Organization of the sections in the chapters on "Technology of semiconductors"

A detailed classification scheme is used so that the information requested can be easily traced by the table of content and the headings of paragraphs. A 5-digit decimal classification is employed to organize the technological sections. Additional unnumbered headings of paragraphs guide the reader to the information requested without the need to read a whole section or paragraph.

The decimal classification continues from the sections on fundamental physical data in the subvolumes III/17a, b. The sections 6 and 7 contain the technology data of tetrahedrally and non-tetrahedrally bonded semiconductors, respectively. The order of semiconductor materials listed is the same as in the sections on fundamental physical data. The order is shown on the inside of the front cover.

For each material, subsections (3. digit in decimal classification) are presented which include the following information:

1. Technological data:

phase diagrams
vapour pressures
solubilities
melt diffusion coefficients
distribution coefficients
viscosity
surface tension
etc.

2. Crystal growth:

synthetization
purification
zone melting
pulling techniques
directional solidification
sheet growth
vapour phase growth
gradient freeze methods
solid state recrystallization
wafer preparation
etc.

3. Characterization of crystal properties:

doping profiles
impurities
defects
polytype verification
surface properties
diagnostic techniques
etc.

4. Device technology:

basic device structures
diffusion
ion implantation
nuclear transmutation doping
epitaxy
fabrication of layers
metallization
lithography
etching processes
final device preparation
etc.

The further classification of the topics listed is not unified because the topics vary according to the applicability of various technological steps or data for each material.

LANDOLT-BÖRNSTEIN

Numerical Data and Functional Relationships in Science and Technology

New Series

Editors in Chief: K.-H. Hellwege · O. Madelung

Group III: Crystal and Solid State Physics

Volume 17

Semiconductors

Editors: O. Madelung · M. Schulz · H. Weiss †

Subvolume c

Technology of Si, Ge and SiC

W. Dietze · E. Doering · P. Glasow · W. Langheinrich
A. Ludsteck · H. Mader · A. Mühlbauer · W. v. Münch
H. Runge · L. Schleicher · M. Schnöller
M. Schulz · E. Sirtl · E. Uden · W. Zulehner

Edited by M. Schulz · H. Weiss †

Springer-Verlag Berlin · Heidelberg · New York · Tokyo 1984

LANDOLT-BÖRNSTEIN

Zahlenwerte und Funktionen
aus Naturwissenschaften und Technik

Neue Serie

Gesamtherausgabe: K.-H. Hellwege · O. Madelung

Gruppe III: Kristall- und Festkörperphysik

Band 17

Halbleiter

Herausgeber: O. Madelung · M. Schulz · H. Weiss †

Teilband c

Technologie von Si, Ge und SiC

W. Dietze · E. Doering · P. Glasow · W. Langheinrich
A. Ludsteck · H. Mader · A. Mühlbauer · W. v. Münch
H. Runge · L. Schleicher · M. Schnöller
M. Schulz · E. Sirtl · E. Uden · W. Zulehner

Herausgegeben von M. Schulz · H. Weiss †

Springer-Verlag Berlin · Heidelberg · New York · Tokyo 1984

CIP-Kurztitelaufnahme der Deutschen Bibliothek

Zahlenwerte und Funktionen aus Naturwissenschaften und Technik/Landolt-Börnstein. - Berlin; Heidelberg; New York; Tokyo: Springer. Teilw. mit d. Erscheinungsorten Berlin, Heidelberg, New York. - Parallelt.: Numerical data and functional relationships in science and technology

NE: Landolt, Hans [Begr.]; PT N.S./Gesamthrsg.: K.-H. Hellwege. Gruppe 3, Kristall- und Festkörperphysik. Bd. 17. Halbleiter/Hrsg.: O. Madelung ... Teilbd. c. Technologie von Si, Ge und SiC/W. Dietze ... Hrsg. von M. Schulz; H. Weiss. 1984.

ISBN 3-540-11474-2 Berlin, Heidelberg, New York, Tokyo
ISBN 0-387-11474-2 New York, Heidelberg, Berlin, Tokyo

NE: Hellwege, Karl-Heinz [Hrsg.]; Madelung, Otfried [Hrsg.]; Schulz, Max [Hrsg.]; Dietze, W. [Mitverf.]

This work is subject to copyright. All rights are reserved, whether the whole or part of the material is concerned specifically those of translation, reprinting, reuse of illustrations, broadcasting, reproduction by photocopying machine or similar means, and storage in data banks.

Under § 54 of the German Copyright Law where copies are made for other than private use a fee is payable to 'Verwertungsgesellschaft Wort' Munich.

© by Springer-Verlag Berlin-Heidelberg 1984

Printed in Germany

The use of registered names, trademarks, etc. in this publication does not imply, even in the absence of a specific statement, that such names are exempt from the relevant protective laws and regulations and therefore free for general use.

Typesetting, printing and bookbinding: Universitätsdruckerei H. Stürtz AG Würzburg

2163/3020 - 543210

Contributors

W. Dietze, Siemens AG, 8000 München 46, FRG

E. Doering, Fondation suisse pour la recherche en microtechnique, 2000 Neuchâtel 7, Switzerland

P. Glasow, Siemens AG, 8520 Erlangen, FRG

W. Langheinrich, Institut für Halbleitertechnik der Technischen Hochschule 6100 Darmstadt, FRG

A. Ludsteck, Siemens AG, 8000 München 46, FRG

H. Mader, Fachbereich Elektrotechnik der Fachhochschule, 8000 München 2, FRG

A. Mühlbauer, Institut für Elektrowärme der Technischen Universität 3000 Hannover, FRG

W. von Münch, Institut für Halbleitertechnik der Universität, 7000 Stuttgart 1, FRG

H. Runge, Siemens AG, 8000 München 83, FRG

L. Schleicher, Siemens AG, 8000 München 80, FRG

M. Schnöller, Siemens AG, 8000 München 83, FRG

M. Schulz, Institut für angewandte Physik der Universität, 8520 Erlangen, FRG

E. Sirtl, Heliotronic GmbH, 8263 Burghausen, FRG

E. Uden, Valvo RHW der Philips GmbH, 2000 Hamburg 54, FRG

W. Zulehner, Wacker-Chemitronic, 8263 Burghausen, FRG

Vorwort

Bereits in der 6. Auflage des Landolt-Börnstein waren Tabellen über die Physik und die Technologie der Halbleiter in verschiedenen 1959, 1962 und 1965 erschienenen Bänden enthalten. In der Neuen Serie bietet jetzt der Band III/17, zu dem der hier vorgelegte Teilband III/17c gehört, erstmals in dieser umfassenden Form, ein detailliertes und vollständiges Daten- und Informationsmaterial über Halbleiter an. Die Daten der physikalischen Eigenschaften füllen sieben, die der Technologie zwei Teilbände. Eine Übersicht über den Inhalt aller Teilbände ist auf der Innenseite des Einbanddeckels abgedruckt.

Die technologischen Daten sollten so schnell wie möglich, d.h. sobald sie geschlossen vorlagen, publiziert werden. Aus diesem eher chronologischen als systematischen Grund erscheinen sie zusammengefaßt in den Teilbänden III/17c und III/17d inmitten der Reihe der Teilbände.

Teilband III/17c behandelt die Technologie von Si, Ge und SiC. Die technologischen Daten der übrigen Halbleiter sind in Teilband III/17d erfaßt, der in Kürze erscheinen wird. Dem Gewicht der Anwendungen entsprechend ist vor allen Dingen der Silicium-Technologie und den III–V Halbleitern ein großes Volumen gewidmet, doch wurden alle technologisch wichtigen Halbleiter, bis hin zu „exotischen" Stoffen berücksichtigt.

Für jeden Halbleiter reicht die Information von den Materialdaten der Halbleiter selbst und der Hilfsstoffe für ihre Technologie über technologische Verfahren, die bereits in großem Umfang eingesetzt werden oder sich noch in der Entwicklung befinden, bis zur Passivierung und Einkapselung der Bauelemente in genormte Gehäuse.

Es war für die Herausgeber und die Autoren nicht leicht, eine einheitliche Darstellungsform und eine angemessene Schranke für den Gesamtumfang so zu finden, daß die wichtige nutzbare Information nicht in unwesentlichen Einzelheiten erstickt. Gerade die Halbleitertechnologie befindet sich derzeit noch in einer rasch fortschreitenden Entwicklung, so daß bei zu vielen technischen Details die Gefahr des Veraltens besteht. Es wurde daher bei speziellen Verfahren und nicht allzu gebräuchlichen Variationen häufig auf die spezifische Fachliteratur verwiesen.

Die Texte nehmen einen etwas größeren Raum ein als in den vorhergehenden Teilbänden, da technologische Fakten der tabellarischen Darstellung nicht so gut angepaßt sind wie physikalische Zahlenwerte. Auf eine Erklärung der Funktionsprinzipien für die Bauelemente wird verzichtet. In den einleitenden Kapiteln zur Bauelement-Technologie wird jedoch auf das Zusammenwirken der verschiedenen Technologieschritte und auf die möglichen Alternativen hingewiesen.

Insgesamt hoffen wir, die Darstellung so gewählt zu haben, daß ein nützliches Nachschlagewerk für die Forschungs- und Entwicklungslaboratorien von Hochschulen und Industrie entstanden ist.

Die Herausgabe von Technologie-Tabellen im Band III/17 der Neuen Serie des Landolt-Börnstein wurde geplant und begonnen von Herbert Weiß, der bereits 1959 zusammen mit H. Welker die ersten Halbleiter-Daten im Landolt-Börnstein publiziert hatte. Infolge eines tragischen Unfalls hat H. Weiß die Vollendung dieser zweiten, wesentlich vollständigeren Bearbeitung nicht mehr erlebt. Als Erlangener Kollege des Verstorbenen hat der Mitherausgeber gern die ehrenvolle Aufgabe übernommen, die Herausgabe der beiden Technologie-Teilbände abzuschließen.

Der erste von ihnen wird hiermit vorgelegt. Seine Fertigstellung war nur möglich dank der guten Kooperation der Autoren, die allen Änderungs- und Ergänzungswünschen durch die Herausgeber in ihren Manuskripten bereitwillig nachgekommen sind. Gedankt sei auch den Mitarbeitern in der Landolt-Börnstein Redaktion, insbesondere Herrn Dr. H. Seemüller und Frau R. Brangs für ihre engagierte Mitarbeit, ferner meiner Sekretärin in Erlangen, Frau G. Loy sowie dem Springer-Verlag für ein weiteres Beispiel hervorragender Buchherstellung.

Dieser Band wird wie alle Bände des Landolt-Börnstein ohne finanzielle Hilfe von anderer Seite veröffentlicht.

Erlangen, im Juli 1983 **Der Herausgeber**

Preface

Several volumes of the 6th edition of Landolt-Börnstein which were published in 1959, 1962, and 1965 already included tables on the physics and technology of semiconductors. A detailed and complete presentation of information on semiconductors is published for the first time in this comprehensive form within the New Series in volume III/17, of which this subvolume III/17c is a part. The data on physical properties are published in seven, the technological information in two subvolumes. A survey of the contents of all subvolumes is printed on the inside of the cover.

The intention was to publish the technological data as soon as possible, i.e. as soon as all the contributions had been completed. For this more chronological than systematical reason, the technology sections are published together in subvolumes III/17c and III/17d in the middle of the series of subvolumes.

The technology of Si, Ge, and SiC is discussed in subvolume III/17c. The technological data of the other semiconductors are covered in subvolume III/17d, which will be published shortly. In accordance with the importance of application, emphasis has been placed especially on the Si-technology and that of the III–V-semiconductors. Nevertheless, all technologically important semiconductors, even exotic ones, are included.

For each semiconductor, the information given ranges from material constants of the semiconductor and of auxiliary chemicals needed during production, via technological methods which are already in use or still under development, to the passivation and encapsulation of the devices into standardized packages. The editor and authors had no easy task to find a standardized form of presentation and an appropriate limit to the total extent in such a way that the important information requested is not drowned in unnecessary details. This is especially so because semiconductor technology is still under rapid development at the moment. Too many technical details would increase the risk of obsolescence. References are therefore often given to specific technical literature for unusual variations and special methods.

The technological subvolumes contain more text than the previous subvolumes. Descriptions of technology cannot as easily be presented in tabular form as physical data. The functional principles of the different devices are not explained. However, the introductory chapters to the device technology discuss the mutual dependence of the different technological steps and mention alternatives.

We hope to have chosen the form of presentation in such a way that a useful reference work for research and development in laboratories, universities, and industry is possible.

The technology tables in volume III/17 were planned and commenced by Herbert Weiß, who had already edited the first semiconductor data together with H. Welker in the Landolt-Börnstein in 1959. Because of a tragic accident, H. Weiß could not see the completion of this second, more comprehensive manuscript. The editor, a colleague of H. Weiß at Erlangen university considered it an honour to continue with the edition of the two technology subvolumes.

The first of these two subvolumes is presented herewith. Its publication was only possible by the good cooperation of the authors, who readily conformed to all requests for changes and additions to their manuscripts. Thanks are also due to the members of the editorial staff of the Landolt-Börnstein, especially to Dr. H. Seemüller and Frau R. Brangs whose competent work was of great assistance, furthermore to my secretary in Erlangen, Frau G. Loy, as well as to the Springer-Verlag for another example of excellent book production.

This volume is published without financial assistance from any other organisation, as are all the other volumes of the Landolt-Börnstein.

Erlangen, July 1983

The Editor

Table of contents

Semiconductors

Subvolume c: Technology of Si, Ge and SiC
(edited by M. Schulz · H. Weiss †)

A. Introduction

1 General remarks

Contents of the technology sections

Data are listed for pure and doped semiconductors. Methods, procedures, and materials and apparatuses employed in the synthesis and purification of materials and for the growth of single crystals are described. Typical crystal properties obtained and diagnostic techniques used for the characterization are outlined. Detailed information is presented on the basic device structures and the processing involved in their fabrication. If fundamental physical semiconductor data are requested, the reader is referred to the subvolumes 17a and 17b. Only those fundamental physical semiconductor data, which are frequently used in device technology are repeated for convenience.

Form of presentation

Technological facts are summarized in the form of tables or are presented in concise statements. Where possible, relations of parameters, flow diagrams of procedures, and outlines of constructions are depicted in figures. Important technological terms or crucial facts and numbers are pointed out for easy finding by using *italic letters.*

Organization of chapters

A detailed classification scheme is used so that the information requested can be easily traced by the table of content and the headings of paragraphs. A 5-digit decimal classification is employed to organize the technological sections. Additional unnumbered headings of paragraphs guide the reader to the information requested without the need to read a whole section or paragraph.

The decimal classification continues from the sections on fundamental physical data in the subvolumes III/17a, b. The sections 6 and 7 contain the technology data of tetrahedrally and non-tetrahedrally bonded semiconductors, respectively. The order of semiconductor materials listed is the same as in the sections on fundamental physical data. The order is shown on the inside of the front cover.

For each material, subsections (3. digit in decimal classification) are presented which include the following information: [1])

1. Technological data:

phase diagrams
vapour pressures
solubilities
melt diffusion coefficients
distribution coefficients
viscosity
surface tension
etc.

2. Crystal growth:

synthetization
purification
zone melting
pulling techniques
directional solidification
sheet growth
vapour phase growth
gradient freeze methods
solid state recrystallization
wafer preparation
etc.

3. Characterization of crystal properties:

doping profiles
impurities
defects
polytype verification
surface properties
diagnostic techniques
etc.

4. Device technology:

basic device structures
diffusion
ion implantation
nuclear transmutation doping
epitaxy
fabrication of layers
metallization
lithography
etching processes
final device preparation
etc.

The further classification of the topics listed is not unified because the topics vary according to the applicability of various technological steps or data for each material.

[1]) This survey is also shown on the fly-leaf.

2 Frequently used symbols

The symbols used are consistent throughout all the subvolumes III/17a···i of the Landolt-Börnstein volume III/17 on semiconductor data. The units given are most frequently used in the figures and tables. For the conversion of equivalent units see Tables 4 and 5.

Table 1. Alphabetical list of symbols.[2])

Symbol	Property	Unit
a	thermal diffusivity $\left(a=\frac{\kappa}{d\cdot C}\right)$	$cm^2 s^{-1}$
a	lattice constant of wurtzite crystals	$Å=10^{-10}$ m
a	radioactivity	s^{-1}
A	area	cm^2
b	line width	$nm=10^{-9}$ m
$\boldsymbol{b}$	Burgers vector	$Å=10^{-10}$ m
B	brightness (of ion sources)	$A cm^{-2} s^{-1}$
$\boldsymbol{B}$	magnetic induction	$T=V s m^{-2}$
c_{ij}	elastic stiffness coefficients	$N m^{-2}$
c	velocity of light	$cm s^{-1}$
c	lattice constant of wurtzite crystals	$Å=10^{-10}$ m
c	concentration, solubility	cm^{-3}, at %, atoms cm^{-3}
C	capacitance	$F=A s V^{-1}$
C	specific heat	$J mol^{-1} K^{-1}$, $cal g^{-1} K^{-1}$
$CSDG$	critical stress of dislocation generation	$g mm^{-2}$
$CRSS$	critical resolved shear stress	$g mm^{-2}$
d	depth, layer thickness, diameter	$Å=10^{-10}$ m
d	specific density	$g cm^{-3}$
d	lattice constant of cubic crystals	$Å=10^{-10}$ m
D	diffusion coefficient	$cm^2 s^{-1}$
D^*	detectivity of ir-detectors	$cm Hz^{-\frac{1}{2}} W^{-\frac{1}{2}}$
e	elementary charge	$C=As$
$\boldsymbol{E}$	electrical field strength	$V cm^{-1}$
E	energy	$V A s=J$, eV
f	focal length	$mm=10^{-3}$ m
f	frequency	Hz
f	misfit parameter	–
f	factor	–
F	Gibbs free energy (free enthalpy)	$cal mol^{-1}$, $J mol^{-1}$
FF	fill factor	–
g	gain factor	–
g	generation/evaporation rate	s^{-1}
	per volume	$cm^{-3} s^{-1}$
	per area	$cm^{-2} s^{-1}$
g	gravitational acceleration	$cm s^{-2}$
G	Gibbs free energy (free enthalpy)	$cal mol^{-1}$, $J mol^{-1}$
$\hbar$	Planck constant, $\hbar=h/2\pi$	$J s$
h	height	m
H	enthalpy	$cal mol^{-1}$, $J mol^{-1}$
H	hardness	$Pa=N m^{-2}$
	empirical scale	–
I	flux density	$cm^{-2} s^{-1}$
I	electrical current	A
j	electrical current density	$A cm^{-2}$
k	distribution or segregation coefficient	–

[2]) Please also consult Table 3.

Symbol	Property	Unit
$\boldsymbol{k}$	wave vector	cm^{-1}
k	electromechanical coupling factor	–
k	Boltzmann constant	$J K^{-1}$
k	reaction rate	s^{-1}
K	equilibrium constant	–
K	absorption coefficient	cm^{-1}
$K (\equiv a)$	thermal diffusivity	$cm^2 s^{-1}$
L, l	length	m
m	mass	g
M	molecular weight	–
	total mass, weight	kg
n	ideality factor	–
n	electron concentration	cm^{-3}
n	refractive index	–
N	concentration of dopant	cm^{-3}
NA	numerical aperture $NA = n \sin \alpha$	–
p	hole concentration	cm^{-3}
p	pressure	$Pa = 10^{-5}$ bar
P	power	$J s^{-1}$
Pr	Prandtl number	–
Q	charge	A s
Q	activation energy of diffusion coefficients	eV
r	rate	s^{-1}
r	ratio	–
r	Hall scattering factor	–
r	radial distance	m
R	radius	m
R	resistance	$\Omega = V A^{-1}$
R	reflectivity	–
R_0	gas constant	$J mol^{-1} K^{-1}$
R_H	Hall coefficient	cm^3
R_p	projected range	μm
s	solidified fraction	–
s	spacing, proximity distance	μm
S	selectivity	–
S	sensitivity	$J cm^{-2}$, $C cm^{-2}$
S	entropy	$cal mol^{-1} K^{-1}$, $J mol^{-1} K^{-1}$
t	time	s
T	temperature	K
u	undercutting	μm
v	velocity	$cm s^{-1}$
V	voltage	V
V	volume	m^3
w	concentration	wt %
w	width	m
x	mole fraction	–
x	variable distance	m
X	stress	$Pa = N m^{-2}$
x	electronegativity of metals	–
z	figure of merit	–
Z	atomic number	–
⟨ ⟩ or (s)	indicates solid material	
{ } or (l)	indicates liquid material	
( ) or (g)	indicates gaseous material	
$[\,]_x$	indicates dissolved in X	

Table 2. Greek symbols.

Symbol	Property	Unit
α	angle	°, rad
α	power coefficient	–
α	thermoelectric power	$mV K^{-1}$
α	linear thermal expansion coefficient	K^{-1}
β	absorption coefficient	cm^{-1}
β	angle	°, rad
δ	small thickness	µm
γ	contrast	–
γ	disorder coefficient	–
γ	surface tension	$N m^{-1}$
Δ	difference, change of quantity	
ε	dielectric constant	–
ε_0	dielectric permittivity of free space	$F m^{-1}$
η	efficiency, yield	–
η	viscosity	$Pa s = 10 P$
θ	angle of incidence	°, rad
κ	compressibility	Pa^{-1}
κ	thermal conductivity	$W cm^{-1} K^{-1}$
λ	wavelength	$nm = 10^{-9} m$
μ	mobility	$cm^2 V^{-1} s^{-1}$
$\tilde{\nu}$	wave number $\tilde{\nu} = 1/\lambda$	cm^{-1}
ν	frequency	$Hz = s^{-1}$
ν	kinematic viscosity $\nu = \eta/d$	$m^2 s^{-1} = 10^{-4} St$
ϱ	resistivity	Ωcm
ϱ	space charge	$C cm^{-3}$
σ	electrical conductivity	$\Omega^{-1} cm^{-1}$
σ	cross-section	cm^2, $barn = 10^{-24} cm^2$
φ	azimut, phase angle	°, rad
Φ	barrier energy, work function	eV
χ	electron affinity	eV
χ	susceptibility	–
ω	angular frequency	$rad s^{-1}$

Table 3. Meaning of indices used.
The following indices are frequently used to further specify the symbols of Tables 1 and 2. Where the same letter is used, different meanings can be usually distinguished from the context in which they appear.

Index	Meaning	Index	Meaning
a	anneal	d	decay
a	acceptor	D	diffusion
abs	absolute value	D	drain
A	activation	e	electron
b	boiling	e	etch
b	bulk material value	E	emitter
B	barrier	eff	effective
B	base electrode of transistor	ex	external
c(rit)	critical value	f	forward
calc	calculated	f	formation
cw	continuous wave	f	facet
C	conduction band	F	Fermi
C	contact	FB	flatband
C	collector	g	gaseous
d	donor	g	growth
d	dopant	g	gap

Index	Meaning	Index	Meaning
g	generation	p	constant pressure
G	gate	p	electron-hole pair
h	hole	puls	pulse
H	Hall effect	r	rise
i	intrinsic, initial	r	reverse
in	incident, input	R	recombination
inc	incorporation (in lattice)	rel	relative
int	internal	s	saturation
ir	infrared	s	solid
it	interface trap	s	sublimation
I	implantation	s	sputtering
j	junction	S	source
l	liquid	sc	space charge
m	metal	sc	short circuit
m	melting	sur	surface
m(ax)	maximum or peak value	t	trap
min	minimum value	tet	tetrahedral
n	electron	th	thermal
o	equilibrium, vacuum, initial or reference value	t(h)	threshold
		tot	total
oc	open circuit	v	vapour or vaporization
out	output	V	valence band
ox	oxide	x	partial component, fraction
p	plasma potential	1/2	value at half life time
p	hole	□	per square, sheet value
p	pinch-off		

3 Conversion of units

Table 4. Conversion of pressure units.

	$Pa = Nm^{-2}$	$at = kp\,cm^{-2}$	atm	bar	Torr	$mm\,H_2O = kp\,m^{-2}$
1 Pa	$=1$	$1.02 \cdot 10^{-5}$	$9.87 \cdot 10^{-6}$	10^{-5}	$7.5 \cdot 10^{-3}$	0.102
1 at	$=9.81 \cdot 10^{4}$	1	0.968	0.981	736	10^{4}
1 atm	$=1.013 \cdot 10^{5}$	1.033	1	1.013	760	$1.033 \cdot 10^{4}$
1 bar	$=10^{5}$	1.02	0.987	1	750	$1.02 \cdot 10^{4}$
1 Torr	$=133$	$1.36 \cdot 10^{-3}$	$1.32 \cdot 10^{-3}$	$1.33 \cdot 10^{-3}$	1	13.6
1 mm H_2O	$=9.81$	10^{-4}	$9.68 \cdot 10^{-5}$	$9.81 \cdot 10^{-5}$	$7.36 \cdot 10^{-2}$	1

Table 5. Conversion of energy units.

Quantity	*E*	*E*	*E*	*E*	*E*
Unit	J	kpm	kWh	kcal	erg
1 J	$\hat{=}1$	0.101972	$2.777778 \cdot 10^{-7}$	$2.38846 \cdot 10^{-4}$	10^{7}
1 kpm	$\hat{=}9.8067$	1	$2.72407 \cdot 10^{-6}$	$2.34228 \cdot 10^{-3}$	$9.8066 \cdot 10^{7}$
1 kWh	$\hat{=}3.60000 \cdot 10^{6}$	$3.67098 \cdot 10^{5}$	1	859.85	$3.60000 \cdot 10^{13}$
1 kcal	$\hat{=}4.18680 \cdot 10^{3}$	426.9348	$1.16300 \cdot 10^{-3}$	1	$4.18680 \cdot 10^{10}$
1 erg	$\hat{=}10^{-7}$	$1.019716 \cdot 10^{-8}$	$2.78 \cdot 10^{-14}$	$2.388459 \cdot 10^{-11}$	1
1 eV	$\hat{=}1.60219 \cdot 10^{-19}$	$1.63377 \cdot 10^{-20}$	$4.4506 \cdot 10^{-26}$	$3.8268 \cdot 10^{-23}$	$1.60219 \cdot 10^{-12}$
1 cm^{-1}	$\hat{=}1.98618 \cdot 10^{-23}$	$2.02534 \cdot 10^{-24}$	$0.55172 \cdot 10^{-29}$	$4.7439 \cdot 10^{27}$	$1.9862 \cdot 10^{-16}$
1 K	$\hat{=}1.38041 \cdot 10^{-23}$	$1.40763 \cdot 10^{-24}$	$3.8345 \cdot 10^{-30}$	$3.2971 \cdot 10^{-27}$	$1.38041 \cdot 10^{-16}$
1 Ry	$\hat{=}2.1784 \cdot 10^{-18}$	$2.2213 \cdot 10^{-19}$	$0.60511 \cdot 10^{-24}$	$0.52030 \cdot 10^{-21}$	$2.1784 \cdot 10^{-11}$

continued

Table 5 (continued)

Quantity	$V=E/e$	$\tilde{\nu}=E/hc$	$T=E/k$	E
Unit	V	cm^{-1}	K	Ry
1 J	$\hat{=}6.2415\cdot10^{18}$	$0.50348\cdot10^{23}$	$0.72442\cdot10^{23}$	$4.5905\cdot10^{17}$
1 kpm	$\hat{=}6.1208\cdot10^{19}$	$4.9374\cdot10^{23}$	$0.71042\cdot10^{24}$	$4.5018\cdot10^{18}$
1 kWh	$\hat{=}2.2469\cdot10^{25}$	$1.8125\cdot10^{29}$	$2.6079\cdot10^{29}$	$1.6526\cdot10^{24}$
1 kcal	$\hat{=}2.61317\cdot10^{22}$	$2.10797\cdot10^{26}$	$3.0330\cdot10^{26}$	$1.9220\cdot10^{21}$
1 erg	$\hat{=}6.2415\cdot10^{11}$	$0.50348\cdot10^{16}$	$0.72442\cdot10^{16}$	$4.5905\cdot10^{10}$
1 eV	$\hat{=}1$	$0.80667\cdot10^{4}$	$1.16066\cdot10^{4}$	$0.73549\cdot10^{-1}$
1 cm^{-1}	$\hat{=}1.23967\cdot10^{-4}$	1	1.43883	$0.911763\cdot10^{-5}$
1 K	$\hat{=}0.86158\cdot10^{-4}$	0.69501	1	$0.63368\cdot10^{-5}$
1 Ry	$\hat{=}1.3596\cdot10$	$1.096776\cdot10^{5}$	$1.5781\cdot10^{5}$	1

4 Abbreviations frequently used in semiconductor technology

It is common practice in semiconductor technology to use abbreviations for standard devices, methods, conditions, diagnostic techniques, or specific apparatus designs. The following abbreviations are frequently used:

AAS atomic absorption spectroscopy
AES auger electron spectroscopy
AM air mass
AMO solar spectrum outer space
AM1 solar spectrum at earth's surface for optimum conditions at sea level, sun at zenith
AM2 solar spectrum at earth's surface for average weather conditions
APCVD atmospheric pressure CVD
AsSG arsenosilicate glass
BARRITT barrier injection transit time
BCCD buried channel CCD
BH buried-heterostructure
BJT bipolar junction transistor
BP boiling point
BSF back surface field (solar cell)
BSG borosilicate glass
BSR back surface reflection (in solar cells)
CAD computer-aided design
CAM computer aided manufacture
CAST capillary action shaping technique
CCD charge coupled device
CCDLTS constant-capacitance DLTS
CCPD charge coupled photodiode
CDE chemical dry etching
CDI collector diffusion isolation
CERDIL ceramic dual in line
CERDIP ceramic dual in line package
CID charge injection device
CIE complete island etch
CMOS complementary MOS
CMSR conductor-semiconductor-resistor-metal structure
CMT cadmium mercury telluride
CP chemical polish
CP critical point (in phase diagrams)
CPAA charged particle activation analysis
CRC chemical reaction in colloidal solution
CRSS critical resolved shear stress
CSDG critical stress of dislocation generation
CTE coefficient of thermal expansion
CV capacitance voltage

CVD	chemical vapour deposition
CW	continuous wave
CZ	Czochralski (growth)
DBR	distributed Bragg reflection
DC	direct current
DD	double drift
DDC	dual dielectric charge storage device
DDLTS	double DLTS
DFB	distributed feedback
DH	double-heterostructure
DIAC	diode ac switch
DIL	dual in line
DIMOS	double implanted MOS
DIP	dual in line package
DLTS	deep level transient spectroscopy
DMOS	double diffused MOS
DMZ	dimethyl zinc
DO	diode outline (package)
DOS	density of states
DOVETT	double velocity transit time diode
DRAM	dynamic RAM
DSPS	derivative surface photovoltage spectroscopy
DTA	differential thermal analysis
EAROM	electrically alterable ROM
EBIC	electron beam induced current
ECDE	electron-induced chemical dry etching
EDAX	energy dispersive analysis of X-rays
EDRI	evaporation-diffusion en regime isotherme (isothermal evaporation and interdiffusion)
EELS	electron energy loss spectroscopy
EEPROM	electrically erasable PROM
EFG	edge-defined film-fed growth
ELO	epitaxial lateral overgrowth
EP	etch pit
EPD	etch pit density
EPROM	erasable programmable read only memory
ESCA	electron spectroscopy for chemical analysis
ESFI	epitaxial silicon film on insulators
ESP	edge-supported pulling
EXAFS	extended x-ray absorption fine structure
FAMOS	floating gate avalanche injection MOS
FCC	face-centered cubic (crystal structure)
FCT	field-controlled thyristor
FED	field effect diode
FET	field effect transistor
FTS	Fourier transform spectroscopy
FWHM	full width half maximum
FZ	floating zone
GCM	glycidyl metacrylate-co-methyl methacrylate (resist)
GGL	gain-guided laser
GTO	gate turn-off thyristor
HCP	hexagonal close-packed (crystal structure)
HCT	mercury (Hg) cadmium telluride
HD	homojunction diode
HEC	high efficiency (solar) cell
HEED	high energy electron diffraction
HEM	heat exchanger method
HEMT	high electron mobility transistor
HIC	high ionic current (ion implantation)

HEXFET	hexagonal MOSFET
HMOS	high-performance MOS
HP	high-purity
HRG	horizontal ribbon growth
HTCVD	high-temperature CVD
I^2	ion implantation
I^2L	integrated injection logic
IC	integrated circuit
ICC	interface-controlled crystallization
ICP	inductively coupled plasma (in AES spectroscopy)
ID	input diode
IEC	international electrical commission
IG	input gate
IGL	index-guided laser
IGFET	insulated gate field effect transistor
IMPATT	impact avalanche transit time
IR	infrared
IRED	infrared emitting diode
IS	inverted Stepanov (growth method)
ITO	indium tin oxide
JEDEC	Joint Electron Device Engineering Councils
JFET	junction field effect transistor
JGFET	junction gate field effect transistor
KDP	potassium (K) dihydrogen phosphate
LAS	light-activated SCR (THYRISTOR)
LASER	light amplification by stimulated emission of radiation
LCRT	laser cathode ray tube
LEC	liquid encapsulation Czochralski (growth)
LED	light emitting diode
LEED	low energy electron diffraction
LESS	lateral epitaxy seeded solidification
LISA	light switching array
LOC	large optical cavity
LOCOS	local oxidation
LPCVD	low-pressure CVD
LPE	liquid phase epitaxy
LSA	limited space charge accumulation
LSI	large scale integration
LSS	Lindhard, Scharff, and Schiott (ion implantation range computations)
LSS	light spot scanning
LSSA	low-angle silicon sheet growth
LSSD	level sensitive scan design
LTCVD	low-temperature CVD
LTT	lead tin telluride
MAOS	metal-Al_2O_3-SiO_2-semiconductor
MBE	molecular beam epitaxy
MBS	multiple blade slurry (saw)
MCT	mercury cadmium telluride
MEK	methyl ethyl ketone
MESFET	metal-semiconductor barrier FET
MG	metallurgical grade
MIBK	methyl isobutyl ketone
MIM	metal-insulator-metal
MIMIM	metal-insulator-metal-insulator structure
MIOS	metal-insulator-oxide-semiconductor
MIS	metal-insulator-semiconductor
MMIC	monolithic microwave integrated circuit
MNOS	metal-Si_3N_4-SiO_2-semiconductor

MO	microelectronic outline (package)
MOM	metal-oxide-metal
MOS	metal-oxide-semiconductor
MOSFET	MOS field effect transistor
MO-VPE	metalorganic VPE
MP	melting point
MTBF	mean time between failures
MTF	medean time to failure
MW	microwave
NAA	neutron activation analysis
NDR	negative differential resistivity
NEA	negative electron affinity
NEP	noise equivalent power
NF	noise figure
NMOS	n-channel MOS
NTC	negative temperature coefficient
NTD	neutron transmutation doping
NVRAM	non-volatile RAM
OED	oxidation-enhanced diffusion
OG	output gate
OM VPE	organometallic VPE
OISF, OSF	oxidation-induced stacking fault
ORD	oxidation-reduced diffusion
PB	plasma barrel reactor
PC	photoconductive, printed circuit
PCD	photoconductive decay
PCDE	photon-induced chemical dry etching
PDG	pendant drop growth
PDO	plasma-deposited oxide
PECVD	plasma-enhanced CVD
PIXE	particle-induced X-ray emission
PL	photoluminescence
PLANOX	planar oxide
PMIPK	polymethyl isopropenyl ketone
PMMA	polymethyl methacrylate
PP	parallel plate (reactor)
POCT	periodically oscillating crystal temperature
POST	periodically oscillating source temperature
PP	parallel-plate reactor
PPB	parts per billion 10^{-9} (a) by atoms or (w) by weight
PPM	parts per million 10^{-6} (a) by atoms or (w) by weight
PROM	programmable read only memory
PSD	position-sensitive radiation detector
PSG	phosphorus silicate glass
PT	plasma triode (etching)
PTC	positive temperature coefficient
PTIS	photothermal infrared spectroscopy
PUT	programmable unijunction transistor
PV	photovoltaic
PVC	poly-n-vinylcarbazole
PVD	physical vapour deposition
RAD	ribbon-against-drop (pulling)
RAM	random access memory
RBS	Rutherford backscattering
RCT	reverse conducting thyristor
RE	rare earth (metals)
REM	raster electron microscope
RF (rf)	radio frequency

RHEED	reflection high energy electron diffraction
RIBE	reactive ion beam etching
RIE	reactive ion etching
ROM	read only memory
RPM	rotations per minute
RPS	rotations per second
RT	room temperature
RTR	ribbon-to-ribbon (growth)
SAM	scanning Auger microscopy
SAMOS	stacked gate avalanche injection MOS
SATO	selfaligned thick oxide
SAW	surface acoustic wave
SB	Schottky surface barrier
SBD	Schottky barrier diode
SBIGFET	Schottky barrier insulated gate field effect transistor
SBS	silicon bilateral switch
SCCD	surface channel CCD
SCH	separate confinement heterostructure
SCIM	silicon coating by inverted meniscus
SCLC	space charge limited current
SCR	silicon-controlled rectifier
SD	single drift
SEM	scanning electron microscope
SEXAFS	surface EXAFS
SH	single heterostructure
S.I.	semiinsulating
SIL	single in line
SILO	sealed-interface oxidation
SIMOS	stacked-gate injection MOS
SIMS	secondary ion mass spectroscopy
SIS	semiconductor-insulator-semiconductor (structure)
SIPOS	semi-insulating polycrystalline oxygen-doped Si
SIT	static induction transistor
SNR	signal to noise ratio
SO	small outline (package), standard outlines
SOA	safety operating area
SOC	silicon on ceramics
SOD	standard outlines of diodes
SOI	silicon on insulator
SOS	silicon on sapphire
SOT	standard outlines of transistors
SPE	solid-phase epitaxy
SPRITE	signal processing in the element
SRAM	static RAM
SRH	Shockley Read Hall (recombination statistics)
SSD	synthesis solid diffusion
SSMS	spark source mass spectroscopy
SSPD	selfscanned photodiode
SSR	solid state recrystallization
SUS	silicon unilateral switch
TAB	tape automated bonding
TC	thermocompression (bonding)
TCA	trichloroethane
TCE	trichloroethylene
TDI	time delay integration
TDMS	thermal desorption mass spectroscopy
TE	transverse electric (mode); triethyl (in MO-VPE)
TED	transferred electron device

TEGFET	two-dimensional electron gas FET
TEM	transmission electron microscope
TFT	thin film transistor
THM	travelling heater method
TIBA	tri-isobutyl aluminum
TJC	tandem junction solar cell
TJS	transverse junction stripe
TM	transverse magnetic; trimethyl (in MO-VPE)
TMS	packages for chip carriers
TO	transistor outline (package)
TOM	temperature oscillating method
TRAPPATT	trapped plasma avalanche triggered transit
TRIAC	triode ac switch
TSC	thermally stimulated current
TSCAP	thermally stimulated capacitance
TSG	terrestrial solar grade
TSIC	thermally stimulated ionic current
TSM	travelling solvent method
TTL	transistor-transistor logic
TUNNETT	tunnel transit time diode
TV	television
TVS	triangular voltage sweep
UHF	ultra-high frequency
UHV	ultra-high vacuum
UJT	unijunction transistor
UMG	upgraded metallurgical grade
UMOS	u-shaped grooved MOS
UP	ultrapure (germanium)
UPS	ultraviolett photoemission spectroscopy
UV	ultraviolett (radiation)
VARACTOR	variable capacitor
VARISTOR	variable resistor
VDR	voltage-dependent resistor
VHSIC	very high speed integrated circuit
VJFET	vertical junction FET
VLSI	very large scale integration
VMOS	vertical MOS
VPE	vapour phase epitaxy
VUV	vacuum UV
WEB	dentritic web growth
XPS	x-ray photoemission spectroscopy

B. Technology of semiconductors

6 Tetrahedrally bonded semiconductors

6.1 Silicon and germanium

6.1.1 Technological data

6.1.1.0 Introduction

The incorporation of impurities into the silicon and germanium lattice is mainly determined by their solid solubility and the respective *distribution coefficient* which are both basically derived from the binary *phase diagram.* The actual incorporation of solute atoms can be described by an effective distribution coefficient that is considerably influenced by the solute *diffusion coefficient* in the melt. The diffusion behaviour of the impurity atoms in the liquid phase depends to a certain degree on their size, for which the *tetrahedral covalent radius* is a reasonable measure.

If solidification takes place under reduced pressure or vacuum, solvent as well as solute atoms can evaporate partly into the surroundings, which leads to a reduced solute concentration in the melt. The amount of evaporation of impurity atoms during solidification is described by the *evaporation rate.*

In order to investigate the heat flow in a growing crystal and in the respective liquid phase, the *thermophysical data* of the crystal as well as the physical *properties of the melt* have to be known.

6.1.1.1 Phase diagrams and liquidus curves

The appropriate representation of the equilibrium between the solid and the liquid state of impure Si or Ge for different temperatures is given by the *binary phase diagram.* It indicates whether an alloyed crystal, i.e. a solid solution of impurities, exists and how much of the solute can be dissolved.

For the growth of semiconductor crystals, the equilibrium between the solid and the liquid phase plays a major role and the *liquidus* (Figs. 1 and 2) *and solidus curves* (Figs. 3 and 4) are the most important parts of the phase diagram. They have to be established much more precisely and reliably than is usual for metallic systems. *Thermal analysis* is a suitable method to establish the liquidus curves in Si and Ge systems [53H, 60T1, 70G]. The solidus curves can be obtained through *controlled crystallization experiments* and a thorough analytical investigation of the solidified samples [56T1, 60T2, 70G].

The solidus and liquidus curves for semiconductor impurity systems are often separately plotted because of the very *small solid solubility of impurity atoms* which makes it nearly impossible to include both curves into one diagram using a linear scale for the composition. For this reason, it is often better to split each system into *two diagrams* with a joint temperature scale but different scales for the solidus (logarithmic) and the liquidus curve (linear) [70G]. The liquidus curves can be described by a simple equation including two constants, a and b, for each system [60T1, 70G]:

$$T = \frac{\Delta H_f + a(1-x_l)^2}{\Delta S_f - R \cdot \ln x_l + b(1-x_l)^2}$$

where ΔH_f is the latent heat of fusion of Si or Ge; ΔS_f the alteration of entropy during phase conversion, i.e. melting, equal to the latent heat of fusion divided by the melting temperature in Kelvin units; R the gas constant; x_l the mole ratio of Si or Ge in the liquid binary system.

Table 1. Coefficients a and b of the liquidus curve equation [60T1].

Impurity	Silicon		Germanium	
	a [cal mol^{-1}]	b [cal mol^{-1} K^{-1}]	a [cal mol^{-1}]	b [cal mol^{-1} K^{-1}]
Aluminum	−4140	−1.22	−5360	−3.16
Gallium	3250	0.83	−150	0
Indium	11450	3.37	1570	0.56
Thallium	(16600)	(3.80)	(5700)	(1.90)

continued

Table 1 (continued)

Impurity	Silicon		Germanium	
	a [cal mol^{-1}]	b [cal mol^{-1} K^{-1}]	a [cal mol^{-1}]	b [cal mol^{-1} K^{-1}]
Tin	8145	1.50	1680	1.08
Lead	19830	4.58	8780	4.08
Arsenic	−49990	−32.40	−5600	−4.16
Antimony	3290	−1.61	2640	1.98
Bismuth	14840	2.06	5505	1.49
Copper	−11910	−7.19	−7360	−7.67
Silver	−7910	−7.63	−5500	−7.13
Gold	−19540	−10.28	−4865	−1.02
Zinc	4280	1.14		
Cadmium			4110	1.75

Figs. 5···7 show phase diagrams for the Si-impurity systems: Si—Ge, Si—B, Si—C, Si—N, and Si—O.

6.1.1.2 Solubility in solid Si and Ge

The solid solubility c_s of impurity atoms in Si and Ge is given by the solidus curve of the *respective phase diagram.* For most impurities, a retrograde solubility can be observed (Figs. 3 and 4) [60T2]. After solidification, the solubility *first increases* with decreasing temperature, *reaches a maximum* at a certain temperature, and *decreases after* that with further dropping temperature. As a result of the retrograde solubility in a *supersaturated solution,* the impurities have the tendency to *precipitate* during cooling down and form their own phase in the host lattice. Precipitated doping elements lose their doping effect.

Table 2. Maximum solid solubility c_s^m of impurities in Si and Ge.

Impurity	c_s^m [atoms cm^{-3}]	T [°C], Remarks	Ref.
		Silicon	
H	$1.7 \cdot 10^{19}$		56P, 72H
Li	$6.5 \cdot 10^{19}$	1200, Figs. 3, 9	57P1, 60T2
Cu	$1.5 \cdot 10^{18}$	1300, Figs. 3, 9	56S1, 57C2, 60T2
	$1.0 \cdot 10^{18}$	1250	58B2
Ag	$2.0 \cdot 10^{17}$	1350, Fig. 3	60B, 70G
Au	$1.2 \cdot 10^{17}$	1300, Figs. 3, 9	57C3, 60T2
Zn	$6.0 \cdot 10^{16}$	1325, Figs. 3, 9	57F1, 57C1, 60T2
	$1.5 \cdot 10^{16}$	1270	70B
Cd	$0.1 \cdots 3.0 \cdot 10^{16}$		62F, 67S
B	$6.0 \cdot 10^{20}$	1420, Fig. 9	60T2
	$1.0 \cdot 10^{21}$	1420, Fig. 3	68H, 73S
Al	$2.0 \cdot 10^{19}$	1100, Figs. 3, 9	60T2
	$5.0 \cdot 10^{20}$	1030	60G, 70G
Ga	$4.0 \cdot 10^{19}$	1250, Figs. 3, 9	60T2
In	$0.4 \cdots 2.0 \cdot 10^{18}$		67S
Tl	$8.8 \cdot 10^{17}$	$c_s^m = 5.2 \cdot 10^{21} \cdot k_0$	62F
C	$3.0 \cdot 10^{17}$	1420	70N, 74N
	$5.0 \cdot 10^{17}$	1420, Fig. 3	71B
Sn	$5.0 \cdot 10^{19}$	1200, Figs. 3, 9	59T1, 60T2
N	$4.5 \cdot 10^{15}$	1420	74N
P	$1.3 \cdot 10^{21}$	1200, Figs. 3, 9	60T2, 70G
As	$1.8 \cdot 10^{21}$	1150, Figs. 3, 9	60T2, 70G
Sb	$7.0 \cdot 10^{19}$	1325, Figs. 3, 9	59R, 60T2
Bi	$8.0 \cdot 10^{17}$	1325, Figs. 3, 9	60T2
O	$3.0 \cdot 10^{18}$	1420, Fig. 3	71B, 74N

continued

Table 2 (continued)

Impurity	c_s^m [atoms cm^{-3}]	T [°C], Remarks	Ref.
		Silicon (continued)	
S	$3.0 \cdot 10^{16}$	1325, Fig. 3	59C, 60T2
Cr	$1.0 \cdot 10^{14}$	1200···1250	70L
Se	$7.0 \cdot 10^{16}$	1200, Fig. 3	78V, 79V
Mn	$3.0 \cdot 10^{16}$	1325, Fig. 3	56C, 60T2
Fe	$3.0 \cdot 10^{16}$	1325, Figs. 3, 9	56S1, 57C2, 60T2
	$>5.0 \cdot 10^{16}$	$c_s = 5 \cdot 10^{22} \cdot \exp(7.3 \cdots 2.87\ \mathrm{eV/kT})$	80W
Co	$2.0 \cdot 10^{16}$	1325, Figs. 3, 9	57C2, 60T2
	$>2.0 \cdot 10^{16}$	$c_s = 1.80 \cdot 10^{21} \cdot \exp(-1.45\ \mathrm{eV/kT})$	70B, 78E
Ni	$8.0 \cdot 10^{17}$	1300, Fig. 3	62A1, 69S
Pd	$2.9 \cdot 10^{16}$	1200	72A, 67S
Ir	$9.0 \cdot 10^{16}$	1250	76A
Pt	$>4.0 \cdot 10^{17}$	Fig. 3	69B
		Germanium	
H	$3.0 \cdot 10^{18}$	937	56T3, 70G
Li	$7.5 \cdot 10^{18}$	825, Figs. 4, 9	57P2, 58R, 60T2
Cu	$6.8 \cdot 10^{16}$		80G
	$3.5 \cdot 10^{16}$	875, Figs. 4, 9	58W, 59T3, 60T2
Ag	$9.0 \cdot 10^{14}$	875, Figs. 4, 9	60T2
	$4.0 \cdot 10^{18}$	750	62K, 70G
	$6.8 \cdot 10^{15}$	$c_s^m = 4.4 \cdot 10^{21} \cdot k_0$	62F, 79G
Au	$3.0 \cdot 10^{16}$	900, Figs. 4, 9	59T3, 62S, 70G
Be	$4.0 \cdot 10^{20}$	$c_s^m = 4.4 \cdot 10^{21} \cdot k_0$	62F, 78G2
Zn	$2.5 \cdot 10^{18}$	750, Figs. 4, 9	59T3, 60T2
Cd	$2.5 \cdot 10^{18}$	825, Fig. 4	59K1, 70G
Ba	$8.8 \cdot 10^{18}$	$c_s^m = 4.4 \cdot 10^{21} \cdot k_0$	62F, 75G
B	$1.0 \cdot 10^{18}$		56B2, 70G
Al	$4.3 \cdot 10^{20}$	700, Figs. 4, 9	59T2, 60T2
Ga	$5.0 \cdot 10^{20}$	650, Figs. 4, 9	59T2, 60T2
In	$4.0 \cdot 10^{18}$	800, Figs. 4, 9	60L, 60T2
Tl	$9.0 \cdot 10^{18}$	910, Fig. 4	62T2, 70G
Sn	$5.0 \cdot 10^{20}$	400···232, Figs. 4, 9	59T1, 59T3, 60T2
Pb	$6.0 \cdot 10^{17}$	870, Figs. 4, 9	60T2
P	$2.0 \cdot 10^{20}$	560, Fig. 4	62A2, 70G
As	$7.0 \cdot 10^{19}$	800, Figs. 4, 9	61S, 62T1, 70G
Sb	$1.2 \cdot 10^{19}$	800, Figs. 4, 9	56T1, 60T2
Bi	$6.0 \cdot 10^{16}$	910, Figs. 4, 9	62T1, 70G
S	$5.0 \cdot 10^{15}$		59T3, 60T2
Te	$7.0 \cdot 10^{18}$	850, Fig. 4	62I, 70G
V	$1.3 \cdot 10^{15}$	$c_s^m = 4.4 \cdot 10^{21} \cdot k_0$	62F
Mn	$4.4 \cdot 10^{15}$	$c_s^m = 4.4 \cdot 10^{21} \cdot k_0$	62F
Fe	$1.3 \cdot 10^{15}$	875, Figs. 4, 9	57B, 60T2
Co	$2.0 \cdot 10^{15}$		59T3, 60T2
Ni	$8.0 \cdot 10^{15}$	875, Figs. 4, 9	59T3
Pt	$2.2 \cdot 10^{16}$	$c_s^m = 4.4 \cdot 10^{21} \cdot k_0$	62F

6.1.1.3 Equilibrium distribution coefficient

The equilibrium distribution coefficient k_0 of a solute is the *ratio* of the *equilibrium concentration in the solid* c_s and in the *liquid phase* c_l *at the melting point* taken from the phase diagram: $k_0 = c_s/c_l = x_s d_s / x_l d_l$, where x_s, x_l and d_s, d_l are the corresponding mole ratios and densities.

In general k_0 will depend on concentration except for very dilute solutions that are typical for semiconductors. Liquidus and solidus curves are effectively straight lines. *For* $k_0 < 1$, the liquidus and solidus lines slope *downward* (Fig. 8a), whereas for the corresponding diagram for a solute which *raises* the *melting point* of the solvent k_0 is greater than unity, i.e., $k_0 > 1$ (Fig. 8b).

Most impurities of interest have distribution coefficients $k_0 < 1$ in Si and Ge which vary considerably for different elements.

The fact that many of the elements have $k_0 < 1$ can be used in refining the starting material, because the solute will be enriched in the liquid due to $k_0 < 1$ during solidification (section 6.1.2.3).

Table 3. Distribution coefficient k_0 of impurities in Si and Ge.

Impurity	k_0	Remarks	Ref.
		Silicon	
Li	0.01	Fig. 9	57P1, 60T2
Cu	$4.0 \cdot 10^{-4}$	Figs. 9, 10, 11	53T2, 60T2
Ag	$1.0 \cdot 10^{-6}$	$k_0 = 1.92 \cdot 10^{-22} \cdot c_s^m$	62F
Au	$2.5 \cdot 10^{-5}$	Figs. 9, 10, 11	57C3, 60T2
Zn	$1.0 \cdot 10^{-5}$		57H1, 60T2
Cd	$1.0 \cdot 10^{-6}$	$k_0 = 1.92 \cdot 10^{-22} \cdot c_s^m$	62F
B	0.8	Figs. 9, 10, 11, 12	56T2, 60T2
Al	$2.0 \cdot 10^{-3}$	Figs. 9, 10, 11, 12	59H, 60T2
Ga	$8.0 \cdot 10^{-3}$	Figs. 9, 10, 12	59H, 60T2
In	$4.0 \cdot 10^{-4}$	Figs. 10, 11, 12	59H, 60T2
Tl	$1.7 \cdot 10^{-4}$	Fig. 12	81K
C	0.07		70N, 74N
	0.06		82K
Ti	$3.6 \cdot 10^{-6}$		76H
Ge	0.33	Figs. 10, 11	53T1, 60T2
Sn	0.016	Figs. 9, 10, 11	59T1, 60T2
N	$7.0 \cdot 10^{-4}$		74N
P	0.35	Figs. 9, 10, 11, 12	54B, 59H, 60T2
As	0.3	Figs. 9, 10, 11, 12	54B, 60T2
Sb	0.023	Figs. 9, 10, 11, 12	59H, 60T2
Bi	$7.0 \cdot 10^{-4}$	Figs. 9, 10, 11, 12	60T2
O	1.4		74N
S	$1.0 \cdot 10^{-5}$		59C, 60T2
Cr	$1.1 \cdot 10^{-5}$		76H
Mn	$1.0 \cdot 10^{-5}$	Figs. 9, 11	56C, 60T2
Fe	$8.0 \cdot 10^{-6}$	Figs. 9, 11	57C2, 60T2
	$6.0 \cdot 10^{-5}$		72P
Co	$8.0 \cdot 10^{-6}$	Figs. 9, 11	57C2, 60T2
Ni	$3.0 \cdot 10^{-5}$	$k_0 = 1.92 \cdot 10^{-22} \cdot c_s^m$	62F
	$8.0 \cdot 10^{-6}$		72P
Ta	$1.0 \cdot 10^{-7}$		54B, 60T2
		Germanium	
Li	$2.0 \cdot 10^{-3}$	Fig. 9	57P1, 60T2
Cu	$1.5 \cdot 10^{-5}$	Figs. 9, 10, 11	53B, 60T2
	$1.3 \cdot 10^{-5}$		80G
		Germanium (continued)	
Ag	$4.0 \cdot 10^{-7}$	Figs. 9, 10, 11	59T3, 60T2
	$9.0 \cdot 10^{-4}$	$k_0 = 2.27 \cdot 10^{-22} \cdot c_s^m$	62F
	$1.6 \cdot 10^{-6}$		79G
Au	$1.3 \cdot 10^{-5}$	Figs. 9, 10, 11	59T3, 60T2
Be	0.081		78G2
Zn	$4.0 \cdot 10^{-4}$	Fig. 9	56T4, 59T3, 60T2
Cd	$1.0 \cdot 10^{-5}$		56W, 59T3, 60T2
B	12.2		70E
	17.0	Figs. 10, 11	56B2, 60T2
	21.0		78G1
Al	0.073	Figs. 9, 10, 11	59T2, 60T2
Ga	0.087	Figs. 9, 10, 11	58L, 59T2, 60T2
In	$1.0 \cdot 10^{-3}$	Figs. 9, 10, 11	60L, 60T2
Tl	$4.0 \cdot 10^{-5}$	Figs. 10, 11	53B, 60T2
Si	5.5	Figs. 10, 11	60T2
Sn	0.02	Figs. 9, 10, 11	54B, 59T1, 60T2
Pb	$1.7 \cdot 10^{-4}$	Figs. 9, 10, 11	60T2
P	0.08	Figs. 10, 11	59H, 60T2
As	0.02	Figs. 9, 10, 11	55J, 60T2
Sb	$3.0 \cdot 10^{-3}$	Figs. 9, 10, 11	53B, 59H, 60T2
Bi	$4.5 \cdot 10^{-5}$	Figs. 9, 10, 11	53B, 58M, 60T2
S	$1.1 \cdot 10^{-6}$	$k_0 = 2.27 \cdot 10^{-22} \cdot c_s^m$	62F
Te	$1.0 \cdot 10^{-6}$		59T3, 60T2
V	$3.0 \cdot 10^{-7}$	Fig. 11	55W, 60T2
Mn	$1.0 \cdot 10^{-6}$	Fig. 11	55W, 59T3, 60T2
Fe	$3.0 \cdot 10^{-5}$	Figs. 9, 11	57B, 60T2
Co	$1.0 \cdot 10^{-6}$	Fig. 11	54T, 60T2
Ni	$3.0 \cdot 10^{-6}$	Figs. 9, 11	57T, 60T2
Pt	$5.0 \cdot 10^{-6}$		54D, 60T2

The values of the equilibrium distribution coefficients can be *correlated* with *various parameters*, e.g., *maximum solid solubilities* (Fig. 9), *tetrahedral covalent radii* (Fig. 10), *heat of sublimation* (Fig. 11), and *melt diffusion coefficients* (Fig. 12). These relations are useful for estimating unknown values of k_0.

The correlation between maximum solid solubility c_s^m and the distribution coefficient for Si is according to [62F]: $c_s^m = 5.2 \cdot 10^{21} \cdot k_0$ atoms cm^{-3} and for Ge: $c_s^m = 4.4 \cdot 10^{21} \cdot k_0$ atoms cm^{-3}. In case of silicon this correlation is *not valid* for *boron, carbon, nitrogen* and *oxygen* [73S, 74N], whereas for germanium the correlation is not fulfilled for *boron* and probably not for *carbon, nitrogen* and *oxygen.*

Methods of determining the distribution coefficients are described in [66P].

6.1.1.4 Melt diffusion coefficient

The melt diffusion coefficient D_l is a measure for the ability of impurity atoms that are repelled at the freezing interface during solidification due to $k_0 < 1$, to diffuse back into the bulk liquid. The melt diffusion coefficient influences the effective incorporation of impurities into the solid (see section 6.1.2.3). Its value can be *correlated* with the *tetrahedral covalent radii* r_{tet}. For Si the melt diffusion coefficients of *group V impurities decrease* with r_{tet}, whereas for *group III* elements a *reverse trend* is found [63K] (see Fig. 13). The correlation between the melt diffusion, coefficient D_l and the equilibrium distribution coefficient k_0 shown in Fig. 12 is a consequence of the interdependence between D_l and r_{tet} of Fig. 13. These relations permit the *determination of unknown* k_0 of group III and V impurities.

Table 4. Diffusion coefficient D_l in the melt for impurities in Si and Ge.

Impurity	D_l [$cm^2 s^{-1}$]	Remarks	Ref.
	Silicon		
B	$2.4 \cdot 10^{-4}$	Figs. 12, 13	63K
	$1.8 \cdot 10^{-4}$	recalculated [1])	
	$3.3 \cdot 10^{-4}$		68S
Al	$7.0 \cdot 10^{-4}$	Figs. 12, 13	63K
	$5.3 \cdot 10^{-4}$	recalculated [1])	
	$2.3 \cdot 10^{-5}$		62T3
Ga	$4.8 \cdot 10^{-4}$	Figs. 12, 13	63K
	$3.6 \cdot 10^{-4}$	recalculated [1])	
	$6.6 \cdot 10^{-5}$		68S
In	$6.9 \cdot 10^{-4}$	Figs. 12, 13	63K
	$5.2 \cdot 10^{-4}$	recalculated [1])	
	$1.7 \cdot 10^{-5}$		68S
Tl	$7.8 \cdot 10^{-4}$	Figs. 12, 13	81K
	$5.9 \cdot 10^{-4}$	recalculated [1])	
C	$2.0 \cdot 10^{-4}$		82K
	$1.5 \cdot 10^{-4}$	recalculated [1])	
	$4.8 \cdot 10^{-4}$		73G
P	$5.1 \cdot 10^{-4}$	Figs. 12, 13	63K
	$3.9 \cdot 10^{-4}$	recalculated [1])	
	$2.3 \cdot 10^{-4}$		62T3
	$2.7 \cdot 10^{-4}$		68S
	Silicon (continued)		
As	$3.3 \cdot 10^{-4}$	Figs. 12, 13	63K
	$2.5 \cdot 10^{-4}$	recalculated [1])	
	$2.4 \cdot 10^{-4}$		62T3
Sb	$1.5 \cdot 10^{-4}$	Figs. 12, 13	63K
	$1.1 \cdot 10^{-4}$	recalculated [1])	
	$1.4 \cdot 10^{-4}$		68S
Bi	$1.0 \cdot 10^{-4}$	Figs. 12, 13	81K
	$7.6 \cdot 10^{-5}$	recalculated [1])	
	Germanium		
Cu	$1.0 \cdot 10^{-5}$		80G
Ag	$8.3 \cdot 10^{-6}$		79G
Be	$1.4 \cdot 10^{-4}$		78G2
B	$3.0 \cdot 10^{-4}$		56B2
	$1.4 \cdot 10^{-4}$		78G1
Ga	$1.0 \cdot 10^{-4}$		63K
	$7.2 \cdot 10^{-5}$	recalculated [1])	
As	$1.3 \cdot 10^{-4}$		63K
	$8.9 \cdot 10^{-5}$	recalculated [1])	
Sb	$6.7 \cdot 10^{-5}$		63K
	$4.7 \cdot 10^{-5}$	recalculated [1])	

[1]) The reported literature values in [63K], [81K] and [82K] have been recalculated using the kinematic viscosity given in [69G] for Si: $\nu = 0.0035\ cm^2\ s^{-1}$ and for Ge: $\nu = 0.00135\ cm^2\ s^{-1}$.

6.1.1.5 Evaporation rate

The evaporation rate g permits the *calculation of dopant losses* during solidification by evaporation from the free surface of the Si or Ge melt: $V \cdot dc_l = A \cdot g \cdot c_l \cdot dt$, where c_l is the dopant concentration in the liquid, V is the volume, and A the free surface of the melt [55V, 58Z, 81K].

Table 5. Evaporation rate g of impurities in Si and Ge melts.

Impurity	g [$cm\ s^{-1}$]	Ref.
	Silicon	
Cu	$4.0 \cdot 10^{-5}$	67H
	$8.0 \cdot 10^{-5}$	56B1
Au	$1.5 \cdot 10^{-5}$	67H
B	$8.0 \cdot 10^{-6}$	56B1
Al	$1.6 \cdot 10^{-4}$	56B1
Ga	$2.0 \cdot 10^{-3}$	56B1
In	$8.0 \cdot 10^{-3}$	56B1
P	$1.6 \cdot 10^{-4}$	56B1
	$4.0 \cdot 10^{-4}$	61H1, 64K, 81K
	Silicon (continued)	
As	$8.0 \cdot 10^{-3}$	56B1
Sb	0.13	56B1
Mn	$4.0 \cdot 10^{-4}$	56B1
Fe	$4.0 \cdot 10^{-5}$	56B1
	$2.1 \cdot 10^{-5}$	67H
	Germanium	
P	$3.4 \cdot 10^{-5}$	61G2
As	$1.5 \cdot 10^{-4}$	61G2
Sb	$1.3 \cdot 10^{-3}$	61G2

Some of the values listed are experimentally determined [61H1, 67H, 81K]; others are calculated from the reciprocal evaporation coefficient E according to [56B1] using the relation: $g = 2\,E/d_l$, where d_l is the melt density. The evaporation of the dopant *influences* to a certain degree the *impurity distribution along a* grown *crystal.*

6.1.1.6 Tetrahedral covalent radius

In Fig. 10 the distribution coefficient k_0 in Si and Ge is plotted versus the tetrahedral covalent radius r_{tet} of the impurity elements.

Table 6. Tetrahedral covalent radius r_{tet} for impurities [45P].

Impurity	r_{tet} [Å]	Impurity	r_{tet} [Å]	Impurity	r_{tet} [Å]	Impurity	r_{tet} [Å]
H	0.03	B	0.88	Ge	1.22	Bi	1.46
Li	1.23	Al	1.26	Sn	1.40	O	0.66
Cu	1.35	Ga	1.26	Pb	1.46	S	1.04
Ag	1.53	In	1.44	N	0.70	Te	1.32
Au	1.50	Tl	1.47	P	1.10		
Zn	1.31	C	0.77	As	1.18		
Cd	1.48	Si	1.17	Sb	1.36		

6.1.1.7 Thermophysical properties

The calculation of *heat flow* and *temperature distribution* in growing Si and Ge crystals requires the knowledge of the *thermal conductivity* (Fig. 14), *specific heat* (Fig. 15), and *emissivity* as a function of the temperature at elevated values. The electrical resistivity versus temperature (see Fig. 16) is of interest to the crystal grower as far as zone melting by means of induction heating is considered. The *thermal expansion* coefficient is useful to estimate the shrinkage behaviour of silicon and germanium crystals during the cooling-down period.

Table 7. Thermal conductivity κ of high-purity Si and Ge crystals [64G].

T K	κ W cm^{-1}K^{-1}		T K	κ W cm^{-1}K^{-1}	
	Si	Ge		Si	Ge
20		15.0	900	0.356	0.177
30		10.5	1000	0.310	0.171
40		7.7	1100	0.280	0.169
50	26.0	5.9	1200	0.261	0.173
60	21.0	4.7	1210		0.173
70	17.0	3.7	1300	0.248	
80	13.9	3.1	1400	0.237	
90	11.4	2.55	1500	0.227	
100	9.5	2.25	1600	0.219	
125	6.0	1.66	1681	0.216	
150	4.20	1.30			
175	3.25	1.10			
200	2.66	0.95			
250	1.95	0.73			
300	1.56	0.60			
400	1.05	0.44			
500	0.80	0.338			
600	0.64	0.269			
700	0.52	0.219			
800	0.43	0.193			

Table 8. Specific heat C_p of high-purity Si and Ge.

T K	C_p Ws g^{-1} K^{-1}		
	Si		Ge
	[63S2]	[60K]	[66S1]
273	0.6909		
298.15			0.3220
373	0.7704		
400			0.3375
473	0.8248		
500			0.3433
573	0.8478		
600			0.3479
673	0.8646		
700			0.3517
773	0.8813		
800			0.3571
873	0.8981		
900			0.3651
973	0.9127		
1000			0.3747
1073	0.9274		
1100			0.3852
1173	0.9420		
1200		0.9835	0.3957
1210			0.3961 (s)
1210			0.3806 (l)
1273	0.9588		
1300		0.9948	0.3806
1373	0.9818		
1400		1.0057	0.3806
1500		1.0166	0.3806
1600		1.0274	
1690	(MP)	1.0367	
1700		1.0480	
1800		1.0572	
1900		1.0660	

Table 9. Spectral emissivity at $\lambda = 0.65$ μm of Si and Ge [56A].

T K	Spectral emissivity	
	Si	Ge
1000	0.64	0.56
1100	0.62	0.55
1200	0.60	0.53
1300	0.57	
1400	0.54	
1500	0.50	
1600	0.48	
1688	0.46	

Table 10. Normal total emissivity of Ge [64B1].

T K	Normal total emissivity	T K	Normal total emissivity
703	0.535	959	0.571
707	0.543	972	0.578
710	0.530	986	0.583
727	0.541	988	0.569
746	0.531	1014	0.566
755	0.542	1021	0.569
762	0.535	1030	0.587
768	0.537	1031	0.565
790	0.547	1055	0.578
810	0.554	1069	0.591
818	0.550	1073	0.566
833	0.558	1103	0.580
865	0.566	1107	0.566
871	0.570	1121	0.570
901	0.574	1129	0.570
905	0.566	1143	0.566
910	0.577	1163	0.562
929	0.577	1171	0.543
935	0.571	1191	0.548

Table 11. Thermal linear expansion α of Si [77T] and Ge [75T].

T K	α [% K^{-1}]	
	Si	Ge
20	−0.022	
25	−0.022	−0.094
50	−0.022	−0.095
75	−0.023	
100	−0.024	−0.089
150	−0.024	
200	−0.019	−0.050
250	−0.010	
293 RT	0.000	0.000
400	0.033	0.064
500	0.066	0.127
600	0.102	0.193
700	0.141	0.262
800	0.181	0.333
900	0.223	0.406
1000	0.266	0.481
1200	0.356	0.636
1400	0.448	
1600	0.540	

6.1.1.8 Properties of liquid Si and Ge

The following physical properties of liquid Si and Ge are of interest to the crystal grower.

Table 12. Physical properties of liquid Si and Ge.

Property	Si Data	Si Ref.	Ge Data	Ge Ref.	Unit
Melting point (MP)	1410	37H, 50B, 59G, 61H2	936	52G, 58G	°C
	1420	58B1, 61O	937	56S2, 62H, 57H2	°C
Boiling point (BP) at 1 bar	2477	50B, 59G	2830	62H	°C
	2480	61H2	2980	50B, 58G	°C
Heat of fusion	$50.7 \cdot 10^3$	57O, 62H	$33.9 \cdot 10^3$	52G, 58G	$Ws\,mol^{-1}$
Heat of vapourization at BP	$297 \cdot 10^3$	50B, 61H2, 59G	$284 \cdot 10^3$	50B, 58G	$Ws\,mol^{-1}$
Vapour pressure at MP	$4.8 \cdot 10^{-5}$	50B, 59G	$10^{-9} \cdots 10^{-10}$	50B, 57H2	bar
	$1.4 \cdot 10^{-3}$	57H2			
Density at MP, liquid	2.53	69G	5.51	69G	$g\,cm^{-3}$
solid	2.30	69G	5.26	69G	$g\,cm^{-3}$
Volume contraction on melting	10.0	69G	4.7	69G	%
Specific heat at MP, liquid	1.037	60K, 70T1	3.79	66S1, 70T1	$Ws\,g^{-1}K^{-1}$
solid	1.037	60K, 70T1	3.96	66S1, 70T1	$Ws\,g^{-1}K^{-1}$
			2.22	65G	$Ws\,g^{-1}K^{-1}$
Surface tension at MP	$7.2 \cdot 10^{-3}$	53K, 65R	$6.0 \cdot 10^{-3}$	53K	$N\,cm^{-1}$
Kinematic viscosity at MP	0.0035	69G	0.00135	69G	$cm^2\,s^{-1}$
	0.0106	63K	0.0055	63K	$cm^2\,s^{-1}$
Thermal conductivity at MP, liquid	0.67	66S2	0.712	64B2	$W\,cm^{-1}K^{-1}$
solid	0.31	66S2	0.243	64B2	$W\,cm^{-1}K^{-1}$
	0.216	64G	0.173	64G	$W\,cm^{-1}K^{-1}$
Electrical resistivity at MP, liquid	$0.081 \cdot 10^{-3}$	53M, 69G	$0.056 \cdot 10^{-3}$	61G1	Ω cm
			$0.067 \cdot 10^{-3}$	69G	Ω cm
solid	$2.35 \cdot 10^{-3}$	53M	$0.165 \cdot 10^{-3}$	61G1	Ω cm
	$1.7 \cdot 10^{-3}$	69G	0.77	69G	Ω cm
Peltier coefficient at solid/liquid interface	0.100	61O, 64D	0.085	61O	V
Spectral emissivity at 0.65 μm near MP, liquid	0.27	81S	–		
Spectral emissivity at 0.65 μm near MP, solid	0.46	56A, 81S	–		

References for 6.1.1

11K Koenigsberger, J., Weiss, J.: Ann. Phys. **35** (1911) 1.
35S Stöhr, H., Klemm, W.Z.: Z. Anorg. Allgem. Chem. **241** (1935) 305.
37H Hoffmann, F., Schulze, A.: Phys. Z. **38** (1937) 901.
45P Pauling, L.: The Nature of the Chemical Bond; Ithaca, N.Y.: Cornell Univ. Press **1945**.
49P Pearson, G.L., Bardeen, J.: Phys. Rev. **75** (1949) 865.
50B Brewer, L., in: The Chemistry and Metallurgy of Miscellaneous Materials; Thermodynamics, L.L. Quill (ed.), New York: McGraw-Hill **1950**.
52G Greiner, E.S.: J. Metals **4** (1952) 1044.
53B Burton, J.A., Kolb, E.D., Slichter, W.P., Struthers, J.D.: J. Chem. Phys. **21** (1953) 1991.
53H Hume-Rothery, W.J., Christian, J.W., Pearson, W.B.: Equilibrium Diagrams of Metallic Systems; New York: Reinhold Publishers **1953**.
53K Keck, P.H., van Horn, W.: Phys. Rev. **91** (1953) 512.
53M Mokrovskij, N.P., Regel, A.R.: Zh. Tekh. Fiz. **23** (1953) 779.
53T1 Thurmond, C.D.: J. Phys. Chem. **57** (1953) 827.
53T2 Thurmond, C.D., Struthers, J.D.: J. Phys. Chem. **57** (1953) 831.
54B Burton, J.A.: Physica **20** (1954) 845.
54D Dunlap, W.C.: Phys. Rev. **96** (1954) 50.
54T Tyler, W.W., Newman, R., Woodbury, H.H.: Phys. Rev. **96** (1954) 874.
54V Van der Maesen, F., Brenkman, J.A.: Philips Res. Rept. **9** (1954) 225.
55J Jillson, D.C., Sheckler, A.C.: Phys. Rev. **98** (1955) 229.
55V Van den Boomgaard, J.: Philips Res. Rept. **10** (1955) 319.
55W Woodbury, H.H., Tyler, W.W.: Phys. Rev. **100** (1955) 659.
56A Allen, F.G.: J. Appl. Phys. **101** (1956) 1676.
56B1 Bradshaw, S.E., Mlavsky, A.I.: J. Electron. **2** (1956) 134.
56B2 Bridgers, H.E., Kolb, E.D.: J. Chem. Phys. **25** (1956) 648.
56C Carlson, R.O.: Phys. Rev. **104** (1956) 937.
56I Ioffe, A.F.: Can. J. Phys. **34** (1956) 1342.
56K Kuprovski, B.B., Geld, P.V.: Fiz. Met. Metalloved. **3** (1956) 182.
56M Miller, R.C., Savage, A.: J. Appl. Phys. **27** (1956) 1430.
56P Papazian, H.A., Wolski, S.P.: J. Appl. Phys. **27** (1956) 1561.
56S1 Struthers, J.D.: J. Appl. Phys. **27** (1956) 1560.
56S2 Stull, D.R., Sinke, G.C.: Advances in Chemistry, Washington DC.: Am. Chem. Soc. **1956**.
56T1 Thurmond, C.D., Trumbore, F.A., Kowalchik, M.: J. Chem. Phys. **25** (1956) 799.
56T2 Theuerer, H.C.: Trans. A.I.M.E. **206** (1956) 1316.
56T3 Thurmond, C.D., Guldner, W.G., Beach, A.L.: J. Electrochem. Soc. **103** (1956) 603.
56T4 Tyler, W.W., Woodbury, H.H.: Phys. Rev. **102** (1956) 647.
56W Woodbury, H.H., Tyler, W.W.: Bull. Am. Phys. Soc. **1** (1956) 127.
57B Bugay, A.A., Kossenko, W.W., Miseliuk, E.G.: Zh. Tekh. Fiz. **27** (1957) 210; Sov. Phys. Tech. Phys. (English Transl.) **2** (1957) 183.
57C1 Carlson, R.O.: Phys. Rev. **108** (1957) 1390.
57C2 Collins, C.B., Carlson, R.O.: Phys. Rev. **108** (1957) 1409.
57C3 Collins, C.B., Carlson, R.O., Gallagher, C.J.: Phys. Rev. **105** (1957) 1168.
57F1 Fuller, C.S., Morin, F.J.: Phys. Rev. **105** (1957) 379.
57F2 Fuller, C.S., Reiss, H.: J. Chem. Phys. **27** (1957) 318.
57G Gudmundsen, R.A., Maserjian jr., J.: J. Appl. Phys. **28** (1957) 1308.
57H1 Hall, R.N.: J. Phys. Chem. Solids **3** (1957) 63.
57H2 Honig, R.E.: RCA Rev. **18** (1957) 195.
57N Navon, D., Chernyshov, V.: J. Appl. Phys. **28** (1957) 823.
57O Ollete, M.: C.R. Acad. Sci. Paris **244** (1957) 1033.
57P1 Pell, E.M.: J. Phys. Chem. Solids **3** (1957) 77.
57P2 Pell, E.M.: J. Phys. Chem. Solids **3** (1957) 74.
57T Tyler, W.W., Woodbury, H.H.: Bull. Am. Phys. Soc. **2** (1957) 135.
57W Woodbury, H.H., Tyler, W.W.: Phys. Rev. **105** (1957) 84.
58B1 Billig, E., Gasson, D.B.: J. Sci. Instr. **35** (1958) 360.
58B2 Boltaks, B.I., Sozinov, I.I.: Zh. Tekh. Fiz. **28** (1958) 679; Sov. Phys. Techn. Phys. (English Transl.) **3** (1958) 636.

58G GMELINS Handbuch der anorganischen Chemie, 8th ed., Syst. No. 45, Germanium (supplement Vol.); Weinheim/Bergstraße: Verlag Chemie **1958**, p. 89.
58J John, H.F.: J. Electrochem. Soc. **105** (1958) 741.
58L Leverton, W.F.: J. Appl. Phys. **29** (1958) 1241.
58M Mortimer, G.: J. Electrochem. Soc. **105** (1958) 731.
58R Reiss, H., Fuller, C.S.: J. Phys. Chem. Solids **4** (1958) 58.
58W Wolfstirn, K.B., Fuller, C.S.: J. Phys. Chem. Solids **7** (1958) 141.
58Z Ziegler, G.: Z. Metallkunde **49** (1958) 491.
59A Abeles, B.: J. Phys. Chem. Solids **8** (1959) 340.
59C Carlson, R.O., Hall, R.N., Pell, E.M.: J. Phys. Chem. Solids **8** (1959) 81.
59G GMELINS Handbuch der anorganischen Chemie, 8th ed., Syst. No. 15, Silicon, Part B; Weinheim/Bergstraße: Verlag Chemie **1959**.
59H Hall, R.N., in: Fortschritte der Hochfrequenztechnik; Frankfurt/Main: Akademische Verlagsgesellschaft **1959**, p. 128.
59K1 Kosenko, V.V.: Fiz. Tverd. Tela. **1** (1959) 1622; Sov. Phys. Solid State (English Transl.) **1** (1960) 1481.
59K2 Kettel, F.: J. Phys. Chem. Solids **10** (1959) 52.
59P Pankove, J.I.: Rev. Sci. Instr. **30** (1959) 495.
59R Rohan, J.J., Pickering, N.E., Kennedy, J.: J. Electrochem. Soc. **106** (1959) 705.
59T1 Trumbore, F.A., Isenberg, C.R., Porbansky, E.M.: J. Phys. Chem. Solids **9** (1959) 60.
59T2 Trumbore, F.A., Porbansky, E.M., Tartaglia, A.A.: J. Phys. Chem. Solids **11** (1959) 239.
59T3 Tyler, W.W.: J. Phys. Chem. Solids **8** (1959) 59.
60B Boltaks, B.I., Hsüeh, Shih-yin: Fiz. Tverd. Tela. **2** (1960) 2677; Sov. Phys. Solid State (English Transl.) **2** (1961) 2383.
60G Glasow, W.M., Lju, Tschen-Juan: Nachr. Akad. Wiss. UdSSR, Abt. techn. Wiss.-Metallurgie Brennstoffe Nr. 4 (1960) 150 (russ.).
60K Kantor, P.B., Kisel, O.M., Fomichev, E.M.: Ukrain. Fiz. Zh. **5** (1960) 358.
60L Lee, M.A.: Solid-State Electron. **1** (1960) 194.
60S1 Stuckes, A.D.: Phil. Mag. **5** (1960) 84.
60S2 Slack, G.A., Glassbrenner, C.J.: Phys. Rev. **120** (1960) 782.
60T1 Thurmond, C.D., Kowalchik, M.: Bell Syst. Tech. J. **39** (1960) 169.
60T2 Trumbore, F.A.: Bell Syst. Tech. J. **39** (1960) 205.
61G1 Glasov, V.M., Chizhevskaya, S.N.: Fiz. Tverd. Tela. **3** (1961) 2694; Sov. Phys. Solid State (English Transl.) **3** (1962) 1964.
61G2 Gould, J.R.: Trans. A.I.M.E. **221** (1961) 1154.
61H1 Hoffmann, A., in: Halbleiterprobleme, Vol. 6; Braunschweig: Vieweg **1961**, p. 152.
61H2 Hampel, C.A.: Rare Metals Handbook, 2nd ed.; New York: Reinhold **1961.**
61M Morris, R.G., Hust, J.G.: Phys. Rev. **124** (1961) 1426.
61O O'Connor, J.R.: J. Electrochem. Soc. **108** (1961) 713.
61S Spitzer, W.G., Trumbore, F.A., Logan, R.A.: J. Appl. Phys. **32** (1961) 1822.
62A1 Aalberts, J.H., Verheijke, M.L.: Appl. Phys. Lett. **1** (1962) 19.
62A2 Abrikosov, N.K., Glasov, V.M., Liu, Chen-Yüan: Zh. Neorg. Khim. **7** (1962) 831; J. Inorg. Chem. (English Transl.) **7** (1962) 429.
62A3 Abeles, B., Beers, D.S., Cody, G.D., Dismukes, J.P.: Phys. Rev. **125** (1962) 44.
62F Fischler, S.: J. Appl. Phys. **33** (1962) 1615.
62H Honig, R.E.: RCA Rev. **23** (1962) 567.
62I Ignatov, V.D., Kosenko, V.E.: Fiz. Tverd. Tela. **4** (1962) 1627; Sov. Phys. Solid State (English Transl.) **4** (1962) 1193.
62K Kosenko, V.E.: Fiz. Tverd. Tela. **4** (1962) 59; Sov. Phys. Solid State (English Transl.) **4** (1962) 42.
62S Syed, A.S.: Can. J. Phys. **40** (1962) 286.
62T1 Trumbore, F.A., Spitzer, W.G., Logan, R.A., Luke, C.L.: J. Electrochem. Soc. **109** (1962) 734.
62T2 Tagirov, V.I., Kuliev, A.A.: Fiz. Tverd. Tela. **4** (1962) 272; Sov. Phys. Solid State (English Transl.) **4** (1962) 196.
62T3 Turovskii, B.M.: Zh. Fiz. Khim. **36** (1962) 1815; Sov. J. Phys. Chem. (English Transl.) **36** (1962) 983.
63C Chapmann, P.W., Tufte, O.N., Zook, J.D., Long, D.: J. Appl. Phys. **34** (1963) 3291.
63K Kodera, H.: Jpn. J. Appl. Phys. **2** (1963) 212.

63M Morris, R.G., Martin, J.L.: J. Appl. Phys. **34** (1963) 2388.
63S1 Statz, H.: J. Phys. Chem. Solids **24** (1963) 699.
63S2 Shanks, H.R., Maycock, P.D., Sidles, P.H., Danielson, G.C.: Phys. Rev. **130** (1963) 1743.
64B1 Brekhovskikh, V.F.: Inzhen.-Fiz. Zhur. **7 (5)** (1964) 66.
64B2 Brice, J.C., Whiffin, P.A.: Solid-State Electron. **7** (1964) 183.
64D Dorner, J.: Inter. Z. Elektrowärme **22** (1964) 331.
64G Glassbrenner, C.J., Slack, G.A.: Phys. Rev. A **134** (1964) 1058.
64K Kadănka, J., Mička, J.: Internationale Halbleiterkonferenz, Prague, **1964**.
65G Gerlich, D., Abeles, B., Miller, R.E.: Appl. Phys. **36** (1965) 76.
65R Runyan, W.R.: Silicon Semiconductor Technology; New York: McGraw-Hill **1965**.
66P Pfann, W.G.: Zone Melting, 2nd ed.; New York: Wiley **1966**.
66S1 Sommelet, P., Orr, R.L.: J. Chem. Eng. Data **11** (1966) 64.
66S2 Shashkov, Y.M., Grishin, V.P.: Fiz. Tverd. Tela. **8** (1966) 567; Sov. Phys. Solid State (English Transl.) **8** (1966) 447.
67H Hadamovsky, H.-F.: Habilitationsschrift, Bergakademie Freiberg/Sa; Freiberg **1967**.
67S Schibli, E., Milnes, A.G.: Mat. Sci. Eng. **2** (1967) 173.
68H Hesse, J.: Z. Metallkunde **59** (1968) 499.
68S Shashkov, Y.M., Gurevich, V.M.: Zh. Fiz. Khim. **42** (1968) 2058; Russ. J. Phys. Chem. (English Transl.) **42** (1968) 1082.
69B Bailey, R.F., Mills, T.G., in: Semiconductor Silicon, E.L. Kern and R.R. Haberecht (eds.); New York: The Electrochemical Society **1969**, p. 481.
69G Glasov, V.M., Chizhevskaya, S.N., Glagoleva, N.N.: Liquid Semiconductors; New York: Plenum Press **1969**.
69S Shunk, F.A.: Constitution of Binary Alloys, Sec. Suppl.; New York: McGraw-Hill **1969**.
70B Bakhadyrkhanov, M.K., Boltaks, B.I., Kulikov, G.S., Pedyash, E.M.: Fiz. Tekh. Poluprovodn. **4** (1970) 873; Sov. Phys. Semicond. (English Transl.) **4** (1970) 739.
70E Edwards, W.D.: J. Electrochem. Soc. **117** (1970) 1062.
70G Glasov, V.M., Semskov, V.S.: Die Germanium- und Silizium-Zweistofflegierungen; Berlin: Deutscher Verlag der Wissenschaften **1970**.
70L Lebedev, A.A., Sultanov, N.A.: Fiz. Tekh. Poluprovodn. **4** (1970) 2208; Sov. Phys. Semicond. (English Transl.) **4** (1970) 1900.
70N Nozaki, T., Yatsurugi, Y., Akiyama, N.: J. Electrochem. Soc. **117** (1970) 1566.
70T1 Touloukian, Y.S., Buyco, E.H.: Thermophysical Properties of Matter, Vol. 5: SPECIFIC HEAT; New York: Plenum Press **1970**.
70T2 Touloukian, Y.S., De Witt, D.P.: Thermophysical Properties of Matter, Vol. 7: THERMAL RADIATIVE PROPERTIES; New York: Plenum Press **1970**.
71B Bean, A.R., Newman, R.C.: J. Phys. Chem. Solids **32** (1971) 1211.
72A Azimov, S.A., Yunusov, M.S., Tusunov, N.A., Sultanov, N.A.: Fiz. Tekh. Poluprovodn. **6** (1972) 1438; Sov. Phys. Semicond. (English Transl.) **6** (1972) 1252.
72B Bean, A.R., Newman, R.C.: J. Phys. Chem. Solids **33** (1972) 255.
72H Hadamovsky, H.-F.: Halbleiterwerkstoffe; Leipzig: VEB Deutscher Verlag für Grundstoffindustrie **1972**.
72P Potemkin, A.Y.: Izv. Akad. Nauk SSSR, Neorg. Mater **8** (1972) 1353; Inorg. Mater (English Transl.) **8** (1972) 1197.
73G Gnesin, G.G., Raichenko, A.I.: Poroshkovaya Metallurg. **12** (1973) 35; Sov. Powder Metallurg. (English Transl.) **12** (1973) 383.
73S Sirtl, E., in: Semiconductor Silicon, H.R. Huff and R.R. Burgess (eds.); Princeton: The Electrochemical Society, **1973**, p. 54.
73Y Yatsurugi, Y., Akiyama, N., Endo, Y.: J. Electrochem. Soc. **120** (1973) 975.
74N Nozaki, T., Yatsurugi, Y., Akiyama, N., Endo, Y.; Makide, Y.: J. Radioanal. Chem. **19** (1974) 109.
75G Goncharov, L.A., Leonov, P.A., Khorvat, A.M.: Izv. Akad. Nauk SSSR, Neorg. Mater. **11** (1975) 1169; Inorg. Mater. (English Transl.) **11** (1975) 997.
75T Touloukian, Y.S.; Kirby, R.K., Taylor, R.E., Desai, P.D.: Thermophysical Properties of Matter, Vol. 12: THERMAL EXPANSION, Metallic Elements and Alloys; New York: Plenum Press **1975**
76A Azimov, S.A., Umarov, B.V., Yunusov, M.S.: Fiz. Tekh. Poluprovodn. **10** (1976) 1418, Sov. Phys. Semicond. (English Transl.) **10** (1976) 842.

76H Hopkins, R.H., Davis, J.R., Rai-Choudhury, P. Blais, P.D., McHugh, J.P., Seidensticker, R.G., McCormick, J.R.: Fifth Quarterly Report, ERDA/JPL 954 331/77-1 (December **1976**).
77T Touloukian, Y.S., Kirby, R.K., Taylor, R.E., Lee, T.Y.R.: Thermophysical Properties of Matter, Vol. 13: THERMAL EXPANSION, Nonmetallic Solids; New York: Plenum Press **1977**
78E Evwaraye, A.: J. Electron. Mater. **7** (1978) 383.
78G1 Goncharov, L.A., Egorov, K.G., Kervalishvili, P.D., Leonov, P.A., Orlof, P.B., Khorvat, A.M.: Izv. Akad. Nauk SSSR, Neorg. Mater. **14** (1978) 985; Inorg. Mater (English Transl.) **14** (1978) 772.
78G2 Goncharov, L.A., Kervalishvili, P.D.: Izv. Akad. Nauk SSSR, Neorg. Mater. **14** (1978) 989; Inorg. Mater (English Transl.) **14** (1978) 775.
78V Vydyanath, H.R., Lorenzo, J.S., Kröger, F.A.: J. Appl. Phys. **49** (1978) 5928.
79G Goncharov, L.A., Kervalishvili, P.D.: Izv. Akad. Nauk SSSR, Neorg. Mater. **15** (1979) 2101; Inorg. Mater. (English Transl.) **15** (1979) 1653.
79V Vydyanath, H.R., Lorenzo, J.S., Kröger, F.A.: Infrared Physics **19** (1979) 93.
80G Goncharov, L.A., Kervalishvili, P.D., Ryabykina, L.V., Chudakov, V.S.: Izv. Akad. Nauk SSSR, Neorg. Mater. **16** (1980) 590; Inorg. Mater (English Transl.) **16** (1980) 394.
80W Weber, E., Riotte, H.G.: J. Appl. Phys. **51** (1980) 1484.
81K Keller, W., Mühlbauer, A.: Floating-Zone Silicon; New York: Marcel Dekker **1981**.
81S Sirtl, E.: private communication; unpublished data (1960).
82K Kolbesen, B.O., Mühlbauer, A.: Solid State Electron. **25** (1982) 759.

6.1.2 Crystal growth

6.1.2.1 Deposition of polycrystalline silicon

6.1.2.1.0 Introduction

Polycrystalline silicon of semiconductor quality, i.e. of highest purity, is needed as starting material for the production of silicon crystals by the crucible pulling or the float zone process. This so-called polysilicon is worldwide mainly manufactured by the Siemens-process. The feature of this process is that polycrystalline silicon is deposited from a gaseous mixture of a silicon compound, especially trichlorosilane, and hydrogen on silicon slimrods at temperatures exceeding 1000 °C. The advantage of this procedure is that both starting materials, i.e. trichlorosilane as well as hydrogen, can be produced at a very high purity, and that silicon deposition takes place on a silicon substrate. It is therefore impossible to introduce impurities into the silicon by wall reactions. Polysilicon is obtained in the form of compact rods and can be easily handled during the following crystal growth processes.

6.1.2.1.1 Deposition technology

Survey of techniques

The technology of the high-purity preparation of metals by chemical vapour phase deposition (CVD) from the metal halides was developed by van Arkel and de Boer [25V] and first applied to silicon by Hölbling [27H]. The characteristic feature of this technique is the decomposition of highly purified metal halides on wires of the same metal heated to high temperatures. The metal is deposited on the hot wire, when the metal vapour pressure is low enough. This process is one kind of metal preparation without a contact to hot walls and therefore avoiding impurities introduced by them. The resulting metal purity may be very high.

Such a CVD process was taken into consideration for the preparation of semiconductor-grade silicon, but proved to be impracticable because of the brittleness and poor electrical conductivity of silicon wires. Amberger therefore prepared pure silicon by decomposition of a silicon halide - hydrogen gaseous mixture on graphite and silicon rods heated by a glow discharge [56A]. Bischoff developed a process [54B], which proved to be a break-through for the semiconductor silicon. This process was lateron improved and completed to the so-called *Siemens-process* [56S], which is characterized by the decomposition of gaseous mixtures of trichlorosilane and hydrogen. Silicon tetrachloride or dichlorosilane instead of trichlorosilane is scarcely used. Silicon is deposited on silicon slimrods, which are connected by a silicon bridge and heated by a direct current. Slim-rods are prepared by a float zone process (slim-rod pulling) of the polysilicon rods produced [81S1].

Reactor design

A CVD-reactor used for the deposition of polysilicon rods is shown in Fig. 1. The reaction chamber is surrounded by a quartz bell jar made of fused or sintered quartz. The bell jar is gas-tight mounted on a water-cooled base plate. For high demands on silicon purity, a silver layer is used to protect the base plate against chemical attack of the reaction products. Silver electrodes in the base plate are connected to the silicon slim-rods (cores) by graphite fittings. The electric connection of two cores is made by a silicon core bridge. A gaseous mixture of trichlorosilane ($SiHCl_3$) containing a large fraction of hydrogen, which has to be decomposed on the heated cores, is blown through a gas nozzle into the reaction chamber. After decomposition, the wasted gas mixture leaves the reactor through the gas outlet. The reaction chamber is surrounded by a water-cooled metal hood, which absorbs radiation heat during the deposition process and ensures a safe processing.

The electric power supply for heating the silicon rods is designed for different tasks:

Because the semiconductor silicon exhibits a negative temperature coefficient for the resistivity, a constant temperature of the rods can only be maintained by regulating the current to a constant value rather than the voltage. Thyristor-controlled power supplies are in use.

The deposition process starts on thin silicon rods. A high-voltage and low-current bias is needed in the initial phase. At the end of the process, the rods have grown to a large diameter, so that a low-voltage and a high-current bias has to be used. It is common to install a power supply, which is adjusted in steps during the deposition process [73B3].

Undoped silicon exhibits a high resistivity at room temperature. In order to start the heating process of the cores, a very high voltage has to be applied. Normally, it is easier to preheat the cores by radiation heat to a temperature of more than 300 °C, where further heating is possible by a reduced voltage. Preheating elements are shown in Fig. 1 inside the metal hood of the reactor.

Decomposition process

During silicon deposition, a temperature between 1000 °C and 1200 °C is preferred. The composition of the gaseous mixture, which is fed into the reaction chamber, is expressed by the mole ratio (moles $SiHCl_3$/moles H_2) and is in the range between 0.05 and 0.5. The gas flow rate depends on the dimensions of the decomposition reactor. It is regulated to an optimum of silicon deposition rates. The radial growth of silicon may reach a growth rate of up to 1 mm/h. The basic chemical reaction is

$$SiHCl_3 + H_2 \rightleftarrows Si + 3\,HCl.$$

This reaction operates in both directions towards the equilibrium point, which is temperature-dependent. In practice, silicon yields of between 10 % and 30 %, respective to the employed amount of trichlorosilane, are obtained. Low mole ratios and long residence times are favourable for high silicon yields. Nevertheless, the loss of un-reacted trichlorosilane and especially hydrogen is very high because of the large hydrogen excess in the gaseous mixture. It is therefore important and usual for an economical process to have a recycling system for the waste gases leaving the decomposition process.

The typical construction of the deposition reactor as described is designed so that more than 90 % of the radiation energy from the heated rods is removed by the water-cooling system of the whole reactor. Replacement of the transparent quartz glas bell jar by an opaque sintered quartz bell jar reduces the heat loss only insignificantly. Thermal insulation outside the bell jar causes an unacceptably high temperature of the bell jar. The deposition process is therefore strongly energy-consuming and a large through-put at a high deposition rate is of greatest interest to save energy.

For a high through-put, one has to consider the resulting low silicon yield related to the trichlorosilane consumption and the additional silicon deposition on the bell jar. A silicon deposit destroys the quartz of the bell jar slowly, but lowers the radiation heat loss during the process.

The number of silicon rods, that are arranged in a deposition reactor, can be two or more, i.e. 4, 6, 8, or 12, if there is space enough. The space needed depends on the rod diameter produced. A deposition reactor of this type cannot be larger than the largest quartz bell jars that can be produced. Bell jars are not larger than 1200 mm in diameter and 2500 mm in length. This limitation may be ignored, if the quartz bell jar is replaced by a metal bell jar [78K2]. Metal bell jars lead to problems in maintaining the purity of the polysilicon.

The growth rate of silicon during the deposition process is small, so that the production of silicon rods is a very slow procedure. It is still possible and usual to produce polysilicon rods of more than 200 mm diameter. These rods are the starting material for the crucible pulling processes.

Purity requirements

The purity of polysilicon depends on the purity of the starting products trichlorosilane and hydrogen, and on the quality of the materials coming into contact with the hot reaction gases, i.e. silver, quartz, graphite, and teflon. The purification of the reaction chamber between the deposition processes has to be performed carefully. The content of boron, which is one of the most undesired impurities in polysilicon, is considerably reduced during the process itself compared to the content of boron halides in trichlorosilane [63S]. Similar reduction is not generally found for other impurities. The requirement for float-zone polysilicon is now to obtain a more than 500 Ω cm n-type material, which is compensated by boron only to a small extent. Electrical measurement for purity control is usually performed after a one-pass float-zone process in pure argon atmosphere. A survey for typical impurity contents of polysilicon is given under section 6.1.3.2.3. A 100 Ω cm polysilicon material is pure enough for the crucible pulling process, where additional impurities are introduced into the silicon, which cannot be eliminated afterwards. Lower resistivities are obtained by introducing boron or phosphorus halides into the gaseous mixture during the deposition process (polysilicon doping).

6.1.2.1.2 Processes connected to deposition

Purification of starting materials

The deposition process cannot be seen apart from the whole silicon technology. Fig. 2 gives a short and rough imagination of all the production steps beginning with quartz and carbon as natural products and ending with the semiconductor-grade silicon crystal. The first step, the arc furnace reduction of quartz yields metallurgical-grade silicon. This silicon is by far too impure for semiconductor purposes, because the silicon content is not more than 99.5% (see section 6.1.2.5). The silicon reduced from quartz therefore has to be purified by converting it to a silicon compound, e.g. trichlorosilane. This is a low-boiling fluid ($T_b = +32$ °C) and can be purified by a highly efficient fractionated distillation technique. Fig. 3 shows the vapour pressure of trichlorosilane compared to some other chlorides of impurity elements which have to be separated. The pure trichlorosilane then is converted back to pure silicon by the decomposition process. The chemical reaction for the trichlorosilane and the silicon preparation is used in both directions.

$$\mathrm{Si} + 3\,\mathrm{HCl} \underset{\text{deposition reactor}}{\overset{\text{fluid bed reactor}}{\rightleftarrows}} \mathrm{SiHCl_3} + \mathrm{H_2}.$$

The process temperature for the fluid bed reactor is low, about 300 °C, and exothermic in contrast to the conditions of the decomposition process.

The waste gases of the decomposition process are recycled for economical reasons. The main chemical products contained are hydrogen (H_2), hydrogen chloride (HCl), trichlorosilane ($SiHCl_3$) and silicon tetrachloride ($SiCl_4$). The silicon compounds are condensated at a low temperature (about −70 °C), HCl is rinsed, and hydrogen may be fed back to the deposition reactor after being dried. The mixture of the silicon compounds is distilled again to separate the $SiHCl_3$ and to recycle it. $SiCl_4$ is a by-product, which usually is employed for other chemical processes.

Slim rod pulling and doping

Among the crystal pulling techniques, the slim-rod pulling is important to obtain long silicon cores for the deposition process. The advantage of these cores is that they may be monocrystalline, so that the resistivity can be determined. Resistivity measurement is important not only for high-ohmic polysilicon, but also for predoped silicon which is prepared this way. The important doping elements are boron and phosphorus. Boron and phosphorus halides can be added to the gaseous mixture of trichlorosilane and hydrogen. They are decomposed during the deposition process of silicon to about the same extent as trichlorosilane. Slim-rods of 5 to 8 mm diameter are pulled from this doped material and, after resistivity measuring, are introduced to the silicon deposition process of high ohmic silicon. The final polysilicon resistivity is predetermined by the core resistivity and the polysilicon diameter [64H2].

Polysilicon for crucible pulling

Materials for crucible pulling need not contain monocrystalline silicon cores. It is possible to use polycrystalline cores having a square cross-section made by sawing a polysilicon rod into many slim-rods. Moreover, this polysilicon may be introduced to the crucible pulling process not only as rod pieces but also as small chips prepared from the larger rods.

6.1.2.1.3 Deposition conditions

Reactant gas composition

It is a very simplified assumption, that the deposition process only follows the mentioned basic chemical reaction. Many chemical reactions are involved because of the high temperature conditions inside the reaction chamber. Besides the main products trichlorosilane, hydrogen, hydrogen chloride and silicon (gaseous and solid) other silicon compounds are found like silicon tetrachloride ($SiCl_4$), silicon dichloride ($SiCl_2$), dichlorosilane (SiH_2Cl_2), monochlorosilane (SiH_3Cl) and silane (SiH_4). Fig. 4 shows the gas composition as a function of temperature under thermodynamic equilibrium conditions [74S2]. Silicon dichloride is stable only at high temperatures, but it is found as a polymeric compound ($[SiCl_2]_x$) on cooled parts of the deposition reactor, where it is formed by a fast cooling process.

Deposition yield

Silicon deposition cannot be performed very near to equilibrium conditions, because the silicon deposition rate is too low. Moreover, uniform gas temperature and composition throughout the reaction chamber is not achievable. However, conclusions of practical importance can be made by assuming equilibrium conditions. At deposition temperatures exceeding 1250 °C, formation of silicon dichloride increases, so that the yield of silicon deposited decreases with a rising temperature. The maximum silicon yield under equilibrium conditions is determined for different temperatures and gas composition by Fig. 5 [72H2]. These yields are not reached in practice, because high silicon deposition rates are preferred rather than high silicon yields. For high deposition rates, it is advantageous to have a large silicon surface, which increases the silicon yield.

The following parameters, which are partly interconnected are important for the silicon deposition process. They are nearly independent of the reactor dimension:

composition of gas mixture employed (mole ratio)
residence time of gas mixture inside the reactor (reactor volume/flow rate)
deposition temperature (silicon rod temperature)
silicon growth rate (deposition rate/silicon surface)
geometrical size and surface quality of the silicon rods produced are important for the float zone process
purity of the deposited silicon

Regarding the recycling processes of the waste gases, the composition of the gases which varies in a large range has to be taken into account additionally.

6.1.2.1.4 Alternative processes

Alternative Si compounds

Among silicon compounds trichlorosilane is preferred for the deposition process, because it can be easily prepared at high yields. Its boiling point (+32 °C) is low and advantageous for purification by distillation techniques. It may be handled without danger. Other compounds considered, the chemical vapour phase deposition reactions, and the boiling points are described as follows:

Table 1. Compounds used for polysilicon deposition, boiling temperature T_b, and chemical reaction.

Compound	T_b °C	Reaction	Ref.
Chlorosilane compounds			
$SiCl_4$	58	$SiCl_4 + 2H_2 \rightleftarrows Si + 4HCl$	
$SiHCl_3$	32	$SiHCl_3 + H_2 \rightleftarrows Si + 3HCl$	56S
SiH_2Cl_2	8	$SiH_2Cl_2 \rightleftarrows Si + 2HCl$	
SiH_3Cl	−30	$SiH_3Cl \rightleftarrows Si + HCl + H_2$	
SiH_4	−112	$SiH_4 \rightleftarrows Si + 2H_2$	75Y

continued

Table 1 (continued)

Compound	T_b °C	Reaction	Ref.
Other halide compounds			
$SiBr_4$	153	$SiBr_4 + 2H_2 \rightleftarrows Si + 4HBr$	
SiI_4	288	$SiI_4 \rightleftarrows Si + 2I_2$	
SiF_2		$2SiF_2 \rightleftarrows Si + SiF_4$	74W

$SiCl_4$ (silicon tetrachloride) may be employed for the deposition process similarly to $SiHCl_3$, however, the silicon yield which is achieved is lower, unless the hydrogen consumption is increased considerably. SiH_2Cl_2 (dichlorosilane) seems better compared to $SiHCl_3$. The silicon yield is high and the hydrogen consumption low, however, SiH_2Cl_2 is more difficult to prepare and to handle, the deposition process causes problems.

SiH_3Cl (monochlorosilane) cannot be seriously considered because of its difficult preparation and handling. SiH_4 (monosilane) is important for silicon epitaxy processes and is also applied to the polysilicon production by the Komatsu process [75Y]. The advantage of the SiH_4 deposition process is a lower deposition temperature (about 700···800 °C), which saves energy, and a high silicon yield. The production of SiH_4, however, is more expensive and far more dangerous than the $SiHCl_3$ production.

Silicon bromides and iodides also have to be considered for the deposition process. Both compounds, silicon tetrabromide and silicon tetraiodide, yield silicon deposition at temperatures below 1000 °C; similar deposition is also possible with hydrogen compounds like $SiHBr_3$ or $SiHI_3$, however, their high boiling points and the high cost of bromine and iodine are great disadvantages. Reduction of SiF_4 (silicon tetrafluoride), which is a gaseous compound, by hydrogen is possible only at very high temperatures and therefore cannot be utilized. An alternative process is possible by SiF_2 (silicon difluoride) [74W].

The problem is, that SiF_2 is a gaseous compound, which is only stable at high temperatures and therefore has to be purified under high temperature conditions. This requires a complicated technology. The deposition process itself is practiced at lower temperatures.

Alternative deposition substrates

Heating of silicon rods is complicated because of the low electric conductivity of silicon at room temperature. The silicon deposition process is occasionally performed on metal wires, especially on tantalum, or carbon. These substrates are a source of impurities for the deposited silicon. At least the metal wire has to be removed after the deposition process. Another silicon deposition often practiced is the deposition inside a heated quartz tube. This technique, however, is not economical for large production rates [49L].

It is a principal disadvantage of the common deposition technique that the reaction takes place at the relatively small silicon surface region. Another technique developed is to obtain spontaneous nucleation and following deposition of silicon in the vapour phase, if silane is blown into a heated reaction chamber [81L1]. In this case the deposition surface is enlarged remarkably. A gaseous mixture of a silicon compound and hydrogen can be decomposed on small silicon particles in a fluid bed reactor [76W1], [81I].

Replacement of hydrogen

Hydrogen as reducing agent for silicon compounds may be replaced by some metals:

$$SiCl_4 + 2\,Zn \longrightarrow Si + 2\,ZnCl_2$$
$$SiCl_4 + 4\,Na \longrightarrow Si + 4\,NaCl$$
$$SiF_4 + 4\,Na \longrightarrow Si + 4\,NaF$$

These chemical reactions are usually practiced in heated tubes or similar systems and yield at least two solids, which have to be separated. This complicates the silicon preparation and results in a silicon product containing additional impurities.

The first process, reduction by zinc, is known as the *DuPont process* [49L]. It was studied again for a production of low-cost silicon [76B], similarly to the sodium reduction of SiF_4 [79K1], and $SiCl_4$ [78R], processes, which are known for a long time.

6.1.2.2 Preparation and purification methods of Ge

Germanium is found at low concentrations all over the world in many ores, but rarely at high concentrations as for instance in the form of Germanite or Reniérite which are both copper sulfides containing large amounts of Ge, Fe, Zn, As etc., or in the form of Argyrodite which is a silver-germaniumsulfide. The germanium content of these minerals is in the range of 2 to 10% [58g].

The main production of germanium today follows other processes having germanium as a by-product, e.g. in the zinc production from zinc sulfide ores or in the coal combustion process which has a remarkable germanium content in the remaining ash. In both cases, the germanium content may exceed 0.1%.

Concentrated products containing sulfides have to be roasted first, to convert sulfide to sulfate. The GeO_2, that has been formed, is leached out by hydrochloric acid. Germanium is distilled from this solution as $GeCl_4$ and separated. In order to purify $GeCl_4$, several distillation steps are applied until impurities cannot be detected by emission spectroscopy. $GeCl_4$ then is hydrolized in purest water and the resulting GeO_2 deposit is dried. GeO_2 reacts with hydrogen in two steps at temperatures not higher than 700 °C in a graphite crucible yielding a germanium powder, which afterwards is melted to germanium ingots at 1000 °C [51P].

$$GeCl_4 + 2H_2O \rightleftarrows GeO_2 + 4HCl$$
$$GeO_2 + H_2 \rightarrow GeO + H_2O$$
$$GeO + H_2 \rightarrow Ge + H_2O$$

The impurity content of this purified germanium is now in the range of part per billion (ppb) and can be further reduced by the following horizontal float zone purification techniques.

Graphite of highest purity is employed as a crucible material for zone melting, because liquid germanium does not wet graphite and because the solubility of carbon in germanium is very low. Zone melting is performed in a nitrogen atmosphere by inductive heating systems. Several kilograms of germanium are purified by this technique in one step. After removing the impure end of the ingot, germanium of highest purity is yielded containing impurities of only $4 \cdot 10^{13}$ cm^{-3} which corresponds to a resistivity of 50 Ω cm [53F, 53P].

6.1.2.3 Czochralski growth of Si and Ge

6.1.2.3.0 Introduction

From the whole production of single crystal silicon, about 80% is produced by the Czochralski method. In the year 1980, this implied about 2000 metric tons of dislocation-free, workable Czochralski silicon single crystals. In contrast to silicon, germanium as a semiconductor material has lost very much of its importance.

6.1.2.3.1 Czochralski pulling technique

The pulling of mono- or polycrystals against gravity out of a melt which is held in a crucible is called after J. Czochralski [17C], who determined the crystallization velocity of metals by this method in 1916. The basic principle for the pulling of single crystals by the Czochralski method (CZ-pulling) is shown in Fig. 6. The method was first applied to the growth of Si- and Ge-single crystals by Teal, Little, and Bühler [50T1, 50T2]. Chips of polycrystalline or monocrystalline (remelt) Si are molten in a crucible under inert gas ambient or vacuum. After reaching the proper temperature in the melt, a slim seed crystal is dipped into the melt and slowly withdrawn upward under rotation, so that melt crystallizes at the seed. With the right pulling speed, crystal rotation, temperature, and crucible rotation, a single crystal of the wanted shape grows at the seed crystal. The orientation of the seed crystal and of the grown crystal is exactly parallel.

Many environmental conditions have to be controlled.

Growth techniques of Si and Ge are similar. Differences of Si and Ge are based on the chemical nature and on the level of the melting temperature. Liquid Si reacts more aggressively with nearly all materials than liquid Ge. The problems in the pulling of Si-single crystals are therefore substantially greater with respect to crucible materials, furnace construction and furnace atmosphere.

The pulling process of the Czochralski method is described in detail for the case of silicon and the *dislocation-free crystal pulling according to Dash* [59D].

Starting procedure

Chips of polycrystalline Si are molten in a silica crucible under inert gas, mostly argon, at atmospheric pressure or at reduced pressure, e.g. 10 to 30 mbar [75C2]. While this inert gas is permanently lead downwards through the pulling chamber it carries away the reaction and evaporation products which are produced at considerable quantities.

After the Si has molten completely, one has to wait for some time until the temperature of the melt has reached the value for crystal pulling and the whole furnace is in a thermal balance. In the ideal case, this means that the temperature does not fluctuate at any place in the furnace, that the temperature gradients stabilize, and that the heat-supply and heat-loss are in balance. At melting, especially for rapid melting, a substantially higher temperature gradient is needed between the heating element and the crucible than lateron during crystal pulling because the melt heat must be supplied in the shortest possible time whereas during crystal growth heat is released again.

In order to avoid that the temperature of the melt overshoots too much, the heating power is reduced before final melting to about the value which is needed for crystal pulling. The control of the heating power can be performed either empirically or by measuring the temperature at a suitable place. It is possible to test the proper temperature by dipping the seed into the melt by way of trial. When reaching the proper temperature the seed crystal, a slim monocrystal of mostly cylindrical shape, begins to melt. The temperature of the melt is then lowered slightly. After this initial phase the seed is withdrawn from the melt in such a manner that the seed grows without expansion of its diameter (Fig. 7).

Growth of dislocation-free crystals

For reasons which are discussed later, the Si-single crystals for todays applications have to be grown dislocation free. This aim can be reached even by a *dislocated seed crystal* by applying a pulling technique which was developed by Dash [59D]. This is important because during each dip, slip dislocations are produced in the seed crystal by the large temperature differences and by the surface tension phenomena at the transition to the melt.

Dislocations normally propagate into the growing crystal, particularly in crystals of large diameters. The propagation of dislocations is affected by cooling strains and faulty crystal growth. The strain which occurs as a result of the different cooling speeds in the inner and outer parts of the crystal is probably the main reason in the case of large crystals. Because of the high tensions and the high temperatures in the crystals, dislocations are able to spread into adjacent slip planes by cross slip, by multiplication processes, and by diffusion-controlled climbing.

For the usual crystal orientations, i.e. [111]-, [100]-directions, all {111}-lattice planes, and with these all the normal slip planes are oblique to the crystal axis with respect to the pulling direction. All dislocation motions which run only over one slip plane, are therefore conducted out of the crystal. For propagation in the growth direction, dislocations have to move downwards zigzag by using the four equivalent {111}-slip planes.

A propagation without slip and climbing is possible in the case of a screw dislocation which penetrates the freezing interface. A screw dislocation can be built up directly by adding of Si-atoms from the melt [51B]. However, after some time, the screw dislocation also arrives at the free crystal surface because its dislocation axis is [110] [58H2] and is therefore also oblique to the [100]- or [111]-crystal axis. Only for crystal pulling in the [110]-direction the screw dislocation can propagate parallel to the crystal axis, and thus grow infinitely far in principle. In this case, it is possible to suppress the screw dislocation by fast growing.

Dislocations which are always present or generated in the seed, may be eliminated according to Dash [59D] by the following procedure (*Dash-Technique*):

The crystal diameter is temporarily reduced during growth to about 2 to 4 mm and the growth velocity is raised to about 6 mm/min.

The latter velocity is especially necessary for critical orientations such as the [111]-orientation.

With these two procedures, the crystal, even a [110]-oriented crystal, will be dislocation-free after a few centimeters growth.

For small crystal diameters and rapid growth, the cooling strain is too low to drive dislocations sufficiently fast in growth direction or to cause cross slip and dislocation multiplication. Climbing of dislocations by diffusion processes is slower than the slip motion. The [110]-crystals can be made dislocation free as easily as the [100]-crystals. A growth velocity of 2 mm/min in the thin crystal neck is sufficient for conversion to dislocation free growth.

Widening of the crystal diameter

The dislocation free state of the crystal can be watched in the development of nodes and grooves on the crystal surface. When the dislocation free state is reached the cross section of the crystal is enlarged by reducing the pulling speed until the desired diameter of the crystal is reached. During widening a more or less flat or tapered-off cone grows.

Shortly before the desired diameter is reached the pulling velocity is raised again to a value for which the crystal grows at a constant diameter. Possible forms of the seed cone are shown in Fig. 8. Because of the seed rotation the crystal section grows nearly circular in most cases. Instead of varying the pulling velocity, it is possible to change the temperature of the melt. However, this procedure is too slow, particularly for large melt volumes of up to

50 kg Si. In spite of the large diameters which may lead to high cooling strains, the crystals stay dislocation-free, because it is difficult for a crystal to generate a first dislocation in a perfect lattice.

Normally, a flat seed cone is grown (Fig. 8). A flat seed cone is most economical because of a short growth time and a small material loss. For some crystal specifications, e.g. for highly antimony-doped crystals, it is very difficult, sometimes impossible to grow a flat seed cone. One is forced in these cases to grow a more or less tapered-off seed cone. The loss in material and time with tapered-off cones is high for large crystal diameters because ingots are shorter than those of smaller diameters. For example, a crystal of 154 mm diameter grown from a 50 kg-melt can be grown to a length of about 1 m. A crystal of 80 mm diameter from a 24 kg-melt may reach a length of about 2 m. A tapered-off seed cone like that in Fig. 8 consumes about 20% of the pulling time and about 6% of the material which is required for the cylindrical part of the 154 mm diameter crystal. In the case of the 80 mm ∅-crystal, the corresponding values are only 5% and 1.7%.

Advanced pulling stage

The pulling velocity during the growth of the cylindrical crystal part is generally not constant and has to be reduced towards the bottom end of the crystal. The reduction is mainly caused by the increasing heat radiation of the crucible wall due to the sinking melt level (Fig. 9). Because of the increasing heat loss in the crucible part which moves out of the thermal shield, the heating power must be raised. The heating power must be further increased because the heat receiving area at the outside of the crucible below the thermal shield becomes smaller due to the lifting crucible. The maximum possible growth velocity is therefore high at the beginning and low at the end of the crystal pulling.

After transition from the seed cone to the cylindrical part, the heat power has to be reduced for a period of time due to the large solidification heat which is generated and cannot be fully eradiated by the short crystal.

Final pulling stage

In order to complete the crystal growth dislocation free, the crystal diameter has to be reduced gradually to a small size. An end cone like that in Fig. 8 is grown. The crystal can then be lifted out of the melt slowly.

If the crystal is lifted out of the melt in its cylindrical part, numerous slip dislocations spread into the crystal from its bottom end as it is shown schematically in Fig. 10 and in reality in section 6.1.3, Fig. 35. Dislocations are generated by thermal shocks, by surface tension phenomena, and by freezing melt drops at the bottom of the crystal. Once generated, dislocations spread into the crystal and multiply driven by the cooling strain in the crystal. Depending on the temperature distribution in the crystal, dislocations run more or less far upwards, normally as far as the crystal diameter.

Closing down of pulling

After finishing the bottom end cone, a residual melt is left in the crucible which is the more the larger the crystal diameter is. With skill, it is possible to pull out the melt completely by the crystal growth; however, this procedure is not profitable for several reasons: high impurity content of the residual melt, loss of time, deterioration of the crystal quality in its lower part by formation of precipitates.

The difficulties in the complete emptying of the crucible arise from surface tension effects which tend to tear off the melt from the crystal (Fig. 11).

Remelt of a crystal

An important aspect of Czochralski growth is the profitable possibility to remelt the crystal if it has lost its dislocation free structure or another important quality.

For remelting, the crystal is simply dipped into the melt and the heat power is raised so that the crystal remelts. There are two possibilities: the first possibility is to remelt the crystal totally and to restart the growth from the beginning. This is always done for short crystals. The second possibility is to remelt only the faulty part of the crystal and to start a new crystal growth on the residual melt by the Dash-technique. The restarting procedure includes diminishing the crystal diameter, forming a new neck and a new shoulder. The Dash-technique has to be applied again because for every remelting process slip dislocations propagate from the liquid-solid interface into the formerly dislocation free crystal.

The possibility for remelting is an essential part of the economy of CZ-crystal pulling. Remelting is not possible for float zone or pedestal pulling.

Size of apparatus for Czochralski growth

In the beginning of crucible pulling, Si-single crystals had weights of about 200 g and the crucibles had diameters of about 60 mm. Today Si-melts up to 50 kg and crucible diameters up to 350 mm are employed. The Table 2 presents typical data ranges of pulling apparatus' used for Czochalski crystal growing of silicon.

Table 2. Typical data for pulling apparatus' for Czochralski growth of Si.

Crucible diameter:	from 180 to 350 mm
heigth:	from 160 to 280 mm
capacity:	from 6 to 50 kg Si
crystal diameter:	from 50 to 155 mm
length:	from 500 to 2200 mm
weight:	from 5 to 48 kg Si
seed shaft travel:	up to 2500 mm
seed shaft rotation:	from ≈ 0 to 50 RPM (reversible)
seed shaft speed:	from ≈ 0 to 10 mm/min (slow),
	from 20 to 800 mm/min (fast)
crucible shaft travel:	up to 500 mm
crucible shaft rotation:	from ≈ 0 to 20 RPM (reversible)
crucible shaft speed:	from ≈ 0 to 1 mm/min (slow),
	from ≈ 0 to 200 mm/min (fast)
power control:	up to 150 kW
overall height:	up to 9 m
overall weight:	up to 9500 kg
pressure range:	10^{-5} to 2 bar
gas flow (argon):	0.4 to 3 m^3/h
vacuum pumps (mechanical):	up to 200 m^3/h

Table 3. Typical data for Czochralski growth of Si.

Parameter	Values	Unit	Remarks
crystal rotation	2 ···50	RPM	depending on crystal specification, melt and furnace assembly
pull rate	0.4··· 3.2	mm/min	
crucible rotation	0 ···20	RPM	
crucible lift	0 ··· 0.5	mm/min	
argon flow	0.4··· 1.2	m^3/h	pulling at 5···50 mbar
argon flow	2.0··· 4.0	m^3/h	pulling at 1 bar
power consumption	20 ···80	kW	
process times:			
loading, evacuation	0.2··· 0.6	h	
melting	2 ··· 4	h	
seed and cone	0.5··· 4	h	
cylindrical part	4 ···24	h	
end cone	0.2··· 2	h	
furnace cooling	2 ··· 3	h	
removal	0.2	h	
yield, dislocation-free	80 ···95	%	

6.1.2.3.2 Starting materials

Poly-silicon

The single crystals for the production of electronic devices must be of very high purity, particularly with respect to electrically active impurities. For this reason, the polycrystalline Si or Ge used for crystal pulling has to be of about the same purity as the single crystal. For example, the boron and phosphorus contents have to be less than 10^{13} atoms/cm^3 Si = 0.2 ppba.

Such a pure poly-Si is obtained by the trichlorosilane ($SiHCl_3$) process (see section 6.1.2.1). The following Table 4 presents a typical analysis of impurities in poly-Si for crystal pulling.

Table 4. Typical impurity content {ppba} in poly-silicon. The symbol "<" means below detection limit, which is given behind it. The elements were detected by repeated neutron activation analysis (NAA) for different samples at KWU, Erlangen, Germany, and at AERE, Harwell, Great Britain [82Z].

Ag	As	Au	Cd	Ce	Co	Cr	Cs	Cu
0.01	0.01	10^{-5}	<0.006	<0.006	0.002	<0.008	$<10^{-4}$	0.1
Fe	**Ga**	**Hf**	**Hg**	**In**	**K**	**Mn**	**Mo**	**Na**
<0.5	<0.02	$<2\cdot10^{-4}$	<0.002	<0.006	<0.1	<0.2	<0.004	<0.4
Ni	**Pt**	**Sb**	**Se**	**Sn**	**Ta**	**Ti**	**Zn**	**Zr**
0.4	<0.004	0.0008	<0.004	<0.1	$<4\cdot10^{-4}$	<0.5	0.1	<0.2

Poly-Ge

The purification of Ge in the course of its conversion to high purity polycrystalline Ge is more complicated than purification of Si. For Ge, the distillation in the form of $GeCl_4$ is an important purifying step. For high resistance (>20 Ω cm) crystal qualities, Ge is purified finally by zone melting (see section 6.1.2.2), sometimes with additions of Mg or Be to reduce the oxygen content which may reach values of up to $2\cdot10^{18}$ cm^{-3} [72H1].

Filling of crucibles

The poly-Si is crushed or sawn and then cleaned. It is filled into the crucible, under cleanroom conditions, it is piled up on the crucible to achieve a higher filling with melt (Fig. 12). In most cases, the dopant is filled into the crucible together with the poly-Si. In the case of volatile dopants, they are added directly to the melt.

For the recharging of poly-Si into the hot crucible which is empty or partly filled with melt, different chip or grain sizes are needed in dependence of the technique applied. The largest poly-Si pieces used are long rods like those shown in Fig. 22, the smallest grains have diameters of a few millimeters.

Doping of starting materials

As dopants for silicon, B, P, Sb, and As are mostly used in this sequence of importance, for Ge, the elements In, Ga, Sb, and As are predominant.

For high dopant concentrations, the doping element can be added simply to the poly-material or to the melt because their quantities can be still well handled.

In the case of medium or low dopant concentrations, this method is hardly or not at all practicable, because the element quantities are too small. Quantities are in the order of $10^{-3}\cdots10^{-6}$ g.

For a Si-crystal of 10 Ω cm p-type, a quantity of 0.1 mg boron must be weighed out if a 10 kg-melt is to be prepared. Such small quantities are too difficult to weigh out and to handle in the rough industrial practice.

For low doping, diluted alloys of Si or Ge are prepared which contain the dopant elements only in concentration ratios between 10^{-2} and 10^{-5}. In this form the dopant can easily be weighed out and handled.

Seed crystals

Seed crystals are normally elongated, cylindrical single crystals. The diameter used is between 5 and 20 mm and the length is between 100 and 300 mm. Seeds are grown dislocation free in special crystal pullers either by the pedestal pulling method (Fig. 13) or by the Czochralski method. The slim, long single crystals are carefully ground and their crystal orientation exactly pinpointed. The angle between the cylindrical axis and the crystal orientation must be close to zero.

The exact orientation of the seed crystal is not sufficient for an exact orientation of the grown crystal. The seed holder and the crystal shaft must guarantee a precise guide. The axis of the seed crystal must be moved exactly vertical to the melt surface. The misorientation allowed in CZ-crystals is less than 1°.

6.1.2.3.3 Crucible materials

Because of its high affinity to most elements, the high solubility of its melt to many elements, and its high melting point, liquid Si sets higher requirements to the crucible material and the growing ambient than Ge, which melts at about 500 degrees lower than Si.

Liquid Si reacts with *every* crucible material to a considerable amount. The question is with which crucible material it reacts sufficient slowly for crystal pulling. From the binary phase diagrams it may be seen that no metal or metal alloy is suitable because all metals are quickly dissolved by liquid Si in any ratio. The solubility limits in liquid and solid Si and the equilibrium distribution coefficients of C, N, and O are listed in Table 5 (see also sections 6.1.1.2 and 6.1.1.3).

Table 5. Equilibrium distribution coefficient k and solubility in liquid Si, c_l^m and solid Si, c_s^m [74N].

Impurity	k	c_l^m		c_s^m		Ref.
		10^{17} cm^{-3}	ppma	10^{17} cm^{-3}	ppma	
C	0.07 ± 0.01	40	81	3.5	7	74N
N	$7 \cdot 10^{-4}$	60	121	0.045	0.09	74N
O	1.4 ± 0.3	22	44	31	63	74N

Carbon compounds

The only elemental material which dissolves slowly in liquid Si is vitreous carbon [76C3]. Dense SiC, especially vitreous SiC is of nearly the same resistance. Other carbides cannot be used because the second component is usually solved in liquid Si in a high concentration thus deteriorating the melt for growth of crystals of semiconductor quality. In the case of TiC, for instance, too high a concentration of Ti is contained in the melt.

All carbon containing crucible materials have to be excluded because C cannot evaporate out of the melt like O in the form of SiO. Because of the reactions of the liquid Si with vitreous C or with SiC, the Si-melt soon reaches the solubility limit for C which is 81 ppma at the melting point [74N] and small SiC-crystals develop in the melt. Especially at the freezing interface of the crystal, SiC-crystals develop first because of the higher carbon content in the melt due to the segregation of C during freezing. SiC-particles terminate the dislocation-free crystal growth. C and SiC therefore cannot be used as crucible material.

Nitrogen compounds

The situation is similar for the nitrides [74N, 76C3]. Si_3N_4-crystallites develop in the melt and impede a dislocation-free crystal growth.

Oxides

In the case of oxides the conditions are more favourable than for C and N because the resulting silicon-oxygen compound SiO can easily evaporate out of the melt.

Fig. 14 gives a SiO-vapour pressure of about 12 mbar at 1420 °C [74K]. Growth conditions for which oxygen does not reach its solubility limit neither in the melt nor in the crystal near the freezing interface can easily be realized. The oxygen contents observed in normal CZ-Si-crystals is about 20 ppma. The values are far below the maximum solubility at the melting point $c_s^m = 63$ ppma [74N] (see Table 5).

Besides vitreous silica, no other oxide can be used as crucible material for silicon crystal pulling because other oxides contaminate the Si-melt with their compound metals, e.g. Al_2O_3 enriches the melt with Al.

Silica crucibles are always made of vitreous silica. Crystalline SiO_2 cannot be used because it cracks due to its phase transitions during cooling or heating. Vitreous silica has no fixed melting point and softens gradually with raising temperature. Near the melting temperature of Si, it is already so soft that the silica crucible must be supported by an outer crucible made of heat-resistant material. As the best material for this support crucible or susceptor graphite has been proven to be the best material. Graphite reacts with silica according to

$$2\langle SiO_2\rangle + 4\langle C\rangle \rightleftarrows \langle SiC\rangle + 3(CO) + (SiO)$$ [1].

The reaction is rather slow because both components are solids and the reaction forms an isolating gas layer between them.

[1]) ⟨ ⟩ indicate the solid and ( ) the gaseous state.

Crucible fabrication

There are several raw materials and several manufacturing methods for the silica crucibles. The purest raw material is produced synthetically from Si-compounds, for example by the reaction

$$(SiCl_4) + 2\,(H_2O) \rightleftarrows \langle SiO_2 \rangle + 4\,(HCl).$$

This synthetic silica is too expensive and, because of its high OH-content, too weak for large crucibles. A survey of crucible materials and their impurity content is given in the Table 6.

Table 6. Impurity analysis of synthetic and natural silica. The data (in [ppmw]) are given as in the literature cited or as quoted by various manufacturers. The marks "<" and "≦" mean below the detection limit as given and maximum measured value, respectively.

Source	Ag	Al	As	Au	B	Ca	Co	Cr
Synthetic								
Ref. [69D]		<0.25	<0.02	<0.1	0.1	<0.1	10^{-4}	0.03
Heraeus		0.1	0.03		<0.01	0.1		
TSL	<0.04	0.1	<0.001	<0.01	0.007	<0.7		<0.02
Natural								
Ref. [69D]		74			4	16		0.1
Ref. [69D]		68			3	0.4		
Heraeus		≦10	0.08	$3\cdot10^{-4}$	<0.1	≦3		
Quartz + Silice		21	<1		≦3	1.5	<1	<1
TSL	<0.04	17	<0.01	<0.03	0.22	3		<0.09
Toshiba		30			0.3	–		
GE		28			<1	3		3
US Fused Q.		12			0.06	1.3		
Corning		35	0.02		1.5		0.2	

Source	Cu	Fe	Ga	K	Li	Mg	Mn	Na
Synthetic								
Ref. [69D]	<1	<0.2	<0.02	0.1			<0.02	<0.1
Heraeus	0.004	0.2		<0.001	<0.05	0.1	<0.01	0.04
TSL	<0.001	<0.3		0.02	<0.004	<0.1	$<5\cdot10^{-4}$	0.04
Natural								
Ref. [69D]	1	7		6	7	4	1	9
Ref. [69D]	1	1.5		<1	1		0.2	5
Heraeus	0.07	0.8		0.8	2	0.2	0.01	1
Quartz + Silice	<1	0.3			1.4	0.2	<1	2.9
TSL	0.16	1		3	2	0.1	<0.09	3
Toshiba	0.5	2		2				2
GE	<1	4		3	0.5	1	2	3
US Fused Q.	0.04	1		1				1.1
Corning	0.2		0.1	6		1.8	0.17	1.5

Source	Ni	OH	P	Sb	Ti	U	Zn	Zr
Synthetic								
Ref. [69D]		1200	<0.001	0.1			<0.1	
Heraeus		1200	0.05	0.002	<0.1		<0.001	
TSL	<0.4	1200	<0.02	<0.001	<0.03	<0.01		<0.2

continued

Table 6 (continued)

Source	Ni	OH	P	Sb	Ti	U	Zn	Zr
Natural								
Ref. [69D]		60	0.01	0.3	3			3
Ref. [69D]		450	0.005	0.1	2	$6\cdot10^{-4}$		
Heraeus		130	0.1	0.15	0.8	$3\cdot10^{-4}$		<0.1
Quartz + Silice	<1			<1	0.8			
TSL	0.4	5	<0.05	<0.06	1.4	<0.04		<1.5
Toshiba								
GE					2			2
US Fused Q.			0.2		0.5			
Corning	0.5		0.4		<1.5		0.6	<0.2

The classical and mostly used raw material is natural quartz in the form of large and pure rock-crystals which are found in several deposits in the world, especially in Brazil. Crystals are cleaned and crushed to fine grit or powder.

In recent years the very expensive raw material rock-crystal has been substituted more and more by the cheaper "pure" quartz sands which can be found more frequently and in larger deposits. These sands are not so pure as the rock-crystals and have to be purified more or less expensively. Recycling of vitreous silica is also partly in use.

Transparent crucible

There are mainly two methods in use for the transformation of the grit into a crucible. The oldest and more expensive method is the melting of the quartz grit to a rod of vitreous silica by an arc. By several forming steps which all include processing by a very hot flame ($T > 1800\,°C$), the rod is transformed to a large tube and then closed to a crucible. The material obtained is completely transparent and free of bubbles. It is the most heat-resistant and chemically resistant type of all vitreous silica qualities.

Snowball crucible

The more recent, shorter and cheaper method to produce a crucible is to directly melt the quartz grit to a crucible in a rotating, water-cooled metal mould by an arc. In this case, the vitreous silica is not transparent because many bubbles develope during melting. Because of the incompletely molten exterior the crucible looks white. This crucible type is therefore called "*snowball-crucible*". Because of the incompletely molten exterior, this crucible type is less resistant against temperature shocks and mechanical stress. The snowball crucible is mostly used because its price is lower than that of the transparent crucible.

Slip-cast crucible

A third, more rarely used way to manufacture silica crucibles is to use a hot mould made of graphite. This crucible type is called "*slip-cast crucible*".

Crucibles for Ge

Much easier than for Si, with respect to crucible material are the conditions for Ge. In contrast to Si, crucibles of graphite can be used, because Ge practically does not react with graphite. The graphite, however, must be of very high purity.

For the system Ge-graphite, the following advantages result compared to the system Si-silica:

Graphite is more rigid at high temperatures and cheaper than vitreous silica;

Graphite has a much better thermal conductivity than silica. Heat is better transmitted from the heater to the melt;

Graphite crucibles need no support crucibles;

complicated and dimensionally accurate shapes can easily be realized;

the floating crucible method can be applied.

6.1.2.3.4 Heating and furnace equipment

Silicon growth

Due to the physical and chemical properties of Si only two heating methods are suitable for CZ-Si-pulling:

Resistance heating,

induction heating.

All other possibilities like electron beam melting, arc melting, melting by focussed light, etc. are excluded.

Resistance heating

The heating of the crucible and the susceptor by a graphite heating element according to the arrangement in Figs. 6 or 16 is simple and inexpensive. It leads to a smooth melt for crystal pulling. All modern crystal pullers are equipped with resistance heaters for the crucible and the susceptor.

Induction heating

In the beginning of Si-crystal crucible pulling, induction heating (Fig. 15) competed with resistance heating as long as the charge weights were not greater than about 1 kg. For higher weights induction heating is no longer in use.

The principle of construction for induction heating is shown in Fig. 15. In order to avoid sublimation of SiO on the water-cooled induction coil, it is mounted outside of the silica tube, which isolates the pulling space from the air. In this method, the silica crucible is held by a graphite crucible. Distances between the single winding induction coil and the graphite crucible must be kept narrow for good energy transmission. Because of the good electric conductivity of graphite, the susceptor absorbs a large fraction of the induction energy. At low frequencies, the induction field penetrates the graphite wall and heats directly the melt to a large extent. Direct heating of the melt leads to strong convection which may be disadvantageous for crystal growth. High frequencies are therefore applied so that only the graphite crucible is heated and the melt remains unaffected. Due to the high frequencies which have to be used, this heating technique becomes expensive and difficult to realize for large crucible loads where a high heating power is needed.

Furnace materials

In resistance-heated CZ-pullers the heating element, the crucible susceptor, and most other furnace parts are made by *graphite.*

For heat insulation, mostly *carbon felt*, *graphite foil*, and *carbon foam* are used. Above the melt, where no dust particles, carbon fibres or similar particles are acceptable because they may drop into the melt, *molybdenum* is often used for the construction of the thermal shield.

Insulation materials containing oxides cannot be employed for the thermal shield, because hot carbon reduces these oxides.

As a result of the reaction with the passing gaseous SiO, the heater and the susceptor are subjected to strong alterations. The hot graphite ⟨C⟩ reacts with gaseous (SiO) according to the reaction

$$2\langle C\rangle + (SiO) \rightleftarrows \langle SiC\rangle + (CO)$$

to form gaseous carbon monoxide and solid silicon carbide.

The porous graphite is gradually converted into SiC not only at its surface but also in its interior. Both its thermal and its electrical conductivity are modified by the change. Disadvantageous changes result in the temperature distribution around the crucible. The susceptor therefore has to be replaced frequently, the heater less frequently, and the other parts rarely.

The CO generated must be kept away from the Si-melt under all circumstances because it is instantly absorbed by the melt. A suitable guiding of the inert gas streams has to be applied so that no CO can flow back to the melt.

The lower part of the crucible shaft consists of a water-cooled steel shaft which can be rotated, lifted and lowered, and is sealed against air.

Germanium growth

For the furnace construction of Ge-crystal pullers far more materials can be used than in the case of Si. As graphite is a low-priced material with very advantageous properties, it is also used in the Ge growth as material for the heater and the furnace construction. CZ-crystal pullers for Si and Ge therefore look very similar.

6.1.2.3.5 Growth ambients

Ambients for Si-growth

Due to the reaction of the Si-melt with the silica crucible which causes a high SiO-vapour pressure, crystal growth in vacuum is unfavourable or hardly possible. SiO evaporates violently in vacuum with boiling of the melt. Si and the dopant elements (except B) also evaporate more or less violently. Antimony as dopant evaporates so rapidly that a reasonably low resistivity of the crystal cannot be reached. (See section 6.1.3.2.1 and Fig. 6 of section 6.1.3).

Ambient pressure

Pulling is generally performed in two pressure ranges with large throughputs of inert gas:
Near atmospheric pressure,
between 5 and 50 mbar.

Growing under atmospheric pressure results in low evaporation rates but also in high oxygen contents in the melt and the crystal, and in disturbing sublimation phenomena of SiO above the melt. Dislocation rates in the crystals are therefore higher than those in crystals grown at low pressure.

For growing under reduced pressure, low oxygen contents are achieved. Enrichment of Sb by segregation in the melt may be compensated by pressure control. This compensation is not possible for P and As, except at very low pressures. Sublimation of SiO takes place only at cold spots in the apparatus which are more distant from the melt.

Gas flow

In both pressure ranges, the evaporated substances and also the products of reactions, e.g. CO, must be carried away by the inert gas, usually argon, to the space below the crucible and from there out of the crystal puller. The velocity of flow of the gas must be high enough to avoid backdiffusion or backstreaming of the gaseous components. For growing at reduced pressures, only small quantities of inert gas have to be applied. The gas throughput is therefore less expensive than at high pressure.

For pulling at reduced pressure, the apparatus, however, is more expensive because vacuum systems including mechanical pumps, vacuum valves, instruments etc. have to be used.

The main design rule for the inert gas flow is to avoid the sublimation of SiO, in the case of high Sb-concentrations also the sublimation of Sb in the space above the melt, i.e. the upper part of the thermal shield, on the seed shaft and the holder, on the seed itself, on the crystal, and in the gas, and to avoid the backstreaming of CO to the melt. Deposits of SiO are usually not compact and particles may easily fall down or be carried away by the streaming gas into the melt, where they disturb the dislocation-free crystal growth to a large probability.

With correct guiding of the gas flow, slight deposits may grow near the crucible edge, on the upper side of the thermal shield, and on the lateral wall of the growth chamber. Strong deposits are formed only in the space below the crucible and the heater, where they do not disturb the crystal growth.

Ambients for Ge-growth

Besides some dopants, there are no substances having a high vapour pressure in the case of the pulling of Ge-single crystals out of graphite crucibles. The vapour pressure of Ge at its melting point which is between 10^{-6} and 10^{-7} mbar [57H], and that of graphite are extremely low. Ge-crystals can therefore be grown both in vacuum as well as under protecting gas (H_2, $Ar+H_2$, Ar, etc.). Because of the lack of a compound like SiO, there are less problems with the growth ambient than in the case of Si.

6.1.2.3.6 Automatic control

There are several possibilities to monitor the pulling process by a physical measuring method and to control the growth automatically on the base of the measured values. In the Figs. 17···20, several possibilities are shown. The optical and the infrared-optical observing techniques are prevailing. These methods measure the crystal diameter immediately at the freezing interface and can therefore respond rapidly to alterations in the growth of the crystal.

6.1.2.3.7 Recharging methods

The usual method in CZ-crystal pulling is the solitary charging and emptying of a crucible by growing one crystal. After completing the crystal growth, the furnace and the crucible must be cooled down to room temperature, the crucible with its frozen residual melt is removed, and another charged crucible is installed in the crystal puller. The furnace is heated again for a new run.

Cooling, flooding, crucible handling, evacuating, and reheating needs several hours of machine time, and in the case of Si also consumes an expensive silica crucible which is destroyed by the freezing residual melt. A method for recharging the hot crucible in the furnace is therefore profitable. The requirements for recharging can be fulfilled either by a continuous, semicontinuous, or discontinuous recharging of the crucible which is partly filled with melt or empty, by new poly-Si or poly-Ge, and dopant.

A continuous and also semi-continuous pulling may have the further advantage of nearly constant pulling conditions i.e. constant melt level, constant temperature distribution and constant dopant concentration, which leads to single crystal material having nearly constant parameters.

Segregation of impurities limits the recharging of the same melt. By continuous addition of new poly-Si or poly -Ge, which contains always a certain amount of impurities, the melt accumulates all those elements for which the distribution coefficient is smaller than 1.

In the case of Si, the silica crucible also supplies impurities, especially Al (see section 6.1.2.3.3) to the melt. The recharging must therefore be terminated after some time because of too high an impurity concentration in the melt.

The continuous pulling process seems also impossible because it requires a continuous removing of grown crystal. This is not imaginable in view of the brittleness and size of the crystals and of the susceptibility to problems in the crystal growth.

Only semi- and discontinuous recharging methods are under consideration. Semi-continuous means a temporary recharging of the crucible by solid or liquid Si or Ge during crystal growth (see Fig. 21). This procedure is only profitable for great seed shaft travels (>1,5 m), but has the advantage, that only a small amount of melt and a small crucible are needed. This recharging method still could not be successfully applied to the case of Si because of the enormous technical difficulties.

Refilling of Si, solid or liquid, during crystal pulling is a very critical action because the dislocation-free crystal growth immediately changes to dislocated crystal growth when slight mechanical vibrations occur in the melt and when crystalline particles of Si emerge in the melt at the freezing interface. Such particles are always formed when chips of Si are filled into a melt. The material problems and problems of temperature stability, solidification of Si at immersed recharging devices, still prevent the application of this recharging method. Figs. 21···23 show three possible recharging methods.

6.1.2.3.8 Czochralski pulling apparatus

Apparatus for silicon growth

Pulling apparatus for Czochralski silicon crystals have water-cooled walls of stainless steel and are all equipped with resistance heaters and furnace parts mainly made of graphite. In most cases, single-phase, in some cases also three-phase alternating-current is used for heating. If a three-phase alternating-current is used, a more or less strong stirring effect may occur in the silicon melt in dependence of the construction of the heater. All machines are built like vacuum systems with vacuum flanges, mechanical pumps, vacuummeters, gas distribution systems, and flowmeters. Many machines are equipped with a front opening chamber and a valve between the chamber and the furnace to permit the removal of the crystal from the hot furnace and the recharging of the crucible by silicon and dopant (Figs. 16 and 22).

The water-cooled seed and crucible shafts made of stainless steel can be exactly lifted and rotated in both directions by motor-driven precision mechanisms. For pressure or vacuum operation, the shafts are sealed to the air mostly by lip seals, both grease-types and dry-types are available. The whole setup is constructed in a manner to provide a minimum of vibrations during operation because the dislocation-free crystal growth depends strongly on vibration-free pulling.

Besides the pulling of the crystal by a shaft, rope or chain pulling is sometimes employed. A rope or a chain is wound by a pulling mechanism which can be rotated on the whole (Fig. 22). This mechanism for crystal pulling is cheaper and its construction is lower in height than for shaft pulling. The reduced construction height allows lower room heights and thus leads to better economy in building costs.

All modern pulling apparatuses are supplied with systems for automatic operation. In most cases, temperature control systems for the hot zone or the melt or both are built-in; in all cases an automatic diameter control system exists. Pulling apparatuses are commercially available which run completely automatically from the shutting of the furnace chamber after loading until the reopening after the cooling down of the furnace.

Apparatus for germanium growth

The pulling apparatuses for Czochralski germanium crystals have in principle the same construction as those used for silicon growth but they can be built simpler and thus cheaper. Ge-growth is simpler because the melting point of Ge is nearly 500 degrees lower than that of Si, the vapour pressure of Ge at its melting point is lower, and especially because there is no strongly evaporating substance like SiO as in the case of Si crucible pulling.

A special feature for Ge crucible pulling is given which cannot be realized with Si because of material problems. This is the *floating crucible system* [58L]. Its principle is shown in the Figs. 24···26. A small inner crucible floats on the melt of a large outer crucible. Their melts are connected by a small hole in the bottom of the inner crucible. If a crystal is pulled out of the inner crucible then melt from the outer crucible flows through the hole into the inner crucible thus preventing the enrichment or depletion of the inner melt with dopants of which the distribution coefficient k is smaller or greater, respectively, than one. By this method, the dopant concentration in the crystal may be kept nearly constant over a greater length than for normal pulling if the melt volumes and concentrations in both crucibles are designed in the right manner [65L, 72H1, 73B1].

The floating crucible technique is possible in the case of Ge because the graphite crucibles do not react with the Ge melt, keep their shape, are not wetted by the melt, and have a good thermal conductivity. In the case of Si the silica reacts with the melt, changes its shape, is partly wetted by the melt, and has a low thermal conductivity so that the floating crucible technique is not applicable.

6.1.2.3.9 Impurity incorporation

Normal freezing

For Czochralski growth, the law of normal freezing is valid. This law has to be applied if at the beginning of the freezing the whole substance is liquid and freezes step by step, for example by pulling out a crystal. After Pfann [52P, 57P], the axial distribution of impurities in the grown crystal is given by

$$\frac{c}{c_0} = k(1-s)^{k-1},$$

where c is the impurity concentration in the crystal at the freezing interface where the fraction s of the starting melt volume has been frozen, c_0 is the impurity concentration in the melt at the beginning, and k is the distribution coefficient.

In Fig. 27 axial impurity distributions in Czochralski grown crystals for several distribution coefficients are plotted.

Influence of mechanical and thermal parameters

The incorporation of impurities at the freezing interface strongly depends on the distribution coefficient k, which is a constant for every impurity only for very slow freezing in a stagnant melt. In this case, it is called equilibrium distribution coefficient k_0. For fast freezing velocities, k always tends towards 1 because in this case the impurity concentration in the melt cannot be distributed fast enough. In a melt layer at the freezing interface, the impurity concentration usually is higher for $k<1$ or lower for $k>1$, respectively, with respect to the residual melt. If the redistribution of the concentration at the freezing interface is only controlled by diffusion of the impurity in a stagnant melt, then the following equation of Burton, Prim, and Slichter [53B] is valid for k:

$$k = \frac{k_0}{k_0 + (1-k_0)\exp\left(-\frac{vd}{D}\right)}$$

with k = effective distribution coefficient,
k_0 = equilibrium distribution coefficient,
v = freezing velocity,
d = thickness of the diffusion-controlled layer in the melt in front of the freezing interface,
D = diffusion coefficient of the impurity in this layer.

However, during real crystal pulling the melt is always more or less strongly stirred by thermal convection, by crystal rotation, by crucible rotation, and sometimes also by gas streams above the melt, Figs. 28a–f. The melt flows destroy or diminish the diffusion layer in the melt at the freezing interface. The effective distribution coefficient k can no longer be calculated by the above equation. It has to be determined experimentally [78W].

Attempts were made to investigate and to calculate the flows in the crucible during crystal pulling [60G3, 65W, 67C, 68C, 71W, 72S1, 75C1, 75K, 76C1, 76W2, 77L1, 77L2, 77B2, 77S1, 77C, 77L3, 77B1, 78K1, 80K1, 80M]. Only simplified environmental conditions were used in these calculations. They therefore describe the reality only approximately. It is extremely difficult to calculate the melt flows exactly in a real Czochralski crucible for silicon crystal pulling because first the environmental conditions cannot be determined exactly and secondly an accurate calculation needs an immense expenditure.

Influence of vapour pressure

In the case of volatile impurities the axial and radial impurity distributions in Czochralski grown crystals are further influenced by the evaporation of the impurities from the melt during the pulling process. The evaporation mechanisms were investigated by Hoffmann [61H]. Bradshav and Mlavsky [56B] performed a thermodynamic analysis of the evaporation during Czochralski growth of Si. In this case, the real conditions are also more complicated than it is taken into account in the computational analysis.

6.1.2.3.10 Oxygen incorporation

During Czochralski growth liquid Si reacts with the silica of the crucible according to

$$\{\mathrm{Si}\} + \langle \mathrm{SiO_2} \rangle \rightleftarrows 2\,(\mathrm{SiO}).$$

It is not clear in which manner the oxygen is incorporated into the Si-melt, as solved $[\mathrm{SiO}]_{\mathrm{Si}}$ or as solved $[\mathrm{O}]_{\mathrm{Si}}$. However, it is known that after transport through the melt the oxygen evaporates easily and rapidly at the melt surface in form of silicon monoxide, even for high inert gas pressures.

As it is shown in Fig. 14, the vapour pressure of SiO at the melting temperature of Si is 12 mbar [74K]. It increases rapidly at higher temperatures.

In order to avoid the sublimation of SiO near the crystallization front, i.e. at the crystal surface or in the gas above the melt, a constant stream of argon is lead over the melt to carry away the gaseous SiO. The SiO partial pressure above the melt is therefore permanently kept low and the evaporation of SiO is little restrained.

According to the measurements of Chaney and Varker [76C2] the silica is dissolved by the Si melt at a velocity of 1.5 mg $SiO_2/h \cdot cm^2 = 6.9$ µm/h. This result is confirmed by the investigations in [82Z] which show that about a hundred times more oxygen is delivered into the melt by the silica during a crystal pulling process than what is found in the crystal eventually. About every 7 minutes the whole oxygen content of the melt is exchanged. This is estimated by consideration of the distribution coefficient of oxygen in silicon and of the solubility of oxygen in liquid silicon near the melting point.

According to Nozaki et al. [73Y, 74N] who investigated the binary phase system Si-O near the side of Si (Fig. 29), the equilibrium distribution coefficient of oxygen in silicon is $k_0 = 1.4$, the solubility of oxygen in the melt 44 ppma = $= 2.2 \cdot 10^{18}$ cm^{-3}, and in the solid 62 ppma $= 3.1 \cdot 10^{18}$ cm^{-3}, at the melting point, respectively.

The oxygen content in the silicon is very much dependent on the melt flow conditions. The highest oxygen content is expected when oxygen is directly transported from the crucible bottom to the growth interface e.g. for the rotation conditions of Fig. 28f.

For a low oxygen content in the crystal the following conditions in the melt should be met (Fig. 30):

An extra, isolated eddy current B should exist beneath the freezing interface which can evaporate SiO on the side of the crystal and which is fed with SiO by a melt of low oxygen contents or by a slow transfer mechanism (for example by the transversal diffusion of SiO in a laminar stream).

The SiO feeding eddy current A should evaporate a large fraction of its SiO content before it is in contact with current B. (Typical results are shown in section 6.1.3, Fig. 17.)

The oxygen content of the melt is further influenced by the thickness of the diffusion layer at the silica-melt interface which may be altered mechanically, by the temperature at this interface and by the impurity concentration in the melt and in the silica.

Sources of impurities in silicon

There are several possible sources of impurities in a CZ-silicon crystal. Each source may be the most important one at any time.

The first possible is the *polycrystalline Si.* Normally polysilicon is very pure because of the perfection of the processing technique.

The second possible source is the *crushing and handling* of the poly-Si until it is charged into the crucible. One has to take care that the poly-Si does not come into contact with critical materials, liquids and dust. Clean room working conditions have to be applied. The possibilities of pollution are very numerous and different everywhere. Because of this, it is not reasonable to discuss it in detail.

The third possible source is the *silica crucible* of which about 10 to 70 g are dissolved by the Si melt during crystal growth, depending on time, temperature, and the wetted silica surface. With this quantity of silica dissolved, also its impurities (see section 6.1.2.3.3) are incorporated into the melt. It is obvious that the impurity concentration in the melt caused by the silica is less at the beginning of the pulling process and increases continuously until its end.

The fourth possible source is the *pulling ambient.* A large variety of impurities may be introduced by the furnace parts which become hot during crystal pulling and may therefore evaporate impurities. In order to avoid the transfer of these evaporating impurities to the melt the inert gas streams have to be conducted in the right manner so that back-diffusion of impurities to the melt is impossible. The modern pulling apparatus and furnace constructions accomplish this requirement and the evaporation of impurities from the hot parts of the furnace is no problem. The purity of the pulling ambient is further influenced by the purity of the inert gas which, for example, may now and then contain gases like CO, CO_2, CH_3, etc. in ppm-amounts, thus influencing the carbon content of the melt (Fig. 31 a, b).

Cool parts of the apparatus in the space above the melt can also supply impurities if they are not clean or are defective. For example a leaking seal of the crystal shaft can emit oil into the melt.

Sources of impurities in germanium

In contrast to silicon there are less problems with contamination for crucible pulling of germanium crystals. Here, the crucible material is very pure graphite with which the germanium melt does not react to a measurable amount.

For germanium, the limiting factors for obtaining high resistivities are given by the purity of the starting polycrystalline Ge and by the cleanliness of the furnace and the handling.

6.1.2.4 Zone melting

6.1.2.4.0 Introduction

The zone-melting process was introduced in 1952 by W.G. Pfann [52P, 66P]. It is a combination of purification (zone refining) and crystal growing. The first important application was the purification of germanium for use in transistors [55P, 62P, 66P]. Besides its use for purification purposes of semiconductors like Ge and Si, zone melting is mainly used to grow silicon crystals by the float-zone technique on an industrial scale [75H1, 81D3, 81S1].

The principal arrangement of the zone-melting process is shown in Fig. 32. A charge placed in a horizontal boat has a liquid zone generated by a short heater surrounding the boat. The heater and thus the liquid zone is slowly moved from one end of the boat to the other. Melting occurs at the front end of the molten zone and freezing takes place at the trail end. The zone-refining effect takes place at the trailing liquid-solid interface. It is based on the higher solubility of impurities in the liquid. This corresponds to a distribution coefficient smaller than one, which is in effect for most impurities in Si and Ge. Impurities travel with the zone and become concentrated in the last part of the charge. The process may be repeated many times, thus improving the purity with the number of zone passes. If the final pass starts with a single-crystal seed, a single crystal grows down the boat. Such an arrangement works very well for Ge but not for Si, because all presently known crucible materials react with the silicon melt and increase its impurity level. Consequently, a crucible-free technique for Si growth was developed from the zone-refining method. This so-called floating-zone technique [52T, 53K, 54E], for which the freely floating zone suspended between the freezing and the melting interface is typical, permits the growth of high-purity Si crystals. Fig. 33 shows the basic setup of the floating-zone process.

After a single zone pass through a uniformly doped starting material, the solute concentration along the ingot is given by the relation

$$\frac{c_s}{c_0} = 1 - (1-k)\exp\left(-k\frac{x}{L}\right)$$

[66P, 60P], where c_0 is the starting concentration, k the distribution coefficient, L the zone length and x the distance along the ingot. Calculated values of c_s/c_0 are plotted in Fig. 34 for various values of k. The ultimate distribution after many passes is shown in Fig. 35. The smaller the distribution coefficient and the zone length the more effective is the zone refining [64S, 66P].

6.1.2.4.1 Silicon float-zone technology

Basic technology

The float-zone method is the only technique that allows the growth of silicon single crystals of high enough purity for the use in high-voltage and high-power devices [65H, 69S1, 72H1, 75H1]. The basic features of silicon float-zone growth are shown in Fig. 33. A comparison of various modifications of the basic technique is given in the following Table 7.

Table 7. Properties of different float-zone growth techniques.

Method	Maximum crystal diameter mm	Ratio crystal diameter / coil diameter	Radial dopant homogeneity	Ref.
Original float-zone technique	30	<1	good	75H1, 72H1, 81D3, 81K
Top-seeded method	35	<1	very good	81K, 81D3
Standing pedestal (e.g. slim pulling)	40		very good	81K, 81D3
Bottom-seeded needle-eye technique: concentric mode	>100	>1	good	59K, 81K, 81D3
Bottom-seeded needle-eye technique: eccentric mode	60	>1	very good	81K, 81D3

Induction heating

The floating zone is generated by means of a *watercooled induction coil* fed by radio frequency power in the megahertz range [81K]. Because semiconductor-grade silicon has only a small electrical conductivity at room temperature, it must be *preheated by radiation* heating. When the supply rod is hot enough, it may pick up energy directly from the electromagnetic field of the coil.

Growth conditions

Growth is usually started at the bottom end so that the silicon seed crystal can be introduced into the melt. Growth from top to bottom leads to better uniformity (*Top-Seeded Method*), but this technique is limited to smaller crystal diameters.

By means of relative motion between the coil and the crystal, the floating zone passes through the feed rod from the seed end and a single crystal grows under the floating zone. Both ends, particularly the seed end, are *rotated*, usually in *opposite directions*. This ensures growth of straight cylindrical single crystals. Growth may take place either in *inert gas*, generally argon is used, or in *vacuum*. The process may be repeated in order to increase the purity by the segregation effect, and additionally by *evaporation of volatile impurities* if vacuum is applied as an ambient. The diameter of the resulting crystal is given by the growth rate and the rate of material supply which is fed into the floating zone. The purification effect of float-zone melting for most impurities is considerable. Exceptions are boron and phosphorus, two important dopants.

For obtaining undoped silicon, it is essential to have these two elements in sufficiently low concentration in the starting material. Even with the zone-refining effect, a maximum purity is only achievable with very pure starting silicon. Because of the purity of polycrystalline material achieved today, the main emphasis of the float-zone process is given to single crystal growth. With the modern bottom-seeded *needle-eye method* [59K, 81K], characterized by the internal diameter of the induction coil which is less than the diameters of the feed rod and that of the growing crystal, single crystals having diameters of up to 100 mm can be grown on an industrial scale (Fig. 36).

Apparatus requirements

Modern float-zone machines must meet stringent specifications of rather different kind. They are listed in the Table 8.

Table 8. Apparatus design requirements.

Growth requirements	Machine specifications	Ref.
Smooth, vibration-free growth	High mechanical stability with smoothly running drives	
Vacuum operation	High vacuum-tight feed-through sleeves	
Large crystal diameter growth (stable floating zone)	Fixed coil zoner with precisely movable crystal and feed rod	
	Reliable high frequency (rf) current feeding connection	
	Frequency between 2 and 3 MHz	81K
	Rf power required between 4 kW (ca. 50 mm) and 11 kW (ca. 100 mm)	81K
Avoiding arcing in argon ambient	Low-inductance one-turn coils for currents well above 1000 A with inner diameters between 30 and 38 mm	81K
Large crystal weights (>4 kg)	Crystal supporting system	
Constant crystal diameter	Automatic diameter control unit	
Large crystal lengths (up to 1500 mm)	Overall height up to 9 meters (fixed-coil zoner)	81D3, 81K

Dislocation-free crystals

In order to ensure an appropriate crystal quality, the single crystal grown should be dislocation-free. According to the basic work by Dash [58D, 59D, 60D2] and its technical realization by Ziegler [60Z, 61Z] dislocations may be avoided by necking-in the used seed [58K2] for a length of 20 to 30 mm at its fusion point. This bottle-neck portion has a diameter of about 3 mm. Dislocation-free float-zone growth has been obtained under argon atmosphere with slight overpressure. Typical dislocation-free silicon single crystals obtained by float-zone growth are shown in Fig. 37.

6.1.2.4.2 Forces in a floating zone

The stability of the freely floating zone is an essential prerequisite for the float-zone process. The most important stabilizing force is the surface tension. There exists a maximum zone length, L_{max}, above which the liquid zone is no longer stable:

$$L_{max} = K(\gamma/d_l g)^{\frac{1}{2}}$$

where K is a dimensionless factor varying from 2.62 [64G, 73B2], 2.84 [56H, 73B2] to 3.18 [54K, 56H, 73B2]. γ is the surface tension, d_l is the density of molten silicon, and g the gravitational acceleration. The calculated maximum stable zone length lies between 14 and 17 mm, whereas experimental values observed are higher by a factor of almost two (Fig. 38). The stable zone length depends on the growth method applied. The reason for the discrepancy between calculated and experimental values may be the large electromagnetic supporting force which is neglected in the calculation. This force has the same order of magnitude as the surface tension force [67M, 81K]. Further stabilizing forces as well as destabilizing and stirring forces which simultaneously influence the floating zone are summarized in the Table 9 [81K].

Table 9. Survey on forces in a floating zone.

Character	Forces
Stabilizing forces:	Surface tension of melt Solid-liquid interface tension Electrodynamic forces Gasflow
Destabilizing forces:	Gravity Centrifugal forces by rotation Vibrations
Stirring forces:	Density gradients cause thermal flow Field gradients cause electro-dynamic flow Surface tension gradients cause Marangoni flow Centrifugal forces cause centrifugal flow

6.1.2.4.3 Doping of float-zone grown crystals

Silicon single crystals doped with group III or group V elements are usually required. The dopant concentration introduced must meet the desired concentrations as exactly as possible over the whole length of the crystal. In addition to this uniform axial distribution a homogeneous radial dopant distribution is desired. Doping methods which are usually applied in float-zone crystal growth are listed in the Table 10.

Table 10. Doping methods for float-zone Si crystals.

Method	Remarks	Ref.
Doping of polycrystalline Si		
Core doping	used for B and P	81K
Generation method (multistage core doping)	suitable for B and P, mainly used for P at low concentration levels	58H1, 81K
Doping during deposition	suitable for B, As, P at high concentrations with limited accuracy	60S, 81K
Doping during float-zone growth		
Pill doping	used for dopants having low k_0 (In, Ga, Sb etc.), good uniformity	66P, 81K
Gas doping	argon ambient for P	55V, 59G, 66H2, 74M2
	vacuum ambient for P	81K
Doping of single crystals		
Neutron transmutation doping (NTD)	P only, very good uniformity, see section 6.1.4.3	74S1, 75H2, 76H1, 81K

Doping of polycrystalline Si

Dopants can be introduced *during growth of the polycrystal* [60S] or by using a *doped core* upon which undoped polycrystalline silicon is deposited [58H1, 81K]. The latter method is particularly suited for *boron* doping where segregation ($k_0 = 0.8$) and evaporation effects may be neglected during float-zone growth. A good axial homogeneity generally may be expected. For *phosphorus* ($k_0 = 0.35$) the homogeneity near the seed end is worse, and for *antimony* ($k_0 = 0.023$) no homogeneity is reached over the actual growth length.

Very low dopant concentrations may be achieved by means of the *generation method* [58H1, 81K] (Fig. 39). The starting rod is produced by deposition doping. Then, by successive and repeated steps of slim pulling, poly-deposition and float-zone growth, the desired dopant level in the single crystal is obtained.

Doping during float-zone growth

For dopants having extremely small distribution coefficients such as indium, gallium or antimony, the *pill doping* [66P] is the most suitable method. In this technique, the doping pill is inserted into the floating zone shortly after seeding. Because of the small distribution coefficient of the dopant, the dopant content of the melt practically does not change during growth [81K]. This yields uniformly doped crystals.

The *gas doping* during single crystal growth is especially suited for dopants having high distribution coefficients such as phosphorus [55V, 59G, 66H2, 74M2, 81K]. Gaseous compounds of the dopant are blown on the melt during the float-zone process. The method may be adapted to growth under vacuum and in inert gas. For gas doping of phosphorus, resistivities ranging from 5···200 Ω cm can be achieved in vacuum at a reproducibility of better than ±10%.

Doping in the single crystal

The neutron transmutation doping of the undoped dislocation-free single crystal yields an excellent phosphorus homogeneity in as-grown crystals of any diameter [51L, 61T1, 74S1, 75H2, 76H1, 76J, 77H]. A doping precision of better than ±5% on macroscopic and microscopic scales and in axial and radial directions is state of the art. For details see section 6.1.4.3.

Radial doping uniformity

The radial dopant inhomogeneity in the growing crystal may have several causes, which are interlinked. Two of these causes are *convective flow* in the melt and *crystallographic orientation* of the growing interface. The most important growth parameters which influence the shape of the floating zone and the radial dopant distribution are given in the Table 11 [81K]. Doping profiles, microscopic inhomogeneities (striations) and defects are discussed together with characterization methods in section 6.1.3.

Table 11. Growth parameters influencing Si crystal properties.

Growth parameter	Growth situation	Impact on dopant distribution	Radial resistivity profile
Crystal orientation [111]			
Convex growing interface	Facet formation	major (−)	central dips
Concave growing interface	Secondary facets possible	minor (−)	small lateral dips
Seed rotation			
Low rotation rate (≈2 rpm)	Concave growing interface maintained	no	normal
Fast rotation rate	Both interfaces grow more and more convex, i.e., for [111] facet formation	major (−)	central dips
	Forced centrifugal flow acting on diffusion layer of growing interface	major (−)	central and peripheral dips
Eccentric rotation (offset seed axis)	Growing interface remains concave even for fast rotation rates (up to ca. 60 mm possible)	major (+)	reduced variations
	Large temperature variations at growing interface	major (−)	striations

continued

Table 11 (continued)

Growth parameter	Growth situation	Impact on dopant distribution	Radial resistivity profile
Induction coil			
Inner diameter large	Large external zone length		
	Convex interfaces, [111]	major (−)	central dips
	Reduced zone stability		
Inner diameter small	Stable zone, low external zone length; concave growing interface possible	minor (−)	no marked dips for low rotation rates
Growth rate	Influence on interface shape, will be enhanced by rotational effects	minor (−) to major (−)	no marked dips (low rotation) marked dips (fast rotation)
Crystal diameter			
up to 60 mm	As discussed above	as above	as above
above 60 mm	No systematic results available as yet, fast rotation rates not possible, eccentric operation not effective		
Dopant	Growth rate dependence of k different for group III and V elements	major (−)	striations
	Different diffusion behaviour too	major (−)	inhomogeneous

6.1.2.4.4 Redistribution of impurities

Impurities having an equilibrium distribution coefficient $k_0<1$ are constantly repelled into the melt during crystal growth.

Effective distribution coefficient

If the repelling rate is higher than the rate of impurity transportation back into the melt by diffusion or convection, a concentration pile-up occurs at the freezing interface (Fig. 40). The concentration profile and thickness δ of the diffusion layer depend to a certain extent on the growth rate and also on the fluid flow in the melt and the diffusion behaviour of the impurities [53B, 66P]. The ratio of the impurity concentration in the solid, c_s, and in the bulk liquid, c_1, are described by an effective distribution coefficient

$$k=c_s/c_1=k_0\cdot c_1'/c_1$$

where c_1' is the impurity concentration in the liquid at the freezing interface.

A quantitative analysis of segregation under non-equilibrium conditions [53B, 63K] results in the relation

$$k=\frac{k_0}{k_0+(1-k_0)\exp(-v_g\,\delta/D_1)}$$

where v_g is the growth velocity (assuming equal densities of solid and liquid), δ the diffusion layer thickness, and D_1 the diffusion coefficient of the impurity in the melt (see section 6.1.1.4). The validity of this formula has been experimentally proven [63K, 81K] and is widely accepted. The theoretical deviation of k from k_0 as a function of the growth velocity is shown in Fig. 41 for group III and group V impurities in silicon and in Fig. 42 for germanium. Figs. 43⋯49 present experimental data for the variation of k for the dopants B, Al, Ga, In, P, As, and Sb, respectively, in Si.

Diffusion layer

For a constant growth rate, the melt flow caused by thermal convection or crystal rotation influences the thickness δ of the diffusion layer at the growing interface (Fig. 50) [53B, 75C3, 81K]. According to the equation describing k, the dopant incorporation increases with increasing diffusion layer thickness δ. With the variation of the diffusion layer thickness across the interface, a radial impurity variation results (Fig. 50) [67C, 75C3, 81K].

The change of dopant concentration in the crystal resulting from steady-state changes in diffusion layer thickness is [82K]:

$$\frac{\Delta c_s}{c_s} = \frac{1}{k}\frac{dk}{d\delta}\Delta\delta = \frac{d(\ln k)}{d\delta}\Delta\delta.$$

Thus, the change of dopant concentration caused by the change of δ is proportional to the slope of the semi-logarithmic curves k versus δ (Fig. 51). The thickness of the diffusion layer is in the range of several hundred micrometers for technical growth conditions [63K, 81K].

A comparison of calculated and experimental values of the effective distribution coefficient and diffusion layer thickness in silicon is given for several dopants in the Tables 12 and 13. The diffusion layer thickness has been calculated according to [53B]:

$$\delta = 1.6\, D_l^{\frac{1}{3}}\, \nu^{\frac{1}{6}}\, \omega^{-\frac{1}{2}}$$

where D_l is the melt diffusion coefficient, ν the kinematic viscosity, and ω the angular velocity of the rotated crystal.

Table 12. Effective distribution coefficients k of dopants for float-zone growth at given growth velocity v_g and rotation rate $\omega/(2\pi/60)$

k_0 = equilibrium distribution coefficient
k_{calc} = calculated effective distribution coefficient
k_{exp} = experimentally observed effective distribution coefficient

Dopant	v_g mm min^{-1}	$\omega/(2\pi/60)$ rpm	k_0	k_{calc}	k_{exp}	Ref.
P	3	8	0.35	0.51	0.56	81K
	4	20	0.35	0.48		
Sb	4	20	0.023	0.074	0.061	
Ga	4	20	0.008	0.014	0.017	
In	4	20	0.0004	0.0006	(0.0047)	

Table 13. Diffusion layer thickness δ of dopants for float-zone growth at given growth velocity v_g and rotation rate $\omega/(2\pi/60)$

δ_{calc} = calculated diffusion layer thickness
δ_{exp} = diffusion layer thickness derived from the experimentally observed effective distribution coefficient

Dopant	v_g mm min^{-1}	$\omega/(2\pi/60)$ rpm	δ_{calc} µm	δ_{exp} µm	Ref.
P	3	8	655	877	81K
Sb	4	20	275	228	
Ga	4	20	406	549	
In	4	20	458	(2592)	

6.1.2.4.5 Faceting

The dopant incorporation in growing crystals is determined not only by the normal segregation effect but also by additional influences such as faceting. Especially dislocation-free Si and Ge crystals grown in the [111] direction may exhibit growth facets, which usually occur when the growing interface has a *convex curvature* towards the melt. Such facets are completely flat regions of the interface, which coincide with *(111) faces* of the silicon lattice. Fig. 52a–f illustrates the different growth mechanisms inside and outside the faceted areas. On a facet, a two-dimensional nucleus can form only after a certain critical amount of supercooling ΔT has been reached (Fig. 52b). When it is formed, rapid lateral layer growth begins forming a new (111)-plane terminating near the solidification isotherm T_m (Fig. 52c). Outside the facet, growth of the crystal can follow the moving isotherm much better than inside, because here many nucleation sites are always available, due to the presence of the rough surface [69C, 58J]. The critical supercooling temperature for two-dimensional nucleation lies between 5 °C and 9 °C for silicon [69C, 74A].

Dopant incorporation on facets

Dopant incorporation in faceted areas is controlled by an additional mechanism [61T2] besides the diffusion mechanism [53B], which dominates in the rough region of the interface. According to [61T2, 74M1], the increased dopant concentration observed in faceted regions can be explained by assuming that the dopant in the smooth interface is built-in into the lattice so rapidly by the fast-growing facet layers that no back-diffusion into the melt can take place. This results in an enhanced dopant concentration in the area of (111) facets. For crystal growth with a convex growing interface, *central dopant cores* may be otained (see Fig. 53, and section 6.1.3.3.1). In such cores, the high dopant concentration leads to resistivity dips [63D, 69C, 73M, 81K].

The coring effect may be explained in terms of an increased effective distribution coefficient k_f in faceted areas [60D1, 63M, 73B2]. The Table 14 gives the ratios of distribution coefficients k_f/k_0 found in Si and Ge crystals.

Table 14. Ratio of distribution coefficients k_f/k_0 on facets.

Material	Dopant	k_0	k_f/k_0	Ref.
Si	P	0.35	1.2 ···1.3	81K
			1.07···1.12	63M
	As	0.30	1.13···1.35	63M
	Sb	0.02	1.3 ···1.45	
	Ga	$8 \cdot 10^{-3}$	1.26	
	In	$4 \cdot 10^{-4}$	1.44	

Material	Dopant	k_0	k_f/k_0	Ref.
Ge	P	0.08	2.5	60D1
	As	0.02	1.8	
	Sb	$3 \cdot 10^{-3}$	1.45	
	Bi	$4.5 \cdot 10^{-5}$	1.65	
	Ga	$8.7 \cdot 10^{-2}$	0.85	
	In	$1 \cdot 10^{-3}$	1.4	
	Tl	$4 \cdot 10^{-5}$	1.2	

These ratios are closely connected to the individual growth situation of each crystal and do not necessarily reflect the influence of the dopant element considered.

Small (111) facets have also been found *on concave interfaces* in Si [73M, 74M1], and GaSb [77K1] crystals. These secondary facets are always located near the deepest region of the curved interface and show increased dopant concentration (Fig. 53). The dopant increase is less pronounced in secondary than in primary central facets.

6.1.2.4.6 Growth of high purity germanium

High purity (HP) germanium probably is the purest semiconductor; it is used as a base material for X- and γ-ray detectors (see section 6.1.4.0). An extremely low net doping concentration $|N_D - N_A|$ of around 10^{10} cm^{-3} (1 electrically active atom in $4 \cdot 10^{12}$ germanium atoms) is required for 1 cm thick field regions at ca. 600 V reverse bias voltage. The concentration of deep levels has to be as low as possible, to avoid charge trapping.

A first feasibility study to purify Germanium to netto impurity concentrations lower than 10^{12} cm^{-3} was prepared by R. N. Hall in 1966 [66H1]. The first crystals having net-impurity concentration between 10^{10} and 10^{11} cm^{-3} were grown in 1970 [70H]. The specialised technology required resulted in a few laboratories, which have succeeded in producing HP-Ge crystals (General Electric, Lawrence Berkeley Laboratories, Siemens Central Laboratories, Hoboken, Ortec). The fundamentals and surveys on Ge technology are given in [71H1, 71H3, 74H1, 78H3, 78G1].

Starting material

It is quite easy to purify Ge to the point, where it is intrinsic at room temperature. As starting material for high-purification, polycrystalline germanium is used which is routinely zone-refined in solid graphite boats by the Eagle-Pitcher process to a shallow impurity concentration below the intrinsic carrier concentration of $2.4 \cdot 10^{13}$ cm^{-3} at 20 °C. Analysis shows, that the material normally contains P, B, and Al at levels of $10^{11} \cdots 10^{13}$ cm^{-3} [81H] as electrically active impurities. The concentration of neutral impurities (e.g. Si, O, H) has not been investigated, since their concentration in the final crystal is determined by the crystal growing technique. Material suppliers are mainly Metallurgie Hoboken and Eagle-Pitcher.

Methods for high purification and growing single crystals

Two processes are commonly used to prepare HP-Ge for high-energetic radiation detector purposes:

a) Purification by horizontal zone refining of the polycrystalline material, followed by growing the single crystal by one Czochralsky process.

b) Subsequent crystal generation growth by the Czochralsky technique using the pure seed ends of previous generations thus achieving purification.

Fig. 54 shows the flow diagrams of the 2 processes.

Zone purification

Because of its high density 5.3 g cm^{-3} and small surface tension, $\sigma_{938\,°C}=600$ dyn cm^{-1}, Ge cannot be purified in diameters larger than 30 mm by the vertical crucible-free technique (floating zone process) [54H, 56H]. Purification is usually performed by the horizontal one-zone refining process (Fig. 55). Fig. 56a, b shows a schematic representation of the impurity distribution obtained after zone refining a), and of impurity profiles of crystals (I···III) b), grown from sections of the refined bar.

The segregation coefficients k_0 and k of impurities in Ge as given in Fig. 42 can only be applied to impurity concentrations down to the intrinsic level at 20 °C ($\approx 10^{13}$ cm^{-3}) [66P].

At lower impurity concentrations, interactions between Ge, the container, and the ambient in the refiner terminate the purification process, where – by formation of nonsegregating complexes – the effective segregation coefficient of a number of impurities (Al, Ga) may approach unity [81H]. The most important fundamental premises for a successful refining process are:

The smallest possible concentration of B and Al in the starting material;

Pure boat material with a separating layer of the purest, most impermeable material between the charge and the boat;

Avoidance of impurities from the quartz envelope, which can be taken up by the boat during the refining process;

Pure gas atmosphere.

Effect of Ge container and container coatings on purification

Ge wets quartz and will stick upon freezing. Solid graphite boats or quartz boats covered by thin graphite layers are in use.

The importance of the container for high purification has been demonstrated by Hall [74H1], Haller [76H2], Edwards [73E], Doremus [73D], and Hubbard [78H3]. The interaction of impurities in the Ge-crystal and the effect of the surrounding boat on the segregation of impurities and on the formation of mainly binary and ternary Si-, O-, B-, and Al-complexes suggest 4 methods for refining commercial 40 Ω cm Ge to residual impurity concentrations $|N_A - N_D| \leqq 2 \cdot 10^{10}$ cm^{-3}:

a) Refining in a solid graphite boat under low O- and Si-conditions, so that Al segregates, followed by refining by amorphous carbon smoke on the quartz boat.

Silicon and oxygen contaminations from the quartz boat through the porous carbon smoke getter boron by the formation of Si—O (B)-complexes.

b) Pyrolytic carbon/carbon smoke refining followed by refining in a Silica-coated quartz boat.

The technological difficulty of this process is, that molten germanium wets pyrolytic carbon-coated quartz and that solidified germanium adheres to the walls of the boat, which may crack.

c) Repeated silica smoke refining.

This process has the advantages of yielding large crystalites in polycrystalline ingots and of high purity, when single-step zone refining is not sufficient [78H3].

d) Pyrolytic carbon/silica smoke refining which combines the aluminum segregating environment of carbon refining with the boron gettering property of silica.

Glasow [77G2] succeeded by using quartz boats coated with suitable thin silica layers in a single-step purification.

Zone refining procedure

The composition of the atmosphere controls the equilibrium between impurity complexes and their dissociation products in liquid Ge [79D1]. Forming gas (90% N_2, 10% H_2) is normally used as the growth ambient. The melt zone is created by induction heating, which exhibits the following advantages over ohmic heating:

– The quartz tube in which the boat is contained remains substantially cooler than the melt. The danger of impurities caused by diffusion from the quartz envelope is therefore significantly reduced.

– The stirring within the liquid zone caused by RF heating has a positive effect upon the refining process.

The melt zone length should be as small as possible (~a few cm). The ingot to zone length ratio should be about 20:1. Typical zone refined bars have a trapezoidal or half circular, but also rectangular cross section. The speed of zone travel is in the order of 10 cm/h at 60 cm bar length as recommended by Pfann for best resistivity uniformity and crystal perfection.

Single-crystal growth

The Czochralski growth method is used.

The same conditions regarding container material as for zone purification are also relevant for the single crystal growth process. A highly purified hydrogen atmosphere is the only ambient which produces good detector material at high yield.

The principal conditions for the gas ambient are:

High purity.

No interaction with apparatus material which liberates impurities.

No complexing reactions with Ge defects e.g. vacancies, interstitials, or dislocations to generate electrically active centers.

Proper thermal enviroment.

Compatibility with heating by radio frequency.

The ambients, which have been tried, are vacuum, inert gases, nitrogen, and hydrogen [82H].

Vacuum-grown crystals have generally low purity. Trapping occurs in detectors made from such crystals.

In *inert gases* and *nitrogen grown* crystals have high oxygen-concentrations. Detectors made from such crystals show charge trapping. The residual oxygen shows smooth pits when crystals are preferentially etched.

Radiofrequency heating cannot be used with high-purity inert gases because of their low breakdown voltage. N_2 and H_2 have high breakdown voltages.

High purity *hydrogen* as protective gas, purified by passing through a palladium diffusion cell, leads to crystals which produce the best detectors. However, H_2 is soluble in Ge and forms complexes with various point defects – as for example the divacancy – hydrogen center (V_2H), causing trapping of holes and degradation of detector performance. The concentration of vacancies can be kept small if dislocations are present at a concentration $>100\ cm^{-2}$ at uniform distribution in the crystal cross section.

Because dislocations also produce charge trapping, the dislocation density in the crystal has to be below $10^4\ cm^{-2}$.

Growth apparatus

All other growth conditions and the process itself are described in section 6.1.2.3.1. The pulling apparatus is similar to that of Fig. 15.

A single crystal puller is shown in Fig. 57 [77G2]. The outer envelope with the graphite susceptor and the crucible is sealed by vacuum-tight, water-cooled metal flanges. The RF coil has conducting fins to create a uniform temperature distribution and is situated outside the envelope in a tank of deionized water which cools the parts of the container heated by the radiation from the graphite. In this manner, the transfer of impurities caused by outgassing or diffusion along the walls to the semiconductor melt is avoided. High-purity hydrogen, which was further purified by passing through a palladium diffusion cell is used as the protective gas. The pulling machine is subjected to a laminar air flow, so that loading of the system can take place under dust-free conditions.

Careful treatment of the zone-refined material, before it is loaded into the pulling apparatus, is of great importance for maintaining its purity. This applies also to the crucible. While it is sufficient to etch the germanium surface by the usual etching mixtures of superpure chemicals, new crucibles have to be repeatedly leached by pulling several preliminary loads of germanium. In order to enable the crucible to be used as often as possible, all the germanium must be removed at the end of each run.

Pulling speed and the rotation speed of the crystal and the crucible influence the crystal properties similarly to Si as discussed in section 6.1.2.3. The thermal symmetry caused by the rotation and the intensive stirring of the melt are advantageous for the radial uniformity of the doping and the effective distribution coefficient. The optimum growing parameters for single crystals having a diameter of approximately 40 mm are e.g.: Pulling speed 6.5···7 cm/h, seed rotation 50 turns per min, crucible rotation 5 turns per min. These conditions are held constant throughout the crystal growth process. Lower pulling speeds increased the density of dislocations which are also non-uniformly distributed. Among the crystal orientations [111], [100], and [113], the last proved to be the most suitable also for detector material at dislocation densities above $10^4\ cm^{-2}$ [79H].

Single crystal generation growing

Ge purification by growing succesive generations of Czochralski crystals, each time removing the sections of the crystal, where most of the impurities accumulate, is described in [71H1]. Fig. 58 shows the Czochralski furnace used by R.N. Hall [70H].

No separate crucible is used. The crucible made of synthetic quartz (spectrosil) is constructed on the lower part of the furnace envelope of fused quartz. The germanium is molten by thermal radiation from a susceptor consisting of a hollow cylinder and a disc of silicon, which are rf-heated. [100]-oriented seed crystals are mainly used. Crystal growth rates are typically 10 cm/h, the seed rotation rate 2 rps. Before loading, the charge and the crucible are etched in white etch (3 HNO_3:1 HF), rinsed in deionized water and finally in isopropanol. The atmosphere is high-purity hydrogen. The removal of all Ge at the end of the pulling is again required.

6.1.2.5 Unconventional Si crystallization techniques

Unconventional crystallization techniques are employed to produce silicon of *solar grade* quality in large quantities and at low cost [81D2]. The material may be monocrystalline or polycrystalline. Information on the quality ranges and their terminology is given in Table 15. Refining techniques of Si are discussed in section 6.1.2.1 and in [81D1].

Table 15. Quality ranges of Si [81D2].

Silicon category	Concentration of el. active impurities at. %
poly-Si hyperpure	$10^{-7} \cdots 10^{-8}$
CZ-monocrystal	$3 \cdot 10^{-6} \cdots 3 \cdot 10^{-7}$
TSG	$10^{-3} \cdots 10^{-5}$ (or less)
HP2	$5 \cdot 10^{-3} \cdots 10^{-4}$
HP1	$2 \cdot 10^{-1} \cdots 5 \cdot 10^{-3}$
MG	$2 \cdots 3 \cdot 10^{-1}$

HP1 = high purity, basic stage of refining
HP2 = high purity, advanced stage of refining
TSG = terrestrial solar grade, final stage of refining
MG = metallurgical grade

Mould-based processes, e.g. *directional solidification* and *casting techniques* are well suited to large scale-production of Si. In the early stages of semiconductor technology, various ways for making mould-based ingots have been tested. Three different methods are described by Runyan [65R].

Directional solidification or Bridgman-Stockbarger crystal growth is an effective and simple technique used e.g. to grow InP (see section 6.3.2.2.4). Directional solidification has become an economically attractive approach to large-scale production of Si ingots for terrestrial solar cell applications. Directional solidification may be combined with the refining procedure to purify metallurgical grade silicon [78H1, 78H2, 80K3].

Mould casting consists of pouring Si into a preheated or cooled mould and subsequent (directional) solidification. The casting methods can be practised batchwise or continuously. Mould casting is also well suited for large scale production.

Ambit casting – in this process, a cold plug is immersed into a silicon melt where a silicon disk immediately starts to be formed according to heat extraction conditions. This process shows some characteristics of liquid phase epitaxy LPE and is today applied in a modified form for sheet growth of Si.

The most common variations of casting and directional solidification techniques are listed in Table 16.

Table 16. Development in casting and related techniques.

Die material	Process	Cristallinity of ingot	Company	Ref.
quartz	one way mould	mono	Crystal systems	76S2
ceramics		poly	Union carbide	81D1
(quartz?)		poly	Solarex	79L, 81L2
graphite		poly	IBM	79C
graphite	reusable mould	poly	CGE	81F
graphite		poly	Wacker	78A, 80H1
graphite	continuous casting	poly	Wacker	81D1

6.1.2.5.1 Directional solidification

In directional solidification techniques, the high reactivity of molten Si by which almost every material is wetted, causes severe problems by contamination of the Si melt. Because the solidifying melt adheres to the container wall, the silicon ingot and the container usually are shattered by the different thermal contraction.

Container materials

Container materials used for directional solidification must exhibit a *high degree of inertness* against molten silicon because of the considerable residence time of the melt. Dissolution of the container material can hardly be completely suppressed. High-density graphite [79C, 81F], quartz ceramics [76S2], and mullite [77W2] can be used as container materials. Quartz and graphite are commercially available in high-purity grades so that only oxygen and carbon cause serious problems with respect to contamination. A SIMS analysis of impurities in the silicon ingots shows spectra which are characteristic for the container materials [80K2].

Cracking of the container during the cooling process may be reduced by adjustment of the thermal expansion coefficient [79C, 77W2]. Quartz crucibles may be designed mechanically unstable so that they fracture during solidification before the ingot starts to crack [81L2]. Cracking of the container may also be reduced by employing protection of the inner container wall by powders of Si_3N_4 [78A], SiO_2, Si, and graphite [65R].

Solidification

Two basic techniques may be differentiated for directional solidification [81D1]:

The container is moved within a fixed temperature profile causing growth of solidified silicon material from bottom to top [79C, 81F] (Fig. 59). Crystallization may be initiated either by seeding or spontaneous nucleation. Monocrystalline growth usually cannot be maintained throughout the total solidification process [79C].

The temperature profile is varied by reducing the heating energy while both the container and the heating system are kept at a fixed position [79C] (Fig. 60). In the *heat exchanger method* (HEM); the gradual reduction of the heating current is combined with a simultaneous localised heat extraction by a helium jet stream [76S2] (Fig. 61). The control of different stages of monocrystalline growth may additionally be facilitated by a seed monocrystal in direct contact with the gas-cooled heat exchanger [81D1].

Solidification rates range from 0.7⋯10 mm/min at cross-sections (50 × 50) and (300 × 300) mm^2 [79C]. The weight of ingots may reach 50 kg. Crystallization is obtained at low rates 0.5⋯6.0 kg/h depending on the cross-section.

Largely monocrystalline ingots may be grown by HEM. A typical feature of directionally solidified silicon is a rather coarse grain structure. A more or less columnar pattern depending on the crystallization conditions is observed along the main vector of solidification. Typical crystallite diameters range from 1⋯20 mm. Etch pit densities (EPD) reported are $10^2 \cdots 10^4$ cm^{-2}. They may reach values of 10^6 cm^{-2} [79C, 81F].

6.1.2.5.2 Mould casting

Casting is the filling of a melt into a mould or die by the aid of gravity, centrifugation or pressure and the subsequent solidification process within the container. For large scale production, reusable containers and continuous casting are advantageous. Containers mostly consist of graphite. Cold moulds reduce the reaction of the silicon melt with the mould. However, crack-free ingots are difficult to achieve. Preheating of the mould reduces the probability of cracking [78A, 80H1]. The Wacker process which is based on mould casting is described in detail.

Wacker process

The basic outline for the casting of multicrystalline ingots by the Wacker process is shown in Fig. 62. High-purity semiconductor-grade silicon or metallurgical-grade silicon is melted in a quartz crucible with a graphite support around it. The silicon melt is poured through a graphite-supported quartz funnel and fed into a preheated graphite mould. Rotating the mould ensures an even distribution of the heat which helps to develop a homogeneous chill layer of solid silicon before the silicon reacts with the mould walls.

The silicon melt can also be directly poured into the mould without using a funnel. A thermally stabilised funnel homogenises the silicon melt. The subsequent solidification process is carried out in a vertical temperature gradient which is adjusted by a water-cooled rotating shaft. For controlled heat extraction, the rotating shift is thermally insulated from the mould. In order to avoid melt inclusion which leads to a highly stressed zone in the ingot or to cracking, the formation of a solidified crust at the melt surface due to heat loss must be prevented by an appropriate temperature gradient. Temperature profiles inside the mould are shown in Fig. 63 for the different processing stages.

Among Si_3N_4, SiC, SiO_2, or high-melting ceramics, e.g. mullite, graphite has been proven to be the most suitable mould material because of the following properties [81D1]:

good heat conduction

small expansion coefficient close to that of Si so that cracking of silicon ingots during the cooling period is prevented

high temperature shock resistance

good workability; complex mould structures can be fabricated and easily be repaired for reuse

a protective reducing atmosphere is automatically established during casting and solidification; a gas bolster between the mould wall and the ingot balances the expansion differences between graphite and silicon and thus inhibits the formation of stressed zones.

Initial *nucleation* at the bottom and on the side walls of the mould is usually randomly oriented. A preferred orientation develops during solidification toward the heat flow vector [64C]. A columnar structure proceeds to grow when the same thermal gradient persists throughout the solidification process.

The ingot solidifies at a rate of 85 g/min. Cast ingots have a square-shaped cross-section of $(106 \times 106)\,mm^2$ or $(162 \times 162)\,mm^2$ and a weight of up to 8 kg.

Mould casting may be performed on a *semicontinuous* (Fig. 64) or *continuous* basis (Fig. 65). Additional components for semicontinuous production are a recharging chamber, a cooling chamber and a loading/unloading chamber for withdrawing the mould-containing solidified ingots and introducing an empty mould. In continuous casting, it is possible to keep the solidification front inside the mould. Continuous casting, therefore, is less sensitive to temperature fluctuations. The temperature gradients perpendicular to the growth direction are rather shallow and the crystallization front is quite planar. An increase in growth velocity $v = 10$ mm/min requires steeper temperature gradients which result in a more disturbed crystal structure [81D].

6.1.2.5.3 Sheet growth

Many efforts have been made to create and stabilize direct growth processes for silicon sheet to avoid material losses and expensive developments of high-efficiency slicing [80S1]. The growth of sheet crystals from the melt rests on the exact control of the crystallization front by control of the temperature field in the growth system and by control of the solid-liquid interface. Silicon sheets may be obtained without any shaping guide and by more or less complex interaction of the solidifying guide with the guiding system. Four categories of interactions can be named [81D1]:

very weak interactions of the guiding system and the melt to be crystallized exist when the solid material is converted into a sheet e.g. by zone melting. Such techniques are the ribbon-to-ribbon (RTR) [78G2, 80B3] process, the horizontal ribbon growth (HRG) [79K3, 81D1], and the low-angle silicon sheet (LSSA) [80J] growth.

weak interactions of the guiding system and the melt to be crystallised occur when the guiding system is applied only to a certain part of the molten source. Such techniques are the dendritic web growth (WEB) [80D3, 80D1, 80S3, 79D2], the edge-supported pulling (ESP) [66T, 80C2], the interface-controlled crystallization (ICC) [80D1] or the roller quenching methods.

strong interactions occur when the melt is forced through a corset-like guiding system e.g. as in the Stepanow method [69S2, 59S], the edge-defined film growth (EFG) [77R, 80K5], and the capillary action shaping technique (CAST) [72C].

very strong interactions occur when the guiding system remains in close contact after solidification of the melt e.g. as in silicon on ceramics (SOC) [78B, 80B2] in ribbon against drop (RAD) [78M, 80Z], and silicon coating by inverted meniscus (SCIM) [80H2, 80B3].

When the solid/liquid interface is oriented perpendicular to the crystal pulling direction, heat dissipation can only occur via the ribbon cross-section (RTR, EFG, CAST). The low heat dissipation limits the pulling rates to 10···100 mm/min. In the cases of a biased crystallization front (wedge type), heat extraction can be much more efficient thus providing increased pulling rates >100 mm/min [77B3, 80C1].

A survey of various sheet growth techniques is given in Table 17.

Table 17. Properties of sheet growth techniques.

Technique	Interaction	Geometry, process control	Remarks	Fig.	Ref.
RTR (ribbon-to-ribbon growth)	very weak	feed material stock and surface tension control; heating by laser scanning or electron beam	high-purity ribbons obtained low segregation	66	78G2, 80B1, 80B3
			Mo-contamination by feed deposition; after-heating required to relieve strain; dislocations $10^6 \cdots 10^7\,cm^{-2}$ microprecipitates $10^{13}\,cm^{-3}$.		80S2

continued

Table 17 (continued)

Technique	Interaction	Geometry, process control	Remarks	Fig.	Ref.
HRG (horizontal ribbon growth)	very weak	ribbon pulled off the melt; meniscus horizontally; uniform gas cooling	100···145 mm/min monocrystalline growth 850 mm/min polcrystalline growth	67	81D1, 79K3, 69B
			precise control of T and melt level required; approx. quality of CZ, profile hard to control		80S3, 80K3
LSSA (low angle silicon growth)	very weak	ribbon pulled off the melt meniscus at small (4°···10°) angle	properties similar to HRG	67	80J
ICC (interface controlled crystallization)	weak	shaping of one ribbon surface by cooled ramp protected by molten slag	coarse-structured multicrystalline foil growth rate 120 mm/min foil 80 × 50 × 0.4 mm	68	81D1
		cold ramp variation	continuous ribbon growth	69	81D1
roller quenching method	weak	molten silicon ejected through nozzle onto cold rotating cylinder	rapid high growth rates 10···40 m/s; ribbons 20···200 μm thick; 1···50 mm wide; small grain size, mechanical stress, uncontrolled contamination	70	80T, 79T
WEB (dendritic web growth)	weak	dendritic silicon needles replace heterogeneous filaments in ESP needles formed by appropriate seeding	web-like ribbon; favoured for low-cost growth; coarse multicrystalline structure 0.2 mm thick, 30···50 mm wide, 450 cm long	71	80D3, 80S3, 79D2
ESP (edge-supported pulling)	weak	edges of ribbon stabilized by two tungsten rods, grooved along axis; act as frame for liquid film by capillary action and wetting,	web-like ribbon, but coarse multicrystalline structure	71	66T
		pair of quartz or graphite filaments act as nucleation aids			80C2
EFG (edge-defined film-fed growth)	strong	ribbon shaped by a die	good growth stability; automated continuous growth; low-cost; 50···100 mm wide ribbons	72	77R, 80K5
		CVD-coated die	contamination risk because $k_{eff} \approx 1$ crystal quality limited		80D2
CAST (capillary action shaping technique)	strong	very similar to EFG shaping die		72	72C
Stepanov method	strong	external force to extrude melt through shaping die			69S2, 59S

continued

Table 17 (continued)

Technique	Interaction	Geometry, process control	Remarks	Fig.	Ref.
IS inverted Stepanov process	strong	melt extrusion through shaping die by gravity		73	77K2, 80K6
RAD (ribbon against drop)	very strong	graphite remains as substrate support, sticking of melt to substrate	thin polycrystalline films at good stabilities; batch process and continuous; limited material quality	74	78M, 80Z
SOC (silicon on ceramic)	very strong	spreading of liquid thin Si film on mullite substrate	preferred by oriented polycrystalline Si layer sticking to substrate, material quality limited by interaction with mullite		78B, 80B2
SCIM (silicon coating by inverted meniscus)	very strong	graphitized ceramic substrate moving over trough containing molten Si	improved modification of SOC	75	80H2, 80B3

6.1.2.6 Wafer preparation

6.1.2.6.1 Ingot preparation

Annealing

After growth Czochralski Si-crystals of high resistivity must be annealed at temperatures between 600 and 700 °C in air or nitrogen with subsequent rapid cooling. This process is necessary to destroy the oxygen donors which may be present in the as-grown CZ crystal at concentrations of up to 10^{15} cm$^{-3} \triangleq 5\ \Omega$ cm.

Testing

The crystals are preferentially etched e.g. by the Sirtl etch [61S1] to remove the oxide layer for electrical measurement and to reveal possible crystal defects like dislocations, twins, and grain boundaries. The seed cone and the bottom end cone, and occasionally faulty parts are cut away. From each end, a test slice is prepared for further crystal defect examination. The measurement of the axial and radial resistivity distribution and a control of the conductivity type are standard tests. According to the axial resistivity distribution, the crystal or parts of the crystal are assorted with respect to different resistivity specifications.

Grinding

For the manufacturing of electronic devices, thin and round silicon or germanium wafers having thicknesses from 150⋯700 µm and radii from 25⋯125 mm are needed. Because of tight diameter tolerances for the wafers, the ingot must also have the same diameter tolerance.

In spite of the high perfection and microprocessor-controlled automatization of the modern crystal pullers which deliver cylindrical single crystals with very good shape control and with tight diameter tolerance, there must be an ingot machining before sawing.

The crystal rods are either centerless or cylindrically ground, Fig. 76a, b. Each technique has its advantages and disadvantages; both are in use.

Orientation

After grinding, the crystal orientation is pinpointed for the following flat grinding. There are two methods in use – optical orientation, x-ray orientation.

For optical orientation, sections of the crystal are etched by KOH. By this etch, etch-pits with (111)-surfaces inclined to the cutting plane develop in the rough as-sawn surfaces. A focused laser beam perpendicularly pointed on the crystal section is mainly reflected by (111)-planes. Reflection patterns are observed in 4 directions for (100)-

sections, in 3 directions for (111)-sections, and in 2 directions for (110)-sections. Fig. 77a, b shows the patterns observed for (100)- and (111)-sections. Fig. 78 shows the apparatus arrangement of the measurement.

More exact is the orientation of Si crystals by the X-ray method. The arrangement for the X-ray orientation is shown in Fig. 79.

Flats

At the perimeter of the ingot, spots are marked where flats are to be ground. For standard wafers, a wide (110)-flat, the so-called "primary flat" is ground. It marks the direction for preferential breaking i.e. parallel or perpendicular to the flat. This direction is needed for wafer partitioning into chips in device processing. The width of the flat depends on the wafer diameter.

The orientation and conductivity type is marked in standard wafers by a second smaller flat (Fig. 80).

Ingot etching

For standard applications, ingots are etched down about 200 μm in diameter to remove the surface damage which was generated by the diameter and flat grinding. For this purpose typical polish etches are used, which are given in the Table 18.

Table 18. Etches for Si single crystals.

Name	Purpose	Composition, procedure	Ref.
	Facetting, cleaning, damage removing	KOH (or NaOH) in H_2O	
Kendall	Vertical groove etching	KOH (44%) in H_2O	79K2
	Chemical polishing, damage removing	HNO_3 (65%) : HF (49%) : CH_3OOH 3 : 2 : 2	
Dash	Revealing of crystal defects	HNO_3 (65%) : HF (49%) : CH_3OOH 3 : 1 : 10	56D
Sirtl	Revealing of crystal defects	HF (49%) : CrO_3 (5 M) 1 : 1	61S
Secco d'Arag.	Revealing of crystal defects	HF (49%) : $Na_2Cr_2O_7$(0.15 M) + ultrasonic agitation 2 : 1	72S2
Schimmel I	Revealing of crystal defects	HNO_3 (65%) : HF (49%) 1 : 155	76S1
Wright-Jenkins	Revealing of crystal defects	60 ml HF (49%) : 30 ml HNO_3 (69%) : 30 ml CrO_3 (5 M) : 2 g $Cu(NO_3)_2 \cdot 3H_2O$: 60 ml CH_3OOH : 60 ml H_2O	77W1
Seiter	Revealing of crystal defects at [100]	HF (49%) : CrO_3 (12 g CrO_3 + 10 ml H_2O) 1 : 9	77S2
Schimmel II	Revealing of crystal defects at [100]	HF (49%) : CrO_3 (1 M) (: H_2O heavy / 1.5 doping) 2 : 1	79S2

6.1.2.6.2 Slicing, lapping, and edge rounding

Slicing

Because of its unsurpassed sawing quality and material economy only the "*inside diameter slicing*" is used today. The essential part of such a machine is its sawblade which is a thin, round stainless steel sheet having an aperture in its centre. The edge of the aperture is coated by very small diamond particles. The blade is clamped on its outer periphery and then tensioned radially until it is rigid enough to be used for slicing. For sawing, the ingot is located in the aperture of the rotating blade and moved towards its edge, (Fig. 81). Because of the peripheral, clamping of the blade, the sawing loss (about 0.3 mm) can be reduced by using very thin foils. For special applications e.g. for growing epitaxial layers, the wafers are cut at well-defined off-orientations of some degrees from the crystal axis.

Lapping

For the tight thickness tolerances which are requested for certain discrete devices, and for the flatness requirements of modern mask printing techniques, the wafers are lapped after an intermediate cleaning process. The lapping machine consists of two plane-parallel rotating steel discs, between which the wafers are sandwiched. The abrasion of the wafer surface is obtained by a well defined lapping powder e.g. SiC or Al_2O_3 which is fed in a suspension between the rotating discs.

The depth of the surface damage obtained depends on the grain size of the lapping powder and therefore can be controlled. Normally the damage depth of lapped wafers is smaller than that of as-cut wafers, and does not exceed a depth of 25 μm. As-cut or lapped wafers are very often used in discrete device applications. For other electronic device applications, especially if photolithographic masking steps are involved, a smooth polished surface is needed.

After each abrasive processing the thickness of the wafers is measured.

Edge rounding

A very important step to increase the wafer yield in device processing is the edge rounding. Fig. 82 shows three types of possible edge roundings.

The advantages of edge rounding in device manufacturing are:
The rounded edge prevents the "epi-crown" in epitaxial deposition processes which is given by the enhanced deposition rates at the edge of the wafers, due to crystal orientation irregularities.
Edge-rounded wafers give smoother photoresistive films on the whole wafer surface. There is no more photoresist built-up at the edge.

The edge-rounding improves the mechanical strength of the wafer circumference and results in less edge chipping due to handling in the line and in reduced slippage during high-temperature process steps.

6.1.2.6.3 Etching, polishing, wafer finishing

Etching

After edge rounding and cleaning, most of the wafers are passed through an etching step to remove any mechanical damage. In this process special care has to be taken that the already good plane-parallel surfaces of the wafer are not damaged or distorted. Both acid etching and caustic etching are in use for this purpose.

Backside damage

Some device manufacturers apply a well-defined backside damage on the wafers to improve the yield in the line. A damage at the wafer backside which is restricted to a certain depth below the surface, generates a dislocation network or stacking faults near the surface during the first high-temperature process, normally an oxidation, without penetrating the wafer. These crystal defects getter fast diffusing elements during high temperature processing similar to intrinsic gettering. There are many methods in use to produce such a damage, some of them are described in the literature or in patents [62S, 64P, 65M, 67P, 75L, 76R, 80R].

Polishing

Polishing of Si-wafers basically consists of the combined actions of chemical and mechanical forces. Here "mechanical" does not mean any mechanical abrasion by hard materials like diamond paste or aluminum-oxide polishing as it was done in the past. Such hard materials cause damage in the polished surface and lead to stacking fault growth during oxidation.

Two different methods of polishing are under consideration in the wafer manufacturing industry: The wax-free or "free polishing" process and the process based on wax-mounted wafers. The latter method is mostly applied. Fig. 83 shows a schematical drawing of a polishing machine for wax-mounted wafers.

After the polishing process, the polished surface of the wafer has to be free of any residual mechanical damage which causes stacking faults in subsequent oxidations. A further very stringent requirement is set by the photolithographic process. The polishing surface has to be in the focal plane of the optical system which focusses the structure of the mask into the photoresist covering the surface of the wafer. Practical values of the flatness of wafers are below ± 1.5 microns.

Cleaning, packaging

Finally, the polished wafers are carefully cleaned in a multistep process, giving a wafer which is practically free of particles and residues. Before packaging each wafer is thoroughly inspected by many random tests. The wafers are vacuum-tight packed so that they are protected against contamination and damage.

References for 6.1.2

Handbooks

58g Gmelins Handbuch der Anorganischen Chemie, 8. Auflage: Germanium, Ergänzungsband, Verlag Chemie GmbH, Weinheim/Bergstr., **1958**.

Bibliography

17C Czochralski, J.: Z. Phys. Chemie **92** (1917) 219.
25V Van Arkel, A.E., de Boer, J.H.: Z. Anorg. Allg. Chem. **148** (1925) 345.
27H Hölbling, R.: Z. Angew. Chem. **40** (1927) 655.
49L Lyon, D.W., Olsen, C.M., Lewis, E.D.: J. Electrochem. Soc. **96** (1949) 359.
50T1 Teal, G.K., Little, J.B.: Phys. Rev. **77** (1950) 809.
50T2 Teal, G.K., Little, J.B.: Phys. Rev. **78** (1950) 647.
51B Burton, W.K., Cabrera, N., Frank, F.C.: Phil. Trans. Roy. Soc. **243A** (1951) 299.
51L Lark-Horowitz, K., in: "Proceedings of the Conference at the University Reading", London, Butterworth **1951**, 47.
51P Powell, A.R., Lever, F.M., Walpole, R.E.: J. Appl. Chem. **1** (1951) 541.
52P Pfann, W.G.: Trans. AIME **194** (1952) 747.
52T Theuerer, H.C.: USP 3, 060, 123, filed Dec. 17, **1952**, patented Oct. 23, 1962.
53B Burton, J.A., Prim, R.C., Slichter, W.P.: J. Chem. Phys. **21** (1953) 1987.
53F Fahnestock, J.D.: Electronics **26** (1953) 131.
53K Keck, P.H., Golay, J.E.: Phys. Rev. **89** (1953) 1297.
53P Pfann, W.G., Olsen, K.M.: Phys. Rev. **89** (1953) 322.
54B Bischoff, F.: DBP 1.061.117 (**1954**).
54E Emeis, R.: Z. Naturforsch. **9a** (1954) 67.
54H Heywang, W., Ziegler, G.: Z. Naturforsch. **9a** (1954) 561.
54K Keck, P.H., van Horne, W., Soled, J., McDonald, A.: Rev. Sci. Instr. **25** (1954) 331.
55P Pearson, G.L., Brattain, W.H.: Proc. I.R.E. **43** (1955) 1794.
55V Van den Boomgaard, J.: Philips Res. Rep. **10** (1955) 319.
56A Amberger, E.: Doctoral Thesis, Univ. München (1956).
56B Bradshav, S.E., Mlavsky, A.J.: J. Electron. **2** (1956) 143.
56D Dash, W.C.: J. Appl. Phys. **27** (1956) 1193.
56H Heywang, W.: Z. Naturforsch. **11a** (1956) 238.
56S Schweickert, H., Reuschel, K., Gutsche, H.: DBP 1.061.593 (**1956**).
57H Honig, R.E.: RCA Review **18** (1957) 195.
57P Pfann, W.G.: Metall. Rev. **2** (1957) 297.
58D Dash, W.C.: J. Appl. Phys. **29** (1958) 736.
58H1 Hoffmann, A., Keller, W., Reuschel, K., Rummel, T.: DBP 1.153.540, filed Sept. 10, **1958**, patented March 12, 1964; USP 2,970,111, filed Sept. 21, 1959, patented Jan. 31, 1961.
58H2 Hornstra, J.: J. Phys. Chem. Solids **5** (1958) 129.
58J Jackson, K.A.: Liquid Metals and Solidification, Cleveland, American Society for Metals **1958**, p. 174.
58K1 Kaiser, W., Frisch, H.L., Reiss, H.: Phys. Rev. **112** (1958) 1546.
58K2 Keller, W: DBP 1.094.710, filed Febr. 19, **1958**, patented Feb. 20, 1969; USP 3,159,459, filed Feb. 18, 1959, patented Dec. 1, 1964.
58L Leverton, W.F.: J. Appl. Phys. **29** (1958) 1241 (and in Engl. Patent 754767, London, 15.8.1956).
59D Dash, W.C.: J. Appl. Phys. **30** (1959) 459.
59G Goorissen, J., van Run, A.M.J.G.: Proc. IEE **106**, Pt.B, Suppl. **17** (1959) 858.
59K Keller, W.: DBP 1.148.525 filed May 29, **1959**, patented Dec. 27, 1963.
59S Stepanov A.V.: Sov. Phys. Tech. Phys. **29** (1959) 339.
60D1 Dikhoff, J.A.M.: Solid State Electron. **1** (1960) 202.
60D2 Dash, W.C.: J. Appl. Phys. **31** (1960) 736.
60G1 Goorissen, J., Karstensen, F., Okkerse, B., in: Solid State Physics, New York, Academic Press **1960**, p. 23.
60G2 Goorissen, J.: Philips Tech. Rev. **21** (1960) 185.
60G3 Goss, A.J., Adlington, R.E.: Solid State Phys. **1** (1960) 28.
60P Parr, N.L.: Zone Refining and Allied Techniques, London, George Newnes **1960**.

60S Sirtl, E.: DBP 1.193.918, filed June 14, **1960**, patented June 22, 1967; USP 3,172,857, filed June 9, 1961, patented March 9, 1965.
60Z Ziegler, G.: Internal Siemens report 1–108, Munich, **1960**.
61H Hoffmann, A.: „Halbleiterprobleme VI", Friedrich Vieweg & Sohn, Braunschweig, **1961**, 159.
61S Sirtl, E., Adler, A.: Z. Metallk. **52** (1961) 529.
61T1 Tanenbaum, M., Mills, A.D.: J. Electrochem. Soc. **108** (1961) 171.
61T2 Trainor, A.; Bartlett, B.E.: Solid State Electron. **2** (1961) 106.
61Z Ziegler, G.: Z. Naturforsch. **16a** (1961) 219.
62P Petritz, R.L.: Proc. IRE **50** (1962) 1025.
62S Stickler, R., Booker, G.R.: J. Electrochem. Soc. **109** (1962) 743.
63D Dikhoff, J.A.M.: Philips Tech. Rev. **25** (1963/64) 195.
63H Hirth, J.P., Pound, G.M.: "Progress in Materials Science 11", Mac Millan, New York, **1963**, 15.
63K Kodera, H.: Jpn. J. Appl. Phys. **2** (1963) 212.
63M Milvidskij, M.G.; Berkova, A.V.: Sov. Phys. Solid State **5** (1963) 517.
63S Sirtl, E.: GDCh-Meeting, Heidelberg (**1963**).
64C Chalmers B.: "Principles of Solidification", John Wiley, **1969**.
64G Green, R.E.: J. Appl. Phys. **35** (1964) 1297.
64H1 Hall, R.N., Racette, J.H.: J. Appl. Phys. **35** (1964) 379.
64H2 Hoffmann, A., Keller, W., Reuschel, K., Rummel, T.: DBP 1.153.540 (**1964**).
64P Pugh, E.N., Samuels, L.E.: J. Electrochem. Soc. **111** (1964) 1429.
64S Schildknecht, H.: „Zonenschmelzen", Weinheim/Bergstr., Verlag Chemie, **1964**.
65H Herlet, A.: Solid State Electron. **8** (1965) 655.
65L Leung, C.W.: Solid State Electron. **8** (1965) 571.
65M Mets, E.J.: J. Electrochem. Soc. **112** (1965) 420.
65R Runyan, W.R.: "Silicon Technology", McGraw Hill N.Y. **1965**, 33.
65W Wilcox, W.R., Fullmer, L.D.: J. Appl. Phys. **36** (1965) 2201.
66H1 Hall, R.N., in: "Proc. of the meeting on semiconductor materials for gamma-ray detectors" eds. W.L. Brown and S. Wagner **1966**, 27.
66H2 Hadamovsky, H.-F.: Wolf, E., Hässner, E., Spiegler, G., in: "Second International Symposium Reinststoffe in Wissenschaft und Technik", Berlin, Akademie-Verlag **1966**, 591.
66P Pfann, W.G.: "Zone Melting", New York, Wiley **1966**.
66T Tsivinsky, S.V., Koptev, Y.I., Stepanov, A.V.: Fiz. Tverd. Tela **8** (1966) (English Transl.: Sov. Phys. Solid State **8** (1966) 449.
66Z Zwanenburg, G.: Deutsches Patent 1.519.850 of 4.10.**1966**.
67C Carruthers, J.R.: J. Electrochem. Soc. **114** (1967) 959.
67M Mühlbauer, A.: Int. Z. Elektrowärme **25** (1967) 461.
67P Pomerantz, D.: J. Appl. Phys. **38** (1967) 5020.
68C Carruthers, J.R., Nassau, K.: J. Appl. Phys. **39** (1968) 5205.
69B Bleil, C.E.: J. Cryst. Growth 5 (1969) 99.
69C Ciszek, T.F., in: "Semiconductor Silicon", RR. Haberecht and E.L. Kern, eds., Princeton, The Electrochemical Society **1969**, p. 156.
69D Dumbaugh, W.H., Schultz, P.C.: Encyclopedia of Chemical Technology **18** (1969) 73.
69K Kendall, D.L., Vries, de D.B.: "Semiconductor Silicon", R.R. Haberecht and E.L. Kern, eds., Princeton, The Electrochemical Society **1969,** 358.
69N Newman, R.C., Smith, R.S.: J. Phys. Chem. Solids **30** (1969) 1493.
69S1 Spenke, E., in: "Semiconductor Silicon", R.R. Haberecht and E.L. Kern, eds., Princeton, The Electrochemical Society **1969,** 1.
69S2 Stepanov, A.V.: Bull. Acad. Sci USSR, Phys. Ser. **33** (1969) 1826.
69Y Young, R.C., Westhead, J.W., Corelli, J.C.: J. Appl. Phys. **40** (1969) 271.
70G Goorissen, J., in: "Handbook of Semiconductor Electronics", 3rd ed., L.P. Hunter, ed., New York, McGraw Hill **1970**, p. 6 ... 14.
70H Hall, R.N., Baertsch, R.D., Soltys,T.J., Petrucco L.J.: General Electric Annual Rpt. NYO-3870-4 (**1970**).
71H1 Hall, R.N., Soltys, T.J.: IEEE Trans. Nucl. Sci., **NS-18/1** (1971) 160.
71H2 Hansen, W.L.: Nucl. Instr. Methods **94** (1971) 377.
71W Whiffin, P.A.C., Brice, J.C.: J. Cryst. Growth **10** (1971) 91.
72A Andrychuk, D.: US-Patent 3,692,499 of Sept. 19 **1972**.

72C Ciszek, T.F.: Mater. Res. Bull. **7** (1972) 73.
72H1 Hadamovsky, H.-F.: "Halbleiterwerkstoffe", Leipzig VEB Deutscher Verlag für Grundstoffindustrie, **1972**.
72H2 Hunt, L.P., Sirtl, E.: J. Electrochem. Soc. **119** (1972) 1741.
72S1 Schulz-DuBois, E.O.: J. Cryst. Growth **12** (1972) 81.
72S2 Secco d'Aragona F.: J. Electrochem. Soc. **119** (1972) 948.
73B1 Beeftink, F.M., Vriezen, H.: DOS 2.301.148 of 11.1.**1973**.
73B2 Brice, J.C.: "The Growth of Crystals from Liquids", New York, North-Holland/American Elsevier **1973**, p. 268.
73B3 Bardahl, N.: Siemens-Z. **47** (1973) 160.
73D Doremus R.H.: "Glass Science", J. Wiley, N.Y. (**1973**).
73E Edwards, J.: J. Appl. Phys. **34** (1973) 2497.
73H Hu, S.M.: Phys. Status Solidi **(b) 60** (1973) 595.
73L Leskoschek, W., Feichtinger, H., Vidrich, G.: Phys. Status Solidi **(a) 20** (1973) 601.
73M Mühlbauer, A. in: "Semiconductor Silicon", H.R. Huff and R.R. Burgess, eds., Princeton, The Electrochemical Society **1973**, p. 107.
73P Patel, J.R.: J. Appl. Phys. **44** (1973) 3903.
73Y Yatsurugi, Y., Akiyama, N., Endo, Y., Nozaki, T.: J. Electrochem. Soc. **120** (1973) 975
74A Abe, T.: J. Cryst. Growth **24/25** (1974) 463.
74H1 Hall, R.N.: IEEE-Trans. Nucl. Sci. **NS-21/1** (1974) 260.
74H2 Hansen, W.L., Haller, E.E.: IEEE Trans. Nucl. Sci. **NS-22/1** (1974) 251.
74K Kubaschewski, G., Chart, T.G.: J. Chem. Thermodyn. **6** (1974) 467.
74M1 Mühlbauer, A., Sirtl, E.: Phys. Status Solidi **(a) 23** (1974) 555.
74M2 Mühlbauer, A., Watanabe, K.: Internal Siemens report 74.L.22, Munich, **1974**.
74N Nozaki, T., Yatsurugi, Y., Akiyama, N., Endo, Y., Makide, Y.: J. Radioanal. Chem. **19**. (1974) 109.
74S1 Schnöller, M.: IEEE Trans. Electron. Devices **ED-21** (1974) 313.
74S2 Sirtl, E., Hunt, L.P., Sawyer, D.H.: J. Electrochem. Soc. **121** (1974) 919.
74W Wolf, M.: Proc. Int. Conf. on Photovoltaics Power Generation, Hamburg, **1974**, 699.
75C1 Capper, P., Elwell, D.: J. Cryst. Growth **30** (1975) 352.
75C2 Chartier, C.P., Sibley, C.B.: Solid State Technol. (1975) 31.
75C3 Carruthers, J.R., Witt, A.F., in: "Crystals Growth and Characterization", R. Ueda and J.B. Mullin, eds., Amsterdam, North Holland **1975,** p. 107.
75H1 Herrmann, H., Herzer, H., Sirtl, E., in: "Festkörperprobleme 15", H.J. Queisser ed., Braunschweig, Pergamon/Vieweg **1975,** p. 279.
75H2 Herrmann, H.A., Herzer, H.: J. Electrochem. Soc. **122** (1975) 1568.
75K Kobayashi, N., Arizumi, T.: J. Cryst. Growth **30** (1975) 177.
75L Lawrence, J.E., Santoro, J.C.: US Patent 3,905,162 from Sept. 16, **1975**.
75M Matsumoto, S., Arai, E., Nakamura, H., Niimi, T.: Jpn. J. Appl. Phys. **14** (1975) 1665.
75P1 Patel, J.R., Authier, A.: J. Appl. Phys. **46** (1975) 118.
75P2 Petroff, P.M., Kock, de A.J.R.: J. Cryst. Growth **30** (1975) 117.
75S Suzuki, T.: Canadian Patent No. 971085, 15. July **1975**.
75Y Yusa, A., Yatsurugi, Y, Takaishi, T.: J. Electrochem. Soc. **122** (1975) 1700.
76B Blocher, J.M., Browning, M.F.: Quarterly Techn. Rep., JPL Contr. 954339, **1976**.
76C1 Carruthers, J.R.: J. Cryst. Growth **32** (1976) 13.
76C2 Chaney, R.E., Varker, Ch.J.: J. Cryst. Growth **33** (1976) 188.
76C Carruthers, J.R.: J. Cryst. Growth **42** (1977) 379.
76C3 Chaney, R.E., Varker, Ch.J.: J. Electrochem. Soc. **123**. (1976) 846.
76H1 Haas, E., Schnöller, M.: J. Electron. Mater. **5** (1976) 57.
76H2 Haller, E.E., Hansen, W.L., Hubbard, G.S., Goulding, E.S.: IEEE-Trans. Nucl. Sci. **NS-23/1** (1976) 81.
76J Janus, H.M., Malmros, O.: IEEE Trans. Electron. Devices **ED-23** (1976) 797.
76R Rozgonyi, G.A., Deysher, R.P., Pearce, C.W.: J. Electrochem. Soc. **123** (1976) 1910.
76S1 Schimmel, D.G.: J. Electrochem. Soc. **123**, (1976) 734.
76S2 Schmid F.: Proc. 12 IEEE Photovoltaic Specialists Conf., Batton Rouge, IEEE New York, **1976** p. 146.
76W1 Wakefield, G.F.: USP 3.998.659 (**1976**).
76W2 Whiffin, P.A.C., Bruton, T.M., Brice, J.C.: J. Cryst. Growth **32** (1976) 205.

77B1 Brandle, C.D.: J. Cryst. Growth **42** (1977) 400.

77B2 Brice, J.C., Whiffin, P.A.C.: J. Cryst. Growth **38** (1977) 245.

77B3 Brissot, J.J., in: "Curr. Top. Mat. Sci. **2**", eds. Kaldis E., Scheel H.J., North Holland Publ., **1977**, 796.

77G1 Glasow, P.A.: IEEE Trans. Nucl. Sci. **NS-23** (1976) 92.

77G2 Glasow, P.A., Raab, G.: Siemens Forschungsbericht SN 0015 Strahlentechnik, **1977**.

77H Herzer, H., in: "Semiconductor Silicon", H.R. Huff and E. Sirtl, eds., Princeton, The Electrochemical Society **1977**, p. 106.

77J Joyce, G.Ch.: US-Patent 4,032,389 of June 28, **1977**.

77K1 Kumagawa, M., Asaba, Y., Yamada, S.: J. Cryst. Growth **41** (1977) 245.

77K2 Kim, K.M.: Final Report DOE/JPL 954 465, June **1977**.

77L1 Langlois, W.E.: "Appl. Math. Modelling"1977, 196.

77L2 Langlois, W.E., Shir, C.C.: "Computer Methods in Applied Mechanics and Engeneering 12", **1977**, 145.

77L3 Langlois, W.E.: J. Cryst. Growth **42** (1977) 386.

77L4 Lorenzini, R.E., et al.: US-Patent 4,036,595 of July 19, **1977**.

77R Ravi, K.V.: J. Cryst. growth **39** (1977) 1.

77S1 Shiroki, K.: J. Cryst. Growth **40** (1977) 129.

77S2 Seiter, H., in: "Semiconductor Silicon" eds. Huff H.R., Sirtl E., The Electrochem. Soc. Inc., Princeton N.J. **1977**, p. 187.

77W1 Wright-Jenkins, M.: J. Electrochem. Soc. **124** (1977) 757.

77W2 Wirth, D.G., Sibold, J.D.: Quarterly Rept. DOE/JPL 954878-77/1, Oct. **1977**.

78A Authier, B.: Adv. Solid. State. Phys. **18** (1978) 1.

78B Belouet, C.: Proc. 1 E.C. Photovoltaic Solar Energy Conf., Luxemburg, D. Reidel Publ. Dordrecht, **1978**, 164.

78E Electronics, June **1978**, 44.

78G1 Glasow, P., Müller, A., Raab, G., Wolf, H.J.: Siemens Forsch. u. Entw. Berichte **7**. (1978) 4.

78G2 Gurtler, R.W.: Proc. 13 IEEE Photovoltaic Specialist Conf., Washington D.C., IEEE N.Y., **1978**, 363.

78H1 Hanoka, J.I., Strock, H.B., Kotval P.S.: same as [78G2], p. 485.

78H2 Hunt, L.P.: same as [78G2], p. 333.

78H3 Hubbard, G.S., Haller, E.E., Hansen, W.L.: IEEE-Trans. Nucl. Sci. **NS-25** (1978) 362.

78K1 Kobayashi, N.: J. Cryst. Growth **43** (1978) 357.

78K2 Köppl, F., Hamster, H., Grießhammer, R., Lorenz, H.: DOS 2854707 (**1978**).

78M Maciolek, R.B., Heaps, J.D., Zook, J.D.: J. Electron. Mater. **7** (1978) 441.

78R Reed, W.H., Meyer, T.N., Fey, M.G., Harvey, T.J., Arcalla, F.G.: Proc. 13th IEEE PSC, Washington D.C., **1978**, 370.

78W Wilson, L.O.: J. Cryst. Growth **44** (1978) 371.

79C Ciszek, T.F., Schwuttke, G.H., Yang, K.H.: J. Cryst. Growth **46** (1979) 527.

79D1 Darken Jr., L.S.: IEEE-Trans. Nucl. Sci. **NS-22/1** (1979) 324.

79D2 Duncan, C.S.: DOE/JPL Rpts. 954 654-79/2 and 79/3, **1979**.

79K1 Kapur, V.K., Nanis, L., Sanjurjo, A.: Ext. Abstr., Spring Meeting, Electrochem. Soc., Boston **1979**, 164.

79K2 Kendall, D.L.: Ann. Rev. Mater. Sci. **9** (1979) 373.

79K3 Kudo, B.: 11. Conf. on Solid State Devices, Tokyo, **1979**, 151.

79L Lindmayer, J., in: "Proc. 3. E.C. Photovoltaic Solar Energy Conf. Cannes", D. Reidel Publ., Dordrecht, **1981**, 1096.

79S1 Spenke, E.: Z. Werkstofftech. **10** (1979) 262.

79S2 Schimmel, D.G.: J. Electrochem. Soc. **126** (1979) 479.

79T Tsuya, N.: 11. Conf. on Solid State Devices, Tokyo, **1979**, 153.

80B1 Baghdadi, A., Gurtler, R.W.: Proc. 14. IEEE Specialists Conf., San Diego, IEEE N.Y., **1980**, 236.

80B2 Belouet, C., in: "Shaped Crystal Growth" eds. G. Cullen and T. Surek, Suppl. Issue J. Cryst. Growth **50** (1980) 279.

80B3 Belouet, C., in: "Proc. of Symposium Electronic and Optical Properties of Polycrystalline or Impure Semiconductors and Novel Silicon Growth Methods", eds. X.V. Ravi and B. O'Mara, The Electrochem. Soc. Inc., Pennington, **1980**, 195.

80C1 Cullen, G., Surek, T.: eds. of "Shaped Crystal Growth" Suppl. Issue J. Cryst. Growth **50** (1980).

80C2 Ciszek, T.F., Hurd, J.L.: same as [80B3], p. 213.
80D1 Dietze, W.: Metall **34** (1980) 676.
80D2 Duffy, M.T.: same as [80B1], p. 347.
80D3 Duncan, C.S.: same as [80B1], p. 25.
80H1 Helmreich, D.: same as [80B3], p. 182.
80H2 Heaps, J.D.: same as [80B1], p. 39.
80J Jewett, D.N., Bates, H.E.: same as [80B1], p. 1404.
80K1 Kobayashi, N., Arizumi, T.: J. Cryst. Growth **49** (1980) 419.
80K2 Kazmerski, L.L., Ireland, P.J., Ciszek, T.F.: Appl. Phys. Lett. **36** (1980) 323.
80K3 Khattak, C.P., Schmidt, F., Hunt, L.P.: same as [80B3], p. 223.
80K4 Kudo, B.: same as [80B1], p. 247.
80K5 Kalejs, J.B.: same as [80B1], p. 13.
80K6 Kim, K.M.: same as [80B1], p. 212.
80L Lane, R.L., Kachare, A.H.: J. Cryst. Growth **50** (1980) 437.
80M Miyazawa, Sh.: J. Cryst. Growth **49** (1980) 515.
80R Reed, C.L., Mar, K.M.: J. Electrochem. Soc. **127** (1980) 2058.
80S1 Surek, T.: same as [80B3], p. 173.
80S2 Strunk, H.: Ast D.: Techn. Rept DOE/JPL 954852, **1980**.
80S3 Seidensticker, R.G., Hopkins, R.H.: same as [80B1], p. 221.
80T Tsuya, N.: J. Electron. Mater. **9** (1980) 111.
80Z Zook, J.D.: same as [80B1], p. 260.
81D1 Dietl, J., Helmreich, D., Sirtl, E., in: "Crystals Growth, Properties and Applications, Vol. 5.", Springer Verlag, Berlin, **1981**, p. 43.
81D2 Dietl, J., in: "Proc. of Symp. Mater. and New Processing Technologies for Photovoltaics", eds. J.A. Amick The Electrochem Soc. Inc., Pennington, **1981**, 48.
81D3 Dietze, W., Keller, W., Mühlbauer, A., In: "Crystals, Growth, Properties and Applications, Vol. 5", Springer Verlag, Berlin, **1981**, p. 1.
81F Fally, J., Guenel, C.: "Proc. 3. E.C. Photovoltaic Solar Energy Conf.", Cannes, D. Reidel Publ. Dordrecht, **1981**, 598.
81H Haller, E.E., Hansen, W.L., Goulding, F.S.: Adv. Phys. **30/1** (1981) 93.
81I Iya, S.K., Flagella, R.N., Di Paolo, F.S.: Proc. Symp. on Materials and New Processing Technologies for Photovoltaics, Orlando, **1981**, 80.
81K Keller, W., Mühlbauer, A.: "Float-Zone Silicon", New York, Marcel Dekker, **1981**.
81L1 Levin, H.: "Proc. Symp. on Materials and New Processing Technologies for Photovoltaics", Orlando **1981**, 68.
81L2 Lindmayer, J.: same as [81F], p. 178.
81S1 Spenke, E., Heywang, W.: Phys. Status Solidi **(a) 64** (1981) 11.
81S2 Sirtl, E.: same as [81F], p. 236.
82H Hansen, W.L., Haller, E.E., Luke, W.N.: IEEE-Trans. Nucl. Sci. **NS-29/1** (1982) 738.
82K Kolbesen, B.O., Mühlbauer, A.: Solid-State Electron. **25** (1982) 759.
82Z Zulehner, W., Huber, D., in: "Crystals, Growth, Properties and Applications, Vol. 8: Silicon/Chemical Etching, Czochralski-Grown Silicon", Springer-Verlag, Berlin, Heidelberg, New York, **1982**.

6.1.3 Characterization of crystal properties

6.1.3.1 Properties of polycrystalline silicon

The macroscopic properties of polysilicon are nearly identical to those of monocrystalline silicon. Deviations are mainly caused by the grain boundaries of the polycrystalline material.

Physical data of silicon [81D3]

atomic number	14
atomic weight	28.086
atomic radius	$1.33 \cdot 10^{-10}$ m
crystal structure	diamond structure
lattice spacing	$5.43 \cdot 10^{-10}$ m
interatomic distance	$2.35 \cdot 10^{-10}$ m
atoms/cm^3	$4.96 \cdot 10^{22}$
density at 20 °C	2.33 g/cm^3
density at melting point	2.55 g/cm^3
melting point	1420 °C
boiling point	2630 °C
specific heat 80···120 °C	0.754 J g^{-1} K^{-1}
linear thermal expansion coefficient at 20 °C	$2.33 \cdot 10^{-6}$ K^{-1}

The typical cross-section of polycrystalline silicon rods is shown in Figs. 1a and b. Fig. 1a shows a silicon rod having a monocrystalline core, on which silicon deposition started at about 1000 °C core temperature. The crystallites are growing in radial direction with preferred ⟨110⟩ orientation. The size of crystallites depends on the growth temperature. Deposition temperatures exceeding 1100 °C lead to very rough silicon surfaces.

The silicon rod shown in Fig. 1b presents a cross-section through polysilicon, the deposition of which started at about 1100 °C core temperature. In this case, a monocrystalline silicon deposition takes place in six directions, when started with a slim-rod of ⟨111⟩-orientation. Besides these directions polycrystalline growth appears and predominates with decreasing decomposition temperatures. It is possible to prepare completely monocrystalline silicon rods at a temperature of about 1150 °C and slow deposition rates [63S2]. Polycrystalline material without monocrystalline areas is preferred for the float zone process.

Polysilicon rods having large diameters (>100 mm) usually contain microcracks and holes, which cannot be seen from the surface. It is not explained how cracks can appear. They are possibly caused by thermal strain during the deposition process. Microcracks are disadvantageous for the float-zone process, but not for crucible pulling techniques. Measurements by electrical methods cannot be performed on polysilicon because the resistivity is strongly influenced by grain boundaries. Only doped polysilicon of resistivity less than 1 Ωcm may be measured without a large error. In all other material, high resistivities are found. It is common practice first to pull a monocrystal by the float zone technique with argon ambient and then to measure the resistivity. The same procedure is performed to determine impurities by infrared or photoluminescence methods. The neutron activation method is the only method to detect low contents of all impurities in polysilicon. A typical result is presented in section 6.1.3.2.3. Impurities may be located in the crystal lattice or at the grain boundaries. Because of the lack of direct measurements it is often impossible to distinguish both sources of impurities. The carbon content was measured in a cross-section of a rod [73S3]. It was found that the content decreased from the core to the surface. The carbon and oxygen content of polysilicon are of the order 10^{16} atoms cm^{-3}. Both elements are the dominant but not the most important impurities in polysilicon.

6.1.3.2 Properties of Czochralski silicon

Table 1. Typical data on Czochralski grown Si-single crystals.

Quality	Producible	Commercial	Unit
Diameter	up to 180	50···125	mm
Length	up to 2200	500···1600	mm
Weight	up to 48	5···35	kg
Orientations	⟨100⟩, ⟨111⟩, ⟨110⟩, ⟨511⟩, and others	⟨100⟩, ⟨111⟩	–

continued

Table 1 (continued)

Quality	Producible	Commercial	Unit
Dopants			
Boron	0.001···100	0.001···80	Ω cm
	10^{20}···10^{14}	10^{20}···$2\cdot10^{14}$	cm^{-3}
Phosphorus	0.002···50	0.03···50	Ω cm
	$6\cdot10^{19}$···10^{14}	$7\cdot10^{17}$···10^{14}	cm^{-3}
Antimony	0.004···50	0.006···0.02	Ω cm
	$2\cdot10^{19}$···10^{14}	10^{19}···10^{18}	cm^{-3}
Arsenic	0.001···50	0.001···0.01	Ω cm
	$8\cdot10^{19}$···10^{14}	$8\cdot10^{19}$···$5\cdot10^{18}$	cm^{-3}
Aluminum	0.1···100	not standard	Ω cm
	$5\cdot10^{17}$···10^{14}	not standard	cm^{-3}
Gallium	0.06···100	not standard	Ω cm
	10^{18}···10^{14}	not standard	cm^{-3}
Indium	1···100	not standard	Ω cm
	$1.5\cdot10^{16}$···10^{14}	not standard	cm^{-3}
Impurities			
Carbon	$<10^{16}$ by selection, up to	$<5\cdot10^{15}$···$5\cdot10^{16}$	cm^{-3}
	$4\cdot10^{17}$ by doping		cm^{-3}
Oxygen	$2\cdot10^{17}$···$2\cdot10^{18}$	$5\cdot10^{17}$···$14\cdot10^{17}$	cm^{-3}
Crystal structure			
Dislocations	free [1])	free	
Precipitates	free – high	free – few	

[1]) For optical applications also polycrystalline rods are grown by the Czochralski method.

A detailed discussion of the properties is given in the following sections.

6.1.3.2.1 Doping profiles

Axial distribution

The incorporation of impurities or dopants is influenced by the growth conditions of Czochralski growth. For dopants having small distribution coefficients and high vapour pressures a large variation in the distribution is possible. If the distribution coefficient is close to 1 the variation is small. Among the usual dopants for silicon, boron shows the smallest difference and antimony the largest difference in the axial and radial distribution. Boron is an example for a distribution coefficient close to 1 ($k_0=0.8$) and a very low vapour pressure. Antimony shows a small distribution coefficient ($k_0=0.023$) and a high vapour pressure at the melting point of silicon.

The Figs. 2···9 show some typical and some extraordinary axial resistivity distributions observed in CZ-Si-crystals for the dopants: B, Al, Ga, In, P, As, and Sb.

Radial distribution

Far more than the axial distribution, the radial impurity distribution is affected by the stirring of the melt, which again is affected by the crystal and crucible rotation, by the dimensions, and by temperature distributions in the melt. In the case of volatile dopants like antimony also the evaporation plays an important role, (Fig. 10).

A further strong influence is given by the crystal orientation. The greatest difference may be found between the orientations ⟨100⟩ and ⟨111⟩. The ⟨111⟩-orientation shows the largest radial resistivity variations, about double the magnitude observed for ⟨100⟩ orientation. As shown in [69C, 74M2, 74A1, 74B3, 74R1] large radial concentration differences are caused by facetted growth at the freezing interface of the ⟨111⟩-crystals. The {111}-crystal layers are built up very rapidly by starting at the edge of the crystal and growing towards its centre, thus causing a segregation and an enrichment of impurities in the centre of the crystal. Normally the {111}-layers do not grow over the whole radius in one go because of the solidification heat emitted and of the mostly higher temperature in the centre of the growing {111}-plane (convex freezing interface and isotherme).

Figs. 11···14 show several examples for radial resistivity variations for different dopants and pulling conditions.

Striations

The fluctuating freezing in the microscopic scale is influenced by crystal and crucible rotation, thermal convection, and crystal orientation. The microscopic resistivity variations (striations), both radial and axial therefore react in nearly the same manner as the macroscopic radial resistivity variation.

Especially the influence of the distribution coefficient k_0 and of the crystal orientation is strong.

In contrast to float-zone pulling there are only weak remelt phenomena during Czochralski crystal growth. During float zone pulling for each crystal revolution, a certain amount of the just frozen crystal layer is remelted by passing through the high-power zone of the induction coil. For crucible pulling, remelt phenomena are not caused directly by heater unsymmetries but by irregular flows in the melt which cause temperature fluctuations at the freezing interface. Depending on melt geometry and furnace construction, the temperature differences may be large in such a melt, temperature differences of 30 °C or more between the hot and the cold point (= the freezing crystal) in the melt are not seldom. However, by certain combinations of crystal and crucible rotation, the temperature fluctuations near the freezing interface are reduced. In most cases, the microscopic resistivity variations (= striations) in crucible-pulled crystals are substantially smaller than in float zone crystals.

Figs. 15 and 16 show examples for microscopic resistivity variations measured by the spreading resistance method.

Compensation

Due to the reaction between the silicon melt and the silica crucible, the highest resistivities which can be made reproducibly by the Czochralski technique are limited to about 100 Ω cm. Higher resistivities may be obtained, but then the degree of compensation is uncertain. Widely varying resistivities are the result.

The main limiting factor is the generation of *oxygen donors* (or thermal donors) which are formed by the oxygen in the temperature region between 500 and 300 °C [54F1, 57F] in the cooling period during crystal growth. Oxygen donors are destroyed by a special annealing after crystal growth, however a residual amount always remains (section 6.1.3.2.2).

The second limiting factor is the dopant content of the silica (section 6.1.3.3.3). Especially boron with its high distribution coefficient ($k_0 = 0.8$) is a critical impurity in silica.

Because of its high purity, polycrystalline Si is not the limiting factor for obtaining resistivities greater than 100 Ω cm.

6.1.3.2.2 Oxygen in CZ silicon

Axial profile

Fig. 17 shows axial oxygen distributions in Czochralski grown Si single crystals, all pulled under the same pressure and argon flow. Their pulling processes differed in crystal and crucible rotation, in the geometry and temperature distribution of the melt, and in the guiding of the argon above the melt.

The strongest influence on the oxygen content is excerted by the crystal and crucible rotation. For example the crystals of curves *1* and curve *7* in Fig. 17 are pulled under the same conditions, only the crystal and crucible rotations were different. The oxygen reduction due to an additional eddy current as discussed in section 6.1.2.3.10 is clearly visible.

Radial profile

Not only the axial but also the radial distribution of oxygen in Si crystals is affected by the above current conditions in the melt. Figs. 18a···d show radial oxygen distributions of four crystals, grown at different pulling conditions.

Solubility and diffusivity of oxygen

In most of the Czochralski silicon crystals, oxygen is the strongest impurity present at contents of about 10^{18} cm^{-3} or 20 ppma. Only in heavily doped crystals, the oxygen contents are exceeded by the dopant concentrations. Next to oxygen, the second strongest impurity carbon shows only contents 10···200 times less than oxygen.

Normally in CZ crystals, the greatest fraction of the oxygen is in the dissolved state where it occupies interstitial lattice sites [56K, 57H, 57F, 72B1, 73S1, 73T, 74M1, 74V, 77C1, 77G2, 77T2, 77P3, 77H4, 78G, 79C1, 80O1]. Figs. 19 and 20 show the result of several investigations on the solubility and diffusivity of oxygen in silicon.

Interstitial impurity oxygen exhibits a relatively high diffusivity whereas on the other hand its solubility decreases strongly to low temperatures. Because of this behaviour, oxygen is the most important precipitation forming element in CZ-silicon. The mobility and the precipitation behaviour of oxygen is to be seen in Fig. 21 which shows the result of 20 h-annealing at different temperatures for samples of an oxygen-rich CZ-crystal.

Oxygen donors

An inconvenient property of oxygen in silicon with respect to the manufacturing of electronic devices is its ability to form donors especially in the temperature range from 300 to 500 °C, most violently at about 450 °C. This effect was first reported by [55F, 56F]. [57K] assumed an SiO_4-complex for the donor. He concluded this from the fact that the oxygen donor concentration is only a small fraction of the oxygen concentration but grows rapidly with increasing oxygen content proportional to the fourth power of the oxygen concentration. This conclusion is not generally accepted and numerous investigations were made on this matter [54F1, 55F, 56K, 57K, 58K, 73V1, 79M, 79G]. None of the investigations supports a model of common consent.

In a recent investigation [81R2], three different oxygen donors were found after prolonged annealing.

The 450 °C-oxygen donors may be easily and rapidly annihilated by an annealing above 500 °C with a following rapid cooling to temperatures below 300 °C.

A residual concentration of the oxygen donors may be left if the quenching is not fast enough, especially for high oxygen concentrations. In critical cases it may be necessary to anneal and quench the silicon material in wafer form instead of the whole crystal.

Depending on the annealing time at about 450 °C (Fig. 22) and on the oxygen content in the crystal, the concentration of oxygen or thermal donors may reach values of up to $5 \cdot 10^{16}\,cm^{-3}$ corresponding to a resistivity of 0.15 Ω cm. Because of the usually high oxygen concentrations and the long period near 450 °C during crystal growth [57F, 82Z], the seed end of the crystals contains more oxygen donors than the bottom end. Figs. 23···25 show typical axial distributions of oxygen donors in CZ-crystals after crystal pulling.

The oxygen donors formed between 300 °C and 500 °C ("450 °C-donors") are not the only oxygen-related donors in silicon; however, they are the most important ones because of the high concentrations which they can reach in many cases.

Further oxygen donors are formed at about 550 °C [79K1, 81R2], 650 °C [79K1, 81R2] and 750 to 800 °C [79K1, 81L, 81R2]. The formation of these donors strongly depends on preannealing and of course on the oxygen concentration. The carbon content also seems to play a role in the formation of these donors [79K1, 81L, 81R2]. According to [81L], the donor generated at 750 °C is a complex involving carbon and oxygen.

Investigations in [82Z] show the influence of preannealing (partly already introduced by the crystal pulling) upon the formation of oxygen donors above 500 °C and give a formation maximum at 770 °C (Figs. 26 and 27).

For the donors formed between 550 and 800 °C, the structure (i.e. the number of oxygen atoms, carbon atoms, vacancies, self-interstitials, etc. involved and the configuration) remains unknown until today.

Influence of oxygen on crystal properties

Besides its negative peculiarity to form donors oxygen has some further negative but also positive characteristics. Both the negative and the positive properties of oxygen depend on the absolute oxygen concentration and on its solute or precipitation state in the crystal.

Oxygen shows negative effects if during device preparation oxygen precipitates are present at or very near to the polished wafer surface. In this case large oxidation-induced stacking faults are formed in the active areas of electronic devices at the surface which may cause device failure by leakage currents, filamentary shorts, reduced lifetimes and microplasma generation sites [77K3, 73R3].

Generation of oxygen precipitates near the wafer surface may be prevented by suitable device processing by which oxygen precipitates, small dislocation loops and stacking faults are formed only in the inside of the wafer, whereas a sufficiently deep defect-free layer is grown at the polished surface, in which no stacking faults develop during oxidation or diffusion steps. In this case the lattice defects in the inside of the wafer show beneficial effects:

First they act as getter centres ("intrinsic gettering") for fast diffusing impurity elements like Cu, Fe, Na, Ag, etc. which are deleterious in electronic devices. Secondly they cause precipitation or particle hardening of the lattice which restrains the motion of dislocations leading to less slippage in the wafers in high temperature processes.

The possibility of intrinsic gettering is the greatest advantage of CZ-silicon for the manufacturing of highly, and more so for very highly integrated circuits. Very highly integrated circuits of the future probably cannot be made without intrinsic gettering.

Because of the lack of these two features – intrinsic gettering and precipitation hardening – float-zone silicon single crystals are less used for integrated circuits than CZ-silicon.

Under some circumstances the precipitation of oxygen in the interior of the wafer may deteriorate the wafer. This occurs when oxygen in very high concentrations is precipitated at high temperatures and under lattice strain. The high temperature and the high oxygen content generate large oxide precipitates and large dislocation loops around them. By the strain, the dislocation loops grow and move through the lattice over large distances to the wafer surface.

6.1.3.2.3 Impurities in CZ silicon

Concentrations and distributions

Besides oxygen and the dopant, *carbon* is the next important impurity in a CZ-crystal. In a normal CZ-crystal, the carbon content of the first 65 % of the crystal length is lower than $1 \cdot 10^{16}\ cm^{-3} = 0{,}2$ ppma (Fig. 28). Because of the normal freezing, the carbon content increases in the last 35 % of the crystal length to about $5 \cdot 10^{16}\ cm^{-3} = 1{,}0$ ppma at the extrem end of the crystal. A large fraction of the whole carbon content originally present in the poly-Si remains in the melt which is left in the crucible. With an effective distribution coefficient of $k \approx 0{,}1$ for carbon and the average carbon concentration of less than $4 \cdot 10^{15}\ cm^{-3} = 0{,}08$ ppma at the seed end of the crystals, the average carbon content of the starting polycrystalline silicon produced by chemical vapor deposition can be calculated to about $4 \cdot 10^{16}\ cm^{-3} = 0{,}8$ ppma.

The concentrations of the *other elements* are normally again lower than that of carbon, especially, because they have in general very low distribution coefficients (example: Fe with $k_0 = 8 \cdot 10^{-6}$). They accumulate according to their distribution coefficient at the bottom end of the crystal, however, the largest fraction remains in the melt which is left in the crucible.

Table 2. Impurity concentration c in Cz-silicon. The analysis was usually made by neutron activation analysis NAA. The analysis was repeated on different samples at various labs: KWU, Erlangen Germany and AERE, Harwell Great Britain.

Impurity	c ppba	Remark	Ref.
Ag	<0.008	NAA below detection limit	
As	0.015	NAA	
Au	$<10^{-5}$	NAA below detection limit	
Cd	0.008	NAA	
Cr	<0.03	NAA below detection limit	
Cu	<0.01	NAA	
Fe	≤0.1	ESR	82Z
Ga	<0.002	NAA below detection limit	
In	<0.03	NAA below detection limit	
K	<0.3	NAA below detection limit	
Mn	<1	NAA below detection limit	
Mo	<0.03	NAA below detection limit	
Na	<0.3	NAA below detection limit	
Ni	<0.2	NAA below detection limit	
Pt	<0.01	NAA below detection limit	
Sb	0.002	NAA	
Se	<0.01	NAA below detection limit	
Sn	<0.3	NAA below detection limit	
Sr	<0.5	NAA below detection limit	
Ta	<0.001	NAA below detection limit	
W	<0.001	NAA below detection limit	
Zn	<0.05	NAA below detection limit	
Zr	<1	NAA below detection limit	

Influence of impurities on crystal properties

Many of the impurities like carbon are electrically neutral and do not directly affect the crystals with respect to their applicability in electronic devices. But they may influence the precipitation behaviour of Si-self-interstitials, vacancies, oxygen, and other fast diffusing elements by forming complexes with them or by forming small precipitates which act as nucleation centres for the precipitation of the fast diffusing point defects. In float zone silicon, for example, [75F1] found a strong correlation between the so-called "swirl"-defects and the carbon content of the crystals.

Electronic levels for donor and acceptor centers are listed in section 1.2.2 of volume III/17a. Diffusion coefficients and their temperature dependence are shown in Fig. 29 and discussed in section 6.1.4.1 in more detail.

Of great importance are some fast diffusing elements like Cu or Fe. Under certain circumstances they are electrically active, especially as "life time killers". They are feared because they are able to diffuse nearly everywhere in a wafer during device processing and tend to accumulate in the electrically active areas. They are particularly gettered at or around crystal defects which are present. In order to avoid or to keep the negative electrical effects at a low level, the concentration of these elements must be kept very low in the crystal.

6.1.3.2.4 Lattice defects

Pointlike defects

As pointlike defects, all precipitates (incoherent, semicoherent, coherent), the stacking faults and the small dislocation loops which are generated at precipitations, are included in this section.

Effect of cooling rate on precipitates

In a slowly cooled dislocation-free crystal, precipitates are always present, at least those of self-point defects. Self-point defects are at the freezing point present in a higher equilibrium concentration than the amount which is solvable in the crystal at low temperatures. In CZ-crystals, supersaturated impurities are always additionally present, above all oxygen, which may act as nuclei for precipitations.

For rapid cooling, especially if the impurity concentrations are not high, the formation of precipitates is difficult because it needs time. Self-point defects and impurities have to diffuse a certain distance before they reach the precipitation site. Precipitation is slower for the slowly diffusing point defects than for the fast diffusing species. For rapid and normal cooling e.g. for crystal pulling, only small, in many cases, very small precipitates below the detection limit grow. For very rapid cooling or quenching precipitates cannot grow. Rapid cooling is impossible for crucible-pulled large crystals.

Precipitation is also slowed down in the formation of the first precipitation nuclei by homogeneous nucleation. The attracting forces between a few isolated point defects are weak and are mostly exceeded by their kinetic energy. Attracting forces are different for different point defect types. Each type needs another supercooling condition for the formation of a stable nucleus which contains a number of point defects.

In reality, crucible-pulled Si-crystals contain less precipitates, particularly oxygen precipitates, than are expected considering the high impurity concentrations. In many cases, dependent on doping, detectable precipitates in a large number are only generated by annealing after crystal pulling, especially for repeated annealings. The formation of precipitates and pointlike defects in Czochralski-grown Si depends strongly on the annealing conditions during device processing.

Nucleation of precipitates

For the first cooling during crystal growth, the point defect type which passes first through its solubility limit should precipitate at first in principle. However, depending on supercooling conditions and diffusivity, some point defect types exceed their solubility limit so far that another type forms the first stable precipitation nuclei.

In Czochralski crystals, the oxygen atoms are the point defects of the highest concentration, except some dopant concentrations. Particularly for the high concentrations, oxygen often forms the first precipitation nuclei. After the classical nucleation theory [63H, 65B], there exists a temperature-dependent *critical radius* r_c for each impurity or self-point defect. Below r_c the nuclei are not stable and tend to decompose. Above r_c they are stable and tend to grow. For the homogeneous nucleation of oxygen in silicon, [77F3] found a critical radius of $r_c \approx 10$ Å (about 200 oxygen atoms) at 1150 °C. For formation of precipitates, a supercooling of about 80 °C was required. [80O2] found at 1050 °C the same radius $r_c \approx 10$ Å but a necessary supercooling of about 300 °C.

When exceeding the critical radius, the oxygen precipitates grow to crystalline, mostly square-shaped oxide platelets (cristobalite) parallel to {100}-lattice planes. Particle sizes up to 5000 Å were reported [76M1, 76T, 78Y, 79T2, 79P2, 80S1].

The SiO_2-precipitates need about double the volume of the Si. They induce a compressive strain on the surrounding lattice. This strain may be released by

a) punching of prismatic dislocation loops, leading to plastic deformation,
b) emission of Si-interstitials,
c) absorption of vacancies (less probable).

Case a) is confirmed several times by TEM-investigations [76M1, 76T, 78Y, 79T2, 79P2, 80S1]. The decision between case b) and c) cannot be made conclusively. Case b) is favoured by the majority of investigators because of its good consistency with the diffusion and precipitation phenomena observed in silicon.

Case b), the emission of Si-interstitials is diffusion-controlled and is effective for a high diffusivity, i.e. at high temperatures. For high temperatures, the emission of self-interstitials is predominant. The formation of large extrinsic stacking faults around some of the precipitates is an indication of the emission of Si-self interstitials [73D3, 74R1, 75T, 76T, 77P1, 78P, 78S3, 78Y, 79P1, 79T1, 79T2, 79W, 79P2, 80W1, 80S3, 80H3, 80S1, 80O2, 80F2, 80T, 80H4, 80U, 72O, 74J, 75A, 77M1, 77R3, 77K3, 80S2, 64Q, 66J1, 66B, 66J2, 71S, 72P1, 72P2, 72M, 72R, 73H2, 74R2, 74H2, 74P, 75H3, 76P2, 76R3, 77M3, 77A1, 77P2, 77S5, 77R3, 77M4, 78T3, 78M2, 78K2, 78S1, 78K1, 78C] (Fig. 30). The self-interstitials emitted agglomerate around an oxide particle between two regular {111}-lattice planes forming an additional {111}-lattice plane which interrupts the normal stacking sequence of the lattice atoms perpendicular to this plane. This layer of Si-interstitials is called extrinsic stacking fault.

The edge of the interstitial layer forms a partial dislocation loop, a so-called *Frank partial dislocation* [49F], which is sessile, i.e. it cannot slip. It can move only by climbing, i.e. by emitting or absorbing of vacancies or interstitials.

In many cases oxygen is not the first precipitating point defect [79D, 80D3], especially in crystals with low oxygen contents (see also float zone crystals). [76M1] found that all types of crystal defects which are known from oxygen-free float-zone crystals, are also present in CZ-crystals, beginning from the smallest precipitates up to stacking faults (single, double), perfect dislocation loops and combinations of all the three. They are all of interstitial type, thus they are probably precipitates of Si-self-interstitials.

As confirmed by the industrial practice, the formation of pointlike crystal defects is strongly influenced not only by impurities but also by the dopants. The experience shows that boron suppresses the formation of self-point precipitates beginning at boron contents of about 10^{15} cm^{-3}. After [79D, 80D3], the suppression does not take place up to boron concentrations of about 10^{17} cm^{-3}. This effect is not confirmed by the industrial experience. The oxygen precipitation is not measurably influenced by the boron contents.

The n-type dopants also suppress the formation of some self-point precipitates. After [79D, 80D3], the large self-interstitial precipitates (extrinsic stacking faults and perfect dislocation loops) do not grow in presence of donor concentrations above 10^{17} cm^{-3}. Only precipitates of vacancy type are found.

An example for self-point defect precipitates in a low oxygen CZ-Si-crystal is shown in Fig. 31. The large defects which are easily made visible by etching, show no arrangement in layers as it is typical for oxygen precipitates (Fig. 32). Their distribution is more homogeneous having a high density near the crystal surface and a decreasing density towards the crystal centre.

The content of pointlike crystal defects in CZ-Si-crystals depends on the contents of all other impurities. High contents of most impurities, e.g.: C, Fe or Cu, immediately result in high defect numbers.

All impurities in Si-single crystals are incorporated into the crystal in more or less oscillating concentrations fluctuating in the direction of the freezing. The magnitude of the fluctuation depends on the distribution coefficients, variations in the freezing conditions, and is affected by irregular flows in the melt beneath the freezing interface.

Precipitation takes first place in layers of high concentrations. If precipitates are once formed in a layer then in most cases, the supersaturated point defects of the adjacent layers having lower concentrations diffuse to the precipitates just formed. The formation of precipitates in the layers of lower concentrations is therefore further restrained.

An arrangement of oxygen precipitates in layers may be seen in Figs. 32 and 33.

Swirls

Similar to the freezing interface, the layers are mostly bent like spherical segments and their sections result in rings of crystal defects (Fig. 34a, b). This ring-shaped arrangement of crystal defects is called "swirl". The expression "swirl" or "swirls" originates from the spiral patterns of pointlike defects in float-zone crystals where the defect layers wind up in the crystal like a helix. In float-zone crystals and in Czochralski crystals, the pointlike crystal defects are often called "swirl defects".

An important difference exists between the mobility of the typical swirl defects, i.e. precipitations of self-point defects, and the mobility of oxygen precipitates. The diffusion coefficients of vacancies and self-interstitials differ from the diffusion coefficient of oxygen by a factor a about 10^5 and 10^3, respectively, at 1200 °C (Fig. 29).

Table 3. Fractional concentration c_v and diffusion coefficient D_v of Si-vacancies [78M1].

T K	c_v	D_v $cm^2 s^{-1}$
1693	$1.3 \cdot 10^{-8}$	$4.2 \cdot 10^{-4}$
1500	$5.0 \cdot 10^{-10}$	$1.2 \cdot 10^{-4}$
1300	$6.5 \cdot 10^{-12}$	$2.0 \cdot 10^{-5}$
1100	$1.7 \cdot 10^{-14}$	$1.8 \cdot 10^{-6}$

Table 4. Fractional concentration c_I and diffusion coefficient D_I of Si-self-interstitials [77S4].

T K	c_I	D_I $cm^2 s^{-1}$
1693	$4.0 \cdot 10^{-7}$	$4.3 \cdot 10^{-6}$
1500	$2.7 \cdot 10^{-8}$	$7.2 \cdot 10^{-7}$
1300	$8.6 \cdot 10^{-10}$	$7.6 \cdot 10^{-8}$
1100	$7.8 \cdot 10^{-12}$	$3.7 \cdot 10^{-9}$

The state of the swirl defects is always strongly changed during device processing. Their most important characteristic is that they tend to dissolve by diffusion of self-interstitials and/or vacancies to stable sinks, especially to the wafer surface where they are eliminated. Swirl defects can diffuse out to the wafer surface nearly completely during a usual oxidation or diffusion step for device processing.

The diffusion path $\bar{x}=\sqrt{2Dt}$ is about 230 μm after 1 hour at 1300 K.

Dislocations, twins, grain boundaries

When Si-single crystals contain a dense network of dislocations, the swirl defects cannot be observed because dislocations absorb the point defects, particularly the self-point defects, and swirl defects cannot grow. This fact can be seen in Fig. 35, where slip dislocations originating from the lower end of the crystal have dissolved the swirl defects.

In crystals for electronic devices, no dislocations, no twins, and no grain boundaries are allowed because they totally degrade the devices which are made of such crystals. A discussion on the types and properties of these defects is not necessary. During device processing slip dislocations are introduced into the crystal in many cases. This problem is discussed in section 6.1.4.

6.1.3.3 Properties of float-zone silicon

6.1.3.3.1 Doping profiles

Axial profiles

The incorporation of dopants and impurities into silicon crystals during float-zone growth is mainly influenced by the growth conditions, by the impurity itself, by the doping method used, and by the desired concentration level of the dopant. For dopants having extremely small distribution coefficients as e.g. indium ($k_0=4 \cdot 10^{-4}$) or gallium ($k_0=8 \cdot 10^{-3}$), only the pill-doping method yields acceptable axial homogeneity (Fig. 36). For phosphorus, the most important dopant in FZ crystal growth having a distribution coefficient of $k_0=0.35$, the axial resistivity profile strongly depends on the doping method used and on the resistivity level required. Figs. 37 and 38 show some axial resistivity profiles of phosphorus-doped crystals. Boron-doped silicon crystals grown from poly-doped feed rods generally show good axial homogeneity because of the high distribution coefficient ($k_0=0.8$) and the very small evaporation rate of boron. The axiale profile of a high-resistivity boron-doped crystal is shown in Fig. 39.

Fig. 40 shows a typical axial profile of the carbon concentration observed in float-zone crystals.

Radial Profiles

The radial resistivity distribution is mainly affected by the stirring of the melt and the crystal rotation, and by the crystal orientation. Figs. 41···46 show typical radial resistivity profiles of the usual dopants boron and phosphorus in medium- and high-resistivity float-zone silicon. Radial profiles of carbon and phosphorus in a float-zone crystal are compared in Fig. 47.

Enhanced dopant concentrations are observed in areas of (111) facets (section 6.1.2.4.5). For crystal growth exhibiting a convex-growing interface, *central dopant cores* may be observed after etching crystal sections cut along the growth direction (Fig. 48). The corresponding resistivity profile shows a marked dip in the central region. The enhanced dopant incorporation in crystal regions having concave growing interfaces leads to lateral resistivity dips which are also visible in Fig. 48. The lateral inhomogeneity of phosphorus-doped crystals can also be revealed by autoradiography (Fig. 49). The radial resistivity profile observed in relation to the interface shape from variable growth speed showing faceting growth is shown in Fig. 50.

Fig. 51 gives examples for the radial resistivity profiles of the dopants P, Sb, Ga, and In. In order to avoid the influence of faceting on the dopant distribution, all crystals were grown in the ⟨100⟩ direction, both in the concentric and the eccentric growth method. The homogenizing effect of the eccentric method is demonstrated.

Dopant compensation

The electrical resistivity of silicon crystals is determined by the uncompensated centers, i.e. the difference $|N_a - N_d|$ of the total concentration of acceptors and donors, and by the mobility of the majority carriers. The compensated centers in the polycrystalline feed rods are mainly phosphorus and boron. The donor phosphorus usually prevails. During float-zone processing under vacuum, phosphorus segregates and evaporates much more than boron, the net phosphorus concentration is reduced, and an increase of resistivity results with the number of zone passes (Figs. 52 and 53). Compensation occurs after a certain number of zone passes depending on the initial concentrations and species of donor and acceptor impurities, and the corresponding resistivity reaches a maximum. Further zone refining leads to a transition to p-type conductivity because of the residual boron content, which cannot be removed. The final resistivity is determined by the residual boron concentration of float-zone crystals.

Microscopic profiles (striations)

Diagnostic techniques having high spatial resolution reveal resistivity fluctuations on a microscale in silicon single crystals. Fig. 54 shows radial microscopic resistivity profiles of phosphorus-doped float-zone crystals measured by the spreading resistance technique. The profiles were taken from the peripheral part of 100 mm-diameter slices. One crystal (top) was doped by neutron transmutation, (see section 6.1.4.3), the other two crystals were doped by conventional methods. Fig. 55 shows radial microscopic resistivity profiles of boron-doped FZ crystals at quite different resistivity levels. The amplitude of the microscopic resistivity variations lies between $\pm 5\%$ and about $\pm 50\%$ for conventionally doped crystals, strongly depending on type of dopant, dopant concentration, crystal diameter, and growth conditions used [74K4, 75H2, 75C, 81K].

6.1.3.3.2 Impurity striations

Microscopic impurity variations have been revealed in Si and Ge crystals by various diagnostic techniques, such as chemical etching [60E, 63D, 69V, 70K2, 73M2], electrical resistivity measurements [65M, 67M, 73B3, 73W, 76M2], X-ray topography [63S1, 65R, 73A, 77G2], infrared breakdown radiation measurement [73B3, 73V2, 75M2], electroreflectance [73S2], SEM-EBIC analysis [73R1, 75D, 76K], and autoradiography [73M1]. For the diagnostic techniques see section 6.1.3.5.

Impurity variations are caused by temperature variations at the freezing interface, which cause locally varying growth rates and distribution coefficient fluctuations [54C, 62T, 63C, 73B1, 75C, 81K]. Variations in convection, as in non-steady flow, similarly lead to a fluctuating distribution coefficient k via a fluctuating diffusion layer thickness δ next to the growing interface. As a consequence, the as-grown crystals show layers having axially and laterally varying impurity concentrations which reflect the local conditions at the moment of their growth [64M, 75C].

Cross sections or lengthwise cuts of crystals show spiral patterns after a special chemical treatment or parallel striae that closely follow the shape of the growing interface (Fig. 56). The contrast lines revealed are called *striations*. Striations may show different appearances, depending on the diagnostic method used (Fig. 57), because various types of impurities are involved. Whereas shallow donors and acceptors are responsible for resistivity fluctuations which may be detected by spreading resistance measurements [67M, 73B3, 76M2] or be made visible by striation etching, impurities such as carbon, oxygen, or heavy metals may considerably influence the lattice parameter. This may be revealed by X-ray topography (Fig. 58 a) or also by chemical etching (Fig. 58 b).

According to the segregation behavior of carbon, it is obvious [82K] that concentration fluctuations of carbon may be the main reason for lattice parameter variations associated with striations in homogeneously float-zone silicon containing low oxygen concentrations as shown in Fig. 58.

It is important to distinguish between *rotational* and *non-rotational* striations. Rotational striations are periodical resistivity variations showing a period λ which is equal to the distance by which the crystal grows during one revolution: $\lambda = v/f$, where v is the macroscopic growth velocity (zone travelling rate) and $f = 60\omega/2\pi$ the rotation rate. These rotational striations correspond to the dark lines in the micrograph of Fig. 59. Non-rotational striations of various origins have been detected in zone-molten Ge and in float-zone Si crystals.

Dopant striations cause inhomogeneous breakdown across the active area of power devices [75M2]. In carbon-rich float-zone silicon having a carbon content exceeding $c_C = 5 \cdot 10^{16}\ \mathrm{cm}^{-3}$ harmful crystalline defects preferentially form at the carbon concentration peaks of the striations during power device wafer processing. These defects strongly deteriorate the device performance [82K].

The amplitude of dopant striations can gradually be reduced by choosing growth conditions which minimize fluctuations of the growth velocity v and the diffusion layer thickness δ [74K1, 75H2]. Practically complete elemination of dopant striations (Fig. 54a) can be achieved by applying the neutron transmutation doping technique [74S1, 77H5]. This doping method is suitable for silicon and can be applied in general for phosphorus doping only (section 6.1.4.3).

6.1.3.3.3 Impurities

The final impurity content in undoped, high-purity float-zone silicon crystals is mainly determined by the purity of the polycrystalline starting material and by the growth ambient as well as by the cleanness of the growth chamber. The concentrations of the *oxygen* and the *carbon*, the main residual impurities, are generally between 10^{15} and $5 \cdot 10^{16}$ cm^{-3} [70N, 74N, 82K] (Fig. 60). *Carbon* is already present in the polycrystalline material, whereas the oxygen content will be strongly influenced by the growth ambient (Fig. 60).

In high-purity silicon crystals the residual quantities of the main dopants *boron* and *phosphorus* are in the range of $10^{12} \cdots 10^{13}$ cm^{-3} (Fig. 61), whereas *arsenic* and *antimony* are detectable at concentrations which are less by one to two orders of magnitude.

Metallic impurities such as *gold*, *copper*, and *iron* are typically present at concentrations below 10^{13} cm^{-3} and they are often below their specific detection limit of neutron activation analysis [74B3, 80H5, 81H1]. These deep-level impurities act as efficient recombination centers and reduce the minority carrier lifetime even if their concentrations are very small. They must be removed as completely as possible.

For certain applications, e.g. of some types of infrared radiation detectors, ultrapure silicon having residual boron concentrations of about 10^{12} cm^{-3} ($10000 \cdots 20000\ \Omega$ cm) is required. This very low boron level can be achieved by applying many zone passes under vacuum in order to eliminate residual quantities of boron and phosphorus mainly by segregation [59H3, 77K1]. *Phosphorus*, *arsenic*, and *antimony* are also reduced in concentration as a result of evaporation (see section 6.1.1.5) [58Z, 81K].

The most important impurity introduced intentionally is *phosphorus*. Doping levels of $10^{13} \cdots 10^{15}$ cm^{-3} ($500 \cdots 5\ \Omega$ cm) are usual for power devices. *Boron*-doped float-zone silicon containing concentrations of about 10^{15} cm^{-3} will be used for the fabrication of integrated circuits. For special applications such as infrared radiation detectors, *indium* and *gallium* have to be incorporated intentionally at concentrations of approximately $2 \cdot 10^{17}$ cm^{-3}.

Most of the impurities present in the starting polycrystalline silicon can be removed by zone refining because of their small distribution coefficients which range between 0.35 for *phosphorus* and $2.5 \cdot 10^{-5}$ for *gold* (see section 6.1.1.3). An exception is boron having a distribution coefficient close to unity ($k_0 = 0.8$). Because it is practically impossible to remove *boron* economically by zone refining, it is absolutely necessary to keep the boron level as low as possible in the starting polycrystalline silicon. This can be achieved by careful chemical purification of the trichlorosilane used. The extremely low distribution coefficients of *metallic impurities* help to remove these elements by segregation during float-zone processing.

Further purification by evaporation occurs if crystal growth is carried out under vacuum, where the purification effect for certain impurities is determined by their respective evaporation coefficients (see section 6.1.1.5).

Typical impurity concentrations in float-zone single crystals and their specific detection limits are given in Table 5.

Table 5. Impurity concentrations in FZ-silicon and detection limits. The impurity concentration was usually detected by neutron activation analysis (NAA) and for oxygen, carbon, boron and phosphorus by infrared spectroscopy (IR) as well as by charged particle activation analysis (CPA) for oxygen, carbon, boron, and nitrogen.

Element	c cm^{-3}	Detection limit cm^{-3}	Remarks	Ref.
Ag	$\leqq 5 \cdot 10^{10}$	$5 \cdot 10^{10}$	NAA	81H1
As	$< 3 \cdot 10^{10}$	$1 \cdot 10^{10}$	NAA	
Au	$< 1 \cdot 10^{9}$	$1 \cdot 10^{8}$	NAA	
B	$< 1 \cdot 10^{12}$	$5 \cdot 10^{11}$	IR, ultra-pure Si	80B
C	$< 5 \cdot 10^{16}$	$5 \cdot 10^{15}$	IR, standard Si	74N, 82K
Cd	$< 3 \cdot 10^{11}$	$3 \cdot 10^{10}$	NAA	81H1
Co	$< 3 \cdot 10^{11}$	$2 \cdot 10^{10}$	NAA	81H1
Cr	$< 5 \cdot 10^{11}$	$8 \cdot 10^{10}$	NAA	
Cu	$< 5 \cdot 10^{10}$	$2 \cdot 10^{10}$	NAA	
Fe	$\leqq 4 \cdot 10^{12}$	$4 \cdot 10^{12}$	NAA	
Ga	$\leqq 7 \cdot 10^{10}$	$7 \cdot 10^{10}$	NAA	
In	$< 5 \cdot 10^{11}$	$1 \cdot 10^{11}$	NAA	
K	$\leqq 1 \cdot 10^{13}$	$1 \cdot 10^{13}$	NAA	
Mn	$\leqq 4 \cdot 10^{13}$	$4 \cdot 10^{13}$	NAA	
Mo	$\leqq 1 \cdot 10^{12}$	$1 \cdot 10^{12}$	NAA	
N	$< 5 \cdot 10^{14}$	$5 \cdot 10^{13}$	CPA	71E, 74N
Na	$\leqq 2 \cdot 10^{13}$	$2 \cdot 10^{13}$	NAA	81H1
Ni	$< 3 \cdot 10^{12}$	$1 \cdot 10^{12}$	NAA	

Table 5 (continued)

Element	c cm^{-3}	Detection limit cm^{-3}	Remarks	Ref.
O	$<5\cdot10^{15}$	$1\cdot10^{15}$	IR, standard Si	80D4
P	$<5\cdot10^{12}$	$5\cdot10^{11}$	IR, ultrapure Si	80B
Pd	$\leqq7\cdot10^{12}$	$7\cdot10^{12}$	NAA	81H1
Pt	$\leqq2\cdot10^{11}$	$2\cdot10^{11}$	NAA	
Sb	$<5\cdot10^{9}$	$2\cdot10^{9}$	NAA	
Sn	$<1\cdot10^{13}$	$3\cdot10^{12}$	NAA	
Sr	$<1\cdot10^{13}$	$5\cdot10^{12}$	NAA	
Ta	$<2\cdot10^{10}$	$5\cdot10^{9}$	NAA	
W	$\leqq1\cdot10^{10}$	$1\cdot10^{10}$	NAA	
Zn	$<2\cdot10^{12}$	$3\cdot10^{11}$	NAA	

6.1.3.3.4 Defects

At present, float-zone single crystals are generally grown free of line dislocations and macroscopic defects, like twins and stacking faults. The main residual defects which may occur in such crystals are *swirl defects*, but also *etching depressions* or *hydrogen defects* may be present in dislocation-free FZ crystals.

Swirl defects

In dislocation-free float-zone silicon crystals microdefects arranged in a spiral or striated pattern in planes perpendicular or parallel to the growth direction may often be revealed by preferential etching (Fig. 62) [65P, 67A]. At least two types of these microdefects which are usually called *swirl defects* have to be distinguished in size, density, and spatial distribution [73D1, 77D2, 80D2]. These are the larger "*A* swirl defects" and the smaller "*B* swirl defects" (Fig. 63a, b). Apart from preferential etching, swirl defects may be revealed by X-ray topography [76R1, 77D2], and by autoradiography [73M1, 77H1], in particular in combination with decoration methods [73C, 73D2]. The influence of swirl defects on electrical properties of devices has been studied by the SEM-EBIC method [75D]. For the diagnostic techniques see section 6.1.3.5.

Swirl defects may be easily distinguished from striations which also form spiral patterns on etched slices but appear as lines and not as dots. Pattern of swirl defects consist of rows of etch pits showing a flat bottom in contrast to the deep dislocation etch pits (Fig. 64a–c). A single swirl defect disappears on etching, forms a shallow etch pit which also disappears on further etching, and new ones form at new spots indicating the point defect character of swirl defects.

The average concentration of *A* swirl defects was found to be typically between $10^6\cdots10^7$ cm^{-3} [73D1, 80D2]. The *B* swirl defect concentration strongly depends on the carbon content of the crystals. At carbon concentrations of 10^{17} cm^{-3}, *B* defect concentrations up to 10^{11} cm^{-3} have been observed [77F2], whereas at carbon concentrations below 10^{16} cm^{-3} the *B* defect concentration approaches zero [77F2]. The typical concentrations are $10^7\cdots10^8$ cm^{-3} [73D1, 80D2]. A peripheral crystal region about 2 mm in width is generally free of *A* swirl defects but *B* swirl defects still often occur. Around dislocations in a zone of about 1 mm, no swirl defects are found [73D1]. The relationship between the concentration of *A* and *B* swirl defects and the crystal growth rate is shown in Fig. 65. Swirl defects vanish at growth velocities higher than 5 mm/min and at extremely low growth velocities of lower than 0.2 mm/min [80D2]. A strong influence of carbon is observed. With increasing carbon concentration, the elimination of *B* swirl defects becomes more difficult [80D2]. The addition of hydrogen to the growth ambient, generally argon, prevents the formation of swirl defects [73D1, 77D1], but may cause other large harmful defects during the float-zone growth which are called "*hydrogen defects*" [81K].

The microscopic structure of swirl defects has extensively been studied mainly by high-voltage TEM. It was established that the *A* swirl defects consist of perfect dislocation loops, showing a typical size in the order of $1\cdot10^{-6}$ m and occuring as single loops, loop clusters, or complicated loop arrangements (Fig. 66a–c) [74B1, 75F1]. The complicated *A* swirl defects mainly occur in crystals of low carbon concentration ($<10^{16}$ cm^{-3}). With an increasing carbon content, the structure is simpler, single loops dominate which often contain a stacking fault at carbon concentrations above $5\cdot10^{16}$ cm^{-3} [75F1]. These dislocation loops are of interstitial type [75F1, 75F2, 75P]. The nature of *B* swirl defects could not be detected by TEM [75F1, 76P1].

In large diameter (50 mm···100 mm) float-zone silicon single crystals grown at velocities above 3 mm/min generally no microdefects arranged in a swirl pattern may be revealed by the usual diagnostic techniques listed in section 6.1.3.5. The microdefect density in such crystals in terms of shallow etch pits is typically below 10^3 cm^{-2}.

Nucleation and formation of swirl defects

For the nucleation and formation of swirl defects in float-zone silicon, a number of models have been proposed [73D1, 75F1, 76P1, 77F1, 77H6, 77C2, 79C2, 80D3], and reviewed in [81D1, 81F]. All the models confirm that the driving force in the formation of swirl defects is a supersaturation of thermal point defects and that the heterogeneous nucleation involving most likely carbon prevails. The recent models [75F1, 76P1, 77F1, 80D3] consider it well established that for the interstitial-type *A* defect dislocation loops are formed by condensation of silicon self-interstitials. They state that the *B* defects are prestages of the *A* defects. According to [77F1], the *B* defects consist of three-dimensional droplet-like agglomerates of silicon interstitials and carbon atoms which collapse into the interstitial-type dislocation loops, the *A* defects, when exceeding a critical size. Other explanations of the swirl defect formation are given in [76P1, 77C2, 77H6].

A comprehensive model covering all experimental results on swirl defects in FZ as well as in CZ silicon crystals is still lacking.

Etching depressions

After polish-etching of dislocation-free float-zone silicon crystals of diameters above 45 mm marked depressions may be found in the crystal interior (Fig. 67) [75K1, 81K]. This serious-looking defect has not yet shown any influence on device performance. The etching depressions may be caused by the thermomechanical stress which occurs during the cooling phase after solidification, whereby the presence of point defects is again important. Gold decoration of slices containing the etching depression show a decoration pattern which closely resembles the depression geometry. Since the gold distribution is similar to that in silicon without depression, the defect has basically to be present in any large-diameter dislocation-free FZ crystal. The formation of a depression depends only on the content of defects.

Hydrogen defects

Float-zone crystal growth in the presence of hydrogen may lead to very strong defects [81K]. The hydrogen defects can be revealed by etching of wafers or cross-sections of crystals by acids or hydroxides (Fig. 68). Deep holes, grooves, or craters are formed in the inner regions of the crystal. How the defect is caused is not yet clear. Devices made from crystals containing hydrogen defects have no blocking capability. For this reason, hydrogen is no longer used as a growth ambient, neither pure nor as a mixture. Only pure argon is employed for dislocation-free crystal growth.

6.1.3.4 Properties of high purity germanium (HP-Ge)

Impurities and defects in HP-Ge are reviewed in [81H2], where also a numerous literature is compiled.

6.1.3.4.1 Impurities in the starting material

Commercial "intrinsic Ge" which is used as starting material for HP-Ge-crystals normally contains P, B, and Al as electrically active impurities at concentrations of $10^{12}\cdots10^{13}$ cm^{-3} [68S, 81H2]. Type and concentration of neutral impurities are not investigated, because their concentration in the final crystals are dependent on the HP-Ge growing technique.

6.1.3.4.2 Impurities in purified Ge

Shallow impurities

The major shallow impurities are the acceptors B and Al [70H, 74H1, 76H1] and the donor P. Fig. 69 shows a selection of typical net impurity profiles of zone refined Ge-crystals [74H3]. For zone purification and refining by multiple Czochralski growth technique, interactions between the melt, the container, the container coatings, and the ambient atmosphere cause the formation of nonsegregating B- and Al-compounds and determine the residual impurities [76H1]. The Table 6 lists boat materials and possible interactions which may occur during zone purification [78H4].

Table 6. Possible contaminations of Ge during purification.

Source	Interaction	c cm^{-3}	Remarks	Ref.
Boat-material				
Pure graphite	contamination P, B	$10^{11} \cdots 10^{13}$	AL segregates	
Synthetic quartz [1])				
Uncovered	contamination P	$<2 \cdot 10^{10}$	formation of ternary or higher order compounds with Si, O, and Al from the container; partly releasing electrically active Al [76H1]; no effective segregation	81H2
Carbon-coated	formation of complexes involving B, O, and Si (coating seems porous)		certain purifications	78H4
Pyrolytic carbon-coated	no interaction melt-quartz		SiO-complexes cannot form, B is not gettered; Al segregates	78H4
Silica smoke-coated	gettering of impurities by complex formation		major residual impurity Al with small amounts of B and Ga. Presence of excess Si allows some segregation (Si is a strong scavenger of oxygen).	78H4, 77G1
Pyrolytic carbon, Silica-coated	no interaction		Al segregates, B gettered by silica	78H4
Ambient				
Vacuum			low purification	81H2, 82H2
N_2 (and silica crucible)	high oxygen concentration	$(>10^{14})$		
H_2	complexes with various point defects			81H2, 82H2
$90\%N_2+10\%H_2$			usual technique	78H4

[1]) Synthetic quartz "Suprasil", by Amersil Inc. Sayreville, N.J., USA and "Spectrosil" by Thermal American Fused Quartz Co., Montville, N.J., USA have successfully been used to purify and grow ultra-pure germanium.

Fig. 70 shows the net doping concentration versus the length of ingots purified in graphite and quartz boats having various carbon coatings [78H4]. During purification, forming gas (90% N_2, 10% H_2), is generally used. Carbon is a source of B, Al segregates because of the low Si- and O-environment (Ingot *90* in Fig. 70). Quartz-coated boats partially lead to interaction of the melt with quartz under formation of electrically inactive borosilicates (Ingot *91* in Fig. 70).

Pyrolytic carbon separates Ge from quartz.

Fig. 71 shows the axial concentration profile of a crystal purified by this method refining in silica smoke-covered boats. The major impurity is aluminum, which does not segregate, and the segregating boron, which is gettered by the silica.

Fig. 72 shows a typical axial concentration profile of an ingot grown by pyrolytic carbon/silica smoke refining which combines the aluminium segregating environment of carbon refining with the boron gettering property of silica.

A typical axial dependence of the net shallow level concentration of a *single* crystal grown by the Czochralski technique from a silica-coated quartz crucible under H_2 atmosphere is shown in Fig. 73 [81H2]. Near the head end, the crystal is uniformly p-type. In a transition region, the type changes first near the axis and more slowly at the perifery ("coring"). In the tail end region, the crystal is mostly n-type with the possibility of a p-type skin [73H1].

Deep impurities

Detector applications of HP-Ge require the lowest possible concentration of deep levels. Most deep impurities in Ge show a low solubility and a small segregation coefficient. They can therefore in principle be removed by zone purification. The elemental deep impurity investigated is *copper* [64H, 54F2]. Copper may be easily introduced into a Ge-crystal in its interstitial form at low temperature by accident. It exists in two forms: interstitial and substitutional. Its solubility in Ge containing excess vacancies is greatly enhanced [81H2]. Other deep impurities are Be [74H1] and Zn [81H2].

The optical transitions of neutral elemental acceptors [64R] and donors are surveyed in [81H2].

Neutral Impurities

Hydrogen

The most complete study of hydrogen in Ge is given by [60F], which presents the diffusion coefficient D and the solubility s of H at temperatures between 800 °C and 910 °C:

$$D=2.7\cdot10^{-3}\exp\left(-\frac{0.38\ \text{eV}}{kT}\right)\ [\text{cm}^2\ \text{s}^{-1}].$$

The solubility of hydrogen at 1,013 bar is:

$$s=1.6\cdot10^{24}\exp\left(-\frac{2.3\ \text{eV}}{kT}\right)\ [\text{cm}^{-3}].$$

The solubility is proportional to the square root of the partial hydrogen pressure. At the melting point of germanium (935 °C), the solid solubility is between 10^{14} and 10^{15} cm^{-3} at 1.013 bar. An accurate value is difficult to obtain and it is not clear whether the hydrogen solubility is retro-grade as are the solubilities of many elemental impurities [60T]. [82H2] has determined a hydrogen concentration of about $2\cdot10^{15}$ cm^{-3} by growing crystals in hydrogen spiked with tritium and counting the tritium β-decays in detectors made from these crystals. Annealing studies show that the hydrogen is strongly bound either to defects or as H_2 with a dissociation energy >3 eV. This energy is lowered to 1.8 eV when copper is present. [82H2] reviews the benefits and problems encountered in using a hydrogen ambient.

Smooth pits on preferentially etched dislocation-free crystals have been found [82H2] to be due to hydrogen precipitates estimated to contain 10^8 H atoms each.

Oxygen

Elemental interstitial oxygen is electrically inactive in Ge at low concentrations; however, the lithium-oxygen complex produces a shallow donor level. By its tendency to bond two neighboring germanium atoms, oxygen behaves primarily as stationary impurity in complex formation [81H2].

Silicon

Silicon replaces a Ge-atom in its substitutional position almost perfectly [81H2]. Additions of silicon to a germanium melt contained in a quartz boat suppresses "free" oxygen so that normal segregation of aluminum is achieved [76H].

Carbon

Carbon is believed to form a shallow acceptor complex with hydrogen. Using ^{14}C-spiked pyrolytic graphite-coated quartz crucibles for the growth of HP-Ge-crystals, an average value of the total carbon concentration of $[^{14}\text{C}+^{12}\text{C}]\approx2\cdot10^{14}$ cm^{-3} is found [82H1].

Complexes

Acceptor (A)-type and donor (D)-type complexes [81H2] are:

Hydrogen-related:	A(H, Si), A(H, C), D(H, O)
Lithium-related:	D(Li) [56R], D(Li, O)
Divacancy-Hydrogen:	A(V_2H)
Copper-related:	A(H, Li, Cu)

The hydrogen-related so-called "fast" acceptors and the "fast" donors have been discovered by heating HP-Ge grown in hydrogen atmosphere at 700 K and rapidly quenching it to room temperature, and after annealing [75H1]. Fast impurities are not observed, when the hydrogen is removed by outdiffusion or in materials crystallized in a nitrogen atmosphere [78H3].

A single H-atom is involved in the complexes [79H2]. Si is also involved in the formation of the fast centres [76H2].

The donor *lithium* is technically used for the Li-ion-drift process [e.g. 60P1]. Li and Li-oxygen donors are investigated by IR-transmission [65A], photo thermal spectroscopy (PTS) [72S2, 79S2, 78H2, 75B] and are surveyed in [81H2].

The *divacancy-hydrogen acceptor* (V_2H) shows a level $E_v = 0.071$ eV [81H2] and is found in dislocation-free material. V_2H-acceptors can be created in dislocated crystals grown in hydrogen atmosphere by irradiation with 1 MeV γ-rays [80V].

Copper-related acceptors are discussed in [81H2] and [77H2].

Dislocations

Dislocations play an important role in HP-Ge-detectors (e.g. [76G, 79H4]). They can act as nucleation centers for excess hydrogen or sinks for vacancies; the total absence of dislocations leads to precipitates in hydrogen ambient grown crystals and to formation of the divacancy-hydrogen complex [82H2]. [77R1] used dislocation-free HP-Ge crystals in the study of large, strainfield confined electron hole drops which showed ten times longer lifetimes than in dislocated Ge.

6.1.3.5 Diagnostic techniques

6.1.3.5.1 Doping profiling

The dopant concentration and distribution in Si and Ge crystals may be analysed by measuring the electrical resistivity which is roughly inversely proportional to the dopant concentration, by means of simple methods which give reliable results. The Table 7 lists the standard techniques in use.

Table 7. Standard measurement techniques for doping profiles.

Method	Resistivity profile	Sampling volume [cm^3]	Ref.
Two-point probe	macroscopic axial (along a crystal)	44 (75 mm crystal, probe distance 10 mm)	71D
Four-point probe	macroscopic radial (across a wafer)	$\approx 10^{-4}$	54V, 74B2
Spreading-resistance probe	microscopic axial and radial	$\approx 10^{-9}$	67M, 74E

Apart from the standard measurement techniques routinely used for resistivity profiling, a number of modified methods not requiring the attachment of electrical contacts have been established but have not received wide application. These techniques include radio frequency resistivity measurement of high-resistivity silicon rods using capacitive coupling [59K], non-destructive photovaltaic measurement of radial resistivity profiles of high-resistivity slices [78B], contactless radio frequency radial resistivity measurement system [74K2], spreading-resistance method using non-blocking aluminum-silicon contacts [73B3].

6.1.3.5.2 Impurity and defect analysis

The most important impurities which have to be characterized in electronic-grade silicon crystals are the *dopants* boron, phosphorus, antimony, and arsenic, the light-element impurities *oxygen* and *carbon* present at relatively high concentrations, and the deep level impurities such as the *heavy metals* gold, copper, iron, and others.

The dopants intentionally introduced are acceptors or donors providing the desired *conductivity type* (p- or n-type), and a certain *concentration level* defining the required *electrical conductivity* of the crystal. Carbon and oxygen are of particular interest since they may cause the formation of defects during wafer processing degrading the electrical properties of devices. Finally, the fast diffusing heavy metals may precipitate at crystal defects and they can act as recombination centers reducing the *lifetime* (diffusion length) of minority carriers. Thus it is important to identify the impurities and determine their respective concentrations in silicon single crystals. This may be achieved by *electrical techniques*, *physical analysis methods*, and *chemical analysis techniques* as well.

Electrical techniques

Resistivity measurements are routinely used for concentration analysis of dopants. The conversion between resistivity and dopant concentration (Fig. 74) [62I, 78T2] is of great importance because in device design the calculation of various parameters such as breakdown voltage involves the dopant concentration which is very difficult to measure directly. Carrier concentration and mobility describe the electrical conductivity and consequently the resistivity of a crystal. The determination of the Hall mobility is a principal task in impurity characterization. Carrier lifetime and diffusion length measurements are one way to probe the presence of non-dopant impurities (and defects) which act as recombination centers.

Various electrical techniques used for impurity analysis [70K1, 77B, 80D1] are listed in Table 8.

Table 8. Electrical measurement techniques for impurity analysis.

Impurity	Property	Diagnostic technique	Remarks	Ref.
Dopants	resistivity	two-point probe	large probe distance: macroscopic profiles	71D
			small probe distance: microscopic profiles (striations)	65M
		four-point probe	macroscopic lateral profiles of silicon	54V, 58S 61L, 68G
			sheet resistance	74B2
		one-point probe (bridge-method)	microscopic profiles	70K3
		spreading-resistance probe	microscopic profiles	67M, 73B3, 74E
		van der Pauw method	average resistivity of samples	58V
		contactless RF method	macroscopic axial profiles of rods	59K
			macroscopic lateral profiles of slices	74K2
	conductivity type	diode probe and thermal probe	p- or n-type	
	carrier concentration	Hall measurement	carrier concentration vs. temperature; compensation	58V, 60P2, 77A2, 78L
		junction CV-measurement	net dopant density	75M1, 78T2
	mobility	Hall measurement	Hall mobility of electrons or holes	60P2, 70K1
		combination of Hall and resistivity measurements	calculated mobility	
		drift measurements	drift mobility	54L
Non-dopants	carrier lifetime	photoconductive-decay method (PCD)	minority carrier lifetime	71A, 73G, 74G1
		photocurrent technique	minority carrier lifetime profiles of samples	75R1, 77S2
	carrier diffusion length	surface photovoltaic method	minority carrier diffusion length	74A2

Physical analysis methods

Various physical methods suitable for impurity identification and characterization are well established. They are used preferably for laboratory investigations. The most common analysis methods for the examination of dopant impurities and other shallow level impurities, such as oxygen and carbon, and the deep level impurities including impurity-induced deep level active centers (e.g. microdefects) are given in Table 9. For electronic levels of impurities see section 1.2.2 of volume III/17a.

Table 9. Physical methods for impurity analysis.

Diagnostic technique	Impurity	Sensitivity cm^{-3}	Remarks	Ref.
Infrared absorption spectroscopy	oxygen $[O_i]$	$1 \cdot 10^{15}$	concentration measurement at room temperature	80D4, 79A, 81P
			macroscopic and microscopic profiling	81H3, 81M
Scanning IR absorption			microprofiling (resolution approx. 30 μm)	81R1
	carbon	$5 \cdot 10^{15}$	concentration measurement at room temperature	80D5, 69B
			macro- and microprofiling	73A, 82K, 74N
	dopants: B, P As, Al, Ga, In [O]-donors	$5 \cdot 10^{11}$	simultaneous identification of dopant impurities at low temperatures (10···20 K); determination of impurity concentrations; detection of net dopant concentration which determine the electrical conductivity, and of total dopant concentration; applicable to $\varrho > 30\ \Omega$ cm; restricted in size and shape of the specimens used	67W, 64P, 75K2, 80B
Photoluminescence Analysis (PL)	B P Al As [O]-donors	$1 \cdot 10^{11}$ $5 \cdot 10^{10}$ $2 \cdot 10^{11}$ $5 \cdot 10^{11}$	simultaneous identification of dopant impurities at low temperatures; detection of net and total dopant concentration; applicable to $\varrho > 10\ \Omega$ cm; not restricted in size and shape of the specimen	77T1, 78T1, 81T
Deep level transient spectroscopy (DLTS and DDLTS)	deep level impurities (e.g. heavy metals)	10^{10}	impurity-induced active centers (defects) are revealed; activation energies, capture cross-sections, and concentrations are detectable and their spatial distribution	74L, 77S1 80W2, 81G, 81J
Derivative surface photovoltage spectroscopy (DSPS)	deep level impurities		type and concentration of deep centers (impurity induced microdefects) as well as their spatial resolution	80J1, 81J
Electron beam induced current measurement (EBIC)	shallow and deep level impurities (dopants and recombination centers)		spatial distribution on a macro- and microscale (e.g. resistivity striations and swirl defects)	73R1, 75D, 81J

Chemical analysis techniques

A range of instrumental methods of chemical analysis has been developed to reach the requirements for local impurity analysis in semiconductor materials. The analysis of electronic-grade bulk silicon generally requires extreme analytical sensitivity of ppb-range and spatial resolution. Conventional wet chemical methods do not meet these requirements. Typical characterization techniques used are reviewed in [69G, 70K1], and [80M], and the most relevant analysis methods are summarized in Table 10.

Table 10. Chemical methods for impurity analysis.

Diagnostic technique	Applications	Sensitivity	Remarks	Ref.
Atomic absorption spectroscopy (AAS) (flame)	impurities (survey)	100 ppb	sensitivity depends on impurity to be analysed and analytical treatment; requires samples in solution	80M
(carbon furnace)		10 ppb		80M
Atomic emission spectroscopy (AES) (flame and plasma)	impurities (survey)	10···100 ppb	preconcentration steps normally required; sensitivity depends on impurity and treatment	70K1, 75L, 80M, 69G
(inductively coupled plasmamode (ICP)	impurities	1 ppb	no preconcentration; information on practically all elements of interest in silicon; solution of samples required	
Spark-source mass spectroscopy (SSMS)	impurities	5 ppb	information on practically all elements simultaneously; solid samples can be used; adequate tool for analysing solar-grade silicon	69M1, 80M, 81D2
Charged particle activation analysis (CPAA)	light impurities (e.g. B, C, N, O)	1···10 ppb	sensitivity depends strongly on impurity; chemical separation is seldom necessary; restricted to light element impurities	70N, 74N
Neutron activation analysis (NAA)	impurities	10^{-3}···1 ppb	information on nearly all elements; chemical separation after irradiation is advantageous; for sensitivity see section 6.1.3.3.3.	55M, 64G, 66H, 69M2, 70K1, 80M
Autoradiography	localization of dopants and impurities		reveals distribution of the analysed elements, shows the decoration of crystal defects by dopants or impurities	69M2

Crystal defects in silicon single crystals are distinguished by their dimensionality which is revealed by different diagnostic techniques. These techniques are summarized in Table 11.

Table 11. Diagnostic techniques for defect detection.

Defect	Dimensionality	Diagnostic technique	Ref.
Dislocations	one-dimensional or line defects	**etching**	67B
Lineage		Sirtl etch for {111} faces	61S
Slippage		Secco etch for {100} faces	72S1
		anodic etching	80F1
		X-ray topography	64A
		IR transmission topography (after decoration)	56D
		transmission electron microscopy (TEM)	65H
		autoradiography (after decoration)	77H1
Stacking faults	two-dimensional defects	cf. dislocations	
Striations	line defects	**etching**	
		Kämper-Mayer etch, applicable for n- and p-type silicon from 0.001···1000 Ω cm	70K2, 73M2
		Sirtl etch for highly doped silicon	61S, 69V
		anodic etching	80F1

continued

Table 11 (continued)

Defect	Dimensionality	Diagnostic technique	Ref.
		electrical resistivity measurement spreading-resistance technique reveals resistivity fluctuations at high spatial resolution	67M 73B3 74V 76M2
		X-ray topography double crystal reflection topography reveals lattice parameter variations caused, e.g. by carbon	65R 74B3 77D2
		EBIC electron beam-induced current analysis exhibits carrier concentration variations	73R2 75D
		autoradiography (after decoration)	77H1
Swirl defects *A*-defect *B*-defect	atomic point-defect agglomerates	**etching** Sirtl etch reveals shallow etch pits, modified version after Bernewitz and Mayer reveals hillocks, anodic etching reveals all kinds of defects	61S 73B2 80F1
		TEM sample preparation after Kolbesen et al., TEM exhibits dislocation loops	73R2, 74B3 74K3 77D2
		X-ray topography	70C 73D2 74D
		EBIC	75D
		autoradiography (after decoration)	77H1
Hydrogen defects	dislocation loop (line defect) agglomerates	**etching** polish etch	81K 67B
Etching depressions	atomic point-defect agglomerates	**etching**	75K1 81K
		polish etch	67B
		X-ray topography	73D1
		autoradiography (after decoration)	77H1

For the most important etchants, the mixtures as well as the obtainable results are listed in Table 12.

Table 12. Etchants for polishing and revealing crystal imperfections.

Name	Mixture vol % if nothing else mentioned		Etching rate and condition	Effect	Ref.
CP4 etch	HNO_3 (conc.)	45.45	(0.15⋯10) μm/min at 23 °C	polish etch	59H1 67B
	HF (40 %)	27.27			
	acetic acid (98 %)	27.27			
	bromine 0.5 ml for 100 ml etchant				
CP6 etch	HNO_3 (conc.)	45.45	10 μm/min at 23 °C	polish etch	59H1 67B
	HF (40 %)	27.27			
	acetic acid (98 %)	27.27			
7:7:5 etch	HNO_3 (conc.)	36.84		polish etch	
	HF (40 %)	36.84	at 23 °C		
	acetic acid	26.32			

continued

Table 12 (continued)

Name	Mixture vol % if nothing else mentioned	Etching rate and condition	Effect	Ref.
Dash etch	HNO_3 (conc.) 21.4 HF (48 %) 7.2 acetic acid (98 %) 71.4		preferential etch	56D 67B
Staining etch	HNO_3 (conc.) 0.5 HF (48 %) 99.5	illumination	reveals p-n junctions	56F 67B
Sirtl etch	CrO_3 50 g in 100 ml dest. H_2O } standard solution standard solution 50 HF (40 %) 50	1.3 µm/min at 23 °C	preferential etch, especially for {111} faces	61S 67B
Modified Sirtl etch	as above	12···18 °C throughout reaction time	preferential etch, reveals swirl hillocks	73B2
Secco etch	$K_2Cr_2O_7$ 4.4 g in 100 ml dest H_2O } standard solution standard solution 50 HF (49 %) 100	1.5 µm/min at 25 to 30 °C	preferential etch, especially for {100} faces	72S1
Mayer etch	$NaNO_2$ 200–1000 mg (depending on resistivity) HF (48 %) 100 ml wetting agent 2–5 drops	illumination by 100 W lamp, etching time 15···20 min	striation etch	73M2
Alcaline etch	NaOH (10 %) KOH (10 %)	ca. 100 °C ca. 100 °C	preferential etch used for seed orientation	67B

6.1.3.5.3 Characterization-techniques for high purity germanium (HP-Ge)

Because the net impurity concentration in HP-Ge-crystals is smaller than the intrinsic concentration at 20 °C, the electrical properties of HP-Ge have to be measured by cooling the sample to the extrinsic range. For net-impurity concentrations of 10^{11} cm^{-3}, this implies using temperatures $T<200$ K.

Typical measurements are given in Table 13.

Table 13. Characterisation techniques of HPGe.

Technique	Analysed property	Remarks	Ref.
Two-point probe	conduction type net impurity concentration	Pt-probe on n-type Ge In-probe on p-type Ge	71H, 72H 73H1, 77G1
Ingot resistance	conductivity	Fig. 75 In-soldered contacts to seed and tail end of ingot, equidistant probes on ingot surface by Ga – In eutectic	73H1, 77G1

continued

Table 13 (continued)

Technique	Analysed property	Remarks	Ref.
Hall effect	concentrations	cleaning procedure of surface required: 1. etch $4HNO_3+1HF$ 2. methanol rinsing 3. N_2 – gas drying liquid Ga-In eutectic contacts for p-type Ge, In (15 wt. %)-Hg contacts alloyed at 250 °C for n-type Ge	58P, 60M, 58B, 73H1
CV analysis	impurity concentrations	pn-diode required	
Photothermal ir-spectroscopy (PTIS)	impurity spectrum	Fig. 76	81H2
	impurity incorporation	two-step process for photo-excitation of impurities	65L, 66K, 68L
		Fourier transform spectroscopy (FTS) used at long wave-lengths (Fig. 77)	81H2
		preferred technique for analysis of Ge	72S2, 74S2, 74H1, 78H1, 79S2
		temperature dependence of PTIS	79J
		impurity in magnetic field	77J
		uniaxial studies	77K1, 78H1, 80H2, 80J2
Deep level transient spectroscopy (DLTS)	level spectrum	direct measurement of detector diode	77M2, 79H3
Electron paramagnetic resonance (EPR)	paramagnetic impurity centers	high Q-factor required	62L, 81H2
Radio tracer method	neutral impurities		78H1
	hydrogen properties	tritium doping	80H1
	carbon properties	C^{14}, β-decay	82H2
Li-precipitation	oxygen impurity	resolution $N_0 \geqq 10^{12}$ cm^{-3} precipitation of oxygen to Li – O clusters, Li evaporated at $T=400$ °C followed by rapid quenching to room temperature, 4-point probe measurement of Li-donors	82H1, 66F
Spark source mass spectroscopy (SSMS)	non-gaseous impurities	sensitivity in ppb range detection of Si in Ge, Na and K always observed	81H2

References for 6.1.3

49F Frank, F.C.: Proc. Phys. Soc. Lond. **A62** (1949) 202.
54C Camp, P.R.: J. Appl. Phys. **25** (1954) 459.
54F1 Fuller, C.S., Ditzenberger, J.A., Hannay, N.B., Buchler, E.: Phys. Rev. **96** (1954) 833.
54F2 Fuller, C.S., Severiens J.C.: Phys. Rev. **96** (1954) 21.
54L Ludwig, G.L., Watters, R.L.: Phys. Rev. **101** (1954) 1699.
54V Valdes, L.B.: Proc. IRE **42** (1954) 420.
55F Fuller, C.S., Ditzenberger, J.A., Hannay, N.B., Buchler, E.: Acta Met. **3** (1955) 97.
55M Morrison, G.H., Cosgrove, J.F.: Anal. Chem. **27** (1955) 810.
56D Dash, W.C.: J. Appl. Phys. **27** (1956) 1193.
56F Fuller, C.S., Ditzenberger, J.A.: J. Appl. Phys. **27** (1956) 544.

56K Kaiser, W., Keck, P.H., Lange, C.F.: Phys. Rev. **101** (1956) 1264.
56R Reiss, H., Fuller C.S., Morin F.J.: Bell Syst. Tech. J. **35** (1956) 535.
57F Fuller, C.S., Logan, R.A.: J. Appl. Phys. **28** (1957) 1427.
57H Hrostowski, H.J., Kaiser, R.H.: Phys. Rev. **107** (1957) 966.
57K Kaiser, W.: Phys. Rev. **105** (1957) 1751.
57S1 Southgate, P.D.: Proc. Phys. Soc. Lond. **B70** (1957) 804.
57S2 Struthers, J.D.: J. Appl. Phys. **27** (1957) 1409 and **28** (1957) 516.
58B Beer, A.C., Willardson, R.K.: Phys. Rev. **110** (1958) 1286. Also, Yee, J.H., Swierkowski, S.P., Armantrout, G.A., Wichner, R.: J. Appl. Phys. **45** (1974) 3949.
58K Kaiser, W., Frisch, H.L., Reiss, H.: Phys. Rev. **112** (1958) 1546.
58S Smits, F.M.: Bell Syst. Techn. J. **37** (1958) 711.
58V Van der Pauw, L.J.: Philips Res. Rep. **13** (1958) 1.
58Z Ziegler, G.: Z. Metallk. **49** (1958) 491.
59H1 Holmes, P.J.: Proc. IEE **106**, Pt. B, Suppl. 10/18 (1959) 861.
59H2 Hrostowski, H.J., Kaiser, R.H.: J. Phys. Chem. Solids **9** (1959) 214.
59H3 Hoffmann, A., Reuschel, K., Rupprecht, H.: J. Phys. Chem. Solids **11** (1959) 284.
59K Keller, W.: Z. angew. Phys. **11** (1959) 346.
59L Logan, R.A., Peters, A.J.: J. Appl. Phys. **30** (1959) 1627.
60E Edwards, W.D.: Can. J. Phys. **38** (1960) 439.
60F Frank, R.C., Thomas, J.E.: J. Phys. Chem. Solids **16** (1960) 144.
60H Haas, C.: J. Phys. Chem. Solids **15** (1960) 108.
60M Miyazawa, H., Maeda, H.: J. Phys. Soc. Jap. **15** (1960) 1924.
60P1 Pell, E.M.: J. Appl. Phys. **31** (1960) 291.
60P2 Putley, E.H.: The Hall Effect and Related Phenomena, London, Butterworths 1960.
60T Trumbore, F.A.: Bell Syst. Techn. J. **39** (1960) 205.
61L Logan, M.A.: Bell Syst. Techn. J. **40** (1961) 885.
61N Newman, R.C., Wakefield, J.: J. Phys. Chem. Solids **19** (1961) 230.
61S Sirtl, E., Adler, A.: Z. Metallk. **52** (1961) 529.
62I Irvin, J.C.: Bell Syst. Techn. J. **41** (1962) 387.
62L Ludwig, G.W., Woodburg, H.H.: *Solid State Physics,* **13**, 223 ed. F. Seitz and D. Turnbull, Academic Press Inc., New York (1962). Also Wilson, D.K.: Phys. Rev. **134** (1964) A265.
62T Turovski, B.M., Milvidskij, M.G.: Sov. Phys. Solid State **3** (1962) 1834.
63C Carruthers, J.R., Benson, K.E.: Appl. Phys. Lett. **3** (1963) 100.
63D Dikhoff, J.A.M.: Philips Techn. Rev. **25** (1963) 195.
63H Hirth, J.P., Pound, G.M.: Progress in Materials Science **11**. Mac Millan N.Y. (1963) 15.
63S1 Schwuttke, G.H.: J. Appl. Phys. **34** (1963) 1662.
63S2 Sirtl, E., Spielmann, V.: Z. Angew. Phys. **15** (1963) 295.
64A Amelinckx, S.: The Direct Observation of Dislocations, New York, Academic Press 1964.
64G Gebauhr, W., Martin, J.: Z. Anal. Chem. **200** (1964) 266.
64H Hall, R.N., Racette, J.H.: J. Appl. Phys. **35** (1964) 379.
64M Müller, A., Wilhelm, M.: Z. Naturforsch. **19a** (1964) 254.
64P Pajot, B.: J. Phys. Chem. Solids **25** (1964) 613.
64Q Queisser, H.J., Loon, van P.G.G.: J. Appl. Phys. **35** (1964) 3066.
64R For donors see: Reuszer, J.H., Fisher, P.: Phys. Rev. **135** (1964) A1125. For acceptors: Jones R.L., Fisher P.: J. Phys. Chem. Solids **26** (1965) 1125.
65A Aggarwal R.L., Fisher, P., Mourzine, V., Ramdas, A.K.: Phys. Rev. **138** (1965) A882.
65B Burke, J.: The Kinetics of Phase Transformations of Metals, Pergamon, London (1965) Chaps. 6, 7.
65H Hirsch, P.B., Howie, A., Nicholson, R.B., Pashley, D.W., Whelan, M.J.: Electron Microscopy of Thin Crystals, London, Butterworth 1965.
65L Lifshits, T.M., Nad, F.Ya.: Dokl. Akad. Nauk SSSR **162** (1965) 801; Soviet Phys. Doklady **10** (1965) 532. For an extensive review see: Kogan, Sh.M., Lifshits, T.M.: Phys. Status Solidi (a) **39** (1977) 11.
65M Mühlbauer, A., Kappelmeyer, R., Keiner, F.: Z. Naturforsch. **20a** (1965) 1089.
65P Plaskett, T.S.: Trans. AIME **233** (1965) 809.
65R Renninger, M.: Z. Angew. Phys. **19** (1965) 20.
66B Booker, G.R., Tunstall, W.J.: Phil. Mag. **13** (1966) 71.
66F Fox, R.J.: IEEE Trans. Nucl. Sci. NS-**13** (1966) 367.

66H Heinen, K.G., Larrabee, G.: Anal. Chem. **38** (1966) 1853.
66J1 Jaccodine, R.J., Drum, C.M.: Appl. Phys. Lett. **8** (1966) 29.
66J2 Joshi, M.L.: Acta Met. **14** (1966) 1157.
66K Kogan, Sh.M., Sedunov, B.I.: Fiz. Tverd. Tela **8** (1966) 2382; Sov. Phys. Solid State **8** (1967) 1898.
67A Abe, T., Maruyama, S.: Denki Kagaku **35** (1967) 149.
67B Bogenschütz, A.F.: Ätzpraxis für Halbleiter, Munich, Hanser 1967.
67M Mazur, R.G.: J. Electrochem. Soc. **114** (1967) 255.
67W White, J.J.: Can. J. Phys. **45** (1967) 2797.
68G Gegenwarth, H.: Solid State Electron. **11** (1968) 787.
68L Lifshits, T.M., Likhtman, N.P., Sidorov, V.I.: Fiz. Tekh. Poluprov. **2** (1968) 782; Sov. Phys. Semicond. (English Transl.) **2** (1968) 652.
68S Sze, S.M., Irvin, J.C.: Solid State Electron. **11** (1968) 599.
69B Baker, J.A.: in Semiconductor Silicon, R.R. Haberecht and E.L. Kern, eds., New York, The Electrochem. Society 1969, p. 566.
69C Ciszek, T.F.: in ref. [69 B], p. 156.
69G Gebauhr, W.C.J.: in ref. [69B], p. 517
69M1 Malm, D.L.: in ref. [69B], p. 534.
69M2 Martin, J.A.: in ref. [69B], p. 547.
69V Vieweg-Gutberlet, F.: Solid-State Electron. **12** (1969) 731.
70C Chikawa, J., Asaeda, Y., Fujimoto, I.: J. Appl. Phys. **41** (1970) 1922.
70H Hall, R.N., Baertsch, R.D., Soltys, T.J., Petrucco, L.J.: Annual Report GE Nr. 3, U.S. AEC Contract Nr. AT (30-1) 3870, March 1970.
70K1 Kane, P.F., Larrabee, G.B.: Characterization of Semiconductor Materials, New York, McGraw-Hill 1970.
70K2 Kämper, M.: J. Electrochem. Soc. **117** (1970) 261.
70K3 Krausse, J.: Paper at the 2. DFG Kolloq., Burghausen 1970.
70N Nozaki, T., Yatsurugi, Y., Akiyama, N.: J. Radioanal. Chem. **4** (1970) 87.
71A ASTM Method F 28, Annual Book of ASTM Standards, (1971), p. 486.
71B Bean, A.R., Newman, R.C.: J. Phys. Chem. Solids **32** (1971) 1211.
71D DIN 50 430 (German standard for two-point probe measurement), Berlin 1971.
71E Engelmann, C.: J. Radioanal. Chem. **7** (1971) 89, 281.
71H Hall, R.N., Soltys, T.J.: IEEE Trans. on Nuclear Science Vol. NS-18, Nr. 1, Feb. 1971.
71S Sugita, Y., Kato, T.: J. Appl. Phys. **42** (1971) 5847.
72B1 Bean, A.R., Newman, R.C.: J. Phys. Chem. Solids **33** (1972) 255.
72B2 Bell, R.J.: Introductory Fourier Transform Spectroscopy, Academic Press, New York (1972).
72H Hall, R.N., Baertsch, R.D., Soltys, T.J., Petrucco, L.J.: General Electric Annual Report 1972, Contract No. AT (11-1)-3193.
72K Kushner, R.A.: Electrochemical Society Fall Meeting Abstract (1972) 643.
72M Matsui, J., Kawamura, T.: Jpn. J. Appl. Phys. **11** (1972) 197.
72O Odgen, R.: Phys. Status Solidi (a) **14** (1972) K 101.
72P1 Prussin, S.: J. Appl. Phys. **43** (1972) 733.
72P2 Prussin, S.: J. Appl. Phys. **43** (1972) 2850.
72R Ravi, K.V.: J. Appl. Phys. **43** (1972) 1785.
72S1 Secco d'Aragona, F.: J. Electrochem. Soc. **119** (1972) 948.
72S2 Seccombe, S.D., Korn, D.M.: Solid State Commun. **11** (1972) 1539.
73A Abe, T., Abe, Y., Chikawa, J.: in Semiconductor Silicon, H.R. Huff and R.R. Burgess, eds., Princeton, The Electrochem. Society 1973, p. 95.
73B1 Barthel, J., Jurisch, M.: Kristall Techn. **8** (1973) 199.
73B2 Bernewitz, L.I., Mayer, K.R.: Phys. Status Solidi (a) **16** (1973) 579.
73B3 Burtscher, J., Krausse, J., Voss, P.: in ref. [73 A], p. 581.
73C Ciszek, T.F.: in ref. [73 A], p. 150.
73D1 De Kock, A.J.R., Roksnoer, P.J., Boonen, P.G.T.: in ref [73 A], p. 83.
73D2 De Kock, A.J.R.: Philips Res. Rep. Suppl. 1 (1973).
73D3 Dyer, L.D., Voltmer, F.W.: J. Electrochem. Soc. **120** (1973) 812.
73G Graff, K., Pieper, H., Goldbach, G.: in ref. [73 A], p. 170.
73H1 Haller, E.E., Hansen, W.L., Goulding, F.S.: IEEE Trans. on Nucl. Science **20**/1 (1973) 481.
73H2 Hsieh, C.M., Maher, D.M.: J. Appl. Phys. **44** (1973) 1302.

73M1 Martin, J.A., Haas, W.E.: in ref. [73A], p. 161.
73M2 Mayer, K.R.: J. Electrochem. Soc. **120** (1973) 1780.
73R1 Ravi, K.V., Varker, C.J.: in ref. [73A], p. 136.
73R2 Ravi, K.V., Varker, C.J.: in ref. [73A], p. 83.
73R3 Ravi, K.V., Varker, C.J., Volk, C.E.: J. Electrochem. Soc. **120** (1973) 533.
73S1 Sirtl, E.: in ref. [73A], p. 54.
73S2 Sittig, R., Zimmermann, W.: in ref. [73A], p. 590.
73S3 Spenke, E.: BMFT-NT74, „Physik und Chemie des Siliziums“ Final Report 1973.
73T Takano, Y., Maki, M.: in ref. [73A], p. 469.
73V1 Voltmer, F.W., Digges, T.G., Jr.: J. Crystal Growth **19** (1973) 215.
73V2 Voss, P.: IEEE Trans. Electron. Devices **ED-20** (1973) 299.
73W Witt, A.F., Lichtensteiger, M., Gatos, H.C.: J. Electrochem. Soc. **120** (1973) 1119.
74A1 Abe, T.: J. Crystal Growth **24/25** (1974) 463.
74A2 ASTM Method F 391-737, February 1974.
74B1 Bernewitz, L.I., Kolbesen, B.O., Mayer, K.R., Schuh, G.E.: Appl. Phys. Lett. **25** (1974) 277.
74B2 Bullis, W.M.: NBSIR 74-496 (August 1974).
74B3 Burtscher, J.: in Proc. European Summer School, "Scientific Principles of Semiconductor Technology", Bad Boll (1974), p. 63.
74B4 Butler, N.R., Fisher, P.: Bull. Am. Phys. Soc. Series II, **19** No. 1, 92 (1974).
74D De Kock, A.J.R., Roksnoer, P.J., Boonen, P.G.T.: J. Cryst. Growth **22** (1974) 311.
74E Ehrstein, J.R., ed.: NBS Special Public. 400-10 (December 1974).
74G1 Graff, K., Pieper, H.: Electron. Mat. Conf., Metallurg. Soc. AIME, Boston 1974.
74G2 Gruzin, P.L., Zemskii, S.V., Bulkin, A.D., Makarov, N.M.: Sov. Phys. Semicond. **7** (1974) 1241.
74H1 Haller, E.E., Hansen, W.L.: Solid. State Commun. **15** (1974) 687; Haller, E.E., Hansen, W.L.: IEEE Trans. Nucl. Sci. NS-**21** (1974) 279 and Haller, E.E., Hansen, W.L., Goulding, F.S.: IEEE Trans. Nucl. Sci. NS-**22** (1975) 127.
74H2 Hu, S.M.: J. Appl. Phys. **45** (1974) 1567.
74H3 Hausen, W.L., Haller, E.E.: IEEE Trans. on Nucl. Science NS **21** (1974) 251.
74J Jenkins, M., Ravi, K.V.: ECS Spring Meeting (1974) 79.
74K1 Keller, W., Mühlbauer, A.: Phys. Status Solidi (a) **25** (1974) 149.
74K2 Keller, W.: Int. Elektron. Rdsch. **28** (1974) 87.
74K3 Kolbesen, B.O., Mayer, K.R., Schuh, G.E.: J. Phys. E (Sci. Instrum.) **8** (1974) 197.
74K4 Krausse, J.: in NBS Special Public. 400-10 (1974), p. 109.
74L Lang, D.V.: J. Appl. Phys. **45** (1974) 3023.
74M1 Malyshev, V.A.: Sov. Phys. Semicond. **8** (1974) 92.
74M2 Mühlbauer, A., Sirtl, E.: Phys. Status Solidi (a) **23** (1974) 555.
74N Nozaki, T., Yatsurugi, Y., Akiyama, N., Endo, Y., Makide, Y.: J. Radioanalyt. Chem. **19** (1974) 109.
74P Prussin, S.: J. Appl. Phys. **45** (1974) 1635.
74R1 Ravi, K.V.: J. Electrochem. Soc. **121** (1974) 1090.
74R2 Ravi, K.V., Varker, C.J.: J. Appl. Phys. **45** (1974) 263.
74S1 Schnöller, M.: IEEE Trans. Electron. Devices **ED-21** (1974) 313.
74S2 Skolnick, M.S., Eaves, L., Stradling, R.A., Portal J.C., Askenazy S.: Solid State Commun. **15** (1974) 1403.
74V Vieweg-Gutberlet, F.G.: in NBS Special Public. 400–10 (1974), p. 185.
75A Authier, A., Patel, J.R.: Phys. Status Solidi (a) **27** (1975) 213.
75B Bykova, E.M., Goncharov, L.A., Lifshits, T.M., Sidorov, V.I., Hall, R.N.: Fiz. Tekh. Poluprov. **9** (1975) 1853; Sov. Phys. Semic. **9** (1976) 1288.
75C Carruthers, J.R., Witt, A.F.: in Crystals Growth and Characterization, R. Ueda and J.B. Mullin, eds., Amsterdam, North-Holland 1975, p. 107.
75D De Kock, A.J.R., Ferris, S.D., Kimerling, L.C., Leamy, H.J.: Appl. Phys. Lett. **27** (1975) 313.
75F1 Föll, H., Kolbesen, B.O.: Appl. Phys. **8** (1975) 319.
75F2 Föll, H., Kolbesen, B.O., Frank, W.: Phys. Status Solidi (a) **29** (1975) K83.
75H1 Hall, R.N.: Inst. Phys. Conf. Ser. **23** (1975) 190; also, Hall, R.N.: IEEE Trans. Nucl. Sci. NS-**21**, No. 1, (1974) 260.
75H2 Herrmann, H., Herzer, H., Sirtl, E.: in Festkörperprobleme, Vol. 15, H.J. Queisser, ed., Braunschweig, Pergamon/Vieweg 1975, p. 279.

75H3 Hu, S.M.: Appl. Phys. Lett. **27** (1975) 165.
75H4 Hubbard, G.S., Haller, E.E., Hansen, W.L.: Nucl. Instr. and Meth. **130** (1975) 481.
75K1 Keller, W., Mühlbauer, A.: Inst. Phys. Conf. Ser. No. 23 (1975) 538.
75K2 Kolbesen, B.O.: Appl. Phys. Lett. **27** (1975) 353.
75L Laudise, R.A.: in ref. [75C], p. 255.
75M1 Mattis, R.L., Buehler, M.G.: in NBS Spec. Public. 400-17 (1975), p. 27.
75M2 Mühlbauer, A., Sedlak, F., Voss, P.: J. Electrochem. Soc. **122** (1975) 1113.
75P Petroff, P.M., de Kock, A.J.R.: J. Cryst. Growth **30** (1975) 117.
75R1 Reichl, H., Bernt, H.: Solid-State Electron. **18** (1975) 453.
75R2 Ruge, I.: „Halbleiter-Technologie", Volume 4 in „Halbleiter-Elektronik", Editors Haywang, W. and Müller, R., Springer-Verlag, Berlin, Heidelberg, New York (1975) 116.
75T Tice, W.K., Huana, T.C.: Appl. Phys. Lett. **24** (1975) 157.
76G Glasow, P.A., Hayller, W.L.: IEEE Trans. Nucl. Sci. NS-23, Nr. 1, (1976) 92.
76H1 Haller, E.E., Hansen, W.L., Hubbard, G.S., Goulding, F.S.: IEEE Trans Nucl. Sci. NS-**23**, No. 1, (1976) 81.
76H2 Haller, E.E., Hubbard, G.S., Hansen, W.L., Seeger, A.: Radiation Effects in Semiconductors, 1976. Inst. Phys. Conf. Ser. **31** (1977) 309.
76K Kamm, J.D.: Solid-State Electron. **19** (1976) 921.
76M1 Maher, D.M., Staudinger, A., Patel, J.R.: J. Appl. Phys. **47** (1976) 3813.
76M2 Murgai, A., Gatos, H.C., Witt, A.F.: J. Electrochem. Soc. **123** (1976) 224.
76P1 Petroff, P.M., de Kock, A.J.R.: J. Cryst. Growth **35** (1976) 4.
76P2 Petroff, P.M., Rozgonyi, G.A., Sheng, T.T.: J. Electrochem. Soc. **123** (1976) 565.
76R1 Renninger, M.: J. Appl. Crystallogr. **9** (1976) 178.
76R2 Roksnoer, P.J., Bartels, W.J., Bulle, C.W.: J. Cryst. Growth **35** (1976) 245.
76R3 Rozgonyi, G.A., Kushner, R.A.: J. Electrochem. Soc. **123** (1976) 570.
76S Schnöller, M.S.: IEEE Trans. Electron. Devices, **ED-23** (1976) 803.
76T Tan, T.Y., Tice, W.K.: Phil. Mag. **34** (1976) 615.
77A1 Armigliato, A., Servodori, M., Solmi, S., Vecchi, I.: J. Appl. Phys. **48** (1977) 1806.
77A2 ASTM Method F 76, Annual Book of ASTM Standards, Part 43 (November 1977).
77B Bullis, W.M., Vieweg-Gutberlet, F.G.: in ref [77D1], p. 360.
77C1 Capper, P., Jones, A.W., Wallhouse, E.J., Wilkes, J.G.: J. Appl. Phys. **48** (1977) 1646.
77C2 Chikawa, J., Shirai, S.: J. Cryst. Growth **39** (1977) 328.
77D1 De Kock, A.J.R.: in ref. [77D1], p. 568.
77D2 De Kock, A.J.R.: in Crystal Growth and Materials, E. Kaldis and H.J. Scheel, eds., Amsterdam, North-Holland 1977, p. 662.
77F1 Föll, H., Gösele, U., Kolbesen, B.O.: J. Cryst. Growth **40** (1977) 90.
77F2 Föll, H., Gösele, U., Kolbesen, B.O.: in Semiconductor Silicon, H.R. Huff and E. Sirtl, eds., Princeton, The Electrochemical Society 1977, p. 568.
77F3 Freeland, P.E., Jackson, K.A., Lowe, C.W., Patel, J.R.: Appl. Phys. Lett. **30** (1977) 31.
77G1 Glasow, P., Raab, G.: Forschungsbericht BMFT-FB-SN 0015.
77G2 Graff, K., Hilgarth, J., Neubrand, H.: in ref. [77D1], p. 575.
77H1 Haas, E.W.: Internal Technical Report KWU R 53-3/77, Erlangen, 1977.
77H2 Haller, E.E., Hubbard, G.S., Hansen, W.L.: IEEE Trans. Nucl. Sci. NS-**24** No. 1, (1977) 48.
77H3 Haller, E.E., Hubbard, G.S., Hansen, W.L., Seeger, A.: Inst. Phys. Conf. Ser. No. **31** (1977) 309.
77H4 Helmreich, D., Sirtl, E.: in ref. [77D1], p. 626.
77H5 Herzer, H.: in ref. [77D1], p. 106.
77H6 Hu, S.M.: J. Vacuum Sci. Techn. **14** (1977) 17.
77H7 Haller, E.E.: 1st Seminar on Photoelectric Spectroscopy of Semiconductors, Moscow **(1977)**.
77J Jonglbloets, H.W.H.M., Stoelinga, J.H.M., van de Steeg, M.J.H., Wyder, P.: Physica **89B** (1977) 18.
77K Kimerling, L.C., Leamy, H.J., Patel, J.R.: Appl. Phys. Lett. **30** (1977) 73.
77K1 Kern, E.L., Yaggy, L.S.: in ref. [77D1], p. 52.
77M1 Mahajan, S., Rozgonyi, G.A., Brasen, D.: Appl. Phys. Lett. **30** (1977) 73.
77M2 Miller, G.L., Lang, D.V., Kimerling, L.C.: Annual Review of Material Science, 377, Annual Review Inc. (1977).
77M3 Murarka, S.P., Quintana, G.: J. Appl. Phys. **48** (1977) 46.
77M4 Murarka, S.P.: J. Appl. Phys. **48** (1977) 5020.

77P1 Patel, J.R., Jackson, K.A., Reiss, H.: J. Appl. Phys. **48** (1977) 5279.

77P2 Patel, J.R.: in ref. [77D1], p. 521.

77P3 Pearce, C.W., Rozgonyi, G.A.: in ref. [77D1], p. 606.

77R1 Rice, T.M.: Solid State Phys. **32** (1977) 1 eds., H. Ehrenreich, F. Seitz and D. Turnbull, Academic Press, N.Y.; J.C. Hensel, T.G. Phillips and G.A. Thomas, ibid 88; Jeffries, C.D.: Science **189** (1975) 955.

77R2 Rozgonyi, G.A., Seidel, T.E.: J. Crystal Growth **38** (1977) 359.

77R3 Rozgonyi, G.A., Seidel, T.E.: in ref. [77D1], p. 616.

77S1 Schulz, M., Lefèvre, H.: in ref. [77D1], p. 142.

77S2 Schwab, G.: in ref. [77D1], p. 481.

77S3 Seeger, A., Föll, H., Frank, W.: Inst. Phys. Conf. Ser. 31 (1977) "Radiation Effects in Semiconductors 1976", 12.

77S4 Seiter, H.: in ref. [77D1], p. 187.

77S5 Shiraki, H.: in ref. [77D1], p. 546.

77T1 Tajima, M.: Jpn. J. Appl. Phys. **16** (1977), 2263, 2265.

77T2 Tempelhoff, K., Spielberg, F.: in ref. [77D1], p. 585.

78B Blackburn, D.L., Larrabee, R.D.: in Semiconductor Characterization Techniques, P.A. Barnes and G.A. Rozgonyi, eds., Princeton, The Electrochem. Society 1978, p. 168.

78C Claeys, C.L., Declerck, G.J., Overstraeten, van R.J.: Rev. Phys. Appliquée **13** (1978) 797.

78G Groza, A.A., K̄uts, V.I.: Sov. Phys. Semicond. **12** (1978) 562.

78H1 Haller, E.E.: Izv. Akad. Nauk SSSR, Phys. Ser. **42**, No. 6, 1131 (1978); Bull. Acad. Sci. USSR Phys. Ser. 42 No. 6, 8 (1979).

78H2 Haller, E.E., Falicov, L.M.: Phys. Rev. Lett. **41** (1978) 1192; also: Inst. Phys. Conf. Ser. **43** (1979) 1039.

78H3 Hall, R.N., Soltys, T.J.: IEEE Trans. Nucl. Sci. **25/1** (1978) 385.

78H4 Hubbard, G.S., Haller, E.E., Hansen, W.L.: IEEE Transaction on Nuclear Science, Vol. NS **25**, N° 1, (1978) 362.

78K1 Katz, L.E., Kimerling, L.C.: J. Electrochem. Soc. **125** (1978) 1680.

78K2 Krivanek, O.L., Maher, D.M.: Appl. Phys. Lett. **32** (1978) 451.

78L Larrabee, R.D.: in ref. [78B], p. 71.

78M1 Masters, B.J., Gorey, E.F.: J. Appl. Phys. **49** (1978) 2717.

78M2 Murarka, S.P.: J. Appl. Phys. **49** (1978) 2513.

78P Plougonven, C., Leroy, B., Arhan, J., Lecuiller, A.: J. Appl. Phys. **49** (1978) 2711.

78S1 Shimizu, H., Yoshinaka, A., Sugita, Y.: Jpn. J. Appl. Phys. **17** (1978) 767.

78S2 Shimizu, H., et al.: United States Patent 4, 116, 719 Sept. 26 (1978).

78S3 Staudinger, A.: J. Appl. Phys. **49** (1978) 3870.

78T1 Tajima, M.: in ref [78B], p. 159.

78T2 Thurber, W.R., Mattis, R.L., Liu, J.M.: in ref [78B], p. 81.

78T3 Tsubouchi, N., Miyoshi, H., Abe, H.: Jpn. J. Appl. Phys. **17** (1978) Suppl. 17-1, 223.

78Y Yang, K.H., Anderson, R., Kappert, H.F.: Appl. Phys. Lett. **33** (1978) 225.

79A ASTM F-121-1979.

79C1 Cazcarra, V.: Inst. Phys. Conf. Ser. **46** (1979) 303.

79C2 Chikawa, J., Shirai, S.: Jpn. J. Appl. Phys. **5** (1979) 153.

79D De Kock, A.J.R., Staca, W.T., Wijgert, van de W.M.: Appl. Phys. Lett. **34** (1979) 611.

79G Gaworzewski, P., Schmalz, K.: Phys. Status Solidi (a) **55** (1979) 699.

79H1 Hall, R.N., Soltys, T.J., Euwaraye, A.O.: Annual Report Nr. 10, U.S. AEC Contract Nr. CH-3193-6) Nr. 10.

79H2 Haller, E.E.: Inst. Phys. Conf. Ser. **46** (1979) 205.

79H3 Haller, E.E., Li, P.P., Hubbard, G.S., Hansen, W.L.: IEEE Trans. Nucl. Sci. NS-**26**, No. 1, (1979) 265.

79H4 Hubbard, G.S., Haller, E.E., Hansen, W.L.: IEEE Trans. Nucl. Sci. **26** (1979) 303.

79J Jongbloets, H.W.H.M., Stoelinga, J.H.M., van de Steen, M.J.H., Wyder, P.: Phys. Rev. B**20** (1979) 3328.

79K1 Kanamori, A., Kanamori, M.: J. Appl. Phys. **50** (1979) 8095.

79K2 Kock, de A.J.R., Stacy, W.T., Wijgert, van de W.M.: Appl. Phys. Lett. **34** (1979) 611.

79M Muller, S.H., Sieverts, E.G., Ammerlaan, C.A.J.: Inst. Phys. Conf. Ser. **46** (1979) 297.

79P1 Patrick, W.J., Hu, S.M., Westdorn, W.A.: J. Appl. Phys. **50** (1979) 1399.

79P2 Patrick, W.J., Hearn, E., Westdorn, W., Bohg, A.: J. Appl. Phys. **50** (1979) 7156.

79S1 Schimmel, D.G.: J. Electrochem. Soc. **126** (1979) 479.
79S2 Schoenmaekers, W.K.H., Clauws, P., van den Steen, K., Broeck, J. Henck, R.: IEEE Trans. Nucl. Sci. NS-**26**, No. 1, (1979) 256; Clauws, P., van den Steen, K., Broeckx, J., Schoenmaekers, W.: Inst. Phys. Conf. Ser. **46** (1979) 218.
79T1 Takaoka, H., Oosaka, J., Inoue, N.: Jpn. J. Appl. Phys. **18** (1979) Supplement 18-1, 179.
79T2 Tempelhoff, K., Spielberg, F., Gleichmann, R., Wruck, D.: Phys. Status Solidi (a) **56** (1979) 213.
79W Wada, K., Takaoka, H., Inoue, N., Kohra, K.: Jpn. J. Appl. Phys. **18** (1979) 1629.
80B Baber, S.C.: Thin Solid Films **72** (1980) 201.
80D1 Dean, P.J., Cullis, A.G., White, A.M.: in ref [80D2], p. 113.
80D2 De Kock, A.J.R.: in Handbook on Semiconductors, T.S. Moss, ed., Vol. **3**, S.P. Keller, ed., Amsterdam, North-Holland 1980, p. 247.
80D3 De Kock, A.J.R., van de Wijgert, W.M.: J. Cryst. Growth **49** (1980) 718.
80D4 DIN 50 438, Pt. 1 (German standard for oxygen determination), Berlin 1980.
80D5 DIN 50 438, Pt. 2 (German standard for carbon determination), Berlin 1980.
80F1 Föll, H.: J. Electrochem. Soc. **127** (1980) 1925.
80F2 Freeland, P.E.: J. Electrochem. Soc. **127** (1980) 754.
80G Gass, J., Müller, H.H., Stüssi, H., Schweitzer, S.: J. Appl. Phys. **51** (1980) 2030.
80H1 Haller E.E., Goulding, F.S.: Handbook on Semiconductors, Vol. 4, Ch. 6C, ed. C. Hilsum, North-Holland Publ. Co. (1980) in print.
80H2 Haller, E.E., Joós, B., Falicov, L.M.: Phys. Rev. B**21** (1980) 4729.
80H3 Hu, S.M.: Appl. Phys. Lett. **36** (1980) 561.
80H4 Hu, S.M.: J. Appl. Phys. **51** (1980) 3666.
80H5 Huber, H., Sirtl, E.: Jpn. J. Appl. Phys. **19**, Suppl. 19-1 (1980) 615.
80J1 Jastrzebski, L., Lagowski, J.: RCA Rev. **41** (1980) 181.
80J2 Joós, B., Haller, E.E., Falicov, L.M.: Phys. Rev. B, **22** (1980) 832.
80K Kock, de A.J.R., Wijgert, van de W.M.: J. Cryst. Growth **49** (1980) 718.
80M Millett, E.J.: J. Cryst. Growth **48** (1980) 666.
80O1 Ohsawa, A., Honda, K., Ohkawa, S., Ueda, R.: Appl. Phys. Lett. **36** (1980) 147.
80O2 Osaka, J., Inoue, N., Wada, K.: Appl. Phys. Lett. **36** (1980) 288.
80S1 Shimura, F., Tsuya, H., Kawamura, T.: J. Appl. Phys. **51** (1980) 269.
80S2 Shimura, F., Tsuya, H., Kawamura, T.: Appl. Phys. Lett. **37** (1980) 483.
80S3 Shirai, S.: Appl. Phys. Lett. **36** (1980) 156.
80T Tajima, M.: J. Appl. Phys. **51** (1980) 2247.
80U Umeno, M., Hildebrandt, G.: Z. Naturforsch. **35a** (1980) 342.
80V Vassilyeva, E.D., Emtsev, V.V., Haller, E.E., Mashovets, T.V.: Sov. Phys. Semic. (1980), submitted.
80W1 Wada, K., Inoue, N.: J. Crystal Growth **49** (1980) 749.
80W2 Weber, J., Wagner, P.: 15th Int. Conf. Phys. Semicond., Tokyo, 1980.
81D1 De Kock, A.J.R.: in Defects in Semiconductors, J. Narayan and T.Y. Tan, eds., Amsterdam, North-Holland 1981, p. 309.
81D2 Dietl, J., Helmreich, D., Sirtl, E.: in Crystals, Growth, Properties and Application, Vol. 5, Berlin, Heidelberg, New York: Springer 1981, p. 43.
81D3 Dietze, W., Keller, W., Mühlbauer, A.: in ref. [81D2], p. 1.
81F Föll, H., Gösele, U., Kolbesen, B.O.: J. Cryst. Growth **52** (1981) 907.
81G Graff, K., Pieper, H.: in Semiconductor Silicon, H.R. Huff and R.J. Kriegler, eds., Pennington, The Electrochemical Society 1981, p. 331.
81H1 Haas, E.W.: Private communication by the Radiochemical Laboratory of the Kraftwerk Union AG, Erlangen 1981, data established in 1972–1979.
81H2 Haller, E.E., Hansen, W.L., Goulding, F.S.: Adv. Phys. **30/1** (1981) 93.
81H3 Hoshikawa, K., Hirata, H., Nakanishi, H., Ikuta, K.: in ref [81G], p. 101.
81J Jastrzebski, L., Zanzucchi, P.: in ref [81G], p. 138.
81K Keller, W., Mühlbauer, A.: Floating-Zone Silicon, New York, Marcel Dekker 1981.
81L Leroueille, J.: Phys. Status Solidi (a) **67** (1981) 177.
81M Murgai, A.: in ref. [81G], p. 113.
81P Patel, J.R.: in ref [81G], p. 189.
81R1 Rava, P., Gatos, H.C., Lagowski, J.: in ref [81G], p. 232.
81R2 Reichel J.: Phys. Status Solidi (a) **66** (1981) 277.
81T Tajima, M., Masui, T., Abe, T., Iizuka, T.: in ref [81G], p. 72.

82H1 Haller, E.E., Hansen, W.L., Luke, P., Murray, R.Mc, Jarrett, B.: IEEE Trans. Nucl. Sci. NS 29/1 (1982) 745.
82H2 Hansen, W.L., Haller, E.E., Luke, P.N.: IEEE Trans. Nucl. Sci. NS-**29**/1 (1982) 738.
82K Kolbesen, B.O., Mühlbauer, A.: Solid State Electron. **25** (1982) 775.
82Z Zulehner, W., Huber, D.: Crystals, Growth, Properties and Application, Vol. 8; Silicon; Czochralski-Grown Silicon. Berlin, Heidelberg, New York: Springer 1982.

6.1.4 Device technology

6.1.4.0 Basic device structures

6.1.4.0.0 Introduction

This section gives a short description of the basic mechanisms, the main properties, and the structure of the most important silicon and germanium devices as well as a detailed description of their technology. Adequate to their importance more silicon devices than germanium devices are treated.

The semiconductor silicon exhibits two essential advantages in comparison to germanium. These are:

the larger bandgap of Si than that of Ge results in a low reverse current of pn-junctions

SiO_2 is a stable and high-performance insulator on silicon. GeO_2 is water-soluable and therefore not suited for the fabrication of devices. The planar technology frequently employed in the fabrication of discrete devices and integrated circuits can only be used with silicon.

The smaller bandgap and the higher mobility of germanium than those of silicon are of importance for photodetectors and microwave diodes.

The manufacture of silicon and germanium devices includes wafer processing as well as wafer probe and chip assembly. Wafer processing can be described in terms of the following process steps:

fabrication of layers
lithography
etching
doping

An example of wafer processing is shown in Fig. 1. Representative process steps for the fabrication of an MOS transistor having a polysilicon gate are illustrated. All these steps are discussed in detail in the following sections. The fabrication of layers is described in the sections 6.1.4.4 and 6.1.4.5. Lithography is covered by the section 6.1.4.6. The focus of section 6.1.4.7 is on etching. Various doping methods are discussed in the sections 6.1.4.1, 6.1.4.2, and 6.1.4.3. Figure 2 illustrates the final device preparation stage, which is described at length in section 6.1.4.8.

Acknowledgements

The author would like to thank D. Widmann, H. Patalong, E. Krimmel, C. Weyrich, G. Winstl, M. Zerbst, H. Friedrich, W. Beinvogl, and W. Müller of the Siemens Research Laboratory, W. Spaeth, J. Müller, G. Olk of the Siemens Device Division, and R. Müller of the Technische Universität München, for many helpful discussions.

6.1.4.0.1 Diodes

Table 1. Basic structures of the most important Si and Ge diodes and short descriptions of their operational principle, their characteristic, and their applications.

No.	Diode	Semiconductor	Symbols	Structure	Doping profile	Characteristics	Operational principle	Applications	Refs.
1	p-n junction diode	Si		metal, SiO_2, p^+, n, Si, n^+, metal diffused planar junction on epitaxial substrate	N_a, N_{d1}, N_{d2}, N, x, 0 N_d : donator density N_a : acceptor density	I, V, 0 current voltage characteristic	forward current voltage characteristic determined by minority carrier injection, reverse voltage limited by avalanche breakdown or tunneling	rectifier, demodulator, switch diode	81 S1 81 N1 80 M1 79 M2
2	Zener diode	Si		metal, SiO, p^+, n, Si, n^+, metal diffused planar junction on epitaxial substrate	N_a, N_{d1}, N_{d2}, N, x, 0	I, V_z, V, 0 current voltage characteristic	breakdown voltage V_z determined by avalanche or tunnel effect	reference diode (reference voltage = V_z)	81 S1 81 N1 80 M1 79 M2 80 L1 82 01
3	psn diode	Si		metal, p^+, s(π,ν), Si, n^+, metal s : lightly doped π : lightly doped on p-type ν : lightly doped on n-type	N_{a1}, N_{a2} or N_{d1}, N_{d2}, N, x, 0	I, V, 0 current voltage characteristic	high breakdown voltage obtained by a lightly doped middle layer	high-power rectifier	81 S1 79 M2
4	Shockley diode	Si		metal (anode), p, n, p, n, Si, metal (cathode)	N_{a1}, N_{d1}, N_{a2}, N_{d2}, N, x, 0	I, "on", "off", V_{BR}, V, 0 current voltage characteristic	the diode is in the off state until the breakover voltage V_{BR} is reached, at which avalanche conditions develop and the device turns on	trigger, switch	81 N1 81 S1 79 M2 80 H2 80 M1 80 L1 78 T1 75 M1
5	Diac	Si		metal (anode1), n, p, n, p, n, metal (anode 2)	N_{d1}, N_{a1}, N_{d2}, N_{a2}, N_{d3}, N, x, 0	I, V, 0 current voltage characteristic	the current voltage characteristic demonstrates that there is a breakover voltage in either direction	trigger, switch in ac applications	81 N1 81 S1 79 M2 80 H2 80 M1 80 L1 78 T1 75 M1
6	Schottky-barrier diode	Si		top metal (Ti-Pt-Au), barrier metal (Mo, Ti), Si_3N_4, SiO_2, p^+, guard-ring, n, n^+, Si, ohmic contact guard-ring diode	N_a (guard-ring), N_{d1}, N_{d2}, N, x, 0	I, V, 0 current voltage characteristic	majority-carrier transport across an energy barrier at the metal-semiconductor interface	detector, mixer, switch, Schottky-barrier clamped transistor, photodiode, solarcell	81 S1 80 M1 79 M2 78 Y1 82 M2

(Error in No. 2: SiO should read SiO_2).

continued

Table 1 (continued)

No.	Diode	Semiconductor	Symbols	Structure	Doping profile	Characteristics	Operational principle	Applications	Refs.
7	Bulk-barrier diode	Si	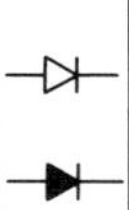	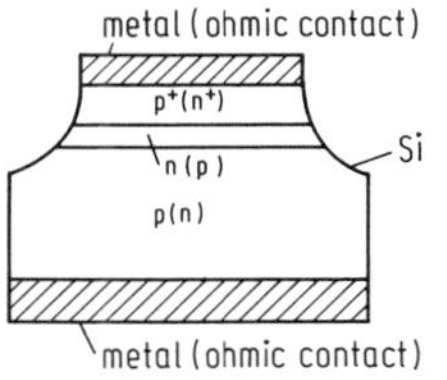 metal (ohmic contact); $p^+(n^+)$; n (p); p(n); Si; metal (ohmic contact); double implanted diode	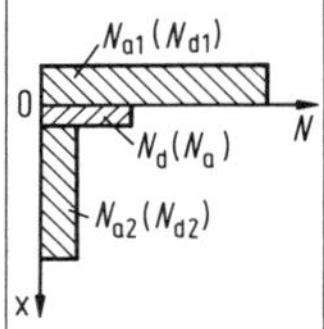 $N_{a1}(N_{d1})$; $N_d(N_a)$; $N_{a2}(N_{d2})$	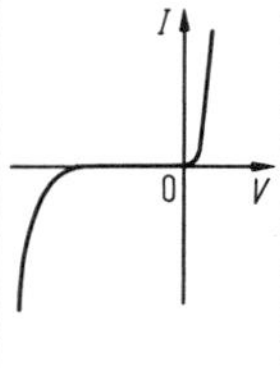 current voltage characteristic	majority-carrier transport across an energy barrier in the semiconductor bulk	detector of small microwave signals, microwave mixer, switch, clamped transistor, photodiode for IR, UV and visible light, thermistor	79 M3, 81 M1, 82 L1, 82 M3
8	Point-contact diode	Ge	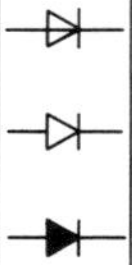	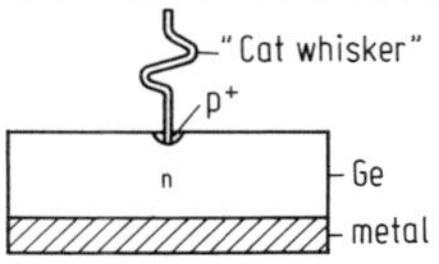 "Cat whisker"; p^+; n; Ge; metal; p-n junction generated by a high current passed through the whisker	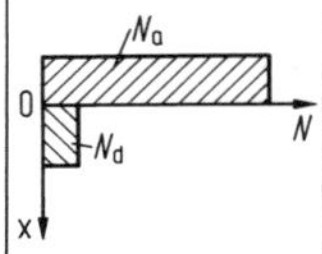 N_a; N_d	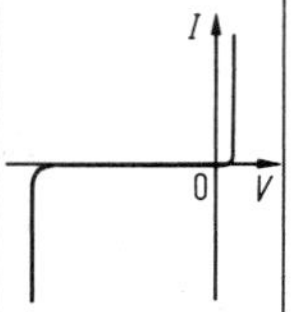 current voltage characteristic	minority-carrier transport, small capacitance due to the small contact area	microwave detector, microwave mixer	81 S1, 81 N1, 82 M2
9	Tunnel diode	Ge		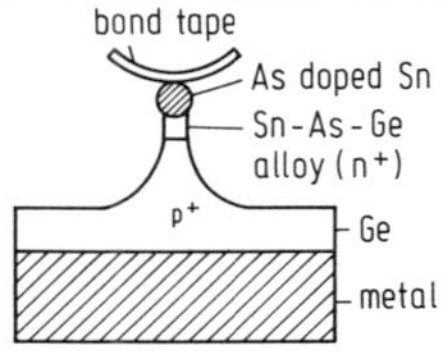 bond tape; As doped Sn; Sn-As-Ge alloy (n^+); p^+; Ge; metal; alloyed mesa junction	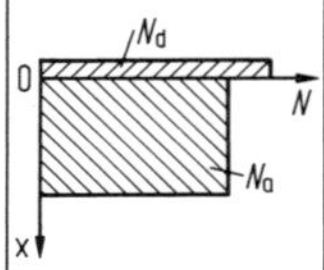 N_d; N_a	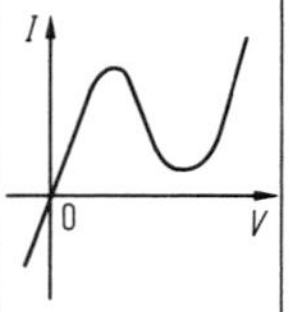 current voltage characteristic	tunneling in forward-biased p^+n^+ junction, negative differential resistance, p^+ and n^+ region degenerated	oscillator, parametric amplifier, switch	81 S1, 81 N1, 81 H3, 79 M2, 80 L1, 82 M2
10	Backward diode	Ge		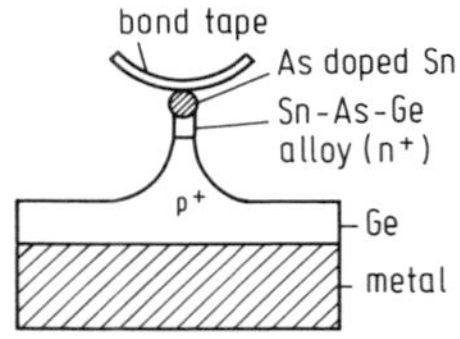 bond tape; As doped Sn; Sn-As-Ge alloy (n^+); p^+; Ge; metal; alloyed mesa junction	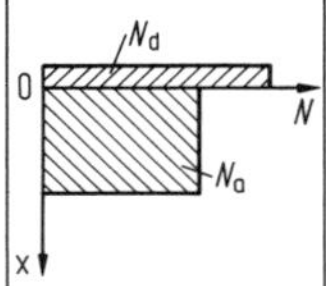 N_d; N_a	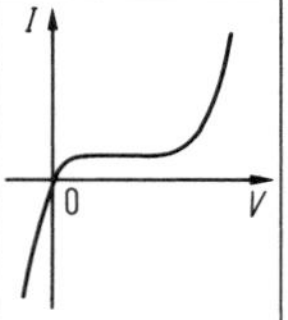 current voltage characteristic	tunneling in reverse-biased junction or near zero bias, high nonlinearity	rectifier of small signals, microwave mixer, microwave detector	81 S1, 81 H3, 79 M2, 82 M2
11	Pin diode	Si	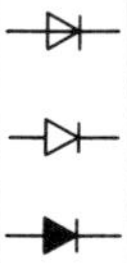	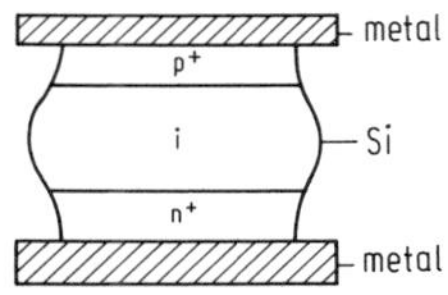 metal; p^+; i; Si; n^+; metal; point-on technology; point-on diffusion of n^+ and p^+ in intrinsic (i) silicon wafer	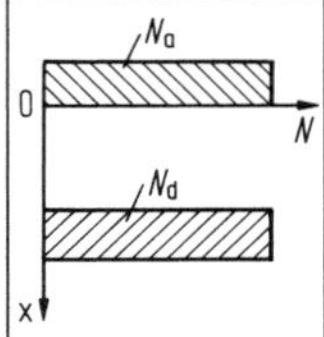 N_a; N_d	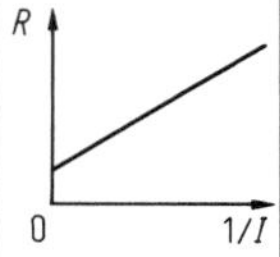 series resistance vs. reciprocal of forward current	charge carrier injection into the i-region decreases the resistance, nearly constant capacitance	attenuator, switch	81 S1, 80 M1, 79 M2, 82 M2
12	Varactor diode (varicap diode)	Si	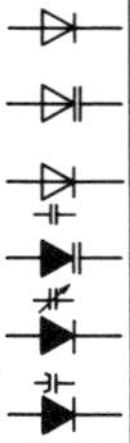	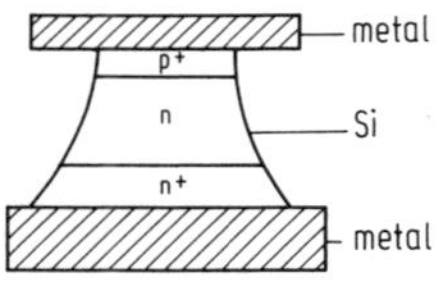 metal; p^+; n; Si; n^+; metal; diffused mesa junction	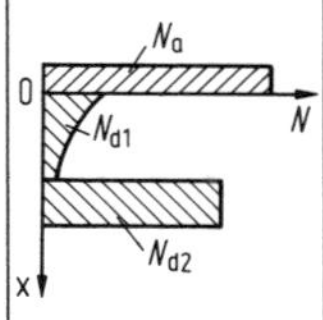 N_a; N_{d1}; N_{d2}; $N_{d1} = Bx^{-m}$ hyper-abrupt	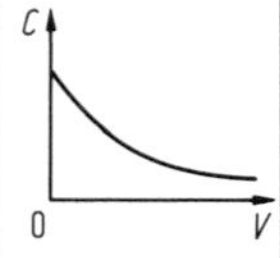 capacitance voltage characteristic	reactance varies with bias voltage	parametric amplifier, harmonic generator, mixer, detector, voltage-variable tuner, frequency multiplier	81 S1, 81 N1, 79 M2, 80 O1

(Error in No. 11: point-on technology should read paint-on technology).

continued

Table 1 (continued)

No.	Diode	Semiconductor	Symbols	Structure	Doping profile	Characteristics	Operational principle	Applications	Refs.
13	MIS diode	Si		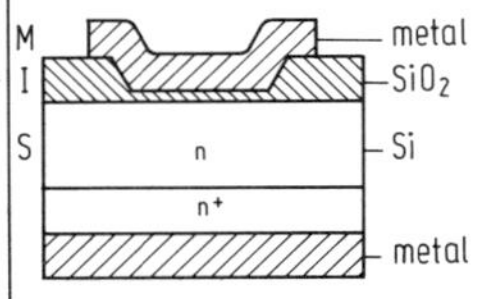 M, I, S; metal, SiO_2, n, Si, n^+, metal oxidation on epitaxial substrate	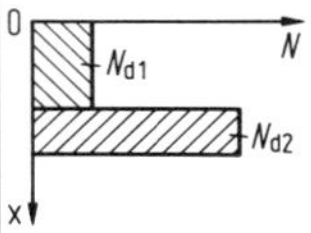 0, N_{d1}, N, N_{d2}, x	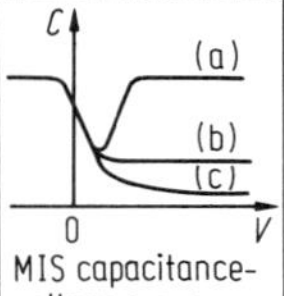 C, (a), (b), (c), 0, V MIS capacitance-voltage curves (a) low-frequency (b) high-frequency (c) deep depletion	capacitance depends on the applied voltage and the frequency	varactor, charge-coupled device, MIS tunnel diode, MIS switch diode	81 S1 79 M2 82 M2
14	Step-recovery diode	Si	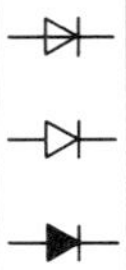	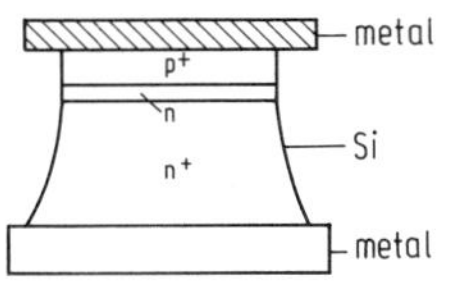 metal, p^+, n, Si, n^+, metal implanted junction on epitaxial substrate	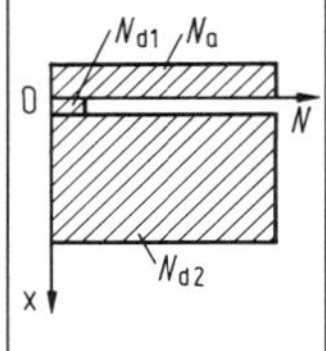 N_{d1}, N_a, 0, N, N_{d2}, x	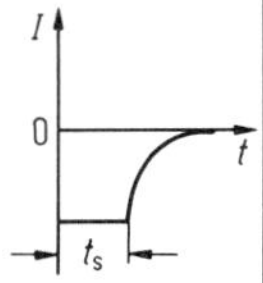 I, 0, t, t_s t_s : storage time I : current	ultrahigh switching speed due to the thin n-doped layer	generator for subnanoseconds pulses, generator for microwave harmonics	81 S1 76 T1 75 M1 82 M2
15	IMPATT diode impact ionization avalanche transit time	Si		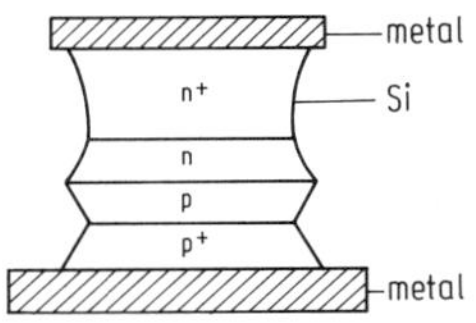 metal, n^+, Si, n, p, p^+, metal double-drift IMPATT diode using ion implantation	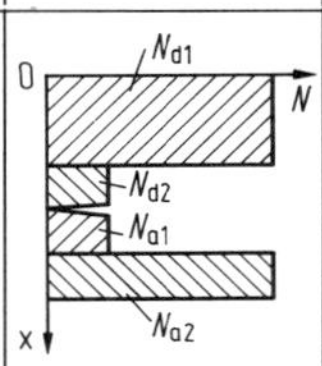 0, N_{d1}, N, N_{d2}, N_{a1}, N_{a2}, x	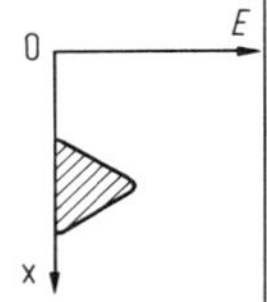 0, E, x electric field distribution	avalanche and transit-time effects	microwave oscillator, microwave amplifier	81 S1 81 H3 80 M1 79 M2 82 M2
16	TRAPATT diode trapped plasma avalanche-triggered transit	Si		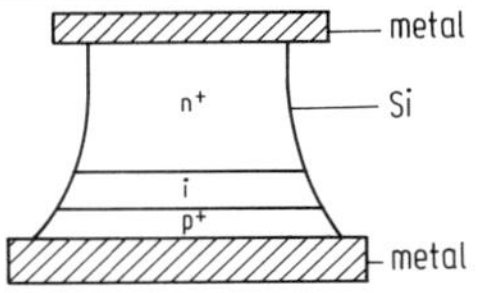 metal, n^+, Si, i, p^+, metal diffused mesa junction on epitaxial substrate	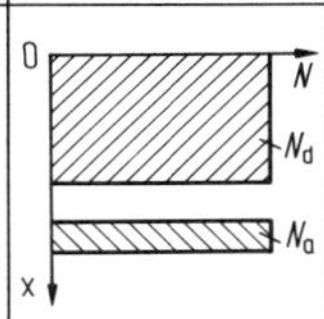 0, N, N_d, N_a, x	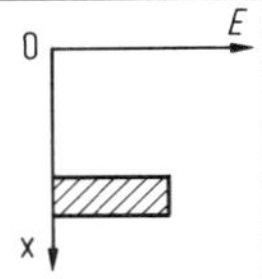 0, E, x electric field distribution	trapped plasmas, avalanche triggered transit of charge carriers	microwave oscillator, microwaves amplifier	81 S1 79 M2 82 M2
17	BARRITT diode barrier injection and transit time	Si		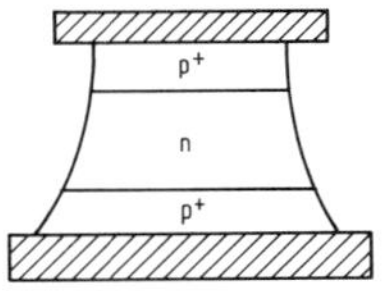 p^+, n, p^+ diffused mesa junction on epitaxial substrate	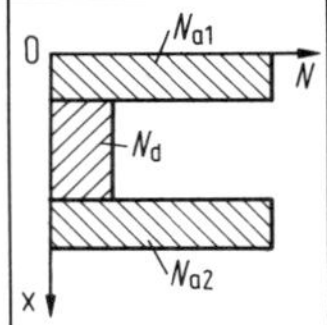 0, N_{a1}, N, N_d, N_{a2}, x	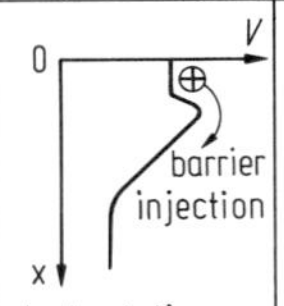 0, V, ⊕, barrier injection, x electrostatic potential distribution under bias condition	barrier injection and transit-time effects, low noise	microwave oscillators, microwave mixer	81 S1 81 H3 82 M2
18	pn-photodiode	Si, Ge	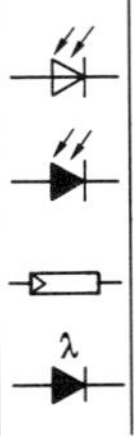 λ	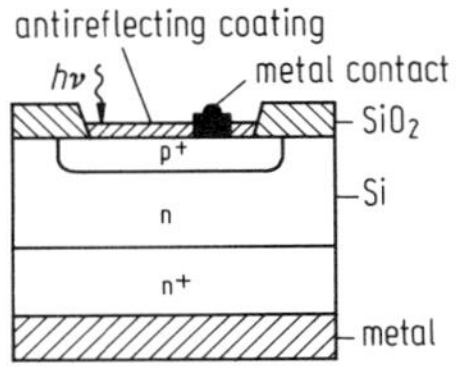 antireflecting coating, $h\nu$, metal contact, SiO_2, p^+, n, Si, n^+, metal diffused planar junction on epitaxial substrate	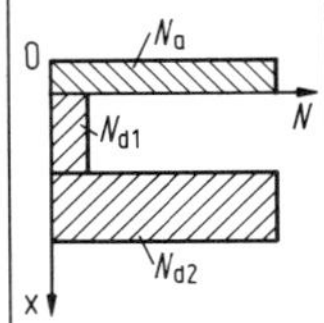 0, N_a, N_{d1}, N, N_{d2}, x	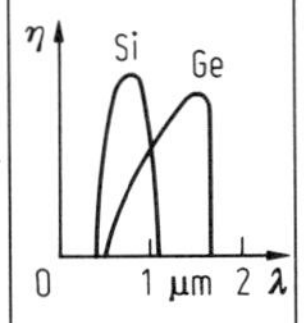 η, Si, Ge, 0, 1 µm, 2, λ quantum efficiency as a function of wavelength	generation of electron hole pairs by photons	photodetector, solarcell	81 S1 79 M2 80 L1 82 S1

continued

Table 1 (continued)

No.	Diode	Semiconductor	Symbols	Structure	Doping profile	Characteristics	Operational principle	Applications	Refs.
19	pin-photodiode	Si		antireflecting coating, $h\nu$, metal contact, SiO_2, p^+, i, Si, n^+, metal; diffused planar junction on epitaxial substrate	N_a, N_d, 0, N, x	η, Si, Ge, 0, 1 µm, 2, λ; quantum efficiency as a function of wavelength	generation of electron hole pairs by photons, for high frequency	photodetector, solarcell	81 S1, 79 M2, 80 L1, 82 S1
20	Avalanche photodiode	Si		antireflecting coating, n^+, metal, SiO_2, n, p, Si, π, p^+, metal; guard-ring n^+-p-π-p^+ structure	N_d, N_{a1}, N_{a2}, 0, N, x	0, E, x, avalanche multiplication; electric field distribution	avalanche multiplication gives rise to internal current gain, operated at high reverse-bias voltages	photodetector for microwaves at high internal current gain	81 S1, 80 L1, 82 S1

Representative for all diodes listed in Table 1 is the following sequence of production steps, typical for Si pn-diodes. These production steps are illustrated by Fig. 3a···m. The technology processes involved are discussed in detail in separate sections, the numbers of which are given in parentheses:

Deposition of an n-type silicon layer on an n^+-type silicon substrate by epitaxy; Fig. 3b (section 6.1.4.4).

Fabrication of an SiO_2-layer by oxidation; Fig. 3c (section 6.1.4.5).

Depositing a photoresist film and exposing it to light through a photomask 1; Fig. 3d (section 6.1.4.6).

Removal of the exposed positive photoresist by a development solvent; Fig. 3e (section 6.1.4.6).

Wet chemical or dry etching of SiO_2; Fig. 3f. The photoresist pattern serves as an etch mask (section 6.1.4.7).

Photoresist stripping; Fig. 3g (section 6.1.4.6).

p-type doping through the SiO_2 window by diffusion; Fig. 3h. SiO_2 growth during diffusion (section 6.1.4.1).

Deposition of photoresist and exposure through a photomask 2. Photoresist pattern generated by development. Transfer of the photoresist mask to the SiO_2 layer by etching. Photoresist stripping; Fig. 3i (sections 6.1.4.6 and 6.1.4.7).

Deposition of a metal layer (mostly aluminium); Fig. 3k (section 6.1.4.5).

Deposition of photoresist and exposure through a photomask 3. Developed photoresist serves as a mask for etching the metal layer. Photoresist stripping Fig. 3l (sections 6.1.4.6 and 6.1.4.7).

Wafer probe for selection of the working diodes and die assembly; Fig. 3m (section 6.1.4.8).

6.1.4.0.2 Transistors

Table 2. Basic structures of the most important Si and Ge transistors and short descriptions of their characteristic, their operational principle, and their applications.

No.	Transistor	Semiconductor	Symbols	Structure	Doping profile	Characteristics	Operational principle	Applications	Refs.
1	pnp bipolar transistor	Si	B, C, E	Al (emitter), Al (base), B, E, B, SiO_2, p^+, n, p, Si, p^+, metal (collector), ground plate, C; double-diffused planar-epitaxial transistor on epitaxial substrate	N_{a1}, N_d, N_{a2}, N_{a3}, 0, N, x; N_a: acceptor density, N_d: donator density	$-I_C$, $-I_B$, 0, $-V_{CE}$; collector current $-I_C$ vs. collector emitter voltage $-V_{CE}$	$-I_C$, 0, $-I_B$; injection of minority carriers from emitter to base, collector current $-I_C$ controlled by base current $-I_B$	amplifier, switch, mixer, oscillator	81 S1, 78 Y1, 79 M2, 81 H4, 75 M1

Note: see also 83M1.

continued

Table 2 (continued)

No.	Transistor	Semiconductor	Symbols	Structure	Doping profile	Characteristics	Operational principle	Applications	Refs.
2	npn bipolar transistor	Si	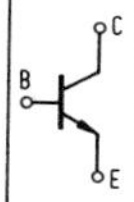 C B E	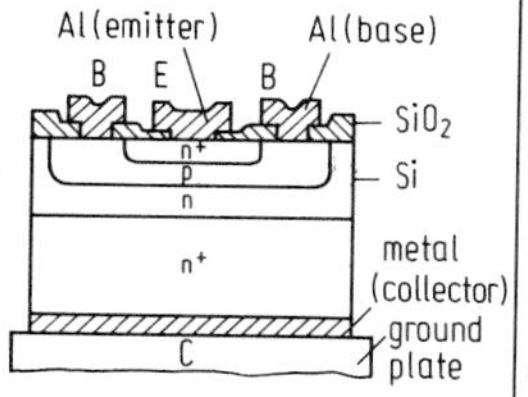 Al(emitter) Al(base) B E B n+ p n n+ SiO2 Si metal (collector) ground plate C double-diffused planar-epitaxial transistor on epitaxial substrate	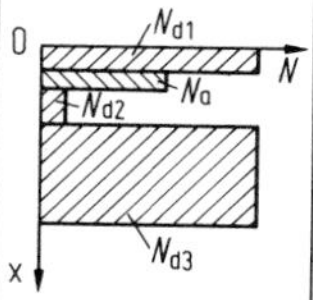 0 Nd1 Nd2 Na N Nd3 x	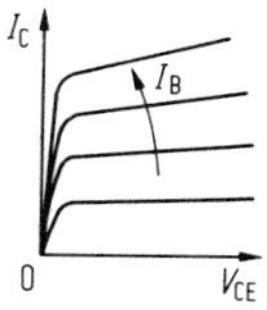 IC IB 0 VCE collector current I_C vs. collector emitter voltage V_{CE}	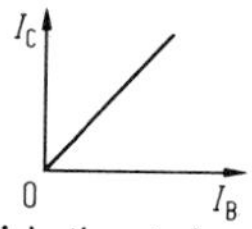 IC 0 IB injection of minority carriers from emitter to base, collector current I_C controlled by base current I_B	amplifier, switch, mixer, oscillator	81 S1 81 H4 78 Y1 79 M2 75 M1
3	pnp bipolar transistor	Ge	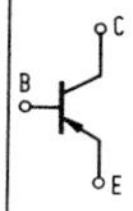 C B E	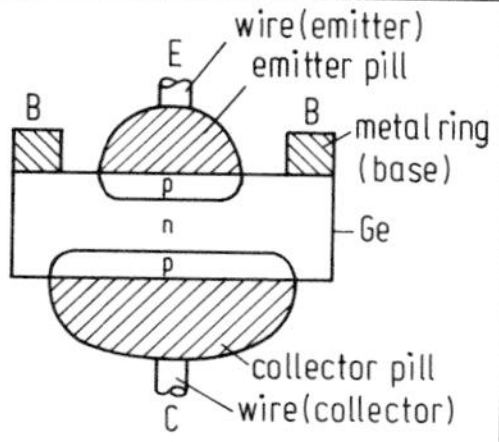 wire (emitter) E emitter pill B B metal ring (base) p n p Ge collector pill C wire (collector) alloyed transistor	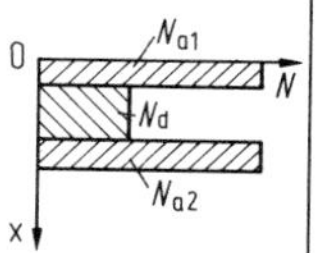 0 Na1 N Nd Na2 x	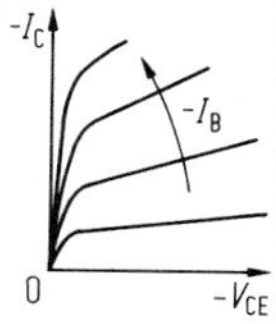 -IC -IB 0 -VCE collector current $-I_C$ vs. collector emitter voltage	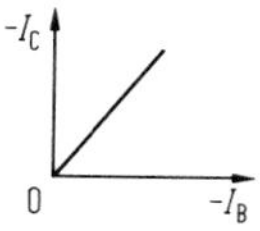 -IC 0 -IB injection of minority carriers from emitter to base, collector current $-I_C$ controlled by base current $-I_B$	amplifier, switch, mixer, oscillator	81 S1 79 M2 75 M1
4	Bipolar photo-transistor	Si	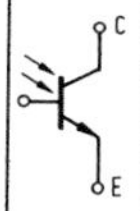 C E	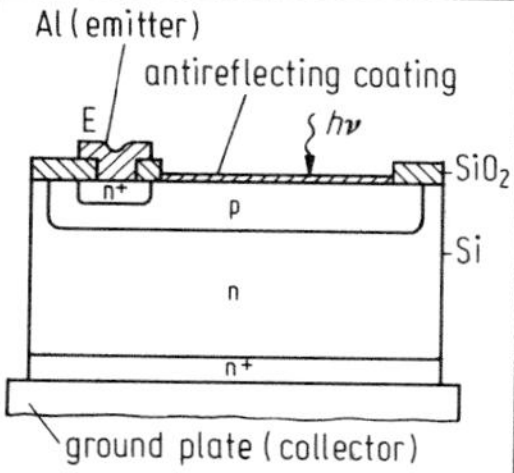 Al (emitter) antireflecting coating E hν SiO2 n+ p Si n n+ ground plate (collector)	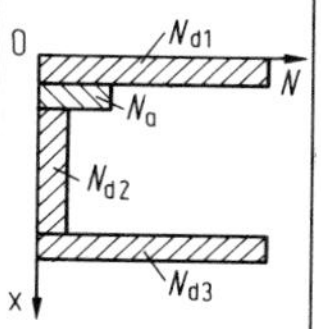 0 Nd1 N Na Nd2 Nd3 x	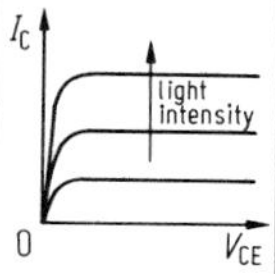 IC light intensity 0 VCE current voltage characteristic	generation of electron-hole pairs in the space charge region due to impinging photons, amplification by the transistor	photodetector	81 S1 82 S1 79 M2 80 L1 78 T1 75 M1 80 M1
5	Unijunction-transistor (UJT)	Si	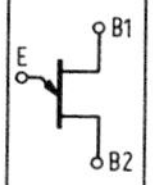 E B1 B2	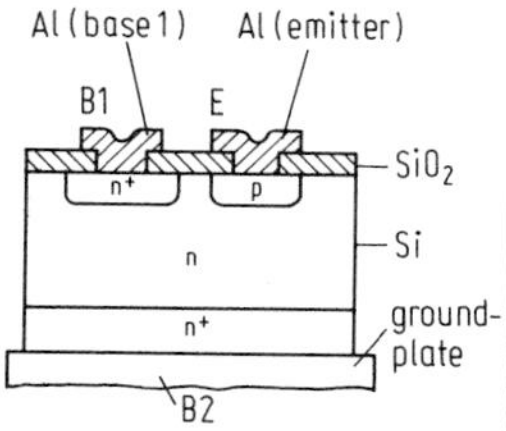 Al (base 1) Al (emitter) B1 E n+ p SiO2 Si n n+ ground-plate B2 diffused base 1 and emitter	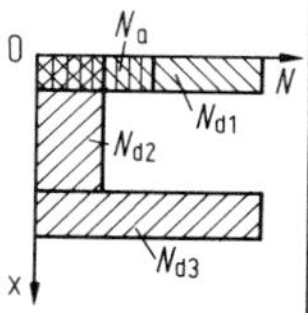 0 Na N Nd1 Nd2 Nd3 x	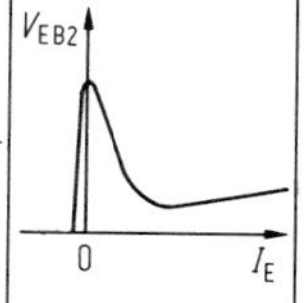 VEB2 0 IE emitter voltage V_{EB2} vs. emitter current I_E	negative differential resistance, modulation of the conductivity	oscillator, inductivity, pulse generator, ramp generator	81 S1 80 L1 78 T1 75 M1 81 N1

Note: in No. 2 see also 83M1.

continued

Table 2 (continued)

No.	Transistor	Semiconductor	Symbols	Structure	Doping profile	Characteristics	Operational principle	Applications	Refs.
6	JFET (junction field-effect transistor) n-channel	Si	normally-on JFET; normally-off JFET	Al (source), Al (gate), Al (drain), S, G, D, SiO$_2$, p^+, n, Si, p^+, ground-plate; fabricated by epitaxial-diffused process	N_{a1}, N_d, N_{a2}, N, x; N_d depends on threshold voltage V_{th} for "normally-on" or "normally-off" type transistors	I_D, V_{GS}, V_{DS}; drain current I_D vs. drain source voltage V_{DS}	lateral current flow is controlled by an externally applied vertical electric field; I_D, V_{th} 0 V_{GS} "normally-on" JFET; I_D, 0 V_{th} V_{GS} "normally-off" JFET	microwave amplifier, fast switch, microwave mixer, microwave oscillator	81 S1, 78 Y1, 80 M1, 79 M2, 80 L1, 81 M2, 81 N1
7	JFET (junction field-effect transistor) p-channel	Si	normally-on JFET; normally-off JFET	Al (source), Al (gate), Al (drain), S, G, D, SiO$_2$, n^+, p, Si, n^+, ground-plate; fabricated by epitaxial-diffused process	N_{d1}, N_a, N_{d2}, N, x; N_a depends on threshold voltage V_{th} for "normally-on" or "normally-off" type transistors	$-I_D$, $-V_{GS}$, $-V_{DS}$; drain current $-I_D$ vs. drain source voltage $-V_{DS}$	lateral current flow is controlled by an externally applied vertical electric field; $-I_D$, V_{th} 0 $-V_{GS}$ "normally-on" JFET; $-I_D$, 0 V_{th} $-V_{GS}$ "normally-off" JFET	microwave amplifier, fast switch, microwave mixer, microwave oscillator	81 S1, 78 Y1, 80 M1, 79 M2, 80 L1, 81 M2, 81 N1
8	JGFET (junction gate field-effect transistor); SIT (static induction transistor)	Si		Al (gate), Al (source), Al (gate), G, S, G, SiO$_2$, p^+, n^+, p^+, n, Si, n^+, D, metal (drain)	N_{d1}, N_a, N_{d2}, N_{d3}, N, x	I_D, V_{GS}, V_{DS}; drain current I_D vs. drain source voltage V_{DS}	I_D, 0 V_{GS}; drain current I_D controlled by the gate voltage V_{GS}	power switching transistor, power amplifier	81 K1, 75 N1

continued

Error in No. 6 and 7: the lateral current flow is controlled by pn-junction depletion layers and not by an external field; in No. 8 see also 78N1.

Table 2 (continued)

No.	Transistor	Semiconductor	Symbols	Structure	Doping profile	Characteristics	Operational principle	Applications	Refs.
9	VJFET (vertical junction field-effect transistor) multi-channel FET SIT (static induction transistor)	Si	G, D, S	Al (source), Al (gate), n^+, SiO_2, p^+, n^-, Si, n^+, metal (drain)	0, N_{d1}, N, N_{d2}, N_a, N_{d3}, x	I_D, V_{GS}, 0, V_{DS}; drain current I_D vs. drain source voltage V_{DS}	I_D, 0, V_{GS}; drain current I_D controlled by the gate voltage V_{GS}	power switching transistor, power amplifier	75 N1 81 S1 81 K1 80 M1 80 L1
10	MES FET (metal on semiconductor field-effect transistor) n-channel	Si	normally-on MESFET; normally-off MESFET; G, D, S	Al (source), Schottky-contact (gate), Al (drain), S, G, D, n, π or semi-insulating, B; Schottky-contact (gate) on n-type epitaxial layer	0, N, N_d, x; N_d depends on threshold voltage V_{th} for "normally-on" or "normally-off" mode	I_D, V_{GS}, 0, V_{DS}; drain current I_D vs. drain source voltage V_{DS}	lateral current flow is controlled by a Schottky contact depletion layer; I_D, V_T, 0, V_{GS} "normally-on" MESFET; I_D, 0, V_T, V_{GS} "normally-off" MESFET	microwave amplifier, fast switch, microwave mixer, microwave oscillator	81 S1 75 M1 80 L1
11	MOS FET n-channel enhancement-type (metal oxide semiconductor field-effect transistor)	Si	G, D, B, S	Al (source), Al (gate), Al (drain), S, G, D, n^+, SiO_2, p, Si, ground-plate, B; diffused or ion-implanted source and drain	0, N_{a1}, N, N_{a2}, x; N_{a1} depends on threshold voltage V_{th}	I_D, V_{GS}, 0, V_{DS}; drain current I_D vs. drain source voltage V_{DS}	I_D, 0, V_{th}, V_{GS}; drain current I_D controlled by gate voltage V_{GS}	amplifier, switch, mixer, oscillator	81 S1 79 M2 81 H4 78 Y1 80 L1 81 M2 81 N1
12	MOS FET p-channel enhancement-type (metal oxid semiconductor field-effect transistor)	Si	G, D, B, S	Al (source), Al (gate), Al (drain), S, G, D, p^+, SiO_2, n, Si, ground-plate, B; diffused or ion-implanted source and drain	0, N_{d1}, N, N_{d2}, x; N_{d1} depends on threshold voltage V_{th}	$-I_D$, $-V_{GS}$, 0, $-V_{DS}$; drain current $-I_D$ vs. drain source voltage $-V_{DS}$	$-I_D$, 0, V_{th}, $-V_{GS}$; drain current $-I_D$ controlled by gate voltage $-V_{GS}$	amplifier, switch, mixer, oscillator	81 S1 79 M2 81 H4 78 Y1 80 L1 81 M2 81 N1

Note: in No. 9 see also 78N1.

continued

Table 2 (continued)

No.	Transistor	Semiconductor	Symbols	Structure	Doping profile	Characteristics	Operational principle	Applications	Refs.
13	MOS FET n-channel depletion-type (metal oxide semiconductor field-effect transistor)	Si		Al (source), Al (gate), Al (drain), S, G, D, n^+, n, n^+, SiO_2, Si, p, ground-plate, B; diffused or ion-implanted source and channel	0, N_d, N_a, N, x; N_d depends on threshold voltage V_{th}	I_D, V_{GS}, 0, V_{DS}; drain current I_D vs. drain source voltage V_{DS}	I_D, V_{th} 0 V_{GS}; drain current-I_D controlled by gate voltage-V_{GS}	amplifier, switch, mixer, oscillator	81 S1, 79 M2, 81 H4, 78 Y1, 80 L1, 81 M2, 81 N1
14	MOS FET p-channel depletion-type (metal oxide semiconductor field-effect transistor)	Si		Al (source), Al (gate), Al (drain), S, G, D, p^+, p, p^+, SiO_2, Si, n, ground-plate, B	0, N_a, N_d, N, x; N_a depends on threshold voltage V_{th}	$-I_D$, $-V_{GS}$, 0, $-V_{DS}$; drain current-I_D vs. drain source voltage $-V_{DS}$	$-I_D$, V_{th} 0 $-V_{GS}$; drain current -I_D controlled by gate voltage -V_{GS}	amplifier, switch, mixer, oscillator	81 S1, 79 M2, 81 H4, 78 Y1, 80 L1, 81 M2, 81 N1
15	HMOS FET (high performance metal oxide semiconductor field-effect transistor)	Si		Al (source), Al (gate), polysilicon, Al (drain), G, S, D, n^+, p, n^+, SiO_2, p^-, Si, ground-plate, B; polysilicon gate technology, implanted source and drain, p-type implantation	0, N_{a1}, N_{a2}, N, x	I_D, V_{GS}, 0, V_{DS}; drain current I_D vs. drain source voltage V_{DS}	I_D, 0, V_{th}, V_{GS}; the p-implatation controls the threshold voltage V_{th} and increases the punch-through voltage	MOS transistor in integrated circuits	81 S1
16	DMOS FET (double-diffused metal oxide semiconductor field effect transistor)	Si		Al (source), Al (gate), polysilicon, Al (drain), G, S, D, n^+, n^-, n^+, p, p^-, L, SiO_2, Si, B, ground plate; polysilicon gate technology, double diffusion of p and n^+	0, N_d, N_a, N, x	I_D, V_{GS}, 0, V_{DS}; drain current I_D vs. drain source voltage V_{DS}	I_D, 0, V_{th}, V_{GS}; very short channel (L) determined by the p-doped region; the channel is followed by a lightly doped drift region	power MOS-transistor, amplifier for high frequency, fast switch	81 S1, 81 H4, 80 L1, 78 Y1, 80 M1
17	DIMOS FET (double-implanted metal oxide semiconductor field-effect transistor)	Si		Al (source), Al (gate), polysilicon, Al (drain), G, S, D, n^+, n^-, n^+, p, p^-, L, SiO_2, Si, B, ground plate; polysilicon gate technology, double implantation of p and n^+	0, N_d, N_a, N, x	I_D, V_{GS}, 0, V_{DS}; drain current I_D vs. drain source voltage V_{DS}	I_D, 0, V_{th}, V_{GS}; very short channel (L) determined by the p-doped region; the channel is followed by a lightly doped drift region	power MOS-transistor, amplifier for high frequency, fast switch	77 T1, 81 S1

continued

Table 2 (continued)

No.	Transistor	Semiconductor	Symbols	Structure	Doping profile	Characteristics	Operational principle	Applications	Refs.
18	VMOS FET (vertical metal oxide semiconductor field-effect transistor)	Si		V-structure is fabricated on <100>-oriented silicon substrate, using a nonisotropic etch		drain current I_D vs. drain source voltage V_{DS}	very short channel (L) determined by the V-shaped grooved MOS structure	power MOS-transistor, amplifier for high frequency, fast switch	81 S1, 81 H4, 80 L1, 78 Y1, 80 M1, 81 N1
19	CMOS FET (complementary metal oxide semiconductor field-effect transistor)	Si				load curves	when the input voltage V_i is high, transistor Q_1 is turned on and Q_2 is turned off, when the input voltage is low, Q_1 is turned off and Q_2 is turned on, under either input condition, very little current is drawn in the steady state.	inverter for integrated circuits, integrated logic, integrated memory	80 M1, 78 Y1, 80 L1, 78 T1, 80 H1
20	SOS MOS FET n-channel depletion-type (silicon on sapphire metal oxide semiconductor field-effect transistor)	Si				drain current I_D vs. drain source voltage V_{DS}	drain current I_D controlled by gate voltage V_{GS}; V_{th} threshold voltage	integrated logic, integrated memory, CMOS integrated circuits	81 H4, 81 S1, 80 M1, 78 Y1, 80 H1
21	SOS MOS FET p-channel depletion-type (silicon on sapphire metal oxide semiconductor field-effect transistor)	Si				drain current $-I_D$ vs. drain source voltage $-V_{DS}$	drain current $-I_D$ controlled by gate voltage $-V_{GS}$	integrated logic, integrated memory, CMOS integrated circuits	81 H4, 81 S1, 80 M1, 78 Y1, 80 H1

continued

Note: in No. 19 see also 83R1.

Table 2 (continued)

No.	Transistor	Semiconductor	Symbols	Structure	Doping profile	Characteristics	Operational principle	Applications	Refs.
22	SOS MOS FET n-channel enhancement-type (silicon on sapphire metal oxide semiconductor field-effect transistor)	Si		Al (source), Al (gate), Al (drain), S, G, D, SiO_2, Si, n^+, p, n^+, sapphire	0, N, N_a, N_d, x	I_D vs. V_{DS}, V_{GS}; drain current I_D vs. drain source voltage V_{DS}	I_D, 0, V_{th}, V_{GS}; drain current I_D controlled by gate voltage V_{GS}; V_{th} threshold voltage	integrated logic, integrated memory, CMOS integrated circuits	81 H4, 81 S1, 80 M1, 78 Y1, 80 H1
23	SOS MOS FET p-channel enhancement-type (silicon on sapphire metal oxide semiconductor field-effect transistor)	Si		Al (source), Al (gate), Al (drain), S, G, D, SiO_2, Si, p^+, n, p^+, sapphire	0, N, N_d, N_a, x	$-I_D$, 0, $-V_{DS}$, $-V_{GS}$; drain current $-I_D$ vs. drain source voltage $-V_{DS}$	$-I_D$, 0, V_{th}, $-V_{GS}$; drain current $-I_D$ controlled by gate voltage $-V_{GS}$; V_{th} threshold voltage	integrated logic, integrated memory, CMOS integrated circuits	81 H4, 81 S1, 80 M1, 78 Y1, 80 H1
24	MNOS (MIOS) memory transistor (metal nitride (insulator) oxide semiconductor)	Si		Al (source), Al (gate), polysilicon, Al (drain), S, G, D, SiO_2, p^+, p^+, Si_3N_4, SiO_2, n, Si, ground plate	0, N, N_a, N_d, x	I_D, stored charges, without charges, 0, U_{GS}; current voltage characteristic	charges are injected from the silicon across the SiO_2 and stored at the SiO_2-Si_3N_4 interface	nonvolatile memory cell in integrated circuits	81 S1, 81 H4, 80 L1, 78 Y1, 75 M1
25	FAMOS memory transistor (floating-gate avalanche-injection metal oxide semiconductor)	Si		Al (source), Al (gate), polysilicon (floating gate), Al (drain), S, G, D, SiO_2, p^+, p^+, n, Si, ground plate	0, N, N_a, N_d, x	I_D, negatively charged floating gate, uncharged floating gate, 0, U_{DS}; current voltage characteristic	charges are injected from the silicon across the SiO_2 and stored on the floating polysilicon gate	nonvolatile memory cell in integrated circuits	81 S1, 81 H4, 78 Y1, 80 L1, 75 M1

continued

Table 2 (continued)

No.	Transistor	Semiconductor	Symbols	Structure	Doping profile	Characteristics	Operational principle	Applications	Refs.
26	SAMOS (SIMOS) memory transistor (stacked-gate avalanche (injection) metal oxide semiconductor)	Si	D, G, S	Al (source), 2nd polysilicon (control gate), Al (gate), 1st polysilicon (floating gate), G, Al (drain), S, D, SiO_2, p^+, p^+, Si, n, ground plate	0, N, N_a, N_d, x	V_{th}, V_D, 0, V_G; control gate voltage V_G; threshold voltage (V_{th}) shift of a SAMOS (SIMOS) memory transistor as a function of control gate voltage V_G and various drain biases V_D [81 S1]	charges are injected from the silicon across the SiO_2 and stored on the floating polysilicon gate	nonvolatile memory cell in integrated circuits	81 S1 81 H4 80 L1

6.1.4.0.3 Bipolar transistors

In the following the fabrication steps of a planar epitaxial bipolar transistor are given. These steps are illustrated in Fig. 4a···o. The technological processes involved are described in detail in seperate sections, the numbers of which are given in parentheses:

Deposition of an n-type silicon layer on an n^+-type silicon substrate by epitaxy; Fig. 4b (section 6.1.4.4).

Fabrication of an SiO_2 layer by oxidation; Fig. 4c (section 6.1.4.5).

Depositing a photoresist film and exposing it to light through a photomask 1; Fig. 4d (section 6.1.4.6).

Removal of the exposed positive resist by a development solvent; Fig. 4e (section 6.1.4.6).

Transfer of the photoresist pattern to the SiO_2 layer by wet chemical or dry etching; Fig. 4f (section 6.1.4.7). Photoresist stripping; Fig. 4g (section 6.1.4.6).

Diffusion of a dopant for the p-type base of the bipolar transistor. SiO_2 growth during diffusion; Fig. 4h (section 6.1.4.1).

Deposition of a photoresist film and exposure through a photomask 2. Generation of a photoresist pattern by development. Transfer of the photoresist mask to the SiO_2 layer by etching. Photoresist stripping; Fig. 4i (sections 6.1.4.6 and 6.1.4.7).

Diffusion of a dopant for the n^+-type emitter. SiO_2 growth during diffusion; Fig. 4k (section 6.1.4.1).

Etching of contact windows in the SiO_2 layer by using a photoresist mask 3; Fig. 4l. Photoresist stripping (sections 6.1.4.6 and 6.1.4.7).

Deposition of a metal layer (mostly aluminum); Fig. 4m (section 6.1.4.5).

Etching of the metal layer using a photoresist mask 4. Photoresist stripping; Fig. 4n (sections 6.1.4.6 and 6.1.4.7).

Wafer probe for selection of the properly working transistors and die assembly; Fig. 4o (section 6.1.4.8).

6.1.4.0.4 MOS-transistors

In the following the process flow for the fabrication of an n-channel metal-gate MOS-transistor is given. This process flow is illustrated in Fig. 5a···o. The technological processes involved are described in detail in separate sections, the numbers of which are given in parentheses:

Fabrication of an SiO_2 layer by oxidation on a p-type silicon wafer; Fig. 5b (section 6.1.4.5).

Depositing a photoresist film and exposing it to light through a photomask 1; Fig. 5c (section 6.1.4.6).

Removal of the exposed positive photoresist by a development solvent; Fig. 5d (section 6.1.4.6).

Transfer of the photoresist pattern to the SiO_2 layer by wet chemical or dry etching; Fig. 5e (section 6.1.4.7).

Photoresist stripping; Fig. 5f (section 6.1.4.6).

Diffusion of a dopant for the n^+-type source and drain regions of the MOS-transistor. SiO_2 growth during diffusion; Fig. 5g (section 6.1.4.1).

Deposition of a photoresist film and exposure through a photomask 2. Etching of SiO_2 for gate, source, and drain using the photoresist mask. Photoresist stripping; Fig. 5h (section 6.1.4.6 and 6.1.4.7).

Fabrication of the gate oxide by oxidation; Fig. 5i (section 6.1.4.5).

Etching of contact windows in the SiO_2 film down to source and drain using a photoresist mask 3. Resist stripping; Fig. 5k (sections 6.1.4.6 and 6.1.4.7).

Deposition of a metal layer, mostly Al/Cu/Si; Fig. 5e (section 6.1.4.5).

Etching of the metal layer using a photoresist mask 4. Resist stripping and annealing of the metal layer; Fig. 5m (sections 6.1.4.5, 6.1.4.6 and 6.1.4.7).

Deposition of a passivation layer, SiO_2, Si_3N_4 or organic materials; Fig. 5n (section 6.1.4.5).

Etching of via-holes in the passivation layer to lead bonding using a photoresist mask 5; Fig. 5o (section 6.1.4.7 and 6.1.4.6).

Photoresist stripping (section 6.1.4.6).

Wafer probe for selection of the properly working transistors and die assembly (section 6.1.4.8).

6.1.4.0.5 Thyristors

In the following the necessary process steps used in the fabrication of thyristors are given. These steps are illustrated in Fig. 6a···m. The technological processes involved are described in detail in separate sections, the numbers of which are given in parentheses:

Diffusion of p-doping atoms (Al+Ga or Al+B) in the n-type silicon wafer; Fig. 6b (section 6.1.4.1).

Fabrication of an SiO_2-layer by oxidation; Fig. 6c (section 6.1.4.5).

Depositing a photoresist film and exposing it to light through a photomask; Fig. 6d (section 6.1.4.6).

Removal of the exposed positive photoresist by a development solvent; Fig. 6e (section 6.1.4.6).

Transfer of the photoresist pattern to the SiO_2 layer by wet chemical etching (section 6.1.4.7). Photoresist stripping; Fig. 6f (section 6.1.4.6).

Diffusion of a dopant for the n^+-type cathode of the thyristor; Fig. 6g (section 6.1.4.1).

Etching of the SiO_2 layer; Fig. 6h (section 6.1.4.7).

Alloying of an aluminum-coated molybdenum plate an the back of the wafer; Fig. 6i.

Evaporation of an aluminum-silver double layer through a mask used as cathode and gate, respectively; Fig. 6k (section 6.1.4.5).

Shaping of the wafer by lapping; Fig. 6l.

Deposition of a passivation layer and assembly; Fig. 6m (section 6.1.4.8).

6.1.4.0.6 Bipolar integrated circuits

In the following the sequence of the production steps for bipolar integrated circuits using junction isolation is given. These steps are illustrated in Fig. 7a···v. The technological processes involved are described in detail in separate sections, the numbers of which are given in parentheses.

Fabrication of an SiO_2 layer by oxidation on a p-type silicon wafer; Fig. 7b (section 6.1.4.5).

Depositing a photoresist film and exposing it to light through a photomask 1; Fig. 7c (section 6.1.4.6).

Removal of the exposed positive photoresist by a development solvent; Fig. 7d (section 6.1.4.6).

Transfer of the photoresist pattern to the SiO_2-layer by wet chemical or dry etching; Fig. 7e (section 6.1.4.7).

Photoresist stripping; Fig. 7f (section 6.1.4.6).

Diffusion of n-doping atoms (mostly As) used in a buried layer; Fig. 7g (section 6.1.4.1).

Oxide etching (section 6.1.4.7).

Deposition of an n-type silicon layer on the buried layer by epitaxy; Fig. 7h (section 6.1.4.4).

Fabrication of an SiO_2 layer by oxidation; Fig. 7i (section 6.1.4.5).

Etching of windows in the SiO_2 layer using a photoresist mask 2. Photoresist stripping; Fig. 7k (sections 6.1.4.6 and 6.1.4.7).

Diffusion of p-doping atoms for the isolation of neighboring devices; Fig. 7l (section 6.1.4.1).

Growth of an SiO_2 layer during diffusion; Fig. 7m (sections 6.1.4.1 and 6.1.4.5).

Etching of windows in the SiO_2 layer using a photoresist mask 3. Photoresist stripping; Fig. 7n (sections 6.1.4.6 and 6.1.4.7).

Diffusion of p-doping atoms for the base of the bipolar transistors; Fig. 7o (section 6.1.4.1).

Growth of an SiO_2 layer during diffusion; Fig. 7p (sections 6.1.4.1 and 6.1.4.5).

Etching of windows in the SiO_2 layer using a photoresist mask 4. Photoresist stripping; Fig. 7q (sections 6.1.4.6 and 6.1.4.7).

Diffusion of n-doping atoms for the emitters and collectors of the bipolar transistors; Fig. 7r (section 6.1.4.1).

Growth of an SiO_2-layer during diffusion; Fig. 7s (sections 6.1.4.1 and 6.1.4.5).

Etching of contact windows in the SiO_2 layer using a photoresist mask 5; Fig. 7t (sections 6.1.4.6 and 6.1.4.7).

Deposition of a metal layer; Fig. 7u (mostly Al; section 6.1.4.5).

Etching of the metal layer using a photoresist mask 6 (sections 6.1.4.6 and 6.1.4.7).
Photoresist stripping (section 6.1.4.6).
Wafer probe for selection of the properly working circuits and die assembly; Fig. 7v (section 6.1.4.8).

6.1.4.0.7 MOS-integrated circuits

In the following the fabrication steps of an MOS integrated circuit manufactured by the LOCOS- and polysilicon gate technology are given. The steps are illustrated in Fig. 8a···v. The technological processes involved are described in detail in separate sections, the numbers of which are given in parentheses:

Fabrication of an SiO_2 layer by oxidation of a p-type silicon wafer; Fig. 8b (section 6.1.4.5).

Fabrication of an Si_3N_4 layer by chemical vapour deposition (CVD); Fig. 8c (section 6.1.4.5).

Deposition of a resist film and exposing it to radiation (photons, electrons, X-rays, or ions); Fig. 8d (section 6.1.4.6).

Removal of the exposed positive resist; Fig. 8e (section 6.1.4.6).

Etching of the Si_3N_4 layer using a resist mask 1; Fig. 8f (section 6.1.4.7).

Implantation of p-doping ions for the isolation of neighboring devices in the MOS circuit; Fig. 8g (section 6.1.4.2).

Resist stripping; Fig. 8h (section 6.1.4.6).

Local oxidation of silicon (LOCOS process). The Si_3N_4 film prevents oxide growth; Fig. 8i (section 6.1.4.5).

Etching of the Si_3N_4 film (section 6.1.4.7). Implantation of p-doping ions for the channel of the transfer transistor and the varactor of the memory cell of a dynamic random access memory (DRAM); Fig. 8k (section 6.1.4.1).

Oxide growth, etching of the thin SiO_2 layer and reoxidation to prevent the "white ribbon effect" (section 6.1.4.5).

Deposition of a polycrystalline silicon film (polysilicon 1); Fig. 8e (section 6.1.4.5).

Phosphorus or arsenic doping by diffusion (section 6.1.4.1) or by implantation of ions (section 6.1.4.2).

Etching of the polysilicon film 1 using a resist mask 2; Fig. 8m (sections 6.1.4.6 and 6.1.4.7).

Resist stripping (section 6.1.4.6).

Fabrication of an SiO_2 isolation by oxidation; Fig. 8n (section 6.1.4.5).

Deposition of a polycrystalline silicon film (polysilicon 2); Fig. 8o (section 6.1.4.5).

Phosphorus or arsenic doping by diffusion (section 6.1.4.1) or by ion implantation (section 6.1.4.2).

Etching of the polysilicon film 2 using a resist mask 3; Fig. 8p (sections 6.1.4.6 and 6.1.4.7). Resist stripping (section 6.1.4.6). Implantation of n-doping ions (As or P) and annealing of the silicon crystal (section 6.1.4.2).

Fabrication of an SiO_2 layer by oxidation and subsequent chemical vapor deposition (CVD); Fig. 8q (section 6.1.4.5).

Etching of contact windows in the SiO_2 layer using a resist mask 4; Fig. 8r (sections 6.1.4.6 and 6.1.4.7). Resist stripping (section 6.1.4.6).

Deposition of a metal layer (Al/Si, Al/Si/Cu, Al/Si/Ti ...); Fig. 8s; (section 6.1.4.5).

Etching of the metal layer using a resist mask 5; Fig. 8t (sections 6.1.4.6 and 6.1.4.7).
Resist stripping (section 6.1.4.6).

Deposition of a passivation layer (SiO_2, Si_3N_4 or organic materials); Fig. 8u (section 6.1.4.5).

Etching of via-holes in the passivation layer using a resist mask 6; Fig. 8v (sections 6.1.4.6 and 6.1.4.7).
Resist stripping (section 6.1.4.6).
Wafer probe for selecting the properly working circuits and die assembly (section 6.1.4.8).

6.1.4.0.8 CMOS integrated circuits

In the following, the sequence of the production steps for CMOS (complementary MOS) integrated circuits is given. The steps are illustrated in Fig. 9a···s. The technological processes involved are described in detail in separate sections, the numbers of which are given in parentheses:

Fabrication of an SiO_2 layer by oxidation of an n-type silicon wafer; Fig. 9b (section 6.1.4.5).

Deposition of a resist film and exposing it to radiation (photons, electrons, X-rays, or ions); Fig. 9c (section 6.1.4.6).

Removal of the exposed positive resist; Fig. 9d (section 6.1.4.6).

Etching of the SiO_2 layer using a resist mask 1; Fig. 9e (section 6.1.4.7).

Resist stripping; Fig. 9f (section 6.1.4.6).

Diffusion of dopants for the p-well. SiO_2 growth during diffusion; Fig. 9g (section 6.1.4.1).

Etching of windows in the SiO_2 layer using a resist mask 2; Fig. 9h (sections 6.1.4.6 and 6.1.4.7).
Resist stripping (section 6.1.4.6).

Diffusion of p-type doping atoms for the source and the drain of the p-channel transistor. SiO_2 growth during diffusion; Fig. 9i (section 6.1.4.1).

Etching of windows in the SiO_2 layer using a resist mask 3; Fig. 9k (sections 6.1.4.6 and 6.1.4.7).
Resist stripping (section 6.1.4.6).

Diffusion of n-type doping atoms for the source and the drain of the n-channel transistor. SiO_2 growth during diffusion; Fig. 9l (section 6.1.4.1).

Etching of windows in the SiO_2 layer using a resist mask 4; Fig. 9m (sections 6.1.4.6 and 6.1.4.7).
Resist stripping (section 6.1.4.6).

Fabrication of the gate oxides by oxidation; Fig. 9n (section 6.1.4.5).

Etching of contact windows in the SiO_2 layer, using a resist mask 5; Fig. 9o (sections 6.1.4.6 and 6.1.4.7).
Resist stripping (6.1.4.6).

Deposition of a metal layer (Al/Si; Al/Si/Cu); Fig. 9p (section 6.1.4.5).

Etching of the metal layer using a resist mask 6; Fig. 9q (sections 6.1.4.6 and 6.1.4.7).

Resist stripping (section 6.1.4.6).

Deposition of a passivation layer (SiO_2, Si_3N_4, or organic materials); Fig. 9r (section 6.1.4.5).

Etching of via-holes in the passivation layer using a resist mask 7; Fig. 9s (sections 6.1.4.6 and 6.1.4.7).
Resist stripping (6.1.4.6).
Wafer probe for selecting the properly working circuits and die assembly (section 6.1.4.8).

6.1.4.0.9 Nuclear radiation detectors

Semiconductor radiation detectors are generally slice-shaped, cylindrical, sometimes ring-shaped devices, fabricated from silicon or germanium sometimes from cadmium telluride or mercury iodide, occasionally from gallium arsenide. They are used to detect individual high-energetic, ionizing particles, e.g. electrons, protons, α-particles, heavy ions, or fission fragments, as well as X- and γ-rays, and indirectly neutrons and to measure energy at very high accuracy in the range from 1 keV to many GeV.

Semiconductor radiation detectors differ from other semiconductor devices by 4 unique features:

The use of *electric field zones* as radiation-sensitive regions which are up to several centimeters thick for sufficient absorption of radiation. The sensitive area may reach sizes of up to 70 cm^2, the volumes of up to 200 cm^3. The volume of nuclear radiation detectors in semiconductor material can thus be as large as 10^8-times that of a transistor.

Very pure material having *extremely low concentration of dopants*, *recombination* and *trapping centres*, is necessary for obtaining deep radiation-sensitive regions and almost perfect charge carrier transport ($\sim$99.9% or more) over distances ranging up to several cm.

The large surfaces require *passivation procedures* different to those used in other semiconductor devices.

The frequent need to operate detectors at *low temperatures* ($\sim$77 K) in order to reduce the leakage current and the noise.

Nuclear radiation detectors based on semiconductors are reviewed in [61C1, 63D1, 68B1, 69K1, 71B1, 74G1, 81H1]. For applications of semiconductor nuclear radiation detectors see the annual IEEE Transactions on Nuclear Science, published by The Institute of Electrical and Electronics Engineers, Inc., N.Y.. The technical applications are reviewed in [82G1].

Basic mechanisms and detector structures

The following semiconductors are commonly used as detector bulk materials: silicon (Si), germanium (Ge), cadmium-telluride (CdTe), mercury-iodide (HgI_2).

This section is focussed on Si- and Ge-detectors. The properties of CdTe- and HgI_2-detectors are discussed in the sections 6.4.4.0.2 and 7.2, respectively.

Radiation detectors may be classified in different ways according to their structure and to their operation:
The structure types are: a) the diode type detector, and b) the resistance type detector (see Fig. 10).
Si- and Ge-detectors usually are of the diode type. CdTe- and HgI_2-detectors are of the resistance type.
The operation may be as: charged particle detector or photon (X- or γ-ray) detector.
Bulk material used for charged particle detectors is mainly Si, sometimes Ge; bulk material for photon detectors is sometimes Si (low energetic X-rays), mainly Ge, but also CdTe and HgI_2.

In contrast to other radiation-sensitive semiconductor devices, e.g. photocells, nuclear radiation detectors detect individual high-energetic photons (X- and γ-rays) or high-energetic ionising particles and measure their energy at high accuracy. The detector operates analogous to an ionisation chamber in which the gas is replaced by a solid. The detector consists of a semiconductor crystal between two conducting electrodes. An external voltage is applied to the electrodes producing the electric field of the barrier in the diode-type detector or the constant field between the electrodes in the resistance-type detector. When photons are absorbed or charged particles penetrate the crystal, free charge carriers are generated, whose number is proportional to the energy loss in the material, which is collected by the electric field to the detector electrodes. The current pulse induced in the external circuit represents the basic signal information obtained.

Table 3 gives a survey of semiconductor radiation detectors. Table 4 shows the essential applications.

For Table 3, see next page.

Table 4. Applications of semiconductor nuclear detectors [82G1].

Application	Detectors	Radiation measured
Physics research	Si, Si(Li), Ge, CdTe, HgI_2	particles, γ, X-rays
Space research, astrophysics	Si, Si(Li), Ge	particles, γ, X-rays
Nuclear technology		
monitoring neutron flux distribution in PWR's	Si	γ
monitoring operating systems	Ge	γ
fuel element integrity	Ge	γ
burn-up determination	Ge	γ
high activity measurement in the stack	Si	β
effluent monitoring	Ge	γ
environmental surveillance	Ge, Si	γ, α, β
radiation protection	Ge, Si	γ, α
mineral prospecting (U, Th)	Si(Li), Ge	γ, X-rays
Chemical analysis		
instrumental neutron activation analysis	Ge	γ
X-ray fluorescence analysis	Si(Li), Ge	X-rays
diffractometry	Si(Li)	X-rays
Medical diagnostic		
radioisotope imaging	Ge, CdTe, HgI_2	γ
computer tomography	Si, Ge, CdTe, HgI_2	X-rays
X-ray fluorescence analysis	Si(Li), Ge	X-rays
catheter probes	Si, CdTe	γ

Table 3. Semiconductor nuclear radiation detector data summary.

Starting material	Detector type	Main application	Useful energy range	Range of active area mm^2	Range of absorbing thickness mm	Volume cm^3	Operating temperature (range)
Silicon	surface barrier	charged particle spectroscopy, identification and detection	see Fig. 11⋯14	1⋯7000	$2\cdot10^{-3}\cdots2$		RT (−30 to +50 °C)
Silicon	diffused or implanted	charged particle spectroscopy identification and detection in adverse environment	see Fig. 11⋯14	10⋯1000	0.1⋯0.5		RT
Silicon	Si(Li)	charged particle spectroscopy and identification	see Fig. 11⋯13	up to 7000	up to 10		RT
Silicon	Si(Li)	X-ray spectroscopy	1⋯30 keV	12⋯100	up to 5		78K
Germanium (High Purity)	planar	X-ray spectroscopy	3⋯100 keV	12⋯100	up to 10		78K
	planar	γ-ray and p-spectroscopy		up to 3000	up to 20		78K
	coaxial	γ-ray spectroscopy		up to 3000	up to 100	up to 100	78K
Germanium	Ge(Li) coaxial	γ-spectroscopy	50 keV⋯ 10 MeV	up to 3000	up to 100	up to 200	78K
Cadmium-telluride		X- and γ-spectroscopy and detection	3 keV⋯ 1 MeV	up to 50	up to 1 for spectrometers up to 7 for counters		RT
Mercury-iodide		X- and γ-spectroscopy and detection	3 keV⋯ 1 MeV	up to 100	up to 0.75 for spectrometers up to a few mm for counters		RT

*) If not stated otherwise.
**) i = intrinsic.

Structure**)		Contact material and thickness in μg cm^{-2} entrance	exit	Energy resolution keV					Operating voltage V*)
Me—n or p—Me	gold—n, p-aluminum (partially or fully depleted)	Au, 40	Al, 40	14···50 (α-particles) 5···40 (β-particles)					up to several 100
p$^+$—n or p—n$^+$	implanted (200···550 Å) or diffused (~2000 Å) phosphorus, arsenic, boron	Al, 20	Al, 20	12···50 (α-particles) 3···40 (β-particles)					up to several 100
Me—i—n$^+$	gold—i—lithium-diffused	Au, 40		20···60 (α-particles) 12···70 (β-particles)					100 V/mm
p$^+$—i—n$^+$	aluminum—i—lithium-diffused	Al, 40		0.150					100 V/mm
				X- or γ-Energy [keV] 5.9	60	122	661	1330	
Me—n or p—n$^+$ or p$^+$—n or p—n$^+$	palladium—n or p—lithium-diffused			0.15		0.5			up to 3000
	or boron—implanted—n, p—lithium-diffused			0.3··· 0.15		0.5··· 0.8	1.0	1.8	up to 5000
	or boron—implanted—n, p—phosphorus implanted					0.8··· 1.2	1.2	2	up to 5000
p$^+$—i—n$^+$	Me—i—lithium-diffused					0.8··· 1.2	1.2	1.8··· 2.2	up to 5000
Me—n or p—Me				1.1	1.7	3.5	8.0	14	up to 100 up to 500 (depending on material type and contact)
Me—n or p—Me				0.65	1.2	2.5	4.5	22	1000···5000

Detector properties — material requirements

The most important parameters of nuclear radiation detectors are the efficiency η and the energy resolution ΔE.

Table 5 shows the relationsship of the bulk material and the detector parameters to the efficiency and the energy resolution requirements.

Table 5. Relation of detector operation requirements to material and device requirements.

Operation requirements	Physical requirements	Material and device requirements
High efficiency	large sensitive area	large single crystal slice
	thick sensitive region	low impurity doping concentration high bias voltage
	high stopping power	high atomic number
High energy resolution	small energy of e, h-pair generation	small bandgap E_g
	low noise	small leakage current
	large drift length	low trap concentration in bulk

The *efficiency* for a monoenergetic radiation is defined as the ratio of the total number of events observed to the total number of photons or charged particles incident on the active detector volume during the time interval.

The *energy resolution* in full width at half maximum (FWHM) is defined by the detector contribution to the FWHM of a pulse-height distribution corresponding to an energy spectrum. *Full width at half maximum* (FWHM) is the full width of a distribution measured at half the peak maximum. For a normal distribution, FWHM is equal to $2(2 \ln 2)^{1/2}$ times the standard deviation σ_x, given in eV or in % of the energy or number of channels, respectively.

Table 6 lists the material data of Si, Ge, CdTe, and HgI_2, which are essential for detectors. The detection sensitivity is dependent on the stopping power and the absorption of radiation. For high-energetic X- and γ-rays for which the Compton interaction is dominant, the detection is additionally dependent on the detector volume.

Table 6. Comparison of semiconductor material data essential to (planar) semiconductor nuclear detectors.

Property	Unit	Si	Ge data at 78 K	CdTe	HgI_2
Dielectric constant ε		12	16	10 [78Z1]	7.4
Density d at RT	$g\,cm^{-3}$	2.33	5.32	6.06	6.40
Number of atoms N	cm^{-3}	$5.02 \cdot 10^{22}$ [73B1]	$4.44 \cdot 10^{22}$ [73B1]	$3.04 \cdot 10^{22}$ [73B1]	$2.54 \cdot 10^{22}$ [73B1]
Atomic number Z		14	32	48.5	80.5
Atomic weight M		28.086	72.59	112.4 127.6	200.6 2.126.9
E_g (RT)	eV	1.16	0.74	1.45	2.15
Energy per electron-hole $E_{h,e}$ at RT	eV	3.72	2.96	4.43	4.15
Fano-factor F		~0.1	~0.1	0.4 calc [79W1]	0.46 exp [78D]
Electron mobility μ_n at RT	$cm^2\,V^{-1}\,s^{-1}$	1350	$4.5 \cdot 10^4$	1000	100
Hole mobility μ_p at RT	$cm^2\,V^{-1}\,s^{-1}$	480	$4.5 \cdot 10^4$	80···100	4
Saturation velocity v_{sat} at RT	$cm\,s^{-1}$	$1 \cdot 10^7$ [81S1]	$6 \cdot 10^6$ [81S1]	$1.5 \cdot 10^7$	$6 \cdot 10^6$
Electric field of v_{sat}	$kV\,cm^{-1}$	100 [81S1]	3(p), 40(n) [81S1]	11.5	80
Electron mobility, lifetime product $\mu_n \tau$	$cm^2\,s^{-1}$	2	> 500	$1.5 \cdots 3 \cdot 10^{-3}$	10^{-4}
Hole mobility, lifetime product $\mu_p \tau$	$cm^2\,s^{-1}$	1	> 500	$3 \cdot 10^{-4}$	10^{-5}

continued

Table 6 (continued)

Property	Unit	Si	Ge data at 78 K	CdTe	HgI_2
Drift length L					
at field corresponding					
to maximum velocity					
(electrons)	mm	10^3	$>10^5$	15	8
(holes)	mm	500	$>10^5$	0.1	0.03
Resistivity ϱ	Ωcm	$10 \cdots 10^5$	$\sim 10^4$	$10^2 \cdots 10^9$	10^{13}

Absorption of charged particles

Charged particle detectors are usually made of silicon, however, germanium may also be used. Detectors made of CdTe, HgI_2 or other compound semiconductors are not in use for *particle* spectroscopy. The high density of germanium results in a reduced range of energetic particles. For a given detector thickness, particles of significantly higher energy may be detected by a germanium detector rather than by a silicon detector. However, silicon detectors are often more convenient to use because they may be operated at room temperature while germanium detectors must be operated at cryogenic temperature. In addition, silicon detectors typically have thinner dead layers than Ge detectors.

For spectroscopy, the trace of the particle must be completely located within the sensitive region so that all charge carriers generated are collected by the electrodes and a correct energy measurement is obtained. The detector is applicable to particles having a penetration range up to the sensitive thickness, for irradiation perpendicular to the sensitive area, and equal to, or less than the detector geometry, for parallel irradiation.

Penetration ranges R of protons (p), deuterons (d), tritons (t), helium 3 (^{3}He), and alpha particles (α) are given in Fig. 11. The range of heavy particles (α, p, d etc.) at energies 0.1 to 500 MeV is 0.85 to 0.55 times smaller in Ge than in Si. However, the range per mass coverage is larger for Ge than for Si (Fig. 12). The range of electrons is not well defined. Extrapolated range values in Si and Ge are shown in Fig. 13. For electrons at 0.1 MeV, the range is about 25% larger in Ge than in Si. At 1 MeV, the difference is about 10%. At 10 MeV, the range is about the same in Si and Ge. For 100 MeV electrons, the range is slightly less in Ge than in Si.

For identification of charged particles, telescopes of thin (down to a few μm) ΔE-detectors and thick E-detectors are used in combination. The ΔE-detector (thickness $d \ll R$) is sensitive to the differential energy loss $\mathrm{d}E/\mathrm{d}x$. The differential energy loss in Si and Ge is shown for various particles in Figs. 14 and 15.

Absorption of X- and γ-rays

For the spectroscopy of X- and γ-rays, detectors are made of Si (1···30 keV range) and Ge. The absorption-coefficients of X- and γ-rays in Si and Ge vs. energy are shown in Figs. 16 and 17. Photo absorption, Compton absorption, Compton scattering, and pair production contribute to the absorption.

When Compton-scattered photons are not absorbed in the detector, the total absorption coefficient is given by $\mu_a = \tau + \sigma_a + \kappa$. When all Compton-scattered photons are absorbed in the detector, the total absorption is given by $\mu_0 = \mu_a + \sigma_s$. The γ-ray absorption including all secondary processes involved depends ultimately on the shape and the size of the detector.

Sensitive layer thickness and capacitance

The structure of semiconductor radiation detectors made of Si and Ge is commonly an abrupt p^+n- or n^+p-junction (or a Schottky barrier contact on n- or p-type material) with a very thin p^+- or n^+-layer, respectively, which is used as radiation entrance window.

The radiation sensitive, electric field zone of width w lies mainly in the low-doped n- or p-layer. In the case of a pn-structure, the width w is approximately given by

$$w=0.5\cdot(\varrho_n V)^{1/2} \quad [\mu m] \text{ n-type Si, } T=300\text{ K}$$

$$w=0.3\cdot(\varrho_p V)^{1/2} \quad [\mu m] \text{ p-type Si, } T=300\text{ K}$$

$$w=4.21\cdot 10^3\left[\frac{V}{|N_d-N_a|}\right]^{1/2} \quad [cm] \text{ n- or p-type Ge, } T=78\text{ K}$$

ϱ_n and ϱ_p are the specific resistivity values of the Si used. Since the mobility of holes and electrons is the same for Ge at 78 K, the net doping concentration $|N_d-N_a|$ in cm^{-3} may be used to describe w. The applied voltage V is given in volts. The depletion capacitance $C=\varepsilon\varepsilon_0 A/w$ is frequently used to determine the width w of the sensitive layer. The nomographs of Figs. 18 and 19 relate ϱ_n, ϱ_p, and C to w for Si and Ge, respectively.

For cylindrical coaxial (HPGe-) detectors, the depletion voltage is given by [75S1]:

$$V_d=\frac{e|N_d-N_a|}{2\varepsilon}\left[R_1^2\ln\frac{R_2}{R_1}-\frac{1}{2}(R_2^2-R_1^2)\right],$$

where R_2 and R_1 are the outer and inner radius, respectively.

Windows and dead layers of the cryostat or of the detector elements can cause a *loss of efficiency* at low energies. Cryostat-windows made of material having low atomic number Z are used to minimize this effect. Fig. 20 gives experimentally determined and calculated detection efficiencies of Si-detectors and the influence of a 7.5 μm beryllium window in the cryostat. [79K1]. Fig. 21 gives the efficiency of 3 and 5 mm detectors made of Si and Ge in the energy range up to 1000 keV. The advantages and disadvantages of Ge and Si as X-ray-detector-material are summarized as follows:

Germanium:

Advantages

The high atomic number results in good efficiency at high energies.

The small value of the e, h-pair generation energy reduces the importance of noise and possibly the statistical contribution to the energy resolution.

Disadvantages

The high efficiency at high energy generates background problems from high-energy sources.

The dead layer (entrance window) corresponds to a high efficiency loss at low energy.

The low-energy tailing on low-energy peaks.

The high efficiency at high energy increases the probability of an amplifier overload.

The high Z results in an increased spectral complication by escape peak problems and a complex relation between efficiency and energy.

The small band-gap increases the detector leakage current noise.

Silicon:

Advantages

Dead layer effects are small.

Si exhibits relative freedom of escape peak complications in complex spectra.

Si is not sensitive to a high-energy background.

Si can be used at high temperatures with low detector noise contribution.

Disadvantages

The efficiency is low at high energy (> 30 keV).

The average energy required to form an electron-hole pair is large.

Energy resolution

For a given detector, 4 factors determine the energy resolution obtainable

- Statistical fluctuations in the ionisation process
- Charge trapping effects
- Noise associated with detector leakage current
- Additional noise of the electronic system.

Charge generation and statistical limit of resolution

The average energy $E_{h,e}$ consumed for the generation of a free electron-hole pair is approximately given by [61S1, 66K1, 65K1].

$$E_{h,e} \sim 3 \cdot E_g.$$

Experimental results are reported in [68B1] (Fig. 22). The ionisation energy varies with temperature in the same manner as the bandgap (Fig. 23). The ionisation energy $E_{h,e}$ is independent of the kind and energy of the incident particle or photon.

The statistical broadening of a monoenergetic particle or photon peak is given by the expression

$$\mathrm{FWHM} = 2.35 \sqrt{E \cdot F \cdot E_{h,e}};$$

where E is the energy of the incident radiation and F is the Fano Factor [46F1], a correction factor, which corrects for the fact, that the observed broadening is substantially less than predicted by Poisson statistics. For Si and Ge, F is not accurately known, but measurements have indicated that it should be between 0.05 and 0.15 [70P1, 70E1].

For low-noise detector-systems (FWHM $\approx$ 150 eV), the statistical broadening limits the energy resolution for γ- or X-ray energies above 10 keV. At lower energies the resolution is determined by the system noise.

Charge trapping effects

Loss of free charge carriers by trapping in the semiconductor crystal can be a significant source of spectral distortion, frequently resulting in low energy tails on spectral distributions of monoenergetic radiation.

The geometrical effect of carrier collection from different distances in the field region is the primary factor causing a spread of the standard deviation σ of the charge pulse height Q [67D1]. The resolution as a function of the drift length L_n/w and L_p/w for electrons and holes, respectively, relative to the field zone width w, is shown in Fig. 24. Effects of trapping on the spectral line shape are reported in [69T1, 69A1, 70M1, 68W1]. Ranges of the drift lengths for electrons and holes in various materials are listed in Table 6.

Noise due to leakage currents

The resolution due to leakage currents is given by the shot noise

$$\mathrm{FWHM} = 2.7 \cdot 10^7 \cdot (I_s t)^{1/2}\ [\mathrm{keV}]$$

with I_s in [A] the saturation leakage current, and t in [s] the amplifier time constants. Leakage currents less than 10^{-9} A at $t \sim 1$ μs are required for FWHM $\leq$ 1 keV. Minority carrier diffusion, bulk and surface generation, and for cooled detector operation thermal photoexcitation contribute to the leakage current.

The *diffusion current density* is for an assymetric, abrupt n^+ p-junction:

$$j_D = 4.4 \cdot 10^{-16}\, \tau^{-1/2}\, \varrho_p \quad [\mathrm{A\,cm^{-2}}]$$

with τ [s] the minority carrier lifetime and ϱ_p [Ωcm] the specific resistivity. The current density of 1000 Ωcm p-type Si having a lifetime at room temperature $\tau = 1$ ms is $j_D = 0.4$ nA cm^{-2}.

The *generation current density* is related to recombination centers in the bulk material

$$j_g = \frac{q n_i}{2\tau} w.$$

τ is the effective lifetime, w is the width of the depletion layer, and n_i is the intrinsic carrier concentration. The flow direction of the generation current is opposite to the diffusion current. In deep depletion layers, the generation of carriers may be dominant. For Si having $\varrho_n = 10000$ Ω cm, $V = 200$ V and $\tau = 1$ ms $j_g = 60$ nA cm^{-2}, for HPGe having $\tau = 10^{-2}$ s and $w = 1$ cm $j_g = 2 \cdot 10^{-4}$ A cm^{-2}. Cooling is required for Ge to freeze out the intrinsic carrier generation n_i. The temperature dependence of n_i for various materials is shown in Fig. 25.

Especially for surface barrier silicon-detectors but also for diffused detectors with an unprotected surface, the diode current measured is much greater than the sum of the two components j_D and j_g. This is due to the *surface generation current density* j_{sur}, which strongly depends on the surface treatment, on the atmosphere in which the detector was fabricated, and on the surface coating.

Fig. 26 shows a comparison between reverse currents of a surface barrier detector and a passivated ion-implanted silicon detector [82K1].

The leakage current contribution j_{th} by *electron-hole pair generation due to thermal radiation* from hot parts of the cryostat is important for cooled Ge detectors. Besides on optical parameters, j_{th} essentially depends on the size of the detector surface area. j_{th} can be dominant especially for large coaxial germanium-detectors, if they are not thermally shielded. Estimates are given in [68M1, 70G1]. An experimental result is shown in Fig. 27.

Shielding of both planar and coaxial germanium detectors is performed by surrounding the detectors by an aluminum case, which is in thermal contact with the detector or with the cooling detector holder [70G1].

Detector structures and fabrication

The structure of radiation detectors can have the following forms:

a) n^+—p—p^+ (entrance window: n^+-layer)
b) p^+—n—n^+ (entrance window: p^+-layer)
c) n^+—i—p^+ (entrance window: p^+-layer)
d) n^+—i—Me (entrance window: Me-layer)
e) Me—n—Me
f) Me—p—Me

The structures are fabricated from either low-doped (e.g. $10^2 \leqq \varrho \leqq 10^5$ Ωcm for Si) p- or n-type or by lithium drift compensated bulk semiconductor material. The junction is formed by a thin heavily doped surface layer or by forming a Schottky barrier.

The detectors are operated in 2 different modes:

a) Partially depleted *(E-detectors)*: The depletion layer penetrates into, but not through the bulk material. The wafer remains partly undepleted.

b) Totally depleted *(ΔE-detectors)*: The applied voltage is higher than its reach-through-value. The thickness of the depletion layer of the junction is nearly equal to the thickness of the semiconductor.

Differential *dE/dx-Detectors* are transmission detectors whose thickness including the entrance and exit windows is small compared to the range of the incident particle. Fig. 28 shows the structure, the charge- and the potential distribution of E-detectors (n^+—p—p^+) and of ΔE-detectors (n^+—p—p^+ and n^+—i—p^+) [74G1].

The technological procedures used to prepare pn-junctions in other semiconductor devices are also common techniques to produce Si—pn—radiation detectors. Solid phase epitaxy is also employed for the manufacture of nuclear radiation detectors. Table 7 gives a survey of contact materials and the technologies employed in silicon and germanium detectors.

Table 7. Contact materials and technology currently employed in silicon and germanium radiation detectors [81W1].

Technology	Ge n^+	Ge p^+	Si n^+	Si p^+
Diffusion	lithium		lithium, phosphorus, arsenic	boron
Implantation	phosphorus	boron	phosphorus, arsenic	boron
Schottky barrier		nickel, chromium gold, palladium	aluminum	palladium, gold
Solid phase epitaxy		aluminum, palladium, platinum		palladium, platinum

Diffused and ion-implanted silicon detectors

Phosphorus or arsenic are generally used for n^+-diffusion or n^+-implantation. Boron is used as p^+-dopant [64H1, 72Z1, 74P1, 82K1]. The diffusion- and implantation parameters are similar to those used for the fabrication of silicon solar cells (section 6.1.4.0.1). Diffusion and implantation are commonly used in combination with the standard planar process (oxide passivation, photo engraving) (section 6.1.4.0.3). The advantage of ion implantation is the thin dead layer (~500 Å in comparison to 0.2 μ for diffusion) resulting in a low energy straggling. Fig. 29 shows the successive steps of the manufacturing process of passivated ion-implanted silicon detectors. The appropriate fabrication data are for example [82K1]:

n-bulk material: $\varrho = 4000$ Ω cm

oxide passivation: 1030 °C with dry oxygen, leading to SiO_2 layers of 1600 to 6000 Å thickness (addition of HCl).

B-implantation front: energy: 10···15 keV
dose: $5 \cdot 10^{14}$ ions cm^{-2} (standard)
$1 \cdot 10^{13}$ ions cm^{-2} (position sensitive)
As-implantation back: energy: 30 keV
dose: $5 \cdot 10^{15}$ ions cm^{-2}
Annealing: at 600 °C in dry nitrogen ambient

Surface barrier silicon detectors

A surface-barrier (SB) is formed by depositing (vacuum evaporation) a thin film of metal on one face of the semiconductor wafer (Schottky barrier contact). Its rectifying properties are influenced by the difference in material work functions and by surface states [66W1, 73E1, 74I1]. Table 8a, b [75R] lists values of work functions obtained for various essential metals and for doped Si and Ge. SB-contacts frequently also utilize an interfacial metal/semiconductor compound e.g. a silicide [81H1].

Table 8a, b. Work-functions of various metals (a) and of the semiconductors Si and Ge at various doping concentrations (b) [75R2, 81H1].

a)

Metal	Ag	Al	Au	Cr	Cs	Cu	Mo	Ni
Φ_B [eV]	4.31	4.20	4.70	4.51	1.79	4.52	4.3	4.74

Metal	Pb	Pd	Pt	Ta	Ti	V	W	Mg
Φ_B [eV]	4.20	4.82	5.28	4.12	3.95	4.12	4.5	3.70

b)

		Doping concentration [cm^{-3}]					
		n-Type			*p-Type*		
		10^{14}	10^{15}	10^{16}	10^{14}	10^{15}	10^{16}
Φ_B [eV]	Si	4.32	4.26	4.20	4.82	4.88	4.94
	Ge	4.43	4.38	4.33	4.51	4.56	4.61

Commonly used surface barrier contacts are:

Pd, Au on high resistivity n—Si base material as blocking contact
Al on high resistivity n—Si base material as non-injecting, ohmic contact
Al on high resistivity p—Si base material as blocking contact
Au on high resistivity p—Si base material as non-injecting, ohmic contact

During fabrication of the surface barrier detector [62F1], the Si wafer is embedded in a ring of epoxy resin, in order to adjust the field distribution at the boundary for high bias voltages. Fig. 30 schematically shows the preparation steps for a surface barrier detector [62F1, 72G1].

High purity germanium detectors

(Fig. 31 A)

The procedures employed to fabricate pn-junctions in HPGe are partly based on the technology to produce lithium-drifted detectors. Lithium is generally used in diffused n^+-type layers (see below for the fabrication of Si(Li)- and Ge(Li)-detectors). Thin radiation entrance windows are commonly fabricated by using a Au- or Pd-surface barrier layer as p^+-contact. However, implantation of boron in p^+, of phosphorous and antimony in thin n^+-layers (thickness about 0.02 μm) is also used [70M2, 73D1, 72H1]. The implantation energies are relatively low in comparison to those used in the production of other semiconductor devices. Typical implantation parameters are given in [77H1, 77H2]:

P-implantation: energy: 25 keV; dose: 10^{14}…10^{16} cm^{-2}, at beam current 1 μA cm^{-2} to reduce sample heating; angle of incidence: 8° of the crystal growth axis ⟨100⟩ to avoid channeling; sample temperature: $T = 77$ K.

annealing: Pre-anneal for 24···60 h at 150 °C, then increasing the temperature 10 °C/10 min to a maximum of 330 °C, then slow cooling (2 °C/min).

B-implantation: energy: 25 keV; dose: 10^{14} cm^{-2}.

Other implantation parameters are [77P1]:

P-implantation: energy: 20 keV; dose: 10^{13}···10^{14} cm^{-2} at 77 K,

annealing: T = 400 °C,

B-implantation: energy: 20 keV; dose: 10^{14} cm^{-2}.

Boron does not require annealing up to doses of 10^{16} cm^{-2}. The lattice disorder is so small, that no amorphous layer can be observed [70M2, 73D1].

The steps for fabricating a planar detector with a diffused n$^+$-contact and a Pd-surface barrier layer are schematically shown in Fig. 31 B [78G1].

Analogous techniques are developed for making both planar and coaxial detectors.

The advantages of HPGe-detectors compared with lithium-drifted detectors are:

- room temperature handling and storage possibility,
- since no undepleted material is present in these detectors, slow pulses produced by diffusion of carriers from undrifted material into the drifted region as with Li-drifted detectors are absent.
- the lack of undepleted material in the case of coaxial detectors improves the peak/Compton ratio,
- repair of radiation damage in high-purity germanium detectors can be accomplished by a simple thermal annealing cycle whereas redrifting is necessary in Li-drifted detectors,
- coaxial detectors with p$^+$ on the outside and n$^+$ on the inside can be produced. The predominance of hole trapping in radiation-damaged germanium makes such detectors less sensitive to radiation damage than detectors made with n$^+$ on the outside and p$^+$ on the inside.

Lithium-drifted detectors

Li-drift is used to reduce the net doping of p-type crystals down to values $|N_a - N_d| = 10^{10}$ cm^{-3} by compensation. The field zones in the n$^+$—i—p$^+$ structures achieved by this technology are

in Si up to 10 mm
in Ge up to 30 mm

The process steps may be divided into *Li-diffusion* and *Li-drift*. Although the basic principles of both steps are the same in Si and Ge, the practical differences are substantial.

Lithium-diffusion

Evaporated Li is diffused into the p-type material up to the solubility limit to form the pn-junction and to create the Li-source for the drift process.

Fig. 32a and b show the solubility of lithium in Si and Ge, respectively. Fig. 33 shows the diffusion constant and Fig. 34 the drift mobility of lithium ions in Si and Ge, respectively.

Calculated lithium distributions in Si and Ge are represented in Figs. 35···38 [68B1].

Lithium-drift process

The lithium-drift process is used to produce a p—i—n-structure i.e. a sensitive region deeper than that of a p—n—diode.

At room temperature, the lithium donors are in interstitial sites. An electrostatic interaction appears to occur between the lithium and boron ions in p-type material, leading to a pairing phenomenon represented by the reaction

$$\begin{array}{ccccc} \mathrm{Li} & \rightleftarrows & \mathrm{Li}^+ & + & e^- \\ & & + & & + \\ \mathrm{B} & \rightleftarrows & \mathrm{B}^- & + & e^+ \\ & & \downarrow\uparrow & & \downarrow\uparrow \\ & & (\mathrm{Li}^+\,\mathrm{B}^-) & & e^-e^+, \end{array}$$

forming a neutral, immobile (Li$^+$ B$^-$)-pair.

At high reverse bias voltage and at high temperature, Li ions drift through the n$^+$ p-junction into the p-type region to the boundary where the electric field vanishes. At this boundary, the Li ions compensate the boron acceptors thus shifting the boundary of the p-type zone continuously through the p-type wafer.

The thickness d_p of the compensated intrinsic region for planar detectors is proportional to the square root of the bias voltage V and the drifting time:

$$d_p = \sqrt{2\mu V t_p}$$

μ is the mobility of the Li-ions.

Figs. 39 and 40 give nomographs for the determination of the i-region thickness at given drift parameters V, t, and T.

Usual drift conditions are for

silicon: $T = 120$ °C and $V = 500 \cdots 1000$ V;
germanium: $T = 40$ °C and $V = 1000$ V.

The time t_c of lithium-drift to obtain a given depth d_c in a *coaxial cylindrical* Ge(Li)-diode in comparison to the drift time t_p in a *planar diode* is shown in Fig. 41.

Figs. 42 and 43 show various detector structures fabricated by lithium drift in Si and Ge.

Figs. 44 A, B gives the manufacturing steps of a Si(Li)-detector and Fig. 45 shows the experimental setup of drifting lithium in germanium or silicon.

Position sensitive detectors (PSD)

PSD's are position- and energy-sensitive devices, employing either

- a linear array of separate detectors in the one-dimensional PSD, or
- a mosaic of separate detectors in the two-dimensional PSD, or
- an array or mosaic of separate detectors on the same substrate material,
- resistive electrodes as charge dividers [68O1, 79W1].

Fig. 46 shows a one-dimensional multielement silicon detector array with oxide passivation [74G1]. Figs. 47 and 48 show various kinds of two-dimensional PDS's fabricated on germanium substrates.

The detector element Fig. 48 is called orthogonal-strip two-dimensional Ge detector [69P1, 73P1, 74K1, 74S1, 78K1].

Figs. 49 and 50 show the principle of resistive charge deviders. This special form of a silicon PSD determines the entry point of the particle or photon from a measurement of the magnitude or duration of currents which flow in the resistive layer. Evaporated metal films as well as diffused or implanted layers are used as resistive electrodes.

Detector cryostats

Detectors for X-rays and γ-rays are usually operated at liquid nitrogen temperature in order to reduce the thermally generated leakage current and the noise associated. Additionally, the charge carrier mobility is increased at liquid nitrogen temperatures. Fig. 51 shows the 3 typical detector housings and liquid nitrogen reservoirs used.

References for 6.1.4.0

46F1 Fano, V.: Phys. Rev. **70** (1946) 44.
61C1 Czulius, W., Engler, H.D., Kuckuck, H.: Ergeb. Exakten Naturwiss. **34** (1961) 236.
61S1 Shockley, W.: Czech. J. Phys. **B 11** (1961) 81.
62F1 Fox, R.J., Borkowski, C.J.: IRE Trans. Nucl. Sci. NS **9/3** (1962) 213.
63D1 Dearnaley, G., Northrop, D.C.: Semiconductor Counters for Nuclear Radiations, **1963** E. u. F.N. Spon Limited, 22 Henrietta Street, London, W. C. 2.
64H1 Hansen, W.L., Goulding, F.S.: Nucl. Instrum. Methods **29** (1964) 345.
65G1 Goulding, F.S.: Semiconductor Detectors for Nuclear Spectrometry, **1965** UCRL-report 16231, Lawrence Radiation Lab., Berkeley, California.
65M1 Malm, H.L., Tavendale, A.J., Fowler I.L.: Can. J. Phys. **43** (1965) 1173.
65R1 Van Roesbroeck, W.: Phys. Rev. **139** (1965) A1 702.
66A1 Andersson-Lindström, G., Zansig, B.: Nucl. Instrum. Methods **40** (1966) 277.
66K1 Klein, C.A.: Phys. Soc. Jpn. Suppl. **21** (1966) 307.
66M1 Malm, H.L.: IEEE Trans. Nucl. Sci. NS **13/3** (1966) 285.
66T1 Tavendale, A.J.: Proceedings of the International Atomic Energy Agency Panel, Vienna, June **1966** 4.
66T2 Tavendale, A.J.: IEEE Trans. Nucl. Sci. NS **13/3** (1966) 315.

66W1 Walter, F.J., Boshart, R.R.: IEEE Trans. Nucl. Sci. NS **13/3** (1966) 189.
66W2 Williamson, C.F., Boujot, J.-P., Picard, J.: Centre D'Etudes Nucleaires de Saclay Rapport CEA · R 3042.
67D1 Day, R.B., Dearnaley, G., Palms, J.M.: IEEE Trans. Nucl. Sci. NS **14** (1967) 487.
68B1 Bertolini, G., Coche, A.: Semiconductor Detectors, North Holland Publishing Company-Amsterdam, **1968**.
68K1 Klein, C.A.: IEEE Trans. Nucl. Sci. NS **15** (1968) 214. Klein, C.A., J. Appl. Phys. **39** (1968) 2029.
68L1 Lehmann, E., Martin, H., Raithel, K.: Siemens Z. **42** (4) (1968) 248.
68M1 Mc Intyre, R.J.: IEEE Trans. Nucl. Sci. NS **15** (1968) 6.
68O1 Owen, R.B., Awcock, M.L.: IEEE Trans. Nucl. Sci. NS **15/3** (1968) 290.
68P1 Pehl, R.H., Goulding, F.S., Landis, D.A., Lenzlinger, M.M.: Nucl. Instrum. Methods **59** (1968) 45.
68W1 Webb, P.P., Malm, H.L., Chartrand, M.C., Green, R.M., Sakai, E., Fowler, I.L.: Nucl. Instrum. Methods **63** (1968) 125.
69A1 Armantrout, G.A.: UCRL (University of California Lawrence Radiation Laboratory) report No 50485, **1969**.
69K1 Kuhn, A.: Halbleiter- und Kristallzähler, Akademische Verlagsgesellschaft Geest & Portic KG, Leipzig, **1969**.
69L1 Lothrop, R.P.: University of California, Lawrence Berkely Laboratory report, No. UCRL–19413, UC–37 Instruments TID-4500 (54th Ed.) **1969**.
69P1 Parker, R.P., Gemmersen, E.M., Wankling, J.L., Ellis, R.: Medical Radioisotope Scintigraphy, **I** (1969) 71, Vienna, IAEA.
69T1 Trammel, R., Walter, F.J.: Nucl. Instrum. Methods **76** (1969) 317.
70B1 Baertsch, R.D., Hall, R.N.: IEEE Trans. Nucl. Sci. NS **17/3** (1970) 235.
70C1 Czulius, W., Reiss, B., Schupp, C.: Forschungsbericht St. Sch. 0572 Bundesministerium für wissenschaftliche Forschung, **1971**.
70E1 Eberhardt, J.E.: Nucl. Instrum. Methods **80** (1970) 291.
70G1 Glasow, P.: Forschungsbericht f. das Bundesministerium f. Bildung u. Wissenschaft Nr. III A2-5891-St. Sch. 0540, **1970**.
70G2 Glasow, P.: Nucl. Instrum. Methods **80** (1970) 141.
70M1 Mayer, J.W., Martini, M., Zanio, K.R., Fowler, I.L.: IEEE Trans. Nucl. Sci. NS **17** (1970) 221.
70M2 Mayer, I.W., Ericsson, L., Davies, J.A.: Ionimplantation in Semiconductors, (Silicon and Germanium), Academic Press, New York and London, **1970**.
70M3 Moody, N.F., Paul, W., Joy, M.L.G.: Proc. IEEE **58/2** (1970) 217.
70P1 Pehl, R.H., Goulding, F.S.: Nucl. Instrum. Methods **81** (1970) 329.
71B1 Bücker, H.: Theorie u. Praxis der Halbleiterdetektoren, Springer, Berlin-Heidelberg-New York **1971**.
71G1 Glasow, P., Roth, S.: Forschungsbericht K 71-16, Bundesministerium für Bildung und Wissenschaft, **1971**.
71H1 Herlet, A.: Solid State Devices, Conference Series No. 12, Page 9 (1971) 123.
71J2 Jaclevic, I.M., Goulding, F.S.: IEEE Trans. Nucl. Sci. NS **18/1** (1971) 127.
72G1 Glasow, P., Spillekothen, H.G.: Siemens Forsch. Entwicklungsber. **1** (1972) 287.
72H1 Herzer, H., Kalbitzer, S., Ponpon, J.P., Stuck, B., Siffert, P.: Nucl. Instrum. Methods **101** (1972) 31.
72H2 Harth, W.: Halbleitertechnologie, B.G. Teubner, Stuttgart, W-Germany, **1972**.
72J1 Jaclevic, J.M., Goulding, F.S.: IEEE Trans. Nucl. Sci. NS **19/3** (1972) 384.
72P1 Pehl, R.H., Cordi, R.C., Goulding, F.S.: IEEE Trans. Nucl. Sci. NS **19/1** (1972) 265.
72Z1 Zulliger, H.R., Drummond, W.E., Middleman, L.M.: IEEE Trans. Nucl. Sci. NS **19/3** (1972) 306.
73B1 Bertolini, G., Cappelani, F., Restelli, G.: Nucl. Instrum. Methods **112** (1973) 219.
73D1 Dearnaley, G., Freeman, J.H., Nelson, R.S., Stephen, J.: Ionimplantation, North-Holland Publishing Company, Amsterdam, **1973**.
73E1 Elad, E., Inskeep, C.N., Sareen, R.A., Nestor, P.: IEEE Trans. Nucl. Sci. **20/1** (1973) 534.
73M1 Müller, R.: Bauelemente der Halbleiter-Elektronik, Springer, Berlin-Heidelberg-New York, **1973**.
73P1 Parker, R.P., Gemmersen, E.M., Ellis, R., Bell, I.: Medical Radioisotope Scintigraphy, **I** (1973) 193, Vienna, IAEA.
73S1 Stuck, R., Ponpon, J.P., Berger, R., Siffert, P.: Radiat. Eff. **20** (1973) 75.
74G1 Goulding, F.S., Pehl, R.H.: Semiconductor Radiation Detectors Nuclear Spectroscopy and Reactions, Part A, S. 289–343, Academic Press, Inc. **1974**.
74I1 Inskeep, C., Elad, E., Sareen, R.A.: IEEE Trans. Nucl. Sci. NS **21** (1974) No 1.
74K1 Kaufman, L., Camp, D.C., McQuaid, J.H., Armantrout, G.A., Swierkowski, S.P., Lee, K.: IEEE Trans. Nucl. Sci. **21/1** (1974) 652.
74M1 Marler, I.M., Hewka, P.: IEEE Trans. Nucl. Sci. NS **21/1** (1974) 297.

74P1 Peterström, S., Holmen, G.: Nucl. Instrum. Methods **119** (1974) 151.
74S1 Schlosser, P.A., Miller, D.W., Gerber, M.S., Redmond, R.F., Harpster, I.W., Collins, W.J., Hunter, W.W.: Trans. Nucl. Sci. **21/1** (1974) 658.
75M1 Moeschwitzer, A.: Halbleiterelektronik Wissenspeicher, Huethig, Heidelberg, W-Germany, **1975**.
75N1 Nishizawa, J.I., Terasaki, T., Shibata, J.: IEEE Transact. Electron Devices **ED-22** (1975) 185.
75P1 Pehl, R.H., Cordi, R.C.: Trans. Nucl. Sci. NS **22/1** (1975) 177.
75R2 Ruge, I.: "Halbleiter-Technologie" Series on "Halbleiter-Elektronik", Springer, **1975** 168.
75S1 Siffert, P., Ponpon, J.P., Cornet, A.: L'oude electrique **55/5** (1975) 281.
76H1 Hoffmann, A., Stocker, K.: Thyristor-Handbuch, Siemens AG, Munich, Germany **1981**.
76T1 Tholl, H.: Bauelemente der Halbleiterelektronik, **1** Teubner, Stuttgart, W-Germany (1976).
77H1 Hubbard, G.S., Haller, E.E., Hansen, W.L.: IEEE Trans. Nucl. Sci NS **24** No. 1 (1977) 161.
77P1 Protic, D., Riepe, G.: IEEE Trans. Nucl. Sci. NS **24/1** (1977) 65.
77T1 Tihany, J., Widmann, D.: IEEE Tech. Dig., Int. Electron Device Meeting, (1977) 399.
78B1 Brauer, F.P., Mitzlaff, W.A.: IEEE Trans. Nucl. Sci. NS **25/1** (1979) 398.
78G1 Glasow, P., Müller, A., Raab, G., Wolf, H.J.: Siemens Forsch. Entwicklungsber. **7** (1978) 181.
78I1 Ipavich, F.M., Lundgren, R.A., Lambird, B.A., Gloeckler, G.: Nucl. Instrum. Methods **154** (1978) 291.
78K1 Kaufman, L., Hosier, K., Lorenz, V., Shosa, D., Hoenninger, J., Cheng, A., Okerlund, M., Hattner, R.S., Price, D.C., Williams, S., Ewins, J.H., Armantrout, G.A., Camp, D.C., Lee, K.: Investigative Radiology **13/3** (1978) 223.
78N1 Nishizawa, J., Ohmi, T., Mochida, Y., Matsuyama, T., Tida, S.: IEDM Tech. Digest (1978) 676.
78S1 Schrenk, H.: Bipolare Transistoren, Springer, Berlin-Heidelberg-New York, **1978**.
78T1 Tholl, H.: Bauelemente der Halbleiterelektronik, **2** Teubner, Stuttgart, W-Germany (1978).
78W Walton, J.T., Sommer, H.A., Greiner, D.E., Bieser, F.S.: IEEE Trans. Nucl. Sci. NS **25/1** (1978) 391.
78Y1 Yang, E.S.: Fundamentals of Semiconductor Devices, Mc. Graw Hill, N.Y., **1978**.
78Z1 Zanio, K.: Semiconductor and Semimetals, Academic Press. **13** (1978).
79G1 Gerlach, W.: Thyristoren, Springer, Berlin-Heidelberg-New York, **1979**.
79H1 Hubbard, G.S., Haller, E.E.: Nucl. Instrum. Methods **164** (1979) 121.
79K1 Kuniluvo, S.: Nucl. Instrum. Methods **165** (1979) 21.
79L1 Laesgaard, E.: Nucl. Instrum. Methods **162** (1979) 93.
79M1 Müller, R.: Grundlagen der Halbleiterelektronik, Springer, Berlin-Heidelberg-New York, **1979**.
79M2 Müller, R.: Bauelemente der Halbleiterelektronik, Springer, Berlin-Heidelberg-New York, **1979**.
79M3 Mader, H.: European Patent 0003130, **1979**.
79R1 Raudorf, T.W., Trammel, R.C., Darken, L.S.: IEEE Trans. Nucl. Sci. NS **26/1** (1979) 297.
79R2 Riepe, G., Protic, D.: Nucl. Instrum. Methods **165** (1979) 31.
79W1 Walton, J.T., Hubbard, G.S., Haller, E.E.: IEEE Trans. Nucl. Sci. NS **26/1** (1979) 334.
80C1 Colclaser, R.A.: Microelectronics: Processing and Device Design, Ch. 1, J. Wiley, New York, **1980**.
80F1 Fagin, F., Klein, T.: Large and Medium Scale Integration, McGraw Hill, New York (1980) 24.
80H1 Hoffmeister, E.: Halbleiterbauelemente für die Elektronik, Siemens AG, München, **1980**.
80L1 Lacour, H.: Elektronische Bauelemente II, Verlag Berliner Union Stuttgart **1980**.
80M1 Milnes, A.G.: Semiconductor Devices and Integrated Electronics, Van Nostrand Reinhold Company, New York, **1980**.
80R1 Rein, H.M., Ranfft, R.: Integrierte Bipolarschaltungen, Springer, Berlin-Heidelberg-New York, **1980**.
80W1 Walton, J.T., Goulding, F.S., Haller, E.E., Pehl, R.H.: 1981 INS International Symposium on Nuclear Radiation Detectors, Tokyo, 23–26 March **1981**.
81H1 Haller, E.E., Goulding, F.S.: Nuclear, Radiation Detectors, Handbook on Semiconductor Nr. 4 C. Hilsum, North Holland Publishing Company, **1981**.
81H2 Hansen, W.L., Haller, E.E.: IEEE Trans. Nucl. Sci. NS **28/1** (1981) 541.
81H3 Harth, W., Claassen, M.: Aktive Mikrowellendioden, Springer, Berlin-Heidelberg-New York, **1981**
81H4 Hilsum, S.: Handbook on Semiconductors, Vol. 4, Device Physics, North Holland, Amsterdam-New York-Oxford, **1981**.
81K1 Kahng, D.: Silicon Integrated Circuits, Part B, Academic Press New York, **1981**.
81M1 Mader, H.: United States Patent 4,278,986, **1981**.
81M2 Müseler, H., Schneider, T.: Elektronik, Carl Hanser-Verlag, München-Wien, **1981**.
81N1 Nashelsky, L., Boylestad, R.: Devices: Discrete and Integrated, Prentice-Hall, N.J., **1981**.
81P1 Patalong, G.: Private communications, **1981**.
81S1 Sze, S.M.: Physics of Semiconductor Devices, John Wiley and Sons, N.Y., **1981**.
81V1 Voltmer, F.W.: VLSI Electronics Microstructure Science (Einspruch, N.G. ed.) Ch. 1, Academic Press, New York (1981) 1.

81W1 Walton, J.T., Goulding, F.S., Haller, E.E., Pehl, R.H.: Proceedings of the 1981 INS International Symposium on Nuclear Radiation Detectors, Tokio, 23–26 March **1981** 125.
82C1 Czulius, W.: Private communication 1982.
82G1 Glasow, P.A.: Trans. Nucl. Sci. NS **29/3** (1982) 1159.
82K1 Kemmer, J.: IEEE Trans. Nucl. Sci. NS **29** (1982) 733.
82L1 Langer, E., Selberherr, S., Mader, H.: Solid State Electronics **25** (1982) 317.
82M2 Müller, J.: Siemens AG., Device Department; private communications, **1982**.
82M3 Mader, H.: IEEE Transact. Electron Devices, **ED-29(11)** (1982) 1766.
82O1 Olk, G.: Siemens AG., Device Department; private communications, **1982**.
82P1 Pehl, R.H.: IEEE Trans. Nucl. Sci. NS **29/3** (1982) 1101.
82S1 Spaeth, W.: Siemens AG., Device Department; private communications, **1982**.
83M1 Mader, H., Müller, R., Beinvogl, W.: IEEE Transact. Electron Devices ED-30 (1983).
83R1 Ruge, I.: „Halbleiter-Technologie“, Series on „Halbleiter-Elektronik“, Springer, **1983**.

6.1.4.1 Diffusion

6.1.4.1.0 Introduction

Fick's first law

The law governing the macroscopic phenomenon of the redistribution of concentrations, independent of the diffusing species and the medium, is Fick's [1855F, 60J] first law.

$$J = -D \cdot \frac{\partial c}{\partial x} = -D \text{ grad } c. \tag{1}$$

The flux J is the quantity of substance passing through a plane of one cm^2 in one second in the direction x normal to the plane. The negative sign indicates the opposite direction of the flux compared to the concentration gradient. D is a factor of proportionality, the diffusion coefficent. The unit of D is [$cm^2\,s^{-1}$].

For the semiconductor technology, the diffusion of substitutional lattice constituents and of interstitial atoms is considered. Such species may be silicon itself, vacancies or impurities which are incorporated in the silicon lattice. The most detailed understanding of the diffusion process is obtained by treating the problem as a random walk process. A good survey is given in [63S]. The statistical description of random walk diffusion in the silicon lattice is discussed in [68G1].

Interstitial diffusion

Interstitial diffusion takes place by jumping of atoms from one interstitial void to the next adjacent one. Due to the squeezing through the lattice, the jumping atom has to overcome a potential barrier of height E_{AI}. Assuming a Boltzmann energy distribution, the frequency of jumping ν_I related to the frequency of lattice vibrations ν_0 is

$$\nu_I = 4\,\nu_0 \cdot e^{-\frac{E_{AI}}{kT}} \tag{2}$$

taking into account that jumping from one interstitial void to the next is possible in four different ways. Interstitial diffusion is therefore thermally activated according to

$$D_I = D_0 \cdot e^{-\frac{E_{AI}}{kT}} \tag{3}$$

Substitutional diffusion

When an atom jumps from one lattice site to the next, chemical bonds must be broken and new bonds are formed. For the jump process an energy barrier of height E_{AS} must be overcome. A similar reasoning for the jumping frequency applies as in the case of interstitial jumping. Additionally, the number of vacancies (Schottky defects) has to be taken into account. The frequency for substitutional jumping is

$$\nu_S = 4\nu_0 \cdot e^{-\frac{(E_{AS}+E_S)}{kT}} \tag{4}$$

where E_S is the energy for the formation of a Schottky defect. Substitutional diffusion is thermally activated by a larger energy than interstitial diffusion according to

$$D_S = D_0 \cdot e^{-\frac{(E_{AS}+E_S)}{kT}}. \tag{5}$$

Fick's second law

A steady state is never established in diffusion. The concentration and the flux are related according to

$$\frac{\partial c}{\partial t} = -\frac{\partial J}{\partial x} \tag{6}$$

the transport equation in one dimension. Combining the transport and diffusion equations yields Fick's second law

$$\frac{\partial c}{\partial t} = D \cdot \frac{\partial^2 c}{\partial x^2}. \tag{7}$$

6.1.4.1.1 Diffusion coefficients

In single crystal silicon, the diffusion coefficient is the same in all directions of the lattice. Differences may be caused by imperfections in real crystals.

The temperature dependence of the diffusion coefficient usually is given in the form

$$D = D_0 \cdot e^{-\frac{E}{kT}}. \tag{8}$$

Both, the prefactor D_0 and the exponential term are temperature dependent. However, D_0 usually is considered to be constant within a temperature range of some hundred degrees, because the temperature dependence of D is mainly controlled by the exponential factor.

The Table 1 lists diffusion coefficients of defects and impurities in single-crystal silicon, polycrystalline silicon, and germanium. The impurities are listed in the order of the position in the periodic table. Properties and diffusion mechanisms of specific impurities are discussed at the end of the table. Diffusion coefficients as function of temperature are plotted in Figs. 1⋯4. The values are believed to be valid for intrinsic diffusion.

Table 1. Diffusion coefficients. Abbreviations used in the table: c_{sur}: range of surface concentration investigated, T [°C]: range of temperature investigated, T: tracer measurements, pn: pn junction depth measurements, distribution assumed, SR: sheet resistance measurements, distribution assumed, ISR: incremental sheet resistance measurements, CV: capacitance voltage method, CD: concentration dependence not taken into account, $\bar{D}$: average D, typical of high concentration, I: interstitial, S: substitutional, L: limited.
Selected data are given in Figs. 1···4.

Dopant	D_0 $cm^2 s^{-1}$	E_A eV	c_{sur} cm^{-3}	T °C	D (1100 °C) $cm^2 s^{-1}$	Remarks	Fig.	Ref.
Silicon (single crystal)								
H	$9.4 \cdot 10^{-3}$	0.48		967···1207	$1.6 \cdot 10^{-4}$	mass spectrometry	1	56W
	$4.2 \cdot 10^{-5}$	0.56			$3.7 \cdot 10^{-7}$	T, evolution of gas		68I1
Li	0.0094	0.78		450···1000	$1.3 \cdot 10^{-5}$	pn		53F
	0.0023	0.66		360··· 877	$8.7 \cdot 10^{-6}$	ion drift, mobility		54F
	0.0023	0.72			$5.2 \cdot 10^{-6}$	low temperature ion pairing relaxation method		58M
	0.0022	0.70			$5.9 \cdot 10^{-6}$	pn		59S2
	0.0025	0.66			$9.4 \cdot 10^{-6}$	outdiffusion at high temperature and ion drift at low temperature		60P1, 60P2
	0.00265	0.63			$1.3 \cdot 10^{-5}$	ISR		66P3
	0.0038	0.66		300··· 550	$1.4 \cdot 10^{-5}$	ISR	1	71L
Na	0.00165	0.72		800···1100	$3.8 \cdot 10^{-6}$	pn	1	67S2
K	0.0011	0.76		800···1100	$1.8 \cdot 10^{-6}$	pn	1	67S2
B	10.5	3.69	$1.0 \cdot 10^{21} \cdots 1.0 \cdot 10^{22}$	950···1275	$3.0 \cdot 10^{-13}$	ISR, CD, $\bar{D}$		56F
	1.4	3.51	$1.6 \cdot 10^{17} \cdots 7.0 \cdot 10^{18}$	1050···1350	$1.8 \cdot 10^{-13}$	pn		60K2
	17.1	3.68	$2.0 \cdot 10^{20} \cdots 5.0 \cdot 10^{20}$	1120···1335	$5.3 \cdot 10^{-13}$	modified ISR		60Y
	16.0	3.69	$\approx 10^{21}$	1050···1350	$4.5 \cdot 10^{-13}$	pn, CD, $\bar{D}$		61W3
	2.02	3.52			$2.4 \cdot 10^{-13}$	ISR, redistribution during oxidation		64K1
	106.0	4.25	$1.0 \cdot 10^{15} \cdots 1.0 \cdot 10^{17}$		$2.6 \cdot 10^{-14}$	ISR, diffusion from a sputtered source		68N
	5.1	3.7	$3.0 \cdot 10^{18} \cdots 8.2 \cdot 10^{19}$	1100···1250	$1.3 \cdot 10^{-13}$	pn, SR, vacuum diffusion	2	69O
Al	2800	3.8		1200···1400	$3.1 \cdot 10^{-11}$	pn, CD		56G
	8.0	3.47	$1.0 \cdot 10^{16} \cdots 4.0 \cdot 10^{17}$	1085···1375	$1.5 \cdot 10^{-12}$	pn	2	56F
	4.8	3.36	$1.0 \cdot 10^{18} \cdots 2.7 \cdot 10^{19}$	1050···1380	$2.2 \cdot 10^{-12}$	capacitance method (CV), CD		56M
	0.5	3.0	$3.0 \cdot 10^{18} \cdots 1.3 \cdot 10^{19}$		$4.8 \cdot 10^{-12}$	SR, CD		67K
	1.38	3.41	$1.2 \cdot 10^{18} \cdots 1.5 \cdot 10^{18}$	1120···1390	$4.2 \cdot 10^{-13}$	ISR, source epilayer; diffusion in hydrogen		71G3

Table 1 (continued)

Dopant	D_0 $cm^2 s^{-1}$	E_A eV	c_{sur} cm^{-3}	T °C	D (1100 °C) $cm^2 s^{-1}$	Remarks	Fig.	Ref.
Ga	3.6	3.51	$1.4 \cdot 10^{19} \cdots 2.1 \cdot 10^{20}$	1105⋯1360	$4.7 \cdot 10^{-13}$	pn, CD		56F
	225	4.12	$1.4 \cdot 10^{17} \cdots 4.4 \cdot 10^{18}$	1130⋯1258	$1.7 \cdot 10^{-13}$	pn, ISR		58K
	0.347	3.39	$1.4 \cdot 10^{18} \cdots 4.2 \cdot 10^{18}$	1120⋯1390	$1.2 \cdot 10^{-13}$	ISR, source epilayer; diffusion in hydrogen		71G3
	60	3.89	$1.0 \cdot 10^{18} \cdots 5 \cdot 10^{18}$	900⋯1050	$3.1 \cdot 10^{-13}$	T, "isoconcentration" studies with boron; D concentration dependent, vacuum diffusion	2	71M
	124	3.96	intr. ⋯$8 \cdot 10^{20}$	900⋯1050	$3.6 \cdot 10^{-13}$	D^0, neutral defect		75S2
	0.716	3.46	intr. ⋯$8 \cdot 10^{20}$	900⋯1050	$1.4 \cdot 10^{-13}$	D^+, positive defect, $D = D^0 + D^+ \cdot (n_i/n)$		
	0.54	3.25	$5 \cdot 10^{15} \cdots 1 \cdot 10^{19}$	1100⋯1250	$6.2 \cdot 10^{-13}$	ISR, CV		79G
In	16.5	3.9	$2.8 \cdot 10^{17} \cdots 6.7 \cdot 10^{19}$	1105⋯1360	$7.9 \cdot 10^{-14}$	pn, CD,	2	56F
						T, data consistent with Fuller [56F]		66M2
	0.79	3.63	$7.0 \cdot 10^{15} \cdots 1.5 \cdot 10^{16}$	1120⋯1390	$3.7 \cdot 10^{-14}$	ISR, source epilayer; diffusion in hydrogen		71G3
Tl	16.5	3.9	$9.0 \cdot 10^{16} \cdots 1.2 \cdot 10^{17}$	1105⋯1360	$7.9 \cdot 10^{-14}$	pn, CD	2	56F
	1.37	3.7	$6 \cdot 10^{15}$	1120⋯1390	$3.6 \cdot 10^{-14}$	ISR, source epilayer, diffusion in hydrogen, only two temperatures investigated	2	71G3
C	0.33	2.92		1070⋯1400	$6.3 \cdot 10^{-12}$	T		61N1
	1.9	3.1		1070⋯1400	$7.9 \cdot 10^{-12}$	reanalyzed data of Newman [61N1]	2	62N
	33.2	2.94		950⋯1100	$5.3 \cdot 10^{-10}$	T, carbide formation		73G4
Si	1800	4.86		1220⋯1400	$2.6 \cdot 10^{-15}$	T		66P2
direct	9000	5.14		1100⋯1300	$1.2 \cdot 10^{-15}$	T	2	66M1
meas.	1460	5.02		1047⋯1387	$5.4 \cdot 10^{-16}$	T		77M2
	31100	5.25		1100⋯1400	$1.7 \cdot 10^{-15}$	recalculated from data of Masters [66M], Peart [66P]		73H
Si	18100	4.86		900⋯1300	$2.6 \cdot 10^{-14}$	T, from dissociative Au-diff-mechanism		64W1, 64W2
indirect meas.	1000	4.20		700⋯ 900	$3.8 \cdot 10^{-13}$	T, from dissociative mechanism of Ni-diffusion		67B
	230	5.00		700⋯ 900	$1.0 \cdot 10^{-16}$	calculated from Bonzel's [67B] data		68S1
	15700	5.23			$9.9 \cdot 10^{-16}$	D^0 neutral defect		75S2
	148	4.84			$2.5 \cdot 10^{-16}$	D^- negative defect		
	0.019	3.91			$8.4 \cdot 10^{-17}$	D^+ positive defect, $D = D^0 + D^+ \cdot (n_i/n) + D^- \cdot (n/n_i)$		
Ge	$6.26 \cdot 10^5$	5.28		1150⋯1350	$2.6 \cdot 10^{-15}$	T		57P
	1540	4.7		1250⋯1350	$8.6 \cdot 10^{-15}$	T		73V
	2500	4.97		>1000	$1.4 \cdot 10^{-15}$	T	2	79H
	0.35	3.93		<1000		T		79H

Table 1 (continued)

Dopant	D_0 $cm^2 s^{-1}$	E_A eV	c_{sur} cm^{-3}	T °C	D (1100 °C) $cm^2 s^{-1}$	Remarks	Fig.	Ref.
Sn	$2.16 \cdot 10^5$	5.39			$3.5 \cdot 10^{-15}$	T, enhanced in p-type, same in n-type		66M2
	32	4.25		1050…1200	$8.0 \cdot 10^{-15}$	T, enhanced in p-type, same in n-type	2	68Y
	0.054	3.5		1100…1200	$7.6 \cdot 10^{-15}$	doped oxide source, backscattering technique		74A
						T, $D = 2 \cdot 10^{-13}$ at 1200 °C		75S1
N		4.55				based on entropy of formation of NI_3		68P
P	10.5	3.69	$6.0 \cdot 10^{20} \cdots 5.0 \cdot 10^{22}$	950…1235	$3.0 \cdot 10^{-13}$	ISR, CD, $\overline{D}$		56F
	0.0032	2.6	$\approx 10^{20}$		$9.1 \cdot 10^{-13}$	SR, pn, prediffusion		62M1
	4.9	3.7	$3 \cdot 10^{18}$	1200…1300	$1.3 \cdot 10^{-13}$	SR, pn, drive in (Gaussian distribution)		62M1
	2.73	3.58	$2.2 \cdot 10^{18} \cdots 7.5 \cdot 10^{18}$		$1.9 \cdot 10^{-13}$	ISR, T; low concentration T only		62M2
	0.074	3.3	below intrinsic concentration at diffusion temperature	1150…1350	$5.6 \cdot 10^{-14}$	ISR		71G1
	1.1	3.4	$1.0 \cdot 10^{18} \cdots 1.0 \cdot 10^{20}$	900…1200	$3.6 \cdot 10^{-13}$	T, vacuum diffusion		71F2
	5.3	3.69	below intrinsic concentration at diffusion temperature	950…1200	$1.5 \cdot 10^{-13}$	T, vacuum diffusion, isoconcentration studies with P and P^{32}	2	73M
	0.39	3.12	extrinsic at diffusion temperature	950…1200	$1.4 \cdot 10^{-12}$	T, vacuum diffusion, isoconcentration studies with P and P^{32}		73M
	0.045	3.19			$8.8 \cdot 10^{-14}$	D^0 neutral defect		75S2
	92	4.14			$5.8 \cdot 10^{-14}$	D^- negative defect, $D = D^0 + D^- \cdot (n/n_i)$		
As	0.32	3.56	$5.8 \cdot 10^{17} \cdots 4.5 \cdot 10^{18}$	1095…1380	$2.7 \cdot 10^{-14}$	pn, CD		56F
	68.6	4.23	$\approx 10^{17} \cdots \approx 10^{19}$	1100…1350	$2.0 \cdot 10^{-14}$	SR, CD		62A
	2.564	3.88	$7.0 \cdot 10^{17} \cdots 7.0 \cdot 10^{18}$	1125…1312	$1.5 \cdot 10^{-14}$	SR		64R
	83000	5.2	$1.0 \cdot 10^{19} \cdots 2.0 \cdot 10^{19}$	1164…1280	$6.7 \cdot 10^{-15}$	SR		68H1
	60	4.20	$7.1 \cdot 10^{19} \cdots 2.2 \cdot 10^{21}$	850…1150	$2.3 \cdot 10^{-14}$	T, vacuum diffusion, isoconcentration studies with As, As^{76}	2	69M
	60	4.20	below intrinsic concentration at diffusion temperature		$2.3 \cdot 10^{-14}$	D^- negative defect, $D = D^- \cdot (n/n_i)$		75S2
	0.065	3.44	$2.0 \cdot 10^{17} \cdots 1.5 \cdot 10^{19}$	1160…1390	$1.5 \cdot 10^{-14}$	ISR, source epilayer; diffusion in hydrogen		71G2
Sb	5.6	3.95	$1.4 \cdot 10^{18} \cdots 9.2 \cdot 10^{21}$	1095…1380	$1.8 \cdot 10^{-14}$	pn, CD, $\overline{D}$		56F
	0.112	2.86		940…1300	$3.5 \cdot 10^{-12}$	T		58P
	12.9	3.98	$1.9 \cdot 10^{19} \cdots 5.0 \cdot 10^{19}$	1190…1398	$3.2 \cdot 10^{-14}$	T	2	59R2
	0.214	3.65	$2.5 \cdot 10^{17} \cdots 3.5 \cdot 10^{18}$	1190…1405	$8.5 \cdot 10^{-15}$	ISR, source epilayer; diffusion in hydrogen		71G2

Table 1 (continued)

Dopant	D_0 $cm^2\ s^{-1}$	E_A eV	c_{sur} cm^{-3}	T °C	D (1100 °C) $cm^2\ s^{-1}$	Remarks	Fig.	Ref.
Bi	1030	4.64	$1.0 \cdot 10^{17} \cdots 2.4 \cdot 10^{18}$	1220⋯1380	$9.5 \cdot 10^{-15}$	pn, CD, $\overline{D}$	2	56F
	896	4.12			$6.7 \cdot 10^{-13}$	pn		65P2
	1.08	3.85	$3.0 \cdot 10^{15} \cdots 2.0 \cdot 10^{16}$	1190⋯1390	$7.9 \cdot 10^{-15}$	ISR, source epilayer; diffusion in hydrogen		71G2
O	1.35	3.5		1250⋯1405	$1.9 \cdot 10^{-11}$	pn, after 450 °C heat treatment		59L
	0.21	2.44			$2.3 \cdot 10^{-10}$	internal friction calculation		60H
	0.23	2.56			$9.2 \cdot 10^{-11}$	comparison of annealing data with internal friction measurements	2	64C
S	0.92	2.2		1050⋯1370	$7.7 \cdot 10^{-9}$	resistivity and Hall measurements		59C
	0.0059	1.82		1000⋯1200	$1.2 \cdot 10^{-9}$	T	2	73G
Se	2.47	2.48		800⋯1250	$1.9 \cdot 10^{-9}$	angle lap, surface conductance		78V
	0.11	2.42		1000⋯1250	$1.4 \cdot 10^{-10}$	ISR	2	79K2
Cu	0.04	1.0		800⋯1100	$8.5 \cdot 10^{-6}$	T, I—S-diffusion		58B
	0.0047	0.43		300⋯ 700	$1.2 \cdot 10^{-4}$	T, I-diffusion, probably more accurate	1	64H
Ag	0.002	1.60		1100⋯1350	$2.7 \cdot 10^{-9}$	T, D (1200 °C) = $6.7 \cdot 10^{-9}\ cm^2\ s^{-1}$	2	61B2
						T, outdiffusion, D (1200 °C) = $6 \cdot 10^{-4}\ cm^2\ s^{-1}$		75U
Au	0.011	1.12		800⋯1200	$8.5 \cdot 10^{-7}$	T, IL, I—S-mechanism	1	56S, 57S
	11500	3.11			$4.4 \cdot 10^{-8}$	T, VL, I—S-mechanism		61B3
	$2.44 \cdot 10^{-4}$	0.39		700⋯1300	$9.0 \cdot 10^{-6}$	T, IL		64W1
	$2.75 \cdot 10^{-3}$	2.04		700⋯1300	$8.9 \cdot 10^{-11}$	T, SL		64W2
				900⋯1100		T, IL, I—S-mechanism, $D = 3 \cdot 10^{-7}\ cm^2\ s^{-1}$, no clear temperature dependence		73H
	0.0178	1.13		700⋯1300	$1.3 \cdot 10^{-6}$	T, IL, I—S-mechanism, reanalyzed data of [64W1]	2	73H
	$1.94 \cdot 10^{-7}$	0.61		900⋯1200	$1.1 \cdot 10^{-9}$	T, VL, I—S-mechanism		73H
	1150	3.12		700⋯1300	$4.0 \cdot 10^{-9}$	T, VL, I—S-mechanism, reanalyzed data of [64W2]	2	73H
Zn	0.1	1.4			$7.3 \cdot 10^{-7}$	electric field dependence studies	1	63M
				900⋯1360		$D = 10^{-6} \cdots 10^{-7}\ cm^2\ s^{-1}$		57F
						T, outdiffusion D (1200 °C) = $2 \cdot 10^{-6}\ cm^2\ s^{-1}$		75U
Cd				1200		D (1200 °C) = $1 \cdot 10^{-8}\ cm^2\ s^{-1}$		57C

Table 1 (continued)

Dopant	D_0 $cm^2\,s^{-1}$	E_A eV	c_{sur} cm^{-3}	T °C	D (1100 °C) $cm^2\,s^{-1}$	Remarks	Fig.	Ref.
Ti	$2.0\cdot10^{-5}$	1.50		1000⋯1250	$6.2\cdot10^{-11}$		2	77B
Ta				1215⋯1294		$D=1\cdot10^{-13}\cdots1\cdot10^{-12}\,cm^2\,s^{-1}$		67S1
Cr	0.01	1.0		1100⋯1250	$2.1\cdot10^{-6}$	pn	1	74W2
						D (1200 °C) $<10^{-8}\,cm^2\cdot s^{-1}$		57C
Mn				1200		T, $D>2\cdot10^{-7}\,cm^2\,s^{-1}$		56C
		1.3		1000⋯1350		T, $D=1.1\cdot10^{-6}\cdots2.1\cdot10^{-5}\,cm^2\,s^{-1}$		72B1
Fe	$6.2\cdot10^{-3}$	0.87		1100⋯1250	$4.0\cdot10^{-6}$	T	1	56S
				1000⋯1115		T, $D>5\cdot10^{-6}\,cm^2\,s^{-1}$ fast		57C
				880⋯1115		$D=1.3\cdot10^{-7}\cdots7\cdot10^{-7}\,cm^2\,s^{-1}$ slow		57C
	$6.3\cdot10^{-4}$	0.58		100⋯ 500	$4.7\cdot10^{-6}$	SR, outdiffusion		72B2
						T, outdiffusion, D (1200 °C) $=4\cdot10^{-6}\,cm^2\,s^{-1}$		75U
Co	$9.2\cdot10^{4}$	2.8		900⋯1200	$4.8\cdot10^{-14}$	T	1	77K
				600⋯ 800		T, $D=10^{-12}\cdots4\cdot10^{-11}\,cm^2\,s^{-1}$		77M1
Ni	0.013	1.4			$9.4\cdot10^{-8}$	apparent diffusion coefficient from decay of electrically active centers (subst. Ni) assumed to be vacancy diffusion-controlled		67Y
	0.1	1.91		450⋯ 800	$9.7\cdot10^{-9}$	T, apparent diffusion coefficient vacancy controlled I—S-mechanism assumed	1	67B
	$\approx10^{-1}$	1.88		325⋯ 400	$1.3\cdot10^{-8}$	sputter-etch technique with Auger-analysis		78B
	$2\cdot10^{-3}$	0.47		800⋯1300	$3.8\cdot10^{-5}$	T, IL, I—S-mechanism	1	80B
Ru				1000⋯1280		ISR, $D=5\cdot10^{-7}\cdots5\cdot10^{-6}\,cm^2\,s^{-1}$		74Y
Os				1280		ISR, $D_1=2\cdot10^{-6}\cdots2.8\cdot10^{-7}\,cm^2\,s^{-1}$, $D_2=4.5\cdot10^{-8}\,cm^2\,s^{-1}$, near surface and in bulk material		78A3
Ir	0.042	1.3		950⋯1250	$7.1\cdot10^{-7}$	ISR, neutron activation	1	76A
Pt	0.1	1.9		450⋯ 800	$1.1\cdot10^{-8}$		1	67B
He	0.11	1.26		960⋯1200	$2.6\cdot10^{-6}$	mass spectometry		56W
	$5.1\cdot10^{-4}$	0.58		500⋯1100	$3.7\cdot10^{-6}$	rate of evolution	1	64L
	$1.28\cdot10^{-3}$	1.8			$3.2\cdot10^{-10}$	calculated		79K1

Table 1 (continued)

Dopant	D_0 $cm^2\,s^{-1}$	E_A eV	c_{sur} cm^{-3}	T °C	D (1100 °C) $cm^2\,s^{-1}$	Remarks	Fig.	Ref.
silicon (polycrystalline)								
B				1000···1200		$D=10^{-12}...10^{-11}$ $cm^2\,s^{-1}$		72K
	$6.01\cdot10^{-3}$	2.39		900···1050	$1.01\cdot10^{-11}$	BN-source		75H2
	$1.51\cdot10^{-3}$	2.51		900···1050	$9.2\cdot10^{-13}$	B_2H_6-source		75H2
Al	$1.3\cdot10^{7}$	2.64		350··· 425	D (500 °C)= $1.61\cdot10^{-17}$	evaporated Al-source		80H
P				1000···1200		$D=10^{-12}...10^{-11}$ $cm^2\,s^{-1}$		72K

Table 1 (continued)

Dopant	D_0 $cm^2\,s^{-1}$	E_A eV	c_{sur} cm^{-3}	T °C	D (800 °C) $cm^2\,s^{-1}$	Remarks	Fig.	Ref.
Germanium (single crystal)								
H						mass spectrometry, D (800 °C) $>5\cdot10^{-5}$ $cm^2\,s^{-1}$		56W
Li	$1.3\cdot10^{-3}$	0.46		350···800	$9.0\cdot10^{-6}$	pn		53F
	$2.5\cdot10^{-3}$	0.52		150···600	$9.0\cdot10^{-6}$	ion drift, mobility		54F
	$9.1\cdot10^{-3}$	0.57		300···500	$1.9\cdot10^{-5}$	ISR	3	66P3
Be	0.5	2.5		720···900	$9.0\cdot10^{-13}$	pn	4	61B1
B	$1.8\cdot10^{9}$	4.55		600···900	$7.6\cdot10^{-13}$	pn		54D
	$1.1\cdot10^{7}$	4.54		770···850	$5.2\cdot10^{-15}$	pn	4	67M
Al	$1.6\cdot10^{2}$	3.24		750···850	$9.7\cdot10^{-14}$	pn	4	67M
Ga	40	3.15		300···600	$6.4\cdot10^{-14}$	pn	4	54D
	20	3.03			$1.2\cdot10^{-13}$	method unknown		67G
In	20	3.0			$1.6\cdot10^{-13}$	concentration profiles from impedance measurements		55B
	33	3.03			$1.9\cdot10^{-13}$	T		66P1
	10	2.77		510···880	$9.8\cdot10^{-13}$	T		65P1
	$1.2\cdot10^{8}$	4.54		800···850	$5.68\cdot10^{-14}$	D^0, neutral defect	4	75S2
	$1.4\cdot10^{17}$	6.45		800···850	$7.07\cdot10^{-14}$	D^-, negative defect		
					$1.3\cdot10^{-13}$	$D=D^0+D^-\cdot(n/n_i)$, $(n/n_i=1)$		

Table 1 (continued)

Dopant	D_0 $cm^2 s^{-1}$	E_A eV	c_{sur} cm^{-3}	T °C	D (800 °C) $cm^2 s^{-1}$	Remarks	Fig.	Ref.
Tl	$6 \cdot 10^{-2}$	2.69			$1.4 \cdot 10^{-14}$	estimated value	4	59R1
	$1.7 \cdot 10^{3}$	3.42		800···930	$1.5 \cdot 10^{-13}$	T		62T1
Ge	7.8	2.98		766···928	$7.8 \cdot 10^{-14}$	T		56L
	44	3.15		730···930	$7.0 \cdot 10^{-14}$	T Steigman-technique		61W2
	10.8	3.02		730···930	$7.1 \cdot 10^{-14}$	T Gruzin-technique		
	3.53	3.00		730···930	$2.9 \cdot 10^{-14}$	D^0 neutral defect	4	75S2
	1.48	2.88		730···930	$4.4 \cdot 10^{-14}$	D^- negative defect		
					$7.3 \cdot 10^{-14}$	$D = D^0 + D^- \cdot (n/n_i)$, $(n/n_i = 1)$		
Sn	$1.2 \cdot 10^{-10}$	0.45		800···850	$9.2 \cdot 10^{-13}$	D^0	4	75S2
	$1.6 \cdot 10^{3}$	3.29		800···850	$5.6 \cdot 10^{-13}$	D^-		
					$1.5 \cdot 10^{-12}$	$D = D^0 + D^- \cdot (n/n_i)$, $(n/n_i = 1)$		
Pb						D (800 °C) $= 2 \cdot 10^{-14}$ $cm^2 s^{-1}$		63B
P	3.3	2.5		600···900	$6 \cdot 10^{-12}$	pn	4	54D
As	0.71	2.21		700···900	$3 \cdot 10^{-11}$	pn		52F
	12.7	2.3		600···900	$2 \cdot 10^{-10}$	pn		54D
	2.1	2.39		700···900	$1.2 \cdot 10^{-11}$	concentration profiles from impedance measurements		55B
	$5.4 \cdot 10^{2}$	2.9		700···900	$1.3 \cdot 10^{-11}$	pn		59K
	3	2.43		700···900	$1.2 \cdot 10^{-11}$			61A
	1.5	2.39		580···900	$8.9 \cdot 10^{-12}$	pn		62W
	10.3	2.51		700···790	$1.7 \cdot 10^{-11}$	pn		68I1
	$9.4 \cdot 10^{11}$	5.10		800···850	$1.0 \cdot 10^{-12}$	D^0 neutral defect	4	75S2
	$3.9 \cdot 10^{-4}$	1.62		800···850	$9.6 \cdot 10^{-12}$	D^- negative defect		
					$1.1 \cdot 10^{-11}$	$D = D^0 + D^- \cdot (n/n_i)$, $(n/n_i = 1)$		
Sb	10	2.5		600···900	$1.8 \cdot 10^{-11}$	pn		54D
	1.2	2.3		700···900	$1.9 \cdot 10^{-11}$	concentration profiles from impedance measurements		55B
	1.3	2.26		800···900	$3.2 \cdot 10^{-11}$	T		57M
	17	2.7		730···900	$3.5 \cdot 10^{-12}$	pn		59K
	3.2	2.42		700···855	$1.4 \cdot 10^{-11}$	T		66P1

Table 1 (continued)

Dopant	D_0 $cm^2\,s^{-1}$	E_A eV	c_{sur} cm^{-3}	T °C	D (800 °C) $cm^2\,s^{-1}$	Remarks	Fig.	Ref.
	6.5	2.57		700···800	$5.5\cdot10^{-12}$			68S2
	$1.6\cdot10^{8}$	4.20		767···875	$3.0\cdot10^{-12}$	D^0 neutral defect	4	75S2
	$6.9\cdot10^{-2}$	2.03		767···875	$2.0\cdot10^{-11}$	D^- negative defect		
					$2.3\cdot10^{-11}$	$D=D^0+D^-\cdot(n/n_i)$, $(n/n_i=1)$		
Bi						$D(850\,°C)=3.0\cdot10^{-11}\,cm^2\,s^{-1}$, estimated value		59R1
	3.3	2.47		650···850	$8.3\cdot10^{-12}$		4	67G
O	0.17	2.02			$5.5\cdot10^{-11}$	internal friction, calculation		60H
	0.4	2.08			$6.8\cdot10^{-11}$	comparison of annealing data with internal friction measurements	4	64C
S						$D\,(920\,°C)\approx10^{-9}\,cm^2\,s^{-1}$		59T
Se						$D\,(920)\,°C)\approx10^{-10}\,cm^2\,s^{-1}$		59T
Te	5.6	2.43		750···900	$2.2\cdot10^{-11}$	T	4	62I
Cu	$1.9\cdot10^{-4}$	0.18		750···900	$2.7\cdot10^{-5}$	I-diffusion		63B
	$4\cdot10^{-2}$	0.99		600···700	$9\cdot10^{-7}$	S-diffusion, T		63B
	$4\cdot10^{-3}$	0.33		350···750	$1.1\cdot10^{-4}$	T	3	64H
Ag	$4.4\cdot10^{-2}$	1.0		700···900	$8.8\cdot10^{-7}$	T,	3	57B
						$D\,(710\,°C)=2\cdot10^{-6}\,cm^2\,s^{-1}$, I-diffusion		61W1
	$4\cdot10^{-2}$	2.23		800···900	$1.3\cdot10^{-12}$	T		62K
Au	2.25	2.5		600···900	$4.1\cdot10^{-12}$	T		55D
	$3.5\cdot10^{-6}$	0.63			$3.8\cdot10^{-9}$	I-diffusion	3	68G2
Zn	5	2.7		600···900	$1.0\cdot10^{-12}$	T, pn	4	54D
	0.65	2.54		600···900	$7.6\cdot10^{-13}$	T		56K
Cd	$1.75\cdot10^{9}$	4.4		760···915	$3.8\cdot10^{-12}$	T	4	60K1
Fe	0.13	1.1		750···900	$8.9\cdot10^{-7}$	T	3	57B
						T, $D\,(800\,°C)=2.3\cdot10^{-7}\,cm^2\,s^{-1}$		61W1
Co	0.16	1.12		750···850	$8.8\cdot10^{-7}$	T	3	61W1
Ni	0.8	0.9		670···900	$4.7\cdot10^{-5}$	T	3	54M
He	$6.1\cdot10^{-3}$	0.69		750···900	$3.5\cdot10^{-6}$	mass spectrometry	3	56W

Substitutional diffusion

Substitutional impurities, e.g. the group III and the group V elements of the periodic table, which are frequently used to adjust conductivity type and resistivity, diffuse by a substitutional mechanism in the silicon lattice. The activation energy for forming a vacancy is required to brake the chemical bonds to the adjacent silicon atoms. The activation energy is in the range of 3 to 4 eV. This energy is less than the activation energy for self diffusion in silicon which is approximately 5 eV.

Interstitial diffusion

Interstitial diffusion is much faster than substitutional diffusion due to the small activation energies in the range of 0.5···1.6 eV. Besides hydrogen and helium, the alkali metals diffuse by the interstitial mechanism. These elements are interstitial impurities in silicon.

Dissociative diffusion

Substitutional and interstitial diffusion processes are involved in the dissociative mechanism. The species located on a lattice site having the concentration c_S dissociates into a vacancy having the concentration c_V by forming an interstitial defect having concentration c_I or interstitial $[Si]_I$ may kick out [81G1] substitutional impurities from a lattice site.

The reactions are:

$$\begin{aligned} c_S &\rightleftharpoons c_I + c_V, \\ c_S + [Si]_I &\rightleftharpoons c_I. \end{aligned} \tag{9}$$

Interstitial defects diffuse more rapidly and thus determine the diffusion process. The equilibrium is influenced through the vacancy or interstitial generation by other processes. The diffusion constant of the dissociative process is an apparent diffusivity, it may be vacancy-limited or interstitial-limited.

Elements which may be substitutional and interstitial impurities are Cu, Ag, Au, Fe, Co, Ni, Pt and probably other elements having activation energies for diffusion in the range of 1.6···3 eV. A clear separation of the vacancy-limited diffusion from the interstitial-limited diffusion could be achieved for gold, which is frequently used to adjust the charge carrier lifetime in silicon.

Effects of concentration

Donor and acceptor impurities (substitutional impurities) in the silicon lattice are ionized. If a concentration gradient of ions is present a concentration gradient of charge carriers, i.e. electrons or holes, is also present. Because the charge carriers diffuse much faster than ions, they build up a space charge which causes an electric field. Ions drift in the electric field in the direction of the concentration gradient, donor impurities as well as acceptor impurities. It is difficult to separate the diffusion from the drift. An effective diffusion constant D_{eff} is defined in [61L]:

$$D_{eff} = D\left(1 + \frac{1}{\sqrt{1+4(n_i/c)^2}}\right) \tag{10}$$

assuming the two limiting values

$$\begin{aligned} c \ll n_i;\ D_{eff} &= D, \\ c \gg n_i;\ D_{eff} &= 2D, \end{aligned} \tag{11}$$

n_i is the intrinsic charge carrier concentration.

The effect of drift fields is noticed at concentrations c above the intrinsic concentrations at the diffusion temperatures ($n_i \approx 10^{19}$ cm^{-3} at 1100 °C) i.e. at doping concentrations $c > 5 \cdot 10^{19}$ cm^{-3}. Makris and Masters [73M] have shown for phosphorus and Masters and Fairfield [69M] for arsenic that in the presence of high homogeneous doping concentrations, the diffusion constant is raised by a factor of ten. In the notation of Shaw [75S2], this concentration dependence can be described by

$$D = D^0 + D^+ \cdot \frac{n_i}{n} + D^- \cdot \frac{n}{n_i} \tag{12}$$

where D^0, D^+ and D^- are diffusion coefficients of the diffusing species combined with a neutral, monovalent positive or negative defect of the native lattice. This description holds for impurity diffusion and self diffusion. In Table 1 the notation of Shaw is used for the concentration dependence. The mechanism of diffusion at high concentrations is not yet understood.

Effects of oxidation

In silicon, diffusion as a doping procedure is frequently performed under oxidizing conditions. Oxidation perturbs the point defect concentrations, causing the intrinsic diffusion of substitutional impurities to be oxidation enhanced or reduced (OED = oxidation enhanced diffusion or ORD = oxidation reduced diffusion). The amount of the observed deviation from intrinsic diffusion depends on the time and the temperature of the diffusion in dry oxygen ambient and thus is oxidation rate determined. Figs. 5 and 6 show results of the time dependence, Figs. 7⋯9 show results of the temperature dependence.

In Fig. 10, plots are given which are based on theoretical calculations and measurements. The diffusion enhancement of boron and phosphorus in dry oxygen ambient may be calculated as a function of time and temperature. Other types of oxidizing ambients, e.g. wet oxygen [81M, 81L] and mixtures of oxygen and nitrogen [82M] seem to produce values as expected from oxidation in dry oxygen according to the oxidation rates.

Diffusion in polycrystalline silicon

Diffusion in polycrystalline silicon occurs in the silicon gate MOS-technology when

thin polycrystalline silicon films are doped by diffusion,

aluminum/poly-silicon contacts are annealed.

Data on diffusion of impurities in polycrystalline thin films of silicon are included in Table 1. In these cases, grain boundary diffusion is prevalent. The diffusion properties vary with the type of the material structure. The material structure depends on deposition conditions and changes during annealing processes [75H1] and hence depends on the diffusion conditions itself [75H2].

6.1.4.1.2 Introduction of impurities by diffusion

Impurities are usually introduced into silicon by annealing the silicon wafer [68G1, 75R]

- in an ambient containing the dopant to be introduced. The dopant may be present as the vapour of the chemical element to be introduced or the vapour of a chemical compound usually an oxide of the element to be introduced.
- in direct contact with the dopant. The dopant may be the chemical element to be introduced or a chemical compound of the element to be introduced, usually an oxide in the form of a silicate glass (*phosphorus silicate glass* PSG, *boron silicate glass* BSG, *arseno silicate glass* AsSG).

A survey of diffusion methods is given in Table 2.

Table 2. Diffusion methods (BSG: boron silicate glass, PSG: phosphorus silicate glass, AsSG: arseno silicate glass,

Doping element	Source				Type of ambient	Composition of dopant in ambient	Number of furnace zones	Type of tube used
	material	inside of furnace	outside of furnace	T [°C]				
Boron	B_2O_3	×		600···1200	ox	B_2O_3	2	open
	B_2O_3	×		600···1200	red	?	1	semi-closed box
	B_2O_3	×		DT	ox		1	open
	BBr_3		×	10···30	ox	B_2O_3	1	open
	BCl_3		×	RT	ox	B_2O_3	1	open
	B_2H_6		×	RT	ox	B_2O_3	1	open
	BN	×		DT	ox	B_2O_3	1	open
	BSG	×		DT	ox, inert	B_2O_3	1	open
	elemental B-powder	×		DT	argon, vacuum	elemental boron	1	sealed
	B-doped silicon	×		DT	argon, vacuum	elemental boron	1	sealed
	B-implant	×		DT	ox		1	open
	B-doped poly-Si	×		DT	ox		1	open
Aluminum	Al	×		DT	red	Al	1	closed
	Al			DT	red		1	open
Gallium	Ga_2O_3	×			red	Ga_2O	2	open
	Ga	×		DT	argon	Ga	1	sealed
Indium	In_2O_3	×			red	In_2O	2	open
	In	×		DT	red		1	sealed
Phosphorus	P_2O_5	×		200···400	ox	P_2O_5	2	open
	P_2O_5	×		1000···1200	ox		1	open
	$POCl_3$		×	2···40	ox	P_2O_5	1	open
	PBr_3		×	0···30	ox	P_2O_5	1	open
	PCl_3		×	170	ox	P_2O_5	1	open
	PH_3		×	RT	ox	P_2O_5	1	open
	PSG	×		DT	ox, inert		1	open
	$(NH_4)_3\ PO_4$	×		DT	ox	P_2O_5	1	semi-closed box

RT: room temperature, DT: diffusion temperature, ox: oxidizing ambient, red: reducing ambient.

Type of diffusion		Typically adjusted surface concentration	Mask		Remarks	Standard method
prede-position	one step		SiO_2	Si_3N_4		
×		max.	×	×		
			poor	×	no usual process no useful diffusion mask	
	×	max.			"paint-on" source; in contact with wafer; for unmasked diffusions only	×
×		max.	×	×		×
×		max.	×	×		×
×		max.	×	×		×
×		max.	×	×		×
	×	depends on composition of BSG	×	×	source in contact with wafer; prepared as "spin-on" film or CVD-film, "solid to solid" diffusion	
	×	max	poor	×		
	×	depends on source concentration and surface ratio	poor	×		
	×		photo-resist		especially for shallow layers below solubility	×
	×	max.	poor		"source" in contact with oxide and silicon; this case occurs in the silicon gate process	
	×	max.			Al reacts with SiO_2 evaporated Al on Si. In contact formation below melting point of Al; diffusion of Si into Al is the dominant process	
	×				not used, no proper diffusion mask	
	×	max.	fails	×	for deep unmasked diffusions	×
	×	max.	fails		not in use; no proper diffusion mask	
	×		fails		not in use	
×		max.		×		
	×	max.			"paint-on" source; in contact with wafer; for unmasked diffusions only	×
×		max.	×	×		×
×		max.	×	×		×
×		max.	×	×		
×		max.	×	×		×
	×	depends on composition of PSG	×	×	source in contact with wafer; prepared as "spin-on"-or CVD-film; "solid to solid" diffusion	
	×	max.	×	×	rarely used	

Table 2 (continued)

Doping element	Source material	inside of furnace	outside of furnace	T [°C]	Type of ambient	Composition of dopant in ambient	Number of furnace zones	Type of tube used
	P red	×		200···300	red	P	2	open
	P-doped Si	×		DT	inert, vacuum	P	1	sealed
	P-implant	×		DT	ox		1	open
	P-doped poly-Si	×		DT	ox		1	open
Arsenic	As_2O_3	×		150···250	ox	oxide of As	2	open
	AsH_3		×	RT	ox	oxide of As	1	open
	AsSG	×		DT	inert, ox		1	open
	As	×		DT	inert	As	1	sealed
Antimony	Sb_2O_4	×		≈900	ox	oxide of Sb	2	open
	SbH_3		×	RT	ox	oxide of Sb	1	open
Gold	Au	×		DT	inert		1	open

Diffusion from the vapour phase

Vapour of elemental dopant

When the vapour of the elemental dopant is in equilibrium with its solution in silicon (mixed phase), Henry's law may be applied to determine the surface equilibrium concentration.

$$c_{sur} = \mathrm{H} \cdot p \qquad (13)$$

where c_{sur} is the concentration in the silicon, p the partial pressure of the dopant in the vapour phase, and H a constant of proportionality. c_{sur} is limited at a given temperature by the maximum solid solubility. Values of H are not given in the literature. Below the maximum solubility, diffusion methods suffer in reproducibility.

Diffusions from an elemental vapour dopant may be performed in an open tube system by using a reducing (hydrogen) ambient as shown in Fig. 11 or in evacuated sealed tubes, sometimes backfilled with argon prior to the sealing. The latter method is commonly used for gallium diffusions, (Fig. 12). A method to adjust surface concentrations below the maximum solid solubility at the diffusion temperature uses doped silicon as a source material in a sealed tube (Fig. 13), [71F, 73G]. Data of boron and phosphorus are given in Fig. 14.

Table 2 (continued)

Type of diffusion: predeposition	Type of diffusion: one step	Typically adjusted surface concentration	Mask: SiO_2	Mask: Si_3N_4	Remarks	Standard method
	×	≦max.	×	×	not in use	
	×	depends on source concentration and surface ratio	×	×		
	×		photoresist		especially for shallow layers below solubility	×
	×	max.	×		"source" in contact with oxide and silicon; this case occurs in the silicon gate process	
×		max.	×	×		
×		max.	×	×		
	×	max.	×	×	source in contact with wafer; prepared as "spin on"-or CVD-film; "solid to solid"-diffusion	×
	×	max.	×	×		
×		max.	×			×
×		max.	×			
	×				source evaporated on back of wafer; used to adjust charge carrier lifetime	×

Vapour of doping compounds

The diffusion process is rather complicated when compounds are applied. This is explained for the case of phosphorus. The ambient contains the oxide of phosphorus P_2O_5 and oxygen. In the initial stages at the interface gas/silicon or in later stages at the interface PSG/silicon, the following chemical reaction takes place:

$$5\,Si + 2\,P_2O_5 \rightarrow 5\,SiO_2 + 4\,P. \tag{14}$$

The phosphorus diffuses into the silicon. Silicon dioxide and phosphorus pentoxide form phosphorus silicate glass (PSG) on the surface of the silicon. The transition of the phosphorus from the vapour phase or from the PSG phase to the silicon, the chemical reaction, seems not to be rate determining. Dependence of surface concentration on partial pressure of the phosphorus in analogy to Henry's law is observed. Below the maximum solubility at the diffusion temperature, the doping concentrations obtained are not reproducible.

Diffusions of this type are usually performed in open tube systems in oxidizing ambient using a solid source (Fig. 15a), liquid source (Fig. 15b), or gaseous source (Fig. 15c). This process is the typical "predeposition" diffusion process followed by the "drive-in" diffusion process, applied for boron, phosphorus and arsenic. The "closed box" diffusion method (Fig. 16) is also an open tube process. It is operated at atmospheric pressure. Prior to the drive-in diffusions, which are performed in an open tube system without source, the PSG-films are removed in order to avoid mask failures. A special diffusion method is based on a boron nitride source. Boron nitride wafers covered by boron oxide serve as the diffusion source. Wafers to be diffused and source wafers are in alternating positions in an open tube system (Fig. 17).

Diffusion from a surface layer in contact with silicon

Elemental dopants

Two-phase systems have to be considered for diffusion of elemental dopants deposited in a surface layer.

In practice, only gold is diffused into silicon for doping purposes from an evaporated source layer. Because of the very complex distribution mechanism of gold (see 6.1.4.1.5), only as much gold is evaporated as is necessary to adjust the bulk concentration by consumption of the whole gold source.

When aluminum is annealed in contact with silicon for contact formation diffusion of silicon into the aluminum layer occurs and leads to the formation of spikes in silicon [76L]. Data of the diffusion and solubility of the Al/Si-system are presented in Table 3.

Table 3. Diffusion and solubility data of the system polycrystalline thin film Al/monocrystalline Si (melting temperatures: 1412 °C (Si), 660 °C (Al)).

	Si in Al	Ref.	Al in Si	Ref.
Solubility at 500 °C in [cm^{-3}]	$\approx 3 \cdot 10^{20}$		$\leqq 10^{19}$	
D (500 °C) in [$cm^2 s^{-1}$]	$2 \cdot 10^{-8}$	71C1	$8 \cdot 10^{-23}$	71G3
Prefactor D_0 in [$cm^2 s^{-1}$]	$2.59 \cdot 10^{-3}$	71C1	1.38	71G3
Activation energy E_A in [eV]	0.79	71C1	3.41	71G3

Compound surface layer

The dopant silicate glass layers (BSG, PSG, AsSG) are deposited from chemically reacting liquids by spinning techniques [69G] or by chemical vapour deposition (see section 6.1.4.4).

The relation of the surface concentration obtained in silicon to the source concentration in PSG or BSG is presented in Figs. 18 and 19. Diffusions using PSG or BSG sources are not reproducible. The AsSG source is applied for buried layer diffusions aiming at maximum concentrations.

Sources for solid to solid diffusions produced by "paint-on" techniques usually consist of the oxide of the doping element. They are painted on from a slurry of B_2O_3 or P_2O_5 in alcohol. Paint-on processes are used for deep unmasked diffusions only.

Ion implantation

Because the solubility of doping materials (see section 6.1.1.2) decreases with decreasing temperature from its maximum value at a certain temperature, phase segregation occurs during cooling down from the diffusion temperature when diffusion is performed under conditions giving maximum solubility at the surface. Because most diffusions giving surface concentrations less than the solubility are not reproducible, predeposition diffusions are substituted by ion implantations. At present, ion implantation is frequently applied for low-level dopings, or shallow doping layers with high concentrations where the total dose can be implanted in a reasonable time (see section 6.1.4.2).

6.1.4.1.3 Properties of diffusion source materials

Various diffusion source materials are in use depending on the diffusion technique applied as discussed in section 6.1.4.1.2. Table 4 lists the melting points, the boiling points and the temperatures for 10 Torr ($\triangleq 1.33 \cdot 10^3$ Pa) vapour pressure of commonly used diffusion source material.

Table 4. Properties of source materials [77W1]. (BP: boiling point, MP: melting point, T_{p10}: temperature at which vapour pressure reaches 10 Torr $\triangleq 1.33 \cdot 10^3$ Pa).

Compound	MP °C	BP °C	T_{p10} °C	Remarks
B	2300	2550	3030	
B_2O_3	450	1860		transforms to glass; vapour pressure unstable
BBr_3	−46	91.3	−10	fuming liquid
BCl_3	−107	12.5	−66.9	
B_2H_6	−16.5	−92.5		
BN				sublimes at $\approx$ 3000 °C; is converted to oxide on top for use as source
BSG				see phase diagram B_2O_3—SiO_2, Fig. 29
B-implant				replaces diffusion source

continued

Table 4 (continued)

Compound	MP °C	BP °C	T_{p10} °C	Remarks
Al	660	2467	1780	
Ga	29.8	2403	1570	
$Ga_2O_3(\alpha)$	1900			above 600 °C transition to β-Ga_2O_3
In	156.6	2080		
In_2O_3	1910			volatile
P_4 violet	590		127	
P_2O_5	569		384	sublimes ≧300 °C, stable form
			189	metastable form
$POCl_3$	2	105.3	2.0	fuming liquid
PCl_3	−112	75.5	−21.3	fuming liquid
PBr_3	−40	172.9	47.8	fuming liquid
PH_3	−133	−187.7		poisonous
$(NH_4)_3PO_4$				decomposes
PSG				see phase diagram SiO_2—P_2O_5, Fig. 30, decomposes
P-implant				replaces diffusion source
As gray		613	437	sublimes
As_2O_3		312.8	259.7	sublimes
AsH_3	−116.3	−55		
AsSG				decomposes
As-implant				replaces diffusion source
Sb	630.7	1750	1033	
Sb_2O_3	656	1550	666	sublimes
SbH_3	−88	−17		
Sb-implant				replaces diffusion source
Au	1064	2807	2160	

6.1.4.1.4 Diffusion in the system SiO_2/Si

Diffusion in the system SiO_2/Si is of interest

when silicon dioxide on silicon is used as a diffusion mask.

when dopants incorporated in silicon are redistributed during thermal oxidation,

when dopants introduced into thermally oxidized silicon by ion implantation are redistributed by diffusion.

The diffusion coefficient of the dopant in the oxide and its temperature dependence are also described by exponential activation

$$D = D_0 \cdot \exp(-E_A/kT). \tag{14}$$

The diffusion coefficient depends on the concentration and additionally on the valence state of the dopant (oxidizing or reducing ambient). Table 5 gives a survey of properties on diffusion in silicon dioxide and some data on diffusion in Si_3N_4. A plot of diffusion coefficients versus temperature is depicted in Fig. 20.

Besides the diffusion coefficients of the dopants in silicon and silicon dioxide the distribution constant (segregation constant)

$$k = c_{Si}/c_{SiO_2} \tag{15}$$

is important for the redistribution of dopants. c_{Si} and c_{SiO_2} are the concentrations in the silicon and the silicon dioxide at the interface. Values of k are presented in Table 6. The temperature dependence of k for boron in the system Si/SiO_2 is given in Fig. 21.

Table 5. Diffusion coefficients of impurities in silicon dioxide (SiO_2) and silicon nitride (Si_3N_4). D(1100 °C) values are calculated when D_0 and E_A are given (BSG: boron silicate glass, PSG: phosphorus silicate glass, AsSG: arseno silicate glass, c_0: range of source concentration; c_{sur}: range of surface concentration).

Dopant	D_0 $cm^2 s^{-1}$	E_A eV	D(1100 °C) $cm^2 s^{-1}$	T-range °C	c_0 mol %
Silicon dioxide (SiO_2)					
B	$7.23 \cdot 10^{-6}$	2.38	$1.3 \cdot 10^{-14}$		0.001···0.1
	$1.23 \cdot 10^{-4}$	3.39	$4.4 \cdot 10^{-17}$		
	$7.38 \cdot 10^{-4}$	3.58	$5.3 \cdot 10^{-17}$		3.5···7.0
	$3.16 \cdot 10^{-4}$	3.53	$3.5 \cdot 10^{-17}$		<18
					>18
	$1.61 \cdot 10^{-5}$	2.82	$7.1 \cdot 10^{-16}$		9.1···9.4
	$3.01 \cdot 10^{-2}$	3.56	$2.6 \cdot 10^{-15}$		14.9···15.8
	$5.16 \cdot 10^{-2}$	4.06	$6.4 \cdot 10^{-17}$	1050···1250	
	$6.2 \cdot 10^{-2}$	4.06	$7.7 \cdot 10^{-17}$	1050···1250	
	$2.3 \cdot 10^{-5}$	2.56	$9.2 \cdot 10^{-15}$		
	$3.6 \cdot 10^{-10}$	1.56	$6.8 \cdot 10^{-16}$		
Ga	$1.04 \cdot 10^{5}$	4.17	$5.1 \cdot 10^{-11}$	1100···1250	
	0.73	2.46	$6.8 \cdot 10^{-10}$	800··· 900	
	$5.2 \cdot 10^{-4}$	1.77	$1.7 \cdot 10^{-10}$	800···1250	
P	$5.73 \cdot 10^{-5}$	2.30	$2.1 \cdot 10^{-13}$		
	$6.39 \cdot 10^{-11}$	1.27	$1.4 \cdot 10^{-15}$		
	4.72	4.21	$1.7 \cdot 10^{-15}$		3.0
	$1.86 \cdot 10^{-1}$	4.03	$3.0 \cdot 10^{-16}$		0.1···3.5
	$1.9 \cdot 10^{-9}$	1.1	$1.7 \cdot 10^{-13}$		
	7.23	4.44	$3.6 \cdot 10^{-16}$	1100···1250	> solid solubility
	$2.0 \cdot 10^{-10}$	1.7	$1.1 \cdot 10^{-16}$	770···1095	
	$7.4 \cdot 10^{-14}$	1.7	$4.2 \cdot 10^{-20}$	770···1095	
	$4.0 \cdot 10^{-10}$	1.62	$4.5 \cdot 10^{-16}$	1100···1250	> solid solubility
	$8.8 \cdot 10^{-10}$	1.63	$9.1 \cdot 10^{-16}$	1100···1250	> solid solubility
	$3.9 \cdot 10^{-9}$	1.63	$4.0 \cdot 10^{-15}$	800···1250	liquidus composition at diffusion-temperature
As	$9.82 \cdot 10^{1}$	4.88	$1.2 \cdot 10^{-16}$		
	$2.48 \cdot 10^{2}$	4.90	$2.5 \cdot 10^{-16}$		0.8
	1.63	4.04	$2.4 \cdot 10^{-15}$		2.5
			$3.5 \cdot 10^{-15}$		5.8
			$1.5 \cdot 10^{-16}$		5.8
			$6.5 \cdot 10^{-18}$		
Sb	$1.31 \cdot 10^{16}$	8.75	$9.9 \cdot 10^{-17}$		3.0
Au	$1.5 \cdot 10^{-7}$	2.14	$2.1 \cdot 10^{-15}$	1000···1300	
Na	5.0	1.50		180··· 400	
H_2	$5.65 \cdot 10^{-4}$	0.45	$1.3 \cdot 10^{-5}$	400··· 900	
D_2	$5.01 \cdot 10^{-4}$	0.46	$1.0 \cdot 10^{-5}$	400··· 900	
He	$3.04 \cdot 10^{-4}$	0.24	$4.0 \cdot 10^{-5}$	24··· 300	
	$7.40 \cdot 10^{-4}$	0.29	$6.4 \cdot 10^{-5}$	300···1000	
^{20}Ne	$2.21 \cdot 10^{-4}$	0.49	$3.5 \cdot 10^{-6}$	400···1200	
^{22}Ne	$2.08 \cdot 10^{-4}$	0.49	$3.3 \cdot 10^{-6}$	400···1200	

continued

c_{sur} cm^{-3}	Source	Ambient	Remarks	Ref.
$1\cdot10^{19}\cdots2\cdot10^{20}$	B_2O_3 vapour	O_2+N_2	[1])	61T
$6\cdot10^{18}$	B_2O_3 vapour	Ar	[1])	62H
$>1\cdot10^{20}$	B_2O_3 vapour	Ar	$D(1200\,°C)=6.2\cdot10^{-15}$ cm$^2\cdot$s^{-1}	62H
$4\cdot10^{19}\cdots8\cdot10^{19}$	BSG	N_2	[1])	69B
$<3\cdot10^{20}$	BSG		[1])	71B
$>3\cdot10^{20}$	BSG		$D(1100\,°C)\simeq2\cdot10^{-11}$ cm$^2\cdot$s^{-1}	71B
$7\cdot10^{19}$	BSG	Ar	[1])	71S
$2\cdot10^{20}$	BSG	Ar	[1])	71S
$1\cdot10^{18}\cdots8\cdot10^{18}$	B_2O_3 vapour	Ar		72W2 [4])
$1.5\cdot10^{19}\cdots3\cdot10^{20}$	B_2O_3 vapour	Ar		72W2
			CVD-oxide	72M
$1.8\cdot10^{19}\cdots5\cdot10^{19}$	B-powder	Ar	sealed tube	69O
	Ga_2O_3 vapour	H_2+N_2	[1])	64G2
	Ga_2O_3 vapour	H_2+H_2O		74W1
	Ga_2O_3 vapour		combined from 74W1 and 64G2	74W1 [4])
$8\cdot10^{20}\cdots1\cdot10^{21}$	P_2O_5 vapour	N_2	[1])	59S1
$3\cdot10^{19}\cdots2\cdot10^{20}$	P_2O_5 vapour	N_2	[1])	60A
$5\cdot10^{19}\cdots1\cdot10^{20}$	P_2O_5 vapour	O_2+N_2	[1])	61T
$8\cdot10^{17}\cdots8\cdot10^{19}$	PSG	N_2	[1])	70B
	$NaH^{32}PO_4$			65L
	^{32}PSG	Ar	outdiffusion from PSG	75G [4])
	^{32}P-powder	sealed tube	oxide remained amorphous	74C
	^{32}P-powder	sealed tube	oxide crystallized	74C
	^{32}PSG	Ar	diffusion into dry oxide [2])	75G
	^{32}PSG	Ar	diffusion into wet oxide [2])	75G
	$POCl_3$	O_2	"diffusion-const." of the PSG-liquid [3])	68E
$1\cdots2\cdot10^{19}$	AsH_3	O_2+N_2	[1])	68H1
$1\cdots6\cdot10^{19}$	AsSG	Ar	[1])	72W1 [4])
	AsSG	Ar	[1])	72W1
$5\cdot10^{19}$	AsSG	Ar		73G3
$4\cdot10^{20}$	AsSG	O_2		73G3
	AsSG		$D(1150\,°C)=1.04\cdot10^{-15}$ cm^2 s^{-1}	73C
			sealed tube	79S
$5\cdot10^{19}$	Sb_2O_5 vapour	O_2+N_2	[1])	61T [4])
				66C [4])
			diffusion in. vitreous silica	59O [4])
			permeation through fused quartz	62L [4])
			permeation through fused quartz	62L
			permeation through fused quartz	61S
			permeation through fused quartz	61S [4])
			permeation through fused quartz	61F
			permeation through fused quartz	61F [4])

Footnotes see p. 139.

continued

Table 5 (continued)

Dopant	D_0 $cm^2 s^{-1}$	E_A eV	D (1100 °C) $cm^2 s^{-1}$	T-range °C	c_0 mol %
O_2	$1.5 \cdot 10^{-2}$	3.09	$6.8 \cdot 10^{-14}$	1000···1100	
Silicon nitride (Si_3N_4)					
Al				450··· 530	
Ga			$5 \cdot 10^{-17}$		

Table 6. Distribution coefficients $k = c_{Si}/c_{SiO}$ of various elements.

Element	k	T °C	Remarks	Ref.
B	0.32		redistribution	64G2, 65D
	$\approx 10^{-2}$	1150···1250	oxide masking	62H
	0.68	1100	oxide masking; argon, closed capsule, el. B source	69O
	0.38	1150		
	0.10	1200		
	0.07	1250		
	0.55		redistribution	67C
	0.06	1200	diffusion from doped oxide	69B
	0.2	1100		
	0.1	1100···1250		70H
	0.52	1050	diffusion; argon ambient; B_2O_3 source	72W2
	0.60	1100		
	0.65	1150		
	0.76	1200		
	0.84	1250		
Al	$< 10^{-3}$			71W
Ga	≈ 20	1100···1250	oxide masking	64G1
	$> 10^3$		thermodyn. estimate	60T
In	$> 10^3$		thermodyn. estimate	60T
P	≈ 10		redistribution	65D
	≈ 10		redistribution	64G2
	$> 10^3$		thermodyn. estimate	60T
	7···25	1100···1260	$H_3{}^{32}PO_4$ diffusion; no clear temperature dependence	76G
As	≈ 10		redistribution	64G2, 65D
	10	1150	diffusion from AsSG	73C
Sb	≈ 10		redistribution	64G2, 65D

Table 5 (continued)

c_{sur} cm^{-3}	Source	Ambient	Remarks	Ref.
			permeation through vitreous silica	61N2 [4]
			$D(600\,°C)=1\cdot10^{-16}\,cm^2\,s^{-1}$	65W
			$D(1000\,°C)=2\cdot10^{-14}\,cm^2\,s^{-1}$	65W
			infused silica	
		dry O_2	$D(1000\,°C)=2.3\cdot10^{-13}\,cm^2\,s^{-1}$	82I
		1000 ppm H_2O in O_2	$D(1000\,°C)=2.4\cdot10^{-12}\,cm^2\,s^{-1}$ (from thermal oxidation)	82I
			$D\approx7.3\cdot10^{-3}\,cm^2\,s^{-1}$	78O
	vapour deposited Ga			75L

[1]) Values calculated or recalculated by Ghezzo [73G1].
[2]) The diffusion coefficient refers to one mol % P_2O_5 in the source. The calculated diffusion coefficient is to be multiplied with the mol % P_2O_5 concentration in the source.
[3]) This is not a true diffusion coefficient. It describes the movement of the phase boundary PSG liquid/SiO_2. D_0 and E_A were calculated from "Diffusion-Coefficients" measured at discrete temperatures to which definite liquidus concentrations are related.
[4]) See Fig. 20.

6.1.4.1.5 Diffusion profiles

Concentration profiles may be derived from the solution of Fick's law for a number of situations (intrinsic diffusion) or from computer modelling taking additionally into account field enhancement, concentration enhancement and oxidation enhancement of the special situations (extrinsic diffusion). For the first case, literature exists giving solutions for special situations [63B]. For the latter case, concentration profiles are derived by process simulators e.g. SUPREM [79A] or ICECREM [80R]. Real chemical concentration profiles can be obtained from radiotracer measurements or secondary ion mass spectrometry. Electrical measurements yield charge carrier concentration profiles.

Intrinsic diffusion (Fick's law)

Constant current source

This situation occurs when an undoped silicon wafer is exposed to a source of constant concentration during diffusion. The impurity concentration $c(x,t)$ is given by

$$c(x,t)=c_{sur}\cdot\operatorname{erfc}\frac{x}{2\sqrt{D\cdot t}},\qquad(16)$$

with c_{sur} [cm^{-3}] concentration at surface ($x=0$)
D [$cm^2 s^{-1}$] intrinsic diffusion coefficient
x [cm] distance from surface,
t [s] diffusion time.

The errorfunction (erfc $=1-$erf) is tabulated [60J]. A schematic plot of concentration profiles in linear scales is given in Fig. 22.

Instantaneous source

Instantaneous source diffusion occurs when a finite quantity is placed as a source in the surface of an undoped wafer and is consumed by diffusion into the silicon only.

The impurity concentration $c(x, t)$ is given by

$$c(x,t)=\frac{Q}{\sqrt{\pi D t}}\exp\left[-\left(\frac{x}{2\sqrt{Dt}}\right)^2\right], \tag{17}$$

with Q [cm^{-2}] amount of source material in the surface,
D [$cm^2 s^{-1}$] intrinsic diffusion constant,
x [cm] distance from surface,
t [s] diffusion time.

A schematic plot of the Gaussiantype of concentration profiles in linear scales is shown in Fig. 23. Both types of profiles in a semilogarithmic plot as usual for experimental concentration profiles are presented in Fig. 24. The exact mathematical description may be found in [63B].

Successive diffusions

A successive diffusion profile occurs in its basic form when a first, from a constant current source, diffused layer is driven into the silicon wafer in a second heating cycle. This is the ideal case of the two step diffusion consisting of a predeposition diffusion and a drive-in diffusion. The problem has been solved by Smith [58S] and may be represented in the form:

$$c(x,t_1,t_2)=\frac{2c_{\mathrm{sur}\,1}}{\pi}\int_0^{\alpha}\frac{\exp[-\beta(1+\xi^2)]}{1+\xi^2}\,d\xi, \tag{18}$$

with

$$\alpha=\left(\frac{D_1 t_1}{D_2 t_2}\right)^{1/2},$$

$$\beta=\left(\frac{x}{2\sqrt{D_1 t_1+D_2 t_2}}\right)^2.$$

The integral function is tabulated [58S]. The indices 1 and 2 indicate parameters of the first and second diffusion steps, respectively. In the plot given in Fig. 24 the erfc-profile and the Gaussian profile are the limiting envelopes of the Smith-profiles, $\alpha=\infty$ and $\alpha\simeq 0$, respectively.

The fabrication of an npn- or pnp-structure is based on successive diffusions. The starting material is homogeneously doped ($\approx 10^{16}$ cm^{-3}). In a first heating cycle a constant current source diffusion is performed with the first impurity. This impurity is driven in in a second heating cycle resulting in a Gaussian type of profile. In a third heating cycle, the second impurity is diffused from a constant current source. The resulting concentration profiles are schematically presented in Fig. 25.

These profiles are observed and diffusion coefficients may be derived from the results (Fig. 26).

Extrinsic diffusion

In the practice of device fabrication, extrinsic diffusion normally occurs but it is difficult to separate the different effects caused by high concentrations or field enhancements.

Concentration effects

In comparison to profiles observed after intrinsic diffusion, profiles having surface concentrations above the intrinsic concentration at the diffusion temperature are flattened. Isoconcentration studies performed with radiotracer doping elements in specimens with the same inactive diffused doping element showed that at concentrations above the intrinsic concentration near the surface, profiles are observed which are shaped according to the ideal source conditions and which may be described by the extrinsic diffusion coefficient [73M] (Fig. 26). However, the entire profile can not be described by a single value of the diffusion coefficient.

Field enhancement effect

In Fig. 27 concentration profiles are shown which result from a successive diffusion of arsenic after a gallium diffusion. The dip in the gallium profile is attributed to diffusion of gallium against the concentration gradient due to the field effect caused by the arsenic diffusion [81M1].

The profile can be constructed by computer modelling taking into account the effects of concentration, of defects created and of the electric field [81M1]. The profile is also explained without the electric field effect contribution [81F]. The emitter push [77W2] and base retardation [77W2] phenomena are caused by the interaction of diffusing dopants. At present, no consistent theory exists which can explain the observed phenomena.

Concentration profile of gold

In device fabrication, gold is diffused into silicon wafers from the back of the wafer under conditions in which the diffusion length $\sqrt{D \cdot t}$ exceeds the wafer thickness. A typical concentration profile which results from such conditions, but is obtained in a homogeneously doped 500 Ω cm p-type wafer, is shown in Fig. 28. By use of a limited amount of gold (implanted gold as diffusion source) the bulk concentration may be adjusted by the implantation dose [74S]. Gold is the element in which the influence of defects on the redistribution was first observed. There is still a lack of understanding of the Au-diffusion mechanism [83W] and even more so for other doping elements which diffuse by the dissociative mechanism.

Profiles in two dimensions

At mask edges diffusion profiles are two-dimensional. These two-dimensional profiles are very complex because of the non-ideality of the edge and the differences in diffusion coefficients between the oxidized and the nonoxidized areas. The horizontal diffusion is about 80% of the vertical diffusion. Investigations on this problem were first presented in [65K]. In [80T], implantation with subsequent diffusion is studied at the mask edge. The influence of the oxide thickness at the mask edge is reported in [81L].

6.1.4.1.6 Properties of masking layers

Masking layers are used for the local introduction of dopants into silicon. Requirements for masking layers are

resistance against chemical attack and indiffusion of dopants,
availability of selective etchants to structure the masking layer by lithography,
stability and good adhesion to the substrate during the high-temperature processing of diffusion,
sufficient purity in order to avoid contamination of the substrate,
no effect on the diffusion process.

Frequently used masking layers are SiO_2 and Si_3N_4. The properties of SiO_2-films and nitride films as masking layers are surveyed in Tables 7 and 8. Etching properties of the films are discussed in section 6.1.4.7. The effects of SiO_2 on the diffusivity of various dopants has been reviewed in section 6.1.4.1.4.

Table 7. Oxide masking of dopants for silicon in oxidizing (ox) or reducing (red) ambient. (+) indicates that SiO_2 acts as a mask.

B		Al		Ga		In		P		As		Sb	
ox	red	ox	red	ox	red	ox	red	ox	red	ox	red	ox	red
+	fails	no diffusion system	reacts with SiO_2	no diffusion system	fails	no diffusion system	fails	+	+	+		+	

Phosphorus and boron are always diffused into silicon under oxidizing conditions where a high concentration of B_2O_3 or P_2O_5 is formed. The silicate glasses BSG and PSG show low viscosities and start to flow at high temperatures. Silicon dioxide dissolves into the glass. The masking properties are limited by the "melt-through" process.

Phase-diagrams of the SiO_2—B_2O_3 and the SiO_2—P_2O_5 systems are presented in Figs. 29···31, Table 9 lists further data. Empirical data on the melt-through process which are used to estimate the stability against pre-deposition processes are presented in Figs. 32 and 33.

The kinetics of PSG formation from P_2O_5 vapour (using a $POCl_3$ source) were established by Eldrige and Balk [68E] and presented by [75G] in the form

$$l_{PSG} = 9.032 \cdot 10^6 \cdot \sqrt{c_p \cdot t} \cdot \exp(-0.815\,\text{eV}/kT)\ [\text{Å}] \tag{19}$$

with l_{PSG} the thickness of PSG formed, t the annealing time in [s], T the annealing temperature, and c_p the phosphorus concentration in [vol. %] in the vapour phase. Calculated data and experimental values for a phosphine source published by Ghoshtagore [75G] are given in Fig. 34. Ghoshtagore [75G] derived a masking relationship for the diffusion from a CVD-PSG-layer with a phosphorus concentration below liquidus composition

$$l_{PSG} \approx 2\sqrt{Dt}\ \text{arg erfc}\,(c_s/c_l), \tag{20}$$

with l_{PSG} the thickness of the newly formed PSG layer, c_s the solid solubility of P_2O_5 in SiO_2, and c_l the liquidus composition (Table 9). D is the "diffusion coefficient", valid for the "diffusion of the liquid" through the SiO_2, which is also given in Table 9. Calculated and experimental values are given in Fig. 35.

Si_3N_4 is a good diffusion mask, but is not commonly used because of the poor properties of the interface Si_3N_4/Si. The masking properties compared to the masking properties of SiO_2 are given in Table 8 and Fig. 36. Additional data are given in [66H] and [66D]. In some cases, the system Si/SiO_2/Si_3N_4 may be used as a mask. This system plays an important role in the local oxidation of silicon. In Figs. 37 and 38, data on Si_3N_4 as oxidation mask are presented.

Table 8. Masking properties of silicon nitride [68H2]. CVD—Si_3N_4 prepared at 1000 °C. Film compositions indicated by $Si_xO_yN_z$ contain approx. 20% SiO_2. In sealed tube diffusions, the elemental dopant was used, the tube was evacuated and back filled with argon prior to sealing. In box diffusions, a semisealed box was used containing BSG or PSG. Where mask failure is noted, but no masked junction depth is given, mask failure was detected by using a thermoelectric probe. In all cases, 1 Ω cm Si wafers were used.

Dopant	Diffusion system	Type of film	Thickness Å	T °C	t h	Masked	Masked junct. depth µm	Unmasked junct. depth µm	Comments
B	sealed tube	SiO_2	2000	1100	4	no	6	6	oxide control
B	sealed tube	Si_3N_4	≧ 300	1100	4	yes		6	
B	sealed tube	Si_3N_4	≧ 400	1200	20	yes		25	
B	sealed tube	Si_3N_4	≧ 500	1200	30	yes		25	film conversion
B	sealed tube	Si_3N_4	1000	1250	10	yes		30	
B	sealed tube	Si_3N_4	≦ 600	1250	10	no	spikes	30	spiking failure
B	sealed tube	$Si_xO_yN_z$	500	1100	4	no	6	6	
B	sealed tube	$Si_xO_yN_z$	1000	1100	4	yes		6	
B	box	Si_3N_4	500…1500	1100	5	yes		4	
B	box	Si_3N_4	1500	1200	5	no			film conversion
P	sealed tube	Si_3N_4	≧1500	1100	3	yes			
P	sealed tube	Si_3N_4	≦1000	1100	3	no			
P	sealed tube	Si_3N_4	≧1500	1100	5	yes		5	
P	sealed tube	Si_3N_4	≧1500	1100	10	yes		9	
P	sealed tube	Si_3N_4	≦1100	1100	10	no		9	
P	sealed tube	Si_3N_4	1500	1200	4	no	3	12	
P	sealed tube	$Si_xO_yN_z$	≧1500	1100	4	yes		5	
P	box	Si_3N_4	1000	1100	1	no		6	film conversion
Ga	sealed tube	SiO_2	2000	1100	4	no	6	6	oxide control
Ga	sealed tube	Si_3N_4	1000	1100	4	yes		6	
Ga	sealed tube	$Si_xO_yN_z$	≦1000	1100	4	no	3.4	6	
Ga	sealed tube	$Si_xO_yN_z$	1500	1100	4	yes		6	
Ga	sealed tube	Si_3N_4	300	1185	15	no	10	18	
Ga	sealed tube	Si_3N_4	1200	1185	15	yes	spikes		localized spiking
Ga	sealed tube	Si_3N_4	1500	1200	4	yes	spikes	14	localized spiking
As	sealed tube	Si_3N_4	1500	1100	4	yes		0.6	
As	sealed tube	$Si_xO_yN_z$	1500	1100	4	yes		0.6	
As	sealed tube	Si_3N_4	1000	1150	6	yes		1.8	
As	sealed tube	Si_3N_4	1500	1150	6	yes	spikes	1.8	films cracked
As	sealed tube	$Si_xO_yN_z$	≧1000	1150	6	yes	spikes	1.8	films cracked
As	sealed tube	Si_3N_4	1500	1200	20	no	6	11	

Table 9. Data on the PSG system [68E, 75G].

T °C	Solid solubility c_s of P_2O_5 in SiO_2		Liquidus composition c_l mol %	"Diffusion coefficient" D of PSG-liquid at liquidus composition $cm^2 s^{-1}$
	P_2O_5 mol %	P_2O_5 cm^{-3}		
1100	$1.09 \cdot 10^{-2}$	$2.50 \cdot 10^{18}$	10.0	$4.04 \cdot 10^{-15}$
1150	$1.37 \cdot 10^{-2}$	$3.15 \cdot 10^{18}$	8.3	$5.45 \cdot 10^{-15}$
1200	$1.70 \cdot 10^{-2}$	$3.90 \cdot 10^{18}$	6.7	$6.91 \cdot 10^{-15}$
1250	$2.09 \cdot 10^{-2}$	$4.80 \cdot 10^{18}$	5.2	$8.19 \cdot 10^{-15}$

6.1.4.1.7 Gettering

Noble and transition metals and native lattice defects may form electrically active trap centers in silicon which deteriorate the charge carrier lifetime. Precipitated impurities deteriorate the reverse current voltage characteristics. Gettering processes are applied to remove all types of defects and precipitations from a silicon wafer or from the zones in which pn-junctions of devices are formed. The methods of gettering are surveyed in Table 10.

Backside damage gettering

Backside damage of the crystal lattice may be achieved by heavy doping, e.g. with phosphorus as a standard method with or without abrasion, by abrasion alone, by laser irradiation, or by ion implantation. A typical result of gettering gold and copper by a phosphorus diffused layer is shown in Fig. 39. Similar gettering behaviour as gold and copper show Ti, V, Cr, Mn, Fe, Co, Ni and Ag [81G2]. It is assumed that these elements are also removed from active device regions in gettering processes. A typical result on the reverse current characteristics of a diode is given in Fig. 40.

Oxygen precipitation

In recent years, the role of oxygen and carbon in silicon was intensively studied. The precipitation of oxygen may be used in intrinsic gettering. Suitable multiple heating cycles of the CZ wafer result in oxygen precipitates in the center of the wafer and on top and rear surfaces of the wafer in oxygen "denuded zones" having a thickness of about 25 μm. In the zone of precipitation between the denuded zones, the precipitates of oxygen act as nucleation centers for metal impurity precipitates. Results on oxygen precipitation are shown in Fig. 41.

Mechanism

Gettering is still an empirical process in semiconductor technology which is not well understood. Defects diffuse to the gettering zone. Impurities precipitate there and thus do not further contribute to the concentration gradient in lattice diffusion. The impurities may diffuse quantitatively to the sink. When the annealing is performed in an ambient containing halides, impurities may be removed as volatile or chemically inactive halide compounds into the ambient. For gettering, annealing procedures must be performed at temperatures high enough that defects diffuse sufficient rapidly, sink regions are formed to remove the defect from active zones of devices. These sink regions must not be annealed out by the high-temperature treatment. Development of suitable gettering processes is still an active field of research and development.

Table 10. Gettering methods.

Method	Type	Conditions	Gettering zone	Remarks	Ref.
Backside damage gettering	diffusion	predeposition diffusion $T \geqq 900\,°C$ $t \approx 1\,h$	diffused layer, together with PSG film formed	standard method	68L, 82A2
	grinding, abrasion, sandblasting	particle size $\geqq 5\,\mu m$	damaged layer, particle size-dependent	additional to diffusion	81T, 82S
	laser irradiation	YAG Q-switched	molten region		81P
	ion implantation	Ar ($10^{14}\,cm^{-2}$)	implanted region		81P, 79R
Oxygen precipitation (intrinsic gettering)	multistep annealing	annealing cycles with different temperatures 750···1100 °C	precipitation zone in center of wafer; denuded zones at surfaces	under investigation	82S
Halide oxidation	thermal oxidation using dry O_2, halide	$T \geqq 1150\,°C$ $t \approx 1\,h$	chemical defects are lost to ambient; reduces oxidation stacking faults	used in formation of thermal oxides	81C

References for 6.1.4.1

1855F Fick, A.: Pogg. Ann. **94**(1955) 59.
52F Fuller, C.S.: Phys. Rev., **86** (1952) 136.
53F Fuller, C.S., Ditzenberger, J.A.: Phys. Rev. **91** (1953) 193.
54D Dunlap, W.C., jr.: Phys. Rev. **94** (1954) 1531.
54F Fuller, C.S., Severiens, J.C.: Phys. Rev. **96** (1954) 225.
54M Van der Maesen, F., Brenkman, J.A.: Philips Res. Rept. **9** (1954) 225.
55B Bösenberg, W.: Z. Naturf. **10a**. (1955) 285.
55D Dunlap, W.C., jr.: Phys. Rev. **97** (1955) 614.
56C Carlson, R.O.: Phys. Rev. **104** (1956) 937.
56F Fuller, C.S., Ditzenberger, J.A.: J. Appl. Phys. **27** (1956) 544.
56G Goldstein, B.: Bull. Am. Phys. Soc., Ser. II, **1** (1956) 145.
56K Kosenko, V.E.: Izv. Akad. Nauk. S.S.S.R., Ser. Fiz. **20** (1956) 1527.
56L Letaw, H., jr., Portnoy, W.M., Slifkin, L.: Phys. Rev. **102** (1956) 636.
56M Miller, R.C., Savage, A.: J. Appl. Phys. **27** (1956) 1430.
56S Struthers, J.D.: J. Appl. Phys. **27** (1956) 1560.
56W van Wieringen, A., Warmolitz, N.: Physica **22** (1956) 849.
57B Bugai, A.A., Kosenko, V.E., Miselyuk, E.G.: Sov. Phys. Techn. Phys. (English Transl.) **2** (1957) 183.
57C Collins, C.B., Carlson, R.O.: Phys. Rev. **108** (1957) 1409.
57F Fuller, C.S., Morin, F.J.: Phys. Rev. **105** (1957) 379.
57M Miller, R.C., Smits, F.M.: Phys. Rev. **107** (1957) 65.
57P Petrov, D.A., Shashkov, Yu.M., Akimchenko, I.P.: Voprosy Met. I. Fiz. Poluprovod. (Moscow, Akad. Nauk SSSR) 130, Sbornik, 1957 (Chem. Abst. 54, 1960, 17190c.)
57S Struthers, J.D.: J. Appl. Phys. **28** (1957) 516.
58B Boltaks, B.I., Sosinov, I.I.: Th. Tekh. Fiz. **28** (1958) 679.
58K Kurtz, A.D., Gravel, C.L.: J. Appl. Phys. **29** (1958) 1456.
58M Maita, J.P.: J. Phys. Chem. Solids **4** (1958) 1546.
58P Petrov, D.A., Schaschkov, J.M., Belanovski, A.S.: "Die Diffusion der Beimengungen im Silizium", in: Semiconductors and Phosphors, eds. M. Shon, H. Welker, Interscience Publishers, Inc. New York, **1958**, pp. 652···655.
58S Smith, R.C.T.: Australian J. Phys. **6** (1958) 127.
58V Valenta, M.W.: Ph. D. Thesis Univ. Illinois 1958 (Univ. Microfilm 58-5509).
59C Carlson, R.O., Pell, E.M.: J. Phys. Chem. Solids **8** (1959) 81.
59K Karstensen, F.: Z. Naturforsch. **14a** (1959) 1031.
59L Logan, R.A., Peters, A.S.: J. Appl. Phys. **30** (1959) 1627.
59O Owen, A.E., Douglas, R.W.: J. Soc. Glass Tech. **43** (1959) 159.
59R1 Reiss, H., Fuller, C.S.: Semiconductors (editor N.B. Hannay) (Reinhold Publ. Corp., N. Y. 222 (1959)).
59R2 Rohan, J.J., Pickering, N.E., Kennedy, J.: J. Electrochem. Soc. **106** (1959) 705.
59S1 Sah, C.T., Sello, H., Tremere, D.A.: J. Phys. Chem. Solids **11** (1959) 288.
59S2 Shashkov, M., Akimchenko, I.P.: Sov. Phys. Dokl. (English Transl.) **4** (1959) 1115.
59T Tyler, W.W.: J. Phys. Chem. Solids **8** (1959) 59.
60A Allen, R.B., Bernstein, H., Kurtz, A.D.: J. Appl. Phys. **31** (1960) 334.
60H Haas, C.: J. Phys. Chem. Solids **15** (1960) 108.
60J Jost, W.: Diffusion in Solids, Liquids, Gases. Academic Press, New York **1960**.
60K1 Kosenko, V.E.: Sov. Phys. Solid State (English Transl.) **1** (1960) 1481.
60K2 Kurtz, A.D., Yee, R.: J. Appl. Phys. **31** (1960) 303.
60P1 Pell, E.M.: Phys. Rev. **119** (1960) 1014.
60P2 Pell, E.M.: Phys. Rev. **119** (1960) 1222.
60T Thurmond, C.D.: Distribution Coefficients of Impurities distributed between Ge or Si Crystals and Ternary Alloys or Surface Oxides in: Properties of Elemental and Compound Semiconductors, H.C. Gatos, ed., Interscience **1960**, p. 121.
60Y Yamaguchi, J., Horiuchi, S., Matsumura, K.: J. Phys. Soc. Jpn. **15** (1960) 1541.
61A Albers, W.: Solid State Electron. **2** (1961) 85.
61B1 Belyaev, Yu.I., Zhidkov, V.A.: Sov. Phys. Solid State (English Transl.) **3** (1961) 133.
61B2 Boltaks, B.I., Shih-Yin, H.: Sov. Phys. Solid State (English Transl.) **2** (1961) 2283.

61B3 Boltaks, B.I., Kulikov, G.S., Malkovich, R.Sh.: Fiz. Tekh. Poluprovod. **2** (1961) Sov. Phys. Solid State (English Transl.) **2** (1961) 2134.
61F Frank, R.C., Swets, D.E., Lee, R.W.: J. Chem. Phys. **35** (1961) 1451.
61L Lehovec, K., Slobodskoy, A.: Solid State Electron **3** (1961) 45.
61N1 Newman, R.C., Wakefield, J.: J. Phys. Chem. Solids **19** (1961) 230.
61N2 Norton, F.J.: Nature **171** (1961) 701.
61S Swets, D.E., Lee, R.W., Frank, R.C.: J. Chem. Phys. **34** (1961) 17.
61T Thurston, M.O., Tsai, J.C.C., Kang, K.D.: Ohio State University Research Foundation Final Report AD 261201, Contract DA-36-039-50-83874, Columbus Ohio **1961**.
61W1 Wei, L.Y.: J. Phys. Chem. Solids **18** (1961) 162.
61W2 Widmer, H., Gunther-Mohr, G.R.: Helv. Phys. Acta **34** (1961) 635.
61W3 Williams, E.L.: J. Electrochem. Soc. **108** (1961) 795.
62A Armstrong, W.J.: J. Electrochem. Soc. **109** (1962) 1065.
62H Horiuchi, S., Yamaguchi, J.: Jpn. J. Appl. Phys. **1** (1962) 314.
62I Ignatkov, V.D., Kosenko, V.E.: Sov. Phys. Solid State (English Transl.) **4** (1962) 1193.
62K Kosenko, V.E.: Sov. Phys. Solid State (English Transl.) **4** (1962) 42.
62L Lee, R.W., Frank, R.C., Swets, D.E.: J. Chem. Phys. **36** (1962) 1062.
62M1 Mackintosh, I.M.: J. Electrochem. Soc. **109** (1962) 392.
62M2 Maekawa, S.: J. Phys. Soc. Jpn. **17** (1962) 1592.
62N Newman, R.C., Wakefield, J.: "Diffusion and Precipitation of Carbon in Silicon", in Metallurgy of Semiconductor Materials, Vol. 15, J.B. Schroeder, ed., Interscience Publishers, New York, 1962, pp. 201.
62T1 Tagirov, V.I., Kuliev, A.A.: Sov. Phys. Solid State (English Transl.) **4** (1962) 196.
62T2 Tien, T.Y., Hummel, R.A.: J. Am. Ceram. Soc. **45** (1962) 422.
62W Wölfle, R., Dorendorf, H.: Solid State Electron. **5** (1962) 98.
63B Boltaks, B.I.: Diffusion in Semiconductors, Infosearch Ltd. London **1963**, 162.
63M Malkovich, R.Sh., Alimbarashvili, N.A.: Sov. Phys.-Solid State (English Transl.) **4** (1963) 1725.
63S Shewmon, P.G.: Diffusion in Solids, New York: McGraw Hill Book Co., **1963**.
64C Corbett, J.W., McDonald, R.S., Watkins, G.D.: J. Phys. Chem. Solids **25** (1964) 873.
64G1 Grove, A.S., Leistiko, O., Sah, C.T.: J. Phys. Chem. Solids **25** (1964) 985.
64G2 Grove, A.S., Leistiko, O., Sah, C.T.: J. Appl. Phys. **35** (1964) 2695.
64H Hall, R.N., Racette, J.H.: J. Appl. Phys. **35** (1964) 379.
64K1 Kato, T., Nishi, Y.: Jpn. J. Appl. Phys. **3** (1964) 377.
64K2 Kooi, E.: J. Electrochem. Soc. **111** (1964) 1383.
64L Luther, L.C., Moore, W.J.: J. Chem. Phys. **41** (1964) 1018.
64R Raju, P.S., Rao, N.R.K., Rao, E.V.K.: Indian J. Pure Appl. Phys. **2** (1964) 353.
64W1 Wilcox, W.R., La Chappelle, T.J.: J. Appl. Phys. **35** (1964) 240.
64W2 Wilcox, W.R., La Chappelle, T.J., Forbes, D.H.: J. Electrochem. Soc. **111** (1964) 1377.
65D Deal, B.E., Grove, A.S., Snow, E.H., Sah, C.T.: J. Electrochem. Soc. **112** (1965) 308.
65K Kennedy, D.P., O'Brien, R.R.: IBM J. Res. Dev. **9** (1965) 179.
65L Li, K.C., Xue, S.Y., Zhu, S., Huang, Y.: Acta Phys. Sinica **20** (1965) 496.
65P1 Pantaleev, V.A.: Soviet Phys. Solid State (English Transl.) **7** (1965) 734.
65P2 Pomerrenig, D.: Acta. Phys. Austriaca **20** (1965) 338.
65R Rockett, T.J., Foster, W.R.: J. Am. Ceram. Soc. **48** (1965) 75.
65W Williams, E.L.: J. Am. Ceram. Soc. **48** (1965) 190.
66C Collins, D.R.: J. Appl. Phys. Lett. **8** (1966) 323.
66D Doo, V.Y.: IEEE Trans. ED **13** (1966) 561.
66H Hu, S.M.: J. Electrochem. Soc. **113** (1966) 693.
66M1 Masters, B.J., Fairfield, J.M.: Appl. Phys. Lett. **8** (1966) 280.
66M2 Millea, M.F.: J. Phys. Chem. Solids **27** (1966) 315.
66P1 Pavlov, P.V.: Sov. Phys. Solid State (English Transl.) **8** (1966) 2377.
66P2 Peart, R.F.: Phys. Status Solidi **15** (1966) K119.
66P3 Pratt, B., Friedman, F.: J. Appl. Phys. **37** (1966) 1893.
66S Snow, E.H., Deal, B.E.: J. Electrochem. Soc. **113** (1966) 263.
67B Bonzel, H.P.: Phys. Status Solidi **20** (1967) 493.
67C Chen, W.H., Chen, W.S.: J. Electrochem Soc. **114** (1967) 1297.
67G Glazov, V.M., Zemskov, V.S.: Fiz. Khem. Osnovi Lagirovaniya Poluprovodnikov (Isdatelstvo 'Nauka' Moscow 1967) 98.

67K Kao, Y.C.: Electrochem. Techn. **5** (1967) 90.
67M Meer, W., Pommerrenig, D.: Z. Angew. Phys. **23** (1967) 370.
67S1 Smith, A.M.: "Fundamentals of Silicon Integrated Devices Technology", Vol. 1, R.M. Burger, R.P. Donovan, eds., Prentice Hall, Inc. N. J. 1967, 204.
67S2 Svob, L.: Solid State Electron. **10** (1967) 991.
67Y Yoshida, M., Saito, K.: Jpn. J. Appl. Phys. **6** (1967) 573.
68E Eldrige, J.M., Balk, P.: Trans. AIME **242** (1968) 539.
68G1 Ghandi, S.K.: The Theory and Practice of Microelectronics J. Wiley & Sons Inc., New York **1968**.
68G2 Gromova, O.N., Khodunova, K.M.: Fiz. Khim. Obr. Mater, No. **5** (1968) 150.
68H1 Hsueh, Y.W.: Electrochem. Techn. **6** (1968) 361.
68H2 Heumann, F.K., Brown, D.M., Mets, E.: J. Electrochem. Soc. **115** (1968) 99.
68I1 Ichimiya, T., Furuichi, : International J. Appl. Radiation Isotopes **19** (1968) 573.
68I2 Isawa, N.: Jpn. J. Appl. Phys. **7** (1968) 81.
68L Lambert, J.L., Reese, M.: Solid State Electron. **11** (1968) 365.
68N Nagano, K., Iwauchi, S., Tanaka, T.: Jpn. J. Appl. Phys. **7** (1968) 1361.
68P Panteleev, V.A., Akinkina, E.I.: Zh. Fiz. Khim. **42** (1968) 922.
68S1 Seeger, A., Chik, K.P.: Phys. Status Solidi **29** (1968) 455.
68S2 Sharma, B.L., Mukerjee, S.N.: Proc. 55th Indian Science Congress (1968) p. 71.
68Y Yeh, T.H., Hu, S.M., Kastl, R.H.: J. Appl. Phys. **39** (1968) 4266.
69B Barry, M.L., Olofsen, P.: J. Electrochem. Soc. **116** (1969) 854.
69G Genser, M.: US Patent No. 3084079, November 25, **1969**.
69K Kendall, D.L., DeVries, D.B.: Semiconductor Silicon, R.R. Haberecht, E.L. Kern, eds., The Electrochem. Soc., Symp. Ser. **1969**.
69M Masters, B.I., Fairfield, I.M.: J. Appl. Phys. **40** (1969) 23290.
69O Okamura, M.: Jpn. J. Appl. Phys. **8** (1969) 1440.
70B Barry, M.L.: Electrochem. Soc. **117** (1970) 1405.
70H Huang, J.S.T., Welliver, L.C.: J. Electrochem. Soc. **117** (1970) 1577.
70S Sharma, B.L.: Diffusion in Semiconductors. Trans. Tech. Publications, Clausthal-Zellerfeld, West Germany **1970**.
71B Brown, D.M., Kennicott, P.R.: J. Electrochem. Soc. **118** (1971) 293.
71C1 McCaldin, J.O., Sankur, H.: Appl. Phys. Lett. **19** (1971) 524.
71C2 Chiu, T.L., Gosh, H.N.: IBM J. Res. Dev. **15** (1971) 472.
71F1 Fränz, I., Langheinrich, W.: Solid State Electron. **14** (1971) 499.
71F2 Fränz, I., Langheinrich, W.: Solid State Electron. **14** (1971) 835.
71G1 Ghoshtagore, R.N.: Phys. Rev. **B, 3** (1971) 389.
71G2 Ghoshtagore, R.N.: Phys. Rev. **B, 3** (1971) 397.
71G3 Ghoshtagore, R.N.: Phys. Rev. **B, 3** (1971) 2507.
71L Larne, I.C.: Phys. Status Solid **6** (1971) 143.
71M Makris, J., Masters, B.J.: J. Appl. Phys. **42** (1971) 3750.
71S Schwenker, R.O.: J. Electrochem. Soc. **118** (1971) 313.
71W Wolf, H.F.: Semiconductors, John Wiley & Sons Inc. **1971**, 361.
72B1 Bakhadyrkhanov, M.K., Boltaks, B.I., Kulikov, G.S.: Sov. Phys. Solid State (English Transl.) **14** (1972) 1441.
72B2 Boltaks, B.I., Bakhadyrkhanov, M.K., Kulikov, G.S.: Sov. Phys. Solid State (English Transl.) **13** (1972) 2240.
72K Kamins, T.I., Manoliu, J., Tucker, R.N.: J. Appl. Phys. **43** (1972) 83.
72M Mukherjee, S.P., Evans, P.E.: Thin Solid Films **14** (1972) 299.
72S Spitsyn, V., Arakelian, V.S., Rezuikov, A.G., Menshutin, L.N.: Dokl. Akad. Nauk. SSSR **205** (1972) 82.
72W1 Wong, J., Ghezzo, M.: J. Electrochem. Soc. **119** (1972) 1413.
72W2 Wilson, P.R.: Solid State Electron. **15** (1972) 961.
73C Chaudhari, P.K., Frey, W.J., Quinn, R.M.: J. Electrochem. Soc. **120** (1973) 910.
73G1 Ghezzo, M., Brown, D.M.: J. Electrochem. Soc. **120** (1973) 146.
73G2 Gereth, R., Kostka, A., Kreuzer, K.: J. Electrochem. Soc. **120** (1973) 966.
73G3 Ghezzo, M., Brown, D.M.: J. Electrochem. Soc. **120** (1973) 110.
73G4 Gruzin, P.L., Zemskii, S.W., Bulkin, A.D., Makarov, N.M.: Fiz. Tekh. Poluprovodn. **7** (1973) 1853. Sov. Phys. Semiconduct. (English Transl.) **7** (1973).

73H Huntley, F.H., Willoughby, A.F.W.: Phil. Mag. **28** (1973) 1319.
73M Makris, I.S., Masters, B.I.: J. Electrochem. Soc. **120** (1973) 1252.
73S Shaw, D.: Atomic Diffusion in Semiconductors, London: Plenum Press **1973**.
73V McVay, G.L., Ducharme, A.R.: J. Appl. Phys. **44** (1973) 1409.
74A Asaka, Y., Horie, K., Nakamura, G.: Jpn. J. Appl. Phys. **13** (1974) 1533.
74C Campbell, D.R., Alessandrini, E.I., Tu, K.N., Lew, J.E.: J. Electrochem. Soc. **121** (1974) 275.
74P Prince, J.L., Schwettmann, F.N.: J. Electrochem. Soc. **121** (1974) 705.
74S Schulz, M., Goetzberger, A., Fränz, I., Langheinrich, W.: Appl. Phys. **3** (1974) 275.
74W1 Wagner, S., Povilonis, E.I.: J. Electrochem. Soc. **121** (1974) 1487.
74W2 Wurker, W., Roy, K., Hesse, K.: Mater. Res. Bull. **9** (1974) 971.
74Y Yanusov, M.S., Tursunov, N.A.: Fiz. Tekh. Poluprovodn. **8** (1974) 1145.
75F Fränz. I., Langheinrich, W.: Solid State Electron. **18** (1975) 209.
75G Ghoshtagore, R.N.: Solid State Electron. **18** (1975) 399.
75H1 Horiuchi, S.: Solid State Electron. **18** (1975) 1111.
75H2 Horiuchi, S., Blanchard, R.: Solid State Electron. **18** (1975) 529.
75L Lodding, A., Lundkvist, L.: Thin Solid Films **25** (1975) 491.
75M Murarka, S.P.: Phys. Rev. B **12** (1975) 2502.
75R Ruge, I.: Halbleitertechnologie, Berlin, Heidelberg, New York: Springer **1975**.
75S1 Seregin, P.P., Nistiryuk, I.V., Nasredinov, F.S.: Fiz. Tverd. Tela **17** (1975) 2310.
75S2 Shaw, D.: Phys. Status Solidi (b) **72** (1975) 11.
75U Uskov, V.A.: Inorg. Mater. **11** (1975) 848.
76A Azimov, S.A., Umarov, B.V., Yunusov, M.S.: Fiz. Tekh. Poluprovdn. **10** (1976) 1418.
76C Colby, J.W., Katz, L.E.: J. Electrochem. Soc. **123** (1976) 409.
76G Goncharov, E.E., Lifirenko, V.D., Makarov, N.M., Ryabova, G.G.: Izv. Vyss. Kcheb. Zar. Fiz. **5** (1976) 29. Sov. Phys. J.: **19** (1976) 566.
76L Learn, A.J.: J. Electrochem. Soc. **23** (1976) 894.
77B Boldyrev, V.P., Prokovskii, F.I., Romanovskaya, S.G., Tkach, A.V., Shimanovich, I.E.: Fiz. Tekh. Poluprovodn. **11** (1977); Sov. Phys. Semicond. (English Transl.) **11** (1977) 709.
77J Jones, C.L., Willoughby, A.F.W.: Semiconductor silicon 1977, H.R. Huff, E. Sirtl, eds., The Electrochemical Society **1977**, p. 684.
77K Kitagawa, H., Hashimoto, K.: Jpn. J. Appl. Phys. **16** (1977) 173.
77M1 Malkovich, R.Sh., Pokoeva, V.A.: Fiz. Tverd. Tela **19** (1977) 2731.
77M2 Mayer, H.J., Mehrer, H., Maier, K.: Radiation Effects in Seminconductors, Inst. of Physics Conf. Series No. 31, Bristol, Inst. of Phys. (1977) p. 186.
77W1 Weast, R.C.: Handbook of Chemistry and Physics 57th edition CRC-Press Cleveland Ohio.
77W2 Willoughby, A.F.W.: J. Phys. D.: Appl. Phys. **10** (1977) 455.
78A1 Antoniadis, D.A., Lin, A.M., Dutton, R.W.: Appl. Phys. Lett. **33** (1978) 1030.
78A2 Antoniadis, D.A., Gonzalez, A.G., Dutton, R.W.: J. Electrochem. Soc. **125** (1978) 813.
78A3 Azimov, S.A., Yunusov, M.S., Nurkuziev, G., Karimor, F.R.: Fiz. and Tekh. Poluprovod. **12** (1978) 1655.
78B Berning, G.L.P., Levenson, L.L.: Thin Solid Films **55** (1978) 473.
78O Ogata, H., Kanayama, K., Ohtani, M., Fujiwara, K., Abe, H., Nakayama, H.: Thin Solid Films **48** (1978) 333.
78V Vydianat, H.R., Lorenzo, I.S., Kröger, F.: J. Appl. Phys. **49** (1978) 5928.
79A Antoniadis, D.A., Dutton, R.W.: IEEE Trans. ED **26** (1979) 490.
79G Ghoshtagore, R.N.: Solid State Electron. **22** (1979) 877.
79H Hettich, G., Mehrer, H., Maier, K.: Int. Conf. on Phys. of Semiconductors: Proc. Inst. Physics London **1979**, p. 500.
79K1 Kaplan, D.R., Weigel, C., Corbett, I.W.: Phys. Status Solidi **94** (1979) 359.
79K2 Kim, C.S., Sakata, M.: Jpn. J. Appl. Phys. **18** (1979) 247.
79R Ryssel, H., Schmiedt, B.: The Electrochem. Soc. Meeting Ext. Abstracts **79–2** (1979) 1260.
79S Shabde, S.N., Helliwell, K.: J. Electrochem. Soc. **126** (1979) 2279.
80B Bakhadyrkhanov, M.K., Zainabidinov, S., Khamidov, A.: Sov. Phys. Semicond. (English Transl.) **14** (1980) 243.
80H Hwang, J.C.M., Ho, P.S., Lewis, J.E., Campbell, D.R.: J. Appl. Phys. **51** (1980) 1576.
80R Ryssel, H., Haberger, K., Hoffmann, K., Prinke, G., Dümcke, R., Sachs, A.: IEEE Trans. ED **27** (1980) 1484.
80T Tielert, R.: IEEE Trans. ED **27** (1980) 1479.

81C Claeys, C., Declerck, G., van Overstraeten, R., Bender, H., van Landuyt, J., Amelinckx, S.: Semiconductor Silicon 1981, H.R. Huff, R.J. Kriegler, eds., The Electrochemical Society **81–5** (1981) 731.

81F Fair, R.B.: Semiconductor Silicon 1981, H.R. Huff, R.J. Kriegler, eds., The Electrochemical Society **81–5** (1981) 963.

81G1 Goesele, U., Morehead, F., Föll, H., Frank, W., Strunk, H.: Semiconductor Silicon 1981, H.R. Huff, R.J. Kriegler. eds., The Electrochemical Soc. **81–5** (1981) 766.

81G2 Graff, K., Pieper, H.: Semiconductor Silicon 1981, H.R. Huff, R.J. Kriegler, eds., The Electrochemical Society **81–5** (1981) 331.

81L Lin, A.M., Antoniadis, D.A., Dutton, R.W.: J. Electrochem. Soc. **128** (1981) 1131.

81M1 Mallam, N., Jones, C.L., Willoughby, A.F.W.: Semiconductor Silicon 1981, H.R. Huff, R.J. Kriegler, eds., The Electrochemical Society **81–5** (1981) 979.

81M2 Mizuo, S., Higuchi, H.J.: Jpn. J. Appl. Phys. **20** (1981) 739.

81P Pearce, C.W., Katz, L.E., Seidel, T.E.: Semiconductor Silicon 1981, H.R. Huff, R.J. Kriegler, eds., The Electrochemical Society **81–5** (1981) 705.

81T Takano, Y., Kozuka, H., Ogirima, M., Maki, M.: Semiconductor Silicon 1981, H.R. Huff, R.J. Kriegler, eds., The Electrochemical Society **81–5** (1981) 743.

82A1 Antoniadis, D.A., Moskowitz, I.: Proc, 1st Int. Symp. on VLSI Science and Technology 1982, C.J. Dell'Oca, W.M. Bullis, eds. **82-7** (1982) 5. The Electrochem. Society.

82A2 Andrews, J.M., Heimann, P.A., Kushner, R.A.: VLSI Science and Technology 1982, C.J. Dell'Oca, W.M. Bullis, eds., The Electrochem. Society **82–7** (1982) 73.

82A3 Antoniadis, D.A.: J. Electrochem. Soc. **129** (1982) 1093.

82I Irene, E.A.: J. Electrochem. Soc. **129** (1982) 413.

82M Miyake, M., Harada, H.: J. Electrochem. Soc. **129** (1982) 1097.

82S Shimura, F.: VLSI Science and Technology 1982, C.J. Dell'Oca, W.M. Bullis, eds., The Electrochem. Soc. **82–7** (1982) 17.

83W Will, N., Hofmann, K., Schulz, M.: in Proc. Int. Conf. Insulating Films on Semiconductors INFOS 83, J. Verwey, ed. North Holland Publ. Corp., in press (1983).

6.1.4.2 Ion implantation

6.1.4.2.0 Introduction

Ion implantation is a method of doping semiconductor materials by first ionizing the dopant, accelerating it in an electric field to a high energy, and then firing the ion into the semiconductor.

Ion implantation has steadily surpassed all the other methods of doping used in semiconductor technology. With the advent of high-current implantation systems, it is now possible to achieve high doping concentrations as required, e.g. in source-drain regions of MOS transistors and in emitter regions of bipolar transistors.

The typical features of ion implantation are:

The concentration profile exhibits an approximately Gaussian peak in the interior of the semiconductor material. Other shapes of the doping profile can be obtained by a sequence of implants performed at different implantation energies [77F1, 77Z1].

The result of a combination of ion implants and subsequent annealing steps can be accurately calculated in advance.

Implants through thin passivation layers (e.g. SiO_2 or Si_3N_4) are possible.

Ion implantation is not dependent on the chemical solubility of the implanted ions, i.e. any desired ion-substrate combination may be chosen.

The desired doping concentration can be monitored with high accuracy by measuring the charge introduced by the implanted ions into the semiconductor material.

The homogeneity of the implantation on a wafer as well as the reproducibility from chip to chip and from batch to batch are of the order of a few percent.

Irrespective of the purity of the starting materials, doping by ion implantation remains almost free of other impurities because the ion beam passes through a mass separator prior to acceleration. Isotopically pure doping, e.g. by only ^{10}B or only ^{11}B is possible.

Ion implantation is a low-temperature process in which the chip temperature usually remains below about 200 °C.

Masking against ion implantation is possible by materials having a low melting point, e.g. by Al or by photo-resist.

The lateral spread of the doping profile under the edge of the mask is small. High packing densities and high accuracy can be achieved by self-aligning of the masking structures.

Radiation damage is caused in the semiconductor crystal when the ion is decelerated. Usually the implanted ion does not come to rest at a lattice site. In order to be able to act as a dopant, the ion must lodge at a lattice site and, in addition, the lattice defects must be annealed. Especially in the case of compound semiconductors, this is only possible to a limited extent.

The range of the implanted ions is limited by the maximum acceleration voltage of the equipment used (approx. 200···400 kV). Doping at greater depth is achieved by a drive-in diffusion step following the ion implantation.

A large amount of review literature exists on ion implantation [70M2, 71M1, 72M1, 73D1, 73S1, 73W1, 78R1]. Ion implantation is the topic of a conference series. The proceedings of the conferences are published in [67G1, 70M1, 71E1, 71R1, 72N1, 73C1, 74R1, 75N1, 76C1, 76G1, 77C1, 80G1]. Special monographs deal with the ion-solid interaction [80G3], with the implantation-technology [75P1] and special applications, e.g. to metals [74P1].

6.1.4.2.1 Depth distribution of implanted ions

The following parameters are of significance in ion implantation technology:

Definition of implantation parameters

Parameter	Definition	Symbol	Unit
fluence or dose	number of ions per unit surface that have passed through the surface plain (of a semiconductor during implantation)	$N_{\square}$ or N_I	cm^{-2}
flux or dose rate	number of ions per unit surface and per unit time (which penetrate the surface of the semiconductor)	$N_{\square,t}$ or $N_{I,t}$ (not standardized)	$cm^{-2}s^{-1}$

continued

(continued)

Parameter	Definition	Symbol	Unit
target current	ion current applied by the accelerator to the target into which ions are to be implanted	I_T	μA
implanted area		A	cm^2
duration of implantation		t	s
charge		q	As

The target current is related to the dose by

$$I_T = N_{I,t} \cdot q \cdot A. \tag{1}$$

The range distribution of the implanted ions is approximately of Gaussian shape and can be characterized by a mean range. The ion trajectory of a single ion is usually neither straight nor normal to the surface.

The projected range is a useful parameter (Fig. 1). The mean projected range R_p, calculated from many ion trajectories, is used for calculations of the doping profile.

The following terminology is generally accepted:

Implantation terminology

Parameter	Definition	Symbol	Unit
depth in semiconductor		x	μm
implantation profile or concentration profile	density distribution of ions implanted into the substrate	$N(x)$	cm^{-3}
mean projected range of implantations	(see Figs. 1 and 2)	R_p	μm
standard deviation of projected range or straggling	(see equation (3))	ΔR_p	μm
concentration distribution of charge carriers:			
for electrons		$n(x)$	cm^{-3}
for holes		$p(x)$	cm^{-3}
peak density of implanted ions		N_{max}	cm^{-3}

$N(x)$ must not be confused with N_I!

The following formulas relate the quantities:

$$N_I = \int_0^\infty N(x)\,dx, \tag{2}$$

$$N(x) = \frac{N_I}{\sqrt{2\pi}\,\Delta R_p} \exp\left[-\frac{(x-R_p)^2}{2\Delta R_p^2}\right], \tag{3}$$

$$N_{max} = \frac{N_I}{\sqrt{2\pi}\,\Delta R_p}. \tag{4}$$

Since relations of general validity for R_p and ΔR_p were first introduced by Lindhard, Scharff and Schiott [63L1], the theory outlined here is often referred to as the LSS theory and R_p is called the LSS range.

Figure 2 shows an example of concentration profiles. R_p and ΔR_p data are tabulated for numerous ion-substrate combinations (see sections 6.1.4.2 and 6.3.4.2).

A reduced range is frequently used:

$$x' = \frac{x-R_p}{\Delta R_p}. \tag{5}$$

With (4) and (5), equation (3) becomes

$$N(x) = N_{max} \exp\left(-\frac{x'^2}{2}\right). \tag{6}$$

Deviations from the Gaussian distribution

Equation (3) describes the concentration profile in the region of the profile peak with high accuracy. An improved description of the overall profile is possible by introducing higher moments of the Gaussian distribution or other distributions developed from mathematical models (e.g. Pearson IV) [75W1, 76D1, 77W3, 78M1, 78W1, 78W2, 78Y1, 80I1, 80W3, 80W4, 80Z1, 82R1].

Exact profile calculations are of essential importance for the computer simulation of ion implantation.

A very comprehensive comparison of the mathematical description of concentration profiles with experimental results is presented in [80K1].

Channelling

The Gaussian distribution of the implanted ions given in equation (3) applies only to amorphous substrates. Ions fired into a single crystal along its principal axes (e.g. ⟨110⟩, ⟨100⟩, ⟨111⟩) are guided along by the rows of atoms in the crystal as if they were travelling through open channels. Because these channeled ions are slowed down far less than those that are not guided, they penetrate deeper into the crystal. This channeling effect continues for as long as the angle of trajectory of the guided ions is smaller than a critical angle Ψ_c relative to the crystal axis (Fig. 3). Ψ_c is a function of the ion energy, the ion mass, the substrate mass, and the lattice constant of the substrate [65L1].

When ions are fired into the crystal at angles greater than Ψ_c a small portion is scattered into open channels and, since they achieve large ranges, the concentration profile usually shows an ion concentration that falls off with an exponential distribution tail in the region of deep penetration. Such tails are observed especially in the case of P^+ implantation in silicon (Fig. 11).

Detailed information on ion channeling is given in [77H1, 77W2, 78M2, 80C1, 80D1, 80M2].

Multilayer systems

For the fabrication of semiconductor devices, ions are often implanted through a masking layer (SiO_2, Si_3N_4, or photoresist). In such cases, the resulting profile can be calculated by using a combination of two profiles (Fig. 4) [75I1].

The following relations hold

$$N_1(x)=\frac{N_I}{\sqrt{2\pi}\,\Delta R_{p1}}\exp\left[-\frac{(R_{p1}-x)^2}{2\Delta R_{p1}^2}\right]\qquad 0\leqq x\leqq d, \tag{7}$$

$$N_2(x)=\frac{N_I}{\sqrt{2\pi}\,\Delta R_{p2}}\exp\left[-\frac{\{d+[((R_{p1}-d)\,R_{p2}/R_{p1})-x]\}^2}{2\Delta R_{p2}^2}\right]\qquad x\geqq d \tag{8}$$

(d is the thickness of layer 1).

Further references treating the multilayer problem are [72F1, 72G1, 75S1, 77W1].

Knock-on implantation

Ions implanted through a masking layer collide with atoms of the masking layer. In this case, they transfer kinetic energy to the atoms of the masking layer. As a result, atoms of the masking layer are implanted into the substrate leading to a so-called knock-on or recoil implantation. Recoil-implanted atoms, especially oxygen, cause problems by inducing defect clusters during annealing.

Detailed information is found in the following references: [73C1, 73C2, 74C1, 75G1, 75M1, 75M2, 75S2, 76G3, 76M1, 78F1, 78G1, 79H1, 80H1, 80H2, 80W1, 80W2].

Sputtering

Implanted ions sputter off atoms from the surface of the substrate. After implantation of the dose N_I, a substrate layer of thickness δ has been removed.

$$\delta=\frac{r_s}{N}\cdot N_I \tag{9}$$

where r_s is the sputter rate, i.e. the number of target atoms sputtered per implanted ion and N denotes the atomic density of the target. r_s strongly depends on the energy [71E2, 69S1] and on the angle of incidence of the sputtering ion. Fig. 5 shows the sputter rate as a function of the mass of the implanted ion. In Table 1, the thickness of the silicon layer removed is given for several important dopants [75A1].

In the case of high ion doses, the concentration profile will be changed considerably. In the case of saturation, the concentration profile is

$$N(x)=\frac{d}{M \cdot m_a \cdot 2r_s}\,\mathrm{erfc}\left[\frac{x-R_p}{2\Delta R_p}\right] \tag{10}$$

d = density of target material (g cm^{-3})
M = mass number of target material
m_a = atomic unit mass ($m_a = 1.66 \cdot 10^{-27}$ kg).

In contrast to the normal implantation profile (equation (3)), the peak concentration appears at the surface. This effect is particularly pronounced in the case of implantation in GaAs (section 6.3.4.2).

Table 1. Thickness δ of silicon removed by sputtering after implantation of various doses N_I of the ions listed [75A1].

Ion	δ nm		
	$N_I = 10^{15}$ cm^{-2}	$N_I = 10^{16}$ cm^{-2}	$N_I = 10^{17}$ cm^{-2}
B	$6 \cdot 10^{-2}$	0.6	6
Ar	0.35	3.5	35
As	0.6	6	60
Sb	0.78	7.8	78

Detailed information on sputtering is given in the following references: [62C1, 68T1, 69S1, 71E2, 71T1, 72C1, 72S1, 72W1, 73K1, 74T1, 75F1, 78B1, 78R1, 79I1, 79L1, 82R1].

Lateral spread

During deceleration, the ions follow a zigzag course (Fig. 1). A lateral displacement therefore occurs. A notable concentration of ions implanted through a masking layer is collected under the edges of the mask. The lateral spread can be approximated by a Gaussian distribution. Its mean square deviation, usually denoted by ΔX or ΔY, has been computed for various ion-substrate combinations [72F2, 75G2].

If the edge contour of the masking layer is described by $d_{ox}(x)$ and the energy lost by the ions in the masking layer is assumed to be approximately equal to that lost in the substrate material (e.g. SiO_2 masking layer on Si), the following relation [77R1] is obtained for the three-dimensional distribution of the ions in the semiconductor material (Fig. 6)

$$N(x,y,z)=\frac{N_I}{(2\pi)\Delta X\,\Delta R_p}\int_{-\infty}^{\infty}\exp\left(-\frac{(x-\xi)^2}{2\Delta X^2}-\frac{(z-d_{ox}(\xi)-R_p)^2}{2\Delta R_p^2}\right)\mathrm{d}\xi. \tag{12}$$

Detailed information on the lateral spread is found in the following references: [72A1, 72F2, 73F1, 73I1, 73O1, 74P2, 76G2, 77S1, 78K1, 78K2, 79S1, 79S2].

Amorphous layers

Each ion leaves behind a track of displaced lattice atoms when it is slowed down by a series of collisions in the substrate. After implantation of a sufficient number of ions, the separate tracks overlap and an amorphous layer develops. The dose necessary to create such a layer depends on ion mass, ion energy, and target temperature. The result of the theoretical and experimental investigations for the most important dopants in Si is shown in Fig. 7.

The annealing behaviour of amorphous layers is different from that of non-amorphous regions. Amorphous layers usually anneal at much lower temperatures. By diffusion of defects and subsequent creation of defect clusters, however, some very stable "defect islands" may remain after annealing. These defect clusters normally can only be dissolved at high temperatures.

Simulation programs

Ion implantation among other technological processes is included in computer simulation programs for device fabrication processes. The program SUPREM of the Stanford University is widely used [78A1]. Another simulation program is given in [79G1]. The concentration profiles are computed as composite Gauss functions (equations (7) and (8)) or by a Pearson IV distribution [80K1].

6.1.4.2.2 Annealing

Implanted ions usually lodge at interstitial positions of the target crystal lattice, which has suffered radiation damage. For high ion doses, the entire implanted layer is rendered amorphous.

No comprehensive theory of the radiation damage has as yet been published. Similar to the concentration profile, the radiation damage profile can be described in a simplified form by a three-dimensional Gaussian distribution. The mean range of the radiation damage is smaller than that of the ions by which it was caused.

References on the causes of radiation damage are:
[69V1, 70B2, 70M2, 70V1, 71G2, 72G2, 72G3, 72V1, 73V1, 75B4, 75K1, 76C2, 76M3, 76M4, 77F2, 77G1, 77N1, 77S3, 78C2, 78D1, 78K2, 78R1, 78S1].

When ions are implanted for the purpose of doping a semiconductor, the radiation damage must be removed and the implanted ions lodged at lattice sites. The process by which this is accomplished is called annealing.

Annealing is influenced by the following factors:

a) The type and number of lattice defects observed after annealing depend on the ion mass, the ion energy, and the implantation temperature.

After formation of an amorphous layer by ion implantation, Si recrystallizes completely at about 550 to 600 °C; Ge recrystallizes at about 420 °C.

b) The starting material may have variable orientations, doping concentrations or other properties related to floating zone refining or crucible pulling which all affect the material properties observed after annealing.

c) Formation of complexes is possible, e.g. vacancy + oxygen, vacancy + impurity atom, etc. during annealing.

d) Annealing can be performed in one or more steps.

e) The environment during annealing (e.g. Ar, N_2, O_2, or vacuum) is important.

The majority carrier concentration $p(x)$ or $n(x)$ and/or the carrier mobility are usually measured as a function of the annealing temperature. If after annealing $p(x) \equiv N(x)$ or $n(x) \equiv N(x)$, the annealing is complete. This usually does *not* imply that all the point defects have vanished. Particularly in devices where characteristic parameters are influenced by minority carriers, it is often necessary to organize the annealing process in a way different to the case where only the majority carrier concentration is important.

In many cases, annealing times in the order of ms or even ns are sufficient. The implanted layer may be heated by a high power laser, an electron beam or a flash lamp, either in one flash or by scanning the implanted area at a high speed. The "pulse annealing" is still in an exploratory state. It is not yet introduced to production [80A2].

Profile variation during annealing

The elevated temperature during annealing causes implanted ions to diffuse. In the simplest case (constant diffusion coefficient D), the Gaussian distribution (equation (3)) broadens during the annealing period t [78R1].

$$N(x) = \frac{N_{max}}{\sqrt{1 + \frac{2Dt}{\Delta R_p^2}}} \exp - \left[\frac{(R_p - x)^2}{2\Delta R_p^2 + 4Dt}\right]. \tag{13}$$

Additionally, surface effects, segregation effects at interfaces and the effects of the dopant concentration and the radiation damage concentration on the diffusion coefficient have to be taken into account. Satisfactory approximations of the profile variation are obtained by simulation programs (cf. section 6.1.4.2.1).

References on diffusion given in the literature are:

a) for profiles: [61L1, 65M1, 66D1, 68H1, 68K1, 68S1, 69D1, 70S1, 71C1, 73F2, 73P1, 73R1, 73S2, 74B1, 74D1, 74F1, 74S1, 74S2, 75B1, 75D1, 75D2, 75F2, 75F3, 75J1, 78R1].

b) for segregation effects: [64K1, 67C1, 73A1, 74P3, 75M3, 76G1, 76M2].

c) for radiation damage effects: [69N1, 70G1, 70T1, 71G1, 72M2, 72N1, 72O1, 72T1, 74T1, 74Z1, 75B2].

6.1.4.2.3 Implantation of dopants in silicon

Arsenic

Arsenic implantation is widely used in source-drain and emitter implantations. High ion doses ($N_I \geqq 5 \cdot 10^{15}$ cm^{-2}) are used in both cases. The implanted layer is rendered amorphous. The optimum annealing temperatures are above 900 °C (Fig. 8).

A typical effect of arsenic implantation at high implantation doses is the appearance of an activation limit. Although the peak ion concentration may be considerably higher, a carrier concentration of only about 10^{20} cm^{-3} is reached, i.e. the carrier profile deviates from the actual doping profile (Fig. 9). When an As-implanted layer is annealed, the activation limit remains unchanged while the doping broadens because of diffusion (Fig. 10). This effect results in an extraordinarily sharp drop in the carrier concentration as is required, e.g. for emitter implantations.

Since arsenic is often implanted through a very thin oxide layer, e.g. the screening oxide for source-drain implantation, or through the flat edges of oxide masks, knock-on implantation of oxygen occurs (section 6.1.4.2.1).

Detailed information on As-implants is found in the references: [67M1, 67P1, 69B1, 69C1, 69E1, 69F1, 69J1, 72R1, 73B1, 73F2, 73I2, 73S3, 74M1, 75R1, 75R2, 75S3, 77E1, 78C1, 78M3, 78O1, 78R1, 80M1, 80S2, 80T1].

Phosphorus

Around 1970, a number of implantations were performed at elevated temperatures. It was, however, found that in this case defect clusters appear that are almost impossible to anneal out. When annealing these implants reverse annealing occurs. Phosphorus implants typically does not show a Gaussian shape. In [77B1], three different regions of the profile are identified. In [79I2], asymmetrical profiles are described by joint Gaussian profiles. In [79H1] higher moments of the Gaussian distribution are taken into account to describe the phosphorus profiles (cf. section 6.1.4.2.1).

Above the amorphous dose ($N_I = 3 \cdot 10^{14}\ \mathrm{cm}^{-2}$), phosphorus implantations can be annealed almost completely at around 600 °C (Fig. 11).

The most prominent phenomenon in phosphorus-implantations is the large channeling tail that falls off exponentially with a characteristic length of 93 nm [79H1] and therefore extends to great depths (Fig. 12). This exponential tail has to be accounted for in almost all device applications.

By implantation with doses N_I above $N_I = 10^{16}\ \mathrm{cm}^{-2}$, the solid solubility is exceeded and the carrier profile does not correspond to the doping concentration (activation limit).

When annealing P^+-implantations in wet oxygen, a considerable reduction of carriers was observed near the SiO_2—Si interface [79Y1].

Further information is found in the following references:

a) on activation: [65M2, 66G1, 67G2, 67G3, 67K1, 70S3, 72W1, 72Y1, 75H7, 76M5, 77G3, 77M1, 78S2, 78W3, 80M3],

b) on range distribution: [69F2, 70D1, 71C2, 72R1, 73B2],

c) on phosphorus channeling and channeling distribution tails: [68D1, 68G1, 71G3, 71M2, 72R2, 73B2, 74B1, 74B2, 74C2, 75C1, 77B3, 79I2].

Boron

The implantation of boron does not normally render the implanted layer amorphous ($N_{I,\,\mathrm{amorphous}} > 2 \cdot 10^{16}\ \mathrm{cm}^{-2}$). B^+ implantation is mainly used for shifting the threshold voltage ($N_I \approx 5 \cdot 10^{11}\ \mathrm{cm}^{-2}$) and in field implantations ($N_I \approx 1 \cdot 10^{12}\ \mathrm{cm}^{-2}$) of MOS transistors and for base doping of transistors ($N_I \approx 10^{12} \cdots 10^{14}\ \mathrm{cm}^{-2}$).

The annealing characteristics of B^+ implantations are shown in Fig. 13. Reverse annealing occurs at temperatures between 500 and 600 °C.

The ion sources employed for the generation of B^+ ions generally use BF_3 as a feed gas which decomposes preferentially into BF_2^+ during ionization. The B^+ ion current is lower than the BF_2^+ current. For this reason frequently BF_2^+ is implanted when high B^+ ion doses are required. The compound decomposes when hitting the semiconductor surface. The energy of the boron ion can be calculated from

$$E_{B^+} = \frac{m_{B^+}}{m_{BF_2^+}} \cdot E_0 \tag{14}$$

m_{B^+} denotes the mass of the boron ion and $m_{BF_2^+}$ the mass of the BF_2^+ molecule. Charge and energy exchange between the various decomposition products of BF_3 during preacceleration and separation may, however, cause B^+ ions of a different acceleration energy to strike the target resulting in implantation profiles of BF_2^+ implants which exhibit double peaks.

Detailed information on B^+ implantation is found in the following references:

a) on the annealing behaviour: [61R1, 65K1, 66K1, 66M1, 66P1, 67K1, 70B1, 70N1, 71S2, 74H1, 74M2, 74S3, 75P2, 76A1, 77S2, 80A1],

b) on the concentration profiles: [68R1, 70D1, 71S1, 73H1, 73K2, 73R2, 75B3, 75H2, 75H3, 75O2, 75W2, 78R1],

c) on the BF_2^+-implantation: [72M3, 75P3, 77B2, 78R1, 78T1, 79T1, 79T2, 79W1, 79W2, 80S1],

d) on the B^{++}-implantation: [75H1, 75H2, 77B1],

e) on B^{10}- vs. B^{11}-implantation: [75O1, 75V1, 77R2].

Antimony

Comparatively few papers are published on Sb implantation in Si. Sb has only a short range in Si and, therefore, it is usually driven in after implantation. Since high Sb^+-currents are available in modern implantation machines, Sb is widely used in bipolar applications, and for shallow implantations, e.g. for controlling the barrier height of Schottky diodes. Figs. 14 and 15 show the annealing behaviour of Sb^+-implants. Like As and P, Sb implanted at concentrations exceeding the amorphous dose ($N_I \geqq 10^{14}$ cm^{-2}) exhibits a rapid increase in activation at about 550 °C.

Sb^+ implantations above the amorphous dose are almost completely annealed. Rather poor annealing is obtained, when the implanted layer was not amorphized during implantation (e.g. at elevated temperatures).

Like phosphorus, Sb implantations show an exponential tail caused by channeling [71C2]. When implantation takes place at elevated temperatures, considerable diffusion occurs. During annealing outdiffusion was observed [75M3].

Detailed information on Sb-implantation is found in the following references: [67D1, 67M3, 69B1, 69C1, 70C1, 70J1, 70N2, 75D3, 75M3, 80C2, 80C3, 80G2, 80K2].

Gettering by Ar-implantation

For impurity gettering, the back of a wafer is heavily damaged by ion bombardment, preferably by Ar^+. Appropriate annealing results in a dislocation network that remains stable during subsequent anneal steps and serves in the gettering of impurities in silicon, e.g. of Cu and Au. The impurity concentration on the unimplanted front side of the chip is reduced, and the lifetime of minority carriers is considerably increased in this region.

Details may be found in the following references: [72B1, 73B3, 73H2, 73P2, 73S4, 76S1, 77G2, 78R2].

6.1.4.2.4 Implantation of dopants in Ge

In comparison to Si and GaAs, very few reports are published on implantation into Ge. Doping of Ge is performed by implanting the same ions as in Si. Amorphous Ge layers anneal out at about 420 °C. The implantation temperature is a significant factor, because many defects already anneal out when high current implantations are employed. Local temperature differences will cause nonuniform doping of the wafer. Low-dose B^+ implants already anneal at about 200 °C. For higher ion doses, as well as in the case of all other ions, an anneal temperature of about 500 to 550 °C is required to assure complete annealing.

In the range tables, Ge is usually listed together with GaAs because the mean density and mass of both semiconductors are almost identical.

More information is found in the following references:

on p-type doping: [67M2, 68A1, 68B1, 69F1, 69J2, 70S2, 70T2, 71H1, 74A1, 74P4, 74S4, 75H4, 75H5, 75H6, 75O3, 75W3, 77A2],

on n-type-doping: [70T2, 71H1, 74S4, 77A2],

on radiation damage: [68B1].

6.1.4.2.5 Useful formulas

Sheet resistance of implanted layers

The sheet resistance of an implanted layer is given by

$$R_\square = \frac{1}{q} \int_0^\infty \frac{\mathrm{d}x}{N(x) \cdot \mu(N(x))}, \tag{15}$$

where $\mu(N(x))$ denotes the carrier mobility, depending on doping concentration and -profile. $N(x)$ can be obtained e.g. from equation (3). The sheet resistance of implantations in silicon may be determined from Figs. 16 and 17. The curves are calculated by assuming the bulk mobility as given by the Irvin curves [62I1] and complete annealing.

Implantation through masking layer

For practical applications, the knowledge of the fraction X of the implanted ions that penetrate a masking layer of the thickness d is essential.

The following relation holds for X:

$$X = \frac{N_{Si}}{N_{\square}} = \frac{1}{\sqrt{2\pi}} \int_{0}^{\infty} \exp\left[-\frac{1}{2}\left(\frac{x-R_p}{\Delta R_p}\right)^2\right] dx. \tag{16}$$

N_{Si} denotes the number of ions that reach the substrate silicon through the masking layer. Fig. 18 shows X as a function of the renormalized parameter

$$x' = \frac{d-R_p}{\Delta R_p}. \tag{17}$$

For R_p and ΔR_p, the data of the masking layer have to be inserted.

Threshold voltage shift of MOS transistors

The threshold voltage of MOS transistors can be shifted by implanting small doses ($N_{\square} \leqq 10^{12}$ cm^{-2}) of a suitable ion into the channel region. If the concentration profile is of an extent less than the width of the space charge region at the onset of inversion, the implanted ions may be treated as a surface charge. Using the capacitor equation, the threshold voltage shift ΔV_{th} of the transistor is calculated to

$$\Delta V_{th} = \frac{q \cdot N_{Si}}{C_{ox}} \tag{18}$$

where N_{Si} is the function of ions penetrating into silicon (equation (16)). If the annealing of defects remains incomplete, the ions which are not lodged at lattice sites usually form deep levels. Due to the sharp bending of the conduction and valence bands, these deep levels are charged at the onset of inversion and therefore add to the threshold voltage shift. At low anneal temperatures (approx. 500 °C), it is always necessary to insert the total N_{Si} in equation (18) and not only the annealed fraction [75R3, 75R4, 75R5, 76R1].

Nomographs for B^+ and P^+ implantation allow the determination of the threshold voltage shift to be expected for a given implantation energy and oxide thickness (Fig. 19···22).

6.1.4.2.6 Range tables

The range tables calculated for B, P, As, Sb, and Ar implants in Si are taken from [78R1]. All calculations are based on a computer programme devised by J. Biersack. Further range tables can be found in [73D1, 75G2, 77S4, 80R1]. The ranges of implantation into Ge are identical to those in GaAs given in section 6.3.4.2.

When using range distribution tables it is essential to choose the correct density of the substrate. Apart from photoresists, this applies in particular to SiO_2 and Si_3N_4, where the densities actually obtained by the semiconductor technology differ from the densities of the respective single crystal on which e.g. the calculations in [75G2] are based.

If the actual densities differ from those in the tables, the range can be scaled with sufficient accuracy by

$$R_{p,\,actual} = \frac{d_{table}}{d_{actual}} R_{p,\,table}. \tag{19}$$

All range calculations refer to the most abundant isotope. Table 2 lists the parameters of the implanted layer and Table 3 the parameters of the ions. Tables 4···10 contain the range data.

Table 2. Basic data for the computation of ion implantation ranges in Table 4.

Application	Designation	Chemical composition	Density d g cm^{-3}	Components	Atomic number Z	Atomic weight M	Ratio of components
substrate	silicon	Si	2.33	Si	14	28.086	1
	germanium	Ge	5.35	Ge	32	72.590	1
masking layer	silicon dioxide	SiO_2	2.27	Si	14	28.086	1
				O	8	15.999	2
	silicon nitride	Si_3N_4	2.90	Si	14	28.086	3
				N	7	14.007	4

continued

Table 2 (continued)

Application	Designation	Chemical composition	Density d g cm^{-3}	Components	Atomic number Z	Atomic weight M	Ratio of components
photo resist	AZ111	unknown	1.20	C	6	12.010	68.80
				H	1	1.008	7.17
				O	8	15.999	24.03
	KTFR	C_9H_{13}	1.05	C	6	12.010	9
				H	1	1.008	13
	PMMA	$C_5H_8O_2$	1.19	C	6	12.010	5
				H	1	1.008	8
				O	8	15.999	2
	AZ1350	unknown	1.30	C	6	12.010	74.00
				H	1	1.008	6.10
				O	8	15.999	19.90

Table 3. Data of ions used in implantation of Si and Ge.

Ions	Isotope	Atomic number Z	Atomic weight M	Application
Boron	^{11}B	5	11.009	doping of Si/Ge
Phosphorus	^{31}P	15	30.974	
Arsenic	^{75}As	33	74.921	
Antimony	^{121}Sb	51	120.900	
Hydrogen	^{1}H	1	1.008	radiation damage, gettering
Silicon	^{28}Si	14	28.086	
Argon	^{40}Ar	18	39.962	

Table 4. Si. Projected range R_p and straggling ΔR_p for implantation of various ions in Si.

E keV	Sb R_p µm	Sb ΔR_p µm	Ar R_p µm	Ar ΔR_p µm	As R_p µm	As ΔR_p µm	B R_p µm	B ΔR_p µm	P R_p µm	P ΔR_p µm	Si R_p µm	Si ΔR_p µm	H R_p µm	H ΔR_p µm
10	0.0092	0.0030	0.0118	0.0060	0.0099	0.0037	0.0310	0.0193	0.0128	0.0074	0.0132	0.0079	0.1757	0.0701
20	0.0147	0.0045	0.0215	0.0102	0.0164	0.0060	0.0634	0.0318	0.0243	0.0129	0.0253	0.0139	0.3155	0.0906
30	0.0195	0.0059	0.0312	0.0143	0.0225	0.0081	0.0956	0.0414	0.0359	0.0181	0.0378	0.0196	0.4330	0.1019
40	0.0241	0.0072	0.0410	0.0181	0.0284	0.0101	0.1270	0.0492	0.0479	0.0232	0.0507	0.0250	0.5362	0.1095
50	0.0284	0.0084	0.0509	0.0219	0.0343	0.0121	0.1574	0.0556	0.0601	0.0280	0.0638	0.0302	0.6295	0.1150
60	0.0327	0.0097	0.0609	0.0256	0.0400	0.0140	0.1868	0.0610	0.0726	0.0327	0.0773	0.0352	0.7152	0.1192
70	0.0369	0.0108	0.0711	0.0291	0.0458	0.0159	0.2153	0.0657	0.0852	0.0372	0.0909	0.0401	0.7950	0.1226
80	0.0410	0.0120	0.0813	0.0326	0.0516	0.0177	0.2429	0.0699	0.0979	0.0415	0.1047	0.0448	0.8701	0.1254
90	0.0451	0.0131	0.0916	0.0360	0.0573	0.0196	0,2696	0.0735	0.1108	0.0458	0.1187	0.0493	0.9411	0.1278
100	0.0491	0.0143	0.1020	0.0393	0.0631	0.0214	0.2956	0.0768	0.1238	0.0498	0.1327	0.0536	1.0088	0.1299
110	0.0532	0.0154	0.1124	0.0426	0.0689	0.0232	0.3209	0.0798	0.1368	0.0538	0.1469	0.0578	1.0736	0.1317
120	0.0572	0.0165	0.1229	0.0457	0.0747	0.0250	0.3456	0.0825	0.1499	0.0575	0.1611	0.0618	1.1359	0.1332
130	0.0611	0.0176	0.1335	0.0488	0.0805	0.0268	0.3697	0.0849	0.1631	0.0612	0.1753	0.0657	1.1959	0.1347
140	0.0651	0.0187	0.1441	0.0518	0.0863	0.0286	0.3933	0.0872	0.1762	0.0648	0.1896	0.0694	1.2540	0.1359
150	0.0691	0.0198	0.1547	0.0547	0.0921	0.0304	0.4163	0.0893	0.1894	0.0682	0.2038	0.0731	1.3103	0.1371
160	0.0730	0.0209	0.1654	0.0576	0.0980	0.0321	0.4389	0.0913	0.2026	0.0715	0.2181	0.0766	1.3650	0.1382
170	0.0769	0.0219	0.1760	0.0604	0.1039	0.0339	0.4610	0.0931	0.2157	0.0747	0.2324	0.0799	1.4182	0.1391
180	0.0809	0.0230	0.1867	0.0631	0.1097	0.0356	0.4828	0.0948	0.2289	0.0778	0.2466	0.0832	1.4701	0.1400
190	0.0848	0.0241	0.1974	0.0658	0.1157	0.0373	0.5041	0.0964	0.2420	0.0808	0.2608	0.0864	1.5208	0.1409
200	0.0887	0.0251	0.2081	0.0684	0.1216	0.0391	0.5251	0.0979	0.2550	0.0837	0.2750	0.0894	1.5704	0.1417
220	0.0966	0.0272	0.2294	0.0734	0.1335	0.0425	0.5661	0.1007	0.2811	0.0893	0.3032	0.0953	1.6666	0.1431
240	0.1044	0.0293	0.2507	0.0782	0.1454	0.0458	0.6059	0.1032	0.3070	0.0945	0.3312	0.1007	1.7592	0.1443
260	0.1123	0.0314	0.2720	0.0828	0.1575	0.0491	0.6446	0.1055	0.3326	0.0994	0.3590	0.1059	1.8487	0.1455
280	0.1202	0.0335	0.2931	0.0871	0.1696	0.0524	0.6824	0.1076	0.3581	0.1041	0.3866	0.1108	1.9355	0.1465
300	0.1280	0.0355	0.3142	0.0913	0.1817	0.0557	0.7193	0.1095	0.3833	0.1084	0.4139	0.1153	2.0199	0.1474
320	0.1359	0.0375	0.3352	0.0953	0.1939	0.0589	0.7553	0.1113	0.4083	0.1126	0.4410	0.1197	2.1022	0.1483
340	0.1438	0.0396	0.3561	0.0991	0.2061	0.0620	0.7906	0.1130	0.4330	0.1165	0.4678	0.1238	2.1825	0.1490
360	0.1517	0.0416	0.3768	0.1028	0.2184	0.0652	0.8253	0.1145	0.4575	0.1202	0.4944	0.1277	2.2611	0.1498
380	0.1596	0.0436	0.3974	0.1063	0.2307	0.0683	0.8593	0.1160	0.4817	0.1238	0.5206	0.1314	2.3382	0.1504
400	0.1675	0.0456	0.4179	0.1096	0.2431	0.0713	0.8928	0.1173	0.5057	0.1272	0.5467	0.1350	2.4138	0.1510
420	0.1755	0.0476	0.4382	0.1129	0.2555	0.0743	0.9257	0.1186	0.5294	0.1304	0.5724	0.1383	2.4881	0.1517
440	0.1835	0.0496	0.4584	0.1160	0.2679	0.0773	0.9581	0.1198	0.5529	0.1334	0.5979	0.1415	2.5612	0.1522
460	0.1914	0.0515	0.4784	0.1190	0.2804	0.0802	0.9901	0.1210	0.5762	0.1364	0.6232	0.1446	2.6333	0.1527
480	0.1994	0.0535	0.4983	0.1219	0.2928	0.0831	1.0216	0.1221	0.5992	0.1392	0.6482	0.1476	2.7043	0.1533
500	0.2074	0.0554	0.5181	0.1246	0.3053	0.0860	1.0527	0.1231	0.6220	0.1419	0.6729	0.1504	2.7744	0.1538

Table 5. SiO_2. Projected range R_p and straggling ΔR_p for implantation of various ions in SiO_2.

E keV	B R_p μm	B ΔR_p μm	P R_p μm	P ΔR_p μm	As R_p μm	As ΔR_p μm	Sb R_p μm	Sb ΔR_p μm	H R_p μm	H ΔR_p μm	Si R_p μm	Si ΔR_p μm	Ar R_p μm	Ar ΔR_p μm
10	0.0325	0.0189	0.0127	0.0058	0.0097	0.0031	0.0098	0.0025	0.1579	0.0514	0.0132	0.0063	0.0117	0.0048
20	0.0698	0.0331	0.0240	0.0104	0.0161	0.0050	0.0151	0.0039	0.2749	0.0640	0.0252	0.0112	0.0213	0.0084
30	0.1084	0.0445	0.0355	0.0147	0.0220	0.0068	0.0198	0.0050	0.3711	0.0707	0.0376	0.0160	0.0310	0.0118
40	0.1468	0.0539	0.0473	0.0189	0.0278	0.0085	0.0242	0.0061	0.4549	0.0750	0.0503	0.0205	0.0407	0.0152
50	0.1844	0.0617	0.0594	0.0229	0.0335	0.0102	0.0284	0.0072	0.5303	0.0782	0.0633	0.0248	0.0506	0.0185
60	0.2210	0.0684	0.0717	0.0268	0.0392	0.0118	0.0326	0.0082	0.5996	0.0806	0.0766	0.0290	0.0607	0.0217
70	0.2566	0.0742	0.0841	0.0306	0.0448	0.0135	0.0367	0.0093	0.6640	0.0825	0.0900	0.0330	0.0708	0.0248
80	0.2912	0.0792	0.0967	0.0343	0.0504	0.0151	0.0407	0.0103	0.7246	0.0841	0.1035	0.0369	0.0811	0.0279
90	0.3248	0.0837	0.1093	0.0378	0.0561	0.0167	0.0447	0.0113	0.7820	0.0854	0.1172	0.0406	0.0915	0.0310
100	0.3574	0.0877	0.1221	0.0412	0.0617	0.0182	0.0486	0.0122	0.8368	0.0866	0.1309	0.0441	0.1020	0.0339
110	0.3893	0.0913	0.1349	0.0445	0.0673	0.0198	0.0526	0.0132	0.8894	0.0876	0.1447	0.0475	0.1125	0.0368
120	0.4202	0.0945	0.1478	0.0477	0.0730	0.0214	0.0565	0.0142	0.9400	0.0885	0.1585	0.0508	0.1231	0.0397
130	0.4505	0.0975	0.1607	0.0508	0.0787	0.0229	0.0603	0.0151	0.9888	0.0893	0.1722	0.0540	0.1338	0.0424
140	0.4800	0.1002	0.1736	0.0537	0.0844	0.0245	0.0642	0.0161	1.0362	0.0901	0.1860	0.0570	0.1445	0.0451
150	0.5088	0.1027	0.1865	0.0566	0.0901	0.0260	0.0681	0.0170	1.0823	0.0907	0.1998	0.0599	0.1553	0.0478
160	0.5370	0.1050	0.1994	0.0594	0.0958	0.0275	0.0720	0.0179	1.1272	0.0913	0.2135	0.0627	0.1661	0.0504
170	0.5646	0.1072	0.2123	0.0620	0.1016	0.0290	0.0758	0.0189	1.1710	0.0919	0.2272	0.0654	0.1769	0.0529
180	0.5917	0.1092	0.2252	0.0646	0.1074	0.0305	0.0797	0.0198	1.2138	0.0924	0.2408	0.0680	0.1877	0.0554
190	0.6182	0.1110	0.2380	0.0671	0.1131	0.0320	0.0835	0.0207	1.2557	0.0929	0.2544	0.0705	0.1986	0.0578
200	0.6442	0.1128	0.2509	0.0696	0.1189	0.0335	0.0874	0.0216	1.2968	0.0934	0.2680	0.0729	0.2094	0.0602
220	0.6948	0.1160	0.2763	0.0742	0.1306	0.0365	0.0951	0.0235	1.3769	0.0942	0.2949	0.0774	0.2312	0.0647
240	0.7438	0.1188	0.3017	0.0785	0.1423	0.0394	0.1027	0.0253	1.4545	0.0950	0.3215	0.0817	0.2529	0.0691
260	0.7911	0.1214	0.3268	0.0826	0.1541	0.0422	0.1104	0.0271	1.5299	0.0956	0.3478	0.0857	0.2745	0.0733
280	0.8371	0.1237	0.3517	0.0865	0.1659	0.0451	0.1182	0.0289	1.6035	0.0962	0.3739	0.0894	0.2961	0.0773
300	0.8818	0.1259	0.3763	0.0901	0.1778	0.0479	0.1259	0.0307	1.6754	0.0968	0.3996	0.0929	0.3177	0.0811
320	0.9253	0.1278	0.4008	0.0936	0.1898	0.0507	0.1336	0.0324	1.7459	0.0973	0.4251	0.0962	0.3391	0.0848
340	0.9678	0.1296	0.4249	0.0968	0.2017	0.0534	0.1413	0.0342	1.8152	0.0978	0.4502	0.0993	0.3604	0.0883
360	1.0092	0.1312	0.4489	0.0999	0.2138	0.0561	0.1491	0.0360	1.8834	0.0983	0.4751	0.1023	0.3817	0.0916
380	1.0497	0.1328	0.4726	0.1028	0.2258	0.0588	0.1569	0.0377	1.9505	0.0987	0.4997	0.1051	0.4028	0.0949
400	1.0893	0.1342	0.4961	0.1056	0.2379	0.0614	0.1647	0.0394	2.0169	0.0991	0.5239	0.1077	0.4238	0.0980
420	1.1281	0.1356	0.5193	0.1083	0.2500	0.0640	0.1725	0.0412	2.0824	0.0995	0.5479	0.1102	0.4446	0.1009
440	1.1662	0.1368	0.5423	0.1108	0.2621	0.0665	0.1803	0.0429	2.1473	0.0999	0.5717	0.1126	0.4653	0.1038
460	1.2035	0.1380	0.5651	0.1133	0.2743	0.0691	0.1881	0.0446	2.2115	0.1002	0.5951	0.1149	0.4059	0.1066
480	1.2402	0.1391	0.5876	0.1156	0.2865	0.0715	0.1960	0.0463	2.2752	0.1006	0.6183	0.1170	0.5064	0.1092
500	1.2763	0.1402	0.6099	0.1178	0.2987	0.0740	0.2038	0.0480	2.3384	0.1008	0.6413	0.1191	0.5267	0.1118

Table 6. Si_3N_4. Projected range R_p and straggling ΔR_p for implantation of various ions into Si_3N_4.

E keV	B		P		As		Sb		H		Si		Ar	
	R_p µm	ΔR_p µm	R_p µm	ΔR_p µm	R_p µm	ΔR_p µm	R_p µm	ΔR_p µm	R_p µm	ΔR_p µm	R_p µm	ΔR_p µm	R_p µm	ΔR_p µm
10	0.0258	0.0149	0.0101	0.0046	0.0077	0.0024	0.0078	0.0020	0.1246	0.0406	0.0105	0.0050	0.0093	0.0038
20	0.0554	0.0261	0.0190	0.0082	0.0127	0.0040	0.0119	0.0031	0.2168	0.0505	0.0200	0.0089	0.0169	0.0066
30	0.0859	0.0351	0.0282	0.0116	0.0174	0.0054	0.0156	0.0040	0.2927	0.0558	0.0298	0.0126	0.0245	0.0094
40	0.1163	0.0425	0.0375	0.0149	0.0220	0.0067	0.0191	0.0049	0.3588	0.0593	0.0399	0.0162	0.0323	0.0120
50	0.1460	0.0486	0.0471	0.0181	0.0265	0.0081	0.0225	0.0057	0.4184	0.0617	0.0502	0.0196	0.0401	0.0146
60	0.1749	0.0539	0.0568	0.0212	0.0310	0.0094	0.0258	0.0066	0.4730	0.0637	0.0607	0.0229	0.0480	0.0172
70	0.2031	0.0584	0.0666	0.0242	0.0355	0.0107	0.0290	0.0074	0.5239	0.0652	0.0713	0.0261	0.0561	0.0196
80	0.2304	0.0624	0.0765	0.0270	0.0399	0.0119	0.0322	0.0082	0.5718	0.0664	0.0820	0.0291	0.0642	0.0221
90	0.2569	0.0659	0.0866	0.0298	0.0444	0.0132	0.0353	0.0090	0.6173	0.0675	0.0928	0.0320	0.0724	0.0245
100	0.2828	0.0690	0.0967	0.0325	0.0488	0.0145	0.0384	0.0097	0.6606	0.0684	0.1036	0.0348	0.0807	0.0268
110	0.3079	0.0719	0.1068	0.0351	0.0533	0.0157	0.0415	0.0105	0.7022	0.0692	0.1145	0.0375	0.0890	0.0291
120	0.3324	0.0744	0.1170	0.0376	0.0578	0.0169	0.0446	0.0113	0.7423	0.0700	0.1254	0.0401	0.0974	0.0313
130	0.3563	0.0768	0.1272	0.0401	0.0622	0.0182	0.0477	0.0120	0.7811	0.0706	0.1363	0.0426	0.1059	0.0335
140	0.3796	0.0789	0.1374	0.0424	0.0667	0.0194	0.0508	0.0128	0.8186	0.0712	0.1472	0.0450	0.1143	0.0357
150	0.4024	0.0809	0.1476	0.0447	0.0713	0.0206	0.0538	0.0135	0.8552	0.0717	0.1581	0.0473	0.1228	0.0378
160	0.4247	0.0827	0.1578	0.0469	0.0758	0.0218	0.0569	0.0143	0.8908	0.0722	0.1689	0.0495	0.1314	0.0398
170	0.4465	0.0844	0.1680	0.0490	0.0803	0.0230	0.0599	0.0150	0.9255	0.0727	0.1797	0.0516	0.1399	0.0418
180	0.4679	0.0860	0.1781	0.0510	0.0849	0.0242	0.0629	0.0157	0.9595	0.0731	0.1905	0.0536	0.1485	0.0438
190	0.4888	0.0874	0.1883	0.0530	0.0895	0.0254	0.0660	0.0165	0.9928	0.0735	0.2012	0.0556	0.1570	0.0457
200	0.5094	0.0888	0.1984	0.0549	0.0940	0.0266	0.0690	0.0172	1.0255	0.0738	0.2119	0.0575	0.1656	0.0476
220	0.5494	0.0913	0.2185	0.0586	0.1032	0.0289	0.0751	0.0187	1.0892	0.0745	0.2332	0.0611	0.1828	0.0512
240	0.5881	0.0936	0.2385	0.0620	0.1125	0.0312	0.0812	0.0201	1.1509	0.0751	0.2542	0.0645	0.1999	0.0546
260	0.6255	0.0956	0.2584	0.0653	0.1218	0.0335	0.0872	0.0215	1.2110	0.0757	0.2750	0.0676	0.2170	0.0579
280	0.6619	0.0975	0.2780	0.0683	0.1311	0.0358	0.0933	0.0230	1.2696	0.0762	0.2956	0.0706	0.2341	0.0611
300	0.6972	0.0991	0.2975	0.0712	0.1405	0.0380	0.0994	0.0244	1.3269	0.0766	0.3159	0.0734	0.2511	0.0641
320	0.7316	0.1007	0.3168	0.0739	0.1499	0.0402	0.1055	0.0258	1.3832	0.0770	0.3360	0.0760	0.2680	0.0670
340	0.7652	0.1021	0.3359	0.0765	0.1594	0.0423	0.1116	0.0272	1.4385	0.0775	0.3559	0.0784	0.2849	0.0698
360	0.7979	0.1034	0.3548	0.0789	0.1689	0.0445	0.1177	0.0286	1.4929	0.0778	0.3756	0.0808	0.3016	0.0724
380	0.8300	0.1046	0.3735	0.0812	0.1784	0.0466	0.1238	0.0300	1.5465	0.0782	0.3950	0.0830	0.3183	0.0750
400	0.8613	0.1057	0.3921	0.0834	0.1879	0.0487	0.1300	0.0314	1.5995	0.0785	0.4141	0.0850	0.3349	0.0775
420	0.8920	0.1068	0.4104	0.0855	0.1975	0.0507	0.1361	0.0327	1.6519	0.0788	0.4331	0.0870	0.3513	0.0798
440	0.9221	0.1078	0.4286	0.0876	0.2071	0.0528	0.1423	0.0341	1.7037	0.0791	0.4519	0.0889	0.3677	0.0821
460	0.9516	0.1087	0.4466	0.0895	0.2166	0.0548	0.1485	0.0355	1.7551	0.0794	0.4704	0.0907	0.3840	0.0843
480	0.9806	0.1096	0.4644	0.0913	0.2262	0.0567	0.1547	0.0368	1.8060	0.0797	0.4887	0.0924	0.4001	0.0864
500	1.0092	0.1104	0.4820	0.0931	0.2359	0.0587	0.1609	0.0382	1.8565	0.0800	0.5068	0.0941	0.4162	0.0884

Table 7. AZ111. Projected range R_p and straggling ΔR_p for implantation of various ions in photoresist AZ 111.

E keV	B R_p µm	B ΔR_p µm	P R_p µm	P ΔR_p µm	As R_p µm	As ΔR_p µm	Sb R_p µm	Sb ΔR_p µm	H R_p µm	H ΔR_p µm	Si R_p µm	Si ΔR_p µm	Ar R_p µm	Ar ΔR_p µm
10	0.0564	0.0249	0.0219	0.0079	0.0163	0.0043	0.0165	0.0035	0.2386	0.0590	0.0228	0.0085	0.0199	0.0066
20	0.1196	0.0433	0.0414	0.0143	0.0272	0.0070	0.0256	0.0054	0.4009	0.0703	0.0437	0.0155	0.0365	0.0117
30	0.1834	0.0574	0.0614	0.0205	0.0373	0.0095	0.0337	0.0071	0.5325	0.0761	0.0652	0.0222	0.0530	0.0166
40	0.2456	0.0684	0.0818	0.0264	0.0473	0.0120	0.0413	0.0086	0.6468	0.0799	0.0873	0.0286	0.0699	0.0215
50	0.3057	0.0774	0.1027	0.0322	0.0571	0.0144	0.0487	0.0102	0.7498	0.0826	0.1097	0.0348	0.0869	0.0262
60	0.3636	0.0848	0.1238	0.0378	0.0668	0.0168	0.0559	0.0116	0.8446	0.0847	0.1326	0.0407	0.1042	0.0308
70	0.4193	0.0911	0.1452	0.0432	0.0765	0.0191	0.0630	0.0131	0.9332	0.0863	0.1556	0.0463	0.1218	0.0351
80	0.4730	0.0965	0.1668	0.0483	0.0862	0.0215	0.0700	0.0145	1.0169	0.0877	0.1789	0.0517	0.1395	0.0398
90	0.5248	0.1011	0.1886	0.0533	0.0960	0.0238	0.0769	0.0159	1.0966	0.0889	0.2022	0.0569	0.1573	0.0442
100	0.5749	0.1052	0.2104	0.0581	0.1057	0.0260	0.0838	0.0173	1.1730	0.0900	0.2256	0.0618	0.1754	0.0484
110	0.6235	0.1089	0.2323	0.0627	0.1155	0.0283	0.0906	0.0187	1.2466	0.0909	0.2490	0.0665	0.1935	0.0525
120	0.6706	0.1122	0.2543	0.0671	0.1253	0.0306	0.0974	0.0201	1.3179	0.0917	0.2724	0.0710	0.2117	0.0566
130	0.7164	0.1151	0.2762	0.0714	0.1351	0.0328	0.1042	0.0215	1.3871	0.0924	0.2957	0.0753	0.2300	0.0605
140	0.7610	0.1178	0.2981	0.0755	0.1450	0.0350	0.1109	0.0228	1.4545	0.0931	0.3190	0.0794	0.2484	0.0644
150	0.8045	0.1202	0.3200	0.0794	0.1549	0.0373	0.1177	0.0242	1.5204	0.0937	0.3422	0.0833	0.2668	0.0681
160	0.8469	0.1225	0.3418	0.0832	0.1648	0.0395	0.1244	0.0255	1.5849	0.0943	0.3652	0.0870	0.2852	0.0718
170	0.8883	0.1245	0.3635	0.0868	0.1748	0.0416	0.1311	0.0269	1.6481	0.0948	0.3882	0.0906	0.3037	0.0753
180	0.9288	0.1264	0.3852	0.0903	0.1848	0.0438	0.1379	0.0282	1.7103	0.0953	0.4110	0.0941	0.3221	0.0788
190	0.9684	0.1282	0.4068	0.0937	0.1948	0.0460	0.1446	0.0295	1.7715	0.0957	0.4337	0.0974	0.3406	0.0822
200	1.0073	0.1299	0.4283	0.0970	0.2049	0.0481	0.1513	0.0308	1.8318	0.0962	0.4563	0.1005	0.3591	0.0854
220	1.0827	0.1329	0.4710	0.1031	0.2251	0.0524	0.1647	0.0335	1.9502	0.0970	0.5010	0.1065	0.3960	0.0918
240	1.1555	0.1355	0.5132	0.1089	0.2455	0.0566	0.1782	0.0361	2.0660	0.0977	0.5452	0.1120	0.4328	0.0978
260	1.2259	0.1379	0.5550	0.1143	0.2659	0.0607	0.1916	0.0387	2.1797	0.0984	0.5887	0.1172	0.4694	0.1035
280	1.2941	0.1400	0.5963	0.1193	0.2865	0.0648	0.2051	0.0412	2.2917	0.0990	0.6316	0.1219	0.5058	0.1090
300	1.3604	0.1420	0.6372	0.1240	0.3071	0.0688	0.2186	0.0438	2.4023	0.0996	0.6740	0.1264	0.5421	0.1141
320	1.4249	0.1437	0.6775	0.1285	0.3278	0.0728	0.2322	0.0464	2.5117	0.1001	0.7157	0.1306	0.5781	0.1191
340	1.4878	0.1453	0.7174	0.1326	0.3486	0.0767	0.2457	0.0489	2.6202	0.1007	0.7569	0.1345	0.6139	0.1238
360	1.5491	0.1468	0.7568	0.1366	0.3694	0.0806	0.2593	0.0514	2.7280	0.1012	0.7975	0.1382	0.6495	0.1283
380	1.6091	0.1482	0.7958	0.1403	0.3903	0.0844	0.2729	0.0539	2.8352	0.1016	0.8375	0.1416	0.6848	0.1326
400	1.6678	0.1495	0.8343	0.1439	0.4112	0.0882	0.2866	0.0564	2.9419	0.1021	0.8771	0.1449	0.7199	0.1367
420	1.7254	0.1507	0.8723	0.1472	0.4322	0.0919	0.3003	0.0589	3.0483	0.1026	0.9161	0.1480	0.7546	0.1407
440	1.7818	0.1518	0.9099	0.1504	0.4532	0.0955	0.3140	0.0614	3.1545	0.1030	0.9546	0.1509	0.7891	0.1445
460	1.8372	0.1529	0.9470	0.1534	0.4743	0.0991	0.3277	0.0639	3.2606	0.1035	0.9926	0.1537	0.8234	0.1481
480	1.8916	0.1539	0.9838	0.1563	0.4954	0.1026	0.3415	0.0663	3.3667	0.1039	1.0301	0.1564	0.8574	0.1516
500	1.9452	0.1548	1.0201	0.1591	0.5165	0.1061	0.3553	0.0687	3.4727	0.1043	1.0672	0.1589	0.8911	0.1550

Table 8. KTFR. Projected range R_p and straggling ΔR_p for implantation of various ions into photoresist KTFR.

E keV	B		P		As		Sb		H		Si		Ar	
	R_p µm	ΔR_p µm	R_p µm	ΔR_p µm	R_p µm	ΔR_p µm	R_p µm	ΔR_p µm	R_p µm	ΔR_p µm	R_p µm	ΔR_p µm	R_p µm	ΔR_p µm
10	0.0570	0.0167	0.0226	0.0064	0.0170	0.0038	0.0175	0.0032	0.1944	0.0351	0.0236	0.0067	0.0205	0.0055
20	0.1172	0.0287	0.0422	0.0114	0.0281	0.0061	0.0269	0.0049	0.3212	0.0406	0.0445	0.0122	0.0371	0.0096
30	0.1763	0.0377	0.0620	0.0163	0.0385	0.0083	0.0352	0.0063	0.4280	0.0436	0.0658	0.0175	0.0537	0.0136
40	0.2330	0.0445	0.0820	0.0210	0.0485	0.0103	0.0431	0.0077	0.5243	0.0457	0.0874	0.0225	0.0703	0.0175
50	0.2871	0.0499	0.1024	0.0255	0.0583	0.0124	0.0506	0.0090	0.6138	0.0473	0.1093	0.0273	0.0870	0.0213
60	0.3388	0.0543	0.1229	0.0299	0.0681	0.0144	0.0580	0.0103	0.6984	0.0486	0.1314	0.0318	0.1039	0.0250
70	0.3883	0.0579	0.1436	0.0341	0.0778	0.0163	0.0652	0.0115	0.7794	0.0496	0.1536	0.0362	0.1210	0.0286
80	0.4358	0.0610	0.1644	0.0381	0.0875	0.0182	0.0723	0.0128	0.8575	0.0506	0.1759	0.0403	0.1382	0.0321
90	0.4815	0.0637	0.1853	0.0419	0.0972	0.0202	0.0793	0.0140	0.9331	0.0514	0.1982	0.0443	0.1555	0.0355
100	0.5255	0.0660	0.2062	0.0456	0.1069	0.0221	0.0863	0.0152	1.0068	0.0521	0.2206	0.0481	0.1729	0.0389
110	0.5682	0.0680	0.2272	0.0492	0.1166	0.0239	0.0932	0.0163	1.0788	0.0528	0.2429	0.0516	0.1904	0.0422
120	0.6095	0.0698	0.2481	0.0526	0.1263	0.0258	0.1000	0.0175	1.1494	0.0534	0.2651	0.0551	0.2079	0.0453
130	0.6496	0.0714	0.2690	0.0559	0.1360	0.0277	0.1069	0.0187	1.2187	0.0540	0.2873	0.0583	0.2255	0.0484
140	0.6886	0.0729	0.2898	0.0590	0.1457	0.0295	0.1137	0.0198	1.2869	0.0545	0.3093	0.0614	0.2431	0.0515
150	0.7266	0.0742	0.3106	0.0620	0.1555	0.0313	0.1205	0.0210	1.3542	0.0550	0.3313	0.0644	0.2607	0.0544
160	0.7637	0.0754	0.3313	0.0649	0.1653	0.0331	0.1272	0.0221	1.4207	0.0555	0.3531	0.0672	0.2784	0.0572
170	0.7999	0.0766	0.3519	0.0677	0.1751	0.0349	0.1340	0.0233	1.4864	0.0560	0.3748	0.0699	0.2960	0.0600
180	0.8354	0.0776	0.3724	0.0704	0.1849	0.0367	0.1407	0.0244	1.5515	0.0564	0.3963	0.0725	0.3137	0.0627
190	0.8701	0.0786	0.3928	0.0729	0.1948	0.0385	0.1475	0.0255	1.6160	0.0568	0.4177	0.0750	0.3313	0.0654
200	0.9042	0.0795	0.4131	0.0754	0.2046	0.0402	0.1542	0.0266	1.6800	0.0572	0.4390	0.0774	0.3489	0.0679
220	0.9705	0.0811	0.4534	0.0801	0.2245	0.0437	0.1676	0.0288	1.8066	0.0580	0.4810	0.0819	0.3840	0.0729
240	1.0346	0.0826	0.4932	0.0845	0.2443	0.0471	0.1810	0.0310	1.9319	0.0587	0.5225	0.0861	0.4190	0.0775
260	1.0967	0.0839	0.5326	0.0886	0.2643	0.0505	0.1945	0.0332	2.0560	0.0593	0.5633	0.0899	0.4538	0.0820
280	1.1572	0.0851	0.5715	0.0924	0.2843	0.0538	0.2079	0.0354	2.1794	0.0600	0.6036	0.0935	0.4884	0.0862
300	1.2162	0.0861	0.6099	0.0960	0.3044	0.0571	0.2213	0.0375	2.3021	0.0606	0.6432	0.0968	0.5227	0.0902
320	1.2738	0.0871	0.6478	0.0993	0.3246	0.0603	0.2348	0.0396	2.4244	0.0612	0.6823	0.0999	0.5569	0.0941
340	1.3302	0.0880	0.6852	0.1025	0.3448	0.0635	0.2483	0.0418	2.5465	0.0617	0.7208	0.1029	0.5908	0.0977
360	1.3855	0.0889	0.7222	0.1055	0.3650	0.0666	0.2617	0.0439	2.6685	0.0622	0.7588	0.1056	0.6244	0.1012
380	1.4397	0.0897	0.7587	0.1083	0.3853	0.0697	0.2752	0.0460	2.7905	0.0629	0.7962	0.1082	0.6578	0.1046
400	1.4931	0.0904	0.7948	0.1110	0.4056	0.0727	0.2888	0.0480	2.9126	0.0634	0.8331	0.1106	0.6909	0.1078
420	1.5456	0.0911	0.8304	0.1135	0.4260	0.0757	0.3023	0.0501	3.0349	0.0638	0.8696	0.1129	0.7238	0.1108
440	1.5974	0.0917	0.8656	0.1159	0.4464	0.0786	0.3159	0.0522	3.1576	0.0643	0.9055	0.1151	0.7564	0.1138
460	1.6484	0.0924	0.9004	0.1182	0.4667	0.0815	0.3294	0.0542	3.2806	0.0646	0.9410	0.1172	0.7887	0.1166
480	1.6987	0.0930	0.9348	0.1204	0.4871	0.0844	0.3430	0.0562	3.4041	0.0653	0.9761	0.1192	0.8208	0.1193
500	1.7485	0.0936	0.9689	0.1225	0.5075	0.0872	0.3566	0.0582	3.5282	0.0655	1.0107	0.1211	0.8526	0.1219

Table 9. PMMA. Projected range R_p and straggling ΔR_p for implantation of various ions into PMMA.

E keV	B		P		As		Sb		H		Si		Ar	
	R_p µm	ΔR_p µm	R_p µm	ΔR_p µm	R_p µm	ΔR_p µm	R_p µm	ΔR_p µm	R_p µm	ΔR_p µm	R_p µm	ΔR_p µm	R_p µm	ΔR_p µm
10	0.0540	0.0171	0.0213	0.0063	0.0158	0.0037	0.0161	0.0032	0.1917	0.0385	0.0222	0.0067	0.0192	0.0054
20	0.1113	0.0294	0.0397	0.0114	0.0261	0.0060	0.0248	0.0048	0.3184	0.0450	0.0419	0.0122	0.0349	0.0096
30	0.1678	0.0386	0.0583	0.0163	0.0358	0.0082	0.0325	0.0062	0.4244	0.0485	0.0619	0.0174	0.0504	0.0135
40	0.2222	0.0458	0.0772	0.0209	0.0451	0.0102	0.0397	0.0076	0.5194	0.0509	0.0822	0.0224	0.0659	0.0174
50	0.2743	0.0514	0.0963	0.0255	0.0542	0.0123	0.0467	0.0089	0.6071	0.0527	0.1029	0.0272	0.0817	0.0212
60	0.3241	0.0560	0.1156	0.0298	0.0633	0.0142	0.0535	0.0101	0.6895	0.0541	0.1237	0.0318	0.0975	0.0249
70	0.3718	0.0599	0.1351	0.0340	0.0724	0.0162	0.0601	0.0114	0.7680	0.0553	0.1446	0.0361	0.1136	0.0285
80	0.4177	0.0632	0.1547	0.0380	0.0814	0.0181	0.0667	0.0126	0.8432	0.0564	0.1656	0.0403	0.1297	0.0320
90	0.4619	0.0660	0.1744	0.0419	0.0904	0.0200	0.0732	0.0138	0.9158	0.0573	0.1867	0.0443	0.1460	0.0355
100	0.5045	0.0685	0.1941	0.0456	0.0994	0.0219	0.0796	0.0150	0.9863	0.0581	0.2078	0.0480	0.1623	0.0388
110	0.5457	0.0706	0.2138	0.0492	0.1084	0.0238	0.0860	0.0162	1.0548	0.0588	0.2288	0.0516	0.1787	0.0421
120	0.5857	0.0726	0.2335	0.0526	0.1175	0.0256	0.0923	0.0173	1.1218	0.0595	0.2498	0.0551	0.1952	0.0453
130	0.6246	0.0743	0.2532	0.0559	0.1265	0.0275	0.0986	0.0185	1.1873	0.0601	0.2707	0.0583	0.2117	0.0484
140	0.6623	0.0759	0.2729	0.0590	0.1356	0.0293	0.1049	0.0196	1.2517	0.0607	0.2915	0.0615	0.2283	0.0514
150	0.6992	0.0773	0.2924	0.0620	0.1447	0.0311	0.1112	0.0208	1.3149	0.0612	0.3122	0.0645	0.2449	0.0543
160	0.7351	0.0786	0.3120	0.0650	0.1538	0.0329	0.1175	0.0219	1.3772	0.0617	0.3328	0.0673	0.2614	0.0572
170	0.7702	0.0798	0.3314	0.0678	0.1629	0.0347	0.1237	0.0230	1.4386	0.0622	0.3533	0.0701	0.2780	0.0600
180	0.8045	0.0809	0.3508	0.0705	0.1721	0.0365	0.1300	0.0241	1.4993	0.0626	0.3737	0.0727	0.2946	0.0627
190	0.8382	0.0820	0.3700	0.0731	0.1812	0.0383	0.1362	0.0253	1.5593	0.0630	0.3939	0.0752	0.3112	0.0654
200	0.8712	0.0829	0.3892	0.0756	0.1904	0.0400	0.1424	0.0264	1.6186	0.0634	0.4140	0.0776	0.3278	0.0680
220	0.9353	0.0847	0.4272	0.0803	0.2089	0.0435	0.1548	0.0286	1.7357	0.0642	0.4538	0.0822	0.3608	0.0729
240	0.9974	0.0862	0.4649	0.0847	0.2274	0.0469	0.1673	0.0307	1.8510	0.0649	0.4930	0.0863	0.3937	0.0776
260	1.0575	0.0876	0.5020	0.0888	0.2460	0.0503	0.1797	0.0329	1.9648	0.0655	0.5316	0.0902	0.4264	0.0821
280	1.1159	0.0889	0.5387	0.0927	0.2647	0.0536	0.1921	0.0351	2.0775	0.0662	0.5697	0.0939	0.4590	0.0864
300	1.1728	0.0900	0.5750	0.0963	0.2834	0.0569	0.2046	0.0372	2.1891	0.0667	0.6072	0.0973	0.4914	0.0904
320	1.2284	0.0911	0.6109	0.0997	0.3022	0.0601	0.2170	0.0393	2.3000	0.0673	0.6442	0.1004	0.5235	0.0943
340	1.2828	0.0920	0.6462	0.1029	0.3210	0.0633	0.2295	0.0414	2.4103	0.0679	0.6806	0.1034	0.5554	0.0980
360	1.3360	0.0929	0.6812	0.1060	0.3399	0.0664	0.2420	0.0435	2.5201	0.0684	0.7166	0.1062	0.5871	0.1015
380	1.3882	0.0938	0.7157	0.1088	0.3588	0.0695	0.2545	0.0456	2.6296	0.0688	0.7520	0.1088	0.6186	0.1049
400	1.4394	0.0945	0.7498	0.1115	0.3778	0.0726	0.2670	0.0477	2.7389	0.0693	0.7870	0.1113	0.6498	0.1081
420	1.4899	0.0953	0.7835	0.1141	0.3968	0.0755	0.2795	0.0498	2.8481	0.0699	0.8215	0.1136	0.6807	0.1112
440	1.5395	0.0960	0.8168	0.1165	0.4158	0.0785	0.2921	0.0518	2.9572	0.0704	0.8556	0.1158	0.7115	0.1142
460	1.5883	0.0966	0.8498	0.1189	0.4348	0.0814	0.3047	0.0538	3.0664	0.0708	0.8892	0.1179	0.7419	0.1170
480	1.6365	0.0972	0.8823	0.1211	0.4538	0.0843	0.3173	0.0559	3.1757	0.0711	0.9224	0.1199	0.7722	0.1198
500	1.6841	0.0978	0.9145	0.1232	0.4728	0.0871	0.3299	0.0579	3.2852	0.0715	0.9552	0.1219	0.8022	0.1224

Table 10. AZ 1350. Projected range R_p and straggling ΔR_p for implantation of various ions into photoresist AZ 1350.

E keV	B		P		As		Sb		Si		Ar		H	
	R_p µm	ΔR_p µm	R_p µm	ΔR_p µm	R_p µm	ΔR_p µm	R_p µm	ΔR_p µm	R_p µm	ΔR_p µm	R_p µm	ΔR_p µm	R_p µm	ΔR_p µm
10	0.0518	0.0229	0.0201	0.0072	0.0150	0.0039	0.0152	0.0032	0.0210	0.0078	0.0183	0.0060	0.2193	0.0538
20	0.1100	0.0398	0.0381	0.0131	0.0250	0.0064	0.0236	0.0049	0.0402	0.0142	0.0335	0.0107	0.3681	0.0641
30	0.1687	0.0527	0.0565	0.0188	0.0344	0.0087	0.0310	0.0065	0.0600	0.0203	0.0488	0.0153	0.4887	0.0694
40	0.2260	0.0629	0.0753	0.0243	0.0435	0.0110	0.0381	0.0079	0.0803	0.0263	0.0643	0.0197	0.5935	0.0728
50	0.2812	0.0711	0.0945	0.0296	0.0525	0.0132	0.0449	0.0093	0.1010	0.0319	0.0800	0.0240	0.6878	0.0752
60	0.3344	0.0779	0.1140	0.0347	0.0615	0.0154	0.0515	0.0107	0.1220	0.0374	0.0959	0.0283	0.7746	0.0771
70	0.3856	0.0836	0.1337	0.0396	0.0704	0.0175	0.0580	0.0120	0.1433	0.0426	0.1121	0.0325	0.8557	0.0786
80	0.4350	0.0885	0.1536	0.0444	0.0794	0.0197	0.0644	0.0133	0.1647	0.0475	0.1284	0.0365	0.9324	0.0799
90	0.4826	0.0928	0.1736	0.0490	0.0884	0.0210	0.0708	0.0146	0.1862	0.0523	0.1449	0.0405	1.0054	0.0809
100	0.5287	0.0966	0.1937	0.0534	0.0973	0.0239	0.0772	0.0159	0.2077	0.0568	0.1614	0.0444	1.0753	0.0819
110	0.5733	0.0999	0.2139	0.0576	0.1064	0.0260	0.0834	0.0172	0.2293	0.0611	0.1782	0.0482	1.1427	0.0827
120	0.6166	0.1029	0.2341	0.0617	0.1154	0.0280	0.0897	0.0184	0.2508	0.0652	0.1949	0.0519	1.2080	0.0834
130	0.6587	0.1056	0.2543	0.0656	0.1244	0.0301	0.0960	0.0197	0.2723	0.0691	0.2118	0.0556	1.2714	0.0841
140	0.6997	0.1080	0.2745	0.0693	0.1335	0.0321	0.1022	0.0209	0.2937	0.0729	0.2287	0.0591	1.3331	0.0847
150	0.7396	0.1102	0.2946	0.0729	0.1427	0.0342	0.1084	0.0222	0.3151	0.0765	0.2457	0.0625	1.3935	0.0853
160	0.7785	0.1123	0.3147	0.0764	0.1518	0.0362	0.1146	0.0234	0.3363	0.0799	0.2627	0.0659	1.4525	0.0858
170	0.8166	0.1142	0.3348	0.0797	0.1610	0.0382	0.1208	0.0246	0.3575	0.0832	0.2797	0.0691	1.5105	0.0862
180	0.8538	0.1159	0.3547	0.0829	0.1702	0.0402	0.1270	0.0258	0.3785	0.0864	0.2967	0.0723	1.5675	0.0867
190	0.8902	0.1176	0.3746	0.0860	0.1795	0.0422	0.1332	0.0271	0.3994	0.0894	0.3137	0.0754	1.6236	0.0871
200	0.9259	0.1191	0.3944	0.0890	0.1888	0.0441	0.1394	0.0283	0.4202	0.0923	0.3307	0.0784	1.6789	0.0875
220	0.9952	0.1218	0.4337	0.0947	0.2074	0.0480	0.1518	0.0307	0.4614	0.0978	0.3647	0.0842	1.7874	0.0882
240	1.0620	0.1242	0.4726	0.1000	0.2261	0.0519	0.1642	0.0331	0.5020	0.1028	0.3986	0.0898	1.8937	0.0889
260	1.1267	0.1264	0.5111	0.1049	0.2450	0.0557	0.1766	0.0355	0.5421	0.1075	0.4323	0.0950	1.9980	0.0895
280	1.1893	0.1283	0.5492	0.1095	0.2639	0.0594	0.1890	0.0378	0.5816	0.1119	0.4659	0.1000	2.1008	0.0901
300	1.2502	0.1301	0.5868	0.1138	0.2829	0.0631	0.2015	0.0402	0.6206	0.1160	0.4993	0.1048	2.2024	0.0906
320	1.3094	0.1317	0.6239	0.1179	0.3020	0.0668	0.2139	0.0425	0.6590	0.1198	0.5325	0.1093	2.3029	0.0911
340	1.3671	0.1332	0.6606	0.1217	0.3212	0.0704	0.2265	0.0448	0.6969	0.1234	0.5655	0.1136	2.4026	0.0916
360	1.4235	0.1346	0.6969	0.1254	0.3404	0.0739	0.2390	0.0471	0.7343	0.1268	0.5983	0.1177	2.5017	0.0921
380	1.4786	0.1358	0.7328	0.1288	0.3597	0.0774	0.2515	0.0495	0.7711	0.1300	0.6308	0.1217	2.6003	0.0924
400	1.5325	0.1370	0.7602	0.1320	0.3790	0.0809	0.2641	0.0517	0.8075	0.1330	0.6630	0.1255	2.6985	0.0930
420	1.5853	0.1381	0.8032	0.1351	0.3983	0.0843	0.2767	0.0540	0.8434	0.1358	0.6951	0.1291	2.7964	0.0934
440	1.6371	0.1391	0.8378	0.1380	0.4177	0.0876	0.2894	0.0563	0.8789	0.1385	0.7269	0.1326	2.8942	0.0938
460	1.6879	0.1401	0.8720	0.1408	0.4371	0.0909	0.3021	0.0585	0.9130	0.1410	0.7584	0.1359	2.9919	0.0941
480	1.7379	0.1410	0.9058	0.1434	0.4565	0.0941	0.3147	0.0608	0.9484	0.1434	0.7897	0.1391	3.0896	0.0944
500	1.7870	0.1419	0.9393	0.1460	0.4760	0.0973	0.3275	0.0630	0.9825	0.1458	0.8208	0.1422	3.1873	0.0947

6.1.4.2.7 Evaluation of ion implanted layers

Table 11. Comparison of widely used measuring techniques [78R1].

Measuring technique	Measured quantity	Measuring depth *) μm	Diameter of measured point *) μm	Accuracy *) at best cm^{-3}	Accuracy *) dynamic range	Lateral resolution μm
CV-measurement	doping concentration	$10^{-2}\cdots10$ (1)	$100\cdots10^4$	10^{13}	$1:10^4$	
Sheet resistivity	resistivity	(3)	$20\cdots2\cdot10^4$ (4)	10^{12}	$1:10^8$	
Spreading resistance	resistivity	several μm	≈10	10^{12}	$1:10^8$	
Hall effect	doping concentration, mobility	(3)	$20\cdots2\cdot10^4$ (6)	10^{12}	$1:10^8$	
Rutherford-back-scattering	distribution of implanted atoms; radiation defects; localization of implanted atoms	several μm	$10\cdots10^4$	$10^{18}\cdots10^{20}$ (7)	$1:10^4$	$10\cdots10^4$
Nuclear reaction	distribution and localization of implanted atoms	$\approx1\cdots10$	$10\cdots10^4$	10^{15} (8)	(8)	$10\cdots10^4$

*) Figures in parentheses refer to the column "Remarks".

The measurement techniques used for the evaluation of ion implanted layers in principle are the same as those used in all other semiconductor applications. However, ion implanted layers have some particular properties, which give rise to the following problems:

a) Ion implanted layers are extremely thin. The extension of ion implanted doping profiles usually is much less than 1 μm, so that high local resolution is needed.

b) The gradient of the doping profile is extremely steep. In some cases, the Debye-length limits the free carrier gradient.

c) On account of the complex annealing behaviour of ion implanted layers, one cannot assume the doping profile and the profile of the carrier concentration to be identical.

The most important measurement techniques used in ion implantation can be grouped into two categories:

Measurement of the carrier concentration:
Most suited are van der Pauw-Hall effect measurements combined with layer removal and measurements of capacitance voltage-curves (CV-measurements). Spreading resistance measurements also yield information on carrier concentration profiles.

Measurement of doping concentration:
Suitable are backscattering techniques, SIMS, and activation analysis combined with layer removal techniques.

The different methods are compared in Table 11. In many cases, only a combination of several methods can give reliable results, particularly when complex defect structures are created during the anneal. A review of measurement techniques is given in [82R1].

Table 11 from left page (continued)

Depth profiling *)		Identification of implanted species *)	Typical time of measurement	Reproducibility %	Remarks
resolution µm	range µm				
$10^{-2}\cdots10$ (1)	$10^{-1}\cdots10^{3}$ (2)		1···10 min	±2	(1) depending on Debye-length; (2) depending on substrate doping; also in combination with layer removal techniques
≈0.01 (5)	$10^{-1}\cdots10^{5}$ (5)		1 min; a profile takes several h	±2	(3) thickness of layer; (4) with dedicated structures 20 µm, otherwise 2 mm; (5) in combination with layer removal techniques
	$10^{-1}\cdots10^{5}$		10 min for homogeneous material; 20 min for depth profile (without the time necessary for the preparation of the wedge)	±5	depth profile by measuring wedged samples; profile obtained by complicated calculation with an error of approx. ±20%
≈0.01 (5)	$10^{-1}\cdots10^{5}$ (5)		10 min; profile several h	±5	(6) dedicated structures required
0.3	several µm	good (7)	10 min; measurement of radiation defects and localization of lattice sites 1 h	±10	does not damage the sample; (7) measuring accuracy depends on mass of implanted atoms
$10^{-2}\cdots3$ (9)	1···10	good (8)	1 h		(8) depending on reaction cross section; (9) in combination with layer removal techniques high resolution, otherwise depending on reaction

*) Figures in parentheses refer to the column "Remarks".

Van der Pauw-Hall effect measurement

The Hall-effect is used to determine the carrier mobility and concentration in implanted layers. The most reliable results are obtained by the van der Pauw-method. For high resolution measurements, the implanted layer is contacted by four diffused regions. Current is passed through two contacts (1, 2), and in the presence of a magnetic field B, the voltage is measured across the other two contacts (3, 4).

One obtains:

sheet resistivity
$$\varrho_{\square}=\frac{\pi}{\ln 2}\cdot\frac{V_{3,4}}{I_{1,2}}, \tag{20}$$

sheet Hall coefficient
$$R_{\mathrm{H}\square}=\frac{\Delta V_{\mathrm{H}}}{2I\cdot B}, \tag{21}$$

sheet carrier concentration
$$N_{\mathrm{I,eff}}=\frac{r}{q\cdot R_{\mathrm{H}\square}}, \tag{22}$$

sheet carrier mobility
$$\mu_{\square,\mathrm{eff}}=\frac{R_{\mathrm{H}\square}}{r\cdot\varrho_{\square}}. \tag{23}$$

The Hall scattering factor r is usually assumed to be $r=1$. After the measurement, a thin layer of the semiconductor can be removed. When the van der Pauw-measurement is repeated, the carrier concentration in the removed layer can be determined from the two measurements. By repeating this process a complete carrier concentration profile is obtained. The measurement can be automatized almost completely. In this case, measurement and layer removal are performed in the same wafer holder [78R1]. After optimization of the process (electrolyte, current density, probe geometry etc.) 10 nm-layers can be reproducibly removed.

Spreading resistance

The spreading resistance method can be used for comparatively fast measurements. However, on account of the difficult physical effects that influence the evaluation of the results, only limited accuracy is obtained.

Capacitance-voltage measurements

The CV-method is the classical method for all low-dose MOS-implantations. A pn-junction or an isolating oxide is necessary on top of the layer that is to be measured. By applying a sufficient voltage of correct polarity at the gate on top of the semiconductor, the semiconductor is depleted of mobile carriers to depth x. By varying the gate voltage, the width of the space charge region can be controlled. Using the Poisson equation and the capacitor formula one obtains:

$$N(x) - \frac{C^3}{q \varepsilon_0 \varepsilon_r A^2} \left(\frac{\mathrm{d}C}{\mathrm{d}V}\right)^{-1} \tag{24}$$

and

$$x - \frac{\varepsilon_0 \varepsilon_r A}{C}. \tag{25}$$

This calculation is valid if the Debye-length can be neglected. Under the assumption of a Gaussian implantation profile, a critical implantation dose N_c can be calculated

$$N_c = 100 \cdot \frac{K T \varepsilon_0 \varepsilon_r}{q^2 \Delta R_p^2}. \tag{26}$$

When this dose is exceeded, the doping concentration and the majority carrier concentration will differ by more than one percent [78R1].

Backscattering

A method particularly adapted to profile measurements of ion implanted layers is the measurement of the energy loss of backscattered light ions. A high energy beam (>200 keV) of H^+ ions is collimated by a series of apertures, so that the beam divergence is less than 0.1° at a beam current of some nA. When this beam hits the semiconductor surface, ions are backscattered from the surface and from atoms deep in the crystal. By measuring the energy of the backscattered ions, the depth, where the ions were backscattered, can be determined (scatter events at the surface give the highest energy). After measuring a large number of ions and sorting them with respect to their energy with the help of a multichannel analyzer, a "backscattering spectrum" is obtained (Fig. 23). When the primary ion beam is aligned parallel to a low index crystal orientation, channeling occurs (cf. section 6.1.2.1). In this case an interstitial atom will be hit with a much enhanced probability. By turning the crystal and investigating different low index axes, the exact location of an interstitial atom can be determined (Fig. 24).

These measurements, often referred to as Rutherford backscattering (RBS), are easy to perform when the implanted impurity atom is much heavier than the host lattice (e.g. As in Si). Great difficulties are encountered, when very light atoms are to be investigated (e.g. B in Si).

Backscattering techniques are the subject of many publications. The most comprehensive review is found in [78C3].

Secondary ion mass spectroscopy (SIMS)

The sputter effect of ion implantation (cf. section 6.1.4.2.1) can be intentionally used to investigate implanted layers. A low energy ion beam (e.g. Ar^+) removes atoms of the implanted layer which are mass analyzed. When the removal rate is constant, a depth profile of the implanted impurity can be recorded. In order to obtain reproducible results, it is important that the sputtering process is uniform. A number of peripheral effects, e.g. changes in the sputtering coefficient of the silicon due to orientation etc., or changes in the ion yield have to be accounted for. A detailed review on SIMS techniques is given in [82W1].

Activation analysis

In many cases, implanted atoms can be activated by irradiation with thermal neutrons. By measuring the decay products of the activated ions and in combination with a layer removal technique, doping profiles can be obtained. In Table 12 the most important radioactive processes are listed. Boron distributions can be measured by using the (n, α-)reaction of the ^{10}B isotope.

$$^{10}B + n \rightarrow {}^{7}Li(859\ keV) + \alpha(1471\ keV), \quad (27)$$

$$^{10}B + n \rightarrow {}^{7}Li(1010\ keV) + \alpha(1777\ keV). \quad (28)$$

The α-particles emitted have a very small energy spread. By measuring the energy distribution of the α-particles emitted from a neutron-activated B^+ implanted sample, the ^{10}B-distribution in the semiconductor can be determined from the energy loss [73C3, 73M1, 75M4].

Table 12. Nuclear reactions suited for profile measurements. β^-: electron, β^+: positron, γ: γ-particle, IT: internal conversion, EC: electron capture.

Reaction	Half life	Type of reaction	main energy keV	Decay end product	Activation cross section $10^{-24}\ cm^2$	Decay rate *) $s^{-1}\ mg^{-1}$
$^{69}Ga(n,\gamma)^{70}Ga$	20 m	β^-	1650	Ge	1.4	$1.85 \cdot 10^8$
		γ	170			
$^{71}Ga(n,\gamma)^{72}Ga$	14.1 h	β^-	960	Ge	5.0	$1.52 \cdot 10^8$
		γ	630, 840			
$^{69}Ga(n,2n)^{68}Ga$	68 m	β^+	1900	Zn	0.55	$2.8 \cdot 10^4$
		γ	810···1880			
$^{69}Ga(n,p)^{69}Zn$	14 h	IT	440	Ga	0.024	$1.1 \cdot 10^2$
$^{71}Ga(n,p)^{71}Zn$	3.9 h	β^-	1500	Ga	0.0005	$2.6 \cdot 10^1$
		γ	390, 490, 610			
$^{69}Ga(n,\alpha)^{66}Cu$	5.1 m	β^-	2630	Zn	0.105	$4.0 \cdot 10^3$
		γ	830, 1040			
$^{71}Ga(n,\alpha)^{68}Cu$	30 s	β^-	3500	Zn		
		γ	810···1880			
$^{75}As(n,\gamma)^{76}As$	26.5 h	β^-	2970	Se	4.5	$2.0 \cdot 10^8$
		γ	560···2700			
$^{75}As(n,2n)^{74}As$	18 d	β^+	900	Ge	0.55	$2.8 \cdot 10^4$
		β^-	1360	Se		
		γ	600, 640···2530			
$^{75}As(n,p)^{75}Ge$	82 m	β^-	1200	As	0.118	$5.0 \cdot 10^1$
		γ	70···630			
$^{75}As(n,\alpha)^{72}Ga$	14.1 h	β^-	960	Ge	0.123	$5.2 \cdot 10^1$
		γ	630, 840			
		γ	560···2700			
$^{197}Au(n,\gamma)^{198}Au$	2.7 d	β^-	1370	Hg	98.8	$5.7 \cdot 10^{12}$
		γ	410			
$^{50}Cr(n,\gamma)^{51}Cr$	27.8 d	γ	320	Mn	17	$2.8 \cdot 10^7$
$^{70}Ge(n,\gamma)^{71}Ge$	11.4 d	EC		As	3.2	$4.6 \cdot 10^7$
$^{113}In(n,\gamma)^{114}In$	50.1 d	IT		Sn	8	$2.5 \cdot 10^7$
		γ	190			
$^{31}P(n,\gamma)^{32}P$	14.3 d	β^-	1710	S	0.19	$2.3 \cdot 10^7$
$^{30}Si(n,\gamma)^{31}Si$	2.7 h	β^-	1470	P	0.4	$1.85 \cdot 10^6$
$^{64}Zn(n,\gamma)^{65}Zn$	245 d	EC		Ga	0.46	$8.1 \cdot 10^5$
		γ	1110			

*) After 12 hours irradiation with fast neutrons of a flux density of $2 \cdot 10^{10}\ cm^{-2}s^{-1}$ and thermal neutrons of a flux density of $2.5 \cdot 10^{13}\ cm^{-2}s^{-1}$.

6.1.4.2.8 Implantation apparatus

Table 13. Data survey of commercial implantation equipment. Only machines used in large scale semiconductor device production are included in the table. Implanters used in research and development are more versatile. The data vary to the specific design; see text).

Manufacturer	Type of machine	Source type	Analysis	Scan system	Wafer size and chamber capacity		Max. beam currents	Energy range	Target chamber type
					∅ mm	pieces			
Eaton (Nova)	high current	Freeman with integral vaporizer	magnetic	fully mechanical scan	75 100 125 150	18 13 10 8	$^{11}B^+$: 4.0 mA $^{31}P^+$: 12.5 mA $^{75}As^+$: 12.5 mA $^{121}Sb^+$: 6.0 mA	20···80 keV variable extraction energy	spinning disc single end station with dual discs
Varian-Extrion		Freeman with integral vaporizer	magnetic	magnetic and angle corrector	50 75 100 125 150	25, 50 25, 50 25 13 10	$^{11}B^+$: 5 mA $^{31}P^+$: 12.5 mA $^{75}As^+$: 12.5 mA $^{121}Sb^+$: 6 mA	10···80 keV and 10···120 keV variable	spinning disc with dual end station
Balzers		Rene Bernas gas and solid	magnetic	hybrid electrostat. 1000 Hz mech. 120 rpm	125 100 75	24 26 52	B: 2 mA P, As: 4 mA Sb: 1.5 mA	40···200 keV full current 10···30 keV reduced current	carousel with wafers on the inner side
Lintott (applied materials)		Freeman dual vaporizer	magnetic	fully mechanical	125 100	24 40	B: 4 mA P, As: 12.5 mA Sb: 6 mA	10···120 keV	race track single chamber
Eaton	medium current	Freeman with integral vaporizer	magnetic	electrostatic	50···150 mm fragments on special wafer holders, single wafer clamp with cassette-to-cassette continuous feed		$^{11}B^+$: 600 μA $^{31}P^+$: 1600 μA $^{75}As^+$: 1500 μA $^{121}Sb^+$: 800 μA	20···200 keV post acceleration	single wafer optional tooling for implants at high temperatures or cryogenic temperatures
Varian-Extrion		Freeman with integral vaporizer	magnetic	electrostatic	50···150 mm		$^{11}B^+$: 600 μA $^{31}P^+$: 1500 μA $^{75}As^+$: 1500 μA maximum total beam current: $^{75}As^+$: 3000 μA	10···200 keV variable	dual end station, single wafer processing

Table 13 from left page (continued)

Wafer handling	Wafer orientation	Wafer cooling	Wafer tilt	Dose measurement	Dose uniformity (standard deviation)
automatic loading optional	provided	wafers clamped to conformable heat conducting pads	0° or 7°	magnetically suppressed Faraday assembly located behind spinning disc to measure beam current pulses transmitted by precision slot in disc. Signal also used to control radial scan rate in real time. system geometry permits transport of fully neutralized beam through field-free region to wafer surface	$\sigma \leqq 0.5\%$
auto-loader cassette-to-cassette	flat orientation	1) radiation 2) partly DI-water 3) fully DI-water	0° or 7°	high precision, self-calibrating dose control	in the range $10^{12} \cdots 10^{17}$ ion/cm² $\sigma \leqq 2\%$ for all wafer sizes
automatic loading	possible by hand	good thermal contact by centrifugal force; wafer plates used as heat absorber	7°	direct measurement of carousel current; magnetic electron suppression	$\sigma \leqq 1\%$
manual	optional	heat sink plates	0°···10°	carousel and plates inside Faraday cup; magnetic suppression; direct collection of beam; vanes maintain beam constant	$\sigma \leqq 0.5\%$
automatic cassette-to-cassette gravity feed	optional	gas cooling system	0°···15° continuously adjustable	electrostatically suppressed Faraday assembly, with separate magnetically suppressed striking plate for beam set-up and machine maintenance	$\sigma \leqq 0.75\%$
vertical wafer handling system cassette-to-cassette	flat orientation optional	gas contact cooling	0° and 7° fixed	electrostatically suppressed Faraday assembly	$\sigma \leqq 0.75\%$

Ion implantation can be performed by a wide variety of machines ranging from large linear accelerators to highly sophisticated isotope separators. For the use of ion implantation as a doping technology, a standard machine configuration has been developed.

It consists of a preacceleration stage, magnetic analysis at high potential, and postacceleration to the final energy. The wafer-handling is completely automatized and the machines are designed for maximum throughput of standard Si-wafers. All parameters (ion current, beam energy etc.) are controlled to an accuracy of a few percent. The high energy power supply is stabilized to at least 10^{-5}. The vacuum is in the range of 10^{-6} mbar or better. Cryopumping is used throughout the system, with the exception of the pump at the ion source, which usually is a diffusion pump. In Table 13 of the previous pages, the machine parameters of a number of commercial implantation systems used in device technology are shown.

Modern implantation equipment is very close to the ideal of a "one button"-machine. The development in the last years has shown a steady increase in complexity of the machines, particularly due to electronic control systems.

An extensive review of implantation equipment and all related questions can be found in [73D1]. A number of conferences were devoted to implantation machines [80W5, 82R1].

Ion source

A hot-cathode Freeman-type source is used in most implanters. It produces high currents, is reliable, and easy to service (Fig. 25).

Although highly toxic, BF_3, AsH_3, and PH_3 are today's standard source gases. With strong and monitored ventilation, most hazards can be controlled. Care is necessary when servicing source or analyzer parts. They may be highly contaminated by poisonous gases and/or deposits. Properties of important source materials are listed in Table 14.

The choice of the correct source material is not easy. Because of better control of source pressure, fluid or gaseous substances are generally preferred. Complex molecules are not recommended, because they dissociate into many different ionized fragments and the useful ion current is low. The composition of the source material is important. For instance, SO_2 is not a suitable material for either S^+ or O^+ because the S^+ beam is contaminated by O_2^+ and O^+ by S^{++} which are not separated by the mass analyzer.

Table 14. Source materials used in ion implantation [73D1].

Ion	Z	M	Source material	Source temperature [°C]					Maximum ionization cross section 10^{-16} cm^2	Ionization potential eV	Remarks
				Pressure [Torr]				Other vapour pressures			
				10^{-4}	10^{-3}	10^{-2}	760				
Al	13	26.9815	Al	972	1082	1217	2056		6.2	6.0	take care to avoid metallisation of insulators
			$AlCl_3$				180.2	2 Torr at 100 °C			
Sb	51	121.75	Sb	425	475	533	1440		7.2	8.5	toxic
			Sb_2O_3		399	450	1425				
			Sb_2S_3								
			Sb_2I_3				401.0	1 Torr at 163.6 °C			
Ar	18	39.948	Ar				−185.6		2.8	15.7	no particular problems
As	33	74.9216	As	204	237	247	610		5.0	10.5	toxic. GaAs decomposes conveniently and controllably in source furnace
			AsH_3				−62.1				
			As_2O_3				457.2	1 Torr at 212.5 °C			
			Cd_3As_2		314	359					
			GaAs								
			$AsCl_3$				130.3	5 Torr at 11.4 °C			
Be	4	9.01218	Be		1395	1552			3.2	9.3	toxic (BeO)
				997	1097	1227					
			$BeCl_2$		176	209	487				
			$BeBr_2$		174	207	474				
			BeF_2		562	623					
			BeI_2		160	194	487				
B	5	10.81	B	1707	1867	2027			2.6	8.3	BCl_3 convenient, but deactivates filament (use tungsten not tantalum), avoid excess BCl_3 in discharge to minimize reaction with filament. Boron halides hydrolyze readily, clean dry apparatus must be used B_2H_6 spontaneously inflammable – not recommended. Diluted stabilized mixtures with inert gases are commercially available, but boron beam fraction low
			BF_3				−110.7				
			BCl_3				12.7				
			B_2H_6				−86.5				

continued

Table 14 (continued)

Ion	Z	M	Source material	Source temperature [°C] Pressure [Torr] 10^{-4}	10^{-3}	10^{-2}	760	Other vapour pressures	Maximum ionization cross section 10^{-16} cm^2	Ionization potential eV	Remarks
Cd	48	112.40	Cd	177	217	265	765		6.3	9.0	no particular problems
			$CdCl_2$		386	437	967				
			$CdBr_2$		341	390					
			CdI_2		289	340	796				
			CdO		691	772					
			CdS		610	686					
			CdSe		588	661					
			CdTe		505	574					
Ga	31	69.72	Ga	907	1007	1132			5.9	6.0	no problem as element
			$GaCl_3$				200	1 Torr at 48 °C			
			GaN		715	825					
			$Ga_2O_3 + CCl_4$								
Ge	32	72.59	Ge	1137	1257	1397			5.7	8.1	Element convenient but requires high temperature source; other volatile materials may be suitable. $GeCl_4$ similar to $SiCl_4$
			GeH_4				−88.9				
			$GeCl_4$				84.0	60 Torr at 16.2 °C [3]			
			GeS_2	sublimes							
Au	79	196.9665	Au	1132	1252	1397			5.9	9.2	Element convenient but requires high temperature source; volatile gold compounds are thermally unstable. Readily ionized in sputtering source
He	2	4.00260	He				−268.6		0.21	24.5	no particular problems
H	1	1.0080	H_2				−252.5		0.22	13.5	no particular operational problems; element forms explosive mixture with air. Water vapour can also be used
			H_2O				100	17.5 Torr at 20 °C			
			H_2S				−60.4				
In	49	114.82	In	742	837	947			7.7	5.8	no particular problems
			$InCl_3$		224	256					
			$InBr_3$		113	141					

continued

Table 14 (continued)

Ion	Z	M	Source material	Source temperature [°C]					Maximum ionization cross section 10^{-16} cm^2	Ionization potential eV	Remarks
				Pressure [Torr]				Other vapour pressures			
				10^{-4}	10^{-3}	10^{-2}	760				
Mg	12	24.305	Mg	327	377	439	1107		5.4	7.6	if the element is used, take care to avoid metallization of insulators
			MgF_2		1058	1158					
			$MgCl_2$		≈500						
			$MgBr_2$		482	552					
N	7	14.0067	N_2				−195.8		1.5	14.5	no particular problem
O	8	15.9994	O_2				−183.1		1.3	13.6	no particular problem. CO_2 convenient and minimizes source and filament oxidation
			CO_2								
P	15	30.9738	P (yellow)				280.0	1 Torr at 76.6 °C	4.5	10.9	PCl_3 convenient; some filament deactivation if excess vapour used. Phosphorus halides hydrolyze readily, clean dry apparatus must be used. Phosphine (toxic) can also be used with large PH^+ fraction
			P (violet)				417	1 Torr at 237 °C			
			PH_3				−87.5				
			PCl_3				74.2	60 Torr at 10.2 °C			
			PBr_3				175.3	1 Torr at 7.8 °C			
Si	14	28.086	Si	1337	1472	1632	2287		5.4	8.1	no particular problems, similar precautions as for PCl_3 (see phosphorus)
			SiH_4				−111.5				
			$SiCl_4$				56.8	100 Torr at 5.4 °C			
			SiS_2								
S	16	32.06	S	55	80	109	444.6		3.9	10.3	no particular problems, if ZnS is used Zn^{2+} may contaminate S^+, all sulphur compounds tend to corrode copper source components
			H_2S				−10.0				
			CS_2				46.5	200 Torr at 10.4 °C			
			COS				−49.9				
			S_2Cl_2				138.0	5 Torr at 15.7 °C			
			Sb_2S_5								
			CdS		610	686					
			ZnS		774	863					
Te	52	127.60	Te	280	323	374			7.1	9.0	toxic, cadmium telluride preferred if machine resolution is adequate to separate Cd^+ and Te^+
			TeF_6				−38.6				
			$TeCl_2$					100 Torr at 240 °C			
			$TeCl_4$				392	100 Torr at 306 °C			

continued

Table 14 (continued)

Ion	Z	M	Source material	Source temperature [°C]					Maximum ionization cross section 10^{-16} cm^2	Ionization potential eV	Remarks
				Pressure [Torr]				Other vapour pressures			
				10^{-4}	10^{-3}	10^{-2}	760				
Te (continued)			TeO_2		590	651					
			CdTe		505	574					
			ZnTe								
Sn	50	118.69	Sn	997	1107	1247			7.7	7.3	no particular problems; precautions for $SnCl_4$ similar to PCl_3
			$SnCl_4$				113	10 Torr at 3.5 °C			
			SnS		534	602					
			SnSe								
			SnH_4				−25.3				
Zn	30	65.37	Zn	247	292	344			4.9	9.4	no particular problems
			ZnS		774	863					
			ZnSe								
			ZnTe								
			$ZnCl_2$		273	317					

Current

Ion currents used in standard implantation systems range from some 10 nA to approx. 20 mA. The implantation machines are roughly divided into 3 categories:

a) up to approx. 50 μA: low current
b) up to approx. 1 mA: medium current
c) 1 mA and more: high current.

Measurement of the ion current is far from being trivial [82R1]. In addition to the positively charged ions of the beam itself, secondary electrons and ions from target and beam apertures have to be accounted for. Further problems arise from atoms of the residual gas colliding with the beam, causing these atoms to become ionized and neutralizing ions of the beam. In order to avoid these problems one has to observe the following points

optimum vacuum conditions

the Faraday cup for measuring the ion beam has to be as long and as wide as possible

a very careful selection and isolation of the bias voltages of the Faraday cup and apertures

apertures must not be hit by the beam

in order to avoid problems of the neutral beam, the beam line has to be set off approximately 7° before reaching the target. When using mechanical scanning, neutral beam effects are distributed homogeneously across the wafer.

Beam analysis

In most machines, a sector magnet is used for analyzing the beam. By shaping the pole pieces in a suitable way or by utilizing movable pole pieces the fringing fields can be used for imaging and focussing.

Usually the current through the magnet can be controlled and monitored at the control desk. In this case, mass spectra can be obtained by plotting the ion current on the y- and magnet current on the x-axis of an x−y plotter. If the value x on the x-axis is proportional to the current I through the magnet and the magnetisation is not saturated, the following relation holds:

$$\sqrt{\frac{M}{n}} = \text{const} \cdot x, \qquad (29)$$

where M is the mass number of the ion recorded at position x and n its charge state.

The constant is specific for each implantation apparatus. It depends on the preacceleration voltage and can be determined by applying equation (29) to well known ion groups, e.g. $^{11}B/^{10}B$. By using equation (29) any unknown ion in the mass spectrum can be identified. In Fig. 26, the natural isotope distribution of the elements is shown as it appears at infinite resolution.

The mass resolution $M/\Delta M$ of a commercial implanter is at least $M/\Delta M = 100$, i.e. mass 100 and 101 can just be discriminated. Care is necessary when using a low intensity ion beam, which is in close vicinity of a mass with high beam intensity. As the mass spectrum has only finite mass resolution, an implantation with a low intensity beam can be heavily contaminated by ions of a neighbouring mass.

An important factor in ion implantation is the beam purity. Apart from Aston-bands, that are discussed below, every ion beam contains impurities that stem from collisions with atoms of the residual gas, from apertures, from source materials such as the filament etc.These contaminations are dragged along by the electric field of the beam. Usually not all of them are diverted in the mass separator.

From a practical point of view, the most critical beam contamination is the cross contamination. When after a high dose implantation (e.g. As^+ for the source-drain implantation of a MOS transistor) a small dose of another dopant (e.g. B^+ for the threshold voltage shift) is implanted, the B^+-beam may contain several percent of As^+. As these ions have only been dragged along, they have very low kinetic energy. The best way to get rid of this beam contamination is a *thin screening oxide* by which the contaminant ions are stopped.

In many ion sources, ions with higher charge states are also produced, i.e. not only As^+ but also As^{++}, As^{+++} etc. Usually the ratio of single, double and triple charged ions is approximately 1:10 and 1:100. Reasonable currents can usually only be obtained for doubly charged ions. When using a doubly charged ion beam, care has to be taken to avoid contamination by "Aston-bands". When e.g. the vacuum in the preacceleration field is poor, charge exchange processes occur between ions and molecules of the same species, e.g.

$$As_2 + As^{++} \rightarrow As_2^+ + As^+.$$

In the mass spectrum, the As_2^+ shows up at exactly the same position of the mass spectrum as As^{++}, but its components have only 1/4 of the energy of the As^{++} ion. Implantations with Aston-bands show a doping concentration exhibiting a second peak closer to the surface.

An Aston-band will appear at a mass position M′ after an inelastic collision changing both mass and charge of the ion:

$$M_1^{n+} \rightarrow M_2^{m+}, \tag{30}$$

$$M' = \frac{M_2^2}{M_1} \cdot \frac{n}{m^2}. \tag{31}$$

The presence of Aston-bands can be detected by varying the pressure in the preacceleration region. If the ion current at the interesting mass is not dependent on pressure, the absence of the Aston-band contamination is likely.

Target chamber

One of the most difficult problems in ion implantation is the wafer handling. Two systems have become widely accepted [82R1]:

a) For small and medium implantation doses each wafer is implanted separately. Cassette systems are used. Transport and implantation of the wafer without using holders is also possible. In both cases, the wafer can be implanted at a definite angle to the beam in order to avoid channeling. Heating of the wafer has to be kept to a minimum because local overheating, mechanical stress or even rupture and mechanical failures may occur on account of poor heat dissipation.

b) For medium and high doses, implantation in batches is used. Ferris-wheel, carousels and rotating disk systems are available. In order to spread the beam power over a large area. mechanical scanning is necessary. In addition, the disk itself can be water-cooled. In high-throughput machines either two target chambers, which are alternatively used for implantation and pumping, or high speed pumping systems are provided.

Scanning

The equipment available today shows a wide variety of scanning systems. For low and medium current applications, electrostatic scanning, and for high current implantations, mechanical scanning is used. Although electrostatic scanning seems to be much simpler and more adapt to implantation systems, it has two major drawbacks:

a) One has to be extremely careful that no standing (Lissajous') pattern is obtained by incidental coupling of the x- and y-scanners or by coupling one scanner with plasma oscillation of the source or the high voltage power supply [80G2, 82R1].

b) Scanning the beam across the wafer changes the angle between the surface normal and the beam continuously. Channeling directions may be hit and the concentration profile, therefore, may be locally altered.

Adjustments and process control

Mechanical adjustment

Although many machines are laser-adjusted during installation, an additional check of the beam position after installation is necessary, because the optical axis and the ion beam axis usually do not coincide. The beam position on actual wafers can be checked conveniently by using silicon on sapphire (SOS-)wafers. These wafers have the same size as standard Si-wafers, but are transparent. After implantation of a dose $N_I \approx 10^{14}\ \mathrm{cm}^{-2}$, the thin silicon film becomes opaque. By applying only x-scanning followed by only y-scanning, a cross pattern is written on the SOS-wafer, from which the exact position of the beam as well as the beam scan height and width can be determined. The SOS-wafer can be used again after annealing [80B1].

Dose adjustment

Several implantation machines are usually installed at one implantation facility. Because implantations on these machines have to be identical, some sort of "reference implantation" has to be devised. For implantation into Si, Si-wafers covered with e.g. 120 nm-SiO_2 (MOS-quality) are best suited. The shift of the threshold voltage or the flatband voltage shift [75R3···5] is measured after implanting one half of the wafer with approximately $5 \cdot 10^{11}\ \mathrm{cm}^{-2}$ B^+ or P^+ and annealing at 900 °C in N_2 and 500 °C in H_2. The dose actually implanted can be determined with good accuracy by using equation (18).

Process control

As all processes in modern Si-technology, ion implantation has to be controlled continuously. One method is described in the previous section. In order to obtain good local resolution, it is necessary to measure several 10^3 to 10^4 points on one wafer. The time-consuming and difficult measurement of threshold voltage shift is better substituted by measuring sheet resistivity with the help of van der Pauw structures. By applying automatic measuring techniques, a spatial resolution of approximately 20 μm can be obtained.

References for 6.1.4.2

61L1 Lehovec, K., Slobodskoy, A.: Solid State Electron. **3** (1961) 45.
61R1 Rourke, F.M., Sheffield, J.L., White, F.A.: Rev. Sci. Instrum. **32** (1961) 455.
62C1 Carter, G., Colligon, J.S., Leck, J.H.: Proc. Phys. Soc. **79** (1962) 299.
62I1 Irvin, J.C.: Bell Syst. Tech. J. **41** (1962) 387.
63L1 Lindhard, J., Scharff, M., Schiott, H.E.: K. Dan. Vidensk. Selsk., Mat. Fys. Medd. **33** (1963) No. 14.
64K1 Kato, T.K., Nishi, Y.: Jpn. J. Appl. Phys. **3** (1964) 377.
65K1 King, W.J., Burrel, J.T., Harrison, S., Martin, F., Kellett, C.M.: Nucl. Instrum. Methods **38** (1965) 178.
65L1 Lindhard, J.: K. Dan. Vidensk. Selskab. Mat. Fys. Medd. **34** (1965) No. 14.
65M1 McCaldin, J.O.: Nucl. Instrum. Methods **38** (1965) 153.
65M2 Manchester, K.E., Silbey, C.B., Alton, G.: Nucl. Instrum. Methods **38** (1965) 169.
66D1 Davies, J.A., Jespergard, P.: Can. J. Phys. **44** (1966) 1631.
66G1 Gusev, V.M., Titov, V.V., Guseva, M.I., Kurimiyi, V.I.: Sov. Phys. Solid State **7** (1966) 1673.
66K1 Kellett, C.M., King, W.J., Leith, F.A.: US Air Force Report AF 19 (628)-4970 AD 635267 (1966).
66M1 Martin, F.W., King, W.J., Harrison, S.: IEEE Trans. Nucl. Sci. **NS-13** (1966) 22.
66P1 Pavlov, P.V., Vasil'yev, K., Zorin, E.I., Tetelbaum, I., Tulovchikov, V.S., Chigirinskaya, T.Yu.: Izv. S.S.S.R. Radioelektron. **13** (1966) 493.
66S1 Schiott, H.E.: K. Dan. Vidensk. Selsk. Mat. Fys. Medd. **35** (1966).
67C1 Chen, W.H., Chen, W.I.: J. Electrochem. Soc. **114** (1967) 1297.
67D1 Davies, J.A., Denhartog, J., Eriksson, L., Mayer, J.W.: Can. J. Phys. **45** (1967) 4053.
67G1 Proc. Int. Conf. Appl. Ion Beams Semiconductor Technology (Ed. Glotin, P.) Editions OPHRYS, Gap (France), (1967).
67G2 Glotin, P., Grapa, J., Monfret, H.: In: [67G1], 619.
67G3 Gibbons, J.F.: In: [67G1], 561.
67K1 Kerr, J.A., Large, L.N.: In: [67G1], 601.
67M1 Mayer, J.W., Marsh, O.J., Shifrin, G.A., Baron, R.: Can. J. Phys. **45** (1967) 4073.
67M2 Mayer, J.W., Davies, J.A., Eriksson, E.: Appl. Phys. Lett. **11** (1967) 365.
67M3 Marsh, O.J., Mayer, J.W., Shifrin, G.A.: In: [67G1], 513.
67P1 Pavlov, P.V., Tetelbaum, D.I., Zorin, E.I., Alekseev, V.I.: Sov. Phys. Solid State **8** (1967) 2141.
68A1 Alton, G.D., Love, L.O.: Can. J. Phys. **46** (1968) 695.
68B1 Björkquist, K., Domeij, B., Eriksson, L., Fladda, G., Fontell, A., Mayer, J.W.: Appl. Phys. Lett. **13** (1968) 379.
68D1 Dearnaley, G., Freeman, J.H., Gard, G.A., Wilkins, M.A.: Can. J. Phys. **4** (1968) 587.
68G1 Galaktionova, I.A., Gusev, V.M., Naumenko, V.G., Titov, V.V.: Sov. Phys. Semicond. **2** (1968) 656.
68H1 Hu, S.M.: J. Appl. Phys. **39** (1968) 3844.
68K1 Kennedy, D.P., Murley, P.C., Kleinfelder, W.: IBM J. Res. Dev. **12** (1968) 399.
68R1 Roosild, S., Dolan, R., Brickmanan, B.: J. Electrochem. Soc. **115** (1968) 307.
68S1 Seeger, A., Chik, K.P.: Phys. Status Solidi **29** (1968) 455.
68T1 Thompson, H.W.: Philos. Mag. **18** (1968) 377.
69B1 Baron, R., Shifrin, G.A., Marsh, O.J., Mayer, J.W.: J. Appl. Phys. **40** (1969) 3702.
69C1 Crowder, B.L., Morehead, jr., F.F.: Appl. Phys. Lett. **14** (1969) 313.
69D1 Dearnaley, G., Wilkins, M.A., Goode, P.D., Freeman, J.H., Gard, G.A.: report AERE-R 6179, **1969**.
69E1 Eriksson, L., Davies, J.A., Johansson, N.G.E., Mayer, J.W.: J. Appl. Phys. **40** (1969) 842.
69F1 Fistul', V.J.: Heavily Doped Semiconductors. New York (1969).
69F2 Fairfield, J.M., Crowder, B.L.: Trans-Met. Soc. AIME **24** (1969) 469.
69J1 Johnson, W.S.: Thesis SU-SEL-69-014. Stanford University, **1969**.
69J2 Jones, K.C., Stevens, P.R.C.: Electron. Lett. **5** (1969) 499.
69N1 Nelson, D.G., Gibbons, J.F., Johnson, W.S.: Appl. Phys. Lett **15** (1969) 246.
69S1 Sigmund, P.: Phys. Rev. **184** (1969) 383.
69V1 Vook, F.L., Stein, H.J., Borders, J.A.: Appl. Phys. Lett. **14** (1969) 328.
70B1 Blamires, N.G.: In: [70M1], 52.
70B2 Bicknell, R.W., Allen, R.M.: Radiat. Eff. **6** (1970) 45.
70C1 Crowder, B.L., Fairfield, J.M.: Electrochem. Soc. **112** (1970) 363.
70D1 Davies, D.E.: Solid State Electron. **13** (1970) 229.
70G1 Gamo, K., Masuda, K., Namba, S., Ishihara, S., Kimura, I.: Appl. Phys. Lett. **17** (1970) 391.
70J1 Johansson, N.G.E., Mayer, J.W.: Solid State Electron. **13** (1970) 123.

70M1 European Conference on Ion Implantation (Ed. Mitchell, E.W.J) Reading (1970) Stevenage, England.
70M2 Mayer, J.W., Erikson, L., Davies, J.A.: Ion Implantation in Semiconductors. New York (1970).
70N1 North, J.C., Gibson, W.M.: Appl. Phys. Lett. **16** (1970) 126.
70N2 Nelson, R.S., Cairns, J.A., Blamires, N.: Radiat. Eff. **6** (1970) 131.
70S1 Schneider, I., Marrone, M., Kabler, M.N.: Appl. Opt. **9** (1970) 1163.
70S2 Sigurd, D., Fladd, G., Erikson, L., Björkquist, K.: Radiat. Eff. **3** (1970) 145.
70S3 Shannon, J.M., Ford, R.A., Gard, G.A.: Radiat. Eff. **6** (1970) 217.
70T1 Tsuchimoto, T., Tokuyama, T.: Radiat. Eff. **6** (1970) 121.
70T2 Tinsley, A.W., Jones, K.C.: In: [70M1], 187.
70V1 Vook, F.L., Stein, H.J.: Radiat. Eff. **6** (1970) 11.
71C1 Chiu, T.L., Ghosh, H.N.: IBM J. Res. Dev. **15** (1971) 472.
71C2 Crowder, B.L.: J. Electrochem. Soc. **118** (1971) 943.
71E1 Ion Implantation (Eds. Eisen, F.H., Chadderton, C.S.) London (1971).
71E2 EerNisse, E.P.: J. Appl. Phys. **42** (1971) 480.
71G1 Gamo, K., Iwaki, H., Masuda, K., Namba, S., Ishihara, S., Kimura, I.: In: [71R1], 459.
71G2 Gibbons, J.F.: Proc. IEEE **60** (1971) 1062.
71G3 Goode, P.D., Wilkins, M.A., Dearnaley, G.: In: [71E1], 187.
71H1 Herzer, H., Kalbitzer, S.: In: [71R1], 307.
71M1 Mazzio, J.: Ion Implantation: A Selective Bibliography. Sandia Laboratories, Albuquerque, N. Mex. NTIS Report No. SC-B-71 0148.
71M2 Moline, R.A.: J. Appl. Phys. **42** (1971) 3553.
71R1 Ion Implantation in Semiconductors (Eds. Ruge, I.; Graul, J.) Berlin-Heidelberg-New York, Springer **1971**.
71S1 Seidel, T.E.: In: [71R1], 47.
71S2 Stephen, J., Grimshaw, G.A.: Radiat. Eff. **7** (1971) 73.
71T1 Tinsley, A.W., Grant. W.A., Carter, G., Nobles, M.J.: in: [71R1], 199.
72A1 Akasaka, Y., Horie, K., Kawazu, S.: Appl. Phys. Lett. **21** (1972) 128.
72B1 Buck, T.M., Pickar, K.A., Poate, I.M., Hsieh, C.M.: Appl. Phys. Lett. **21** (1972) 485.
72C1 Carter, G., Baruch, J.N., Grant, W.A.: Radiat. Eff. **16** (1972) 107.
72F1 Furukawa, S., Ishihara, H.: J. Appl. Phys. **43** (1972) 1268.
72F2 Furukawa, S., Matsumura, H., Ishiwara, H.: Jpn. J. Appl. Phys. **11** (1972) 134.
72G1 Gibbons, J.F.: In: [72N1], 79.
72G2 Gibbons, J.F.: Proc. IEEE **60** (1972) 1062.
72G3 Gerasimenko, N.N., Dvurechensky, A.V., Romanov, S.I., Smirnov, L.S.: Radiation Damage and Defects in Semiconductors, Conf. Series No. **16**, Inst. of Phys., London and Bristol (1972).
72M1 Morgan, R., Greenhalgh, K.R.: Ion Implantation: A Bibliography. AERE Harwell, England, AERE-Bib-176 (1972).
72M2 Minear, R.L., Nelson, D.G., Gibbons, J.F.: J. Appl. Phys. **43** (1972).
72M3 Müller, H., Ryssel, H., Schmid, K.: J. Appl. Phys. **43** (1972) 2006.
72N1 Proc. US-Japan Seminar Ion Implantation in Semiconductors (Ed. Namba, S.) Kyoto (August 1971) Jap. Soc. for the Promotion of Science, **1972**.
72O1 Ohmura, Y., Mimura, S., Kanazawa, M., Abe, T., Konaka, M.: Radiat. Eff. **15** (1972).
72R1 Reddi, V.G.K., Yu, A.Y.C.: Solid State Technol. **15** (1972) 35.
72R2 Reddi, V.G.K., Sansbury, J.D.: Appl. Phys. Lett. **20** (1972) 30.
72S1 Sigmund, P.: Rev. Roum. Phys. **17** (1972) 823, 969, 1079.
72T1 Tarui, Y., Komiga, Y., Teshima, H., Takahashi, R.: In: [72N1].
72V1 Vook, F.L.: Radiation Damage and Defects in Semiconductors. Conf. Series No. **16**, Inst. of Physics, London and Bristol (1972).
72W1 Whitton, I.L., Carter, G., Baruch, J.N., Grant, W.A.: Radiat. Eff. **16** (1972) 101.
72Y1 Yoshihiro, N., Ikeda, T., Tamura, M., Tokuyama, T., Tsuchimoto, T.: Proc. 3rd Conf. on Solid State Devices, Tokyo (1971) Suppl. Oyo Buturi **41** (1972) 225.
73A1 Allen, W.G., Atkinson, C.: Solid State Electron. **16** (1973) 1283.
73B1 Baldo, E., Cappellani, F., Restelli, G.: Radiat. Eff. **19** (1973) 271.
73B2 Blood, P., Dearnaley, G., Wilkins, M.A.: In: [73C1], 75.
73B3 Buck, T.M., Poate, J.M., Pickar, K.A., Hsieh, C.M.: Surf. Sci. **35** (1973) 362.
73C1 Ion Implantation in Semiconductors and other Materials (Ed. Crowder, B.L.) New York (1973).
73C2 Cass, T.C., Reddi, V.G.K.: Appl. Phys. Lett. **23** (1973) 268.
73C3 Crowder, B.L., Ziegler, J.F., Cole, G.W.: In: [73C1].

73D1 Dearnaley, G., Freeman, J.H., Nelson, R.S., Stephen, J.: Ion Implantation, North-Holland Publ. Comp. Amsterdam (1973).
73F1 Furukawa, S., Matsumura, H.: Appl. Phys. Lett. **22** (1973) 97.
73F2 Fair, R.B., Weber, G.R.: J. Appl. Phys. **44** (1973) 273.
73H1 Hofker, W.K., Werner, H.W., Oosthoek, D.P., de Grefte, H.A.: in: [73C1], 133.
73H2 Hsieh, C.M., Mathews, J.R., Seidel, H.D., Pickar, K.A., Drum, C.M.: Appl. Phys. Lett. **22** (1973) 238.
73I1 Ishiwara, H., Furukawa, S., Matsumura, H., Natsuaki, N.: Jpn. J. Appl. Phys. **42** (1973) 124.
73I2 Iwaki, M., Gamo, K., Masuda, K., Namba, S., Ishihara, S., Kimura, I.: In: [73C1], 111.
73K1 Krimmel, E.F., Pfleiderer, H.: Radiat. Eff. **19** (1973) 83.
73K2 Kostka, A., Kalbitzer, S.: Radiat. Eff. **19** (1973) 77.
73M1 Mezey, G., Szökefalvi-Nagy, Z., Badinka, Cs.: Thin Solid Films **19** (1973) 173.
73O1 Okabayashi, H., Shinoda, D.: J. Appl. Phys. **44** (1973) 4220.
73P1 Perloff, D.S.: J. Electrochem. Soc. **120** (1973) 1135.
73P2 Poate, J.M., Seidel, T.E.: in: [73C1], 317.
73R1 Ryssel, H.: Thesis, TU München (1973).
73R2 Ryssel, H., Müller, H., Schmid, K., Ruge, I.: in: [73C1], 215.
73S1 Seager, D.K.: Ion Implantation: A Bibliography. Sandia Laboratories, Albuquerque, N.Mex. NTIS Report No. SC-B-71048.
73S2 Shaw, D. (Ed.): Atomic Diffusion in Semiconductors. London (1973).
73S3 Schwettmann, F.N.: Appl. Phys. Lett. **22** (1973) 570.
73S4 Seidel, T.E., Meek, R.L.: in: [73C1], 305.
73V1 Vook, F.L.: Radiation Damage and Defects in Semiconductors (Ed. Whitehouse, J.E.). Inst. of Phys. Conf. Ser. **16** (1973) 60.
73W1 Wilson, R.G., Brewer, G.R.: Ion Beams, New York (1973).
74A1 Anderson, G.W., Davey, J.E., Comas, J., Saks, N.S., Lucke, W.H.: J. Appl. Phys. **45** (1974) 4528.
74B1 Blood, P., Dearnaley, G., Wilkins, M.A.: J. Appl. Phys. **45** (1974).
74B2 Blood, P., Dearnaley, G., Wilkins, M.A.: Radiat. Eff. **21** (1974) 245.
74C1 Chu, W.K., Müller, H., Mayer, J.W., Sigmon, T.N.: Appl. Phys. Lett. **25** (1974) 297.
74C2 Cembali, F., Galloni, R., Mousty, F., Rosa, R., Zagnani, F.: Radiat. Eff. **21** (1974) 255.
74D1 Douglas, E.C., Dingwall, A.G.F.: IEEE Trans. Electron Devices **ED-21** (1974) 324.
74F1 Fair, R.B.: Solid State Electron. **17** (1974) 17.
74H1 Hofker, W.K., Werner, H.W., Oosthoek, D.P., Koeman, N.J.: Appl. Phys. **4** (1974) 125.
74M1 Müller, H., Kranz, H., Ryssel, H., Schmid, K.: Appl. Phys. **4** (1974) 115.
74M2 Michel, A.E., Fang, F.F., Pan, E.S.: J. Appl. Phys. **45** (1974) 2991.
74P1 Application of Ion Beam to Metals (Eds. Picraux, S.T., EerNisse, E.P., Vook, F.L.) New York (1974).
74P2 Pan, E., Fang, F.F.: J. Appl. Phys. **45** (1974) 2801.
74P3 Prince, J.L., Schwettmann, F.N.: J. Electrochem. Soc. **121** (1974) 705.
74P4 Pompon, J.P., Grob, J.J., Stuck, R., Burger, P., Siffert, P.: in: [74P1], 420.
74R1 Proc. Int. Conference on Ion Implantation in Semiconductors (Eds. Rzewuski, H., Fiderkiewicz, A., Werner, Z., Tom, M., Zaki, C., Lada, A.) Inst. of Nuclear Research, Swierk, Polen (1974).
74S1 Schroen, W.: in: [74E1], 235.
74S2 Schneider, J.: Diplomarbeit, TU München (1974).
74S3 Schwettmann, F.N.: J. Appl. Phys. **45** (1974) 1919.
74S4 Schmid, K., Kranz, H., Ryssel, H., Müller, W., Dathe, J.: Phys. Status Solidi (a) **23** (1974) 323.
74T1 Tsai, J.C.C., Marabito, J.M.: Surf. Sci. **44** (1974) 247.
74Z1 Zelevinskaya, V.M., Kachurin, G.A., Pridachin, N.B.: Sov. Phys. Semicond. **8** (1974) 252.
75A1 Andersen, H.H., Bay, H.L.: J. Appl. Phys. **46** (1975) 1919.
75B1 Bourgoin, J.C., Corbett, J.W.: in: [75H1], 149.
75B2 Baruch, P., Monnier, J., Blanchard, B., Castaing, C.: Appl. Phys. Lett. **26** (1975) 77.
75B3 Barz, A., Rosinski, W., Jurov, A., Wielinski, L., Wojtowiez-Natanson, B.: Phys. Status Solidi (a) **27** (1975) K65.
75B4 Baranova, E.C., Gusev, V.M., Martynenko, Y.U., Kaibullin, I.B.: Radiat. Eff. **25** (1975) 157.
75C1 Cembali, F., Galloni, R., Zignani, Z.: Radiat. Eff. **26** (1975) 61.
75D1 Dearnaley, G., Gard, G.A., Temple, W., Wilkins, M.A.: Appl. Phys. Lett. **27** (1975) 17.
75D2 Davies, J.A., Foti, G., Howe, L.M., Mitchell, J.B., Winterbon, K.B.: Phys. Rev. Lett. **34** (1975) 1441.
75D3 Drum, C.M.: Electrochem. Soc. Meeting (1975).
75F1 Fritsche, C.R., Rothemund, W.: Appl. Phys. **7** (1975) 39.
75F2 Fair, R.B., Tsai, J.C.C.: J. Electrochem. Soc. **122** (1975) 1689.

75F3 Fair, R.B.: J. Electrochem. Soc. **122** (1975) 800.
75G1 Goetzberger, A.: Int. Electron Devices Meeting, Washington (1975).
75G2 Gibbons, J.F., Johnson, W.S., Mylroie, S.M.: Projected Range Statistics, 2nd edition, Dowden, Hutchinson and Ross Inc., (1975) Stroudsbourg, USA.
75H1 Hidetoshi, N., Takashi, S., Isuneo, F.: Fujitsu Sci. Tech. J. **11** (1975) 113.
75H2 Hofker, W.K.: Philips Res. Rep. Suppl. **8** (1975).
75H3 Hofker, W.K., Oosthoek, D.P., Koeman, N.J., de Grefte, H.A.M.: Radiat. Eff. **24** (1975) 223.
75H4 Holmén, G., Burén, A., Högberg, P.: Radiat. Eff. **24** (1975) 51.
75H5 Holmén, G., Burén, A.: Radiat. Eff. **24** (1975) 45.
75H6 Holmén, G.: Radiat. Eff. **24** (1975) 7.
75H7 Hasegawa, S., Shimizu, T.: J. Phys. Soc. Jpn. **38** (1975) 766.
75I1 Ishihara, H., Furukawa, S., Yamada, J., Kawamura, M.: in: [75N1], 423.
75J1 Jain, R.K., Van Overstraeten, R.J.: J. Electrochem. Soc. **122** (1975) 552.
75K1 Kimerling, L.C., Poate, J.M.: in: [75H1], 126.
75M1 Moline, R.A., Reutlinger, G.W., North, J.C.: in: Atomic Collisions in Solids **1** (Eds. Datz, J., Appleton, B.R., Meak, C.D.) New York (1975).
75M2 Moline, R.A., Cullis, A.G.: Appl. Phys. Lett. **26** (1975) 551.
75M3 Müller, H., Gyulai, J., Chu, W.K., Mayer, J.W., Sigmon, T.W.: J. Electrochem. Soc. **122** (1975) 1234.
75M4 Müller, K., Henkelmann, R., Boroffka, H.: Nucl. Instrum. Methods **128** (1975) 417.
75N1 Ion Implantation in Semiconductors (Ed. Namba, S.) New York (1975).
75O1 Ohmura, Y., Koike, K.: Appl. Phys. Lett. **26** (1975) 221.
75O2 Ohmura, Y., Koike, K., Kobayashi, H.: in: [75N1], 183.
75O3 Ohdomari, I., Itoh, T.: Jpn. J. Appl. Phys. **11** (1975) 458.
75P1 Plunkett, J.C., Stone, J.L.: A Selected Bibliography on Ion Implantation in Solid State Technology. Solid State Technol. (1975)
75P2 Prussin, S., Fern, A.M.: J. Electrochem. Soc. **122** (1975) 830.
75P3 Prussin, S.: in: [75N1], 449.
75R1 Ryssel, H., Schmid, K., Ruge, I.: in: [75N1], 169.
75R2 Ryssel, H., Kranz, H.: Appl. Phys. **7** (1975) 11.
75R3 Runge, H.: in: [75N1], 703.
75R4 Runge, H.: Appl. Phys. **8** (1975) 43.
75R5 Runge, H.: Phys. Status Solidi. (a) **30** (1975) 147.
75S1 Satya, A.V.S., Palanki, H.R.: in: [75N1], 405.
75S2 Sigmon, T.W., Chu, W.K., Müller, H., Mayer, J.W.: in: [75N1], 633.
75S3 Sigmon, T.W., Chu, W.K., Müller, H., Mayer, J.W.: Appl. Phys. **5** (1975) 347.
75V1 Vaidyanathan, K.V., Chatterjee, P.K., Streetman, B.G.: Appl. Phys. Lett. **27** (1975) 648.
75W1 Winterbon, K.B.: Ion Implantation Range and Energy Deposition Distributions **2** (1975) Plenum Press New York.
75W2 Wittmaack, K., Schulz, F., Hietl, B.: in: [75N1], 193.
75W3 Wang, K.L., Gray, P.V.: IEEE ED (June 1975) 353.
76A1 Anand, K.V., Sobhy, M.I., El-Dhaher, A.H.G.: Int. J. Electron. **40** (1976) 169.
76C1 Application of Ion Beams to Materials 1975 (Eds. Carter, G., Colligon, J.S., Grant, W.A.) Inst. of Phys. Conf. Ser. **28** (1976).
76C2 Csepregi, L., Chu, W.K., Muller, H., Mayer, J.W., Sigmon, T.W.: Radiat. Eff. **28** (1976) 227.
76D1 Desalvo, A., Rosa, R.: Radiat. Eff. **31** (1976) 41.
76G1 Proc. Int. Conference on Ion Implantation in Semiconductors (Ed. Gyulai, J.) Ungar. Acad. d. Wissensch., Budapest (1976).
76G2 Grant, W.A., Williams, J.S., Dodds, D.: Radiat. Eff. **29** (1976) 189.
76G3 Goetzberger, A., Bartelink, D.J., McVittie, J.P., Gibbons, J.F.: Appl. Phys. Lett. **29** (1976) 259.
76M1 Mader, S., Michel, A.: J. Vac. Sci. Technol. **13** (1976) 391.
76M2 Mayer, J.W., Csepregi, L., Gyulai, J., Nagy, T., Mezey, G., Revesz, P., Kotai, E.: Thin Solid Films **32** (1976) 303.
76M3 Matsumura, H., Furukawa, S.: Jpn. J. Appl. Phys. **14** (1976) 1983.
76M4 Matsumura, H., Furukawa, S.: J. Appl. Phys. **47** (1976) 1746.
76M5 Miyao, M., Natsuaki, N., Yoshihiro, N., Tamura, M., Tokuyama, T.: Jpn. J. Appl. Phys. **15** (1976) 57.
76R1 Runge, H.: Electron. Eng. (1976) 41.
76S1 Sigmon, T.W., Csepregi, L., Mayer, J.W.: J. Electrochem. Soc. **123** (1976) 1116.
77A2 Axmann, A., Schulz, M., Fritzsche, C.R.: Appl. Phys. **12** (1977) 175.

77B1 Blamires, N.G., Smith, B.J.: J. Phys. D: Appl. Phys. **10** (1977) 799.
77B2 Beanland, D.G.: in: [77C1], 31.
77B3 Burkhardt, F., Wagner, C.: Phys. Status Solidi (a) **39** (1977) K63.
77C1 Ion Implantation in Semiconductors and other Materials (Eds. Chernov, F., Borders, J.A., Brice, D.W.) New York (1977).
77E1 Evwaraye, A.O.: J. Appl. Phys. **48** (1977) 1840.
77F1 Freeman, J.H., Booker, D.V.: Nucl. Instrum. Methods **144** (1977) 144.
77F2 Fritzsche, C.R.: Appl. Phys. **12** (1977) 347.
77G1 Gerasimenko, N.N.: Radiation Effects in Semiconductors. Inst. Phys. Conf. Ser. **31**, Bristol-London (1977) 164.
77G2 Geipel, H.J., Tice, W.K.: Appl. Phys. Lett. **30** (1977) 325.
77G3 Galloni, R., Pedulli, L., Zignani, F.: Radiat. Eff. **32** (1977) 223.
77H1 Huez, M., Cappellani, F., Restelli, G.: Radiat. Eff. **32** (1977) 147.
77M1 Mitsuishi, T., Sasaki, Y., Asami, H.: Jpn. J. Appl. Phys. **16** (1977) 367.
77N1 Nelson, R.S.: Radiat. Eff. **32** (1977) 19.
77R1 Runge, H.: Phys. Status Solidi (a) **39** (1977) 595.
77R2 Ryssel, H., Kranz, H., Müller, K., Henkelmann, R.A., Biersack, J.: Appl. Phys. Lett. **30** (1977) 399.
77S1 Shimizu, R., Kang, S.T., Koshikawa, T., Ogata, H., Kanayama, K., Ogata, Y., Akasaka, Y., Horie, K.: J. Appl. Phys. **48** (1977) 1745.
77S2 Seidel, T.E., Payne, R.S., Moline, R.A., Costello, W.R., Tsai, J.C.C., Gardner, K.R.: IEEE Transactions on Electron Devices **ED-24** (1977) 717.
77S3 Smirnov, L.S.: Physical Process in Irradiated Semiconductors. ("Nauka", Novosibirsk, 1977).
77S4 Smith, B.: Ion implantation range data for silicon and germanium device technologies. Learned Information (Europe) Ltd., Oxford (1977).
77W1 Winterbon, K.B.: Appl. Phys. Lett. **31** (1977) 649.
77W2 Wilson R.G., Dunlap, H.L., Jamba, D.M.: Proceedings of the 3rd International Symposium on Silicon Materials Science and Technology. Proceedings Vol. **77-2**, The Electrochemical Society, Inc., Princeton (1977).
77W3 Williams, M.M.R.: Radiat. Eff. **31** (1977) 233.
77Z1 Zaremba, A.J., Marcyk, G.T., Streetman, B.G.: IEEE Trans. **ED-24** (1977) 163.
78A1 Antoniadis, D.A., Hansen, S.E., Dutton, R.W.: SUPREM II - A Program for IC Process Modelling and Simulation. Technical Report No. 5019-2, Stanford University, Stanford, **1978**.
78B1 Bilyus, K.I., Pranyavichyus, L.I.: Sov. Phys. Semicond. **12** (1978) 801.
78C1 Christodoulides, C.E., Baragiola, R.A., Chivers, D., Grant, W.A., Williams, J.S.: Radiat. Eff. **36** (1978) 73.
78C2 Cembali, F., Dori, L., Galloni. R., Servidori, M., Zignani, F.: Radiat. Eff. **36** (1978) 111.
78C3 Chu, W.K., Mayer, J.W., Nicolet, M.A.: Backscattering Spectroscopy, Academic Press New York, **1978**.
78D1 Dvurechensky, A.V., Kachurin, G.A., Antonenko, A.Kh.: Radiat. Eff. **37** (1978) 179.
78F1 Fischer, G., Carter, G., Webb, R.: Radiat. Eff. **38** (1978) 41.
78G1 Grötzschel, R., Klabes, R., Kreissig, U., Schmidt, M.: Radiat. Eff. **36** (1978) 129.
78K1 Kawata, H., Sato, T., Sakurai, T.: Fujitsu Sci. Tech. J. **14** (1978) 83.
78K2 Krimmel, E.F., Oppolzer, H., Runge, H.: Rev. Phys. **13** (1978) 791.
78M1 More, R.M., Venskytis, F.J.: Radiat. Eff. **35** (1978) 129.
78M2 Marcovich, A., Bahir, G., Bernstein, T., Kalish, R.: Radiat. Eff. **39** (1978) 65.
78M3 Mende, G.: Phys. Status Solidi (a) **45** (1978) K121.
78O1 Ohmura, Y., Inoue, T.: J. Appl. Phys. **49** (1978) 3597.
78R1 Ryssel, H., Ruge, I.: Ionenimplantation, Stuttgart (1978) Teubner-Verlag.
78R2 Runge, H., Wörl, H.: Phys. Status Solidi (a) **45** (1978) 509.
78S1 Seshan, K., Washburn, J.: Radiat. Eff. **37** (1978) 147.
78S2 Suski, J., Krynicki, J., Rzewuski, H., Gyulai, J., Loferski, J.J.: Radiat. Eff. **35** (1978) 13.
78T1 Tsai, M.Y., Streetman, B.G., Williams, P., Evans, Jr., C.A.: Appl. Phys. Lett. **32** (1978) 144.
78W1 Winterbon, K.B.: Radiat. Eff. **39** (1978) 31.
78W2 Winterbon, K.B., Sanders, J.B.: Radiat. Eff. **39** (1978) 39.
78W3 Wagner, C., Burkhardt, F.: Phys. Status Solidi (a) **47** (1978) 131.
78Y1 Yudin, V.V.: Appl. Phys. **15** (1978) 223.
79G1 Geipel, H.J., Fortino, A.G.: IEEE Trans. Solid-State circuits and Electron Devices **SC-14** (1979) 430.
79H1 Hirao, T., Inoue, K., Yaegashi, Y., Takayanagi, S.: Jpn. J. Appl. Phys. **18** (1979) 647.
79I1 Ishitani, T., Tamura, H.: Radiat. Eff. Lett. **43** (1979) 149.

79I2 Inoue, K., Hirao, T., Yeagashi, Y., Takayanagi, S.: Jpn. J. Appl. Phys. **18** (1979) 367.
79L1 Lee, Hee Yong: Radiat. Eff. Lett. **43** (1979) 161.
79S1 Sakurai, T., Kawata, H., Sato, T., Hisatsuga, T., Hashimoto, H., Furuya, T.: J. Appl. Phys. **50** (1979) 1287.
79S2 Seidowski, E., Mamaev, G.I., Ledovskoj , W.P.: Krist. Tech. **14** (1979) 825.
79T1 Tsai, M.Y., Streetman, B.G.: J. Appl. Phys. **50** (1979) 183.
79T2 Tsai, M.Y., Day, D.S., Streetman, B.G.: J. Appl. Phys. **50** (1979) 188.
79W1 Wada, Y., Hashimoto, N.: J. Appl. Phys. **50** (1979) 6257.
79W2 Wada, Y., Hashimoto, N.: J. Appl. Phys. **50** (1979) 5720.
79Y1 Yagi, K., Oyu, K., Tamura, M., Tokuyama, T.: Appl. Phys. Lett **35** (1979) 275.
80A1 Alestig, G., Holmén, G., Martenson, M., Peterström, S.: Radiat. Eff. **49** (1980) 7.
80A2 Anderson, C.L., Celler, G.K., Rozgonyi, G.A.: Laser and electron beam processing of electronic materials. The Electrochemical Society, Proc. Vol. **80-1**, Princeton, N.J. (1980).
80B1 Badalec, R., Runge, H.: in: [80W5] 91.
80C1 Comas, J., Wilson, R.G.: J. Appl. Phys. **51** (1980) 3697.
80C2 Chu, W.K., Sullivan, M.J., Ku, S.M., Shatzkes, M.: Radiat. Eff. **47** (1980) 7.
80C3 Chu, W.K., Kastl, R.H., Murley, P.C.: Radiat. Eff. **47** (1980) 1.
80D1 Desalvo, A., Galloni, R., Rosa, R., Zignani, F.: J. Appl. Phys. **51** (1980) 1994.
80G1 Gyulai, J., Lohner, T.: Radiat. Eff. **47**, **48**, **49** (1980).
80G2 Glawischnig, H.: in: [80W5] 60.
80G3 Gibson, W.N., Teitelbaum, H.: Ion Solid Interaction. Institute of Electrical Engineers, Hersts, England, (1980).
80H1 Hirao, T., Fuse, G., Inoue, K., Takayanagi, S.T., Yaegashi, Y., Ichikawa, S., Izumi, T.: J. Appl. Phys. **51** (1980) 262.
80H2 Hirao, T., Inoue, K., Fuse, G., Takayanagi, S., Yaegashi, Y.: Radiat. Eff. **47** (1980) 99.
80I1 Iafrate, G.I., Ziegler, J.F., Nass, M.J.: J. Appl. Phys. **51** (1980) 984.
80K1 Kalbitzer, S., Oetzmann, H.: Radiat. Eff. **47** (1980) 57.
80K2 Kótai, E., Nagy, T., Meyer, O., Gyulai, J., Révész, P., Mezey, G., Lohner, T., Manuaba, A.: Radiat. Eff. **47** (1980) 27.
80M1 Mandurah, M.M., Saraswat, K.C., Kamins, T.I.: Appl. Phys. Lett. **36** (1980) 683.
80M2 Myers, D.R., Wilson, R.G.: Radiat. Eff. **47** (1980) 91.
80M3 Miyao, M., Yoshihiro, N., Tokuyama, T.: Radiat. Eff. **47** (1980) 85.
80R1 Ryssel, H.: Private Communication.
80S1 Sigmon, T.W., Deline, V.R., Evans, Jr., C.A., Katz, W.M.: J. Electrochem. Soc.: Solid State Sci. Technol. **127** (1980) 981.
80S2 Scovell, P.D., Young, J.M.: Electron. Lett. **16** (1980) 614.
80T1 Tsai, M.Y., Morehead, F.F., Baglin, J.E.E., Michel, A.E.: J. Appl. Phys. **51** (1980) 3230.
80W1 Winterbon, K.B.: Radiat. Eff. **48** (1980) 97.
80W2 Winterbon, K.B.: Radiat. Eff. **46** (1980) 181.
80W3 Wilson, R.G.: Radiat. Eff. **46** (1980) 141.
80W4 Winterbon, K.B.: Radiat. Eff. **46** (1980) 15.
80W5 Wilson, I.H., Stephens, K.G.: Inst. Phys. Conf. Ser. **54**, Bristol - London (1980).
80Z1 Ziegler, J.F., Iafrate, G.J.: Radiat. Eff. **46** (1980) 199.
82R1 Ryssel, H., Glawischnig, H.: Ion Implantation Techniques: Springer Verlag Berlin **1982**.
82W1 Wittmaack, K.: Vacuum **32** (1982) 65.

6.1.4.3 Nuclear transmutation doping

6.1.4.3.0 Introduction

High power devices, e.g. thyristors for high voltage D.C. transmission, require silicon crystals having a phosphorus doping within tight limits and good uniformity.

Conventional doping methods, e.g. the adding of phosphorus before or during the growth of the silicon single crystal lead to a nonuniform distribution of the dopant in the silicon crystal [74b1]. (See sections 6.1.3.2 and 6.1.3.3.). Macroscopic variations of the resistivity are observed along the axis of a silicon rod (Fig. 1) and across a silicon wafer (Fig. 3a). The deviation from the average resistivity is of the order of ±20%. Additional microscopic variations of the resistivity are superimposed on the macroscopic variations at a period of about 100 μm and an amplitude of about ±25% (Figs. 2 and 4a).

Phosphorus doping of Si independent of the growth conditions of the crystal is possible by the nuclear transmutation doping (NTD-doping). In NTD, the dopant is generated in the crystal volume after the growth of the single crystal by the aid of neutron irradiation [51l1, 61T1, 71K1, 73S1, 74S1, 75H1, 76H1, 76H2, 79m1, 81g1]. This doping method leads to a very exact target value of doping and to a homogeneous dopant distribution (Figs. 3b and 4b).

NTD doping of Ge is so far only of scientific interest [60F1].

6.1.4.3.1 Nuclear transmutation

Silicon consists of the three isotopes which are present in the following natural abundances [81s2]: ^{28}Si (92.23%), ^{29}Si (4.67%), ^{30}Si (3.10%).

By the capture of thermal neutrons, the isotopes ^{29}Si, ^{30}Si, and ^{31}Si are formed. The isotopes ^{29}Si and ^{30}Si are stable in contrast to the isotope ^{31}Si which decays to the stable phosphorus ^{31}P by emission of β^- particles:

$$^{30}\mathrm{Si}(n,\gamma)^{31}\mathrm{Si} \xrightarrow{2.6\,\mathrm{h}} {}^{31}\mathrm{P}+\beta^-.$$

Because of the low absorption cross section of the neutron reaction ($\sigma_{abs}=0.16$ barn [81s2]), the distribution of the transmutation-grown phosphorus is nearly independent of the diameter and the size of the silicon ingot. The distribution of the phosphorus generated only depends on the spatial distribution of the neutron flux in the irradiation facility during the irradiation time [76H1, 76H2].

The concentration of the phosphorus generated depends linearly on the neutron dose and can be adjusted accurately.

6.1.4.3.2 Radiation defects

During the neutron irradiation, lattice defects are generated in the silicon crystals [61b1, 61C1, 63W1, 69W1, 67B1, 70L1, 71C1, 75N1, 76T1, 77T1, 78G1]. Defects are caused by fast neutrons, γ-radiation present in the reactor, γ-radiation emitted during the reaction of the neutrons, and γ- and β^--radiation originating from the decay of the intermediate silicon isotope ^{31}Si to stable phosphorus ^{31}P.

A list of defects identified in neutron-irradiated float-zone silicon containing a low concentration of oxygen and carbon after 30 min. N_2-anneal by the aid of deep-level transient spectroscopy (DLTS) is shown in Table 1.

Defects may be completely annealed by a suitable temperature treatment [61T1, 77H1, 77S1, 78Y1, 79C1], e.g. by an annealing for about 1 hour at 750 °C. All the defects not due to fast neutron damage can be annealed at temperatures lower than 750 °C. For technological details see section 6.1.4.3.8.

Table 1. Si. Electron traps identified in neutron-irradiated float-zone silicon containing a low concentration of oxygen and carbon after 30 min. N_2 anneal by the aid of deep-level transient spectroscopy (DLTS). Energy depth E_C-E_t, electron capture cross-section σ_n, and temperature range of existence $T_{min} \cdots T_{max}$ are given. (HOB = electronic higher-order band) [73c1].

E_C-E_t eV	σ_n cm^2	T_{min} °C	T_{max} °C	Identification	Ref.
0.15	$2\cdot10^{-16}$		450	HOB	73c1, 77G1, 78G1, 79s1
0.17	$4\cdot10^{-16}$	475	550	HOB	73c1, 77G1, 78G1, 79s1
0.19	$2\cdot10^{-16}$	450	625		77G1, 78G1, 79s1
0.20	$4\cdot10^{-17}$	450	725		77G1, 78G1, 79s1
0.32	$1\cdot10^{-15}$		550	vacancy complex	77G1, 78G1, 79s1
0.40	$3\cdot10^{-16}$	450	525	vacancy complex	77G1, 78G1, 79s1
0.46	$3\cdot10^{-15}$		425	vacancy (5-fold)	77G1, 78G1, 79s1
0.13	$8\cdot10^{-17}$	625	725	surface	77G1, 78G1
0.50	$2\cdot10^{-15}$	450	525	surface	77G1, 78G1, 79s1

6.1.4.3.3 Side reactions of neutron irradiation

Two essential kinds of side reactions occur during the neutron irradiation of pure silicon:

reactions of fast neutrons with the silicon isotopes (Table 2) [74i1],

reaction of the phosphorus ^{31}P forming the unstable phosphorus isotope ^{32}P which decays to stable sulfur ^{32}S at a half-life of 14 days:

$$^{31}P(n,\gamma)^{32}P \xrightarrow{14\,d} {}^{32}S+\beta^-.$$

Only the generation of the unstable phosphorus isotope ^{32}P has to be considered in practical cases because of its relatively long half-life of 14 days. The concentration of ^{32}P causes the waiting time from the end of the neutron irradiation to the delivery from the irradiation facility. After the decay of ^{32}P, the irradiated silicon can be handled as non-radioactive material [76H1, 76H2].

Table 2. Si. The main reactions of silicon isotopes with fast neutrons [74i1]. $\bar{\sigma}$ = average cross section for the reaction in a fission neutron spectrum. E_{th} = threshold energy for the reaction.

Reaction	$\bar{\sigma}$ mb	E_{th} MeV
(n, p)-reactions:		
$^{28}Si(n,p)^{28}Al \xrightarrow{2.25\,min} {}^{28}Si+\beta^-$ (stable)	2	4
$^{29}Si(n,p)^{29}Al \xrightarrow{6.6\,min} {}^{29}Si+\beta^-$ (stable)	0.56	3
$^{30}Si(n,p)^{30}Al \xrightarrow{3.3\,sec} {}^{30}Si+\beta^-$ (stable)	0.05	8
(n, α)-reactions:		
$^{28}Si(n,\alpha)^{25}Mg$ (stable);	0.56	2.75
$^{29}Si(n,\alpha)^{26}Mg$ (stable);	7.9	0.03
$^{30}Si(n,\alpha)^{27}Mg \xrightarrow{9.5\,min} {}^{27}Al+\beta^-$ (stable)	0.07	4.3

If silicon contains dopants or other impurities, the reactions of these elements with neutrons have to be taken into account especially with respect to the formation of unstable isotopes with long half-lives. Thermal neutron capture cross-sections and half-lives of activation products are listed for the usual doping elements in Table 3 [81s2].

Table 3. Si. Reactions of doping elements with thermal neutrons [81s2].

Doping element	Natural abundance %	Thermal neutron capture cross-section σ [mb]	Activation product	Half-live of activation product $t_{1/2}$	Resulting stable isotope
(n, γ)-reactions					
^{6}Li	7.5	28	^{7}Li		^{7}Li
^{7}Li	92.5	37	^{8}Li	842 ms	^{4}He
^{10}B	20	500	^{11}B		^{11}B
^{11}B	80	5.5	^{12}B	0.02 s	^{12}C
^{12}C	98.9	3.4	^{13}C		^{13}C
^{13}C	1.1	0.9	^{14}C	5730 y	^{14}N
^{14}N	99.63	75	^{15}N		^{15}N
^{15}N	0.37	$24\cdot10^{-3}$	^{16}N	7.13 s	^{16}O
^{16}O	99.762	0.178	^{17}O		^{17}O
^{17}O	0.038		^{18}O		^{18}O
^{18}O	0.2	0.16	^{19}O	27.1 s	^{19}F
^{27}Al	100	230	^{28}Al	2.246 m	^{28}Si
^{31}P	100	180	^{32}P	14.3 d	^{32}S
^{32}S	95.02	530	^{33}S		^{33}S
^{33}S	0.75		^{34}S		^{34}S
^{34}S	4.21	240	^{35}S	87.5 d	^{35}Cl

continued

Table 3 (continued)

Doping element	Natural abundance %	Thermal neutron capture cross-section σ [mb]	Activation product	Half-live of activation product $t_{1/2}$	Resulting stable isotope
^{36}S	0.02	150	^{37}S	5.0 m	^{37}Cl
^{69}Ga	60.1	1680	^{70}Ga	21.15 m	^{70}Ge
^{71}Ga	39.9	4710	^{72}Ga	14.1 h	^{72}Ge
^{70}Ge	20.5	3200	^{71}Ge	11.2 d	^{71}Ga
^{72}Ge	27.4	980	^{73}Ge		^{73}Ge
^{73}Ge	7.8	15000	^{74}Ge		^{74}Ge
^{74}Ge	36.5	143, 240	^{75m}Ge, ^{75}Ge	47 s, 83 m	^{75}As
^{76}Ge	7.8	90, 50	^{77m}Ge, ^{77}Ge$\rightarrow$^{77}As	53 s, 11.3 h, 38.8 h	^{77}Se
^{75}As	100	4300	^{76}As	26.4 h	^{76}Se
^{74}Se	0.9	51800	^{75}Se	120 d	^{75}As
^{76}Se	9.0	21000, 64000	^{77m}Se, ^{77}Se	17.5 s	^{77}Se
^{77}Se	7.6	42000	^{78}Se		^{78}Se
^{78}Se	23.5	330, 100	^{79m}Se, ^{79}Se	3.9 m, $7 \cdot 10^{4}$ y	^{79}Br
^{80}Se	49.6	80, 530	^{81m}Se, ^{81}Se	57.3 m, 18 m	^{81}Br
^{82}Se	9.4	39, 6	^{83m}Se, ^{83}Se$\rightarrow$^{83}Br	69 s, 22.4 m, 2.4 h	^{83}Kr
^{113}In	4.3	7500, 3900	^{114m}In, ^{114}In	49.5 d, 71.9 s	^{114}Cd, ^{114}Sn
^{115}In	95.7	92000, 65000 45000	^{116m_2}In, ^{116m_1}In, ^{116}In	2.2 m, 54 m, 14 s	^{116}Sn
^{112}Sn	1	350, 800	^{113m}Sn, ^{113}Sn	21.4 m, 115.1 d	^{113}In
^{114}Sn	0.7	1200	^{115}Sn		^{115}Sn
^{115}Sn	0.4	50000	^{116}Sn		^{116}Sn
^{116}Sn	14.7	6	^{117m}Sn	13.6 d	^{117}Sn
^{117}Sn	7.7	2600	^{118}Sn		^{118}Sn
^{118}Sn	24.3	16	^{119m}Sn	293 d	^{119}Sn
^{119}Sn	8.6	2300	^{120}Sn		^{120}Sn
^{120}Sn	32.4	1, 140	^{121m}Sn, ^{121}Sn	50 y, 27 h	^{121}Sb
^{122}Sn	4.6	180, 1	^{123m}Sn, ^{123}Sn	40.1 m, 129.2 d	^{123}Sb
^{124}Sn	5.6	130, 4	^{125m}Sn, ^{125}Sn$\rightarrow$^{125}Sb	9.5 m, 9.46 d, 2.77 y	^{125}Te
^{121}Sb	57.3	55, 6200	^{122m}Sb, ^{122}Sb	4.2 m, 2.7 d	^{122}Sn, ^{122}Te
^{123}Sb	42.7	11, 35, 4280	^{124m_2}Sb, ^{124m_1}Sb, ^{124}Sb	20 m, 1.6 m, 60.3 d	^{124}Te
^{120}Te	0.096	340, 2000	^{121m}Te, ^{121}Te	154 d, 16.8 d	^{121}Sb
^{122}Te	2.6	1100, 1700	^{123m}Te, ^{123}Te	119.7 d, $1.2 \cdot 10^{13}$ y	^{123}Sb
^{123}Te	0.908	406000	^{124}Te		^{124}Te
^{124}Te	4.816	40, 6800	^{125m}Te, ^{125}Te	57.4 d	^{125}Te
^{125}Te	7.14	1560	^{126}Te		^{126}Te
^{126}Te	18.95	125, 900	^{127m}Te, ^{127}Te	109 d, 9.35 h	^{127}I
^{128}Te	31.69	15, 200	^{129m}Te, ^{129}Te$\rightarrow$^{129}I	33.6 d, 69.9 m, $1.57 \cdot 10^{7}$ y	^{129}Xe
^{130}Te	33.8	20, 270	^{131m}Te, ^{131}Te$\rightarrow$^{131}I $\rightarrow$^{131m}Xe	30 h, 25 m, 8.02 d, 11.9 d	^{131}Xe
^{190}Pt	0.01	150000	^{191}Pt	2.8 d	^{191}Ir
^{192}Pt	0.79	2200, < 14000	^{193m}Pt, ^{193}Pt$\rightarrow$^{193m}Ir	4.33 d, ~50 y, 10.6 d	^{193}Ir
^{194}Pt	32.9	90, 1200	^{195m}Pt, ^{195}Pt	4.02 d	^{195}Pt
^{195}Pt	33.8	27000	^{196}Pt		^{196}Pt
^{196}Pt	25.3	50, 700	^{197m}Pt, ^{197}Pt$\rightarrow$^{197m}Au	94.4 m, 18.3 h, 7.8 s	^{197}Au
^{198}Pt	7.2	27, 3700	^{199m}Pt ^{199}Pt$\rightarrow$^{199}Au$\rightarrow$^{199m}Hg	13.6 s, 30.8 m, 3.14 d, 42.6 m	^{199}Hg

continued

Table 3 (continued)

Doping element	Natural abundance %	Thermal neutron capture cross-section σ [mb]	Activation product	Half-live of activation product $t_{1/2}$	Resulting stable isotope
^{197}Au	100	98800	^{198}Au	2.6935 d	^{198}Hg
^{209}Bi	100	14, 19	^{210m}Bi, ^{210}Bi$\rightarrow$^{210}Po	$3\cdot10^6$ y, 5 d, 138.4 d	^{206}Pb
(n, α)-reactions					
^{6}Li	7.5	940000	^{3}H	12.323 y	^{3}He
^{10}B	20.0	3837000	^{7}Li		^{7}Li
^{17}O	0.038	235	^{14}C	5730 y	^{14}N
^{32}S	95.02	4	^{29}Si		^{29}Si
^{33}S	0.75	140	^{30}Si		^{30}Si
(n, p)-reactions					
^{14}N	99.63	1810	^{14}C	5730 y	^{14}N
^{33}S	0.75	2	^{33}P	25.3 d	^{33}S

6.1.4.3.4 Irradiation exposure time

The phosphorus concentration obtained in Si is with sufficient accuracy given by the equation:

$$c_P = c_{Si} \cdot \sigma \cdot \Phi \cdot t = 1.65 \cdot 10^{-4}\,[\mathrm{cm}^{-1}] \cdot \Phi \cdot t$$

where c_P is the phosphorus concentration, c_{Si} is the concentration of ^{30}Si, σ is the cross section (for the reaction of the silicon isotope ^{30}Si with thermal neutrons $\sigma = 0.107$ b, 1 b$=10^{-24}$ cm^2), Φ is the thermal neutron flux density, and t is the irradiation time.

The effective cross section of the phosphorus-generating reaction must be determined for each irradiation facility because the cross section depends on the type of the reactor [76H2, 77B1] due to differing neutron energy spectra in the various types of reactors.

The relation between the irradiation time t, the resistivity ϱ and the number of phosphorus atoms n_d generated after an irradiation by a thermal neutron flux density of 10^{13} cm^{-2} s^{-1}, is depicted in Fig. 5 [62B1].

6.1.4.3.5 Selection and preparation of the starting silicon

Semiconductor grade silicon is pure so that no special pretreatment before the neutron irradiation is necessary. Care has only to be taken that the surface is free of metallic impurities. This may be reached, e.g. by etching the silicon surface with an HF—HNO_3 mixture (1:10) and rinsing by pure water.

In order to obtain a sufficient homogeneity, the dopant concentration of the starting material should be as low as possible. Fluctuations in the background doping are superimposed on the uniform NTD-doping and thus determine the residual non-uniformity. An example of the uniformity improvement obtained is shown in Fig. 6.

6.1.4.3.6 Neutron irradiation

Desired resistivity values usually range between $\varrho = 300\,\Omega$cm and $\varrho = 5\,\Omega$cm. A neutron irradiation dose of $\Phi \cdot t = (1\cdot10^{17} \cdots 6\cdot10^{18})$ neutrons/cm^2 is therefore required. This high neutron irradiation dose can be obtained in nuclear reactors in a reasonable exposure time of a few hours up to several days.

It is advantageous to perform the neutron irradiation in reactors having a high ratio of thermal to fast neutrons because then the annealing of the lattice defects (section 6.1.4.3.8) may be performed at a lower temperature [78Y1].

Figs. 7 and 8 show a possible arrangement of silicon samples during the irradiation in a nuclear reactor (Reactor OSIRIS) [81b2].

The time that must elapse before the irradiated pure silicon can be handled in the same manner as a non-irradiated material are depicted in Fig. 10 for various resistivities.

If the irradiated silicon contains dopants or other impurities, the radioactivity of the neutron-activated elements has to be taken into account. The radioactivity a of the elements at the time t can be calculated by the following equation:

$$a = a_0 \exp\left[-\frac{t \ln 2}{t_{1/2}}\right]$$

where $t_{1/2}$ is the half-life period of the radioactive decay, and a_0 is the initial radioactivity.

6.1.4.3.7 Radioactivity of the irradiated silicon

The radioactivity of the irradiated silicon requires special precautions for the handling of the material [79h1]. The activity may originate from the surface and from the volume of the samples. The radioactive impurities on the surface originating, e.g. from corrosion products or fusion products in the reactor can be etched away. Usually the surfaces are totally free of any radio-activity [76H1, 76H2].

The main radio-activity originates from the activation of the phosphorus produced by nuclear transmutation to the unstable isotope ^{32}P. The relationship of the neutron flux density Φ to the ^{32}P radioactivity a for various resistivities is depicted in Fig. 9 [79h1]. Other radioactivities may be neglected because of the short half-life of the intermediate silicon isotope ^{31}Si and the high purity of the commercial semiconductor silicon.

The irradiated silicon has to be stored in a safety area until the radioactivity of the phosphorus ^{32}P has decayed and the crystals may be handled without special care [81s1].

6.1.4.3.8 Annealing

The resistivity of the irradiated silicon is in the range of 10^5 Ωcm because of radiation-induced lattice defects. A proposed method to anneal the defects [78Y1] is a heat treatment at 750 °C for 1 hour in a vacuum furnace [79C1] or a heating with the help of infrared light [81y1]. The exact resistivity value may be determined on an off-cut of the silicon wafer after a phosphorus diffusion at 1250 °C. After removing the diffused layer by lapping, the exact resistivity may be measured well [77S1].

The completeness of the annealing treatment may be controlled by infrared absorption (e.g. [75K1, 81p1, 79J1]), by photoluminescence methods (e.g. [81h1, 81t1]) or with the aid of deep level transient spectroscopy (e.g. [77G1, 77s1, 78G1, 79s1]).

Minority carrier life-time measurements of annealed samples mostly lead to values of about $\tau = 50 \cdots 100$ μs. The values are dependent on the heat treatment. It has been shown that conventionally doped silicon samples after having received the same annealing treatment show the same minority carrier lifetime as the irradiated and annealed silicon samples [76V1].

The annealing behavior of different samples (Table 5) irradiated in nuclear reactors by different fluxes of fast and thermal neutrons (Table 4) is shown in Figs. 11⋯15 [78Y1].

Fig. 16 shows trap spectra of irradiated float-zone silicon as obtained after 30 min. anneals in N_2 at various temperatures [78G1].

Table 4. Si. Characteristics of irradiation facilities for irradiated silicon samples as listed in Table 5 [78Y1]. (Φ = neutron flux density).

Reactor	Irradiation location	Sample environment	Fast neutrons: Φ cm^{-2} s^{-1}	Thermal neutrons: Φ cm^{-2} s^{-1}
Oak Ridge Bulk Shielding Reactor (BSR)	in core CP 15	H_2O	$3.1 \cdot 10^{12}$	$7.2 \cdot 10^{12}$
	SW, NW holes	D_2O	$2.8 \cdot 10^{9}$	$2.8 \cdot 10^{12}$
	center line (CL)	D_2O	$1.4 \cdot 10^{8}$	$8.9 \cdot 10^{11}$
	N, S holes	D_2O	$1.0 \cdot 10^{8}$	$7.1 \cdot 10^{11}$
National Bureau of Standards Reactor (NBSR)	G-2 hole in core (C)	D_2O	$2.2 \cdot 10^{13}$	$1.1 \cdot 10^{14}$
Oak Ridge Research Reactor (ORR)	in core (C)	H_2O	$1.1 \cdot 10^{14}$	$3.5 \cdot 10^{14}$

Table 5. Si. Characteristics of silicon samples irradiated in various reactors as listed in Table 4 [78Y1]. ($\Phi \cdot t$ = neutron fluence).

Sample No.	Initial resistivity Ωcm, type (n or p)	Reactor (irradiation location), irradiation time in h	Fast neutrons (>1 MeV) $\Phi \cdot t$ cm^{-2}	Thermal neutrons $\Phi \cdot t$ cm^{-2}
1	30, n	BSR (S), 1	$3.6 \cdot 10^{11}$	$2.5 \cdot 10^{15}$
2	30, n	BSR (S), 2	$7.2 \cdot 10^{11}$	$5.0 \cdot 10^{15}$
3	1000, p	BSR (CL), 69	$3.8 \cdot 10^{13}$	$2.2 \cdot 10^{17}$
4	700, n	BSR (SW), 25	$2.5 \cdot 10^{14}$	$2.5 \cdot 10^{17}$
5	700, n	BSR (SW), 76	$7.6 \cdot 10^{14}$	$7.6 \cdot 10^{17}$
6	1000, p	BSR (SW), 984	$1.0 \cdot 10^{16}$	$1.0 \cdot 10^{19}$
7	700, n	BSR (CP15), 10	$1.1 \cdot 10^{17}$	$2.6 \cdot 10^{17}$
8	700, n	NBSR (G-2), 504	$3.2 \cdot 10^{19}$	$1.6 \cdot 10^{20}$
9	700, n	NBSR (G-2), 1008	$6.4 \cdot 10^{19}$	$3.2 \cdot 10^{20}$
10	700, n	ORR (C), 264	$1.0 \cdot 10^{20}$	$3.0 \cdot 10^{20}$
11	100, p	BSR (SW), 112	$1.1 \cdot 10^{15}$	$1.1 \cdot 10^{18}$
12	100, p	BSR (SW), 76	$7.6 \cdot 10^{14}$	$7.6 \cdot 10^{17}$
13	10, n	BSR (SW), 112	$1.1 \cdot 10^{15}$	$1.1 \cdot 10^{18}$
14	10, n	BSR (SW), 76	$7.6 \cdot 10^{14}$	$7.6 \cdot 10^{17}$
15	700, n	BSR (SW), 76	$7.6 \cdot 10^{14}$	*)
16	1000, p	BSR (SW), 76	$7.6 \cdot 10^{14}$	*)

*) Shielded against thermal neutrons by Cd foil.

The annealing treatment of irradiated silicon crystals is not detrimental to the production of silicon devices. Various authors have shown that irradiated silicon applied to device manufacturing shows the same electrical properties as conventionally doped silicon [61T1, 64M1, 71K1, 73S1, 74S1, 76H1, 76H2, 76H3, 76P1, 77K1, 78P1, 79C1, 81b1, 81y1].

References for 6.1.4.3

Bibliography

60F1 Fritsch, H., Cuevas, M.: Phys. Rev. **119** (1960) 1238.

61C1 Corbett, J.W., Watkins, G.D.: Phys. Rev. Lett. **7** (1961) 314.

61T1 Tanenbaum, M., Mills, A.D.: J. Electrochem. Soc. **108** (1961) 171.

62B1 Irvin, J.C.: Bell Syst. Tech. J. **41** (1962) 387.

63W1 Watkins, G.D.: J. Phys. Soc. Jpn. Suppl. (2) **18** (1963) 22.

64M1 Messier, J., Le Coroller, Y., Merlo Flores, J.: IEEE Trans. Nucl. Sci. **11** (1964) 276.

67B1 Bertoletti, M., Papa, T., Sette, D., Vitali, G.: J. Appl. Phys. **38** (1967) 2645.

69W1 Watkins, G.D.: IEEE Trans. Nucl. Sci. NS-**16** (No. 6) (1969) 13.

70L1 Lappo, M.T., Tkachev, V.D.: Fiz. Tekh. Poluprovodn. **4** (1970) 2192; Sov. Phys. Semicond. (English Transl.) **4** (1971) 1882.

71C1 Coy, R.A., Thomas, J.E. jr.: J. Appl. Phys. **42** (1971) 1236.

71K1 Kharchenko, V.A., Solov'ev, S.P.: Fiz. Tekh. Poluprovodn. **5** (1971) 1641; Sov. Phys. Semicond. (English Transl.) **5** (1972) 1437.

73S1 Schnöller, M.: "Gleichrichter und Thyristoren mit homogenem Breakdown", Kolloquium "Halbleiterleistungsbauelemente" Freiburg (FRG), 14., 15. Nov. **1973**.

74S1 Schnöller, M.S.: IEEE Trans. Electron Devices **21** (1974) 313.

75H1 Herrmann, H.A., Herzer, H.: J. Electrochem. Soc. **122** (1975) 1568.

75K1 Kolbesen, B.O.: Appl. Phys. Lett. **27** (1975) 353.

75N1 Newman, R.C., Totterdell, D.H.J.: J. Phys. C **8** (1975) 3944.

76H1 Haas, E., Schnöller, M.S.: IEEE Trans. Electron Devices **23** (1976) 803.

76H2 Haas, E.W., Schnöller, M.S.: J. Electron. Mater. **5** (1976) 57.

76H3 Hill, M.J., van Isegham, P.M., Zimmermann, W.: IEEE Trans. Electron Devices **23** (1976) 809.

76P1 Platzöder, K., Loch, K.: IEEE Trans. Electron Devices **23** (1976) 805.

76T1 Tokuta, Y., Usami, A.: J. Appl. Phys. **47** (1976) 4952.
76V1 Voss, P.: Private communication **1976**.
77B1 Bartel, M., Haas, E., Schnöller, M. (Siemens AG): Deutsche Offenlegungsschrift 2617320 (German patent application, 3. Nov. 1977).
77G1 Guldberg, J.: Appl. Phys. Lett. **31** (1977) 578.
77H1 Herzer, H.: in [77h1, p. 106].
77K1 Konorova, L.F., Litovskii, M.A., Malkovich, R.Sh., Nikitina, J.P., Savin, E.P.: Fiz. Tekh. Poluprovodn. **11** (1977) 2036; Sov. Phys. Semicond. (English transl.) **11** (1977) 1195.
77S1 Schnöller, M. (Siemens AG): Deutsche Offenlegungsschrift 2607414 (German patent application) 25. Aug. 1977.
77T1 Tokuta, Y., Usami, A,: J. Appl. Phys. **48** (1977) 1668.
78G1 Guldberg, J.: J. Phys. D **11** (1978) 2043.
78P1 Prussin, S., Cleland, J.W.: J. Electrochem. Soc. **125** (1978) 350.
78Y1 Young, R.T., Cleland, J.W., Wood, R.F., Abraham, M.M.: J. Appl. Phys. **49** (1978) 4752.
79C1 Corish, J., Bénière, F., Agrawal, V.K., Haridoss, S., Defeux, C.: J. Appl. Phys. **50** (1979) 6838.
79J1 Jagannath, G., Grabowski, Z.W., Ramdas, A.K.: Solid State Commun. **29** (1979) 355.

Review articles and books

51l1 Lark-Horovitz, K.: Semiconducting Materials. Proc. of Conf. at Univ. Reading (Butterworth, London) **1951**, pp. 47···69.
61b1 Billington, D.S., Crawford, J.H.: Radiation Damage in Solids, Princeton, New Jersey, Princeton University Press **1961**.
73c1 Chen, C.S., Lowell, R., Corelli, J.C.: Radiation Damage and Defect Centers in Semiconductors. The Institute of Physics, London, **1973**, p. 210.
74b1 Burtscher, J.: Resistivity Fluctuations, Growth Striations and Swirls in Silicon Single Crystals. Proceedings of the European Summer School: Scientific Principles of Semiconductor Technology. July 8–12, **1974** at Bad Boll, Ed. H. Weiß.
74i1 Handbook on Nuclear Activation Cross-sections. IAEA Wien, **1974**. Technical Report No. 156.
74k1 Krausse, J.: National Bureau of Standards Special Publication 400–10, Spreading Resistance Symposium, Proceedings of a Symposium held at NBS, Gaithersburg, MD, June 13–14, **1974**.
77h1 Semiconductor Silicon 1977, Proc. of the Third International Symposium on Silicon Materials Science and Technology. Edited by Huff, H.R., and Sirtl, E.: "Electronics and Electrothermics and Metallurgy Divisions" Proceedings Vol. **77-2**. The Electrochemical Society, Inc., P.O. Box 2071, Princeton, N.J. 08540.
77s1 Schulz, M., Lefèvre, H.: in [77h1, p. 142···151].
79h1 Haas, E.W., Martin, J.A.: in [79m1, p. 27···36].
79m1 Neutron Transmutation Doping in Semiconductors **1979**. Edited by Meese, J.M., Plenum Publ. Corp. 227 West, 17th Street, New York N.J. 10011.
79s1 Schulz, M., Lefèvre, H., Pohl, F.: Forschungsbericht T 79-117, Technologische Forschung und Entwicklung. Fachinformationszentrum Energie, Physik, Mathematik GmbH, Kernforschungszentrum, 7514 Eggenstein-Leopoldshafen 2.
81b1 Braggins, T.T., Thomas, R.N.: in [81g1, p. 437···446].
81b2 Bréant, P., Cherruan, F., Genthon, J.-P.: in [81g1, p. 287···303].
81g1 Neutron Transmutation - Doped Silicon. Proceedings of the Third International Conference on Neutron Transmutation Doping of Silicon, held Aug. 27–29, 1980, in Copenhagen, Denmark. Edited by Jens Guldberg. **1981** Plenum Press, New York.
81h1 Hammond, R,B., Meese, J.M.: in [81g1, p. 417···421].
81p1 Pajot, B., Débarre, D.: in [81g1, p. 423···435].
81s1 Smith, T.G.G.: in [81g1, p. 183···191].
81s2 Seelmann-Eggebert, W., Pfennig, G., Münzel, H., Klewe-Nebenius, H.: Karlsruher Nuklidkarte, 5. Auflage **1981**. Kernforschungszentrum Karlsruhe GmbH.
81t1 Tajima, M., Yusa, A.: in [81g1, p. 377···394].
81y1 Yusa, A., Itoh, D., Kim, C., Kim, H., Hushimi, K., Ohkawa, S.: in [81g1, p. 473···485].

6.1.4.4 Silicon epitaxy

6.1.4.4.1 Introduction

Definitions

The word "epitaxy" is derived from Greek. Its meaning is "to arrange upon". In the field of crystal growth it is defined as the growth of a single crystalline layer on a single crystalline substrate under conditions where the lattice structure and the surface orientation of the substrate crystal determine crystal structure, orientation, and quality of the growing single crystal layer.

If the growing layer and the substrate are basically identical crystals, we speak of *homoepitaxy*, e.g. silicon on silicon. Both crystals may differ in parameters which are unimportant to crystal growth, e.g. different dopant elements and dopant concentrations.

The other case, where the growing film and the substrate are different crystals, is denoted as *heteroepitaxy*, e.g. silicon on sapphire.

Application of epitaxial silicon layers

In contrast to diffusion or ion implantation, where the dopant concentration of the crystal is increased, the epitaxial deposition results in the growth of a single crystalline layer on the surface of a crystal, leaving its dopant level nearly unaffected. Since layers are doped during growth, almost any kind of vertical (i.e. perpendicular to the surface) doping profile can be achieved. Laterally (i.e. in a plane parallel to the substrate surface) the layer is always homogeneously doped. Neglecting selective epitaxial growth, which is usually not used in production lines, all epitaxially grown layers cover the whole surface of the substrate. The main advantage of epitaxial growth is its ability to produce thin silicon films with a vertically constant dopant level. In practice, nearly all epitaxial layers are required with a vertically constant dopant concentration. Typical applications of silicon epitaxial films are given in Table 1.

Table 1. Application of epitaxial silicon layers.

Features of epitaxial layers			Application					
Layer structure	Thickness range µm	Dopant concentration range cm^{-3}	Diodes	Transistors	Sensors	Bipolar IC	MOS-IC	Power MOS
Homoepitaxy								
Vertically constant dopant concentration								
nn^+-structure	0.1···100	$5\cdot10^{12}\cdots1\cdot10^{17}$	×	×	×		×	×
pp^+-structure	1···100	$1\cdot10^{14}\cdots1\cdot10^{17}$		×			×	
pn^+-structure	10···100			×				
np^+-structure	10···100			×				
pp-structure	1··· 5	$1\cdot10^{15}\cdots5\cdot10^{16}$					×	
$np(n^+)$-structure	1··· 20	$1\cdot10^{14}\cdots1\cdot10^{17}$				×		
Intentionally produced vertical doping profile								
nn^+-structure	0.5···100	$5\cdot10^{12}\cdots5\cdot10^{17}$	×	×				×
Heteroepitaxy								
Si on sapphire	0.5··· 1.2	intrinsic					×	

6.1.4.4.2 Basic technology

Epitaxial silicon growth methods

Table 2. Survey of the different growth methods for silicon epitaxy.

Growth method	Growth conditions			Homo-epitaxial growth	Hetero-epitaxial growth	Special features	Ref.
	p Pa	T K	r µm/min				
Molecular beam epitaxy (MBE)	$\leqq 1 \cdot 10^{-7}$	≈ 1120	$10^{-4} \cdots 10^{-3}$	+	+		78O2
Vapour-phase epitaxy (VPE)	$101.3 \cdot 10^{3}$ $\approx 13 \cdot 10^{3}$	1220···1500 1150···1320	0.1···5.0 0.1···1.0	+ +	+ +	see section 6.1.4.4.2	
Solid-phase-epitaxy (SPE)	$\leqq 1.3 \cdot 10^{-4}$	600··· 850	$10^{-3} \cdots 10^{-2}$	+		Growth via a transport medium, such as a metal	75C, 76S, 77M2, 78L, 78M
Electro-crystallization epitaxy	$101.3 \cdot 10^{3}$	370···1700	≈ 0.1	+		Growth by electro-deposition from molten salts	76C2, 81E
Grapho-epitaxy	$101.3 \cdot 10^{3}$	$\geqq 1570$			+	Crystallization of an amorphous silicon layer deposited on an amorphous substrate with an artificial surface relief structure having long-range order	79G, 80F
Ionized-cluster beam epitaxy	$1 \cdot 10^{-5} \cdots 1 \cdot 10^{-2}$	670···1020	$\approx 5 \cdot 10^{-2}$	+	+	Silicon clusters are formed by super-condensation phenomena following adiabatic expansion after injection through small nozzles in a vacuum chamber	77T, 79T2

From the epitaxial growth methods summarized in the Table 2, only two are of practical interest today, the *molecular-beam epitaxy* and the *vapour-phase epitaxy*.

The very small growth rates and the ultra high vacuum needed in the molecular-beam process are advantages for basic research work and the manufacturing of multilayer epitaxy or very thin layers as used for Schottky diodes. Because of these limited applications this process is not further discussed in this paper.

Vapour-phase epitaxy (VPE) is the only process which is commercially used on a large scale. The following data are therefore restricted to VPE.

In VPE gaseous or vaporized compounds of silicon (e.g. SiH_4 or $SiHCl_3$) and dopants (mostly hydrides such as PH_3 or B_2H_6) heavily diluted by an inert gas (usually H_2) are fed into an open-tube type reaction chamber in which heated substrates are located. The reaction species decompose on or very close to the hot silicon surface leading to an epitaxial silicon crystal growth.

Fundamentals of VPE

The VPE crystal growth is empirically well under control. The fundamental processes, vapour-phase mass transport, and surface reactions are theoretically understood, but the relevant theories are not suited to predict the properties of a grown layer with the accuracy which is required in device technology.

Gas dynamics and mass transport have been studied by different authors, for horizontal reactors mainly by Eversteijn et al., Takahashi et al. and Ban et al. [65S1, 68R1, 69A, 70E1, 70E2, 71E, 71R, 72T, 74E, 75B1, 78B1, 78B2] and for cylinder reactors by Fuji et al., Curtis et al. and Manke et al. [72F, 73C, 74D, 76C1, 77M1]. Gas reactions and the composition of hot gas have been theoretically investigated by Sirtl et al. [72H2, 74S1] whereas Sedgwick et al., Ban et al. and Nishizawa et al. measured the gas composition by different methods [75B1, 77S, 78N]. Duchemin [76D1, 77D2] investigated gas compositions theoretically and experimentally. Surface reaction and growth kinetics have been studied by Burton et al., Shaw and Blanc [51B, 74S2, 78B5]. The doping mechanism in VPE has been investigated thoroughly by different authors [68S1, 69S, 70R1, 70R2, 71B, 72B1, 72H1, 74B, 75G1, 77D1, 78K1, 78K2, 78K3, 78R, 79K, 79R1, 79R2, 80K1, 80K2].

For a survey on silicon epitaxy, the reader is referred to recently published review articles which all emphasize different aspects of epitaxy [78B3, 78B6, 78H, 79H1, 80P].

Special features of VPE

Advantages and problems

In semiconductor technology, VPE is almost exclusively used for manufacturing epitaxial silicon films. The enormous flexibility of the technique, the high production rate, and the good crystal quality of silicon layers which are usually achieved by this method, make VPE very favourable for application in silicon technology. The widespread ranges of thickness and dopant concentration which are used for device production are shown in Fig. 1. These dopant concentration profiles are valid for the most commonly used dopant atoms, i.e. boron B, arsenic As, and phosphorus P. Apart from silane SiH_4, which is restricted to the growth of thin layers (<1 μm), films having thicknesses of more than 0.3 μm can be grown from all the other silicon sources in use, i.e. dichlorosilane SiH_2Cl_2, trichlorsilane $SiHCl_3$, and silicon tetrachloride $SiCl_4$.

A further advantage of VPE is its qualification for the growth of layers showing narrow tolerance ranges of thickness and dopant concentration.

The density of crystal defects is very low in vapour-phase epitaxial films, but the defects themselves are still an unsolved problem because of the continuously increasing requirements for high yield in device technology. The crystal and surface quality of epitaxially grown layers cannot be better than that of the polished substrate slice. A large number of different defects can grow or can be created during epitaxial deposition, e.g. stacking faults, slip, etch crown, protruding defects and haze. For a detailed discussion of crystal and surface defects, see for instance [70S, 73W, 76R, 78B5].

Other typical problems of VPE are the autodoping effect [68S2, 69G, 70P, 72L, 73K, 73S, 74L, 75B2, 75I, 75S, 77D1, 77L, 77O, 78H, 78S, 79T1, 80S1, 80S2, 80S3], diffusion of dopant atoms from the substrate into the growing layer and the memory effect, i.e. undesired doping by species remaining inside the reactor from previous runs. The feed gases like N_2, H_2, HCl, etc. may be contaminated by dopant compounds and other impurities (e.g. metals) leading to additional sources of unexpected impurity species.

Typical deposition data:

The consecutive steps of a vapour-phase epitaxial growth process are described in Table 3.

Temperature and duration are usually kept constant for each process step except in-situ etching and deposition. Values are chosen within the given ranges. These values are fed into the reactor control unit. The etch and growth rates have to be adapted to other parameters influencing the epitaxial growth, e.g. buried layer shift or autodoping. The parameters are usually reproducible for each kind of epitaxial layer.

The fine tuning of the growth process is performed by varying the etch and deposition duration in accordance with the result of the second last or last run.

Table 3. Steps of a standard vapour-phase epitaxial growth process.

Process step		Gases fed into the reaction chamber				T	r	t	p
No.	Purpose	Main stream	Etch gas	Dopant source	Silicon source	K	µm/min	min	bar
1	Loading	low N_2 flow				≈300			1 (atmospheric pressure)
2a	Purging	high N_2 flow				≈300		3···5	
2b	Purging	high H_2 flow				≈300		3···5	
3	Heating	normal or high H_2 flow				>300		5···10	
4	In-situ etching	normal H_2 flow	HCl			1370···1500	0.1···0.5	variable	≦1 (atmospheric pressure or less)
5	Purging	normal H_2 flow				1220···1500		1···2	
6	De-position	normal H_2 flow	PH_3 or AsH_3 or B_2H_6		SiH_4	1220···1320	≈0.1	variable	
					SiH_2Cl_2	1320···1420	0.1···3.0		
					$SiHCl_3$	1400···1470	0.2···1.5		
					$SiCl_4$	1450···1500	0.2···1.0		
7	Cooling down	normal or high H_2 flow				>300		5···10	
8	Purging	high N_2 flow				≈300		3···5	
9	Un-loading	low N_2 flow				≈300			1 (atmospheric pressure)

Reduced pressure operation

Reduced pressure operation improves the epitaxial deposition properties considerably. Low-pressure deposition results in an improved growth uniformity across the slice as well as from slice to slice [78O1, 79H2], in a reduction of autodoping effects [76O, 77L, 77O, 78D, 79H2] and in the disappearance of a buried layer shift and wash-out [79H2].

Reactors for VPE

For vapour-phase growth of epitaxial films three different types of reactors are in use. The oldest type is called *vertical reactor* (Fig. 2). The name indicates that the main direction of gas flow in the reaction chamber is vertical. Typically, the reaction chamber of a vertical reactor has a high volume leading to a mixture of feed gases and gaseous reaction products ("mixed gas reactor"), making the transient response time longer than that of all other types of reactors. In all vertical reactors, the susceptor or the nozzle(s) of the gas inlet rotate around the symmetry axis of the reaction chamber, thus increasing the uniformity of the growth.

The next generation of reactors, *horizontal reactors*, is characterized by a mainly horizontal gas flow (Fig. 3) where the feed gases do not mix with the gaseous reaction products. The advantages of horizontal reactors are a higher production capacity and an easier control of thickness and resistivity of the growing film by tilting the susceptor [70E1, 70E2, 74E, 78B2], and adjusting a proper temperature profile. The uniformity of layers grown in a horizontal reactor is limited by flow disturbances at the leading edge of the susceptor, by edge effects on both sides of the susceptor, and by the decrease of the concentration of the reactant species along the susceptor.

Improved uniformity of epitaxial layers is guaranteed by the use of *cylinder reactors* (=*barrel reactors*). The reaction chamber has the shape of a straight cylinder (Fig. 4) and the susceptor is a truncated pyramid. In cylinder reactors the slices stand nearly vertical in shallow pockets on the sides of the susceptor pyramid. By these shapes of the reaction chamber and the susceptor, edge effects are minimized and the concentration decrease along the susceptor is smaller than that in horizontal reactors. As with most types of vertical reactors, the susceptor of a cylinder reactor rotates.

In principle, all these types of reactors can be operated at reduced pressure. Barrel-type reactors with radiation heating are mostly used for this purpose.

Usually, reactors are heated by induction or radiation. In induction-heated systems, graphite plates are used for susceptors which are heated by eddy-currents. Induction heating results in a steep temperature gradient across the thickness of the slices [71H, 72B2], causing stresses in the slices which produce slips in the silicon crystal particularly at large diameters (>5 cm). Slip-free slices can be manufactured by radiation heating because an extremely small temperature gradient across the thickness of the slices is obtained [78H].

For temperature control, optical pyrometers, semiconductor detectors, or thermocouples are in use. The latter are not suitable for induction heated systems. These temperature control methods determine a reference temperature rather than the actual temperature of the growing silicon surface. This usually causes no problems in practice, but the temperature has to be corrected for a comparison of different reactors and for theoretical work, e.g. for the use of thermodynamic data.

Quality requirements for epitaxial slices

The quality standards of epitaxial films are defined for

layer thickness
concentration of deliberately added dopant atoms
levels of impurities (including undesired doping)
crystallographic parameters and surface properties,

according to the requirements of the finally produced electronic device type. The growth rate and consequently the layer thickness is exclusively controlled by the deposition parameters, whereas dopant and impurity concentrations as well as surface and crystallographic properties are determined by the deposition process, the properties of the substrate, and the condition of the reactor.

The standard specification of epitaxial layers is given in the "Book of SEMI Standards", whereas the test methods have been developed by DIN (Deutsches Institut für Normung e.V.) and ASTM (The american society for testing and materials). Usually the standards of DIN and ASTM are very similar or even identical. These standards can be obtained through:

Book of SEMI Standards:
Semiconductor Equipment and Materials
Institute, Incorporated
625 Ellis Street
Mountain View, CA 94043
USA

DIN Standards:
Beuth-Vertrieb GmbH
D-1000 Berlin 30

Annual Book for ASTM Standards, Part 43 (1983):

ASTM, The American Society for Testing and Materials
1916 Race Street
Philadelphia, PA 19103
USA

6.1.4.4.3 Scope of the data collection

As mentioned before, VPE is an empirical technology which is too complex and depends too much on the individual reactor and its conditions to make reliable theoretical predictions of the parameters of the growing layer, within a few percent deviation from the experimental result. The theoretical predictions are therefore not suitable in practice. There are, however, a variety of different problems like the comparison of depositions from different silicon sources or necessary changes in the growth conditions which require calculations, based on physico-chemical data of the reacting compounds, the properties of the semiconductor crystals, and the gas flow conditions. These data and the basic relationships are important for research and theoretical work.

The physico-chemical data and the basic relationships are collected here if they seem to be reliable and if needed for problems as mentioned above. Only the properties of the materials of the reaction chamber, e.g. quartz and graphite are not reviewed and collected here for two reasons: Firstly, these materials are not involved in the epitaxial growth process, but they may be a source of undesired impurities, and secondly the properties of these materials depend strongly on the manufacturing conditions. Making predictions of their influence is almost impossible. This influence has to be determined experimentally.

6.1.4.4.4 Properties of materials used in Si epitaxy

Silicon source materials

For vapour-phase epitaxial growth, the following silicon source materials are in use: Silane SiH_4, dichlorosilane SiH_2Cl_2, trichlorosilane $SiHCl_3$ and silicon tetrachloride $SiCl_4$. Some physical data of monochlorosilane SiH_3Cl, which may be a by-product of epitaxial growth, is added for completeness.

Basic properties and critical constants of these compounds are collected in the Tables 4···8.

In the field of VPE, those physical and thermodynamic properties are of interest which relate to vapourization, gas dynamics and formation or decomposition. The factors important to vaporization of SiH_2Cl_2, $SiHCl_3$, and $SiCl_4$ are vapour pressure (Figs. 5 and 6), heat of vaporization (Figs. 7, 8), liquid heat capacity (Figs. 9, 10), and liquid density (Figs. 11, 12). The temperature dependence of the thermal conductivity of the liquid is published only for $SiCl_4$ (Fig. 13). Gas-flow dynamics and vapour-phase mass transport are based on gas heat capacity (Figs. 14, 15 and 16), gas viscosity (Figs. 17, 18, and 19), and gas thermal conductivity (Figs. 20, 21, and 22). These properties are presented also for SiH_4 and SiH_3Cl. The heat of formation and Gibbs' free energy of formation of SiH_4 and $SiCl_4$ are plotted versus temperature in Figs. 23 and 24, and Figs. 25 and 26, respectively, For SiH_2Cl_2, these values are given in the Tables 7 and 8.

Table 4. Molecular weight M and properties of Si source materials used in VPE.

Material	M	State and density d at p=1.013 bar	Ref.
SiH_4	32.12	colourless gas at T=300 K	78B4
		d=1.342 kgm^{-3} at T=293.2 K	80B
SiH_3Cl	66.55	colourless gas at T=300 K	73D
SiH_2Cl_2	100.98	colourless gas at T=300 K	73D
		d=4.168 kgm^{-3} at T=298.2 K	80B
$SiHCl_3$	135.43	colourless liquid	73D
$SiCl_4$	169.9	colourless liquid	79Y

Table 5. Melting temperature T_m and boiling temperature T_b of Si source materials used in VPE.

Material	T_m K	T_b K	Ref.
SiH_4	88.5	161.3	78B4
SiH_3Cl	155.1	242.8	73D
SiH_2Cl_2	151.2	281.5	73D
$SiHCl_3$	146.6	305.0	73D
$SiCl_4$	203.8	330.5	79Y

Table 6. Critical temperature T_{crit}, pressure p_{crit}, and density d_{crit} of silicon source materials for VPE.

Material	T_{crit} K	p_{crit} bar	d_{crit} kg m^{-3}	Remarks	Ref.
SiH_4	269.7	48.43		questionable values	78B4
			247	estimated value	78B4
SiH_3Cl	396.7	48.43	382	estimated values	73D
SiH_2Cl_2	449.5	44.99	479	estimated values	73D
$SiHCl_3$	479.2		505		73D
		41.14		estimated value	73D
$SiCl_4$	507.2	37.49	520.7		79Y

Table 7. Heat of formation ΔH_f of gaseous SiH_2Cl_2 for low and high temperatures T, $p=1.013$ bar, [71S].

T K	ΔH_f kJ mol^{-1}	T K	ΔH_f kJ mol^{-1}
0	−306.352	800	−318.733
100	−309.214	900	−318.867
200	−311.566	1000	−318.850
298.15	−313.800	1500	−317.536
300	−313.842	2000	−365.757
400	−315.628	2500	−363.058
500	−316.934	3000	−360.648
600	−317.833	4000	−753.342
700	−318.415	5000	−748.928

Table 8. SiH_2Cl_2. Gibbs' free energy of formation ΔF_f for low and high temperatures T, $p=1.013$ bar, [71S].

T K	ΔF_f kJ mol^{-1}	T K	ΔF_f kJ mol^{-1}
0	−306.352	800	−241.375
100	−303.194	900	−231.693
200	−296.294	1000	−222.007
298.15	−288.315	1500	−173.816
300	−288.156	2000	−116.759
400	−279.311	2500	−54.827
500	−270.073	3000	6.577
600	−260.613	4000	174.686
700	−251.027	5000	406.145

Etch gases

Besides anhydrous hydrogen chloride HCl which is usually used for in-situ etching of silicon, sometimes etching by sulfur hexafluoride SF_6 is preferred for obtaining very smooth surfaces. Both compounds are shipped in cylinders as a liquified gas under their own pressure. Data important to vaporization, such as vapour pressure (Figs. 27 and 28), heat of vaporization (Figs. 29 and 30), heat capacity of the liquid (Figs. 31 and 32), liquid density (Figs. 33 and 34), and liquid thermal conductivity (Figs. 35 and 36) have been added to the properties which are related to mass transport and gas dynamics, i.e. gas heat capacity (Figs. 37 and 38), gas viscosity (Figs. 39 and 40), and gas thermal conductivity (Figs. 41 and 42). The heat of formation and Gibbs' free energy of formation are plotted versus temperature in Figs. 43 and 44, and Figs. 45 and 46, respectively. Basic physical properties and critical data are summarized in the Tables 9⋯11.

Table 9. Molecular weight M and properties of etch gases used in VPE.

Material	M	State and density d at $p=1.013$ bar	Ref.
HCl	36.46	colourless gas at $T=300$ K	76A
		$d=1.500$ kg m^{-3} at $T=298.2$ K	80B
SF_6	146.06	colourless gas at $T=300$ K	75A
		$d=6.162$ kg m^{-3} at $T=193.2$ K	80B

Table 10. Melting temperature T_m and boiling temperature T_b of etch gases used in VPE.

Material	T_m K	T_b K	Remarks	Ref.
HCl	158.74	188.13		76A
SF_6	222.35		at $p=2.24$ bar	80B
		209.45	sublimation at $p=1.013$ bar	80B

Table 11. Critical temperature T_{crit}, pressure p_{crit}, and density d_{crit} of etch gases used in VPE.

Material	T_{crit} K	p_{crit} bar	d_{crit} kg m^{-3}	Ref.
HCl	324.7	83.09	418	76A
SF_6	318.71	37.59	753	75A

Gas-phase dopants

Doping atoms are incorporated in the covalent lattice of silicon on regular sites, that is by substitution of silicon atoms. This substitution is possible only for elements which have a covalent radius close to that of silicon. The tetrahedral covalent radii are summarized in section 6.1.1. Boron B, aluminum Al, and gallium Ga from the third group of the periodic table, and phosphorus P, arsenic As, and antimony Sb from the fifth group fit this condition. For practical reasons, only B, P, As, and sometimes Sb are used for doping of silicon layers grown by vapour-phase epitaxy.

In principle different kinds of compounds of the dopant elements, e.g. halogens and hydrides, can be applied to VPE. Since compounds which are gaseous at room temperature offer several advantages, they are universally used in modern reactors. These gaseous compounds are arsine AsH_3, phosphine PH_3 and stibine SbH_3 for n-type doping, and diborane B_2H_6 for p-type doping. The critical constants and other basic physical data are collected in the Tables 12···14. The temperature dependence of the heat capacity, the viscosity, and the thermal conductivity in the gas phase are plotted in Figs. 47···52. For the thermodynamic data of formation of these hydrides, see Figs. 53···56.

Table 12. Molecular weight M and properties of gas-phase dopants used in Si VPE.

Material	M	State and density d at p=1.013 bar	Ref.
PH_3	34.00	colourless gas at T=300 K	74Y1
		d=1.402 kg m^{-3} at T=298.2 K	80B
AsH_3	77.95	colourless gas at T=300 K	74Y1
		d=3.243 kg m^{-3} at T=293.2 K	80B
SbH_3	124.77	colourless gas at T=300 K	74Y1
B_2H_6	26.67	colourless gas at T=300 K	74Y2
		d=1.247 kg m^{-3} at T=273.2 K	80B

Table 13. Melting temperature T_m, and boiling temperature T_b of gas-phase dopants used in Si VPE.

Material	T_m K	T_b K	Ref.
PH_3	140.2	185.42	74Y1
AsH_3	156.3	210.68	74Y1
SbH_3	185.2	255.2	74Y1
B_2H_6	107.7	180.7	74Y2

Table 14. Critical temperature T_{crit}, pressure p_{crit}, and density d_{crit} of gas-phase dopants used in Si VPE.

Material	T_{crit} K	p_{crit} bar	d_{crit} kg m^{-3}	Ref.
PH_3	327.31	65.35	218	74Y1
AsH_3	364.5	71.23	588	74Y1
SbH_3	440.4	73.06	794	74Y1
B_2H_6	290.1	41.39	160	74Y2

Main stream gases

The gas used for purging the reactor has to be an inert gas, which is very pure and dry. Furthermore, it must be available at large quantities at low cost. Nitrogen N_2 fulfills all these requirements. Additional requirements are valid for the main stream gas (H_2) used for the dilution of the reacting species. Even at high temperatures it must not react with silicon or the materials of the reaction chamber. This gas and some noble gases are suitable, but the latter must be rejected in consideration of their high cost.

Physical data of N_2 and H_2 are given in the Tables 15···21. In addition to purging, nitrogen is also used for the final cooling of the reactor after an epitaxial deposition. Relevant physical properties for the two gases hydrogen and nitrogen are also compiled in the Tables 15···17.

Table 15. Molecular weight M and properties of main stream gases used in Si VPE.

Material	M	State and density d at p=1.013 bar	Ref.
H_2	2.016	colourless gas at T=300 K d=0.08235 kg m^{-3} at T=298.2 K	80B
N_2	28.013	colourless gas at T=300 K d=1.145 kg m^{-3} at T=298.2 K	80B

Table 16. Melting temperature T_m, and boiling temperature T_b of main stream gases used in Si VPE.

Material	T_m K	T_b K	Ref.
H_2	14.02	20.38	80B
N_2	63.30	77.35	80B

Table 17. Critical temperature T_{crit}, pressure p_{crit}, and density d_{crit} of main stream gases used in Si VPE.

Material	T_{crit} K	p_{crit} bar	d_{crit} kg m^{-3}	Ref.
H_2	33.24	12.97	31	80B
N_2	126.26	34.00	321.6	80B

Table 18. Density d at p=1.013 bar of main stream gases used in Si VPE [75B3].

T [K]	d [kg m^{-3}]	T [K]	d [kg m^{-3}]
N_2		H_2	
248.15	1.3765	50	0.4954
273.15	1.25052	100	0.2457
293.15	1.1652	150	0.1637
298.15	1.1455	200	0.1227
313.15	1.0908	250	0.0982
333.15	1.0253	273.15	0.0898879
353.15	0.9672	293.15	0.08376
373.15	0.9154	298.15	0.08235
423.15	0.8072	350	0.0701
473.15	0.7219	400	0.0614
573.15	0.5960	450	0.0545
673.15	0.5074	500	0.0491
		550	0.0446
		600	0.0409

Table 19. Heat capacity C_p for main stream gases used in Si VPE at p=1.013 bar, [75B3].

T K	C_p J mol^{-1} K^{-1}	T K	C_p J mol^{-1} K^{-1}
N_2		H_2	
0	0.000	0	0.000
100	29.104	100	22.564
200	29.108	200	27.271
298	29.125	298	28.836
300	29.125	300	28.844
400	29.246	400	29.183
500	29.577	500	29.259
600	30.108	600	29.326
700	30.752	700	29.439
800	31.430	800	29.652
900	32.091	900	29.907
1000	32.698	1000	30.204
1500	34.853	1500	32.300
2000	35.987	2000	34.288
2500	36.635	2500	35.878
3000	37.049	3000	37.066
3500	37.351	3500	38.116
4000	37.585	4000	39.087
4500	37.782	4500	39.978
5000	37.966	5000	40.786
		5500	41.547
		6000	42.258

Table 20. Viscosity η at p=1.013 bar of main stream gases used in Si VPE [75B3].

T [K]	η [cP]	T [K]	η [cP]
N_2		H_2	
100	0.00692	20	0.001093
140	0.00948	40	0.002067
180	0.01182	60	0.002876
220	0.01398	80	0.003579
260	0.01599	100	0.004211
300	0.01787	150	0.005598
400	0.02217	200	0.006813
500	0.02602	250	0.007923
600	0.02955	273.15	0.008411
700	0.03284	300	0.008957
800	0.03593	400	0.010867
900	0.03886	500	0.012642
1000	0.04165	600	0.014903
1200	0.04691	700	0.015846
1400	0.05180	800	0.017335
1600	0.05641	900	0.018757
1800	0.06078	1000	0.020128
2000	0.06495	1100	0.021440

Table 21. Thermal conductivity κ at p=1.013 bar of main stream gases used in Si VPE [75B3].

T [K]	κ [$\cdot 10^{-6}$ J cm^{-1} s^{-1} K^{-1}]	T [K]	κ [$\cdot 10^{-5}$ J cm^{-1} s^{-1} K^{-1}]
N_2		H_2	
100	96	10	7.402
150	139	20	15.478
200	183	40	29.778
250	222	60	42.228
300	259	80	54.173
350	293	100	66.454
400	327	150	98.083
450	359	200	128.198
500	389	250	156.126

continued

Table 21 (continued)

T [K]	κ [$\cdot 10^{-6}$ J cm^{-1} s^{-1} K^{-1}]
N_2	
600	446
700	498
800	548
900	597
1000	647
1200	760
1400	858
1600	962
1800	1067
2000	1163

T [K]	κ [$\cdot 10^{-5}$ J cm^{-1} s^{-1} K^{-1}]
H_2	
273.15	168.239
300	181.699
350	205.589
400	228.133
450	250.341
500	271.875
550	293.242
600	314.777
650	336.648
700	358.014

Substrate materials

All substrates have to be highly pure single crystalline material, with an exact orientation according to the requirements of nucleation and lattice match. The surface on which epitaxial silicon growth will occur must be polished thoroughly and cleaned effectively prior to epitaxial deposition.

Silicon

For homoepitaxial growth by VPE slices with [111]- or [100]-orientation are used as substrates cut from monocrystalline rods, which have been grown by the float zone or the Czochralski-method. The position of the different crystallographic directions with respect to the [111]- and [100]-directions, respectively, can be seen from the stereographic projections in Figs. 57 and 58.

Sapphire and spinel

For heteroepitaxial growth (SOS technique) mostly sapphire, α-Al_2O_3, is used as substrate material, whereas spinel $MgAl_2O_4$ is of minor importance. The physical properties which are important to the SOS technique and/or device technology are collected in Table 22 [78C], where silicon data are repeated for comparison.

Table 22. Physical properties of the substrate materials Si, α-Al_2O_3, and $MgAl_2O_4$.

Parameter	Silicon Si	Sapphire α-Al_2O_3	Spinel $MgAl_2O_4$
Crystal structure	Face-centered cubic	Rhombohedral	Face-centered cubic
Unit cell dimension [Å]	a=5.4301	a=4.758, c=12.991	a=8.083
Density [g cm^{-3}]	2.33	3.98	3.58
Hardness [mohs]	7	9	8
Indentation No. [kg mm^{-2}]	1000	1525···2000	1320
Melting point [°C]	1412	2030	2105
Vapourization process: BP [°C]	3145,	2980,	Unknown,
species	Si-Si_7	Al_2O and O_2	Unknown (Mg-rich)
Dielectric constant	11.7 (500 Hz···30 MHz)	9.4 (‖c-axis) (100 Hz···100 kHz)	8.4 (100 Hz···100 kHz)
Dissipation factor, tan δ		10^{-3}···10^{-4}	10^{-3}···10^{-4}
Refractive index	3.4975 (at 1.357 μm)	1.7707 (at 5461 Å)	1.7202 (at 5461 Å)
Optical transmission	Transparent in infrared	80% at minimum (λ=0.24···6.0 μm, d=0.04445 cm)	80% at minimum (λ=0.31···5.1 μm, d=0.04445 cm)
Thermal conductivity [cal/cm · s · K] at 25 °C	0.30	0.065 (60° to c-axis)	0.035
Thermal expansion coefficient [1/K] at 25···800 °C	$3.59 \cdot 10^{-6}$	$8.40 \cdot 10^{-6}$ (60° to c-axis)	$7.45 \cdot 10^{-6}$

The usual orientation of sapphire slices is [1012] but epitaxial silicon growth is also possible on other orientations as shown below [67C].

Table 23. Heteroepitaxial silicon on sapphire-orientation relationships.

Deposition plane		Parallel directions			Mismatch along [110] rows		Mismatch between equivalent [110] rows	
Silicon	Sapphire	Silicon		Sapphire	%	Site ratio	%	Site ratio
1. (001) on	$(\bar{1}012)$	[110]	‖	$[2\bar{2}01]$	8.5	Al:Si=1:1	3.8	Al:Si=1:1
2. (111) on	$(11\bar{2}4)$	$[1\bar{1}0]$	‖	$[1\bar{1}00]$	7.1	Al:Si=1:2	1.7	Al:Si=2:3
3. (111) on	$(11\bar{2}0)$	$[1\bar{1}0]$	‖	$[\bar{2}201]$	8.4	Al:Si=1:1	3.8	Al:Si=2:1
4. (111) on	(0001)	$[11\bar{2}]$	‖	$[1\bar{1}00]$	19.4	Al:Si=1:1	19.4	Al:Si=1:1
5. (111) on	(0001)	$[11\bar{2}]$	‖	$[\bar{1}\bar{2}30]$	8.7	Al:Si=1:1	8.7	Al:Si=1:1

The mutual position of the crystallographic orientations of the rhombohedral sapphire structure is given in Fig. 59.

6.1.4.4.5 Epitaxial silicon growth

Requirements for single crystal growth

Epitaxial monocrystalline growth by vapour-phase deposition is possible only if the deposition temperature is sufficiently high and the growth rate small enough. Typical values are given in Table 2 in section 6.1.4.4.2. In all other cases, the growing film will be polycrystalline, independent of the monocrystalline or polycrystalline nature of the substrate (Fig. 60). At high-temperature growth, the rate limiting step of the overall reaction is the gas phase diffusion of the reacting species to the growing surface [74E]. The transfer of material occurs by diffusion through a region of laminar flow just above the susceptor. As in flow dynamics, this region is called stagnant boundary layer.

Mass transport

Horizontal reactors

Models have been published which simplify in different ways the fields of temperature, gas velocity and concentration of the reacting species in the reaction chamber [65S1, 68R, 69A, 70E1, 70E2, 71E, 71R, 72T, 74E, 75B1, 78B2]. None of these models is applicable for calculations of the overall growth rate and the growth rate variations along and across the susceptor with adequate precission necessary for avoiding test runs, i.e. within a few per cent, but some of them are very helpful for understanding the mass transport mechanisms and for estimating how strongly they are influenced by each of the different parameters important to mass transport.

Mass transport can be best understood by means of the stagnant layer model, which is based on the boundary layer problem of flow dynamics [65S2] and which was first introduced by Everstejin et al [70E1, 70E2, 71E, 74E], and developed further by Berkman et al [78B2]. This model simplifies the gas flow as description of two flow regions (Fig. 61): Laminar flow inside the stagnant layer, where the gas velocity and the concentration of the reacting species increase continuously from zero with increasing distance from the hot surface of the susceptor and where the gas temperature decreases linearly. Above the stagnant boundary layer a region of mixed flow (forced convection spiral flow [70E1, 72T]) exists, where the temperature, the gas velocity, and the concentration of the reacting species are constant. At the interface of the two regions the values of the parameters are equal.

Berkman's [78B2] value of the thickness of the stagnant boundary layer $\delta_T(x)$ for a tilted susceptor is given in Fig. 62 and by the relation:

$$\delta_T(x)=1.226\left(\frac{D_T x}{v_T(x)}\right)^{\frac{1}{2}}-\frac{0.564}{b(x)}\left(\frac{D_T x}{v_T(x)}\right)-\frac{0.0768}{b^2(x)}\left(\frac{D_T x}{v_T(x)}\right)^{\frac{3}{2}}$$
$$-\frac{0.0265}{b^3(x)}\left(\frac{D_T x}{v_T(x)}\right)^2-\frac{0.01298}{b^4(x)}\left(\frac{D_T x}{v_T(x)}\right)^{\frac{5}{2}}\approx\tfrac{1}{2}b(x)\ [\mathrm{cm}],$$

where x [cm] is the distance from the leading edge of the susceptor as defined in Fig. 62. The temperature dependence of the self-diffusion coefficient D_T [$\mathrm{cm^2\,s^{-1}}$] of the main stream gas and of the mean gas velocity $v_T(x)$[$\mathrm{cm\,s^{-1}}$] in the mixed flow region obey to the relationship

$$D_T = D_0 \left(\frac{T_A}{T_0}\right)^m, \quad v_T(x) = v_0(x)\left(\frac{T_A}{T_0}\right),$$

respectively, where the value of m is $1.75 \leqq m \leqq 2$. Usually the average value $m = 1.88$ is used. D_0 and v_0 are the self-diffusion coefficients and the mean gas velocity at a reference temperature T_0 [K], which is usually room temperature. $v_0(x)$ depends on tiltangle θ as

$$v_T(x) = v_0(0)\frac{b_0}{b(x)}$$

where

$$b(x) = b_0 - x \sin\theta$$

b_0 [cm] and θ are defined in Fig. 62. T_A [K] is defined as

$$T_A = \frac{\delta_T(x)\left[T_s - \frac{T_s - T_m}{2}\right] + [b(x) - \delta_T(x)]T_m}{b(x)}$$

where T_s [K] and T_m [K] are the temperature of the susceptor surface and the gas in the mixed flow region, respectively. With $\delta_T(x) \approx \frac{1}{2}b(x)$ T_A can be simply expressed by

$$T_A \approx \tfrac{1}{4}(T_s + 3T_m).$$

The epitaxial growth rate $r(x)$ in µm min^{-1} is given by the following relationship [78B2]:

$$r(x) = \frac{(1.76 \cdot 10^5) D_T^{(Si)} p_0}{T_A \cdot b(x)} \exp\left\{\frac{2D_T^{(Si)}}{b_0 v_T \sin\theta}\left[\ln\left(\frac{b_0}{\sin\theta} - x\right) - \ln\frac{b_0}{\sin\theta}\right]\right\}.$$

The partial pressure of the reactive species at $x = 0$ is denoted by p_0 [atm]. Contrary to D_T, which is the selfdiffusion coefficient, $D_T^{(Si)}$ is defined as the diffusion coefficient of the reacting species in the host carrier gas. Reliable experimental values of binary diffusion coefficients important to silicon VPE have not been published, but various equations for calculating them are known. Some of these equations are based on empirical relations. A detailed treatment of this problem is given in [77R].

Cylinder reactors

Calculations of the growth rate in barrel type reactors have been made by Fujii et al. [72F] and Manke et al. [77M1]. Both assume a parabolic gas velocity profile (not a stagnant layer model) inside the annular passage between the susceptor and the quartz bell jar. Furthermore, they assume a decrease of the concentration of the reacting species along the axis of the reactor due to the deposition of silicon. Contrary to Fujii, who based his calculations on a constant temperature profile which depends only on the radius, Manke takes into account developing effects of the mean gas temperature. Manke's approach seems to be more adapted to the practice. Since the calculation of the growth rate in this case has to be done in cylinder coordinates and is therefore much more complex than for a horizontal reactor, the interested reader is referred to Manke's original paper [77M1].

Surface reactions

Epitaxial growth from silane, SiH_4

The growth of epitaxial silicon layers from SiH_4 is described by the overall relation

$$SiH_4(g) \rightarrow Si(s) + 2H_2(g)$$

(g) and (s) denote the gaseous and solid state, respectively. This pyrolysis of silane is expected to consist of the following different reaction steps [78B3]:

1. Diffusion of SiH_4 from the mixed flow region through the stagnant layer to the silicon surface:

$$SiH_4(m) \xrightarrow{k_{D_0}} SiH_4(g).$$

2. Adsorption of silane at the silicon surface:

$$SiH_4(g) + * \underset{k_{-1}}{\overset{k_1}{\rightleftarrows}} SiH_4(ad).$$

3. Decomposition of silane:

$$SiH_4(ad) \underset{k_{-2}}{\overset{k_2}{\rightleftarrows}} Si(ad) + 2H_2.$$

4. Adsorption of hydrogen at the silicon surface:

$$\tfrac{1}{2}H_2 + * \underset{k_{-3}}{\overset{k_3}{\rightleftarrows}} H(ad).$$

5. Evaporation of adsorbed silicon:

$$Si(ad) \xrightarrow{k_4} Si(g) + *.$$

6. Surface diffusion and incorporation of silicon at a kink site:

$$Si(ad) \xrightarrow{k_5} Si(s).$$

7. Gas phase decomposition

$$SiH_4(g) \xrightarrow{k_6} Si(g) + 2H_2(g).$$

Reactant species in the mixed flow region are denoted by (m). (ad) and "*" denote an adsorbed atom or molecule and a free site on the silicon surface, respectively. k_i is the reaction rate.

In a steady state situation and under the assumption of a small surface coverage and the absence of gas phase reactions ($k_1 \gg k_6$), the diffusion limited growth rate is given by

$$r \sim k_{D_0} p_{SiH_4}(m)$$

where p_{SiH_4} (m) is the partial pressure of silane in the mixed flow region.

Epitaxial growth from chlorosilanes

The overall deposition reactions for the different silicon sources are:

a) Dichlorosilane: $SiH_2Cl_2(g) \rightleftarrows Si(s) + 2HCl(g)$
b) Trichlorosilane: $SiHCl_3(g) + H_2(g) \rightleftarrows Si(s) + 3HCl(g)$
c) Silicon tetrachloride: $SiCl_4(g) + 2H_2(g) \rightleftarrows Si(s) + 4HCl(g)$

In the hot gas, conversion reactions in the silicon-hydrogen-chlorine system take place, leading to a number of silicon compounds which are involved in the epitaxial silicon growth. Consequently the system of the detailed growth reactions is much more complex than in the case of SiH_4 epitaxy. The consecutive steps are expected to be the same as in the case of SiH_4, but no detailed investigation of these reactions has been published.

As in the case of epitaxial growth from SiH_4 the deposition rate is mainly determined by the diffusion of the reacting species to the growing surface, but it is not proportional to the partial pressure of the used chlorosilane in the gas mixture fed to the reactor, since HCl, the main by-product of the growth reaction, acts as an etchant.

Gas-phase equilibrium

Since the diffusion of the reacting species to the growing surface is the rate limiting step, all other steps of the growth mechanism are near equilibrium. For equilibrium data of the gas phase, a problem which has been treated mainly by Hunt and Sirtl [72H2, 74S1], the reader is referred to the chapter on polycrystalline silicon growth section 6.1.2.1., where the relevant data are summarized.

Nucleation

Nucleation may be hampered by impurity atoms occupying nucleation sites and by defects. Nucleation difficulties can be overcome by

a) using defect-free single crystalline substrates,
b) effectively cleaning of the substrates,
c) providing substrates with nucleation sites.

Single crystalline substrates

Nearly defect-free substrates are obtained by cutting slices from single-crystalline rods with subsequent lapping, etching, and polishing.

Cleaning

A wet cleaning procedure is necessary to remove all surface contaminations such as mounting wax from polishing, grease, dust, and oxide layers. Further cleaning may be achieved by a heat treatment of the substrates in hydrogen H_2 prior to deposition. In the case of silicon substrates, the natural oxide layer is removed at temperatures above 1100 °C according to the reaction:

$$SiO_2(s) + Si(s) \rightleftarrows 2\,SiO(g).$$

In the SOS technique, hydrogen acts as an etchant:

$$2H_2(g) + Al_2O_3(s) \rightleftarrows Al_2O(g) + 2H_2O(g),$$
$$H_2(g) + MgO(s) \rightleftarrows Mg(g) + H_2O(g).$$

In vapour-phase homoepitaxy, in-situ etching is achieved by hydrogen chloride HCl, which is usually used, or by sulfur hexafluoride SF_6, which is of minor importance, according to the relations:

$$Si(s) + 2HCl(g) \rightleftarrows SiCl_2(g) + H_2(g)$$
$$4Si(s) + 2SF_6(g) \rightleftarrows SiS_2(g) + 3SiF_4(g).$$

The in-situ etching has to be performed at high temperature and at a low etching rate, in order to obtain a smooth surface. Otherwise the etched surface will be bunched or even pitted (Fig. 63).

Nucleation sites

Atomic steps at the surface of the substrate act as nucleation sites. Epitaxial growth on perfect surfaces, e.g. on a (111)-surface, results in very imperfect silicon layers [65T, 71L1]. A slight misorientation from the (111)-orientation leads to a homogeneous density of atomic steps all over the surface. The result is a homogeneous silicon growth. The misorientation and its effect on the surface structure is shown in Fig. 64. The vector normal to the surface of the slice is tilted out of the [111]-direction towards [110] by an angle ε. For discrete devices ε has to be 2.0°···2.5° and for bipolar devices 4°, with a tolerance range of ±0.5° in both cases. The higher value for bipolar devices is necessary for reducing the buried layer pattern shift, distortion and wash-out [68D, 69E, 70D, 71L, 73J]. For silicon substrates with (100)-orientation often no misorientation is required.

The usual orientation of sapphire substrates is $(10\bar{1}2)$ within ±2°, since with this orientation the best single crystalline silicon layers with (100)-orientation can be obtained. Other possible orientations are given in section 6.1.4.4.4. Spinel and silicon have both a face-centered cubic-structure. The orientation of the silicon overgrowth therefore matches the orientation of the spinel substrate.

6.1.4.4.6 Doping during epitaxy

Intentional doping

The doping of epitaxially growing layers is controlled by the same mechanisms as the epitaxial growth, i.e. the mass transport to the surface and the incorporation of the dopant in the growing crystal layer. The dopant incorporation rate is also a function of the growth rate, making the doping theory more complex than the growth theory. Duchemin [76D1, 77D1] published a theory of doping epitaxial layers under steady state conditions whereas Reif et al. [78R, 79R1, 79R2] presented a physico-chemical model describing the incorporation of dopant atoms under transient and steady state conditions. They describe the doping mechanism by 8 consecutive steps (Fig. 65):

1. Forced convection mass transport of the dopant hydride (e.g. AsH_3) inside the mixed flow region.
2. Boundary layer mass transport of the dopant compound from the mixed flow region through the stagnant layer to the growing surface.
3. Gas phase chemical reactions (e.g. $2AsH_3 \rightleftarrows 2AsH_2 + H_2$).
4. Adsorption of the dopant-containing species at adsorption sites on the silicon surface.
5. Chemical dissociation of the dopant-containing compound into monoatomic dopant and hydrogen (e.g. $AsH_2 \rightleftarrows As + 2H$).
6. Surface diffusion of the dopant atoms and their incorporation at kink and step sites on the surface.
7. Burying of the dopant atoms by subsequently arriving silicon atoms.
8. Desorption of hydrogen from the surface.

This model is described by a set of first-order linear differential equations. For steady-state conditions, i.e. for a constant dopant molar fraction x_d in the main stream gas, the model gives a dopant concentration N_d inside the growing layer of

$$N_d = \frac{x_d}{R_d r + K_d}$$

where r is the growth rate and R_d and K_d are thermodynamic parameters. R_d is the sum of the "resistances" of steps 1···6, whereas K_d is mainly determined by the equilibrium constants of the steps 4···7 and the desorption of H_2. The calculations of Duchemin [76D1, 77D1] give a similar result. The above equation shows that for low growth rates the dopant concentration is independent of the growth rate, whereas at high deposition rates the dopant level is inversely proportional to the growth rate (Fig. 66).

The dependence of the dopant concentration on the deposition temperature is difficult to predict by each of the existing models for two reasons: Firstly, each of the consecutive steps depend individually on temperature. The reaction rates of endothermic and exothermic mechanisms increase or decrease, respectively, with increasing temperature. Secondly, the temperature dependence of the whole doping mechanism is equal to that of the rate limiting step, which is usually not known. From the temperature dependence of the dopant concentration of As at high growth rates, Reif et al. [79R2] concluded that the rate limiting step is one of the surface mechanisms (steps 4···6) for As doping.

The experimental result is, that for n-type dopants the concentration decreases with increasing temperature while p-type doping shows the opposite temperature dependence. In Table 24, values of the enthalpy of incorporation ΔH_{inc} of P, As, Sb, and B are summarized, which have been calculated from the temperature dependence of the doping level N_d measured by different authors under different deposition conditions. ΔH_{inc} [kcal mol^{-1}] has been calculated from

$$N_d(T) = N\,e^{-\frac{\Delta H_{inc}}{RT}}$$

where R is the gas constant [kcal mol^{-1} K^{-1}] and N [cm^{-3}] the (fictitious) dopant concentration at an infinitely high deposition temperature.

For Table 24, see p. 209.

Autodoping

In homoepitaxial growth, autodoping is caused by dopant atoms evaporated from heavily doped substrates or diffused areas prior to the deposition (evaporation from the front and back of the substrate), or during the whole growth process (evaporation from the back only). These evaporated atoms contribute to autodoping if they are incorporated into the growing layer instead of diffusing into the mixed gas region, where they are transported out of the reactor with a probability close to one.

In the case of heteroepitaxy, i.e. the SOS technique, chemical reactions are responsible for autodoping. Besides the etching of sapphire and spinel by hydrogen (see section 6.1.4.4.5), the following reaction also results in an undesired autodoping:

$$2\,Si(s) + Al_2O_3(s) \rightarrow 2\,SiO(g) + Al_2O(g).$$

Aluminum may be incorporated into the growing layer, leading to an undesired p-type doping of the silicon film [76D2].

6.1.4.4.7 Low-pressure deposition

The pressure dependence of the diffusion coefficient D_T is given by

$$D_{T,R} = D_{T,A} \frac{p_A}{p_R}$$

where the subscripts "R" and "A" denote the reduced and the atmospheric total pressure, respectively.

Usually a reactor is operated under low pressure with the same mass flow rates as those used at atmospheric pressure, i.e. the molar fractions of the species in the gas mixture are kept constant, and the total pressure p is reduced. Under these conditions the mean gas velocity $v_T(x)$ in the mixed flow region varies with total pressure according to

$$v_{T,R}(x) = v_{T,A}(x) \frac{p_A}{p_R}$$

and the thickness of the stagnant layer (see section 6.1.4.4.5) is therefore independent of the total pressure p. The growth rate is independent of p as far as only gas flow dynamics and mass transport are considered, but Duchemin et al. [78D] showed that at pressures below 0.13 bar the growth rate is controlled by hydrogen adsorbed at the silicon surface. Consequently, the deposition rate increases with increasing temperature. At high total pressures, the growth is mass transfer-limited and not dependent on temperature. This change of the rate limiting step is responsible for a change of the growth rate between atmospheric and reduced pressure operation. The burying rate of dopant atoms (see section 6.1.4.4.6) also changes, resulting in different dopant concentrations at atmospheric and reduced pressures.

The fact, that the diffusion coefficient increases inversely with the pressure, whereas the thickness of the stagnant boundary layer is independent of pressure, is the explanation for the strongly reduced autodoping at low pressure epitaxial growth. The evaporated dopant atoms can escape more easily through the stagnant layer into the mixed flow region, where they are transported out of the reactor.

6.1.4.4.8 Defects in epitaxial layers

Blanc [78B5] showed that the correlation between the misfit f, and the critical thickness of the epitaxial film h_c, above which *misfit dislocations* are expected to occur, is given by

$$h_c \approx \frac{|\boldsymbol{b}|}{2|f|}$$

where $|\boldsymbol{b}|$ is the magnitude of the Burgers vector and f is defined as:

$$f = \frac{a_1 - a_2}{(a_1 \cdot a_2)^{\frac{1}{2}}}$$

where a_1 and a_2 are the lattice parameters of the substrate and the epitaxial layer, respectively. Misfit dislocations are revealed by X-ray topography.

Slip dislocations, which can be seen by the same technique, are created by thermal gradients in the slice [71H, 72B2]. Slip occurs always between (111)-orientated planes.

Epitaxial stacking faults usually nucleate at the interface substrate-epitaxial film (Fig. 67). Their surfaces are parallel to (111)-planes; i.e. they have the shape of a tetrahedron or a square-shaped pyramid in the case of [111] and [100] orientated substrates, respectively. They can be used for determining the thickness of an epitaxial layer (see section 6.1.4.4.9).

6.1.4.4.9 Evaluation of epitaxial silicon films

Most of the evaluation methods are (nearly) identical for bulk silicon and epitaxial silicon films. Only those methods which are specific for the evaluation of epitaxial layers and are used routinely in production lines are considered here. Special methods necessary for research and trouble-shooting are cited in the literature [70S, 73W, 76R, 78B5].

The thickness of epitaxially grown layers can be determined by

a) etching with Sirtl-etch [61S] and measuring the *stacking fault size* by optical microscopy [64W]. The thickness h of the epitaxial layer is related to the length L of the base of the stacking fault (Fig. 67) by

$$h_{[111]} = \sqrt{\tfrac{2}{3}}L = 0.816 \cdot L \quad \text{for [111]-orientation, or by}$$
$$h_{[100]} = \sqrt{\tfrac{2}{3}}L = 0.707 \cdot L \quad \text{for [100]-orientation.}$$

This method is always applicable if (undesired) stacking faults are present but it is a destructive one. Its accuracy is limited to $> \pm 10\%$,

b) *interference of infrared light*. This non-destructive method is limited to epitaxial layers grown on heavily doped silicon substrates [62A]. Automized IR spectrometers using Fourier analysis are preferred to interferometers. The accuracy and reproducibility of this method is very good (up to ± 0.05 μm) but the measured thickness is usually not exactly the crystallographic thickness because of the diffusion of dopant atoms from the substrate into the epitaxial layer,

c) *bevelling and staining* [64W]. This destructive method is restricted to slices where the epitaxial layer and the substrate silicon have different conductivity types. The thickness can be determined within roughly ± 0.3 μm by means of an interference microscope.

The resistivity of epitaxially grown layers is measured by the *four-point probe* method [58S], together with a determination of the layer thickness. The four-point probe measurement gives an average resistivity value according to

$$\varrho = R_4 \cdot h \cdot F$$

where R_4 and F are the resistance measured by the four-point probe and a geometrical correction factor, respectively. This resistance measurement depends on the isolation of the film by junction formation at the substrate-layer interface. The method is restricted to p-on-n or n-on-p type structures.

Resistivity profiles can be measured by the *spreading resistance* method on bevelled specimens [66M]. The resistivity of a small volume ($\approx 10^{-12}$ cm^3) under the probe is measured effectively, giving a good resolution of resistivity profiles. Large corrections must be made when the probe approaches a pn-junction or a large doping gradient within less than about 1 μm. The corrections are usually introduced automatically by the computerized equipment. Special care must be taken to maintain the quality of the bevelled surface.

Concentration profiles are obtained by CV *measurement* [62T] using a mercury [72S] or an evaporated probe. Using a mercury probe, this method becomes non-destructive. Evaporated probes give more reliable concentration values and result in a higher break-through voltage, i.e. a deeper profile. Different types of profile plotters which determine the concentration profiles in somewhat different ways are in use. The basic equations for all these types are

$$x = \frac{\varepsilon \cdot \varepsilon_0 A}{C},$$

$$N(x) = \frac{-C^3}{e \cdot \varepsilon \cdot \varepsilon_0 A^2 (\mathrm{d}C/\mathrm{d}V)}$$

where A is the contact area and x the depth of the depletion layer. ε_0 and ε are the permittivities of free space ($\varepsilon_0 = 8.859 \cdot 10^{-14}$ AsV^{-1} cm^{-1}) and silicon ($\varepsilon = 11.75$), respectively. e is the elementary charge of electron ($e = 1.602 \cdot 10^{-19}$ As). V and C are the blocking voltage and the capacitance of the diode, respectively. Examples for CV profiles are presented in Fig. 1.

Surface characterization is performed either by *visual inspection* using a bright collimated light or by inspection with an optical microscope using *interference contrast* [55N]. The former gives quick information of the film quality of the whole slice but cannot distinguish between different defects. The inspection by interference contrast gives detailed information concerning the nature of the defects but is limited to smaller areas.

Many of the evaluation methods are covered by DIN or ASTM standards.

Table 24. Enthalpy of incorporation ΔH_{inc} of P, As, Sb, B.

Si source	Growth temperature range K	Dopant concentration range cm^{-3}	P Dopant source	P ΔH_{inc} $kcal\,mol^{-1}$	As Dopant source	As ΔH_{inc} $kcal\,mol^{-1}$	Sb Dopant source	Sb ΔH_{inc} $kcal\,mol^{-1}$	B Dopant source	B ΔH_{inc} $kcal\,mol^{-1}$	Ref.
$SiCl_4$	1350···1650	$1\cdot10^{16}\cdots1\cdot10^{17}$	PCl_3	−11							68S1
$SiCl_4$	1350···1650	$5\cdot10^{15}\cdots5\cdot10^{17}$			$AsCl_3$	−62					68S1
$SiCl_4$	1350···1550	$1\cdot10^{15}\cdots5\cdot10^{16}$					$SbCl_3$	−62			68S1
$SiCl_4$	1350···1650	$1\cdot10^{17}\cdots5\cdot10^{18}$	PH_3	−13	AsH_3	−32					70R1
$SiCl_4$	1350···1650	$5\cdot10^{17}\cdots5\cdot10^{18}$							B_2H_6	+19	70R1
$SiCl_4$	1250···1550	$5\cdot10^{15}\cdots1\cdot10^{17}$			AsH_3	−34					77D1
SiH_4	1300···1500	$1\cdot10^{16}\cdots1\cdot10^{17}$							B_2H_6	+18	71B
SiH_4	1300···1350	$1\cdot10^{15}\cdots5\cdot10^{15}$			AsH_3	−34					78R
SiH_4	1400···1600	$1\cdot10^{16}\cdots5\cdot10^{17}$	PH_3	−41							75G
SiH_2Cl_2; $SiHCl_3$	1250···1650	$1\cdot10^{19}\cdots1\cdot10^{20}$	PH_3	−15							77D1
SiH_2Cl_2; $SiHCl_3$	1150···1550								B_2H_6	+8	77D1
SiH_2Cl_2	1100···1700	$\approx1\cdot10^{19}$					$SbCl_5$	−1			77D1

References for 6.1.4.4

51B Burton, W.K., Cabrera, N., Frank, F.C.: Phil. Trans. Roy. Soc. London **243** (1951) 299.
55N Nomarski, G., Weill, A.R.: Rev. Metall. (Paris) **52** (1955) 121.
56B Bond, W.L., Smits, F.M.: Bell Syst. Techn. J. **35** (1956) 1209.
58S Smits, F.M.: Bell Syst. Techn. J. **37** (1958) 711.
61S Sirtl, E., Adler, A.: Z. Metallkunde **52** (1961) 529.
62A Alber, M.P., Combs, J.F.: J. Electrochem. Soc. **109** (1962) 709.
62T Thomas, C.O., Kahng, D., Manz, R.C.: J. Electrochem. Soc. **109** (1962) 1055.
64W Walsh, R.J.: Solid State Technol. **7** (1964) 23.
65S1 Shepherd, W.H.: J. Electrochem. Soc. **112** (1965) 988.
65S2 Schlichting, H.: "Grenzschichttheorie", Karlsruhe: Verlag G. Braun, **1965**, Chapter VII, p. 108.
65T Tung, S.K.: J. Electrochem. Soc. **112** (1965) 436.
66M Mazur, R.G., Dickey, D.H.: J. Electrochem. Soc. **113** (1966) 255.
67C La Chapelle, T.I., Miller, A., Morritz, F.L., in: Progress in Solid State Chemistry; Riess, H. (ed.); New York: Pergamon Press **1967**, 1.
68D Drum, C.M., Clark, C.A.: J. Electrochem. Soc. **115** (1968) 664.
68R Rundle, P.C.: Int. J. Electron. **24** (1968) 405.
68S1 Shepherd, W.H.: J. Electrochem. Soc. **115** (1968) 541.
68S2 Shepherd, W.H.: J. Electrochem. Soc. **115** (1968) 652.
69A Andrews, R.W., Rynne, D.M., Wright, E.G.: Solid State Technol. **12** (1969) 61.
69E Enomoto, T., Yukawa, K., Iwata, Y., Ohkubo, T.: Jpn. J. Appl. Phys. **8** (1969) 1301.
69G Gupta, D.C., Yee, R.: J. Electrochem. Soc. **116** (1969) 1561.
69S Swanson, T.B., Tucker, R.N.: J. Electrochem. Soc. **116** (1969) 1271.
70D Drum, C.M., Clark, C.A.: J. Electrochem. Soc. **117** (1970) 1401.
70E1 Eversteijn, F.C., Severin, P.J.W., v.d. Brekel, C.H.J., Peek, H.L.: J. Electrochem. Soc. **117** (1970) 925.
70E2 Eversteijn, F.C., Peek, H.L.: Philips Res. Rept. **25** (1970) 472.
70P Pogge, H.B., Boss, D.W., Ebert, E., in: Chemical Vapor Deposition, Second Int. Conf.; Blocher, J.M.; Whithers, J.C. (eds.), New York: The Electrochem. Soc. Inc., **1970**, 768.
70R1 Rai-Choudhury, P., Salkovitz, E.I.: J. Cryst. Growth **7** (1970) 353.
70R2 Rai-Choudhury, P., Salkovitz, E.I.: J. Cryst. Growth **7** (1970) 361.
70S Schwuttke, G.H.: Microelectron. Reliab. **9** (1970) 397.
71B Bloem, J.: J. Electrochem. Soc. **118** (1971) 1837.
71E Eversteijn, F.C.: Philips Res. Rept. **26** (1971) 134.
71H Huff, H.R., Braken, R.C., Rea, S.N.: J. Electrochem. Soc. **118** (1971) 143.
71L Ludsteck, A.: Internal Siemens Report **1971**.
71R Rundle, P.C.: J. Cryst. Growth **11** (1971) 6.
71S Stull, D.R., Prophet, H.: "JANAF Thermochemical Tables", 2nd ed.; NSRDS-NBS 37; Natl. Bur. Stds., Washington, D.C., **1971**.
72B1 Bloem, J.: J. Cryst. Growth **13/14** (1972) 302.
72B2 Bloem, J., Goemans, A.H.: J. Appl. Phys. **43** (1972) 1281.
72F Fujii, E., Nakamura, H., Haruna, K., Koga, Y.: J. Electrochem. Soc. **119** (1972) 1106.
72H1 Hurle, D.T.J., Logan, R.M., Farrow, R.F.C.: J. Cryst. Growth **12** (1972) 73.
72H2 Hunt, L.P., Sirtl, E.: J. Electrochem. Soc. **119** (1972) 1741.
72L Lekholm, A.: J. Electrochem. Soc. **119** (1972) 1122.
72S Severin, P.J., Poodt, G.: J. Electrochem. Soc. **119** (1972) 1384.
72T Takahashi, R., Koga, Y., Sugawara, K.: J. Electrochem. Soc. **119** (1972) 1406.
73B Bloem, J.: J. Cryst. Growth **18** (1973) 70.
73C Curtis, B.J., Dismukes, J.P., in: Chemical Vapor Deposition, Fourth Int. Conf.; Wakefield, G.F., Blocher jr., J.W. (eds.); The Electrochem. Soc. Softbound Symp. Series, Princeton, N.J., **1973**, 218.
73D Drews, M.A., Schenck jr., W.L., Smith, J.D., Walker, J., Yaws, C.L.: Solid State Technol. **16** (1973) 39.
73J Juleff, E.M.: Microelectron. **4** (1973) 18.
73K Kondo, A., Ishii, T., Takahashi, K., Fujibayashi, K., in: Electrochem. Soc. Extended Abstracts, Fall Meeting, Boston, Oct. 7–11, **1973**, Abstract No. 179, p. 451.
73S Skelly, G., Adams, A.C.: J. Electrochem. Soc. **120** (1973) 116.

73W Watts, B.E.: Thin Solid Films **18** (1973) 1.
74B Bloem, J., Giling, L.J., Graef, M.W.M.: J. Electrochem. Soc. **121** (1974) 1354.
74D Dittman, F.W., in: Chemical Reaction Engineering II; Hulburt, H.M. (ed.); Advances in Chemistry Series 133; Washington, D.C.: The American Chemical Society, **1974**, 463.
74E Eversteijn, F.C.: Philips Res. Rept. **29** (1974) 45.
74L Langer, P.H., Goldstein, J.I.: J. Electrochem. Soc. **121** (1974) 563.
74S1 Sirtl, E., Hunt, L.P., Sawyer, D.H.: J. Electrochem. Soc. **121** (1974) 919.
74S2 Shaw, D.W., in: Crystal Growth, Theory and Techniques, Vol. 1, Goodman, C.H.L. (ed.), Plenum Press **1974**, p. 1
74Y1 Yaws, C.L., Setty, H.S.N., Hopper, J.R., Swinderman, E.M.: Solid State Technol. **17** (1974) 47.
74Y2 Yaws, C.L., Hopper, J.R., Swinderman, E.M.: Solid State Technol. **17** (1974) 31.
75A Adler, L.S., Yaws, C.L.: Solid State Technol. **18** (1975) 35.
75B1 Ban, V.S., Gilbert, S.L.: J. Cryst. Growth **31** (1975) 284.
75B2 Bozler, C.O.: J. Electrochem. Soc. **122** (1975) 1705.
75B3 Braker, W., Mossman, A.L.: The Matheson Unabridged Gas Data Book, East Rutherford, N.J.: Matheson, **1975**
75C Canali, C., Campisano, S.U., Lau, S.S., Liau, Z.L., Mayer, J.W.: J. Appl. Phys. **46** (1975) 2831.
75G Giling, L.J., Bloem, J.: J. Cryst. Growth **31** (1975) 317.
75I Ishii, T., Takahashi, K., Kondo, A., Shirahata, K.: J. Electrochem. Soc. **122** (1975) 1523.
75S Speth, M.: Internal Siemens Report **1975**.
76A Adler, L.S., Yaws, C.L.: Solid State Technol. **19** (1976) 35.
76C1 Curtis, B.J.: J. Electrochem. Soc. **123** (1976) 437.
76C2 Cohen, U., Huggins, R.A.: J. Electrochem. Soc. **123** (1976) 382.
76D1 Duchemin, J.-P.: Ph.D. Thesis, L'université de Caen **1976**.
76D2 Druminski, M.: Electrochem. Soc. Meeting Las Vegas, Oct. 17–22, **1976**, Extended Abstract No. 183, p. 490.
76O Ogirima, M., Saida, H., Suzuki, M., Maki, M.: Electrochem. Soc. Spring Meeting, Washington, May **1976**, Abstract No. 102, p. 273.
76R Ravi, K.V.: Thin Solid Films **31** (1976) 171.
76S Saito, T., Sek, Y.: Appl. Phys. Lett. **29** (1976) 600.
77D1 Duchemin, J.-P.: Rev. Tech. Thomson-CSF **9** (1977) 411.
77D2 Duchemin, J.-P.: Rev. Tech. Thomson-CSF **9** (1977) 33.
77L Lee, P.H., Wank, M.T., Benzing, W.C.: IEDM (1977) 482.
77M1 Manke, C.W., Donaghey, L.F.: J. Electrochem. Soc. **124** (1977) 561.
77M2 Majni, G., Ottaviani, G.: Appl. Phys. Lett. **31** (1977) 125.
77O Ogarima, M., Saida, H., Suzuki, M., Maki, M.: J. Electrochem. Soc. **124** (1977) 903.
77R Reid, C.R., Prausnitz, J.M., Sherwood, T.K.: The Properties of Gases and Liquids, Third edition, McGraw-Hill **1977**.
77S Smith jr., J.E., Sedgwick, T.O.: Thin Solid Films **40** (1977) 1.
77T Tagaki, T., Yamada, I., Sasaki, A.: Thin Solid Films **45** (1977) 569.
78B1 Ban, V.S.: J. Electrochem. Soc. **125** (1978) 317.
78B2 Berkman, S., Ban, V.S., Goldsmith, N., in: Heteroepitaxial Semiconductors for Electronic Devices; Cullen, G.W., Wang, C.C. (eds.), Springer Verlag **1978**, 264.
78B3 Bloem, J., Giling, L.J., in: Current Topics in Material Science; Vol. 1; Kaldis, E. (ed.); North Holland Publishing Company **1978**, 147.
78B4 Borreson, R.W., Yaws, C.L., Hsu, G., Lutwack, R.: Solid State Technol. **21** (1978) 43.
78B5 Blanc, J., in: Heteroepitaxial Semiconductors for Electronic Devices; Cullen, G.W., Wang, C.C. (eds), Springer Verlag **1978**, 282.
78B6 Bollen, L.J.M.: Acta Electron. **21** (1978) 185.
78C Cullen, G.W., in: Heteroepitaxial Semiconductors for Electronic Devices; Cullen, G.W., Wang, C.C. (eds), Springer Verlag **1978**, 6.
78D Duchemin, J.-P., Bonnet, M., Koelsch, F.: J. Electrochem. Soc. **125** (1978) 637.
78H Hammond, M.L.: Solid State Technol. **21** (1978) 68.
78K1 Kühne, H.: Krist. Tech. **13** (1978) 841.
78K2 Kühne, H.: Krist. Tech. **13** (1978) 939.
78K3 Kühne, H.: Krist. Tech. **13** (1978) 1059.
78L Lau, S.S.: Proc. of the Sympos. on Thin Film Phenomena – Interfaces and Interactions, Atlanta, Georgia, Oct. **1977** (publ. 1978) 269.

78M Majni, G., Ottaviani, G.: Proc. of the Sympos. on Thin Film Phenomena – Interfaces and Interactions, Atlanta, Georgia, Oct. **1977** (publ. 1978) 293.
78N Nishizawa, J., Nihira, H.: J. Cryst. Growth **45** (1978) 82.
78O1 Ogirama, M., Saida, H., Suzuki, M., Maki, M.: J. Electrochem. Soc. **125** (1978) 1879.
78O2 Ota, Y.: Proc. of the Sympos. on Thin Film Phenomena – Interfaces and Interactions, Atlanta, Georgia, Oct. **1977** (publ. 1978) 309.
78R Reif, R., Kamins, T.I., Saraswat, K.: J. Electrochem. Soc. **125** (1978) 1860.
78S Srinivasan, G.R.: J. Electrochem. Soc. **125** (1978) 146.
79G Geis, M.W., Flanders, D.C., Antoniadis, D.A., Smith, H.I.: 25th Int. Electron. Devices Meeting, Technical Digest, Washington, D.C., USA, December 3–5, **1979**, 210.
79H1 Hammond, M.L.: Microelectron. J. **10** (1979) 4.
79H2 Herring, R.B.: Solid State Technol. **22** (1979)75.
79K Kühne, H.: Krist. Tech. **14** (1979) 333.
79R1 Reif, R., Kamins, T.I., Saraswat, K.: J. Electrochem. Soc. **126** (1979) 644.
79R2 Reif, R., Kamins, T.I., Saraswat, K.: J. Electrochem. Soc. **126** (1979) 653.
79T1 Tabe, M., Nakamura, W.: J. Electrochem. Soc. **126** (1979) 822.
79T2 Tagaki, T., Yamada, I., Matsubara, K.: Thin Solid Films **58** (1979) 9.
79Y Yaws, C.L., Shah, P.N., Patel, P.M., Hsu, G., Lutwack, R.: Solid State Technol. **22** (1979) 65.
80B Braker, W., Mossman, A.L.: Matheson Gas Data Book, **1980**. East Rutherford, N.J.: Matheson.
80F Flanders, D.C.: J. Vac. Sci. Technol. (USA) **17** (1980) 1195.
80K1 Kühne, H.: Krist. Tech. **15** (1980) 177.
80K2 Kühne, H., Sperling, R.: Krist. Tech. **15** (1980) 403.
80P Pogge, H.B., in: Handbook of Semiconductors; Moss, T.S. (ed.), Vol. 3, Kelles, S.P. (ed.), North Holland Publishing Company **1980**, 335.
80S1 Srinivasan, G.R.: J. Electrochem. Soc. **127** (1980) 1334.
80S2 Srinivasan, G.R.: J. Appl. Phys. **51** (1980) 4824.
80S3 Srinivasan, G.R.: J. Electrochem. Soc. **127** (1980) 2305.
81E Ellwell, D.: J. Cryst. Growth **52** (1981) 741.

6.1.4.5 Fabrication of layers

6.1.4.5.0 Introduction to insulating films

During the fabrication of semiconductor devices, various insulating films and fabrication methods are used. Besides the thermal oxidation of silicon to grow an insulating film directly on a silicon wafer, other preparation methods are used which deposit dielectrics to complement the properties of thermal SiO_2 layers. The deposited dielectrics used for semiconductor applications are primarily oxides or glasses. The individual deposition method is not related to a specific dielectric film. One can use several methods to prepare the same film. The choice of the deposition method will always be influenced by the specific situation related to the complete fabrication process for semiconductor devices.

Table 1 shows an overview of the preparation and uses of various dielectric films in modern semiconductor technology.

Table 1. Deposition methods and uses of dielectric films in semiconductor technology [79D1].

Types of dielectric film		Deposition methods [1])	Uses of dielectric films [2])
Silicon dioxide	SiO_2	A, B, C, D, G	1, 2, 3, 5, 7
Phosphosilicate glass	$SiO_2 \cdot P_2O_5$	A, F	1, 3, 4, 6, 7, 8, 10, 11
Borosilicate glass	$SiO_2 \cdot B_2O_3$	A, F	3
Silicon nitride	Si_3N_4	A, C, G	1, 5, 6, 7, 8, 9, 10
Aluminum oxide	Al_2O_3	A, G	1, 5, 6, 8, 9, 10
Other glasses		E	1, 2, 5, 6, 10, 11
Organic films		A, F	1, 2, 6, 10

[1]) A: Chemical vapour deposition
B: Evaporation
C: Plasma
D: Pyrolysis
E: Sedimentation
F: Spin-on, spray, etc.
G: Sputter

[2]) 1: Corrosion protection
2: Device isolation
3: Dopant diffusion source
4: Getter impurities
5: Increase breakdown voltage
6: Insulate metal layers
7: Mask against dopants
8: Mask against impurities
9: Mask against oxidation
10: Mechanical protection
11: Smooth out topography

For the thermal oxidation of silicon the fabrication method is described in section 6.1.4.5.1. Due to the good reproducibility of the published data, the influence of various sample parameters and preparation conditions on the growth of the silicon dioxide layers may be reviewed in 6.1.4.5.2. For the deposition techniques mentioned in Table 1, only a general description of the more important methods are given in sections 6.1.4.5.3 and 4, because the literature is not very consistent.

Books and general review articles on the fabrication of insulating layers are: [68B4, 69c1, 69v1, 69w1, 70m1, 73C2, 74B1, 75S3, 77A3, 77K1, 78p, 78v1, 82n].

6.1.4.5.1 Thermal oxidation of silicon

Thermal oxidation of silicon wafers to form silicon dioxide layers is the most important process in fabricating dielectric films for semiconductor devices. Thermal oxides form stable insulating layers on silicon. They are the basis of modern integrated circuits.

Oxidation kinetics

A thermal oxide grows on silicon wafers when the wafers are heated in a gas stream containing oxygen or water. A model describing the growth kinetics of thermal oxides was developed many years ago [65D3]. This first-order model is based on the diffusion and surface reaction of the oxidizing species. The model depicted in Fig. 1 relates the flux of the oxidant from the gas ambient through the oxide to the concentration at the silicon surface. The three fluxes I_1, I_2, and I_3 can be described as follows:

The gas phase flux I_1 is given by

$$I_1 = h(c^* - c_0)$$

where h is the gas phase mass transfer coefficient related to the concentration in the solid, c^* is the equilibrium concentration of the oxidant in the oxide, and c_0 is the concentration of the oxidant at the outer surface of the oxide.

The flux across the oxide I_2 is given by the diffusion of the oxidant through the oxide

$$I_2 = D_{\text{eff}} \frac{c_0 - c_i}{d_{\text{ox}}},$$

where D_{eff} is the diffusion coefficient of the oxidant in the oxide, and $\frac{c_0 - c_i}{d_{\text{ox}}}$ the oxidant gradient in the oxide.

Finally, the oxidizing species react at the silicon surface to form a layer of SiO_2. Assuming that the oxidation rate is proportional to the concentration of the oxidant at the silicon surface, the corresponding flux I_3 can be expressed by

$$I_3 = r_s c_i,$$

where r_s is the chemical surface reaction rate constant for oxidation.

Since at steady state $I_1 = I_2 = I_3$, c_i and c_0 can be expressed in terms of c^* and with N_1, the number of oxidant molecules incorporated in the oxide per unit volume, the flux I_3 can be written as

$$I_3 = N_1 \frac{d\,d_{\text{ox}}}{dt} = \frac{r_s c^*}{1 + \frac{r_s}{h} + \frac{r_s d_{\text{ox}}}{D_{\text{eff}}}}.$$

With $d_{\text{ox}}(t=0) = d_i$ as initial condition, the oxide growth can be expressed as a function of oxidation time t as follows

$$d_{\text{ox}}^2 + A\, d_{\text{ox}} = B(t + \tau)$$

where

$$A \equiv 2 D_{\text{eff}} \left(\frac{1}{r_s} + \frac{1}{h} \right)$$

$$B \equiv \frac{2 D_{\text{eff}} c^*}{N_1}$$

$$\tau \equiv \frac{d_i^2 + A\, d_i}{B}$$

The oxidation behaviour can be divided into two growth regimes.

For

$$t + \tau \ll \frac{A^2}{4B}: \quad d_{\text{ox}} \approx \frac{B}{A}(t + \tau).$$

This relation is applicable to thin oxides with B/A, the linear rate constant.

For

$$t + \tau \gg \frac{A^2}{4B}: \quad d_{\text{ox}}^2 \approx B(t + \tau).$$

This relation describes the growth of thick oxides with B, the parabolic rate constant.

In the first regime, the oxidation is limited by the surface reaction while in the second regime the diffusion of the oxidizing species through the oxide is the rate limiting process. The two regimes can clearly be seen when plotting the experimental data using normalized coordinates (Fig. 2).

It is found experimentally that the values of d_i differ for wet or dry oxidation. While $d_i = 0$ is valid for wet oxidation, an effective $d_i \approx 200$ Å is found for dry oxidation [65D3] thus, indicating that the growth of thin layers in dry oxygen has to be described by a different growth law. The oxidation rate equation can only be used to describe the growth of oxides thicker than 230 Å. For this oxide thickness regime, the oxidation rate is a function of the oxidizing species (water or dry oxygen), pressure of the oxidizing species, crystal orientation of the silicon surface, the addition of chlorine species to the oxidant, and the silicon substrate doping. The influence of these parameters on B or B/A are discussed in the following.

Very recently Tiller [80T1, 80T2] proposed a model which can be viewed as an extension of the basic Deal-Grove model [65D3] described above. The Tiller model pays particular attention to the reaction at the Si/SiO_2 interface using a mechanism which takes free volume supply into consideration. During the reaction of silicon with the oxidizing species, a free volume at the Si/SiO_2 interface is needed, so that the newly formed SiO_2 molecules can fit into the SiO_2 network. At the interface, the free volume V_f will react with the Si atoms and ionized O^{2-}, which has been formed at the SiO_2/gas interface, in the following manner:

$$V_f + Si + 2O^{2-} + 4p^+ \rightleftharpoons SiO_2.$$

The free volume may be supplied by one of the three mechanisms depicted in Fig. 3. The upper mechanism uses dislocations and strain due to lattice mismatch at the Si/SiO_2 interface. In the middle reaction, silicon vacancies Si_V from the silicon substrate provide the free volume, while for the bottom mechanism, silicon atoms are moved into interstitial sites Si_I to create the volume for the SiO_2 growth. All three of these mechanisms may play a role in the oxidation process with varying dominance depending on specific process conditions.

This model shows that the kinetics of oxidation cannot be described by neutral oxygen diffusion in the SiO_2 alone; the diffusion of ionized species and the flux of vacancies into the Si/SiO_2 interface must be considered. Combining the fluxes of O^{2-} and O_2 at the gas/SiO_2 and SiO_2/Si interface yields a relation for the film thickness d_{ox} which is equal in form to that obtained by the Deal-Grove analysis but uses changed definitions for A, B, and τ [65D3, 66N1, 69L2, 69R2, 75H3, 77D1, 77D2, 77E1, 78B1, 80T1, 80T2].

Oxidation equipment

A system for thermal oxidation of silicon usually consists of a resistance-heated furnace containing an open quartz tube and a gas system regulating gas flows of oxidizing species (Fig. 4). The oxidizing species may be dry oxygen or water vapour. In order to obtain water vapour, one can boil deionized water, which is the simplest way, or drip deionized water on a heated silica surface, or generate steam inside the oxidation tube by burning hydrogen in an oxygen atmosphere. This last process isoften called pyrogenic steam. The silicon wafers pushed or pulled with a boat-puller into or from the furnace, are placed vertically, closely stacked, in a quartz carrier boat.

Oxidation rates

Wet and dry oxydation

The water content of the oxidant is very important and has to be controlled very carefully, since the oxidation rate varies drastically. The variation of the oxidation rate constants A/B and B with temperature for wet and dry oxidation is shown in Figs. 5 and 6 for ⟨111⟩ oriented silicon wafers [65D3].

The enhanced oxidation rate of wet oxidation is due to the absorption of oxidant into the silicon dioxide and not to the speed of diffusion through the oxide. The "wet oxidants" diffuse more slowly than dry oxygen, but the silicon dioxide can adsorb much more water than dry oxygen so that the steam growth rate is higher than that of dry oxygen [65D3, 76J1, 76R2, 77I1, 77I3, 79R4].

Crystal orientation

The effect of crystal orientation on the oxidation rate is due to the density of silicon atoms at the SiO_2/Si interface which are available for the surface reaction. The effect is pronounced for the oxidation regime where the linear rate constant is dominant, i.e. for low temperatures and short oxidation times, except for the initial oxide growth regime. This behaviour can clearly be seen in Fig. 7.

Calculating the linear and parabolic rate constants by means of the general relationship demonstrates that there is almost no difference in the parabolic rate constant B for the two orientations ⟨111⟩ and ⟨100⟩. The variation in oxide growth only appears in the constant A (Fig. 8 and Table 2) [61L1, 77H3, 78D1, 77D5].

Table 2. Rate constants for ⟨111⟩ and ⟨100⟩ silicon [77D5].

T °C	Orien-tation ⟨hkl⟩	τ h	A μm	B $\mu m^2/h$	B/A μm/h
1200	100	0.03	0.0399	0.0453	1.14
	111	0.03	0.0404	0.0458	1.13
1100	100	0.09	0.101	0.0247	0.246
	111	0.09	0.0845	0.0244	0.289
1000	100	0.35	0.195	0.00913	0.0467
	111	0.35	0.120	0.00956	0.0797
900	100	3.2	0.429	0.00332	0.00775
	111	1.2	0.214	0.00381	0.0178
800	100	10.0	0.441	0.000755	0.00171
	111	5.0	0.354	0.00119	0.00335
700	100				0.000222
	111				0.000348

Impurity effects

When growing a thermal oxide on a doped silicon wafer, two effects have to be considered: impurity redistribution and enhanced oxidation.

Impurity redistribution

When a doped silicon wafer is thermally oxidized, the doping profile in the silicon is altered. For phosphorus, arsenic, and antimony, the impurity concentration in silicon near the Si/SiO_2 interface tends to pile up; for boron the surface concentration decreases (Fig. 9).

The change of surface concentration of the impurity is usually described in terms of a segregation coefficient k.

$$k=\frac{\text{silicon impurity concentration } c_{si}}{\text{oxide impurity concentration } c_{ox}}.$$

For $k>1$, the impurity remains in the silicon during oxidation ("pile up"); for $k<1$ the impurity segregates in the SiO_2 layer ("pile down"). The magnitude of the segregation coefficient is a function of crystal orientation, oxidizing species, oxidation temperature and pressure. Values published in the literature are not very reliable: [76C2, 78F3] (see also section 6.1.4.1.4).

Boron: $k=0.15\cdots0.4$,

arsenic, antimony, phosphorus: $k=1.2\cdots10$.

Enhanced oxidation on doped substrates

The oxidation rate B/A for heavily doped silicon wafers ($c_B \geqq 10^{19}$ cm^{-3}), particularly if n-type impurities are involved, is considerably higher than that for lightly or moderately doped substrates (Figs. 10 and 11).

Ho and Plummer [79H8] model this effect by taking into account the electrical influence of the doping on the interface reaction. In accordance with the oxidation model of Tiller [80T1, 80T2], which considers the influence of silicon vacancies Si_V on the surface reaction at the Si/SiO_2 interface, Ho and Plummer state that the enhanced linear oxidation rate B/A of highly doped silicon is due to the increased Si_V concentration present in heavily doped silicon layers. The increase of the Si_V concentration with doping is caused by the shift of the Fermi level in the silicon band gap. This model has been successfully used to describe a number of experimental data (Figs. 12 and 13) [79H3]. Furthermore, it has been observed that the parabolic rate constant B is nearly unchanged [77D5] (Fig. 14).

General literature is found in [65D2, 76R4, 77D5, 78I2, 78O1, 79H8, 79H3, 79H9, 78H7].

Initial oxidation regime

The initial oxidation regime, i.e. up to 200 Å SiO_2 thickness is of importance when for increasing packing densities of very large scale integration circuits the lateral dimensions and the horizontal dimensions, i.e. the SiO_2 layer thickness, are reduced. For a small oxide thickness, the general relation for oxide growth is not valid. The initial oxide growth has been measured in situ to describe the oxidation behaviour just at the beginning of any oxidation cycle (Fig. 15) [77I1, 76I1].

A relation has been given by Hooper and coworkers [75H2], who observed a linear-parabolic law. The rate constants (Table 3) differ from thick-film data. The time t, which is related to the oxide thickness d_{ox} can be expressed by

$$t = a_0 + a_1 d_{ox} + a_2 d_{ox}^2, \quad (d_{ox} < 200\ \text{Å}).$$

Table 3. Temperature dependence of growth parameters of ⟨111⟩ Si at 760 Torr oxygen for films less than 200 Å [75H2].

T °C	a_0 s	a_1 s Å^{-1}	a_2 s Å^{-2}
720	−4386.0	251.7	0.621
780	−576.0	58.5	0.268
832	−216.0	24.0	0.0654
870	−162.6	10.0	0.0396
930	−92.4	3.0	0.0174

An exponential law is reported in recent publications by [82M1, 83H1]. The general literature on this subject is [69R2, 70G2, 75H2, 75F1, 76I1, 77I1, 78B4, 78G5, 78I1, 79L1, 79R2, 72V1, 72G3, 83H1, 82M1, 83F1].

6.1.4.5.2 Modified thermal oxidation processes

Oxidation in chlorine oxygen mixtures

The addition of chlorine species during thermal oxidation of silicon is important to improve the threshold stability of MOS devices, to reduce the pinhole density in the oxide film, and to increase the dielectric strength of the oxides and the minority carrier lifetime in the silicon wafer. All these effects together result in a significant yield improvement which is a prerequisite for an increasing complexity of semiconductor devices.

The most common chlorine species added during wet or dry oxidation is HCl gas, but gases as e.g. trichloroethylene (TCE), trichloroethane (TCA), tetrachlorocarbide (CCl_4) or chlorine (Cl_2) are also used.

The addition of chlorine increases the growth rate of SiO_2 remarkably. The reason is that during oxidation according to the reaction

$$4HCl + O_2 \rightleftharpoons H_2O + 2Cl_2.$$

H_2O and Cl_2 form.

This reaction changes the oxidation constants in the following way: The parabolic rate constant B increases with the HCl concentration if the concentration is higher than 1%. No orientation effect is found for B, while the linear rate constant B/A does depend on the orientation as discussed above. The initial rapid increase for B and B/A is probably caused by an increased solubility of oxidants in the chlorine-containing oxides. The linear increase of the parabolic rate constant is related to the enhanced diffusion rate of the oxidizing species due to the incorporation of the chlorine in the growing oxide (Figs. 16, 17, and 18) [77D5].

The incorporation of the chlorine has been extensively studied and measured by various methods:

TEM, SEM, STEM and IR-spectroscopy [77M7, 78M1, 79M1]

SIMS [74K5, 78D4, 79F2]

Auger electron spectroscopy [73K2, 73C4, 76J2, 77J1, 81R1]

X-ray fluorescence [75V1, 78M1]

electron microprobe [75V1]

nuclear backscattering [75V1, 73M3, 77B3]

Rutherford backscattering [77B3, 78M4, 79R3].

The investigations show that the chlorine is not homogeneously incorporated but is mostly concentrated near the SiO_2/Si interface (Fig. 19) and the concentration does not increase linearly with the partial pressure of the Cl_2 (Figs. 20 and 21).

Not only does the chlorine influence the oxidation kinetics, it also has beneficial effects on the following semiconductor device parameters:

Due to the reduction of sodium concentration during HCl-oxidation (Fig. 22), the threshold voltage stability increases [72K3, 72K4, 72K1, 73K2, 74K5, 75D1, 75V1, 77R2, 78J1, 78S3, 79R7].

The dielectric breakdown voltage increases [74O1, 78H6].

The oxide defect density decreases [74O1].

The interface trap density decreases [72S1, 71F1, 73B6, 74S1, 75D1, 75J2, 78J1, 78S6].

The silicon minority carrier lifetime increases [71R1, 72R1, 73B6, 73Y1, 74G1].

Oxidation-induced stacking faults (OISF) in the silicon wafer shrink during HCl-oxidation or a preoxidation annealing step in an HCl containing ambient (Fig. 23) [75S2, 76S10, 76S4, 76S1, 76H2, 77H11, 77C9, 77S2, 78H11, 78H5, 78C5, 79C2, 79H6, 79H7].

A large amount of literature exists on the effect of chlorine on thermal oxidation and the properties of SiO_2.

General reviews:
[73K2, 77S2, 78H11, 79M3, 79M4, 80M3].

Kinetics of chlorine incorporation:
[73H3, 74V3, 76R4, 77H3, 77T2, 78D1, 78D2, 79H6, 79S2, 78H6].

Gases used in chlorination:

HCl
[71R1, 72R1, 72K3, 72K4, 73K2, 71F1, 72S1, 72K1, 73B6, 73K4, 73K3, 73Y1, 74O1, 75S2, 75V1, 76S10, 76S4, 76S1, 76R7, 77R2, 77T2, 77H3, 77H5, 78D1, 78D2, 78S6, 79M3, 79M4, 79M1, 79R3, 79R7, 81R1].

Trichloroethylene (TCE):
[72C2, 73Y1, 74O1, 75D1, 75J2, 76H2, 76H6, 77H11, 78C2, 78H11, 78H5, 79S2, 79F2, 79H6, 79H7, 78H6].

Trichloroethane (TCA):
[77C9, 78C5, 78J1, 78L2, 79C2].

CCl_4
[73Y1, 74O1, 78S6, 79F2].

Cl_2
[72K3, 72K4, 73K2, 73Y1, 74O1, 78D2].

Oxygen pressure

In characterizing the oxidation process by the Deal-Grove model, the parabolic rate constant B and the linear rate constant B/A are both dependent on the pressure of the oxidizing species. While at the present time most oxidation activities occur in the atmospheric pressure regime, some papers deal with oxidation at reduced pressure, and most recently oxidation at high oxygen pressure has assumed a certain importance because of reduced oxidation times and thus reduced diffusion of dopants in device structures.

Reduced-pressure oxidation

When using O_2/N_2 mixtures instead of pure oxygen, it is possible to oxidize silicon at reduced oxygen pressure. This technique has not been used very often, although it has the advantage of slowing down the growth rate and thus permits to grow very thin oxide layers at high temperatures. Examples of oxidation under reduced pressure are given in Figs. 24, 25, and 26.

High-pressure oxidation

Reviews on high-pressure oxidation are given in [79Z1, 81O1, 81C2]. Increasing the pressure in an oxidation system enhances the oxidation rate. While in an atmospheric system, the oxide growth is controlled by the temperature, in a high-pressure system, the growth rate is determined by temperature and pressure, thus allowing greater process flexibility. The use of high-pressure oxidizing reduces the oxidation time, and permits oxidation at a lower temperature (Fig. 27) which is of great interest with respect to very large scale integration (VLSI) technology [81C2]. The oxidation rates of steam oxidation at a pressure of 9.81 bar are given in Fig. 28.

Equipment for high-pressure oxidation

The first experiments in the high-pressure regime were performed in the 1960···1962 period [60L1, 61S1, 61L1, 62L1], but it was not until 1977 that the first commercial equipment was announced [77C1]. A schematic drawing of an apparatus for high-pressure oxidation, which is suited to semiconductor device fabrication, is shown in Fig. 29a [77M9]. A sealed tube which contains a certain amount of water for pressurization and oxidation has also been used quite often in research work [77M8, 78M5] (Fig. 29b). An experimental setup for high pressure oxidation at up to 500 bar was described recently, but such a system is not yet commercially available.

Details on equipment are found in [81O1, 61L1, 77C1, 73P2, 77M8, 77M9, 77T4, 78M5].

Application

There are two main areas in which high-pressure is applied: in the field of local oxidation or oxide isolation technologies, and in the suppresion of oxidation-induced stacking faults (OSF), which deteriorate the quality of MOS devices.

The selective oxidation, or LOCOS techniques [70A4] use Si_3N_4 layers as an oxidation mask to protect specific locations from being oxidized. The two parameters which characterize these techniques are the oxidation behaviour of the nitride and the properties of the SiO_2/Si_3N_4 edges during the oxidation.

It has been found that the amount of nitride oxidized during a given period of time is proportional to the oxidation pressure; therefore all published data can be normalized to one atmosphere and compared with results reported for atmospheric oxidation. Fig. 30 [81C2] summarizes the published data.

The SiO_2/Si_3N_4 edge behaviour is described by the lateral oxidation, which is the extent to which the oxide moves laterally under the Si_3N_4 layer. Lateral oxidation is affected by the temperature and the ratio of the oxide pad thickness to silicon nitride thickness, but not by the oxidation pressure [78M3]. The data in Figs. 31 and 32 can also be taken for wet oxidation under atmospheric pressure.

The beneficial influence of high-pressure oxidation on the formation of oxidation-induced stacking faults has been shown for oxides up to 5 μm thickness on ⟨100⟩ and ⟨111⟩ n- and p-type doped wafers, grown in dry oxygen or steam at pressures of up to 20 bar and temperatures in the range from 700 °C to 1150 °C. The data presented in Fig. 33 [77T1] for a steam oxidation at 6.5 bar as well as the values reported by different groups can be described by the OSF model of Murarka [77M5]. Applications of high-pressure oxidation are further discussed in [78T8, 78M3, 79T3, 80I1, 78Z1, 80A1, 77T1, 75M11, 74P2, 77M9, 77M3, 78K2, 79Z2, 78T5, 77M8].

Impurity redistribution during high-pressure oxidation

High-pressure oxidation tends to bring the segregation coefficient towards unity for all impurities [80F1]. The pile-down effect for boron and the pile-up effect for arsenic have been measured and found to be less pronounced at high pressures (Fig. 34) [81C2].

Due to these changes in the segregation coefficient, the impurity-enhanced oxidation at high pressure differs from that at atmospheric pressure in such a way that for heavy n^+-doping, the oxidation rate is reduced (Fig. 35) [80F1]. For high boron doping, the enhancement of the oxidation rate is increased [80F1]. More information is found in [80F1, 79K3, 78M5, 78K5].

Further details on the oxidation kinetics are given in [81C2, 78M5, 61L1, 62L1, 77T3, 75Z2, 77T2, 79T1].

6.1.4.5.3 Chemical vapour deposition

Chemical vapour deposition (CVD) is the most widely used technique for depositing dielectric layers in semiconductor technology. Chemical vapour deposition is a deposition technique in which gaseous substances react to form a permanent film on a heated substrate. This technique is very versatile because the deposited films only depend on the gas mixtures passing over the heated substrate surface. A large amount of literature and general reviews exist on this topic [66P1, 69F3, 70C3, 72H3, 75h1, 77B2, 74B1, 77K1, 77R1, 77A3, 78K1, 78h1, 78H2, 79K1, 80c1, 80M1, 80S1, 75K4, 69C4, 69c1, 70A2, 73C2, 73W2, 73B5, 76S8, 79S4, 78H1, 79R1, 75S3, 81D1, 79R5]. The CVD technology is also regularly discussed on symposia and conferences [70b1, 72g1, 73w1, 75b1, 77d1, 79s1, 78k1, 78k2].

Chemical reactions

Several reactions have been utilized in semiconductor device fabrication in order to produce the required layers. These reactions can be classified in the following manner:

Pyrolytics

In the early years, film deposition was frequently affected by means of a pyrolytic decomposition of gases. Typical examples are the decomposition of tetraethylorthosilicate (TEOS) to produce silicon dioxide layers, or the decomposition of silane gas (SiH_4) to form polycrystalline silicon films

$$\text{TEOS} \rightarrow SiO_2 + \text{by-products}$$
$$SiH_4 \rightarrow Si + 2H_2.$$

Oxidation

SiO_2 films can also be produced by oxidation at low temperatures. The following reaction is utilized for this process:

$$SiH_4\,(g) + 2O_2\,(g) \rightarrow SiO_2\,(s) + 2H_2O\,(g).$$

If a doping gas, e.g. phosphine (PH_3), is added to these starting gases, a silicate glass results. Phosphosilicate glass (PSG) is the most important type of glass for the passivation of semiconductor devices:

$$SiH_4\,(g) + 2PH_3\,(g) + 6O_2\,(g) \rightarrow SiO_2P_2O_5\,(s) + 5H_2O\,(g).$$

Hydrolysis

A typical example of a hydrolysis reaction is the deposition of alumina in accordance with the following relation:

$$Al_2Cl_6\,(g) + 3CO_2\,(g) + H_2\,(g) \rightarrow Al_2O_3\,(s) + 3HCl\,(g) + 3CO\,(g).$$

Ammonolysis

The ammonolysis reaction is utilized for the important deposition of silicon nitride. The gases silane and ammonia react by

$$3SiH_4\,(g) + 4NH_3\,(g) \rightarrow Si_3N_4\,(s) + 12H_2\,(g).$$

It is desirable in all cases that the chemical reaction heterogeneously takes place on the surface of the substrate. A homogeneous reaction is undesirable, since it leads to the formation of dust-like deposits. A homogeneous gas-phase nucleation can be prevented to a large extent by selecting suitable conditions of temperature and gas flow.

The various chemical reactions taking place during a chemical vapour deposition can be determined if the composition of the solid phases and the different gas phases has been established. Difficulties arise especially in measuring the vapour phases in situ. The composition of the solid films can be measured easily after the experiment with well-known methods. For in situ measurements the following methods have been successfully applied:

Mass spectrometry [75B4, 75B5, 75B6, 77B1, 79K2].

Raman spectroscopy [75S7, 76S9, 77S3, 79B3].

Absorption spectroscopy [63R1].

Gas chromatography [73L1, 77S8].

These techniques have been summarized recently in [77S8].

Thermodynamics and kinetics

The choice of the deposition temperature and of the concentrations of the reactants depend on the thermodynamics and the kinetics of the specific CVD system.

Calculation of the thermodynamical equilibrium in a multicomponent system helps a great deal in finding the right temperature range, reactor pressure and input concentrations of reactants. Various ways for calculating the thermodynamics in a CVD-system are published [70v1, 74f1, 75J1, 76R6, 80S1].

Knowledge of the thermodynamical properties does not answer the question of how much progress is actually made by the chemical reaction during deposition. This depends on the kinetics. Usually a heterogeneous reaction takes place during deposition. CVD reactions can generally be divided into the following steps:

diffusion of reactants to the surface

adsorption of reactants at the surface

surface events, such as chemical reactions, surface motion, lattice incorporation etc.

desorption of products from the surface

diffusion of products away from the surface.

Examinations of the kinetics during the deposition of SiO_2 and Si_3N_4 films showed, that the reaction mechanism is mainly influenced by the silicon-forming compound [74F2].

Transport phenomena

Besides thermodynamics and kinetics, transport phenomena are of especial importance for optimizing the design of a CVD reactor. The most important parameters influencing the homogeneity and efficiency of a deposition are:

velocity of the gas flow
temperature and temperature distribution
pressure
reactor geometry
gas or vapour characteristics.

Methods for measuring or modelling the transport phenomena have been described by several authors [67B2, 72T2, 74E1, 75B6, 77M6, 77W1, 78B6, 78B5, 78W1]. No standard rules have been established.

CVD systems

Various systems have been developed for CVD at atmospheric pressure (called: APCVD), low pressure (LPCVD) and at low pressure with plasma enhancement (PECVD). While in the past most work has been performed in APCVD systems, the acceptance of LPCVD and PECVD systems is growing rapidly. In addition to the general remarks made above, which hold for all systems, some specific points have to be considered for LPCVD or PECVD systems.

Low-pressure CVD (LPCVD)

The chemical reactions involved in the deposition procedure of films at low pressure are basically the same as those in the APCVD, and the same starting gases can be employed. The essential difference between APCVD and LPCVD lies in the fact that the reduction of pressure changes the deposition rate determining step. At low pressure, the ratio of the mass transfer rate of the gaseous reactants to the surface reaction rate shifts in favour of the surface reaction, while at atmospheric pressure these two rates are approximately of the same order of magnitude [79K1]. The enhanced mass transfer rate under low-pressure conditions is of special importance to the deposition uniformity and economics [77R1]. The advantage of a low reactor pressure on thickness uniformity of the deposited layer has been shown by modelling the diffusion current of the reactants [79B1]. If the emphasis is not only on uniformity but also on a high throughput, better step coverage and conformity, and pinhole density, the improved quality of LPCVD compared with APCVD is especially important in modern very large scale integration (VLSI) processing. Numerous articles have been published [79K1] since the appearance of the pioneering work [62S1, 68O1, 73T1].

A disadvantage of the LPCVD technique is that some reactions require a high deposition temperature. In cases in which the substrates do not permit high-temperature stress, the same chemical reactions can be used at lower substrate temperature when plasma enhancement is employed in place of thermal energy.

Plasma-enhanced CVD (PECVD)

The first work which showed the feasibility of a plasma-enhanced chemical reaction in CVD technology appeared as early as 1963 [63A1]. But it was not until the introduction of the first radial flow reactor [74R1] that interest in PECVD began to grow [79R1].

In principle, use is made of the same gases as those employed for the other CVD processes, but the reaction does not occur through the action of thermal energy but by means of the glow discharge effect. The deposition temperature can be kept relatively low, typically at 300 °C, which is the main reason for the rapid acceptance of the PECVD technique for passivation coatings.

After generation of the chemically reactive species by the plasma, a large variety of reactions can occur [67m1, 69K2, 74h1, 73R1, 76B4, 76h1, 78B2, 78w1, 79B1]. The major inelastic collisions are listed in Table 4 [79R1]. [79R1]. The plasma potentials which exist during deposition are described in [75H1, 77H2]. Processes which can lead to contamination and radiation damage during plasma deposition are discussed in [79V2].

The properties of the usually amorphous films change to a large extent in response to variations of plasma parameters, e.g. RF power or reactant gas flow ratios [78H1, 78S8, 78S9, 78V1, 79V1].

Table 4. Possible inelastic collisions in a low-power RF gas discharge. e = electron, n = any neutral species [molecule (m), atom, radical], $x\cdot$ = atom or radical, i = ion, s = surface, and an * indicates that this reaction is believed to be important for film deposition [79R1].

Reaction	Remarks
I. Electrons + neutrals yield:	
*(1) Excited species + slow e	excitation may be metastable
$\rightarrow i+e$	Penning or autoionization
$\rightarrow 2x\cdot s$	Penning dissociation
*(2) Negative i_a	attachment or capture
$\rightarrow$ negative $i_b + x\cdot$	dissociative attachment
*(3) Positive i_a + 2 slow e	ionization
$\rightarrow$ positive $i_b + x\cdot$	dissociative ionization
$+e \rightarrow 2x\cdot s$	dissociation
$+e \rightarrow$ positive i_b + negative i	
II. Electrons + ions yield:	
(4) Neutral	attachment or recombination
*(5) $2x\cdot s$	dissociative recombination
III. Ions + neutrals yield:	
*(6) Other ions + other neutrals:	metathesis or transfer
$i_a + n \rightarrow i_b + x\cdot$	atom or ion transfer
$i_a + n \rightarrow i_b + m$	ion-molecule transfer
$i_a + n_b \rightarrow i_b + n_a$	charge transfer
(7) Neutral + electron:	
negative $i + n_a \rightarrow n_b + e$	associative detachment
IV. Ions + ions (opposite charge) yield:	
(8) Neutrals	recombination
(9) New $x\cdot s$	dissociative recombination
V. Neutrals + neutrals (usually $x\cdot$) yield:	
*(10) Other neutrals:	metasthesis or transfer
$m_{ab} + x\cdot_c \rightarrow m_{bc} + x\cdot_a$	especially if b = H, H abstraction
$2x\cdot s \rightarrow m$	recombination or addition
2 (m or radical)$_a$ $\rightarrow$ (m or radical)$_b$ + (m or radical)$_c$	disproportionation
(11) Ion + electron:	
$n_a + n_b \rightarrow n_a + i_b + e$	collisional detachment
VI. Surface and miscellaneous reactions	
(12) Excited species + $s \rightarrow n$	deexcitation or positive ion impact
(13) $s + i \rightarrow x\cdot + i_s$	sputtering
(14) $s + x\cdot \rightarrow i + e$	surface contact ionization
(15) $s + e$ or $h\nu \rightarrow i_s + e\,s$	photoelectric effect or electron impact

CVD reactors

The reactor is the most essential part in the CVD system, followed by the gas system configuration (Fig. 36) and the equipment for electrical control. In the case of APCVD, many types of different reactors have been described, while LPCVD and PECVD reactors have a more or less standardized configuration.

Atmospheric-pressure reactors

APCVD reactors are divided into low-temperature- and high-temperature reactors according to the deposition temperature applied. High-temperature reactors are further divided into *hot-wall* and *cold-wall* reactors. Another classification is used with respect to the gas flow. Depending on the flow direction, the reactors are called *vertical* or *horizontal* reactors.

Low-temperature reactors differ mainly in their gas injection systems [72W4, 73B2, 75K4, 77K1, 78K1]. The following reactor configurations have been described

horizontal tube displacement reactor
vertical rotary batch-type reactor
vertical continuous reactor using an extended area dispenser for the premixed gases
vertical continuous reactor using laminar flow nozzles for separate gas streams.

These reactor types are schematically represented in Fig. 37 [75K4].

Almost all the high-temperature reactors use a cold-wall. Heating of the Si-wafers via a susceptor is accomplished by RF-heating or radiation lamps. An example of an RF-heated reactor is shown in Fig. 38 [77G3]. The susceptor, which is made of SiC-coated graphite, is the only part kept at high temperature. This is of advantage for endothermic reactions because the reaction will occur only on the wafer surface. For exothermic reactions, a hot-wall system is preferred. This reactor type is generally tubular in form and uses a resistance element to heat the wafers in a similar manner as a diffusion furnace used for thermal oxidation of Si wafers. Such reactors have been used for research, but are of minor importance for deposition at atmospheric pressure.

Low-pressure reactors

A low-pressure reactor should be optimized with respect to the temperature control, because under low-pressure conditions, the rate of surface reaction controls the deposition. In most systems described, the reactor consists of a diffusion-type hot-wall furnace with a circular tube containing the wafer standing vertically in a boat [77R1, 77H10]. A typical LPCVD system is schematically shown in Fig. 39 [79K1]. Such reactors are used for high-temperature as well as low-temperature depositions. For low-temperature silicon dioxide deposition, special inserts for injecting the gases at different places are necessary [77R1, 77L1, 78F1, 78T1].

Plasma-enhanced reactors

The plasma-enhanced chemical vapour deposition is determined by the RF power density and the plasma distribution. Sufficient control of the uniformity of these parameters over a large area was first achieved with planar flow reactors [74R1, 76R1, 78R1]. The gases can flow either from the center to the edge of the vacuum chamber (Fig. 40) or in the opposite direction. The wafers are placed on a heatable plate at ground potential. The RF-power electrode is located at a distance of about 5 cm. This type of reactor has been discussed by several authors with respect to deposition uniformity affecting parameters, such as residual gas concentration of the controlling species, the electron density, and the residence time of the gases [79R1, 79R5, 80E1].

Vertical or horizontal reactor approaches using inductive coil-electrode geometries or microwave excitation have also been published [73T3, 77H7, 77S1].

A reactor configuration, in which the wafers can be staggered in a diffusion furnace, positioned vertically and closely packed, has been suggested recently [79R6]. This hot-wall resistance-heating concept (Fig. 41) provides the system with a good temperature control, the number of particles on the deposited layers decreases, and the throughput increases compared to the planar radial flow reactor.

6.1.4.5.4 Evaporation

The evaporation techniques can also be used for depositing dielectrics. The material can be evaporated directly from a resistance-heated filament [69H1, 70H1, 73S1], with an electron beam gun [69M1, 71S1, 72H2, 74V2], or in a reactive mode using an oxygen atmosphere, called *reactive evaporation* [74B3]. The latter technique has been studied for Al_2O_3 and SiO_2 layer deposition [74K2]. General reviews on evaporation of insulating films are given in [70G1, 70H2, 71H1, 74B3, 76A1, 76H3, 66p1, 70B1].

Evaporation has the advantage that a deposition is possible at relatively low temperatures (< 300 °C). A very important disadvantage lies in the fact that during evaporation charged particles are generated which will be trapped in the deposited film [69H1, 69M1, 70H1, 71S1, 72H2, 72J2, 73S1, 74V2]. Due to this severe restriction, the evaporated dielectrics are of minor interest in modern integrated-circuit technology [68P1, 70H1, 71H1, 72H2].

6.1.4.5.5 Sputtering

The sputtering technique can be used for depositing dielectrics. Only the parameters which are important for deposition of dielectrics will be discussed. General reviews on sputtering are given in [55W1, 62K1, 66M2, 68c1, 69c1, 69C1, 69W1, 70M1, 70W1, 70J1, 70L2, 70M3, 71V1, 73m1, 76O1, 76T3, 76W1, 77C2, 77L2, 80c1].

Two methods are commonly used: RF sputtering and magnetron sputtering.

RF sputtering

The apparatus for RF sputtering (Fig. 42) is, in principle, the same as that for DC sputtering except for the cathode potential. Instead of DC power, radio frequency (RF) power is needed for sputtering insulators. The substrates can be grounded, floating, or biased. The sputter gas is usually argon or a reactive gas such as oxygen or nitrogen. In the latter case, the sputtering process is called *reactive sputtering*, which is widely used for dielectrics and well documented in the literature [64S2, 66P1, 67P2, 68H2, 69F2, 69P1, 69W1, 70C1, 70F1, 73H1, 75A2, 75G2, 75S9, 76D5, 76G1, 76W4, 79F1].

The sputter action generates not only neutral particles but also charged particles and radiation. This subject has been studied in the following publications:

Secondary electrons: [68H1, 68T1, 69B1, 72B3, 74C2, 75S8];

Ions: [69B2, 71B2, 72C1, 73B3, 75W3, 76K5, 76K6, 77C6, 78C4];

X-rays and photons: [71T4, 73B3, 75B3, 75K1, 75M5, 75W4, 76B7].

The radiation can damage sensitive devices such as MOS transistors [73M2, 74M1].

If the substrates are biased, this sputter mode is called *bias sputtering*. Changing the bias potential influences the stoichiometry of the deposited films [67P1, 69W3, 70M2, 71T5, 75C1, 76D4, 76H9, 76S3, 76W5, 78G2], the step coverage, and other physical film properties [70C2, 70M1, 71V1, 76W1, 77D3].

The *film composition* also depends on the target properties. These are:

a) the target parameters, e.g. heat dissipation, electrical isolation, ground shielding and the material composition [63M1, 66M2, 66D2, 68H1, 68V1, 70M1, 70J1, 73C3];
b) changes of the target composition during the process [67P1, 69A3, 69O1, 71T5, 72N1, 73S3, 75M6, 76D4, 76C4, 76F1, 76M6, 76P1, 76W5, 76W6];
c) chemical dissociation [68V1, 69V1, 70S4, 70T2, 74K3, 75M7, 75N1, 76K7, 76K8, 77H8, 77K2].

A very important parameter for the *film quality* is the amount of sputter gas which is incorporated in the sputtered film. The following sputter parameters influence the film quality: target voltage, substrate bias voltage, sputter gas pressure, substrate temperature, target-to-substrate distance, inert gas atomic volume [67W1, 69W3, 70L3, 70S3, 71B3, 71M1, 71M2, 73B4, 73S2, 74B2, 74F1, 75C1, 75L1, 76R5, 77C7].

Magnetron sputtering

Magnetrons are defined as diode sputtering sources in which magnetic fields are used in combination with the cathode surface to form electron traps which are designed that the $\boldsymbol{E} \times \boldsymbol{B}$ electron drift currents close into themselves [78F2]. The magnetically enhanced sputtering techniques have been developed many years ago [63K1, 65G1, 69W4, 71M3] and introduced by Chapin [74C1] to deposit insulators.

Various magnetron sputtering source configurations have been described by Thornton and Penfold [78T7]. For deposition of dielectrics, the two most common types are inverter magnetrons [76C3, 77F1, 77L2, 77S7, 78F2] and planar magnetrons [76M3, 76V1, 77M2, 77N1, 77S6, 78W3]. Examples are given in the previous section.

Magnetron RF sputtering in comparison to e-beam evaporation and conventional RF sputtering, offers an alternative low-temperature deposition technique due to the following sputter parameters:

High deposition rate [78S3], large substrate areas [78H10], low deposition temperature [76V1], bias sputtering [77F1, 77S6], minimal bombardement of substrates by electrons or ions [76H2], and good step coverage [70L1, 73K1, 74V1, 76K3, 80S2]. It also permits reactive sputtering [77C8, 80M2]. The application for magnetron-sputtered dielectrics is a low-temperature hermetic passivation process employing Al_2O_3 or Si_3N_4 on aluminum metallization.

The important points of magnetron sputtering have been reviewed by several authors [76W7, 78F2, 78T6, 78T7, 78W2, 78W3, 79C1].

6.1.4.5.6 Anodization

Anodization techniques are in very limited use in semiconductor device technology. Anodization has only been used successfully in two fields. The first one is forming Al_2O_3 films for electrolytic capacitors or protective layers on aluminum metallization [71D1, 72D1, 73C1]. The second one is growing thin SiO_2 films with a good thickness accuracy at low temperatures. The latter method has been used to good advantage for stripping very thin layers of the silicon wafers to determine doping profiles in the wafer without changing the concentration and depth distribution [65D1, 78G3].

In *wet anodization*, the metal or semiconductor is connected to the anode in an electrolytic cell. The electrolytes are aqueous solutions containing anions like acetates, citrates, borates or sulfates. When a DC-current is flowing through the electrolyte, the oxide layer grows. During oxide formation, the substrate material is consumed as in thermal oxidation. Because the current decreases as the film thickness increases, oxide growth by an anodization process is self-limiting and can thus be controlled very precisely. The quality of the films is poor with respect to ionic contamination. Reviews on anodization are given in [70C3, 71D2, 76S5].

Gaseous anodization has also been used for preparing oxides of aluminum and silicon [67J3, 70O1, 76H1, 77O1]. This method uses a plasma of oxygen and therefore is also called *plasma oxidation* [65L1]. In order to obtain film growth, the sample is connected to a positive bias.

Other film deposition processes

Other existing deposition processes have no significance to semiconductor device technology because the application is either very limited or the usefulness has not been confirmed. These methods have been reviewed or cited in the following references: [76W4, 77A3, 76C1, 77W1, 78v1, 69v1, 75M4, 78T2].

6.1.4.5.7 Properties of insulating films

The properties of insulating films on semiconductors are still not very well established and strongly dependent on the fabrication technology. Only for the thermal oxide, intrinsic properties which do not depend on the material purity and structure quality may be given. These data are listed in Table 5. Barrier heights are given in Table 6.

Table 5. Intrinsic data of thermal SiO_2. (For the band structure diagram see Fig. 43).

Property	Value	Unit	Ref.
Energy gap E_g	8.8	eV	82n, 71P2
Dielectric permittivity ε	3.84		82n
Electron affinity χ_{SiO_2}	1.0	eV	82n, 65W1
Barrier height to Si($\chi_{Si}-\chi_{SiO_2}$)	3.23	eV	82n, 71P2
Breakdown field	8	MV	81W2

Table 6. Metal work function Φ_M and the work function difference to silicon ($\Phi_M-\chi_{Si}$). (The work function difference $\Phi_M-\chi_{Si}$ is independent of the semiconductor doping. In order to determine the actual work function difference

$$\Phi_{MS}=\Phi_M-\Phi_{Si}=\Phi_M-\chi_{Si}-E_g/2+(E_F-E_i)$$

the correction due to the doping-dependent position of the Fermi energy has to be added (see Fig. 44).

Gate metal	Φ_M eV	$\Phi_M-\chi_{Si}$ eV	Remarks	Ref.
Ag	4.97	0.76		75K6
Al	4.13	−0.06		75K6, 82R1
Au	5.06	0.91		75K6
Cr	4.18	0		75K6
Cu	4.87	0.63		75K6
Mg	3.19	−1.05		75K6
Ni	4.6	0.4		75K6
$NiSi_2$	4.5	0.3		80S4
Sn	3.42	−0.83		75K6
poly-Si (B-degenerate)	5.4	1.2	doping Fig. 45	74W1
poly-Si (P-degenerate)	4.1	−0.1	doping Fig. 45	74W1
Pd	4.82	0.6		66G1
Pd_2Si	5.2	1.0		80S4
Pt	5.29	1.1		66G1
PtSi	5.15	0.95		80S4
RhSi	4.96	0.76		80S4
WSi_2	4.8	0.6		80S4

A large amount of information has been collected on the variation of thermal oxide properties with technology. These properties are still topic of regularly held conferences and are found in books and review papers [82n, 81s, 78p, 79l, 79r]. The following Tables 7···10 give a survey of the information available ordered with respect to the fabrication technology and for various specific properties.

Table 7. Electrical properties of thermal SiO_2.

Property	Remarks	Ref.
Layer		
Oxide trapped charge	trap levels charged by hot carrier injection caused by ion implantation, e-beam evaporation or radiation damage	78D5, 82n
	water related	71N1, 78D6, 82n
Fixed oxide charge	always positive $10^{11}\cdots10^{12}$ cm^{-2}	82n
	Deal triangle for annealing	67D1
Mobile ionic charge	Na, K mobile ions	67H3, 77S9
	$\mu=\mu_0\,e^{-E_A/kT}$	
	Na: $\mu_0=1$ cm^2/Vs, $E_A=0.66$ eV	77S9
	K: $\mu_0=0.03$ cm^2/Vs, $E_A=1.1$ eV	77S9
Oxide conductivity	only electrons mobile, holes trapped	78D5, 82n
Interface		
Interface state density	midgap density $10^8\cdots10^{12}$ cm^{-2} eV^{-1}	83S1, 82n 77C10, 76G7
	continuous increase to E_C	83S1
	U-shaped continuous distribution	76G7, 77C10
	dangling bond $E_V+0.35$ eV	83S1, 81P1
	capture cross-section decreases near E_C	83S1, 76G7
Interfacial layer	15···30 Å strained layer	81H1
	roughness $\lesssim 10$ Å	81H1
Non-uniformities	surface potential fluctuation	83S1, 82n
	gross non-uniformity (Na)	73D1

Table 8. Measurement techniques for the interface characterization [84L].

Method	Determination of	Ref.
Capacitance voltage (CV) technique	interface state density, flat band voltage, oxide charge, doping concentrations	82n, 76G7, 84L1
Conductance technique	interface state density, capture cross-sections, surface potential fluctuations	82n, 76G7, 67N1
Transient spectroscopy DLTS, CC-DLTS	interface state density, capture cross-sections,	83S1, 78J3, 79S5
	gate boundary effects	82Z1
Thermally stimulated ionic current TISC	mobile ions	77S9, 82n
Triangular voltage sweep TVS	mobile ions	71K6, 82n
Avalanche injection	oxide traps, capture cross-sections	78D5, 84L1

continued

Table 8 (continued)

Method	Determination of	Ref.
Electron beam-induced currents EBIC	gross non-uniformities	84L1
Zerbst method	carrier lifetime	66Z1, 71S3, 82n

Table 9. Properties of SiO_2 [82n].

Type of oxide	Density	Index of refraction n	Dielectric constant (at 1 MHz unless otherwise stated)	Resistivity	Dielectric strength	Etch rates r_e		
						1.8 molar HF etch	buffered HF etch [1])	p_O-etch [1])
	$g\,cm^{-3}$	$\lambda = 5460\,Å$		$\Omega\,cm$	$10^6\,V\,cm^{-1}$	$Å\,min^{-1}$	$Å\,min^{-1}$	$Å\,min^{-1}$
Thermal oxide								
Steam ($97\,°C$ H_2O)								
Open tube	2.00···2.20	1.45···1.46	3.9	$10^{15}\cdots10^{17}$	6.8···9.0	162	438	120
High pressure	2.32	1.48			6.8···9.0	156···162		
Dry oxygen	2.24···2.27	1.46···1.47	3.84	$3\cdot10^{15}\cdots 2\cdot10^{16}$	6.8···9.0	198	408	120
Anodic oxide								
Electrolytic	1.80	1.32···1.49			5.2···20	21600	3000···4500	10800···13800
Plasma		1.46···1.47			10	174	540	144
Deposited oxide								
Silane			4.4···4.6 (10 kHz)		5		1200···6000	
Bulk silica glass	2.20	1.46	3.8	$10^{15}\cdots10^{16}$	1···5			

[1]) Buffered HF etch: $10\,cm^3$ HF (48%), $100\,cm^3$ NH_4F solution ($0.45\,kg$ $NH_4F/680\,cm^3$ de-ionized H_2O); p_O-etch: $15\,cm^3$ HF (48%), $10\,cm^3$ HNO_3 (70%), $300\,cm^3$ de-ionized H_2O.

Table 10. Properties of insulating films.

Preparation method	Reactants	Deposition temperature °C	Ref.	Remarks
SiO_2				
Thermal growth at atmospheric pressure	O_2 or H_2O	700···1200	65D3, 69L2	oxidation rates increase with T
	H_2O+O_2+Ar	800···1200	66N1, 74I1, 79R4, 76R2	rates increase with p_{H_2O}
	O	600···1000	78B1	oxidation in atomic oxygen
	O_2+N_2	1200	75H3	reduced oxidation rate by $p_{O_2}<1$ atm
	O_2 or steam	700···1200	71H3, 79H9, 78O1, 78I2, 78H7, 76R4, 69R2, 65D2	increased rate by dopants
	O_2 or H_2	700···1200	80T1, 80T2	volume increase
	O_2+H_2O	600···1000	79R2	initial oxidation regime

continued

Table 10 (continued)

Preparation method	Reactants	Deposition temperature °C	Ref.	Remarks
SiO_2			70G2, 82M1, 83H1	$d_{ox} \leqq 6$ nm;
			72G3	$p_{O_2} = (0.01 \cdots 1.0)$ atm, $d_{ox} \leqq 30$ nm;
			72V1	reduced O_2 partial pressure
			76I1	in situ thickness measurements
			78I1	$d_{ox} \leqq 20$ nm; oxide properties
Thermal growth at high pressure				
$\lesssim 500$ atm	O_2	700···850	75M11, 78Z1, 79Z2, 80I1	growth rates and properties
		900	77M3	local oxidation
(40···150) atm	steam	500···800	61L1	crystal orientation
90 atm	steam	700	74P2	local oxidation
<25 atm	steam	750···900	80F1	effect of substrate doping reduced
20 atm	steam	700	78K5, 79K3	effect of substrate doping reduced
			78K2	formation of stacking faults
15 atm	wet + HCl	600···1000	77M8, 78M5	sealed tube
10 atm	steam	700	79T1	theory, oxidation kinetics
6.6 atm	steam	800···1100	73P2, 77M8, 77M9, 77T1, 77Z2, 78M3	local oxidation, oxide properties
	dry + 14 ppm H_2O	950···1100	75Z2, 77Z2	
Evaporation	SiO_2		64L1, 68P1, 70H1	general properties
			77Z1	electric characterization
Anodization	Si	3···50	65D1, 76B5, 76J1, 76M2, 77G1, 78A1 78G3, 78K6, 79J1	ethylene glucol $+0.04$ n KNO_3 $+(1 \cdots 2)$ g/liter of $Al(NO_3)_3 \cdot 9H_2O$
Ion implantation	$Si + O_2^+$	RT	76D3, 78K3	dose: $1 \cdot 10^{18}$ cm^{-2} energie: 30 keV anneal: (550···800) °C
RF-sputtering	SiO_2	RT···200	76S6	anneal: 1050 °C
			70L1, 71V1, 73K1	step coverage
			70G4, 76H5, 76H8	deposition rate
			67P3, 69T3	low stress conditions
			78P1, 79H4	general properties
			80H1	gate oxide
RF-sputtering with bias	SiO_2	RT···200	77D4,	general properties
			78T3	model
RF-magnetron sputtering	SiO_2	RT···200	77U1, 79H5	deposition rate
			77H9, 80S2	step coverage
reactive sputtering	$Si + O_2$	RT···200	76O1, 77K3, 78B3	general properties
microwave plasma	$Si + O_2$	<600	76B6, 76D1, 76D2, 78H9, 79G1	general properties
	$Si(OC_2H_5)_4$	300	78K4	anneal: 400 °C, H_2

continued

Table 10 (continued)

Preparation method	Reactants	Deposition temperature °C	Ref.	Remarks
SiO_2				
RF-plasma	$Si + O_2$		75B2, 75S4, 76A4, 79H2	general properties
APCVD	$SiH_4 + O_2$	450	67G1, 68K3, 70A2 70B3, 70K3, 73B1, 73B2, 75M1	general properties use for passivation
	$SiH_4 + O_2$	450	68S1, 69S2, 76V2, 78V2	kinetics
			76M5	model
	$SiH_4 + O_2$	600···1000	68C1	general properties
	tetraalkoxysilanes	depends on molecular weight	76B3	general properties
	$Si(OC_2H_5)_4$	765	62J1, 69H2, 70A2, 75W1, 76G5	general properties
	$Si(OC_2H_5)_4 + O_2$	250···500	68N1	$(C_2H_5)_3Sb$ as catalyst
	$C_2H_5Si(OCH_3)_3$	700	61J1, 63E1, 65K1, 76A5	general properties
	$HSi(C_3H_7)_3 + O_2$	750	74A1	general properties
	$SiCl_4 + H_2 + NO$	1150	67R1	general properties
	$SiCl_4 + H_2 + CO_2$	1100	64S1, 65T1	general properties
	$SiBr_4 + H_2 + NO$	850	67R1	general properties
	$SiBr_4 + H_2 + CO_2$	800	66R1	general properties
	$SiH_4 + H_2 + CO_2$	800···1050	69S1, 75K2, 76G2	temp. dependent SiH_4 mole fraction
	$SiH_4 + H_2 + CO_2 + HCl$	800···1050	76G3, 79G2	HCl reduces deposition rate
	$SiH_2Cl_2 + N_2O$	900	76L1	gas flow dependent
LPCVD	$Si(OC_2H_5)_4$	700···800	61K1, 62S1, 68O1, 76B1, 77K1	(0.4···0.7) torr
			79H1	model
	$Si(OC_3H_7)_4$	700···800	73T1, 73T2	general properties
	$SiH_4 + O_2$	425	77R1, 79B2, 79T2	todays standard
	$SiH_4 + N_2O$	700	78D3	Si-content varied
	$SiH_4 + N_2O$	800···900	77R1	general properties
	$SiH_2Cl_2 + N_2O$	900	77R1	general properties todays standard
PECVD	$Si(OC_2H_5)_4 + O_2$	200	64I1, 65I1, 66S2, 69S3, 72M1	general properties
	$SiH_4 + N_2O$	<400	65S1, 69A2, 74K1, 78V1, 79H2, 79V3, 79R5, 80M1, 80R1	todays standard for passivation
	$SiCl_4 + O_2$	950···1000	75K3, 76K2	general properties
	$SiH_4 + CO_2$	<400	79R5	RF power
Arsenosilicate glass (AsSG)				
APCVD	$Si(OC_2H_5)_4 + AsCl_3 + O_2$	500	67L1, 70A3	general properties
	$Si(OC_2H_5)_4 + AsCl_3$	500	69C3	general properties

continued

Table 10 (continued)

Preparation method	Reactants	Deposition temperature °C	Ref.	Remarks
Arsenosilicate glass (AsSG)				
APCVD	$Si(OC_2H_5)_4 + AsCl_3 + CO_2 + H_2$	~500	69T1	general properties
	$SiCl_4 + AsCl_3 + CO_2 + H_2$	~500	69T1	general properties
	$SiH_4 + AsCl_3 + O_2$	500	71W1, 72W1, 72W2, 72W3, 73W1	general properties
	$SiH_4 + AsH_3 + O_2$	500	71P1, 72F1, 73G1	general properties
LPCVD	$SiH_4 + AsH_3 + O_2$	400···500	78H3, 78H4	doping source
Antimony silicate glass (SbSG)				
APCVD	$SiH_4 + (CH_3)_3Sb + O_2$	315	70G3	general properties
Boron silicate glass (BSG)				
APCVD	$Si(OC_2H_5)_4 + B(OCH_3)_3$	700	65S2, 68B1, 68B2, 80P1	general properties
	$Si(OC_2H_5)_4 + B(OC_3H_7)_3$	688	69W2	general properties
	$SiH_4 + B_2H_6 + O_2$	325···450	68F1, 68F2, 69S2, 70K1, 70K2, 71B1, 71K2, 71T2, 71T3, 72B2, 72T1, 75W2	doping source
LPCVD	$Si(O_2C_3H_7)_4 + B(OCH_3)_3 + O_2$	750	73T1, 73T2	general properties
Phosphorus silicate glass (PSG)				
APCVD	$Si(OC_2H_5)_4 + PO(OCH_3)_3$	700···800	65S2, 68F2	general properties
	$Si(OC_3H_7)_4 + PO(OCH_3)_3$	700···800	65S2, 68F2	general properties
	$SiCl_4 + POCl_3 + O_2$	900	66M1	H_2O added
	$SiH_4 + POCl_3 + O_2$	~500	72G1	general properties
	$SiBr_4 + PH_3 + O_2 + H_2$	600···1000	76G6, 77G1	general properties
	$SiH_4 + PH_3 + O_2$	300···500	68F1, 68F2, 69S2, 69T2, 70K1, 70K2, 70K3, 70S1, 70S7, 71S2, 73B1, 75M1, 75S5, 75S6, 75W2, 76K1	passivation and doping source
LPCVD	$Si(OC_2H_5)_4 + PO(OCH_3)_3$	700···800	75S1	general properties
	$Si(OC_3H_7)_4 + PO(OCH_3)_3$	700···800	73T1, 73T2	general properties
	$SiH_4 + PH_3 + O_2$	350···450	74V1, 76K1, 77K1, 77L1, 77R1, 78F1	todays standard
	$SiH_2Cl_2 + PH_3 + O_2$	600···700	78T1	general properties

continued

Table 10 (continued)

Preparation method	Reactants	Deposition temperature °C	Ref.	Remarks
Al_2O_3				
Oxidation	$Al+O_2$		76A2	general properties
	$Al+H_2O$		76H4, 78C1	general properties
Wet anodizing	Al	RT···90	68A1, 71D1, 72D1, 73C1, 76J1, 77A2, 77A4, 77I2, 78G1, 79T4, 80B1, 80K1, 81C1	for capacitors mainly
Gaseous anodizing	$Al+O_2$	RT	63M2, 68L1, 68W1, 77C3, 80S3	plasma
Evaporation	Al_2O_3	RT···600	64L1, 70A4, 71H1, 75E1, 75M10, 78L3	e-gun
Reactive-evapor.	$Al+O_2$		69F1	general properties
RF-sputtering	Al_2O_3	RT···500	69K4, 69P2, 70G4, 70S5, 70S6, 74K4	general properties
RF-magnetron sputtering	Al_2O_3	RT···500	77N1, 78K4	general properties
DC-reactive-sputtering	$Al+O_2$	RT···500	62D1, 63P1, 64S3, 65K2, 66F1, 68T2, 69F1, 71H3, 71I1, 72B4, 74B3, 77B5, 81N1	general properties
DC-magnetron reactive sputter.	$Al+O_2$	RT	80M2	general properties
APCVD	$Al(OC_2H_5)_3+O_2$	300···500	67A1, 68M1, 69K3, 70D2, 71H2, 74P1	general properties
		400···750	69R1	general properties
	$Al(OC_3H_7)_3+O_2$	420	67A1, 70D2, 70D3, 71C1, 75M8	general properties
		250···500	66E1	general properties
	Al-alkoxide		68S2	general properties
	$Al(CH_3)_3+NO$	>650	67A1, 70D2, 70H3, 71H2	general properties
	$AlCl_3$ or $AlBr_3+CO_2+H_2$ or $NO+H_2$	400···1200 (typ. 800···900)	65T1, 67T1, 69D1, 69T1, 70B2, 70T1, 70T3, 70W2, 71K3, 71K4, 71M4, 72J1, 72K2, 73A1, 73M1, 76A3, 76S7, 76T1, 78S4, 79N1	most frequently used, standard
LPCVD	$AlCl_3+CO_2+H_2$	700···900	73T2	7 torr
	$Al(OC_3H_7)_3+O_2$	700···800	72H1	10 torr
PECVD	$AlCl_3+O_2$	480	71K1	general properties
Si_3N_4				
Thermal nitridation	N_2	1000···1300	69L1, 73G2, 75A1, 75R1, 76A6, 76G4, 77S4, 78I3	Very sensitive to O_2-impurities
			73H2	UHV
	$N_2+3\%\ NH_3$	900···1100	76K9	general properties
	$Ar+5\%\ NH_3$	900···1200	79M2	general properties

continued

Table 10 (continued)

Preparation method	Reactants	Deposition temperature °C	Ref.	Remarks
Si_3N_4				
	NH_3	950···1300	67Y1, 68Y2, 71K5, 75S10, 78I4, 79M2	general properties
	NO	1020···1100	81W1	UHV
Ion implantation	N_2^+	500	67P4, 70F2, 73F1, 75M9, 75W5, 76W2, 76W3, 78J2, 79Y1, 80R2	anneal: >900 °C energy: 10···200 keV dependent on application
Evaporation	Si_3N_4	200···250	68K2, 69E1	e-beam; pinholes
RF-sputtering	Si_3N_4	300	68V1, 69C5, 71V2, 75K5, 75M3, 76M4	low power
RF reactive sputtering	$Si+N_2$	≦250	66D2, 67C3, 67H1, 67H2, 67J2, 72A1, 73R2, 76R5, 78B3	general properties
	$Si+NH_3$	<250	75M3, 75P1, 76S3, 78K3	general properties
DC magnetron reactive sputtering	$Si+NH_3$	<250	77C5	general properties
APCVD	$SiBr_4+NH_3$	550···800	69A4	general properties
	SiF_4+NH_3	700···1200	69A4, 77L3	general properties
	$SiCl_4+NH_3$	650···1000	67C1, 67S2, 68A2, 68D1, 68G1, 70K4, 72M2, 72W4, 73P1, 76N1, 77H6, 80M4	N_2 or H_2 carrier gas
	SiH_4+NH_3	650···1000	66D1, 66H1, 67B1, 67C2, 68B3, 68B4, 68D2, 68Y1, 69G1, 69L1, 70D1, 70K4, 70S2, 71T1, 74M2, 76G4, 76M1, 77H4, 77M1, 77M4, 78C3	NH_3/SiH_4 >30, carrier gas: N_2: Low dielectric strength H_2: high dielectric strength todays standard
	SiH_4+NH_3	600···900	70Y1	Pt or NiO as catalyst
	$SiH_2Cl_2+NH_3$	700···1100	72D2, 78M2	general properties
	$SiCl_4+N_2H_4$	700···1100	75S10	general properties
	$SiH_4+N_2H_4$	550···1150	67Y2	general properties
LPCVD	$SiCl_4+NH_3$	1100···1400	77B4, 78G4	polycrystalline
	SiH_4+NH_3	750···850	73T2, 77R1	sensitive to wafer spacing
	$SiH_2Cl_2+NH_3$	750···850	75G1, 77K1, 77R1, 78D3, 78M2, 79A1, 79B2	good uniformity todays standard
PECVD	$SiBr_4+N_2$	~500	69A1	general properties
	SiI_4+N_2	~500	77S1	general properties
	SiH_4+N_2	≦450	68D1, 68K1, 69K1, 71T1, 72G1, 75B1, 76T2	≈0.1 torr poor uniformity
	SiH_4+NH_3	≦300	65S1, 66S1, 67J1, 67S1, 68P2, 69A2, 71T1, 73S4, 74R1, 75M2, 76B2, 76R1	0.5···1.0 torr good uniformity, for passivation todays standard,

continued

Table 10 (continued)

Preparation method	Reactants	Deposition temperature °C	Ref.	Remarks
Si_3N_4				
			76S2, 76T2, 76T4, 76W4, 77A1, 77C4, 77H1, 77H7, 77S5, 78H8, 78L1, 78S1, 78S2, 78S7, 78S10, 78Y1, 79S1, 80R1, 81S1	general properties, sensitive to deposition parameters
UV-CVD	$SiH_4+N_2H_4$	200	69C2, 72B1	many pinholes
	$SiH_4+N_2H_4$	550···1150	67Y2	porous layers
	SiH_4+NH_3	<200	80P2	low temp., passivation layer

References for 6.1.4.5.1 ··· 6.1.4.5.7

Books and review articles

66p1 Powell, C.F., Oxley, J.H., Blocher, jr., J.M. (eds.): Vapor Deposition, New York: Wiley **1966**.

67m1 McTaggart, F.K.: Plasma Chemistry in Electrical Discharges, New York: Elsevier **1967**.

68c1 Carter, G., Colligon, J.S.: Ion Bombardment of Solids, New York: Am. Elsevier **1968**.

69c1 Chopra, K.L.: Thin Film Phenomena. New York: Mc Graw Hill **1969**.

69v1 Vratny, F., (ed.): Thin Film Dielectrics, New York: Electrochemical Soc. **1969**.

69w1 Wolf, H.F.: Silicon Semiconductor Data. Oxford: Pergamon Press Inc. **1969**.

70b1 Blocher, jr., J.M., Withers, J.L., (eds.): Chemical Vapor Deposition – Second International Conference, New York: Electrochemical Soc. **1970**.

70m1 Maissel, L.I., Glang, R. (eds.): Handbook of Thin Film Technology, New York: Mc Graw-Hill **1970**.

70v1 Van Zeggeren, F., Storey, S.H.: The Computation of Chemical Equilibria, London: Cambridge Univ. Press **1970**.

71m1 Milek, J.T.: Silicon Nitride for Microelectronic Applications, Part 1, Handbook of Electronic Materials, **3**, IFI/Plenum, New York: Plenum **1971**.

72g1 Glaski, F.A., (ed.): Chemical Vapor Deposition – Third International Conference, Hinsdale, Illinois: Am. Nucl. Soc. **1972**.

72m1 Milek, J.T.: Silicon Nitride for Microelectronic Applications, Part 2, Handbook of Electronic Materials, Vol. 6, IFI/New York: Plenum **1972**.

73m1 Mattox, D.M.: Sputter Deposition and Ion Plating Technology, New York: Thin Film Div. American Vacuum Soc. **1973**.

73w1 Wakefield, G.F., Blocher, J.M., (eds.): Proc. 4th Int. Conf. on Chemical Vapor Deposition, Princeton, N.J.: Electrochemical Soc. **1973**.

74c1 Chapman, B.N., Anderson, J.C., eds.: Science and Technology of Surface Coatings, New York: Academic Press **1974**.

74f1 Faktor, M.M., Garrett, I.: Growth of Crystals from the Vapour, London: Chapman and Hall, **1974**.

74h1 Hollahan, J.R., Bell, A.T. (eds): Techniques and Application of Plasma Chemistry, New York: Wiley **1974**.

75b1 Blocher, J.M., Hintermann, H.E., Hall, L.H. (eds.): Proc. 5th Int. Conf. on Chemical Vapor Deposition, Princeton, N.J.: Electrochemical Soc. **1975**.

75h1 Hastie, J.W.: High Temperature Vapors: Science and Technology Chap. 3, New York: Academic Press **1975**.

76h1 Howatson, A.M.: An Introduction to Gas Discharges, 2nd ed., New York: Pergamon Press **1976**.

77d1 Donaghey, L.F., Rai-Choudhury, P., Tauber, R.N. (eds.): Proc. 6th Int. Conf. on Chemical Vapor Deposition, Princeton, N.J.: Electrochem. Soc. **1977**.

78h1 Hammond, M., Tauber, R.N. (Chairmen): CVD, Electrochem. Soc. Extend. Abstr. **78-2** (1978) 583 ··· 605.

78k1 Kern, W. (Chairman): Symposium on Low-Pressure Chemical Vapor Deposition, Murray Hill, N.J.: Greater New York Chapter, American Vacuum Soc. March 15, **1978**.

78k2 Kern, W. (Session Moderator): Low Pressure and Plasma Chemical Vapor Deposition, 25th Nat. Symp. Amer. Vac. Soc., San Francisco, CA., Dec. 1, **1978**.

78p Pantelides, S., (ed.): The physics of SiO_2 its interfaces Proc. of the Topical Conference, Yorktown Heights N.Y.: Pergamon Press **1978**.

78v1 Vossen, J.L., Kern, W., (eds.): Thin Film Processes, New York: Academic Press **1978**.

78w1 Watel, G., (ed.): Electronic and Atomic Collisions, Amsterdam: North Holland Publishing Company **1978**.

79l Lucowsky, G., Pantelides, S., Galeener, F.L.: The Physics of MOS Insulators, Pergamon Press **1980**.

79r Roberts, G.G., Morant, J., (eds.): Insulating Films on Semiconductors Proceedings of the INFOS 79 Conference, Durham The Institute of Physics Conf. Ser. 50.

79s1 Sedgwick, T.O., Lydtin, H., (eds.): Proc. 7th Intern. Conf. on Chemical Vapor Deposition, Princeton, N.J., Electrochemical Society, **1979**.

80c1 Chapman, B.: Glow Discharge Processes, New York: J. Wiley **1980**.

81s Schulz, M., Pensl, G., (eds.): Insulating Films on Semiconductors. Series in Electrophysics, Berlin Heidelberg New York: Springer **7** (1981).
82n Nicollian, E.H., Brews, J.R.: MOS (Metal Oxide Semiconductor) Physics and Technology, New York: J. Wiley **1982**.
84b Balk, P., (ed.): The SiO_2-Si System Elsevier **1984**.

Bibliography

55W1 Wehner, G.K.: Adv. Electron. Electron. Phys. **7** (1955) 239.
60L1 Ligenza, J.R., Spitzer, W.G.: J. Phys. Chem. Solids **14** (1960) 131.
61J1 Jordan, E.L.: J. Electrochem. Soc. **108** (1961) 478.
61K1 Klerer, J.: J. Electrochem. Soc. **108** (1961) 1070.
61L1 Ligenza, J.R.: J. Phys. Chem. **65** (1961) 2011.
61S1 Spitzer, W.G., Ligenza, J.R.: J. Phys. Chem. Solids **17** (1961) 196.
62D1 Da Silva, E.M., White, P.: J. Electrochem. Soc. **109** (1962) 12.
62J1 Jorgensen, P.L.: J. Chem. Phys. **37** (1962) 874.
62K1 Kay, E.: Adv. Electron. Electron. Phys. **17** (1962) 245.
62L1 Ligenza, J.R.: J. Electrochem. Soc. **109** (1962) 73.
62S1 Sandor, J.: Electrochem. Soc. Extend. Abstr. No. 96 Spring Meeting **1962**, 228.
63A1 Alt, L.L., Ing, jr., S.W., Laendle, W.W.: J. Electrochem. Soc. **110** (1963) 445.
63E1 Egagawa, H., Morita, Y., Maekawa, S.: Jpn. J. Appl. Phys. **2** (1963) 765.
63K1 Kay, E.: J. Appl. Phys. **34** (1963) 760.
63M1 Maissel, L.I., Vaughn, J.H.: Vacuum **13** (1963) 421.
63M2 Miles, J.L., Smith, P.H.: J. Electrochem. Soc. **110** (1963) 1240.
63P1 Pollack, S.R., Freitag, W.D., Morris, C.E.: Electrochem. Technol. **1** (1963) 96.
63R1 Richman, D.: RCA Rev. **24** (1963) 596.
64I1 Ing, S.W., Davern, W.: J. Electrochem. Soc. **111** (1964) 120.
64K1 Kerr, D.R., Logan, J.S., Burkhardt, P.J., Pliskin, W.A.: IBM J. Res. Dev. **8** (1964) 376.
64L1 Lewis, B.: Microelectron. Reliab. **3** (1964) 109.
64P1 Pliskin, W.A., Gnall, R.P.: J. Electrochem. Soc. **111** (1964) 872.
64S1 Steinmaier, W., Bloem, J.: J. Electrochem. Soc. **111** (1964) 206.
64S2 Schwartz, N.: Trans. Natl. Vac. Symp. 10th, Boston, 1963, **1964**, 325.
64S3 Schilling, R.B.: Proc. IEEE **52** (1964) 1350.
65D1 Duffek, E.F., Benjamini, E.A., Mylroie, C.: Electrochem. Technol. **3** (1965) 75.
65D2 Deal, B.E., Skalar, M.: J. Electrochem. Soc. **112** (1965) 430.
65D3 Deal, B.E., Grove, A.S.: J. Appl. Phys. **36** (1965) 3770.
65G1 Gill, W.D., Kay, E.: Rev. Sci. Instrum. **36** (1965) 277.
65I1 Ing, S.W., Davern, W.: J. Electrochem. Soc. **112** (1965) 284.
65K1 Klerer, J.: J. Electrochem. Soc. **112** (1965) 503.
65K2 Kaplan, L.H.: Electrochem. Technol. **3** (1965) 335.
65L1 Ligenza, J.R.: J. Appl. Phys. **36** (1965) 2703.
65S1 Sterling, H.F., Swann, R.C.G.: Solid State Electron. **8** (1965) 653.
65S2 Scott, J., Olmstead, J.: RCA Rev. **26** (1965) 357.
65T1 Tung, S.K., Caffrey, R.E.: Trans. Metall. Soc. AIME **233** (1965) 572.
65W1 Williams, R.: Phys. Rev. **140** (1965) A 569.
66D1 Doo, V.Y., Nichols, D.R., Silvey, G.A.: J. Electrochem. Soc. **113** (1966) 1279.
66D2 Davidse, P.D., Maissel, L.I.: J. Appl. Phys. **37** (1966) 571.
66E1 Eversteijn, F.C.: Philips Res. Rep. **21** (1966) 379.
66F1 Frieser, R.G.: J. Electrochem. Soc. **113** (1966) 357.
66G1 Goodman, A.M.: J. appl. Phys. **37** (1966) 3580.
66H1 Hu, S.M.: J. Electrochem. Soc. **113** (1966) 693.
66M1 Miura, Y., Tanaka, S., Matu Kura, Y., Osafune, H.: J. Electrochem. Soc. **113** (1966) 399.
66M2 Maissel, L.I.: Phys. Thin Films **3** (1966) 61.
66N1 Nakayama, T., Collins, F.C.: J. Electrochem. Soc. **113** (1966) 706.
66P1 Perny, G.: Vide **21** (1966) 106.
66R1 Rand, M.J., Ashworth, J.L.: J. Electrochem. Soc. **113** (1966) 48.
66S1 Sterling, H.F., Joyce, B.A., Alexander, J.: Vide **21** (1966) 80.
66S2 Secrist, D.R., Mac Kenzie, J.D.: J. Electrochem. Soc. **113** (1966) 914.
66Z1 Zerbst, M.: Z. Angew. Phys. **22** (1966) 30.

67A1 Aboaf, J.A.: J. Electrochem. Soc. **114** (1967) 948.
67B1 Bean, K.E., Gleim, P.S., Yearley, R.L., Runyan, W.R.: J. Electrochem. Soc. **114** (1967) 733.
67B2 Bradshaw, S.E.: Int. J. Electron. **23** (1967) 381.
67C1 Chu, T.L., Lee, C.H., Gruber, G.A.: J. Electrochem. Soc. **114** (1967) 717.
67C2 Chu, T.L., Szedon, J.R., Lee, C.H.: Solid State Electron. **10** (1967) 897.
67C3 Cordes, L.F.: Appl. Phys. Lett. **11** (1967) 383.
67D1 Deal, B.E., Sklar, M., Grove, A.S., Snow, E.H.: J. Electrochem. Soc. **114** (1967) 266.
67G1 Goldsmith, N., Kern, W.: RCA Rev. **28** (1967) 153.
67H1 Hu, S.M., Gregor, L.V.: J. Electrochem. Soc. **114** (1967) 827.
67H2 Hu, S.M., Kerr, D.R., Gregor, L.V.: Appl. Phys. Lett. **10** (1967) 97.
67H3 Hofstein, S.R.: IEEE Trans. **ED-14.** (1967) 794.
67J1 Joyce, R.J., Sterling, H.F., Alexander, H.F.: Thin Solid Films **1** (1967) 481.
67J2 James, A.R., Shirn, G.A.: J. Vac. Sci. Technol. **4** (1967) 37.
67J3 Jackson, N.F.: J. Mater. Sci. **2** (1967) 12.
67L1 Lee, D.B.: Solid State Electron. **10** (1967) 623.
67N1 Nicollian, E.H., Goetzberger, A.: Bell Syst. Techn. J. **46** (1967) 1055.
67P1 Patterson, W.L., Shirn, G.A.: J. Vac. Sci. Technol. **4** (1967) 343.
67P2 Pompei, J.: Proc. Symp. Deposition Thin Films Sputter., 2nd, Univ. Rochester, **1967**, 127.
67P3 Pliskin, W.A., Davidse, P.H., Lehman, H.S., Maissel, L.I.: IBM J. Res. Dev. **11** (1967) 461.
67P4 Pavlov, P.V., Shitova, E.V.: Sov. Phys. Dokl. **12** (1967) 11.
67R1 Rand, M.J.: J. Electrochem. Soc. **114** (1967) 274.
67S1 Swann, R.C.G., Mehta, R.R., Cange, T.P.: J. Electrochem. Soc. **114** (1967) 713.
67S2 Seki, H., Moriyama, K.: Jpn. J. Appl. Phys. **6** (1967) 1345.
67T1 Tung, S.K., Caffrey, R.E.: J. Electrochem. Soc. **114** (1967) 275C.
67W1 Winters, H.F., Kay, E.: J. Appl. Phys. **38** (1967) 3928.
67Y1 Yamazaki, S., Kato, Y., Taniguchi, J.: Jpn. J. Appl. Phys. **6** (1967) 408.
67Y2 Yoshioka, S., Takayanagi, S.: J. Electrochem. Soc. **114** (1967) 962.
68A1 Argall, F., Jonscher, A.K.: Thin Solid Films **2** (1968) 185.
68A2 Arizumi, T., Nishinaga, T., Ogawa, H.: Jpn. J. Appl. Phys. **7** (1968) 1021.
68B1 Brown, D.M., Engeler, W.E., Garfinkel, M., Gray, P.V.: Solid State Electron. **11** (1968) 1105.
68B2 Brown, D.M., Engeler, W.E., Garfinkel, M., Gray, P.V.: J. Electrochem. Soc. **115** (1968) 874.
68B3 Brown, D.M., Gray, P.V., Heumann, F.K., Philipp, H.R., Taft, A.E.: J. Electrochem. Soc. **115** (1968) 311.
68B4 Brown, G.A., Robinette, jr., W.C., Calson, H.G.: J. Electrochem. Soc. **115** (1968) 948.
68C1 Chu, T.L., Szedon, J.R., Gruber, G.A.: Trans. Metall. Soc. AIME **242** (1968) 532.
68C2 Chu, T.L., Szedon, J.R., Lee, C.H.: J. Electrochem. Soc. **115** (1968) 318.
68D1 Dalton, J.V., Drobek, J.: J. Electrochem. Soc. **115** (1968) 865.
68D2 Doo, V.Y., Kerr, D.R., Nichols, D.R.: J. Electrochem. Soc. **115** (1968) 61.
68F1 Fisher, A.W., Amick, J.A., Hyman, H., Scott, jr., J.H.: RCA Rev. **29** (1968) 533.
68F2 Fisher, A.W., Amick, J.A.: RCA Rev. **29** (1968) 549.
68G1 Grieco, M.J., Worthing, F.L., Schwartz, B.: J. Electrochem. Soc. **115** (1968) 525.
68H1 Holland, L., Putner, T., Jackson, G.N.: J. Phys. E **1** (1968) 32.
68H2 Hollands, E., Campbell, D.S.: J. Mater. Sci. **3** (1968) 544.
68K1 Kuwano, Y.: Jpn. J. Appl. Phys. **7** (1968) 88.
68K2 Kendall, E.J.M.: J. Phys. D **1** (1968) 1409.
68K3 Kern, W.: RCA Rev. **29** (1968) 525.
68L1 Locker, C.D., Skolnick, L.P.: Appl. Phys. Lett. **12** (1968) 396.
68M1 Matsushita, M., Yoga, Y.: Electrochem. Soc. Extend. Abstr. 90, Spring Meeting **1968**, 230.
68N1 Nakai, Y.: Electrochem. Soc. Extend. Abstr. No. 84, Spring Meeting **1968**, 215.
68O1 Oroshnik, J., Kraitchman, J.: J. Electrochem. Soc. **115** (1968) 649.
68P1 Pliskin, W.A., Castrucci, P.P.: Electrochem. Technol. **6** (1968) 85.
68P2 Parnell, M., Sterling, H.F.: Electron. Commun. **43** (1968) 63.
68S1 Strater, K.: RCA Rev. **29** (1968) 618.
68S2 Sedgwick, T., Aboaf, J.: IEEE Trans. **ED-15** (1968) 1015.
68T1 Toombs, P.A.B.: J. Phys. D **1** (1968) 662.
68T2 Tanaka, T., Iwauchi, S.: Jpn. J. Appl. Phys. **7** (1968) 1420.
68V1 Vossen, J.L., O'Neill, J.L.: RCA Rev. **29** (1968) 149.
68W1 Waxman, A., Zaininger, K.H.: Appl. Phys. Lett. **12** (1968) 109.

68Y1 Yeargan, J.R., Taylor, H.L.: J. Electrochem. Soc. **115** (1968) 273.
68Y2 Yamazaki, S., Kato, Y., Okamoto, T., Taniguchi, I.: Jpn. J. Appl. Phys. **7** (1968) 846.
69A1 Androshuk, A., et al.: U.S. Patent 3424661 (1969).
69A2 Alexander, J.H., et al.: in [69v1] p. 186.
69A3 Anderson, G.S.: J. Appl. Phys. **40** (1969) 2884.
69A4 Aboaf, J.A.: J. Electrochem. Soc. **116** (1969) 1736.
69B1 Brodie, I., Lamont, L.T., Myers, D.O.: J. Vac. Sci. Technol. **6** (1969) 124.
69B2 Benninghoven, A.: Z. Phys. **220** (1969) 159.
69C1 Chittick, R.C., Alexander, J.H., Sterling, H.F.: J. Electrochem. Soc. **116** (1969) 77.
69C2 Collet, M.G.: J. Electrochem. Soc. **116** (1969) 110.
69C3 Cuccia, A., Shrank, G., Queirolo, G.: in [69h1] p. 506.
69C4 Chu, T.L.: J. Vac. Sci. Technol. **6** (1969) 25.
69C5 Červenak, J., Aleksandrov, L.N., Lovyagin, R.N., Krivorotov, E.A.: J. Vac. Sci. Technol. **6** (1969) 938.
69D1 Doo, Y.V., Tsang, P.J.: J. Electrochem. Soc. **116** (1969) 116 C.
69E1 Elliot, E.: Thin Solid Films **3** (1969) R47.
69F1 Ferrieu, E., Pruniax, B.: J. Electrochem. Soc. **116** (1969) 1008.
69F2 Fiedler, O., Reisse, G., Schöneich, B., Weissmantel, C.: Proc. 4th Int. Vacuum Congr. Manchester, 1968, Inst. Phys. London, **1969**, 569.
69F3 Feist, W.F., Steele, S.R., Ready, D.W.: Phys. Thin Films **5** (1969) 237.
69G1 Gregor, L.V.: in [69v1] p. 447.
69H1 Hoenig, S.A., Pope, R.A.: Appl. Phys. Lett. **14** (1969) 271.
69H2 Heunisch, G.W.: Anal. Chim. Acta **48** (1969) 405.
69K1 Kuwano, Y.: Jpn. J. Appl. Phys. **8** (1969) 876.
69K2 Kaufman, F.L.: Chemical Reactions in Electrical Discharges (Advances in Chemistry Series 80) Washington, D.C., Ann. Chem. Soc., **1969**, 29.
69K3 Korzo, V.F., Ibraimov, N.S., Halkin, B.D.: J. Appl. Chem. **42** (1969) 989.
69K4 Krongelb, S.: J. Electrochem. Soc. **116** (1969) 1583.
69L1 Langheinrich, W., Eisbrenner, D.: Metalloberfläche **23** (1969) 129.
69L2 Laverty, S.J., Ryan, W.D.: Int. J. Electron. **26** (1969) 519.
69M1 Meyer, D.E.: Trans. Metall. Soc. AIME **245** (1969) 593.
69O1 Ogar, W.T., Olson, N.T., Smith, H.P.: J. Appl. Phys. **40** (1969) 4997.
69P1 Pompei, J.: Proc. Symp. Deposition Thin Films Sputter., 3rd, Univ. Rochester, **1969**, 165.
69P2 Pratet, T.H.: Thin Solid Films **3** (1969) R 23.
69R1 Ryabova, L.A., Savitskaya, Y.S.: J. Vac. Sci. Technol. **6** (1969) 934.
69R2 Revesz, A.G., Evans, R.J.: J. Phys. Chem. Solids **30** (1969) 551.
69S1 Swan, R., Pyne, A.: J. Electrochem. Soc. **116** (1969) 1014.
69S2 Strater, K., Mayer, A.: in [69h1] p. 469.
69S3 Secrist, D.R.: Adv. Chem. Ser. No. 80 (1969) p. 242.
69T1 Teshima, H., Tarui, Y., Takeda, O.: Denki Shikenjo Iho **33** (1969) 631.
69T2 Tokuyama, T., Miyazaki, T., Horiuchi, M.: in [69v1] p. 297.
69T3 Totta, P.A., Sopher, R.P.: IBM J. Res. Dev. **13** (1969) 226.
69V1 Vossen, J.L.: Proc. Symp. Deposition Thin Films Sputter., 3rd, Univ. Rochester **1969**, 80.
69W1 Wehner, G.K.: in [69v1] p. 117.
69W2 Whittle, K.M., Vick, G.L.: J. Electrochem. Soc. **116** (1969) 645.
69W3 Winters, H.F., Raimondi, D.L., Horne, D.E.: J. Appl. Phys. **40** (1969) 2996.
69W4 Wasa, K., Hayakawa, S.: Rev. Sci. Instrum. **40** (1969) 693.
70A1 Appels, J.A, Kooi, E., Paffen, M.M., Schatorje, J.J.H., Verkuylen, W.H.C.G.: Philips Res. Rep. **25** (1970) 118.
70A2 Amick, J.A., Kern, W.: Symposium on Chemical Vapor Deposition, 2nd Int. Conf. Princeton, N.J.: Electrochemical Soc. **1970**, 551.
70A3 Arai, E., Terunuma, Y.: Jpn. J. Appl. Phys. **9** (1970) 691.
70A4 Abbott, R.A., Kamins, T.I.: Solid State Electronics **13** (1970) 565.
70B1 Blech, J., Sello, H., Gregor, L.V.: in [70m1] Chap. 23.
70B2 Balk, P., Stephany, S.: NTZ-Nachr. Tech. Z. **23** (1970) Nov.
70B3 Barry, M.L.: in [70b1] p. 595.
70C1 Červenǎk, J., Živčáková, A., Buch, J.: Czech. J. Phys. B **20** (1970) 84.
70C2 Christensen, O.: Solid State Technol. **13** (12) (1970) 39.

70C3 Campbell, D.S.: in [70m1] Chap. 5.
70D1 Duffy, M.T., Kern, W.: RCA Rev. **31** (1970) 742.
70D2 Duffy, M.T., Kern, W.: RCA Rev. **31** (1970) 754.
70D3 Duffy, M.T., Revesz, A.G.: J. Electrochem. Soc. **117** (1970) 372.
70F1 Fiedler, O.: Wiss. Z. Tech. Hochsch. Karl-Marx-Stadt **12** (1970) 483.
70F2 Freeman, J.H., Gard, G.A., Mazey, D.J., Stephen, J.H., Whiting, F.B.: Proc. Europ. Conf. on Ion Implantation, Reading **1970**, 74.
70G1 Glang, R.: in [70m1] Chap. 1.
70G2 Goodman, A.M., Breece, J.M.: J. Electrochem. Soc. **117** (1970) 982.
70G3 Gittler, F.L., Porter, R.A.: J. Electrochem. Soc. **117** (1970) 1551.
70G4 Grantham, D.H., Paradis, E.L., Quinn, D.J.: J. Vac. Sci. Technol. **7** (1970) 343.
70H1 Howson, R.P., Taylor, A.: Thin Solid Films **6** (1970) 31.
70H2 Harrop, P.J., Campbell, D.S.: in [70m1] Chap. 16.
70H3 Hall, L., Robinette, B.: in [70b1] p. 637.
70J1 Jackson, G.N.: Thin Solid Films **5** (1970) 209.
70K1 Kern, W., Heim, R.C.: J. Electrochem. Soc. **117** (1970) 562.
70K2 Kern, W., Heim, R.C.: J. Electrochem. Soc. **117** (1970) 568.
70K3 Kern, W., Fisher, A.W.: RCA Rev. **21** (1970) 715.
70K4 Kohler, W.A.: Trans. Metall. AIME **1** (1970) 735.
70L1 Logan, J.S., Maddocks, F.S., Davidse, P.D.: IBM J. Res. Develop. **14** (1970) 182.
70L2 Le Comber, P.G., Spear, W.E.: Phys. Rev. Lett. **25** (1970) 509.
70L3 Lee, W.W.Y., Oblas, D.W.: J. Vac. Sci. Technol. **7** (1970) 129.
70M1 Maissel, L.I.: in [70m1] Chap. 4.
70M2 Maissel, L.I., Jones, R.E., Standley, C.L.: IBM J. Res. Develop. **14** (1970) 176.
70M3 Mac Donald, R.J.: Adv. Phys. **19** (1970) 457.
70O1 O'Hanlon, J.F.: J. Vac. Sci. Technol. **7** (1970) 330.
70P1 Pliskin, W.A., Zanin, S.J.: in [70m1] Chap. 11.
70S1 Schlacter, M.M., Schlegel, E.S., Keen, R.S., Lathlaen, R.A., Schnable, G.L.: IEEE Trans. Electron. Devices **ED-17** (1970) 1077.
70S2 Schaffer, P.S., Swaroop, B.: Am. Ceram. Soc. Bull. **49** (1970) 536.
70S3 Schwartz, G.C., Jones, R.E.: IBM J. Res. Dev. **14** (1970) 52.
70S4 Sachse, H.B., Nichols, G.L.: J. Appl. Phys. **41** (1970) 4237
70S5 Salama, C.A.T.: J. Electrochem. Soc. **117** (1970) 913.
70S6 Sedgwick, T.O., Aboaf, J.A., Krongelb, S.: IBM J. Res. Dev. **14** (1970) 2.
70S7 Sunami, H., Itoh, Y., Sato, K.: J. Appl. Phys. **41** (1970) 5115.
70T1 Tung, S.K., Caffrey, R.E.: J. Electrochem. Soc. **117** (1970) 91.
70T2 Takei, W.J., Formigoni, N.P., Francombe, M.H.: J. Vac. Sci. Technol. **7** (1970) 442.
70T3 Tsujide, T., Nakanuma, S., Ikushima, Y.: J. Electrochem. Soc. **117** (1970) 703.
70W1 Wehner, G.K., Anderson, G.S.: in [70m1] Chap. 3.
70W2 Wong, P., Robinson, Mc D.: J. Am. Ceram. Soc. **53** (1970) 617.
70Y1 Yamazaki, S., Wada, K., Taniguchi, T.: Jpn. J. Appl. Phys. **9** (1970) 1467.
71B1 Brown, D.M., Kennicott, P.R.: J. Electrochem. Soc. **118** (1971) 293.
71B2 Benninghoven, A.: Surf. Sci. **28** (1971) 541.
71B3 Blachman, A.G.: Metall. Trans. **2** (1971) 699.
71C1 Carnes, J.E., Duffy, M.T.: J. Appl. Phys. **42** (1971) 4350.
71D1 Dell'Oca, C.J., Learn, A.J.: Thin Solid Films **8** (1971) R 47.
71D2 Dell'Oca, C.J., Pulfrey, D.L., Young, L.: Phys. of Thin Films, Francombe, M.H., Hoffman, R.W. (eds.), New York: Academic Press **1971**, 1.
71F1 Fogels, E.A., Salama, C.A.T.: J. Electrochem. Soc. **118** (1971) 2002.
71H1 Hoffman, D., Leibowitz, D.: J. Vac. Sci. Technol. **8** (1971) 107.
71H2 Hall, L.H., Robinette, W.C.: J. Electrochem. Soc. **118** (1971) 1624.
71H3 Hattori, T., Iwauchi, S., Nagano, K., Tanaka, T.: Jpn. J. Appl. Phys. **10** (1971) 203.
71I1 Iwauchi, S., Tanaka, T.: Jpn. J. Appl. Phys. **10** (1971) 260.
71K1 Katto, H., Koga, Y.: J. Electrochem. Soc. **118** (1971) 1619.
71K2 Kern, W.: RCA Rev. **32** (1971) 429.
71K3 Kalter, H., Schatorjé, J.J.H., Kooi, E.: Philips Res. Rep. **26** (1971) 181.
71K4 Kamoshida, M., Mitchell, I.V., Mayer, J.W.: Appl. Phys. Lett. **18** (1971) 292.
71K5 Kamchatka, M.I., Ormont, B.F.: Zh. Fiz. Khim. **45** (1971) 2202.

71K6 Kuhn, M., Silversmith, D.J.: J. Electrochem. Soc. **118** (1971) 966.
71M1 Mattox, D.M., Kominiak, G.J.: J. Vac. Sci. Technol. **8** (1971) 194.
71M2 Mitchell, I.V., Maddison, R.C.: Vacuum **21** (1971) 591.
71M3 Mullaly, J.R.: Res./Dev. **22**(2) (1971) 40.
71M4 Mitchell, I.V., Kamoshida, M., Mayer, J.W.: J. App. Phys. **42** (1971) 4378.
71N1 Nicollian, E.H., Berglund, C.N., Schmidt, P.F., Andrews, J.M.: J. Appl. Phys. **42** (1971) 5654.
71P1 Parekh, P.C., Goldstein, D.R., Chan, T.C.: Solid State Electron **14** (1971) 281.
71P2 Powell, R.J., Berglund, C.N.: J. Appl. Phys. **42** (1971) 4390.
71R1 Robinson, P.H., Heiman, F.P.: J. Electrochem. Soc. **118** (1971) 141.
71S1 Shuermeyer, F.L., Chase, W.R., King, E.L.: J. Appl. Phys. **42** (1971) 5856.
71S2 Schnable, G.L.: IEEE Int. Conv. Dig. (1971) 586.
71S3 Schroder, D.K., Guldberg, J.: Solid State Electron. **14** (1971) 1285.
71T1 Taft, E.A.: J. Electrochem. Soc. **118** (1971) 1341.
71T2 Tenney, A.S.: J. Electrochem. Soc. **118** (1971) 1658.
71T3 Taft, E.A.: J. Electrochem. Soc. **118** (1971) 1985.
71T4 Tsong, I.S.T.: Phys. Status Solidi (a) **7** (1971) 451.
71T5 Tarng, M.L., Wehner, G.K.: J. Appl. Phys. **42** (1971) 2449.
71V1 Vossen, J.L.: J. Vac. Sci. Technol. **8** (1971) 512.
71V2 Vossen, J.L.: J. Vac. Sci. Technol. **8** (1971) 751.
71W1 Wong, J., Ghezzo, M.: J. Electrochem. Soc. **118** (1971) 1540.
72A1 Audisio, S.C., Leidheiser, jr., H.: J. Electrochem. Soc. **119** (1972) 408.
72B1 van der Brekel, C.H.J., Severin, P.J.: J. Electrochem. Soc. **119** (1972) 372.
72B2 Brown, D.M., Garfinkel, M., Ghezzo, M., Taft, E.A., Tenney, A., Wong, J.: J. Cryst. Growth **17** (1972) 276.
72B3 Ball, D.J.: J. Appl. Phys. **43** (1972) 3047.
72B4 Bunshah, R.F., Raghuram, A.C.: J. Vac. Sci. Technol. **9** (1972) 1385.
72C1 Coburn, J.W., Kay, E.: J. Appl. Phys. **43** (1972) 4965.
72C2 Chen, M.C., Hile, J.W.: J. Electrochem. Soc. **119** (1972) 223.
72D1 Dell'Oca, C.J., Barry, M.L.: Solid State Electron. **15** (1972) 659.
72D2 Delong, D.J.: Solid State Technol. **15**(10) (1972) 29.
72F1 Fair, R.B.: J. Electrochem. Soc. **119** (1972) 1389.
72G1 Gereth, R., Scherber, W.: J. Electrochem. Soc. **119** (1972) 1248.
72G2 Ghezzo, M.: J. Electrochem. Soc. **119** (1972) 1428.
72G3 Ghez, R., van der Meulen, Y.L.: J. Electrochem. Soc. **119** (1972) 1100.
72H1 Hough, R.L.: in [72g1] p. 232.
72H2 Hoffman, D., Leibowitz, D.: J. Vac. Sci. Technol. **9** (1972) 326.
72H3 Haskell, R.W., Byrne, J.G., Herman, H. (ed.): Treatise on Materials Science and Technology. New York: Academic Press **1** (1972) 293.
72I1 Iida, K., Tsujide, T.: Jpn. J. Appl. Phys. **11** (1972) 840.
72I2 Ionov, N.I.: Prog. Surf. Sci. **1** (1972) 237.
72K1 Kriegler, R.J.: Thin Solid Films **13** (1972) 11.
72K2 Kamoshida, M., Mitchell, I.V., Mayer, J.W.: J. Appl. Phys. **43** (1972) 1717.
72K3 Kriegler, R.J., Cheng, Y.C., Colton, D.R.: J. Electrochem. Soc. **119** (1972) 388.
72K4 Kriegler, R.J.: J. Appl. Phys. Lett. **20** (1972) 449.
72M1 Mukherjee, S.P., Evans, P.E.: Thin Solid Films **14** (1972) 105.
72M2 Mac Kenna, E., Kodama, P.: J. Electrochem. Soc. **119** (1972) 1094.
72N1 Nagnib, H.M., Kelly, R.: J. Phys. Chem. Solids **33** (1972) 1751.
72R1 Ronen, R.S., Robinson, P.H.: J. Electrochem. Soc. **119** (1972) 747.
72S1 Severi, M., Soncini, G.: Electron. Lett. **8** (1972) 402.
72T1 Tenney, A.S., Wong, J.: J. Chem. Phys. **56** (1972) 5516.
72T2 Takahashi, R., Koga, Y., Sugawara, K.: J. Electrochem. Soc. **119** (1972) 1406.
72V1 Van der Meulen, Y.L.: J. Electrochem. Soc. **119** (1972) 530.
72W1 Wong, J., Ghezzo, M.: J. Electrochem. Soc. **119** (1972) 1413.
72W2 Wong, J.: J. Electrochem. Soc. **119** (1972) 1071.
72W3 Wong, J.: J. Electrochem. Soc. **119** (1972) 1080.
72W4 Wohlheiter, V.D., Whitner, R.A.: J. Electrochem. Soc. **119** (1972) 945.
73A1 Aboaf, J.A., Kerr, D.R., Bassous, E.: J. Electrochem. Soc. **120** (1973) 1103.
73B1 Baliga, B.J., Ghandhi, S.K.: J. Appl. Phys. **44** (1973) 990.

73B2 Benzing, W.C., Rosler, R.S., East, R.W.: Solid State Technol. **16**(11) (1973) 37.
73B3 Benninghoven, A.: Surf. Sci. **35** (1973) 427.
73B4 Blachman, A.G.: J. Vac. Sci. Technol. **10** (1973) 299.
73B5 Bloem, J., et al.: "Semiconductor Silicon 1973", Huff, H.R., Burgess, R.R. (eds.), Princeton, N.J.: Electrochem. Soc. **1973** Chap. **2**.
73B6 Baccarani, G., Severi, M., Soncini, G.: J. Electrochem. Soc. **120** (1973) 1436.
73C1 Collins, D.R., Shortes, S.R., Mc Mahon, W.R., Bracken, R.C., Penn, T.C.: J. Electrochem. Soc. **120** (1973) 521.
73C2 Chu, T.L., Smeltzer, R.K.: J. Vac. Sci. Technol. **10** (1973) 1.
73C3 Cambey, L.A.: Proc. Conf. Sputtering Technol. Autom. Prod. Equip.; Materials Research Corp. Orangeburg, N.Y. (1973) p. 129.
73C4 Chou, N.J., Osburn, C.M., van der Meulen, Y.J., Hammer, R.: Appl. Phys. Lett. **22** (1973) 380.
73D1 Di Stefano, T.H.: J. Appl. Phys. **44** (1973) 527.
73F1 Fritsche, C.R., Rothemund, W.: J. Electrochem. Soc. **120** (1973) 1603.
73G1 Ghezzo, M., Brown, D.M.: J. Electrochem. Soc. **120** (1973) 110.
73G2 Guthrie, R.B., Riley, F.L.: Proc. Br. Ceram. Soc. **22** (1973) 275.
73H1 Heller, J.: Thin Solid Films **17** (1973) 163.
73H2 Heckingbottom, R., Wood, P.R.: Surf. Sci. **36** (1973) 594.
73H3 Hirabayashi, K., Iwamura, J.: J. Electrochem. Soc. **120** (1973) 1595.
73K1 Kern, W., Vossen, J.L., Schnable, G.L.: Proc. 11th. Ann. Reliab. Phys. Symp., New York, IEEE **1973**, 214.
73K2 Kriegler, R.J.: "Semiconductor Silicon 1973", Huff, H.R., Burgess, R.R., (eds.), Princeton, N.J.: Electrochem. Soc. **1973**, 363.
73K3 Kriegler, R.J.: Denki Kagaku **41** (1973) 466.
73K4 Kriegler, R.J., Devenyi, T.F.: 11th IEEE Rel. Phys. Symp. 1973, IEEE, New York **1973**, 153.
73L1 Lieberman, M.L., Noles, G.T.: in [73w1].
73M1 Mehta, D.A., Butler, S.R., Feigl, F.J.: J. Electrochem. Soc. **120** (1973) 1707.
73M2 Mc Caughan, D.V., Kushner, R.S., Murphy, V.T.: "Semiconductor Silicon 1973", Huff, H.R., Burgess, R.S., (eds.), The Electrochemical Society Softbound Symposium Series, Princeton, N.J. **1973**, 376.
73M3 Meek, R.L.: J. Electrochem. Soc. **120** (1973) 308.
73P1 Parekh, P.C., Molea, J.R.: Solid State Electron. **16** (1973) 954.
73P2 Panousis, P.T., Schneider, M.: Electrochemical Soc., Spring Meeting, Abstr. No. 53, Chicago, **1973**.
73R1 Rand, M.J., Roberts, J.F.: J. Electrochem. Soc. **120** (1973) 446.
73R2 Rothemund, W., Fritzsche, C.R.: Thin Solid Films **15** (1973) 199.
73S1 Shalabutov, Y.K.: Sov. Phys. Semicond. **7** (1973) 322.
73S2 Schmidt, P.H.: J. Vac. Sci. Technol. **10** (1973) 611.
73S3 Shimizu, H., Ono, M., Nakayamu, K.: Surf. Sci. **36** (1973) 817.
73S4 Shima, Y., Miyazaki, T., Nakamura, N., Adachi, E., Tokuyama, T.: Jpn. J. Appl. Phys. **12** (1973) 309.
73T1 Tanikawa, E., Okabe, T., Maeda, K.: Denki Kagaku **41** (1973) 491.
73T2 Tanikawa, E., Takayama, O., Maeda, K.: in [73w1] p. 261.
73T3 Townsend, W.G., Uddin, M.E.: Solid State Electron. **16** (1973) 39.
73W1 Wong, J.: J. Electrochem. Soc. **120** (1973) 122.
73W2 Watts, B.F.: Thin Solid Films **18** (1973) 1.
73Y1 Young, D.R., Osburn, C.M.: J. Electrochem. Soc. **120** (1973) 1578.
74A1 Avigal, Y., Beinglass, I., Schieber, M.: J. Electrochem. Soc. **121** (1974) 1103.
74B1 Balk, P.: Inst. Phys. Conf. Ser. **19** (1974) 51.
74B2 Berg, R.S., Kominiak, G.J., Mattox, D.M.: J. Vac. Sci. Technol. **11** (1974) 52.
74B3 Bunshah, R.F.: in [74c1] p. 361.
74C1 Chapin, J.S.: Res./Dev. **25** (1974) 37.
74C2 Chapman, B.N., Downer, D., Guimaraes, L.J.M.: J. Appl. Phys. **45** (1974) 2115.
74E1 Eversteijn, F.C.: Philips Res. Rep. **29** (1974) 45.
74F1 Fagen, E.A., Nowicki, R.S., Saguin, R.W.: J. Appl. Phys. **45** (1974) 50.
74F2 Fischer, H.: Z. Phys. Chem (Leipzig) **255** (1974) 773.
74G1 Green, J.M., Osburn, C.M., Sedgwick, T.O.: J. Electron. Mater. **3** (1974) 579.
74I1 Irene, E.A.: J. Electrochem. Soc. **121** (1974) 1613.

74K1 Kirk, R.W.: in [74h1] p. 347.
74K2 Kerner, K.O.T.: Proc. 6th Int. Vac. Congress, Jap. J. Appl. Phys. Suppl. No. 2, Part 1 (1974) 463.
74K3 Kim, K.S., Winograd, N.: Surf. Sci. **43** (1974) 625.
74K4 Kennedy, T.N.: Electron. Packag. Prod. **11** (1974) 136.
74K5 Kriegler, R.J., Aitken, A., Morris, J.D.: Proc. 5th Conf. on Solid State Devices, Tokyo 1973; Suppl. Jpn. J. Appl. Phys. **43** (1974) 341.
74M1 Mc Caughan, D.V., Kushner, R.A.: Proc. IEEE **62** (1974) 1236.
74M2 Mellottée, H., Delbourgo, R.: Bull. Soc. Fr. Céram. **102** (1974) 65.
74O1 Osburn, C.M.: J. Electrochem. Soc. **121** (1974) 809.
74P1 Politycki, A., Hieber, K.: in [74c1] p. 159.
74P2 Powell, R.J., Ligenza, J.R., Schneider, M.S.: IEEE Trans. Electron. Dev. **ED-21** (1974) 636.
74R1 Reinberg, A.R.: Electrochem. Soc. Extend. Abstr. **74-1** (1974) 19.
74S1 Singh, B.R., Tyagi, B.D., Chandorkar, A.N., Marathe, B.R.: Extended Abstracts Electrochem. Soc. Meeting, San Francisco (1974) Paper 42.
74V1 Vossen, J.C., Schnable, G.L., Kern, W.: J. Vac. Sci. Technol. **11** (1974) 60.
74V2 Volland, G., Pagnia, H.: App. Phys. **3** (1974) 77.
74V3 Van der Meulen, Y.J., Lahill, J.G.: J. Electron. Mater. **3** (1974) 371.
74W1 Werner, M.: Solid State Electron. **17** (1974) 769.
75A1 Atkinson, A., Moulson, A.J., Roberts, E.W.: J. Mater. Sci. **10** (1975) 1242.
75A2 Abe, T., Yamashina, T.: Thin Solid Films **30** (1975) 19.
75B1 Bourdon, B., Sifre, G.: in [75b1] p. 281.
75B2 Bardos, L., Loncar, G., Stoll, J., Musil, J., Zacek, F.: J. Phys. D. **8** (1975) L. 195.
75B3 Blaise, G., Bernheim, M.: Surf. Sci. **47** (1975) 324.
74B4 Ban, V.S., Gilbert, S.L.: J. Electrochem. Soc. **123** (1975) 1382.
75B5 Ban, V.S.: J. Electrochem. Soc. **123** (1975) 1389.
75B6 Ban, V.S., Gilbert, S.L.: J. Cryst. Growth **31** (1975) 284.
75C1 Cuomo, J.J., Gambino, R.J.: J. Vac. Sci. Technol. **12** (1975) 79.
75D1 Declerck, G.J., Hattori, T., May, G.A., Beaudoin, J., Meindl, J.D.: J. Electrochem. Soc. **122** (1975) 436.
75E1 Eisele, K.M.: J. Electrochem. Soc. **122** (1975) 148.
75F1 Frantsuzov, A.A., Makrushin, N.J.: Sov. Phys. Tech. Phys. **20** (1975) 372.
75G1 Goldman, J., Mc Millan, J., Price, J.B.: Electrochem. Soc. Extend. Abstr. **75-2** (1975) 340.
75G2 Geraghty, K.G., Donaghey, L.F.: in [75b1] p. 219.
75H1 Holland, L.: Thin Solid Films **27** (1975) 185.
75H2 Hopper, M.A., Clarke, R.A., Young, L.: J. Electrochem. Soc. **122** (1975) 1216.
75H3 Hess, D.W., Deal, B.E.: J. Electrochem. Soc. **122** (1975) 579.
75J1 Jones, M.E., Shaw, D.W.: "Treatise on Solid State Chemistry", (N.B. Hannay) New York: Plenum Press **5** (1975) 283.
75J2 Janssens, E.J., Declerck, G.J.: Proc. ESSDERC 1975 p. 16.
75K1 Kerkdijk, C.B., Kelly, R.: Surf. Sci. **47** (1975) 294.
75K2 Kroll, W.J., Titus, R.L., Wagner, J.B.: J. Electrochem. Soc. **122** (1975) 573.
75K3 Koenigs, J.D., Kuppers, D., Lydtin, H., Wilson, H.: in [75b1] p. 270.
75K4 Kern, W.: Solid State Technol. **18**(12) (1975) 25.
75K5 Kominiak, G.J.: J. Electrochem. Soc. **122** (1975) 1271.
75K6 Kar, S.: Solid State Electron. **18** (1975) 169.
75L1 Lee, W.W.Y., Oblas, D.W.: J. Appl. Phys. **46** (1975) 1728.
75M1 Middlehoek, Y., Klinklhamer, A.Y.: in [75b1] p. 19.
75M2 Morzeck, P., et al.: Wiss. Z. Tech. Hochsch. Karl-Marx-Stadt **17** (1975) 329.
75M3 Mogab, C.J., Petroff, P.M., Sheng, T.T.: J. Electrochem. Soc. **122** (1975) 815.
75M4 Möller, R., Fabian, L., Weise, H., Weissmantel, L.: Thin Solid Films **29** (1975) 349.
75M5 Meriaux, J.P., Goutte, R., Guilland, C.: Appl. Phys. **7** (1975) 313.
75M6 Mathiew, H.J., Landolt, D.: Surf. Sci. **53** (1975) 228.
75M7 Murti, P.K., Kelly, R.: Surf. Sci. **47** (1975) 282.
75M8 Mutoh, M., Mizokami, Y., Matsui, H., Hagiwara, S., Ino, N.: J. Electrochem. Soc. **122** (1975) 987.
75M9 Mitchell, J.B., Pronko, P.P., Shewchun, J., Thompson, D.A., Davies, J.A.: J. Appl. Phys. **46** (1975) 332.

75M10 Munro, P.C., Thompson, H.W.: J. Electrochem. Soc. **122** (1975) 127.
75M11 Marshall, S., Zeto, R.J., Thornton, C.G.: J. Electrochem. Soc. **122** (1975) 1411.
75N1 Naguib, H.M., Kelly, R.: Radiat. Eff. **25** (1975) 79.
75P1 Popova, L.J., Shekerdjiiski, V.J., Lasarova, V.B., Beshkov, G.D., Vitanov, P.K.: Bulg. J. Phys. **2** (1975) 609.
75R1 Raider, S.I., Gdula, R.A., Petrak, J.R.: Appl. Phys. Lett. **27** (1975) 150.
75S1 Sugawara, K., Yoshimi, T., Sakai, H.: in [75b1] p. 407.
75S2 Shiraki, H.: Jpn. J. Appl. Phys. **14** (1975) 747.
75S3 Schnable, G.L., Kern, W., Comizzoli, R.B.: J. Electrochem. Soc. **122** (1975) 1092.
75S4 Sulimin, A.D., Ostashkin, L.P., Neustroev, S.A., Yakovenko, V.G., Lukanov, N.M.: Fiz. Khim. Obrab. Mater. **2** (1975) 57.
75S5 Shibata, M., Yoshimi, T., Sugawara, K.: J. Electrochem. Soc. **122** (1975) 157.
75S6 Shibata, M., Sugawara, K.: J. Electrochem. Soc. **122** (1975) 155.
75S7 Sedgwick, T.O., Smith, J.E., Ghez, R., Cowher, M.E.: J. Cryst. Growth **31** (1975) 264.
75S8 Shintani, Y., Nakanishi, K., Takawaki, T., Tada, O.: Jpn. J. Appl. Phys. **14** (1975) 1875.
75S9 Shinoki, F., Itoh, A.: J. Appl. Phys. **46** (1975) 3381.
75S10 Swaroop, B.: 4th Interamerican Conf. on Materials Technology, Caracas, Venezuela, **1975**, 224.
75V1 Van der Meulen, Y.J., Osburn, C.M., Ziegler, J.F.: J. Electrochem. Soc. **122** (1975) 284.
75W1 Wohlheiter, V.D., Whitner, R.A.: Electrochem. Soc. Extend. Abstr. **75-1** (1975) 424.
75W2 Wahl, G.: in [75b1] p. 391.
75W3 Werner, H.W.: Surf. Sci. **47** (1975) 301.
75W4 White, C.W., Simms, D.C., Toek, N.M., Caughan, D.V.: Surf. Sci. **49** (1975) 657.
75W5 Wada, Y., Usui, H., Ashikawa, M.: Jpn. J. Appl. Phys. **14** (1975) 1351.
75Z1 Zaininger, K.H.: "Surface Physics of Phosphors and Semiconductors", Scott, G.G., Reed, C.E. (eds.) London, Academic 1975, Chap. 9.
75Z2 Zeto, R.J., Thornton, C.G., Hryckowian, E., Bosco, C.D.: J. Electrochem. Soc. **122** (1975) 1409.
76A1 Archibald, P., Parent, E.: Solid State Technol. **19(7)** (1976) 32.
76A2 Alexandrova, J., Atanasova, E., Ivanovich, M., Kamadjiev, P., Kirov, K., Simeonov, S.: Electron. Lett. **12** (1976) 169.
76A3 Atanasova, E., Kamadjiev, P., Kirov, K., Simeonov, S.: Thin Solid Films **32** (1976) 77.
76A4 Abe, H., Emoto, H.: Jpn. J. Appl. Phys. **15** (1976) 925.
76A5 Albella, J.M., Criado, A., Merino, E.M.: Thin Solid Films **36** (1976) 479.
76A6 Atkinson, A., Moulson, A.J., Roberts, E.W.: J. Am. Ceram. Soc. **59** (1976) 285.
76B1 Burt, D., Taraci, R., Zavion, J.: U.S. Patent 3934060, Jan. 20, 1976.
76B2 Bersin, R.L.: Solid State Electron. **19** (1976) 31.
76B3 Belyakova, O.I., Zhagata, L., Kalnach, J., Putnynya, J., Feltyn, J.: Jzv. Akad. Nauk. SSSR, Neorg. Mater. **12(4)** (1976) 631.
76B4 Bell, A.T.: J. Macromol. Sci. Chem. **A 10** (1976) 367.
76B5 Barber, H.D., Lo, H.B., Jones, J.E.: J. Electrochem. Soc. **123** (1976) 1404.
76B6 Bardos, L., Loncar, G., Musil, J., Zacek, F.: J. Microwave Power **11** (1976) 184.
76B7 Blaise, G.: Surf. Sci. **60** (1976) 65.
76C1 Campbel, D.S.: Thin Solid Films **32** (1976) 3.
76C2 Colby, J.W., Katz, L.E.: J. Electrochem. Soc. **123** (1976) 409.
76C3 Clarke, P.J.: Solid State Technol. **19(12)** (1976) 77.
76C4 Coburn, J.W.: J. Vac. Sci. Technol. **13** (1976) 1037.
76D1 Dragila, R., Bardos, L., Loncar, G.: Phys. Status Solidi (a) **35** (1976) 291.
76D2 Dragila, R., Bardos, L., Loncar, G.: Thin Solid Films **34** (1976) 115.
76D3 Dylewski, J., Joshi, M.C.: Thin Solid Films **35** (1976) 327.
76D4 Dove, D.B., Gambino, R.J., Cuomo, J.J., Kobliska, R.J.: J. Vac. Sci. Technol. **13** (1976) 965.
76D5 Donaghey, L.F., Geraghty, K.G.: Thin Solid Films **38** (1976) 271.
76F1 Färber, W., Betz, G., Braun, P.: Nucl. Instrum. Methods **132** (1976) 351.
76G1 Goranchev, B., Orlinov, V., Popova, V.: Thin Solid Films **33** (1976) 173.
76G2 Gaind, A.K., Ackermann, G.K., Lucarini, V.J., Bratter, R.L.: J. Electrochem. Soc. **123**, 1976) 111.
76G3 Gaind, A.K., Ackermann, G.K., Nagarajan, A., Bratter, R.L.: J. Electrochem. Soc. **123** (1976) 238.
76G4 Gebhardt, J.J., Tanzilli, R.A., Harris, T.A.: J. Electrochem. Soc. **123** (1976) 453.
76G5 Gorokhov, E.B., Demyanov, E.A., Neizvestnyi, I.G., Pokrovskaya, J.G.: Inorg. Mater. **12** (1976) 232.

76G6 Gdula, R.A., Li, P.C.: Electrochem. Soc. Extend. Abstr. **76-2** (1976) 634.
76G7 Goetzberger, A., Klausmann, E., Schulz, M.: CRC Critical Reviews in Solid State Sci. **6** (1976) 1.
76H1 Hyder, S.B., Yep, T.O.: J. Electrochem. Soc. **123** (1976) 1721.
76H2 Hattori, T.: J. Electrochem. Soc. **123** (1976) 945.
76H3 Hill, R.J.: Physical Vapor Deposition, Berkeley, Airco-Temescal, **1976**, 15.
76H4 Harada, H., Satoh, S., Yoshida, M.: IEEE Trans. Reliab. **R-25** (1976) 290.
76H5 Huang, P.C., Schaible, P.M.: Electrochem. Soc. Extend. Abstr. **76-2** (1976) 754.
76H6 Heald, D.L., Das, R.M., Khosla, R.P.: J. Electrochem. Soc. **123** (1976) 302.
76H7 Hoffman, V.: Solid State Technol. **19**(12) (1976) 57.
76H8 Hosokawa, N.: J. Vac. Soc. Jap. **19** (1976) 82.
76H9 Hoff, P.K., Switkowski, Z.E.: Appl. Phys. Lett. **29** (1976) 549.
76I1 Irene, E.A., Van der Meulen, Y.J.: J. Electrochem. Soc. **123** (1976) 1380.
76J1 Jaskolska, H., Walis, L., Golkovska, H.: Thin Solid Films, **33** (1976) 281.
76J2 Johannessen, J.S., Spicer, W.E., Strausser, Y.E.: J. Appl. Phys. **47** (1976) 3028.
76K1 Kern, W., Schnable, G.L., Fisher, A.W.: RCA Rev. **37** (1976) 3.
76K2 Kuppers, D., Koenings, J., Wilson, H.: J. Electrochem. Soc. **123** (1976) 1079.
76K3 Kennedy, T.N.: J. Vac. Sci. Technol. **13** (1976) 1135.
76K4 Kern, W.: RCA Rev. **37** (1976) 78.
76K5 Krohn, V.E.: Int. J. Mass. Spectrom. Ion Phys. **22** (1976) 43.
76K6 Kay, E.: Colloq. Int. Pulver. Cathod., 2nd, Nice, 1976.
76K7 Kelly, R., Sanders, J.B.: Nucl. Instrum. Methods **132** (1976) 335.
76K8 Kim, K.S., Baitinger, W.E., Winograd, N.: Surf. Sci. **55** (1976) 285.
76K9 Kooi, E., Van Lierop, J.G., Appels, J.A.: J. Electrochem. Soc. **123** (1976) 1117.
76L1 Lim, M.J.: Electrochemical. Soc. Extend. Abstr. **76-2** (1976) 631.
76M1 Misawa, Y., Yagi, H.: Jpn. J. Appl. Phys. **15** (1976) 1045.
76M2 Mochizuki, H., et al.: Jpn. J. Appl. Phys. **15, suppl. 15-1** (1976) 41.
76M3 Mc Leod, P.S.: U.S. Patent 3956093 (1976).
76M4 Mogab, C.J., Lugujjo, E.: J. Appl. Phys. **47** (1976) 1302.
76M5 Mitra, N.K., Heynen, C.J.H.: Rev. Sci. Instrum. **47** (1976) 757.
76M6 Murti, D.K., Kelly, R.: Thin Solid Films **33** (1976) 149.
76N1 Niihara, K., Hirai, T.: J. Mater. Sci. **11** (1976) 593, 604.
76O1 Orlinov, K., Goranchev, B., Hristov, D., Dimitrov, G., Choubriev, Zh.: Thin Solid Films **36** (1976) 411.
76P1 Pickering, H.W.: J. Vac. Sci. Technol. **13** (1976) 618.
76R1 Rosler, R.S., Benzing, W.C., Baldo, J.: Solid State Technol. **19**(6) (1976) 45.
76R2 Ruzyllo, J., Shiota, I., Miyamoto, N., Nishizawa, J.: J. Electrochem. Soc. **123** (1976) 26.
76R3 Ritchey, P.M., Stach, J., Tressler, R.E.: Electrochem. Soc. Extend. Abstr., Las Vegas 1976, (1976) Abstr. 324.
76R4 Ritchie, K.: Appl. Phys. Lett. **28** (1976) 401.
76R5 Rigo, S., Amsel, G., Croset, M.: J. Appl. Phys. **47** (1976) 2800.
76R6 Reisman, A., Sedgwick, T.O.: in "Phase Diagrams", Alper, A.M. (ed.), New York: Academic Press IV (1976) 1.
76S1 Shibayama, H., Masaki, H., Ishikawa, H., Hashimoto, H.: Appl. Phys. Lett. **29** (1976) 136.
76S2 Sinha, A.K.: Electrochem. Soc. Extend. Abstr. **76-2** (1976) 625 and 629.
76S3 Stephens, A.W., Vossen, J.L., Kern, W.: J. Electrochem. Soc. **123** (1976) 303.
76S4 Shiraki, H.: Jpn. J. Appl. Phys. **15** (1976) 83.
76S5 Schnable, G.L., Schmidt, P.F.: J. Electrochem. Soc. **123** (1976) 310C.
76S6 Schreiber, H.U., Froeschle, E.: J. Electrochem. Soc. **123** (1976) 30.
76S7 Schlesier, K.M., Shaw, J.M. Benyon, jr. C.W.: RCA Rev. **37** (1976) 358.
76S8 Seraphin, B.O.: Thin Solid Films **39** (1976) 87.
76S9 Sedgwick, T.O., Smith, jr., J.E.: J. Electrochem. Soc. **123** (1976) 254.
76S10 Shiraki, H.: Jpn. J. Appl. Phys. **15** (1976) 1.
76T1 Tsang, P.J., Anderson, R.M., Cvikevich, S.: J. Electrochem. Soc. **123** (1976) 57.
76T2 Turban, G., Catherine, Y.: Thin Solid Films **35** (1976) 179.
76T3 Townsend, P.D., Kelly, J.C., Hartley, N.E.W.: "Ion Implantation, Sputtering, and Their Applications", New York, Academic Press, 1976, Chap. 6.
76T4 Turban, G., Catherine, Y.: Vide, Suppl. **182** (1976) 51.

76V1 Van Vorous, T.: Solid State Technol. **19(12)** (1976) 62.
76V2 Vasileva, L.L., Drozdov, V.N., Repinskii, S.M., Svitashev, K.K.: Sov. Microelectron. **5** (1976) 358.
76W1 Westwood, W.D.: Prog. Surf. Sci. **7** (1976) 71.
76W2 Wada, Y., Ashikawa, M.: Jpn. J. Appl. Phys. **15** (1976) 1725.
76W3 Wada, Y., Ashikawa, M.: Jpn. J. Appl. Phys. **15** (1976) 389.
76W4 Weissmantel, C.: Thin Solid Films, **32** (1976) 11.
76W5 Winters, H.F., Coburn, J.W.: Appl. Phys. Lett. **28** (1976) 176.
76W6 West, L.A.: J. Vac. Sci. Technol. **13** (1976) 198.
76W7 Wilson, R.W., Terry, L.E.: J. Vac. Sci. Technol. **13** (1976) 157.
77A1 Anderson, D.A., Spear, W.E.: Phil. Mag. **35** (1977) 1.
77A2 Arslambekov, V.A., Safarov, A.: Sov. Microelectron. **6** (1977) 55.
77A3 Amick, J.A., Schnable, G.L., Vossen, J.L.: J. Vac. Sci. Technol. **14** (1977) 1053.
77A4 Abdulleav, G.B., et al.: Akad. Nauk. AZ SSR **33** (1977) 19.
77B1 Ban, V.S.: in [77d1] p. 59.
77B2 Bryant, W.A.: J. Mater. Sci. **12** (1977) 1285.
77B3 Butler, S.R., Feigl, F.J., Rohatgi, A., Kraner, H.W., Jones, K.W.: Electrochem. Soc. Extend. Abstr. **77-1** (1977) 217.
77B4 Buehler, J., Fitzer, E., Kehr, D.: in [77d1] p. 493.
77B5 Bunshah, R.F., Schramm, R.J.: Thin Solid Films **40** (1977) 211.
77C1 Champagne, R., Toole, M.: Solid State Technol. **20(12)** (1977) 61.
77C2 Currau, J.E.: J. Vac. Sci. Technol. **14** (1977) 108.
77C3 Chang, R.P.H., Chang, C.C., Sheng, T.T.: Appl. Phys. Lett. **30** (1977) 657.
77C4 Catherine, Y., Turban, G.: Thin Solid Films **41** (1977) L 57.
77C5 Clarke, P.J.: Thin Solid Films **45** (1977) 157.
77C6 Cuomo, J.J., Gambino, R.J., Harper, J.M.E., Kuptsis, J.D.: IBM J. Res. Dev. **21** (1977) 580.
77C7 Cuomo, J.J., Gambino, R.J.: J. Vac. Sci. Technol. **14** (1977) 152.
77C8 Clarke, P.J.: J. Vac. Sci. Technol. **14** (1977) 141.
77C9 Claeys, C.L., Laes, E.E., Declerck, G.J., Van Overstraeten, R.J.: in [77h1] p. 773.
77C10 Cheng, Y.C.: Progr. Surf. Sci. **8** (1977) 181.
77D1 Deal, B.E.: Jpn. J. Appl. Phys. **16, suppl. 16-1** (1977) 29.
77D2 Deal, B.E.: in [77h1] p. 276.
77D3 Deppe, H.R., Hieke, E., Sigusch, R.: Electrochem. Soc. Extend. Abstr. **77-1** (1977) 622.
77D4 Davydov, B.B., Mamedov, N.A., Efendiev, K.J., EL-Bakravi, M.D.: Izv. Akad. Nauk. AZ. SSR, Ser. Fiz-Tekh. Mat. Nauk. **4** (1977) 88.
77D5 Dutton, R.W., Antoniadis, D.A.: Nato Advanced Study Institute on Process and Device Modeling for Integrated Circuit Design, Université Catholique de Louvain, July 19–29, **1977.**
77E1 Eer Nisse, E.P.: Appl. Phys. Lett. **30** (1977) 290.
77F1 Fraser, D.B., Cook, H.D.: J. Vac. Sci. Technol. **14** (1977) 147.
77G1 Gdula, R.A., Li, P.C.: J. Electrochem. Soc. **124** (1977) 1927.
77G2 Gal, S., Buzagh-Gere, E., Szatisz, J., Pokol, G.: Proc. 5th Int. Conf. on Therm. Anal. (1977) p. 258.
77G3 Gaind, A.K., Ackermann, G.K., Lucarini, V.J., Bratter, R.L.: J. Electrochem. Soc. **124** (1977) 599.
77H1 Hollahan, J.R., Wauk, M.T., Rosler, R.S.: in [77d1] p. 224.
77H2 Holland, L.: J. Vac. Sci. Technol. **14** (1977) 5.
77H3 Hess, D.W., Deal, B.E.: J. Electrochem. Soc. **123** (1977) 735.
77H4 Hoffmann, G., Hermann, L., Puskas, L., Kovacs, P., Nagy, A.: J. Phys. D. **10** (1977) 1509.
77H5 Hess, D.W., Mc Donald, R.C.: Thin Solid Films **42** (1977) 127.
77H6 Hirai, T., Niikava, K., Goto, T.: J. Mater. Sci. **12** (1977) 631.
77H7 Helix, M.J., Vaidyanathan, K.V., Streetman, B.G.: Electrochem. Soc. Extend. Abstr. **77-2** (1977) 426.
77H8 Holm, R., Storp, S.: Appl. Phys. **12** (1977) 101.
77H9 Hartsough, L.D., Mc Leod, P.S.: J. Vac. Sci. Technol. **14** (1977) 123.
77H10 Hammond, M.L.: in [77d1] p. 144.
77H11 Hattori, T.: Appl. Phys. Lett. **30** (1977) 312.
77I1 Irene, E.A., Ghez, R.: J. Electrochem. Soc. **124** (1977) 1757.
77I2 Iida, K.: J. Electrochem. Soc. **124** (1977) 614.

77I3 Irene, E.A., Ghez, R.: Proc. 3rd Int. Symp. on Silicon Materials Science and Technology, Semiconductor Silicon 1977, Philadelphia, May, 9–13, 1977, (1977) 313.
77J1 Johannessen, J.S., Helms, C.R., Spicer, W.E., Strausser, Y.E.: IEEE Trans. Electron. Dev. **ED-24** (1977) 547.
77K1 Kern, W., Rosler, R.S.: J. Vac. Sci. Technol. **14** (1977) 1082.
77K2 Kelly, R.: Radiat. Eff. **32** (1977) 91.
77K3 Kropman, D., Vinnal, M., Putk, P.: Vacuum **27** (1977) 125.
77L1 Logar, R.E., Wauk, M.T., Rosler, R.S.: in [77d1] p. 195.
77L2 Lamont, L.T.: J. Vac. Sci. Technol. **14** (1977) 122.
77L3 Lin, S.S.: J. Electrochem. Soc. **124** (1977) 1945.
77M1 Misawa, Y., Yagi, H.: Jpn. Appl. Phys. **16** (1977) 1115.
77M2 Mc Leod, P.S., Hartsough, L.D.: J. Vac. Sci. Technol. **14** (1977) 263.
77M3 Marshall, S., Zeto, R.J., Kesperis, J.S.: Electrochem. Soc. Extend. Abstr. **77-1** (1977) Abstr. No. 83.
77M4 Murarka, S.P., Levinstein, H.J., Marcus, R.B., Wagner, R.S.: J. Appl. Phys. **48** (1977) 4001.
77M5 Murarka, J.: Appl. Phys. **48** (1977) 5020.
77M6 Manke, C.W., Donaghey, L.F.: J. Electrochem. Soc. **124** (1977) 561.
77M7 Monkowski, J., Stach, J., Tressler, R.E.: in [77h1] p. 324.
77M8 Maeda, M., Kamioka, H., Shimoda, H., Takagi, M.: 13th Symposium on Semiconductor and IC Technol., Tokyo: 1977.
77M9 Miyoshi, H., Tsubouchi, N., Nishimoto, A., Abe, H.: Electrochem. Soc. Extend. Abstr. **77-2** (1977) 943.
77N1 Nowicki, R.S.: J. Vac. Sci. Technol. **14** (1977) 127.
77O1 O'Hanlon, J.F.: "Oxides and Oxide Films", Vijh, A.K., (ed.), New York: Dekker **5** (1977) 1.
77R1 Rosler, R.S.: Solid State Techn. **20(4)** (1977) 63.
77R2 Rohatgi, A., Butler, S.R., Feigl, F.J., Kraner, H.W., Jones, K.W.: Appl. Phys. Lett. **30** (1977) 104.
77S1 Shiloh, N., Gayer, B., Brinckman, F.E.: J. Electrochem. Soc. **124** (1977) 295.
77S2 Shiraki, H.: in [77h1] p. 546.
77S3 Smith, jr., J.E., Sedgwick, T.O.: Thin Solid Films **40** (1977).
77S4 Shibagaki, M., Horiike, Y., Yamazuki, T.: Jpn. J. Appl. Phys. **17, Suppl. 17-1** (1977) 215.
77S5 Szymanski, A., Podgorski, A., Huczko, A.: Commun.-Symp. Int. Chim. Plasmas, Vol. 3 (1977) Paper S.4.13.
77S6 Schiller, S., Heisig, U., Goedicke, K.: J. Vac. Sci. Technol. **14** (1977) 815.
77S7 Soxman, E.J.: Proc. 7th Intern. Vac. Congr. ed. by R. Dobrozemsky et al. (Vienna, 1977) p. 309.
77S8 Sedgwick, T.O.: in [77d1] p. 59.
77S9 Stagg, J.P.: Appl. Phys. Lett. **31** (1977) 532.
77T1 Tsubouchi, N., Miyoshi, H., Abe, H.: Proc. 9th Conf. on Solid State Devices, Tokyo: 1977, Paper No. A-5-3.
77T2 Tressler, R.E., Stach, J., Metz, D.M.: J. Electrochem. Soc. **124** (1977) 607.
77T3 Tsubouchi, N., Miyoshi, H., Nishimoto, A., Abe, H.: Jpn. J. Appl. Phys. **16** (1977) 855.
77T4 Tsubouchi, N., Miyoshi, H., Nishimoto, A., Abe, H., Sato, J.: Jpn. J. Appl. Phys. **16** (1977) 1055.
77U1 Urbanek, K.: Solid State Technol. **20**(4) (1977) 87.
77W1 Wahl, G.: Thin Solid Films **40** (1977) 13.
77Z1 Zdanowicz, L., Zielinska, K.: Acta Phys. Pol. A **52** (1977) 67.
77Z2 Zeto, R.J., Bosco, C.D., Hryckowian, E., Wilcox, L.L.: Electrochem. Soc. Extend. Abstr. **77-1** (1977) 229.
78A1 Arita, Y.: J. Cryst. Growth. **45** (1978) 383.
78B1 Blanc, J.: Appl. Phys. Lett. **33** (1978) 424.
78B2 Bell, A.T.: Solid State Technol. **21** (1978) 89.
78B3 Buch, J., Cervenak, J.: Thin Solid Films **55** (1978) 185.
78B4 Bauer, R.S., Mc Menamin, J.C., Petersen, H., Bianconi, A.: "Phys. SiO_2 and Its Interfaces" Pantelides, S.T., (ed.), Elmsford, N.Y.: Pergamon **1978**. 401.
78B5 Berkman, S., Ban, V.S., Goldsmith, N.: "Heteroepitaxial Semiconductors for Electronic Devices", Cullen, G.W., Wang, C.C., (eds.), Berlin, Heidelberg New York: Springer **1978**, 264.
78B6 Ban, V.S.: J. Electrochem. Soc. **125** (1978) 317.
78C1 Chang, C.C., et al.: J. Electrochem. Soc. **125** (1978) 787.
78C2 Clark, R.S.: Solid State Technol. **21**(11) (1978) 80.
78C3 Cochet, G., Mellottée, H., Delbourgo, R.: J. Electrochem. Soc. **125** (1978) 487.

78C4 Cuomo, J.J., Gambino, R.J., Harper, J.M.E., Kuptsis, J.D., Weber, J.C.: J. Vac. Sci. Technol. **15** (1978) 281.
78C5 Claeys, C.L., Declerck, G., Van Overstraeten, R.J.: "Semiconductor Characterization Techniques" Barnes, P.A., Rozgonyi, G.A., (eds.), The Electrochemical Soc., Princeton, N.J. **1978**, 366.
78D1 Deal, B.E.: J. Electrochem. Soc. **125** (1978) 576.
78D2 Deal, B.E., Hess, D.W., Plummer, J.D., Ho, C.P.: J. Electrochem. Soc. **125** (1978) 339.
78D3 Dong, D., Irene, E.A., Young, D.R.: J. Electrochem. Soc. **125** (1978) 819.
78D4 Deal, B.E., Hurrle, A., Schulz, M.J.: J. Electrochem. Soc. **125** (1978) 2024.
78D5 DiMaria, D.J. in [78p] p. 160.
78D6 DiMaria, D.J., Young, D.R., Dekeersmaecker, R.F., Hunter, W.R., Serrano, C.M.: J. Appl. Phys. **49** (1978) 5441.
78F1 Frima, H., Goldman, J.: Electrochem. Soc. Extend. Abstr. **78-1** (1978) 704.
78F2 Fraser, D.B.: in [78v1] Chap. II-3.
78F3 Fair, R.B., Tsai, J.C.: J. Electrochem. Soc. **125** (1978) 2050.
78G1 Gould, R.D., Hogarth, C.A.: Thin Solid Films **51** (1978) 237.
78G2 Gambino, R.J., Cuomo, J.J.: J. Vac. Sci. Technol. **15** (1978) 296.
78G3 Guerrero, E., Machek, J., Tobolka, G.: Thin Solid Films **53** (1978) L1.
78G4 Galasso, G.S., Veltri, R.D., Croft, W.J.: Am. Ceram. Soc. Bull. **57** (1978) 453.
78G5 Goldstein, B., Szostak, D.J.: Appl. Phys. Lett. **33** (1978) 85.
78H1 Hollahan, J.R., Rosler, R.S.: in [78v1] p. 335.
78H2 Hitchman, M.C., Kane, J., Widmer, A.E.: in [78h1] p. 589.
78H3 Hochberg, A.K., Dozier, A.: in [78h1] p. 592.
78H4 Hollahan, J.R., Wauk, M., Hammond, M.: Proc. Applied Materials Seminar 1978, Santa Clara, CA., pp. 1–25.
78H5 Hattori, T., Suzuki, T.: Appl. Phys. Lett. **33** (4) (1978) 347.
78H6 Hattori, T.: Jpn. J. Appl. Phys. **17** (1978) 69.
78H7 Ho, C.P., Plumer, J.D., Meindl, J.D., Deal, B.E.: J. Electrochem. Soc. **125** (1978) 665.
78H8 Helix, M.J., Vaidyanathan, K.V., Streetman, B.G., Dietrich, H.B.: Thin Solid Films **55** (1978) 143.
78H9 Hulenyi, L., Harman, R., Csabay, O., Kinder, R.: Int. Wiss. Kolloq.-Tech. Hochsch. Ilmenau **23** (6) (1978) 77.
78H10 Hughes, J.L.: J. Vac. Sci. Technol. **15** (1978) 1576.
78H11 Hattori, T.: Denki Kagaku **46** (1978) 122.
78I1 Irene, E.A.: J. Electrochem. Soc. **125** (1978) 1708.
78I2 Irene, E.A., Dong, D.W.: J. Electrochem. Soc. **125** (1978) 1146.
78I3 Ito, T., Hijiya, S., Nozaki, T., Arakawa, H., Shinoda, M., Fukikawa, Y.: J. Electrochem. Soc. **125** (1978) 448.
78I4 Ito, T., Nozaki, T., Arakawa, H., Shinoda, M.: Appl. Phys. Lett. **32** (1978) 330.
78J1 Janssens, E.J., Declerck, G.J.: J. Electrochem. Soc. **125** (1978) 1696.
78J2 Josquin, W.J.M.J.: Proc. Ion Beam Modification of Materials. Vol. 3, Budapest, (1978) p. 1443.
78J3 Johnson, N.M., Bartelink, D.J., Schulz, M.: in [78p] p. 421.
78K1 Kern, W., Ban, V.S.: in [78v1] p. 257.
78K2 Katz, L.E., Kimerling, L.C.: J. Electrochem. Soc. **125** (1978) 1680.
78K3 Kirov, K., Atanasova, E., Aleksandrova, S.P., Amov, B., Dyakov, A.: Thin Solid Films **48** (1978) 187.
78K4 Kirov, K.I., Georgiev, S.S., Gerova, E.V., Aleksandrova, S.P.: Phys. Status Solidi A **48** (1978) 609.
78K5 Katz, L.E., Howells, jr., B.F.: Electrochem. Soc. Extend. Abstr. **78-2** (1978) 498.
78K6 Kurata, K., Miyata, N., Miyake, K.: Yamaguchi, Daigaku Kogakubu Kenkyu Hokoku **29** (1) (1978) 97.
78L1 Lanford, W.A., Rand, M.J.: J. Appl. Phys. **49** (1978) 2473.
78L2 Linssen, A.J., Peek, H.L.: Philips J. Res. **33** (1978) 281.
78L3 Le Couteile, M., Morin, F.: Thin Solid Films **52** (1978) 63.
78M1 Monkowski, J., Tressler, R.E., Stach, J.: J. Electrochem. Soc. **125** (1978) 1867.
78M2 Morosanu, C.E.: Rev. Roum. Phys. **23** (1978) 595.
78M3 Miyoshi, H., Tsubouchi, N., Nishimoto, A.: J. Electrochem. Soc. **125** (1978) 1824.
78M4 Monkowski, J., Tressler, R.E., Stach, J., Rohatgi, A., Williams, D.B., Butler, S.R.: Electrochem. Soc. Extend. Abstr. **78-2**, Abstr. No. 211.
78M5 Maeda, M., Kamioka, H., Takagi, M.: Electrochem. Soc. Extend. Abstr. **78-1** (1978) 423.

78O1 Ohkawa, S., Nakajima, Y.: J. Electrochem. Soc. **125** (1978) 1997.
78P1 Pudonin, F.A., Seleznev, V.N., Tokarchuk, D.N.: Microelectronics (USA) **7** (1978) 218.
78R1 Reinberg, A.R.: J. Electron. Mater. **7** (1978) 716.
78S1 Sinha, A.K., Levinstein, H.J., Smith, T.E., Quintana, G., Haszko, S.E.: J. Electrochem. Soc. **125** (1978) 601.
78S2 Shibagaki, M., Horiike, Y., Yamazaki, T.: Jpn. J. Appl. Phys. **17, suppl. 17-1** (1978) 215.
78S3 Stagg, J.P., Boudry, M.R.: Rev. Phys. Appl. **13** (1978) 841.
78S4 Silvestri, V.L., Osburn, C.M., Ormond, D.W.: J. Electrochem. Soc. **125** (1978) 902.
78S5 Schiller, S., Heisig, U., Goedicke, K.: Thin Solid Films **54** (1978) 33.
78S6 Singh, B.R., Balk, P.: J. Electrochem. Soc. **125** (1978) 453.
78S7 Sinha, A.K., Lugujjo, E.: Appl. Phys. Lett. **32** (1978) 245.
78S8 Sinha, A.K., Levinstein, H.J., Smith, T.E.: J. Appl. Phys. **49** (1978) 2423.
78S9 Sinha, A.K., Smith, T.E.: J. Appl. Phys. **49** (1978) 2756.
78S10 Stein, H.J., Wells, V.A., Hampy, R.E.: Electrochem. Soc. Extend. Abstr. **78-1** (1978) 695.
78T1 Tobin, P.J., Price, J.B.: Electrochem. Soc. Extend. Abstr. **78-1** (1978) 265.
78T2 Turban, G., Catherine, Y.: Thin Solid Films **48** (1978) 57.
78T3 Ting, C.Y., Vivalda, V.J., Schaefer, H.G.: J. Vac. Sci. Technol. **15** (1978) 1105.
78T4 Tarantor, Y.A., Kasyanenko, E.V., Konorov, P.P., Romanova, A.A.: Sov. Microelectron. **7** (1978) 41.
78T5 Tsubouchi, N., Nishimoto, A., Abe, H., Enomoto, T.: Mitsubishi Denki Giho **52** (1978) 382.
78T6 Thornton, J.A.: J. Vac. Sci. Technol. **15** (1978) 171.
78T7 Thornton, J.A., Penfold, A.S.: in [78v1] Chap. II-2.
78T8 Tsubouchi, N., Miyoshi, H., Abe, H.: Jpn. J. Appl. Phys. **17** (1978) 223.
78V1 Van de Ven, E.P.G.T., Sanders, J.A.M.: Electrochem. Soc. Extend. Abstr. **78-2** (1978) 525.
78V2 Vasilyeva, L.L., Drozdov, V.N., Repinsky, S.M., Svitashev, K.K.: Thin Solid Films **55** (1978) 221.
78W1 Widmer, A.E., Hitchman, M.L.: J. Electrochem. Soc. **125** (1978) 1723.
78W2 Waits, R.K.: J. Vac. Sci. Instrum. **15** (1978) 179.
78W3 Waits, R.K.: in [78v1] Chap. II-4.
78Y1 Yoshimi, T., Tanaka, K.: Electrochem. Soc. Extend. Abstr. **78-2** (1978) 531.
78Z1 Zeto, R.J., Irene, E.A., Dong, D.W.: IEEE Trans. Electron. Dev. **ED-26** (1978) 1359.
79A1 Adams, A.C., Capio, C.D., Haszko, S.E., Parisi, G.I., Povilonis, E.J., Robinson, McD.: J. Electrochem. Soc. **126** (1979) 313.
79A2 Adams, A.C., Capio, C.D.: J. Electrochem. Soc. **126** (1979) 1042.
79A3 Adams, A.C., Murarka, S.P.: J. Electrochem. Soc. **126** (1979) 334.
79B1 Bell, A.E.: J. Vac. Sci. Technol. **16**(2) (1979) 418.
79B2 Brown, W.A., Kamins, T.I.: Solid State Technol. **22**(7) (1979) 51.
79B3 Bäuerle, D.: in [79s1] p. 19.
79C1 Class, W.H.: Solid State Technol. **22**(6) (1979) 61.
79C2 Claeys, C.L., Declerck, G.J., Van Overstraeten, R.J.: Electrochem. Soc. Extended Abstracts. Los Angeles, 1979, **79-2**.
79D1 Deal, B.E., Early, J.M.: J. Electrochem. Soc. **126** (1979) 20C.
79F1 Fukuyama, T., Yanagisawa, S.: Jpn. J. Appl. Phys. **18** (1979) 987.
79F2 Frenzel, H., Singh, B.R., Haberle, K., Balk, P.: Thin Solid Films **58** (1979) 301.
79G1 Goncharov, E.V., Gol'dfarb, V.A., Petrakov, V.I., Surovtsev, L.S.: Fiz. Khim. Obrab. Mater. **3** (1979) 56.
79G2 Gaind, A.K., Kasprzak, L.A.: Solid State Electronics **22** (1979) 303.
79H1 Huppertz, H., Engl, W.L.: IEEE Trans. Electron. Dev. **ED-26** (1979) 658.
79H2 Hollahan, J.R.: J. Electrochem. Soc. **126** (1979) 930.
79H3 Ho, C.P., Plummer, J.D.: J. Electrochem. Soc. **126** (1979) 1523.
79H4 Hara, K., Usami, A., Suzuki, K.: Surf. Sci. **86** (1979) 866.
79H5 Hara, K., Suzuki, K.: Jpn. J. Appl. Phys. **18** (1979) 2027.
79H6 Hattori, T.: J. Electrochem. Soc. **126** (1979) 1789.
79H7 Hattori, T.: Solid State Technol. **22**(11) 1979) 85.
79H8 Ho, C.P., Plummer, J.D.: J. Electrochem. Soc. **126** (1979) 1516.
79H9 Ho, C.P., Plummer, J.D.: IEEE Trans. Electron. Dev. **ED-26** (1979) 623.
79J1 Jain, G.C., Prasad, A., Chakravaty, B.C.: J. Electrochem. Soc. **126** (1979) 89.
79K1 Kern, W., Schnable, G.L.: IEEE Trans. Electron. Dev. **ED-26** (1979) 647.
79K2 Kisker, D., Dun, H., Stevenson, D.A.: in [79s1] 30.

79K3 Katz, L.E., Howells, jr., B.F.: J. Electrochem. Soc. **126** (1979) 1822.
79L1 Lieske, N., Hezel, R.: Thin Solid Films **62** (1979) 197.
79M1 Monkowski, J., Stach, J., Tressler R.E.: J. Electrochem. Soc. **126** (1979) 1129.
79M2 Murarka, S.P., Chang, C.C., Adams, A.C.: J. Electrochem. Soc. **126** (1979) 996.
79M3 Monkowski, J.: Solid State Technol. **22**(7) (1979) 58.
79M4 Monkowski, J.: Solid State Technol. **22**(7) (1979) 113.
79M5 Murarka, S.P.: IEDM Digest (1979) 454.
79N1 Nazar, F.M., Ahmed, N.: Int. J. Electron. **47** (1979) 81.
79R1 Rand, M.J.: J. Vac. Sci. Technol. **16** (1979) 420.
79R2 Revesz, A.G.: J. Electrochem. Soc. **126** (1979) 502.
79R3 Rohatgi, A., Butler, S.R., Feigl, F.J., Kramer, H.W., Jones, K.W.: J. Electrochem. Soc. **126** (1979) 143.
79R4 Revesz, A.G.: J. Electrochem. Soc. **126** (1979) 122.
79R5 Reinberg, A.R.: J. Electron. Mater. **8** (1979) 345.
79R6 Rosler, R.S., Engle, G.M.: Solid State Technol. **22**(12) (1979) 88.
79R7 Rohatgi, A., Butler, S.R., Feigl, F.J.: J. Electrochem. Soc. **126** (1979) 149.
79S1 Stein, H.J., Wells, V.A., Hampy, R.E.: J. Electrochem. Soc. **126** (1979) 1750.
79S2 Singh, B.R., Balk, P.: J. Electrochem. Soc. **126** (1979) 1288.
79S3 Schwartz, S.A., Helms, C.R.: J. Appl. Phys. **50** (1979) 5492.
79S4 Spear, K.E.: in [79s1] 1.
79S5 Schulz, M., Klausmann, E.: Appl. Phys. **18** (1979) 169.
79T1 Thompson, T., Wiesner, J., Carlson, D., Kline, R.: Electrochem. Soc. Extend. Abstr. **79-1** (1979) Abstr. No. 97.
79T2 Taft, E.A.: J. Electrochem. Soc. **126** (1979) 1728.
79T3 Tsubouchi, N., Miyoshi, H., Abe, H., Enomoto, T.: IEEE Trans. Electron. Dev. **ED-26** (1979) 618.
79T4 Thompson, G.E., Wood, G.C.: J. Electrochem. Soc. **126** (1979) 2042.
79V1 Van de Ven, E.P.G.T., Sanders, J.A.M.: Abstr. ESSDERC Munich, Sept. **1979**.
79V2 Vossen, J.L.: J. Electrochem. Soc. **126** (1979) 319.
79V3 Vandenberg, M.: Electrochem. Soc. Extend. Abstr. **79-1** (1979) Abstr. 99.
79Y1 Yadav, A.D., Joshi, M.C.: Thin Solid Films **58** (1979) 300.
79Z1 Zeto, R.J., Korolkoff, N.O., Marshall, S.: Solid State Technol. **22**(7) (1979) 62.
79Z2 Zeto, R.J.: Electrochem. Soc. Extend. Abstr. **79-1** (1979) Abstr. No. 96.
80A1 Agraz-Guerena, J., Panousis, P.T., Morris, B.L.: IEEE Trans. Electron. Dev. **ED-27** (1980) 1397.
80B1 Bernard, W.J., Russel, P.G.: J. Electrochem. Soc. **127** (1980) 1256.
80E1 Egitto, F.D.: J. Electrochem. Soc. **127** (1980) 1354.
80F1 Fuoss, D., Topich, J.A.: Appl. Phys. Lett. **36**(4) (1980) 275.
80H1 Haberle, K., Froeschle, E.: Solid State Electron. **23** (1980) 855.
80I1 Irene, E.A., Dong, D.W., Zeto, R.J.: J. Electrochem. Soc. **127** (1980) 396.
80K1 Klein, N., Moskovici, V., Kadary, V.: J. Electrochem. Soc. **127** (1980) 152.
80M1 Mattson, B.: Solid State Technol. **23**(1) (1980) 60.
80M2 Maniv, S., Westwood, W.D.: J. Vac. Sci. Technol. **17** (1980) 743.
80M3 Monkowski, J., Tressler, R.E., Stach, J.: Thin Solid Films **65** (1980) 153.
80M4 Misawa, Y., Yagi, H.: Jpn. J. Appl. Phys. **19** (1980) 1885.
80P1 Prassad, P.M., Sundarsingh, V.P.: Microelectronic. J. **11**(6) (1980) 21.
80P2 Peters, J.W., Gebhardt, F.L., Hall, T.C.: Solid State Technol. **23**(9) (1980) 121.
80R1 Rosler, R.S., Engle, G.M.: Electrochem. Soc. Extend. Abstr. **80-1** (1980).
80R2 Ramin, M., Ryssel, H., Kranz, H.: Appl. Phys. **22** (1980) 393.
80S1 Spear, K.E., Wang, M.S.: Solid State Technol. **23**(7) (1980) 63.
80S2 Serikawa, T.: Jpn. J. Appl. Phys. **19** (1980) L 259.
80S3 Szczeklik, J., Schabowska, E.: Thin Solid Films **71** (1980) L 17.
80S4 Saraswat, K.C., Mohammadi, F.: IEEE-Trans. **EDL-1** (1980) 18.
80T1 Tiller, W.A.: J. Electrochem. Soc. **127** (1980) 619.
80T2 Tiller, W.A.: J. Electrochem. Soc. **127** (1980) 625.
81B1 Bussmann, E.: Siemens Forsch. Entwicklungsber. **10** (1981) 357.
81C1 Chari, K.S., Mathur, B.: Thin Solid Films **75** (1981) 157.
81C2 Craven, D.R., Stimmel, J.B.: Semiconductor International; June 1981 (1981) 59.

81D1 Doering, E.: in [81s] p. 208.
81H1 Helms, C.R.: in [81s] p. 19.
81N1 Nazar, F.M., Bhatti, M.A.: Int. J. Electron. **50** (1981) 119.
81O1 O'Neill, T.G.: Semiconductor International; June 1981 (1981) 77.
81P1 Poindexter, E.H., Caplan, P.J.: in [81s] p. 150; and Johnson, N.M.: in [81s] p. 35.
81R1 Rouse, J.W., Helms, C.R., Deal, B.E., Razouk, R.R.: J. Vac. Sci. Technol. **18**(3) (1981) 971.
81S1 Segueda, F., Richardson, jr., R.E.: J. Vac. Sci. Technol. **18**(2) (1981) 362.
81W1 Wiggins, M.D., Baird, R.J., Wyablatt, P.: J. Vac. Sci. Technol. **18** (1981) 965.
81W2 Wolters, D.R.: in [81s] p. 180.
82M1 Massoud, H.Z., Ho, C.P., Plummer, J.D.: Stanford Techn. Rept. TR DXG 501-82 (1982).
82R1 Razouk, R., Deal, B.E.: J. Electrochem. Soc. **129** (1982) 808.
82Z1 Zheng, X.Q., Hofmann, K., Schulz, M.: J. Appl. Phys. **53** (1982) 9146.
83F1 Fargeix, A., Ghibaudo, G., Kamarinos, G.: J. Appl. Phys. **54** (1983) 2878.
83H1 Hu, S.M.: Appl. Phys. Lett. **42** (1983) 872.
83S1 Schulz, M.: Surf. Sci. **132** (1983) 422.
84L1 Lefevre, H., Schulz, M.: in [84b].

For subsections 6.1.4.5.8···6.1.4.5.12
(Fabrication of metal layers)
see p. 593ff.

6.1.4.6 Lithography

6.1.4.6.0 Introduction

Lithography processes are used to generate patterns in thin films on a semiconductor substrate or to generate an impurity doping pattern in a semiconductor substrate as is needed in device fabrication. These patterns can be generated by two methods:

i) Patterns are generated by coating the wafer with a *radiation-sensitive resist*. Irradiation of the resist with visible light, UV light, soft X-rays, electrons, or ions changes the solubility of the resist in the developer. A resist pattern is generated when the irradiation is nonuniform, e.g. for illumination through a mask. The resist pattern can be transferred into a thin film by etching (Fig. 1), by a lift-off technique (Fig. 2), or by selective material deposition (Fig. 3), and to the semiconductor substrate by ion implantation of impurities (Fig. 4). The lithographic process involves the deposition, exposing and developing of the radiation-sensitive resist.

ii) Thin-film or impurity-doping patterns are directly generated without the need of a resist (Fig. 5). Exposing the desired areas of a substrate to photon, electron, or ion beams in an appropriate environment, leads to the local deposition of a thin film or the local implantation of impurities. Local exposure can also be used for local etching, oxidation, or inducing a change in conductivity as is shown schematically in Fig. 5. In this class of pattern generation, the lithography process consists solely of exposure to photons, electrons, or ions.

In the fabrication of semiconductor devices, the methods of the first class of pattern generation (i) are mainly used. The methods of the second class (ii) are still under development.

The present section reviews the lithography methods of the first class, which may be divided into four categories:

Optical lithography	contact printing proximity printing 1:1 projection printing reduction projection printing
Electron-beam lithography	projection printing direct writing (scanning)
X-ray lithography	proximity printing
Ion-beam lithography	projection printing direct writing (scanning) proximity printing

The principles of these methods are schematically illustrated in Fig. 6. The various methods by which patterns can be transferred to a wafer are summarized in Fig. 7.

Lithography used in semiconductor device technology is reviewed in a number of excellent books and publications [80B4, 80C3, 80D1, 80W2, 80W6, 81B1, 81B4, 81T2].

Acknowledgements

The author would like to thank D. Widmann, H. Bierhenke, E. Hieke, H.U. Zeitler, K. Anger, P. Tischer, W. Arden and R. Sigusch of the Siemens Research Laboratory, for many helpful discussions.

6.1.4.6.1 Resists*)

Resists are materials which change the solubility in a developer solution when they are exposed to radiation. Resists are classified as either *positive*-working or *negative*-working, depending on whether the solubility in the developer increases (positive) or decreases (negative) upon exposure to radiation, respectively.

Positive resists consist mainly of two components: a film-forming *resin* and a radiation-sensitive *sensitizer*. Both are dissolved in a solvent. The sensitizer is a dissolution inhibiter. When destroyed by exposure to radiation, the solubility of the resin increases in an appropriate developer solution.

Figure 8 shows two mechanisms which change the solubility of a positive resist upon exposure to radiation: side group modification and the scission of polymer chains.

*) Definitions. For substance data see appropriate subsections.

Negative resists consist essentially of two components: a chemically inert *film-forming polymer* and a radiation-sensitive *sensitizer*. Upon exposure to radiation, the sensitizer reacts with the polymer to form cross-links between the polymer molecules (Fig. 9), thus reducing the solubility in the developer solvent.

Resists show two characteristic features: sensitivity and contrast.

i) The *sensitivity* is defined as the incident exposure dose required to induce the desired chemical response in the resist, see Table 1.

Table 1. Definition of the sensitivity S for photo-, X-ray, electron-beam, and ion-beam resist. P is the exposing radiation power, I is the beam current, t is the exposure time required to induce the desired chemical response in the resist, A is the resist area exposed.

Resist	Photoresist	X-ray resist	Electron-beam resist	Ion-beam resist
Sensitivity S = incident exposure dose required to induce desired chemical response in resist	$\frac{P \cdot t}{A}$ J cm^{-2}	$\frac{P \cdot t}{A}$ J cm^{-2}	$\frac{I \cdot t}{A}$ C cm^{-2}	$\frac{I \cdot t}{A}$ C cm^{-2}

ii) The contrast characterizes the response of a resist to development.
For *positive resists* the contrast γ_p is defined as

$$\gamma_p = \left[\log_{10}\left(\frac{D_c}{D_0}\right)\right]^{-1},$$

where D_0 and D_c are extrapolated doses as shown in Fig. 10. For *negative resists* the contrast γ_n is defined as

$$\gamma_n = \left[\log_{10}\left(\frac{D_g^0}{D_g^i}\right)\right]^{-1},$$

where D_g^i and D_g^0 are extrapolated doses as shown in Fig. 10.
The contrast increases with decreasing D_c/D_0 or D_g^0/D_g^i, respectively.

6.1.4.6.2 Photolithography

Photolithography was first employed by the electronics industry for the fabrication of copper printed circuits. At the present time, photolithography is an essential tool in high-density integrated circuit technology. Fig. 11 shows the basic principle of pattern generation in a thin film by photolithography. A photoresist deposited on the thin film is exposed to light incident through a mask. The exposed positive or unexposed negative photoresist is removed in a developing process. The resist pattern can be used as a mask for etching the thin film underneath. The mask can also be transferred to a thin film either by the *lift-off technique* (Fig. 2) or by selective material deposition (Fig. 3). Photolithography is reviewed in a number of excellent publications [76W1, 78C2, 80B3, 80C3, 80D1, 80D2, 80W2, 80W6, 81B1, 81B4].

Exposure techniques

Photolithography is performed using any of the three different exposure techniques: *contact printing, proximity printing,* or *projection printing.* These techniques are described in the following sections.

Contact printing

In contact printing, a photomask is brought into physical contact with the wafer as shown in Fig. 12. For an intimate contact between the wafer and the mask, a high resolution is possible allowing the printing of submicron features.

The main drawbacks of contact printing are:

The contact between mask and wafer produces defects in both the mask and the wafer, thus reducing the device yield.

Particles between the mask and the wafer prevent an intimate contact and thus degrade the resolution.

Wafer deformation and mask bending lead to a misalignment and thus to a reduced resolution.

Proximity printing

In this exposure technique, the wafer is separated from the mask by a small constant distance as shown in Fig. 13. The proximity distance is of the order of 10 µm. Most of the problems associated with the mechanical contact between mask and wafer in contact printing are eliminated.

However, the resolution is not as good as with contact printing. The primary effect limiting the resolution is the Fresnel diffraction of light at the corners of the photomask. The Fresnel diffraction depends on the wavelength of the light and the proximity distance from mask to wafer.

Fig. 14 shows the distribution of the UV light intensity on the photoresist resulting from the irradiation of a photomask having a periodic grid.

The practical minimum period, $2w_{min}$, which can be replicated by proximity printing [76W1] is given by

$$2w_{min} = 3\sqrt{\lambda(s+d/2)},$$

where s is the proximity distance between mask and wafer, λ is the wavelength of the incident light, and d is the photoresist thickness.

Fig. 15 shows the minimum linewidth w_{min} as a function of the wavelength of the light and the proximity distance s. The photoresist thickness is assumed to be 1.5 µm.

The minimum linewidth that can be replicated by the commonly used wavelength of 400 nm and by a proximity distance of 10 µm is about 3 µm. For contact printing, where the spacing s is zero, the minimum linewidth is approximately 0.8 µm.

Projection printing

In projection printing, the image of a mask is projected onto a wafer by either a refractive or a reflective optical system. Two types of optical projection systems have been developed for semiconductor device fabrication: these are 1:1 *reflective scanning projection* and *step-and-repeat reduction projection* systems. The main advantage of projection printing over contact printing is that wafer and mask are completely separated by the projection optic. Problems due to the contact between mask and wafer cannot occur. Projection printing also offers higher resolution than proximity printing.

Fig. 16 schematically shows the optical projection system of a widely used 1:1 reflective scanning projection system. The mask is projected at a ratio 1:1 onto a wafer after three reflections at two spherical mirrors.

Since in this optical system only a small zone is satisfactorily projected, the wafer has to be exposed portion by portion. Full wafer exposure is realized by scanning mask and wafer in parallel. Since the mirror imaging system is used without chromatic aberration, a broad spectrum of radiation, from deep UV to visible light, may be applied for the exposure of the resist.

Fig. 17 schematically shows a step-and-repeat optical reduction projection system incorporating an exposure source, a filter, a condensor lens and a reduction lens. The mask image is reduction-projected onto the wafer. For full coverage the wafer is mechanically stepped between exposure sites.

The *resolution capability* of projection printing depends on the wavelength, the numerical aperture of the objective, the degree of defocussing of mask and wafer surface, and the degree of coherence of the illumination.

The numerical aperture NA is defined by

$$NA = n \sin\alpha,$$

where n is the refractive index in the image space and 2α is the maximum cone angle of rays reaching an image point on the optical axis of the projection system as shown in Fig. 18.

The degree of *coherency* can be quantified by the ratio r

$$r = \frac{(NA)_c}{(NA)_0},$$

where $(NA)_c$ and $(NA)_0$ are the numerical apertures of the condenser and the objective, respectively.

A point exposure source is a coherent source, in which case the numerical aperture of the condenser is zero, resulting in $r=0$.

The *minimum linewidth* which can be reliably replicated by a projection printing system is given by [80W2]

$$w_{min} \simeq \frac{\lambda}{2NA}\left(\frac{1+3r}{1+r}\right)\left(1+0.15\left(\frac{\Delta f}{\Delta f_R}\right)^2\right).$$

where λ is the wavelength, NA the numerical aperture of the objective, r the degree of coherence, Δf the defocus distance, and Δf_R the Rayleigh depth of the focus as expressed by

$$\Delta f_R = \frac{\lambda}{2NA^2}.$$

Fig. 19 shows the minimum linewidth w_{min} as a function of the degree of defocusing Δf and of the numerical aperture of the objective NA.

The finest patterns achieved with projection printing in photoresist films are of $w = 0.5$ μm [82A1].

Besides the limits given by the optical exposure system, the minimum linewidth that can be produced in a photoresist is also determined by standing-wave effects in the photoresist. This effect leads to photoresist edge profiles as shown in Fig. 20 and to linewidth variations [75W1, 75W2, 77B1, 79B3].

Minor changes in resist thickness lead to a linewidth variation of the resist at the interface between resist and substrate.

Photoresists

Photoresists are materials for which the solubility in a developer solution changes when exposed to light. Photoresists are classified as either *positive-* or *negative*-working, depending on whether the solubility increases (positive) or decreases (negative) upon exposure to light.

Photoresists contain a film-forming *resin* dissolved in a *solvent* system, a *sensitizer*, also called a photoinitiator, for increasing the photosensitivity, and *additives* which improve the performance of the resist.

Photoresists are sensitive at wavelengths λ in the range 200···500 nm as shown in Fig. 21. The light absorption occurs in this wavelength range.

In the safelight zone between yellow and red it is possible to work with photoresists without having any radiation-induced change of resist properties.

Positive photoresists

The operation of positive photoresists is based on the photochemical conversion of a light-sensitive insoluble molecule to one that is readily soluble in an alkaline developer solvent [75D1]. Fig. 22 shows reactions obtained by exposing and developing positive photoresists [44S1, 79P1].

The light-sensitive element is converted into a ketene by exposure to light. This ketene in turn reacts with H_2O vapour to form indenecarboxylic acid. During developing, a carboxylic acid group salt is formed.

A large number of sensitizers (light-sensitive elements in the photoresist) are available for positive photoresists. Some are listed in Table 2.

Table 2. Sensitizers for positive photoresists.

No.	Sensitizer	Structure	Remarks	Ref.
1	Naphthoquinone-1,2-diazide-5-sulfochloride	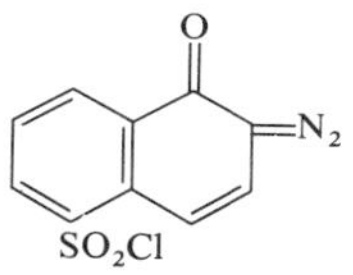 O, $=N_2$, SO_2Cl	naphtalene form	75D1
2	Naphthoquinone-1,2-diazide-4-sulfochloride	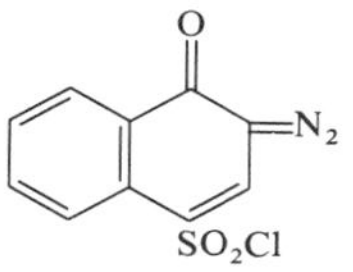 O, $=N_2$, SO_2Cl	naphtalene form	75D1
3	Naphthoquinone-2,1-diazide-4-sulfochloride	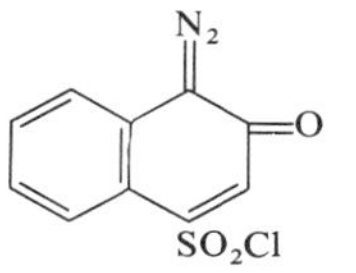 N_2, =O, SO_2Cl	naphtalene form	75D1
4	Naphthoquinone-2,1-diazide-5-sulfochloride	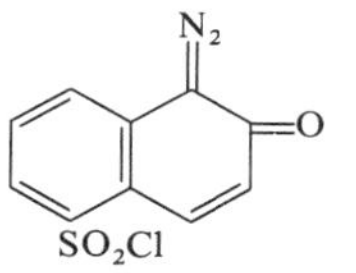 N_2, =O, SO_2Cl	naphtalene form	75D1

continued

Table 2 (continued)

No.	Sensitizer	Structure	Remarks	Ref.
5	Sensitizer with naphthoquinone diazides	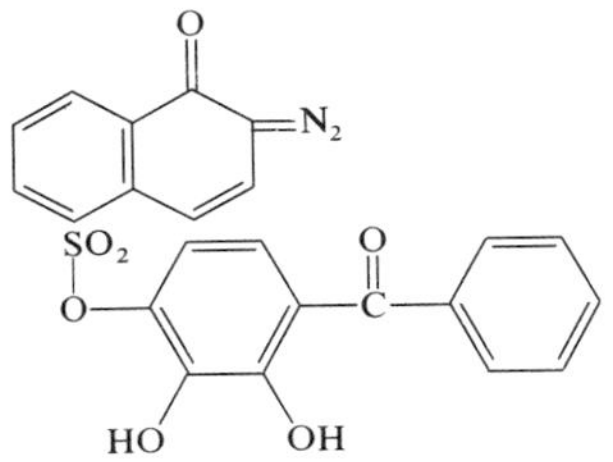 O; $=N_2$; SO_2; O; O; C; HO; OH	central structure	75D1, 75D2
6	Sensitizer with naphthoquinone diazides	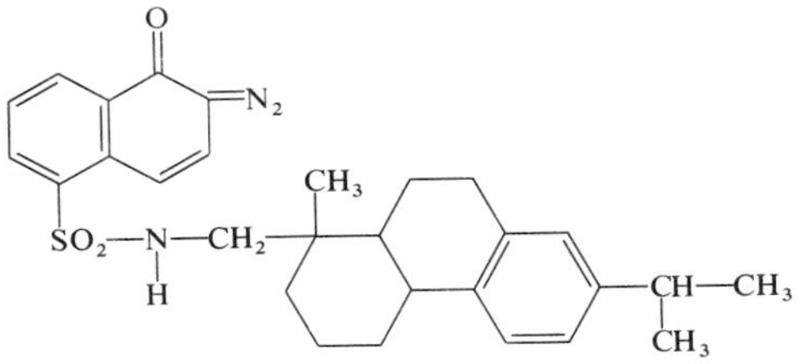 O; $=N_2$; CH_3; SO_2—N—CH_2; H; CH—CH_3; CH_3	mostly used in commercially available positive photoresists	57M1, 78C2

Film-forming *resins* for positive photoresists are listed in Table 3.

Table 3. Resins for positive photoresists.

No.	Resin	Structure	Remarks	Ref.
1	Novolak	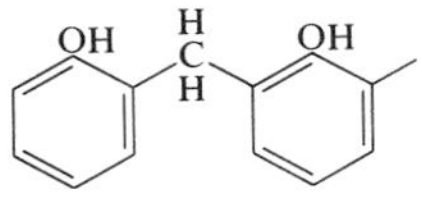 OH; H–C–H; OH	linearly and permanently fusible; does not form cross-linked polymers	75D1
2	Resole	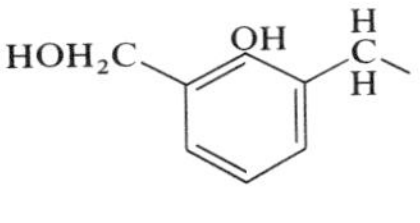 HOH_2C; OH; H–C–H	derivative of phenol, which is more reactive than novolak; resole is often used to prepare novolak resins	75D1
3	Phenol-formaldehyde	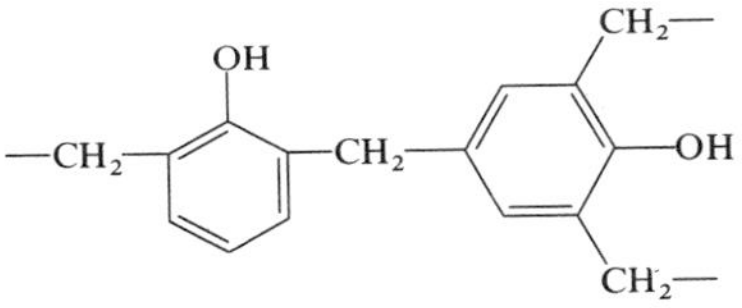 OH; —CH_2—; —CH_2—; CH_2—; —OH; CH_2—	if the reacting medium of phenol formaldehyde mixtures is alkaline in nature and if an excess of formaldehyde is used, the polymerization may be continued by heating	75B3

Table 4 lists developers and strippers for various positive photoresists.

Table 4. Developers and strippers for positive photoresists [75D1].

Developer	Stripper	Resists
Alkaline solutions	acetone DMF, dimethyl formamide MEK, methyl ethyl ketone MIBK, methyl isobutyl ketone dioxine butyl cellosolve butyl carbitol	**Shipley:** AZ-111, AZ-111B, AZ-119, AZ-119B AZ-340, AZ-345, AZ-1350, AZ-1350B, AZ-1350J AZ-1350H, AZ-1450J, AZ-2400 **Waycoat:** positive HR **Dynachem:** PRS-175 **GAF Microline:** PR-102 **Kodak:** KAR-3 **Hunt:** HPR

Positive photoresists usable for deep UV light are listed in Table 5. In general, any electron beam-sensitive resist is also suitable as a photoresist for deep UV light.

Table 5. Positive photoresists for deep UV light.

No.	Resist	λ nm	Developer	Stripper	Ref.
1	Crosslinked GCM, poly (glycidyl methacrylate-comethyl methacrylate)	215	5:1 MIBK:MEK (methyl isobutyl ketone: methyl ethyl ketone)	O_2-plasma	79Y1
2	Poly (methyl methacrylate-co-3-oximino-2-butanone methacrylate) (+methacrylonitrile)	225		O_2-plasma	80W1
3	PMMA (polymethyl methacrylate)	254	MIBK (methyl isobutyl ketone)	O_2-plasma	70M1, 75L1, 76L1, 78M1, 80T7
4	PMIPK (polymethyl isopropenyl ketone)	295 200		O_2-plasma	73L1, 78N1, 79T5
5	Meldrum's acid (5-diazo Meldrum's acid, 2,2-dimethyl-4,6-dioxo-5-diazo-1,3-dioxolane)	254	Shipley AZ developer diluted 1:3 with deionized water	O_2-plasma	80W8, 81G4
6	PBS, poly (1-butene sulfone)	185		O_2-plasma	75A1
7	P (M—OM), poly (methyl methacrylate-co-3-oximino-2-butanone methacrylate)	200···250		O_2-plasma	81C5
8	P (M—OM—CN), poly (methyl methacrylate-co-3-oximino-2-butanone methacrylate-co-methacrylonitrile)	240		O_2-plasma	81C5
9	PMI, poly(methyl methacrylate-co-indenone)	230···300		O_2-plasma	81C5
10	Poly(methacrylonitrile) with a diester of dihydroxybenzophenone with 1,2-diazo-naphthoquinone-5-sulfonic acid as photoactive component	<300	CF_4-plasma	O_2-plasma	81H4

Negative photoresists

The exposure of negative photoresists to light causes cross-linking of polymer molecules. This cross-linking increases the molecular weight thus reducing the solubility in a developer solvent. Fig. 23 shows a probable way of cross-linking polyvinyl cinnamate by exposure.

Table 6 lists important negative photoresists.

Table 6. Negative photoresists.

No.	Resin	Ref.	Structure	Possible resists	Stripper	Developer	Ref.	Sensitizer	Ref.
1	Polyvinyl cinnamate derivates	75D1	$\left[-\overset{H}{\underset{H}{C}}-\overset{H}{\underset{\vert}{C}}-\right]_n$ with side chain $-O-COCH{=}CH-C_6H_5$	Kodak KPR, KPR-2, KPR-3, KPR-4; Waycoat RC and 20	sulfuric acid (concentrated, hot) sulfuric acid saturated with chromic acid commercial stripper	nitrobenzene acetic acid furfural cyclohexanone benzaldehyde methyl glycol acetate xylene isopropanol phthalic acid methanol	52M1, 54M1 52M1 52M1 54M2 52M1, 54M2 52M1, 54M2 52M1, 54M2 54M2 54M2 54M2 54M2	4,4′-diazidodibenzol acetone 2-keto-3-methyl-1,3-diazabenzanthrone (speed : 500) 1-methyl-2-benzoymethylenenaphthothiazoline (speed: 1,400) 4-nitro-2-chloroaniline (speed: 300)	62H1, 62H2 54M1 56R1 52M1
2	Polyvinyl cinnamylidene ester	75D1	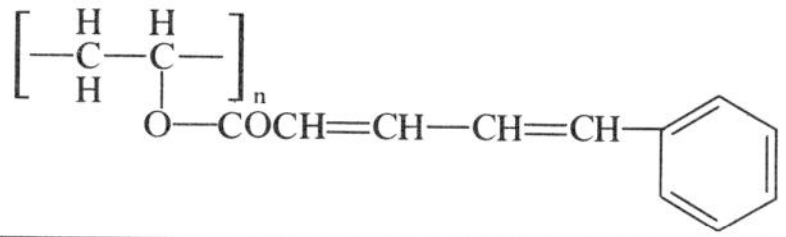 $\left[-\overset{H}{\underset{H}{C}}-\overset{H}{\underset{\vert}{C}}-\right]_n$ with side chain $-O-COCH{=}CH-CH{=}CH-C_6H_5$	Kodak KOR	same as for polyvinyl cinnamate	same as for polyvinyl cinnamate		same as for polyvinyl cinnamate	75D1
3	Allyl ester prepolymer	75D1	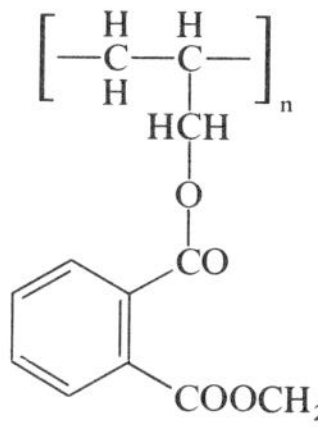 $\left[-\overset{H}{\underset{H}{C}}-\overset{H}{\underset{\vert}{C}}-\right]_n$ (one form) with side chain $HCH-O-CO-C_6H_4-COOCH_2CH{=}CH_2$	DCR wet film resist.	sulfuric acid (concentrated, hot) commercial stripper	methyl isobutyl ketone (MIBK) xylene 1,1,1-trichloroethane	68G1 68G1 68G1	basically the same as those listed for polyvinyl cinnamate	75D1

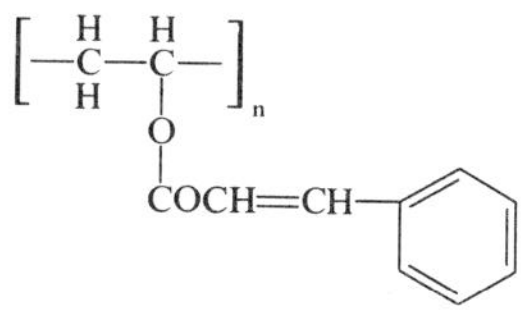

continued

Table 6 (continued)

No.	Resin	Ref.	Structure	Possible resists	Stripper	Developer	Ref.	Sensitizer	Ref.
4	Cyclized-rubber derivates	75D1	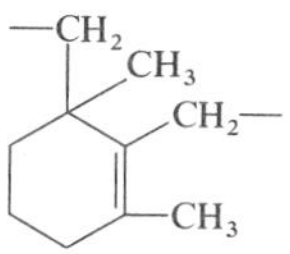 (one form)	Kodak KMER, KTFR; Waycoat IC, SC, and 450	triethylenetetramine (hot) sulfuric acid (concentrated hot) commercial stripper	xylene mixture of the ortho-, meta- and para-xylenes benzene chlorinated hydrocarbon	75D1 75D1 75D1 75D1	(limited to azides) 4,4′-diazido-stilbene 4,4′-diazido-benzo-phenoni 4,4′-diazido-dibenzalacetone	75D1 58H1 58H1 62H1
5	Polyimide precurson	79R1	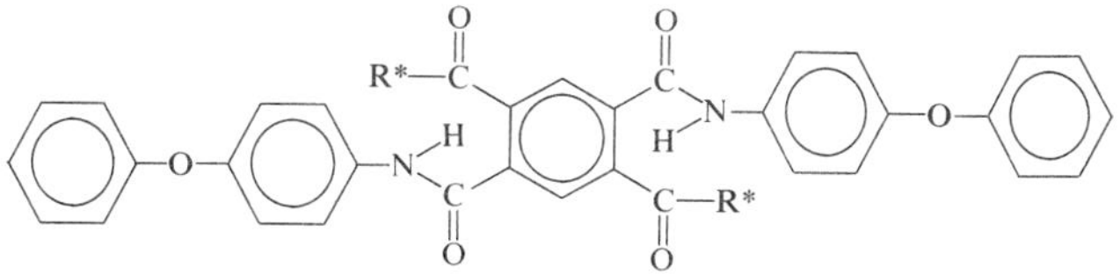		O_2-plasma	PVI-developer		sensitizer R*: allylic group acrylic group methacrylic group	
6	As_2S_3	79C1				0.03 N NaOH; CF_4 plasma			
7	$AgCl/As_2S_3$	79C1				1) fixer solution 2) 0.03 N NaOH, CF_4-plasma			
8	$Ag_2Se/GeSe$	80T4			CF_4—O_2-plasma				

Table 7 lists negative photoresists usable for deep UV light.

Table 7. Negative photoresists for deep UV light.

No.	Resin	Sensitizer	Sensitizer structure	λ_{max} nm	Developer	stripper	Exposure time t [s]	Ref.
1	Cyclized polyisoprene **WR (White Resist)**	diazides	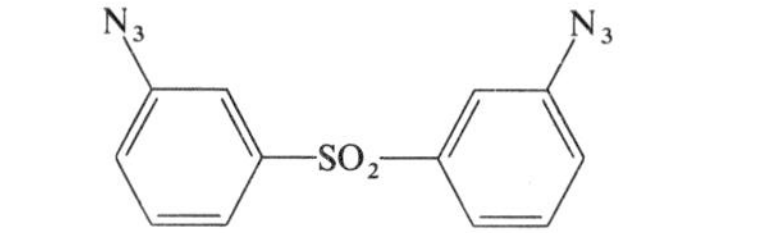 N_3 —SO_2— N_3	240	70% n-heptane		0.6	80I1
			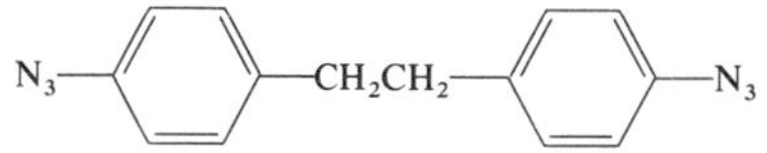 N_3— —CH_2CH_2— —N_3	254	+30% xylene		0.3	
			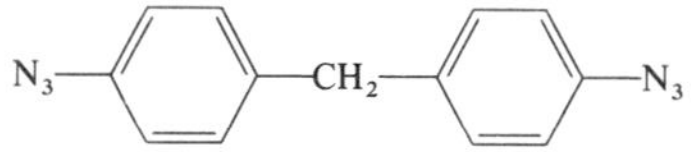 N_3— —CH_2— —N_3	256			0.5	
			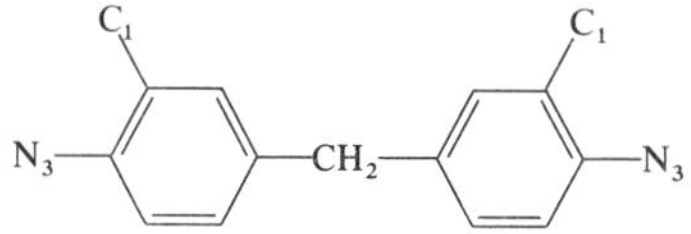 C_1 C_1 N_3— —CH_2— —N_3	258			0.5	
			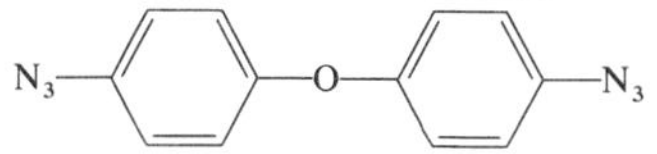 N_3— —O— —N_3	264			0.2	
			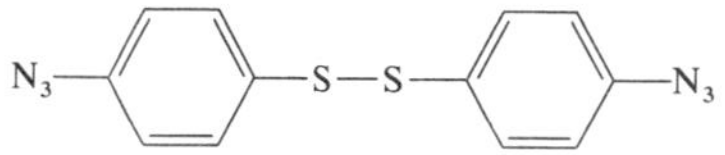 N_3— —S—S— —N_3	266			0.7	
			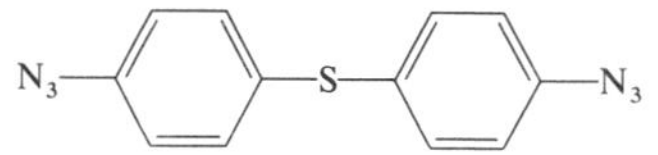 N_3— —S— —N_3	273			0.1	
			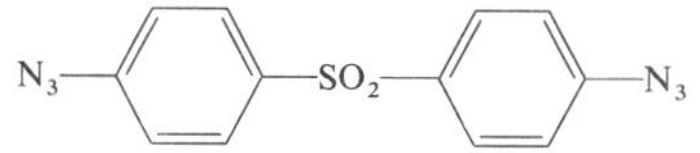 N_3— —SO_2— —N_3	284			0.3	

continued

Table 7 (continued)

No.	Resin	Sensitizer	Sensitizer structure	λ_{max} nm	Developer	stripper	Exposure time t [s]	Ref.
2	DCPA poly(2,3-dichloro-1-propyl-acrylate)	2,3 diphenylindenone	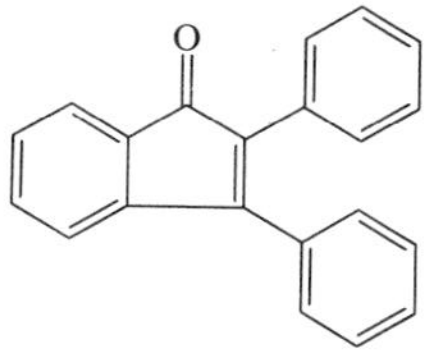	290 350	O_2-plasma			81T1
		acenaphthenequinone	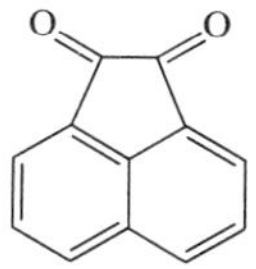					
		perinaphthenone	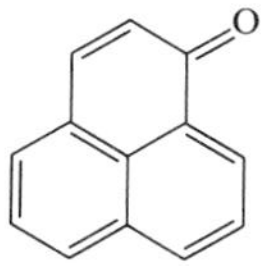					
		phenanthrenequinene	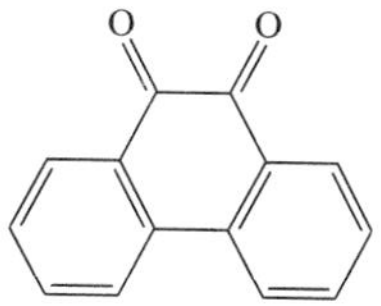					
		ingacure 651	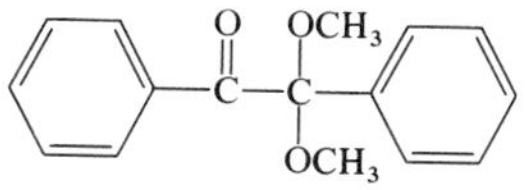					
3	Poly (p-vinylphenol)	3,3′-diazidodiphenyl sulfone		280	aqueous alkaline developers	O_2-plasma		81I2, 81M2
4	AgBr	AgBr	complex		complex development process			81L1

6.1.4.6.3 Electron-beam lithography

The electron beam of a scanning electron system can be used for the *direct generation of patterns* with great pattern flexibility. Fig. 24 shows the principle of generating a pattern in a thin film by means of an electron beam. The chemical bonds of polymer chains in the electron resist are broken by the electron beam irradiation. This leads to chain scission in positive resists and to polymerization by cross-linking in negative resists. The resist pattern remaining after development can be transferred by etching (Fig. 24), lift-off (Fig. 2), or selective deposition (Fig. 3).

Since the electron beam can be focussed to a spot diameter two or three orders of magnetude less than the wavelength of light, very fine structures can be generated. The resolution capability of an electron beam impinging on an electron-sensitive resist is not limited by the beam spot, or by the de Broglie wavelength of electrons, but by scattering of electrons in the resist and the substrate.

Electron-beam lithography systems

Electron-beam lithography systems can be grouped into projection and scanning systems.

Projection systems, also called *parallel-exposure systems*, transfer patterns from a mask to the electron-sensitive resist on a wafer. In a *scanning* system, also called *serial-exposure* or *direct-writing* system, a finely focussed electron beam generates patterns into a resist under the control of a computer. The most important feature of such a system is its great pattern flexibility.

Electron-beam projection systems

Two types of electron-beam projection systems are desribed in this section. These are the *reduction projection* system and the 1:1 *photocathode projection* system (Fig. 25).

The reduction electron-beam projection system [78L1, 79F2, 80F1, 80C1, 80F2, 80W4] is the electron- optical analogon of an optical reduction projection system. In an electron-optical system, the mask is irradiated by electrons to form a demagnified image of the mask on the wafer. Electron-optical systems, which provide electron-beam generation, beam focussing and deflection, are described by [45Z1] and by [80W4]. The mask has to be a freely suspended metal foil [80G3], which causes one of the major problems of the reduction projection.

The photocathode projection system [77S2, 80C1, 80W4] allows the exposure of a complete wafer in a single operation. In this system, the photocathode mask is uniformly illuminated by ultraviolet light. The photocathode mask has regions that emit electrons on exposure to UV illumination. These photoelectrons are accelerated towards the wafer by an electrical potential between the mask and the wafer. A uniform magnetic field focusses the photoelectrons onto the wafer.

Scanning electron-beam systems

In a scanning electron beam system, the electron beam is focussed to a spot of the same size as, or smaller than, the minimum pattern feature. The electron beam is modulated by a deflection unit and is scanned within a definite area of the wafer according to the exposure data stored in the control computer.

Three different spot-forming techniques are used in commercial systems. In the simplest system [80W4], the spot is formed by directly imaging the electron *point source* onto the wafer by means of an electron lens as shown in Fig. 26a. The beam current density distribution $I(x)$ is almost Gaussian. In order to realize high pattern fidelity, the spot size has to be small compared to the minimum feature size.

In *shaped beam* systems [70P1, 74V1, 77M3], an electron-illuminated aperture (typically square or round) is imaged onto the wafer as shown in Fig. 26b. In this system, the beam profile $I(x)$ is more uniform. The spot size is equal to the minimum feature size.

A *variable shaped beam* system [77P1, 77O1, 78T2, 80G1, 80G2] is shown schematically in Fig. 26c. The image of the electron source having aperture 1 is changed to the aperture 2 by the deflection unit 1. The result is a variable shaped beam which is imaged on to the wafer.

The two basic beam-writing techniques are the *raster scan* and the *vector scan* as shown in Fig. 27.

In the raster scan technique, the electron beam scans line by line over the entire area, and is turned on and off as required by the exposure pattern. In the vector scan technique, the beam is directed only to pattern areas that are to be exposed. Raster scanning and vector scanning are both possible in combination with a source beam, a shaped beam, or a variable shaped beam imaging system.

Electron-beam resists

The interaction of high energy electrons with polymeric resists mainly results in two reactions: the *scission of the chain bonds* of the polymeric molecules in positive resists, and the chemical bonding between different polymer molecules, i.e. *cross-linking*, in negative resists.

Chain scission in positive resists results in the molecular weight distribution being shifted in the direction of reduced molecular weight as shown in Fig. 28. The irradiated resist can be removed by a developer, resulting in a *positive image* in the resists as shown in Fig. 24.

Cross-linking in negative resists increases the molecular weight of the polymer. The cross-linked network is insoluble. The dissolution of the non-irradiated regions by a developer leads to a *negative image* of the resist as shown in Fig. 24.

Tables 8 and 9 list various positive and negative electron beam resists, respectively.

Table 8. Positive electron beam resists. (S: sensitivity; γ_p: contrast; w_{min}: resolution, minimum line width.)

No.	Resist	S C cm^{-2}	γ_p	w_{min} μm	Developer	Ref.
1	**PMA polymethacrylates** $(-CH_2-C(CH_3)(C{=}O\,O-\mathbf{R})-)_n$ Monomer **R**:					77G2, 79B1, 79T3, 80H4
1.1	CH_3 PMMA (poly (methyl methacrylate))	$5 \cdots 20 \cdot 10^{-5}$			3:1 IPA: MIBK (isopropanol: methyl isobutyl ketone)	79L1, 75R3, 79T3
		$5 \cdot 10^{-5}$		<0.1		68H1, 69H1, 79B1, 80C1
		$2 \cdot 10^{-4}$	7.59		3:1 IPA: MIBK	80B1
1.2	C_2H_5	$5.5 \cdot 10^{-5}$	2.98		2:1 IPA: MA (isopropanol: methyl alcohol)	80B1
1.3	n-C_3H_7	$5 \cdot 10^{-5}$	11.6		2:1 IPA: MA	80B1
1.4	i-C_3H_7	$4.6 \cdot 10^{-5}$	3.3		2:1 IPA: MA	80B1
1.5	n-C_4H_9	$3 \cdot 10^{-6}$			30:1 IPA: MIBK (isopropanol: methyl isobutyl ketone	80B1
1.6	s-C_4H_9	$2 \cdot 10^{-4}$	~15		MA (methyl alcohol)	80B1
1.7	t-C_4H_9	$1.4 \cdots 3.2 \cdot 10^{-5}$	6⋯11		carb (carbitol or 2-(2-ethoxyethoxy) ethanol)	80B1
1.8	c-C_6H_{11}	$4.6 \cdot 10^{-5}$	1.68		1:1 IPA: MIBK (isopropanol: methyl isobutyl ketone)	80B1
1.9	$C_6H_5CH_2$	$3 \cdot 10^{-4}$	1.68		2:1 MEK: IPA (methyl ethyl ketone: isopropanol)	80B1
1.10	C_6H_5	$3.8 \cdot 10^{-5}$	3.91		dioxane	80B1
1.11	CF_3CH_2	$3.1 \cdot 10^{-5}$	1.34		3:1 IPA: MIBK (isopropanol: methyl isobutyl ketone)	80B1
2	**Copolymers**					79T3
2.1	P[MMA-CO-MAA], copolymer of methyl methacrylate and methacrylic acid	$2 \cdots 5 \cdot 10^{-5}$			ECA: EA (ethyl cellosolve acetate: ethyl alcohol)	79H2

continued

Table 8 (continued)

No.	Resist	S C cm^{-2}	γ_p	w_{min} µm	Developer	Ref.
	Copolymers (continued)					
2.2	P[MMA-CO-IB], copolymer of methyl methacrylate and isobutylene	$5 \cdot 10^{-6}$				75G1, 76G1, 79B1
2.3	P[MMA-CO-AN], copolymer of methyl methacrylate and acrylonitrile	$5 \cdot 10^{-7}$			9:1 AA:EA (amyl acetate; ethyl alcohol)	75H1, 79B1, 79H3
2.4	P[MCA-CO-MCN], copolymer of methyl α-chlorocrylate and methacrylonitrile	$1.8 \cdots 2.4 \cdot 10^{-5}$			BN:MEK (benzonitrile: methyl ethyl ketone)	78C1, 79L1
2.5	P[MFA-CO-MCN], copolymer of methyl α-fluoroacrylate and methacrylonitrile	$6 \cdot 10^{-5}$				81P1
2.6	P[TFEM-CO-MCN], copolymer of 2,2,2-trifluoroethyl methacrylate and methacrylonitrile	$3 \cdots 4 \cdot 10^{-5}$				81P1
2.7	P[TFEM-CO-MMA], copolymer of 2,2,2-trifluoromethyl methacrylate and methyl methacrylate	$2 \cdots 3 \cdot 10^{-5}$				81P1
2.8	P[TFMAN-CO-MCN], copolymer of α-trifluoromethacrylonitrile and methacrylonitrile	$5 \cdot 10^{-5}$				81P1
2.9	P[TFMAN-CO-MMA], copolymer of α-trifluoromethacrylonitrile and methyl methacrylate	$3 \cdot 10^{-5}$				81P1
2.10	P[TFMMA-CO-MMA], copolymer of α-trifluoromethylacrylate and methyl methacrylate	$15 \cdot 10^{-5}$				81P1
3	**Terpolymers**					
3.1	P[MMA-MA-MAN], terpolymer of methyl methacrylate, methacrylic acid and methacrylic anhydride	$1 \cdots 10 \cdot 10^{-6}$			acetone, 2-methoxyethanol, or n-methyl pyrollidone	79M1, 79H4
3.2	P[MMA-MA-ClMA], terpolymer of methyl methacrylate, methacrylic acid and methacryloyl chloride	$1 \cdots 10 \cdot 10^{-5}$	1⋯4	0.2	IPA/MIBK (isopropyl alcohol/ methyl isobutyl ketone), or acetone	79K1
4	**PMCN,** poly (methacrylonitrile)	$1.9 \cdots 7.0 \cdot 10^{-5}$			BN:Tol (benzonitrile: toluene)	79H1, 79L1
5	**PMCA,** poly (methyl α-chloroacrylate)	$2.3 \cdots 2.7 \cdot 10^{-5}$			BN (benzonitrile)	78H1, 79L1

continued

Table 8 (continued)

No.	Resist	S C cm^{-2}	γ_p	w_{min} µm	Developer	Ref.
6	**PBS,** poly (butene-1-sulfone)	$8 \cdot 10^{-7}$ 10^{-6}		0.5		75B1, 79B1 80C1, 79T3
7	**POS,** poly (olefin sulfone)	$\sim 10^{-6}$				73T1, 79B1
8	Poly (cyclopentene sulfone)	$6.7 \cdot 10^{-6}$				75H3, 75H4, 79B1
9	Poly (1-methyl cyclopentene sulfone)	$1.3 \cdot 10^{-6}$				75H3, 75H4, 79B1
10	Poly (methyl-isopropenyl ketone)	$2 \cdot 10^{-6} \cdots 2 \cdot 10^{-5}$				74H1, 79B1
11	**PS,** poly (α-methyl-styrene)	10^{-4}				80C1
12	**Photoresist AZ 1350,** Shipley	$3 \cdots 9 \cdot 10^{-6}$ $5 \cdot 10^{-5}$		 1.0		76T1, 79B1 80C1
13	**FPM,** poly (dimethyltetrafluoropropyl methacrylate)	$3 \cdots 12 \cdot 10^{-6}$		<1.0		77K1, 80S1
14	**NPR** (novolac resin in solid solution with poly (2-methyl-1-pentene sulfone)	$3 \cdot 10^{-6}$ (20 keV)	1.0	<1.0		81B5
15	**Copolymer** (acrylonitrile-methacrylic acid)	$\simeq 10^{-6}$				81H4
16	Poly (n-butyl α-cyanoacrylate)	$5 \cdot 10^{-6}$	$\simeq 3$			81E1
18	**PTFEM,** poly (2,2,2-trifluoroethyl methacrylate)	$2 \cdots 3 \cdot 10^{-5}$				81P1

Table 9. Negative electron beam resists. (S: sensitivity; γ_n: contrast; w_{min}: resolution, minimum line width).

No.	Resist	S C cm^{-2}	γ_n	w_{min} µm	Developer	Ref.
1	**COP,** poly (glycidil meth-acrylate-co-ethyl acrylate)	$5 \cdot 10^{-7}$ $4.2 \cdot 10^{-7}$	0.95 0.72	0.5	 COP developer	71H1, 74T2, 75T1, 78T2, 78Y1, 80C1, 80H3, 80B1, 79T3, 80E1
2	**PGMA,** poly (glycidil methacrylate)	$7 \cdots 8 \cdot 10^{-7}$ $2 \cdot 10^{-8}$ $2 \cdots 20 \cdot 10^{-7}$ (D_g^i)	2 1.0⋯1.24	0.5		79B1, 80H3, 79T1, 75C1, 80E1, 79F1
3	**PDOP,** poly (diallyl phthalate)	10^{-6} $7 \cdot 10^{-6}$ $3 \cdot 10^{-6}$	 1.3 0.83	2 0.5	 1:1 CB:AA (chlorobenzene: amyl acetate)	73B1, 79B1, 80C1 80B1, 80E1
4	**PS,** poly (styrene)	$\approx 2 \cdot 10^{-5}$ $4.6 \cdot 10^{-5}$ $1.87 \cdot 10^{-5}$ (D_g^i)	5 1.43 1.9	0.5	 CB (chlorobenzene)	79W1, 79T2, 80H3, 80B1 79F1, 80E1
5	**P(4Cl-Sty),** poly (4-chlorostyrene)	$1.85 \cdot 10^{-6}$ (D_g^i)	1.82 1.5		CB (chlorobenzene)	80B1 79F1

continued

Table 9 (continued)

No.	Resist	S C cm^{-2}	γ_n	w_{min} µm	Developer	Ref.
7	**P(2Cl-Sty),** poly (2-chloro-styrene)	$3.6 \cdot 10^{-6}$ (D_g^i)	1.6			79F1
8	**P(2&4-ClSty),** poly (2,4-chloro-styrene)	$5.3 \cdot 10^{-6}$ (D_g^i)	1.3			79F1
9	Poly (vinyl acetate)	$4.4 \cdot 10^{-5}$	0.94		Tol (toluene)	80B1
10	Poly (methyl vinyl ketone)	$1.7 \cdot 10^{-4}$	1.29		MIBK (methyl isobutyl ketone)	80B1
11	Poly (vinyl cinnamate)	$3 \cdot 10^{-6}$	1.15		MEK (methyl ethyl ketone)	80B1
12	**P(VC),** poly (vinyl chloride)	$7.4 \cdot 10^{-5}$	0.86		CB (chlorobenzene)	80B1
13	Poly vinyl pyrrolidone	$4.1 \cdot 10^{-5}$	0.66		water	80B1
14	**P(VT),** poly (vinyl toluene)	$4.6 \cdot 10^{-5}$ $7.3 \cdot 10^{-6}$ (D_g^i)	1.43 1.9		CB (chlorobenzone)	80B1 79F1
15	**P(VBCl),** poly (vinylbenzyl chloride)	$2 \cdot 10^{-6}$ $4.6 \cdot 10^{-7}$ (D_g^i)	2.1 1.5		CB (chlorobenzone)	80B1 79F1
16	**P(VBSA),** poly (vinylbenzene sulfonic acid)	$3.5 \cdot 10^{-6}$ (D_g^i)	1.4			79F1
17	Poly (4-iso-pro-polystyrene)	$8.1 \cdot 10^{-5}$	1.2		Tol (toluene)	80B1
18	Poly (4-tert-butylstyrene)	$2.9 \cdot 10^{-4}$	2.88		Tol (toluene)	80B1
19	**P(4Br-Sty),** poly (4-bromo-styrene)	$1.7 \cdot 10^{-6}$ (D_g^i)	1.89		CB (chlorobenzene)	80B1
20	**Copolymers**					
20.1	P(4Cl-Sty-CO-GMA), copolymer of 4-chlorostyrene and glycidyl methacrylate	$3.6 \cdot 10^{-7}$ (D_g^i)	1.1			79F1
20.2	P(3Cl-Sty-CO-GMA), copolymer of 3-chlorostyrene and glycidyl methacrylate	$3.5 \cdot 10^{-7}$ (D_g^i)	1.2			79F1
20.3	P(ClSty-CO-GMA), copolymer of chlorostyrene and glycidyl methacrylate	$3.5 \cdot 10^{-7}$ (D_g^i)	1.1			79F1
20.4	P(VBCl-CO-GMA), copolymer of vinyl benzyl chloride and glycidyl methacrylate	$8.8 \cdot 10^{-7}$ (D_g^i)	1.7			79F1
20.5	P(VT-CO-GMA), copolymer of vinyl toluene and glycidyl methacrylate	$2.4 \cdot 10^{-6}$ (D_g^i)	1.4			79F1
21	**EPB** (epoxidized polybutadiene)	$1 \cdots 2 \cdot 10^{-7}$ $5 \cdot 10^{-8}$ $2 \cdot 10^{-8}$				71H1, 73N1, 73F1, 79B1, 80C1, 80E1, 75C1

continued

Table 9 (continued)

No.	Resist	S C cm^{-2}	γ_n	w_{min} µm	Developer	Ref.
22	**CPB**	$5 \cdot 10^{-8}$	0.9	1.5		80H3
23	**CMS,** chloromethylated polystyrene	$1 \cdots 10 \cdot 10^{-6}$	1.5···2.2	<0.2	acetone/isopropanol	79I1, 80S1, 81K1
24	**αM-CMS,** chloromethylated poly-α-methyl-styrene	$1 \cdots 10 \cdot 10^{-6}$ $1 \cdots 10 \cdot 10^{-6}$	2.2···3	<0.1		81K1
25	**SEL-N**	$4 \cdot 10^{-7}$	1.28	0.5		80H3
26	**OEBR-100**	$3 \cdot 10^{-7}$	0.77	1.0		80H3, 80E1
27	**Photoresist AZ 1350,** Shipley	$1.5 \cdot 10^{-5}$	1.4	0.5		80H3, 80O1
28	**PMCS**					75R3
29	**HUNT WAYCOAT**	$4 \cdot 10^{-7}$ $8 \cdot 10^{-7}$	1.42	1.0		80H3 75C1, 80E1
30	**KTFR-KMER-KPR** (Kodak)	$1 \cdots 9 \cdot 10^{-6}$		1.0		80C1, 75C1, 79T3, 80E1
31	Poly (α-chloroacrylonitrile)	10^{-5}			CF_4-plasma	81H4
32	**Silicones**	10^{-5}		0.5		80C1
33	**Se-Ge**					77Y2
34	**AgBr**	10^{-9}				81L1

Limits of electron-beam lithography

The resolution capability of electron-beam lithography is not limited by the electron-optical system, or by the electron spot diameter, or by the de Broglie wavelength of the electrons, but by the *electron scattering in the resist* and by the *backscattering of electrons* from the substrate to the resist.

The finest patterns normally achieved on solid substrates are of the order of 0.1···0.2 µm [69B1], although beam diameters as small as a few angströms are possible by the conventional electron beam equipment.

Fig. 29 shows Monte Carlo-simulated trajectories of 100 electrons in 400 nm PMMA films on silicon. The simulation described by [74K1], was performed by assuming electrons incident at energies of 10 keV and 20 keV. The electron beam, which strikes the resist normal to its surface, is assumed to be of infinitely small width.

Both, the depth range and the spot diameter of the exposed resist increase with increasing electron energy.

The scattering of electrons leads to the *proximity effect*, i.e. the phenomenon that even if an electron beam system delivers constant irradiation to all points in the pattern, the exposure density depends on size and density of the pattern elements. Due to the scattering of electrons, a dense pattern is more strongly exposed than a sparse pattern. Large pattern elements are more strongly exposed than small elements. This proximity effect leads to *linewidth variations* as shown in Fig. 30.

The proximity effect is also responsible for the effect that the edges and the corners of a pattern element are less exposed than the center parts.

6.1.4.6.4 X-ray lithography

X-ray lithography may be considered as a logically consistent further development of optical lithography [72S1, 72S2].

The essential feature of X-ray lithography is that pattern definition is achieved deep in the resist. The resist is assumed to be exposed by electrons generated by the X-ray photons rather than directly by the X-ray photons themselves. X-ray photons penetrate deep into the resist with neglegible diffraction or scattering. Photoelectrons generated by X-rays exhibit a small penetration range in the resist because their maximum energy is only of the order of the energy of the incoming photons (150···3000 eV). X-ray lithography is therefore suitable for the generation of very fine patterns having extensions of less than 1 µm.

Because it is difficult to deflect or collimate X-ray radiation at sufficient efficiency by using the various systems shown in Fig. 6, only contact and proximity printing are suitable for X-ray lithography. The disadvantage of contact printing is that the mask touches the wafer thus resulting in both mask and resist damage. Proximity printing, for which a small gap exists between the wafer and the mask, is therefore mostly used.

X-ray lithography systems

X-ray lithography can be performed by using the system schematically shown in Fig. 31.

Soft X-ray radiation at wavelengths ranging from 0.4···8.0 nm are used. The X-ray radiation can be generated by a conventional electron-bombardment X-ray source, by a synchrotron, or by means of a plasma discharge.

The mask consists of a pattern of an X-ray absorbing material on a thin substrate membrane of X-ray-transmitting material.

X-ray sources

Electron-bombardment X-ray source

X-ray radiation can be produced by the electron bombardment of an anode (aluminum, copper, etc.) [77Y1, 79N1, 80H1, 80N1]. Two types of radiation occur: the continuum and the characteristic radiation. *Continuum radiation*, which is primarily produced by interactions with the nucleus, can be described by the Kramer's law [23K1, 80N1]:

$$I(E)=P_0 \cdot \mathrm{K} \cdot \mathrm{Z}^{\mathrm{n}} \cdot (E_0-E)/E_0,$$

where $I(E)$ is the total radiation intensity at energy E radiated into all directions, P_0 is the incident electron beam power, Z is the atomic number of the anode metal, n is about 1, and E_0 is the incident electron energy. The value of the constant K for aluminum is $\mathrm{K}=7\cdot10^{-9}\,\mathrm{eV}^{-1}$.

The intensity of the *characteristic radiation* at the energy E_c, which is produced by recombination of excited atom electrons, can be approximately calculated by [73D1]:

$$I(E)=\mathrm{B}\cdot N(\mathrm{Z})\cdot F(x)\cdot(E_0-E_c)^{\mathrm{m}},$$

where $N(\mathrm{Z})$ is a function of the atomic number, $F(x)$ is the target reabsorption factor (which depends on the mass absorption coefficient and the emission angle), B is a factor proportional to the incident electron beam power, E_0 is the incident electron energy, and m = 1.5 (for aluminum).

The characteristic wavelengths $\lambda_c=(1.240\cdot10^{-4})/E_c$ (λ_c in cm, E_c in eV) and generation efficiencies measured at 20 keV are listed in Table 10 for a variety of possible anode materials.

Fig. 32 schematically shows an electron-bombardment X-ray source.

Table 10. Properties of elements used in X-ray sources of interest in X-ray lithography [80N1, 75M1, 67K1, 76S2], λ_{cutoff} is the wavelength at the absorption edge, λ_c is the wavelength of the characteristic radiation of the K or L line as indicated by the index of the element, η is the power efficiency of X-ray generation for bombardment by 20 keV electrons unless otherwise stated.

Element	λ_{cutoff} nm	λ_c nm	η (20 keV) mW/srW
C_K	4.377	4.382	
N_K	3.105	3.160	
O_K	2.337	2.371	
F_K	1.805	1.831	
Ne_K	1.419	1.462	
Cu_L		1.336	0.019 (8 keV)
Na_K	1.148	1.191	
Mg_K	0.951	0.989	
Al_K	0.795	0.834	0.055
Si_K	0.625	0.713	0.060
P_K	0.579	0.616	
Mo_L		0.541	0.053
S_K	0.502	0.538	
Cl_K	0.440	0.473	
Rh_L		0.460	0.065
Pd_L		0.437	0.068
K_K	0.344	0.374	

Synchrotron and electron storage ring radiation

Synchrotrons and electron storage rings are both suitable sources for X-ray lithography applications [76S1, 76F1, 77T1]. In both cases, the motion of electrons at high velocity ($v \simeq c$) in a circular orbit generates intensive electromagnetic radiation emitted tangentially to the orbit. The spectral distribution of the radiation extends over a wide wavelength range. In contrast to an electron bombardment source the synchrotron produces almost parallel radiation in a large area.

Synchrotron radiation was first investigated as an X-ray source for lithography by [76S1]. The radiation was obtained by the German synchrotron ring DESY in Hamburg. [76F1, 77T2] obtained the radiation from the ACO storage ring at Orsay in France.

The radiation power emitted from the total circumference of a storage ring is given by [80H1]:

$$P = \mathrm{K} \cdot \frac{E^4}{R} \cdot I,$$

where E is the electron energy, I is the electron current, R is the storage ring radius, and K is a proportionality constant.

The wavelength of the peak radiation λ_{max} is given by [80H1]:

$$\lambda_{max} = 0.235\, R/E^3$$

(λ_{max} in nm, R in m, E in GeV).

Fig. 33 shows a schematic representation of a synchrotron used as a radiation source for X-ray lithography.

Plasma radiation

Soft X-ray radiation from a plasma can also be used as a source for X-ray lithography [74M1, 78P1, 79M4, 79Y2]. Two methods of producing an appropriate plasma seem to be possible:
a plasma can be generated by a high-power laser beam [74M1, 78P1, 79Y2]; and
a plasma can be generated by an electron beam [77M2, 79M4].

Fig. 34 shows an X-ray lithography system using a *laser-generated* plasma as a soft X-ray source. A laser pulse of high energy ($\approx$100 Joules per shot) is focussed onto a target. As a result the exposed material of the target is vaporized and heated to some millions of degrees Kelvin. This hot plasma emits soft X-rays which can be used as a radiation source for X-ray lithography. Fig. 35 shows an X-ray source in which the plasma is *intensified by an electron beam.*

In this soft X-ray source the plasma is generated by a spark discharge in a capillary. A very intensive electron beam (stored energy $\approx$ 50 Joules) is injected into the plasma thus increasing the intensity of X-ray emission.

X-ray masks

A mask appropriate for X-ray lithography, (Fig. 36) consists of a pattern prepared by X-ray absorbing material on a thin film of X-ray transmitting material, and a substrate ring or grid.

The X-ray absorbing material consists of a high and the transmitting material of a low atomic number Z. The absorbing pattern of the mask frequently is prepared by a thin gold film. For the X-ray transmitting film, any of the various light materials such as Si, Si_3N_4 etc. [80H1] may be chosen.

The X-ray absorption coefficients of materials which are of interest in X-ray lithography, are given in Fig. 37. More detailed absorption data are available in [73B3].

X-ray resists

Absorbed X-ray photons generate photo and Auger electrons [79M5] that break the chemical bonds of the polymer chains, thus leading to chain scission in positive resists and to polymerisation in negative resists. Any electron resist may therefore also be used in X-ray lithography.

Fig. 38 shows the relationship of electron beam and X-ray sensitivity for various resist materials.

The two sensitivities are roughly proportional.

Tables 11 and 12 list various positive and negative X-ray resists.

Table 11. Positive X-ray resists. (λ: wavelength; S: sensitivity; w_{min}: resolution, minimum line width.)

No.	Resist	S mJ cm^{-2}	Target	λ nm	Major abs. element	w_{min} µm	Developer	Ref.
1	PMMA, poly (methyl methacrylate)	500···2000	Al	0.834	O	≦0.1	MIBK: EA (methyl isobutyl ketone: ethyl alcohol)	80T1, 80T2, 80H1, 80H2
2	PMMA, cross-linked	80	Al	0.834			MIBK	74T1
3	PBS, poly (butene-1-sulfone)	~100	Pd	0.436	S	0.5	MEK: 2-PrOH (methyl ethyl ketone: 2-PrOH)	73B1, 78P1, 80H1, 80T2, 74T1, 75B1, 77S1, 80H2
		20	Al	0.834		0.5		
4	P[MMA-co-MAA], copolymer of methyl methacrylate and methacrylic acid	150	Al	0.834	O	<0.1	ECA : EA (ethyl cellosolve acetate: ethyl alcohol)	80T2, 79H2
5.	Tl-P[MMA-co-MAA], Tl-copolymer of methyl methacrylate and methacrylic acid	24	Al	0.834	Tl	<0.5		80T2
6	FBM, poly (hexafluoro butyl methacrylate)	52	Mo	0.541	F	0.3		80T2, 80S1,
		80	Mo	0.541		0.3		78F1, 77M1
		180	Si	0.71			MIBK : IPA (methyl ethyl ketone : isopropanol)	80T1, 80H2
7	FPM, poly (dimethyltetrofluoro propyl methacrylate)	100	Pd	0.436		0.5		80H1, 80S1
8	Poly (2,2-dichloroacetaldehyde)	140	Mo	0.541	Cl			80T2
9	AZ 2400, Shipley	2100	Si	0.71			AZ 2401 + H_2O	80H2
10	Poly (n-butyl α-cyanoacrylate)	40	Cu	1.33				81E1

Table 12. Negative X-ray resists. (λ: wavelength; S: sensitivity; w_{min}: resolution, minimum line width.)

No.	Resist	S mJ cm^{-2}	Target	λ nm	Major abs. element	w_{min} µm	Developer	Ref.
1	COP, poly (glycidyl methacrylate-co-ethylacrylate)	50	Al	0.834		1.0		75T1, 78B1 80T1
		15···20	Al	0.834	O	1···2	MEK : ETCH (methyl ethyl ketone : etch)	80T2, 75B1, 80H2
		175	Pd	0.436	O	1.0		80T2
2	DCPA, poly (2,3-dichloro-1-propylacrylate)	7	Pd	0.436	Cl	1.0		77T1, 80T2,
		10	Pd	0.436	Cl	0.5		77T1, 80T1, 80H1

continued

Table 12 (continued)

No.	Resist	S mJ cm^{-2}	Target	λ nm	Major abs. element	w_{min} µm	Developer	Ref.
3	DCPA+COP (DCOPA)	4⋯22	Pd	0.436		2		79M2, 79M3
		15	Pd	0.436	Cl	0.8		80T2, 80M1
4	PDOP, poly (diallyorthophthalate)	14	Al	0.834			ETOH	74T1, 77S1, 80H2
5	DCPA+N-vinyl carbazole, poly (2,3-dichloro-1-propylacrylate)+N-vinyl carbazole	4.5	Pd	0.436	Cl	0.3		80T2
6	Poly (2,3-dibromo-1-propylacrylate-co-glycidylacrylate)	15	Rh	0.460	Br	1.0		80T2
7	Poly (chloroethylvinylether-co-vinyloxyethyl acrylate)	18	Mo	0.541	Cl, O	1.0		80T2
8	Brominated tetrathiafulvalene-functionalized polystyrene	22	Al	0.834	S, Br	0.2		80T2
9	Hydrocarbon resist	2.0	Al	0.834	C	<0.1		80T2
10	AgBr emulsion	0.4	Pd	0.436	Ag, Br	<0.1		80T2
	Metal acrylates	20⋯25	Al	0.834	Ba, Pd, Ni	1.0		80T2
12	PGMA-EA, poly (glycidyl methacrylate-EA)	50	Pd	0.436		0.5		80H1
13	SEL-N	200	Si	0.71			MeOH : IPA (MeOH : isopropanol)	80H2
14	OEBR	25	Al	0.834		0.5		80H2
15	CER	25	Al	0.834			xylene	74T1, 75S1, 80H2
16	DCPA-Monomer mixtures poly (2,3-dichloro-1-propyl acrylate)-Monomer mixtures						O_2 plasma	80T3
16.1	DCPA-BABTDS (bis-acryloxybutyltetramethyldisiloxane)		Pd	0.436		<0.5	O_2 plasma	81T3
16.2	DCPA-DPDVS (diphenyldivinylsilane)		Pd	0.436			O_2 plasma	81T3
16.3	DCPA-BMBTDS (bis-methacryloxyalkyltetramethyldisiloxane		Pd	0.436			O_2 plasma	81T3
16.4	DCPA-NVC (N-vinyl/carbazole)	3⋯17 ($D^{0.2}$)	Pd	0.436			O_2 plasma	80T3, 80T6, 81T3
16.5	DCPA-NVC-DPDVS		Pd	0.436			O_2 plasma	81T3
16.6	DCPA-(2-(1-naphthyl)-ethyl methacrylate and acrylate)	16⋯30 ($D^{0.2}$)	Pd	0.436			O_2 plasma	80T3

continued

Table 12 (continued)

No.	Resist	S mJ cm^{-2}	Target	λ nm	Major abs. element	w_{min} µm	Developer	Ref.
16.7	DCPA-NVC-DMPSS (dimethylphenylstyrylsilane)		Pd	0.436			O_2 plasma	81T3
16.8	DCPA-NPM (N-phenyl maleimide)	8···11 ($D^{0.2}$)	Pd	0.436			O_2 plasma	80T3, 80T6
16.9	DCPA-NPM-DMPSS (dimethylphenylsilylstyrene)		Pd	0.436			O_2 plasma	81T3
17	Polymer containing N-vinyl carbazole monomer							
17.1	Poly (2,3-dichloro-1-propyl methacrylate)	20.1 ($D^{0.2}$)	Pd	0.436			O_2 plasma	80T3
17.2	Poly (2-chloroethyl acrylate)	13.2 ($D^{0.2}$)	Pd	0.436			O_2 plasma	80T3
17.3	Poly (2,2,2-trichloroethyl acrylate)	5.8 ($D^{0.2}$)	Pd	0.436			O_2 plasma	80T3
17.4	Poly (1,3-dichloro-2-propylacrylate)	4.0 ($D^{0.2}$)	Pd	0.436			O_2 plasma	80T3
17.5	Poly (2,3-dichloro-1-propylacrylate	3.2 ($D^{0.2}$)	Pd	0.436			O_2 plasma	80T3

Limits of X-ray lithography

Using parallel irradiation, the resolution capability of X-ray lithography is limited by the *Fresnel diffraction* [77T2, 78T1] and by the *penetration range* of X-ray radiation-produced photoelectrons into the resist [75F1, 76S2, 77S1]. The Fresnel diffraction depends on the wavelength and the distance from mask to wafer. The arrangement of mask and wafer is shown in Fig. 39.

Fig. 40 shows the X-ray intensity distribution obtained behind a 2 µm wide slit at a proximity distance of 20 µm for three different wavelengths. The resolution capability increases with decreasing photon wavelength.

The penetration range of photoelectrons in the resist increases with decreasing wavelength of the X-ray radiation.

Fig. 41 shows the minimum printable line width w in X-ray proximity printing as a function of the X-ray wavelength.

By using non-parallel irradiation, the resolution is further reduced by two types of *geometrical distortion:* the run-off distortion and the penumbral blurring as shown in Fig. 42.

The geometrical distortion can be reduced by increasing the distance between the mask and the source and by reducing the proximity distance between the mask and the wafer.

6.1.4.6.5 Ion-beam lithography

In ion beam lithography, patterns are generated by the local exposure to an ion beam [73W1, 79K2, 79S2, 80S2, 80W6, 81B4, 81R1, 81R2].

Various methods of pattern generation by using an ion beam are possible (Fig. 43a···f).

The wafer is coated by an ion beam-sensitive resist (Fig. 43a). Exposing desired areas to ions changes the solubility of the resist in a proper developer so that a resist mask is generated. The resist mask can be transferred to the thin film below by etching (Fig. 1), by lift-off (Fig. 2), or by selective material deposition (Fig. 3).

The thin film is patterned directly by local ion etching (Fig. 43b).

For a number of inorganic thin films (SiO_2, Si_3N_4, Si, etc.) the chemical etch rate is increased by irradiation of ions. Wet or dry chemical etching after exposure to ions generates a pattern in the thin film (Fig. 43c).

An impurity pattern is directly generated by ion implantation. p-type conduction is obtained in silicon by implantation of boron, gallium, etc., while n-type conduction is obtained by using phosphorus, arsenic, etc. (Fig. 43d).

In a polymer or inorganic film a mask for reactive ion etching is formed by implantation of low energy ions (Fig. 43e). The implanted ions form nonvolatile compounds which act as a mask for the reactive ion etching.

The resist consists of a thin $GeSe_2$ layer covered by a very thin Ag_2Se layer (Fig. 43f). While exposing the resist to ions, silver from the Ag_2Se layer migrates into the $GeSe_2$ film. The resist is subsequently developed in a solvent in which Ag-doped $GeSe_2$ is insoluble while undoped $GeSe_2$ is soluble. This produces a pattern having a negative relief.

When an ion beam is used for the exposure of resists, higher resolution is possible than with an electron beam because there is less scattering of ions (due to their relatively large mass) and because of the small penetration range of the low-energy secondary electrons produced by ions. The same reasons explain the reduced proximity effect.

Ion-beam resists

For ion beam exposure, all electron-beam resists are well suited. Usually, the resists are polymers, and the exposure process results in cross-linking (negative resist), or scission of molecular chains (positive resist).

Resists are much more sensitive to ions than to electrons and to X-rays, because of the relatively large mass of ions and their small penetration range (typically 0.1 to 1 µm). The greater sensitivity of resists to ions is not entirely beneficial, because the minimum number of ions per focussed spot is limited by statistical fluctuations rather than by the resist sensitivity. Because of the limited penetration range of the ions, very thin resist films can be used.

Tables 13 and 14 list various positive and negative ion beam resists, respectively.

Table 13. Positive ion beam resists. (E: ion energy; S: sensitivity; d_{max}: maximum resist thickness removed; w_{min}: resolution, minimum line width.)

No.	Resist	Ions	E keV	S C cm^{-2}	d_{max} µm	w_{min} µm	Developer	Ref.
1	PMMA, polymethyl methacrylate)	He^+	30	$7 \cdot 10^{-7}$	0.2		MIBK : IPA = 1 : 1	81R2
		He^+	60	$5 \cdot 10^{-7}$	0.2			
		Ar^+	100	10^{-7}	0.2			
		Ar^+	150	10^{-7}	0.2			
		Ga^+	200	10^{-7}	0.2			
		H^+	100	$2 \cdot 10^{-6}$	1.0			
		H^+	60	10^{-6}	0.4	<0.1	MIBK : IPA = 1 : 1	81A1
		Li^+	60	$5 \cdot 10^{-7}$	0.6	<0.5	MIBK : IPA = 1 : 3	
		Si^+	40	$1.6 \cdot 10^{-4}$	1.2	<0.5	O_2 plasma	81A2
2	P[MMA-co-MAA], copolymer of methacrylate with methacrylic and acrylic acids	H^+	50	$9 \cdot 10^{-7}$	0.4		MEK : toluene = 4 : 3	81G1
3	FPM, poly (dimethyltetrofluoropropylmethacrylate)	H^+	50	$3 \cdot 10^{-7}$	0.5	<0.08	MIBK : IPA = 1 : 4	81M3
4	FBM 110	H^+	120	$3 \cdot 10^{-8}$	1.0			81R2
5	FPM 210	H^+	120	$3 \cdot 10^{-7}$	1.0			81R2
6	OEBR 1000	H^+	120	$8 \cdot 10^{-7}$	1.0			81R2
7	EPMF 1	H^+	120	$8 \cdot 10^{-7}$	1.0			81R2
8	OEBR 1010	H^+	120	$1.5 \cdot 10^{-6}$	1.0			81R2
9	AZ 1350 Shipley	H^+	50	$3 \cdot 10^{-6}$	0.5			81M3

Table 14. Negative ion beam resists. (E: ion energy; S: sensitivity; w_{min}: resolution, minimum line width).

No.	Resist	Ions	E keV	S C cm^{-2}	w_{min} µm	Developer	Ref.
1	Cyclized rubber with N-phenyl-maleimide	H^+	50	$1.6 \cdot 10^{-7}$		liquid developer	81G1
2	Novolac	H^+ He^+ O^+	1500 1500 1500	10^{-3} 10^{-5} 10^{-6}			81B6
3	PVC, poly (vinyl chloride)	In^+	20	$3 \cdot 10^{-3}$		O_2 plasma RIE	81T4
4	PVF, poly (vinyl formal)	In^+	20	$3 \cdot 10^{-3}$		O_2 plasma RIE	81T4
5	PMMA, poly (methyl methacrylate)	In^+	20	$3 \cdot 10^{-3}$		O_2 plasma RIE	81T4
6	Poly (2,3-dibromo-1-propylacrylate-co-2,3-dichloro-1-propylacrylate)	In^+	20	$3 \cdot 10^{-3}$		O_2 plasma RIE	81T4
7	Poly (vinyl carbazole)-trinitro fluorenone complex	In^+	20	$3 \cdot 10^{-3}$	<0.3	O_2 plasma RIE	81T4
8	SiO_2	In^+	20	$3 \cdot 10^{-3}$		fluorocarbon plasma RIE	81T4
9	$Ag_2Se/GeSe_2$						79T6, 80B5

*) For definition of S (sensitivity) and w_{min} (resolution or line width) see Table 1 and sections 6.1.4.6.1 and 6.1.4.6.2.

Ion-beam lithography systems

Ion-beam lithography can be performed either by the *ion projection* of a mask image onto the wafer, by *direct writing* with a focussed ion beam, or by *ion-beam proximity printing*.

Ion-beam projection

Fig. 44 schematically shows an ion-beam projection system [80S2]. Ions of low energy emitted by an ion source are shaped into a homogeneous beam by a condensor lens system. The parallel ions irradiate a stencil mask containing the image information in the form of physical holes. The multiple ion beams produced by the holes in the mask are accelerated to an energy sufficient to penetrate the thin film that is to be patterned (typically 60···100 keV for ranges 0.1···1.0 µm). The accelerated ions pass through an inversion lens and a demagnifying lens system to project a demagnified image of the mask onto the wafer. The wafer can be positioned by a step-and-repeat system as shown in Fig. 17 for an optical projection system.

Ion-beam writing

A system for direct writing with a focussed ion beam is shown in Fig. 45 [79S2]. Ions extracted from an ion source by an electrode are focussed by an ion lense system onto the wafer where the pattern is to be generated. The ion beam modulated by a deflection unit scans a definite area of the wafer according to the exposure data in the control computer.

Ion-beam proximity printing

Fig. 46 schematically shows a system for ion-beam proximity printing. It consists of an ion source, an extraction electrode, two apertures, and two electrostatic ion lenses. The image of the mask is directly transferred to the wafer by a parallel ion beam [81B12]. The wafer can be positioned by a step-and-repeat system as shown for an optical projection system in Fig. 17.

Limits of ion-beam lithography

The *resolution* of ion-beam lithography depends on the lithography system used. In ion-beam writing, the limiting factor may be the ion beam spot diameter. In ion-beam projection, the minimal line width may be determined by the lens aberration or by the Rayleigh's diffraction. Independent of the system used, a common limitation of the resolution capability is given by lateral straggling of the exposing ions as a result of nuclear collisions [81R2]. Backscattering and secondary electrons which limit the resolution of electron-beam exposure, or photoelectrons in the case of X-ray exposure, are negligible in ion beam exposure [81R2]. Due to the mass ratio between ions and electrons, the secondary electrons produced by the incident ions have energies of only 5···50 eV, leading to ranges of less than 10 nm.

Fig. 47 shows Monte-Carlo-simulated trajectories of incident ions in PMMA, PMMA on gold, and PMMA on silicon. The simulation, described by [81K3] was performed by assuming ions having an incident energy of 60 keV. The ion beam which strikes the resist normal to its surface, is assumed to be of infinitely small width.

The *minimum line width*, which can be produced by ion-beam lithography also depends on the resist sensitivity and the exposure dose. This is illustrated in Fig. 48. For large exposure areas the minimum ion dose is identical to the resist sensitivity. For constant sensitivity, the minimal ion dose necessary for exposure increases with decreasing line width as a result of the statistical emission of ions from the ion source [81R2].

Limits in ion-beam writing

Beside the limitation by straggling of the exposing ions, the resolution capability of ion-beam writing may also be limited by the ion-beam *spot diameter*. The minimum spot diameter depends on the ion source and the spherical, chromatic, and diffraction aberration of the lens system. The spot diameter d is given by:

$$d=(d_g^2+d_s^2+d_c^2+d_\lambda^2)^{\frac{1}{2}}$$

where $d_g=\sqrt{I/B}/\pi\theta$ is the Gaussian image diameter, $d_s=C_s\theta^3/2$ the spherical aberration, $d_c=C_c\,\theta\,\Delta E/E$ the chromatic aberration and $d_\lambda=C_d/\theta\sqrt{E}$ the spot broadening due to diffraction.

θ is the acceptance half angle of the lense, I is the target current, B is the ion source brightness, ΔE is the axial energy spread of the ion beam entering the lens, E is the energy of the beam entering the lens, C_s is the spherical aberration coefficient, C_c is the chromatical aberration coefficient, and C_d is the diffraction aberration coefficient of the lens.

In Fig. 49, achievable beam diameters are plotted for a typical asymmetric lens, designed to focus hydrogen ion beams of about 50 keV [81R2]. As shown in this figure the spot diameter is limited by the brightness B and by the chromatic or the spherical aberration. The spot broadening due to corpuscular diffraction is negligible. In order to obtain a small spot diameter, the brightness of the ion source must be high and the coefficients of chromatic aberration C_c and spherical aberration C_s have to be small.

In Fig. 50 three types of ion sources are compared under consideration of brightness, energy spread, angular current density, virtual gun size, and possible ions.

Limits in ion-beam projection

In ion-beam projection systems, the resolution is limited by *lens aberration*, by *lateral straggling* of the exposing ions, or by the *Rayleigh's diffraction.*

Fig. 51 shows the relationship between the resolution limited by lens aberration (expressed by the maximum diameter of the aberration figure) and the image field diameter. The aberration is a combination of the geometric and chromatic aberration of the imaging system described by [81S4]. The resolution increases with decreasing image field.

The ultimate limit of an aberration-free imaging system is given by the Rayleigh's diffraction [81R1] which can be expressed by the Rayleigh criterion:

$$d_1=0.61\frac{\lambda}{NA}$$

where d_1 is the minimum distance between two points that can be resolved, λ is the De' Broglie wavelength of the ions, and NA is the numerical aperture of the projection system.

For an image plane that is out of focus, the diameter of the error disk is given by

$$d_2=2\,\Delta z\,\mathrm{tg}(\alpha),$$

where Δz is the distance between image plane and focal plane, and $\alpha=\arcsin(NA)$. d_2 equals the minimum distance between two resolvable points if Δz is the distance between the image plane and the focal plane.

Fig. 52 shows the ultimate limit of resolution of an aberration-free image-forming system.

If the resolution is limited by Fresnel diffraction, the minimum line width obtained by ion-beam lithography is smaller than that obtainable by using electron-beam, X-ray, or photolithography.

Limits in ion-beam proximity printing

In ion-beam proximity printing systems, the resolution capability is limited by lateral *straggling* of the exposing ions, by *lens aberration*, or by *geometrical distortion*. With parallel ion irradiation, the resolution is mainly determined by the lateral straggling of the ions.

With nonparallel ion irradiation, the limiting factor may be the geometrical distortion or the lens aberration.

References for 6.1.4.6

23K1 Kramers, H.A.: Philos. Mag. **46** (1923) 836.
44S1 Sus, O.: Liebig's Annalen der Chemie **556** (1944) 65.
45Z1 Zworkin, V.K., Morton, G.A., Ramberg, E.G., Hillier, J., Vance, A.W.: Electron Optics and the Electron Microscope; New York: J. Wiley 1945.
49S1 Schwinger, J.: Phys. Rev. **75** (1949).
52M1 Minsk, L.M., Deusen, W.P., Robertson, E.M.: U.S. Patent 2, 610, 120 (1952).
54M1 Minsk, L.M., Deusen, W.P., Robertson, E.M.: U.S. Patent 2, 670, 285 (1974).
54M2 Minsk, L.M., Deusen, W.P., Robertson, E.M.: U.S. Patent 2, 670, 286 (1974).
56R1 Robertson, E.M., West, W.: U.S. Patent 2, 732, 301 (1956).
57M1 Moore, R.G.D.: U.S. Patent 2, 797, 213 (1957).
58H1 Hepher, M., Wagner, H.M.: U.S. Patent 2, 852, 379 (1958).
62H1 Hepher, M., Wagner, H.M.: British Patent 892, 811 (1962).
62H2 Hepher, M.: British Patent 892, 812 (1962).
63S1 Starks, C.K., Caroll, N.L.: U.S. Patent No. 3079 502 (1963).
64K1 Knoll, M., Eichmeier, J.: Technische Elektronik, Bd. 1, Berlin: Springer 1964.
66K1 Knoll, M., Eichmeier, J.: Technische Elektronik, Bd. 2, Berlin: Springer 1966.
67K1 Kaelble, E.F. (ed.): Handbook of X-Rays, New York: McGraw-Hill **1967**.
68G1 Gilano, M.N., Martinson, I.W., Ott, L.H.: U.S. Patent 3, 376, 139 (1968).
68K1 Koops, H., Mollenstedt, G., Speidel, R.: Optik **28** (1968) 518.
68H1 Haller, I., Hatzakis, M., Srinivasan, R.: IBM J. Res. Dev. **12** (1968) 251.
68S1 Sokolov, A.A., Ternov, I.M.: Synchrotron Radiation, Pergamon Press **1968**.
69B1 Broers, A.N., Lean, E.G., Hatzakis, M.: Appl. Phys. Lett. **15** (1969) 98.
69H1 Hatzakis, M.: J. Electrochem. Soc. **116** (1969) 1033.
69K1 Koops, H.: Optik **29** (1969) 119.
69L1 Levine, H.A.: Polymer Preprints, Division of Polymer Chemistry, ACS **10(1)** (1969).
70M1 Moreau, W.M., Schmidt, P.R.: Electrochem. Soc. Meeting, Extend. Abstr. No. 187 (1970).
70P1 Pfeiffer, H.C., Loeffler, K.H.: 7th Int. Conf. on Electron Microscopy, Grenoble, France, **1970**, 63.
70S1 Schoenthaler, A.C.: U.S. Patent 3, 418, 295 (1970).
71H1 Hirai, T., Hatano, Y., Nonagaki, S.: J. Electrochem. Soc. **118** (1971) 669.
72B1 Brown, J.R., O'Donnell, J.H.: Macromolecules **5** (1972) 109.
72K1 Koops, H.: Optiks **36** (1972) 93.
72S1 Spears, D.L., Smith, H.I.: Electron. Lett. **8** (1972) 102.
72S2 Spears, D.L., Smith, H.I.: Solid State Technol. **15** (1972) 21.
73B1 Bowden, M.J., Thompson, L.F.: J. Appl. Polym. Sci. **17** (1973) 3211.
73B2 Bartelt, J.L.: Prepr. ACS Div. Org. Coatings Plast. Chem. **33** (1973) 390.
73B3 Bracewell, B.L., Veigele, W.J.: Appl. Spectr. **9** (1973) 375.
73D1 Dyson, N.A.: X-Rays in Atomic and Nuclear Physics, Longmans Group Ltd., London (1973) 203.
73F1 Feit, E.D., Heidenreich, R.D., Thompson, L.F.: Prepr. ACS Div. Org. Coatings Plast. Chem. **33** (1973) 383.
73H1 Harris, R.A.: Solid State Sci. Technol. **120** (1973) 270.
73K1 Koops, H.: J. Vac. Sci. Technol. **10** (1973) 913.
73L1 Levine, A.: Soc. Plast. Eng. (1973) 106.
73N1 Nonagaki, S., Morishita, H., Saitou, N.: Prepr. ACS Div. Org. Coatings Plast. Chem. **33** (1973) 378.

73R1 Roberts, E.D.: Prepr. ACS Div. Org. Coatings Plast. Chem. **33** (1973) 359.
73T1 Thompson, L.F., Bowden, M.J.: Electrochem. Soc. **120** (1973) 1722.
74B1 Brewer, T.L.: Polym. Eng. Sci. **14** (1974) 534.
74B2 Bowden, M.J., Thompson, L.F.: J. Appl. Polym. Symposia **23** (1974) 99.
74K1 Kunz, C.: Proc. IV. Int. Conf. Vac. Ultraviolet Rad. Phys. Hamburg, West Germany, July 22–26 (1974).
74L1 Levine, A.W., Kaplan, M., Poliniak, E.S.: Polym. Eng. Sci. **14** (1974) 518.
74M1 Mallozzi, P.J., Epstein, H.M., Jung, R.G., Applebaum, D.C., Fairand, B.P., Gallagher, W.J., Hecker, R.L., Jour, M.C.: J. Appl. Phys. **45** (1974).
74S1 Seliger, R.L., Fleming, W.P.: J. Appl. Phys. **45** (1974) 1416.
74T1 Thompson, L.F., Feit, E.D., Bowden, M.J., Lanzo, P.V., Spencer, E.G.: J. Electrochem. Soc. Solid State Sci. Technol. **121** (1974).
74T2 Thompson, L.E., Feit, E.D., Heidenreich, R.D.: Polym. Eng. Sci. **14** (1974) 529.
74V1 Varnell, G.L., Spicer, D.F., Rodger, A.C., Holland, R.D.: 6th Int. Conf. on Electron and Ion Beam Science and Technol., Bakish, R. (ed.), **1974**, 97.
75A1 Applebaum, J., Bowden, M.J., Chandross, E.A., Feldmann, M., White, D.L.: Proc. Kod. Microelectronics Seminar-Interface 75, Oct. 1975.
75B1 Bowden, M.J., Thompson, L.E., Ballantyne, J.P.: J. Vac. Sci. Technol. **12** (1975), 1294.
75B2 Bartelt, J.L., Feit, E.D.: J. Electrochem. Soc. **122** (1975) 541.
75B3 Bogenschütz, A.F.: Fotolacktechnik, D-7968 Saulgau: E.G. Leuze Verlag **1975**.
75B4 Ballantyne, J.P.: Prepr. Div. Org. Coatings Plast. Chem. **35** (1975) 235.
75C1 Cole, H.S., Skelly, D.W., Wagner, B.C.: IEEE Trans. Electron Devices **ED-22** (1975) 4.
75D1 DeForest, W.S.: Photoresist, New York: McGraw-Hill **1975**.
75D2 Dill, F.H., Hornberger, W.P., Hauge, P.S., Shaw, J.M.: IEEE **ED-22** (1975) 445.
75D3 Doniach, S., Lindan, I., Spicer, W.E., Winik, H.: J. Vac. Sci. Technol. **12(6)** (1975).
75D4 Dill, F.H., Neureuther, A.R., Tuttle, J.A., Walker, E.J.: IEEE Trans. Electron Devices **22** (1975) 4.
75F1 Feder, R., Spiller, E., Topalian, J.: J. Vac. Sci. Technol. **12** (1975) 1332.
75G1 Gipstein, E., Need, O., Moreau, W.: Prepr. ACS Div. Org. Coatings Plast. Chem. **35** (1975) 246.
75H1 Hatano, Y., Morishita, H., Nonagaki, S.: Prepr. ACS Div. Org. Coatings Plast. Chem. **35** (1975) 258.
75H2 Helbert, J.N., Wagner, B.E., Caplan, P.J., Poindexter, E.H.: J. Appl. Polym. Sci. **19** (1975) 1201.
75H3 Himics, R.J., Desai, N.V., Kaplan, M., Poliniak, E.S.: Prepr. ACS Div. Org. Coatings Plast. Chem. **35** (1975) 266.
75H4 Himics, R.J., Kaplan, M., Desai, N.V., Poliniak, E.S.: Prepr. ACS Div. Org. Coatings Plast. Chem. **35** (1975) 273.
75K1 Kyser, D.F., Viswanathan, N.S.: J. Vac. Sci. Techn. **12** (1975) 1305.
75L1 Lin, B.J.: J. Vac. Sci. Technol. **12** (1975) 1317.
75R1 Roberts, E.D.: Prepr. ACS Div. Org. Coatings Plast. Chem. **35** (1975) 281.
75R2 Rüsbüldt, D., Thimm, K.: Nucl. Instrum. Methods **116** (1974) 125.
75R3 Roberts, E.D.: Philipps Techn. Rdsch. **35(3)** (1975) 72.
75T1 Thompson, L.F., Ballantyne, J.P., Feit, E.D.: J. Vac. Sci. Technol. **12** (1975) 1280.
75W1 Widmann, D.W., Binder, H.: IEEE Transact. Electron Devices **ED-22** (1975) 467.
75W2 Widmann, D.W.: Appl. Optics **14(4)** (1975) 931.
75M1 Maydan, D., Coquin, G.A., Maldonado, J.R., Somekh, S., Lou, D.Y., Taylor, G.N.: IEEE Trans. Electr. Dev. **22** (1975) 429.
76F1 Fay, B., Trotel, J., Petroff, Y., Pinchaux, R., Thiry, P.: Appl. Phys. Lett. **29** (1976) 370.
76G1 Gipstein, E., Moreau, W., Need, O.: Solid State Sci. Technol. **123** (1976) 1105.
76H1 Heritage, M.B.: Electron and Ion Beam Science and Technol., Bakish, R. (ed.), Electrochem. Soc., Princeton, N.J. **1976** 348, 587.
76K1 Kunz, C.: Desy-Bericht F 41-76/12, Dec. **1976**.
76K2 Koch, E.E.: Desy-Bericht F 41-76/08, Sept. **1976**.
76L1 Lin, B.J.: IBM J. Res. Develop. **20** (1976) 213.
76N1 Nakane, Y., Mifune, T.: 11th. Semiconductor Integrated Circuit Symp., Japan, **1976**, 54.
76S1 Spiller, E., Eastman, P.E., Feder, R., Grobmann, W.D., Gudat, W.D., Topalion, J.: Desy Report SR-76/11 (June 1976).
76S2 Sullivan, P.A., McCoy, J.H.: IEEE Trans. Electron Devices **23** (1976) 412.

76T1 Thompson, L.E., Kerwin, R.E.: Annu. Rev. Mater. Sci. **6** (1976) 267.
76W1 Widmann, D., Stein, K.U.: Proc. 2nd Eur. Solid State Circuits Conf., **1976** 29.
77B1 Binder, H., Sigusch, R., Widmann, D.W.: Proc. Intern. Conf. Microlithography, Paris, **1977** 249.
77F1 Feder, R., Spiller, E., Topalian, J.: Polym. Eng. Sci. **17** (1977) 385.
77F2 Friedrich, H., Zeitler, H.V., Bierhenke, H.: J. Electrochem. Soc. **124** (1977) 62.
77F3 Friedrich, H., Bierhenke, H., Zeitler, H.U.: Proc. Int. Conf. Microlith., Paris (1977).
77G1 Gipstein, E., Ouano, A.C., Johnson, D.E., Need, O.V.: III. Polym. Eng. Sci. **17** (1977) 39.
77G2 Gipstein, E., Ouano, A.C., Johnson, D.E., Need, O.U.: IBM J. Res. Devel. (March 1977).
77K1 Kakuchi, M., Sugawara, S., Murase, K., Matsuyama, K.: J. Electrochem. Soc. **124** (1977) 164.
77L1 Lischke, B., Anger, K., Frosien, A., Oelmann, A., Schuster-Woldan, H.: Proc. Int. Conf. on Microlithography, Paris, **1977** 163, 167.
77M1 Murase, K., Kakuchi, M., Sugawara, S.: Proc. Int. Conf. Microlithography, Paris, **1977** 261.
77M2 McCorkle, R.A., Vollmer, H.J.: Rev. Sci. Instrum. **48(8)** (1977) 1055.
77M3 Mauer, J.L., Pfeiffer, H.C., Stickel, W.: IBM J. Res. and Develop. (Nov. 1977) 514.
77O1 Oelmann, A., Lischke, B., Kutzer, E.: Proc. Int. Conf. on Microlithography, Paris (1977) 171.
77P1 Pfeiffer, H.C.: 14th. Symposium on Electron, Ion and Photon Beam Technol. (1977) 887.
77S1 Spiller, E., Feder, R.: Topics in Appl. Physics, H.J. Queisser (ed.), Ch. 3, Berlin, Heidelberg, New York: Springer **1977**.
77S2 Scott, J.P.: Solid State Technol., May **1977** 43.
77T1 Taylor, G.N., Coquin, G.A., Somekh, S.: Polym. Eng. Sci. **17** (1977) 420.
77T2 Trotel, J., Fay, B.: Proc. Int. Conf. Microlithography, Paris, France (June 1977) 201.
77Y1 Yoshingtsu, M., Kozaki, S.: Topics in Applied Physics, Queisser, H.J. ed., Ch. 2, Berlin, Heidelberg, New York: Springer **1977**.
77Y2 Yashikawa, A., Ochi, O., Nagai, H., Mizushima, Y.: Appl. Phys. Lett. **31** (1977)
78B1 Buckley, W.D., Dalle Ave, J.A.: Proc. 8th. Int. Conf. Electron Ion Beam Science and Technology, Seattle, **1978** 458.
78C1 Chen, C.-Y., Pittmann, C.U., Helbert, J.N.: Polymer Preprints **19** (1978) 565.
78C2 Clark, K.G.: Solid State Technol., Aug. **1978** 73.
78F1 Flanders, D.C., Smith, H.I.: J. Vac. Sci. Technol. **15** (1978) 995.
78H1 Helbert, J.N., Chen, C-Y., Pittmann, C.U.: Macromolecules **11** (1978) 1104.
78L1 Lindau, I., Winick, H.: J. Vac. Sci. Technol. **15(3)** (1978).
78L2 Lischke, B., Münchmeyer, W.: Optik **50** (1978) 315.
78M1 Mimira, Y., Ohkubo, T., Takeuchi, T., Sekikawa, K.: Jpn. J. Appl. Phys. **17** (1978) 541.
78N1 Nakane, Y., Tsumori, T., Mifune, T.: Kodac Microelectronics Seminar Proc. Interface-78, **1978**.
78P1 Peckerar, M.C., Greig, J.R., Nagel, D.J., Pechacek, R.E., Whitlock, R.R.: Proc. 8th Int. Conf. Electron Ion Beam Science and Technology, Seattle, **1978** 432.
78T1 Tischer, P.: Proc. 42. Physikertagung der Deutschen Physikalischen Gesellschaft, Berlin, **1978**.
78T2 Thompson, L.F., Stillwagon, L.E., Doerries, E.M.: J. Vac. Sci. Techn. **15** (1978)
78Y1 Yau, L.D., Thibault, L.R.: J. Vac. Sci. Technol. **15** (1978) 960.
79B1 Bowden, M.J.: CRC Critical Reviews in Solid State Sciences, Febr. **1979** 223.
79B2 Bierhenke, H., Zeitler, H.U., Risch, L.: Proc. Microcircuit Eng. 79, Aachen, W-Germany, **1979** 171.
79B3 Binder, H., Lacombat, M.: Transact. Electron Devices **ED-26** (1979) 698.
79C1 Chang, M.S., Hou, T.W., Chen, J.T., Kolwicz, K.D., Zemel, J.N.: J. Vac. Sci. Technol. **16** (1979) 1973.
79F1 Feit, E.D., Thompson, L.F., Wilkins, C.W., Wurtz, M.E., Doerries, E.M., Stillwagon, L.: J. Vac. Sci. Technol. **16** (1979) 1997.
79F2 Frosien, J., Lischke, B., Anger, K.: J. Vac. Sci. Technol. **16(6)** (1979) 182.
79H1 Helbert, J.N., Poindexter, E.H., Stahl, G.A., Chen, C-Y., Pittmann, C.U.: J. Polym. Sci.: Chem. Ed. **17** (1979) 49.
79H2 Hatzakis, M.: J. Vac. Sci. Technol. **16** (1979) 1984.
79H3 Helbert, J.N., Cook, C.F.: Solid State Sci. Technol. **126** (1979) 694.
79H4 Haller, I., Feder, R., Hatzakis, M., Spiller, E.: Solid State Sci. Technol. **126** (1979) 154.
79I1 Imamura, S.: Solid State Sci. Technol. **126(9)** (1979) 1628.
79K1 Kitakohji, T., Yoneda, Y., Kitamura, K., Okuyama, H., Murakawa, K.: J. Electrochem. Soc. **126** (1979) 1881.
79K2 Karapiperis, L., Lee, C.A.: J. Vac. Sci. Technol. **16(6)** (1979) 1625.
79K3 Komuro, M., Atoda, N., Kawakatsu, H.: J. Electrochem. Soc. **126** (1979) 483.

79L1 Lai, J.H., Helbert, J.N., Cook, C.F., Pittmann, C.U.: J. Vac. Sci. Technol. **16** (1979) 199.

79L2 Lischke, B., Frosien, J., Anger, K., Münchmeyer, W.: Optik **54(4)** (1979) 32.

79L3 Long, M., Walker, Ch.: Proc. Kodak Seminar, New Orleans, Oct. **1979** 125.

79M1 Moreau, W., Merritt, D., Moyer, W., Hatzakis, M., Johnson, D., Pederson, L.: J. Vac. Sci. Technol. **16** (1979) 1989.

79M2 Moran, J.M., Taylor, G.N.: J. Vac. Sci. Technol. **16** (1979) 2014.

79M3 Moran, J.M., Taylor, G.N.: J. Vac. Sci. Technol. **16** (1979) 2020.

79M4 McCorkle, R.A.: IBM-Research-Report, IC 7680 (33137), **28** (May 1979).

79M5 Maydan, D., Coquin, G.A., Levinstein, H.J., Sinha, A.K., Wang, D.N.K.: 15th Symp. Electron Ion, Photon Beam Technology, Bosten **1979** paper 1–7.

79M6 Maydan, D., Coquin, G.A., Levinstein, H.J., Shinha, A.K., Wang, D.N.K.: J. Vac. Sci. Technol. **16** (1979) 1959.

79N1 Nakayama, S., Hayusaha, T., Yamazaki, S.: Rev. Electr. Comm. Lab. **27** (Febr. 1979).

79R1 Rubner, R., Ahne, H., Kühn, E., Kolodziej, G.: Photogr. Sci. Eng. **23** (1979) 303.

79P1 Pacansky, J., Lyerla, J.R.: IBM J. Res. Develop. **23** (1979) 42.

79S1 Stengl, G., Kaitna, R., Löschner, H., Wolf, P., Sacher, R.: J. Vac. Sci. Technol. **16(6)** (1979) 1883.

79S2 Seliger, R.L., Kubena, R.L., Olney, R.D., Ward, J.W., Wang, V.: J. Vac. Sci. Technol. **16(6)** (1979) 1610.

79T1 Toniguchi, Y., Hatano, Y., Shiraishi, H., Horigome, S., Nonogaki, S., Naraoka, K.: Jpn. J. Appl. Phys. **18** (1979) 1143.

79T2 Tai, J.H., Shephard, L.T.: Solid State Sci. Technol. **126** (1979) 696.

79T3 Thompson, L.F., Doerris, E.M.: El. Chem. Sci. Technol. **126** (1979) 1699.

79T4 Taniguchi, Y., Hatano, Y., Shiraishi, H., Horigome, S., Nonogak, S., Naraoka, K.: Jpn. J. Appl. Phys. **18** (1979) 1143.

79T5 Tsuda, M., Kawa, S.O., Nakamura, Y., Nagata, H., Yokata, A., Nakane, H., Tsumori, T., Nakane, Y., Mifune, T.: Photographic Sci. Eng. **23** (1979) 290.

79T6 Tai, K.L., Sinclair, W.R., Vadinsky, R.G., Moran, J.M., Rand, M.J.: J. Vac. Sci. Technol. **16** (1979) 1977.

79W1 Wipps, D.W.: Proc. Microcircuit Conf., D 5100 Aachen **1979** 118.

79Y1 Yamashita, Y., Ogura, K., Kunishi, M., Kawazu, R., Ohno, S., Mizokami, Y.: J. Vac. Sci. Technol. **16** (1979) 2026.

79Y2 Yanaguchi, N.: J. Phys. Soc. Jpn. **47** (1979).

79Z1 Zeitler, H.U., Hieke, E.: J. Electrochem. Soc. **126** (1979) 1430.

80B1 Brault, R.G., Miller, L.J.: Proc. 9th. Int. Conf. on Electron and Ion Beam Sci. and Tech., Electrochem. Soc. **1980** 341.

80B2 Broers, A.N.: Proc. IEEE Electron Device Meeting, Washington **1981** 2.

80B3 Bruning, J.H.: J. Vac. Sci. Technol. **17(5)** (1980) 1147.

80B4 Broers, A.N., Chang, T.H.P.: Microcircuit Engineering, Ahmed, H.; Nixon, W.C. eds., New York: Cambridge University Press **1980** 1.

80B5 Brault, G.R., Miller, L.J.: Polymer Engineering & Science **20** (1980) 1064.

80C1 Chang, T.H.P., Hatzakis, M., Broers, A.N.: Microelectronics Interconnectic and Packaging, Lyman, J. (ed.), New York: McGraw-Hill **1980** 3.

80C2 Cohen, Ch.: Electronics, Febr. **1980** 73.

80C3 Colclaser, R.A.: Microelectronics: Processing and Device Design, New York: J. Wiley **1980**.

80D1 Deckert, C.A., Ross, D.L.: J. Electrochem. Soc. **127(3)** (1980) 450.

80D2 Doane, D.A.: Solid State Technol. Aug. **1980** 101.

80E1 Elliot, D.J.: Semiconductor International, May **1980** 61.

80F1 Frosien, J., Anger, K., Lischke, B., Oelmann, A., Münchmeyer, W., Schmitt, R.: Microcircuit Engineering, Ahmed, H. and Nixon, W.C. (eds.), New York: Cambridge University Press, USA **1980** 199.

80F2 Frosien, J.: Dissertation, Technical University, D 1000 Berlin **1980**.

80G1 Goto, E., Soma, T., Idesawa, M., Sasaki, T.: Proc. Electron Ion Beam Sci. Technol., Bakish, R. (ed.), Electrochem. Soc. **80-6** (1980) 92.

80G2 Goto, N., Someya, T., Tanaka, K., Takeuchi, M., Miyauchi, S.: Proc. Electron Ion Beam Sci. Technol., Bakish, R. (ed.), Electrochem. Soc. **80-6** (1980) 98.

80G3 Greschner, J., Bohlen, H., Engelke, H., Nehmiz, P.: Proc. Electron Ion Beam Sci. Technol., Bakish, R. (ed.), Electrochem. Soc. **80-6** (1980) 152.

80H1 Heuberger, A., Betz, H., Pongratz, S.: Advances in Solid State Physics, Treusch, (ed.), D 3300 Braunschweig: Vieweg **1980** 259.
80H2 Hughes, G.P., Fink, R.C.: Microelectronics Interconnection and Packaging, Lyman, J. (ed.), New York: McGraw Hill **1980** 21.
80H3 Hieke, E., Oldham, W.G.: Proc. Microcircuit Engineering 80, Amsterdam, Netherland **1980** 395.
80H4 Horada, K.: Solid State Sci. Technol. **127** (1980) 491.
80I1 Iwayanagi, T., Kohashi, T., Nonogaki, S.: J. Electrochem. Soc. **127** (1980) 2759.
80K1 Koops, H.: Fine Line Lithography, Newman, R. (ed.), Ch. 3, Amsterdam: North-Holland Publishing Company **1980**.
80L1 Lyman, J.: Microelectronics Interconnection and Packaging, Lyman, J. (ed.), New York: McGraw Hill **1980**.
80L2 Lin, B.J.: Fine Line Lithography, Newman, R. (ed.), Ch. 2, Amsterdam: North-Holland Publishing Company **1980**.
80L3 Li, C., Richards, J.: IEDM 80 Technical Digest, Washington (1981) 412.
80M1 Maydan, D.: J. Vac. Sci. Technol. **17** (1980) 1164.
80M2 Moore, R.D.: Proc. Electron Ion Beam Sci. Technol., Bakish, R. (ed.), Electrochem. Soc. **80-6** (1980) 126.
80M3 Moriwaki, K., Aritome, H., Namba, S.: Microcircuit Engineering 80, Amsterdam **1980**.
80N1 Neureuther, A.R.: Synchrotron Radiation Research, Winick, H.; Doniach, S. eds., Ch. Plenum Publishing Corporation **1980**.
80O1 Oldham, W.G., Hieke, E.: IEEE Electron Dev. Lett. **EDL-1** (1980) 217.
80S1 Sugawara, S., Kogure, O., Harada, K., Kakuchi, M., Sukegawa, K., Imamura, S., Miyoshi, K.: Electrochem. Soc. Extend. Abstr. **80-1** (1980) 680.
80S2 Stengl, G., Wolf, P., Kaitna, R., Löschner, H., Sacher, R.: Proc. 3rd Int. Cont. Ionimplantation, Kingston, Canada, July 8–11 **1980**.
80T1 Tischer, P.: From Electronics to Microelectronics, W.A. Kaiser and W.E. Proebster (eds.), North-Holland Publishing Company **1980** 46.
80T2 Taylor, G.N.: Solid State Technol. May **1980** 73.
80T3 Taylor, G.N.: J. Electrochem. Soc. **127** (1980) 2665.
80T4 Tai, K.L., Vadimsky, R.G., Kemmerer, C.T., Wagner, J.S., Lamberti, V.E., Timko, A.G.: J. Vac. Sci. Technol. **17** (1980) 1169.
80T5 Trotel, J.: Proc. Electron Ion Beam Sci. Technol., Bakish, R. (eds.), Electrochem. Soc. **80-6** (1980) 137.
80T6 Taylor, G.N.: European Patent Application No. 0017032 **1980**.
80T7 Tsuda, M.: ACS Symp. Ser. **121** (1980) 281.
80W1 Wilkins, C.W., Reichmanis, E., Chandross, E.A.: Electrochem. Soc. Extend. Abstr. **80-1** (1980) 684.
80W2 Widmann, D.: Device Impact of new Microfabrication Technologies, Summer Course, Katholieke Universiteit Leuven, Belgium **1980**.
80W3 Weber, E.V., Yourke, H.S.: Microelectronics Interconnection and Packaging, Lyman, J. (ed.), New York: McGraw Hill **1980** 13.
80W4 Wittels, N.D.: Fine Line Lithography, Newman, R. (ed.), Ch. 1, Amsterdam: North-Holland Publishing Company **1980**.
80W5 Weber, E.Y.: Fine Line Lithography, Newman, R. (ed.), Ch. 5, Amsterdam: North-Holland Publishing Company **1980**.
80W6 Watts, R.K.: Very Large Scale Integration, Barbe, D.F. (ed.), Ch. 3, Berlin, Heidelberg, New York: Springer **1980** 84.
80W7 Wittekoek, S.: Solid State Technol., June **1980** 80.
80W8 Willson, C.G., Clecak, N.J., Grant, B.D., Twieg, R.J.: Electrochem. Soc. Ext. Abstr. No. 275 (1980).
81A1 Adesida, I., Karapiperis, L., Lee, C.A., Wolf, E.D.: Microcircuit Engineering 81. Abstracts, Lausanne, Switzerland **1981** 56.
81A2 Adesida, I., Chin, J.D., Wolf, E.D., Phillips, J.R.: Microcircuit Engineering 81. Abstracts, Lausanne, Switzerland **1981** 60.
81B1 Bowden, M.J.: J. Electrochem. Soc. **128** (1981) 195C.
81B2 Buckley, W.D., Hughes, G.P.: J. Electrochem. Soc. **128** (1981) 1106.
81B3 Buckley, W.O., Nester, J.F., Windischmann, H.: J. Electrochem. Soc. **128** (1981) 1116.
81B4 Bruning, J.H.: Semiconductor Silicon 1981, Huff, H.R. et al. (eds.) Electrochem. Soc., Pennington, NJ **1981** 666.

81B5 Bowden, M.J., Thompson, L.F., Fahrenholtz, S.R., Doerries, E.M.: J. Electrochem. Soc. **128(6)** (1981) 1304.

81B6 Brown, W.L., Venkatesan, T., Wagner, A.: Solid State Technol., August **1981** 60.

81B7 Broers, A.N.: IEEE Transact. Electr. Dev. **ED-28** (1981) 1268.

81B8 Brodie, I., Westerberg, E.R., Cone, D.R., Muray, J.J., Williams, N., Gasiorek, L.: IEEE Trans. Electron Devices, **ED-28** (1981) 1422.

81B9 Braun, H.: Phys. Bl. **37** (1981) 311.

81B10 Binder, H.J., Hahn, E.: Microcircuit Engineering 81 Abstr., Lausanne, Switzerland **1981** 115.

81B11 Betz, H.: Experimentelle Untersuchungen der Synchrotronstrahlung zur hochauflösenden Lithographie für die Grösstintegration, Dissertation, Technische Universität, D 8000 München, 1981.

81B12 Behringer, U.: Ein Ionenstrahl Proximity Printer für die Halbleitertechnologie, Dissertation, Eberhard-Karls-University, D 7400 Tübingen **1981**.

81C1 Chang, T.S., Kyser, D.F., Ting, C.H.: IEEE Trans. Electron Devices **ED-28** (1981) 1295.

81C2 Chang, T.S., Codella, C.F., Lange, R.C.: IEEE Trans. Electron Devices, **ED-28** (1981) 1428.

81C3 Chatterjee, P.K.: Microcircuit Engineering 81 Abstracts, Lausanne, Switzerland, **1981** 15.

81C4 Chalmeton, V., Esteva, J.M.: Microcircuit Engineering 81 Abstracts, Lausanne, Switzerland **1981** 25.

81C5 Chandross, E.A., Reichmanis, E., Wilkins, C.W., Hartless, R.L.: Solid State Technol., August **1981** 81.

81C6 Chang, M.S., Liu, E.D., O'Toole, M.M.: IEEE Digest of Technical Papers, Symposium on VLSI Technology, Maui, Hawaii **1981** 8.

81D1 Daniel, P.J.: Inst. Phys. Conf. Ser. **57** (1981) 169.

81E1 Eranian, A., Datamanti, E., Dubois, J.C., Serre, B., Abadie, J.M., Schue, F., Montiginoul, C., Giral, L.: Microcircuit Engineering 81 Abstr., Lausanne, Switzerland **1981** 89.

81E2 Ehrlich, D.J., Osgood, R.M., Deutsch, T.F.: J. Electrochem. Soc. **128** (1981) 2039.

81F1 Fujinami, M., Shimazu, N., Takamoto, K., Saitou, N.: IEDM 81 Technical Digest (1981) 566.

81F2 Feuer, M.D., Prober, D.E.: IEEE Trans. Electron Devices, **ED-28** (1981) 1375.

81G1 Grigaitis, P., Pranevicus, L.: Jpn. J. Appl. Phys. **20** (1981) L261.

81G2 Gallagher, R.T.: Electronics, Nov. **1981** 81.

81G3 Griffing, B.F.: IEDM 81 Technical Digest (1981) 562.

81G4 Grant, B.D., Clecak, N.J., Twieg, R.J., Willson, C.G.: IEEE Trans. Electron Devices, **ED-28** (1981) 1300.

81G5 Greeneich, E.W., Tolliver, D.L., Gonzales, A.J.: IEEE Trans. Electron Devices, **ED-28** (1981) 1346.

81H1 Heinrich, K., Betz, H., Heuberger, A., Pongratz, S.: J. Vac. Sci. Technol. **19** (1981) 1254.

81H2 Howard, R.E., Hu, E.L., Jackel, L.D.: IEEE Trans. Electron Devices, **ED-28** (1981) 1378.

81H3 Homma, Y., Nozawa, H., Harada, S.: IEEE Trans. Electron Devices, **ED-28** (1981) 552.

81H4 Hiroaka, H.: J. Electrochem. Soc. **128** (1981) 1065.

81H5 Hatzakis, M.: Solid State Technol., August **1981** 74.

81I1 Iida, Y., Mori, K.: J. Electrochem. Soc. **128** (1981) 2429.

81I2 Iwayanagi, T., Kohashi, T., Nonogaki, S., Matsuzawa, T., Douta, K., Yanazawa, H.: IEEE Trans. Electron Devices, **ED-28** (1981) 1306.

81J1 Jain, P.K., Neureuther, A.R., Oldham, W.G.: IEEE Trans. Electron Devices, **ED-28** (1981) 1410.

81K1 Kogure, O., Sukegawa, K., Imamura, S., Miyoshi, K., Sugawara, S.: Electroch. Soc. Extend. Abstr. **81-1** (1981) 722.

81K2 Krimmel, E.F.: Private communications, **1981**.

81K3 Karapiperis, K., Adesida, L., Lee, S.A., Wolf, E.D.: J. Vac. Sci. Technol. **19** (1981) 1259.

81L1 Lavine, J.M., Masters, J.I., Goldberg, G.M., Das, A.: IEEE Trans. Electron Devices, **ED-28** (1981) 1311.

81L2 Lepselter, M.P., Lynch, W.T.: VLSI Electronics Microstructure Science, Vol. 1, Einspruch, N.G. ed. New York: Academic Press **1981** 83.

81M1 Mader, L., Widmann, D.: Microcircuit Engineering 81 Abstracts, Lausanne, Switzerland **1981** 35.

81M2 Matsuzawa, T., Tomioka, H.: IEEE Trans. Electron Devices, **ED-28** (1981) 1284.

81M3 Moriwaki, K., Aritome, H., Namba, S.: Microcircuit Engineering 81 Abstracts, Lausanne, Switzerland **1981** 58.

81M4 Mori, K., Iida, Y., Suzuki, K., Shinoda, D.: IEEE Digest of Technical Papers, Symposium on VLSI Technology, Maui, Hawaii **1981** 12.

81N1 Nakase, M., Shinozaki, T.: IEEE Trans. Electron Devices, **ED-28** (1981) 1416.
81N2 Nakase, M., Shinozaki, T.: IEEE Digest of Technical Papers, Symposium on VLSI Technology, Maui, Hawaii **1981** 4.
81N3 Nishi, Y.: Microelectronics Journal **12** (1981) 5.
81O1 Oldham, W.G., Neureuther, A.R.: Solid State Technol., May **1981** 106.
81O2 Okazaki, S., Chow, T.P., Steckl, A.J.: IEEE Trans. Electron Devices, **ED-28** (1981) 1364.
81O3 O'Toole, M.M., Liu, E.D., Chang, M.S.: IEEE Trans. Electron Devices, **ED-28** (1981) 1405.
81O4 Obayashi, H., Matsuzawa, T., Iwayanagi, T., Yanazawa, H., Nonogaki, S., Tomioka, H.: IEEE Digest of Technical Papers, Symposium on VLSI Technology, Maui, Hawaii **1981** 6.
81P1 Pittman, C.U., Veda, M., Chen, C.Y., Kwiatkowski, J.H., Cook, C.F., Helbert, J.N.: J. Electrochem. Soc. **128** (1981) 1758.
81R1 Rieder, R., Kaitna, R., Löschner, H., Sacher, R., Stengl, G., Wolf, P.: Proc. 16th. Symp. Electron, Ion and Photon Beam Technol., Dallas, Texas, May 26–29 **1981**.
81R2 Ryssel, H.: Proc. Microcircuit Engineering, Lausanne, Switzerland, September **1981**.
81R3 Rosenfield, M.G., Neureuther, A.R.: IEEE Trans. Electron Devices, **ED-28** (1981) 1289.
81R4 Rammos, E., Chalmeton, V.: Microcircuit Engineering 81 Abstracts, Lausanne, Switzerland **1981** 23.
81R5 Reekstin, J.P., McCoy, J.H.: Solid State Technol., August **1981** 68.
81R6 Rieder, R., Löschner, H., Kaitna, R., Sacher, R., Stengl, G., Wolf, P.: Private Communication, **1981**.
81S1 Sakakibara, Y., Ogawa, T., Komatsu, K., Moriya, S., Kobayashi, M., Kobayashi, T.: IEEE Trans. Electron Devices **ED-28** (1981) 1279.
81S2 Seliger, R.L.: International Patent Classification HO1J 37/317.
81S3 Suzuki, K., Matsui, J., Ono, T., Saito, Y.: J. Electrochem. Soc. **128** (1981) 2434.
81S4 Stengl, G., Kaitna, R., Löschner, H., Rieder, R., Wolf, P., Sacher, R.: J. Vac. Sci. Technol. **19** (1981) 1164.
81T1 Taylor, G.N., Wolf, T.M., Goldrick, M.R.: J. Electrochem. Soc. **128** (1981) 361.
81T2 Tarui, Y., Takeishi, Y.: Semicond. Silicon 1981, Huff, R. (ed.), Electrochem. Soc. (1981) 6.
81T3 Taylor, G.N., Wolf, T.M., Moran, J.M.: J. Vac. Sci. Technol. **19** (1981) 872.
81V1 Venkatesan, T., Taylor, G.N., Wagner, A., Wilkens, B., Barr, D.: J. Vac. Sci. Technol. **19** (1981) 1259.
81W1 Watts, R.K., Bruning, J.H.: Solid State Technol., May **1981** 99.
81W2 Ward, R.: Electronics, Nov. **1981** 144.
81W3 Watts, R.K., Fichtner, W., Fuls, E.N., Thibault, L.R., Johnston, R.L.: IEEE Trans. Electron Devices, **ED-28** (1981) 1338.
81W4 Wilczynski, J.S.: IEEE Digest of Technical Papers, Symposium on VLSI Technology, Maui, Hawaii (1981) 2.
81Y1 Yoshida, K., Kuwano, H., Yamazaki, S.-H.: IEEE Digest of Technical Papers, Symposium on VLSI Technology, Maui, Hawaii **1981** 14.
81Z1 Zacharias, A.: Solid State Technol., August **1981** 57.
82A1 Arden, W.: Siemens Forschungs- und Entwicklungsberichte **11** (1982) 169.

6.1.4.7 Etching processes

6.1.4.7.1 Wet etching

In this section, chemical etchants are reviewed which are used in semiconductor device fabrication. Since it is not possible to cover all etchants, preference will be given to chemical etchants according to their practical usefulness. More information on chemical etching is found in the numerous books, reviews, and treatises specialising in this topic [59D1, 59F1, 59R1, 59T1, 60F1, 60H1, 60K1, 60L1, 60R1, 61S2, 62H1, 62I1, 63E1, 65G1, 67B3, 67M1, 68G1, 69V1, 70G1, 70K1, 70P1, 70R1, 70S1, 71M3, 71W1, 72U1, 73V1, 74H1, 75R2, 76H1, 76P2, 76S2, 76S4, 78K2]. In semiconductor device technology, chemical etching is used for both complete *removal of thin films* and the *generation of patterns* in thin films. Etching techniques for both applications are covered in this section.

Semiconductor devices usually contain thin layers of *semiconductor material, insulators*, and *conductors*. The semiconductor layers may be *single crystals*, or *polycristalline* or *amorphous material*. The principal insulating materials are the *oxides*, the *nitrides*, and the *oxinitrides* of silicon, *silicate glasses*, and *anorganic* and *organic polymers*. Materials used for conductors are *metals*, *metal silicides*, and doped single crystals or *polycristalline silicon* or *germanium*.

An important process step in semiconductor technology is the application of chemical etching for the *structural characterisation* of materials and especially for the detection of lattice defects in semiconductors.

Acknowledgements

The author would like to thank D. Widmann, W. Beinvogl, B. Hasler, A. Koller, W. Kretzig, E. Voith, J. Ivanits, E. Garbac, M. Preloznik, I. Syniawa, Ch. Schmidt, and W. Müller of the Siemens Research Laboratory, for many helpful discussions.

Principle of etching

Wet chemical etching is the removal of material by dissolution in a suitable etchant. Dissolution of the material in a *liquid solvent* without changing the chemical nature of the dissolved material is the simplest way of etching. Only a few of the principal materials used in semiconductor device fabrication are etched in this manner. Dissolution obtained by a chemical change is the appropriate method of etching metals and semiconductors. This type of etching is usually performed in two steps. The first step is an *oxidation-reduction reaction* in which the etched material is converted into a higher oxidation state, the second step leads to the *dissolution of oxidation products* in the form of complex soluble ions. For silicon, the most commonly used oxidizing agent is nitric acid; hydrofluoric acid is used for complexing.

Detailed information on chemical etching mechanisms is given in several treatises [59T1, 60F1, 60K1, 60H1, 60L1, 62H1, 63E1, 68G1, 70G1, 73V1, 78K2].

6.1.4.7.2 Wet etching techniques

Etching techniques for thin films are immersion, spray, electrolytic, and gas-phase etching, mechanical-chemical polishing, and certain fusion techniques [78K2]. The choice of the etching technique depends on the requirements. The most commonly used etching technique for semiconductor device fabrication is *liquid chemical immersion* or *dip etching*, for which the specimen is immersed in the etch solution. *Mechanical agitation* is usually required in order to improve the uniformity and the reproducibility by enhancing the removal of reaction products of the etched surface and the contacting with fresh solution. Gas bubbles produced by the etching reaction often cling to the etched surface and prevent uniform etching. Bubble accumulation can be prevented by adding a suitable surface-active agent to the etch solution.

6.1.4.7.3 Wet etching process for pattern generation

In semiconductor device technology, wet chemical etching is used for pattern generation in thin films. With the aid of appropriate lithography techniques described in section 6.1.4.6, a resist pattern is formed on a thin film deposited on the substrate. This resist pattern acts as a mask for the following etch step, which has to transfer the resist pattern to the film below the resist.

Figure 1 shows a typical edge profile obtained by wet chemical etching. The extent and shape of *undercutting* depend on the etch process. The etch factor f_e defined as the ratio of the etch depth, d, to lateral etching u below the edge of the mask [$f_e = d/u$ (Fig. 1)], is a measure of the undercutting. The etch factor equals unity for isotropic and is unequal to unity for anisotropic etching. Since undercutting decreases with increasing etch factor, the etch factor has to be high in order to assume the accurate transfer of the etch mask pattern to the layer below.

In semiconductor device technology, only the film below the mask has to be etched, not the substrate or the etch mask. Thus the etch process has to be selective with respect to the substrate and the etch mask. The *selectivity* is defined as the ratio of the etch rates of two different materials. It must be high between the substrate and the mask. The selectivity is one of the most important parameters in etching of semiconductor devices. An etch process for pattern generation in thin films can be described by the following properties:

1. Etch rate: $r_e = \dfrac{\text{etch depth } d}{\text{etch time } t}$;
2. Selectivity of etching of material 1 to material 2: $s = \dfrac{\text{etch rate of material 1 } r_{e1}}{\text{etch rate of material 2 } r_{e2}}$;
3. Etch factor: $f_e = \dfrac{\text{etch depth } d}{\text{undercutting } u}$ (Fig. 1).

An ideal etch process for fine-pattern generation in semiconductor device technology requires a high etch factor for assuring the accurate transfer of the etch mask to the thin film below, a high selectivity s of etching the thin film to the substrate and the mask, and a high etch rate r_e for obtaining cost efficiency.

6.1.4.7.4 Wet etchants

This section lists etchants which are useful in semiconductor device technology. The reagents given in the tables are aqueous solutions of the concentrations given in Table 1.

Table 1. Concentration c of etching reagents in aqueous solutions.

Reagents	c [wt %]
HF	49
HNO_3	70
H_3PO_4	85
HCl	37
H_2SO_4	98
CH_3COOH	100
NaOH	>97
KOH	~85
NH_4OH	29
NH_4F	40
Na_2O_2	93
H_2O_2	30
N_2H_2	64

Many etchants are known by special names which usually are abbreviations of the components. The Table 2 lists etchants giving the name and the composition of reagents used.

Table 2. Composition of etchants. The composition ratio is given in volume parts unless otherwise stated.

No.	Etchant	Composition	Ref.
1	A—B-etch	1 ml HF, 2 ml H_2O, 1 g CrO_3, 8 mg $AgNO_3$	75R2
2	Aqua regia	3 vol HCl, 1 vol HNO_3	78K2
3	Bell No. 2 (buffered HF)	54 % H_2O, 36 % NH_4F, 10 % HF	75R2
4	BD-etch ("back door")	2 vol 40 % NH_4F, 1 vol 12···1/2 % $NH_4H_2PO_4$	76P1
5	BHF	Buffered HF prepared by mixing NH_4F, H_2O, and HF; normal buffered HF: 1 vol 49 % HF to 10 vol 40 % NH_4F	73T1, 78K2
6	Copper-etch No. 1	20 ml HF, 10ml HNO_3, 20 ml H_2O, 1 g $Cu(NO_3)_2 \cdot 3H_2O$	52N1, 56J1
7	Copper-etch No. 2	40 ml HF, 35 ml HNO_3, 10 ml CH_3COOH, 25 ml H_2O, 1 g $Cu(NO_3)_2 \cdot 3H_2O$	58F1
8	CP-4	3 HF, 3 HNO_3, 3 CH_3COOH, 0.06 Br_2	75R2
9	CP-4a	3 HF, 3 HNO_3, 3 CH_3COOH	75R2
10	Cyanide-etch	8 g $K_3Fe(CN)_6$, 12 g KOH, 100 ml H_2O	75R2
11	Dash-etch	1 HF, 3 HNO_3, 10 CH_3COOH; (1—3—10)	75R2
12	Dow ("Secco")-etch	44 g $K_2Cr_2O_7$, 1 l H_2O, mix 1 volume of solution with 2 volumes HF	72S1
13	DP-etch	9 vol H_2O to 1 vol P-etch	64P1
14	Iodine-etch	50 ml HF, 100 ml HNO_3, 110 ml CH_3COOH, 0.3 g I_2	63C1
15	Mercury-etch	3 ml HF, 5 ml HNO_3, 3 ml CH_3COOH, 2···3 % $Hg(NO_3)_2$	56V1
16	Modified Sirtl-etch	110 ml of Sirtl-etch mixed with 25 ml HF, 30 ml HNO_3, 100 ml H_2O	75R2
17	P-etch	2 vol HNO_3, 3 vol HF, 60 vol H_2O	78K2
18	Planar-etch	2 HF, 15 HNO_3, 5 CH_3COOH	75R2
19	Richards-Crockers-etch	$2.4 \cdot 10^3$ molar solution of $AgNO_3$ in 2 ml HF, 3 ml HNO_3, 5 ml H_2O	75R2

continued

Table 2 (continued)

No.	Etchant	Composition	Ref.
20	Russian-etch	10 ml HF, 15 ml HNO_3, 5 ml CH_3COOH, 20 ml H_2O, 8 mg I_2, 2 mg KI	75R2
21	Sailor's-etch	60 ml HF, 30 ml HNO_3, 0.2 ml Br_2, 2.3 g $Cu(NO_3)_2$	58A1
22	SBD-etch ("slow back door")	300 vol 40% NH_4F, 100 vol 12···1/2% $NH_4H_2PO_4$, 2 vol 58% NH_4OH	76P1
23	Schell-etch	1 ml HNO_3, 2 ml H_2O	75R2
24	Sirtl-etch	50 g CrO_3, 100 ml H_2O; mix with 100 ml HF just before using	61S3
25	Staining-etch	0.5 HNO_3, 99.5 HF	56F1
26	Superoxol-etch	1 ml HF, 1 ml H_2O_2, 4 ml H_2O	75R2
27	WAg-etch	4 ml HF, 2 ml HNO_3, 4 ml H_2O, 0.2 g $AgNO_3$	75R2
28	White-etch	1 HF, 4 HNO_3	75R2
29	W—R-etch	2 HCl, 1 HNO_3, 2 H_2O	75R2

Table 3. Etchants used for isotropic etching of Si.

No.	Etchant	r_e nm min^{-1}	Material specification	T °C	Remarks	Ref.
1	HF, HNO_3, H_2O or CH_3COOH	depends on composition	(111), (100), (110) n- and p-type Si	RT	most frequently used	59R1, 60R1, 61S1, 72U1, 74R1, 75R3, 75S1, 76S2, 67B3
2	1 HF, 3 HNO_3, 10 CH_3COOH,				for general etching, frequently used	75R3, 67B3
3	2 HF, 15 HNO_3, 5 CH_3COOH (planar etch)				for general etching	75R3
4	HF, HNO_3 (a) 50 HNO_3, 50 HF (b) 75 HNO_3, 25 HF (c) 91 HNO_3, 9 HF	 $2\cdot10^5$ $3\cdot10^4$ $6\cdot10^3$	(100) p-type 12···78 Ω cm n-type 0.05···8 Ω cm	30	for general etching; no difference in etch rates of p- and n-type Si; stirring and sample rotation; rotation 88 rpm	62K1
5	50 ml HF, 100 ml HNO_3, 110 ml CH_3COOH, 0.3 g I_2 (iodine etch)				for general etching	58W1, 75R3, 67B3
6	20 HF, 30 HNO_3, 1 Na_2HPO_4 (2%)				for general etching	61S4
7	100 g NH_4F, 1000 ml H_2O, 2 ml H_2O_2				for fine pattern etching low degree of undercutting below photoresist mask	61C1
8	50 ml HF, 50 ml CH_3COOH, 200 mg $KMnO_4$	$2\cdot10^2$	(100), (111) epi Si n, $p<2\cdot10^{16}$ cm^{-3}	18	for epi Si etching	70T1

continued

Table 3 (continued)

No.	Etchant	r_e nm min^{-1}	Material specification	T °C	Remarks	Ref.
9	108 ml HF, 350 g NH_4F per 1000 ml	 0.043(8) 0.045(8) 0.023(10)	single crystalline Si n-type 0.2···0.6 Ω cm p-type 0.4 Ω cm p-type 15 Ω cm		etching process not selective to SiO_2; for very thin Si films	69H3
10	1 HF, 26 HNO_3, 33 CH_3COOH	 ≈150 <100	poly Si; undoped, B-doped undoped B doped, <0.01 Ω cm		for poly-Si etching	76C1
11	1 HF, 3 HNO_3, 8 CH_3COOH	 (0.7···3)·10^3 no etching occurs	single-crystal Si, n- and p-type doped with As, P, Sb or B <0.01 Ω cm >0.068 Ω cm		for selective etching of n- and p-type Si; selective to SiO_2 and Si_3N_4; etch rate of SiO_2: 0.05 μm/min	73M1, 73S2
12	5 wt % HF	 0	(111), (100) Si n^+-type substrate n-type epi layer		electrochemical etching for thinning of n^+n leaving n epi layer	71M2
13	1 HF, 1 H_2SO_4, 5 H_2O		p^+ bulk Si n type epi layer	25	electrochemical etching preferential etching of p^+ Si	72W1
14	0.5 HNO_3, 99.5 HF		Staining etch		detection of pn-junctions using light	56F1
15	5 g CrO_3, 1 l (5 H_2O, 1 HF) 10 g CrO_3, 1 l (5 H_2O, 1 HF)	80 160	single crystal Si	24 24	etch rate decreases with decreasing temperature	80K1, 80H1
16	45.45 HNO_3, 27.27 HF, 27.27 CH_3COOH, 0.5 ml Br for 100 ml solvent (CP4-etch)	150	single-crystal Si	23		67B3

Table 4. Etchants used for anisotropic etching of Si.

No.	Etchant	r_e μm min^{-1}	Material specification	T °C	Remarks	Ref.
1	KOH solution (3···50 %)		(100) Si; SiO_2 masked	70···90	for V-groove etching	75B1, 67B3
2	65 hydrazine, 35 H_2O	1.6	(100) Si, 3···5 Ω cm	100	for V-groove patterns	75D2
3	100 g KOH in 100 ml H_2O	≈8	(110) Si; SiO_2 masked	boiling	for vertical deep etching in (110) Si	67B3
4	KOH solutions (a) KOH (4N) (b) 6 KOH (4N), 1 isopropanol (c) 6 KOH (6N), 1 isopropanol	 (a) 0.36 (b) 0.162 (c) 0.210	(100) epi Si, n-type 0.01···10 Ω cm	60	for patterning epi Si on sapphire lower etch rates with p-type Si	73C1

continued

Table 4 (continued)

No.	Etchant	r_e µm min^{-1}	Material specification	T °C	Remarks	Ref.
5	50 ml H_2O, 15 g KOH 15 ml isopropanol	 0.9 1.1 0.7	P doped $\leqq 1.5 \cdot 10^{20}$ cm^{-3} B doped $< 5 \cdot 10^{17}$ cm^{-3} (100) epi Si (100) bulk Si poly Silicon		for patterning epi Si on sapphire or spinnel	74R1
6	60 hydrazine, 40 H_2O		(100) Si	110	for diffuse-reflectivity texturing solar cells	75B1
7	Hydrazine-H_2O equimolar (ethylenediamine-pyrocatechol-H_2O)	1…3	(100) Si bulk crystal	100…120	for structural etching; etch rate in (111) minimum, in (211) maximum	58H1, 78B1
8	100 g KOH in 100 ml H_2O	≈8	(110) Si	boiling	for vertical deep etching of moats in (110) Si; selective to SiO_2	70S2
9	4 M NH_4F— 1 M Cu $(NO_3)_2$	 0.185 0.117 0.012	n-type Si, 10…100 Ω cm (100) (110) (111)	22	for very high resolution patterning	73Z1
10	Tetramethylammonium hydroxide, or tri-methyl-2-hydroxyethyl ammonium hydroxid (0.5 wt %)	 0.360 0.00230 0.0163	(100), (111) Si (100) n-type (100) p-type (111) n-type	80…90	alkali-free Si etching; selective to SiO_2; etch rates of thermal SiO_2: 0.3 nm/min, etch rates of CVD SiO_2: 0.7 nm/min	76A1
11	23.4 wt % KOH, 63.3 wt % H_2O 13.3 wt % isopropanol	 0.95…0.99 0.63 0	(100) Si, B doped (100), doped with B, 10^{14}…10^{18} cm^{-3} (100), doped with B, 10^{19} cm^{-3} (100), doped with B, 10^{20} cm^{-3}	80	for thinning of p^+p leaving p^+ layer	

Table 5. Etchants used for etching of germanium.

No.	Etchant	r_e µm min^{-1}	Material specification	T °C	Remarks	Ref.
1	5 HNO_3, 3 HF, 3 CH_3COOH with 0.06 Br_2 (CP-4)		(100), (111) Ge		for general etching	52H1, 53V1, 54K1, 67B3
2	175 ml H_2O_2 (3…4 %), 25 ml 0.2 M KH_2PO_4 containing 12 ml H_3PO_4 per liter	0.020	(100), (111), (201) Ge n-type high resistivity		for thin film etching magnetic stirrer	67P1, 78K2

continued

Table 5 (continued)

No.	Etchant	r_e µm min^{-1}	Material specification	T °C	Remarks	Ref.
3	50 wt % HF, 50 wt % H_2O_2	11.2 21.1 39.3	(100) Ge (111) Ge (110) Ge n-type, 3 Ω cm	25	for general etching	67S1
4	5 H_2O, 1 H_2O_2		(111) Ge	26	for general etching	55C1
5	100 ml H_2O_2 (10 vol %), 8 g NaOH	1.25···5	Ge	70	for controlled etching	62H1, 78K2
6	4 H_2O, 1 HF, 1 H_2O_2 (No. 2 etch; superoxol etch)		(100), (111) Ge		for general etching	54K1, 55C1, 57B1, 67B3
7	5 HNO_3, 3 HF, 3 CH_3COOH (CP-4A, CP-6, CP-8)r		Ge	23···70	much slower than CP-4 at 23 °C	54K1, 62H1, 67B3
8	11 ml CH_3COOH with 30 mg I_2 dissolved, 10 ml HNO_3, 5 ml HF (iodine etch A)		(100), (111) Ge		for general etching	58W1 67B3
9	1 NaOCl(10 %), 10 H_2O		(100), (111) Ge	40	for general etching	57G1

Table 6. Etchants used for etching oxides. CVD=chemical vapor deposition; LTCVD=low temperature CVD; HTCVD=high temperature CVD; LPCVD=low pressure CVD; HPCVD=high pressure CVD.

No.	Etchant	r_e nm min^{-1}	Material	Structure	Formation	Photoresist mask applicable	T °C	Remarks	Ref.
1	$HF—H_2O$ solution:		SiO_2	amorphous	thermal in steam at 1100 °C on (100) Si			etch rate r_e for SiO_2: *) etching selective to Si	71J2, 73H2, 76J1
	95 HF, 5 H_2O	1500					25		
	50 HF, 50 H_2O	360							
	5 HF, 95 H_2O	18							
	95 HF, 5 H_2O	4800					60		
	50 HF, 50 H_2O	1500							
	5 HF, 95 H_2O	90							
2	1 HF, 4 H_2O	300	SiO_2	amorphous	CVD		25		76K6
3	BHF (buffered HF, commonly 40 wt % NH_4F solution with HF)	60···120	SiO_2	amorphous	thermally grown sputtered	yes		The ratio of constituents in BHF varies typically from 9···12 vol % HF and 88···91 vol % buffer solution. BHF is important in pattern etching of SiO_2 films using photoresist masks. Addition of NH_4F prevents depletion of fluoride ions, thus maintaining a stable etch rate. Etching is selective to Si.	65D1, 68H1, 70G1, 78K2
4	BHF, dihydroxyalcohol or glycerol		SiO_2	amorphous	sputtered evaporated LT CVD	yes		for SiO_2 films on Al; addition of dihydroxyalcohol or glycerol improves the selectivity to Al	71H2, 74G1

continued

*)

$r_e = A[HF] + B[HF_2^-] + C$

$A = A_0 \exp(-\Delta E_1/RT)$;
$B = B_0 \exp(-\Delta E_2/RT)$;
$C = -C_0(T/K - 292)$
$[HF]$, $[HF_2^-]$
are molar concentrations

T [°C]	A [nm/min]	B [nm/min]	C [nm/min]
25	0.250	0.966	−0.014
60	1.04	4.86	−0.102

A_0 [nm/min]	B_0 [nm/min]	C_0 [nm/min]
$5 \cdot 10^6$	$2.2 \cdot 10^5$	0.0025

ΔE_1	ΔE_2 (kcal/mol)
9.1	8.1

Table 6 (continued)

No.	Etchant	r_e nm min^{-1}	Material	Structure	Formation	Photoresist mask applicable	T °C	Remarks	Ref.
5	P-etch (3 HF, 2 HNO_3, 60 H_2O)	0.2	SiO_2	amorphous	thermally grown		25	for controllable removal of very thin layers of SiO_2 and for diagnostic of the layers	64P1, 68P1, 70G1
		0.4···1.2			rf sputtered				
		0.6···2			CVD				
		2···7			electron gun evaporated				
		1.8···22.8			anodization				
6	Vapor from aqueous HF		SiO_2	amorphous	different processes			for pattern etching in SiO_2 films	66H1
7	HF, BHF		SiO_2	amorphous	thermally grown				65D1, 65P1, 66M1, 67B2, 68K1, 70K5, 71J2, 73H1, 73H2
					sputtered				72K1, 75G2, 76H2
					evaporated				70P1
					CVD				66M2, 67G2, 68C1, 68D1, 68H2, 69S1, 70B1, 71H1, 71M1, 75G1, 75K3, 76K2, 76K3
					anodization				67K1, 70P1
					plasma				67K2, 69A1, 74K3
8	5 M KOH	~5	SiO_2	amorphous	thermally grown		85		67K1
9	10 wt % NaOH	0.01	SiO_2	amorphous			23		67P1
		0.5					55		
		50					90		
10	HF, HNO_3		SiO	amorphous					70G1
	NH_4F, NH_4OH	500		amorphous		yes	80···90	with usual negativ photoresists line edges without undercutting are obtained	70G1
11	HF, BHF, H_3PO_4		Al_2O_3	amorphous	CVD			CVD below 500 °C	67A1, 69H1, 69K2, 70D1, 75M2
					plasma				71K1, 74K3

continued

Table 6 (continued)

No.	Etchant	r_e nm min^{-1}	Material	Structure	Formation	Photoresist mask applicable	T °C	Remarks	Ref.
11					anodically				69H2, 73L1, 75H1, 77I1
					evaporation				69F1, 69H2
					sputtering				66F1, 74K1, 77N1
					boiling in H_2O				76H1
12	H_3PO_4	10	Al_2O_3	amorphous	$AlCl_3$ hydrolyse at 900···1000 °C	no	180	CVD SiO_2 etch masks are useful for patterning selective to AC	76S1, 78K2
13	H_3PO_4, CrO_3		Al_2O_3	amorphous					75S3
14	HF, H_3PO_4, H_2SO_4, NaOH		GeO_2	hexagonal	> 1000 °C, bulk				78K2
	H_3PO_4, H_2SO_4, NaOH			tetragonal	> 1000 °C, bulk				78K2

Table 7. Etchants used for etching of silicate glasses. PSG: phosphosilicate glass (incorporation of P_2O_5 in SiO_2 network), BSG: borosilicate glass (incorporation of B_2O_3 in SiO_2 network), AsSG: arsenosilicate glass (incorporation of As_2O_3 in SiO_2 network), AlSG: aluminosilicate glass (incorporation of Al_2O_3 in SiO_2 network) (for definitions of CVD, LTCVD, HTCVD, LPCVD and HPCVD see Table 6).

No.	Etchant	r_e nm min^{-1}	Material	Formation	Composition mole % in SiO_2	Photoresist mask applicable	T °C	Remarks	Ref.
1	1 HF, 2 H_2O	$4.5 \cdot 10^3$ $24 \cdot 10^3$	PSG	LTCVD (from SiH_4, PH_3, and O_2 in N_2 at 450 °C)	5% P_2O_5 10% P_2O_5		22	etch rate increases exponentially with P_2O_5 content	66S1, 68E1, 68K2, 70K3, 76K2, 76K3, 76K4, 76K5, 78K1, 78K2, 78K3
2	1 HF, 2 Glycerol	$2.2 \cdot 10^3$ $7.2 \cdot 10^3$	PSG	LTCVD (see No. 1)	5% P_2O_5 10% P_2O_5		22		78K3
3	BHF (48% HF : 40% NH_4F : H_2O, 1:10:11 by volume)	720 $1.14 \cdot 10^3$	PSG	LTCVD (see No. 1)	5% P_2O_5 10% P_2O_5	yes	22	etch rate of PSG in BHF not much affected by layer composition	78K3
4	P-etch (3 HF, 2 HNO_3, 60 H_2O)	480 $1.68 \cdot 10^3$	PSG	LTCVD (see No. 1)	5% P_2O_5 10% P_2O_5	yes	25		78K3
	P-etch	132	PSG	LTCVD	5% P_2O_5	yes	25		78K3

continued

Table 7 (continued)

No.	Etchant	r_e nm min^{-1}	Material	Formation	Composition mole % in SiO_2	Photoresist mask applicable	T °C	Remarks	Ref.
4	P-etch (continued)	720		(450 °C, densified at 1000 °C in N_2 for 60 min)	10% P_2O_5				
	P-etch	360	PSG	thermal	10% P_2O_5	yes	25		78K3
5	1 HF, 25 H_2O	60 540	PSG	thermal (obtained by reacting vapors of $POCl_3$ or P_2O_5 with SiO_2 at 1000 °C)	5% P_2O_5 10% P_2O_5		25		78K3
6	BD ("back door") etch: 2 part 40 wt % NH_4F to 1 part 12···1/2 wt % $NH_4H_2PO_4$	16.4	PSG	thermal (1000···1100 °C SiO_2 films 550···610 nm thick with $POCl_3$ with O_2 in N_2)	5.6% P_2O_5	yes	25	etch rate ratio to SiO_2 2.5:1; for removal of very thin SiO_2 films from contact holes with a minimum removal of PSG	76P1
7	SBD ("slow back door") etch: 300 parts 40 wt % NH_4F, 100 parts 12···1/2 wt % $NH_4H_2PO_4$, 2 part 58% NH_4OH	7.8	PSG	thermal (see No. 6)	5.6% P_2O_5		25	etch rate ratio to SiO_2 2.4:1	76P1
8	DP etch (9 H_2O, 1 P etch)	6.0 24.0	PSG	thermal (see No. 6)	5% P_2O_5 10% P_2O_5		25	application similar to BD etch	64P1, 76P1
9	HF, BHF, P etch, BD etch, DP etch		PSG	LTCVD					67N1, 69T1, 70K4, 71H1, 76P1, 77J1, 77C1
10	HF, BHF, P etch BD etch, DP etch		PSG	thermal					64P1, 66S1, 67N1, 68E1, 68S1, 69B1, 76P1, 78K3

continued

Table 7 (continued)

No.	Etchant	r_e nm min^{-1}	Material	Formation	Composition mole % in SiO_2	Photoresist mask applicable	T °C	Remarks	Ref.
11	BHF (1 HF, 10 NH_4F)	44 21 35	BSG (boro- silicate glass)	LTCVD (reaction of mixtures of O_2, SiH_4, and B_2H_6 diluted in Ar at 300···500 °C)	5 % B_2O_3 17 % B_2O_3 30 % B_2O_3	yes	26(1)	selective to silicon	73T1
	50 % BHF	37 21 47			5 % B_2O_3 17 % B_2O_3 30 % B_2O_3	yes	26(1)	selective to silicon	73T1
	10 % BHF	13 20 73			5 % B_2O_3 17 % B_2O_3 30 % B_2O_3	yes	26(1)	selective to silicon	73T1
	1 % BHF	1 6 52			5 % B_2O_3 17 % B_2O_3 30 % B_2O_3	yes	26(1)	selective to silicon	73T1
12	1 HF, 2 H_2O	600 $18 \cdot 10^3$	BSG	LTCVD (films deposited at 450 °C from SiH_4, B_2H_6, and O_2 in N_2; densified 5 min 770 °C)	5 % B_2O_3 30 % B_2O_3		25(1)	selective to silicon	70K4
13	500 ml HNO_3, 1 ml HF		BSG		5···15 % B_2O_3				71S1
14	BHF, HF		BSG	LTCVD					68K2, 68P1, 69K1, 69I1, 70E1, 70K2, 70K4, 71B2, 72K1, 73R3, 73S1, 73T1, 74C1, 75S2
15	BHF	100 140	AsSG (arseno- silicate glass)	LTCVD (deposited at 500 °C using SiH_4 and AsH_3, each diluted to 1 % in Ar and pure O_2)	2 % As_2O_3 7.5 % AsO_3	yes		densification in argon at 1100 °C for 5 h; etch rate increases exponentially with increasing As_2O_3 content from 0 to 8 mole % As_2O_3	73G1

continued

Table 7 (continued)

No.	Etchant	r_e nm min^{-1}	Material	Formation	Composition mole % in SiO_2	Photoresist mask applicable	T °C	Remarks	Ref.
16	BHF (1 HF, 10 NH_4F)	108	AsSG	LTCVD (reaction of mixture of O_2, SiH_4, and AsH_3 diluted in Ar at 300···500 °C)	2% As_2O_3 5% As_2O_3	yes	26(1)	densification in argon at 1100 °C for 5 hrs	73T1
	50% BHF	80			2% As_2O_3 5% As_2O_3	yes	26(1)	etch rate increases exponentially with increasing As_2O_3 content from 0 to 5 mole % As_2O_3	73T1
	10% BHF	15			2% As_2O_3 5% As_2O_3	yes	26(1)		73T1
	BHF			LTCVD					72W3, 77J1
17	BHF	198	AlSG (alumino-silicate glass)	LTCVD (deposited at 475 °C using SiH_4 trimethyl aluminum, and O_2)	1.5% Al_2O_3	yes	25(1)	densification at 770 °C in argon for 20 min	70K4
18	P-etch (15 HF, 10 HNO_3, 300 H_2O)	$1.08 \cdot 10^3$ 360 180	AlSG	LTCVD	Si/Al = 2 Si/Al = 6 Si/Al = 15		25		66E2
19	HF, BHF, H_3PO_4(180 °C)		AlSG	LTCVD					69K2, 70T2

Table 8. Etchants used for etching of nitrides (CVD = chemical vapor deposition).

No.	Etchant	r_e nm min^{-1}	Material	Structure	Formation	Photoresist mask applicable	T °C	Remarks	Ref.
1	H_3PO_4	10···30	Si_3N_4	amorphous		no	180	selective to SiO_2 and Si: etch rate of SiO_2 is 0···2.5 nm/min and 0.3 nm/min for single-crystal Si	67G1, 71M4
		2.5···20					150	etch rates of Si_3N_4, SiO_2, Si increase with increasing temperature; an increase in water content of phosphoric acid increases the etch rate of silicon nitride	

continued

Table 8 (continued)

No.	Etchant	r_e nm min^{-1}	Material	Structure	Formation	Photo-resist mask applicable	T °C	Remarks	Ref.
2	HF	50···100	Si_3N_4	amorphous	different processes	no	25	the etch rate depends on the method of film preparation	67B2, 70G1
3	1% HF 10% HF	0.4 20 1.5 106	Si_3N_4	amorphous	CVD (from SiH_4, NH_3 in the presence of N_2 at 850 °C)		24.5 90 24.5 90	the activation energy of etch rate is proportional to the square root of the concentration	73H1
4	HF—H_2O solution: 95 HF, 5 H_2O 50 HF, 50 H_2O 5 HF, 95 H_2O 95 HF, 5 H_2O 50 HF, 50 H_2O 5 HF, 95 H_2O	 13.2 6.0 1.0 150 80 11	Si_3N_4	amorphous	CVD			etch rate r_e for Si_3N_4: $r_e = A[HF] + B[HF_2^-] + C$ [HF], $[HF_2^-]$ are molar concentrations T [°C] / A [nm/min] / B [nm/min] / C [nm/min]: 25 / 0.016 / 0.031 / $<10^{-5}$ 60 / 0.19 / 0.37 / 0.002 etching is selective to Si	73H2, 78D1
5	BHF (Bell No. 2: 300 cc H_2O + 200 g NH_4F + 45 cc HF)	0.7 10	Si_3N_4	amorphous	CVD (from SiH_4, NH_3 at 850 °C)	yes	25 60	selective to Si: the etch rate depends on deposition temperature and on the ratio NH_3/SiH_4	67B1
6	P-etch	2.9	Si_3N_4	amorphous	reactively sputtered (sputtering voltage ~1000 V)		23	the etch rate increases with decreasing sputtering voltage	67J1
7	HF BHF H_3PO_4	150···300 20··· 30 60···100	$Si_xN_yH_z$	amorphous	plasma-enhanced CVD (300 °C)	 yes no	23 20···25 180	etch rate very much higher than that of high-temperature CVD nitride	77K2

continued

Table 8 (continued)

No.	Etchant	r_e nm min^{-1}	Material	Structure	Formation	Photo-resist mask applicable	T °C	Remarks	Ref.
8	H_3PO_4, HF, BHF, HF + HNO_3, NaOH, P-etch		Si_3N_4	amorphous	CVD				68W2, 69A1, 69G1, 69K3, 71M4, 71T1, 72G1, 72W2, 73H1, 73H2, 74K3, 75W1, 76K2, 76S5, 77L1, 77R1
9	1 H_2SO_4, 9 H_3PO_4		Si_3N_4	amorphous		no	160		75W1
10	HF, BHF		Si_3N_4	amorphous	sputtering				75K2, 75M1, 76S6
11	HF/Glycerol solution		Si_3N_4	amorphous	CVD	yes		etch rate of Si_3N_4 higher than that of SiO_2	80D1
12	HF, BHF, H_3PO_4, HNO_3		Ge_3N_4	amorphous	annealing at 400···600 °C				78K2

Table 9. Etchants used for etching of oxinitrides.

No.	Etchants	r_e nm min^{-1}	Material	Structure	Formation	Photo-resist mask applicable	T °C	Remarks	Ref.
1	BHF (1 HF, 10 NH_4F)	3 10	$Si_xN_yO_z$	amorphous	CVD (from SiH_4, NH_3, CO_2 at 700···1000 °C)	yes		10^3 ppm oxygen gas 10^4 ppm oxygen gas	80N1
	BHF	1···10			CVD	yes	~25		68B1, 73R1
2	HF	35 150 500	$Si_xN_yO_z$	amorphous	CVD (from SiH_4, NH_3 NO at 1000 °C, 850 °C)		~25	0.25 Vol. % NO 1.25 Vol. % NO 10.00 Vol. % NO	68B1, 73R1
3	H_3PO_4	1···10	$Si_xN_yO_x$	amorphous	CVD (from SiH_4, NH_3 NO at 1000 °C, 850 °C)	no	180	the etch rate of SiO_2 is much higher than that of $Si_xN_yO_z$	68B1, 73B1
4	Dim etch (125 ml HNO_3, 25 ml HF, 110 ml CH_3COOH)	2···20	$Si_xN_yO_z$	amorphous	CVD (from SiH_4, NH_3 NO at 1000 °C, 850 °C)		~23		68B1, 73R1

Table 10. Etchants used for etching of metals.

No.	Etchants	r_e nm min^{-1}	Material	Formation	Photo-resist mask applicable	T °C	Remarks	Ref.
1	76 H_3PO_4, 15 CH_3COOH, 3 HNO_3, 5 H_2O and small amount of NH_4F; 1 vol % NH_4F / 5 vol % NH_4F	160 / 100	aluminum		yes	40	standard aluminum etch; selective to SiO_2; applicable to Al—Si	75A1, 76M1
2	4 H_3PO_4, 4 CH_3COOH, 1 HNO_3, 1 H_2O	33		evaporated	yes	25	small amount of under-cutting, selective to SiO_2	75K1
3	80⋯75 H_3PO_4, 5 HNO_3, 0⋯20 H_2O	150⋯250			yes	40	widely used for Al; selective to SiO_2; applicable to Al—Cu (Cu < 10 %)	70G1
4	74.1 H_3PO_4, 7.4 HNO_3, 18.5 H_2O	900			yes	50	selective to SiO_2, PSG	77K3
5	HCl, 4 H_2O				yes	80	for fine line etching	66E1, 68H1
6	75 g Na_2CO_3, 35 g $Na_3PO_4 \cdot 12H_2O$, 16 g $K_3Fe(CN)_6$, 0.5 liter H_2O	130			yes	24	small amount of under-cutting	75K1
7	0.25 M NaOH	60			yes	25	small amount of under-cutting	66E1, 70G1, 75K1
8	$FeCl_3$, HCl; $HClO_4$, $(CH_3CO)_2O$; HCl, HNO_3, H_2O; $FeCl_3$							59T1, 66E1, 70R1
9	10 HCl, 1 HNO_3, 9 H_2O	25⋯50 · 10^3	Al—Ti			49		70R1
10	dilute HCl or HNO_3		chromium		yes			66E1, 68W2, 70R1, 78K2
11	1 vol (50 g NaOH, 100 ml H_2O), 3 vol (100 g $K_3[Fe(CN)_6]$, 300 ml H_2O) (Kodak EB-5[b] bath)	25⋯100		vacuum-deposited	yes	~25	for etching Cr masks	66E1, 70G1, 70R1, 73N7, 78K2
12	2 g $Ce(SO_4)_2 \cdot 2(NH_4)_2SO_4 \cdot 2H_2O$, 10 ml HNO_3, 50 ml H_2O	8⋯85		sputtered		28	oxide coating etches at 20 nm/min	72J1, 78K2
13	164.5 g $Ce(SO_4)_2 \cdot 2(NH_4)_2SO_4 \cdot 2H_2O$, 43 ml $HClO_4$, H_2O to make 1 liter		chromium			25⋯50	for precision patterning of Cr masks on glass substrates	69E1

continued

Table 10 (continued)

No.	Etchants	r_e nm min^{-1}	Material	Formation	Photo-resist mask applicable	T °C	Remarks	Ref.
14	454 g $AlCl_3 \cdot 6H_2O$, 135 g $ZnCl_2$, 30 ml H_3PO_4, 400 ml H_2O							70G1
15	50 ml glycerin, 50 ml HCl	80						69E1, 70G1
16	9 sat. $Ce(SO_4)_2$ sol., 1 HNO_3	80						69E1, 70G1
17	2 $FeCl_3$ (42° Bé), 1 HCl					80		66E1, 70R1
18	HCl, depassivation with zinc rod	150						66R1, 70G1
19	$FeCl_3$, 36···42° Bé	$50 \cdot 10^3$	copper		yes	49	diluted solutions better for slow etching;	61G1, 61S1, 70G1, 70R1, 78K2
20	5 HNO_3, 5 CH_3COOH, 2 H_2SO_4; dilute with H_2O as desired						etches Cu and Cu-based alloys at the same rate as Ni and Ni-based alloys	72J2, 78K2
21	3 HCl, 1 HNO_3	$(25 \cdots 50) \cdot 10^3$	gold		yes	32···38	the photoresist does not withstand very long; the technique is limited to very thin gold films	70G1, 70R1, 78K2
22	4 g KI, 1 g I_2, 40 ml H_2O (KI—I_2 etch)	$(0.5 \cdots 1) \cdot 10^3$			yes			66Z1, 70G1, 70R1, 78K2
23	1···2% HF		hafnium		yes			59T1, 67H1, 70G1
24	3 HNO_3, 7 HCl, 30 H_2O		iron			60···70		59T1, 74O1, 78K2
25	$FeCl_3$ 36···42° Be		lead			43···54		59T1, 70R1, 78K2
26	1···3 HNO_3, 17···19 H_2O		magnesium					59T1, 66E1
27	1 H_2SO_4, 1 HNO_3, 1···5 H_2O	$\approx 12 \cdot 10^3$ $\approx 25 \cdot 10^3$	molybdenum		yes (for 25 °C)	25 54		66E1, 71B1, 70R1, 75K1
28	22 wt% $Ce(NH_4)_2(NO_3)_6$ in 1% HNO_3	≈ 100				25		75K1
29	92 g $K_3[Fe(CN)_6]$, 20 g KOH, 300 ml H_2O	$\approx 10^3$			yes			66D1, 70G1
30	other etches							59T1, 66B1, 70S1, 75C1, 76B1

continued

Table 10 (continued)

No.	Etchants	r_e nm min^{-1}	Material	Formation	Photo-resist mask applicable	T °C	Remarks	Ref.
31	Fe_3Cl_3, 42⋯49 °Bé	12⋯25 · 10^3	nickel and Ni—Fe (permalloy)		yes	3⋯54		66E1, 70R1
32	other etches							59T1, 66E1, 72J2, 74B1, 75K1, 76H3
33	1 HCl, 10 HNO_3, 10 CH_3COOH	100	palladium		yes	25		76S3, 78K2
34	7 HCl, 1 HNO_3, 8 H_2O	40⋯50	platinum			85		74R2, 75R1
35	other etches							66E1, 70G1, 76F1
36	55 g $Fe(NO_3)_3$ in ethylene glycol to make 100 ml, 25 ml H_2O	≈300	silver		yes	43⋯49		66E1, 70G1
37	4 g KI, 1 g I_2, 40 ml H_2O (KI—I_2 etch)	20⋯60 · 10^3			yes			70G1
38	4 CH_3OH, 1 NH_4OH, 1 H_2O_2	≈6 · 10^3			yes		for pattern etching using a photoresist mask	74O2
39	other etches							59T1, 66E1, 68W2, 70R1
40	2 HNO_3, 1 HF, 1 H_2O		tantalum		yes	~25		70G1
41	9 NaOH or KOH(30 %), 1 H_2O_2	100⋯200			no	90	metal mask must be used; very little undercutting; etches Ta_2O_5 and TaN at same rate as Ta	69G2, 75D1, 78K2
42	other etches							59T1, 70G1, 75T1, 76C2, 78K2
43	2 HNO_3 + 3 H_2O		tellurium					78K2
44	$FeCl_3$, 36⋯42 · Bé		tin			32⋯54		70R1, 78K2
45	1 HF, 9 H_2O	12 · 10^3	titanium		yes	32		66E1, 70G1, 70R1, 78K2
46	1 HF, 2 HNO_3, 7 H_2O	18 · 10^3			yes	32		66E1, 70G1, 70R1 78K2

continued

Table 10 (continued)

No.	Etchants	r_e nm min^{-1}	Material	Formation	Photo-resist mask applicable	T °C	Remarks	Ref.
47	34 g KH_2PO_4, 13.4 g KOH, 33 g $K_3Fe(CN)_6$, H_2O to make 1 liter	~160	tungsten		yes		high resolution can be achieved	75S4, 78K2
48	other etches							59T1, 66B1, 71J1, 75S4, 76H3, 77K1, 78K2
49	2⋯3 HNO_3, 17⋯18 H_2O	$25 \cdot 10^3$	zinc			38⋯49		70R1, 78K2
50	other etchants							59T1, 61E1, 68D2, 69P1, 70A1, 78K2

Table 11. Etchants used for etching of silicides.

No.	Etchants	r_e nm min^{-1}	Material	Formation	Photoresist mask applicable	T °C	Remarks	Ref.
1	1 N NaCl or 1 N Na_2SO_4		CoSi	melted, annealed			electrochemical etch	77K4
2	60 H_3PO_4, 5 HNO_3, 1 HF	60⋯90	CrSi	electron-beam deposited	yes			68W1
3	NaCl		CuSi	melted, annealed			electrochemical etch	77K4
4	1 N NaCl or 1 N Na_2SO_4		NiSi	melted, annealed			electrochemical etch	77K4
5	dilute aqua regia 4 (sat. I_2/CH_3COOH), 3 HNO_3, 1 HF		PtSi	sputtering e-gun evaporation CVD	no	50	SiO_2 mask applicable	73R2, 74R2, 75R1, 78K2 68S2, 78K2
6	HF—HNO_3		$MoSi_2$	magnetron sputtering		~25	etch rates 1⋯3 nm/day for HCl, HNO_3, H_2SO_4, and aqua regia	78M1
7	10:1 BHF solution	20	$TaSi_2$	reactive sintering of sputtered Ta	yes		silicide on a polysilicon layer	80M1
8	10:1 BHF solution	≧150	$TiSi_2$	reactive sintering of sputtered Ti	yes		silicide on a polysilicon layer	80M1
9	98 % HNO_3, 2 % NH_4F		WSi_2	rf diode sputtering	yes		silicide on SiO_2 layer	79S1

Table 12. Etchants used for the detection of lattice defects in Si and Ge (compare section 6.1.3.3.4).

No.	Defect	Etchant	Material	t min	Remarks	Ref.
1	Dislocations	Sirtl etch: 50 g CrO_3, 100 ml H_2O; mix with 100 ml HF just before using	(111), (110) Si	1···7		61S3
2		modified Sirtl etch: made by mixing 110 ml of Sirtl etch with 25 ml HF, 30 ml HNO_3, 100 ml H_2O	(111) Si		better on low-resistivity material than the Sirtl etch	75R2
3		Dow ("Secco") etch: 44 g $K_2Cr_2O_7$, 1 l H_2O, mix 1 volume of solution with 2 volumes HF	(111), (110), (100) Si	5, with ultrasonic agitation	reasonable detection on (100); gives circular pits	72S1
4		mercury etch: 3 ml HF, 5 ml HNO_3, 3 ml CH_3COOH 2···3 % $Hg(NO_3)_2$	(111), (100) Si	2	aged in closed bottle for 6 weeks	56V1
5		dash etch: 1 HF, 3 HNO_3, 10 CH_3COOH	(111), (110) Si	240	works moderately well on (100) faces; few min etching for thin layer	56O1, 75R2
6		copper etch No. 1: 20 ml HF, 10 ml HNO_3, 20 ml H_2O 1 g $Cu(NO_3)_2 \cdot 3H_2O$	(111) Si		most of the metal-ion etchants are developed in an attempt to produce sharp, well defined pits	52N1, 56J1
7		copper etch No. 2: 40 ml HF, 35 mol HNO_3, 25 ml H_2O 10 ml CH_3COOH, 1 g $Cu(NO_3)_2 \cdot 3H_2O$	(111) Si			58F1
8		Sailor's etch: 60 ml HF, 30 ml HNO_3, 0.2 ml Br_2, 2.3 g $Cu(NO_3)_2$ dilute 10:1 with water before using	(111) Si	120	shows Shockley partials but not stair-rod dislocations	58A1
9		ASTM etch: modified copper etch No. 1	(111) Si	240	useful in defining the surface (e.g. (111)) of sampling plans and determination of pit densities	75R2
10		2 HF, 1 (1 M CrO_3)	(100) Si	5	for 0.6···15 Ωcm n- and p-type Si	
11	Stacking faults	dash, Sailor's	(111) Si	15···20 up to 240	enhances delineation of (112) partial dislocations; RT	63C1
12		Sirtl etch	(111) Si	1/4···1/2		68A1

continued

Table 12 (continued)

No.	Defect	Etchant	Material	t min	Remarks	Ref.
13		iodine: 50 ml HF, 100 ml HNO_3, 110 ml CH_3COOH, 0.3 g I_2	(111) Si			63C1
14		Dash, Sirtl	(100) Si			63C1
15		Dash	(110) Si			63C1
16	Twins	CP4A: 3 HF, 5 HNO_3, 3 CH_3COOH	Si			53V1
17	Lineage	etchants for dislocations	Si			75R2
18	Grain boundaries	Sirtl; 1 HF, 3 HNO_3, 6 CH_3COOH 1 HF, 3 HNO_3, 10 CH_3COOH	Si	0.5 6 60	etching at RT	75R2
19	Swirl (A, B)	Sirtl etch	Si		etching at 12···18 °C	73B1
20		Sirtl etch	Si			77G1
21	Micro-defects	Secco etch	Si			77T1
22	"S" pits	Secco etch	Si			77P2
23	Dislocations	CP-4: 3 HF, 5 HNO_3, 3 CH_3COOH with 0.06 Br_2	(100), (110), (111) Ge		yields conical pits	75R2
24		cyanide	(111) Ge	3···4		75R2
25		Dash etch: 1 HF, 3 HNO_3, IOCH_3COOH	Ge			75R2
26	Stacking faults	W Ag	Ge			75R2
27	Twins	CP-4; superoxal; aqueous solution of 10% KOH				75R2
28	Grain boundaries	CP-4 W Ag 1 HF, 1 HNO_3, 1 CH_3COOH	Ge Ge Ge	2 5 1.5	RT	75R2

References for 6.1.4.7.1···6.1.4.7.4

52H1 Heidenreich, R.D.: US Patent 2,619,414 (1952).
52N1 Navon, D., Bray, R., Fan, H.Y.: Proc. IRE **40** (1952) 1342.
53V1 Vogel, F.L., Pfann, W.G., Coney, H.E., Thomas, E.E.: Phys. Rev. **90** (1953) 489.
54K1 Mc Kelvey, J.P., Longini, R.L.: J. Appl. Phys. **25** (1954) 634.
55C1 Camp, P.R.: J. Electrochem. Soc. **102** (1955) 586.
56D1 Dash, W.C.: J. Appl. Phys. **27** (1956) 1193.
56F1 Fuller, C.S., Ditzenberger, J.A.: J. Appl. Phys. **27** (1956) 544.
56J1 Jensen, R.V., Christian, S.M.: RCA Industry Service, Lab. Bull. L 13–1023. Mar. 5 (1956).
56V1 Vogel, F.L., Clarice Lovell, L.: J. Appl. Phys. **27** (1956) 1413.
57B1 Battermann, B.W.: J. Appl. Phys. **28** (1957) 1236.
57G1 Geach, G.A., Irving, B.A., Pillips, R.: Research (London) **10** (1957) 411.
58A1 Allegheny Electric Chemical Co.: Tech. Bull. **6** (June 1958).
58F1 Feuerstein, W.J.: Trans. AIME **212** (1958) 210.
58H1 Harvey, W.W., Gatos, H.C.: J. Electrochem. Soc. **105** (1958) 654.
58W1 Wang, P.: Sylvania Technol. **11** (1958) 50.
59D1 Dewald, J.F.: Semiconductors (1959) 727.
59F1 Faust, J.W. jr.: Methods of Experimental Physics, Vol. 6, New York; Ch. z.B. Academic Press **1959**.
59R1 Robbins, H., Schwartz, B.: J. Electrochem. Soc. **106** (1959) 505, 1020.
59T1 Tegert, W.: The Electrolytic and Chemical Polishing of Metals, Oxford, 2nd ed. Pergamon **1959**.
60F1 Faust, J.W. jr.: The Surface Chemistry of Metals and Semiconductors, Wiley, New York, H.C. Gatos, ed. **1960**.
60H1 Hackerman, N.: The Surface Chemistry of Metals and Semiconductors, Wiley, New York H.C. Gatos, ed. **1960**.
60K1 King, C.V.: The Surface Chemistry of Metals and Semiconductors, Wiley, New York, H.C. Gatos, ed. **1960**.
60L1 Lacombe, P.: The Surface Chemistry of Metals and Semiconductors, Wiley, New York, H.C. Gatos, ed. **1960**.
60R1 Robbins, H., Schwartz, B.: J. Electrochem. Soc. **107** (1960) 108.
60T1 Turner, D.R.: J. Electrochem. Soc. **107** (1960) 810.
61C1 Chappey, M., Meritet, P.: Fr. Patent 1.266.612 (1961).
61E1 Eisenberg, M., Baumann, H.F., Brettner, D.M.: J. Electrochem. Soc. **108** (1961) 909.
61G1 Greer, W.N.: Plating (East Orange, N.J.) **48** (1961) 1095.
61S1 Sayers, J.R., Smit, J.: Plating (East Orange, N.J.) **48** (1961) 789.
61S2 Schwartz, B., Robbins, H.: J. Electrochem. Soc. **108** (1961) 365.
61S3 Sirtl, E., Adler, A.: Z. Metallkd. **52** (1961) 529.
61S4 Stead, R.R.: U.S. Patent 2,973,253 (1961).
62H1 Holms, P.J.: The Electrochemistry of Semiconductors, New York, Ch. 8. Academic Press **1962**.
62I1 Irving, B.A.: The Electrochemistry of Semiconductors, New York, Ch. 6 Academic Press **1962**.
62K1 Klein, D.L., D'Stefan, D.J.: J. Electrochem. Soc. **109** (1962) 37.
63C1 Chu, T.L., Gavaler, J.R.: J. Electrochem. Soc. **110** (1963) 388.
63E1 Efimov, E.A., Erusalimchik, I.G.: Electrochemistry of Germanium and Silicon, Washington, Sigma Press **1963**.
64P1 Pliskin, W.A., Gnall, R.P.: J. Electrochem. Soc. **111** (1964) 872.
65D1 Duffek, E.F., Pilling, D.: Electrochem. Soc. Extend. Abstr. **111** (1965) 244.
65G1 Gatos, H.C., Lavine, M.C.: Prog. Semicond, **9** (1965) 1.
65P1 Pliskin, W.A., Lehman, H.S.: J. Electrochem. Soc. **112** (1965) 103.
66B1 Barnett, G.D., Miller, A.: U.S. Patent 3,232,803 (1966).
66D1 Daltons, J.V.: The Electrochem. Soc. Meeting, Abstr. 23 (1966).
66E1 Eastman Kodak Co.: Rochester, N.Y.; pamphlet p–91 (1966).
66E2 Eversteijn, F.C.: Philips Res. Rep. **21** (1966) 379.
66F1 Frieser, R.G.: J. Electrochem. Soc. **113** (1966) 357.
66H1 Holmes, P.J., Snell, J.E.: Microelectron. Reliab. **5** (1966) 337.
66M1 Mai, C.C., Looney, J.C.: Semicond. Prod. Solid State Technol. **9(1)** (1966) 19.
66M2 Murray, L.A., Goldsmith, N.: J. Electrochem. Soc. **113** (1966) 1297.
66R1 Rogel, A.: Rev. Sci. Instr. **37** (1966) 1416.
66S1 Snow, E.H., Deal, B.E.: J. Electrochem. Soc. **113** (1966) 263.

66Z1 Zyetz, M.C., Despres, A.M.: Am Vac. Soc. Symp. 13th, Extend. Abstr. (1966) 169.
67A1 Aboaf, J.A.: J. Electrochem. Soc. **114** (1967) 948.
67B1 Bean, K.E.: J. Electrochem. Soc. **114** (1967) 733.
67B2 Brown, D.M., Engeler, W.E., Garfinkel, M., Heumann, F.K.: J. Electrochem. Soc. **114** (1967) 730.
67B3 Bogenschütz, A.F.: Aetzpraxis für Halbleiter, Carl Hauser Verlag, München **1967**.
67F1 Finne, R.M., Klein, D.L.: J. Electrochem. Soc. **114** (1967) 965.
67G1 Gelder, W. van, Hauser, V.E.: J. Electrochem. Soc. **114** (1967) 869.
67G2 Goldsmith, N., Kern, W.: RCA Rev. **28** (1967) 153.
67H1 Huber, F., Witt, W., Pratt, I.H.: Proc. Electron. Components Conf., **1967** 66.
67J1 Janus, A.R., Shirn, G.A.: J. Vac. Sci. Technol. **4** (1967) 37.
67K1 Kraitchman, J., Oroshnik, J.: J. Electrochem. Soc. **114** (1967) 405.
67K2 Kraitchman, J.: J. Appl. Phys. **38** (1967) 4323.
67M1 Myamlin, V.A., Pleskov, Y.V.: Electrochemistry of Semiconductors, New York **1967**.
67N1 Nishimatsu, S., Tokuyama, T.: Electrochem. Soc. Extend. Abstr. **170** (1967) 24.
67P1 Primak, W., Kampwirth, R., Dayal, Y.: J. Electrochem. Soc. **114** (1967) 88.
67S1 Schwartz, B.: J. Electrochem. Soc. **114** (1967) 285.
68A1 American Society for Testing and Materials: ASTM F80, Philadelphia, **1968**.
68B1 Brown, D.M., Gray, P.V., Herrmann, F.K., Philipp, H.R., Taft, E.A.: J. Electrochem. Soc. **115** (1968) 311.
68C1 Chu, T.L., Szedon, J.R., Gruber, G.A.: Trans. Metall. Soc. AIME **242** (1968) 532.
68D1 Deal, B.E., Fleming, P.J., Castro, P.L.: J. Electrochem. Soc. **115** (1968) 300.
68D2 Dirksc, T.P., De Witt, D., Shoemaker, R.: J. Electrochem. Soc. **115** (1968) 442.
68E1 Eldridge, J.M., Balk, P.: Trans. Metall. Soc. AIME **242** (1968) 539.
68G1 Ghandhi, S.K.: The Theory and Practice of Microelectronics, Ch. 7, New York, Wiley 1968.
68H1 Haller, I., Hatzakis, M., Srinivasan, R.: IBM J. Res. Develop. **12** (1968) 251.
68H2 Hammond, M.L., Bowers, G.M.: Trans. Metall. Soc. AIME **242** (1968) 546.
68K1 Kern, W.: RCA Rev. **29** (1968) 557.
68K2 Kern, W., Heim, R.C.: Electrochem. Soc. Extend. Abstr. **92** (1968) 234.
68P1 Pliskin, W.A.: Thin Solid Films **2** (1968) 1.
68S1 Schmidt, P.F., Gelder, W. van, Drobek, J.: J. Electrochem. Soc. **115** (1968) 79.
68S2 Shinoda, D.: Electrochem. Soc. Extend. Abstr. **502** (1968) 462.
68W1 Waits, R.K.: Trans. Metall. Soc. AIME **242** (1968) 490.
68W2 Woitsch, F.: Solid State Technol. **11(1)** (1968) 29.
69A1 Alexander, J.H., Joyce, R.J., Sterling, H.F.: Thin Films Dielectrics, New York, Electrochem. Soc. **1969**.
69B1 Balk, P., Eldridge, J.M.: Proc. IEEE **57** (1969) 1558.
69E1 Eastman Kodak Co.: Incidental Intelligence About Kodak Resists **7/1** (1969) 4.
69F1 Ferrien, E., Pruniaux, B.: J. Electrochem. Soc. **116** (1969) 1008.
69G1 Gregor, L.V.: Thin Film Dielectries, New York, Electrochem. Soc. **1969**.
69G2 Grossman, J., Herman, D.S.: J. Electrochem. Soc. **116** (1969) 674.
69H1 Hashimoto, N., Koga, Y., Yamada, E.: Thin Film Dielectrics, New York, Electrochem. Soc. **1969**.
69H2 Hill, B.H.: J. Electrochem. Soc. **116** (1969) 668.
69H3 Hoffmeister, W.: Intern. J. Appl. Radiat. Isot. **2** (1969) 139.
69K1 Kern, W.: J. Electrochem. Soc. **116** (1969) 251 C.
69K2 Koga, Y., Matsushita, M., Kobayashi, K., Nakaido, Y., Toyoshima, S.: Thin Film Dielectrics, New York, Electrochem. Soc. 1969.
69K3 Kuwano, Y.: Jpn. J. Appl. Phys. **8** (1969) 876.
69P1 Powers, R.W., Breiter, M.W.: J. Electrochem. Soc. **116** (1969) 719.
69S1 Swaroop, B.: Thin Film Dielectrics, New York, Electrochem. Soc. 1969.
69T1 Tokuyama, T., Miyazaki, T., Horiuchi, M.: Thin Film Dielectrics, New York, Electrochem. Soc. **1969**.
69V1 Vratny, F.: Thin Film Dielectrics, New York, Electrochem. Soc. **1969**.
70A1 Armstrong, R.D., Bulman, G.M.: J. Electroanal. Chem. **25** (1970) 121.
70B1 Barry, M.L.: Chemical Vapor Deposition, New York, Electrochem. Soc. 1970.
70D1 Duffy, M.T., Kern, W.: RCA Rev. **31** (1970) 754.
70E1 El-Hoshy, A.H.: J. Electrochem. Soc. **117** (1970) 1583.
70G1 Glang, R., Gregor, L.V.: Handbook of Thin Film Technology, New York, McGraw-Hill **1970**.
70K1 Kane, P.F., Larrabee, G.B.: Characterisation of Semiconductor Materials, New York, McGraw-Hill **1970**.
70K2 Kern, W., Fischer, A.W.: RCA Rev. **31** (1970) 715.

70K3 Kern, W., Heim, R.C.: J. Electrochem. Soc. **117** (1970) 562.
70K4 Kern, W., Heim, R.C.: J. Electrochem. Soc. **117** (1970) 568.
70K5 Kern, W., Pnotinen, D.: RCA Rev. **31** (1970) 187.
70P1 Pliskin, W.A., Zanin, S.J.: Handbook of Thin Film Technology, New York, Mc Graw-Hill **1970**.
70R1 Ryan, R.J., Davidson, E.B., Hook, H.O.: Handbook of Materials and Processes for Electronics, New York, Mc Graw-Hill **1970**.
70S1 Shehigolev, P.V.: Electrolytic and Chemical Polishing of Metals, Holon, Israel, Freund. Publ. House **1970**.
70S2 Stoller, A.I.: RCA Rev. **31** (1970) 271.
70T1 Thennissen, M.J.J., Apples, J.A., Verkuylen, W.H.C.G.: J. Electrochem. Soc. **117** (1970) 959.
70T2 Tung, S.K., Caffrey, R.E.: J. Electrochem. Soc. **117** (1970) 91.
71B1 Brown, D.M., Cady, W.R., Spragne, J.W., Salvagni, P.J.: IEEE Trans. Electron Devices **ED-18** (1971) 931.
71B2 Brown, D.M., Kennicott, R.P.: J. Electrochem. Soc. **118** (1971) 293.
71H1 Hall, L.: J. Electrochem. Soc. **118** (1971) 1506.
71H2 Herman, D.S., Schuster, M.A., Oeler, H.G.: Electrochem. Soc. Extend Abstr. **71-1** (1971) 167.
71J1 Johnson, J.W., Wu, C.L.: J. Electrochem. Soc. **118** (1971) 1909.
71J2 Judge, J.S.: J. Electrochem. Soc. **118** (1971) 1772.
71K1 Katto, H., Koga, Y.: J. Electrochem. Soc. **118** (1971) 1619.
71K2 Kern, W., Shaw, J.M.: J. Electrochem. Soc. **118** (1971) 1699.
71M1 Mac Kenna, E.L.: Proc. Semicond. IIC Proc. Conf. Chicago, Illinois, Ind. Sci. Conf. Manage. **1971**.
71M2 Meek, R.L.: J. Electrochem. Soc. **118** (1971) 1240.
71M3 Milek, J.T.: Silicon Nitride for Microelectronic Application, IFI, New York, Plenum **1971**.
71M4 Milek, J.T.: Silicon Nitride for Microelectronic Application, pp. 40–53, IFI, New York, Plenum, **1971**.
71S1 Schwenker, R.O.: J. Electrochem. Soc. **118** (1971) 313.
71T1 Taft, E.A.: J. Electrochem. Soc. **118** (1971) 1341.
71W1 Wolf, H.F.: Semiconductors, New York, Wiley (Interscience) **1971**.
72G1 Gereth, R., Scherber, W.: J. Electrochem. Soc. **119** (1972) 1248.
72J1 Janus, A.R.: J. Electrochem. Soc. **119** (1972) 392.
72J2 Johnston, H.K.: Larson, T.L.: U.S. Patent 3,702,273 (1972).
72K1 Kubota, T.: Jpn. J. Appl. Phys. **11** (1972) 1413.
72P1 Pugacz-Muraszkiewicz, I.J.: IBM J. Res. Develop., Sept. **1972** 523.
72S1 Secco d'Aragona, F.: J. Electrochem. Soc. **119** (1972) 948.
72U1 Unvala, B.A., Holt, D.B., San, A.: J. Electrochem. Soc. **119** (1972) 318.
72W1 Wen, C.P., Weller, K.P.: J. Electrochem. Soc. **119** (1972) 547.
72W2 Wohlheiter, V.D., Whitner, R.A.: J. Electrochem. Soc. **119** (1972) 945.
72W3 Wong, J.: J. Electrochem. **119** (1972) 1071.
73B1 Bernewitz, L.I., Mayer, K.R.: Phys. Status Solidi **16** (1973) 579.
73C1 Clemens, D.P.: Electrochem. Soc. Ext. Abstr. **73-2** (1973) 407.
73D1 Dell'Oca, C.J.: J. Electrochem. Soc. **120** (1973) 1225.
73G1 Ghezzo, M., Brown, D.M.: J. Electrochem. Soc. **120** (1973) 110.
73H1 Harrap, V.: Semiconductor Silicon 1973, Princeton, New Yersey, Electrochem. Soc. **1973**.
73H2 Herring, R., Price, J.B.: Electrochem. Soc. Extend. Abstr. **73-2** (1973) 410.
73L1 Learn, A.J.: J. Appl. Physics **44** (1973) 1251.
73M1 Muraoka, H., Obhashi, T., Sumitomo, Y.: Semiconductor Silicon 1973, Princeton, New Yersey, Electrochem. Soc. **1973**.
73N1 Naraoka, K., Maeda, M.: Jpn. Patent 7308,706 (1973).
73P1 Price, J.B.: Semiconductor Silicon 1973, Princeton, New Yersey, Electrochem. Soc. **1973**.
73R1 Rand, M.J., Roberts, J.F.: J. Electrochem. Soc. **120** (1973) 446.
73R2 Rand, M.J., Roberts, J.F.: Electrochem. Soc. Ext. Abstr. **73-2** (1973) 432.
73R3 Rankel Planger, L.: J. Electrochem. Soc. **120** (1973) 1428.
73S1 Schwettmann, F.N., Dexter, R.J., Cole, D.F.: J. Electrochem. Soc. **120** (1973) 1566.
73S2 Sumitomo, Y., Niwa, K., Sawazaki, H., Sakai, K.: Semiconductor Silicon 1973, Princeton, New Yersey, Electrochem. Soc. **1973**.
73T1 Tenney, A.S., Ghezzo, M.: J. Electrochem. Soc. **120** (1973) 1091.
73V1 Vijh, A.K.: Electrochemistry of Metals and Semiconductors, New York, Decker, **1973**.
73Z1 Zwicker, W.K., Kurtz, S.K.: Semiconductor Silicon 1973, Princeton, New Jersey, Electrochem. Soc. **1973**.

74B1 Baba, K.: Jpn. Patent 7447,225 (1974).
74C1 Chang, S.S.: US Patent 3,784,424 (1974).
74G1 Gajda, J.J.: Annu. Proc. Reliab. Physics, 12th (1974) 30.
74H1 Harman, T.C., Melngailis, I.: Appl. Solid State Sci **4** (1974) 1.
74K1 Kennedy, T.N.: Electron Packag. Prod. **14** (12) (1974) 136.
74K2 Kirk, R.W.: Techniques and Applications of Plasma Chemistry, New York, Wiley **1974**.
74K3 Kirk, R.W.: Techniques and Applications of Plasma Chemistry, pp. 362–374, Wiley, New York 1974.
74O1 Ohno, Y., Hatsuoka, T., Miyaji, K.: Jap. Pat. 7411, 3737 (1974).
74O2 Okamato, F.: Jpn. J. Appl. Phys. **13** (1974) 383.
74R1 Raetzel, C., Schild, S., Schlötterer, H.: Electrochem. Soc. Ext. Abstr. **74-2** (1974) 336.
74R2 Rand, M.J., Roberts, J.F.: Appl. Phys. Lett. **24** (1974) 49.
75A1 Agatsuma, T., Kikuchi, A., Nabada, K., Tomozawa, A.: J. Electrochem. Soc. **122** (1975) 825.
75B1 Baraona, C.R., Brandhorst, H.W.: IEEE Photovoltaic Spec. Conf. Proc., Scottsdale (1975) 44.
75C1 Colom, L.A., Levine, H.A.: U.S. Patent 3,639,185 (1975).
75D1 Day, H.M., Christen, A., Weisenberger, W.H., Hervonen, J.K.: J. Electrochem. Soc. **122** (1975) 769.
75D2 Declerg, M.J., Gerzberg, L., Meindl, J.D.: J. Electrochem. Soc. **122** (1975) 545.
75G1 Gaind, A.K., Ackermann, G.K., Lucarini, V.J., Bratter, R.L.: J. Electrochem. Soc. **122** (1975) 573.
75G2 Grantham, D.H., Swindal, J.: Int. Microelectron. Symp., Alabama, Int. Soc. Hybrid Microelectron. Montgomery **1975**.
75H1 Hirayama, M., Shono, K.: J. Electrochem. Soc. **122** (1975) 1671.
75K1 Kelly, J.J., Minjer, C.H. de: J. Electrochem. Soc. **122** (1975) 931.
75K2 Kominiak, G.J.: J. Electrochem. Soc. **122** (1975) 1272.
75K3 Kroll, W.J., Titus, R.L., Wagner, J.B. jr.: J. Electrochem. Soc. **122** (1975) 573.
75M1 Mogab, C.J., Petroff, P.M., Sheng, T.T.: J. Electrochem. Soc. **122** (1975) 815.
75M2 Mutoh, M., Mizokami, Y., Matsui, H., Hagiwara, S., Ino, M.: J. Electrochem. Soc. **122** (1975) 987.
75R1 Rand, M.J.: J. Electrochem. Soc. **122** (1975) 811.
75R2 Runyan, W.R.: Semiconductor Measurements and Instrumentation, New York, Mc Graw-Hill 1975.
75R3 Runyan, W.R.: Semiconductor Measurement and Instrumentation, Ch. 7, New York, McGraw-Hill 1975.
75S1 Schmidt, C.J., Lenzo, P.V., Spencer, E.G.: J. Appl. Phys. **48** (1975) 4080.
75S2 Schnable, G.A., Kern, W., Comizzoli, R.B.: J. Electrochem. Soc. **122** (1975) 1092.
75S3 Schwartz, G.C., Platter, V.: J. Electrochem. Soc. **122** (1975) 1508.
75S4 Shankoff, T.A., Chandross, E.A.: J. Electrochem. Soc. **122** (1975) 294.
75T1 Takamura, T., Kihara-Morishata, H.: J. Electrochem. Soc. **122** (1975) 386.
75W1 Wohlheiter, V.D.: J. Electrochem. Soc. **122** (1975) 1736.
76A1 Asano, M., Cho, T., Muraoka, H.: Electrochem. Soc. Ext. Abstr. **76-2** (1976) 911.
76B1 Bernardy, R. de, Donaghey, L.F.: Electrochem. Soc. Ext. Abstr. **76-2** (1976) 648.
76C1 Chappelow, R.E., Lin, P.T.: J. Electrochem. Soc. **123** (1976) 913.
76C2 Choo, Y.H., Devereux, O.F.: J. Electrochem. Soc. **123** (1976) 1868.
76F1 Frankenthal, R.F., Eaton, D.H.: J. Electrochem. Soc. **123** (1976) 703.
76H1 Harada, H., Satoh, S., Yoshida, M.: IEEE Trans. Reliab. **R-25** (1976) 290.
76H2 Huang, P.C., Schaible, P.M.: Electrochem. Soc. Ext. Abstr. **76-2** (1976) 754.
76H3 Hughes, H.G., Rand, M.J.: Etching for Pattern Definition, Princeton, New Yersey, Electrochem. Soc. 1976.
76J1 Judge, J.S.: Etching for Pattern Definition, Princeton, New Yersey, Electrochem. Soc. 1976.
76K1 Katz, L.E., Erdman, W.C.: J. Electrochem. Soc. **123** (1976) 1249.
76K2 Kern, W.: RCA Rev. **37** (1976) 78.
76K3 Kern, W.: RCA Rev. **37** (1976) 55.
76K4 Kern, W.: Electrochem. Soc. Ext. Abstr. **76-1** (1976) 119.
76K5 Kern, W., Schnable, G.L., Fischer, A.W.: RCA Rev. **37** (1976) 3.
76K6 Kern, W.: Etching for Pattern Definition, Princeton, New Yersey, Electrochem. Soc. **1976**.
76M1 Mac Arthur, D.: Etching for Pattern Definition, Princeton, New Yersey, Electrochem. Soc. **1976**.
76P1 Pliskin, W.A., Esch, R.P.: Etching for Pattern Definition, Princeton, New Yersey, Electrochem. Soc. **1976**.
76P2 Pryor, H.J., Stachle, R.W.: Treatise, on Solid State Chemistry 4, New York, Plenum **1976**.
76S1 Schlesier, K.M., Shaw, J.M., Benyon, C.W. jr.: RCA Rev. **37** (1976) 358.
76S2 Schwartz, B., Robbins, H.: J. Electrochem. Soc. **123** (1976) 1903.
76S3 Shivaraman, M.S., Svensson, C.M.: J. Electrochem. Soc. **123** (1976) 1258.
76S4 Smithells, C.J.: Metal Reference Book, London, 5th ed. Butterworth **1976**.

76S5 Stein, H.J.: J. Electron. Mater. **5** (1976) 161.
76S6 Stephens, A.W., Vossen, J.L., Kern, W.: J. Electrochem. Soc. **123** (1976) 303.
77C1 Chow, K., Garrison, L.G.: J. Electrochem. Soc. **124** (1977) 1133.
77G1 Graff, K., Hilgarth, J., Neubrandt, H.: Semiconductor Silicon 1977, p. 575, Princeton, Electrochem. Soc. 1977.
77I1 Iida, K.: J. Electrochem. Soc. **124** (1977) 614.
77J1 Jinno, K., Kinoshita, H., Matsumoto, Y.: J. Electrochem. Soc. **124** (1977) 1258.
77K1 Kelsey, G.S.: J. Electrochem. Soc. **124** (1977) 814.
77K2 Kern, W., Roster, R.S.: J. Vac. Sci. Technol. **14** (1977) 1082.
77K3 Kern, W., Comizzdi, R.B.: J. Vac. Sci. Technol. **14** (1977) 32.
77K4 Kuhn, A.T., Shalaby, H., Wakeman, D.W.: Corros. Sci. **17** (1977) 833.
77L1 Lanford, W.A., Rand, M.J.: Electrochem. Soc. Ext. Abstr. **77-2** (1977) 421.
77N1 Nowicki, R.S.: J. Vac. Sci. Technol. **14** (1977) 127.
77P1 Pugacz-Muraszkiewicz, I.J., Hammond, B.R.: J. Vac. Sci. Techn. **14** (1977) 49.
77P2 Petroff, P.M., Katz, L.E., Savage, A.: Semiconductor Silicon 1977, p. 761, Electrochem. Soc. Princeton, 1977.
77R1 Rand, M.J.: Electrochem. Soc. Ext. Abstr. **77-2** (1977) 419.
77T1 Tempelhoff, K., Spiegelberg, F., Gleichmann, R.: Semiconductor Silicon 1977, p. 585, Princeton, Electrochem. Soc. **1977**.
78B1 Bean, K.E.: IEEE Transact. Electr. Devices **ED-25** (1978) 1185.
78D1 Deckert, C.A.: J. Electrochem. Soc. **124** (1978) 320.
78K1 Kern, W.: RCA Rev. **39** (1978) 278.
78K2 Kern, W., Deckert, Ch.: Thin Film Processes, Ch. V-1, New York, Academic Press, **1978**.
78K3 Kern, W., Deckert, Ch.: Thin Film Processes, pp. 417–421, New York, Academic Press **1978**.
78M1 Mochizuki, T., Shibata, K., Inone, T., Ohuchi, K.: Jap. J. Appl. Phys. **17** (1978) 37.
79C1 Crowder, A.L., Zirinsky, S.: IEEE J. Solid State Circuits **SC-14** (1979) 291.
79S1 Saraswat, K.C., Mohammadi, F., Meindl, J.D.: IEDM Technical Digest (1979) 462.
79S2 Schimmel, D.G.: J. Electrochem. Soc. **126** (1979) 479.
80D1 Deckert, Ch.A.: J. Electrochem. Soc. **127** (1980) 2433.
80H1 Hasler, B.: Personal Communication, **1980**.
80K1 Koller, A.: Personal Communication, 1980.
80M1 Murarka, S.P., Fraser, D.B., Sinha, A.K., Levinstein, H.J.: IEEE Trans. Electron. Dev. **ED-27** (1980) 1409.
80N1 Nozaki, T.: European Patent 0010910, (1980).

6.1.4.7.5 Dry etching

Dry etching implies methods by which a substrate surface is etched either physically by ion bombardement, or chemically by a chemical reaction of reactive species generated in a plasma at the surface, or both physically and chemically by an ion-, electron-, or photon-induced chemical reaction at the surface. Most dry etching methods are also known as "*plasma-assisted*" *etching* [78M1]. The various dry etching methods are compared in Table 1.

In *physical etching*, ions are accelerated in an electric field towards the substrate to be etched. The impinging ions, neutral atoms, or neutral molecules erode the surface by momentum transfer. Physical etching techniques are *ion beam etching*, *ion beam milling*, *sputter etching*, and *rf sputter etching*.

In *chemical dry etching*, reactive species generated in a plasma are applied to the substrate to be etched. Etching is affected by a chemical reaction of the reactive species with atoms at the substrate surface. The reaction product has to be a volatile gas, which is removed by a vacuum pump system. Chemical etching techniques are *chemical dry etching* and *plasma etching* in a barrel reactor.

In *physical-chemical dry etching*, the chemical reaction at the surface is induced by impinging ions, electrons, or photons. The reaction product also has to be a volatile gas. Ion-induced chemical dry etching is used for the fabrication of integrated circuits of high packing density. Physical-chemical dry etching techniques are: *plasma etching* in a parallel-plate reactor, *reactive ion* or *reactive sputter etching*, *triode plasma etching*, *reactive ion beam etching* or *reactive ion beam milling*, and *electron- or photon-induced chemical dry etching*.

Table 1. Comparison of dry etching methods.

	Dry etching		
	Ion etching	Reactive etching	
Method	Physical	Chemical	Physical-chemical
Techniques	ion beam etching ion beam milling sputter etching rf sputter etching	plasma etching in a barrel reactor chemical dry etching	plasma etching in a parallel-plate reactor triode plasma etching reactive ion etching - reactive sputter etching reactive ion beam etching - reactive ion beam milling electron-induced chemical dry etching photon-induced chemical dry etching
Principle	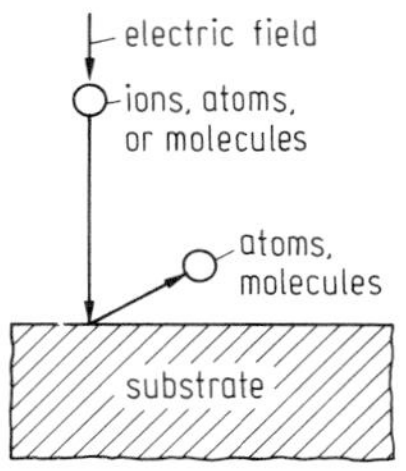 electric field ions, atoms, or molecules atoms, molecules substrate	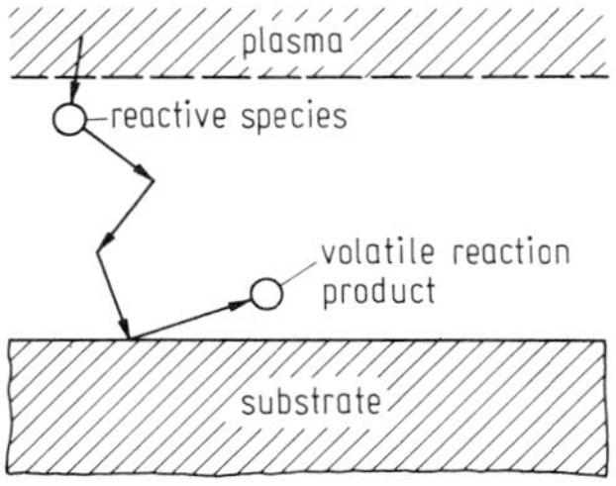 plasma reactive species volatile reaction product substrate	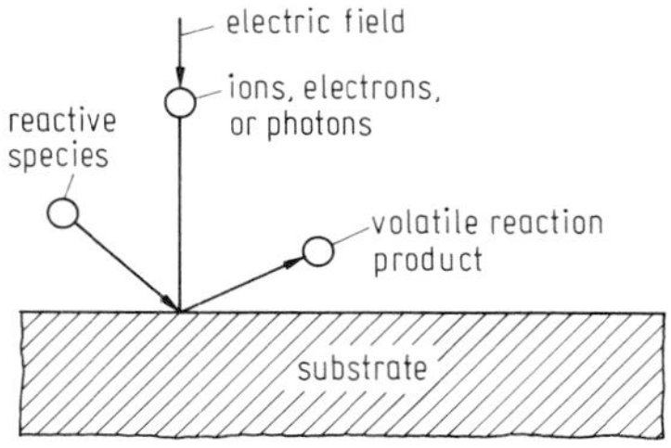 electric field reactive species ions, electrons, or photons volatile reaction product substrate
Etching affected by	momentum transfer of impinging ions, atoms, or molecules.	generation of reactive species in a plasma. motion of reactive species towards substrate to be etched. chemical reaction of reactive species with atoms at substrate surface. desorption of volatile reaction product.	chemical reaction between reactive species and atoms at substrate surface induced by ions, electrons, or photons. desorption of volatile reaction product. the reactive species may be the ion itself or added to the ambient.

In semiconductor device technology, dry etching is more and more used for pattern generation in thin films. With the help of the appropriate lithography techniques described in section 6.1.4.6, the resist pattern is formed in a thin film deposited on a substrate. This resist pattern acts as a mask for the subsequent etching, by which the resist pattern is transferred to the film below the mask. As discussed in section 6.1.4.7.3, an etch process for pattern generation in thin films can be described by the etch factor f_e, for the undercutting, by the etch rate r_e, and by the selectivity S_e.

An ideal etch process for the generation of a very fine pattern in a thin film has to feature a high etch factor for the accurate transfer of the etch mask to the layer below, a high selectivity with respect to the substrate and the mask, and a high etch rate because of economic reasons.

Table 2 shows typical edge profiles produced by various dry etching methods. Physically etched films show no undercutting of the etch mask, however, "trenching" at the boundary of patterns occurs as a result of the forward elastic reflection of incident particles at the slopes.

Most chemical dry etching processes are isotropic and therefore lead to undercutting of the etch mask.

Edge profiles without undercutting and without trenching can be produced by the physical-chemical dry etching. This etching makes possible the accurate transfer of the mask pattern to the film below the mask as required in the generation of fine patterns for integrated circuits of high packing density.

Table 2. Typical edge profiles produced by dry etching methods.

	Ion etching	Reactive dry etching	
Method	Physical	Chemical	Physical-chemical
Techniques	ion beam etching ion beam milling sputter etching rf sputter etching	plasma etching in barrel reactor chemical dry etching	plasma etching in parallel-plate reactor triode plasma etching reactive ion etching - reactive sputter etching reactive ion beam etching - reactive ion beam milling electron-induced chemical dry etching photon-induced chemical dry etching

continued

Table 2 (continued)

Method	Ion etching	Reactive dry etching	
	Physical	Chemical	Physical - chemical
Edge profile	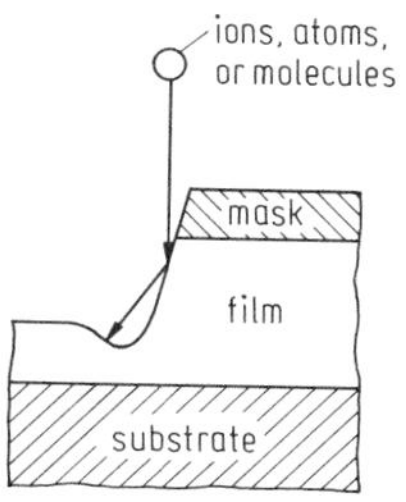 ions, atoms, or molecules; mask; film; substrate	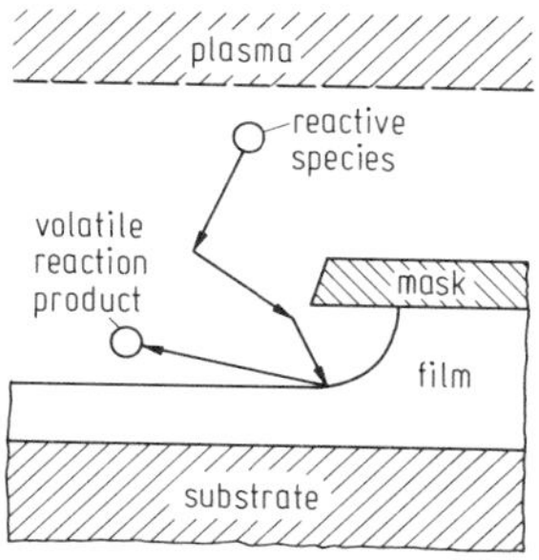 plasma; reactive species; volatile reaction product; mask; film; substrate	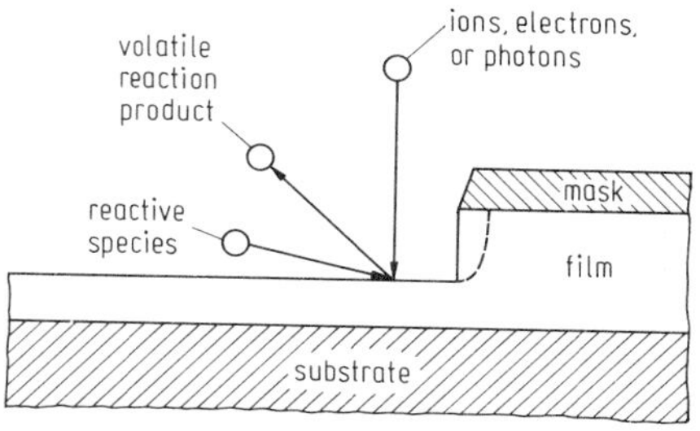 volatile reaction product; ions, electrons, or photons; reactive species; mask; film; substrate

6.1.4.7.6 Ion etching processes

Ion etching can be described as a process by which atoms are removed from a substrate surface (target) by bombardment with energetic ions, neutral atoms, or neutral molecules. Etching results from the momentum transfer between the impinging ions, atoms, or molecules and the atoms of the substrate. A fraction of the energy transferred of an impinging particle is returned from the surface by a particle having a momentum of opposite direction. This fraction of energy must be greater than the lattice binding energy of a surface atom so that sputtering occurs.

Two types of ion etching are generally used; *ion beam etching*, also called *ion beam milling*, where the ion beam is generated in a plasma source separate from the substrate being etched, and *rf sputter etching*, where the substrates are placed on the cathode of a parallel-plate rf discharge.

Ion etching is described in numerous books, reviews and treatises [1852G1, 55W1, 62A1, 61A3, 61R1, 65K1, 67M4, 68C1, 68V1, 69S1, 70J1, 70W1, 72V1, 73O1, 75O1, 76D1, 76M1, 77W1, 78M1, 78V1, 80C1, 80C2, 80C8].

Sputtering yield

The most important material property in ion etching is the sputtering yield, η_s, which is the ratio of the number of sputtered atoms to the number of incident ions, or molecules. The sputtering yield depends on the bombarding atoms and their energy, the material being etched, the angle of incidence, the temperature, and the pressure and composition of the background gas.

Theoretical studies [68A1, 68C2, 68G1, 68H1, 68M1, 68S1, 69S1] lead to the following expression for the sputtering yield, η_s, of a planar target [70W1]:

$$\eta_s = \frac{\alpha S_n(E)}{16\pi^3 a^2 U_0},$$

where α is a function of (M_2/M_1), M_1 is the mass of the target atoms (atoms being etched), M_2 is the mass of incident particles, $S_n(E)$ is the nuclear stopping cross section of the ions at the energy E, $a = 0.0219$ nm is the screening radius proposed by Anderson and Sigmund [65A1] for the Born-Mayer interaction of two atoms, and U_0 is the surface barrier energy, which is taken as the heat of sublimation H_s. For low ion energies (0···1000 eV), the nuclear stopping cross section can be expressed by

$$S_n(E) = 12\pi a^2 \lambda E,$$

where $\lambda = 4\,M_1 M_2/(M_1 + M_2)^2$ is the energy transfer coefficient. The sputtering yield for low ion energies is given by

$$\eta_s = \frac{3\alpha\lambda E}{4\pi^2 H_s}.$$

The sputtering yield at low energies is proportional to the bombarding ion energy E and inversely proportional to the heat of sublimation H_s. At high energies the sputtering yield can be written as [68L1]:

$$\eta_s(E) = 0.042\frac{\alpha S_n(E)}{H_s}.$$

Values of α and $S_n(E)$ are given in ref. [68L1].

Sputtering yield as a function of material sputtered

The inverse heat of sublimation H_s is given as a function of the element atomic number in Fig. 2. [71S1, 76M1]. The sputtering yield η_s observed experimentally by using argon ions is shown in Fig. 3 as a function of the element atomic number [61L1].

The correlation of the sputtering yield η_s and the inverse heat of sublimation (H_s^{-1}) is seen by comparing Figs. 2 and 3.

Both the sputtering yield and the inverse heat of sublimation show a strong periodicity related to the degree of occupation of the 3d, 4d, and 5d electron shells.

Sputtering yield as a function of the atomic number of the bombarding ions

The sputtering yield of silver, copper, and tantalum is shown in Fig. 4 as a function of the atomic number of the bombarding ions [61A1, 61A2]. The sputtering yield varies periodically with the atomic number of the ions similar to the variation with the element atomic number shown in Fig. 3. Noble gas ions give a high yield while elements from the center columns of the periodic table show a low yield. The sputtering yield increases with the atomic number of the ions. At energies lower than 1 keV, the inert gas ions all produce similar yields [69S1].

Sputtering yield as a function of ion energy

The sputtering yield for copper bombarded with Xe, Kr, Ar, and Ne is shown in Fig. 5 as a function of the ion energy [69S1].

For energies less than 1000 eV, the sputtering yield is proportional to the ion energy in accordance with theory. In the energy range $10^4 \cdots 10^6$ eV, the yield reaches a maximum and decreases with increasing energy in accordance with the high-energy approximations.

Sputtering yield as a function of angle of incidence

The sputtering yield depends on the angle of incidence as shown in Fig. 6. With increasing angle of incidence, the sputtering yield of most materials initially increases, reaching a maximum at angles in the range of 40°···60°, and decreases at large angles. The increase in sputtering yield with increasing angle can be visualized by a billiard-game model. The more oblique the ion incidence the smaller is the directional change in momentum required for ejecting atoms in a forward direction. The reduction in sputtering yield observed at high angles of incidence is due to the increased probability of elastic reflection of the incoming ions. The sputtering yield as a function of the angle of incidence has been investigated experimentally [73O1, 75G1, 75O1, 76D1, 77C1] and theoretically [69S1, 75O1]. Fig. 6 shows the sputtering yield of gold, aluminum, and photoresist, Fig. 7 the yield of aluminum, titanium, tantalum, and silver, and Fig. 8 the yield of SiO_2 and photoresist.

A survey of the sputtering yield obtained by using various gases to sputter elemental targets is given in Tables 3···6. All the target materials sputtered are either polycrystalline or amorphous.

Table 3. Sputtering yield η_s of various elements [78V1]. Ion energy $E = 500$ eV.

Target material	η_s[atoms/ion]					Ref.
	He^+	Ne^+	Ar^+	Kr^+	Xe^+	
Be	0.24	0.42	0.51	0.48	0.35	62W1
C	0.07		0.12	0.13	0.17	
Al	0.16	0.73	1.05	0.96	0.82	
Si	0.13	0.48	0.50	0.50	0.42	
Ti	0.07	0.43	0.51	0.48	0.43	
V	0.06	0.48	0.65	0.62	0.63	
Cr	0.17	0.99	1.18	1.39	1.55	
Mn				1.39	1.43	
Mn			1.90			78V1
Bi			6.64			78V1
Fe	0.15	0.88	1.10	1.07	1.00	62W1
Fe		0.63	0.84	0.77	0.88	61W1
Co	0.13	0.90	1.22	1.08	1.08	62W1

continued

Table 3 (continued)

Target material	η_s [atoms/ion]					Ref.
	He⁺	Ne⁺	Ar⁺	Kr⁺	Xe⁺	
Ni	0.16	1.10	1.45	1.30	1.22	
Ni		0.99	1.33	1.06	1.22	61W1
Cu	0.24	1.80	2.35	2.35	2.05	62W1
Cu		1.35	2.0	1.91	1.91	61W1
Cu (111)		2.1		2.50	3.9	62M1
Cu			1.2			55K1
Ge	0.08	0.68	1.1	1.12	1.04	62W1
Y	0.05	0.46	0.68	0.66	0.48	
Zr	0.02	0.38	0.65	0.51	0.58	
Nb	0.03	0.33	0.60	0.55	0.53	
Mo	0.03	0.48	0.80	0.87	0.87	
Mo		0.24	0.64	0.59	0.72	61W1
Ru		0.57	1.15	1.27	1.20	62W1
Rh	0.06	0.70	1.30	1.43	1.38	
Pd	0.13	1.15	2.08	2.22	2.23	
Ag	0.20	1.77	3.12	3.27	3.32	
Ag	1.0	1.70	2.4	3.1		55K1
Ag			3.06			66C1
Sm	0.05	0.69	0.80	1.09	1.28	62W1
Gd	0.03	0.48	0.83	1.12	1.20	
Dy	0.03	0.55	0.88	1.15	1.29	
Er	0.03	0.52	0.77	1.07	1.07	
Hf	0.01	0.32	0.70	0.80		
Ta	0.01	0.28	0.57	0.87	0.88	
W	0.01	0.28	0.57	0.91	1.01	
Re	0.01	0.37	0.87	1.25		
Os	0.01	0.37	0.87	1.27	1.33	
Ir	0.01	0.43	1.01	1.35	1.56	
Pt	0.03	0.63	1.40	1.82	1.93	
Au	0.07	1.08	2.40	3.06	3.01	
Au	0.10	1.3	2.5		7.7	61M1
Pb	1.1		2.7			55K1
Th	0.0	0.28	0.62	0.96	1.05	62W1
U		0.45	0.85	1.30	0.81	
Sb			2.83			78V1
Sn (solid)			1.2			70K1
Sn (liquid)			1.4			

Table 4. Sputtering yield η_s of various elements [78V1, 70M1, 77W1]. Ion energy $E = 1$ keV.

Target material	η_s [atoms/ion]							Ref.
	He⁺	N⁺	Ne⁺	N_2^+	Ar⁺	Kr⁺	Xe⁺	
Be					1.1			75O1
Al					1.9			
Si					1.0			
Si					0.6			61L1, 63S1
Ti					1.1			75O1
Fe					1.3			
Fe			0.85		1.33	1.42	1.82	61W1

continued

Table 4 (continued)

Target material	η_s [atoms/ion]							Ref.
	He^+	N^+	Ne^+	N_2^+	Ar^+	Kr^+	Xe^+	
Fe		0.55		0.78				65K1
Ni					2.1			61L1, 61W2
Ni					2.2			75O1
Ni			1.22		2.21	1.76	2.26	61W1
Ni		0.74		1.05				65K1
Cu					3.2			60Y1, 61L1, 61R1, 63S1
Cu					3.6			75O1
Cu			1.88		2.85	3.42	3.6	61W1
Cu		1.5						65K1
Cu					3.2	2.5		55K1
Cu				1.95				64S1
Ge					1.6			75O1
Ge					1.5			61L1, 63S1
Zr					1.1			75O1
Nb					1.0			
Mo					1.1			
Mo					1.1			61L1, 61P1, 61W2, 68K1
Mo			0.49		1.13	1.27	1.60	61W1
Mo		0.16		0.3				65K1
Pd					3.1			75O1
Ag					4.7			
Ag	1.8		2.4		3.8	4.7		62M1
Cd					11.2			75O1
Sn					0.8			62P1
Ta					0.9			75O1
W					1.1			
W		0.18		0.2				65K1
Pt					2.0			75O1
Au					4.0			
Au					1.0			62P1
Au	0.3		2.1		4.9			61M1
Au					3.6			67R1
Pb	1.5				3.0			55K1
Pb					4.2			75O1
Al_2O_3					0.04			67D1
SiO_2					0.13			

Table 5. Sputtering yield η_s of various elements by argon bombardement at 200 eV and 600 eV [70M1].

Target material	η_s [atoms/ion]		Ref.
	$E=200$ eV	$E=600$ eV	
Ag	1.6	3.4	61L1, 61A1
Al	0.35	1.2	61L1, 65C1
Au	1.1	2.8	61L1, 68R1
Co	0.6	1.4	61L1
Cr	0.7	1.3	
Cu	1.1	2.3	61L1, 60Y1, 61R1, 63S1
Fe	0.5	1.3	61L1, 61W2
Ge	0.5	1.2	61L1, 63S1
Mo	0.4	0.9	61L1, 61P1, 61W2, 68K1
Nb	0.25	0.65	61L1

continued

Table 5 (continued)

Target material	η_s [atoms/ion]		Ref.
	$E=200$ eV	$E=600$ eV	
Ni	0.7	1.5	61L1, 61W2
Os	0.4	0.95	61L1
Pd	1.0	2.4	
Pt	0.6	1.6	
Re	0.4	0.9	
Rh	0.55	1.5	
Si	0.2	0.5	61L1, 63S1
Ta	0.3	0.6	61L1
Th	0.3	0.7	
Ti	0.2	0.6	61L1, 68K1
W	0.3	0.6	61L1, 65C1
Zr	0.3	0.75	61L1

Table 6. Miscellaneous sputtering yield values η_s [70M1, 78V1].

Target material	Bombarding ion	E keV	η_s $\left[\frac{\text{atoms}}{\text{ion}}\right]$	Ref.
Fe	Ar	10	1.0	62P1
Cu	Ne	10	3.2	60R1, 61A3
Cu	Ar	10	6.6	60Y1, 61L1, 61R1, 63S1
Cu	Ar	10	6.25	60R1, 61A3
Cu	Kr	10	8.0	61A3
Cu	Xe	10	10.2	61A3
Ag	Ar	10	8.8	61L1, 61A1
Ag	Ar	10	10.4	59G1
Ag	Kr	10	14.8	59G1
Ag	Xe	10	15.5	60R1, 61A3
Au	Ne	10	3.7	61A3
Au	Ar	10	8.5	61A3, 62P1
Au	Kr	10	14.6	61A3
Au	Xe	10	20.3	61A3
Sn	Ar	10	2.1	62P1
Mo	Ar	10	2.2	61L1, 61P1, 61W1, 68K1
Ti	Ar	10	2.1	61L1, 68K1
Al	Ar	5	2.0	61L1, 65C1
Au	Ar	5	7.9	61L1, 68R1
Cu	Ar	5	5.5	61L1, 60Y1, 61R1, 63S1
Fe	Ar	5	2.5	61L1, 61W2
Ge	Ar	5	3.0	61L1, 63S1
Mo	Ar	5	1.5	61L1, 61P1, 61W2, 68K1
Si	Ar	5	1.4	61L1, 63S1
Ta	Ar	5	1.05	61L1, 65C1
Ti	Ar	5	1.7	61L1, 68K1
W	Ar	5	1.1	61L1, 65C1
Au	Ar	2	5.6	61L1, 68R1
Cu	Ar	2	4.3	61L1, 60Y1, 61R1, 63S1
Fe	Ar	2	2.0	61L1, 61W2
Ge	Ar	2	2.0	61L1, 63S1
Si	Ar	2	0.9	61L1, 63S1
Ti	Ar	2	1.1	61L1, 68K1
SiO_2	Ar	2	0.4	67D1
Al_2O_3	Ar	2	0.11	67D1

rf sputter etching

rf (radio-frequency) sputtering is generally used for etching of thin films. The main advantage of rf sputter etching is that all solid-state films, including dielectrics, can be etched [55W1]. dc sputter etching only permits etching of conductive materials [55W1, 70W1, 78V1].

A schematic representation of an rf sputter etching system is shown in Fig. 9. The system consists of a vacuum chamber, a cooled cathode plate, and mostly an anode parallel to the cathode. The substrate to be etched is placed on the cathode plate. The pressure of the gas (usually Ar) is kept constant in the range of 1···10 Pa. An rf voltage at a frequency in the range of 5···50 MHz is applied to the chamber and the cathode. Electrons oscillating in the rf field gain sufficient energy to ionise gas atoms by collisions. Because electrons are more mobile than ions, a space charge builds up when electrons reach the electrodes and are extracted from the plasma. The cathode plate and the anode are negatively charged with respect to the plasma. The ions are accellerated in the dc field towards the electrodes and cause the ion etching of the substrate placed on the cathode.

The ratio of the voltage developing between the plasma and the capacitively coupled cathode (V_c) to the voltage between the plasma and the directly coupled electrode (V_d) is given by [70K2]:

$$V_c/V_d = (A_d/A_c)^4,$$

where A_c and A_d are the respective surface areas of the capacitively and the directly coupled electrodes. The law is valid at low pressure (<10 Pa) and for $A_c \ll A_d$.

In order to confine the sputter etching to the cathode, the rf generator must be capacitively coupled via a capacitor or an insulator plate on top of the cathode and the surface area of the cathode must be small compared to that of the directly coupled electrode. In this case, the negative potential of the cathode is large compared to that of the anode and the chamber wall. For a large surface area, the anode is connected to the chamber wall. The surface area of the cathode is small compared to the total area of the anode and the chamber walls. Because the ion sheath always is parallel to the surface of the cathode, the incidence angle of the ions is normal to the substrate (target) on the cathode.

Models which describe rf sputtering systems have been developed by Logan et al. [77L1] and Keller, Pennebaker and Simmons [79K1, 79K2, 79P1]. The numerous publications describing the application of sputter etching in semiconductor device technology are also based on the models by Melliar-Smith [76M1], Somekh and Casey [76S1, 77S1], Dimigen and Lüthje [76D1], Vossen and O'Neill [68V1], and Davidse [69D2].

In semiconductor device technology, rf sputter etching is used for pattern generation in thin films. Photoresist etch masks can be used if the films to be etched are thin compared to the resist film.

The drawbacks of rf sputter etching are the *redeposition of the sputtered material* on the side walls, *low selectivity* in etch rates of various materials, *low etch rates*, possible *radiation damage*, and frequent difficulties in the *removal the resist film*. Its advantages are *no undercutting, no chemical attack* of the resist and the dry process.

The Table 7 lists the sputter etch rates for dielectrics, silicon, and germanium. Tables 8 and 9 list sputter etch rates for metals and resist materials.

Ion beam etching

A schematic representation of a system used for ion beam etching, also known as ion beam milling or ion beam sputtering, is shown in Fig. 10. The system comprises a plasma source, an extractor, a neutralizer, and a substrate table. The ion generating plasma source is operated at pressures above 10^{-2} Pa. Ions are extracted from the plasma source by extraction grids and are accelerated towards the substrate. The extraction voltage is typically between 500 and 2000 V. Current densities range up to 2 mA/cm^2. The neutralizer is usually a hot filament which emits electrons in order to keep the substrate electrically neutral. The angle of the substrate table is adjustable. The process chamber is operated at pressures below 10^{-3} Pa.

The advantages and drawbacks of ion beam etching are almost the same as those described for rf sputter etching.

Ion beam etching is described in numerous publications [55K1, 70W1, 71S1, 74B1, 75G1, 76D1, 76D2, 76G1, 76M1, 76S1, 76S2, 77C1, 77S1, 77W1, 78M1, 78R1, 79L1, 80C1].

Ion beam etch rates are listed in Table 10 for dielectrics, in Table 11 for silicon and germanium, in Table 12 for metals, and in Table 13 for resist materials.

Table 7. rf sputter etch rates r_e for dielectrics, silicon, and germanium (V_{rf}: voltage applied to the chamber, f: frequency, P_{rf}/A: power per target area, p_g: gas pressure in the plasma chamber, T_{sur}: temperature of the target surface).

No.	Material	Formation	Gas	r_e nm min^{-1}	f MHz	P_{rf}/A W cm^{-2}	V_{rf} V	T_{sur} °C	p_g Pa	Remarks	Ref.
1	SiO_2	sputtered, thermally grown	Ar	7.5		1		190	1.5	2 cm electrode spacing; dielectric substrate; 100 W	67T1, 70J1
				16		2					
				25		3					
2	SiO_2	rf sputtered thermally grown	Ar	6	13.56	200 [W]			0.3···2		69D1
				12		400 [W]					
				20		600 [W]					
3	Si_3N_4	chemical vapour deposition	Ar	6		1.6	1500 from peak to peak	190	1.5	2 cm electrode spacing; dielectric substrate	67T1 70J1
4	Al_2O_3		Ar	2···5		1.6	1500 from peak to peak	190	1.5	2 cm electrode spacing; dielectric substrate	67T1, 70J1
5	Si		Ar	13.8		1.3	–		2.6		76M1
6	Ge		Ar				500···1200				66W1

Table 8. rf sputter etch rates r_e for metals (for explanation of the symbols see Table 7).

No	Material	Formation	Gas	r_e nm min^{-1}	f MHz	P_{rf}/A W cm^{-2}	V_{rf} V	T_{sur} °C	p_g Pa	Remarks	Ref.
1	Al	evaporated	Ar	12···16		1.6	1500 from peak to peak	190	1.5	2 cm electrode spacing; dielectric substrate	67T1, 70J1
2	Au	evaporated	Ar	20···35		1.6	1500 from peak to peak	190	1.5	2 cm electrode spacing; dielectric substrate	67T1, 70J1
3	Cu	evaporated	Ar	20···35		1.6	1500 from peak to peak	190	1.5	2 cm electrode spacing; dielectric substrate	67T1, 70J1
4	Ni—Cr	sputtered	Ar	10		1.6	1500 from peak to peak	190	1.5	2 cm electrode spacing; dielectric substrate	67T1, 70J1
5	W		Ar	7···7.5		1.6	1500 from peak to peak	190	1.5	2 cm electrode spacing; dielectric substrate	67T1, 70J1
6	Pt		Ar	90		2					74L1
7	Ni		Ar	50		2					74L1
8	Ta		Ar	13		1.3	1000 self-bias		1.3		77S1
9	Ti		Ar	5		2					74L1

Table 9. rf sputter etch rates r_e for resist materials (for explanation of the symbols see Table 7).

No.	Resist	Producer	Gas	r_e nm min^{-1}	f MHz	P_{rf}/A W cm^{-2}	V_{rf} V	T_{sur} °C	p_g Pa	r_e (resist) / r_e (SiO_2)	Remarks	Ref.
1	KTFR (Kodak Thin Film Resist)	Eastman Kodak Chemical Company	Ar	8 18 25	13.56	200 [W] 400 [W] 600 [W]			0.3…2			69D1
2	KTFR	Eastman Kodak	Ar	7…30		1.6	1500 from peak to peak	190	1.5		2 cm electrode spacing; dielectric substrate	67T1 70J1
3	KTFR	Eastman Kodak	Ar	27			−700 (target sheath potential)		0.7	6.0	baked at 150 °C under N_2 for 10 min	72V2
4	KMER	Eastman Kodak	Ar	24			−700 (target sheath potential)		0.7	5.3	baked at 150 °C under N_2 for 10 min	72V2
5	AZ 340	Shipley Company	Ar	10…35		1.6	1500 from peak to peak	190	1.5		2 cm electrode spacing; dielectric substrate	67T1 70J1
6	AZ 1350	Shipley	Ar	25			−700 (target sheath potential)		0.7	8.2	} baked under N_2, 10 min at recommended temperature (Nos. 6…9)	72V2
7	AZ 1350 H	Shipley	Ar	17			−700		0.7	3.8		
8	AZ 111	Shipley	Ar	22			−700		0.7	4.8		
9	Waycote	Hunt Chemical Company	Ar	29			−700		0.7	5.4		
10	RCA	RCA	Ar	30			−700		0.7	5.7	baked under N_2, 10 min at 200 °C	
11	RCA	RCA	Ar	32			−700		0.7	4.7	unbaked	
12	Rezyl 387-5	Koppers Chemical Company	Ar	32			−700		0.7	8.0	} baked under N_2, 10 min at recommended temperature (Nos. 12…16)	
13	Rezyl 837-1	Koppers	Ar	37			−700		0.7	12.3		
14	Rezyl 412-1-50	Koppers	Ar	6			−700		0.7	7.0		
15	Rezyl 1102-5	Koppers	Ar	27			−700		0.7	6.0		
16	Amberlac 80x	Röhm and Haas	Ar	32			−700		0.7	7.0		

continued

Table 9 (continued)

No.	Resist	Producer	Gas	r_e nm min^{-1}	f MHz	P_{rf}/A W cm^{-2}	V_{rf} V	T_{sur} °C	p_g Pa	$\frac{r_e\,(\text{resist})}{r_e\,(SiO_2)}$	Remarks	Ref.
17	Amberlac 292x	Röhm and Haas	Ar	27			−700		0.7	5.1		
18	Beckosol 11-147	Reichhold Chemical Company	Ar	61			−700		0.7	8.5		
19	Beckosol 12-101	Reichhold	Ar	38			−700		0.7	7.6	baked under N_2, 10 min at recommended temperature (17…24)	
20	Alkydol 12-704	Reichhold	Ar	36			−700		0.7	8.0		
21	13-031	Reichhold	Ar	27			−700		0.7	7.2		
22	13-040	Reichhold	Ar	25			−700		0.7	5.0		
23	13-046	Reichhold	Ar	32			−700		0.7	7.0		
24	13-077	Reichhold	Ar	30			−700		0.7	7.2		
25	p-isophthalate		Ar	19			−700		0.7	4.3	baked under N_2, 10 min at 185 °C	
26	p-o-phthalate		Ar	24			−700		0.7	5.3	baked under N_2, 10 min at 185 °C	

Table 10. Ion beam etching rates r_e of dielectrics (E: ion energy, j: ion current density, α: angle of incidence, p_g: gas pressure in the plasma chamber).

No.	Material	Gas	r_e nm min^{-1}	E keV	j mA cm^{-2}	α	p_g Pa	Remarks	Ref.
1	SiO_2	Ar	12	0.3	0.32	0°		maximum etch rate at an angle of incidence $\alpha = 50°$	79L1
			20	0.3		50°			
2	SiO_2	Ar	25	0.5	0.65	0°	<0.013		76S2
3	SiO_2	Ar	28⋯42	0.5					73C1, 73G1, 75G1, 76M1
4	SiO_2	Ar	38⋯67	1.0					73G1, 75L1, 76M1
5	SiO_2	Ar	33	6	≈1	60°⋯75°	0.04		71S1
6	SiO_2	Ar + O_2	15	0.5	0.65	0°		pressure of O_2 0.013 Pa	76S2
7	Al_2O_3	Ar	8.5±1.2	0.5	0.65	0°	<0.013		75G1, 76S2
8	Al_2O_3	Ar	13	1.0					75L1
9	$LiNbO_3$	Ar	28	0.5	0.65	0°	<0.013		76S2
10	$LiNbO_3$	Ar	64	1.0					73G1

[Ref. p. 351

Table 11. Ion beam etching rates r_e of silicon and germanium (for explanation of the symbols see Table 10).

No.	Material	Gas	r_e nm min^{-1}	E keV	j mA cm^{-2}	α	p_g Pa	Remarks	Ref.
1	Si	Ar	28⋯42	0.5					73C1, 73G1, 75G1, 76M1
2	Si (111)	Ar	27±0.6	0.5	0.65	0°			76S2
3	Si (100)	Ar	21.5	0.5	1.0	0°			76G1
4	Si	Ar	36⋯75	1.0					73C1, 75L1
5	Ge	Ar		0.5⋯1.2					66W1

Table 12. Ion beam etching rates r_e of metals (for explanation of the symbols see Table 10).

No.	Material	Gas	r_e nm min^{-1}	E keV	j mA cm^{-2}	α	p_g Pa	Remarks	Ref.
1	Al	Ar	19	0.3	0.32	0°		maximum etch rate at an angle of incidence $\alpha=40°$	79L1
			28			40°			
2	Al	Ar	30⋯70	0.5					73G1, 75G1, 76M1
3	Al	Ar	30	0.5	0.65	0°	<0.013		76S2
4	Al	Ar	42	0.5	1.00	0°	0.013	maximum etch rate at $\alpha=45°$	76S1
			76			45°			
5	Al	Ar	43	1.0	0.85	0°	0.04	beam diameter 5 cm	71S1
6	Al	Ar	45⋯75	1.0					68J1, 75L1
7	Ti	Ar	4	0.3	0.32	0°		maximum etch rate at $\alpha=35°$	79L1
			6			35°			
8	Ti	Ar	10	0.5	0.65	0°	<0.013		76S2
9	Ti	Ar	20	0.5					76M1
10	Ti	Ar	20	1.0					75L1, 76M1
11	Ti	Ar+O_2	1.6	0.5	0.65	0°		pressure of O_2 0.013 Pa	76S1, 76S2
			12		0.60	50°		pressure of O_2 0.026 Pa	
12	V	Ar	22	1.0					73C1, 76M1
13	Cr	Ar	10⋯24	0.5	0.65	0°	<0.013		73C1, 76S2
14	Cr	Ar	20⋯40	1.0					73C1, 75L1
15	Cr	Ar	17	6.0	1.0	60⋯75	0.04		71S1

continued

Table 12 (continued)

No.	Material	Gas	r_e nm min^{-1}	E keV	j mA cm^{-2}	α	p_g Pa	Remarks	Ref.
16	Cr	Ar	2.4	0.5	0.65	0	< 0.013		75C1, 76S2
17	Mn	Ar	27	1.0					73C1, 76M1
18	Fe	Ar	32	1.0					71S1, 75L1, 76M1
19	Cu	Ar	45	0.5					73C1
20	Cu	Ar	45	0.5	1.0	0			76G1
21	Cu	Ar	33	6.0	1.0	60···75	0.04		71S1
22	Zr	Ar	32	1.0					75L1, 76M1
23	Nb	Ar	30	1.0	0.85	0	0.04		71S1, 75L1, 76M1
24	Mo	Ar	23	0.5					75G1, 76G1, 76M1
25	Mo	Ar	40	1.0					75L1, 76M1
26	Ag	Ar	200	1.0					75L1, 76M1
27	Ag	Ar	300	1.0	0.85	0	0.04	beam diameter 5 cm.	71S1
28	Ta	Ar	33	0.5	1.0	0···45	0.013	etch rate decreases for $\alpha > 45°$.	76S1
29	Ta	Ar	15···33	0.5					75G1, 76M1
30	W	Ar	18	0.5	1.0	0			75G1, 76G1, 76M1
31	Au	Ar	30	0.3	0.32	0		etch rate decreases for $\alpha > 0°$.	79L1
32	Au	Ar	100···150	0.5					73G1, 76G1, 76M1
33	Au	Ar	25	0.5	0.65	0	< 0.013		76S2
34	Au	Ar	160···215	1.0					73G1, 75L1, 76M1
35	Permalloy	Ar	33	0.5	1.0	0			76G1
36	Permalloy	Ar	45	0.5	1.0	0	0.013	maximum etch rate at $\alpha = 40°$	76S1
37	Permalloy	Ar	33···45	0.5					76M1, 75G1
38	Permalloy	Ar	17	6.0	1.0	60···75	0.04		71S1

Table 13. Ion beam etching rates r_e of resist materials.

No.	Resist	Producer	Gas	r_e nm min^{-1}	E keV	I mA cm^{-2}	α	p_g Pa	Remarks	Ref.
1	AZ 1350	Shipley Company	Ar	13	0.3	0.32	0° 40°		maximum etch rate at an angle of incidence $\alpha = 40°$	79L1
2	AZ 1350	Shipley	Ar	40 90	0.5	1.0	0° 60°	0.007···0.026	AZ 1350 resist on GaAs; maximum etch rate at $\alpha = 60°$	77S1
3	AZ 1350	Shipley	Ar	15±3	0.5	0.65	0°	<0.013		76S2
4	AZ 1350	Shipley	Ar	20···42	0.5					75G1, 76M1
5	AZ 1350	Shipley	Ar	60	1.0					75L1
6	KTFR	Eastman Kodak Chemical Company	Ar	38	1.0	0.85	0°	0.04	beam diameter 5 cm	71S1
7	KTFR	Eastman Kodak	Ar	39	1.0					75L1
8	Riston 14	du Pont	Ar	25	0.5					75G1
9	PMMA (Polymethyl-methacrylate)		Ar	42±6	0.5	0.65	0°	<0.013		76S2
10	PMMA		Ar	84	1.0	0.85	0°	0.04		71S1, 75L1
11	COP (copolymer electron resist)		Ar	86	0.5					74T1, 76M1

6.1.4.7.7 Reactive dry etching processes

Reactive dry etching exists in a variety of forms and under a variety of names – *plasma etching, plasma-assisted etching, reactive ion etching, reactive sputter etching, triode plasma etching, chemical dry etching, reactive ion beam etching, reactive ion beam milling, electron-induced chemical dry etching,* and *photon-induced chemical dry etching.*

A common feature of all these variants is that the etching results from a chemical reaction of the gaseous reactive species and the surface to be etched.

A volatile reaction product is formed. The chemical reaction may be *spontaneous,* or *ion-, electron-,* or *photon-induced.*

Fig. 11 shows an example of reactive dry etching by a spontaneous chemical reaction. The reactive gas CF_4 dissociates in the plasma to $CF_3^+ + e + F^0$. Fluorine atoms spontaneously react with Si, leading to the formation of the volatile SiF_4. Because of the isotropic velocity distribution of neutral fluorine atoms the etch process is also isotropic.

An example of reactive dry etching by an ion-induced chemical reaction is shown in Fig. 12. The silicon surface is exposed to a XeF_2 gas and to an Ar^+ beam, which induces the chemical reaction [79C4]. The reaction products are the volatile SiF_4 gas and the noble gas Xe. Because of the directional ion beam the etch process is also directional.

In semiconductor device technology, reactive dry etching was first used for the removal of organic photoresist materials and was known as *plasma ashing* or *plasma stripping*. Fig. 13 shows the principle of plasma ashing. In the plasma, a fraction of the molecular oxygen is converted to atomic oxygen, which reacts with the photoresist at room temperature. Reaction products are CO, CO_2, H_2O etc.

The application of reactive dry etching in semiconductor device technology was proposed by Irving [71I1] as "gas plasma vapor etching". Reactive dry etching permits the generation of very fine patterns in the fabrication of integrated circuits of high packing density.

Low-pressure plasma

A low-pressure plasma used in plasma etching is a partially ionized gas consisting of equal numbers of positive and negative charges and a differing number of non-ionized neutral atoms or molecules. A typical degree of ionization in this type of plasma is 10^{-4}. The pressure of such a plasma typically ranges from $10^{-1} \cdots 10^3$ Pa. The average electron density is $10^{10} \cdots 10^{11}$ electrons cm^{-3} at electron temperature of typically $>10^4$ K. The gas temperature ranges from 25⋯300 °C.

The primary function of the plasma is to produce chemically active species which react with a solid surface to form a volatile reaction product. The reaction may occur with or without electron or ion radiation from the plasma. The plasma is produced by collisions of gas molecules or atoms and accelerated electrons which can be generated by an electric or electromagnetic field applied. The discharge reaches a self-sustained steady state when generation and loss processes are in balance.

Tables 14, 15, and 16 list several examples of elementary processes occurring within the plasma and between the plasma and the solid surface.

Table 14. Electron-impact reactions [79B8]. (A: atom; A^+: ion; A^*: excited atom).

Excitation (rotational, vibrational and electronic)	$e + A_2 \rightarrow A_2^* + e$
Dissociative attachment	$e + A_2 \rightarrow A^- + A^+ + e$
Dissociation	$e + A_2 \rightarrow 2A + e$
Ionization	$e + A_2 \rightarrow A_2^+ + 2e$
Dissociative ionization	$e + A_2 \rightarrow A^+ + A + 2e$

Table 16. Heterogeneous reactions [79B8]. (S: surface of solid in contact with the plasma).

Atom recombination	$S - A + A \rightarrow S + A_2$
Metastable deexcitation	$S + M^* \rightarrow S + M$
Atom abstraction	$S - B + A \rightarrow S + AB$
Sputtering	$S - B + M^+ \rightarrow S^+ + B + M$

Table 15. Inelastic collisions of heavy particles [79B8]. (M: heavy metal; M^*: heavy metal in excited state; A, B, C: gas particles).

Penning dissociation	$M^* + A_2 \rightarrow 2A + M$
Penning ionization	$M^* + A_2 \rightarrow A_2^+ + M + e$
Charge transfer	$M^+ + A_2 \rightarrow A_2^+ + M$
	$M^- + A_2 \rightarrow A_2^- + M$
Collisional detachment	$M + A_2^- \rightarrow A_2 + M + e$
Associative detachment	$A^- + A \rightarrow A_2 + e$
Ion-ion recombination	$M^- + A_2^+ \rightarrow A_2 + M$
	$M^- + A_2^+ \rightarrow 2A + M$
Electron-ion recombination	$e^- + A_2^+ \rightarrow 2A$
	$e^- + A_2^+ + M \rightarrow A_2 + M$
Atom recombination	$M + 2A \rightarrow A_2 + M$
Atom abstraction	$A + BC \rightarrow AB + C$
Atom addition	$A + BC + M \rightarrow ABC + M$

Low-pressure plasmas can be produced by a variety of techniques. The principal method used in reactive dry etching is plasma generation induced by an electromagnetic radio frequency (rf) field. Advantages offered by rf low-pressure plasmas are:

Electrons oscillating in the rf field can pick up sufficient energy to cause ionization. The discharge is independent of the yield of secondary electrons from electrodes and walls.

Since the oscillation enhances the efficiency of the electrons for ionizing collisions, the discharge can be operated at pressures down to about 0.1 Pa. A low pressure is essential for directional etching.

The capacitive rf coupling permits that electrodes may be coated by an insulating material and be placed outside the discharge. Etching of insulating material is possible.

Etching mechanisms

Three process steps occur during the reactive dry etching of a surface. In the first step, the gas phase species chemisorb on the surface. The second step is the formation of a volatile reaction product. In the third step, the reaction product is desorbed into the gas phase. These three steps are demonstrated in the following example in which silicon is etched using molecular fluorine [79C5]:

Chemisorption: $(F_2)_{gas} \rightarrow (F_2)_{ads} \rightarrow 2\,F_{ads}$

Reaction: $Si + 4\,F_{ads} \rightarrow (SiF_4)_{ads}$

Desorption: $(SiF_4)_{ads} \rightarrow (SiF_4)_{gas}$

Any of these steps could be rate-limiting in the etching process.

Reactive dry etching may occur with or without the irradiation of ions, electrons, or photons.

Reactive dry etching without radiation

Spontaneous etching without radiation takes place in several surface-gas systems. Molecular fluorine and XeF_2 spontaneously etch silicon at room temperature, but at a very low etch rate [79C4]. In these etching systems, chemisorption is the rate-limiting step [79C5].

The principal radical in reactive dry etching of silicon and silicon compounds is the F atom. In the silicon-fluorine system, the steps for reaction product formation and desorption spontaneously proceed at room temperature at a very high rate [79C5]. High etch rates can be obtained.

F atoms can be generated by a CF_4/O_2 plasma. With increasing oxygen concentration, the fluorine concentration increases together with the etch rate of silicon. This effect is visible in Fig. 14.

Mass spectra and optical spectroscopy studies indicate that CO, CO_2, and COF_2 are the final products in the plasma [76H4, 77C3, 78M3]. The oxygen added to CF_4 produces F atoms and is effective in reducing the carbon contamination of the surface to be etched [78M3, 79C5]. Both effects of the oxygen are necessary for etching silicon at a high rate.

Reactive dry etching with ion irradiation

Coburn and Winters [79C4] found that silicon can be etched at room temperature by exposure to XeF_2 gas together with an Ar^+ beam irradiating. The etch rate obtained is greater than the sum of the etch rates of the XeF_2 gas alone and the etch rate of the ion beam alone. The combined effect is visible in Fig. 15.

The XeF_2 gas can be used as a source of atomic fluorine. XeF_2 is dissociatively chemisorbed onto the surface of silicon. Xe evaporates away, leaving two chemisorbed fluorine atoms. The reaction probability without ion radiation is about 10^{-2}. With ion irradiation, it is increased by more than one order of magnitude. The etching is initiated by the ion beam.

Ion beam-induced etching is of importance in semiconductor device technology because the etch process is directional. The mask pattern can be accurately transferred to the film below as noted in Table 2.

Ion beam-induced chemical dry etching can be performed in numerous commercial systems such as the *parallel-plate reactor* for plasma etching, the *reactive ion etcher*, *reactive ion beam etcher*, and *triode plasma etcher*.

Reactive dry etching with electron irradiation

Electron-induced chemical dry etching was observed by Coburn and Winters [79C4], who found that SiO_2 can be etched by simultaneous exposure to a XeF_2 gas and an electron beam. SiO_2 cannot be etched by the XeF_2 gas alone. The experimental results are shown in Fig. 16.

XeF_2 gas is used as a source of atomic fluorine, which is dissociatively chemisorbed onto the surface of SiO_2.

Electron-induced chemical dry etching is of interest in semiconductor device technology because patterns can be etched into silicon compounds without the use of a photoresist mask by direct-writing using electron beam lithography.

Reactive dry etching with photon irradiation

Steinfeld et al. [80S5] found that reactive neutral fragments generated by the multiple infrared photon dissociation of various molecules using a CO_2 laser are capable of etching silicon dioxide and silicon nitride. Silicon dioxide is etched by CF_3Br at a pressure of 370 Pa at a high etch rate. The etch gases for silicon nitride are CF_2Cl_2 and CDF_3.

Anisotropic etching

If ions or electrons are the etchant species, their directionality produces anisotropic etching as shown in Fig. 17. But it has been observed that the density of ions or electrons is too low to account for the high etch rates obtained by anisotropic etching. The more abundant neutral species are the main etchants. The model of Mogab and Levingstein [80M7] explains anisotropic etching by neutral species.

At the surface to be etched, two different types of chemical reactions occur: recombination and etch reaction with a volatile reaction product. The etch reaction dominates at surfaces which are exposed to ion or electron radiation. The dominant reaction at unbombarded surfaces is recombination. Anisotropic etching as shown in Fig. 17 is obtained.

An example is the anisotropic etching of Si by a gas mixture of C_2F_6 and Cl_2. According to Mogab and Levingstein, the following reactions occur [80M7]:

Dissociation in the plasma:

$$C_2F_6 + e \rightarrow 2\,CF_3 + e \quad (1)$$

$$Cl_2 + e \rightarrow 2\,Cl + e \quad (2)$$

Reactions on the surface:

$$x\,Cl + Si \rightarrow SiCl_x\uparrow \text{ (etching)} \quad (3)$$

$$CF_3 + Cl \rightarrow CF_3Cl\uparrow \text{ (recombination)} \quad (4)$$

If reaction (3) is slow in the absence of ion or electron bombardment, reaction (4) predominates over reaction (3) on the side walls (see Fig. 17) provided that there is a local excess of recombinant species.

Vertical or directional edge profiles as shown in Fig. 17 are obtained.

Vapour pressure of reaction products

The choice of etching gases is limited to reactions where the product of the reactive species and the atoms at the substrate surface is a volatile gas.

Fig. 18 schematically shows the *p-T* phase diagram of a reaction product. The phase diagram determines whether a reaction product at given temperature and pressure is gaseous, liquid, or solid. A survey of vapour pressures at definite temperatures is given for various reaction products in the Tables 17···23.

Table 17. Vapour pressure of silicon compounds [60G2]. (For $p-T$ diagrams see Fig. 19).

No.	Compound	p_{vapour} Pa	T °C	No.	Compound	p_{vapour} Pa	T °C
1	$SiBr_4$	1330	+31	16	SiH_2Br_2	133	−59
2	$SiCl_4$	133	−63	17	$SiHBr_3$	13.3	−62
3	Si_2Cl_6	133	+3	18	SiH_3Cl	133	−168
4	Si_3Cl_8	13.3	+13	19	$SiHCl_3$	13.3	−101
5	SiF_4	13.3	−155	20	SiH_3CN	1330	−17
6	Si_2F_6	13.3	−96	21	SiH_3CNS	133	−39
7	$SiFBr_3$	13.3	−76	22	SiH_3F	133	−152
8	SiF_3Cl	1330	−129	23	SiH_2F_2	133	−146
9	SiF_2Cl_2	133	−126	24	$SiHF_3$	133	−152
10	$SiFCl_3$	13.3	−120	25	SiH_3J	1330	−43
11	SiH_4	133	−175	26	SiH_2J_2	1330	+18
12	Si_2H_6	133	−111	27	SiI_4	1330	+142
13	Si_3H_8	13.3	−91	28	SiO	13.3	+1128
14	Si_4H_{10}	13.3	−53	29	SiO_2	0.013	+1200
15	SiH_3Br	1330	−77				

For Table 18, see next page.

Table 19. Vapour pressure of chlorine compounds [60G2]. (For $p-T$ diagrams see Fig. 21).

No.	Compound	p_{vapour} Pa	T °C
1	AgCl	13.3	+789
2	$AlCl_3$	13.3	+78
3	Al_2Cl_6	13.3	+77
4	$AsCl_3$	1330	+26
5	BCl_3	133	−92
6	B_2Cl_4	133	−56
7	B_4Cl_4	133	+14
8	$BaCl_2$	13.3	+987
9	$BeCl_2$	13.3	+262
10	$BiCl_3$	1330	+264
11	CCl_4	133	−50
12	C_2Cl_4	133	−20
13	C_2Cl_6	13.3	+2
14	CNCl	13.3	−96
15	$COCl_2$	13.3	−120
16	Cl_2O	133	−99
17	Cl_2O_6	13.3	−12
18	Cl_2O_7	13.3	−72
19	$CeCl_3$	13.3	+898
20	$CdCl_2$	13.3	+488
21	$CoCl_2$	13.3	+594
22	$CrCl_2$	13.3	+750
23	$CrCl_3$	13.3	+618
24	CrO_2Cl_2	13.3	−43
25	CsCl	13.3	+638
26	$CuCl_2$	1330	+368
27	Cu_2Cl_2	133	+546
28	$FeCl_2$	1330	+681
29	Fe_2Cl_6	13.3	+175
30	$GaCl_3$	13.3	+23
31	$GeCl_4$	133	−44
32	$GeHCl_3$	13.3	−63
33	HCl	13.3	−165
34	$HgCl_2$	13.3	+100
35	$HfCl_4$	133	+171
36	InCl	133	+304
37	$InCl_2$	133	+341
38	KCl	13.3	+716
39	LiCl	13.3	+674
40	$LaCl_3$	13.3	+919
41	$MgCl_2$	13.3	+776
42	$MnCl_2$	133	+729
43	$MoCl_5$	13.3	+70
44	NOCl	133	−96
45	NH_4Cl	$1.33 \cdot 10^4$	+271
46	NaCl	13.3	+752
47	$NbCl_4$	133	+286
48	$NbCl_5$	1330	+143
49	$NdCl_3$	13.3	+834
50	$NiCl_2$	13.3	+620
51	PCl_3	13.3	−75
52	PCl_5	1330	+78
53	PH_4Cl	133	−91
54	$PbCl_2$	13.3	+474
55	$PrCl_3$	13.3	+869
56	$PuCl_3$	13.3	+937
57	RbCl	13.3	+685
58	SCl_2	133	−64
59	S_2Cl_2	13.3	−34
60	$SOCl_2$	13.3	−81
61	SO_2Cl_2	1330	−24
62	$SbCl_3$	13.3	+18
63	$SbCl_5$	133	+22
64	$ScCl_2$	133	+715
65	$SeCl_4$	1330	+106
66	$SiCl_4$	133	−63
67	Si_2Cl_6	133	+3
68	Si_3Cl_8	13.3	+13
69	$SiHCl_3$	13.3	−101
70	SiH_3Cl	133	−168
71	$SnCl_2$	13.3	+257
72	$SnCl_4$	133	−22
73	$TaCl_5$	13.3	+90
74	$TeCl_4$	1330	+234
75	$ThCl_4$	1330	+697
76	$TiCl_4$	133	−13
77	TlCl	13.3	+357
78	UCl_3	13.3	+895
79	UCl_4	13.3	+457
80	UCl_6	13.3	+73
81	VCl_4	133	−10
82	$VOCl_3$	1330	+14
83	WCl_5	133	+114
84	WCl_6	13.3	+118
85	$WOCl_4$	133	+96
86	$ZnCl_2$	13.3	+361
87	$ZrCl_4$	13.3	+157

Table 18. Vapour pressure of germanium compounds [60G2]. (For $p-T$ diagrams see Fig. 20).

No.	Compound	p_{vapour} Pa	T °C
1	$GeBr_4$	1330	+56
2	$GeCl_4$	133	−44
3	GeF_4	133	−109
4	GeF_3Cl	$1.33 \cdot 10^4$	−45
5	GeF_2Cl_2	$1.33 \cdot 10^4$	−41
6	$GeFCl_3$	1330	−35
7	GeH_4	133	−163
8	Ge_2H_6	13.3	−110
9	Ge_3H_8	13.3	−64
10	$GeHCl_3$	13.3	−63
11	GeJ_4	13.3	+94
12	GeO	133	+614
13	GeS	13.3	+436

Table 20. Vapour pressure of fluorine compounds [60G2]. (For $p-T$ diagrams see Fig. 22).

No.	Compound	p_{vapour} Pa	T °C
1	AlF_3	13.3	+882
2	AsF_3	1330	−16
3	AsF_5	13.3	−131
4	BF_3	13.3	−166
5	BaF_3	13.3	+987
6	BeF_2	13.3	+693
7	CF_4	1330	−169
8	C_2F_4	1330	−133
9	C_2F_6	13.3	−157
10	CHF_3	133	−157
11	CH_3F	133	−147
12	COF_2	1330	−140
13	CaF_2	13.3	+1447
14	CdF_2	1330	+1273
15	CsF_2	133	+710
16	F_2O	13.3	−205
17	F_2O_2	1330	−120
18	GeF_4	133	−109
19	HF	1330	−66
20	IrF_6	$1.33 \cdot 10^4$	+9
21	KF	13.3	+752
22	LiF	13.3	+920
23	MgF_2	1330	+1641
24	MnO_3F	1330	−27
25	MoF_6	13.3	−87
26	NF_3	1330	−171
27	NOF	1330	−114
28	NO_2F	13.3	−156
29	NaF	13.3	+916
30	NbF_5	13.3	+45
31	PF_3	1330	−150
32	PF_5	1330	−123
33	POF_3	1330	−82
34	PbF_2	1330	+904
35	PuF_3	13.3	+1304
36	PuF_6	133	−31
37	RbF	133	+827
38	RuF		
39	RuF_5	1330	+134
40	SF_4	1330	−105
41	SF_6	13.3	−159
42	S_2F_{10}	1330	−53
43	SOF_2	1330	−104
44	SbF_5	1330	+39
45	SeF_4	133	−13
46	SeF_6	13.3	−134
47	SiF_4	13.3	−155
48	Si_2F_6	13.3	−96
49	$SiHF_3$	133	−152
50	SiH_2F_2	133	−147
51	SiH_3F	133	−152
52	SrF_2	133	+1600
53	TaF_5	133	+80
54	TeF_4	13.3	+41
55	TeF_6	13.3	−128
56	TiF_4	1330	+174
57	TlF	13.3	+346
58	UF_4	13.3	+872
59	UF_6	13.3	−50
60	WF_6	13.3	−89
61	ZnF_2	133	+922
62	ZrF_4	13.3	+587

Table 21. Vapour pressure of fluorine-chlorine compounds [60G2]. (For $p-T$ diagrams see Fig. 23).

No.	Compound	p_{vapour} Pa	T °C
1	$CFCl_3$	1330	−61
2	CF_2Cl_2	133	−121
3	CF_3Cl	$1.33 \cdot 10^4$	−113

continued

Table 21 (continued)

No.	Compound	p_{vapour} Pa	T °C
4	$C_2F_2Cl_2$	1330	−59
5	C_2F_3Cl	1330	−98
6	$C_2F_4Cl_2$	1330	−75
7	$COClF$	$1.33 \cdot 10^4$	−86
8	ClF	133	−153
9	ClF_3	1330	−61
10	ClO_3F	13.3	−145
11	$GeFCl_3$	1330	−38
12	GeF_2Cl_2	$1.33 \cdot 10^4$	−42
13	GeF_3Cl	$1.33 \cdot 10^4$	−45
14	$PFCl_2$	$1.33 \cdot 10^4$	−32
15	PF_2Cl	$1.33 \cdot 10^4$	−87
16	$POFCl_2$	$1.33 \cdot 10^4$	+3
17	POF_2Cl	$1.33 \cdot 10^4$	−40
18	$SOClF$	133	−99
19	SO_2ClF	133	−96
20	S_2O_5ClF	133	−21
21	$SiFCl_3$	1330	−72
22	SiF_2Cl_2	$1.33 \cdot 10^4$	−71
23	SiF_3Cl	$1.33 \cdot 10^4$	−103
24	$Si_2OF_3Cl_3$	133	−72

Table 22. Vapour pressure of bromine compounds [60G2]. (For $p-T$ diagrams see Fig. 24).

No.	Compound	p_{vapour} Pa	T °C
1	$AgBr$	13.3	+653
2	$AlBr_3$	$1.01 \cdot 10^5$	+466
3	Al_2Br_6	13.3	+56
4	$Al_2Br_2Cl_4$	1.33	+37
5	$AsBr_3$	1330	+93
6	BBr_3	133	−41
7	B_2H_5Br	133	−94
8	$BeBr_2$	133	+288
9	$BiBr_3$	1330	+280
10	BrF	$1.33 \cdot 10^4$	−26
11	BrF_3	1330	+29
12	BrF_5	13.3	−89
13	CBr_4	13.3	−2
14	C_2Br_4	133	−28
15	CCl_3Br	1330	−2
16	$CHBr_3$	1330	+34
17	CH_2Br_2	133	−35
18	CH_3Br	133	−96
19	$CNBr$	1330	−11
20	$COBrF$	$1.33 \cdot 10^4$	−60
21	$CeBr$	13.3	+757
22	$CdBr_2$	13.3	+450

Table 22 (continued)

No.	Compound	p_{vapour} Pa	T °C
23	$CrBr_3$	13.3	+626
24	$CsBr$	13.3	+642
25	$CuBr$	133	+717
26	Cu_2Br_2	133	+570
27	Cu_3Br_3	133	+592
28	$FeBr_2$	13.3	+516
29	$GaBr_3$	1330	+141
30	$GeBr_4$	1330	+56
31	HBr	133	−139
32	$HgBr_2$	13.3	+100
33	$InBr$	133	+312
34	$InBr_2$	133	+298
35	$InBr_3$	133	+212
36	IBr	13.3	−25
37	KBr	13.3	+674
38	$LaBr$	13.3	+784
39	$LiBr$	13.3	+640
40	$MoBr_4$	1000	+252
41	$MoBr_5$	1000	+169
42	NH_4Br	1330	+249
43	$NaBr$	13.3	+697
44	$NbBr_5$	$1.33 \cdot 10^5$	+284
45	$NdBr$	13.3	+702
46	$NiBr$	13.3	+587
47	PBr_3	1330	+45
48	$PFBr_2$	1330	−33
49	PF_2Br	1330	−89
50	PH_4Br	1330	−20
51	POF_2Br	1330	−51
52	$PSBr_3$	133	+50
53	$PbBr_2$	13.3	+438
54	$PuBr_3$	13.3	+828
55	$RbBr$	13.3	+668
56	$SOBr_2$	13.3	−36
57	SO_2BrF	133	−72
58	$SbBr_3$	1330	+141
59	$SbBr_5$	133	+94
60	$ScBr_2$	1330	+761
61	$SiBr_4$	1330	+31
62	$SiCl_2BrF$	133	−90
63	$SiClBr_2F$	13.3	−91
64	$SiFBr_3$	1330	−17
65	SiF_2Br_2	$1.33 \cdot 10^4$	−33
66	$SiHBr_3$	13.3	−62
67	SiH_2Br_2	133	−59

continued

Table 22 (continued)

No.	Compound	p_{vapour} Pa	T °C
68	SiH_3Br	1330	−77
69	SmBr	13.3	+667
70	$SnBr_2$	13.3	+284
71	$SnBr_4$	13.3	+6
72	$TaBr_4$	1000	+211
73	$TaBr_5$	1330	+215
74	$TiBr_4$	13.3	+16
75	$ThBr_4$	1330	+624
76	TlBr	13.3	+367
77	UBr_3	13.3	+860
78	UBr_4	13.3	+428
79	WBr_4	1000	+216
80	WBr_5	1000	+187
81	WBr_6	1000	+230
82	$ZnBr_2$	13.3	+340
83	$ZrBr_4$	13.3	+172

Table 23. Vapour pressure of iodine compounds [60G2]. (For $p-T$ diagrams see Fig. 25).

No.	Compound	p_{vapour} Pa	T °C
1	AgI	13.3	+697
2	Al_2I_6	13.3	+147
3	AsI_3	1330	+220
4	BI_3	1330	+77
5	BeI_2	133	+282
6	CI_4	1330	+172
7	CNI	13.3	0
8	COIF	$1.33 \cdot 10^4$	−24
9	CdI_2	13.3	+402
10	CrI_2	13.3	+710
11	Cu_2I_2	1330	+654
12	CsI	13.3	+633
13	FeI_2	$1.33 \cdot 10^4$	+666
14	GaI_3	$1.33 \cdot 10^4$	+263

Table 23 (continued)

No.	Compound	p_{vapour} Pa	T °C
15	GeI_4	13.3	+94
16	HI	13.3	−136
17	HgI_2	13.3	+120
18	InI	1330	+433
19	IBr	1330	+29
20	ICl	1330	+8
21	IF_5	13.3	−35
22	IF_7	13.3	−107
23	KI	13.3	+623
24	LiI	13.3	+631
25	MoI_4	133	+261
26	NH_4I	1330	+269
27	NaI	13.3	+597
28	PI_3	1330	+82
29	PH_4I	133	−25
30	PbI_2	13.3	+404
31	PuI_3	133	+832
32	RbI	13.3	+643
33	SbI_3	133	+164
34	ScI_3	1330	+741
35	SiI_4	1330	+142
36	SiH_2I_2	1330	+18
37	SiH_3I	1330	−44
38	SnI_2	133	+388
39	SnI_4	13.3	+87
40	TaI_5	13.3	+213
41	ThI_4	1330	+579
42	TiI_4	1330	+191
43	TlI	13.3	+369
44	UI_3	1.33	+748
45	UI_4	13.3	+428
46	WI_4	133	+292
47	ZnI_2	13.3	+323
48	ZrI_4	13.3	+226

6.1.4.7.8 Systems used for reactive dry etching

Reactive dry etching can be performed in numerous different systems. Table 24 lists the principal reactive dry etching systems, provides a schematic representation of the reactors, and gives information on the edge profiles obtained by the etch processes *). Each system noted in Table 24 is described in the following sections.

*) See section 6.1.4.7.9.

Table 24. Reactive dry etching systems.

No.	Etching	Reactor	Etch process	Etch profile	Ref.
1	Barrel reactor plasma etching (PB) (also called tube or tunnel reactor)	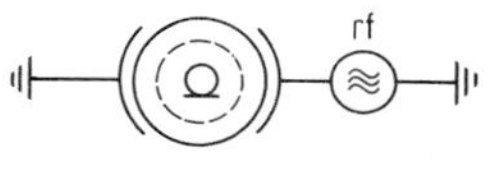 rf	isotropic	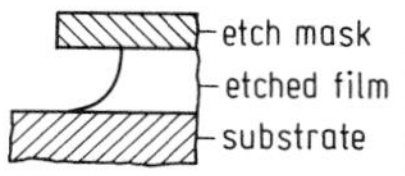 etch mask, etched film, substrate	[67I1] [76B4] [78M1] [80C3]
2	Parallel-plate reactor plasma etching (PP) (also called planar, planar diode, high-pressure diode, anodically-coupled-diode, or Reinberg reactor plasma etching)	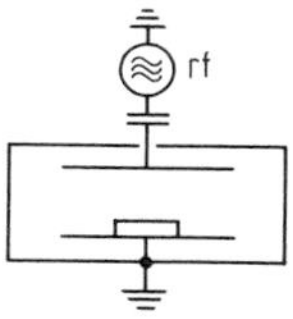 rf	vertical (**a**) anisotropic (**b**) isotropic (**c**)	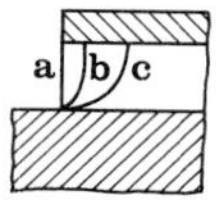 a b c	[73R1] [79R1]
3	Reactive ion etching (RIE) (also called reactive sputter etching and low-pressure-diode or cathodically-coupled-diode etching)	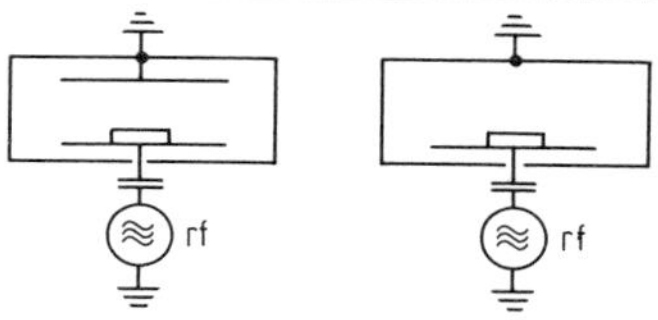 rf rf	vertical (**a**) anisotropic (**b**)	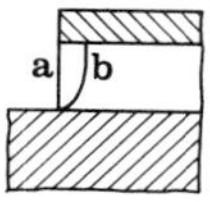 a b	[74H1] [76B2]
4	Triode plasma etching (PT)	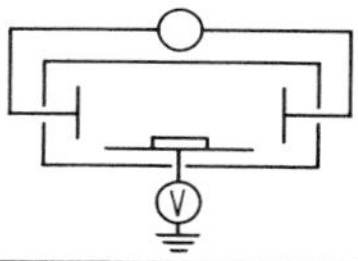 V	vertical (**a**) anisotropic (**b**) isotropic (**c**)	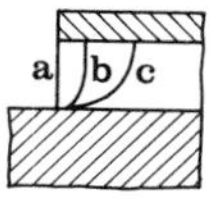 a b c	[79C1] [79C2] [80C3]
5	Chemical dry etching (CDE)	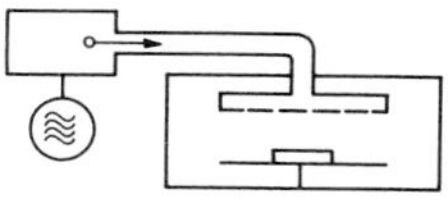	anisotropic (**b**) isotropic (**c**)	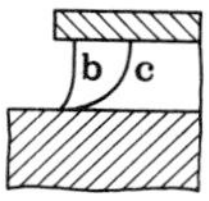 b c	[76H2] [80C3]
6	Reactive ion beam etching (RIBE)	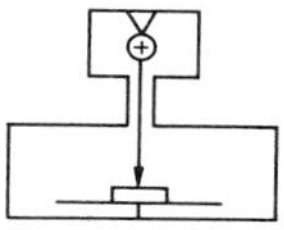 +	vertical (**a**) anisotropic (**b**)	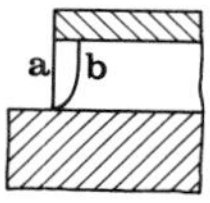 a b	[77C3] [79C4] [79H7] [79V4] [80O1] [80B9]
7	Electron-induced chemical dry etching (ECDE)	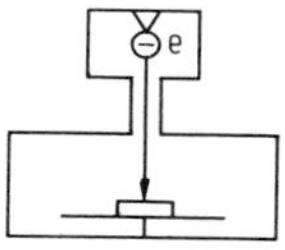 e	vertical (**a**) anisotropic (**b**)	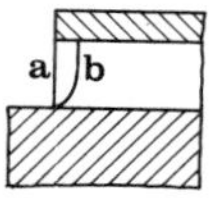 a b	[79C4] [79C5] [79W1]
8	Photon-induced chemical dry etching (PCDE)	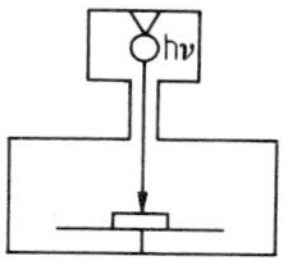 hν	vertical (**a**) anisotropic (**b**) isotropic (**c**)	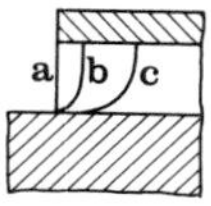 a b c	[80S5]

Note: No. 6 also called reactive ion beam milling.

Plasma etching in the barrel reactor (PB)

The barrel reactor, also known as tunnel or tube reactor, was the earliest commercial reactor for reactive dry etching. It was developed for stripping organic photoresists in oxygen plasma [65H1]. Fig. 26 shows the setup of a barrel reactor. It consists of a quartz chamber, electrodes for rf supply, a gas inlet and an output for the vacuum pump. A boat containing the wafers to be etched is placed inside the quartz chamber. A perforated shield known as etch tunnel is placed between the wafers and the chamber in order to improve the etch uniformity.

Besides photoresist stripping the barrel reactor can be used for the plasma etching of silicon, silicondioxide, siliconnitride, and several other materials. For plasma etching, the evacuated reactor is filled by a reactive gas to a pressure of the order of 100 Pa. The gas is brought to a glow discharge by an rf voltage applied. The radicals generated in the glow discharge diffuse to the wafers, where they chemically react with the surface atoms of the wafers. When the reaction product is a volatile gas, etching occurs.

Etching in the barrel reactor using an etch tunnel is purely chemical. Because of the isotropic velocity distribution of the radicals, the etch process is also isotropic.

Chemical dry etching (CDE)

Chemical dry etching is performed in a reactor in which the discharge region is separated from the etching location as shown in Fig. 27. The discharge is initiated by an rf voltage [76H2], or by microwaves [79T5]. The excited atoms and molecules diffuse from the discharge region through a transport tube to the wafers, which are shielded from the radiation and the bombardment associated with the glow discharge.

The etch process is purely chemical. It can be expected that the resulting edge profiles will be isotropic at high gas pressures and anisotropic at low pressures.

Plasma etching in the parallel-plate reactor (PP)

Plasma etching in the parallel plate reactor (also called planar, planar-diode, high-pressure diode, anodically-coupled-diode, or Reinberg reactor) is performed in the system shown in Fig. 28. It was introduced by A. Reinberg [73R1] initially for plasma deposition. The wafers to be etched are placed on the lower of two parallel electrodes arranged in the recipient. The lower electrode is grounded while the upper electrode is rf powered. A reactive gas fed into the recipient is brought to a glow discharge by the application of an rf voltage, usually at a frequency of 13.56 MHz.

The high mobility of the electrons causes the two electrodes to be negatively charged relative to the plasma. The positive ions of the plasma are accelerated towards the electrodes. The accelerated ions initiate chemical etching of the wafers. The product of the chemical reaction is a volatile gas, which is removed by the vacuum pump.

The edge profile of a masked film etched in a parallel-plate reactor depends on the etching parameters and may be vertical, anisotropic, or isotropic. The gas pressure usually ranges from 10 to 100 Pa.

Reactive ion etching (RIE)-reactive sputter etching

In reactors for reactive ion etching which is also called reactive sputter etching, or low-pressure-diode, or cathodically coupled-diode etching, the wafers to be etched are placed on the cathode electrode to which the rf voltage is applied [74H1, 76B2] as shown in Fig. 29. The upper anode electrode is grounded. It has a larger surface area than the lower electrode. The lower electrode therefore develops a high negative voltage. A complete explanation of this effect was presented by Koenig and Maissel [70K2]. Due to the high negative voltage of the lower electrode, positive ions bombarding the wafers gain a higher kinetic energy than the ions bombarding the upper electrode. The kinetic energy of the ions responsible for etching is usually several hundred eVs.

The pressure of the reactive gas is usually in the range of 1 Pa···10 Pa, one order of magnitude lower than the pressure in a parallel-plate plasma reactor. Depending on the etch parameters, the etch process is vertical or anisotropic.

Fig. 30 shows a schematic of another type of reactor for reactive ion etching. In order to realize a large surface area, the entire recipient is grounded and used as a back electrode.

A vertical arrangement for reactive ion etching is shown in Fig. 31. The concentric hexagonal setup leads to a high etch uniformity [81E6].

Figure 32 shows a reactor for reactive ion etching using a permanent magnet which increases the etch rate [81O1, 82H1].

Triode plasma etching (PT)

In a triode plasma etching system, the discharge is sustained by an applied dc, rf, or microwave supply [79C1, 77S3]. The wafers are placed on a third electrode, which can be dc-biased. The energy and the flux of the ions to the wafer are quasi-independently controllable, which is in contrast to reactive ion etching and plasma etching in a parallel-plate reactor. Triode plasma etching having excitation by microwaves is also called microwave plasma etching [77S3]. Fig. 33 schematically shows a system for triode plasma etching.

Depending on the gas pressure, the flux, and the energy of the bombarding ions, the etch process may be isotropic, anisotropic, or vertical. The undercutting below the etch mask decreases with decreasing gas pressure and increasing ion energy.

Reactive ion beam etching (RIBE)

Reactors for ion beam etching using noble gases and those for reactive ion beam etching which is also known as reactive ion milling, are very similar. The reactor contains a plasma source, an extractor, a neutralizer, and a wafer table as shown in Fig. 34. The materials used inside the reactor have to be selected for reactive gases. The plasma source which generates the ions, is operated at pressures above 10^{-2} Pa. Some of the ions generated are extracted from the plasma source by extraction grids and accelerated towards the wafer. The extraction voltage is typically in the range of 500 V···2000 V. The maximum current density obtained is 2 mA cm^{-2}. The neutralizer usually is a hot filament which emits electrons to keep the wafer neutral. The angle of the wafer table is adjustable. The process chamber is operated at pressures below 10^{-3} Pa.

Reactive ion beam etching can be performed by using the following combination of gases:

Ions of a noble gas (usually Ar) together with a reactive gas in the environment of the wafer to be etched [79C4],

Ions of a reactive gas together with a reactive gas in the environment of the wafer to be etched [77C3, 80O1].

Reactive ion beam etching leads to anisotropic, vertical, or tapered edge profiles. The shape of the edge can be controlled by the angle of incidence of the ions.

Electron-induced chemical dry etching (ECDE)

Winters and Coburn [79C4] found that SiO_2 can be etched by XeF_2 under electron bombardment. Neither exposure to XeF_2 alone nor an electron beam alone produces etching by itself. The etching is performed in the setup shown in Fig. 35, which comprises an electron gun, a reaction chamber having a gas input and an output to a vacuum pump, and a wafer stage.

The resulting edge profile is anisotropic or vertical.

Photon-induced chemical dry etching (PCDE)

Reactive neutral fragments generated by multiple infrared photon dissociation of various molecules in a CO_2 laser etch silicon dioxide and silicon nitride [80S5]. Silicon dioxide is etched by CF_3Br at a pressure of 730 Pa at a high etch rate. The etch gases for silicon nitride are CF_2Cl_2 and CDF_3. The etching is performed in the experimental setup shown in Fig. 36, which comprises a laser, a reaction chamber with a gas input and an output to a vacuum pump, and a wafer stage.

When the reactive radicals are generated at a distance from the wafer surface, the etch process is purely chemical and isotropic.

When the photons induce the chemical etch reaction at the wafer surface, the etch process is anisotropic or vertical.

6.1.4.7.9 Gases used in reactive dry etching

This section presents tables on gases used in reactive dry etching (Table 25) and gases used for etching of silicon (Table 26), of oxides (Table 27), of nitrides (Table 28), of metals (Table 29), of metal silicides (Table 30), and of organic films (Table 31). The reactor systems have been described in the previous section. The films to be etched are usually described by the following short form:

Si: single crystal silicon,
poly-Si: polycrystalline silicon,
Ge: single crystal germanium,
SiO_2: amorphous silicon dioxide,
PSG: phosphorus silicate glass,
Si_3N_4: amorphous silicon nitride,
metals: amorphous metals.

The edge profiles are defined as follows:

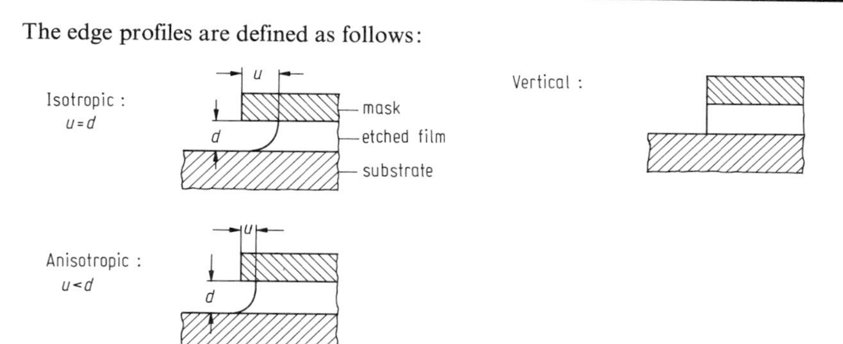

Table 25. Gases used for reactive dry etching.

No.	Gases	Material	Selective to	Technique	Ref.
1	Br_2	Al		RIE	76H3
2	$BrCl_3$	Al	Al_2O_3	PP	80A3
3	$B Br_3$	Al		RIE	80N2
4	BCl_3	Al	SiO_2, photoresists Shipley AZ 1350 and Waycoat WR 300	PP	75H1, 76P1, 81C5, 81D6, 81T6, 82H3
	BCl_3	Al	SiO_2	RIE	77L3, 80N2, 81L2, 82H3
	BCl_3	Al_2O_3			78C1
5	BCl_3/Cl_2	poly-Si	SiO_2	PP, RIE	79H4
	BCl_3/Cl_2	Al, Al/Cu, Al/Cu/Si	SiO_2	PP	79L3, 79M9, 80H11, 81D6, 82H3
	BCl_3/Cl_2	Al, Al/Si/Ti	SiO_2, Si, photoresist	RIE	82B1, 82A1, 83B1, 83S1
6	$BCl_3/Cl_2/He$	Al, Al/Si	SiO_2, Si	PP	82R2, 83B3
7	$BCl_3/Cl_2/CHCl_3/He$	Al, Al/Si	SiO_2, Si	PP	82R2
8	$BCl_3/Cl_2/SiCl_4/He$	Al, Al/Si	SiO_2, Si	PP	82R2
9	BCl_3/CCl_4	Al	SiO_2, Si	PP, RIE	81T12
10	BCl_3/PCl_3	Al, Al/Cu	Si, SiO_2	RIE	80N2, 81T12
11	$BCl_3/CF_4/O_2$	Al/Si, Al/Si/Cu	SiO_2, Si, resist Shipley AZ 1350 J	RIE	81M13
12	BF_n	Si		RIBE	81T13
13	BF_3	poly-Si	SiO_2	RIBE	82O2
14	Cl_2	Al		RIBE	81S1
	Cl_2	Si	SiO_2	RIBE	81M16, 81S1, 81M22
	Cl_2	Al	Al_2O_3	PP	81B21, 81D6
	Cl_2	Si, poly-Si	SiO_2	PP	80B10, 81B4, 81B16, 81B21, 81D6, 81Y1, 82B1, 82B2, 83Z1
	Cl_2	Al		RIE	80N2, 76H3, 81D6
	Cl_2	poly-Si, Si	SiO_2	RIE	81K1, 82F1, 82O3, 81M23, 82B1, 83Z1
	Cl_2	poly-Si	Si_3N_4	PCDE	81S8
	Cl_2	Cr, Sn, Au		PP, RIE	78C1, 78B4, 82H3, 81Z2
15	Cl^+	Si		RIBE	81T13

continued

Table 25 (continued)

No.	Gases	Material	Selective to	Technique	Ref.
16	Cl_2/Ar	Si	SiO_2	RIE	78B8, 80L4, 81H13, 81B24, 81F7, 83P1
	Cl_2/Ar	Al_2O_3		RIE	80H3
	Cl_2/Ar	Si	SiO_2	RIBE	79C4, 81M20
	Cl_2/Ar	Cr_2O_3			78C1
17	Cl_2/Ar, He	Si	SiO_2, Al_2O_3, MgO	RIE	79S2, 79S8, 80H3, 83Z1
18	Cl_2/Ar, Ne, He	Si		RIBE	80C11, 81C7
19	Cl_2/ClF_3	Si, poly-Si	SiO_2	RIE	81F6
20	Cl_2/H_2	Al		RIE	81O5, 82O1, 82O3
	Cl_2/H_2	poly-Si, Si	SiO_2	RIE	81O5, 82O1
21	Cl_2/O_2	Si	SiO_2, Al_2O_3	RIE	81H11
	Cl_2/O_2	Cr		RIE, PB	77Z1, 77P1
22	$Cl_2/O_2/Ar$	Cr		PB	76A2, 80N3
23	$Cl_2/O_2/He$	Cr		PB	76A2, 80N3
24	ClF_3/Cl_2	Si		RIE	81F1
25	$CBrF_3$	Ti, Pt			78M1, 82H3
	$CBrF_3$	Si, poly-Si, Mo	Si_3N_4		
		Si_3N_4	SiO_2, photoresist Shipley AZ 1350 J	RIE	80M3
	$CBrF_3$	SiO_2		PCDE	80S5
26	$CBrF_3/He$	Si	SiO_2	PP	79M9, 79M10
27	$CBrF_3/Cl_2$	poly-Si	SiO_2	RIE	81S11, 82S1
28	$CBrF_3/He/O_2$	Ti	Si_3N_4, Au, Pd	PP	77M4
29	$CClF_3$	poly-Si, Si	SiO_2	PP	79M8, 79M9, 79M10, 80H8, 80H10, 81A11, 81D6
	$CClF_3$	Au, Ti			78M1, 82H3
30	$CClF_3/He$	Si	SiO_2	PP	79M3, 79M9
31	$CClF_3/C_2F_6$	poly-Si	SiO_2	PP	79H4, 79M8
32	CCl_2F_2	Si, poly-Si	SiO_2, photoresist CMS	PP	80H8, 80H10, 81H12
	CCl_2F_2	Si, SiO_2, Si_3N_4, Mo, W, TaN		RIE	76M2
	CCl_2F_2	Si	SiO_2	RIE	76B2, 76M2
	CCl_2F_2	Al		RIE	80N2
	CCl_2F_2	GaAs		RIE	78C1, 80S7
	CCl_2F_2	Si_3N_4		PCDE	80S5
	CCl_2F_2	W		PB	77F1
33	CCl_2F_2/C_2F_6	poly-Si	photoresist	RIE	82W3
34	CCl_2F_4	Au		PB	77P1, 78C1
35	CCl_2F_4/O_2	Pt			78C1
36	CCl_3F	Si, poly-Si	SiO_2, photoresist CMS	PP	80H8, 81M5
37	CCl_3F/O_2	Si, poly-Si	Si_3N_4, SiO_2	RIE	80E1
38	C_2ClF_5	SiO_2, Si, poly-Si		PP	78B4, 81D6, 82K3
39	$C_2Cl_2F_4$	poly-Si	SiO_2, photoresist Shipley AZ 1470	PP, RIE	81Y5
	$C_2Cl_2F_4$	Au		PP, RIE	82H3

continued

Table 25 (continued)

No.	Gases	Material	Selective to	Technique	Ref.
40	$C_2Cl_3F_3/O_2$	Pt			78C1
41	CCl_4	$TiSi_2$	SiO_2	PP	81L7
	CCl_4	Al, Al/Si	SiO_2, Al_2O_3, photoresist Shipley AZ 1350 J	PP	76P1, 78B4, 79H5, 79T2, 80M6, 80T1, 80B7, 80F1, 80R2, 80W2, 80A3, 80H3, 81D6, 81H12, 81K8, 81P1, 81R3, 81T6, 82H3
	CCl_4	Si, poly-Si	SiO_2, photoresist CMS	PP	80B7, 80H8, 81H16, 82B3, 82B4
	CCl_4	$MoSi_2$	SiO_2	PP	82B3
	CCl_4	Al, Al/Si	SiO_2, Al_2O_3, photoresist, PSG	RIE	78S1, 81A5, 79H2, 80N2, 81P1, 81T2
	CCl_4	Al		RIE	76H3, 82I2
	CCl_4	Si		CDE	81N5
	CCl_4	Al/Cu/Si, Al	SiO_2, photoresist Shipley AZ 1350 J	RIBE	80S6, 81D2, 81D9
	CCl_4	Cr			78C1, 82H3
42	CCl_4/Ar	Cr_2O_3		PB	76A2, 77N1, 78C1
	CCl_4/Ar	Si	SiO_2	RIE	79S2, 79S7, 80S2
	CCl_4/Ar	Al, Al/Cu	SiO_2, Al_2O_3, MgO, photoresist Shipley AZ 1350 J	RIE, PP	78S1, 79S1, 81W4, 82R2, 81L9
	CCl_4/Ar	Al_2O_3			78C1
	CCl_4/Ar	Cr			78C1
43	CCl_4/Air	Cr		PB	76A2, 77N1, 80Y2, 80N3
44	CCl_4/Cl_2	Al		PP	79Z2
	CCl_4/Cl_2	Al	Si, SiO_2	RIE	80K3, 81H6
45	$CCl_4/Cl_2/N_2$	Si		PP	78R4, 78R5
46	$CCl_4/Cl_2/He$	Al	SiO_2	PP	82R2
47	CCl_4/He	Al	SiO_2, Si(100), Al_2O_3 photoresist Shipley AZ 1350 J	PP	78B1, 80M6, 81M15, 81T12, 82R2
	CCl_4/He	Al	SiO_2	RIE	81C1, 81T12
48	CCl_4/O_2	Si	SiO_2, Si_3N_4	RIE	81K7
	CCl_4/O_2	Cr		PB	80N3
	CCl_4/O_2	Mo		PP, RIE	80H8, 81H12, 82K1
49	$CCl_4/O_2/Ar$	Cr		PB	80N3
50	$CCl_4/O_2/N_2$	Cr		PB	80N3
51	CF_4	Si, poly-Si	SiO_2	PB	73A1, 77B7, 76M2, 81G5, 81T1
	CF_4	Si	Si_3N_4	PB	79E2, 81G5
		Si_3N_4	SiO_2		
	CF_4	Si, poly-Si	SiO_2, photoresist	PP	80H5, 80M4, 80Z1, 80H8, 80H10, 80N4, 81T1
	CF_4	Si_3N_4	Si	PP	76R1, 78C1
		Si	SiO_2, photoresist		
	CF_4	SiO_2	Si	PP	81T8
	CF_4	polymers		PP	82P1

continued

Table 25 (continued)

No.	Gases	Material	Selective to	Technique	Ref.
51	CF_4	Si	SiO_2, photoresist Shipley AZ 1350 J	RIE	79S5, 80M3, 81B15, 81A10, 81I5, 81M17, 81T1
	CF_4	Si_3N_4	Si	RIE	76B2, 79B4, 80M9
	CF_4	SiO_2	Si	RIE	76S3, 77B4, 77S3, 80M2
	CF_4	Si_3N_4	SiO_2	RIE	79M6
	CF_4	Si_3N_4	SiO_2	RIE	77K1
		SiO_2	Si, photoresist Shipley AZ 2400		
	CF_4	Si, SiO_2, Si_3N_4, Mo, W, TaN		RIE	76M2
	CF_4	Si, SiO_2	Au	RIE	78L2
	CF_4	Mo, W	SiO_2, Si_3N_4		77L3
	CF_4	Mo, W, Ta, Ti			78C1, 78M1, 82H3
	CF_4	Nb, Au, Ti, Cr, Al, Si		PB, RIE, RIBE	81H7
	CF_4	SiO_2		PT	79C1, 80H6
	CF_4	Si		CDE	81N5
	CF_4	Si			80N5
	CF_4	Si_3N_4	Au, GaAs	RIBE	81Y4
	CF_4	Si		RIBE	81D2
	CF_4	SiO_2	Si	RIBE	81H14, 81M8, 81M16, 82H2, 81M22
52	CF_n^+	Si		RIBE	81T13
53	CF_4/Cl_2	Si, poly-Si, $TaSi_2$	SiO_2	RIE	80S1, 83B2, 83M1
54	$CF_4/Cl_2/O_2$	Si		CDE	81H1
55	CF_4/CO_2	Si	SiO_2	PB	80S3, 81S9
56	CF_4/C_2H_4	SiO_2	Si	RIE	81O4
57	CF_4/H_2	SiO_2	Si	PP	79C6, 79C10, 79E4
	CF_4/H_2	Si		RIE	80K6
	CF_4/H_2	SiO_2, PSG, Si_3N_4	Si, photoresist Shipley AZ 1350 B, electron beam resist PMMA, photoresist FPM	RIE	77E1, 79C5, 79C7, 79E1, 79E3, 79E4, 80A1, 80H1, 80E3, 80H8, 80K5, 80M9, 80O2, 81E1, 81E4, 81H2, 81H12, 81M6, 81N1, 81C9, 82I2, 82R1, 83B1
	CF_4/H_2	SiN_4, SiO_2	Au, GaAs	RIBE	81Y4
58	CF_4/HF	Si	SiO_2	RIE	80S4
59	CF_4/He	SiO_2		DC-Plasma	80G1
60	CF_4/NO	Si	SiO_2	PB	81S9
61	CF_4/O_2	Si, poly-Si	Si_3N_4, SiO_2	PB	76K1, 77B7, 77P1, 78B1, 78C1, 78M7, 79D2, 79E2, 79K3,
		Si_3N_4	SiO_2		80B5, 80V1, 81A1, 81M15, 81S9, 81S12
	CF_4/O_2	W, Mo	SiO_2, Al	PB	76B1, 76K1, 80O3
	CF_4/O_2	$MoSi_2$	SiO_2	PB	80C4, 80C6, 80G4, 80M8

continued

Table 25 (continued)

No.	Gases	Material	Selective to	Technique	Ref.
61	CF_4/O_2	Si_3N_4	SiO_2	PP	81M3
	CF_4/O_2	$TaSi_2$	SiO_2	PP	80S10, 81F5
	CF_4/O_2	Si, poly-Si	SiO_2	PP	76R1, 77M3, 78M3, 79C10, 79M8, 79E4, 81B21, 82K3, 81A11
	CF_4/O_2	polymers		PP	82P1
	CF_4/O_2	W Si_2	SiO_2	PP, RIE	80E3, 80G4, 81E4, 81T5
	CF_4/O_2	Si, poly-Si, SiO_2		RIE	79C13, 80E3, 81B1, 81C1, 81E4, 82K2, 82W3
	CF_4/O_2	$MoSi_2$	SiO_2	RIE	82W3
	CF_4/O_2	Si		RIBE	81S1
	CF_4/O_2	SiO_2, Si(100)	photoresist Shipley 1350 H	RIBE	79M12, 81S1
	CF_4/O_2	SiO_2, poly-Si		RIBE	79H7
	CF_4/O_2	Au	Si_3N_4, GaAs	RIBE	81Y4
	CF_4/O_2	poly-Si	SiO_2	CDE	76H2, 81A9, 81H1, 81M18
	CF_4/O_2	Si	Al	PT	79S8, 81B20
	CF_4/O_2	TiS_2			81G6
	CF_4/O_2	Mo W, Pt	SiO_2, Al		78C1
62	$CF_4/O_2/He$	Si_3N_4 Si	Si SiO_2	PB	78J1
63	$CF_4/O_2/Ar$	Si	SiO_2	PP	76R1
64	CF_4/wet air	Si Si_3N_4, Ti	Si_3N_4, Ti SiO_2, COP resist	PB	78Y1
65	C_2F_4	Si, SiO_2		PP	79C6, 79C9
66	C_2F_4/CF_4	SiO_2	Si	PP	79C6, 79C9
67	C_2F_4/O_2	Si	SiO_2	PP	79C6, 79C9
68	C_2F_6	SiO_2	Si		77L3, 78M1
	C_2F_6	SiO_2	Si	PP	76H1, 78B1, 78B4
	C_2F_6	SiO_2, Si_3N_4	Si, photoresist Shipley AZ 1350 J	RIE	80M2
	C_2F_6	SiO_2	Si, Ti, photoresist Waycoat IC	RIE	78M5, 80M2, 81O4, 81Y2
	C_2F_6	TiO_2, V_2O_5	Ti, V	RIE	78M5, 80M2
	C_2F_6	Si		–	80N5
	C_2F_6	SiO_2	Si, poly-Si, Al	RIBE	79H7, 79V4, 80B9, 80O1, 81G2, 81H3, 81H17, 82H2, 82O2
	C_2F_6	Si		CDE	81N5
69	C_2F_6/Ar	SiO_2	Si	RIE	81Y2
70	C_2F_6/Cl_2	poly-Si, Si	SiO_2	PP	79H4, 79M9, 79M10, 80M7, 81D6, 81A11
71	C_2F_6/C_2H_4	Si_3N_4 SiO_2	SiO_2 poly-Si, photoresist Shipley AZ 1350	RIE	78M8
	C_2F_6/C_2H_4	SiO_2	Si, photoresist Shipley AZ 1350 J	RIE	78M5, 80M2, 81Y2
72	C_2F_6/CF_3Cl	poly-Si		PP	81A11

continued

Table 25 (continued)

No.	Gases	Material	Selective to	Technique	Ref.
73	C_2F_6/HF	Si, SiO_2		RIE	80S4
74	C_2F_6/O_2	SiO_2		RIE	81Y2
	C_2F_6/O_2	Si_3N_4	SiO_2, poly-Si	RIE	78M8, 81O3
	C_2F_6/O_2	Si	SiO_2	RIBE	80O1
75	C_3F_8	SiO_2	Si	PP	75H2, 76R1, 76H1
	C_3F_8	Si_3N_4 SiO_2	SiO_2, resist PMMA Si	PP	75H2, 76B2
	C_3F_8	SiO_2	Si	RIE	77I1, 80Y3, 81I5
	C_3F_8	SiO_2	poly-Si	RIBE	80B9, 78C1, 82H2
	C_3F_8	Si			80N5
76	$C_3F_8/CF_4/Ar$	SiO_2	Si	PP	81R1
77	C_4F_8	SiO_2, Si_3N_4	Si	PP	76H1
	C_4F_8	SiO_2	Si, PMMA, photoresist	RIBE	82M1
78	C_4F_8/He	Si		RIE	80K6
79	C_2HCl_3	Al		RIE	80N2
80	$CHCl_3/Cl_2$	Al	SiO_2	PP	81B7, 81B16, 83B3
81	CHF_3	SiO_2	Si		77L3
	CHF_3	SiO_2	Si	PP	76H1, 76R1, 77D1
	CHF_3	SiO_2	Si	RIE	82H1, 82O1, 81O5, 82O3, 82B1, 78L1, 78F1, 79M13, 80E5, 81M9, 81N1, 81T9, 81S2, 82N1, 81C9
	CHF_3	Si_3N_4 SiO_2	SiO_2 Si, photoresist Shipley AZ 1350	RIE	79M13, 80B4
	CHF_3	SiO_2	Si	PT	80K4
	CHF_3	Si_3N_4, SiO_2	Si, Cr, Au, photoresists Shipley AZ 1350 and Waycoat HR 100	RIE	77I1, 78L1, 78L2, 80B1, 80B6, 83B1
	CHF_3	PSG	Si	RIE	81P3
	CHF_3	SiO_2	Si, photoresist Shipley AZ 1350 H	RIBE	79M12
82	CHF_3/H_2	SiO_2	Si	RIE	80O1, 81O5, 82O1
83	CHF_3/NH_3	SiO_2, PSG	Si, photoresist Shipley AZ 2400	RIE	81S2, 82S5
84	CHF_3/N_2	SiO_2	Si	RIE	81S2
85	CHF_3/O_2	SiO_2	Si	RIE	80B6
	CHF_3/O_2	SiO_2, Si			78C6
86	$CHF_3/CF_4/Ar$	SiO_2	Si	PP	81B16
87	$CHF_3/C_2F_6/He$	SiO_2	Si	RIE	81C6
88	CHF_3/SF_6	SiO_2, PSG	Si, photoresist	RIE	82W1
89	F, F_2	Si, Ge		RIBE	81V1
	F, F_2	Si, SiO_2			81M4
90	F^+	Si		RIBE	81T11, 81T13
91	F_2/Ar	Si		RIBE	79C4, 79C5
92	H_2	Si, GaAs, GaSb, InP		PB	82C2, 82C3
93	HCl	Al, poly-Si		RIE	76H3, 83Z1

continued

Table 25 (continued)

No.	Gases	Material	Selective to	Technique	Ref.
94	H_2Cl_2	SnO_3			78B4
95	NF_3	Si_3N_4	SiO_2	RIE	82S4
	NF_3	poly-Si	SiO_2, photoresist	PP	82B4
	NF_3	$MoSi_2$	SiO_2	PP	80C7
96	NF_3/He	SiO_2		DC-plasma	80I1, 81I4
	NF_3/He	poly-Si	SiO_2	PP	81M15
97	NF_3/CF_4	SiO_2 Si	Si SiO_2	PP, RIE	80N6
98	NF_3/CF_4/Ar	polystyrene		PP, RIE	80N6
99	NF_3/O_2/Ar	Si	SiO_2	PP, RIE	80E2
100	O_2	photoresist	Si, SiO_2, Si_3N_4, Al, other metals	PB	65H1, 68I1, 70B1, 74H2, 76K3, 76V1, 77B5, 77B6, 77L3, 77W4, 78M1, 80C3
	O_2	graphite	Si, SiO_2, Si_3N_4, Al, other metals	PB	69G2
	O_2	photoresist epoxy, polyester, organic resists	Si, SiO_2, Si_3N_4, Al, other metals	RIE	81G1, 83B4
101	O_2/Ar	NiFe	Ti	RIBE	80D2
102	O_2/CF_4	photoresist	Si, SiO_2, Si_3N_4, Al, other metals	PB	77B6, 81A6
	O_2/CF_4	photoresist	Si, SiO_2, Si_3N_4, Al, other metals	CDE	81W5
103	O_2/N_2	resist	Si, SiO_2, Si_3N_4, Al, other metals		81B6
104	SF_n^+	Si		RIBE	81T13
105	SF_6	Si(100)		PP	79E4, 79W2
	SF_6	poly-Si, $MoSi_2$, Si_3N_4	SiO_2, photoresist Shipley AZ 1370	PP	80B2, 81B2, 81E5, 82B1
	SF_6	Si(100)		RIE	79E4
	SF_6	poly-Si, $MoSi_2$	SiO_2, photoresist Shipley AZ 1450 J	RIE	81B1
	SF_6	poly-Si	SiO_2, photoresist Shipley AZ 1350 J	RIE	80E1, 79G1, 81E5, 81P2, 81P3, 81T9, 82I2, 82B1
	SF_6	Si, SiO_2, Si_3N_4, Mo, W, TaN		RIE	76M2
	SF_6	Si_3N_4	SiO_2	RIE	81P3
	SF_6	Si		RIE	80P2
	SF_6	W			82H3
	SF_6	Au	GaAs	RIE	82C1
106	SF_6/Ar	poly-Si	SiO_2	PP	77D3
107	SF_6/Cl_2	Si, poly-Si, $TaSi_2$	SiO_2, Si_3N_4	RIE	80F2, 82A1, 83B2, 83M1
108	SF_6/He	Si, poly-Si, $MoSi_2$	SiO_2	RIE	79G1, 81B1
109	SF_6/H_2	Si		PP	79E4, 79W2
110	SF_6/N_2	Si	SiO_2	RIE	80L3
111	SF_6/O_2	Si, poly-Si	SiO_2	PP, RIE	79E4, 83L1
	SF_6/O_2	Si	SiO_2	PB	81A8

continued

Table 25 (continued)

No.	Gases	Material	Selective to	Technique	Ref.
112	$SiCl_4$	Al	SiO_2, PSG, poly-Si	PP	81H4, 82H3
	$SiCl_4$	Al	SiO_2	RIE	80N2, 80S12, 82H3, 82S2
	$SiCl_4$	Al	Si, SiO_2, PMMA, photoresist	RIBE	82M1
113	$SiCl_4/O_2$	Al, Si	SiO_2, Cr, MgF_2	RIE	81H11
114	SiF_4	Si_3N_4	Si, poly-Si	PB	78B4, 79B2
		Si	SiO_2		
	SiF_4	poly-Si	Si_3N_4	PB	79V3
		Si_3N_4	SiO_2		
	SiF_4	Si	SiO_2, Si_3N_4	RIE	81M17, 81M19
	SiF_4	Si	SiO_2	RIBE	80O1, 82O2
115	SiF_4/Cl_2	Si	SiO_2, Si_3N_4	RIE	81M19
116	SiF_4/He	Si	SiO_2	RIE	81F7
	SiF_4/Cl_2/Ar/He	Si	SiO_2	RIE	81F7
117	SiF_4/O_2	Si_3N_4	Si, poly-Si	PB	78B1, 79B2, 79M11, 80V1
		Si	SiO_2		
118	XeF_2	Si		RIBE	79C4, 80C11
	XeF_2	SiO_2, Si_3N_4, SiC		ECDE	79C4, 79C5, 79W1
119	XeF_2/He, Ne, Ar	Si		RIBE	80C11, 81C7

Table 26. Gases used for etching of silicon.

No.	Gases	Structure	Selective to	Technique	Etching profile	Remarks	Refs.
1	BCl_3/Cl_2	poly-Si	SiO_2	PP, RIE			79H4
2	BCl_3/PCl_3	Si	SiO_2	RIE			80N2
3	BF_n^+	Si		RIBE		investigation of etching mechanism	81T13
4	BF_3	poly-Si	SiO_2	RIBE	vertical, tapered		82O2
5	Cl_2	Si, poly-Si	SiO_2	PP	anisotropic, vertical		80B10, 81B4, 81B16, 81B21, 81Y1, 81D6, 82B2, 82B1, 83Z1
	Cl_2	poly-Si, Si	SiO_2	RIE	anisotropic, vertical	loading effect, contamination	81K1, 82F1, 82O3, 81M23, 82B1
	Cl_2	Si	SiO_2	RIBE		investigation of etching mechanism	81M16, 81S1, 81M22
	Cl_2	poly-Si	Si_3N_4	PCDE			81S8
6	Cl^+	Si		RIBE		investigation of etching mechanism	81T13
7	Cl_2/Ar	Si	SiO_2	RIE	tapered	trench etching	78B8, 80L4, 81H13, 81B24, 81F7, 83P1
	Cl_2/Ar	Si	SiO_2	RIBE		Ar^+-ions	79C4, 81M20
8	Cl_2/Ar, He	Si	SiO_2, Al_2O_3, MgO	RIE	vertical, tapered		79S2, 79S8, 80H3, 83Z1
9	Cl_2/Ar, He, Ne	Si		RIBE		etch yield is given	80C11, 81C7
10	Cl_2/ClF_3	Si, poly-Si	SiO_2	RIE	isotropic, anisotropic	undercutting depends on ClF_3 content	81F6
11	Cl_2/H_2	Si, poly-Si	SiO_2	RIE	anisotropic	RIE with magnetic field	81O5, 82O1
12	Cl_2/O_2	Si	SiO_2, Al_2O_3	RIE	anisotropic		81H11
13	ClF_3/Cl_2	Si		RIE	anisotropic	undercutting depends on Cl_3F	81F1
14	$CBrF_3$	Si, poly-Si	Si_3N_4, SiO_2, resist	RIE	vertical		80M3
15	$CBrF_3/Cl_2$	poly-Si	SiO_2	RIE	vertical	high selectivity	81S11, 82S1
16	$CBrF_3/He$	Si	SiO_2	PP			79M9, 79M10
17	$CClF_3$	Si, poly-Si	SiO_2	PP	vertical, overhang		79M8, 79M9, 79M10, 80H8, 80H10, 81D6, 81A11
18	$CClF_3/He$	Si	SiO_2	PP			79M3, 79M9
19	$CClF_3/C_2F_6$	poly-Si	SiO_2	PP	vertical, overhang		79M8, 79H4

continued

Table 26 (continued)

No.	Gases	Structure	Selective to	Technique	Etching profile	Remarks	Refs.
20	CCl_2F_2	Si	SiO_2	RIE			76B2, 76M2
	CCl_2F_2	Si, poly-Si	SiO_2	PP	vertical		80H8, 80H10, 81H12
21	CCl_2F_2/C_2F_6	poly-Si	photoresist	RIE	vertical		82W3
22	CCl_3F	Si, poly-Si	SiO_2, photoresist	PP	vertical		80H8, 81M5
23	CCl_3F/O_2	Si, poly-Si	Si_3N_4, SiO_2	RIE			80E1
24	C_2ClF_5	Si, poly-Si		PP	anisotropic		81D6, 78B4, 82K3
25	$C_2Cl_2F_4$	poly-Si	SiO_2, photoresist	PP, RIE	anisotropic		81Y5
26	CCl_4	Si, poly-Si	SiO_2, photoresist	PP	anisotropic		80B7, 80H8, 81H16, 82B3, 82B4
	CCl_4	Si		CDE		results of IR-absorption spectroscopy	81N5
27	CCl_4/Ar	Si	SiO_2	RIE	vertical		79S2, 79S7, 80S2
28	$CCl_4/Cl_2/N_2$	Si		PP	anisotropic		78R4, 78R5
29	CCl_4/O_2	Si	SiO_2, Si_3N_4	RIE	vertical, tapered	V-groove isolation	81K7
30	CF_4	Si, poly-Si	SiO_2	PB	isotropic		73A1, 77B7, 81G5, 81T1, 76M2
	CF_4	Si	Si_3N_4, SiO_2	PB	isotropic		79E2, 81G5
	CF_4	Si, poly-Si	SiO_2	PP	isotropic, anisotropic		76R1, 78C1, 80H5, 80Z1, 80H8, 80H10, 80H11, 81T1, 81T8
	CF_4	poly-Si	SiO_2, photoresist Shipley AZ 1350 H	PP	vertical, anisotropic	radial flow reactor	80M4
	CF_4	Si	SiO_2, photoresist Shipley AZ 1350 J	RIE	isotropic, anisotropic	etching profile depends on etching parameters, mask, and cathode material	79S5, 80M3, 81M17, 81T1
	CF_4	Si	Au	RIE			78L2
	CF_4	Si		CDE		results of IR absorption spectroscopy	81N5
	CF_4	Si		RIE		comparable etch rates of Si, SiO_2, Si_3N_4; etch rate of photoresist Shipley AZ 1350 J high; etching profile depends on process; study of reaction process	76S3, 79B4, 81B15, 81A10, 81I5
	CF_4	Si					80N5
	CF_4	Si		RIBE	anisotropic		81D2
31	CF_4/CO_2, NO	Si	SiO_2	PB	isotropic		80S3, 81S9
32	CF_4/Cl_2	Si	SiO_2	RIE	anisotropic		80S1, 83B2, 83M1
33	CF_4/H_2	Si		RIE		RIE-induced traps in Si	80K6

continued

Table 26 (continued)

No.	Gases	Structure	Selective to	Technique	Etching profile	Remarks	Refs.
34	CF_4/HF	Si	SiO_2	RIE			80S4
35	CF_4/O_2	Si, poly-Si	Si_3N_4, SiO_2	PB	isotropic		76K1, 77B7, 77P1, 78B1, 78C1, 78M7, 79E2, 79D2, 79K3, 80B5, 80V1, 81A1, 81S9
	CF_4/O_2	Si, poly-Si	SiO_2	PP	isotropic		76R1, 77M2, 78M3, 79C10, 79M8, 79E4, 81B21, 82K3
	CF_4/O_2	Si, poly-Si, SiO_2		RIE	anisotropic, vertical	chrome mask; vertical edge profile at low gas pressure	79C13, 80E3, 81B1, 81C1, 81E4, 82K2, 82W3
	CF_4/O_2	poly-Si		CDE	anisotropic	low etch rate, aerodynamic nozzle	81A9
	CF_4/O_2	Si, poly-Si	SiO_2, Al	CDE	isotropic, anisotropic	etching profile anisotropic at pressures lower than 10^{-1} Pa	76H2, 81H1, 81M18
	CF_4/O_2	Si	Al	PT	anisotropic, vertical	etching profile dependence on angle of incidence	79S8, 81B20
	CF_4/O_2	Si		RIBE		product detection	81S1
36	$CF_4/O_2/Ar$	Si	SiO_2	PP			76R1
37	$CF_4/O_2/Cl_2$	Si	SiO_2	RIE	anisotropic		80S1
38	$CF_4/O_2/He$	Si	SiO_2	PB	isotropic		78J1
39	$CF_4/O_2/Cl_2$	Si		CDE			81H1
40	CF_4/wet air	Si	SiO_2, Si_3N_4, Ti	PB	isotropic		78Y1
41	CF_n^+	Si		RIBE		investigation of etch mechanism	81T13
42	C_2F_4	Si		PP			79C6, 79C9
43	C_2F_4/O_2	Si	SiO_2	PP			79C6, 79C9
44	C_2F_6	Si		CDE		results of IR absorption spectroscopy	81N5
	C_2F_6	Si				study of reaction process	80N5
45	C_2F_6/Cl_2	poly-Si	SiO_2	PP			79H4, 79M9, 79M10, 80M7, 81D6, 81A11
46	C_2F_6/CF_3Cl	poly-Si		PP	anisotropic	overhang for over etching	81A11
47	C_2F_6/O_2	Si	SiO_2	RIBE			80O1
48	C_2F_6/HF	Si		RIE			80S4

continued

Table 26 (continued)

No.	Gases	Structure	Selective to	Technique	Etching profile	Remarks	Refs.
49	C_3F_8	Si				study of reaction process	80N5
50	C_4F_8/He	Si		RIE		RIE-induced traps in Si	80K6
51	F, F_2	Si		RIBE	isotropic	study of kinetics and mechanism	81V1, 81M4
52	F^+	Si				chemical and physical sputtering	81T11, 81T13
53	F_2/Ar	Si		RIBE	anisotropic		79C4, 79C5
54	H_2	Si		PB			82C2, 82C3
55	NF_3	poly-Si	SiO_2	PP	isotropic		82B4
56	$NF_3/Ar/O_2$	Si	SiO_2	PP, RIE			80E2
57	NF_3/CF_4	Si	SiO_2	PP, RIE			80N6
58	NF_3/He	poly-Si	SiO_2	PP, DC-Plasma		end point detection	81M15, 81I4
59	SF_n	Si		RIBE			81T13
60	SF_6	Si, poly-Si	SiO_2, photoresists Shipley AZ 1350 J and AZ 1450 J	RIE	anisotropic, vertical	etching profile depends on process	79G1, 80E1, 81B1, 79E4, 76M2, 80P2, 81E5, 81P2, 81P3, 81T9, 82I2, 82B1, 83B1
	SF_6	Si, poly-Si	SiO_2, photoresist Shipley AZ 1370	PP	anisotropic	large undercutting for overetched poly-Si	80B2, 79W2, 79E4, 81B2, 81E5, 83B1
61	SF_6/Ar	poly-Si	SiO_2	PP	tapered		77D3
62	SF_6/Cl_2	Si, poly-Si	SiO_2, Si_3N_4	RIE	vertical	high selectivity to SiO_2, Si_3N_4	80F2, 83B2, 83M1
63	SF_6/He	Si	SiO_2	RIE	isotropic, vertical	etching profile depends on process	79G1, 81B1
64	SF_6/H_2	Si		PP			79E4, 79W2
65	SF_6/N_2	Si	SiO_2	RIE	isotropic, anisotropic	low selectivity Si:SiO_2	80L3
66	SF_6/O_2	Si, poly-Si	SiO_2	PP, RIE	vertical, tapered		79E4, 83L1
	SF_6/O_2	Si	SiO_2	PB		etch reaction study	81A8
67	$SiCl_4/O_2$	Si	SiO_2, Cr, MgF_2	RIE	anisotropic		81H11
68	SiF_4	poly-Si, Si	Si_3N_4, SiO_2	PB	isotropic		79V3, 78B4, 79B2
	SiF_4	Si	SiO_2	RIBE			80O1, 82O2
	SiF_4	Si	SiO_2, Si_3N_4	RIE	anisotropic	plasma diagnostic	81M17, 81M19

continued

Table 26 (continued)

No.	Gases	Structure	Selective to	Technique	Etching profile	Remarks	Refs.
69	SiF_4/He	Si	SiO_2	RIE	anisotropic		81F7
70	SiF_4/O_2	Si	SiO_2	PB	isotropic		78B1, 79B2, 79M11, 80V1
71	SiF_4/Cl_2	Si	SiO_2, Si_3N_4	RIE	anisotropic	plasma diagnostic	81M19
72	$SiF_4/Cl_2/Ar/He$	Si	SiO_2	RIE	anisotropic	variation of gas ratios	81F7
73	XeF_2	Si		RIBE		XeF_2 toxic	79C4, 80C11
74	XeF_2/Ar, Ne, He	Si		RIBE			81C7, 80C11

Table 27. Gases used for etching of oxides.

No.	Gases	Material	Selective to	Technique	Etching profile	Remarks	Refs.
1	$CBrF_3$	SiO_2		PCDE			80S5
2	CCl_2F_2	SiO_2		RIE			76M2
3	C_2ClF_5	SiO_2		PP	anisotropic		78B4
4	CF_4	SiO_2	Si	PP		Auger spectroscopy	81T8
	CF_4	SiO_2		RIE	vertical		76S5, 77B4, 77K1, 77S3, 80M2, 81B15
	CF_4	SiO_2	Au	RIE	vertical		78L2
	CF_4	SiO_2		PT	vertical		79C1, 80H6
	CF_4	SiO_2	Si	RIBE	tapered	comparison of Ar and CF_4	81H14, 81M8, 81M16, 82H2, 81M22
5	CF_4/C_2H_4	SiO_2	Si	RIE		mass spectrum analysis	81O4
6	CF_4/H_2	SiO_2	Si	PP			79C6, 79C10, 79E4
	CF_4/H_2	SiO_2	Au, Ti, Pt, GaAs, AuGe	RIBE	tapered	sputter-SiO_2	81Y4
	CF_4/H_2	SiO_2, PSG	Si, resists Shipley AZ 1350 B and PMMA, Mo	RIE	vertical	frequently used for vertical SiO_2 etching	77E1, 79C5, 79C7, 79E1, 79E3, 79E4, 80A1, 80H1, 80E3, 80H8, 80K5, 81E1, 81E4, 81H2, 81H12, 81M6, 81N1, 82I2, 82R1, 81C9, 83B1 continued

Table 27 (continued)

No.	Gases	Material	Selective to	Technique	Etching profile	Remarks	Refs.
6	CF_4/H_2	SiO_2	Si	RIE		investigation of contamination and damage induced by RIE	80O2
7	CF_4/He	SiO_2		DC-plasma		no commercial equipment	80G1
8	CF_4/O_2	SiO_2		PB	isotropic		81S12
	CF_4/O_2	SiO_2		RIBE	vertical, tapered		79H7, 79M12
9	C_2F_4	SiO_2		PP			79C6, 79C9
10	C_2F_4/CF_4	SiO_2	Si	PP			79C6, 79C9
11	C_2F_6	SiO_2	Si				77L3, 78M1
	C_2F_6	SiO_2	Si	PP	anisotropic		76H1, 78B1, 78B4
	C_2F_6	SiO_2	Si, resist Shipley AZ 1350 J	RIE	vertical	mass spectra	78M5, 80M2, 81O4, 81Y2
	C_2F_6	SiO_2	Si, Ti, resist Waycoat IC	RIE	vertical		80M2
	C_2F_6	SiO_2	Si, poly-Si, Al	RIBE	vertical, tapered	etching profile depends on process	79H7, 79V4, 80B9, 80O1, 81G2, 81H3, 81H17, 82H2, 82O2
12	C_2F_6/Ar, HF	SiO_2	Si	RIE		investigation of Si-damage	81Y2, 80S4
13	C_2F_6/C_2H_4	SiO_2	Si, resist Shipley AZ 1350 J	RIE			78M5, 80M2, 81Y2
	C_2F_6/C_2H_4	SiO_2	poly-Si, resist Shipley AZ 1350	RIE			78M8
14	C_2F_6/O_2	SiO_2	Si	RIE		investigation of Si-damage	81Y2
15	C_3F_8	SiO_2	Si	PP			75H2, 76H1, 76R1, 76B2
	C_3F_8	SiO_2, PSG	Si, resist Shipley AZ 1370	RIE	vertical		77I1, 80Y3, 81I5
	C_3F_8	SiO_2	poly-Si	RIBE	vertical, tapered	etching profile depends on process	80B9, 78C1, 82H2
16	$C_3F_8/CF_4/Ar$	SiO_2	Si	PP	vertical		81R1
17	C_4F_8	SiO_2	Si	PP			76H1
	C_4F_8	SiO_2	Si, PMMA, photoresist	RIBE	vertical		82M1
18	CHF_3	SiO_2	Si	RIE	vertical	high etch rate RIE with magnetic field	81O5, 82H1, 82O1

continued

Table 27 (continued)

No.	Gases	Material	Selective to	Technique	Etching profile	Remarks	Refs.
18	CHF_3	SiO_2	Si				77L3
	CHF_3	SiO_2	Si	PP			77D1, 76H1, 76R1
	CHF_3	SiO_2	Si, Cr, Au, resists Shipley AZ 1350 and Waycoat HR 100	RIE	vertical	frequently used for SiO_2	77I1, 78F1, 78L1, 78L2, 81C9, 79M13, 80B1, 80B4, 80B6, 80E5, 81M9, 81N1, 81P3, 81S2, 81T9, 82O3, 82O1, 82N1, 82B1, 79M12, 83B1
	CHF_3	SiO_2	Si, resist Shipley AZ 1350 H	RIBE	vertical		
	CHF_3	SiO_2	Si	PT			80K4
19	CHF_3/O_2	SiO_2	Si	RIE	tapered	lateral etching of photoresist produces tapered edge	80B6
	CHF_3/O_2	SiO_2					78C6
20	$CHF_3/CF_4/Ar$	SiO_2	Si	PP	vertical		81B16
21	CHF_3/SF_6	SiO_2	Si, photoresist	RIE	vertical, tapered		82W1
22	$CHF_3/C_2F_6/He$	SiO_2	Si	RIE	vertical	high rf power density	81C6
23	CHF_3/H_2	SiO_2	Si	RIE	anisotropic, vertical	high etch rate, magnetic field	81O1, 81O5, 82O1
24	CHF_3/N_2, NH_3	SiO_2	Si	RIE			81S2, 82S5
25	NF_3	SiO_2		DC-plasma		no commercial equipment	80I1
26	NF_3/CF_4	SiO_2	Si	RIE			80N6
27	SF_6	SiO_2		RIE			76M2
28	XeF_2	SiO_2		ECDE		XeF_2 toxic	79C4, 79C5, 79W1
29	XeF_2/Ar, Ne, He	SiO_2		RIBE		XeF_2 toxic	81C7
30	BCl_3	Al_2O_3				no selectivity to Al	78C1
31	CCl_4/Ar	Al_2O_3				no selectivity to Al	78C1
32	Cl_2/Ar	Al_2O_3		RIE		no selectivity to Al	80H3
	Cl_2/Ar	Cr_2O_3					78C1
33	CCl_4/Ar	Cr_2O_3, Cr_3O_4		PB	isotropic		76A2, 77N1, 78C1
34	H_2Cl_2	SnO_3					78B4
35	C_2F_6	TiO_2	Ti, V	RIE			80M2
	C_2F_6	V_2O_2	V, Ti	RIE			80M2

Table 28. Gases used for etching of nitrides.

No.	Gases	Material	Selective to	Technique	Etching profile	Remarks	Refs.
1	CF_4	Si_3N_4	SiO_2	PB	isotropic		79E2
	CF_4	Si_3N_4	Si, SiO_2, photoresist	PP			76R1, 78C1
	CF_4	Si_3N_4	Si, SiO_2, photoresist	RIE			77K1
	CF_4	Si_3N_4	SiO_2	RIE			79M6
	CF_4	Si_3N_4	Si	RIE		high etch rate on photoresist Shipley AZ 1350 J	76B2, 79B4
	CF_4	Si_3N_4		RIE	anisotropic		80M9
	CF_4	Si_3N_4	Au, GaAs	RIBE	tapered		81Y4
2	CF_4/H_2	Si_3N_4	Si, photoresist CMS	RIE			80H8, 80K5, 80M9, 81H12
	CF_4/H_2	Si_3N_4	Au, Ti, Pt, GaAs	RIBE	tapered	PCVD-Si_3N_4	81Y4
3	CF_4/O_2	Si_3N_4	GaAs	RIBE	tapered		81Y4
	CF_4/O_2	Si_3N_4	SiO_2	PP	anisotropic	endpoint detection	81M15, 81M3
	CF_4/O_2	Si_3N_4	SiO_2	PB	isotropic	frequently used	76K1, 77B7, 77P1, 78B1, 78C1, 78M7, 79D2, 79E2, 79K3, 80B5, 80V1
4	$CF_4/O_2/He$	Si_3N_4	Si, SiO_2	PB	isotropic		78J1
5	CF_4/Wet air	Si_3N_4	SiO_2	PB	isotropic		78Y1
6	C_2F_6	Si_3N_4	Si, poly-Si, SiO_2, photoresist Shipley AZ 1350 J	RIE			78M5, 80M2
7	C_2F_6/C_2H_4	Si_3N_4	SiO_2, poly-Si, photoresist	RIE		used in MOS LSI process	78M8
8	C_2F_6/O_2	Si_3N_4	SiO_2, poly-Si photoresist	RIE		used in MOS LSI process	78M8, 81O3
9	C_3F_8	Si_3N_4	SiO_2, resist PMMA	PP			75H2, 76B2
10	C_4F_8	Si_3N_4	Si	PP			76H1
11	CHF_3	Si_3N_4	SiO_2, Si, photoresist Shipley AZ 1350 H/J	RIE	vertical		79M13, 80B4
	CHF_3	Si_3N_4	Si, Cr, Au, photoresists Shipley AZ 1350 and waycoat HR 100	RIE			77I1, 78L1, 78L2, 80B1, 80B6, 83B1
12	$CBrF_3$	Si_3N_4	SiO_2, photoresist	RIE			80M3

continued

Table 28 (continued)

No.	Gases	Material	Selective to	Technique	Etching profile	Remarks	Refs.
13	CCl_2F_2	Si_3N_4		RIE			76M2
14	NF_3	Si_3N_4	SiO_2, photoresist	RIE	anisotropic	selectivity $Si_3N_4 : SiO_2 \simeq 10:1$	82S4
15	SF_6	Si_3N_4	SiO_2, photoresist	RIE	anisotropic		76M2, 81B2, 81P3
16	SiF_4	Si_3N_4	Si, poly-Si, SiO_2	PB	isotropic		78B4, 79B2
	SiF_4	Si_3N_4	SiO_2	PB	isotropic		79V3
17	SiF_4/O_2	Si_3N_4	Si, poly-Si, SiO_2	PB	isotropic		78B1, 79B2, 79M11, 80V1
18	XeF_2	Si_3N_4		ECDE			79C4, 79C5, 79W1
19	CCl_2F_2	TaN		RIE			76M2
20	CF_4	TaN		RIE			76M2
21	SF_6	TaN		RIE			76M2

Table 29. Gases used for etching metals.

No.	Gases	Material	Selective to	Technique	Etching profile	Remarks	Refs.
1	Br_2	Al		RIE			76H3
2	$BrCl_3$	Al	Al_2O_3	PP			80A3
3	BBr_3	Al		RIE			80N2
4	BCl_3	Al, Al/Si	SiO_2, photoresist Al_2O_3	PP	anisotropic		75H1, 76P1, 81C5, 81D6, 81T6, 82H3
	BCl_3	Al	SiO_2, Al_2O_3	RIE	anisotropic		77L3, 80N2, 81L1, 82H3
5	BCl_3/Cl_2	Al, Al/Si; Al/Cu	SiO_2	PP	anisotropic, vertical	Al/Cu: 4 %Cu; Al/Si: 2 %Si; Al/Si/Cu: 0.75 %Si, 0.5 % Cu	79L3, 79M9, 80H11, 81D6, 82H3, 83B3
	BCl_3/Cl_2	Al/Ti/Si, Al	SiO_2, Si	RIE	anisotropic, vertical		82B1, 82A1, 83B1, 83S1
6	$BCl_3/Cl_2/He$	Al, Al/Si	SiO_2, Si	PP	isotropic		82R2
7	$BCl_3/Cl_2/CHCl_3/$ He	Al, Al/Si	SiO_2	PP	anisotropic		82R2
8	$BCl_3/Cl_2/SiCl_4/$ He	Al, Al/Si	SiO_2	PP	vertical		82R2
9	BCl_3/CCl_4	Al	SiO_2, Si	PP, RIE	anisotropic		81T12

continued

Table 29 (continued)

No.	Gases	Material	Selective to	Technique	Etching profile	Remarks	Refs.
10	BCl_3/PCl_3	Al, Al/Cu	SiO_2, Si	RIE	vertical	2 %Cu	80N2, 80S12
11	$BCl_3/CF_4/O_2$	Al/Si, Al/Si/Cu	SiO_2, Si, photoresist	RIE	anisotropic		81M13
12	Cl_2	Al	Al_2O_3	PP	anisotropic		81B21, 81D6
	Cl_2	Al		RIE			76H3, 80N2, 81D6
	Cl_2	Al		RIBE		detection of reaction products	81S1
13	Cl_2/H_2	Al		RIE	anisotropic	high etch rate; magnetic field	81O5, 82O1, 82O3
14	CCl_2F_2	Al		RIE			80N2
15	CCl_4	Al, Al/Si	SiO_2, Al_2O_3, photoresist, PSG	PP	anisotropic, vertical	etching profile depends on process	76P1, 78B4, 79H5, 79T2, 80M6, 80T1, 80B7, 80F1, 80R2, 80W2, 80A3, 80H8, 81D6, 81H12, 81K8, 81P1, 81R3, 81T6, 82H3
	CCl_4	Al, Al/Si	SiO_2, Al_2O_3, photoresist, PSG	RIE	anisotropic, vertical		78S1, 79H2, 80N2, 81A5, 81P1, 81T2, 82H3, 82I2
	CCl_4	Al, Al/Si/Cu	SiO_2, photoresist	RIBE	tapered, vertical		80S6, 81D2, 81D9
16	CCl_4/Ar	Al, Al/Cu	SiO_2, Al_2O_3, MgO, photoresist	RIE	vertical, anisotropic		78S1, 79S1, 81W4, 81L9
	CCl_4/Ar	Al	SiO_2	PP	anisotropic		82R2
17	CCl_4/Cl_2	Al		PP	anisotropic	two-step process: CCl_4 ; CCl_4/Cl_2	79Z2
	CCl_4/Cl_2	Al	SiO_2, Si	RIE	anisotropic, vertical		80K3, 81H6
18	$CCl_4/Cl_2/He$	Al	SiO_2	PP	isotropic, anisotropic	etching profile depends on process	82R2
19	CCl_4/He	Al	SiO_2, Si, Al_2O_3, photoresist	PP	vertical, anisotropic		78B1, 80M6, 81M15, 81T12, 82R2
	CCl_4/He	Al	SiO_2	RIE	anisotropic, vertical		81C1, 81T12
20	C_2HCl_3	Al		RIE			80N2
21	$CHCl_3/Cl_2$	Al	SiO_2	PP	isotropic, anisotropic, vertical		81B7, 81B16, 83B3

continued

Table 29 (continued)

No.	Gases	Material	Selective to	Technique	Etching profile	Remarks	Refs.
22	HCl	Al		RIE			76H3
23	$SiCl_4$	Al	SiO_2, PSG, poly-Si	PP	anisotropic		81H4, 82H3
	$SiCl_4$	Al	SiO_2	RIE	anisotropic		80N2, 80S12, 82H3, 82S2
	$SiCl_4$	Al	Si, SiO_2, PMMA, photoresist	RIBE	vertical, tapered		82M1
24	$SiCl_4/O_2$	Al	SiO_2, Cr, MgF_2	RIE	anisotropic	etch rate of Al depends on O_2-content	81H11
25	SF_6	Au	GaAs	RIE	anisotropic		82C1
26	$CClF_3$	Au					78M1, 82H3
27	$C_2Cl_2F_4$	Au		PB, PP, RIE	isotropic, anisotropic		77P1, 78C1, 82H3
28	CF_4	Au		PB, RIE, RIBE			81H7
29	Cl_2	Au		RIE			81Z2
30	Cl_2	Cr					78C1, 82H3
31	CCl_4	Cr					78M1, 82H3
32	CCl_4/Ar	Cr					78C1
33	$CCl_4/Ar/O_2$	Cr		PB	isotropic		80N3
34	CCl_4/Air	Cr		PB	isotropic		76A2, 77N1, 80Y2, 80N3
35	$CCl_4/N_2/O_2$	Cr		PB	isotropic		80N3
36	CCl_4/O_2	Cr		PB	isotropic		80N3
37	Cl_2/O_2/Ar, He	Cr		PB	isotropic		80N3
38	$CBrF_3$	Mo	Si_3N_4, SiO_2, resist	RIE			80M3
39	CCl_2F_2	Mo		RIE			76M2
40	CCl_4/O_2	Mo	SiO_2, photoresist	PP, RIE			80H8, 81H12, 82K1
41	CF_4	Mo					76M2, 77L3, 78C1, 78M1
42	CF_4/O_2	Mo	SiO_2	PB	isotropic		76B1, 76K1, 80O3
43	NF_3	Mo	SiO_2	PP			80C7
44	SF_6	Mo		RIE			76M2

continued

Table 29 (continued)

No.	Gases	Material	Selective to	Technique	Etching profile	Remarks	Refs.
45	O_2/Ar	NiFe	Ti	RIBE			80D2
46	CF_4	Nb		PB, RIE, RIBE	isotropic, anisotropic	usable with resists: PMMA, Shipley AZ 1350	81H7
47	$CBrF_3$	Pt					78M1
48	CCl_2F_4/O_2	Pt					78C1
49	$C_2Cl_3F_3/O_2$	Pt					78C1
50	CF_4/O_2	Pt	SiO_2				78C1
51	Cl_2	Sn					78B4
52	CF_4	Ta					78C1, 78M1
53	Ar/CF_4, O_2, CO, CO_2	Ti		RIBE		reactive species generated by ion bombardment of solid material	81A7
54	$CBrF_3$	Ti					78M1, 82H3
55	$CBrF_3/He/O_2$	Ti	Si_3N_4, Au, Pd	PP			77M4
56	$CClF_3$	Ti					78M1, 82H3
57	CF_4	Ti		PB, RIE, RIBE	isotropic, anisotropic		81H7, 82H3
	CF_4	Ti					78C1, 78M1
58	CCl_2F_2	W		PB			77F1
	CCl_2F_2	W		RIE			76M2
59	CF_4	W					76M2, 77L3, 78C1, 78M1, 82H3
60	CF_4/O_2	W		PB	isotropic		76B1, 76K1
	CF_4/O_2	W	SiO_2	PP		endpoint detection	81M15
	CF_4/O_2	W					78C1
61	SF_6	W		RIE			76M2, 82H3

Table 30. Gases used for etching of metal silicides.

No.	Gases	Material	Selective to	Technique	Etching Profile	Remarks	Refs.
1	CF_4/O_2	$MoSi_2$	SiO_2, photoresist Shipley AZ 1470	PB	isotropic	sputtered $MoSi_2$	80C4, 80C6
	CF_4/O_2	$MoSi_2$	SiO_2	PB, RIE	anisotropic, isotropic	sputtered or evaporated $MoSi_2$	80M8, 81B1, 82W3
	CF_4/O_2	$MoSi_2$	SiO_2		isotropic	sputtered or evaporated $MoSi_2$	80G4
2	NF_3	$MoSi_2$	SiO_2	PP	anisotropic	sputtered $MoSi_2$	80C7
3	SF_6	$MoSi_2$	SiO_2	PP	anisotropic	sputtered $MoSi_2$ on poly-Si, photoresist Shipley AZ 1370	80B2
4	SF_6/He	$MoSi_2$	SiO_2	RIE	anisotropic, vertical	sputtered or evaporated $MoSi_2$ on poly-Si photoresist Shipley AZ 1450	81B1, 82B1
5	CCl_4	$MoSi_2$	SiO_2	PP	anisotropic	$MoSi_2$ on poly-Si	82B3
6	CCl_4/O_2	$MoSi_2$	SiO_2	PP	isotropic	$MoSi_2$ on poly-Si	82B3
7	CF_4/O_2	$TaSi_2$	SiO_2	PP	anisotropic	$TaSi_2$, evaporated or sputtered	80S10
	CF_4/O_2	$TaSi_2$			anisotropic	high gas flow required	81F5
8	CF_4/Cl_2	$TaSi_2$	SiO_2	RIE	anisotropic	$TaSi_2$ on poly-Si	83B2, 83M1
9	SF_6/Cl_2	$TaSi_2$	SiO_2	RIE	anisotropic, vertical	$TaSi_2$ on poly-Si	82A1, 83B2, 83M1
10	CCl_4	$TiSi_2$	SiO_2	PP	vertical		81L7
11	CF_4/O_2	$TiSi_2$					81G6
12	CF_4/O_2	$W Si_2$	SiO_2		isotropic	$W Si_2$, sputtered or evaporated	80G4
	CF_4/O_2	$W Si_2$	SiO_2	PP, RIE	anisotropic	$W Si_2$ on poly-Si	80E3, 81E4
	CF_4/O_2	$W Si_2$	SiO_2	RIE	anisotropic	$W Si_2$ on poly-Si	81T5

Table 31. Gases used for etching of organic films.

No.	Gases	Material	Selective to	Technique	Remarks	Refs
1	CF_4/CO_2	photoresist, HPR 204		PB	not selective to Si, SiO_2, Si_3N_4	81S9
2	CF_4/NO	photoresist, HPR 204		PB	not selective to Si, SiO_2, Si_3N_4	81S9
3	CF_4/O_2	polymers		PP	large number of polymers	82P1
4	$NF_3/CF_4/Ar$	polystyrene		RIE, PP		80N6

continued

Table 31 (continued)

No.	Gases	Material	Selective to	Technique	Remarks	Refs.
5	O_2	photoresist	Si, SiO_2, Si_3N_4, Al, other metals	PB	sodium contamination possible	65H1, 68I1, 70B1, 74H2, 76K3, 76V1, 77B5, 77B6, 77L3, 77W4, 78M1, 80C3
	O_2	photoresist, epoxy, polyester, organic resists	Si, SiO_2, Si_3N_4, Al, other metals	RIE	vertical etching profile, anisotropic	81G1, 83B4
	O_2	graphite	Si, SiO_2, Si_3N_4, Al, other metals			69G2
6	O_2/CF_4	photoresist	Si, SiO_2, Si_3N_4, Al, other metals	PB	no sodium contamination	77B6, 77B7, 81A6
	O_2/CF_4	photoresist	Si, SiO_2, Si_3N_4, Al, other metals	CDE	chemiluminescence used for endpoint detection	81W5
7	O_2/N_2	resist	Si, SiO_2, Si_3N_4, Al, other metals	PB	investigation of degradation of IC-devices	81B6

References for 6.1.4.7.5···6.1.4.7.9

1852G1 Grove, W.R.: Philos. Trans. R. Soc. London **142** (1852) 87.
54C1 Castaign, R., Laborie, P.: C.R. Acad. Sci. **238** (1954) 1885.
55K1 Keywell, F.: Phys. Rev. **97** (1955) 1611.
55W1 Wehner, G.K.: Adv. Electron. Phys. **7** (1955) 239.
55W2 Wehner, G.K.: J. Appl. Phys. **26** (1955) 1056.
57M1 Morgulis, N.D., Tischenko, V.D.: Bull. Acad. Sci. USSR, Phys. Ser. **20** (1957) 1082.
59G1 Guseva, M.I.: Fiz. Tverd. Tela **1** (1959) 1540. Sov. Phys. Solid State **1** (English transl.) (1959) 1410.
59W1 Wehner, G.K.: J. Appl. Phys. **30** (1959) 1762.
60B1 Boudart, M.: The Surface Chemistry of Metals and Semiconductors (H.C. Gatas, ed.), New York, Wiley **1960** 409.
60G2 Grau, G.G.: Landolt Börnstein, 6th Edition **II/2a** (1960).
60K1 Kaufman, F., Kelso, J.R.: J. Chem. Phys. **32** (1960) 301.
60K2 Kaufman, F., Kelso, J.R.: Int. Combust. Symp. 8th, **1960** 230.
60R1 Rol, P.K., Fluit, J.M., Kistemaker, J.: Electromagnetic Separation of Radioactive Isotopes, p. 207, Berlin and New York, Springer Verlag **1960**.
60Y1 Yonts, O.C., Normand, C.E., Harrison, D.E.: J. Appl. Phys. **31** (1960) 447.
61A1 Almén, O., Bruce, G.: Nucl. Instrum. Methods **11** (1961) 257.
61A2 Almén, O., Bruce, G.: Nucl. Instrum. Metals **11** (1961) 279.
61A3 Almén, O., Bruce, G.: Trans. Natl. Vac. Symp. 8th, Washington, D.C., **1961** 245.
61F1 Fluit, J.M.: Colloq. Intern. Centre Natl. Rech. Sci., Bellevue **1961**
61L1 Laegreid, N., Wehner, G.K.: J. Appl. Phys. **32** (1961) 365.
61M1 Mc Keown, D., Cabezas, A., Mackenzie, E.T.: Annu. Rep. Low Energy Sputtering Stud.; Space Sci. Lab., General Dynamics, July **1961**.
61P1 Pitkin, E.T.: Electrostatic Propulsion (Langmuir, D.B., Stuhlinger, E., Sellen, J.M., eds.) Academic Press, New York (1961) 195.
61R1 Rol, P.K., Fluit, J.M., Kistemaker, J.: Electrostatic Propulsion (Langmuir, D.B., Stuhlinger, E., Sellen, J.M., eds.), New York, Academic Press, **1961** 203.
61W1 Weysenfeld, C.H., Hoogendorn, A.: Proc. Conf. Ion. Phenom. Gases, 5th Munich. **1** (1961) 124.
61W2 Weysenfeld, C.H., Hoogendorn, A., Koedam, M.: Physica **27** (1961) 963.
62A1 Anderson, G.S., Mayer, W.N., Wehner, G.K.: J. Appl. Phys. **33** (1962) 2991.
62K1 Kay,, E.: Adv. Electron. Phys. **17** (1962) 245.
62M1 Mc Keown, D., Cabezas, A.Y.: Annu. Rep. Space Sci. Lab., General Dynamics, July **1962**.
62P1 Patterson, H., Tomlin, D.H.: Proc. R. Soc. **A 265** (1962) 474.
62R1 Rosenberg, D., Wehner, G.K.: J. Appl. Phys. **33** (1962) 1842.
62W1 Wehner, G.K.: Rep. No. 2309. Minneapolis, General Mills **1962**.
63B1 Buttler, H.S., Kino, G.S.: Phys. Fluids **6** (1963) 1346.
63S1 Southern, A.L., Willis, W.R., Robinson, M.T.: J. Appl. Phys. **34** (1963) 153.
64S1 Snouse, T.W.: NASA Tech. Note **D 2235** (1964).
65A1 Anderson, H.H., Sigmund, P.: Danish Atomic Energy Commission, Risö Report 103, **1965**.
65B1 Broers, A.: Microelectron. Reliab. **4** (1965) 103.
65C1 Carlston, C.E., Magnuson, G.D., Comeaux, A., Mahadevan, P.: Phys. Rev. **138** (1965) A 759.
65D1 Davidse, P.D., Maissel, L.I.: Advances in Vac. Sci. Techn. **2** (1965) 651.
65H1 Hansen, H.R., Pascale, J.V., De Benedictis, T., Rentzepis, P.M.: J. Polymer Sci. **P 4 A 3** (1965) 2205.
65K1 Kaminsky, M.: Atomic and Ionic Impact Phenomena on Metal Surfaces, New York: Academic Press **1965**.
65W1 Weast, R.C., Selby, S.M., Hodgman, C.D.: Handbook of Chemistry and Physics **46** (1965).
66C1 Comas, J., Cooper, C.B.: J. Appl. Phys. **37** (1966) 2820.
66D1 Davidse, P.D.: 13th National Vac. Symp. (American Vacuum Society) Extend. Abstr., San Francisco 1966.
66D2 Davidse, P.D.: SCP and Solid State Technol., **Dec.** (1966) 30.
66F1 Francombe, M.H.: Basic Problems in Thin Film Physics (Niedermeyer, R.; Mayer, H., eds.) p. 52, Göttingen, Vanderhoeck & Ruprecht, **1966**.
66H1 Huss, W.N.: SCP and Solid State Technol., **Dec.** (1966) 50.
66W1 Wolsky, S.P.: SCP and Solid State Technol., **Dec.** (1966) 44.
66Z1 Zaininger, K.H.: Appl. Phys. Lett. **8** (1966) 140.

67B1 Beguin, C.P., Ezell, J.B., Salvemini, J.C., Thomson, J.C., Vickroy, D.G., Margrave, J.L.: The Applications of Plasma to Chemical Processing (Baddour, R.F., Timmins, R.S. eds.) MIT Press, Cambridge, Mass. p. 35, **1967**.
67D1 Davidse, P.D., Maissel, L.I.: J. Vac. Sci. Technol. **4** (1967) 33.
67I1 Irving, S.M.: Electrochem. Soc. Extend. Abstr. No. **67-2** (1967) 460.
67K1 Kloss, F., Herte, L.: SCP and Solid State Technol., Dec. **1967**, 45.
67M1 Manasevit, H.M., Morritz, F.L.: J. Electrochem. Soc. **114** (1967) 204.
67M2 Mc Taggart, F.K.: Plasma Chemistry in Electrical Discharges, Amsterdam: Elsevier **1967**.
67M3 Mitchell, J.P.: IEEE Trans. Electron. Devices **ED-14** (1967) 764.
67M4 Mitchell, J.P., Wilson, D.K.: Bell Syst. Tech. J. **46** (1967) 1.
67R1 Robinson, M.T., Southern, A.L.: J. Appl. Phys. **38** (1967) 2969.
67S1 Snow, E.H., Grove, A.S., Fitzgerald, D.J.: Proc. IEEE **55** (1967) 1168.
67T1 Tsui, R.T.C.: SCP and Solid State Technol., (Dec. 1967) 33.
67Z1 Zaininger, K.H., Holmes-Siedle, A.G.: RCA Rev. **28** (1967) 208.
68A1 Anderson, H.H.: Appl. Phys. Lett. **13** (1968) 85.
68B1 Berry, R.W., Hall, P.M., Harris, M.T.: Thin Film Technology, D. van Nostrand, Princeton, New Jersey **1968**.
68C1 Carter, G., Colligon, J.S.: Ion Bombardement of Solids, New York: Am. Elsevier **1968**.
68C2 Cudermann, J.F., Brady, J.J.: Surface Sci. **10** (1968) 410.
68G1 Gurmin, B.M., Martynenko, T.P., Ryzhov, Yu.A.: Fiz. Tverd. Tela **10** (1968) 411; Sov. Phys. Solid State (English transl.) **10** (1968).
68H1 Higgins, T.B., Olson, N.T., Smith, H.P.: J. App. Phys. **39** (1968) 4849.
68I1 Irving, S.M.: Kodak Photoresist Seminar **2** (1968) 26.
68J1 Jones, R.E., Winters, H.F., Maissel, L.I.: J. Vac. Sci. Technol. **5 (3)** (1968) 84.
68K1 Kurbatov, O.K.: Zh. Tekh. Fiz. **37** (1967) 1814. Soviet Phys. Techn. Phys. (English transl.) **12** (1968) 1328.
68K2 Kern, W.: RCA Review **29** (1968) 557.
68L1 Lindhard, J., Nielsen, V., Scharff, M.: Mat. Fys. Medd. Dan. Vid. Selsk. **36** (1968) 1.
68M1 Musket, R.G., Smith, H.P.: J. Appl. Phys. **39** (1968) 3579.
68M2 Manasevit, H.M.: J. Electrochem. Soc. **115** (1968) 434.
68M3 Muschlitz, E.E.: Science **159** (1968) 599.
68R1 Robinson, M.T., Southern, A.L.: J. Appl. Phys. **39** (1968) 3463.
68S1 Sigmund, P.: Can. J. Phys. **46** (1968) 731.
68S2 Simmons, M.: IEEE Trans. Electron. Devices **ED-15** (1968) 966.
68V1 Vossen, J.L., O'Neill, J.J. Jr.: RCA Rev. **29** (1968) 149.
68V2 Vossen, J.L., O'Neill, J.J. Jr.: RCA Rev. **29** (1968) 566.
69C1 Chopra, K.L.: Thin Film Phenomena, New York: Mc Graw-Hill **1969**.
69C2 Cormia, R.L.: Solid State Technology, **Dec.** (1969) 58.
69D1 Davids, P.D.: Thin Film Dielectrics (F. Vratny, ed.) p. 130, New York: Electrochem. Society **1969**.
69D2 Davids, P.D.: J. Electrochem. Soc. **116** (1969) 100.
69D3 Dugdale, R.A., Maskrey, J.T., Ford, S.D., Harmer, P.R., Lee, R.E.: J. Mater. Sci. **4** (1969) 323.
69G1 Gould, R.F.: Chemical Reactions in Electric Discharges, Advances in Chemistry Series, No. 80. Am. Chem. Soc. Publ. Washington, D.C. 1969.
69G2 Gleit, C.E.: Adv. Chem. Ser. No. **80** (1969) 232.
69H1 Hatzakis, M.: J. Electrochem. Soc. **116** (1969) 1033.
69S1 Sigmund, P.: Phys. Rev. **184** (1969) 383.
69S2 Stewart, A.G.D., Thompson, M.W.: J. Mater. Sci. **4** (1969) 56.
70B1 Bersin, R.L.: Solid State Technol., **June** (1970) 39.
70C1 Christensen, O.: Solid State Technol., **Dec.** (1970) 39.
70F1 Furr, A.K., Finfgeld, C.R.: J. Appl. Phys. **41** (1970) 1739.
70J1 Jackson, G.N.: Thin Solid Films, **5** (1970) 209.
70K1 Krutenat, R.C.; Panzera, C.: J. Appl. Phys. **41** (1970) 4953.
70K2 Koenig, H.R., Maissel, L.I.: IBM J. Res. Develop. **14** (1970) 168.
70M1 Maissel, I.I.: Handbook of Thin Film Technol., Ch. 4, New York: Mc Graw-Hill **1970**.
70M2 Mac Donalds, R.J.: Adv. Phys. **19** (1970) 457.
70V1 Vossen, J.L., O'Neill, J.J., Finlayson, K.M., Royer, L.J.: RCA Review, June **1970** 293.
70W1 Wehner, G.K., Anderson, G.S.: Handbook of Thin Film Technology (Maissel, L.I., Glang, R., eds.), Ch. 3, New York: Mc Graw-Hill **1970**.

71I1 Irving, S.M., Lemons, K.E., Bobos, G.E.: US Patent 3615956 (1971).
71I2 Irving, S.M.: Solid State Technol. **14** (1971) 47.
71O1 Orlinov, V., Mladenov, G., Goranchev, B.: Int. J. Electronics **30** (1971) 233.
71P1 Powell, R.I., Derbenwick, G.F.: IEEE Trans. Nucl. Sci. **NS-18** (1971) 99.
71S1 Spencer, E.G., Schmidt, P.H.: J. Vac. Sci. Technol. **8** (1971) 552.
72H1 Holland, L., Priestland, C.R.D.: Vacuum **22** (1972) 133.
72L1 Lau, S.S., Mills, R.H.: J. Vac. Sci. Technol. **9** (1972) 1196.
72M1 Mc Caughan, D.V., Murphy, V.T.: IEEE Trans. Nucl. Sci. **NS-19** (1972) 249.
72V1 Vogel, J.: Feinwerktechnik + micronic **76** (1972) 338.
72V2 Vossen, J.L., Davidson, E.B.: J. Electrochem. Soc. **119** (1972) 1708.
73A1 Abe, H. et al.: Jpn. J. Appl. Phys. **12** (1973) 154.
73A2 Andersen, H.H., Bay, H.L.: Radiat. Eff. **19** (1973) 139.
73B1 Brandess, R.G., Dudley, R.H.: J. Electrochem. Soc. **120** (1973) 140.
73B2 Bernstein, T., Labuda, E.F.: J. Vac. Sci. Technol. **10** (1973) 108.
73C1 Cantagrel, M., Marchal, M.: J. Mater. Sci. **8** (1973) 1711.
73C2 Chapman, B.: Conference and School on Sputtering Technology and Automated Production Sputtering, MRC, Pebble Beach, California, June 24–26, **1973**, 28.
73G1 Garvin, H.L.: Solid State Technol. **Nov.** (1973) 31.
73H1 Hughes, H.G., Hunter, W.L., Ritchie, K.: J. Electrochem. Soc. **120** (1973) 99.
73H2 Holland, L.: Vacuum **23** (1973) 175.
73K1 Krumme, J.P., Dimigen, H.: IEEE Trans. **MAG-9** (1973) 405.
73M1 Mc Caughan, D.V., Murphy, V.T.: J. Appl. Phys. **44** (1973) 3182.
73M2 Mc Caughan, D.V., Murphy, V.T.: J. Appl. Phys. **44** (1973) 2008.
73M3 Mc Caughan, D.V., Kushner, R.A., Murphy, V.T.: Phys. Rev. Lett. **30** (1973) 614.
73O1 Oechsner, H.: Z. Physik **261** (1973) 37.
73P1 Pickar, K.A., Thibault, L.R.: J. Vac. Sci. Technol. **10** (1973) 1074.
73R1 Reinberg, A.R.: U.S. Patent 3 757 733 (1973).
73S1 Smith, H.I., Melngailis, J., Williamson, R.C., Brogan, W.T.: IEEE Ultrasonics Symp. Proc. (Monterey, Calif. Nov. 5–7, 1973), IEEE Cat. No. 73 CHO 807-85, **1973**, 558.
74A1 Abe, H.: Proc. of the 6th Conf. on Solid State Devices, Tokyo **1974**.
74B1 Bailey, R.F.: Transactions of Production Sputtering, MRC, Raucho La Costa, California **1974**.
74B2 Bauer, H.J., Bogardus, E.H.: J. Vac. Sci. Technol. **11** (1974) 1144.
74C1 Circuits Manufact. **14** (October 1974) 72.
74G1 Ganguli, P.S., Kaufmann, M.: Chem. Phys. Lett. **25** (1974) 221.
74H1 Hosokawa, N., Matsuzaki, R., Asamaki, T.: Jpn. J. Appl. Phys. Suppl. **2**, Pt. 1 (1974) 435.
74H2 Hollahan, J.R., Bell, A.T.: Techniques and Applications of Plasma Chemistry, Ch. 9, New York: J. Wiley **1974**.
74H3 Hasokawa, N., Matsuzaki, R., Asamaki, T.: Jpn. J. Appl. Phys. Suppl. **2** (1974) 1.
74J1 Jakob, A.: U.S. Patent 3, 806, 365 (1974).
74L1 Labuda, E.F., Herb, G.K., Ryder, W.D., Fritzinger, L.B., Szaba, J.M.: Electrochem. Soc. Extend. Abstr. **74-1** (1974) 195.
74L2 Lamont, L.T., Turner, F.T.: J. Vac Sci. Technol. **11** (1974) 47.
74M1 Mc Coughan, D.V., Kushner, R.A.: Proc. IEEE **62** (1974) 1236.
74M2 van der Meulen, Y.J., Hien, N.C.: J. Opt. Soc. **64** (1974) 804.
74S1 Sosnowski, T.P.: Electrochem. Soc. Extend. Abstr. No. **74-2** (1974) 390.
74T1 Thomson, L.F., Feit, E.D., Heidenreich, R.D.: Polym. Eng. Sci. **14** (1974) 529.
74Y1 Yamamoto, Y., Shinada, K., Itoh, T., Yada, K.: Jpn. J. Appl. Phys. **13** (1974) 551.
75A1 Abe, H.: Jpn. J. Appl. Phys. **14** (1975) 287.
75A2 Abe, H.: Jpn. J. Appl. Phys. **14** (1975) 1825.
75A3 Abe, H.: US Patent 3, 880, 684 (1975).
75C1 Cantagrel, M.: J. Vac. Sci. Tech. **12** (1975) 1340.
75C2 Cantagrel, M.: IEEE Trans. **ED-22** (1975) 483.
75C3 Clark, H.A.: Electrochem. Soc. Extend. Abstr. **75-1** (1975) 51.
75D1 Dimigen, H., Luthje, H.: Philips Techn. Rev. **35** (1975) 199.
75F1 Foon, R., Kaufman, M.: Progr. React. Kinet. **8** (1975) 81.
75G1 Glöersen, P.G.: J. Vac. Sci. Technol. **12** (1975) 28.
75H1 Fundamentals of plasma etching, Product Bulletin 7101, IBM Hayward, California.
75H2 Heinecke, R.A.H.: Solid State Electron. **18** (1975) 1146.

75H3 Hamamoto, M.: Electrochem. Soc. Extend. Abstr. **75-2** (1975) 335.
75H4 Holland, L.: Thin Solid Films **27** (1975) 185.
75H5 Harvilchuck, I. M., Logan, I. S., et al.: German Patent, Offenlegungschrift 2617483 (1975).
75J1 Jakob, A.: Electrochem. Soc. Extend. Abstr. No. **75-1** (1975) 457.
75K1 Keller, J., Weston, D.: Photoresist Loss and Etch Rates of Silicon Dioxide and Silicon Nitride During Plasma Processing, Kodak Microelectronics Seminar Proceedings, Interface 75, p. 54. California, October **1975**.
75K2 Konnerth, K. L., Dill, F. H.: IEEE Trans. Electron Devices **ED-22** (1975) 452.
75K3 Kumar, R., Ladas, C.: 1975 IEEE IEDM Tech. Digest, Abstract **2.5** (1975) 27.
75L1 Laznovsky, W.: Vacuum Technology Res./Dev. (Aug. 1975) 47.
75M1 Mader, L., Widmann, D., Höpfner, J.: Electrochem. Soc. Extend. Abstr. **75-1** (1975) 432.
75M2 Maeda, K., Fujino, K.: Denki Kagagu **43** (1975) 22.
75O1 Oechsner, H.: Appl. Phys. **8** (1975) 185.
75S1 Sachse, G. W., Miller, W. E., Gross, Ch.: Solid State Electron. **18** (1975) 431.
75T1 Tisone, T. C., Cruzan, P. D.: J. Vac. Sci. Technol. **12** (1975) 1058.
75V1 Voschenskov, A. M., Bartelt, J. L.: Electrochem. Soc. Extend. Abstr. No. **75-2** (1975) 333.
75Z1 Zielinski, L., Schwartz, G. C.: Electrochem. Soc. Extend. Abstr. **75-1** (1975) 117.
76A1 Agajanian, A. K.: Semiconducting Devices. A Bibliography of Fabrication Technology, Properties, and Applications, IFI/Plenum New York and London **1976**.
76A2 Abe, H., Nishioka, K., Tamura, S., Nishimoto, A.: Jpn. J. Appl. Phys. **15**. Suppl. (1976) 25.
76B1 Bersin, R. L.: Solid State Technol. **19** (1976) 31.
76B2 Bondur, J. A.: J. Vac. Sci. Techn. **13** (1976) 1023.
76B3 Broers, A.: Appl. Phys. Lett. **29** (1976) 596.
76B4 Bersin, R., Singleton, M.: U.S. Patent 3, 879, 597 (1976).
76B6 Bell, G.: Electrochem. Soc. Extend. Abstr. **76-I** (1976) 128.
76C1 Clark, H. A.: Solid State Technol. **19** (1976) 51.
76D1 Dimigen, H., Lüthje, H.: Philips Techn. Rdsch. **35** (1975/76) 217.
76D2 Dimigen, H., et al.: J. Vac. Sci. Technol. **13** (1976) 976.
76D3 Degenkolb, E. O., Mogab, C. J., Goldrick, M. R., Griffiths, J. E.: Appl. Spectroscopy **30** (1976) 520.
76G1 Gloersen, P.: Solid State Technol. **April** (1976) 68.
76H1 Heinecke, R. A. H.: Solid State Electron. **19** (1976) 1039.
76H2 Horiike, Y., Shibagaki, M.: Proc. 7th Conference on Solid State Devices, Tokyo 1975. Supplement to Jpn. J. Appl. Phys. **15** (1976) 13.
76H3 Harvilchuck, J. M., Logan, J. S., Metzger, W. C., Schaible, P. M.: U.S. Patent 3,994,793 (1976).
76J1 Jacob, A.: U.S. Patent 3, 951 843 (1976).
76J2 Jones, W. E., Skolnik, E. G.: Chemical Reviews **76** (1976) 563.
76J3 Jacob, A.: Solid State Technol. **19 (9)** (1976) 70.
76K1 Kumar, R., Ladas, Ch., Hudson, G.: Solid State Technol., **Oct.** (1976) 54.
76K2 Komiya, H., Toyoda, H., Kato, T., Inaba, K.: Jpn. J. Appl. Phys. **15** (suppl.) (1976) 19.
76K3 Kalter, H., van de Ven, E. P. G. T.: Electrochem. Soc. Extend. Abstr. No. **76-1** (1976) 335.
76K4 Kay, E., Coburn, J. W., Kruppa, G.: Le Vide **183** (1976) 89.
76M1 Melliar-Smith, C. M.: J. Vac. Sci. Technol. **13** (1976) 1008.
76M2 Muto, S. Y.: US Patent 3,971,684 (1976).
76N1 Nordine, P. C., Le Grange, J. D.: AIAA J. **14** (1976) 644.
76P1 Poulsen, R. G., Neutwich, H., Ingrey, S.: IEEE, IEDM, Washington, **1976**, 205.
76P2 Poulsen, R. G., Brochu, M.: Etching for Pattern Definition (Hughes, H. G., Rand, M. J. eds.) p. 111, Electrochem. Soc. Princeton, New Jersey **1976**.
76P3 Poulsen, R. G., Brochu, M.: Electrochem. Soc. Ext. Abstr. **76-1** (1976) 143.
76R1 Reinberg, A. R.: Etching for Pattern Definition (Hughes, H. G., Rand, M. J. eds.) Electrochem. Society, p. 91 Princeton New Jersey 1976.
76R2 Romankiw, L. T.: Etching for Pattern Definition (Hughes, H. C., and Rand, M. J. eds.) Electrochem. Soc. Princeton, New Jersey (1976) 161.
76R3 Ryden, W. D., Labuda, E. F., Clemens, J. T.: Etching for Pattern Definition (Hughes, H. G., Rand, M. J. eds.) p. 144, Electrochem. Soc. Princeton, New Jersey 1976.
76R4 Reinberg, A. R.: Electrochem. Soc. Ext. Abstr. **76-1** (1976) 142.
76S1 Somekh, S.: J. Vac. Sci. Technol. **13** (1976) 1003.
76S2 Smith, H. I.: Etching for Pattern Definition (Hughes, H. C., Rand, M. J. eds.) p. 133 Electrochem. Soc. Princeton, New Jersey 1976.

76S3 Schwartz, G.C., Zielinski, L.B., Schopen, T.: Etching for Pattern Definition (Hughes, H.G., Rand, M.I. eds.) Electrochem. Soc. p. 122 Princeton, New Jersey 1970.
76V1 Van de Ven, E.P.G.T., Kalter, H.: Electrochem. Soc. Extend. Abstr. No. **76-1** (1976) 332.
76Z1 Zafiropoulo, A.: Circuits Manufacturing, **April** (1976) 42.
76Z2 Zielinski, L.B.: Electrochem. Soc. Extend. Abstr. No. **76-1** (1976) 200.
77B1 Bunyard, G.B., Raby, B.A.: Solid State Technol. **20** (1977) 53.
77B2 Bell, G., Hasler, B.: Electrochem. Soc. Extend. Abstr. **77-2** (1977) 411.
77B3 Battey, J.F.: J. Electrochem. Soc. **124** (1977) 147.
77B4 Bondur, J.A.: Electrochem. Soc. Extend. Abstr. **77-2** (1977) 371.
77B5 Battey, J.F.: Electrochem. Soc. **124** (1977) 437.
77B6 Bell, G., Stokan, R.: Electrochem. Soc. Extend. Abstr. **77-2** (1977) 383.
77B7 Bell, G., Hasler, B.: Electrochem. Soc. Extend. Abstr. **77-2** (1977) 411.
77C1 Chapman, R.E.: J. Mater. Sci. **12** (1977) 1125.
77C2 Cobum, J.W., Kay, E.: Proc. 7th Int. Vac. Congr. & 3rd Int. Conf. Solid Surfaces, Vienna (1977) 1257.
77C3 Coburn, J.W., Winters, H.F., Chuang, T.J.: J. Appl. Phys. **48** (1977) 3532.
77C4 Clark, H.A., Purcell, E.D.: Electrochem. Soc. Extend. Abstr. **77-2** (1977) 408.
77D1 Degenkolb, E.O., Griffiths, J.E.: Appl. Spectrosc. **31** (1977) 40.
77D3 Darwall, E.P.D.: Electrochem. Soc. Ext. Abstr. **77-2** (1977) 400.
77D4 Deppe, H.-R., Hieke, E., Sigusch, R.: Semiconductor Silicon, p. 1082, Electrochem. Soc. Princeton, New Jersey **1977**.
77D5 Deppe, H.-R., Hasler, B., Höpfner, J.: Solid State Electronics **20** (1977) 51.
77E1 Ephrath, L.M.: Electrochem. Soc. Extend. Abstr. **77-2** (1977) 376.
77E2 Eschwei, M., Gottfried, S.: J. Vac. Sci. Technol. **14** (1977) 1214.
77F1 Fujino, K.: US Patent 4,026,742 (1977).
77G1 Griffiths, J.E., Degenkolb, E.O.: Appl. Spectrosc. **31** (1977) 134.
77H1 Horriike, Y., Shibagaki, M.: Semiconductor Silicon 1977, p. 1071, Electrochem. Soc., Princeton, New Jersey **1977**.
77H2 Harshbarger, W.R., Porter, R.A., Miller, T.A., Norton, P.: Appl. Spectrosc. **31** (1977) 201.
77H3 Herndon, T.O., Burke, R.L.: Interface 77, Kodak Microelectronics Symposium, Montcrey, California (1977) 33.
77H4 Holland, L.: J. Vac. Sci. Technol. **14** (1977) 5.
77H5 Horriike, Y., Shibagaki, M.: Electrochem. Soc. Extend. Abstr. **77-2** (1977) 243.
77I1 Itoga, M., Inoue, M., Kitihara, Y., Ban, Y.: Electrochem. Soc. Extend. Abstr. **77-2** (1977) 378.
77J1 Jacob, A.: Solid State Electron. **20** (1977) 479.
77J2 Jacob, A.: U.S. Patent 4,028,155 (June 1977).
77J3 Jinno, K., Kinoshita, H., Matsumato, Y.: J. Electrochem. Soc. **124** (1977) 1258.
77K1 Kurogi, Y., Tajima, M., Mori, K., Sugibuchi, K.: Electrochem. Soc. Extend. Abstr. No. **77-2** (1977) 373.
77K2 Kinoshita, H., Jinno, K.: Jpn. Appl. Phys. **16** (1977) 381.
77L1 Logan, J.S., Keller, J.H., Simmons, R.G.: J. Vac. Sci. Technol. **14** (1977) 92.
77L2 Lehmann, H.W., Krausbauer, L., Widmer, R.: J. Vac. Sci. Technol. **14** (1977) 281.
77L3 Lassus, M., Founand, J.-P.: Microelectronics and Reliability **16** (1977) 367.
77L4 Lam, D.K.: Electrochem. Soc. Extend. Abstr. **77-2** (1977) 404.
77M1 Matsno, S., Takehara, Y.: Jpn. J. Appl. Phys. **16** (1977) 175.
77M2 Mentall, J.E., Guyon, P.M.: J. Chem. Phys. **67** (1977) 3845.
77M3 Mogab, C.J.: J. Electrochem. Soc. **124** (1977) 1262.
77M4 Mogab, C.J., Shankoff, T.A.: J. Electrochem. Soc. **124** (1977) 1766.
77M5 Mauer, J.L., Carruthers, R., Zielinski, L.B.: Electrochem. Soc. Extend. Abstr. **77-2** (1977) 386.
77M6 Mogab, C.J.: Electrochem. Soc. Extend. Abstr. **77-2** (1977) 402.
77N1 Nishioka, K., Abe, H.: Electrochem. Soc. Extend. Abstr. **77-2** (1977) 406.
77P1 Poulsen, R.G.: Electrochem. Soc. Extend. Abstr. **77-2** (1977) 242.
77P2 Poulsen, R.G., Smith, G.M.: Semiconductor Silicon, p. 1058, Electrochem. Soc. Princeton, New Jersey 1977.
77P3 Porter, R.A., Harshbarger, W.R.: Electrochem. Soc. Extend. Abstr. **77-2** (1977) 388.
77P4 Poulsen, R.G.: J. Vac. Sci. Technol. **14** (1977) 266.
77P5 Pandey, K.C., Sakurai, T., Hagstrum, H.D.: Phys. **B 16** (1977) 3648.
77R1 Reichelderfer, R.F., Welty, J.M., Battey, J.F.: J. Electrochem. Soc. **124** (1977) 1926.
77R2 Reichelderfer, R.F., Vogel, D., Bersin, R.L.: Electrochem. Soc. Extend. Abstr. **77-2** (1977) 414.

77R3 Rowe, I.E., Margaritondo, G., Christmas, S.B.: Phys. Rev. **B 16** (1977) 1581.
77S1 Somekh, S., Casey, H.C. jr.: Appl. Optics **16** (1977) 126.
77S2 Schwartz, G.C., Zielinski, L.B., Schopen, T.S.: Electrochem. Soc. Extend. Abstr. **77-2** (1977) 391.
77S3 Suzuki, K., Okudaira, S., Sakudo, N., Kanomata, I.: Jpn. J. Appl. Phys. **16** (1977) 1979.
77T1 Tsuchimoto, T.: Electrochem. Soc. Extend. Abstr. **77-2** (1977) 423.
77W1 Wechsung, R.: Vakuum-Technik **26** (1977) 227.
77W2 Winters, H.F., Coburn, J.W., Kay, E.: J. Appl. Phys. **48** (1977) 4973.
77W3 Winters, H.F., Coborn, J.W., Kay, E.: Electrochem. Soc. Extend. Abstr. **77-2** (1977) 393.
77W4 Weston, D., Keller, J.: Electrochem. Soc. Extend. Abstr. **77-2** (1977) 381.
77W5 Weitzel, Ch., E.: US Patent 4,052,251 (1977).
77Z1 Zarowin, C.B., Allessandrini, E.I.: Electrochem. Soc. Extend. Abstr. **77-2** (1977) 395.
78A1 Alcorn, G.E., Hamaker, R.W., Stephens, G.B.: European Patent 0002185 (1978).
78B1 Bersin, R.L.: Solid State Technol. **21** (1978) 117.
78B2 Baron, M., Zelez, J.: Solid State Technol., **Dec.** (1978) 61.
78B3 Bell, A.T.: Solid State Technol., **April** (1978) 89.
78B4 Bersin, R.L.: Programmed Plasma Processing, Kodak Microelectronics Seminar Proceedings, San Diego Calif., Oct. 1978, 21.
78B5 Bondur, J.A., Case, W.R., Clark, H.A.: Electrochem. Soc. Extend. Abstr. **78-1** (1978) 760.
78B6 Blech, I., Fraser, D.B., Haszka, S.E.: J. Vac. Sci. Technol. **15** (1978) 13.
78B7 Bondur, J.A., Pogge, H.P.: European Patent 0001100 (1978).
78B8 Bondur, J.A., Pogge, H.P.: European Patent 0000897 (1978).
78B9 Brown, H.L., Bunyard, G.B., Lin, K.C.: Solid State Technol., **July** (1978) 35.
78C1 Circuits Manufacturing **18** (1978) 22.
78C2 Chapman, B.N., Minkiewicz, V.J.: J. Vac. Sci. Tech. **15** (1978) 329.
78C3 Clark, H.A., Bondur, J.A.: Electrochem. Soc. Extend. Abstr. **78-1** (1978) 762.
78C4 Coburn, J.W., Winters, H.F.: J. Vac. Sci. Tech. **15** (1978) 327.
78C5 Crabtree, P.N., Gorin, G., Thomas, R.S.: Scan. Electron. Microscop. **1** (1978) 543.
78C6 Clark, H.A.: European Patent 0002503 (1978).
78C7 Clements, R.M.: J. Vac. Sci. Technol. **15 (2)** (1978) 193.
78C8 Chang, M.S., Chen, J.T.: Appl. Phys. Lett. **33** (1978) 892.
78C9 Chung, S.: Solid State Technol. (April 1978) 114.
78C10 Cuomo, J.J., Gambino, R.J.: Offenlegungsschrift 2845074 Germany, **1978**.
78E1 Eisele, K.M.: Rev. Phys. Appl. **13** (1978) 701.
78E2 Ephrath, L.: J. Electron. Mater. **7** (1978) 415.
78E3 Eser, E., Ogilvie, R.E., Taylor, K.A.: J. Vac. Sci. Technol. **15** (1978) 199.
78F1 Flanders, D.C., Smith, H.I., Lehmann, H.W., Widmer, R., Shaver, D.: Appl. Phys. Lett. **32** (1978) 112.
78F2 Flamm, D.L.: AVS Meeting, Danvers, Massachusetts, **1978**.
78H1 Harper, J.M.E.: Thin Film Processes (Vossen, J.L., Kern, W. eds.) Ch. II-5, New York: Academic Press 1978.
78H2 Heinecke, R.A.H.: Solid State Technol. **21. April** (1978) 104.
78H3 Hirobe, K., Tsuchimoto, T.: Electrochem. Soc. Extend. Abstr. **78-1** (1978) 757.
78H4 Harshbarger, W.R., Porter, R.A.: Solid State Technol. **April** (1978) 99.
78J1 Jacob, A.: Solid State Technol., **April** (1978) 95.
78K1 Kleinknecht, H.P., Meier, H.: J. Electrochem. Soc. **125** (1978) 798.
78K2 Keller, J.H., Mc Kenna, Ch.M.: European Patent 0002726A2 (1978).
78K3 Ketterer, G.: MRC-Symposium, Lindau, West Germany, May 1978.
78L1 Lehmann, H.W., Widmer, R.: J. Vac. Sci. Tech. **15** (1978) 319.
78L2 Lehmann, H.W., Widmer, R.: Appl. Phys. Lett. **32** (1978) 163.
78M1 Melliar-Smith, C.M., Mogab, C.J.: Thin Film Processes (Vossen, J.L., Kern, W. eds.) Ch. V-2, New York: Academic Press **1978**.
78M2 Mauer, J.L., Logan, J.S., Zielinski, L.B., Schwartz, G.C.: J. Vac. Sci. Tech. **15** (1978) 1734.
78M3 Mogab, C.J., Adams, A.C., Flamm, D.L.: J. Appl. Phys. **49** (1978) 3796.
78M4 Morosoff, N., Newton, W., Yasuda, H.: J. Vac. Sci. Tech. **15** (1978) 1815.
78M5 Matsuo, S.: Jpn. J. Appl. Phys. **17** (1978) 235.
78M6 Mc Caughan, D.V., White, C.W., Tolk, N., Murphy, V.T.: J. Appl. Phys. (1978).
78M7 Maddox, R.L., Parker, H.L.: Solid State Technol. **April** (1978) 107.
78M8 Muramato, S., Hasoya, T., Matsuo, S.: The Technical Digest of the 1978 IEDM (1978) 185.
78M9 Mogab, C.J., Harshbarger, W.R.: Electronics, **Aug.** (1978) 117.

78M10 Mader, L.: Res. Rep. NT 595, BMFT-FB, West-Germany Governement, June **1978**.
78R1 Robertson, D.D.: Solid State Technol. **Dec.** (1978) 57.
78R2 Raby, B.A.: J. Vac. Sci. Technol. **15 (2)** (1978) 205.
78R3 Reinberg, A.R.: IEDM 1978, Washington D.C. (1978) 441.
78R4 Reinberg, A.R.: US Patent 4,069,096 (1978).
78R5 Reinberg, A.R.: US Patent 4,094,732 (1978).
78S1 Schaible, P.M., Metzger, W.C., Anderson, J.P.: J. Vac. Sci. Tech. **15** (1978) 334.
78T1 Takahashi, S., Murai, F., Kodera, H.: IEEE Trans. Electron Devices **ED-25** (1978) 1213.
78U1 Ukai, K., Hanazawa, K.: J. Vac. Sci. Technol. **15 (2)** (1978) 338.
78V1 Vossen, J.L., Cuomo, J.J.: Thin Film Processes (eds. Vossen, J.L., Kern, W.) Ch. II-1, Academic Press, New York: **1978**.
78V2 Vossen, J.L.: Electrochem. Soc. Extend. Abstr. No. 386, **78-1** (1978) 960.
78W1 Winters, H.F.: J. Appl. Phys. **49** (1978) 5165.
78Y1 Yan, L.D., Thibault, L.R.: J. Vac. Sci. Technol. **15 (3)** (1978) 960.
78Y2 Yano, H., Hashimoto, H., Toyama, Y.: Electrochem. Soc. Extend. Abstr. **78-1** (1978) 965.
79B1 Busta, H.H., Lajos, R.E., Kiewit, D.A.: Solid State Technol. **Febr.** (1979) 61.
79B2 Boyd, H., Tang, M.S.: Solid State Technol. **22 April** (1979) 133.
79B3 Bogen, P., Hintz, E., Rusbüldt, D.: Proc. 4th Intern. Symp. Plasma Chemistry, Zurich (1979) 111.
79B4 Bondur, J.A.: J. Electrochem. Soc. **126 (2)** (1979) 226.
79B5 Beenakker, C.I.M., van de Poll, R.P.I.: Proc. 4th Intern. Symp. Plasma Chemistry, Zurich (1979) 125.
79B6 Beinvogl, W.: Proc. Microcircuit Conf., Aachen, Germany, **1979**.
79B7 Bruno, G., Capezzuto, P., Cromarossa, F., d'Agostino, R., Latrofa, G.: Proc. 4th Intern. Symp. Plasma Chemistry, Zurich (1979) 460.
79B8 Bell, A.I.: J. Vac. Sci. Technol. **16** (1979) 418.
79B9 Burggraaf, P.S.: Semiconductor International, December (1979) 49.
79C1 Chapman, B.N., Minkiewicz, V.J.: Appl. Phys. Lett. **34** (1979) 192.
79C2 Chapman, B.N.: IBM Tech. Discl. Bull. **21** (1979) 5006.
79C3 Chuang, T.J.: J. Vac. Sci. Tech. **16** (1979) 496.
79C4 Coburn, J.W., Winters, H.F.: J. Appl. Phys. **50** (1979) 3189.
79C5 Coburn, J.W., Winters, H.F.: J. Vac. Sci. Tech. **16** (1979) 391.
79C6 Coburn, J.W., Kay, E.: IBM J. Res. Develop. **23** (1979) 33.
79C7 Coburn, J.W.: J. Appl. Phys. **50** (1979) 5210.
79C8 Coburn, J.W., Winters, H.F.: J. Vac. Sci. Technol. **16** (1979) 1613.
79C9 Coburn, J.W., Kay, E.: Solid State Technol. **April** (1979) 117.
79C10 Coburn, J.W., Kay, E.: J. Vac. Sci. Technol. **16 (2)** (1979) 407.
79C11 Chen, M., Minkiewicz, V.J., Lee, K.: J. Electrochem. Soc. **126** (1979) 1946.
79C12 Curran, I.E.: Proc. 4th. Intern. Symp. Plasma Chem., Zurich, **1979**, 146.
79C13 Cetronio, A., Jannuzzi, G.: 4th Intern. Symp. Plasma Chem. Zurich **1** (1979) 131.
79C14 Curtis, B.I., Brunner, H.R.: Proc. 4th. Intern. Symp. Plasma Chem. Zurich (1979) 139.
79D1 Di Maria, D.J., Ephrath, L.M., Young, D.R.: J. Appl. Phys. **50** (1979) 4015.
79D2 Doken, M., Miyata, I.: J. Electrochem. Soc. **126 (12)** (1979) 2235.
79E1 Ephrath, L.M.: J. Electrochem. Soc. **126** (1979) 1419.
79E2 Enomoto, T., Denda, M., Yasuoka, A., Nakata, I.: Jpn. J. Appl. Phys. **18** (1979) 155.
79E3 Ephrath, L.M.: European Patent 0001 538 (1979).
79E4 Eisele, K.: 4th Symp. Solid State Device Technol. Munich, Germany, **1979**, 74.
79F1 Flamm, D.L., Mogab, C.J.: Proc. 4th Intern. Symp. Plasma Chem., Zurich (1979) 119.
79F2 Flamm, D.L.: Solid State Technol. **April** (1979) 109.
79F3 Flamm, D.L.: Proc. 4th. Intern. Symp. Plasma Chemistry, Zurich (1979) 466.
79G1 Gdula, R.A.: Electrochem. Soc. Extend. Abstr. **79-2** (1979) 1524.
79G2 Gdula, R.A.: IEEE Trans. Electron. Devices **ED-26** (1979) 644.
79G3 Goodner, W.R., Wood, T.E., Hughes, H.G., Smith, J.N., Keller, J.V.: Kodak Interface Sem. Proc. (1979).
79H1 Hiraiwa, A., Mukai, K., Harada, S., Yoshimi, T., Itch, S.: Jpn. J. Appl. Phys. **18** (1979) 191.
79H2 Hirobe, K., Kureishi, Y., Tsuchimoto, T.: Electrochem. Soc. Extend. Abstr. No. 610, **79-2** (1979) 1529.
79H3 Hom-ma, Y., Harada, S.: J. Electrochem. Soc. **126 (9)** (1979) 1531.
79H4 Harsbarger, W.R., Levinstein, H.J., Mogab, C.J.: German Patent Offenlegungsschrift DE 29 30 290 A1 (1979).
79H5 Hoffmann, N.: Proc. NTG Tagung, Baden-Baden, Germany, April **1979** 54.

79H6 Hom-ma, Y., Nozawa, H., Harada, S.: Proc. IEEE IEDM (1979) 54.
79H7 Horiike, Y., Shibagaki, M., Kadono, K.: Jpn. J. Appl. Phys. **18** (1979) 2309.
79H8 Hofmann, D., Wechsung, R.: Proc. 4th Intern. Symp. Plasma Chem., Zurich (1979) 622.
79H9 Hosaka, S., Sakudo, N., Hashimoto, S.: J. Vac. Sci. Technol. **16** (1979) 913.
79H10 Höthker, K., Koizlik, K., Hintz, E.: Proc. 4th Intern. Symp. Plasma Chemistry, Zurich (1979) 109.
79J1 Jakob, A.: Electrochem. Soc. Ext. Abstr. No. **79-2** (1979) 1512.
79K1 Keller, J.H., Pennebaker, W.B.: IBM J. Res. Develop. **23** (1979) 3.
79K2 Keller, J.H., Simmons, R.G.: IBM J. Res. Develop. **23** (1979) 24.
79K3 Kalter, H., van de Ven, E.P.G.T.: Philips Tech. Rundsch. **38** (1979) 203.
79K4 Knop, K., Lehmann, H.W., Widmer, R.: J. Appl. Phys. **50** (1979) 3841.
79K5 Kawamoto, Y., Hashimoto, N.: Jpn. J. Appl. Phys. **18** Suppl. (1979) 277.
79K6 Kay, E.: Proc. 4th Intern. Symp. Plasma Chem., Zurich (1979) 30.
79L1 Lee, R.E.: J. Vac. Sci. Technol. **16** (1979) 164.
79L2 LeClaire, R.: Solid State Technol. **April** (1979) 139.
79L3 Levinstein, H.J.: German Patent Offenlegungsschrift DE 29 30 291 A1 (1979).
79L4 Lamont, L.T.: Solid State Technol. **Sept.** (1979) 107.
79M1 Mauer, J.L., Carruthers, R.A.: Proc. 21st Electron. Matters Conf., Boulder, Colorado 1979.
79M2 Moran, J.M., Maydan, D.: Bell System Tech. J. **58** (1979) 1027.
79M3 Meusemann, B.: 15th Symp. Electr. Ion Photon Beam Technol. Boston **1979**.
79M4 Mauer, J.L., Logan, J.S.: J. Vac. Sci. Technol. **16 (2)** (1979) 404.
79M5 Mogab, C.J.: J. Vac. Sci. Technol. **16** (2) (1979) 408.
79M6 Mauer, J.L., Logan, J.S.: Electrochem. Soc. Ext. Abstr. No. 607, **79-2** (1979) 1521.
79M7 Mathad, G.S., Patnaik, B.: Electrochem. Soc. Ext. Abstr. No. 603, **79-2** (1979) 1510.
79M8 Mayer, T.M., McConville, J.H.: Proc. IEEE Electron. Device Meeting (1979) 44.
79M9 Mogab, C.J.: German Patent Offenlegungsschrift DE 29 30 292 A1 (1979).
79M10 Mogab, C.J.: German Patent Offenlegungsschrift DE 29 30 293 A1 (1979).
79M11 Matheson, Fa.: Techn. Brief TB-151 (1979).
79M12 Meusemann, B.: J. Vac. Sci. Technol. **16 (6)** (1979) 1886.
79M13 Mader, H.: 4th Symp. Solid State Device Technol. Munich, Germany, **1979** 73.
79M14 McLeod, P.S., Hughes, J.L.: J. Vac. Sci. Technol. **16** (1979) 369.
79M15 Murarka, S.P.: Proc. IEDM, Washington, **1979** 454.
79M16 Mogab, C.J.: 4th Symp. Solid State Device Technol., Munich, Germany **1979**.
79M17 Maddox, R.L.: Kodak Interface Seminar Proc., **1979**.
79M18 Mogab, C.J.: Ion Beam, Plasma and Reactive Ion Etching in Solid State Dev. Conf. Series 53, The Institute of Physics, Bristol, England (1979).
79M19 Minkiewicz, V.J., Chen, M., Coburn, J.W., Chapman, B.N., Lee, K.: Appl. Phys. Lett. **35** (1979) 393.
79N1 Niggebrügge, U., Frenzel, H., Bolsen, M., Geelen, H.J.: Proc. Microcircuit Conf. Aachen, Germany (1979) 120.
79N2 Nozawa, H., Nishimura, S., Horriike, Y., Okumura, K., Iizuki, H., Kohyama, S.: Proc. IEDM, Washington, **1979** 366.
79N3 Noble, Jr., Wendell, Ph.: European Patent EP 0 012 863 A3 (1979).
79P1 Pennebaker, W.B.: BM J. Res. Develop. **23** (1979) 16.
79P2 Parry, P.D., Rodde, A.F.: Solid State Technol. **22** (1979) 125.
79P3 Pogge, H.B., Lechoton, J.S., Burkhardt, P.J.: 4th Symp. Solid State Device Technol. Munich, Germany, **1979** 71.
79P4 Pogge, H.B.: European Patent EP 0010596 A1 (1979).
79R1 Reinberg, A.R.: Circuits Manufacturing **April** (1979) 25.
79R2 Ranadive, D.K., Losee, D.L.: Kodak Interface Seminar Proc. (1979).
79R3 Riseman, J.: European Patent EP 0010623 A1.
79S1 Schaible, P.M., Schwartz, G.C.: J. Vac. Sci. Tech. **16** (1979) 377.
79S2 Schwartz, G.C., Schaible, P.M.: J. Vac. Sci. Tech. **16** (1979) 410.
79S3 Smolinsky, G., Flamm, D.L.: J. Appl. Phys. **50** (1979) 4982.
79S4 Stephani, D., Kratschmer, E.: Microcircuit Engineering Conf. Aachen (1979).
79S5 Schwartz, G.C., Rothman, L.B., Schopen, T.J.: J. Electrochem. Soc. **126 (3)** (1979) 464.
79S6 Suzuki, K., Okudaira, S., Nishimatsu, S., Usami, K., Kanomata, J.: Electrochem. Soc. Extend. Abstr. No. 606, **79-2** (1979) 1518.
79S7 Schwartz, G.C., Schaible, P.M.: Electrochem. Soc. Extend. Abstr. No. 612, **79-2** (1979) 1535.
79S8 Suzuki, K., Okudaira, S., Kanomata, J.: J. Electrochem. Soc. **126 (6)** (1979) 1024.

79S9 Shibayama, H., Ogawa, T., Kobayashi, K., Hisatsuga, T.: Digest of Tech. Papers, The 11th Conf. on Solid State Devices, Tokyo (1979) 19.
79S10 Stein, L.: European Patent Application EP 0010657 A1 (1979).
79T1 Taylor, G.N., Wolf, T.M.: Photopolymers Conf., Ellenville, New York (1979) 174.
79T2 Tokunaga, K., Hess, D.W.: Electrochem. Soc. Extend. Abstr. No. 609, **79-2** (1979) 1527.
79T3 Taillet, J.: Proc. 4th Intern. Symp. Plasma Chem., Zurich (1979) 113.
79T4 Takacs, M., Viswanathan, N.S.: Electrochem. Soc. Extend. Abstr. No. **79-2** (1979) 1500.
79T5 Tokuda Seisakusho Ltd., commercial literature (1979).
79T6 Taglauer, E.: Proc. 4th Intern. Symp. Plasma Chemistry, Zurich (1981).
79U1 Ukai, K., Hanazawa, K.: J. Vac. Sci. Tech. **16** (1979) 385.
79V1 Vossen, J.L.: J. Electrochem. Soc. **126** (1979) 319.
79V2 Vossen, J.L.: 4th Internat. Symp. Plasma Chem., Zurich (1979) 344.
79V3 Van de Ven, E.P.G.T., Zijlstra, P.A.: ESSDERC Abstracts, Munich, Germany, Sept. **1979**
79V4 Veeco Catalogue. **1979** 34.
79V5 Viswanathan, N.S.: J. Vac. Sci. Technol. **16 (2)** (1979) 388.
79W1 Winters, H.F., Chapman, B.N.: European Patent 0008348 A1 (1979).
79W2 Wagner, J.J., Brandt, W.W.: Proc. 4th Intern. Symp. Plasma Chem., Zurich (1979) 120.
79W3 Winters, H.F.: 4th Intern. Symp. Plasma Chem. Zürich, **1979** 28.
79Y1 Yaws, C.L., Shah, P.N., Patel, P.M., Hsu, G., Lutwack, R.: Solid State Technol. **Febr.** (1979) 65.
79Y2 Yamazaki, T., Suzuki, Y., Uno, J., Nakate, H.: J. Electrochem. Soc. **126** (1979) 1794.
79Z1 Zelley, A.: Developments in Semiconductor Microlithography IV, **SPIE 174** (1979) 173.
79Z2 Zajac, J.: German Patent Offenlegungsschrift DE 29 30 360 A1 (1979).
79Z3 Zarowin, C.B.: Proc. 4th. Intern. Symp. Plasma Chem., Zürich, **1979** 56.
80A1 Arikado, T., Horiuchi, S.: Electrochem. Soc. Ext. Abstr. No. **80-1** (1980) 263.
80A2 Abe, H., Nakata, H.: Electrochem. Soc. Ext. Abstr. **80-1** (1980) 230.
80A3 Abe, H., Harada, H., Mashiko, Y.: German Patent, Offenlegungsschrift 3037876 (1980).
80B1 Boyd, G.D., Coldren, L.A., Storz, F.G.: Appl. Phys. Lett. **36 (7)** (1980) 583.
80B2 Blash, A., Chang, P., Hsueh, Y.W.: Kodak Microelectronics Seminar Proceedings (Oct. 1980).
80B3 Bennakker, C.J.M., van Dommelen, J.H.J., Dieleman, J.: Electrochem. Soc. Ext. Abstr. No. **80-1** (1980) 330.
80B4 Beinvogl, W.: ESSDERC '80, Conf. Abstr., Brighton, UN (1980) S 39.
80B5 Bell, G.: Electrochem. Soc. Extend. Abstr. No. **80-1** (1980) 337.
80B6 Bondur, J.A., Frieser, R.G.: Electrochem. Soc. Ext. Abstr. No. **80-1** (1980) 288.
80B7 Bruce, R.H., Gelernt, B.: Electrochem. Soc. Ext. Abstr. No. **80-1** (1980) 307.
80B8 Bruno, G., Capezzuto, P.: Electrochem. Soc. Ext. Abstr. No. **80-1** (1980) 315.
80B9 Brown, D.M., Heath, B.A., Coutumas, T., Thompson, G.R.: Appl. Phys. Lett. **37** (1980) 159.
80B10 Bruce, R.H., Reinberg, A.R.: Proc. Sec. Symp. Dry Processing, Japan, Oct. (1980) 131.
80B11 Bruce, R.H., Reinberg, A.R.: Proc. Int. Conf. Microlithography, Amsterdam, Netherland (1980) 533.
80C1 Chapman, B.: Glow Discharge Processes, Ch. 6, New York: J. Wiley **1980**.
80C2 Chapman, B.: Glow Discharge Processes, Ch. 3, New York: J. Wiley **1980**.
80C3 Chapman, B.: Glow Discharge Processes, Ch. 7, New York: J. Wiley **1980**.
80C4 Chow, T.P., Steckl, A.J.: Appl. Phys. Lett. **37 (5)** (1980) 466.
80C5 Curtis, B.J., Brunner, H.J.: J. Electrochem. Soc. Accel. Brief Communication (May 1980) 829.
80C6 Chow, T.P., Steckl, A.J.: Electrochem. Soc. Ext. Abstr. No. **80-1** (1980) 313.
80C7 Chow, T.P., Steckl, A.J.: Proc. IEDM 1980, Washington D.C., **1980** 149.
80C8 Chapman, B.: Glow Discharge Processes, Ch. 5, New York: J. Wiley **1980**.
80C9 Czanderna, A.W.: J. Vac. Sci. Technol. **17** (1980) 72.
80C10 Castellano, R.N.: Proc. 8th Int. Vacuum Congr., Cannes, France, Sept. 22–26 (1980) 74.
80C11 Coburn, J.W.: Proc. Sec. Symp. Dry Processes, Japan, **Oct.** (1980) 103.
80D1 Donnelly, V.M., Flamm, D.L.: Electrochem. Soc. Extend. Abstr. No. **80-1** (1980) 323.
80D2 Dennison, R.W.: Solid State Technol. (Sept. 1980) 117.
80D3 DeKeersmaecker, R.F., DiMaria, D.J.: J. Appl. Phys. **51** (1980) 1085.
80D4 DeKeersmaecker, R.F., DiMaria, D.J.: J. Appl. Phys. **51** (1980) 532.
80D5 Dargent, B., Sibuet, H.: Proc. 8th Int. Vac. Congr., Cannes, France, **Sept.** (1980) 78.
80D6 Duval, P.: Proc. 8th Int. Vac. Congr., Cannes, France, **Sept.** (1980) 26.
80D7 Danilin, B.S., Kireev, Yu.: Mikroélektronika **9 (4)** (1980) 302.
80E1 Endo, N., Kurogi, Y.: IEEE Trans. Electr. Dev. **ED-27** (1980) 1346.
80E2 Eisele, K.M.: Electrochem. Soc. Extend. Abstr. No. **80-1** (1980) 285.

80E3 Ephrath, L.M.: Proc. IEDM 1980, Washington D.C. (1980) 402.
80E4 Electronics (Nov. 1980) 76.
80E5 Eisele, K.M., Hofmann, D.: Proc. 8th Int. Vacuum Congr., Cannes, France, Sept. **22–26** (1980) 62.
80F1 Fok, T.Y.: Electrochem. Soc. Ext. Abstr. **80-1** (1980) 301.
80F2 Forget, L.E., Road, R.: European Patent EP 0 015 403 A1 (1980).
80F3 Flamm, D.L.: Electrochem. Soc. Ext. Abstr. **80-1** (1980) 256.
80F4 Flamm, D.L., Cowan, P.L., Golovchenko, J.A.: J. Vac. Sci. Technol. **17** (1980) 1341.
80F5 Flamm, D.L.: J. Appl. Phys. **51** (1980) 5688.
80G1 Griffin, S.T., Verdeyen, J.T.: IEEE Transactions on Electron Devices **ED-27** (1980) 602.
80G2 Gill, M.D.: Solid State Electronics **23** (1980) 995.
80G3 Grusell, E., Berg, S., Andersson, L.P.: I. Electrochem. Soc. **127** (1980) 1573.
80G4 Geipel, H.J., Hsieh, N., et al.: IEEE Trans. Electron. Dev. **ED-27** (1980) 1417.
80H1 Hollahan, J.R.: Perkin-Elmer Seminar on Sputtering, Plasma Etching and Surface Analysis, Munich, Germany April **1980**.
80H2 Hutt, M., Class, W.: Solid State Technol. (March 1980) 92.
80H3 Heiman, N., Minkiewicz, V., Chapman, B.: J. Vac. Sci. Technol. **17 (3)** (1980) 731.
80H4 Hirobe, K., Tsuchimoto, T.: J. Electrochem. Soc. **127** (1980) 234.
80H5 Horwath, R., Zarowin, C.B., Rosenberg, R.: Electrochem. Extend. Abstr. No. **80-1** (1980) 294.
80H6 Hiraoka, H., Welsh, L.W.: Electrochem. Soc. Extend. Abstr. No. **80-1** (1980) 261.
80H7 Harada, K.: J. Electrochem. Soc. **127** (1980) 491.
80H8 Hirata, K., Ozaki, Y., Oda, M., Kimizuka, M.: Proc. IEDM 1980, Washington D.C. (1980) 405.
80H9 Hutt, M.: SPIE Semicond. Microlithography V **221** (1980) 46.
80H10 Horwath, R.S., Zarawin, C.B.: Proc. Sec. Symp. Dry Processes, Japan, **Oct.** (1980) 1.
80H11 Hata, T., Halon, B., Vossen, J.L.: Proc. Sec. Symp. Dry Processes, Japan, **Oct.** (1980) 19.
80H12 Holland, L.: Surface Technol. **11** (1980) 145.
80H13 Hu, E.L., Howard, R.E.: Appl. Phys. Lett. **37 (1)** (1980) 1022.
80H14 Howard, R.E.: Solid State Technol. **Aug.** (1980) 127.
80H15 Hitchman, M.L., Eichenberger, V.: J. Vac. Sci. Technol. **17** (1980) 1378.
80I1 Ianno, N.J., Verdeyen, J.T.: Electrochem. Soc. Extend. Abstr. **80-1** (1980) 283.
80J1 Jones, A.H.: Proc. LFE Plasma Symp., San Francisco (1980) TM3.
80K1 Klinger, R.E., Greene, J.E.: Electrochem. Soc. Extend. Abstr. **80-1** (1980) 321.
80K2 Kay, E.: J. Vac. Sci. Technol. **17** (1980) 658.
80K3 Kurisaki, T., Horiike, Y., Yamazaki, T.: German Patent, Offenlegungsschrift 3030814 (1980).
80K4 Kosugi, M., Ogawa, T., Shibayama, H., Hisatsugu, T.: Proc. Sec. Symp. Dry Processes, Japan, Oct. **1980**, 27.
80K5 Kurosawa, K., Horiike, Y., Okano, H., Okumura, K.: Proc. Sec. Symp. Dry Processes, Japan, Oct. 1980, 43.
80K6 Kawamoto, Y., Hashimoto, N.: Proc. Soc. Symp. Dry Processes, Japan, **Oct.** (1980) 63.
80L1 Lehmann, H.W.: Perkin-Elmer Seminar on Sputtering, Plasma Etching and Surface Analysis, Munich, Germany, April **1980**.
80L2 Lam, D.K., Koch, G.R.: Solid State Technol. **Sept.** (1980) 99.
80L3 Lehmann, H.W., Widmer, R.: J. Vac. Sci. Technol. **17 (5)** (1980) 1177.
80L4 Lechaton, J.S., Mauer, J.L.: Electrochem. Soc. Extend. Abstr. **No. 80-1** (1980) 270.
80L5 Le Beau, B., Wourms, B.: SPIE Semicond. Microlithography V **221** (1980) 61.
80L6 Lehmann, H.W.: Dry Etching Techniques, EUROCON 80, in: "From Electronics to Microelectronics", ed. by W.A. Kaiser and W.E. Proebster, North-Holland Publ. Co (1980).
80M1 Mader, H.: Perkin-Elmer Seminar on Sputtering, Plasma Etching and Surface Analysis, Munich, Germany, April **1980**.
80M2 Matsuo, S.: J. Vac. Sci. Technol. **17 (2)** (1980) 587.
80M3 Matsuo, S.: Appl. Phys. Lett. **36 (9)** (1980) 768.
80M4 Mader, H.: Electrochem. Soc. Extend. Abstr. **No. 80-1** (1980) 274.
80M5 Mauer, J.L., Logan, J.S.: Electrochem. Soc. Extend. Abstr. **No. 80-1** (1980) 268.
80M6 Mundt, R., Patel, K.C., Cowen, K.: Proc. IEDM 1980, Washington D.C. (1980) 409.
80M7 Mogab, C.J., Levinstein, H.L.: J. Vac. Sci. Technol. **17** (1980) 721.
80M8 Mochizuki, T., Tsujimaru, T., Kashiwagi, M., Nishi, Y.: IEEE Trans. Electron Devices ED-27 (1980) 1431.
80M9 Moriya, T., Hazuki, Y., Kashiwagi, M.: Proc. Sec. Symp. Dry Processes, Japan, **Oct.** (1980) 49.
80N1 Nishizawa, J., Hasaka, N.: Electrochem. Soc. Extend. Abstr. No. **80-1** (1980) 326.

80N2 Nakamura, M., Itoga, M., Ban, Y.: Electrochem. Soc. Extend. Abstr. No. **80-1** (1980) 298.
80N3 Nakata, H., Nishioka, K., Abe, H.: J. Vac. Sci. Technol. **17** (1980) 1351.
80N4 Nagatomo, M., Wakamiya, W., Abe, H.: Proc. Sec. Symp. Dry Processes, Japan, **Oct.** (1980) 95.
80N5 Nishizawa, J., Hayasaka, N.: Proc. Sec. Symp. Dry Processes, Japan, **Oct.** (1980) 109.
80N6 Nakayama, S., Tsuneto, K., et al.: Proc. Sec. Symp. Dry Processes, Japan, **Oct.** (1980) 115.
80O1 Okana, H., Horiike, Y.: Electrochem. Soc. Extend. Abstr. No. **80-1** (1980) 291.
80O2 Ozaki, Y., Hirata, K., Yabumoto, N., Oshima, M.: Proc. Sec. Symp. Dry Processes, Japan, **Oct.** (1980) 55.
80O3 Oda, M., Hirata, K.: Proc. Sec. Symp. Dry Processes, Japan, **Oct.** (1980) 87.
80P1 Pandhumsoporn, T., Hayes, J.: Proc. LFE Plasma Symp. San Francisco, **1980**, TA3.
80P2 Peccoud, Mme., Laporte, M.: Proc. 8th Int. Vac. Congr., Cannes, France, **Sept.** (1980) 66.
80R1 Rothman, L.B., Mauer, J., Schwartz, G.C.: Electrochem. Soc. Extend. Abstr. No. **80-1** (1980) 289.
80R2 Ranadive, D.K., Losee, D.L.: Electrochem. Soc. Extend. Abstr. No. **80-1** (1980) 304.
80S1 Shibagaki, M., Horiike, Y.: Jpn. J. Appl. Phys. **19** (1980) 1579.
80S2 Schwartz, G.C., Schaible, P.M.: Electrochem. Soc. Extend. Abstr. No. **80-1** (1980) 277.
80S3 Sanders, F.H.M., Sanders, J.A.M., Beenakker, C.I.M., Dieleman, J.: Electrochem. Soc. Extend. Abstr. No. **80-1** (1980) 280.
80S4 Smolinsky, G., Mayer, T.M.: Electrochem. Soc. Extend. Abstr. No. **80-1** (1980) 272.
80S5 Steinfeld, J.I., Anderson, T.G., Reiser, C., Denison, D.R., Hartsough, L.D., Hollahan, J.R.: J. Electrochem. Soc. **127** (1980) 514.
80S6 Solid State Technol. (Aug. 1980) 54.
80S7 Smolinsky, G.: Electrochem. Soc. Meeting, May 1980, St. Louis, Late News **1980**.
80S8 Selbrede, S.C.: SPIE Semicond. Microlithography V **221** (1980) 53.
80S9 Schwartz, G.C., Schaible, P.M.: Proc. LFE Plasma Symp. San Francisco (1980) FM7.
80S10 Sinha, A.K., Lindenberger, W.S., Fraser, D.B., Murarka, S.P., Fuls, E.N.: IEEE Trans. Electron. Dev. **ED-27** (1980) 1425.
80S11 Smith, J.N.: Proc. Sec. Symp. Dry Processes, Japan, Oct. **1980**, 71.
80S12 Sato, M., Nakamura, H.: Proc. Sec. Symp. Dry Processes, Japan, **Oct.** (1980) 11.
80T1 Tokunaga, K., Hess, D.W.: J. Electrom. Soc. **127** (1980) 928.
80T2 Tsukada, T., Ukai, K.: Electrochem. Soc. Extend. Abstr. No. **80-1** (1980) 328.
80T3 Tolliver, D.L.: Proc. LFE Plasma Symp. San Francisco (1980) TM2.
80V1 van de Ven, E.P.G.T., Zijlstra, P.A.: Electrochem. Soc. Extend. Abstr. No. **80-1** (1980) 253.
80V2 Vossen, J.L.: Proc. 8th Int. Vac. Congr., Cannes, France, Sept. (1980) 21.
80W1 Wang, D.N.K., Maydan, D., Levinstein, H.J.: Solid State Technol. **Aug.** (1980) 122.
80W2 Winkler, U., Schmidt, W.: Electrochem. Soc. Meeting, May 1980, St. Louis, Late News **1980**.
80Y1 Yamazaki, T., Watakabe, Y., Suzuki, Y., Nakata, H.: J. Electrochem. Soc. **127** (1980) 1859.
80Y2 Yamazaki, T., Suzuki, Y., Nakata, H.: J. Vac. Sci. Technol. **17** (1980) 1348.
80Y3 Yoneda, M., Voya, S., Abe, H.: Proc. Sec. Symp. Dry Processes, Japan, **Oct.** (1980) 35.
80Z1 Zarowin, C.B., Rosenberg, R., Horwath, R.: Electrochem. Soc. Extend. Abstr. No. **80-1** (1980) 266.
80Z2 Zelly, A.: Proc. LFE Plasma Symp. San Francisco, **1980**, TA2.
81A1 d'Agostino, R., Cramarossa, F., De Benedicts, S., Ferraro, G.: J. Appl. Phys. **52(3)** (1981) 1259.
81A2 Arikado, T., Horiuchi, S.: Plasma Processing (eds. Frieser, R.G., Mogab, C.J., eds.) Electrochem. Soc. Pennington, N.J. **1981** 66.
81A3 Atamanov, V.M., Ivanov, A.A., Levadnyi, G.B. et al.: Proc. 5th Int. Symp. Plasma Chem., Edinburgh, UK **1981** 336.
81A4 Atamanov, V.M., Ivanov, A.A., Levadnyi, G.B., Logunov, V.I. et al.: Proc. 5th. Int. Symp. Plasma Chem., Edinburgh, UK (1981) 340.
81A5 Abe, H., Yamazaki, T., Yoneda, M., Nishioka, K., Nakata, H.: 1981 Symp. VLSI Technology, Digest of Technical Papers, Mani **1981** 64.
81A6 Akiya, H., Saito, K., Kobayashi, K.: Jpn. J. Appl. Phys. **20(3)** (1981) 647.
81A7 Ahn, K.Y., Cox, D.E.: European Patent Application 0020935 (1981).
81A8 d'Agostino, R.: J. Appl. Phys. **52(1)** (1981) 162.
81A9 Akiya, H.: Proc. Symp. Dry Process, Tokyo, Japan, **Oct.** (1981) 119.
81A10 Arikado, T., Horiike, Y.: Proc. Symp. Dry Process, Tokyo, Japan, Oct. **1981** 55.
81A11 Adams, A.C., Capio, C.D.: J. Electrochem. Soc. **128** (1981) 366.
81B1 Beinvogl, W., Hasler, B.: Semiconductor Silicon 1981 (Huff, H.R., Kriegler, R.J., Takeishi, Y. eds.) Electrochem. Soc., Pennington, N.J. **1981** 649.
81B2 Beinvogl, W., Deppe, H., Stokan, R., Hasler, B.: IEEE Trans. Electron Devices **ED-28(11)** (1981) 1332.

81B3 Beenakker, C.I.M., van de Poll, R.P.J., Dieleman, J.: Electrochem. Soc. Extend. Abstr. **No 81-2** (1981) 618.

81B4 Bruce, R.H., Flamm, D.L., Donnelly, V.M., Duncan, B.S.: Electrochem. Soc. Extend. Abstr. **No 81-2** (1981) 631.

81B5 Burton, R.H., Smolinsky, G.: Electrochem. Soc. Extend. Abstr. **No 81-2** (1981) 645.

81B6 Beguin, A., Grassionot, G.: Electrochem. Soc. Extend. Abstr. **No 81-2** (1981) 696.

81B7 Bruce, R.H., Malafsky, G.: Electrochem. Soc. Extend. Abstr. **No 81-2** (1981) 703.

81B8 Bondur, J.A., Frieser, R.G.: Plasma Processing (Frieser, R.G., Mogab, C.J., eds.) Electrochem. Soc., Pennington, N.J. **1981** 180.

81B9 Bruno, G., Cappezzuto, P., Cramarossa, F., d'Agostino, R.: Plasma Processing (Frieser, R.G., Mogab, C.J., eds.) Electrochem. Soc., Pennington, N.J. **1981** 208.

81B10 Bell, G.: Plasma Processing (Frieser, R.G., Mogab, C.J., eds.) Electrochem. Soc., Pennington, N.J. **1981** 218.

81B11 Bruce, R.H.: Plasma Processing (Frieser, R.G., Mogab, C.J., eds.) Electrochem. Soc., Pennington, N.J. **1981** 243.

81B12 Beenakker, C.I., Van Domelen, J.H.J., Dieleman, J.: Plasma Processing (Frieser, R.G., Mogab, C.J., eds.) Electrochem. Soc., Pennington, N.J. **1981** 302.

81B13 Bresnock, F.J.: Plasma Processing (Frieser, R.G., Mogab, C.J., eds.) Electrochem. Soc., Pennington, N.J. **1981** 313.

81B14 Brandt, W.W., Roselle, P.: Proc. 5th. Int. Symp. Plasma Chem. ISPC-5, Edinburgh, UK **1981** 296.

81B15 Bolsen, M., Stephani, D.: Microcircuit Engineering Conf., Abstracts, Lausanne, Switzerland **1981** 100.

81B16 Bruce, R.H., Reinberg, A.R.: Microcircuit Engineering Conf., Abstracts, Lausanne, Switzerland **1981** 98.

81B17 Bond, R.A., Dzioba, S., Naguib, H.M.: J. Vac. Sci. Technol. **18(2)** (1981) 335.

81B18 Bruce, R.H.: Private Communication **1981**.

81B19 Bell, A.T., Hess, D.W.: Fundamentals and Applications of Plasma Chemistry for IC Fabrication, Electrochem. Soc., Minneapolis, Minnesota, May **1981**.

81B20 Beenakker, C.I.M., Dommelen, van J.H.J., van de Poll, R.P.J.: J. Appl. Phys. **52(1)** (1981) 480.

81B21 Bruce, R.H.: Solid. State Technol. Oct. (1981) 64.

81B22 Birol, K.: European Patent Application 0031704 (1981).

81B23 Bösch, M.A., Coldren, L.A., Good, E.: Appl. Phys. Lett. **38** (1981) 264.

81B24 Bondur, J.A., Pogge, H.B.: European Patent 0000897 (1981).

81B25 Bond, R.A., Dzioba, S., Naguib, H.M.: J. Vac. Sci. Technol. **18** (1981) 335.

81C1 Curran, J.E., McCulloch, D.J.: Electrochem. Soc. Extend. Abstr. **No 81-2** (1981) 639.

81C2 Chang, R.P.H.: Electrochem. Soc. Ext. Abstr. **No 81-2** (1981) 648.

81C3 Curtis, B.J., Brunner, H.R.: Proc. 5th Int. Symp. Plasma Chem., Edinburgh, UK **1981** 318.

81C4 Coldren, L.A., Miller, B.I., Iga, K.., Rentschler, J.A.: Appl. Phys. Lett. **38 (5)** (1981) 315.

81C5 Cabral, S.M., Silversmith, D.J., Mountain, R.W.: Proc. Kodac Interface, **1981**.

81C6 Crockett, A.R., Stark, M.M.: Proc. Kodak Interface, **1981**.

81C7 Coburn, J.W., Winters, H.F.: J. Vac. Sci. Technol. **18(3)** (1981) 825.

81C8 Coburn, J.W., Chen, M.: J. Vac. Sci. Technol. **18(2)** (1981) 353.

81C9 Chinn, J.D., Adesida, I., Wolf, E.D., Tibero, R.C.: J. Vac. Sci. Technol. **19** (1981) 1418.

81C10 Coldren, L.A., Rentschler, J.A.: J. Vac. Sci. Technol. **19** (1981) 225.

81C11 Chang, R.P.H., Darack, S.: Appl. Phys. Lett. **38** (1981) 898.

81D1 Donnelly, V.M., Flamm, D.L., Collins, G.J.: Electrochem. Soc. Ext. Abstr. **No 81-2** (1981) 621.

81D2 Dzioba, S., Este, G., Bond, R.A., Naguib, H.M.: Electrochem. Soc. Ext. Abstr. **No 81-2** (1981) 628.

81D3 Donohoe, K.G.: Electrochem. Soc. Ext. Abstr. **No 81-2** (1981) 706.

81D4 Donnelly, V.M., Flamm, D.L., Mucha, J.A.: Plasma Processing (Frieser, R.G., Mogab, C.J., eds.) Electrochem. Soc., Pennington, N.J. **1981** 270.

81D5 Donohoe, K.G.: Proc. 5th Int. Symp. Plasma Chem. ISPC-5, Edinburgh, UK (1981) 310.

81D6 Donnelly, V.M., Flamm, D.L.: Solid State Technol., April (1981) 161.

81D7 Dieleman, J.: Proc. Symp. Dry Process, Tokyo, Japan, Oct. (1981) 1.

81D8 Danesh, P., Pantchev, B.G.: Thin Solid Films **82** (1981) L 117.

81D9 Downey, D.F., Bottoms, W.R., Hanley, P.R.: Solid State Technol. Febr. (1981) 121.

81E1 Ephrath, L.M., Petrillo, E.J.: Electrochem. Soc. Ext. Abstr. **No 81-2** (1981) 678.

81E2 Eisele, K.M.: Plasma Processing (Frieser, R.G., Mogab, C.J., eds.) Electrochem. Soc., Pennington, N.J. (1981) 174.

81E3 Ephrath, L.M.: Semiconductor Silicon 1981 (Huff, H.R., Kriegler, R.J., Takeishi, Y., eds.) Electrochem. Soc., Pennington, **1981** 627.
81E4 Ephrath, L.M.: IEEE Trans. Electron Devices. **ED-28(11)** (1981) 1315.
81E5 Eisele, K.M.: J. Electrochem. Soc. **128(11)** (1981) 123.
81E6 Egitto, F.D., Wang, D.N.K., Maydan, D.: Solid State Technol. **Dec.** (1981) 71.
81F1 Flamm, D.L., Wang, D.N.K., Maydan, D.: Electrochem. Soc. Ext. Abstr. **No 81-2** (1981) 658.
81F2 Flamm, D.L.: Plasma Processing (Frieser, R.G., Mogab, C.J., eds.) Electrochem. Soc., Pennington, N.J. **1981** 55.
81F3 Flamm, D.L., Donnelly, V.M., Mucha, J.A., Vasile, M.J.: Proc. 5th Int. Symp. Plasma Chem. ISPC-5, Edinburgh, UK **1981** 293.
81F4 Flamm, D.L., Donnelly, V.M., Bruce, R.H., Collins, G.J.: Proc. 5th. Int. Symp. Plasma Chem. ISPC-5, Edinburg, UK **1981** 307.
81F5 Fraser, D.B., Murarka, S.P., Tretola, A.R., Sinha, A.K.: J. Vac. Sci. Technol. **18(2)** (1981) 345.
81F6 Flamm, D.L., Maydan, D.: PCT International Patent Classification HOIL 21/306, 21/312 (1981).
81F7 Forget, L.E., Gdula, R.A., Hollis, J.C.: European Patent Application 0036144 (1981).
81F8 Flamm, D.L., Donnelly, V.M.: Plasma Chemistry and Plasma Processing **1** (1981) 317.
81G1 Goldstein, I.S., Kalk, F.: J. Vac. Sci. Technol. **19(3)** (1981) 743.
81G2 Gildenblat, G., Heath, B.A.: Electrochem. Soc. Ext. Abstr. **No 81-2** (1981) 693.
81G3 Ginley, D.S., Haaland, D.M., Seager, C.H.: Electrochem. Soc. Ext. Abstr. **No 81-2** (1981) 699.
81G4 Gorin, G.J.: German Patent, Offenlegungsschrift 3023591 (1981).
81G5 Götzlich, J., Ryssel, H.: J. Electrochem. Soc. **128(3)** (1981) 617.
81G6 Guldan, A., Schiller, V., Steffen, A., Balk, P.: 6th Symp. Solid State Device Technol. SSSDT 81, Abstracts, Toulouse, France, **Sept.** (1981) 277.
81G7 Gardiner, K.M., Halley, S.R.: Solid State Technol. **Oct.** (1981) 117.
81H1 Hartman, D.C.: Electrochem. Soc. Extend. Abstr. **No 81-2** (1981) 662.
81H2 Hanson, D., Keyser, T., Pierce, J.M.: Electrochem. Soc. Extend. Abstr. **No 81-2** (1981) 683.
81H3 Heath, B.A.: Electrochem. Soc. Extend. Abstr. **No 81-2** (1981) 690.
81H4 Herb, G.K., Porter, R.A., Cruzan, P.D., Agraz-Guerena, J.; Soller, B.R.: Electrochem. Soc. Extend Abstr. **No 81-2** (1981) 710.
81H5 Hiraoka, H., Welsh, L.W.: Plasma Processing (Frieser, R.G., Mogab, C.J., eds.) Electrochem. Soc., Pennington, N.J. **1981** 59.
81H6 Horiike, Y.: Private communication, **1981**.
81H7 Harada, T., Gamo, K., Namba, S.: Jpn. J. Appl. Phys. **20(1)** (1981) 259.
81H8 Hakhu, J.K.: 1981 Symp. VLSI Technol., Digest of Technical Papers, Maui (1981) 66.
81H9 Hosaka, S., Kawamoto, Y., Hashimoto, S.: J. Vac. Sci. Technol. **18(1)** (1981) 17.
81H10 Horng, Ch.T., Schwenker, R.O.: European Patent 0021147 (1981).
81H11 Horwitz, Ch.M.: IEEE Trans. Electron Devices **ED-28(11)** (1981) 1320.
81H12 Hirata, K., Ozaki, Y., Oda, M., Kimizuka, M.: IEEE Trans. Electron Devices **ED-28(11)** (1981) 1323.
81H13 Horiike, Y., Sugawara, T., Okano, H., Shibagaki, M., Ueda, Y.: Jpn. J. Appl. Phys. **20(4)** (1981) 803.
81H14 Harper, J.M.E., Cuomo, J.J., Leary, P.A., Summa, G.M., Kaufman, H.R., Bresnock, F.J.: J. Electrochem. Soc. **128(5)** (1981) 1077.
81H15 Hikosaka, K., Mimura, T., Joshin, K.: Proc. Symp. Dry Process, Tokyo, Japan, Oct. **1981** 97.
81H16 Horwath, R.S.; Zarowin, C.B.: Proc. Symp. Dry Process, Tokyo, Japan, Oct. **1981** 91.
81H17 Heath, B.A.: Solid State Technol., Oct. **1981** 75.
81I1 Ibbotson, D.E., Donnelly, V.M., Flamm, D.L., Duncan, B.S.: Electrochem. Soc. Extend. Abstr. **No 81-2** (1981) 650.
81I2 Ianno, N.J., Verdeyen, J.T.: Plasma Processing (Frieser, R.G.; Mogab, C.J.; eds.) Electrochem. Soc., Pennington, N.J. **1981** 166.
81I3 Irene, E.A., Tierney, E., Blum, J.M., Aliotta, C.F., Lamberti, A.C., Ginsberg, B.J.: J. Electrochem. Soc. **128** (1981) 1971.
81I4 Ianno, N.J., Greenberg, K.E., Verdeyen, J.T.: J. Electroch. Soc. **128 (10)** (1981) 2174.
81I5 Itakura, H., Nishioka, K., Yoneda, M., Abe, H.: Proc. Symp. Dry Process, Tokyo, Japan, Oct. (1981) 75.
81J1 Jackel, J.L., Howard, R.F., Hu, E.L., Lyman, S.P.: Appl. Phys. Lett. **38** (1981) 907.
81K1 Kravitz, S.H., Soller, B.R., Rieger, D.J.: Electrochem. Soc. Ext. Abstr. **No 81-2** (1981) 655.
81K2 Kern, D.P., Zarowin, C.B.; Plasma Processing (Frieser, R.G., Mogab, C.J., eds.) Electrochem. Soc. Pennington, NJ. **1981** 86.
81K3 Klinger, R.E., Greene, J.E.: Plasma Processing (Frieser, R.G., Mogab, C.J., eds.) Electrochem. Soc., Pennington, N.J. **1981** 257.

81K4 Klinger, R.E., Greene, J.E.: Proc. 5th Int. Symp. Plasma Chem. ISPC-5, Edinburgh, UK **1981** 324.
81K5 Keller, J.V.: European Patent Application 0013483 (1981).
81K6 Klinger, R.E., Greene, J.E.: Proc. Symp. Dry Process, Tokyo, Japan, Oct. **1981** 93.
81K7 Kure, T., Tamaki, Y.: Proc. Symp. Dry Process, Tokyo, Japan, Oct. **1981** 83.
81K8 Kammerdiner, L.: Solid State Technol., **Oct.** (1981) 79.
81L1 Lehmann, H.W., Heeb, E., Frick, K.: Electrochem. Soc. Extend Abstr. **No 81-2** (1981) 615.
81L2 Leahy, M.F.: Electrochem. Soc. Extend Abstr. **No 81-2** (1981) 660.
81L3 Lechaton, J.S., Mauer, J.L.: Plasma Processing (Frieser, R.G., Mogab, C.J., eds.) Electrochem. Soc., Pennington, NJ **1981** 75.
81L4 Laporte, Ph., Peccoud, L.: Proc. 5th Int. Symp. Plasma Chem. ISPC-5, Edinburgh, UK **1981** 344.
81L5 Lyman, J.: Electronics, May 5 (1981) 41.
81L6 Lehmann, H.W.: From Electronics to Microelectronics, Kaiser, W.A., Proebster, W.E. (eds.), North-Holland Publishing Company **1980** 57.
81L7 Laporte, Ph., Peccound, L.: Microcircuit Engineering Conf., Abstracts, Lausanne, Switzerland **1981** 102.
81L8 Lehmann, H.W., Heeb, E., Frick, K.: Solid State Technol. **Oct.** (1981) 69.
81L9 Lee, W.J., Eldridge, J.M.: J. Appl. Phys. **52** (1981) 2994.
81M1 Mader, H.: Plasma Processing (Frieser, R.G., Mogab, C.J., eds.) Electrochem. Soc., Pennington, NJ (1981) 125.
81M2 Mader, H.: Perkin Elmer Seminar on Sputtering, Plasma Etching and Surface Analysis, Cambridge, UK April **1981**.
81M3 Mader, H., Syniawa, B., Durner, H.: Proc. 5th. Int. Symp. Plasma Chemistry ISPC-5, Edinburgh, UK **1981** 301.
81M4 Mucha, J.A., Flamm, D.L., Donnelly, V.M.: Electrochem. Soc. Extend. Abstr. **No 81-2** (1981) 610.
81M5 McOmber, J.I., Pender, M.: Electrochem. Soc. Extend. Abstr. **No 81-2** (1981) 634.
81M6 Mauer, J.L., Desilets, K.R.: Electrochem. Soc. Extend. Abstr. **No 81-2** (1981) 674.
81M7 Montillo, F.J., Frieser, R.G., Chu, W.K.: Electrochem. Soc. Extend. Abstr. **No 81-2** (1981) 681.
81M8 Mayer, T.M., Barker, R.A.: Electrochem. Soc. Extend. Abstr. **No 81-2** (1981) 688.
81M9 Mathad, G.S.: Electrochem. Soc. Extend. Abstr. **No 81-2** (1981) 720.
81M10 Martinet, F.: Electrochem. Soc. Extend. Abstr. **No 81-2** (1981) 723.
81M11 Moriwaki, K., Aritome, H., Namba, S.: Jpn. J. Appl. Phys. **20(7)** (1981) 1305.
81M12 Matsui, S., Mizuki, S., Yamato, T., Aritome, H., Namba, S.: Jpn. J. Appl. Phys. **20(1)** (1981) L38.
81M13 Mizutani, T., Komatsu, H., Harada, S.: Proc. IEEE IEDM, Washington (1981).
81M14 Mei, L., Sweetsen, Chen, Dutton, R.W.: IEEE Int. Electron. Device Meeting IEDM 80, Washington **1980** 831.
81M15 Marcoux, P.J., Foo, P.D.: Solid State Technology **April** (1981) 115.
81M16 Mayer, T.M., Barker, R.A., Whitman, L.J.: J. Vac. Sci. Technol. **18(2)** (1981) 349.
81M17 Matsumoto, H., Sugano, T.: SSD 81-**43** (1981) 73.
81M18 Meguro, T., Kurita, H., Itoh, T.: J. Electrochem. Soc. **128(6)** (1981) 1379.
81M19 Matsumoto, H., Sugano, T.: Proc. Symp. Dry Process, Tokyo, Japan, Oct. **1981** 25.
81M20 Meguro, T., Itoh, T., Okano, H., Horiike, Y.: Proc. Symp. Dry Process, Tokyo, Japan, Oct. **1981** 113.
81M21 Makino, T., Nakumura, H., Asano, M.: J. Electrochem. Soc. **128** (1981) 103.
81M22 Mayer, T.M., Barker, R.A., Whitmann, L.J.: J. Vac. Sci. Technol. **18** (1981) 349.
81M23 Maydan, D.: German Patent Offenlegungsschrift DE 3104024 A1 (1981).
81M24 Mucha, J.A., Flamm, D.L., Donnely, V.M.: Electrochem. Soc. Extend. Abstr. **81-2** (1981) 610.
81N1 Nagarajan, A., Sharif, A.: Electrochem. Soc. Extend. Abstr. **No 81-2** (1981) 667.
81N2 Nakamura, M., Itoga, M., Ban, Y.: Plasma Processing (Frieser, R.G., Mogab, C.J., eds.) Electrochem. Soc., Pennington, N.J. **1981** 225.
81N3 Nishizawa, J., Hayasaka, N.: Plasma Processing (Frieser, R.G., Mogab, C.J., eds.) Electrochem. Soc., Pennington, N.J. **1981** 278.
81N4 Norstrom, H., Olaison, R., Berg, S., Andersson, L.P.: Plasma Processing (Frieser, R.G., Mogab, C.J., eds.) Electrochem. Soc., Pennington, N.J. **1981** 313.
81N5 Nishizawa, J., Hayasaka, N., Motoyoshi, M.: Proc. Symp. Dry Process, Tokyo, Japan, Oct. **1981** 9.
81N6 Nagamoto, M., Abe, H.: German Patent Offenlegungsschrift DE 3048441 A1 (1981).
81O1 Okano, H., Horiike, Y.: Electrochem. Soc. Extend. Abstr. **No 81-2** (1981) 672.
81O2 Okano, H., Horiike, Y.: Plasma Processing (Frieser, R.G., Mogab, C.J., eds.) Electrochem. Soc., Pennington, N.J. **1981** 199.
81O3 Oshima, M.: Jpn. J. Appl. Phys. **20(4)** (1981) 683.

81O4 Oshima, M.: Jpn. J. Appl. Phys. **20(7)** (1981) 1255.
81O5 Okano, H., Yamazaki, T., Horiike, Y.: Proc. Symp. Dry Process, Tokyo, Japan, Oct. **1981** 69.
81O6 O'Hanlon, J.F.: Solid State Technol. **Oct.** (1981) 86.
81P1 Peccoud, L., Laporte, Ph., Chevallier, M., Fried, T.: Electrochem. Soc. Extend. Abstr. **No 81-2** (1981) 712.
81P2 Parrens, P.: Proc. Microcircuit Engineering Conf., Lausanne, Switzerland **1981** 96.
81P3 Parrens, P., Raffat, E., Jeuch, P.: 6th Symp. Solid State Device Technol. SSSDT 81, Abstracts, Toulouse, France, **Sept.** (1981) 194.
81R1 Reinberg, A.R., Dalle Ave, J., Steinberg, G., Bruce, R.: Electrochem. Soc. Extend. Abstr. **No 81-2** (1981) 669.
81R2 Rapakoulias, D., Amouroux, J., Bergougnan, M.P.: Electrochem. Soc. Extend. Abstr. **No 81-2** (1981) 701.
81R3 Ranadive, D.K., Losee, D.L., Khosla, R.P.: Electrochem. Soc. Extend. Abstr. **No 81-2** (1981) 707.
81R4 Rothman, L.B., Mauer, J.L., Schwartz, G.C., Logan, J.S.: Plasma Processing (Frieser, R.G., Mogab, C.J., eds.) Electrochem. Soc., Pennington, N.J. **1981** 193.
81R5 Ranadive, D.K., Losee, D.L.: Plasma Processing (Frieser, R.G., Mogab, C.J., eds.) Electrochem. Soc., Pennington, N.J. **1981** 236.
81R6 Reinberg, A.R.: VLSI Electronics Microstructure Science, Vol. **2**, Einspruch, N.G. ed., New York: Academic Press (1981) 2.
81S1 Smith, D.L., Bruce, R.H.: Electrochem. Soc. Extend. Abstr. **No 81-2** (1981) 625.
81S2 Smolinsky, G., Wang, D.N.K., Maydan, D.: Electrochem. Soc. Extend. Abstr. **No 81-2** (1981) 686.
81S3 Sternheim, M., Van Gelder, W.: Electrochem. Soc. Extend. Abstr. **No 81-2** (1981) 717.
81S4 Smolinsky, G., Mayer, T.M., Truesdale, E.A.: Plasma Processing (Frieser, R.G., Mogab, C.J., eds.) Electrochem. Soc., Pennington, N.J. **1981** 120.
81S5 Schwartz, G.C., Schaible, P.M.: Plasma Processing (Frieser, R.G., Mogab, C.J., eds.) Electrochem. Soc., Pennington, N.J. **1981** 133.
81S6 Sanders, F.H.M., Sanders, J.A.M., Beenakker, C.I.M., Dieleman, J.: Plasma Processing (Frieser, R.G., Mogab, C.J., eds.) Electrochem. Soc., Pennington, N.J. **1981** 155.
81S7 Saccocio, E.J., Cobb, C.A., Holycross, M.: Proc. 5th Int. Symp. Plasma Chem. ISPC-5, Edinburgh, UK **1981** 330.
81S8 Silversmith, D.J., Ehrlich, D.J., Osgood, R.M., Deutsch, T.F.: 1981 Symp. VLSI Technol., Digest of Technical Papers, Maui **1981** 70.
81S9 Sanders, F.H.M., Dieleman, J.: Semiconductor Silicon 1981 (Huff, H.R., Kriegler, R.J., Takeishi, Y. eds.) Electrochem. Soc. Pennington, N.J. **1981** 638.
81S10 Suzuki, K., Okudaira, S., Nishimatsu, S., Kanomata, I.: European Patent Application 0017143 (1981).
81S11 Shibagaki, M., Watanabe, T., Takeuchi, H., Horiike, Y.: Proc. Symp. Dry Process, Tokyo, Japan, Oct. **1981** 39.
81S12 Shibata, H., Serikawa, T., Okamoto, A.: Proc. Symp. Dry Process, Tokyo, Japan, Oct. (1981) 61.
81T1 Turban, G., Pasquéreau, J., Rapeaux, M., Catherine, Y., Grolleau, B.: Electrochem. Soc. Extend. Abstr. **No 81-2** (1981) 636.
81T2 Tsukada, T., Mito, H., Nagasaka, M.: Electrochem. Soc. Ext. Abstr. **No 81-2** (1981) 715.
81T3 Tsukada, T., Ukai, K.: Plasma Processing (Frieser, R.G., Mogab, C.J., eds.) Electrochem. Soc., Pennington, N.J. **1981** 288.
81T4 Tretola, A.R.: Plasma Processing (Frieser, R.G., Mogab, C.J., eds.) Electrochem. Soc., Pennington, N.J. **1981** 295.
81T5 Tsai, M.Y., Chao, H.H., Ephrath, L.M.: J. Electrochem. Soc. **128(10)** (1981) 2207.
81T6 Tokunaga, K., Redeker, F.C., Danner, D.A., Hess, D.W.: J. Electrochem. Soc. **128(4)** (1981) 851.
81T7 Tolliver, D.L.: Solid State Technol. (Nov. 1980) 99.
81T8 Toyoda, H., Itakura, H., Komiya, H.: Jpn. J. Appl. Phys. **20** (1981) 667.
81T9 Toyoda, H., Tobinaga, M., Komiya, H.: Jpn. J. Appl. Phys. **20** (1981) 681.
81T10 Taillet, J.: German Patent DE 3031220 A1 (1981).
81T11 Tachi, S., Miyake, K., Tokuyama, T.: Jpn. J. Appl. Phys. **20(6)** (1981) L 411.
81T12 Takada, T., Tokitomo, K., Hoshino, H.: European Patent Application 0023429 (1981).
81T13 Tachi, S., Miyake, K., Tokuyama, T.: Proc. Symp. Dry Process, Tokyo, Japan, Oct. **1981** 17.
81V1 Vasile, M.J., Stevie, F.A.: Electrochem. Soc. Extend. Abstr. **No 81-2** (1981) 613.
81V2 Van de Ven, E.P.G.T., Zijlstra, P.A.: Plasma Proc. (Frieser, R.G., Mogab, C.J., eds.) Electrochem. Soc., Pennington, N.J. **1981** 112.
81W1 Weston, D.F., Tolliver, D.L.: Electrochem. Soc. Extend. Abstr. **No 81-2** (1981) 665.
81W2 Wang, D.N.K., Maydan, D.: Solid State Technology, (May 1981) 121.

81W3 Winkler, V., Schmidt, F., Hoffman, N.: Plasma Processing (Frieser, R.G., Mogab, C.J., eds.) Electrochem. Soc., Pennington, N.J. **1981** 253.
81W4 Wen-Yaung Lee, Eldridge, J.M.: J. Appl. Phys. **52(4)** (1981) 2994.
81W5 Wang, C.W., Gelernt, B.: Proc. Symp. Dry Process, Tokyo, Japan, Oct. **1981** 121.
81Y1 Yao, W.W., Bruce, R.H.: Electrochem. Soc. Extend. Abstr. **No 81-2** (1981) 652.
81Y2 Yabumoto, N., Oshima, M., Michikami, O., Yoshii, S.: Jpn. J. Appl. Phys. **20(5)** (1981) 893.
81Y3 Yamada, M., Tamano, J., Hattori, S., Yoneda, K., Morita, S.: Proc. Symp. Dry Process, Tokyo, Japan, Oct. **1981** 33.
81Y4 Yamasaki, K., Asai, K., Kurumada, K.: Proc. Symp. Dry Process, Tokyo, Japan, Oct. **1981** 105.
81Y5 Yoneda, M., Itakura, H., Nishioka, K., Abe, H.: Proc. Symp. Dry Process, Tokyo, Japan, Oct. **1981** 47.
81Z1 Zarowin, C.B., Horwath, R.S.: Electrochem. Soc. Extend. Abstr. **No 81-2** (1981) 642.
81Z2 Zarowin, C.B.: Thin Solid Films **85** (1981) 33.
82A1 Arden, W., Beinvogl, W., Müller, W.: Proc. IEEE IEDM, Washington (1982) 403.
82B1 Beinvogl, W., Mader, H.: Siemens Forsch. Entwicklungsber. **11** (1982) 180.
82B2 Bruce, R.H., Reinberg, A.R.: J. Electrochem. Soc. **129** (1982) 393.
82B3 Bernacki, S.E.: Electrochem. Soc. Extend. Abstr. **82-1** (1982) 344.
82B4 Bower, D.H.: J. Electrochem. Soc. **129** (1982) 795.
82C1 Cabral, S.M., Elta, M.E., Chu, A., Mahoney, L.J.: Electrochem. Soc. Extend. Abstr. **82-1** (1982) 348.
82C2 Chang, R.P.H., Chang, C.C., Darack, S.: J. Vac. Sci. Technol. **20** (1982) 45.
82C3 Chang, R.P.H., Chang, C.C., Darack, S.: J. Vac. Sci. Technol. **20** (1982) 490.
82F1 Fraser, D.B., Kinsbron, E., Vratny, F., Johnston, R.L.: J. Vac. Sci. Technol. **20** (1982) 491.
82G1 Ginsberg, B.I.: J. Electrochem. Soc. **128** (1982) 1971.
82H1 Horiike, Y., Okano, H., Yamazaki, T., Horie, H.: Jpn. J. Appl. Phys. Lett. to be published (1982).
82H2 Heath, B.A.: J. Electrochem. Soc. **129** (1982) 396.
82H3 Hess, D.W.: Semicon/West Technical Program Proceedings, May 26–28, San Mateo, California **1982**.
82H4 Hikosaka, K., Mimura, T., Joshin, K., Abe, M.: Electrochem. Soc. Extend. Abstr. **82-1** (1982) 264.
82H5 Hargis, P.J., Kushner, M.J.: Appl. Phys. Lett. **40** (1982) 779.
82I1 Ibbotson, D.E., Flamm, D.L., Donnelly, V.M., Duncan, B.S.: J. Vac. Sci. Technol. **20** (1982) 489.
82I2 Itakura, H., Komiya, H., Ukai, K.: Solid State Technol., **April** (1982) 209.
82K1 Kurogi, Y., Kamimura, K.: Jpn. J. Appl. Phys. **21** (1982) 168.
82K2 Kuwano, H., Miyake, S., Kasai, T.: Jpn. J. Appl. Phys. **21** (1982) 529.
82K3 Koike, A., Imai, K., Hosoda, S., Tomozawa, A., Agatsuma, T.: Electrochem. Soc. Extend. Abstr. **82-1** (1981) 343.
82K4 Korman, C.S.: Solid State Technol., **April** (1982) 115.
82L1 Lam, D.K.: Solid State Technol., **April** (1982) 215.
82M1 Matsuo, S., Adadri, Y.: Jpn. J. Appl. Phys. **21** (1982) L 4.
82N1 Niggebrügge, U., Balk, P.: Solid State Electronics, to be published (1982).
82O1 Okano, H., Horiike, Y.: Plasma Processing, Electrochem. Soc., Pennington, N.J. (1982).
82O2 Okano, H., Horiike, Y., Jpn. J. Appl. Phys. **21** (1982) 696.
82O3 Okana, H., Yamazaki, T., Horiike, Y.: Solid State Technol., **April** (1982) 66.
82O4 Ozaki, Y., Hirata, K.: J. Vac. Sci. Technol. **21** (1982) 61.
82P1 Pederson, L.A.: J. Electrochem. Soc. **129** (1982) 205.
82R1 Ransom, C.M., Chappell, T.I., Ephrath, L.M., Bennett, R.S.: Electrochem. Soc. Extend. Abstr. **82-1** (1982) 346.
82R2 Reichelderfer, R.F.: Solid State Technol., **April** (1982) 160.
82S1 Shibagaki, M., Watanabe, T., Takeuchi, H., Horiike, Y.: Plasma Processing, Electrochem. Soc., Pennington, N.J. **1982**.
82S2 Sato, M., Nakamura, H.: J. Vac. Sci. Technol. **20** (1982) 186.
82S3 Sakai, Y., Reynolds, J.L., Neureuther, A.R.: Electrochem. Soc. Extend. Abstr. **82-1** (1982) 270.
82S4 Sirkin, E.R.: Electrochem. Soc. Extend. Abstr. **82-1** (1982) 341.
82S5 Smolinsky, G., Truesdale, E.A., Wang, D.N.K., Maydan, D.: J. Electrochem. Soc. **129** (1982) 1036.
82T1 Tabe, M.: Jpn. J. Appl. Phys. **21** (1982) 534.
82T2 Ting, C.H., Neureuther, A.R.: Solid State Technol., **Febr.** (1982) 115.
82T3 Tu, C.W., Chang, R.P.H., Schlier, A.R.: Appl. Phys. Lett. **41** (1982) 80.
82V1 Vasile, M.J., Stevie, F.A.: J. Appl. Phys. **53** (1982) 3799.
82W1 Whitcomb, E.C.: Electrochem. Soc. Extend. Abstr. **82-1** (1982) 339.
82W2 Winters, H.F.: J. Vac. Sci. Technol. **20** (1982) 493.
82W3 Whitcomb, E.C., Jones, A.B.: Solid State Technol., **April** (1982) 121.

83B1 Beinvogl, W., Mader, H.: ntz Archiv **5** (1983) 3.
83B2 Beinvogl, W., Hasler, B.: Solid State Technol. April (1983).
83B3 Bruce, R.H., Malafsky, G.P.: J. Electrochem. Soc. **130** (1983) 1369.
83B4 Bassous, E., Ephrath, L.M., Pepper, G., Mikalsen, D.J.: J. Electrochem. Soc. **130** (1983) 478.
83L1 Light, R.W., Bell, H.B.: J. Electrochem. Soc. **130** (1983) 1567.
83M1 Mattausch, H.J., Hasler, B., Beinvogl, W.: J. Vac. Sci. Technol. **B1** (1983) 15.
83P1 Pogge, H.B., Bondur, I.A., Burkhardt, P.I.: J. Electrochem. Soc. **130** (1983) 1592.
83S1 Schwarzl, S., Beinvogl, W.: Electrochem. Soc. Extend. Abstr. No. 83-1 (1983)
83Z1 Zarowin, C.B.: J. Electrochem. Soc. **130** (1983) 1144.

6.1.4.8 Final device preparation

The manufacture of semiconductor devices is not finished with the fabrication of the wafers. This chapter will give advice how to prepare devices in adequate packages.

A selection of the most usual packages is discussed in detail. All other packages are more or less derived from the selected ones. In order to give as much information as possible, patent applications are cited among the references since important details are often described there. Information without reference is the personal experience of the author.

6.1.4.8.1 Survey of package groups

Most common *package outlines* are standardized in the United States by JEDEC (Joint Electron Device Engineering Counsils) called:

DO-packages for diodes (*D*iode *O*utline),
TO-packages for transistors and integrated circuits (*T*ransistor *O*utline),
MO-packages for integrated circuits (*M*icroelectronic-*O*utline),
TMS-packages for chip carriers (abbreviation unknown).

Other countries have their national standardization system. In Europe is a multi national system called IEC (International Electrical Commission), but the JEDEC package designations are more usual in practice and will be applied here. Since the outline drawings of JEDEC are area consuming including drawings, tables with dimensions, and tolerances, only outline drawings taken from [79V1] are presented and supplemented by cross-section drawings which give more information on the internal construction. If there is no JEDEC designation, company internal designations are used, e.g. SOD-, SOT- and SO-numbers (*s*tandard *o*utlines of *d*iodes, *t*ransisters etc.). A survey of the packages is given in Table 1 and in the subsections following the table.

Table 1. Survey of packages. Further details on the construction and the outline of each type are given in the subsections following this table and in the appropriate figures.

Type	Outline	Application	Remarks	Fig. No.	Ref.
A. Hermetically sealed packages for discrete devices					
A.1 Glass-metal packages	DO34	small-signal diodes, switching diodes, varicap diodes, temperature sensors		1, 2	69G2
A.2 Glass-metal headers for transistors					
A.2.1 Matched seal packages	TO-5 TO-18 TO-39	small-signal transistors, high-frequency Si-, Ge- transistors	will be replaced by TO-92 or by T-pack transistors	3a, b, 4, 5	
A.2.2 Compression-sealed packages	TO-39 TO-3	large chips and for high power dissipation		6a, b, 7, 8a, b	
A.3 Ceramic packages	TO-120	high-frequency transistors, tetrodes	same technology as bottom-brazed IC-packages: B2.1	9a, b	

continued

Table 1 (continued)

Type	Outline	Application	Remarks	Fig. No.	Ref.
B. Hermetically sealed packages for IC's					
B.1 Glass-metal headers	TO-78		pins: 6	10a, b	
	TO-99		8		
	TO-74/100		10		
	TO-101		12		
			14		
		integrated circuits chip area $<8\ mm^2$ low power consumption	outline and construction identical to TO-39. Will be replaced by DIL		
B.2 Ceramic packages					
B.2.1 Bottom-brazed	MO-001 AB		pins: 14		
	MO-001 AC		16		
	MO-015 AG		24		
	MO-015 AH		28	11, 12	
	MO-015 AJ		40		
		IC's requiring hermetical seal, also called CERDIL, MOS circuits, military circuits, lab testing			
B.2.2 Side-brazed		same as B.2.1			
	MO-001 AB		pins: 14		
	MO-001 AC		16		
			18		
	MO-026		22		
	MO-015 AG		24		
	MO-015 AH		28	13	
	MO-015 AF		36		
	MO-015 AJ		40		
			also available with 42, 48, 50, and 64 pins		
B.2.3 Glass-ceramic-metal packages	MO-001 AD		pins: 14		
	MO-001 AG		16		
			18		
	MO-026		22		
	MO-015 AD		24		
	MO-015 AH		28	14, 15	
	MO-015 AJ		40		
		digital and analog IC's	also called CERDIP, cheapest hermetically sealed DIL package, limited chip size and thermal resistance		

continued

Table 1 (continued)

Type	Outline	Application	Remarks	Fig. No.	Ref.
B.2.4 Chip carriers			new series of micro-miniature packages,	16	79A1
	type A TMS 002		28, 44, 52, 68, 84, 100, 124, 156 pins,		
	type B TMS 003		28, 44, 52, 68, 84, 100, 124, 156 pins,		
	type C TMS 004		16, 20, 28, 44, 52, 68, 84 pins,	17, 18	
	type D TMS 005		28, 44, 52, 68, 84, 100, 124, 156 pins,		79H3
		digital circuitry	very dense packing possible in circuits, cooling towers on lids required for high-power dissipation		80B3 77P2 77L1 79B1 81O1
C. Hermetic hybrid packages					
		circuits requiring short interconnections and dense packing, ultra-high frequency tuners	not standardized	19	
D. Plastic packages for discrete devices					
D.1 Single-ended plastic package	TO-92	transistors	low-production cost	20, 22	
		LED's		21	
			version having low thermal resistance		74U3
D.2 Double-ended plastic packages	DO-14	general-purpose diodes		23, 26	
	SOD-23	UHF varicap and switching diodes	low interelectrode capacitance,	24, 27	
	SOT-23	thick-film circuits	soft soldering possible, high packing density smallest transistor package	25, 28	77K1
D.3 T- and X-packages	TO-119	transistors, high-frequency tuners	also applicable in strip line technique,	29,31	
			alternative design		79H1
	SOT-103	field effect tetrodes (FET's)	no JEDEC standard, 4-lead version of TO-119	30, 32	
D.4 Power packages					
D.4.1 Low-frequency power packages	SOT-93		plastic design of TO-3	33, 36	
	TO-202			34	
	TO-220			35	
		low-frequency power output stages	3 packages differ in size and power handling capability		

continued

Table 1 (continued)

Type	Outline	Application	Remarks	Fig. No.	Ref.
D.4.2 High-frequency power	TO-117	high-frequency power transistors	4 to 5 Al bonding wires are bonded in parallel	37, 38, 39	75E1
E. Plastic packages for integrated circuits					
E.1 Dual-in-line (DIL) packages	MO-001 AB		pins: 14		
	MO-001 AE		16	40, 42	
	MO-026		22		
	MO-015 AD		24		
	MO-015 AH		28	41	
	MO-015 AJ		40		
		integrated circuits, direct soldering into printed boards	6, 8, 18, and 20 pins are also usual, DIL packages in plastic up to 64 pins are possible, low cost package. high-gain DIL		75H1
E.2 Medium-power DIL and single-in-line packages (SIL)		medium power dissipation for vertical deflection and sound circuits			
	FIN-DIL		"batwing" package	43	68B1
	SIL-9			44	
E.3 Power packages	DIL 16	sound and vertical deflection power circuits	16 pins	45	
	SIL 5, 7, 9, and 13		5, 7, 9, and 13 pins	46	
E.4 Small outline packages					
E.4.1 DIL configuration		watch circuits thick- and thin-film substrates and printed circuits where small room consumption is required			
	SO-6		pins: 6 } same row spacing		
	SO-8		8 } same row spacing		
	SO-14		14 } same row spacing		
	SO-16		16 } same row spacing	47	
	SO-16 large		16 } same row spacing		
	SO-20		20 } same row spacing		
	SO-24		24 } same row spacing	48	
	SO-28		28 } same row spacing		
	VSO-40		40 } same row spacing		
E.4.2 Square flat packages		VLSI high-pin counts, low cost	similar TMS-007 28, 44, 52, 68, 84, 100, 124, 156 pins	49	

continued

Table 1 (continued)

Type	Outline	Application	Remarks	Fig. No.	Ref.
F. Naked chips on substrates					
F.1 Ceramic substrates		professional and consumer circuits in narrow space car radios, video recorders, computers	passivated chips on substrates, chips are glued or soldered to substrate; chip and board protected by epoxy resin		
F.2 Printed circuit board		watch circuits same as F.1	lower cost than ceramic circuit board, power capability reduced		
F.3 Chip carrier	mini pack	only applied to MOS chips by one company	glass epoxy board		77L1, 77A1
F.4 Flexible substrates		cameras, identification or check cards	copper-clad polyimide foil, super 8, 16 mm or 35 mm film reel	50	78S1, 79Y1, 80H2, 81F1, 80H3

A. Hermetically sealed packages for discrete devices

A.1 Glass-metal packages for diodes

The main type of package of this category is the whiskerless diode package DO-34 (Figs. 1, and 2). The semiconductor chip is sealed in a glass sleeve between two wires. The thermal coefficient of expansion of the glass is higher than that of the chip and the wires. During the cooling from the sealing temperature of about 680 °C, the glass sleeve contracts more than the wires and the chips, thus forming a reliable pressure contact. Sealing is performed in graphite jigs at large quantities. The Fe- and Fe—Ni-cores in the wires are only required for transport by magnetic fields during fabrication [69G2]. The outer leads are galvanically or dip-tin or lead/tin finished. Package parts are commercially available.

A.2 Glass-metal headers for transistors

A.2.1 Matched-seal packages

The outline of a TO-18 package as an example is shown in Fig. 3a, b.

A header of Fe—Ni—Co-alloy is hermetically sealed with wires of the same material and a hard glass having a very similar (matched) coefficient of thermal expansion. For details see section 6.1.4.8.2. For chips arranged in a planar structure, the collector lead is welded to the header, the emitter and base leads are isolated by the glass (Fig. 4).

For Ge high-frequency transistors all 3 wires are isolated for reduction of the interelectrode capacitances. A fourth ground wire is welded to the header (Fig. 5).

Die attach and wire bonding are achieved by the procedures described in section 6.1.4.8.3. A can fabricated of Ni, brass, German silver, or Ni-plated Fe is welded to the header. The weld is easily fabricated obtaining long life time of the electrodes if the cans are plated by "Brenner-Nickel" [66B1], containing about 7% of phosphorus which reduces the melting point of the layer. The package is normally filled by dry air or nitrogen. In case of surface-sensitive chips, e.g. germanium transistors or for high power consumption, the package can be filled by either a mixture of silicon rubber and silicon oil [65U1] or with alumina and silicon oil [57M1]. Headers and cans are commercially available on the market. Suppliers of header and can may not be the same ones.

A.2.2 Compression-sealed packages

The outline of a TO-39 package as an example is shown in Fig. 6a, b. The header is a disc of Ni-plated steel or Fe—Ni. The wires are Fe—Ni—Co (Fig. 7). The coefficient of thermal expansion of the glass is lower than that of steel. During cooling down from the sealing temperature, the steel contracts more than the glass and makes a radial compression to glass and wire (see section 6.1.4.8.2). The advantage is a lower price than that of a matched sealed package and better heat transport through the header. The disadvantage is a sensitivity to thermal shock.

Assembly, finish, and availability of components is comparable to the matched-seal package. The advantages of the compression-sealed package permit the design of real power packages of the JEDEC TO-3-family [66B3].

The outline of a power package TO-3 is shown in Fig. 8a, b. The construction is the same as the TO-39 package but at larger dimensions. The base disc has a rhomboid shape having two holes to screw it to heat sinks, which is absolutely necessary for high power dissipation. Instead of using a steel disc, copper is also usual.

A.3 Ceramic packages for discrete devices

The technology applied is the same as that for the bottom-brazed IC-packages (see B.2.1 and Fig. 9a, b).

B. Hermetically sealed packages for integrated circuits

B.1 Glass-metal headers

Outline and construction are identical to the TO-39-package as described under A.2.1 except the number of leads. There are 5 different high pin count TO-39-packages, see Table 1. Fig. 4 shows the construction and Fig. 10a, b gives an example of a TO-74-package. The can is lower in height than that of TO-39.

Headers as well as cans are commercially available on the market.

B.2 Ceramic packages

B.2.1 Bottom-brazed

The typical outline is shown in Fig. 11a···c. The pin counts given in Table 1 fit into the JEDEC standards.

3 layers of rectangular ceramic are laminated and cofired with the metal conductors. The outer leads are hard soldered "brazed" to the metallization of the bottom side. In the center of the bottom layer an area is gold-plated for chip soldering. The middle layer has a window for the chip and around this window also gold-plated pads for wire bonding. The top layer has a still larger window to provide room for the wire bonding operation. A metallized ring allows soldering of a ceramic or metal lid to seal the cavity hermetically. The lids are solder-coated at the bottom side. The outer leads are fabricated by an Fe—Ni—Co-alloy gold-plated and arranged horizontally when purchased. They are bent into a dual-in-line (DIL) configuration for the final device. The bar must be cropped. This package is abbreviated called CERDIL (Fig. 12). Components are commercially available.

B.2.2 Side-brazed

The outline is shown in Fig. 13a···c. Various pin counts fit into JEDEC standards as shown in Table 1.

The construction is internally identical with the bottom-brazed family (see Fig. 11a···c); however, the ceramic body is as wide as the row spacing of the dual-in-line rows and the leads are soldered to the side of the body. The name *side-brazed CERDIL* is therefore common. The advantage of this package is that there is more room for large cavities and chips.

The applications and availability are the same as for the bottom-brazed family.

B.2.3 Glass-ceramic-metal packages.

The outline is shown in Fig. 14a···c. The glass-ceramic-metal packages are mostly in dual-in-line (DIL) configuration abbreviated "CERDIP" with pin counts from 14 to 40 in JEDEC designations as given in Table 1. A ceramic block having the length and the width of the package and a thickness of the bottom half contains a gold-plated cavity in the middle where the chip is placed by Au—Si eutectic soldering. The surface around the cavity is coated by a low-melting glass. Embedded in the glass is a punched and selectively Al-plated lead frame of Fe—Ni 42 alloy which is already bent into the DIL configuration. After die attach and wire bonding, a second block of ceramic, which is also coated by low-melting glass, is soldered on top of it. The chip is hermetically sealed. The sealing glass contains small amounts of water which vaporizes out of the glass in traces during sealing. It is most important to keep the level of water vapour in the cavity below −40 °C dew point to prevent corrosion of the Al-interconnection. The outer leads are finished by galvanic Sn-plating. A cross section of the package is shown in Fig. 15 [73L1, 69S1, 79M1].

A typical sealing procedure is as follows: heat-up rate 80···140 K/min, peak temperature 410···450 °C, peak time 8···12 min, cool down rate 30···50 K/min. Top and bottom ceramic parts as well as lead frames are commercially available.

B.2.4 Chip carriers

Chip carriers are a relatively new series of micro-miniature packages. Up to now JEDEC have standardized 4 types of chip carriers in this hermetically sealed series [79A1].

They are distinguished by different sealing planes and index corners, (Fig. 16). An outline of a 28 lead chip carrier type C is shown in Fig. 17.

A square piece of ceramic contains a gold-plated area in the center for the eutectic Au—Si die attach. Around this area several bond pads are located which are connected to the edges of the ceramic piece where the outer connection pads are placed. Connections can be made by clipping on metal leads or by pushing the chip carrier into a socket. The Si-chip and the bond wires are hermetically sealed by a partly glass-covered lid. A cross section is shown in Fig. 18.

When high-power capability is required, "cooling towers" have to be applied on top of the lids [77L1, 79B1, 81O1, 81L3]. Some companies try BeO instead of Al_2O_3 [81L2]. Carriers and lids are commercially available on the market up to 68 pins.

C. Hermetic hybrid packages

Hermetic hybrid packages are not standardized because they are used in limited quantities only. Hybrid package means that active components like transistors and integrated circuits as well as passive components like resistors and capacitors are assembled in one package on thin or thick film ceramic substrates. The package is a rectangular metal box having glass-isolated leads at the edges. A flat lid is soldered or welded to the box (Fig. 19). Hybrid packages are commercially available in different designs and with different pin numbers.

D. Plastic packages for discrete devices

D.1 Single-ended plastic packages

A typical representative package of this family which is produced at large quantities is the TO-92-package for transistors (Figs. 20 and 22). Another single-ended plastic package especially for light emitting diodes, which is not standardized by JEDEC is shown in Fig. 21.

Because the design and layout for a lead frame depends strongly on the bonding methods and machines applied, these parts are mostly produced by the manufacturers of the transistors themselves and are not available on the market.

D.2 Double-ended plastic packages

Similar or the same assembly technology as that of the single-ended packages. Rotationally symmetric or rectangular packages with leads at opposite sides are usual. The construction varies:

DO-14: the chip is soldered between Cu wires having nailheads. The assembly is encapsulated by plastic. One end is rounded to indicate the polarity (Figs. 23 and 26).

SOD-23: lead frame structure as in TO-92. The same assembly technology is applied (Figs. 24 and 27).

SOT-23: lead frame structure as in TO-92 but smaller. Smallest transistor package. Leads bent towards bottom surface of the plastic body (Figs. 25 and 28).

Components are not commercially available as lead frames.

D.3 T- and X-packages

The construction is similar to TO-92 with the special feature that the leads do not protrude from one single side of the plastic but at angles of 90° to reduce the capacitance between the leads (Figs. 29⋯32). For similar design see [79H1].

D.4 Power packages

D.4.1 Low-frequency power packages

The construction of SOT-93, TO-202, and TO-220 are shown in Figs. 33, 34, and 35, respectively.

SOT-93 and TO-220 use copper heat sinks of 1.3⋯2 mm thickness which are connected to a thinner (0.5 mm) lead frame. The chip is soldered to the heat sink for optimal heat transfer. A hole in the heat sink is for screwing it to cooling fins. The middle part with chip and bonding wires is encapsulated in plastic. The heat sinks are Ni-plated and the leads dip-Sn-plated (Fig. 36) [75L1, 78R1]. The TO-202 package is a simplified version where heat sink and leads are prepared from one piece of the same thickness.

The parts are not commercially available.

D.4.2 High-frequency power packages

The outline of TO-117 is shown in Fig. 37 [75E1]. On top of a special screw a beryllia disc is soldered and on that the leads and the semiconductor chip. All elements above the screw are encapsulated in Si-plastic (Figs. 38 and 39). 4 to 5 Al bonding wires are bonded in parallel in order to operate the device at high currents and to achieve a low inductance. The so-called power tower assembly is commercially available on the market.

E. Plastic packages for integrated circuits

E.1 Dual-in-line (DIL) packages

Outlines of the 16 and 28 lead versions are shown in Figs. 40 and 41 as examples. From the design point of view, DIL packages in plastic up to pin counts of 64 are possible.

A flat lead frame punched into endless strips consists of a die pad in the middle supported by two leads to the outer frame and by as many leads placed around the die pad as required for the package. The maximum number from the design point of view is about 64. The leads are also supported by the frame and by two dam bars which additionally act as stoppers for the plastic flow. The die pad is reduced in height to compensate the die thickness. The material is either Fe-alloy or Cu-alloy, normally 0.25 mm thick. The lead frames are selectively plated by gold, silver, or remain unplated. For improvement of the power handling capability, some manufacturers apply heat spreaders of Al or Cu [68W1, 70U2, 71H1] which are not glued or soldered to the bottom of the die pad. The thermal resistance is near 1/2. After die attach by hard or soft soldering or glueing and wire bonding, the device is encapsulated in plastic and Sn or Pb/Sn finished. In most cases, epoxy plastic is applied, in others silicon plastic is used (Figs. 42 and 54). A dual-in-line package for high gain is presented in [75H1]. Punched lead frames for IC-plastic packages are only in some cases available on the market, fabricated by so-called open tools. The device manufacturers usually have their own designs.

E.2 Medium power DIL and single-in-line (SIL) packages

Both designs of the FIN-DIL (Fig. 43) [68B1] and the SIL-9 package (Fig. 44) are derived from the standard DIL package and are fabricated by the same technology. The outer fins are internally connected to the die pad in order to transport the heat at a low resistance to the outside. The fin DIL-package transports the heat into the printed circuit board, the SIL-package to a separate cooling fin.

E.3 Power packages

The DIL 16 power package (Fig. 45) is very similar to the standard DIL package in Fig. 42. Only the heat spreader deviates. It is a copper block which fills the room between the pad and surface of the package. The heat spreader is soldered or glued to the backside of the die pad which results in a very thermal resistance junction-case. The leads are bent in the opposite direction so that the chip is inserted face down and heat sinks can be applied to the top surface of the package.

The SIL-power-packages (Fig. 46) have a construction like TO-220 and SOT-93 but with more leads. The chip is soldered or glued directly to the heat sink [77G2].

E.4 Small outline packages

E.4.1 DIL configuration

This family of packages became known as SO- (*s*mall *o*utline) range and has not yet JEDEC designations. Pin counts of 6, 8, 14, 16, 20, 24, 28 and 40 are usual. The row spacing of the leads and the body width are the same for the SO-6, 8, 14 and 16 packages. The range continues with SO-16 large, 20, 24, 28, and 40 which are 4 mm wider (Figs. 47 and 48). The leads have a center to center distance of 1.27 mm which is one half compared to the standard DIL-packages. The leads are bent horizontally so that no holes are required in the substrate where the package is to be placed.

The construction is analogous to the standard DIL-range but smaller in size.

E.4.2 Square flat packages

From the design point of view the SO-packages are limited to about 44 pins. If a family of higher pin counts is required leads have to be placed on all four sides of the package and its shape has to be square rather than rectangular.

The outline of square flat packages is very similar to the JEDEC leaded chip carrier range TMS-007. Pin counts of 28, 44, 52, 68, 84, 100, 124, 156 are registered but realized are 28 to 68 pins up to now (Fig. 49).

The construction is basically the same as that of the standard DIL- and the SO-package. The leads are bent underneath the plastic body to save substrate area but the inspection of the solder joints is more difficult compared to the SO-range [77L1, 81L1, 81L3, 81O1, 80M3, 80L1].

F. Naked chips on substrates

Assembly of maked chips means that surface-passivated chips are glued or soldered on to ceramic substrates, printed circuits boards, chip carriers, or flexible substrates and are only protected against mechanical damage by a drop of lacquer like epoxy resin.

F.1 Ceramic substrates

The substrates used are mostly thick-film Al_2O_3-substrates [76H1]. Semiconductor chips are glued or soldered to the ceramic substrate and conventionally wire-bonded. Other passive components like resistors and capacitors are placed on the same substrate (Fig. 19). The chip and bond wires are protected by a drop of epoxy resin. The whole circuit is sometimes covered by a thin layer of resin prepared by the fluid bed powder coating process (see section 6.1.4.8.3).

F.2 Printed circuit board

The construction is the same as that of ceramic substrates, but cheaper glass epoxy printed circuit boards are used. Due to the inferior thermal conductivity of the glass epoxy, the power handling capability is reduced.

F.3 Chip carrier

A carrier for naked chips called *mini pack* is registered in the JEDEC chip carrier family. The chip is bonded on a piece of printed circuit glass-epoxy board. A drop of epoxy resin protects the chip [77L1, 77A1].

F.4 Flexible substrates

The substrate used is in most cases copper-clad polyimide foil in super 8, 16, and 35 mm film size [Fig. 50]. Copper leads on the foil are directly connected to the bond pads on the chip. In order to prevent shorts between leads and edges of the chip bumps are applied either to the chip or to the copper (see section 6.1.4.8.3) [71M2, 72B1, 77S1]. The chip is protected by a drop of epoxy resin. There are also constructions using conventional wire bonding.

6.1.4.8.2 Package components and materials

Glas-metal seals

Matched seals

Matched glass-metal seals are applied to hermetic packages as shown in Figs. 4, 5, and 19. Matched seals require that the thermal expansion coefficients of metal and glass are nearly identical over a wide range of temperature.

An example of material combinations is shown in Table 2.

Table 2. Material combination of matched seals (d = thickness, α = thermal expansion coefficient).

Part	Material	d mm	α K^{-1}
Header	Fe54 Ni28 Co18	0.2	$\approx 5.1 \cdot 10^{-6}$
Glass	Schottglass 8250 (hard)		$\approx 5.0 \cdot 10^{-6}$
Wire	Fe54 Ni28 Co18	0.45 diameter	$\approx 5.1 \cdot 10^{-6}$

The metal parts are oxidized before sealing in order to achieve a chemical adhesion between glass and metal. The sealing temperature is about 900···1000 °C. The oxide is chemically removed after sealing before galvanic plating by Ni or Au [79M4].

The properties of the sealed headers are given in Table 3.

Table 3. Properties of a matched-sealed header.

Isolation resistance	$10^{10} \cdots 10^{12}$ Ω
Electrical field strength	800 V/mm
Current density in the wires	2 A/mm^2
Hermeticity	10^{-8} mbar · l/s
Temperature range	−65···250 °C
Difference of pressure	5···10 bar

For a cheap version having a header of Fe58 Ni42 see [79F1].

Compression seals

Compression glass metal seals are applied to hermetic packages as shown in Figs. 7 and 8. A seal is shown in Fig. 51.

Table 4. Material combinations of compression seals (α = thermal expansion coefficient).

Part	Material	α K^{-1}	Remarks
Example A			
Disc/ring	steel	$\approx 13 \cdot 10^{-6}$	compression
Glass	Schottglass 8422	$\approx 9 \cdot 10^{-6}$	matched at high level
Wire or tube	Ni50 Fe49 Cr1	$\approx 9.1 \cdot 10^{-6}$	
Example B			
Disc/ring	Ni50 Fe49 Cr1	$\approx 9.1 \cdot 10^{-6}$	compression
Glass	Schottglass 8250	$\approx 5 \cdot 10^{-6}$	matched at low level
Wire or tube	Fe54 Ni28 Co18	$\approx 5.1 \cdot 10^{-6}$	
Example C			
Disc/ring	steel	$\approx 13 \cdot 10^{-6}$	compression
Glass	Schottglass 8422	$\approx 9 \cdot 10^{-6}$	
Wire or tube	Fe54 Ni28 Co18	$\approx 5.1 \cdot 10^{-6}$	

The three examples differ by the thermal expansion coefficients of the metal-glass-metal layers. As a consequence a compression and matched seal or only a compression seal is obtained.

The examples B and C are applicable to TO-3 and TO-39 headers. They are thermal shock-resistant. The radial compression builds up during cooling down from the sealing temperature because the inner parts have the low and the outer ring or disc the high coefficient of thermal expansion.

Table 5 shows the dimensions of the compression seals.

Table 5. Dimensions of the compression seals (see Fig. 51).

D_0/D_i	1.3
$\frac{D_0 - D_i}{2}$	0.5 mm
$\frac{d_0 - d_i}{2}$	0.2 mm
h	1.5 mm
$\frac{h_{\text{disc ring}} - h_{\text{glass}}}{2}$	0.2···0.4 mm

The properties of compression seals are the same as those of matched seals (Table 3).

Cans and lids

Hermetic packages shown in the Figs. 3, 4, 5, 6, 7, 8, and 19 need cans or lids welded to the headers. The materials used and their plating to prevent corrosion or to simplify the welding are given in Table 6.

Table 6. Materials for cans and lids.

Material	Plating	Ref.	Material	Plating	Ref.
Fe—Ni—Co	Au (for lids)	74H2	Ni	no	
Steel	Ni or NiP	78F1, 80E2, 82E1	Brass (CuZn)	Ni or NiP	
			German silver (CuZnNi)	no	79D2

The material thickness ranges from 0.2 to 0.5 mm. The plating of NiP is performed by a chemical process according to Brenner [66B1]. The layer contains about 7% P which reduces the melting point of Ni and additionally generates a hard soldering joint to the weld.

The design of the flange of the can depends on the shape of the header. For headers having a weld ring (Fig. 52), as applied in TO-39 compression seal headers, a 90° flange is common and for flat headers as applied in TO-3 a 45° flange is used (Fig. 53). Both designs have a guaranteed welding line which reduces the required welding energy and keeps the glass seals free of strain during welding. For details see [70B1].

Lead frames

Lead frames are used in nearly all plastic packages as well as in discrete devices and in integrated circuits. Examples of lead frames are shown in Figs. 22, 27, 28, 31, 32, 36, and 42. In the following the details of a lead frame are described for the 18 lead DIL-package in Fig. 54 (the numbers refer to the locations in the drawing of Fig. 54):

(*1*) The die pad carries the semiconductor chip. A planarity of 2···10 μm is required in the order of the thickness of the solder or the glue joint.

(*2*) The die pad is connected to the outer frame by support leads. The support leads contribute to the heat transport and are therefore made as wide as possible.

(*3*) The die pad is reduced in height in order to adjust for the die or lead frame thickness and to balance the difference in height level for the wire bonding.

(*4*) The die pad support lead is opened at the ends like a fish tail to prevent tilting of the die pad.

(*5*) The lead tips are placed around the die pad. The tips are landing areas for the bond wire and planished to a planarity of 3 to 5 μm per 100 μm (hatched area in Fig. 54).

(*6*) The burr of stamping the lead frame is on the top side. This gives a maximum of plane landing area for wire bonding. The opposite side is rounded off at the edges.

(*7*) The slots between the lead tips and also the width of the tip itself should not be less than the material thickness. The limitation is the mechanical stability of the punches.

(*8*) If possible the finger tips are designed wider than the lead to achieve a certain anchoring in the plastic.

(*9*) The corners have a curvature radius of at least 0.2 up to 1 mm to increase the life time of the stamping tool.

(*10*) The minimum slot width is only applied in the center part of the lead frame. At other places slots have to be larger to anchor top and bottom part of the plastic body.

(*11*) The leads are to be anchored in the plastic. They have to withstand a force of 25 N to avoid electrical degradation or the removal of leads from the plastic.

(*12*) Dam bars prevent the plastic to flow along the leads during moulding. The dam bars also support the leads and prevent their dislocation during handling.

(*13*) Seating shoulders act as stoppers when the device is placed in holes of a printed circuit (p. c.) board. Between the plastic body and the p.c. board a minimum distance of about 0.5 mm is required to allow cleaning.

(*14*) The outer lead tips are pointed to prevent misplaced leads for automatic insertion into p. c. boards.

(*15*) The outer frame gives mechanical stability to the whole structure.

(*16*) Holes are necessary for locating pins in the stamping tool, the plating machines, the bonders, the moulding tool, and in the cropping tools.

(*17*) For the identification of pin 1 of the package a notch or other mark is placed near pin 1.

(*18*) The lead frames are produced in stripes of about 200 mm length consisting of 6 to 14 products for integrated circuits or up to 36 products for discrete devices or in nearly endless ribbons of several thousand products on a reel.

(*19*) The outer leads have a conical part to improve the stiffness.

Fig. 55 shows a computer drawing of a modern and cost saving "interdigitated lead frame" where the outer leads of two products are shifted into each other to save room. Many of the worlds largest manufacturers of integrated circuits are using it. Thermal aspects for lead frame designs are discussed in [79M2], bondability aspects in [80M2]. For other designs see [75L1, 78O1, 69S1]. The layout of lead frames for tape-bonded hybrids is discussed in [78M1].

Lead frame materials

The selection of lead frame materials is influenced by different and sometimes contrary requirements:

If the semiconductor chip is to be hard soldered by Si—Au eutectic the softening temperature of the lead frame material has to be above about 480 °C.

If the chips have areas in excess of 6 to 10 mm^2 a hard soldering on copper alloy lead frames is only possible with special precautions e.g. slow cooling. The high strains involved lead to the danger of chip cracks. Materials matched to the thermal expansion coefficient of Si like Fe—Ni—Co and Fe—Ni are preferred.

If the chips are soft-soldered or glued the thermal expansion coefficient is no longer important but the thermal conductivity or the cost of the material receive priority.

Automatic insertion into p.c. boards requires a certain stiffness of the outer leads which cannot be well defined in numbers.

A weakness of Cu- and Cu-alloys is the work hardening and therefore a mechanical fatigue of leads during bending. A lead should not break for less than 4 bends of 90° loaded with 250 g.

Table 7 gives a survey of lead frame materials and their most important properties.

Table 7. Properties of lead frame material. α = thermal expansion coefficient, κ = thermal conductivity, H_V = hardness (Vickers) at room temperature, T_s = softening temperature.

Material	α $10^{-6}\cdot °C^{-1}$	κ $W m^{-1} K^{-1}$	H_V	T_s °C	Ref.
Fe 54 Ni 28 Co 18	5.1	16.3	160···220	600	57H1
FeNi 42	5.3	15.1	160···220	600	
Fe	13.2	50.2	160···220	≈600	
CuZn 15	18	159	≈150	≈450	75M1
CuFe 1 Co 0.35	16.9	209	≈170	≈450	
CuFe 2 P	16.2	262	≈130	400···550	
CuZr 0.15	17.8	343	≈130	≈500	72D1
CuAg 0.1	17.7	347	≈130	≈450	72D1 68F1
CuSn 0.12	17.7	364	120···130	≈450	72D1
CuCo 0.2	17.7	385	100···165	500···700	
DHP-Cu (0.04 P)	17.7	339	80···110	≈250	
ETP-Cu (0.02 O)	17.7	391	50···110	≈200	

In the past Fe—Ni—Co was mainly applied. Due to the price increase of Co industry switched to Fe—Ni having nearly the same coefficient of thermal expansion as Si of $\alpha = 4.4 \cdot 10^{-6} K^{-1}$.

The common Cu alloys mostly used are CuFe 2 P for integrated circuits and CuSn 0.12 and DHP- and ETP-Cu for discrete devices.

Plating of lead frames

For eutectic die attach (see section 6.1.4.8.3) the lead frames have to be gold-plated. Because of the high cost, gold is only deposited on the die pad and on the internal finger tips around the pad (hatched area in Fig. 54). The surfaces not to be plated are screened from the electrolyte by silicone rubber masks. For a quick exchange of electrolyte on the surfaces to be plated a jet of electrolyte is blown into the mask cavities. For gold electrolytes see [66R1, 74M1]. For plating of TO-5- and TO-74-headers like those in Figs. 6, 7, and 10 see [81E1].

If a gold spot is too expensive, silver [66K1] or electroless nickel plating [66B1] can be applied. The Si—Au eutectic soldering is performed by applying a gold preform 25 µm thick in an area a little larger than the chip area or gold-deposited area and alloyed on the backside of the chip. A layer thickness of 1.5 µm minimum is required. A cheaper solution is to glue the chips on the lead frames. This process is widely employed for integrated circuits and partly for discrete devices.

Gold diffuses into copper very quickly at elevated temperatures. In order to prevent this interdiffusion a Ni- [66S1] or Co- [66O1] barrier layer has to be applied between the copper lead frame and the gold layer. Barrier layers are also required for all copper alloys used. Plating procedures are performed on lead frames from reel to reel at high speed on machines which are commercially available.

The cheapest bonding avoids plating of the lead frames. The gold is applied on the backside of the chips or the chips are glued to the lead frame. The lead frames then have to be degreased and special precaution has to be taken to prevent oxidation of the copper lead frames by a good protecting gas supply or by reducing the lead frame surface by hydrogen flames [79W1, 79W2].

Wire bonding is possible to gold, silver, and electroless nickel plated as well as on bare lead frames under special conditions which are described in section 6.1.4.8.3.

The Table 8 gives a survey of plating of lead frames and the thicknesses required.

Table 8. Plating of lead frames and headers.

Material of lead frame or header (See Table 7)	Alter-natives of plating	Activator strike	Barrier layer thickness μm	Activator strike	Main layer thickness μm	Ref.
Fe 54 Ni 28 Co 18 or	1			Au	Au 2···5	66R1 74M1
FeNi 42 or	2			Ag	Ag 1.8···4*)	66K1
Fe	3				electroless Ni 2···4	66B1
Cu and Cu-alloys except CuZn	1	Ni	Ni 2···4 or Co 2···4	Au	Au 2···5	66R1, 74M1 66O1 66S1, 79M3
	2			Cu	Ag 1.8···4*)	66K1 66F1
	3	Ni			electroless Ni 2···4	66B1 66S1, 79M3
CuZn	1	Ni	Ni 2···4 or Co 2···4	Au	Au 2···5	66R1, 74M1 66O1 66S1, 79M3
	2	Ni	Ni 0.5···4 or Co 0.5···4	Ag	Ag 1.8···4*)	66K1 66O1 66S1, 79M3
	3	Ni			electroless Ni 2···4	66B1 66S1, 79M3

*) For high frequency applications 7···12 μm is desirable.

Adhesives

Adhesives are applied to glue chips on lead frames and headers. The requirements are as follows:

good adhesion to relevant materials like Si, Au, Ag, Ni, Cu,
good thermal conductivity,
in many cases electrical conductivity,
ability to absorb different coefficients of thermal expansion e.g. Si—Cu,
fast curing,
long pot life,
adequate viscosity for application,
high purity (lowest possible content of halogenes),
short time-temperature stability of e.g. 350 °C if thermocompression wire bonding is employed,
long time-temperature stability of e.g. 150 °C which is the maximum chip temperature during operation.

These requirements can be met by three groups of resins:

Epoxies
Silicones
Polyimides

which are mostly filled by 60 to 75% of silver flakes and/or granules to achieve heat and electrical conductivity. Adhesives filled by gold or silica are also available commercially, however, their use is restricted to special applications. Examples for mixing of adhesives are given in [76D1] and [78C1]. Tests are described in [75D3].

Adhesives may be used unfilled; however, in order to achieve a low thermal resistance of the glue joint a layer thickness of less than 2 μm is necessary. This layer is unable to absorb differences of thermal expansion between e.g. silicon and copper which are $4.4 \cdot 10^{-6}\ K^{-1}$ and $17.7 \cdot 10^{-6}\ K^{-1}$ respectively. Only layers from 3 μm thickness onwards do have this property. This thickness limit depends on chip size. In order to maintain a low thermal resistance a filler is necessary which is mostly silver. In practice, the thickness of glue layers is 3 to 12 μm.

If a low ohmic electrical contact between chip and lead frame is required e.g. for the collector of a transistor the back of the chip has to be metallized e.g. by Ti-Au or similar metals because the silver flakes in the glue make many point contacts to the semiconductor material which results in a diode characteristic.

Bonding wires

For the interconnection of the chips and the tips of lead frames thin wires of Au and Al are used. Diameters of 25, 32 and 38 μm are mainly used but for special purposes Au wires of 7, 10, 15, 22, and 50 μm are employed. The price minimum for Au wires is between 20 and 35 μm diameter. Above this diameter the material cost is increasing, below this diameter the effort in drawing the wire is increasing the cost.

In order to influence the properties of gold wires e.g. the tensile strength, the grain growth at higher temperatures etc., additives are used i.e. Be, Pt, and Pd. Details on Au-wires are described in [73R1, 75C3, 80G1, 80M1].

Replacement of the expensive precious metal Au is topic of world-wide investigations [81B1]. Proposed metals or alloys are:

Au 50 Ag 50···Au 10 Ag 90, Ag 90 Pd 10, Pd, Cu alloys [79B2, 80H1].

The cheapest possible solution is the use of Al-wires. The growth of Au—Al intermetallic compounds which cause purple plague [67K2, 67H1] between the Al bond pad on the chip and the Au-wire is no longer possible. The properties of Al-wires are improved by additions of 0.5···1.5% Si or Fe. Al-wires are used in hermetic packages, i.e. in CERDILs and CERDIPs (see section 6.1.4.8.1 and Fig. 14) ultrasonically wedge-wedge bonded (see section 6.1.4.8.3). It is possible to melt balls on an Al-wire in argon atmosphere [79D1, 81P2, 81S3] and to employ nailhead-bonding which allows a high production speed.

Table 9 gives a survey on electrical resistance of bonding wire materials to calculate the series resistance of wires in a semiconductor device.

Table 9. Electrical resistivity ϱ of bonding wire materials.

Material	ϱ [$\Omega\ mm^2\ m^{-1}$]
Au	0.022
Au 50 Ag 50	0.106
Au 10 Ag 90	0.027
Ag 90 Pd 10	0.065
Pd	0.099
Al	0.029

Protection lacquers

Most electronic devices, which are not passivated by a silicon-oxide or -nitride layer, are protected by a thin layer of *pure polymer*. Polymers are normally dissolved in a thinner [67S1, 70S1, 79G2, 81S1]. After evaporation of the solvents the polymers are more or less baked at high temperature, which depends on the type of the polymer employed. Normally, polymers consist of chemically pure silicones which contain reactive groups, epoxies, epoxy-silicones, co-polymers, polyimides, or ultra-violett-light sensitive acrylic compounds.

These resins do not only protect the chip surface against chemical reactions but also act as mechnical protection for the bonding wires during encapsulation. The protective film or a second layer also act as an α-radiation shield [72S1, 72T1, 79S1, 80A1, 80B4, 80S2] on memory integrated circuits. α-radiation may erase information in memory-cells. Besides polymers the shielding system should also contain small amounts of high-atomic number particle fillers i.e. tungsten carbide [72S1]. The thickness of the layers depends on the type of application.

Plastics

Besides the well-known types of hermetically sealed packages for electronic devices like glass or ceramic packages the plastic envelope is today world-wide applied for device protection [76M1]. Different types of plastic are used in numerous tools and presses forming the required package shapes. This encapsulation is performed by using thermosetting compounds in transfer moulding or in some cases by potting, casting, dropping or fluid bed powder coating (see section 6.1.4.8.3). Plastic encapsulation offers numerous advantages over hermetically sealed packages, e.g. much lower cost and reduction of package size. Besides good mechanical protection, the chip is isolated from chemical influences of the environment. The leads are insulated and anchored in the plastic body. Heat transport from the chip to the surface remains unchanged as long as no heat distributors are applied. Plastic also screens the surface of the chip from light. Marking of the plastic body can be achieved in an easy and optimal way.

Requirements

The most important plastic encapsulation method is performed by the transfer moulding process. Plastic materials which are employed in transfer moulding must meet the requirements listed in the following sections:

An optimal flow at low viscosity during the transfer time is important. The curing time depends on the moulding tool temperature and the type of the plastic. By using the Brabender plastograph test, the characteristic data of moulding compounds can be determined (Fig. 65).

The flow properties of moulding compounds depend on the chemical composition, the prewarming conditions of the pellets or granules, the temperature of the tool, and the heat transport from the tool wall into the plastic. The minimum viscosity of the molten plastic is nearly independent of the tool temperature, which on the other hand strongly influences the *time* of minimum viscosity.

Cavities must be filled before the low-viscosity level of the melt rises. Displacements of the wire connections may occur. Such displacements can be detected by X-rays [82V1].

The thermal expansion of the plastic applied must be in a range, which is nearly identical to that of the metallic substrate (lead frame) used.

A good *adhesion of the plastic* to the leads, wires, and chips is required. For poor adhesion, the pull-out strength of the leads out of the plastic body is reduced and water and ionic contaminations may penetrate between the plastic and the metal [80B4, 80M4, 70H1].

The adherence to the mould surface should be as low as possible. This property is partly obtained by release agents as additives.

A further requirement for an optimal moulding material is a *high purity* of all plastic constituents. This is necessary because corrosion of the chip metallization, electrical instabilities, leakage currents between leads etc. may arise.

The effect of *mechanical stress* on the chip and its electrical behaviour is discussed in section 6.1.4.8.4.

The glass *transition temperature* (T_g) of the plastic applied should be higher than or equal to the maximum junction temperature (T_j) of the chip. The thermal expansion is much higher above the glass transition temperature than below. Modern epoxy and silicone plastic types exhibit sufficiently high glass transition temperatures.

Flame *retardant behaviour* of transfer moulding compounds is often a requirement for security reasons. Silicones meet this requirement without any modification of the composition. Epoxies must be modified by adding flame retardants like antimony oxide and brominated resins [67L1, 76R1].

Test methods

Test methods for characterization of transfer moulding compounds are performed with regard to later assignment in factory tools.

Flow and cure properties are important to obtain good workability. In thermosetting compounds, when heated, different processes occur at the same time: i. softening of the material by heat transfer from the mould to the plastic and ii. chemical reactions, which increase the viscosity by cross-linking of the polymer. Various methods exist for characterizing the flow properties:

The *Emmi-spiral test:* Emmi 1–66 [75E2] is used world-wide for epoxy and silicone compounds (Fig. 66). Standard test conditions are:
Emmi 1–66 mould, tool temperature 180 °C or 160 °C, pressure of plunger 70 bar, 25 g granulate.

The Emmi-spiral-test tool is placed in a press instead of a moulding tool. Besides the spiral length the time of plastic flow in the runner can be determined by recording the time and velocity of the plunger movement. The data allow to calculate an average speed of plastic flow in the runner of the mould which roughly correlates to the viscosity of molten plastic.

Flow rate test: The moulding compound is transferred from a transfer pot through a conical nozzle (0.8 mm diameter) under standardized conditions. The amount of plastic which flows through the nozzle in well-defined intervals of time is registered. The speed of plunger movement is also registered. Undesired and coarse particles or agglomerates of glass fibres may block the nozzle and thus can be detected by this test method (Fig. 67): The typical test conditiones are:

mould temperature: 180 °C, plunger pressure: 70 bar, granulate: 50 g, measurement of outflow (time intervals): 3 s.

The spiral flow test allows to determine the flow properties at moulding temperature with the result of a totally cured plastic spiral. Flow rate test results give information of the first softening and flow through the gate at remarkably lower temperatures between 120 °C and 140 °C because the plastic does not reach the tool temperature within the short test time. The same results as found by the flow rate test can be determined by the disc flow degree test. Theoretical aspects of flow are discussed in [71D1, 73D1].

Flash test: Flash test tools exist in different constructions. The idea of all systems is similar: softened plastic flows through runners into grooves, normally 5, 10, 20, 40, and 80 μm in depth at standardized conditions:

tool temperature 180 °C, injection pressure 70 bar, 25 g granulate.

The distance of the plastic flow in the grooves (flash) is recorded. The distance depends on type and prepolymerization of the plastic, composition of the filler (size distribution), kind of the plastic and tool temperature.
In factory moulds plastic flash is an undesired byproduct on lead frames which must be removed by suitable methods (see section 6.1.4.8.3).

Filler content: All thermosetting compounds used for encapsulation of electronic devices contain different types of chemically pure filler like crystalline quartz, fused quartz, E-glass fibres, antimony oxide, and others [67J1, 75M2, 77S3, 79H4, 79K4]. The filler type has a large influence on: flash, flow properties, mechanical strength of the postcured device, thermal expansion behaviour, stress, electrical behaviour, and heat transfer. The manufacturers optimize the plastic system for the applications. This optimization can be achieved by adapting the polymer systems and the filler type and the size of particles. The spreading of these constituents influences the parameters of the ready plastic product. An increase of filler content or fine filler normally decrease the flow properties and the thermal expansion. Thermocycle test results of final devices are also affected by the filler variation and the chemical composition. Glass fibres influence the mechanical stability of large packages (e.g. for IC's). Normally an anisotropic behaviour of thermal expansion is detected [74B1]. Plastic materials for which the crystalline quartz filler is replaced by fused quartz, possess a reduced coefficient of thermal expansion and thermal conductivity. The particle size distribution is an important factor. Coarse filler particles or prepolymerized plastic particles may block the gates of the moulding tool. Unfilled cavities and wire sweep formation are possible consequences. Particles having a large surface area and primers (silane compounds) on the surface influence the flow properties, the water penetration and the thermal behaviour of the plastic remarkably.
X-ray diffraction gives fast information on the kind and the amount of filler in the plastic. Flame retardant additions i.e. antimony oxide and brominated resins can be analytically determined by atomic absorption spectroscopy or qualitatively by the characteristic X-ray diffraction lines. Bromine is also analytically detectable.

Water extraction: The chemical purity of the transfer moulding compound is essential for long-term stability of electronic devices. Sodium and chlorine ions can be leached out of the plastic by boiling the granulate in water. These ions influence the electrical conductivity on the chip surface. The corrosion of aluminum metalization on the chip is accelerated [74K1, 76H2, 78L1, 80D2]. The source for chlorine normally is the polymer system. The filler seems to be the source of sodium.
Standardized recipes exist for the water extract analysis of original or cured transfer moulding compounds. The water extract is analysed by measuring the electrical conductivity, the pH-value, and the ions which are soluble in water.

Thermal expansion: The thermal expansion behaviour of transfer moulding material is important for the assembly of large chips in relatively small packages [59J1, 72W1, 79R2, 80B4]. This property of thermosetting compounds can be checked in commercially available apparatus offered by several suppliers. The evaluation of the elongation of a test specimen measured vs. body temperature allows the determination of the coefficient of linear thermal expansion α in the range below (α_1) and above (α_2) the glass transition temperature T_g (Fig. 68). For high T_g, the rapid increase in the thermal coefficient of expansion (α_2) also occurs at high temperature. A low coefficient of thermal expansion (α_1) leads to reduced shrinkage of the plastic and stress formation on the chip.
This method can be applied to detect the degree of curing of encapsulated devices. A plastic which is not totally cured shows high linear expansion behaviour and a low T_g (e.g. curve A in Fig. 68). A softening as for a thermoplast can be observed.
If the test specimen is not optimally densified during the moulding process, deviations of the thermal expansion behaviour from the standard curve can be detected.

Inflammability test: Flame retardant plastics are applied to chip encapsulation, where the original plastic fulfills the requirement of UL 94-VO [74U4]. This test method allows the determination of the burning behaviour.
A plastic specimen having standardized size is inflamed under reproducible conditions. The tests are performed by "Underwriters Laboratories". Further test methods exist e.g.: DIN- or ISO-tests but they are not accepted world-wide. Inflammability tests are also possible for encapsulated devices. For small packages e.g. diodes and transistors, they are not standardized, but they also follow UL-prescriptions like UL 478.

Silicone plastics

A typical silicone moulding compound for transfer moulding may have the following formulation [68K2, 75M2]:

Silicone resin: 25···30%, filler: 70···73%, hardener or catalyst: traces (0.3%), pigment: less than 1%, release agent: 1%.

These types of moulding compounds are non-burning. A data collection of two different types of silicones is given in Table 16. The resin normally consists of a copolymer of different monomer methylsilicones, phenyl-silicones or phenyl-methyl silicone monomer units. OH-groups, which exist besides silicone-oxygen-silicone bonds in the chain polymer, react during curing by elimination of water molecules.

Other organo-silicone monomers containing hydrogen or vinyl-functional groups are also applied [73A2]. The filler normally consists of fused or crystalline silica and glass fibres of pure quality. A pure filler or a mixture of different types of fillers- sometimes also presilanized – are homogeneously dispersed in the system [74L1]. Particle size and kind of filler affect the mechanical, electrical, and physical properties of the system (Table 16).

The condensation reaction between different OH-groups is catalyzed by many compounds [60N1, 74S1]. Normally, oxides or salts of heavy metals are applied. The water generated penetrates through the thermoset to the surface. If the injection pressure applied during encapsulation is not high enough, foamy devices showing unacceptable performance and mechanical stability are produced. The application of silicones for fluid bed encapsulation of devices therefore is impossible.

Various types of colouring agents are used as pigments, mostly carbon black or soot is applied to a small amount. Higher pigment content affects the electrical properties. Calcium stearate is used as release agent which also influences the curing behaviour and the physical properties of the cured plastic.

Application

Silicone moulding compounds are used for the encapsulation of power transistors because of the good dielectric and insulation properties, their chemical purity, the moisture resistance, and their thermal stability up to more than 200 °C. The adhesion properties to the substrate and the mechanical behaviour, however, is poor. The adhesion may be improved by backfilling of the devices by a reactive silicone.

Under cooled conditions, the shelf life of silicone plastic is very good.

Silicone plastic filled only by silica without glass fibres exhibits a low and temperature-independent dielectric constant which makes it very suitable for high frequency varicap diode applications.

Typical moulding conditions are:
transfer moulding press at tool temperature of 170···190 °C, injection pressure between 50 and 100 bar, cycle time of 60···180 s.

Typical data of two silicone moulding compounds are listed in Table 10.

Table 10. Data of two thermosetting silicone moulding compounds.

Property	Data	Data	Unit
Raw material	**Compound 1**	**Compound 2**	
Filler	α-SiO_2	amorphous SiO_2 and glass fibres	
Resins	phenyl-methyl-silicones		
Shelf life, 5 °C storage, 90% flow retension	8···12	6	months
Processing			
Mould temperature	180···210	150···180	°C
Injection pressure	10···100	min. 28	bar
Cure time	70··· 90	60···300	s
Post-cure time/temperature	2/200	2/200	h/°C
Spiral flow (Emmi 1/66) at 175 °C, 55 bar	76···101	91	cm
Cured specimens			
Specific weight	1.95	1.88	$g\,cm^{-3}$
Coefficient of linear thermal expansion			
30···60 °C	$55\cdots65\cdot10^{-6}$	$22\cdot10^{-6}$	$°C^{-1}$
60···150 °C	$70\cdots80\cdot10^{-6}$	$24\cdot10^{-6}$	$°C^{-1}$

continued

Table 10 (continued)

Property	Data	Data	Unit
Water absorption			
24 hrs boiling in water	+0.4		%
ASTM D 570		+0.08	%
Inflammability	self extinguishing		
Thermal conductivity	0.40···0.44	0.50	$W m^{-1} K^{-1}$
Dielectric dissipation factor at 1 kHz		$23 \cdot 10^{-4}$	
at 300 kHz	$30 \cdot 10^{-4}$		
at. 1.5 MHz	$25 \cdot 10^{-4}$		
Dielectric constant at 1 kHz		3.79	
at 300 kHz	4.0		
at 1.5 MHz	3.4		

Epoxy plastics

This class of thermosetting compounds also consists of a well homogenized premolten mixture of epoxy-polymer systems, a type of hardener – which depends on the polymer –, an accelerator, pigment, filler, release agent, sometimes flame-retardant additives, and coupling agents for the coupling of the polymer to the filler surface.

Various types of epoxy-hardener systems are distinguished for the thermosetting plastics:

The application of bisphenol-A-epoxy polymer compounds and organic acid anhydrides in the thermosets lead to the socalled "*epoxy-anhydride*" *systems.*

By replacing 1. most of the bisphenol-A-epoxy polymers by epoxydized phenolic and cresolic novolacs and 2. the organic anhydride systems by phenol-formaldehydes the whole organic system can be hot-cured by a catalyst. These are mostly organic nitrogen-containing compounds like imidazole derivates or other amines. These systems are called "*epoxy-novolacs*".

A further possibility exists for curing epoxy-bisphenol-A-systems. As curing agent an aromatic diamine is applied, e.g. diaminodiphenylmethane. Because of the cancerogenic behaviour and the short storage time at room temperature, this material is not employed anymore.

In the following only "epoxy-anhydrides" and "epoxy-novolacs" are considered.

Phenolic novolac-cured encapsulants are only satisfactory for small signal transistors and digital integrated circuits i.e. low-voltage applications. For surface-sensitive devices or higher-voltage applications anhydride-cured materials, which have been proven to have excellent electrical compatibility, are now applied world-wide. Anhydride-cured materials must be handled differently because of their shift in chemistry compared to "phenolic novolac" systems. A survey of the "epoxy-cresol-novolac" is given in [79H6].

Storage and handling

Only "anhydrides" show a strong attraction to water. Containers of plastic must be sealed as long as possible when coming from cooled conditions to room temperature. The hydrophylic behaviour may negatively influence the linear thermal expansion, the T_g-value, and the moisture resistance [78W1]. "Novolacs" and "anhydrides" are comparable with respect to shelf life behaviour at 20 °C and below. At higher temperatures anhydrides seem to show reduced shelf life.

Mouldability

Because of the large differences in the chemical curing both types of epoxies can only be compared by the fact, that they need nearly identical mould temperature (170···195 °C) and curing time (60···120 sec) in the tool. "Anhydrides" generally exhibit higher flashing than "novolacs". The curing is from the chemical point of view a condensation reaction of epoxy groups containing more functional anhydrides or OH-groups of phenolic derivates [68L1, 71L2, 72F1, 73C1, 73S1, 74T1, 75C2, 75T1, 76R1, 78Y1, 79H4]. The flow required of an epoxy compound can be obtained by applying a minimum amount of resin in the formulation by varying the filler particle size distribution or by using a resin having low molecular weight. The type of anhydride hardener also has an influence on spiral flow. An increased flow also increases the flash behaviour. "Anhydrides" normally contain 3···5 wt. %

less filler than "novolacs". Long flash or bleeding is mostly caused by low molecular weight polymers or a wrong filler particle size distribution. Thin flash has a strong affinity to lead frames. Deflashing is difficult to achieve.

Post curing of "anhydrides" normally needs more time or higher temperatures than those of "novolacs".

It is not possible to use one factory moulding tool for novolac or anhydride encapsulation of devices alternatively without intensive cleaning of the tool in between by melamine application.

The two types of plastic can also be employed for fluid bed coating. The release agent is not necessary in this case because no tools are used.

Purity of epoxy plastics

Constituents of anhydride plastics are normally very pure. Hydrolyzable chloride and extractable ion content is about 50% lower for anhydrides than that of novolac compounds. The curing process of anhydrides needs no catalyst. Possible side reactions of the nitrogen containing polar groups in catalysts which may degrade device performance are therefore reduced.

Thermal stability

The weight loss versus temperature of anhydrides is a little higher in the temperature range below 150···175 °C and lower above this temperature range than that of novolacs.

A typical composition of epoxy moulding compounds is given in Table 11.

Table 11. Composition of epoxy moulding compounds.

Compound	Remarks	Fraction
Resin	different types of epoxy polymers	20···25%
Hardener system	phenolic OH-containing compounds	5···10%
	or multifunctional anhydrides	3···8%
Filler	α – or fused quartz in anhydride systems	63···67%
	in novolac systems	67···72%
Coupling agents	(silane compounds)	less than 1%
Pigment	soot or carbon black	less than 1%
Catalyst		less than 0.5%
Release agent		0.5···2%
Flame retardant additives	antimony oxide	1···3%
	brominated resins	2···5%

The advantages of special additives or combinations of filler, resin, and hardener systems are described in [68L1, 71L2, 72F1, 73C1, 73S1, 74T1, 75C2, 75T1, 76R1, 78Y1, 79H4]. The intrinsic properties of two different examples of epoxy moulding compounds are given in Table 12.

Table 12. Examples of thermosetting epoxy moulding compounds.

Property	Data	Data	Unit
Raw material	**Compound 1**	**Compound 2**	
Filler	amorphous SiO_2 and glass fibre	amorphous SiO_2	
Resin/hardener	epoxy-bisphenol/ anhydride	epoxy cresolic novolac/ phenolic formaldehyde polymer	
Shelf life, 5 °C storage, 90% flow retension	6	6	month
Processing			
Mould temperature	175···200	150···180	°C
Injection pressure	35···100	35···100	bar
Cure time at 175 °C	60···90	60	s
Post cure time at 175 °C	16	2	h
Spiral flow (Emmi 1/66) at 175 °C, 50 bar	65···90	75	cm

continued

Table 12 (continued)

Property	Data	Data	Unit
Cured specimens			
Specific weight	1.75	1.83	$g\,cm^{-3}$
Coefficient of linear thermal expansion			
below T_g: α_1	$30 \cdot 10^{-6}$	$21 \cdot 10^{-6}$	$°C^{-1}$
above T_g: α_2	$90 \cdot 10^{-6}$	$70 \cdot 10^{-6}$	$°C^{-1}$
Glass transition temperature T_g	175	162	°C
Water absorption ASTM D 570	1.1		%
Inflammability UL 94 [see 74U4]	VO	VO	
Thermal conductivity, κ	0.67	0.63	$W\,m^{-1}\,K^{-1}$
Dielectric dissipation factor at 100 Hz	$40 \cdot 10^{-4}$		
at 1 kHz		$30 \cdot 10^{-4}$	
at 10 kHz	$50 \cdot 10^{-4}$		
at 1 MHz		$90 \cdot 10^{-4}$	
Dielectric permittivity, ε at 100 Hz	4.0		
at 1 kHz		4.3	
at 10 kHz	3.9		
at 1 MHz		4.2	

Flame retardancy

Epoxy plastics are normally not flame retardant. The requirement for flame-retardant epoxy plastics is met by adding antimony oxide compounds and brominated aromatic epoxy resins as synergetics to the epoxies [74F1, 75K1, 76R1, 80F1].

Comparison of epoxies and silicones

By comparing the properties of epoxy compounds and those of silicones, some properties favour epoxies others favour silicones in applications.

Advantages of epoxies: Low price, good adhesion to lead frames, low coefficient of thermal expansion, high mechanical strength, applicable to fluid bed powder coating.

Advantages of silicones: High purity, high electrical stability at high temperatures, low and temperature-independent dielectric constant, high T_g.

Epoxy-Silicone hybrid plastics

A combination of both epoxy resin and silicon in a hybrid plastic may be used to optimize the properties. Only one type of this hybrid plastic is commercially available. The composition is not known except that it contains a copolymer of epoxy and silicone resin, amorphous silica and flame retardant additives. For details of salt spray testing, autoclave testing, 85 °C 85% relative humidity testing, and the physical properties see [78T1, 77A2].

Other encapsulation plastics

Besides chip encapsulation, the two-shot technique is possible. Single-ended devices are encapsulated by epoxy powder in the fluid bed powder coating process. A thermoplast material is then injected into a form around the cured powder to shape the final outline of the device.

For most applications, the viscosity of thermoplastics is too high leading to breakage of bonding wires in this process. Exceptions are noted in [77S4].

Thermosetting polyimide compounds may also be employed for the encapsulation of semiconductor devices. The mouldability and properties of devices moulded by polyimide are not yet known [78N1].

Heat spreaders

The power handling capability of standard dual-in-line packages can be considerably improved by the application of internal heat spreaders, inserted into the bottom cavity of the transfermoulding tool and encapsulated with plastic in the same way as the lead frame. These devices do not need any external heat sinks or cooling fins. A cross section of a 16 lead DIL package having a heat spreader is shown in Fig. 42. A perspective view of a heat spreader is shown in Fig. 56 [68W1, 70U2, 71H1].

Aluminum and copper are the usual materials for heat spreaders. Aluminum is used together with galvanic tin plating of the outer leads. Copper is used together with dip tinning employing an HCl etch as a pretreatment. Aluminum does not withstand HCl.

The reduction in thermal resistance of a package by a heat spreader is discussed in section 6.1.4.8.4.

6.1.4.8.3 Assembly methods

Assembly is usually performed in strips of lead frames from cassette to cassette or from reel to reel by a single machine [79W1, 79W2]. Chip bonder, wire bonder, and transfermoulding machine may be coupled in a line [76Y1]. A general survey of assemblers is given in [80U1].

Wafer processing

Lapping, etching, grinding

After fabrication, wafers have diameters of 2 or 3 inches, or 100 mm. The thicknesses are 380 and 525 μm, respectively. Wafers are used in their original thickness if they are glued to substrates like lead frames, foils etc. rather than soldered.

Reasons to thin wafers are:

Electrical series resistance in the chip, thermal series resistance in the chip, hard or soft soldering to Cu lead frames or heat sinks to reduce mechanical stress in the chips, small chip sizes e.g. 0.35×0.35 mm with subsequent diamond scribing and breaking, wire bonders are not able to accept the difference in height level.

Wafer thinning can be achieved by 3 methods:

Lapping

The wafer is glued e.g. by wax face-down with at least two other wafers on a stainless steel rotor of a lapping machine and lapped down to the required thickness. Lapping machines are commercially available. SiC is the usual lapping compound.

Etching

Etching is more or less a cleaning process for lapped wafers because the lattice is damaged to a depth in the order of 35 μm deep during lapping. 50 μm etching in (40% HF): (65% HNO_3) = 2:5 volume parts, is usual.

Grinding

The most efficient and accurate way to thin the wafers is grinding. The spread in thickness is less than ± 10 μm. The wafers are simply held in position by vacuum chucks. More than 50 wafers (depending on the wafer diameter) can be ground at a time on grinding machines which are commercially available [81S2].

Scribing, lasering, sawing, and breaking

Because chips are simultaneously produced at large quantities on one wafer, they have to be separated for assembly. Between the active areas of chips are separation lanes of 50 to 100 μm width. Three methods are usual for the separation of chips:

Scribing

An edge of a diamond is drawn along the scribing lanes at a pressure of several cN. The result is a V-shaped groove of several μm depth [80A2]:

Lasering

A laser beam is vertically focussed on the separation lane and melts the material in the whole thickness. The necrystallized polycrystalline zone shows high mechanical stress [74G1].

Sawing

A thin diamond sawing blade made of bronze which is about 25 to 35 μm thick, rotates at a speed in the order of 30000 rpm. A slot is cut in the separation lane. A water jet is used for cooling the blade and the wafer and rinsing the silicon dust. If the wafer is held in position by a vacuum chuck a fraction of the wafer thickness remains uncut, e.g. 80 μm. If the wafer is glued on a sticky foil the wafer may be cut through.

In all cases except the last one, a subsequent breaking operation is required. The breaking is usually combined with a stretching to simplify the pick up procedure for die attach. [67K1, 77G1]. For non-passivated p n junctions, an etching process follows the breaking operation [69G1, 74M2].

Die attachment

There are three different methods for die attach (also called chip bonding): Hard soldering by Au—Si- or Au—Ge-eutect, soft soldering by Au—Sn- or Pb—Sn-alloys and glueing. All methods have in common that the chip has to be picked up and placed on the lead frame or substrate. The most simple method is a vacuum pickup having a tube as shown in Fig. 57. When the chip is in position, the vacuum is turned off. An improved version uses a needle which touches only the edges of the chip (Fig. 58). The advantages are:

More precise positioning, no damage of the active chip surface, better mechanical coupling of chip and needle for the horizontal scrubbing operation.

This method requires a special push-up needle to prevent damage of neighbouring chips. For each size a separate pick up needle is necessary.

Hard soldering

Germanium as well as Silicon form an eutectic alloy with gold. The melting point of a Ge—Au-eutect is 356 °C and 370 °C for an Au—Si-eutect (Figs. 59 and 60) [58H1]. When the back of a Ge- or Si-chip is placed on an Au-plated lead frame or substrate at a temperature exceeding the eutectic melting point, an Au—Ge or Au—Si alloy is formed at the interface [61B1, 64A1]. In practice, the soldering temperatures range between 430 and 480 °C in protecting gas $H_2 + N_2$ ambient.

If a gold layer e.g. 0.3 μm thick is deposited on the back of the wafer and alloyed to the silicon in order to prefabricate a good wetting and to avoid hinge effects during separation (Au—Si eutect is brittle, pure Au is not) the Au layer thickness on the lead frame can be reduced to about 0.6 μm. An alternative method is to deposit a minimum of 1.5 μm Au on the wafer and to alloy it. In this case, gold on the lead frame may be replaced by silver or electroless nickel or even bare frames may be soldered in a reducing atmosphere [79W1, 79W2].

Soft soldering

Hard soldering by Au—Ge or Au—Si eutect requires lead frame material having a similar coefficient of thermal expansion as Si. If the chip area exceeds 6···10 mm^2 Fe—Ni—Co- and Fe—Ni-alloys are used (see section 6.1.4.8.2). The thermal conductivity of these alloys is more than a factor of 20 smaller than that of copper. Much effort was taken to find a soft solder which can balance the differences in thermal expansion of silicon and copper without the disadvantage of lead-tin alloys which exhibit thermal fatigue of the joint during on-off cycling of the device. The optimum alloy is [81O2]:

65% Sn + 25% Ag + 10% Sb.

The back of a silicon chip is not wetted by soft solders. In an additional operation it has to be made wettable by deposition of Ni, Ti + Au, Cr + Ag or similar metals. The lead frames are mainly plated by nickel. A good gas protection by $H_2 + N_2$ is necessary to avoid oxidation.

For low thermal resistance, the solder layer should be as thin as possible. This requirement is contradictory to the function of the solder to balance differences in thermal expansion for which a certain thickness is required. Attempts were made to find a compromise by "waffeling" of the copper surface or by solder channels (runners) [71M1, 73A1, 77S2].

Glueing

Three methods are in use to apply the glue (section 6.1.4.8.2) to lead frames and substrates:

Dispensing through a thin tube

For dispensing through a thin tube like a syringe, high pressure is necessary and the quantity dispensed shows a relatively large spread. This method is only applied for laboratory purposes.

Offset printing

A needle preferably cross-shaped is dipped into the reservoir of the glue and is wetted. The needle then is transferred to the lead frame and pushed on the die pad to deposit a certain amount of glue onto it. The Si chip is positioned on the pad. [70L1, 76Z1]. The amount of glue is well controllable. The cross-shaped printing needle deposits a cross-shaped glue pattern which is squeezed to a rectangle by the chip without including air bubbles. This method is applied in mass production machines.

Silk screening

A silk or better metal screen having openings in the size of the glue area is used as a mask. A squeegy squeezes the glue through the mask on the lead frame. This method is also applied in mass production. An alternative using a glue foil is described in [73H1].

Quality control

The in-line inspection for quality control in the assembly department of hard/soft solder and glue joints can be performed in several ways:

Non-destructive methods

Optical inspection of excessive solder or glue around the chip, X-ray photography which shows a non-uniform distribution of solder or glue [75B1].

Measurement of the thermal resistance [79S2]. This method is restricted to finished devices but it can be performed on 100% of the chips if necessary because of the short test time in the order of one second.

Destructive methods

Cross-sections show the layer thickness distribution and voids. Shear force measurement for glued chips. This technique is also applied at elevated temperatures.

Bend test, where the die pad is bent downwards so that the chip cracks or the solder/glue joint is opened for inspection.

Interconnections

The interconnection between bond pads on the chips and the lead frame, header or substrate can be achieved by wires (bonding wires), beam leads or by thin and small lead frames. In most designs the lead frame is produced and applied on polyimide foil (Fig. 50).

Wire bonding

Wire bonders are described in [71A1, 78B1, 78B2, 80U1]. Various processes are described in the following and in [80R1].

Thermocompression bonding

Four methods are used in practice (Fig. 61). Thermocompression bonding is abbreviated TC-bonding.

Thermocompression wedge bonding (Fig. 61, line 1): The wire emerges from a horizontally mounted capillary (C), is positioned on the bond pad on the chip by a micromanipulator. A tool (T) squashes the wire on the bond pad to double its diameter (position 1). The lead frame is placed on a block heated to 320···350 °C. The tool then moves to position 2 and presses the wire on the lead frame finger tip. The tip scissors (Sc) cut the wire (position 3). Before cutting, the capillary is withdrawn to extract the wire length required for the next bonding operation [58C1].

This method is applied to 7 and 10 μm diameter gold wire on very small emitter and base bonding pads of $13 \times 30\ \mu m^2$ in high-frequency germanium transistors for applications up to 850 MHz. Although it is the only method to interconnect such small devices it is slow and relatively expensive. This method is no longer used in mass production.

Thermocompression stitch bonding (Fig. 61, line 2): Stitch bonding requires wire of at least 15 μm diameter. The capillary (C) is also the bonding tool and presses the wire with its bottom ring surface on the bond pad (position 1). The same operation is performed on the lead frame finger tip (position 2) and the wire is cut (stitched) between the edge of the lead frame and the edge of the capillary (position 3). The stitch bonding method requires a micromanipulator and observation through a microscope.

Thermocompression multiple wedge bonding (Fig. 61, line 3): Tweezers (Tw) pick up two small pieces of wires and hold them in position on the bonding (sites). Four tools (T) simultaneously press the wires on the bond pads and lead frame finger tips. This method is automatically performed and applied e.g. to the transistor structures shown in Figs. 22, 28, and 31. The method requires the wires supplied in a line [64M1].

Thermocompression nailhead bonding (Fig. 61, line 4): The nailhead bonding is also called ball bonding, because a ball is formed at the end of the wire by a hydrogen flame or an electric spark (position 4). The diameter of the ball is larger than the hole in the capillary (C) so that the capillary can press the ball on the bond pad (position 1). The second bond is wedge-bonded as in line 2 (position 2). The wire is separated by clamping and pulling (position 3). At the end of the process a ball is again formed by a flame or a spark (position 4). The bond temperature is 320··· ···350 °C. In some cases the capillary is heated to about 150 °C. The advantage of nailhead bonding is that the capillary can move in any direction. A turning of the product to be bonded is not necessary. Nailhead bonding is well suited for integrated circuits.

Bond strength and electrical resistance of thermocompression bonds are discussed in detail in [63M1]. In most cases, an Au-wire is bonded to Al-bond pads on the chip. Au and Al form an intermetallic compound Al_2Au, which starts to form during bonding. Formation of Al_2Au continues during operation of the device. This compound has a purplish color, is quite brittle and cannot withstand any deformation or mechanical shock. The deterioration of devices by formation of Al_2Au is called "*purple plague*" [67H1, 67K2]. Many attempts were made to avoid the purple plague by barrier layers of Mo, W, Ti, and TiPt between the Al-bond pad and the Au-wire. These metals cannot be bonded by Au wire. A layer of Au is additionally necessary [65C1, 70A1].

The purple plague may be avoided by replacement of the Au-wire by an Al-wire, which is employed in the ultrasonic bonding methods.

Ultrasonic bonding

Ultrasonic wedge bonding and ultrasonic nailhead bonding are commonly used (Fig. 61).

Ultrasonic wedge bonding (Fig. 61 line 5): The wire is pressed on the bonding pad of the chip by a nose of the tool (T) which has a feeding hole for the wire. The tool vibrates horizontally at about 40 kHz with an amplitude of some μm (position 1). No temperature increase is necessary. The tool moves to the lead frame finger tip and prepares the second bond. The tool (T) is tilted a little to weaken the wire (position 2). The wire is clamped and pulled for separation (position 3). Al-wire is mostly employed [77J2, 62P1] e.g. in the CERDIP packages as shown in Fig. 15. In this case, the lead frame finger tips are Al-plated so that a monometallic interconnection system is achieved. The application of Au-wire bond is also reported [73W1, 74P1].

Ultrasonic nailhead bonding (Fig. 61, line 6): The movements of bonding are similar to thermocompression nailhead bonding (line 4), however instead of heat a horizontal ultrasonic scrubbing is applied. The formation of Al_2Au compound is then reduced but still visible in cross-sections. An additional increase of the shear strength of the nailhead bonds can be obtained by a postbake of 4 h at 170 °C. An intermetallic diffusion of Au and Al occurs to some extent. For details see [75W1, 77G3]. Non-destructive methods of monitoring and controling ultrasonic wire bonds are described in [77W1, 79W3]. Ultrasonic bonding on thickfilm substrates is discussed in [79W6].

In the past, Au-wire was exclusively applied in ultrasonic nailhead bonding. Since it is possible to form balls on Al-wires [79D1, 81P2] nailhead bonding is a cost effective method especially for CERDIP and CERDIL packaged integrated circuits.

Thermosonic nailhead bonding (Fig. 61, line 6): A combination of thermocompression and ultrasonic nailhead bonding is the thermosonic nailhead bonding. This method improves the bond quality when an Au-wire is bonded to silver- or electroless nickel-plated lead frames or when Al- and Au-wires are bonded to bare copper lead frames. The temperature can be reduced to the range of 150···300 °C.

The following Table 13 summarizes the usual combinations of bond wires, lead frame or substrate platings and bonding methods.

Table 13. Summary of bondwire material usually employed for various platings of lead frames, substrates, and bonding methods. The material of the bond pads on the chips is Al or Au.

Bonding method	Plating of lead frames or substrates				
	Au	Ag	electroless Ni	Al	bare Cu
Thermocompression wedge	Au	Au			
Thermocompression stitch	Au, Al				
Thermocompression multi wedge	Au	Au			
Thermocompression nailhead	Au	Au			Au
Ultrasonic wedge	Al(Au)		Al(Au)	Al	
Ultrasonic nailhead	Au, Al	Au			
Thermosonic nailhead	Au, Al	Au	Au, Al	Al	Au, Al

Other bond wire materials than Au and Al exist, but are not mentioned because of limited experience.

The *quality of a wire bond* can be observed by the following methods:

destructive wire bond loop pull test [78H1]
non-destructive wire bond loop pull test [74H1]
nailhead shear test [75W1]
SEM photographs [80B1]
cross sections [80B1]

Beam lead bonding

Beam leads are gold ribbons which are fabricated together with the interconnections of the chip on the wafer [66L1, 69C1, 71N1]. The beam leads exceed the edges of the chip by about 150 µm. They are about 50 µm wide and 10 µm thick (Fig. 61, line 7). The advantage of beam leads over bonding wires is a reduced series inductance. They are mainly employed in high-frequency circuits at above 1000 MHz or in high-speed digital circuits.

Beam lead bonding is performed by a bell-like bonding tool which simultaneously presses the beams on the thin film conductors at temperatures of about 350 °C. The bonding tool as well as the substrate are heated. The chip is bonded face down (position 1). Thin-film ceramic substrates are mainly used. The top surface of the bonding tool is shaped like a ball and a spring-loaded wire pulls the tool against a tube. This enables the tool to compensate differences in planarity. Other methods, e.g. the compliant bonding and the wobble bonding are described in [68C1, 68C2]. An apparatus for compliant bonding is described in [69R1].

Direct contact concepts

Direct contacts avoid bonding wires or beam leads and directly interconnect the bond pads on the chip with the lead frame or substrate. Various solutions of direct contact concepts are given in Table 14.

For Table 14, see next page.

Most of the concepts apply bumps in order to maintain a certain distance between the chip surface and the interconnection to avoid short circuits (Fig. 61, line 8).

Because in many cases the substrates and in all cases the chips are not transparent, a complicated aligning in the bonders is required [70P1].

The most common bump technique employed is bonding on polyimide film tapes. In the literature it is often called TAB (tape automated bonding). Test methods and test results of bonds and reliability data of TAB products are presented in [81C1].

Table 14. Groups of direct contact concepts.

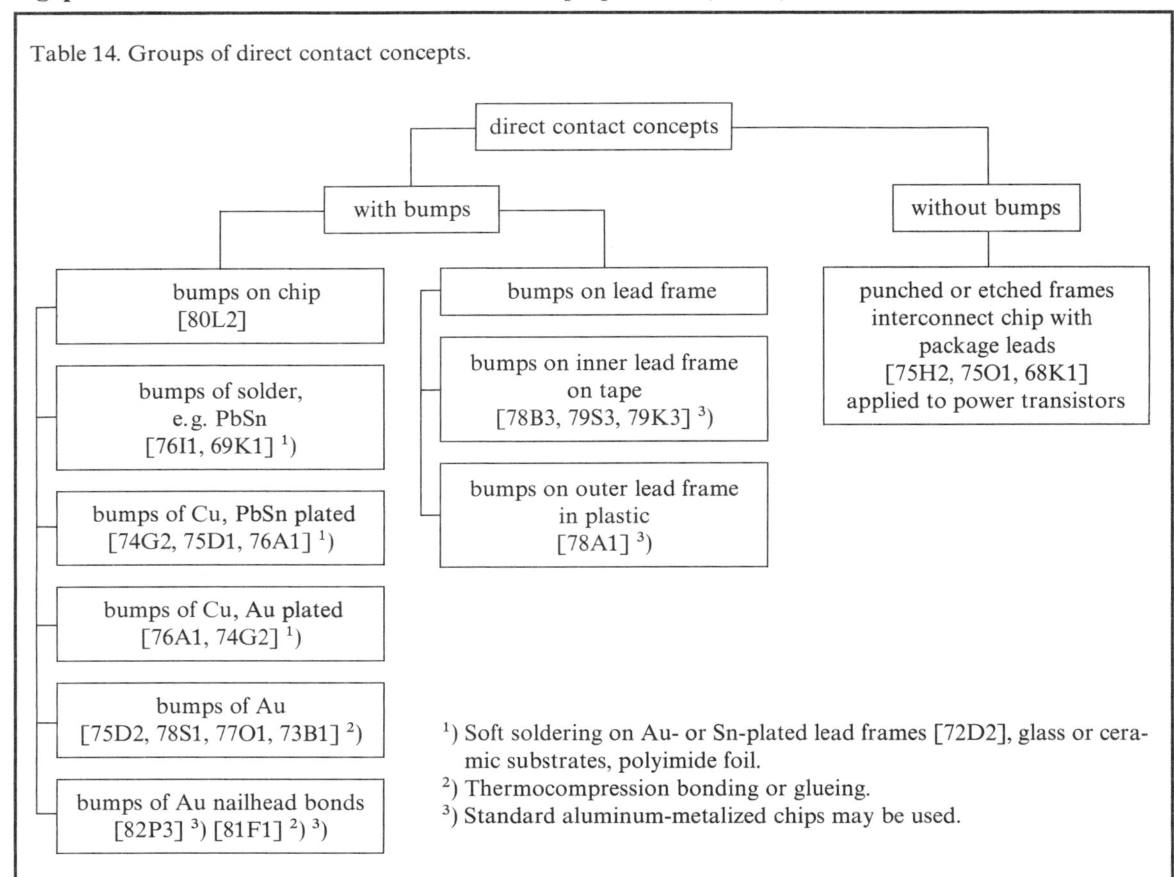

[1]) Soft soldering on Au- or Sn-plated lead frames [72D2], glass or ceramic substrates, polyimide foil.
[2]) Thermocompression bonding or glueing.
[3]) Standard aluminum-metalized chips may be used.

Surface protection

Surface protection is applied in plastic encapsulated devices for the following reasons:

to keep the chip surface free of contaminations from the plastic,

to act as an adhesion promotor between chip and plastic,

to delay the penetration of humidity through the package to the chip,

to screen memory cells from α-radiation.

The application of the protection lacquers described in section 6.1.4.8.2 can be performed by dipping for single-ended lead frames and by dropping the laquer from a syringe for double-ended frames. A subsequent annealing process is necessary. The temperature and the time of annealing depends on the material applied. A comparison of lacquers is described in [81S1].

Since the protection or passivation of chips has been improved e.g. by Si_3N_4 and the contamination of plastics has been reduced considerably in the past years the application of protection lacquers is not so important anymore.

Plastic encapsulation

Transfer moulding

Most of the plastic semiconductor devices are moulded by the transfer moulding process using low-viscosity thermosetting plastics as described in section 6.1.4.8.2. Fig. 62 shows a hydraulic press having a moulding tool in a simplified form. The moulding process is as follows:

The upper and lower half of the moulding tool and the plates are electrically heated to 170···200 °C. The mould is fabricated by chromium steel and highly polished or spark eroded (electro discharge machining).

The lead frames are loaded on the lower half of the mould over the locating pins. Loading is performed with the help of a loading frame carrying 10 to 24 lead frames simultaneously (not shown).

The clamp piston presses the movable table and the lower half of the mould upwards against the upper half. The clamping force depends on the number of lead frames and is in the order of 0.2···1.5 MN. The lower ejector springs retract the lower ejector pins. Just before closing the tool, the lower half of the mould touches the upper driving rods and retracts the upper ejector pins.

Pellets, which are preheated to about 80 °C in a high frequency preheater are loaded through the hole in the upper half of the mould. The pellets melt to a viscosity like honey. The ram transfers the plastic through the runners and gates into the cavities. The transfer time is adjusted in the order of 6···30 s. On the opposite side of the cavities are small grooves (about 5···10 μm deep), the vents (not shown), so that the air can escape from cavities.

When the cavities are filled with plastic the ram force increases to a pressure in the plastic of 30···70 bar to attain a sufficient density.

The plastic cures in 30···120 seconds. After curing the tool opens, the clamp piston retracts, and the upper ejector pins eject the moulded product.

When the tool is fully opened, the lower ejector pins push out the product. After a brief cleaning of the mould surfaces by a brush and by compressed air, the next cycle can be started. In the intervals of 1 shift (e.g. 8 h) an additional preventive cleaning by filling melamine pellets into the cavities is recommended in order to avoid staining.

The runners and the gates are removed from the product and the plastic is postcured at a temperature depending on the type of the plastic for e.g. 4 h at 175 °C in an oven to attain the final properties. Exhausting of the post-curing oven is important, to keep the pins clean for finishing.

The equipment and the tools for transfer moulding are relatively complicated. Attempts to simplify the tools were succesful for e.g. TO-220 and similar packages [68O1] and for integrated circuits [78S2].

For failure analysis, the plastic has to be removed from the device [79W4]. A non-destructive inspection is achieved by X-rays [82V1].

Potting

Potting is a casting process where no tool is required. The outline is given by a mostly rectangular pot prepared by either thermosetting or thermoplastic material. Epoxy and silicon compounds are used as resin [60I1, 65J1, 79P1]. Potting is mainly applied to small rectifiers. Special methods of potting are described in [78A2, 78H2, 79W5, 77F1].

Casting

The casting process is applied e.g. to light emitting diodes (Fig. 21), for which the surface of the package has to be free of burr. A metal tool is used to smoothen the surface. The tool is covered by a release agent like wax. Holes of the size of the package are filled to about 3/4 by an unfilled epoxy resin which contains a transparent dye of the required colour e.g. red, yellow, green, or blue. The lead frame including chip and bondwire is placed into the resin, fixed into position and cured in an oven. Ejection of the cured product out of the tool is relatively easy, because an unfilled resin exhibits a strong shrinkage.

For laboratory purposes, tools can be easily fabricated by castable silicon rubber. A release agent is not required.

Dropping

The dropping technique is employed for the protection of naked chips on thick- and thin-film substrates. Because of the small coefficients of thermal expansion of alumina and glass substrates, the resin used for dropping is highly filled by fused quartz and contains a black dye to prevent light from reaching the chip. The chip and its bonding wires are covered by the drop. The dropping resin as well as the equipment for mixing and dispensing the resin in correct quantities are commercially available.

Fluid bed powder coating

Resin powder is filled into an open box. Air is blown into the box through the bottom so that the powder floats and behaves like a liquid, "fluid bed". The resin powder contains a hardener, filler, dye etc. A thick-or thin-film substrate or a single-ended lead frame heated to a temperature of about 150 °C, is dipped into the fluid bed. The powder particles melt on the surface of the substrate/lead frame and form a dense coating. The process has to be repeated several times gelling the resin in between to obtain the required thickness of some tenth of millimeters [77S4, 79G1].

Injection moulding

Injection moulding is the state of the art for thermoplastic products. (Fig. 63). Plasticized and preheated plastic is injected in a water-cooled tool by a screw. The viscosity of these thermoplastics at moulding temperature is considerably higher than the viscosity of the thermosetting material as is generally used. Bonding wires therefore brake during moulding if not special precautions are taken [79H5].

Plastic flash removal

The moulding tools described are limited with respect to the clamping force, show a certain nonuniformity of the clamping surface at working temperature and the lead frames clamped vary with respect to thickness. Thin layers of plastic therefore bleed out over the surface of the lead frames outside the cavity and form a flash. This flash has to be removed before tin plating of the leads. Several methods are in use:

Soaking of the flash by Dimethylformamide; Methylenchloride mixed with other solvents, trade names are K5, Escaten VC 144; n-Methyl-Pyrrolidon, trade name M-Pyrol.

Removal by blasting by glass beads; blasting by alumina (corund); water jet containing alumina; water jet only; brushing.

The success of the method depends on the type of the plastic, the grade of curing and the kind of equipment used.

Finishing of packages

During finishing of the packages the metal parts are protected against corrosion, preparing the leads for soft soldering in a soldering bath [79K2], and cutting and bending the leads to the length and position required.

The solderability of leads can be tested by dipping the leads into a lead-tin melt of e.g. 260 °C for some seconds and subsequent inspection of the wetted area. 95% wetting is sufficient. This test can be carried out after a storage of the package at 150 °C for 16 h to simulate a burn-in or long time storage.

Galvanic plating of packages or leads

Hermetic metal packages are totally galvanically plated. For ceramic and plastic packages, only the leads are plated. In most cases, pure tin plating is applied [66C1, 70L2]. Deposition of lead-tin is also possible. Fe-alloys are directly plated. Cu-alloys require a Cu-, Co- or Ni-layer as preplating. Co- and Ni-layers prevent the diffusion of Sn into Cu and vice versa at high temperatures. The in-diffusion forms a bronce and decreases the wettability. A Cu-layer improves the wettability and is completely sufficient for most of the applications.

Methods of plating are:

barrel plating for metal and small ceramic and plastic packages,
plating on racks for lead frame packages,
plating on reels for double-ended glass packages.

In lead frame strips, the devices are not separated to prevent currents from bond wires and semiconductor chips. When barrel plating is employed conductive spheres are added to the bath to maintain contact to all parts.

The layer thickness after plating is in the order of 5⋯20 μm.

Glass-isolated packages are never plated by silver because Ag migration over the glass surface leads to shorts between the leads.

Dip tinning of leads

Dip lead-tin plating is employed in competition to galvanic plating. It is still a matter of discussion, whether this dipping method yields layers which show better solderability and cosmetic appearance after burn-in tests. Before dipping into the melt, flux has to be applied, e.g. HCl for copper lead frames.

In practice devices are cropped and bent out of the frame prior to dip tinning. The equipment for dip tinning is commercially available.

Passivation of leads

For high frequency applications it may be of interest that the leads are silver-plated instead of tin-plated. Silver is coloured by some sulfuric compounds and the solderability is reduced. In order to prevent a corrosion passivation of the Ag-surface has to be applied [66W1].

Cutting and bending of leads

The last operations before final electrical testing of a device are cutting the dam bars and frames and bending the leads e.g. in the required dual-in-line-position. For high-volume production, hard-metal tools are required. The devices are automatically transferred in rails for shipping and/or final testing.

6.1.4.8.4 Properties of packages

Thermal resistance

The thermal resistance is expressed in KW^{-1} and gives the difference of temperature between the chip or a distinct location on the chip (junction) and the ambient per power dissipation: R_{th} junction-ambient $= R_{th_{j-a}}$.

For power devices which require cooling fins or heat sinks the thermal resistance between junction and surface of the package is important: R_{th} junction-case $= R_{th_{j-c}}$ [79D3].

For measuring the temperature of a junction, the forward characteristic voltage has to be calibrated as a function of temperature ($V_{forward} = f(T)$) at constant current. The voltage drop over a junction decreases with temperature in the order of 2 mV/K. In practice, a device under test is operated at its operating point at a given power dissipation P in Watt in a box of e.g. 30 liters of volume. Every second, the power is switched off and the voltage drop across the junction is measured for 1 ms. The voltage drop gives the information of the junction temperature and therefore the difference to the ambient temperature. The thermal resistance $R_{th_{j-a}}$ can be calculated by

$$R_{th_{j-a}} = \frac{T_j - T_a}{P} \, KW^{-1}.$$

In order to achieve thermal equilibrium, the test takes 10 to 30 minutes per device. A quick method showing test times of about one second is described in [79S2].

For discrete devices, the thermal resistance is given in the relevant data books. Table 15 summarizes thermal resistance values to show the order of magnitude.

Table 15. Thermal resistance of discrete devices. Differences are due to internal construction.

Package	Outline Fig. No.	$R_{th_{j-a}}$ [KW^{-1}] in printed circuit board	$R_{th_{j-c}}$ [KW^{-1}] on copper block
TO-18	3	450···900	
TO-5	6	200···220	12.5···45
TO-3	8		1.5···2.5
TO-92	20	250···330	
SOT-23	25	400···620	
T-pack	29	230···600	
X-pack	30	335···500	
SOT-93	33		1.2···2.5
TO-202	34		10 ···12.5
TO-220	35		1.4···4.2

For integrated circuits, the thermal resistance values are not in all cases published in data books, because a given type has a function and a typical power dissipation which cannot be influenced by the user [81A1].

Fig. 64 shows the thermal resistance $R_{th_{j-a}}$ of dual-in-line packages as a function of the lead count.

The values vary with:

plastic DIL having lead frames of NiFe,
plastic DIL having lead frames of Cu-alloys,
plastic DIL having lead frames of Cu-alloys including heat spreader,
CERDIL packages, side-brazed,
CERDIP packages,
SO-packages on printed circuit boards,
SO-packages on ceramic substrates.

The values may be influenced by the chip size, the lead frame design, the plastic filler, e.g. α-quartz or fused quartz, and the surface area of the package [79M2].

Table 16 gives approximate values of the thermal resistance of integrated circuit power packages.

Table 16. R_{thj-c} or $R_{thj-fin}$ of IC-power-packages.

Package	Outline Fig.	R_{thj-c} KW^{-1}	$R_{thj-fin}$ KW^{-1}
Fin-DIL 12	43		12
SIL 9	44		10···12.5
DIL 16 power	45	3···5	
SIL 13 power	46	2···5	

Mechanical stress of the chip

Semiconductor wafers and chips are mechanically stressed during manufacture and assembly e.g. by sawing, breaking, wire bonding, and during and after plastic encapsulation at [77D2, 80D1, 82S1, 2]:

closing of the moulding tool and clamping of the lead frame,
injection of the liquid plastic in the cavity,
opening of the tool,
ejection of the product of the mould by the bottom ejector pins,
shrinkage of the plastic due to the chemical reaction,
shrinkage of the plastic due to cooling down to room temperature because of the high coefficient of expansion of the plastic.

The last two items lead to such high static strains in the silicon chip that diffused resistors change their resistance by up to -12%. This piezoresistive effect can be detected by special strain gauge chips having diffused resistors [80D1] and in practice in linear integrated circuits. In some cases this piezoresistive effect has to be taken into account during the design of a circuit.

Technological variations to minimize the stress effects:

glueing of chips instead of soldering,
application of plastics having low thermal expansion coefficient,
layout of the components on a circuit so that all important resistors are located perpendicular to the long axis of the package,
application of other crystal planes than (111) [79K1, 80S1].

The first three items reduce the piezoresistive effect to -2% change for single resistors and to 0.6% difference for two resistors acting e.g. as voltage dividers.

Hermeticity

In order to ensure a reliable function of a sealed device hermeticity is required. The US military standard defines tolerable leak rates as a function of the package volume [79P2] (Table 17).

Table 17. Tolerable leak rates as a function of package volume.

He leak rate $Pa\,m^3\,s^{-1}$	Volume of cavity cm^3
$5 \cdot 10^{-8}$	<0.01
$1 \cdot 10^{-7}$	0.01···0.4
$1 \cdot 10^{-6}$	>0.4

For the test, packages are exposed to 30 bar He pressure for two hours. A mass spectrometer leak tester which is adjusted to the mass of helium is used to sense the leak rate [80R2].

Methods for testing large leakages are:

Bubble test: The device is dipped into hot water. Due to the expansion of the gas in the package bubbles emerge from the leak.

Pressure test: The devices are pressurized to 2···5 bar in aethanol for a certain time and then electrically tested. Leaky devices fail in electrical parameters e.g. I_{CBO}.

The leak testing is described in [77R1]. For testing of packages sealed by adhesives see [78P1].

For hermetically sealed packages, the humidity enclosed during sealing is important [80G2]. The dew point of water vapour enclosed has to be lower than the lowest operating and storage temperature in order to prevent corrosion and deterioration of electrical device functions.

Plastics applied for encapsulation of semiconductor devices do not guarantee a hermetic sealing.

Diffusion of water vapour through the plastic body and along the interface of the plastic and leads is observed. In combination with ionic contamination, corrosion of the metallization of the chip occurs. The quality of plastics employed and the protection of the chips by glass-covers delay this effect beyond the time periods of practical application.

Tests of the water vapour effect are detected by overstress effects e.g.:

boiling in water,
pressure cooker test,
storage at 85 °C at 85% relative humidity,
operating at 85 °C, 85% relative humidity and at bias.

When a device is operated at an elevated power dissipation the package usually dries out.

The above mentioned tests are also part of investigations of the reliability of semiconductor devices [74K1, 76H2, 77R2, 78R2, 78L1, 79P3, 80B2].

Thermal fatigue of package components

In semiconductor mass production of plastic encapsulation, a fatigue of bonding wires during temperature cycling test was observed in the initial phase. The Au-wires show growth of grains especially in and near the ball. The mismatch of thermal expansion between silicon, Au-wire and plastic cause a creep of the wires along the grain boundaries up to rupture.

Fatiguing is suppressed by the following actions and can only be detected by overstress temperature cycling of many hundred times from −65 °C to 150 °C:

use of doped gold wire to maintain fine grain [80B1, 73R1],
use of plastic compounds with coefficients of thermal expansion as low as possible in the practical temperature range, forming the ball on the wire by an electric spark instead of a hydrogen flame in order to obtain quick cooling,
application of ultrasonic or thermosonic wire bonding methods instead of thermocompression to keep the bonding temperatures as low as possible.

Electrical properties

From the design point of view, the inductance of a package lead and the capacitance between leads is important. Table 18 gives relevant data.

Table 18. Comparison of electrical characteristics of dual-in-line packages and chip carriers [77P2].

Number of leads	Dual-in-line			Chip carrier		
	Longest lead	Shortest lead	Long to short ratio	Longest lead	Shortest lead	Long to short ratio
Spurious inductance [nH]						
16	6.4	1.62	3.98	1.13	0.73	1.55
28	14.77	1.62	9.12	1.80	1.15	1.57
40	24.94	1.62	15.4	2.90	1.87	1.55
64	49.14	2.34	21.0	6.44	4.21	1.53
Lead to lead or line to line capacitance [pF]						
16	0.074	0.025	2.96	0.012	0.009	1.44
28	0.148	0.025	5.92	0.019	0.013	1.46
40	0.213	0.025	8.53	0.027	0.019	1.42
64	0.412	0.033	12.48	0.052	0.036	1.44

Acknowledgements

The author thanks U. Weddi, G. Keil and W. Soppa for the procurement and selection of literature and patents, J. Wegener for his contributions to information on plastics and E. Klaue for reviewing of the manuscript.

References for 6.1.4.8

57A1 Andersen, O.L. et al.: J. Appl. Phys. **28** (1957) 923.

57H1 Herrmann, H., Thomas, H.: Über Eisen-Nickel-Kobalt-Legierungen für Glas-Metall-Verbindungen, Metallkunde **48** Nr. 11 (1957) 582···587.

57M1 Milam, D.L.: US patent 3021460, **1957**.

58C1 Christensen, H.: Electrical Contacts with Thermo-Compression Bonds, Bell Laboratories Record **1958**, p. 127···130.

58H1 Hansen, M.: Constitution of Binary Alloys, New York: McGraw Hill Book Company **1958**.

59J1 Jouwersma, C.: Die Diffusion von Wasser in Kunststoffen, Chem. Ing. Tech. **31**, Heft 10, (1959) 652···658.

60I1 Irmler, H., Schierz, W.: German patent 1246888, **1960**.

60N1 Noll, W.: Chemie und Technologie der Silicone, Weinheim/Bergstraße: Verlag Chemie **1960**.

61B1 Bernstein, L.: Gold Alloying to Germanium, Silicon and Aluminum-Silicon Eutectic Surfaces, Semiconductor Products, part 1: July 1961, S. 29···32 part 2: August 1961, S. 35···39.

62P1 Petersen, J.M., McKaig, H.L., DePrisco, C.F.: IRE. Nat. Conv. Rec. **1962** 3···12.

62S1 Sandbank, C.P.: British patent 14602, **1962**; German patent 1282188, 1963.

63M1 McKinnon, M.C., Hoeckelmann, R.F.: IEEE, Int. Conv. Rec. (1963) 93.

64A1 Avila, A.J.: Solid State Technol. November (1964) 22.

64M1 Moesker, G.: Dutch patent 142018, **1964**.

64S1 Schulz, G.: Die Kunststoffe, zweite Auflage, München, **1964**, S. 320···327.

65C1 Cunningham, J.A.: Solid State Electron. **8**, (1965) 735···745.

65J1 Jacobsen, H.: German patent 1489675, **1965**.

65U1 Uden, E., Weiß, H.: German patent 1569317, **1965**.

66B1 Brenner, A.: Vernickelung mittels Reduktionsverfahren, Handbuch der Galvanotechnik, **II**, München: Carl Hanser Verlag 1966, 729.

66B2 Brennen, B.: US patent 3272625, **1966**.

66B3 Boczar, W.J., Dawis, W.C.: US patent 3434018, **1966**.

66C1 Cuthbertson, J.W.: Handbuch der Galvanotechnik, **II**, München: Carl Hanser Verlag: 1966, 338···377.

66F1 Fischer, J., Elze, J.: Handbuch der Galvanotechnik, **II**, München: Carl Hanser Verlag 1966, 23···85.

66K1 Krusenstjern, A. v.: Handbuch der Galvanotechnik, **II**, München: Carl Hanser Verlag 1966, 392···414.

66L1 Lepstelter, M.P.: The Bell System Technical Journal, Vol. **XLV**, No. 2, (1966) 233···253.

66O1 Oosterhout, H.A. v.: Handbuch der Galvanotechnik, **II**, München: Carl Hanser Verlag 1966, 141···148.

66R1 Rinker, E.C., Schumpelt, K.: Handbuch der Galvanotechnik, **II**, München: Carl Hanser Verlag 1966, 414···428.

66S1 Sample, C.H.: Handbuch der Galvanotechnik, **II**, München: Carl Hanser Verlag 1966, 87···141.

66W1 Woldt, G.: Handbuch der Galvanotechnik, **III**, München: Carl Hanser Verlag 1966, 152.

67H1 Electronics **1967**, 236···238.

67J1 Joel, D., Höhne, H., Esch, H., Schubert, M., Kiesewetter, M., Walter, S.: German patent 1615993, **1967**.

67K1 Kielwein, F.: German patent 1652522, **1967**.

67K2 Koschel, H.: Siemens Z. **41**, Heft 3, (1967) 210.

67L1 Lee, H., Neville, K.: Handbook of Epoxy Resins, New York: McGraw Hill Book Company **1967**.

67S1 Schink, N.: German patent 1614703, **1967**.

68B1 Bott, F., Link, W.: German patent 1800213, **1968**.

68C1 Coucoulas, A., Cranston, B.H.: IEEE, Trans. Electronic Devices Vol. ED-**15**, No. 9 (1968) 664···672.

68C2 Clark, J.E.: Wobble Table for Thermocompression Bonding of Beam Lead Silicon Integrated Circuits, Proceedings, Wescon, Paper 2/1, August **1968**.

68F1 Finlay, W.L.: Silver Bearing Copper, New York: Corinthian Editions **1968**.

68K1 Kelley, D.T.: US patent 695015, **1968**; German patent 1817596, 1969.

68K2 Kookootsedes, G.J., Lockhart, F.J.: Mod. Plast. (1968) 150···154, 216.

68L1 Lindner, H.-J.: German patent 1789053, **1968**.

68O1 Overman, J.H.: US patent 752822, **1968**; German patent 1941305, 1969.

68W1 Water, J.T. v. d., Hout, J.J. v., Baelde, A.: Dutch patent 6810761, **1968**; German patent 1937664, 1969.

69C1 Crishal, J.M., Rice, E.J.: US patent 3620932, **1969**.

69G1 Gebrail, S.I.: US patent 812182, **1969**; German patent 2014246, 1970.

69G2 Garceau, W.J.: US patent 3631589, **1969**.

69K1 King, W.J., Wilcox, D.L., Lin, P.T., Winter, E.M.: US patent 850093, **1969**; German patent 2031725, 1970.

69R1 Ramsey, H.J.: US patent 3627190, **1969**.

69S1 Small, L., Campbell, H.J.: US patent 861966, **1969**; German patent 2047458, 1970.
70A1 Anderson, J.H., Maple, T.G., Cox, W.P.: IEEE Trans. Reliab. Vol. R-**19**, No. 1, (1970) 32···34.
70B1 Bower, F.H.: Solid State Technol. August (1970) 56···61.
70H1 Hirsch, H.: Solid State Technol. August (1970) 48···55.
70L1 Lutsch, A., Engelter, A., Fiedeldey, E., Greyvenstein, R.: Solid State Technol. February (1970) 58.
70L2 Langan, J.P.: Electron. Packag. Prod. **6**, No. 6, (1970) 5···11.
70P1 Percival, J.O.: US patent 3667848, **1970**.
70S1 Schimmer, R., Messerschmidt, J.: German patent 2019099, **1970**.
70U2 Uden, E.: German patent 2004768, **1970**.
71A1 Adams, A.L., Yearsby, G.A., Yager, B.P.: US patent 3776447, **1971**.
71D1 Dujardin, G.J.P.: Kunststoffe **61**, Teil 1, Heft 3, (1971) 177···182. Teil 2: [73D1].
71H1 Hupfeld, K.: German patent 2107786, **1971**.
71L2 Lindner, H.J.: German patent 2148690, **1971**.
71M1 Mizukoshi, K., Andou, M., Okano, H.: Japanese patent 59689-71, **1971**; German patent 2238569, 1972.
71M2 Melse, J.L., Groenewegen, M.A.: Dutch patent 7101602, **1971**; German patent 2202801, 1972.
71N1 Nash, M.T.: Us patent 3699010, **1971**.
71U4 Uden, E., Götze, B.: German patent 2124887, **1971**.
72B1 Budde, H.: Dutch patent 7204574, **1972**; German patent 2314247, 1973.
72D1 Niedrig legierte Kupferlegierungen, Deutsches Kupferinstitut, Berlin, ca. **1972**.
72D2 Duffek, E.F., Funk, E.J., Jankowski, A.S., Lane, J.C., Lehner, W.L., Oliver, F.F., Schneider, M.: US patent 3765590, **1972**.
72F1 Fetcher, C.A., Rosso, M.J.: US patent 3849187, **1972**.
72S1 Saddler, I.R.: US patent 3735209, **1972**.
72T1 Tomono, M. et al.: Japanese patent 103588-72; **1972** German patent 2352329, 1973.
72W1 Wieland, H., Unger, P.: Kunststoffe **62**, Heft 8, (1972) 493···496.
73A1 Anders, E.: German patent 2317514, **1973**.
73A2 Antonen, R.C.: US patent 3844992, **1973**.
73B1 Birglechner, G., Botzenhardt, L.: German patent 2315711, **1973**.
73C1 Curtiss, O.E., Tuller, H.W., Nussbaum, R.W.: US patent 3789038, **1973**.
73D1 Dujardin, G.J.P.: Kunststoffe **63**, Heft 6, (1973) 387···390.
73H1 Howarth, G., Jackson, S.: British patent 54644-73, **1973**; German patent 2454578, 1974.
73L1 Ludewig, J.: East German patent Wp 175242, **1973**.
73R1 Ramsey, T.H.: Solid State Technol. October **1973**, 43···47.
73S1 Sellers, R.F., Harvey, J.R.: US patent 3862260, **1973**.
73W1 Wirsing, C.E.: Solid State Technol. October **1973**, 48···50.
74B1 Blaine, R.L.: Thermal Analysis in the Electronics Industry, Du Pont Company, Wilmington, Delaware, June **1974**.
74F1 Factor, A.: J. Chem. Educ. **51**, No. 7, (1974) 453···456.
74G1 Gates, G.A., Ryan, W.J.: US patent 526998, **1974**; German patent 2540430, 1975.
74G2 Grossman, S.E.: Electronics May, **1974**, 89···95.
74H1 Harman, G.G.: A Metallurgical Basis for the Non-destructive Wire Bond Pull Test, 12th Annual Proceedings Reliability Physics, **1974**.
74H2 Hascoe, N.: US patent 3874549, **1974**.
74K1 Koelmans, H.: Metallization Corrosion in Silicon Devices by Moisture-Induced Electrolysis, 12th Annual Proceedings Reliability Physics Symposium, Las Vegas **1974**, p. 168···170.
74L1 Lipinski, B.W.: Dtsch. Farben Z., 28. Jahrgang, Heft **5** (1974) 207···211.
74M1 Mason, D.R., Blair, A., Stevenson, J.S.: British patent 52520-74, **1974**; German patent 2554583, 1975.
74M2 Moore, A.R.: British patent 2541-74, **1974**; German patent 2501532, 1975.
74P1 Pas, H. v. d.: Dutch patent 7406783, **1974**; German patent 2522022, 1975.
74S1 Smith, A.L.: Analysis of Silicones, New York: John Wiley and Sons 1974.
74T1 Tamura, T., Ojima, N., Kondo, S., Jizodo, Y.: Japanese patents 49120829, 49120838, **1974**; German patent 2545471, 1975.
74U3 Uden, E.: German patent 2443988, **1974**.
74U4 "Standard for Tests for Flammability of Plastic Materials, UL 94, second edition", Underwriters Laboratories Inc., Chicago-Northbrook Ill., Melville N.Y., Santa Clara CA, February **1974**.
74W1 Weißert, H. et al.: German patent 2425626, **1974**.
75B1 Bascom, W.D., Bitner, J.L.: Solid State Technol. September **1975**, 37···39.
75C2 Curtis, O.E., Tuller, H.W., O'Neill, C.T., Nussbaum, R.W.: US patent 4034014, **1975**.

75C3 Cohn, E.: Solid State Technol. September **1975**, 31···34.

75D1 Drift, A. v. d. Gelling, W. G., Rademakers, A. Philips Tech. Rundsch. **34**, Nr. 4, S. 73···84, 1974/75.

75D2 Dehaine, G., Kurzweil, K.: Solid State Technol. p. 46···52, October 1975.

75D3 David, R.F.S.: Solid State Technol. p. 40···44, September 1975.

75E1 Euler, G.: Hochfrequenz-Leistungstransistoren, VALVO der Philips GmbH, Hamburg **1975**.

75E2 Method of Test for Spiral Flow, Epoxy Molding Materials Institut, 250 Park Avenue, New York, N.Y. 10017, 212 Mu 7-2675.

75H1 Hintzmann, K., Kaiser, R., Link, W., Minner, W., Mutz, D., Salomon, M., Wingert, A.: German patent 2543968, **1975**.

75H2 Heil, G.: German patent 2541121, **1975**.

75K1 Krevelen, D.W. v.: Chem. Ing. Tech. **47**, Nr. 19, (1975) 793···803.

75L1 Lehner, L.L., Segerson, E.E.: US patent 616456, **1975**; German patent 2636450, 1976.

75M1 Martin, J.P., Gray, T.H.: US patent 596039, **1975**; German patent 2631904, 1976.

75M2 Mink, A.E.: US patent 3948848, **1975**.

75O1 Oppermann, J., Heil, G.: German patent 2543701, **1975**.

75T1 Tuller, H.W., Nussbaum, R.W.: US patent 4042550, **1975**.

75W1 Winkle, R.V.: A Study of the Mechanism of Wire Bonding, Proceedings of the Internepcon in Brighton, UK, **1975**.

76A1 Angelucci, T.: Solid State Technol. July **1976**, 21···25.

76D1 Dietz, L., Tillman, J.J.: US patent 729227, **1976**; German patent 2743773, 1977.

76H1 Holmes, P.S. et al.: Handbook of Thick Film Technology, Electrochemical Publications Ltd., Ayr, Scottland **1976**.

76H2 Hakim, E.B.: Korrosion der Kontakt-Metallisierung bei Halbleitern, Metalloberfläche **30**, Nr. 1, (1976) 15···19.

76I1 Inoue, Y.: Japanese patent 96365-76, **1976**; German patent 2736090, 1977.

76M1 Markstein, H.W.: Electronic Packag. Prod. 51···54, November **1976**.

76R1 Rosler, R.: US patent 648424, **1976**; German patent 2700363, 1977.

76Y1 Yamamoto, M., Makino, K., Suzukawa, K., Nakamura, S., Sakai, K.: Japanese patent 100637-76, **1976**; German patent 2738021, 1977.

76Z1 Zschimmer, G.: German patent 2627178, **1976**.

77A1 Acello, S.; Mini-Pak- A cost Effective Leadless Chip Carrier, Packaging and Production Conference, March **1977**.

77A2 Antonen, R.C., Michael, K.W., Engelmann, J.H.: J. of Electron. Mater. **6**, No. 1, 1977, p. 49···60

77D1 Delorme, R.L.: French patent 7528170; German patent 2704266, 1978.

77D2 Dale, R.J., Oldfield, R.C.: Microelectron. Reliab. **16** (1977) 255···258, Pergamon Press, UK.

77F1 Fierkens, R.H.J.: Dutch patent 7704937, **1977**; German patent 2819287, 1978.

77G1 Gantley, F.C.: US patent 848129, **1977**; German patent 2846398, 1978.

77G2 Greenberg, L.S.: US patent 788330, **1977**; German patent 2815776, 1978.

77J1 Justi, E.: German patent 2712172, **1977**.

77J2 Johnson, K.I., Scott, M.H., Edson, D.A.: Ultrasonic Wire Welding Part I, Solid State Technol. March **1977**, 50.

77J3 Johnson, K.I., Scott, M.H., Edson, D.A.: Ultrasonic Wire Welding, Part II, Solid State Technol. April **1977**, 91.

77K1 Khadpe, S.: 27th Electronic Components Conference Proceedings, May **1977**, 30···33.

77L1 Lyman, J.: Electronics March **1977**, 81···91.

77O1 Oswald, R.G., de Miranda, W.R.: Solid State Technol. March **1977**, 33···38.

77P2 Prokop, J.S., Williams, D.W.: Chip carriers as a means for high-density packaging, Proceedings of the 1977 International Microelectronics Symposium, October 1977.

77R1 Ruthberg, S.: IEEE Trans. Parts, Hybrids and Packaging, Vol. PHP-**13**, No. 2, (1977) 110···115.

77R2 Reich, B.: Reliability of Plastic Encapsulated Semiconductor Devices and Integrated Circuits (PED), Relectronic 77, 4th Symposium on Reliability in Electronics, Hungary, October 1977 [78R2].

77S1 Schermer, G., Breuning, P.: Dutch patent 7713758, **1977**; German patent 2853328, 1978.

77S2 Spanjer, K.G.: US patent 763398, **1977**; German patent 2802439, 1978.

77S3 Strieder, E.: German patent 2728523, **1977**.

77S4 Scholten, G.J., Brandsma, J.: Dutch patent 7701283, **1977**; German patent 2804956, 1978.

77W1 Winkle, R.V.: Non-destructive Monitoring of Ultrasonic Bond Quality, Proceedings International Microelectronic Conference, Brighton **1977**.

78A1 Andrascek, E., Hadersbeck, H., Hacke, H.-J.: German patent 2840973, **1978**.

78A2 Andrascek, E.: German patent 2840972, **1978**.
78B1 Bansemir, M.: East German patent Wp 207161, **1978**; German patent 2929042, 1979.
78B2 Beyerlein, F.W.: Electron. Packag. Prod. January **1978**, 54.
78B3 Burns, C.D., Kanz, J.W.: Solid State Technol. September **1978**, p. 79···81.
78C1 Chandross, E.A., Sharpe, L.H.: US patent 931160, **1978**; German patent 2931633, 1979.
78F1 Fancher, D.R., McDaniel, J.E.: Electron. Packag. Prod. February **1978**, 88···94.
78H1 Harman, G.G., Cannon, C.A.: The Microelectronic Wire Bond Pull Test, Proceedings of the 28th Electronic Components Conference, p. 291, **1978**.
78H2 Hazawa, K., Maeda, Y., Ohta, S.: Japanese patent P 164752-78, **1978**; German patent 2952493, 1979.
78L1 Livesay, B.R.: Solid State Technol. October **1978**, 63···67.
78L2 Ludwig, D.P.: Electron. Packag. Prod. April **1978**, 77···86.
78M1 Miranda, W.R.R. de, Oswald, R.G., Brown, D.: IEEE Trans. Components, Hybrids, and Manufacturing Technology, Vol. CHMT-**1**, No. 4, (1978) 377···383.
78N1 Nishikawa, A., Suzuki, H., Kohkame, H.: Japanese patent P 84004-78, **1978**; German patent 2927995, 1980.
78O1 Otsuki, K., Mochizuki, H., Suzuki, A., Adachi, Y., Kosaka, H., Murakami, G.: Japanese patent P 93607-78, **1978**; German patent 2931449, 1979.
78P1 Perkins, K.L., Licari, J.J., Caruso, S.V.: IEEE Trans. Components, Hybrids, and Manufacturing Technology, Vol. CHMT-**1**, No. 4, (1978) 412···415.
78R1 Romano, L.: Italian patent 23432 A-78, **1978**; German patent 2919540, 1979.
78R2 Reich, B.: Solid State Technol. September **1978** = [77R2].
78S1 Schmitter, D.: Micropack - Integrated Semiconductor Circuits from Strip, Components Report, Siemens, Vol. **XIII**, No. 2, München 1978, p. 38···42.
78S2 Slepcevic, D.: US patent 925295, **1978**; European patent 0007762, 1979.
78T1 Trego, B., Haberer, S.: The development of a new hybrid moulding compound, Electronic Production (GB), **7**, No. 4, April 1978, p. 45···51.
78W1 Woodard, J.B.: Comparison of Phenolic Novolac and Anhydride Cured Semiconductor Moulding Compounds, Proceedings of Hysol Technical Symposium at Semicon/Europe, September **1978**.
78Y1 Yoshizumi, A., Ikeja, H., Wada, M.: Japanese patent P 48776-78; German patent 2916954, 1979.
79A1 Amey, D.: The Jedec Chip Carrier and LSI Package Standard, Sperry Univac, Blue Bell, Pa, USA, **1979**.
79B1 Beall, R.J., Lee, J.C.: US patent 67538, **1979**; German patent 3030763, 1980.
79B2 Bischoff, A.: German patent 2929623, 1979.
79D1 Dawes, C.J.: Ultrasonic ball/wedge bonding of aluminum wires, The Welding Institute Research Bulletin, England, January **1979**.
79D2 Dathe, J., Schmidhuber, E.: German patent 2930760, **1979**.
79D3 Dance, B.: Wärmeableitung bei Halbleitern, Elektronik Industrie, S. 36···39, September **1979**.
79F1 Fussel, L.: US patent 16756, **1979**.
79G1 Gill, M.F., German, J.J., Nelesen, M.A.: Fluidized Bed Packaging for Thick-Film Hybrid Circuits, Proceedings of the 29th Electronic Components Conference, p. 37···39, May 14···16, **1979**.
79G2 Garz, D.P.W.: German patent 2944922, **1979**.
79H1 Horita, K.: Japanese patent P 120631-79, **1979**; German patent 3033516, 1980.
79H2 Horita, K.: Japanese patent 75289-79, **1979**; European patent 0029858, 1980.
79H3 Honda, N., Hayashi, K.: Japanese patent 95327-79, **1979**; European patent 0023400, 1980.
79H4 Hunt, E.R., Rosler, R.K., Peterson, J.O.: US patent 28155, **1979**; German patent 3013470, 1980.
79H5 Hoffmann, H.: German patent 2930760, **1979**.
79H6 Helfant, D., Villanani, T.: Epoxy Resins for Molding Application, 14 Electr./Electrical Insulation Conference, p. 290···311, October **1979**.
79K1 Komatsu, S.: Japanese patent P 165390-79, **1979**; German patent 3047300, 1980.
79K2 Klein-Wassink, R.J.: Philips Tech. Rundsch. **38** 4/5 (1979) 142···150.
79K3 Kanz, J.W., Braun, G.W., Unger, R.F.: IEEE Trans. Components, Hybrids, and Manufacturing Technology, Vol. CHMT-**2**, (1979) 301···308.
79K4 Kishikawa, K.: Japanese patent P 143278-79, **1979**; German patent 3042093, 1980.
79M1 Miura, Y.: Japanese patent P 160245-79, **1979**; German patent 3046375, 1980.
79M2 Mitchel C., Berg, H.M.: IEEE Trans. Components, Hybrids, and Manufacturing Technology, CHMT-**2**, No. 4, December (1979) 500···511.
79M3 Mücke, K.H.: Nickelschichten auf Bodenplatten für Halbleiterbauelemente, Elektronik Produktion und Prüftechnik, November 1979, S. 467···470.
79M4 McCormick, J., Zakrausek, L.: Annu. Proc. Reliab. Phys. April **1979**, 44···50.

79M5 Monnier, M.: Circuiterie sur film polyamide, Toute l'Electronique Août, September 1979, p. 25···30.

79P1 Paquet, R.: Elektronik Nr. **2**, 1979, S. 39···41.

79P2 Petersen, M.E., Sergent, J.: Consideration in the Selection of a Metal Hermetic Package, Proceedings of the 1979 International Microelectronic Conference, **1979**, p. 221···227.

79P3 Peterson, P.W.: The Performance of Plastic Encapsulated CMOS Microcircuits in a Humid Environment, Proceedings of the 29th Electronic Components Conference, May 14···16, **1979**.

79S1 Suzuki, A., Tsubosaki, K., Kato, T.: Japanese patent P 63839-79, **1979**; German patent 3019868, 1980.

79S2 Siegel, B.S.: Electronics (1979) 141, April 12.

79S3 Schramm, K.H.: Das Film-Bonden, Elektronik, Nr. **4**, 39···42, (1979).

79V1 VALVO data books, Hamburg: VALVO der Philips GmbH, 1979···1981.

79W1 Winkler, K., Heitzler, V., Albrecht, A.: German patent 2923440, **1979**.

79W2 Winkler, K., Heitzler, V., Albrecht, A.: European patent 0020857, **1979**.

79W3 Winkle, R.V., Warmer, C.H.: A New Method of Quality Control for Ultrasonic Wire Bonding, Proceedings of Ultrasonics International Conference, Graz, Austria, **1979**, p. 62···68.

79W4 Wensink, B.L.: Solid State Technol. p. 107···111, October **1979**.

79W5 Wakashima, Y.: Japanese patent P 97286-79, **1979**; German patent 3029123, 1980.

79W6 Weissenfelt, M., Collander, P., Järvinen, E., Laurindli, T.: A practical method for evaluating and adjusting ultrasonic wire bonding process, Proceedings of the 29th Electronic Components Conference, 14···16 May, **1979**.

79Y1 Yahya, H.-T., Hoppe, J.: German Patent 2920012, **1979**.

80A1 Abiru, A., Sugimoto, M., Jonomata, J.: Japanese patent 25142/80, **1980**; European patent 0035331, 1981.

80A2 Abe, M., Miyagawa, M., Nakamura, H., Yonezawa, T.: Japanese patent 6079/80, **1980**; European patent 0032801, 1981.

80B1 Bischoff, A., Thiede, H.: Metallkundliche Aspekte bei Verarbeitung und Einsatz von Feinstdrähten in der Halbleitertechnik, Draht, **3** 1980.

80B2 Berg, H.M., Paulson, W.M.: Chip Corrosion in Plastic Packages, Microelectronic Reliability, **20**, p. 247···263, Pergamon Press Ltd., 1980.

80B3 Bauer, J.A.: IEEE Trans. Components, Hybrids, and Manufacturing Technology, Vol. CHMT-**3**, No. 1, March 1980.

80B4 Burggraf, P.S.: Semiconductor Plastic Encapsulants: Molding Compounds, Semiconductor International, p. 47···64, October **1980**.

80D1 Dale, J.R., Oldfield, R.C.: Strain and its Effects in Silicon Microcircuits, Proceedings of the BSSM-International Conference on Product Liability and Reliability, Birmingham **1980**, p. 1.

80D2 Dunkel, W., Wagner, H.: Wassereinfluß auf quarzgefüllte Epoxidharz-Formstoffe, Elektrotechn. Z. Band **101**, Heft 1, (1980), 31···35.

80E1 Exner, K.: European patent 0032728, **1980**.

80E2 Eisentraut, H.: German patent 3010076, **1980**.

80F1 Feinstein, L.G.: Do Flame Retardants Affect the Reliability of Molded Plastic Packages?, Proceedings of the 30th Electronic Component Conference, p. 106···113, April **1980**.

80G1 Gehmann, B.L.: Gold Wire for Automated Bonding in: Solid State Technol. March **1980**.

80G2 Grilletto, C., Vowinkel, F.: Solid State Technol. June **1980**, 71···73.

80H1 Helber, H.: German patent 3011661, **1980**.

80H2 Hoppe, J., Haghiri-Tehrani, Y.: German patent 3019207, **1980**.

80H3 Hoppe, J.: German patent 3029667, **1980**.

80L1 Luis, J.S., Thumm, M., Ribot, P., Gale, G.: Low Cost Chip Carriers for the **1980**'s, Proceedings of the 30th Electronic Components Conference, 1980, p. 436···441.

80L2 Lyman, J.: Tape automated bonding meets VLSI challange, in: Electronics, p. 100···105, December 18, **1980**.

80M1 McGill, G.P., Moore, L.R.: Gold Wire for Automatic Bonders, Semiconductor International, June **1980**.

80M2 McShane, M.B.: Device Design and Lead Frame Layout Techniques for Automatic Wire Bonding, Semiconductor International, June **1980**.

80M3 Masessa, A.J., Mohr, R.G.: IEEE Trans Components, Hybrids, and Manufacturing Technology, Vol. CHMT-**3** No. 3, September **1980**.

80M4 Melliar-Smith, C.M., Matsuoka, S., Hubbauer, P.: Plastic encapsulation of integrated circuits, Plastic and Rubbers: Material and Applications, p. 49···56, May **1980**.

80R1 Rudolf, F., Heinke, G., Labs, J.: Nachrichtentech. Elektron. Jahrgang 30, Heft **3**, (1980) 115···122.

80R2 Ruthberg, S.: A Rapid Cycle Method for Gross Leak Testing with the Helium Leak Detector, Proceedings of the 30th Electronic Components Conference, p. 128···134, April **1980**.

80S1 Suzuki, K., Aoiki, T., Kusanagi, S.: Japanese patent 25142/80, **1980**; European patent 0035331, 1981.
80S2 Suzuki, H., Tanaka, G., Nichikawa, A., Mukai, J., Sato, M., Makino, D., Wakashima, Y.: Japanese patent P 587-80, **1980**; German patent 31 00 303, 1981.
80U1 Ullrich, H.: Weiterentwicklung der IS Montagetechnologie, Bundesministerium für Forschung und Technologie, Forschungsbericht T 80-179, Bonn, Dezember **1980**.
81A1 Andrews, J.A., Mahalingam, L.M., Berg, H.M.: Thermal characteristics of 16- and 40-pin plastic DIP's IEEE-Transactions on Components, Hybrids, and Manufacturing Technology, CHMT-**4**, No. 4 (1981) p. 455···461.
81B1 Bischoff, A., Aldinger, F.: Neue goldsparende Feinstdrahtwerkstoffe, Elektronik Produktion und Prüftechnik, **21**, No. 9, (1981).
81C1 Chaffin, B.: Solid State Technol. September **1981**, 136···138.
81E1 Endicott, D.W., Asher, R.K.: Selective Gold Electroplating of Electronic Multilead Headers, Plating and Surface Finishing, p. 46, January 1981.
81F1 Funk, W., Wijburg, M.A.T.: German registered design, G 8122540. 7, **1981**.
81L1 Lineback, J.R.: TI tries out plastic leaded chip carriers, Electronics, p. 39···40, June **1981**.
81L2 Lyman, L.: Electronics July **1981**.
81L3 Lyman, J.: Electronics, October 20, **1981**
81O1 O'Neil, T.G.: VLSI Packaging Requirements and Trends, Semiconductor International, March **1981**.
81O2 Olsen, D.R., Spanjer, K.G.: Solid State Technol. September **1981**, 121···126.
81P2 Pas, H. v. d.: German patent 31 41 842, **1981**.
81S1 Sinnadurai, N.: An Evaluation of Plastic Coatings for Microcircuits, Third European Hybrid Microelectronics Conference, p. 482, Avignon, **1981**.
81S2 Solid State Technol. September (1981), p. 50, 51.
81S3 Scott, M.H., Johnson, K.J.: Environmental testing of aluminium wire ball bonds, The Welding Institute Research Bulletin, England, p. 218···220, August 1981.
82E1 Erickson, D.: Electron. Packag. Prod. **1982**, 133···138.
82P3 Pas, H.A. v. d., Knobbout, H.A.: German patent 32 09 242, **1982**.
82S1 Schlesier, K.M., et. al.: Piezoresistive Effects in plastic-encapsulated integrated circuits, International Conference on Consumer Electronics (ICCE), Session II, **1982**, p. 28, 29.
82S2 Schlesier, K.M., Keneman, S.A., Mooney, R.T.: Piezoresistivity Effects in Plastic-Encapsulated Integrated Circuits, RCA-Review, **Vol. 43**, December 1982, 590···607
82V1 Voll, H.: Röntgenmikroskopie - Eine neue Methode zur Materialuntersuchung, Mikroskopie - Elektronenmikroskopie, Supplement, January (1982) p. 32, 33.

6.2 Silicon carbide

6.2.0 Introduction

Silicon carbide is considered to be a potentially useful material for electronic devices.

The main advantages are:

high band-gap (depending on polytype),
high breakdown field,
high saturated carrier velocity,
high thermal conductivity,
chemical stability.

The main disadvantages are:

technological problems, mainly with respect to the preparation of deliberately doped, perfect single crystals of adequate size

low carrier mobilities, as compared to elemental group IV semiconductors and most III–V-compounds.

Silicon carbide crystallizes in numerous modification (see Vol. III/17a, section 1.5). The 6 H modification is mainly considered as a material for electronic applications because crystals of this type can be produced by the *Lely technique* with dimensions in the order of 10 × 10 × 3 mm. Large crystals are also found occasionally during the manufacture of SiC abrasive material by the *Acheson process.*

Single-crystal silicon carbide is particularly useful for the following semiconductor devices:

high temperature devices (thermistors, rectifiers, transistors), light-emitting diodes (especially for emission of blue light), uv and particle detectors.

6.2.1 Technological data

6.2.1.1 Figures of merit for devices

Fundamental investigations on the physical limits of semiconductor devices in terms of packing densities, speed of operation, and power handling capability have led to the development of two figures of merit by Johnson [65J]

$$z_J = (4\pi)^{-1} E_b^2 v_s^2$$

and by Keyes [74K]

$$z_K = \kappa (c v_s / 4\pi\varepsilon_r)^{\frac{1}{2}},$$

where E_b is the breakdown field, v_s the saturated drift velocity, κ the thermal conductivity, c the velocity of light, and ε_r the dielectric constant.

As shown in Table 1, silicon carbide compares favourably with conventional semiconductors.

Table 1. Figures of merit for Si, GaAs, and SiC.

	v_s cm s^{-1}	E_b V cm^{-1}	ε_r	κ W cm^{-1}K^{-1}	z_J V^2 s^{-2}	z_K W s^{-1}K^{-1}
Si	10^7	$2 \cdot 10^5$	12	1.5	$3 \cdot 10^{23}$	$7 \cdot 10^7$
GaAs	$2 \cdot 10^7$ [1])	$3 \cdot 10^5$	11	0.5	$3 \cdot 10^{24}$	$3 \cdot 10^7$
SiC	$2 \cdot 10^7$	$3 \cdot 10^6$	9.7	3.5	$3 \cdot 10^{26}$	$2 \cdot 10^8$

[1]) Maximum velocity.

6.2.1.2 SiC phase diagram

The silicon-carbon phase diagram is shown in Fig. 1. The peritectic nature of the system makes growth from a stoichiometric melt unfavourable.

6.2.1.3 Solubility of carbon in silicon

The solubility of carbon in silicon is shown in Fig. 2. The solubility is very low at temperatures below 2000 °C. At 1800 °C, for instance, only 0.1 % carbon can be solved in silicon.

6.2.1.4 Vapour pressure of silane compounds

The vapour pressure of silane compounds containing carbon and chlorine is shown in Fig. 3. These silane compounds are used in the van Arkel process for deposition of polycrystalline SiC.

6.2.2 Crystal growth

6.2.2.0 Introductory remarks

The following methods can be employed for the preparation of silicon carbide crystals:

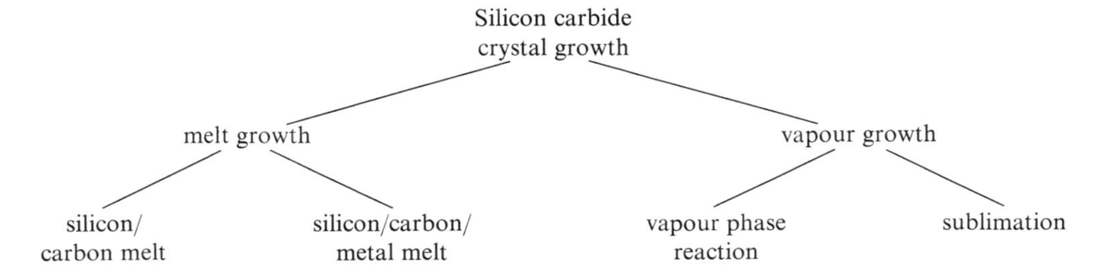

In practice, however, most of these methods have failed to generate SiC crystals which are adequate for an industrial production of electronic devices. A detailed description, therefore, will only be given for the *sublimation process* in section 6.2.2.2. This process requires temperatures in excess of 2500 °C; most crystals grown in this manner are of the 6H polytype.

Small silicon carbide crystals of various polytypes, can be obtained by *thermal decomposition* of certain silane compounds or mixtures of gases containing silicon and carbon in a wide temperature range (starting at about 1200 °C). The size of these crystals, so far, is not adequate for device production. The *van Arkel process*, however, is a useful method for the preparation of extremely pure (or deliberately doped) feed material, which is subsequently converted into 6H crystals by the *Lely technique* (section 6.2.2.2.2).

The principles and problems of *melt growth processes* for silicon carbide are outlined in section 6.2.2.1. Due to the extremely low solubility of carbon in molten silicon, the crystallization of silicon carbide from a silicon/carbon solution takes place with very small growth-rates. This method has been mainly used for the preparation of *high-purity β-SiC* (3C polytype). Other polytypes can be obtained with proper seeding.

The solubility of carbon can be drastically increased by adding a third constituent (transition metal) to the melt. Undesired impurities, however, are introduced at a level determined by the solid solubility of the metal in silicon carbide.

6.2.2.1 Melt growth

Because of the peritectic nature of the Si-C phase diagram (Fig. 1) it is not likely that silicon carbide can be obtained from a stoichiometric melt. All investigations on the growth of silicon carbide from a silicon/carbon solution were performed with a *non-stoichiometric melt* [66N, 67B1, 69M1].

Since the carbon solubility in silicon is very low at temperatures below 2000 °C (Fig. 2) and the silicon vapour pressure at this temperature is in the order of 0.1 Torr, *evaporation of silicon* must be suppressed (or retarded, at least) by a *high-pressure inert atmosphere* (e.g. argon).

The *crucible material* usually is *high-density graphite*. Other crucible materials, including boron nitride, have been used occasionally [69M1]. The crystal growth systems are either of the (vertical) *Bridgman* or of the *Czochralski* type.

Silicon carbide crystals grown from solution are, in general, n-*type*, due to traces of nitrogen in the inert gas. The electron concentration can be deliberately increased by *adding nitrogen* to the ambient. The p-*type* conductivity is obtained by *adding boron, aluminum*, or *gallium* to the melt (see section 6.2.4.3).

The solubility of carbon can be increased by adding *a transition metal* to the melt. The most widely explored system consists of *silicon, carbon* and *chromium* [64G, 65W, 66K1]. Other solvents include *iron* [60H, 60E], *nickel* [60E], *cobalt* [69M1] and *titanium* [74P]. The *travelling solvent method* (TSM) is often employed for silicon carbide growth.

Using the *travelling heater method* (THM) and a sealed-tube arrangement according to Fig. 4 the evaporation of silicon from the melt can be completely compensated by the *iodine transport reactions*

$$SiI_2 \rightarrow Si + 2I \quad \text{and} \quad SiI_4 \rightarrow Si + 4I.$$

These reactions perform a silicon transport from the colder regions of the ampoule to the hot silicon melt. In practice, an *amount of* 10^{-2}g *iodine per* cm^3 *of volume* in the gas phase is adequate to prevent silicon evaporation from a melt at 1800 °C. The whole system, of course, has to be kept at an elevated temperature (e.g. 700 °C) to avoid condensation of iodine [73G].

Although the sealed-tube system according to Fig. 4 has been used only for the deposition of epitaxial layers [74M1], it could well be adapted for the preparation of silicon carbide crystals.

6.2.2.2 Vapour growth

6.2.2.2.1 Deposition of polycrystalline SiC

Polycrystalline silicon carbide can be produced by the *van Arkel method*, i.e. pyrolysis of gaseous compounds at a hot wire or carbon rod. A variety of chemicals may be employed in this process; a selection of these is compiled in Table 2. The carrier gas is usually hydrogen or argon.

Table 2. Gaseous compounds for the van Arkel process.

Compound(s)	Ref.
CH_3SiCl_3 (methyl-trichlorosilane)	59A, 60S, 68N, 69P, 72K
CH_3SiHCl_2 (methyl-dichlorosilane)	63K
$(CH_3)_2SiCl_2$ (dimethyl-dichlorosilane)	60S, 63K, 68N, 78M1
$(CH_3)_3SiCl$ (trimethyl-chlorosilane)	60S, 63K, 68N, 78M1
$(CH_3)_4Si$ (tetramethylsilane)	60S, 68N

continued

Table 2 (continued)

Compound(s)	Ref.
$C_2H_5SiCl_3$ (ethyl-trichlorosilane)	68N
$(C_2H_5)_2SiCl_2$ (diethyl-dichlorosilane)	68N
$SiCl_4$ (silicon tetrachloride) + C_6H_6 (benzene)	68N
$SiCl_4$ (silicon tetrachloride) + $C_6H_5CH_3$ (toluene)	59A, 60S, 68N
$SiHCl_3$ (trichlorosilane) + C_6H_6 (benzene)	63K, 68N
$SiHCl_3$ (trichlorosilane) + $C_6H_5CH_3$ (toluene)	63K
$SiHCl_3$ (trichlorosilane) + CH_2Cl_2 (dichlormethane)	63K

A system for the preparation of pure and boron- or aluminum-doped SiC crystals by pyrolysis of silane compounds is shown in Fig. 5a [78M1]. Details of the reaction chamber can be seen in Fig. 5b. The water-cooled base plate is made of silver-plated copper, the current leads and the gas inlet (nozzles) are of pure silver. The carbon rods (e.g. 6 mm diameter, 150···300 mm length) are heated by AC. It is important that the gas streams ejected from the nozzles are hitting the centre of each rod. The reaction chamber should be carefully evacuated and flushed with pure hydrogen before each run. The deposition of silicon carbide takes place in hydrogen at atmospheric pressure.

The carrier gas is saturated with silane compounds which are kept at a temperature around 20 °C. The vapour pressures of the most common silane compounds in the 200···340 K range are shown in Fig. 3.

The nature and the morphology of the deposit depend on the composition of the gas and the deposition temperature. In general, a stoichiometric (or nearstoichiometric) deposit is obtained in the 1400 to 1600 °C temperature range. A high-density material with small grain size and uniformly distributed excess silicon is obtained at low temperatures. The grain size increases with increasing growth temperature. At temperatures above 1600 °C, excess carbon is formed, which is concentrated mainly between particles of silicon carbide.

Typical dimensions and growth conditions for a laboratory-scale production of silicon carbide are the following:

Quartz bell jar: diameter 150 mm, height 400 mm; carbon rods: diameter 6 mm, length 300 mm; source: dimethyl-dichlorosilane at 18 °C; hydrogen flow rate: 100 l/h; deposition temperature: 1500 °C.

A deposition rate of about 40 g/h is achieved with the above growth conditions.

Doping of the polycrystalline material with boron is accomplished by adding a metered flow of diborane (diluted in argon) during the deposition cycle. A bubbler containing a 10···20% solution of trimethyl-aluminum in n-decane ($C_{10}H_{22}$) serves as a source for aluminum doping; the vapour pressure of n-decane is negligible in the temperature range involved.

6.2.2.2.2 Lely technique

Polycrystalline silicon carbide produced by the van Arkel method or commercial (green or black) silicon carbide grinding powder is converted into 6H single crystals by the *Lely technique*. The basic idea of this process is shown in Fig. 6. The charge of granular SiC is situated in an annular layer between the crucible wall and a cylinder consisting of high-porosity graphite. The assembly is heated in an inert gas atmosphere (argon or helium) to *about 2600 °C*. Silicon carbide decomposes at this temperature and vapour phase species like SiC_2, Si_2C, and silicon are diffusing through the porous graphite into the central cavity. Due to a *slight temperature gradient* between the cavity and the silicon carbide charge, crystallization takes place at the inside of the porous graphite cylinder. Silicon carbide crystals, mostly of the 6H polytype, are grown with the shape of hexagonal platelets. A growth run normally takes about 10···20 hours.

Lely-type furnace design

Some of the Lely-type furnaces are based on *resistance heating*, others utilize *induction heating* (4 kHz or 10 kHz). In order to keep the power input at an acceptable level, an *efficient thermal insulation* is required. Graphite powder, graphite felt, graphite foil and pyrolytic graphite are used for this purpose. Graphite powder and graphite felt are good thermal insulators. Due to the large surface-to-volume ratio of these materials, *prolonged purification* (outgassing) procedures are necessary if high-purity crystals are to be grown. The use of pyrolytic graphite is based on the low conductivity of this material in the direction of the *c*-axis. With a *shield* of 10 mm highly oriented *pyrolytic graphite*, a temperature of 2600 °C can be maintained inside the crucible, while the surface of the shield has a temperature of only 1600 °C. In addition, pyrolytic graphite has a low adsorption of gases, it is dense and impervious to gases and can be made purer than other types of graphite. Large parts consisting of pyrolytic graphite are quite expensive, however. For slightly relaxed specifications on crystal purity, the use of (degassed) *graphite foil* is adequate.

Some of the Lely-type growth systems described in the literature are characterized in the following survey.

Typical Lely systems

Heating method:	Resistance	Inductive (10 kHz)	Inductive (4 kHz)	Inductive (10 kHz)
Insulation:	Graphite powder	Graphite powder	Graphite foil, pyrolytic graphite	Pyrolytic graphite
References:	63K, 66B	72P	77M	70K

Crucible assembly

A *crucible assembly* for an inductively heated (4 kHz, 35 kW) crystal growth system is shown in Fig. 7a. The radial heat shield consists of graphite foil ("Sigraflex" from Sigri Company, Meitingen) whereas the axial insulation is obtained with a stack of discs, which are made from pyrolytic graphite (*c*-axis parallel to the axis of the crucible). Heat is generated mainly in the wall of the graphite crucible. An adjustment of the vertical temperature gradient can be achieved by additional annular heaters at the top and the bottom of the crucible.

The complete arrangement of the crucible assembly, the bell jar, the induction coil, and the water-cooling is shown in Fig. 7b. In a typical growth run, the apparatus is first assembled, evacuated, degassed at about 1800 °C, and flushed with argon (or helium). Finally, power is applied to bring the crucible to the desired growth temperature (2600 °C in general). After the growth run, *silicon carbide platelets* can be recovered from the inner wall of the porous graphite cylinder. Most crystals (more than 80 %) are of the 6 H type. Other polytypes (e.g. 15R, 4 H, and 8 H) are found occasionally.

Doping of crystals

Doping of the crystals can be achieved via the gas phase or by introducing properly doped starting material. For example, n-type samples are easily prepared by *adding nitrogen* to the inert gas.

Variations of the Lely technique

Several variations of the Lely technique have been developed. The *Norton process* is one of these modifications [69K]. The major difference in the Norton process is that it starts with elemental silicon and carbon and synthesizes the silicon carbide in the crystal growth furnace. This process eliminates the need for preparing polycrystalline feed material. The Norton process, however, is very sensitive to temperature gradients, and best results are obtained for very small gradients.

6.2.3 Characterization of crystal properties

6.2.3.1 Polytype verification and crystal defects

As mentioned in section 6.2.2.2.2, the silicon carbide crystals produced by the Lely technique have the shape of hexagonal platelets, with the *c*-axis perpendicular to the two large faces (basal planes), see Fig. 8a. Since a certain percentage of the crystals is not of the desired 6H structure, a polytype verification is necessary. Some crystals are composed of two (or more) layers with different polytypes (Fig. 8b).

Polytype verification is most readily accomplished by generating *Laue transmission patterns* with X-ray diffraction techniques. These patterns are compared with suitable standards [74T]. By checking all spots it is possible to determine if the sample is a single polytype, and if so which one. If the sample is composed of several polytypes, the *relative proportions of the polytypes* can be estimated. Figs. 9a and b show idealized Laue patterns of the most abundant (6H and 15R) polytypes. Distorted spots are indicative for poor crystal perfection.

Alternatively, the *Buerger precession method* [64B] can be employed; in this case the X-ray beam is parallel to an *a*-axis.

In many cases, a simple visual inspection of the *crystal shape* and the colour allows for a discrimination between different polytypes. 15 R crystals, for instance, are often "distorted", i.e. the platelets exhibit alternating long and short edges. Nitrogen doped 6H crystals are green, whereas 15 R crystals (with the same doping) have a yellowish-green appearance. Another technique is to observe the *luminescence* under ultraviolet light at liquid nitrogen temperature and then again as the sample warms up (*thermoluminescence*).

A discrimination between the cubic, hexagonal, and rhombohedral polytypes can also be performed by microscopic inspection of the shape of *dislocation etch-pits*. These etch-pits are readily formed on the (0001) basal plane (Si face) by molten-salt etching, as described in section 6.2.4.5.2. Triangular etch-pits are characteristic for cubic (3C) material. The etch-pits formed on crystals with hexagonal symmetry (e.g. 6H) have a hexagonal shape, whereas "distorted" hexagons (i.e. hexagons with three long and three short sides) are found on 15R crystals [74F] (see Figs. 32 a) and b)).

Crystals which are composed of several polytypes can be detected by the *oxidation method*. As shown in section 6.2.4.4, there are small differences in the thermal oxidation rates of different polytypes. A microscopic inspection of an oxidized $(1\bar{1}00)$ or $(11\bar{2}0)$ face, therefore, reveals regions of different polytypes [75M1].

6.2.3.2 Electrical properties

Conventional *van der Pauw techniques* are employed to obtain information on the electrical properties (resistivity, carrier concentration, mobility) of as-grown SiC crystals. It is recommended, however, to *remove about* 50 μm *of material* from the surfaces. These layers are grown during the cool-down phase of the Lely process; their properties may deviate from those of the bulk material.

Due to high activation energies (see Vol. III/17a, section 1.5) the donors and acceptors in silicon carbide are incompletely ionized at room temperature. Meaningful results, therefore, can only be obtained by measuring the electrical properties over a wide temperature range. Alloyed contacts produced with Au/Ta (for n-type material) or Au/Ta/Al (for p-type material) can be used for measurements up to 1000 °C. Some examples for the temperature dependence of the electrical conductivity of n- and p-type silicon carbide are shown in Fig. 10. This temperature dependence is mainly due to variations of the carrier concentration, as indicated in Figs. 11a, b. (For additional information, see Vol. III/17a, section 1.5).

6.2.3.3 Luminescence

Additional information about the impurity content and the crystal perfection can be obtained from *photoluminescence* and *cathodoluminescence* measurements, especially at low temperatures. The interpretation of such experiments, however, is quite difficult. The incorporation of traces of nitrogen cannot be avoided in practice. Low-temperature spectra of p-type crystals are generally dominated by donor-acceptor pair-recombination. At room temperature, a broad peak resulting from unidentified impurities is frequently observed. Fig. 12 shows an example of photoluminescent spectra of an Al-doped crystal at 2 K and 300 K.

6.2.4 Device technology

6.2.4.0 Introductory remarks

6.2.4.0.1 General processing steps

The realization of silicon carbide devices can be described in terms of the following process steps:

Synthesis of polycrystalline SiC,
SiC crystal growth,
generation of the p—n junctions,
shaping of the p—n junctions and surface protection,
metallization,
encapsulation.

The processes of SiC synthesis and crystal growth for device applications have been treated in section 6.2.2.

The methods of junction formation in silicon carbide are quite similar to those employed with other semiconductor materials, i.e.:

junction formation during crystal growth	junction formation after crystal growth:
	alloying,
	epitaxy (LPE or VPE),
	diffusion,
	ion implantation.

Doping during crystal growth

In order to prepare crystals with grown junctions, p-type and n-type dopants are sequentially introduced into the atmosphere of a Lely-type furnace. In practice, aluminum (or an aluminum compound) is added directly to the charge during the first part of the furnace run. After the p-type dopant has been depleted, nitrogen is introduced into the furnace atmosphere. Thus, a layer of n-type silicon carbide is formed, which surrounds a central core of p-type silicon carbide.

Alloying

Only a few SiC devices have been produced by means of the alloying technique. SiC tunnel diodes, for instance, can be made in the following manner. Heavily Al-doped crystals (with $p > 10^{20}$ cm^{-3}) are fused to tungsten tabs at a temperature around 1900 °C in order to form ohmic contacts. Small pieces of silicon are then alloyed to the exposed (0001) faces of the silicon carbide at temperatures around 2000 °C in a nitrogen or forming gas atmosphere. The excess silicon can be removed by HF/HNO_3 etching after the alloying cycle [64R].

Diffusion

The fabrication of SiC devices using diffusion techniques requires process temperatures in excess of 1800 °C, and special precautions are necessary to prevent the decomposition of silicon carbide in this temperature range (see section 6.2.4.1). It is not possible to use passivating layers to mask against diffusion, making the fabrication of diffused planar devices very difficult.

Ion implantation

Ion implantation appears to be an attractive method to generate p—n junctions in silicon carbide. However, the annealing behaviour of the implanted layers is not well understood at present time.

Epitaxy

Epitaxial deposition techniques (VPE and LPE) are most widely used in the fabrication of silicon carbide devices, as shown in the following sections.

6.2.4.0.2 Diode- (rectifier)- fabrication

The fabrication of silicon carbide diodes with diffused or epitaxial junctions is shown schematically in Fig. 13. Diffusion (boron or aluminum) creates a p-type surface layer that completely encases the (n-type) crystal (Fig. 13a). The p-type layer is partially ground off to expose the n-type bulk (Fig. 13b). A simple and straightforward technique involves dicing and alloying steps only (Fig. 13c).

Small-area diodes can also be produced by means of a selective etching (mesa) technique, as described in Figs. 13c2 to f2). For this technique, it is advantageous to have the junction close to the (000$\bar{1}$) basal plane, since the oxidation and etching rates are higher than on the opposite face (see sections 6.2.4.4. and 6.2.4.5). An oxidation step in wet oxygen at 1070 °C yields an oxide layer of about 4000 Å on the (000$\bar{1}$) face, Fig. 13c2). The mesa regions are defined by conventional photolithographic techniques, Fig. 13d2). Shaping of the silicon carbide is accomplished by gaseous etching as described in section 6.2.4.5.1. (Fig. 13e2). The choice of the ohmic contacts (Fig. 13f2), depends on the maximum operating temperature of the diodes (see section 6.2.4.6).

Special SiC diode structures have been prepared which are capable of detecting α-particles, and with the addition of a conversion layer, thermal neutrons can be counted [64C].

SiC diodes can also be used for the detection of ultraviolet radiation. A peak response near 2800 Å requires a junction depth of less than 1 μm [67C]. The detectors operate at temperature up to 500 °C and are relatively insensitive to visible light.

6.2.4.0.3 Field-effect transistor fabrication

The sequence of the production steps for *Schottky-barrier field-effect* transistors is shown in Fig. 14. A thin layer of n-type silicon carbide is deposited on the (000$\bar{1}$) face of a p-type SiC crystal by *vapour-phase epitaxy* or *liquid-phase epitaxy* (Fig. 14a). The thickness and the doping level depend on the desired device characteristics. The active device area is defined by a groove which is generated by oxidation, photolithography, and gaseous etching, (see Fig. 14b···d), as described in section 6.2.4.0.2. The ohmic (source and drain) contacts consist of a thin nickel layer which is sintered at 1000 °C in vacuum. A contact to the bulk material is made by alloying with eutectic Al/Si (Fig. 14e). The Schottky-gate contact is generated by depositing titanium and gold (or Ti/Pt/Au), with pattern definition by the lift-off technique or by subtractive etching (Fig. 14f).

A method for the realization of SiC field-effect transistors with junction gate (JFET) has been described by Campbell and Berman [69C].

6.2.4.0.4 Light-emitting diode fabrication

The band-gap of 6H silicon carbide is 2.86 eV at room temperature. It is, therefore, possible to fabricate light-emitting diodes from this material with emission peaks in the blue, green, yellow, and red spectral ranges. In view of the well-established production processes utilizing GaP and $GaAs_xP_{1-x}$ for green, yellow, and red diodes it is obvious that silicon carbide will be required only for *blue-emitting diodes*. This type of LEDs will be discussed here.

Due to the indirect nature of the band structure of silicon carbide, band-to-band transitions are not efficient. Consequently, suitable "activator" impurities have to be introduced. At present, only *nitrogen* and *aluminum* are known to enhance the blue emission from 6H silicon carbide. The concentration of these impurities has to be carefully adjusted in the vicinity of the p—n junction.

A p-type (heavily Al-doped) SiC crystal is used as substrate. Since carrier injection into SiC crystals grown by the Lely (sublimation) technique yields a poor quantum efficiency, it is advantageous to deposit a high-quality *p-type epitaxial layer* prior to the generation of the p—n junction (Fig. 15a). The epitaxial deposition processes (VPE and LPE) are discussed in sections 6.2.4.3. At present, the highest quantum efficiencies (around 10^{-4}) are obtained with LPE diodes [78M2].

The further process steps for the definition of the junction area (Figs. 15 b···d) are comprising oxidation, photolithography, and etching, as already discussed. The top contact consists of a *thin layer of nickel*, which is evaporated through a suitable mask and sintered at 1000 °C. The contact to the substrate material is made by evaporating eutectic Al/Si and alloying at 950 °C (Fig. 15e).

Similar fabrication techniques have been described by Brander and Sutton [69B2] and by Suzuki et al. [76S].

SiC diodes *doped with nitrogen and aluminum* are featuring emission peaks at 2.75 and 2.9 eV if the doping level of the active layer is low or moderate. The relative intensities of these peaks depend on the operating temperature and the current level. Diodes with a high doping level ($p \geqq 5 \cdot 10^{18}$ cm^{-3}) exhibit a broad emission band, centered around 2.6 eV (480 nm).

6.2.4.1 Diffusion

Diffusion of electrically active dopants in silicon carbide requires *temperatures exceeding 1800 °C*. In order to avoid decomposition of the SiC crystals it is necessary to provide a "silicon carbide" atmosphere by a crucible design according to Fig. 16. The space between the graphite walls is filled with *SiC powder* which partly decomposes during the diffusion heating cycle, thus protecting the crystals in the inner porous crucible from decomposition. The diffusing species (boron and/or aluminum) are introduced by mixing them with the SiC protective powder. Nitrogen can be introduced via the gas phase.

The temperature dependence of the diffusion coefficients of nitrogen, boron and aluminum is shown in Figs. 17, 18, and 19. There are major discrepancies in the diffusion data reported by different authors. The diffusion of beryllium proceeds via a "fast" and a "slow" mechanism [68M].

The experimental results are summarized in Table 3.

Table 3. Diffusion coefficients and activation energies of impurities in SiC.

Impurity	D_0 cm^2 s^{-1}	E_a eV	Ref.
N	$4.6 \cdots 8.7 \cdot 10^4$	7.6···9.3	66K2
Be (slow)	32	5.2	68M
Be (fast)	0.3	3.1	68M
B	$1.6 \cdots 10^2$	5.6	66V
Al	1.8	4.9	60C
	0.2	4.9	66V
	8.0	6.1	69M2

Simultaneous diffusion of aluminum and boron into n-type SiC crystals is used in the fabrication process of yellow LEDs [69B3].

6.2.4.2 Ion implantation

Ion implantation is an attractive alternative doping technique, provided the associated radiation damage anneals out at reasonable temperatures and the implanted atoms become electrically active. p—n junctions were formed by *donor (N, P, Sb and Bi) implantations* into p-type SiC crystals with subsequent annealing [74M2]. The implantation of acceptor elements was performed with limited success only, i.e. in most cases high-resistivity layers were formed, even with high ion flux and high annealing temperatures.

Annealing

Most annealing experiments were performed with nitrogen implants. A p-type sample, e.g. doped with aluminum with a concentration of $10^{18}\ cm^{-3}$ implanted with nitrogen at a dose of $10^{15}\ cm^{-2}$ at 25 and 85 keV showed a photoresponse to ultraviolet light after *500 °C, 15-min anneal in a nitrogen atmosphere.* Annealing at *750 °C for 15 min* produced an indication of an n-type layer when tested with a thermal probe. After annealing at *1100 °C for 2 min,* ohmic contacts could be obtained and Hall measurements are possible. In general, good junctions are obtained with annealing at 1400 °C. The carrier mobility approaches bulk values with an *annealing step around 1700 °C*; the number of ionized donors is about one-half of the implanted nitrogen dose.

The results of annealing studies on nitrogen-implanted crystals with He^{+} and He^{2+} ions are summarized in Fig. 20.

6.2.4.3 Epitaxy

6.2.4.3.1 Vapour-phase epitaxy (VPE)

The basic design of a reactor for epitaxial growth of silicon carbide differs little from that used in the silicon epitaxy. In essence, a *purified gas supply*, sources of *silicon* and *carbon containing vapours*, metering valves, and a reaction chamber containing a *heated substrate support* are required. Due to the refractory nature of silicon carbide, however, *high growth temperatures* (i.e. in the 1600 to 1850 °C range) are necessary.

Various source compounds, both gaseous and liquid, are used; the most common ones are listed in Table 4. Hydrogen purified by diffusion through a palladium-silver cell, generally serves as the carrier gas as this is necessary for the reduction of many of the source compounds. Some sources, i.e. silane and methyl-trichlorosilane, can be thermally decomposed and can therefore be employed with an inert carrier gas (argon or helium).

Table 4. Compounds used for vapour-phase epitaxy of silicon carbide.

Source compound(s)	T °C	Ref.
CH_3SiCl_3	1600···1800	65S, 69B1
SiH_4/C_3H_8	1600···1800	65S, 69B1, 74W
SiH_4/C_3H_8	1500···1650	71H
$SiHCl_3/C_6H_{14}$	1650	69T
$SiCl_4/CCl_4$	1700···1750	66C
$SiCl_4/C_3H_8$	1570···1630	75M2
$SiCl_4/C_6H_{14}$	1700···1850	72G, 76M, 77M

The reactions involved in the deposition of single-crystal silicon carbide layers vary according to the source species and the ambient gas. The optimum source concentrations are generally established empirically since theoretical predictions are difficult to make.

In practice, the partial pressure of the source compounds is in the $10^{-5}\cdots10^{-3}$ atm range (1 atm hydrogen pressure); silicon to carbon mole ratios ranging from 1:3.5···5:1 have been successfully employed in different VPE systems.

The quality of the epitaxial layers is, in general, improved with decreasing growth-rate. For a given temperature, there exists an upper limit for 6H single crystal growth on 6H substrates as shown in Fig. 21. *Cubic layers* which exhibit a pronounced tendency for *twinning* and formation of *stacking faults* are obtained at high growth rates.

The quality and the growth-rate of epitaxial layers have also been shown to vary according to the *polarity of the substrate surface.* Under suitable conditions layers of good quality can be grown on (0001) and $(000\bar{1})$ faces; the relative rates may differ by up to a factor of two in either direction, depending on the growth system employed. Some experimental results which have been obtained with a *silane/propane epitaxial deposition* system are summarized in Figs. 22 and 23.

An epitaxial deposition system employing *liquid sources* (silicon tetrachloride and hexane) is shown in Fig. 24. *The optimum growth conditions of the epitaxial deposition system are*:

growth temperature	1800 °C
hydrogen flow rate	3 l/h
silicon tetrachloride flow rate	$1.5\cdot10^{-5}\cdots3\cdot10^{-5}$ mol/min
hexane flow rate	$0.8\cdot10^{-5}\cdots1.6\cdot10^{-5}$ mol/min
silicon-to-carbon ratio	1:2.5
growth rate	0.05···0.1 μm/min.

Various sources for the introduction of electrically active impurities have been used. The most important sources are listed in Table 5.

Table 5. Electrically active impurities in vapour growth of SiC.

Donors		Acceptors	
Dopant	Ref.	Dopant	Ref.
N_2	66C, 69T, 74W, 76M	B_2H_6	66C, 69B1, 69T, 72G, 74W, 76M
NH_3	76M	BBr_3	65S
PH_3	66C	$AlCl_3$	69B1, 75M2
PCl_3	65S	$Al(CH_3)_3$	77M

6.2.4.3.2 Liquid-phase epitaxy (LPE)

By analogy to the III–V compounds it may be expected that SiC layers grown from a saturated silicon/carbon melt exhibit superior properties in terms of radiative recombination. It is likely that the silicon solution getters many of the impurities which cannot be eliminated from other systems. On the other hand, it is possible to dope melts with large percentages of elements which are difficult to incorporate in SiC layers grown from the vapour phase or which are difficult to obtain in a convenient volatile form (e.g. aluminum).

Since growth can be performed at temperatures substantially below the diffusion temperature for electrically active dopants, *abrupt p—n junctions can be readily fabricated.* A general disadvantage of the technique is the low solubility of carbon in silicon.

An apparatus for the epitaxial growth of silicon carbide is shown in Fig. 25. The substrate is clamped to the bottom of a graphite crucible containing the silicon melt. The crucible, the heating element and the heat shield are adjusted in such a manner that the crucible wall is maintained at a higher temperature than the seed crystal (temperature gradient about 10···30 °C/cm). Thus, carbon is dissolved from the crucible wall, whereas a silicon carbide layer grows on the substrate. Impurity atoms can be added to the melt or introduced via the gas phase.

A temperature-time program of a growth run for the production of "overcompensated" LEDs is shown in Fig. 26. A *degassing and heat-up cycle* (A) is followed by a *growth period* (B) at 1600 °C; p-type material is grown during this period, since the silicon melt is doped with aluminum. Lateron, the melt is *cooled* at a rate of about 15 °C/min and an *increasing pressure of nitrogen* is established within the system. Thus, a p-type layer which is partially compensated with nitrogen donors is deposited during the growth period (C). Finally an n-type layer is formed, growth period (D). Further slow cooling is recommended. The silicon carbide wafer is recovered by opening the crucible and dissolving the residual silicon in an HF/HNO_3 solution.

A similar technique has been described by Brander and Sutton [69B2] whereas Suzuki et al. developed a *vertical dipping method* [75S].

6.2.4.4 Thermal oxidation

Silicon carbide can be oxidized in a manner similar to that used for the growth of silicon dioxide layers on silicon.

Silicon dioxide layers on silicon carbide are used for various purposes in the technology of silicon carbide devices.

Fig. 27 shown the thickness of the oxide grown on the $(000\bar{1})$ face of 6H silicon carbide by a *dry oxidation process* as a function of the oxidation time. As seen from Fig. 27 a square-root oxidation law is followed, when a long oxidation cycle is applied (i.e. the *oxidation rate is diffusion-limited*).

The time-dependence of the oxide thickness on the (0001) face is shown in Fig. 28. (dry oxidation process). The time-dependence of the oxide thickness on both (0001) and $(000\bar{1})$ faces by a wet oxydation process is shown in Fig. 29.

The oxidation rate depends on the surface polarity, the conductivity (dopant) type, the doping level and the polytype.

Values of oxide thickness obtained with wet oxidation processes (1070 °C, 6 hours; oxygen, saturated with water vapour at 90 °C) are listed in Table 6 (see also Fig. 29).

Table 6. Oxide thickness obtained by wet oxidation [75M1].

	6H, n-type lightly doped	6H, n-type heavily doped	6H, p-type	15R, p-type
$(000\bar{1})$	2560 Å	2730 Å	2420 Å	2200 Å
(0001)	300 Å	310 Å	450 Å	420 Å

The differences of the oxidation rates of opposite crystal faces are less pronounced at high oxidation temperatures (see Figs. 27 and 28).

6.2.4.5 Surface preparation

6.2.4.5.0 Introductory remarks

Several ways of surface preparation are described in the following sections. The main objectives of the surface treatment are:

Polarity determination, i.e. discrimination between the (0001) and $(000\bar{1})$ basal planes of the 6H platelets,
detection of dislocations and grain boundaries,
preparation of smooth and undisturbed surfaces prior to the epitaxial deposition of SiC layers,
junction delineation,
surface structuring (mesa-etching),
surface protection.

Polarity determination

The *polarity discrimination* of 6 H platelets can be performed by X-ray techniques [65B]. It is, however, simpler to use oxidation or etching methods (Figs. 33a and b). As indicated in section 6.2.4.4 the *oxidation rate* of the $(000\bar{1})$ surface (carbon face) is much higher than that of the (0001) surface (silicon face) (see Fig. 29).

Alternatively, the polarity discrimination can be performed by etching in *molten sodium or potassium hydroxide* (e.g. 500 °C, 1 min). With these etching conditions, the (0001) basal plane remains smooth, whereas the $(000\bar{1})$ plane shows a dull ("wormy") appearance (Figs. 33a and b). Similar results are obtained with other molten salt etchants (e.g. K_2CO_3/KOH, $NaOH/Na_2O_2$), and with a saturated aqueous solution of $K_3Fe(CN)_6/NaOH$ (1:1) at 110 °C.

Polishing

Silicon carbide can be *lapped and polished* by means of extremely *hard abrasives*, i.e. *boron carbide*, *cubic boron nitride*, and *diamond*. An *in-situ etching* in *pure hydrogen* is often performed prior to vapour-phase epitaxial deposition processes.

Junction delineation

Most p—n junctions can be revealed by the *oxidation method*, due to small differences in the oxidation rates of p- and n-type material (see section 6.2.4.4). *Electrolytic etching* in a 2:1 mixture of alcohol and 40% HF is used for the delineation of p^+—p and n^+—n junctions and for those p—n junctions in which both sides are lightly doped.

The contrast of the electrolytic delineation can be markedly improved by subsequent oxidation [75M1].

Surface structuring

Surface structuring (mesa formation) is achieved by a combination of oxidation, photolithography and gaseous etching, as described in section 6.2.4.0.2.

6.2.4.5.1 Gaseous etching

Hydrogen etching of silicon carbide in the *1600···1850 °C* temperature range is often used for surface cleaning prior to vapour-phase epitaxial growth. The rate of attack on the (0001) plane is slightly higher than that on the $(000\bar{1})$ plane (see Fig. 30).

Fluorine and *chlorotrifluoride* attack silicon carbide at relatively low temperatures (200···400 °C). However, both these gases are highly reactive to many other materials [69W].

A preferred technique is the use of *gaseous etching* with *chlorine/oxygen* or *chlorine/oxygen/argon mixtures* in the 800···1100 °C temperature range. The optimum concentration for device applications was found to be (by volume): 68% Ar, 26% Cl_2, 6% O_2 [69C]. The etching rate on the (000$\bar{1}$) face is higher than on the (0001) face (Fig. 31).

Silicon dioxide is *not attacked* by the chlorine/oxygen mixture at 1000 °C. By using the *conventional photoresist technique* on an oxidized silicon carbide surface, a precision configuration can be etched into the SiO_2 layer and in turn a corresponding configuration can be formed in silicon carbide by the gaseous chlorine/oxygen etching.

6.2.4.5.2 Molten salt etching

Silicon carbide is a highly refractory material and shows considerable resistance to chemical attack, i.e. it cannot be etched by aqueous (acid, neutral or basic) solutions at room temperature. Reactions of silicon carbide in *molten salts*, therefore, have been widely used for surface treatment.

In general, the solid salt (or salt mixture) is placed in a crucible which is heated in a *vertical muffle furnace* to the required temperature. The SiC crystals are then added. After the required etching time the molten salt is poured off. Finally, the crystals are cleaned in strong acid solutions.

A variety of chemicals can be used for molten salt etching of silicon carbide:

NaOH Na_2CO_3 Na_2SO_4 $NaNO_3$ Na_2O_2 $Na_2B_4O_7$ (Borax) KOH K_2CO_3 K_2SO_4 KNO_3.

The etching reaction requires the *presence of oxygen*. Oxygen is provided by the decomposition of a constituent of the molten salt, e.g. *sodium peroxide*. In other cases, e.g. sodium hydroxide, oxygen from the air dissolved in the molten salt is effective. It has been found advantageous to use a mixture of molten salts to lower the temperature required for etching and to increase the fluidity of the melt.

The etching of silicon carbide in molten salts is performed in the *400···900 °C* temperature range. The attack by molten salts is chiefly on the *(000$\bar{1}$) face*. Table 7 shows the effect of temperature on the etching rate for a 3:1 mixture of sodium hydroxide and sodium peroxide.

Table 7. Etch rate r_e of SiC on (000$\bar{1}$) face by a mixture of [NaOH]:[Na_2O_2]=3:1 [69J].

T °C	r_e µm/min
600	1
700	7
750	14
800	24

Molten salt etching of silicon carbide is mainly used to reveal *dislocations* and *grain boundaries*. As mentioned in section 6.2.3.1, the inspection of etch pits yields information on the polytype character of SiC. (see Figs. 32a and b).

The literature on surface preparation and etching up to 1972 has been compiled by [74F].

6.2.4.5.3 Electrolytic etching

Electrolytic (anodic) dissolution of silicon carbide is the only etching process that may be carried out at *room temperature*. It is mainly used for *junction delineation*, since the etching is *primarily on* the *p-type* material. Other applications include the *cleaning* of junction diodes.

Most electrolytic etching experiments were performed with an electrolyte consisting of a dilute solution of *hydrofluoric acid in water*, *methanol* or *glycol*. For further details and other solutions, see [67B2, 69J, 74F].

6.2.4.6 Contact fabrication

Various materials have been employed to provide electrical (ohmic) contacts to silicon carbide. In view of the desired mechanical bonding strength, the contact material should contain a *transition metal* which is capable of *forming a carbide compound*. The addition of silicon can also serve for this purpose. The choice of the contact material must also take into account the intended temperature limit for device operation.

The most widely used contact material is the *gold-tantalum alloy* (with 2···10% Ta), which is alloyed to silicon carbide *at* approximately *1400 °C*. A small *addition of aluminum* is recommended for contacts to p-type SiC. Gold-based contacts can be *operated up to 1000 °C*.

Most contacts to the p-type regions of SiC devices are fabricated by alloying with the *Al/Si eutectic* (12% Si) in the temperature range of *600···1200 °C*. Operation of these contacts is limited to temperatures *below 550 °C*.

Pure tungsten can also be used as contact material for n- and p-type silicon carbide. The *alloying* is performed *at 1900 °C* [58H]. Tungsten is the only metal whose coefficient of *thermal expansion matches* that of silicon carbide over a wide temperature range. In high-temperature operation, the tungsten contacts *must be protected* from the (oxidizing) ambient.

Ohmic contacts to thin layers of n- or p-type silicon carbide are provided by *evaporation (or sputtering) of tungsten* and a subsequent *sintering process at 1700 °C*. Good ohmic contacts to n-type layers can be formed with *nickel* (e.g. 1500 Å), the required sintering temperature being in the *900···1000 °C range*.

References for 6.2

58H Hall, R.N.: J. Appl. Phys. **29** (1958) 914.

59A Adamski, R.F., Merz, K.M.: Z. Kristallogr. **11** (1959) 350.

59S Scace, R.I., Slack, G.A.: J. Chem. Phys. **30** (1959) 1551.

60C Chang, H.C., Le May, Ch.Z., Wallace, L.F.: Silicon Carbide – A High Temperature Semiconductor (eds. J.R. O'Connor and J. Smiltens), Pergamon Press, Oxford, London, New York, Paris **1960**, p. 496.

60E Ellis, R.C.: Silicon Carbide – A High Temperature Semiconductor (eds. J.R. O'Connor and J. Smiltens), Pergamon Press, Oxford, London, New York, Paris **1960**, p. 124.

60H Halden, F.A.: Silicon Carbide – A High Temperature Semiconductor (eds. J.R. O'Connor and J. Smiltens), Pergamon Press, Oxford, London, New York, Paris **1960**, p. 115.

60L Landolt-Börnstein, Zahlenwerte und Funktionen, 6th Edition, Vol. II, 2a, Springer Verlag, Berlin **1960**, p. 44.

60S Susman, S., Spriggs, R.S., Weber, H.S.: Silicon Carbide – A High Temperature Semiconductor (eds. J.R. O'Connor and J. Smiltens), Pergamon Press, Oxford, London, New York, Paris **1960**, p. 94.

63K Knippenberg, W.F.: Philips Res. Repts. **18** (1963) 161.

64B Buerger, M.J.: The Precession Method in X-ray Crystallography, J. Wiley and Sons, New York, London, Sydney **1964**.

64C Canepa, P.C., Malinaric, P., Campbell, R.B., Ostroski, J.W.: IEEE Trans. Nuclear Sci. **NS-11** (1964) 262.

64G Griffiths, L.B., Mlavski, A.I.: J. Electrochem. Soc. **111** (1964) 805.

64R Rutz, R.F.: IBM J. Res. Dev. **8** (1964) 539.

65B Brack, K.: J. Appl. Phys. **36** (1965) 3560.

65C Chu, T.L., Campbell, R.B.: J. Electrochem. Soc. **112** (1965) 955.

65J Johnson, A.: RCA Rev. **26** (1965) 163.

65S Spielmann, W.: Z. Angew. Phys. **19** (1965) 93.

65W Wright, M.A.: J. Electrochem. Soc. **112** (1965) 1114.

66B Barrett, D.L.: J. Electrochem. Soc. **113** (1966) 1215.

66C Campbell, R.B., Chu, T.C.: J. Electrochem. Soc. **113** (1966) 825.

66K1 Knippenberg, W.F., Verspui, G.: Philips Res. Repts. **21** (1966) 113.

66K2 Kroko, L.J., Milnes, A.G.: Solid State Electron. **9** (1966) 1125.

66N Nelson, W.E., Halden, F.A., Rosengreen, A.: J. Appl. Phys. **37** (1966) 333.

66V Vodakov, Yu.A., Mokhov, E.N., Reifman, M.B.: Fiz. Tverd. Tela Sov. Phys. Solid State (English Transl) **8** (1966) 1040.

67B1 Bartlett, R.W., Nelson, W.E., Halden, F.A.: J. Electrochem. Soc. **114** (1967) 1149.

67B2 Brander, R.W., Boughey, A.L.: Brit. J. Appl. Phys. **18** (1967) 905.

67C Campbell, R.B., Chang, H.C.: Solid-State Electron. **10** (1967) 949.

68M Maslakovets, Yu.P., Mokhov, E.N., Vodakov, Yu.A., Lomakina, G.A.: Fiz. Tverd. Tela **10** (1968) 809; Sov. Phys. Solid State (English Transl.) **10** (1968) 634.

68N Nieberlein, V.A.: SAMPE J., Oct./Nov. **1968**, 72.

69B1 Bartlett, R.W., Mueller, R.A.: Mater. Res. Bull. **4** (1969) 341.

69B2 Brander, R.W., Sutton, R.P.: Brit. J. Appl. Phys. **2** (1969) 309.

69B3 Blank, J.M.: Mater. Res. Bull. **4** (1969) 186.

69C Campbell, R.B., Berman, H.S.: Mater. Res. Bull. **4** (1969) 2.

69J Jennings, V.J.: Mater. Res. Bull. **4** (1969) 199.

69K Kamath, G.S.: Mater. Res. Bull. **4** (1969) 57.
69M1 Marshall, R.C.: Mater. Res. Bull. **4** (1969) 73.
69M2 Mokhov, E.N., Vodakov, Yu.A., Lomakina, G.A.: Soviet Phys. Solid State (English Transl.) **11** (1969) 415. Fiz. Tverd. Tela **11** (1969) 519.
69P Powell, J.A.: J. Appl. Phys. **40** (1969) 4660.
69T Todkill, A., Brander, R.W.: Mater. Res. Bull. **4** (1969) S 293.
69W Wolff, G.A., Das, B.N., Lamport, C.B., Mlavski, A.I.: Mater. Res. Bull. **4** (1969) S 67.
70K Kapteyns, C.J., Knippenberg, W.F.: J. Crystal Growth **7** (1970) 20.
71H Harris, J.M., Gatos, H.C., Witt, A.F.: J. Electrochem. Soc. **118** (1971) 335, 338.
72G Gramberg, G., Königer, M.: J. Crystal Growth **9** (1971) 309.
72K van Kemenade, A.W.C., Stemfoort, C.F.: J. Crystal Growth **12** (1972) 13.
72P Potter, R.M., Sattele, J.H.: J. Crystal Growth **12** (1972) 245.
73G Gillessen, K., v. Münch, W.: J. Crystal Growth **19** (1973) 263.
74C Campbell, R.B., Mitchell, J.B., Shewchun, J., Thompson, D.A., Davies, J.A.: Silicon Carbide – 1973 (eds. R.C. Marshall, J.W. Faust, C.E. Ryan), University of South Carolina Press, Columbia, S.C., **1974**, p. 486.
74F Faust, J.W., Liaw, H.M.: Silicon Carbide – 1973 (eds. R.C. Marshall, J.W. Faust, C.E. Ryan) University of South Carolina Press, Columbia, S.C., **1974**, p. 657.
74H Harris, R.C.A., Call, R.L.: Silicon Carbide – 1973 (eds. R.C. Marshall, J.W. Faust, C.E. Ryan) University of South Carolina Press, Columbia, S.C., **1974**, p. 329.
74K Keyes, R.W.: Silicon Carbide – 1973 (eds. R.C. Marshall, J.W. Faust, C.E. Ryan), University of South Carolina Press, Columbia, S.C., **1974**, p. 534.
74M1 v. Münch, W., Gillessen, K.: Silicon Carbide – 1973 (eds. R.C. Marshall, J.W. Faust, C.E. Ryan), University of South Carolina Press, Columbia, S.C., **1974**, p. 51.
74M2 Marsh, O.J.: Silicon Carbide – 1973 (eds. R.C. Marshall, J.W. Faust, C.E. Ryan), University of South Carolina Press, Columbia, S.C., **1974**, p. 471.
74P Pellegrini, P.W., Feldman, J.M.: Silicon Carbide – 1973 (eds. R.C. Marshall, J.W. Faust, C.E. Ryan), University of South Carolina Press, Columbia, S.C., **1974**, p. 161.
74T Tung, Y., Faust, J.W.: Silicon Carbide – 1973 (eds. R.C. Marshall, J.W. Faust, C.E. Ryan), University of South Carolina Press, Columbia, S.C., **1974**, p. 254.
74W Wessels, B., Gatos, H.C., Witt, A.F.: Silicon Carbide – 1973 (eds. R.C. Marshall, J.W. Faust, C.E. Ryan) University of South Carolina Press, Columbia, S.C., **1974**, p. 25.
75M1 v. Münch, W., Pfaffeneder, I.: J. Electrochem. Soc. **122** (1975) 642.
75M2 Matsunami, H., Nishino, S., Odaka, M., Tanaka, T.: J. Crystal Growth **31** (1975) 72.
75S Suzuki, A., Ikeda, M., Matsunami, H., Tanaka, T.: J. Electrochem. Soc. **122** (1975) 1741.
76M v. Münch, W., Pfaffeneder, I.: Thin Solid Films **31** (1976) 39.
76S Suzuki, A., Ikeda, M., Nagano, N., Matsunami, H., Tanaka, T.: J. Appl. Phys. **47** (1976) 4546.
77M v. Münch, W.: J. Electron. Mater. **6** (1977) 449.
77P Pettenpaul, E.: Dissertation (thesis) Technische Universität Hannover, **1977**, p. 63.
78M1 v. Münch,W., Pettenpaul, E.: J. Electrochem. Soc. **125** (1978) 294.
78M2 v. Münch, W., Kürzinger, W.: Solid-State Electron. **21** (1978) 1129.
79B Burgemeister, E.A., v. Münch, W., Pettenpaul, E.: J. Appl. Phys. **50** (1979) 5790.

Figures

6 Tetrahedrally bonded semiconductors

6.1 Silicon and germanium

6.1.1 Technological data

For Fig. 1, see next page.

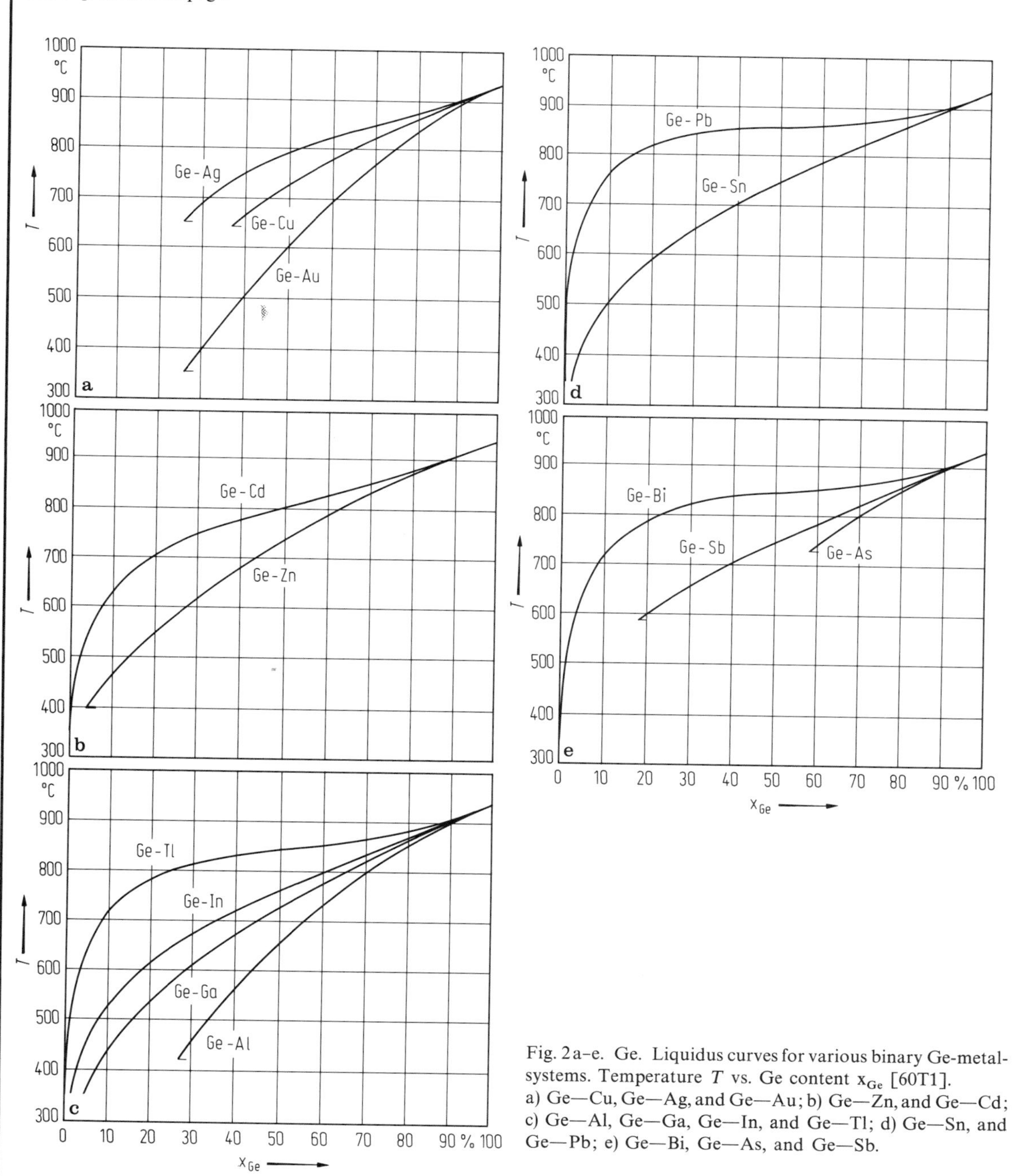

Fig. 2a–e. Ge. Liquidus curves for various binary Ge-metal-systems. Temperature T vs. Ge content x_{Ge} [60T1].
a) Ge—Cu, Ge—Ag, and Ge—Au; b) Ge—Zn, and Ge—Cd; c) Ge—Al, Ge—Ga, Ge—In, and Ge—Tl; d) Ge—Sn, and Ge—Pb; e) Ge—Bi, Ge—As, and Ge—Sb.

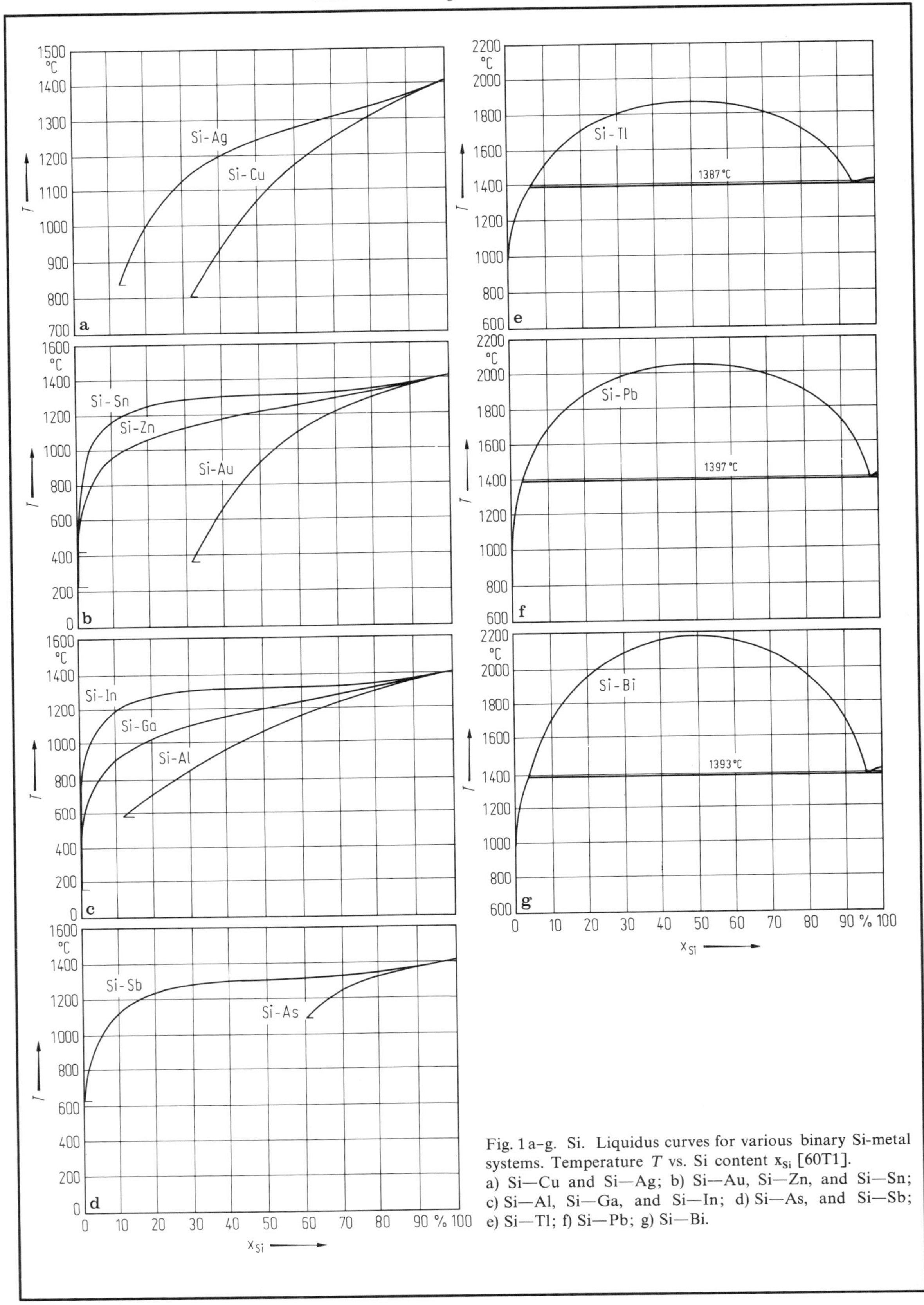

Fig. 1 a–g. Si. Liquidus curves for various binary Si-metal systems. Temperature T vs. Si content x_{Si} [60T1].
a) Si—Cu and Si—Ag; b) Si—Au, Si—Zn, and Si—Sn; c) Si—Al, Si—Ga, and Si—In; d) Si—As, and Si—Sb; e) Si—Tl; f) Si—Pb; g) Si—Bi.

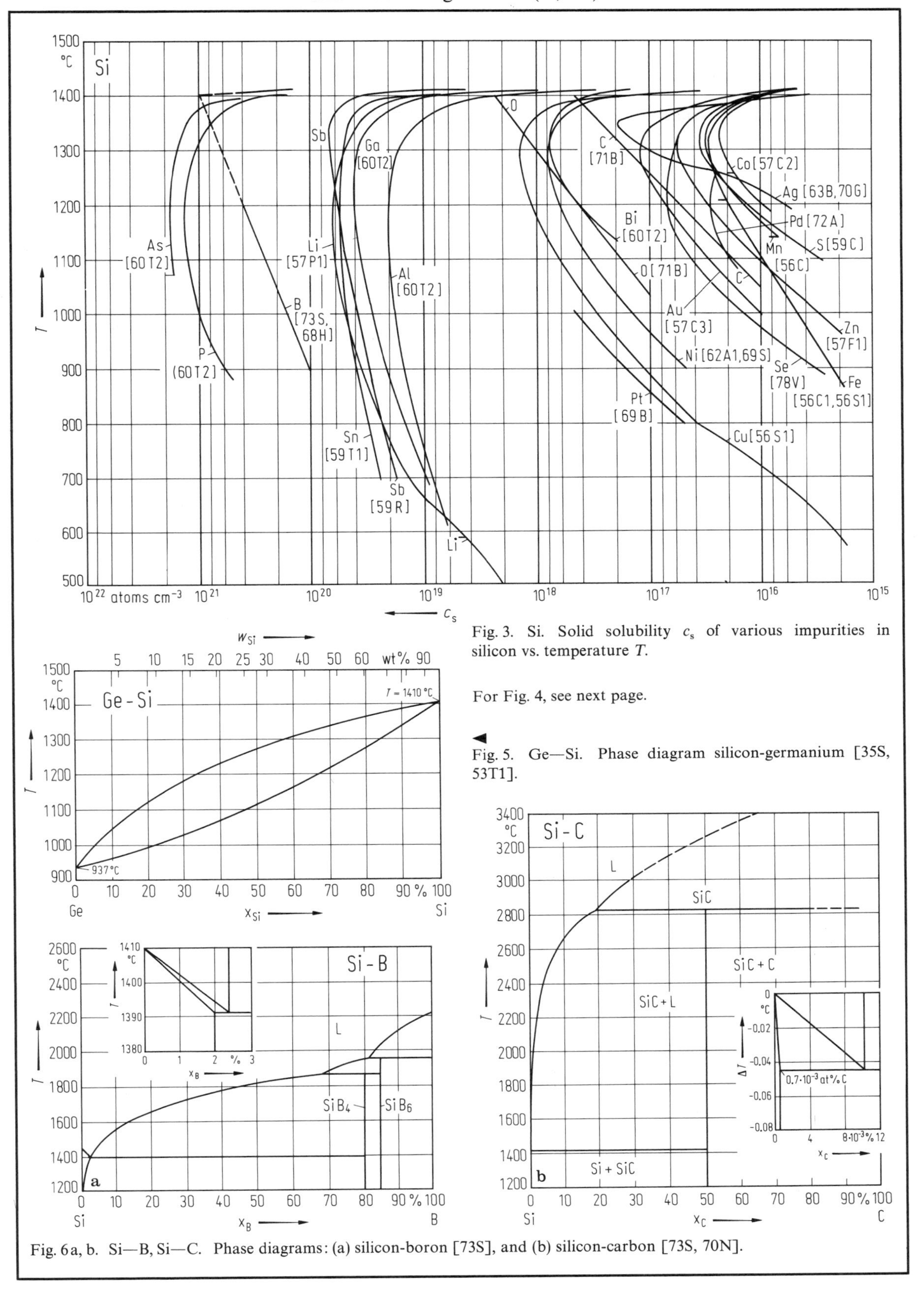

Fig. 3. Si. Solid solubility c_s of various impurities in silicon vs. temperature T.

For Fig. 4, see next page.

◄ Fig. 5. Ge—Si. Phase diagram silicon-germanium [35S, 53T1].

Fig. 6a, b. Si—B, Si—C. Phase diagrams: (a) silicon-boron [73S], and (b) silicon-carbon [73S, 70N].

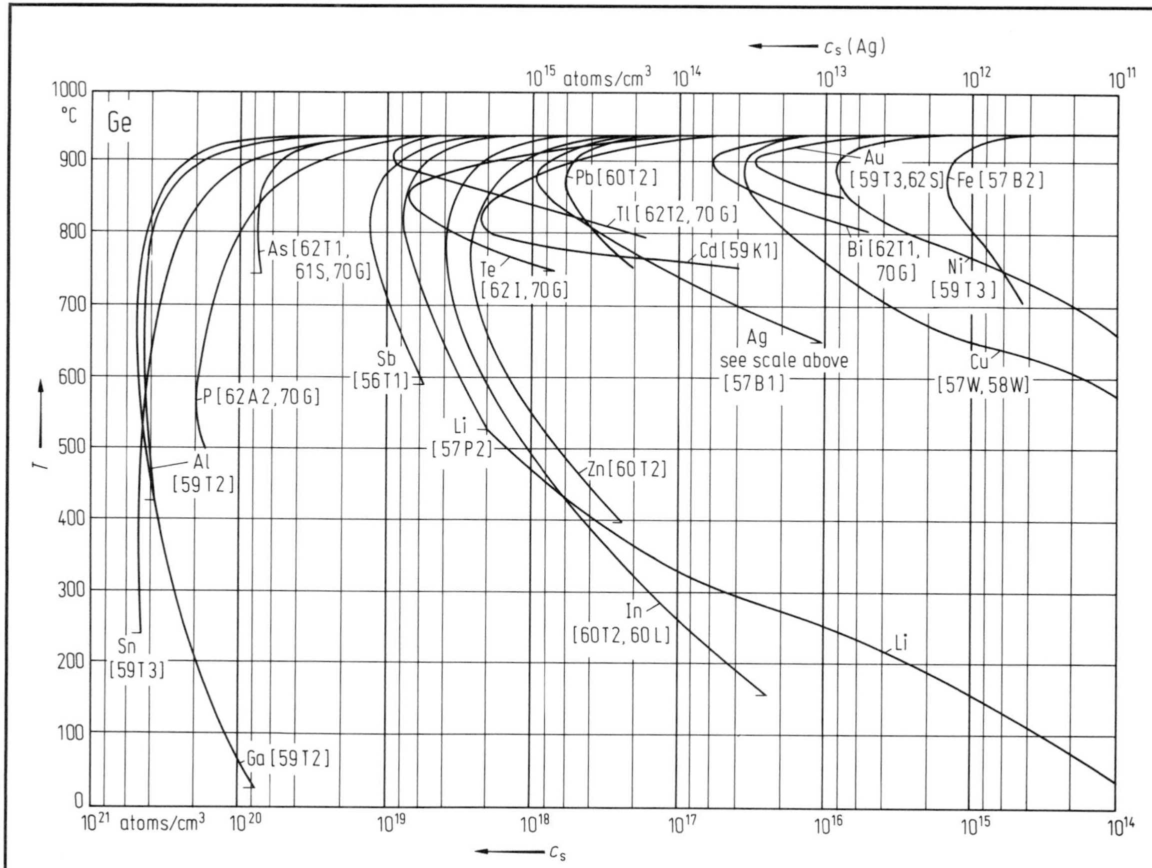

Fig. 4. Ge. Solid solubility c_s of various impurities in germanium vs. temperature T.

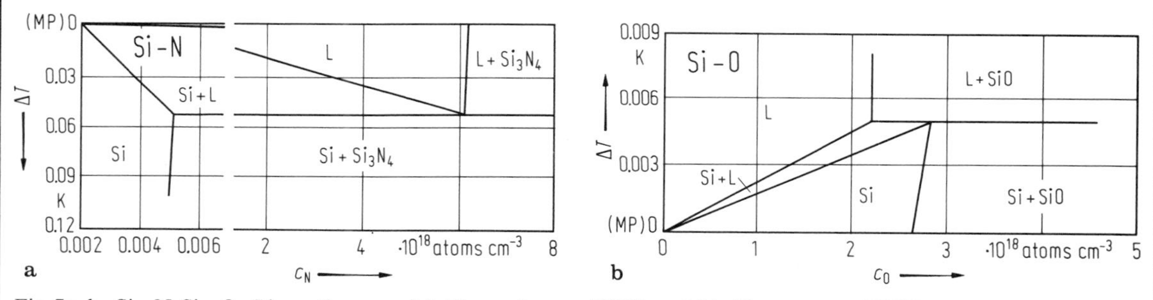

Fig. 7 a, b. Si—N, Si—O. Phase diagrams: (a) silicon-nitrogen [73Y], and (b) silicon-oxygen [73Y].

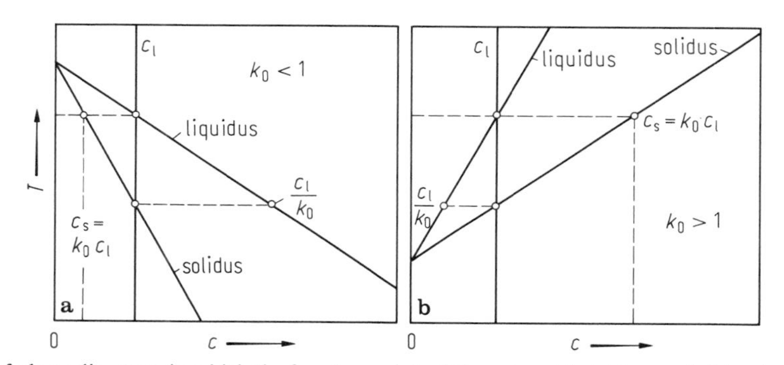

Fig. 8 a, b. Portions of phase diagrams in which the freezing point of the solvent is (a) lowered, (b) raised by the solute [66P].

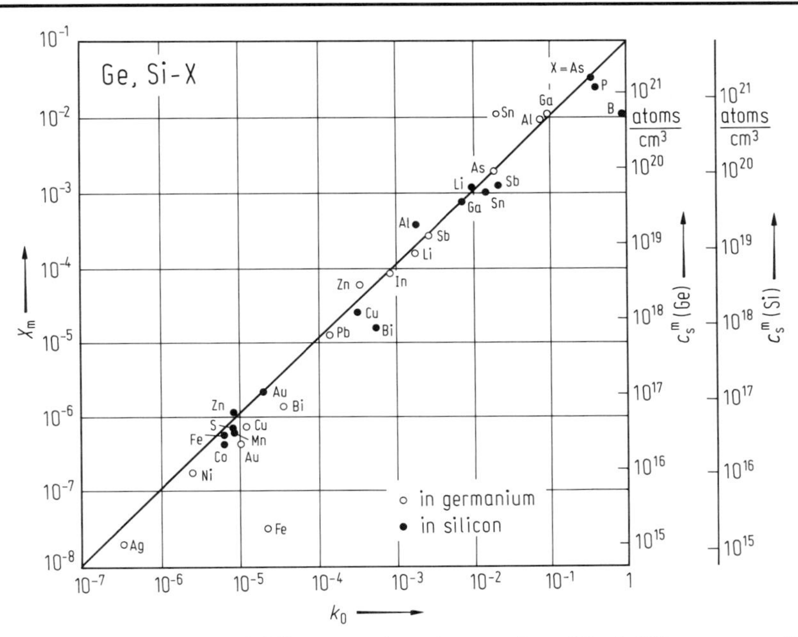

Fig. 9. Ge—X, Si—X. Maximum solid solubility c_s^m, and maximum molar solid solubility x_m, vs. equilibrium distribution coefficient k_0 at the melting point of impurities X in silicon and germanium [62F].

For Figs. 10⋯14, see following pages.

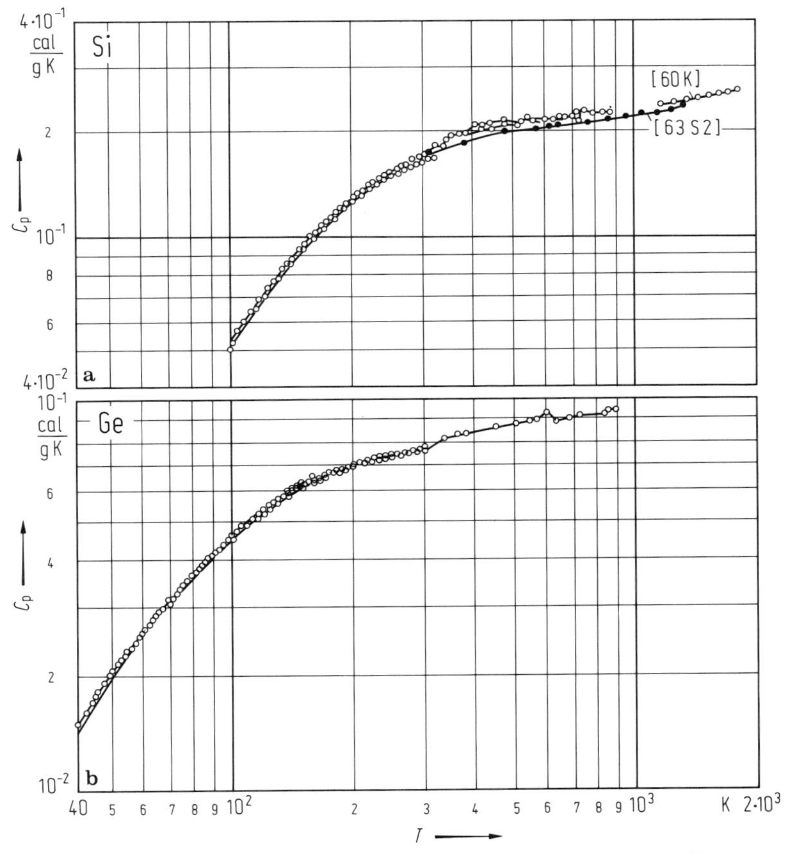

Fig. 15 a, b. Si, Ge. Specific heat C_p vs. temperature T of (a) silicon and (b) germanium [70T1].

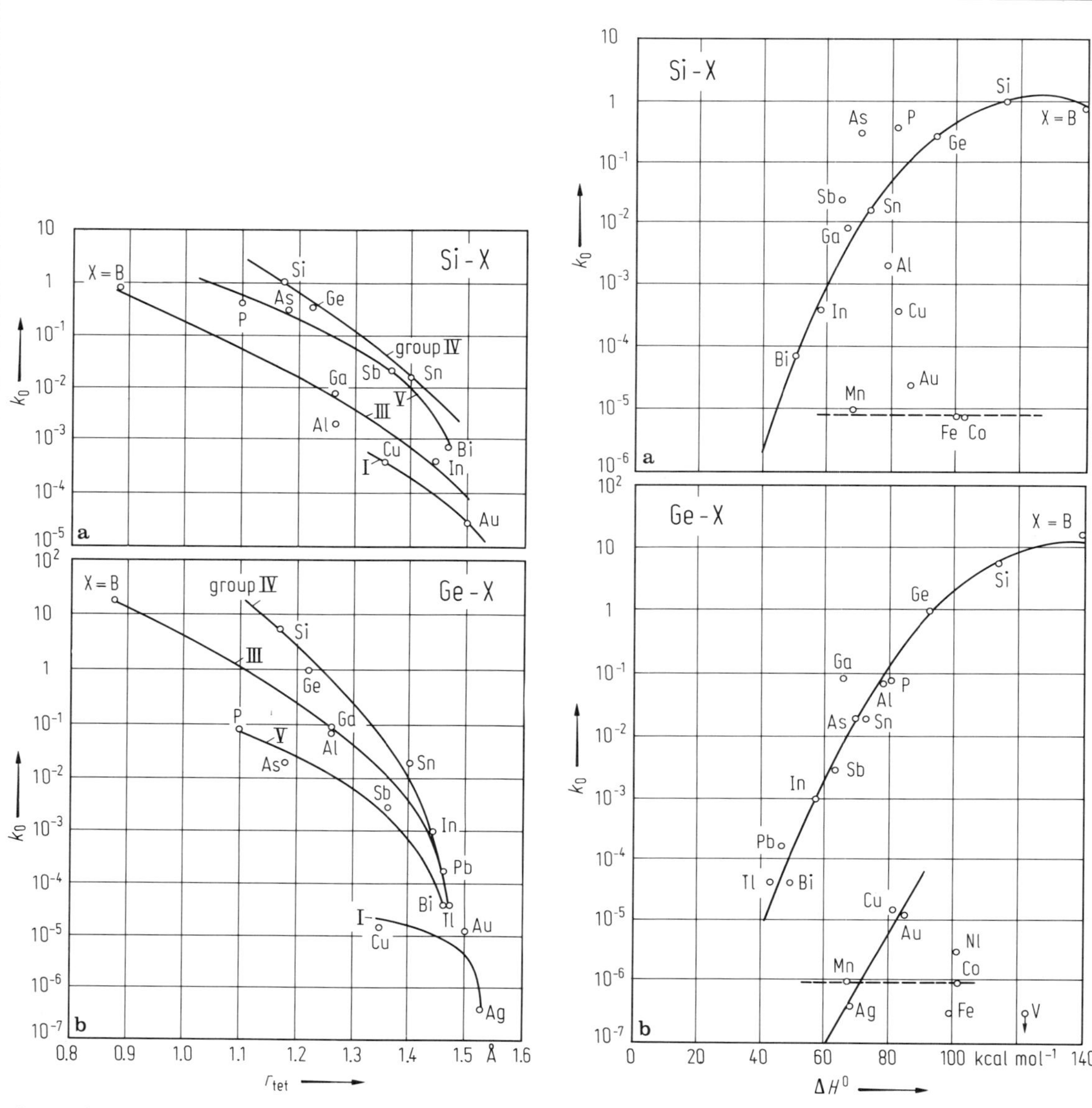

Fig. 10a, b. Si—X, Ge—X. Equilibrium distribution coefficient k_0 at the melting point vs. tetrahedral covalent radius r_{tet} of impurities X in (a) silicon and (b) germanium [60T2].

Fig. 11a, b. Si—X, Ge—X. Equilibrium distribution coefficient k_0 at the melting point vs. heat of sublimation ΔH^0 (at 298 K) for impurities X in (a) silicon and (b) germanium [60T2].

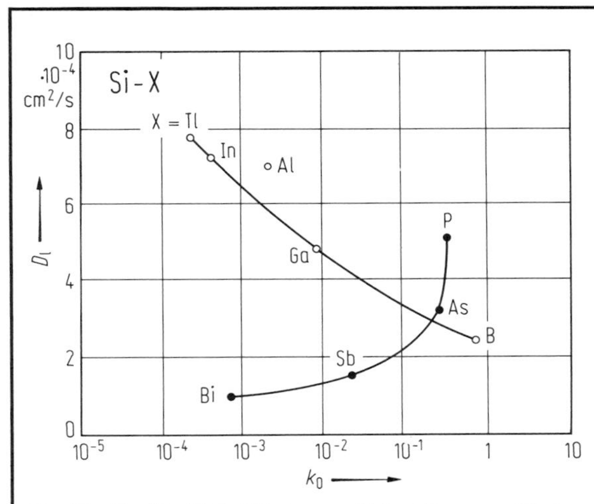

Fig. 12. Si—X. Melt diffusion coefficient D_l vs. equilibrium distribution coefficient k_0 of several impurities X in silicon [81K].

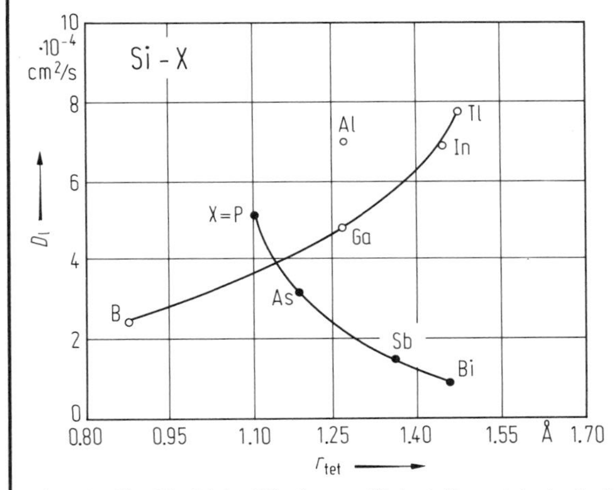

Fig. 13. Si—X. Melt diffusion coefficient D_l vs. tetrahedral covalent radius r_{tet} of several dopants X in silicon [81K].

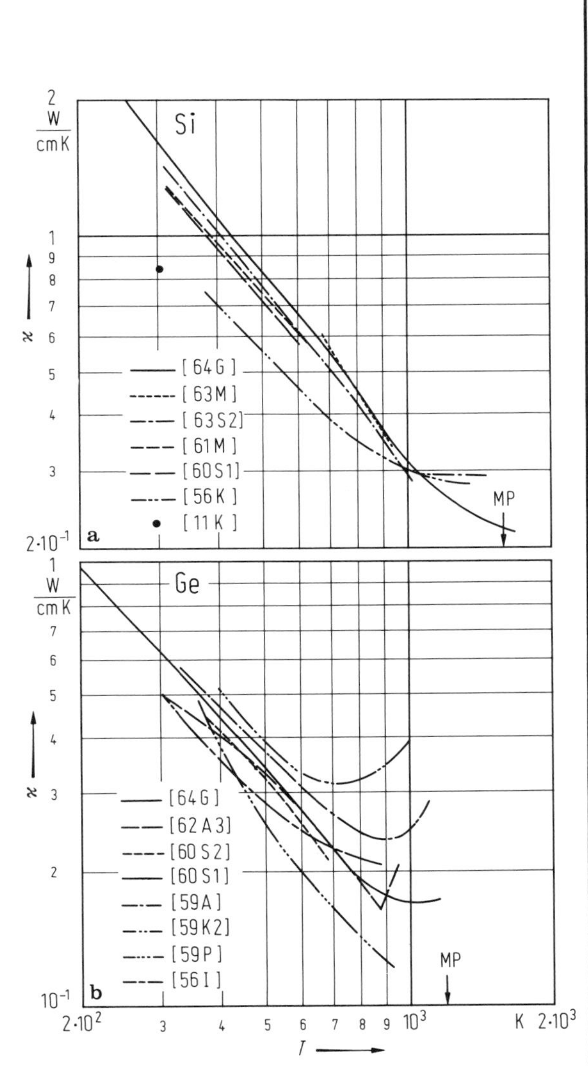

Fig. 14a, b. Si, Ge. High-temperature thermal conductivity κ vs. temperature T of (a) silicon and (b) germanium [64G]. The arrow MP marks the melting temperature.

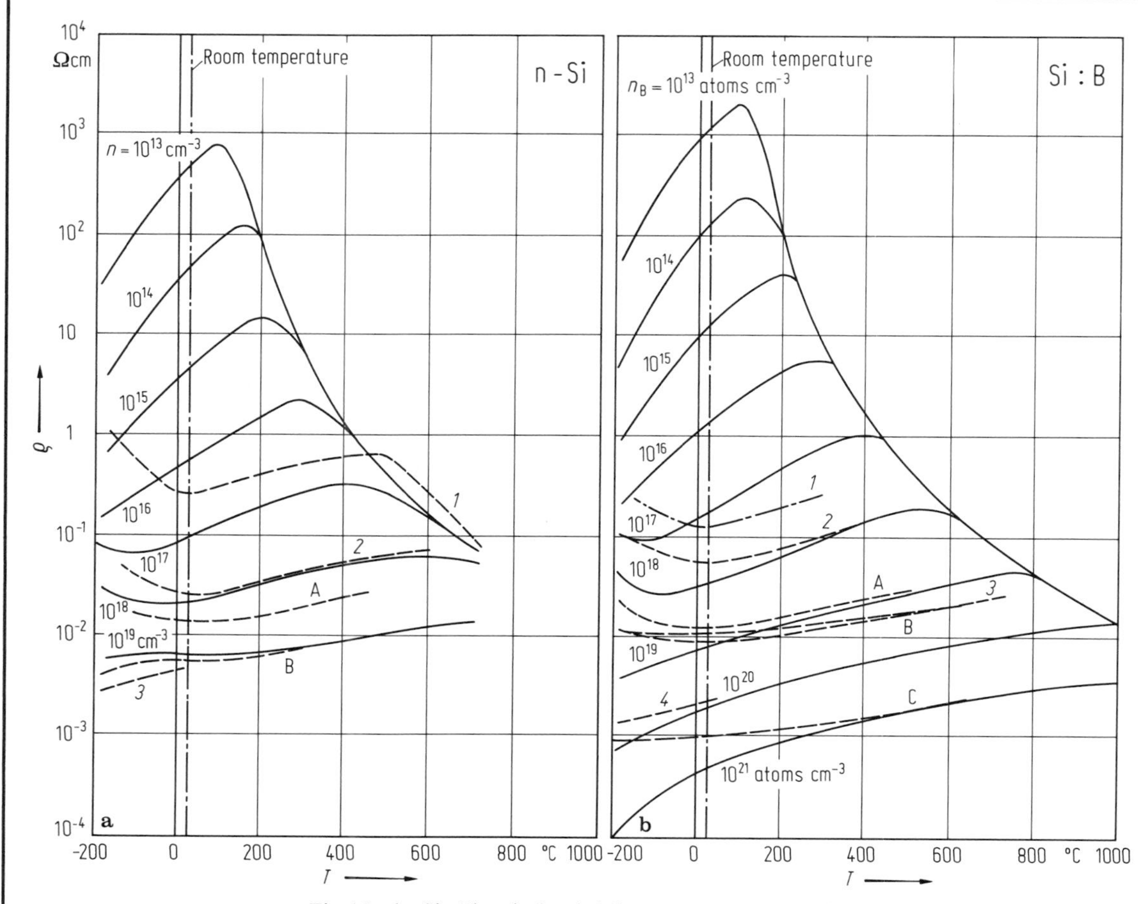

Fig. 16a, b. Si. Electrical resistivity ϱ vs. temperature T for (a) n-type silicon and (b) boron-doped silicon, for various concentrations in atoms cm^{-3} [65R]. Solid lines are calculated for the concentrations indicated. Dashed lines are experimental:

a) A: $1.7 \cdot 10^{18}$ atoms cm^{-3} arsenic; B: $1.1 \cdot 10^{19}$ atoms cm^{-3} phosphorus [63C]; *1*: $1 \cdot 10^{17}$ atoms cm^{-3} phosphorus; *2*: $1.3 \cdot 10^{18}$ atoms cm^{-3} phosphorus; *3*: $1.7 \cdot 10^{19}$ atoms cm^{-3} phosphorus [49P].

b) A: $6 \cdot 10^{18}$ atoms cm^{-3}; B: $1 \cdot 10^{19}$ atoms cm^{-3}; C: $1 \cdot 10^{20}$ atoms cm^{-3} [63C]. *1*: $6 \cdot 10^{17}$ atoms cm^{-3}, *2*: $1.3 \cdot 10^{18}$ atoms cm^{-3}, *3*: $1.4 \cdot 10^{19}$ atoms cm^{-3}, *4*: $1.2 \cdot 10^{20}$ atoms cm^{-3} [49P].

6.1.2 Crystal growth

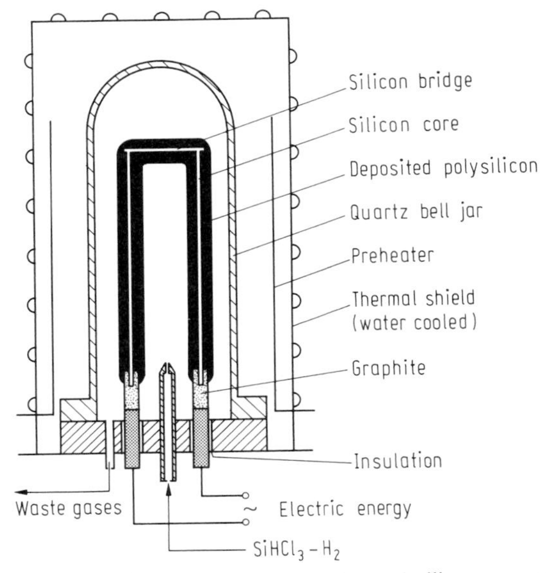

Fig. 1. Si. Schematical drawing of a polysilicon reactor to illustrate the principle of operation [79S1].

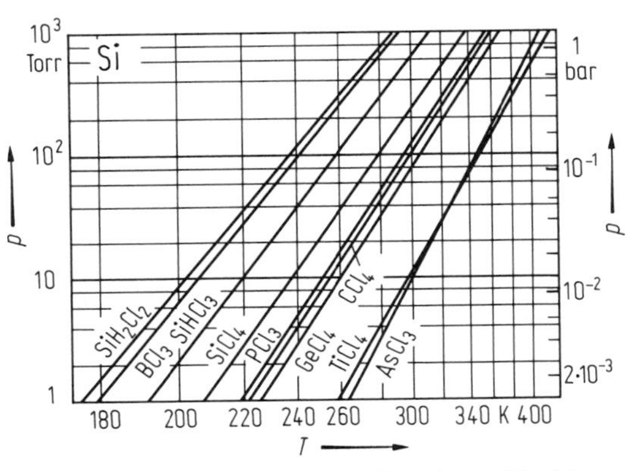

Fig. 3. Si. Vapour pressure p of various chlorides vs. temperature T in comparison to trichlorosilane.

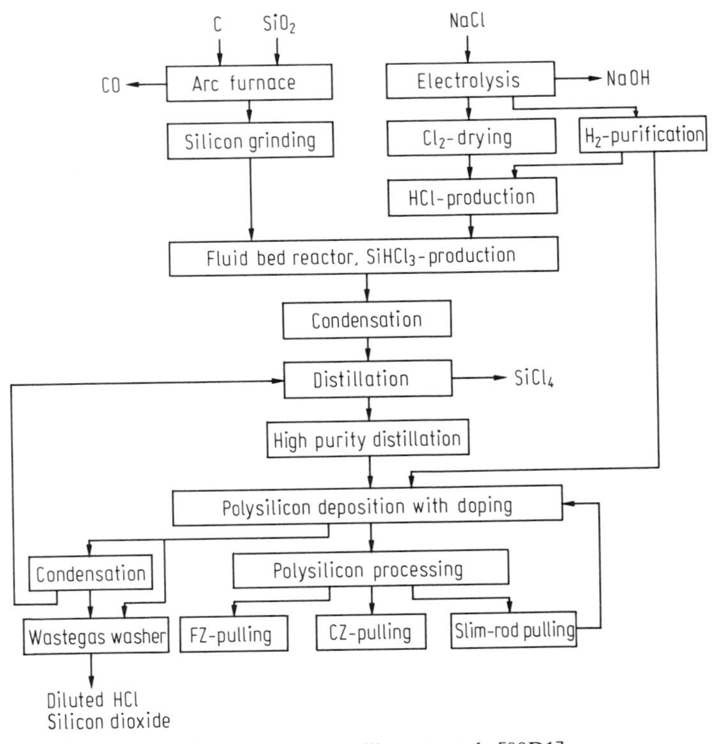

Fig. 2. Si. Production steps leading to semiconductor grade silicon crystals [80D1].

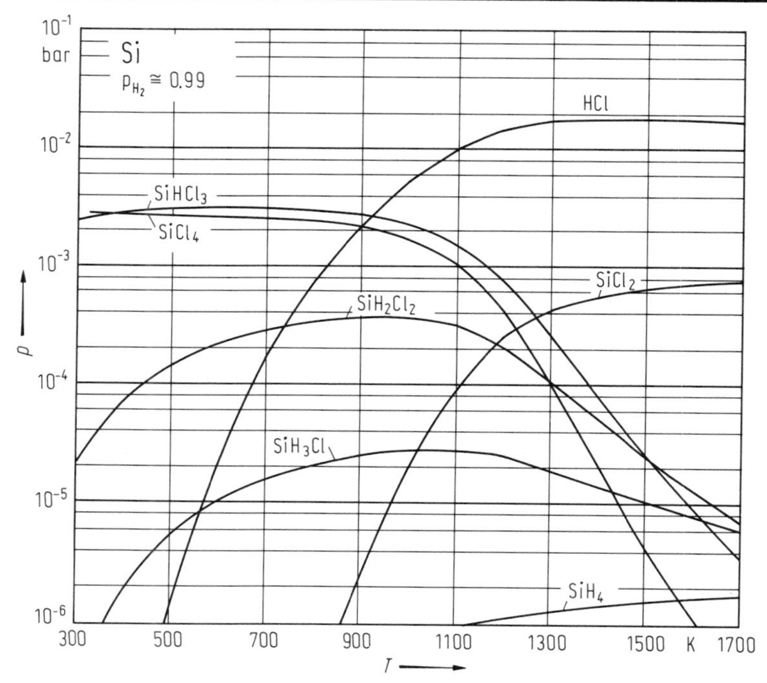

Fig. 4. Si. Partial pressure p of gas phase components at equilibrium vs. temperature T. The composition is determined for 1 bar total pressure and a ratio Cl/H = 0.01 [74S2].

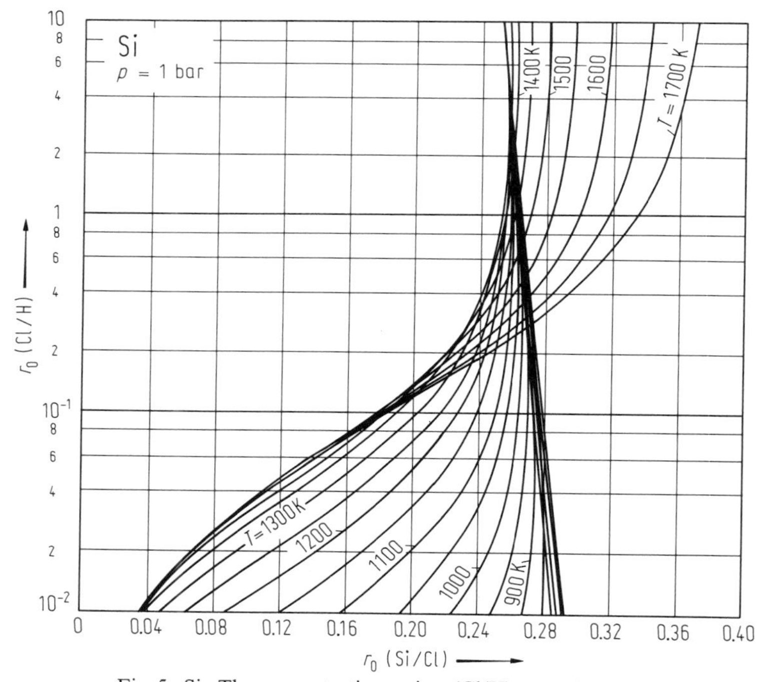

Fig. 5. Si. The concentration ratio r_0(Cl/H) at total pressure $p_t = 1$ bar vs. the concentration ratio r_0(Si/Cl). Parameter is the reaction temperature. The theoretical silicon deposition yield η may be calculated by $\eta = [r(\mathrm{Si/Cl}) - r_0(\mathrm{Si/Cl})]/r(\mathit{Si/Cl})$, where r_0(Si/Cl) is the ratio at equilibrium and r(Si/Cl) the ratio of the starting mixture [72H2].

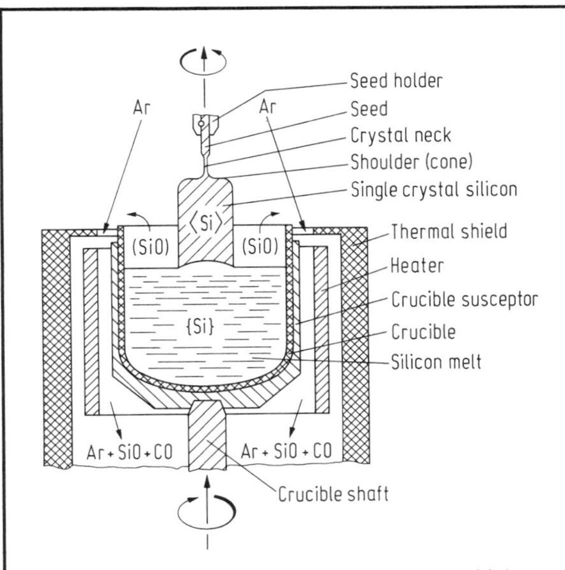

Fig. 6. Si. Czochralski single crystal pulling. Initial stage of the process. ⟨Si⟩ indicates solid, and {Si} indicates liquid silicon. (SiO) is a gaseous reactant.

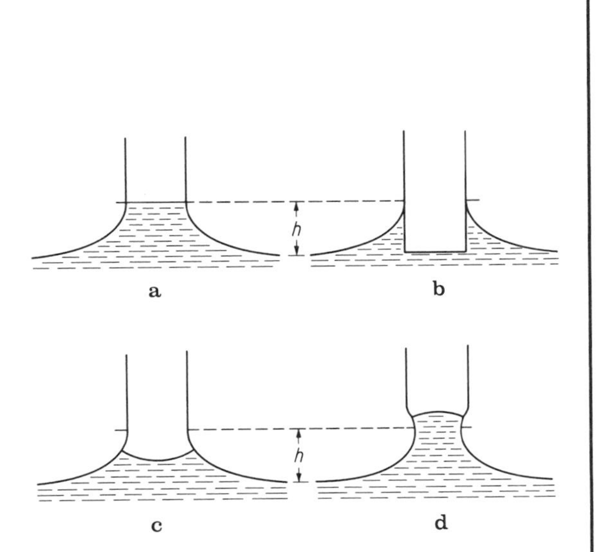

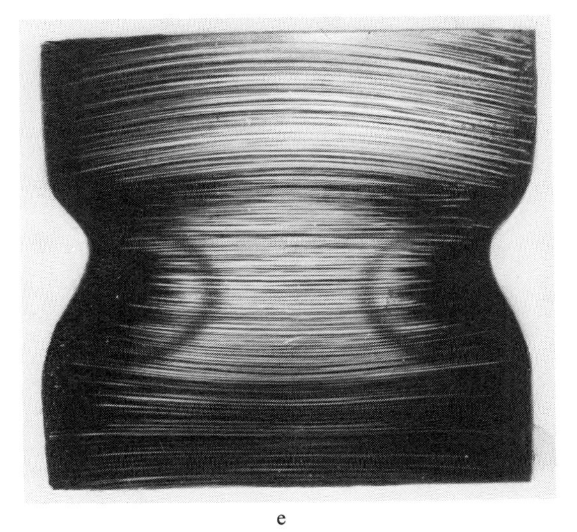

Fig. 7a–e. Si. Wetting conditions and interface shapes for CZ-crystal pulling:

a) Plane freezing isotherm; the crystal diameter remains constant at an angle of 90° between interface and melt surface;
b) a dipped-in crystal;
c) expanding interface is convex, angle between interface and melt surface $>90°$;
d) shrinking interface is concave; angle between interface and melt surface $<90°$;
e) axial section through a real crystal with shrinking (concave interface) and reexpanding (convex interface) diameter; the freezing interfaces are made visible by layers of preferentially etched oxygen precipitates.

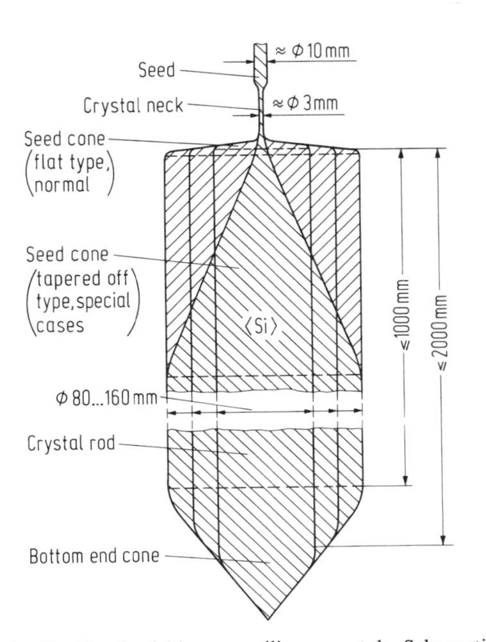

Fig. 8. Si. Czochralski-grown silicon crystals. Schematical axial section of crystals grown to different diameters.

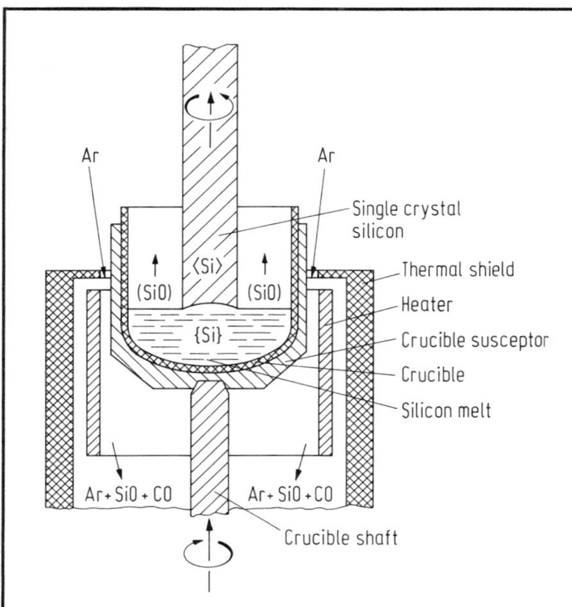

Fig. 9. Si. Advanced stage of Czochralski single crystal pulling. Arrows indicate the gas flow. ⟨Si⟩ is solid, and {Si} is liquid silicon.

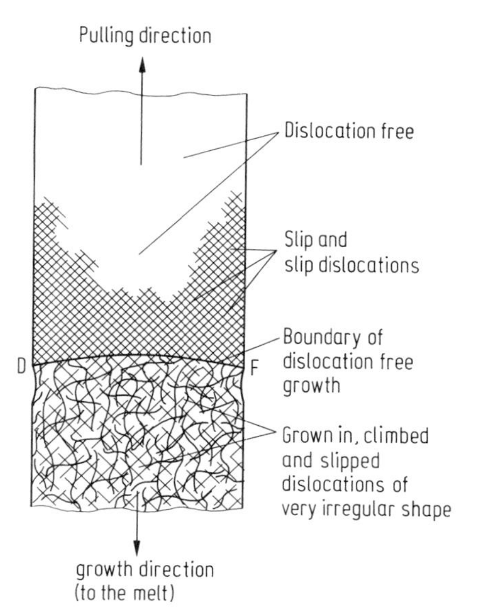

Fig. 10. Longitudinal section through a Czochralski-grown single crystal with a transition from dislocation-free growth to dislocated growth. At the line DF which represents the last solid-liquid interface of the formerly dislocation-free crystal, a disturbance generated the first dislocations. These dislocations were multiplied and spread out by the cooling strain in the crystal. Though the crystal has been hotter and softer in its centre than in the rim region, the dislocations do not reach as far upwards in the center as in the outer regions. This effect is explained by the faster shrinking of the outer parts of the crystal caused by faster cooling which generates tensile strain parallel to the crystal surface. The distribution of dislocations above line DF is a more regular one. The dislocations are mainly located on relatively few slip planes. In the part below line DF, the dislocations are partly grown in and have moved by diffusion-controlled climbing and are partly generated and have moved by slip. They are therefore of very irregular shape and distribution. (See also Fig. 35 in section 6.1.3.)

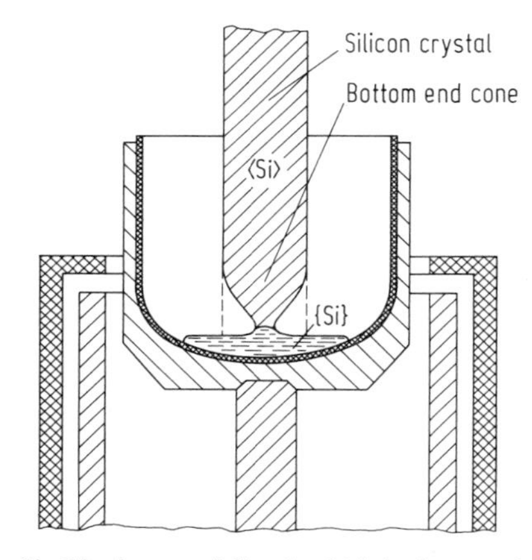

Fig. 11. Si. Final stage of Czochralski single crystal pulling. If the crystal has the shape of the dashed lines, then the crystal cannot be finished dislocation-free. ⟨Si⟩ is solid silicon. {Si} is liquid silicon.

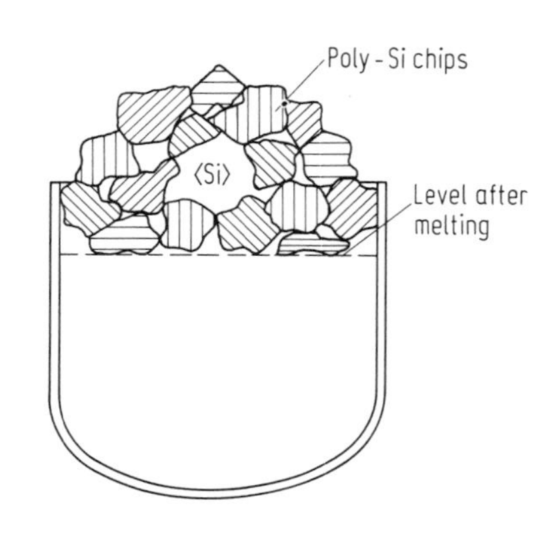

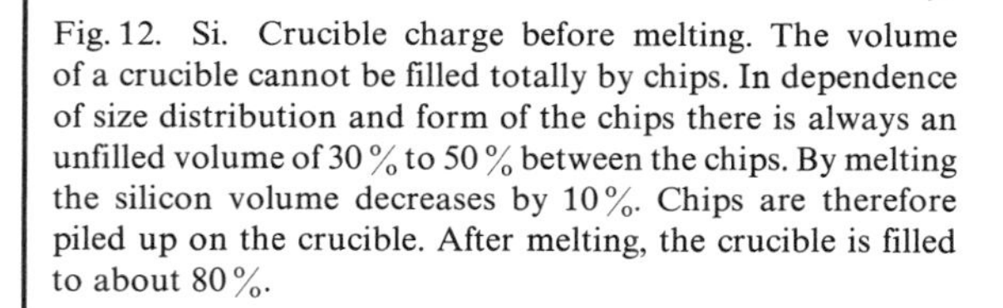

► Fig. 12. Si. Crucible charge before melting. The volume of a crucible cannot be filled totally by chips. In dependence of size distribution and form of the chips there is always an unfilled volume of 30 % to 50 % between the chips. By melting the silicon volume decreases by 10%. Chips are therefore piled up on the crucible. After melting, the crucible is filled to about 80 %.

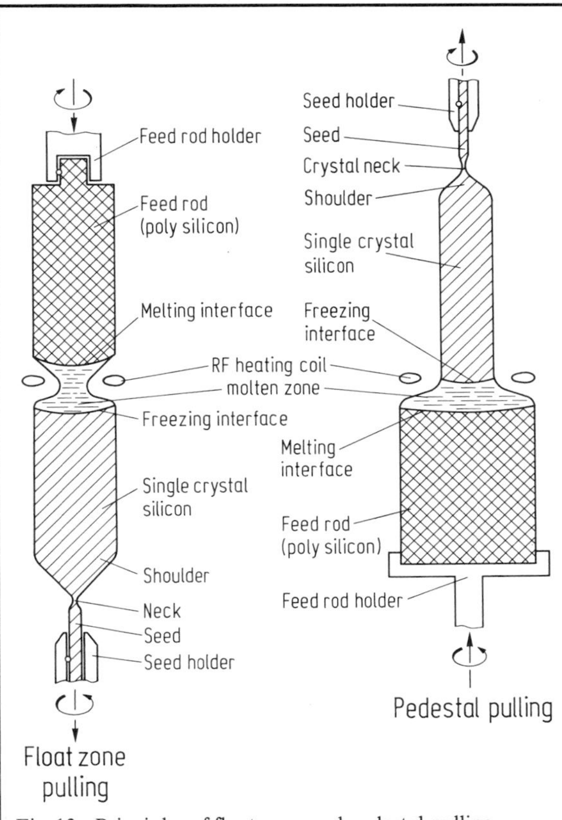

Fig. 13. Principles of float zone and pedestal pulling.

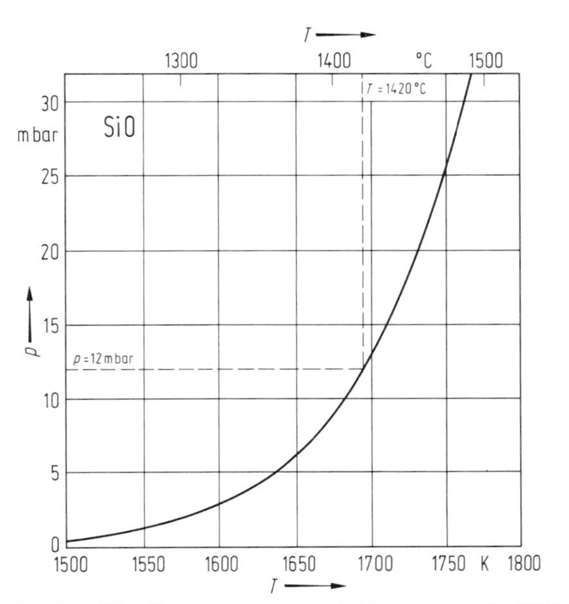

Fig. 14. SiO. Vapour pressure p of silicon monoxide (SiO) vs. temperature T [74K].

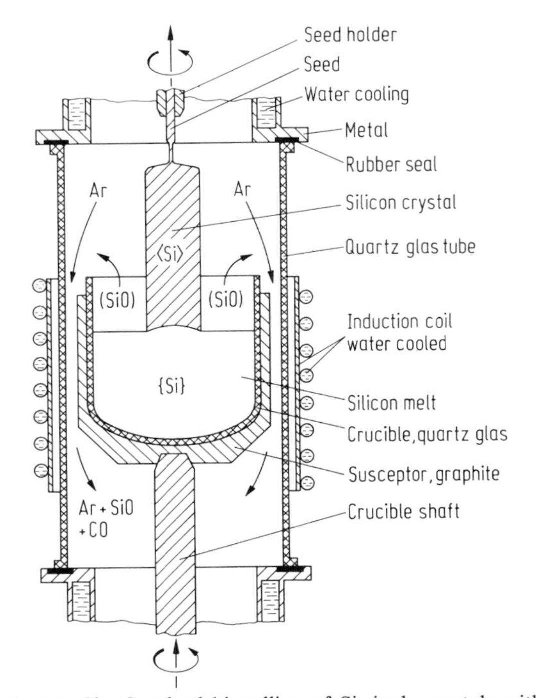

Fig. 15. Si. Czochralski pulling of Si-single crystals with inductive heating, schematically. ⟨Si⟩ and {Si} indicate solid and liquid silicon, respectively.

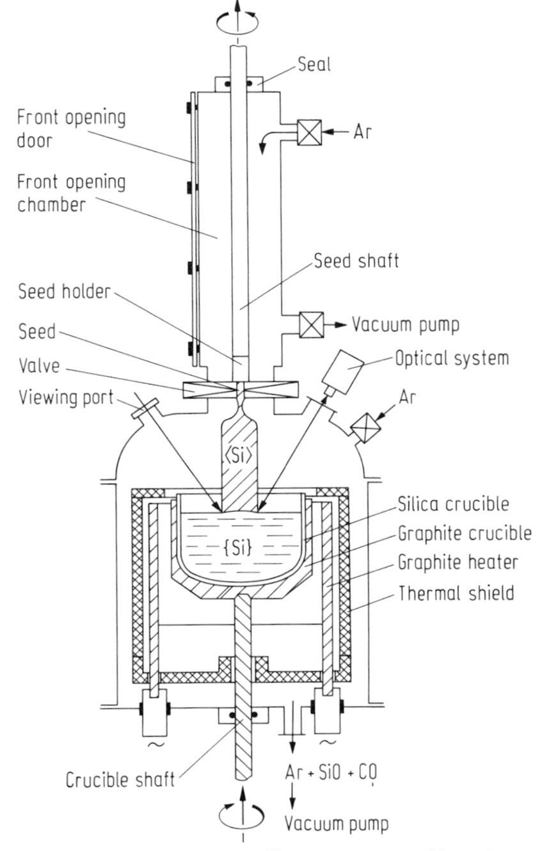

Fig. 16. Si. Czochralski pulling apparatus with resistance heating, automatic optical diameter control, and separating valve between furnace and front opening chamber.

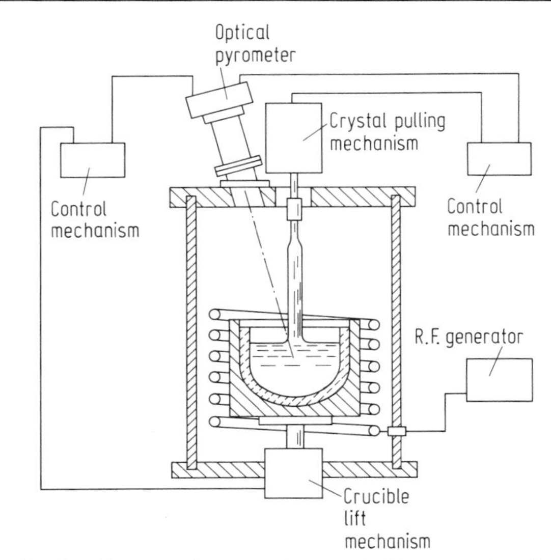

Fig. 17. Diagram of automatic control of CZ-crystal pulling and crucible lifting. The crystal diameter is monitored by an optical pyrometer which looks at the halo between crystal and melt. US-Patent No. 3,692,499 [72A].

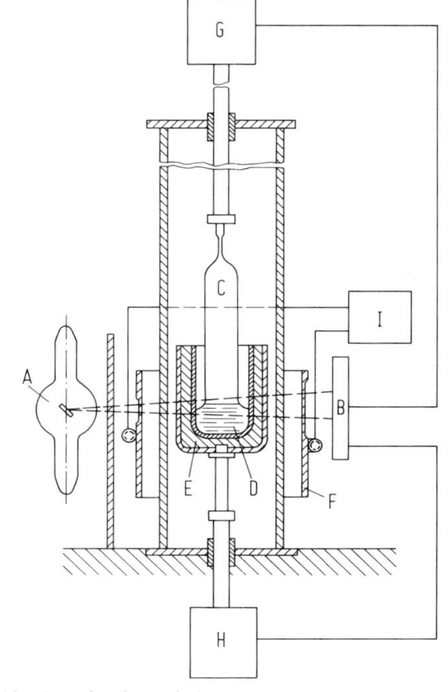

Fig. 18. Monitoring of the crystal diameter and the melt level by X-rays. A: X-ray tube, B: X-ray monitor, C: crystal, D: melt, E: crucibles, F: induction coil, G: crystal pulling mechanism, H: crucible lift mechanism, I: power supply. Deutsches Patent No. 1.519.850 [66Z].

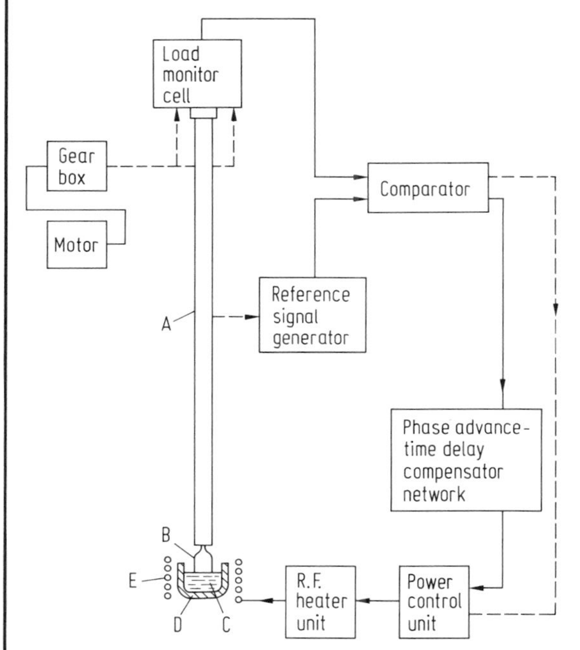

Fig. 19. Block diagram for automatic CZ-crystal pulling by weighing of the crystal; A: crystal shaft, B: crystal, C: melt, D: crucible, E: induction coil. US-Patent No. 4,032,389 [77J].

For Figs. 20, 21, see next page.

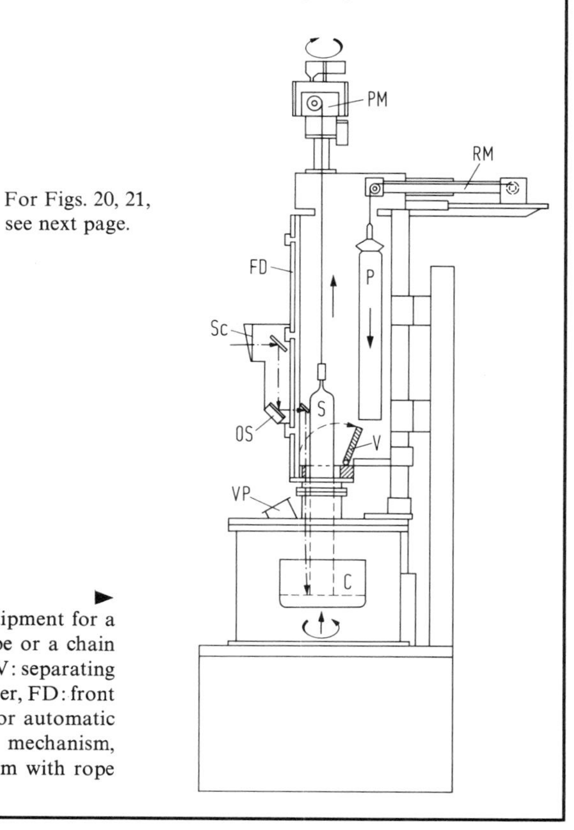

Fig. 22. CZ-crystal puller with recharging equipment for a poly-silicon rod and crystal pulling with a rope or a chain instead of a shaft. S: single crystal, C: crucible, V: separating valve between furnace and front opening chamber, FD: front opening door, Sc: screen, OS: optical system for automatic diameter control, VP: view port, PM: pulling mechanism, P: poly-silicon rod, RM: recharging mechanism with rope or chain [80L].

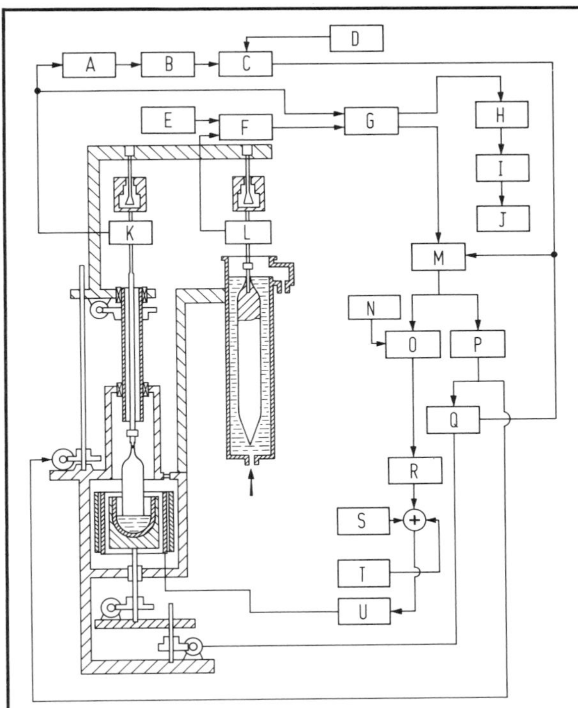

Fig. 20. Automatic crystal pulling by a weighing method. The weight of the crystal is compared with a float which has the same shape as the crystal. By draining off the fluid around the float, the weight of the float increases. The automatic control pulls a crystal of the same weight and shape as that of the float above the melt. A: differential circuit, B: amplifier, C: comparator, D: programmed reference source, E: diameter adjuster, F: zero adjuster, G: polarisation error amplifier, H: differentiation circuit, I: DC amplifier, J: graphic meter, K: first weight detector, L: second weight detector, M: PID controller, N: programmed reference source, O: comparator, P: amplifier, Q: crucible lift controller, R: differentiation circuit, S: programmed control signal source, T: manual power regulator, U: constant power controller, ⊕: adder circuit. Canadian Patent No. 971085 [75S].

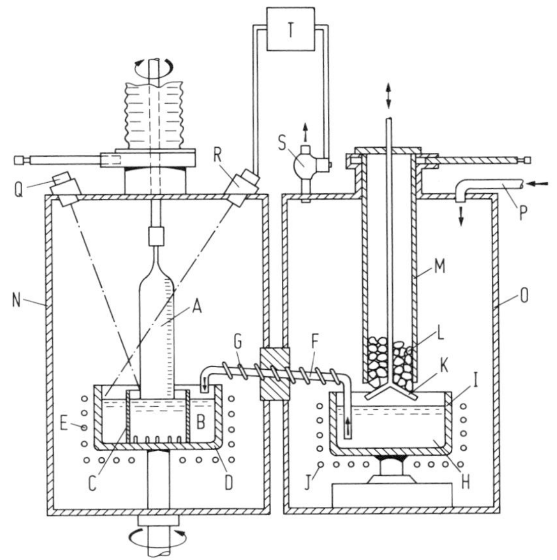

Fig. 21. Automatic recharging of the crucible with melt during CZ-crystal pulling. A: crystal, B: melt for crystal pulling, C: separating wall with holes, D: crucible, E: heating element, F: overflow tube, G: tube heating, H: reserve melt, I: crucible, J: heating element, K: store plug, L: polycrystalline Si or Ge, M: storage bin, N: crystal pulling chamber, O: supply chamber, P: inert gas supply, Q: pyrometer for crystal diameter control, R: pyrometer for melt level control, S: drain valve, T: overflow control. US-Patent No. 4,036,595 [77L4].

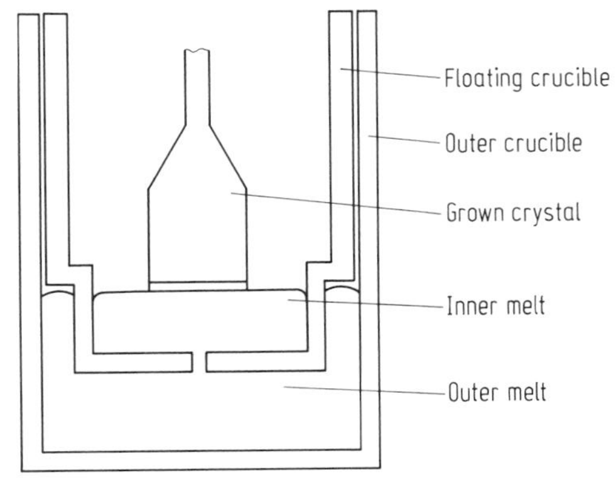

Fig. 24. Ge. Schematic drawing of the floating crucible [65L].

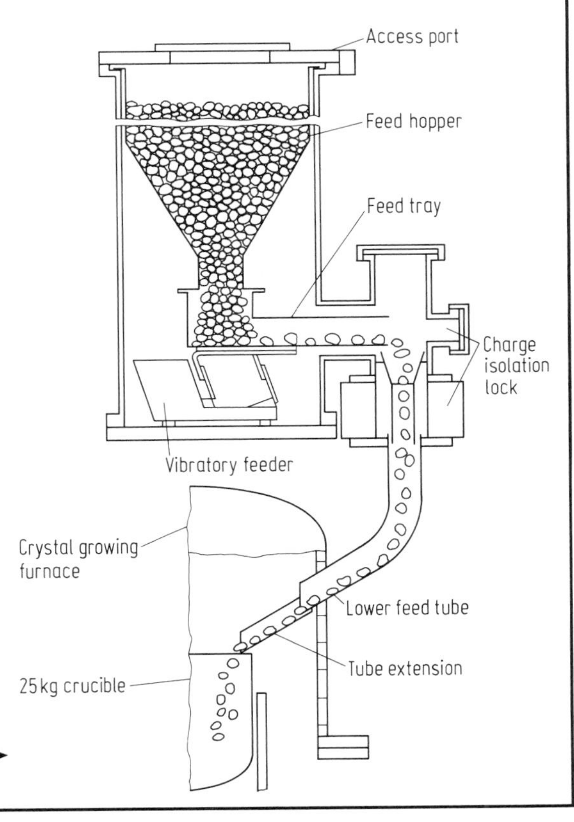

Fig. 23. Recharging of the hot crucible with poly-silicon ► chips [78E].

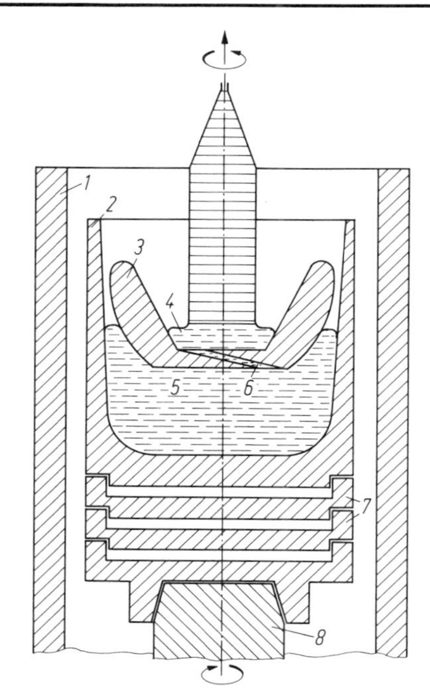

Fig. 25. Ge. Schematic drawing of the floating crucible after [72H1]. *1:* rotating outer wall made of graphite, *2:* outer crucible, *3:* floating crucible, *4:* Ge melt in floating crucible with growing crystal, *5:* Ge melt in outer crucible, *6:* capillary, *7:* heat insulating supports, *8:* rotating crucible holder.

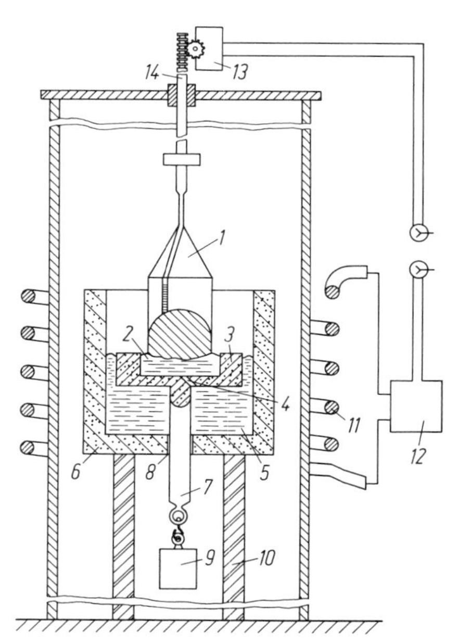

Fig. 26. Ge. Schematic drawing of a floating crucible [73B1]. *1:* crystal, *2:* inner melt, *3:* floating crucible, *4:* hole, *5:* outer melt, *6:* outer crucible, *7:* guide shaft, *8:* guide hole, *9:* load, *10:* support, *11:* induction coil, *12:* power supply, *13:* pulling mechanism, *14:* crystal shaft.

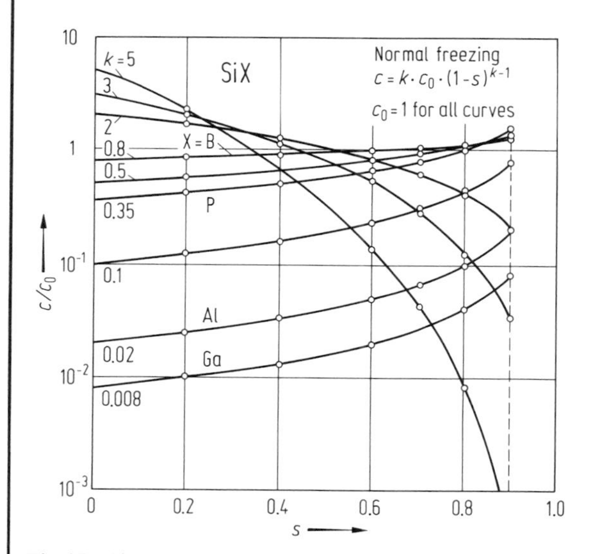

Fig. 27. Si. Fractional concentration c/c_0 of an impurity having distribution coefficient k vs. solidified fraction of silicon s calculated by normal freezing after Pfann [52P, 57P]. The curves show the axial impurity distributions expected for different k values.

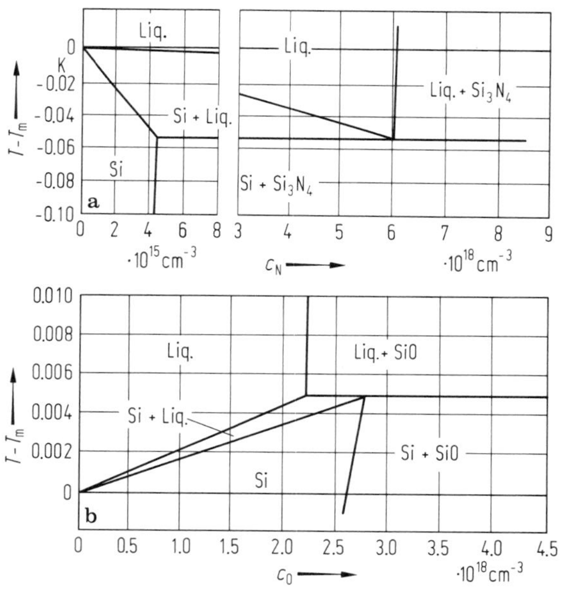

Fig. 29 a, b. Si. Phase diagrams of the Si – N and Si – O systems for low concentrations [73Y].
a) Lowering of the melting temperature $T - T_m$ with increasing nitrogen concentration c_N in silicon.
b) Increase of the melting temperature $T - T_m$ with increasing oxygen concentration c_O.

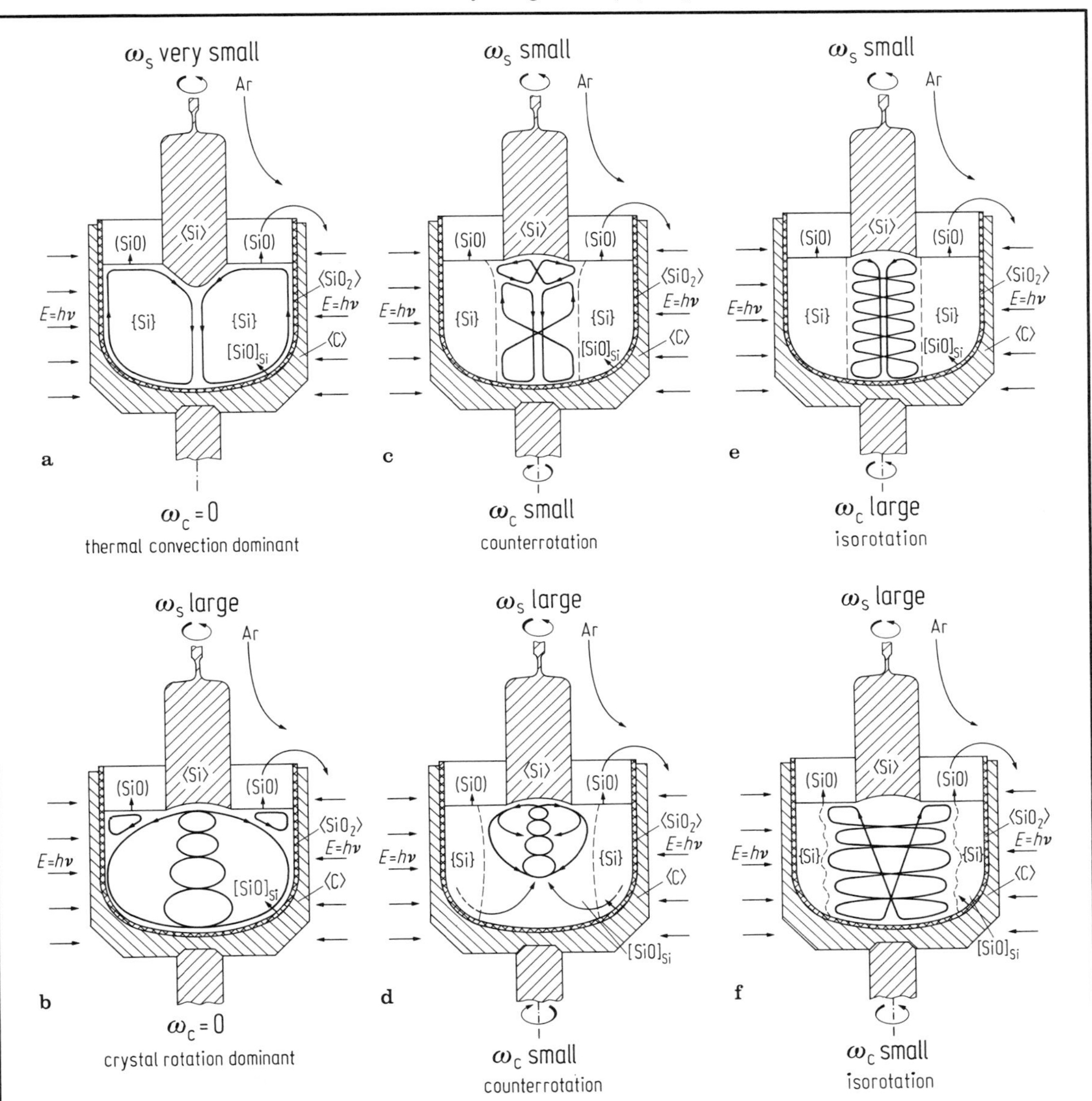

Fig. 28a. Si. Flow characteristics in a CZ-crucible dominated by thermal convection [67C, 68C, 76C1, 77C].

Fig. 28b. Si. Flow characteristics in a CZ-crucible dominated by crystal rotation [67C, 68C, 76C1, 77C].

Fig. 28c. Si. Flow characteristics in a CZ-crucible dominated by counterrotation of the crystal and the crucible at small frequencies [67C, 68C, 76C1, 77C].

Fig. 28d. Si. Flow characteristics in a CZ-crucible dominated by counterrotation with a fast rotation of the crystal [67C, 68C, 76C1, 77C].

Fig. 28e. Si. Flow characteristics in a CZ-crucible dominated by fast rotation of the crucible and slow rotation in the same direction of the crystal [67C, 68C, 76C1, 77C].

Fig. 28f. Si. Flow characteristics in a CZ-crucible dominated by fast rotation of the crystal and slow rotation in the same direction of the crucible [67C, 68C, 76C1, 77C].

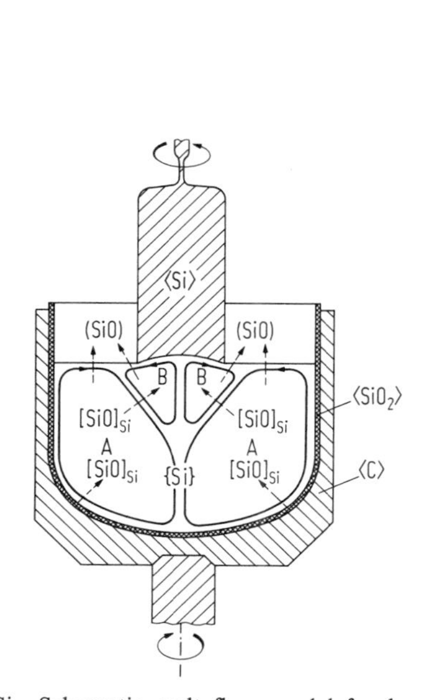

Fig. 30. Si. Schematic melt flow model for low oxygen transfer from the silica to the Si-crystal. Important is the evaporation of SiO from each of the eddy currents A and B and the slight transfer of SiO between A and B due to the low oxygen content of the melt between them.

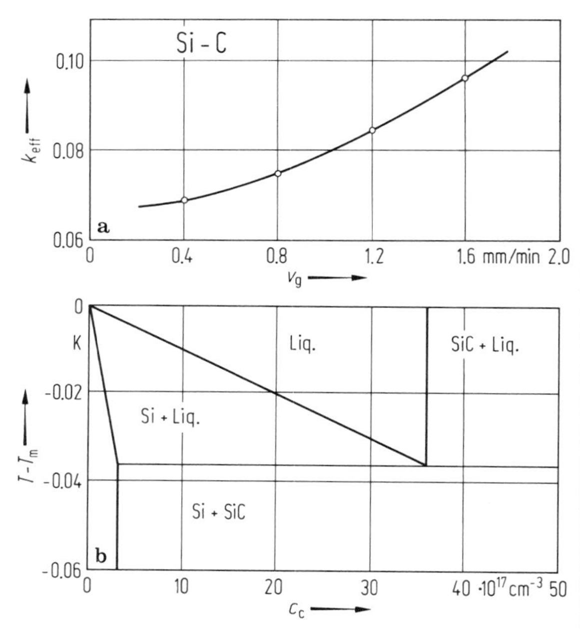

Fig. 31 a, b. Si. Phase diagram of the Si—C system at extremely low carbon concentrations [74N]. a) effective distribution coefficient k_{eff} of carbon vs. growth velocity v_g b) lowering of the melting temperature $T-T_m$ with increasing concentration c_C in silicon.

For Fig. 32, see next page.

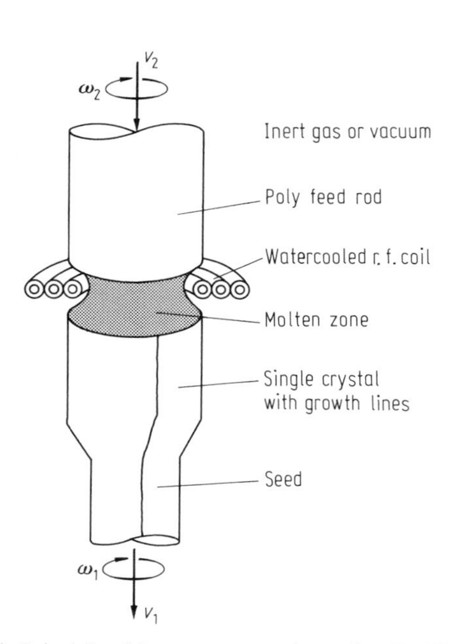

Fig. 33. Principle of float-zone crystal growing [81K].

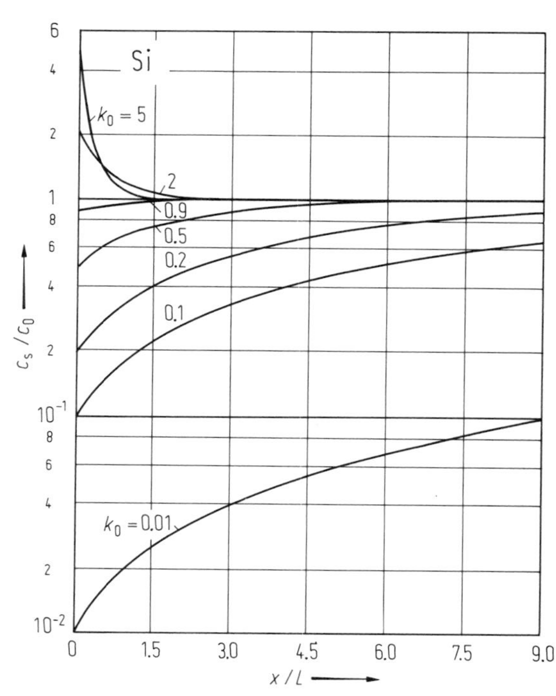

Fig. 34. Calculated single-pass relative solute concentrations c_s/c_0 in the solid vs. distance in zone length x/L for various values of distribution coefficients k_0.

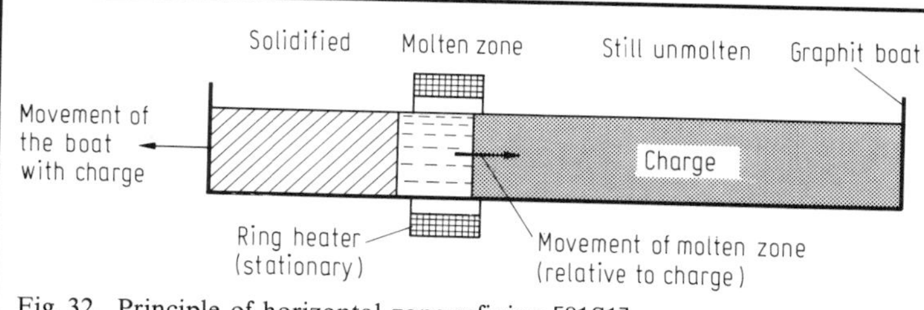

Fig. 32. Principle of horizontal zone refining [81S1].

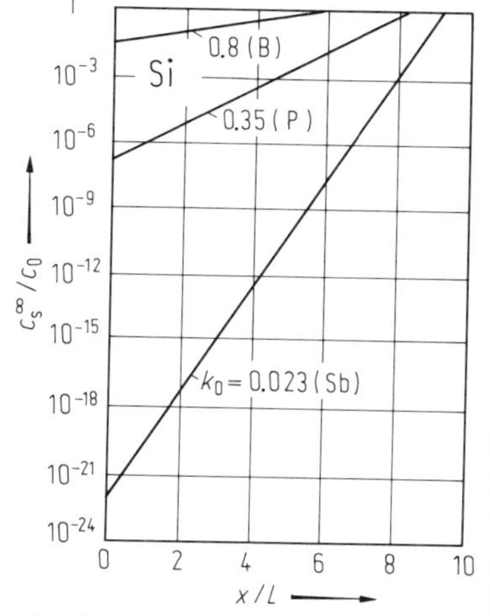

Fig. 35. Si. Calculated ultimate relative solute distribution c_s^∞/c_0 in a zone refined crystal after an infinite number of zone passes vs. distance in zone length x/L. The ratio of zone length to the crystal length is 1:10. Parameter is the distribution coefficient k_0. The values shown are for boron, phosphorus, and antimony, respectively.

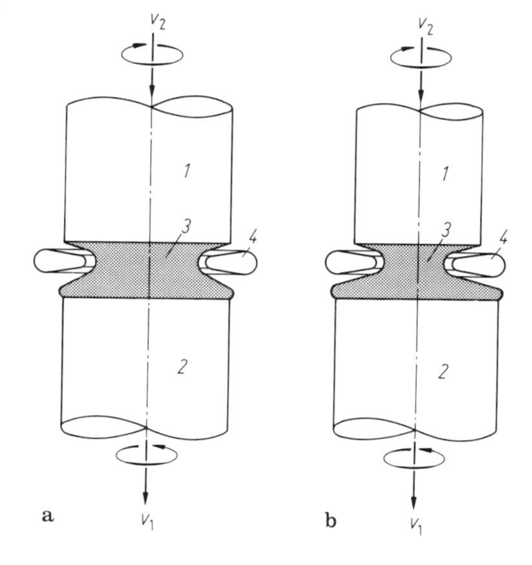

◄ Fig. 36a, b. Si. Needle-eye float-zone growth [59K, 81K]. *1* = feed rod, *2* = single crystal, *3* = melt, *4* = induction coil; v_1 = growth velocity, v_2 = melt down velocity.
a) Feed rod diameter = crystal diameter: $v_1 = v_2$.
b) Feed rod diameter < crystal diameter: $v_1 < v_2$.

Fig. 37. Si. Typical dislocation-free silicon crystals as obtained by float-zone growth.

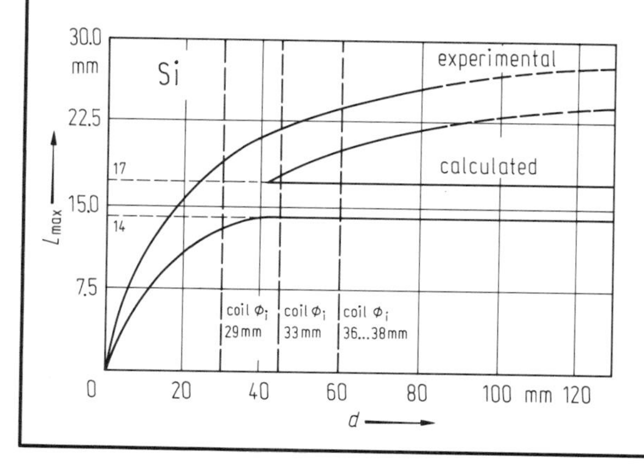

◄ Fig. 38. Si. Maximum zone length L_{max} vs. crystal diameter d [81K]. The ranges required for the inner diameter of the rf coil, ϕ_i is marked by the dashed vertical lines.

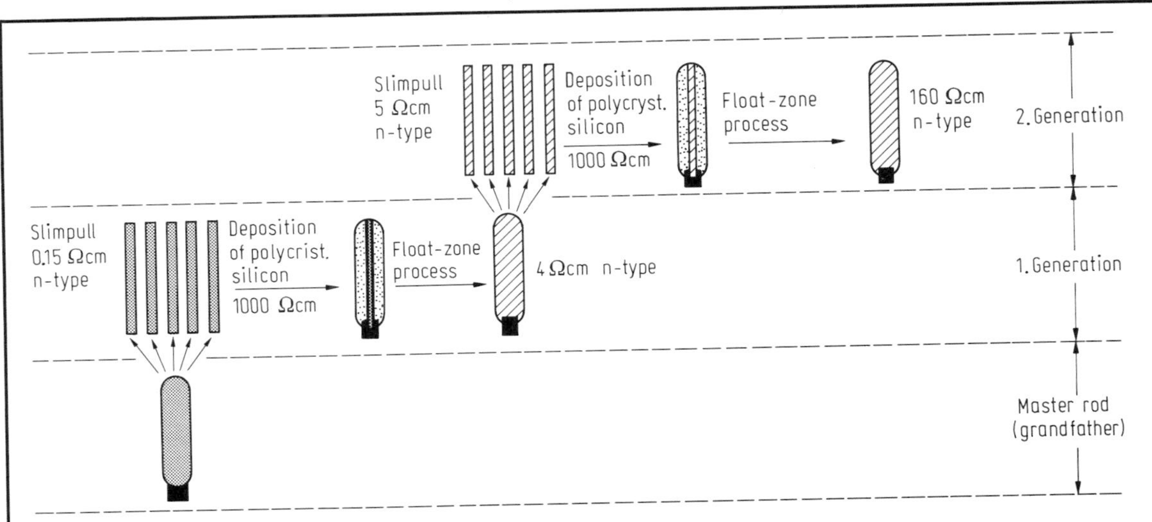

Fig. 39. Si. Schematic outline of the generation method for doping [58H1, 81K].

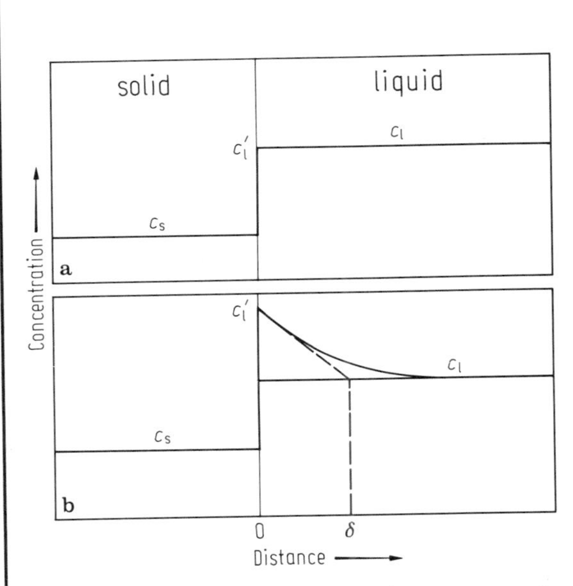

Fig. 40a, b. Solute impurity concentration near a growing interface. c_s and c_l are impurity concentrations in the solid and liquid, respectively, c_l' is the concentration in the liquid at the freezing interface. a) Equilibrium, b) steady-state growth conditions (δ, diffusion layer thickness).

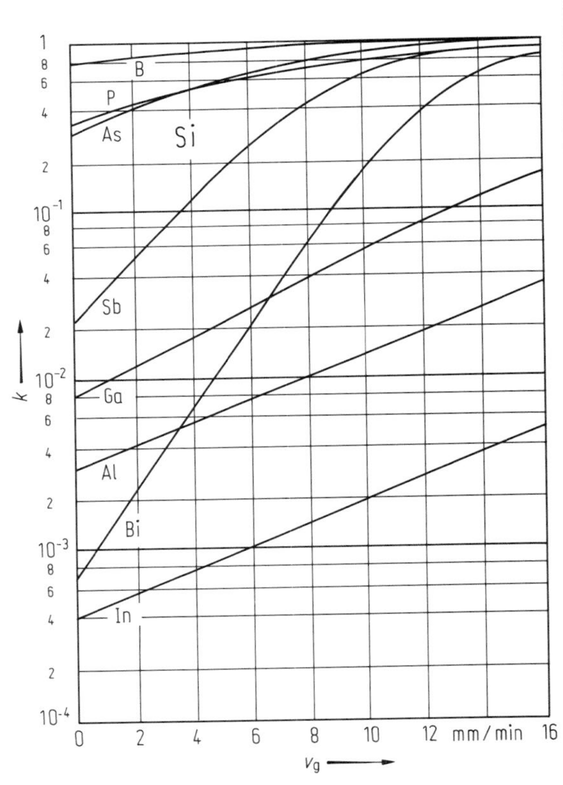

Fig. 41. Si. Calculated effective distribution coefficient k vs. growth velocity v_g of several dopants. The rotation rate used is 8 rpm [81K].

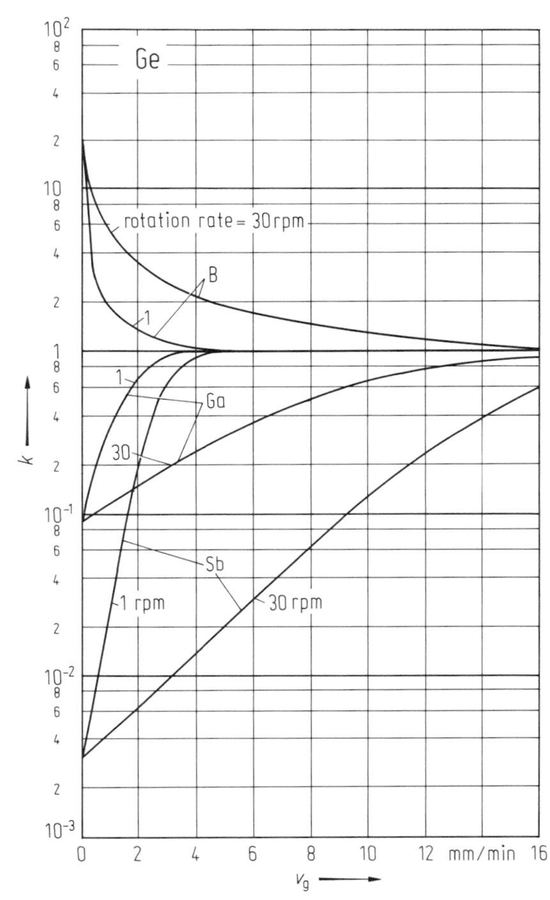

Fig. 42. Ge. Calculated effective distribution coefficient k vs. growth velocity v_g of several dopants. The rotation rate of each dopant is 1 and 30 rpm, respectively.

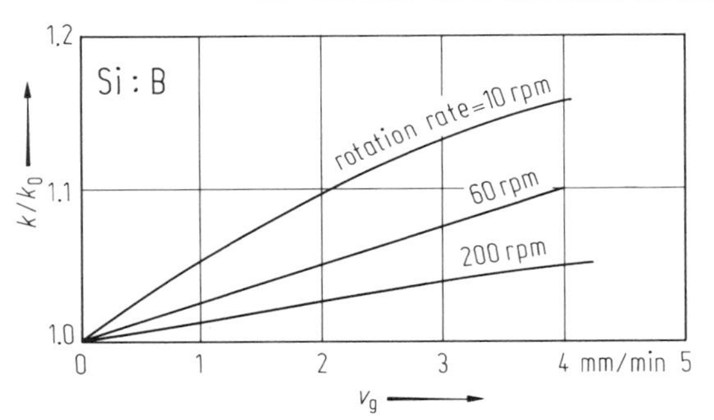

Fig. 43. Si. Relative effective distribution coefficient k/k_0 of B vs. growth velocity v_g. Parameters are the rotation rates 10, 60, and 200 rpm [63K].

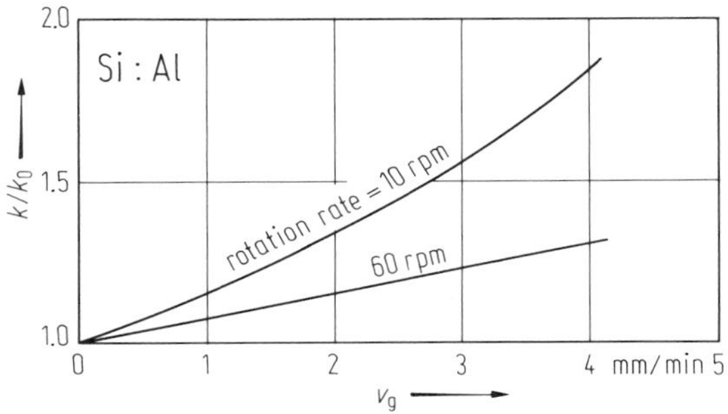

Fig. 44. Si. Relative effective distribution coefficient k/k_0 of Al vs. growth velocity v_g. Parameters are the rotation rates 10 and 60 rpm [63K].

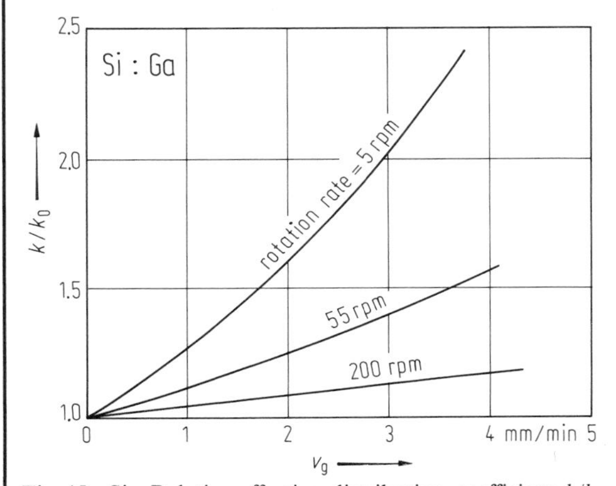

Fig. 45. Si. Relative effective distribution coefficient k/k_0 of Ga vs. growth velocity v_g. Parameters are the rotation rates 5, 55 and 200 rpm [63K].

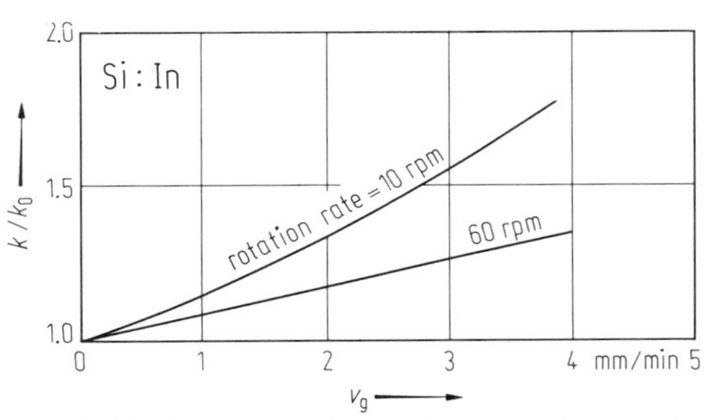

Fig. 46. Si. Relative effective distribution coefficient k/k_0 of In vs. growth velocity v_g. Parameters are the rotation rates 10 and 60 rpm [63K].

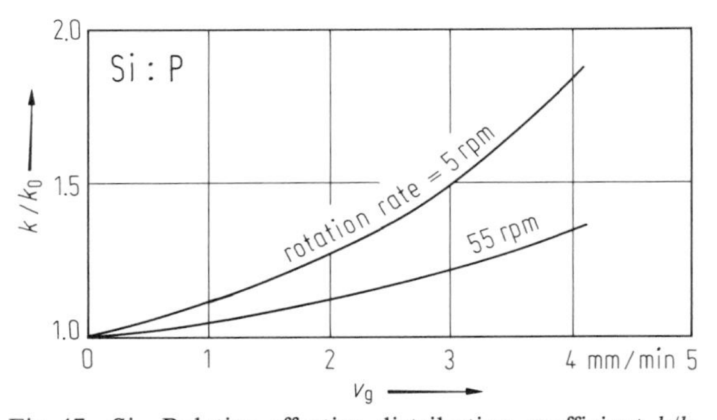

Fig. 47. Si. Relative effective distribution coefficient k/k_0 of P vs. growth velocity v_g. Parameters are the rotation rates 5 and 55 rpm [63K].

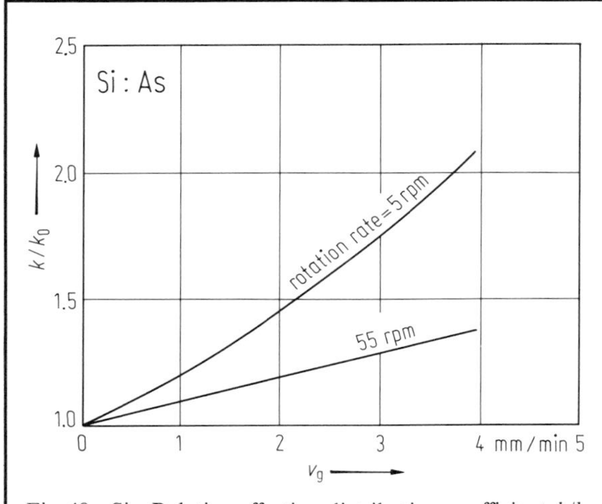

Fig. 48. Si. Relative effective distribution coefficient k/k_0 of As vs. growth velocity v_g. Parameters are the rotation rates 5 and 55 rpm [63K].

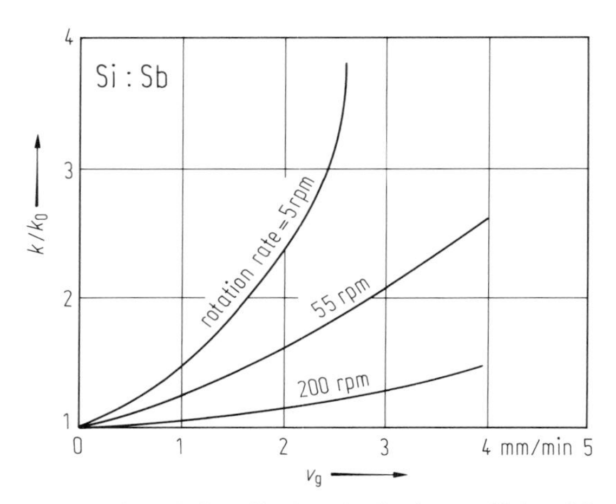

Fig. 49. Si. Relative effective distribution coefficient k/k_0 of Sb vs. growth velocity v_g. Parameters are the rotation rates 5, 55, and 200 rpm [63K].

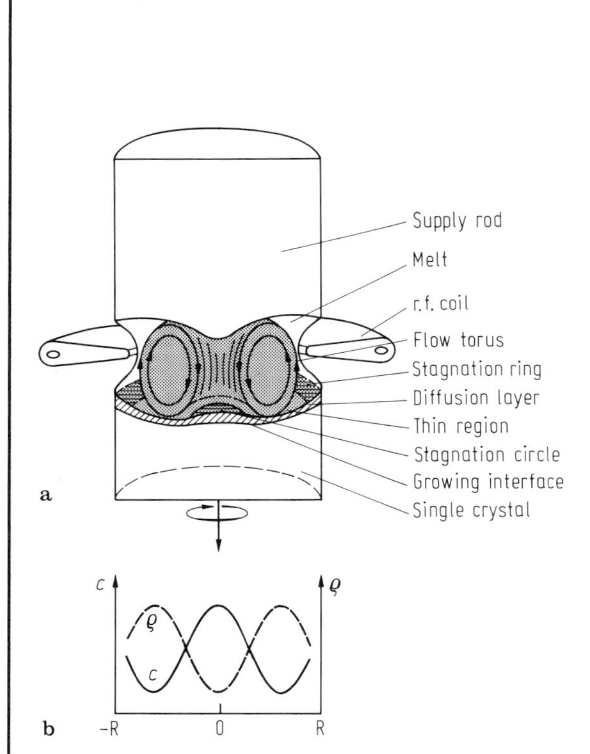

Fig. 50a, b. Si. Model of flow pattern during float-zone growth, and dopant distribution obtained from [81K].
a) Flow torus and diffusion layer thickness.
b) Radial dopant concentration c and resistivity profile ϱ.

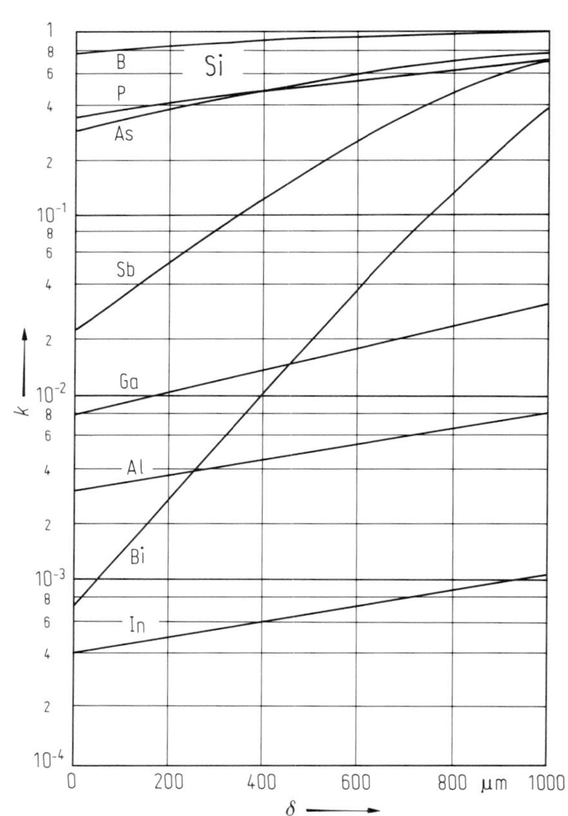

Fig. 51. Si. Calculated effective distribution coefficient k vs. diffusion layer thickness δ of several dopants [81K].

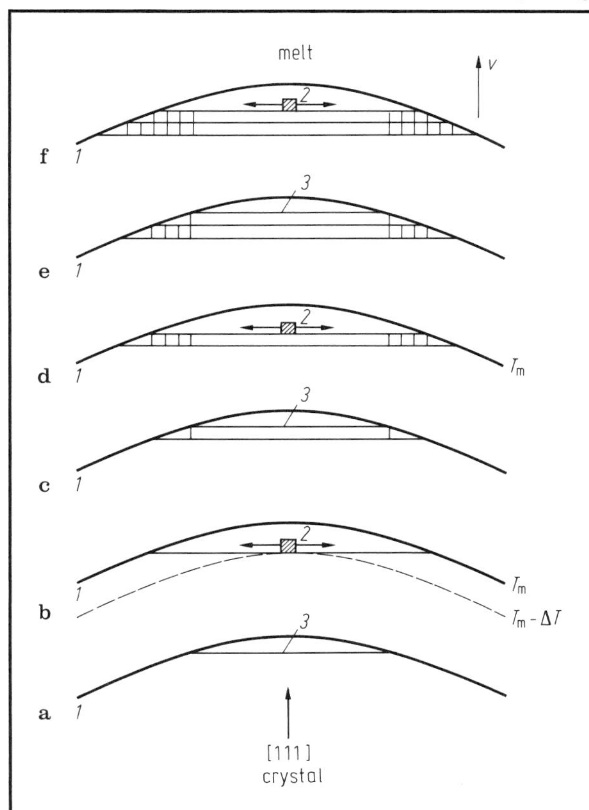

Fig. 52. Schematic drawing of facet nucleation and sheet growth along [111] crystal orientation [81K]: Consecutive growth steps are shown along a)···f).
1 = melting temperature T_m isotherm, 2 = nucleus, 3 = facet, dashed line indicates the critical supercooling $T_m - \Delta T$ isotherm.

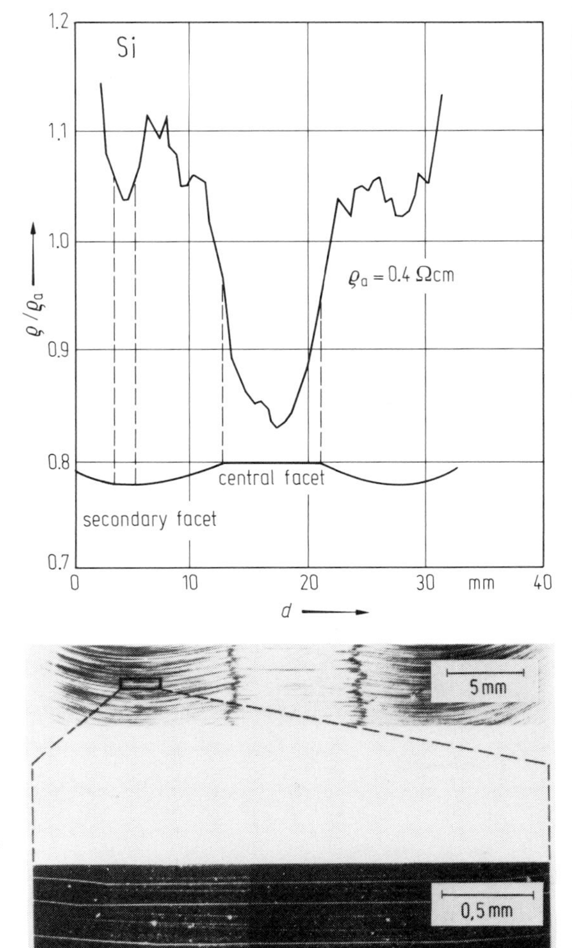

Fig. 53. Si. Primary, central, and secondary facets and resulting coring in a P-doped, dislocation-free [111] silicon crystal with corresponding resistivity profile. A magnified portion of secondary facet is shown in the lower part [81K].

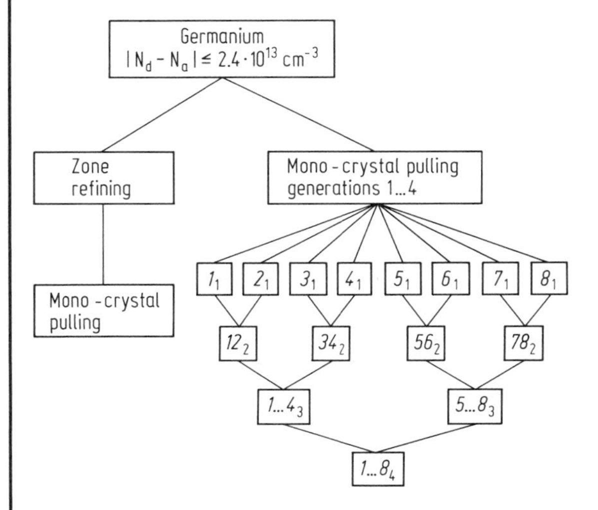

◄

Fig. 54. Ge. Flow diagram for the purification of Germanium [78G1]. The numbers indicate crystals (line 1) and crystal sections (line 2–4), respectively. The indices indicate the sequential generations of single crystals pulled from the subsections.

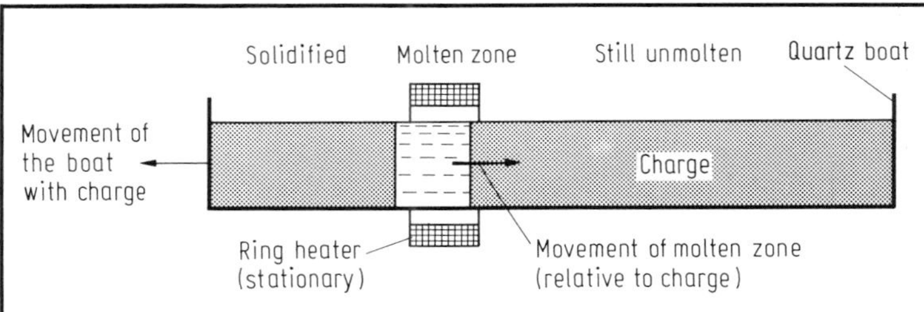

Fig. 55. Ge. Schematic drawing of the (horizontal) single zone refining process as used for germanium.

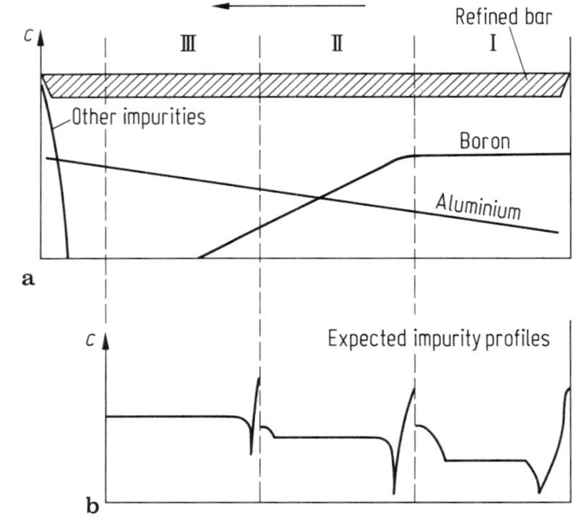

Fig. 56a, b. Ge. Schematic representation of the distribution of impurities during zone refining (a) and of impurity profiles of Ge crystals grown from sections of the refined bar (b). Boron is an occasional impurity while aluminum is almost always dominant [78H3].

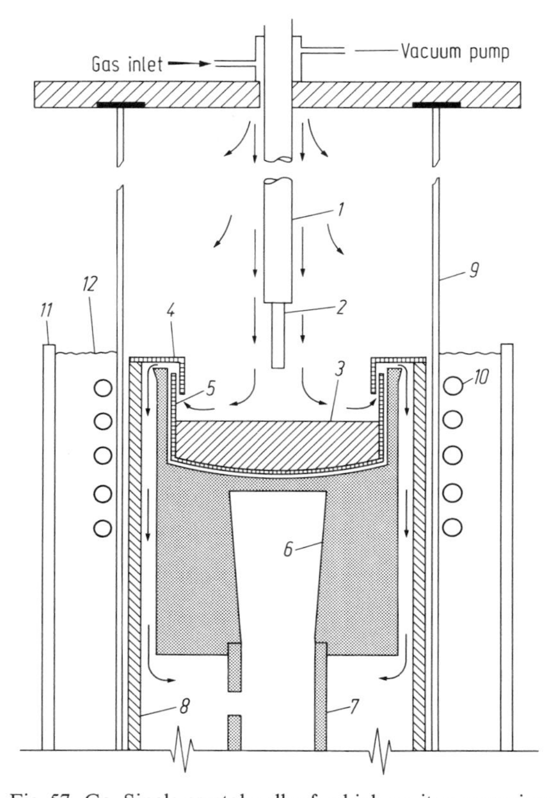

Fig. 57. Ge. Single crystal puller for high purity germanium [77G2]. *1*: Molybdenum seed rod, *2*: seed, *3*: germanium melt, *4*: quartz gas deflector, *5*: quartz crucible, *6*: carbon susceptor, *7*: carbon susceptor pedestal, *8*: sand quartz heat shield, *9*: quartz outer envelope, *10*: RF coil, *11*: pyrex fishbowl, *12*: deionized water.

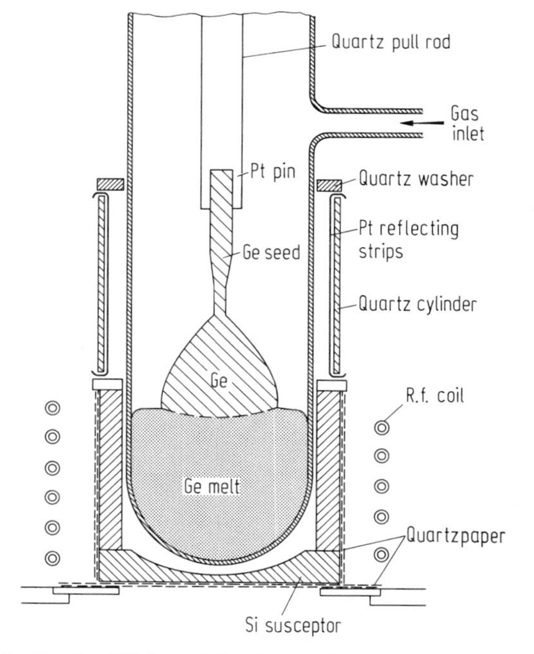

Fig. 58. Ge. RF heated Czochralski furnace [70H].

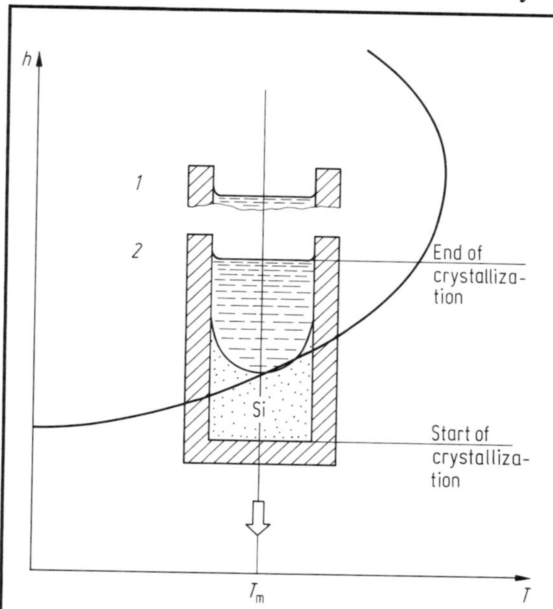

Fig. 59. Si. Schematic graph of ingot growth by directional solidification. The crucible is moved downward in position *h* in a fixed temperature *T* field. Two positions *1* and *2* of the container are indicated. The intersection with the vertical line at the melting temperature T_m marks the solid-melt interface. [81D1, 79C, 81F].

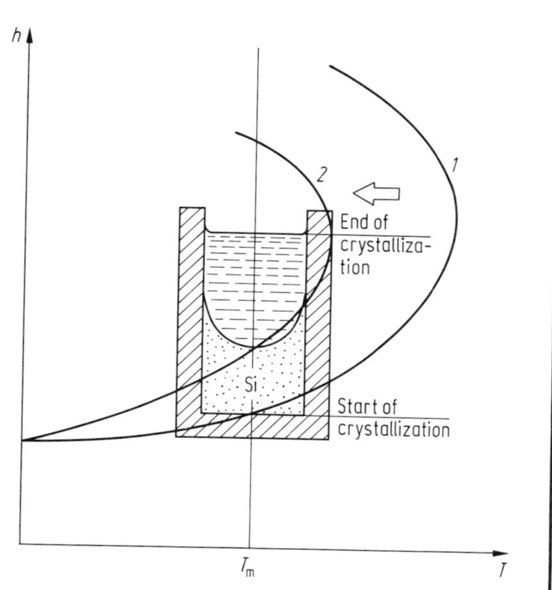

Fig. 60. Si. Schematic graph of ingot growth by directional solidification. The crucible is held fixed while the temperature profile is moved. Two sequential profiles marked *1* and *2* are shown in the diagram of position *h* vs. temperature *T*. The melting temperature T_m is marked by the vertical line [71D1, 79C].

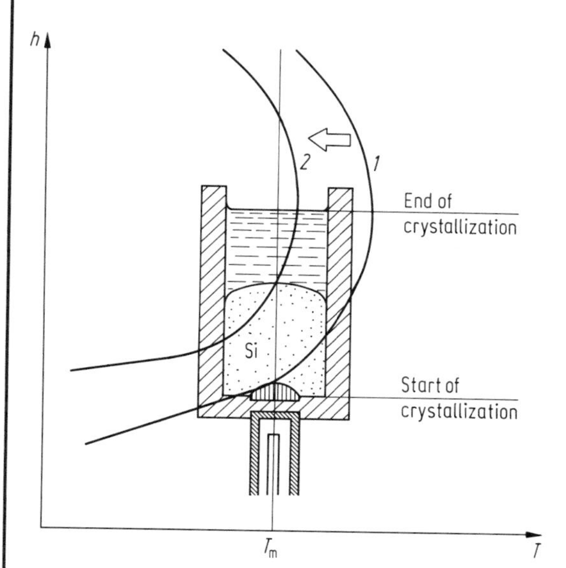

Fig. 61. Si. Schematic graph of ingot growth by the heat exchanger method (HEM) [76S2]. The crucible is attached to the heat exchanger at the bottom. Its position is kept fixed. The temperature profile is moved by adjusting the heat exchanger and the heat power operation. Two sequential profiles marked *1* and *2* are shown. The melting temperature T_m is marked by the vertical line.

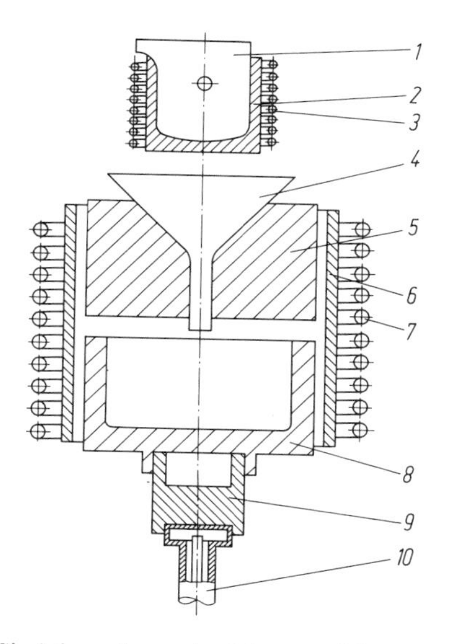

Fig. 62. Si. Schematic graph of the essential parts of mould casting by the Wacker process [80H1, 81D1]. *1*: Quartz crucible, *2*, *5*: graphite supports, *3*, *7*: RF coils, *4*: preheated quartz funnel, *6*: heating element, *8*: graphite mould, *9*: thermal insulation of, *10*: water-cooled rotating shaft.

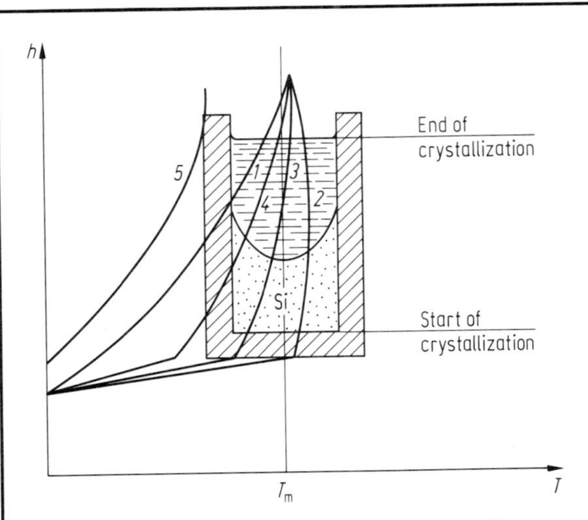

Fig. 63. Si. Schematic graph of the temperature adjustment for the ingot processing in the mould casting technique [81D1]. The sequential stages of temperature distribution (position h vs. temperature T) are given by the curves $1 \cdots 5$; 1: temperature of the preheated mould, 2: temperature after filling by the silicon melt, 3: temperature during crystallisation, 4, 5: temperature during cooling period.

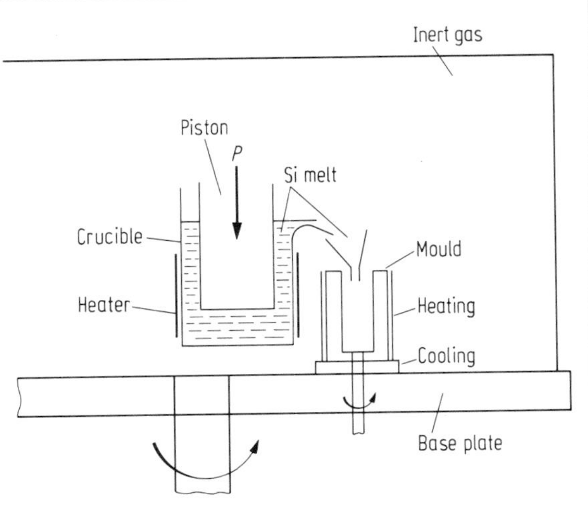

Fig. 64. Schematic view of an equipment for semicontinuous production of silicon ingots by mould casting [81S2].

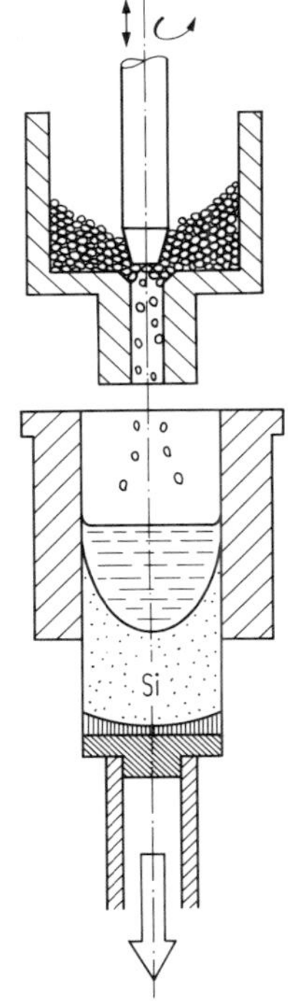

Fig. 65. Si. Schematic graph of ingot preparation by continuous mould casting [81D1]. Silicon is fed from a reservoir to the crystallization chamber. The solidification rate of the ingot is adjusted to the temperature profile to keep the solid/liquid interface at the same level.

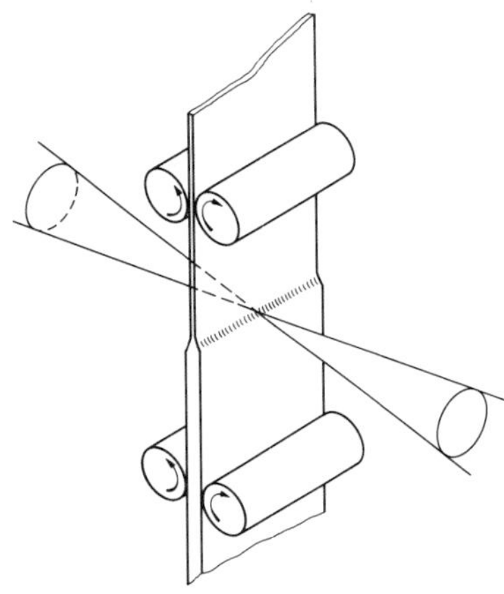

Fig. 66. Schematic graph of the ribbon-to-ribbon (RTR) technology for sheet growth of silicon [78G2, 80B1]. The molten zone between the feed stock (bottom end) and the ribbon (top end) is generated by laser scanning or electron beam action.

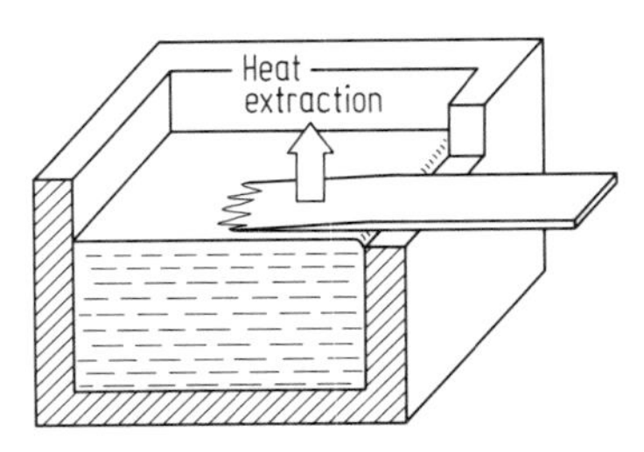

Fig. 67. Schematic graph of horizontal ribbon growth (HRG) [81D1, 79K3] and low-angle silicon sheet growth (LASS). The angle between the ribbon and the surface of the silicon melt ranges between 4° and 10°. The pulling rate is influenced by the length of the wedge-shaped solidification front.

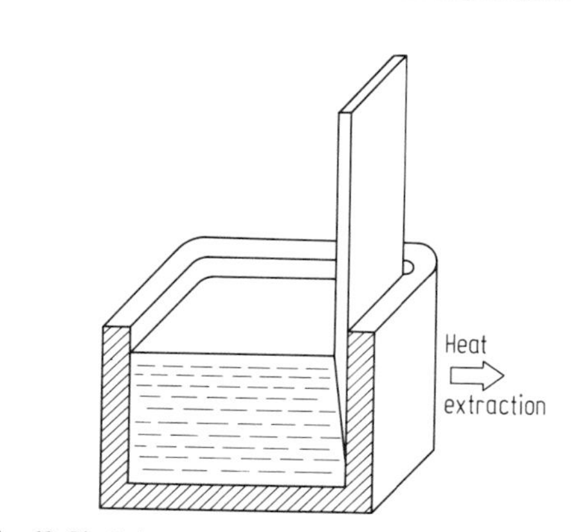

Fig. 68. Si. Schematic graph of the interface-controlled crystallization method (ICC), [81D1]. The ribbon crystallizes at the cooled ramp which is covered by a liquid film of molten slag to avoid sticking of the solidified silicon.

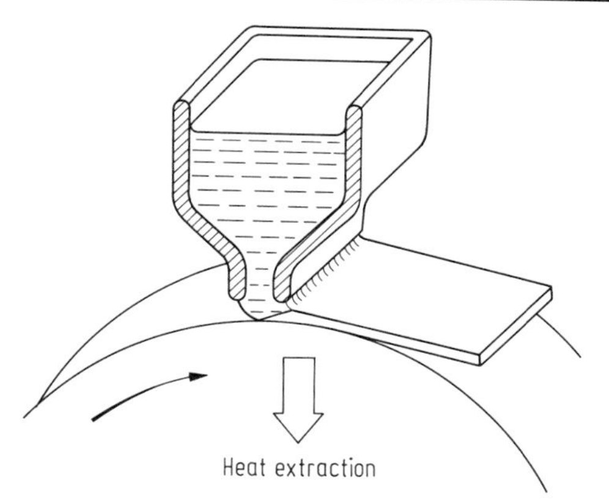

Fig. 70. Si. Schematic graph of the rapid quenching in the roller quenching method [80T, 79T]. A liquid silicon film is quenched in contact with the surface of a very fast rotating drum.

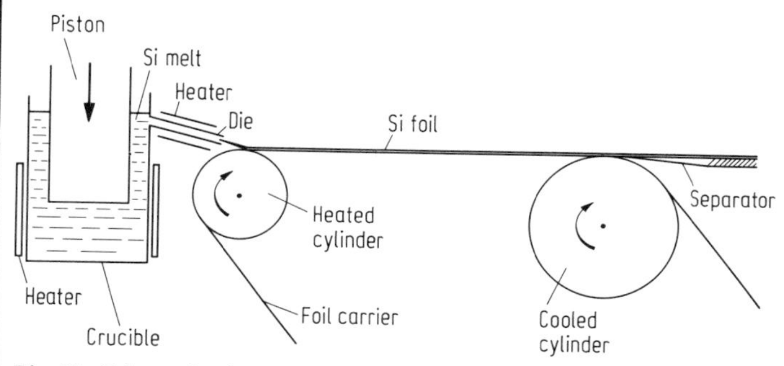

Fig. 69. Schematic view of a set-up for continuous production of ribbons by a modified interface-controlled crystallization (ICC) method [81D1].

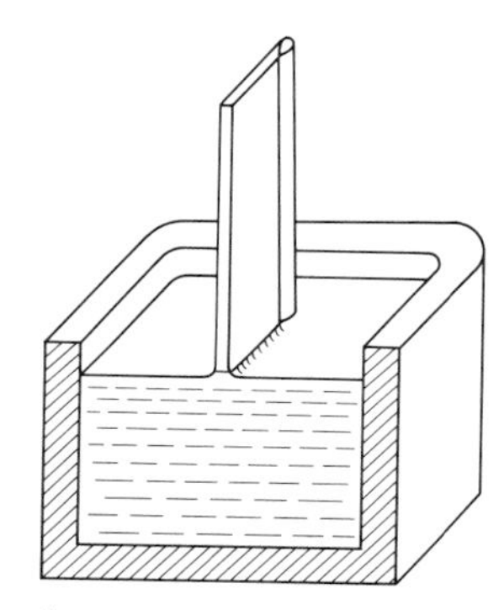

Fig. 71. Si. Schematic graph of the dentritic web growth (WEB) [80D2, 80S3] or the edge-supported pulling (ESP) in general [66T, 80C2]. The edges of the ribbon are stabilized either by dentrites grown simultaneously with the silicon sheet (WEB) or by graphite or quartz filaments fed through the melt to the crystallization front (ESP).

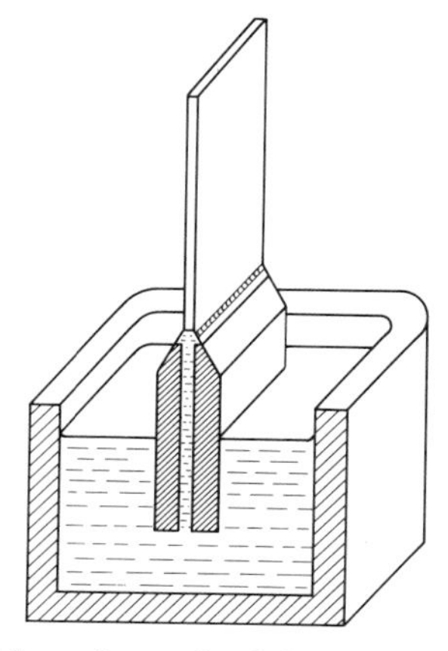

Fig. 72. Si. Schematic graph of the edge-defined film-fed growth (EFG) [77R, 80K5] or the capillary action shaping technique (CAST). The ribbon pulled is shaped by a die. In CAST capillary action helps to extract silicon melt through the die [72C].

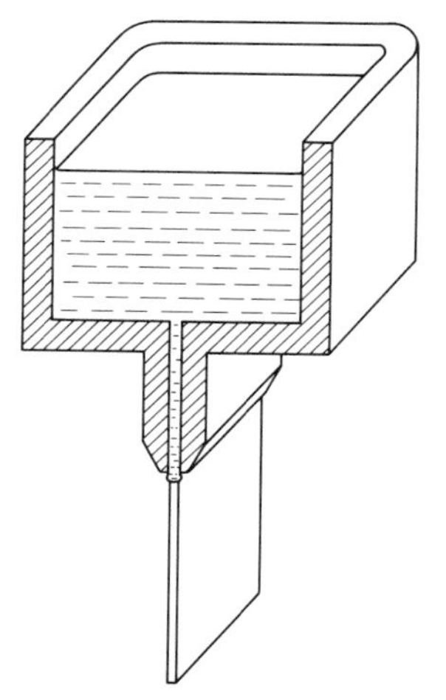

Fig. 73. Si. Schematic graph of the inverted Stepanov (IS) method [77K, 80K6]. The liquid melt is forced through the shaping die by gravity forces.

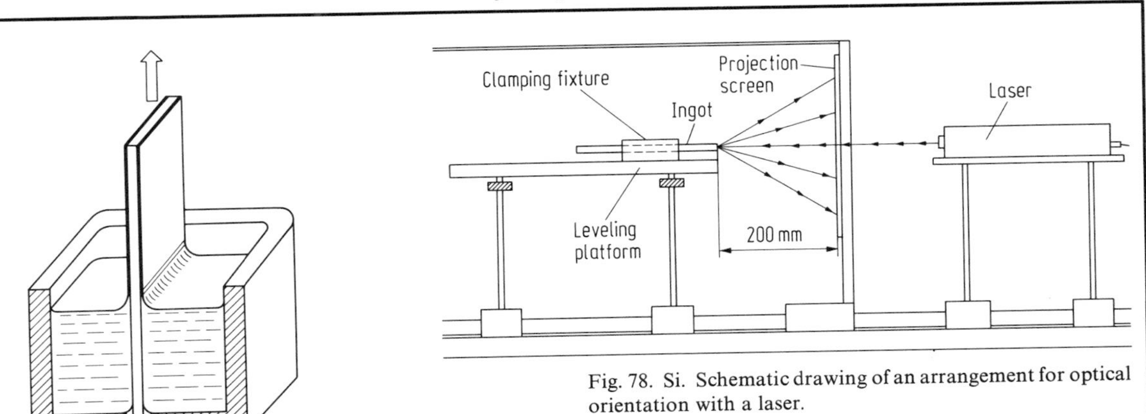

Fig. 78. Si. Schematic drawing of an arrangement for optical orientation with a laser.

Fig. 74. Si. Schematic graph of the ribbon-against-drop (RAD) pulling method [78M]. A graphite foil is fed through the bottom of the melt crucible. Silicon is deposited on both faces of the foil.

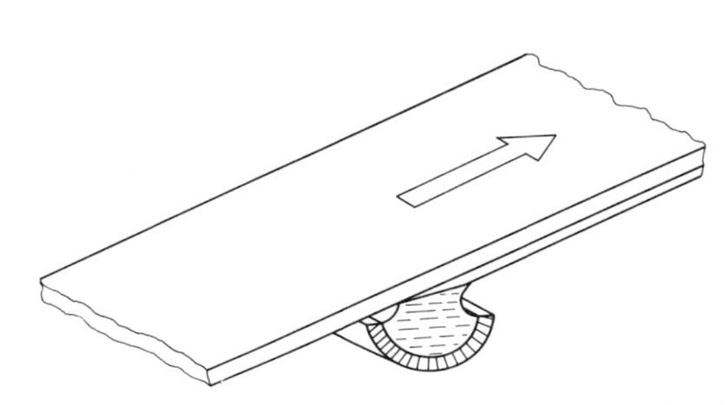

Fig. 75. Si. Schematic graph of the silicon coating by inverted meniscus (SCIM) method [80H2, 80Z]. A supporting graphitized substrate ceramic sheet is coated by moving it over a trough containing molten silicon. The SCIM method is a further development of the original silicon-on-ceramic (SOC) method [78B, 80B2].

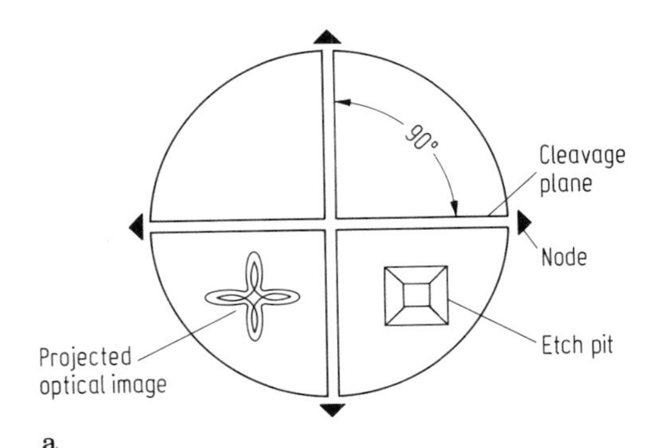

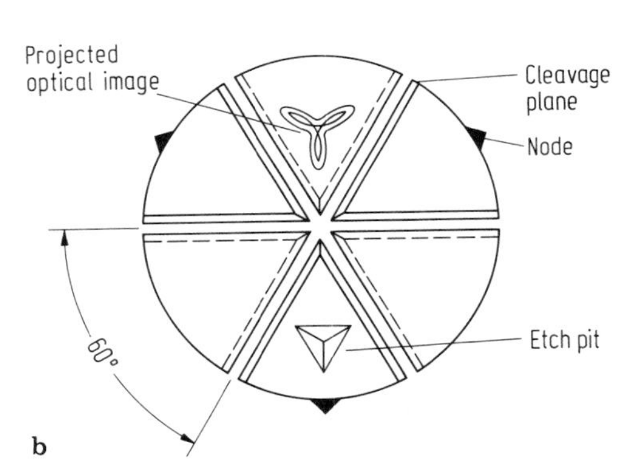

Fig. 77 a, b. Si. Principle of optical orientation of Si ingots. A sawn and KOH-etched cross section of the crystal is irradiated by a laser beam. Due to inclined (111)-surfaces of etch pits or grooves the laser beam is reflected in 4 directions at a (100)-cross section, in 3 directions at a (111)-cross section, and in 2 directions at a (110)-cross section. a) Cross-section and optical image of a (100) plane, b) cross-section and optical image of a (111) plane.

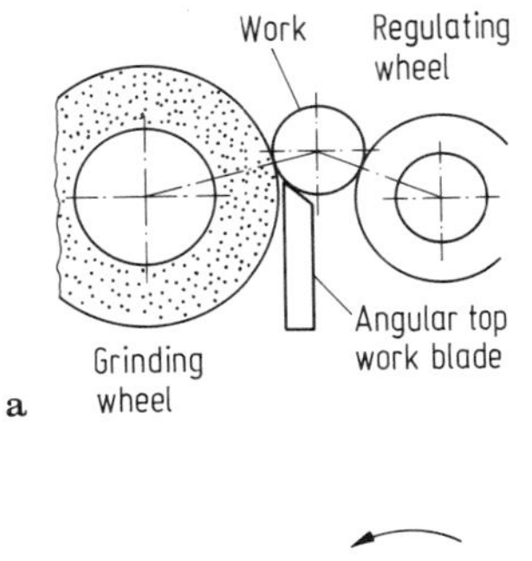

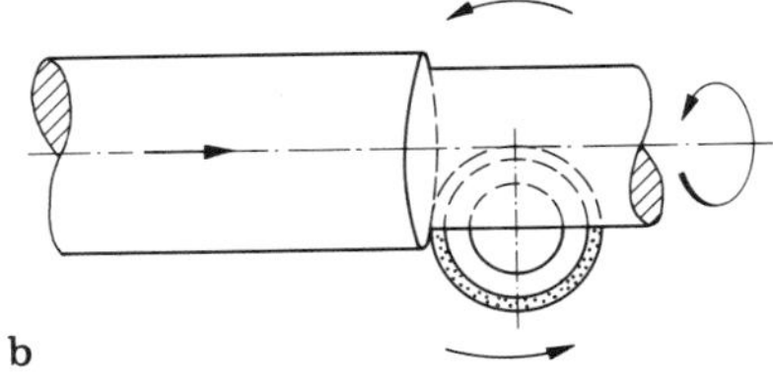

◄

Fig. 76 a, b. Si. Centerless a) and cylindrical b) grinding for silicon crystals.

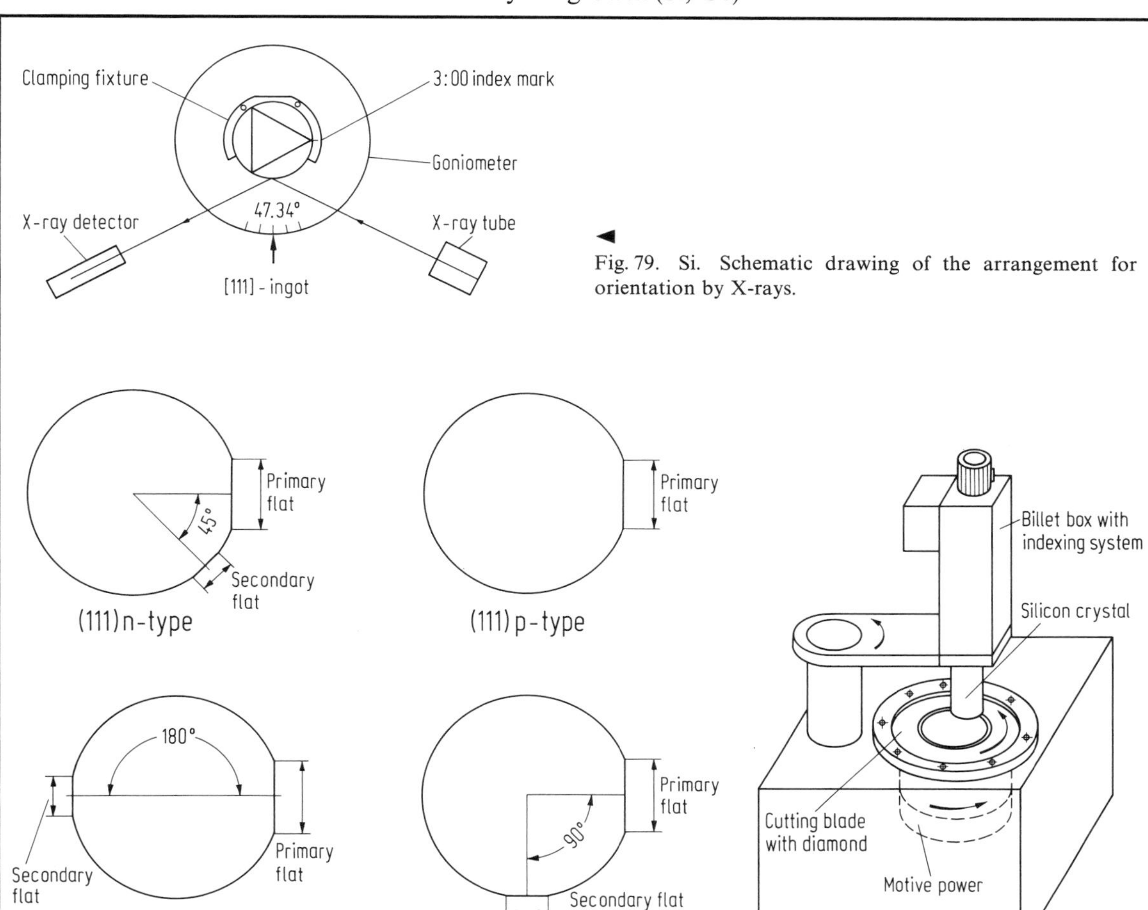

◄ Fig. 79. Si. Schematic drawing of the arrangement for orientation by X-rays.

Fig. 80. Si. Standard silicon single crystal wafers are characterized by flats with respect to orientation and conductivity type.

Fig. 81. Schematic drawing of a cutting machine with an inside diameter sawblade for the slicing of silicon or germanium crystals.

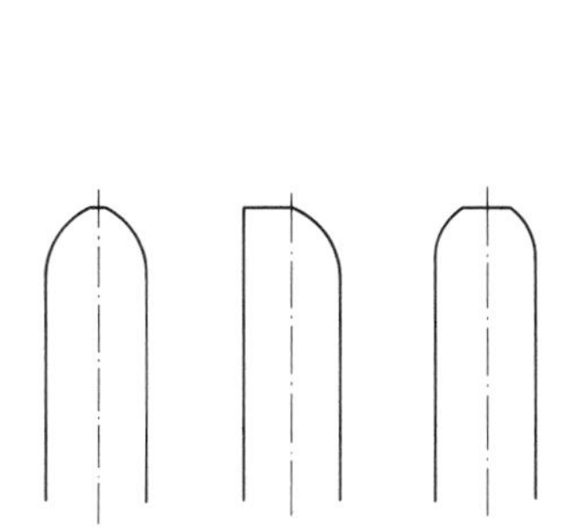

Fig. 82. Si. Schematic drawings of different edge-rounded wafers before etching (section).

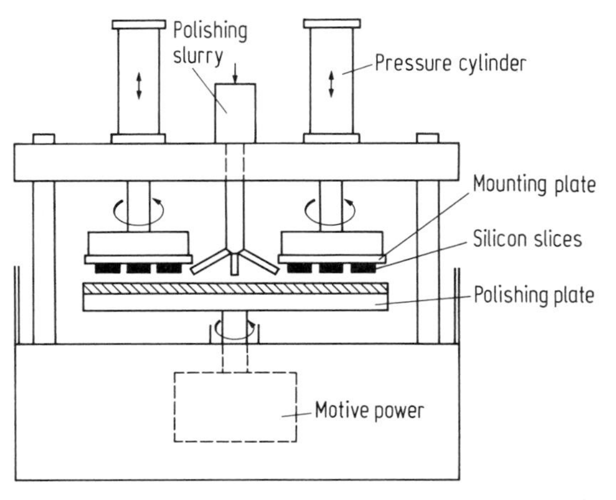

Fig. 83. Schematic drawing of a polishing machine for wax-mounted wafers.

6.1.3 Characterization of crystal properties

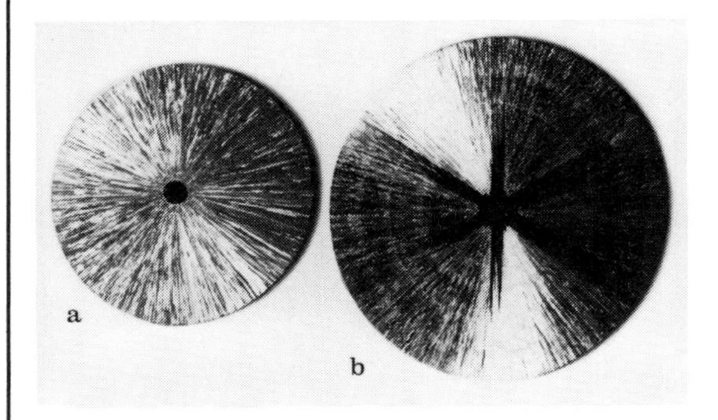

Fig. 1 a, b. Si. Cross-sections taken from polysilicon rods. a) monocrystalline core with polycrystalline deposit; b) monocrystalline core in ⟨111⟩ direction with partly monocrystalline deposit in equivalent ⟨110⟩ directions.

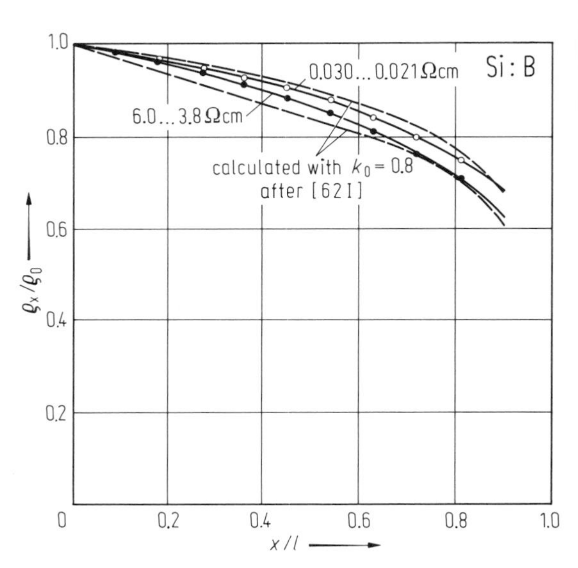

Fig. 3. Si:B. Axial resistivity variation ϱ_x/ϱ_0 in CZ—Si-single-crystals with different boron contents pulled under the same conditions vs. relative crystal length x/l. The broken lines are calculated according to the normal freezing and converted into resistivity values by using Irvin's curve [62I].

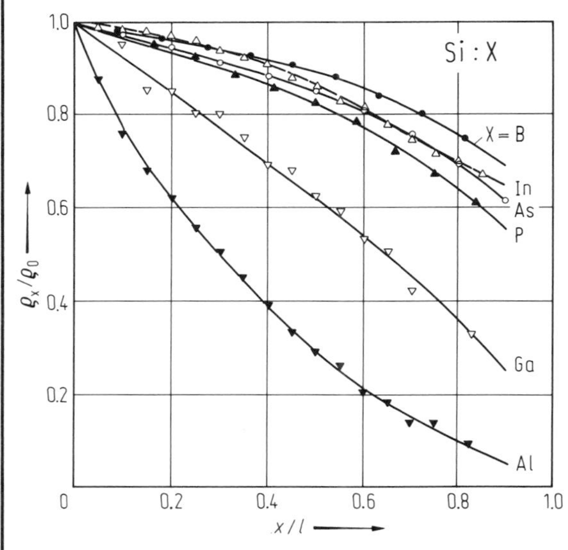

Fig. 2. Si:X (X = B, Al, Ga, In, P, As). Axial resistivity variations ϱ_x/r_0 of different dopants in CZ—Si-single crystals, all grown under the same conditions in reduced pressure vs. relative crystal length x/l. Resistivity ranges: B: $\varrho = 0.03 \cdots 0.021$ Ωcm, Al: $\varrho = 0.3 \cdots 0.021$ Ωcm, Ga: $\varrho = 0.18 \cdots 0.054$ Ωcm, In: $\varrho = 1.05 \cdots 0.36$ Ω cm, P: $\varrho = 0.015 \cdots 0.0086$ Ω cm, As: $\varrho = 0.030 \cdots 0.018$ Ωcm. The axial resistivity variation of Sb pulled under the same pressure of 11 mbar cannot be drawn in this diagram because all values ϱ_x/ϱ_0 are greater than 1, see Fig. 6, curve *B*.

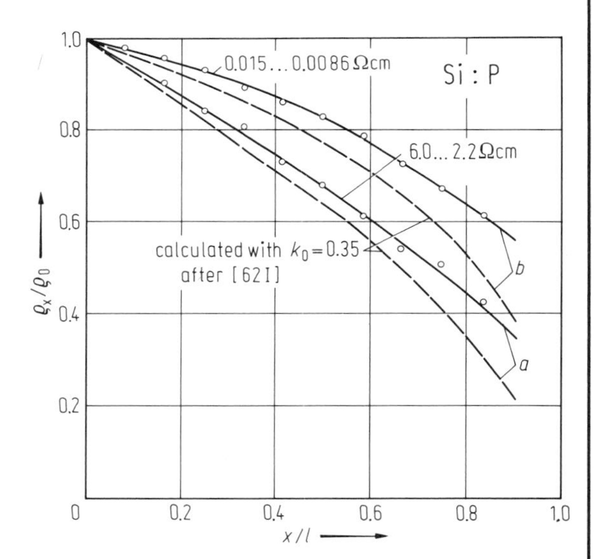

Fig. 4. Si:P. Axial resistivity variation ϱ_x/ϱ_0 in CZ—Si-single-crystals with varying phosphorus content versus relative crystal length x/l. The dashed lines are calculated according to the normal freezing and converted into resistivity values by using Irvin's curve [62I]. The calculated curves are lower because of evaporation of P in the experiment. (Corresponding curves are indicated by a and b.)

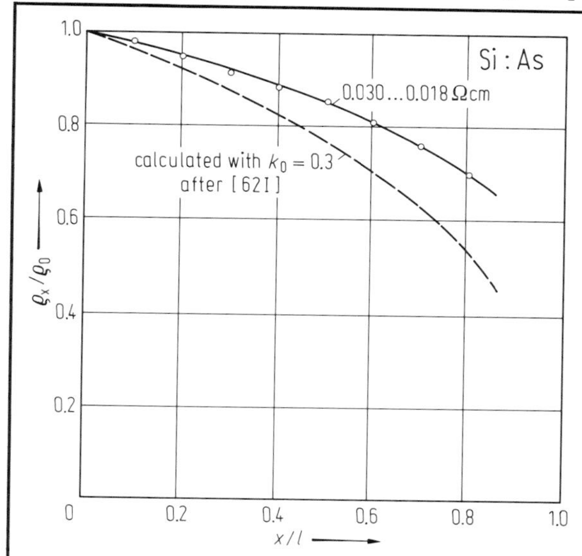

Fig. 5. Si:As. Axial resistivity variation ϱ_x/ϱ_0 in CZ—Si-single-crystals containing As versus relative crystal length x/l. The dashed line is calculated assuming normal freezing and converted into resistivity by using Irvin's curve [62I]. The calculated curve is lower than the experimental curve because of evaporation of As.

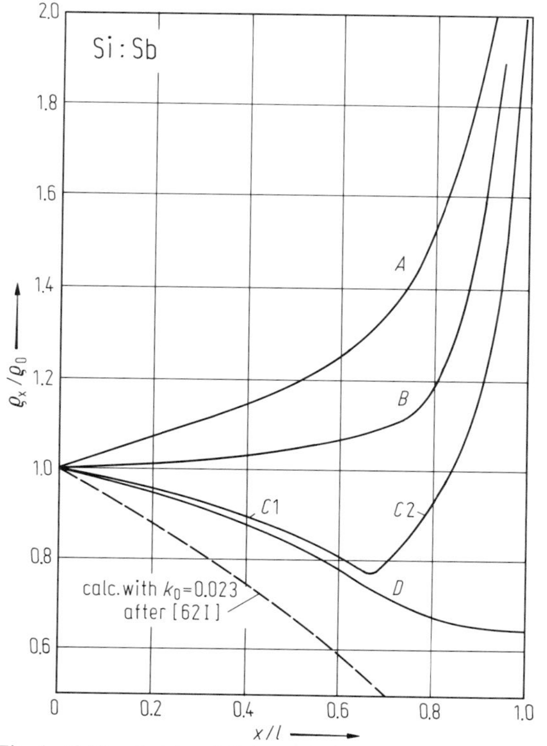

Fig. 6. Si:Sb. Axial resistivity variations ρ_x/ρ_0 in antimony doped CZ—Si-single-crystals vs. relative crystal length x/l. The starting doping concentration (≙0.018 Ω cm) was the same for all curves, also melt volume and crystal diameter. Due to the high vapour pressure of Sb a variety of axial resistivity variations may be obtained by applying different pressures and gas flow characteristics. Curves A and B: Same pressure of 11 mbar but different gas flow characteristics. Curves C: 67 mbar in C 1, 11 mbar in C 2. Curve D: 75 mbar.

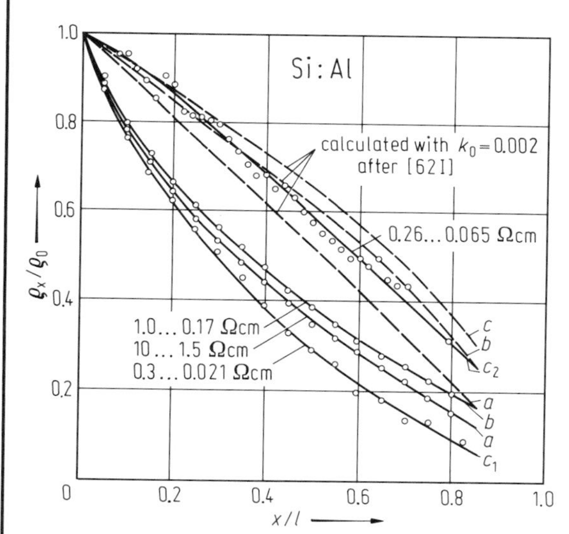

Fig. 7. Si:Al. Axial resistivity variation ϱ_x/ϱ_0 in CZ—Si-single-crystals of different Al-concentrations pulled under the same conditions vs. relative crystal length x/l, (solid lines) in comparison with calculated curves [62I] (dashed lines). For Al-doping of Si, the resulting resistivities are rather uncertain. Especially for high Al-concentrations, resistivities differ sometimes by a factor 2⋯3 for the same Al-concentrations in the melt. In dislocated or polycrystalline Si-rods the values of k_{eff} (calculated according to the resistivity) are often substantially higher than in dislocation-free crystals of lower Al-concentrations. The origin of these differences are unknown. A reaction in the melt between Al and O to aluminium oxide which reduces the number of free Al-atoms for incorporation into the Si-crystal seems possible. (Corresponding curves are indicated).

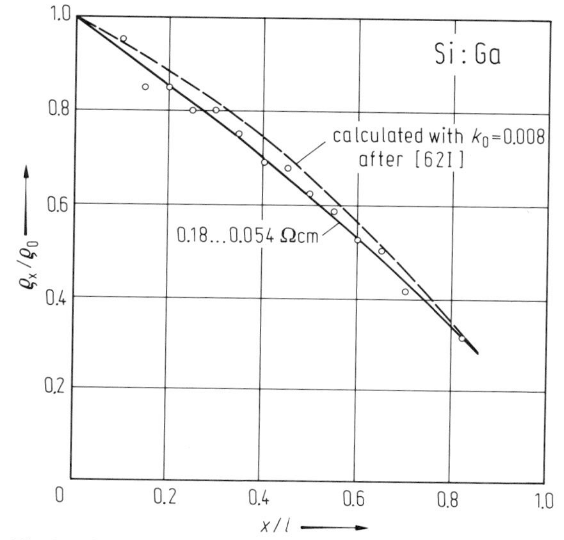

Fig. 8. Si:Ga. Axial resistivity variation ϱ_x/ϱ_0 in Ga-doped CZ—Si-single crystals vs. relative crystal length x/l. The same pulling conditions are used as in Fig. 7. The measured values and the calculated ones are in good agreement.

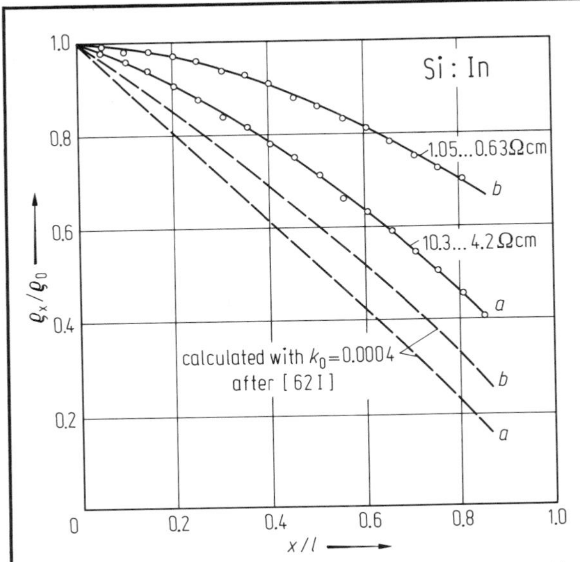

Fig. 9. Si:In. Axial resistivity variation ϱ_x/ϱ_0 in CZ—Si-single-crystals of different In-concentrations vs. relative crystal length x/l. The measured curves are distinctly above the calculated ones because an essential fraction of the In has evaporated out of the melt. After antimony indium has the second highest vapour pressure of the doping elements B, Al, Ga, In, P, As, Sb. (Corresponding curves are indicated).

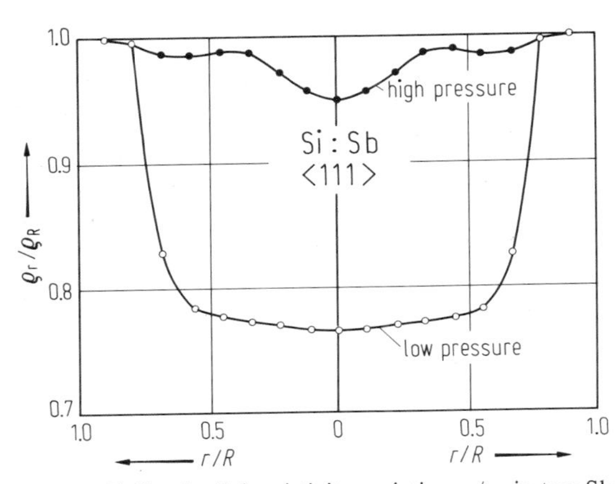

Fig. 10. Si:Sb. Radial resistivity variation ϱ_r/ϱ_0 in two Sb-doped CZ—Si-single-crystals for the same crystal length pulled under different pressures vs. radial position in the crystal r/R. All other crystal and pulling parameters were the same. The crystal which was pulled under the low pressure showed a considerably higher resistivity near the crystal surface caused by evaporation of Sb from the melt near the periphery of the solidification interface.

►

Fig. 13. Si:B, Si:P. Radial resistivity variation ϱ_r/ϱ_R in boron- and phosphorus-doped CZ—Si-single-crystals vs. radial position r/R showing the difference between the incorporation of boron and phosphorus. Although the P-doped crystal was grown with rotation parameters more favourable for uniform radial resistivity it shows a 2.5 times higher radial resistivity variation than the B-doped crystal; both crystals are of ⟨111⟩-orientation.

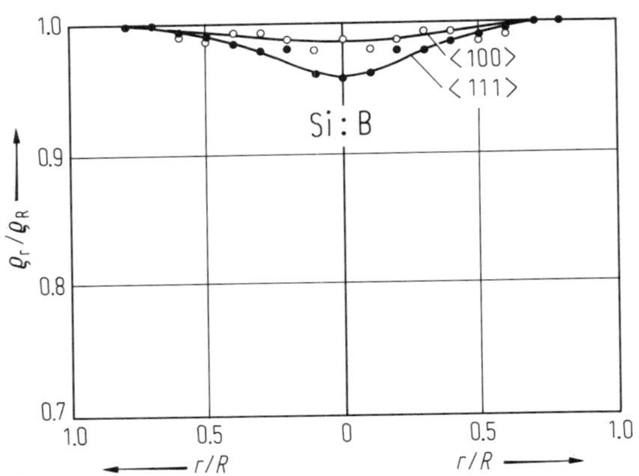

Fig. 11. Si:B. Radial resistivity variation ϱ_r/ϱ_R in boron doped CZ—Si-single-crystals vs. radial position r/R showing the influence of the crystal orientation on the dopant incorporation. For ⟨111⟩ orientation, a stronger enrichment of dopant towards the centre of the crystal takes place than for ⟨100⟩ orientation. The crystals were grown under the same conditions.

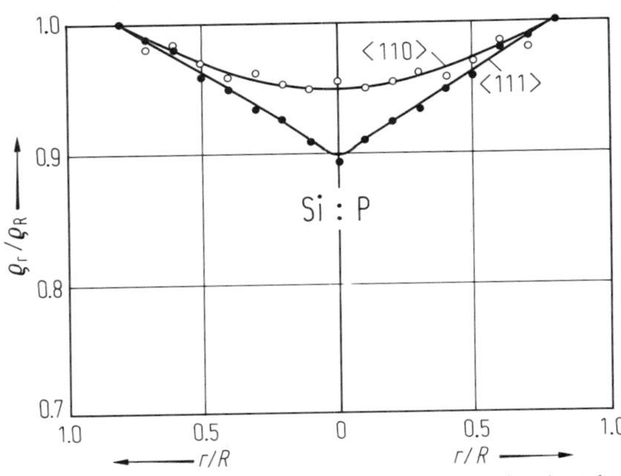

Fig. 12. Si:P. Radial resistivity variation ϱ_r/ϱ_R in phosphorus-doped CZ—Si-single-crystals vs. radial position r/R showing the influence of the crystal orientation on the dopant incorporation. For ⟨111⟩ orientation, a stronger enrichment of dopant towards the centre of the crystal takes place than for ⟨100⟩ orientation. The crystals were grown under the same conditions.

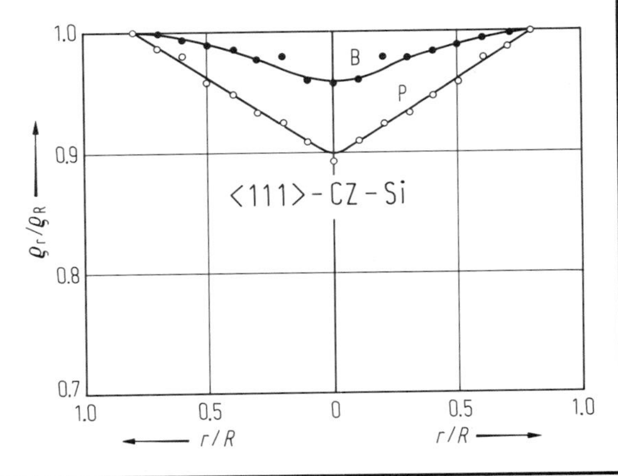

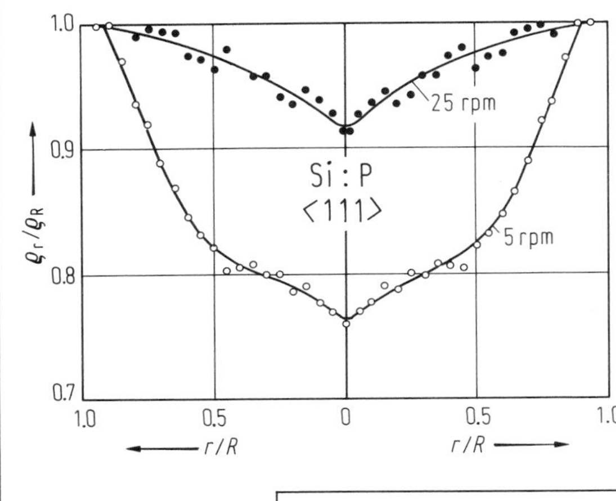

◀ Fig. 14. Si:P. Radial resistivity variation ϱ_r/ϱ_R in a CZ—Si-single crystal vs. radial position r/R. (⟨111⟩, 80 mm ∅, phosphorous, $\varrho = 5\ \Omega$ cm). The pulling conditions for both crystals were the same except the crystal rotation. For low rotation rate (5 rpm), the radial resistivity variation was 24.2% and for high rotation rate (25 rpm) it was 8.8%. The pronounced enrichment of phosphorous in the centre of the crystal is due to the facet growth of the ⟨111⟩-crystal.

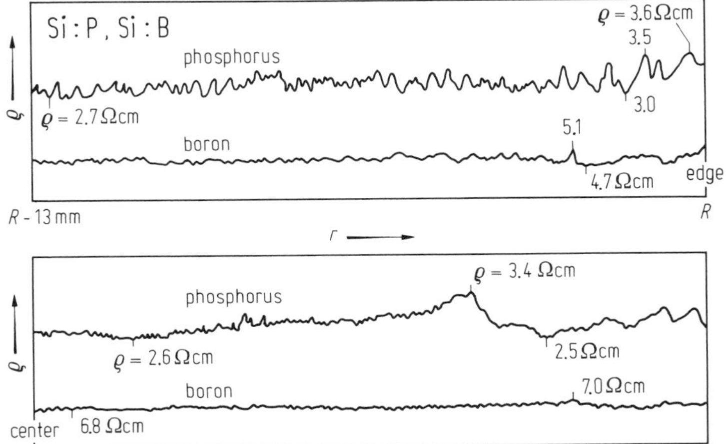

Fig. 15. Si:P, Si:B. Spreading resistivity ρ of boron- and phosphorous-doped CZ—Si-single crystal wafers vs. radial position r showing the influence of the distribution coefficient k on the microscopic resistivity variation (striations). The lower k-value of P causes essentially stronger dopant striations than for the case of B. (7,62 cm diameter wafer, ⟨100⟩ orientation).

For Fig. 16, see next page.

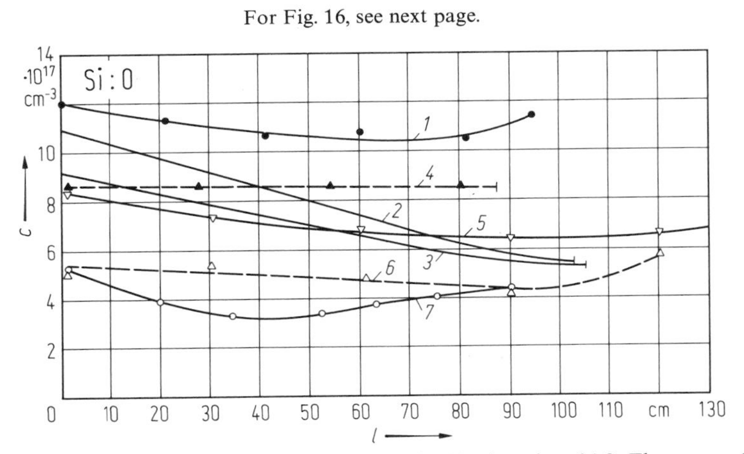

Fig. 17. Si:O. Axial oxygen concentration c in CZ—Si-single crystals grown with different pulling parameters vs. axial position l. The curves *2* and *3* represent typical axial distributions after usual CZ-crystal pulling with constant opposite crystal and crucible rotations in a furnace like that in Fig. 15 of section 6.1.2. The curves *1*, *4*, *5*, *6* and *7* are taken from crystals grown after improved processes. With these processes it is possible to grow crystals with axial more uniform oxygen contents both on high and low levels.

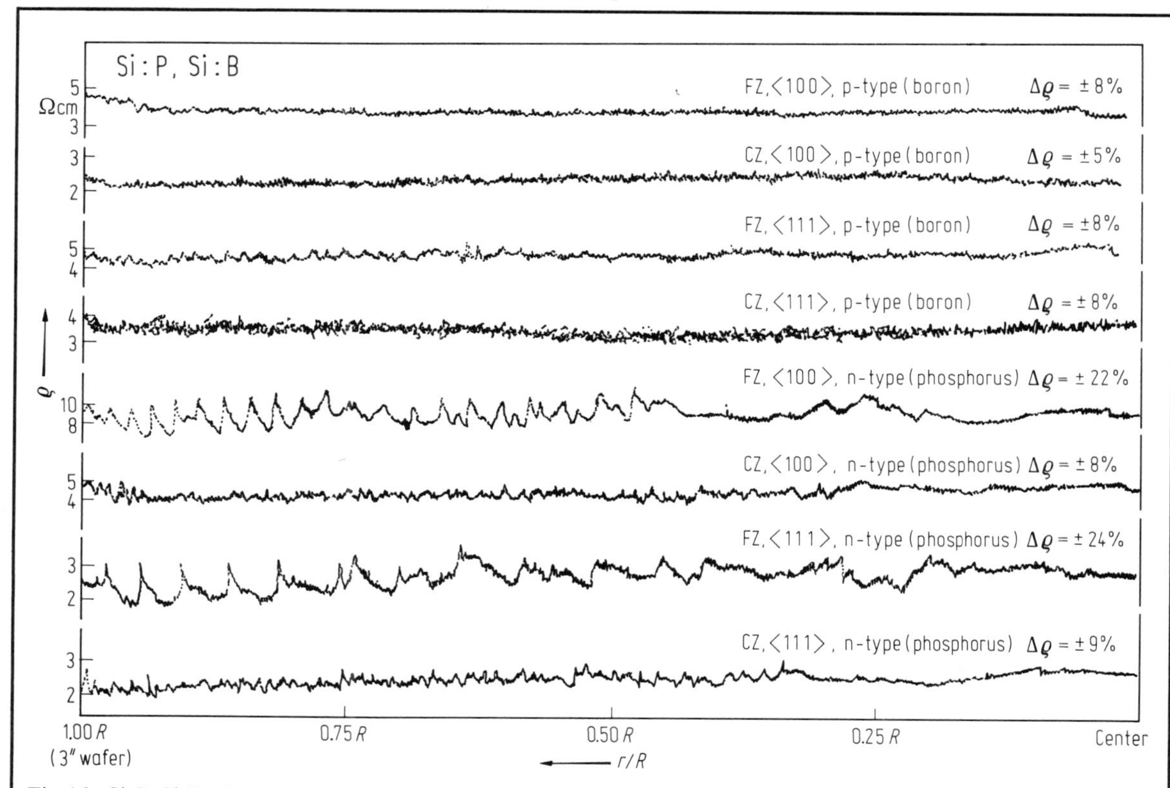

Fig. 16. Si:P, Si:B. Spreading resistivity ϱ vs. radial position r/R. Comparison of Czochralski grown (CZ) with float zone (FZ) silicon crystals of the same specification by spreading resistance measurements. The dopant striations in float zone crystals are greater due to the pronounced remelt phenomena at every crystal rotation. Parameters are the growth type, the conductivity type and dopant, and the measured resistivity variation $\Delta\varrho$.

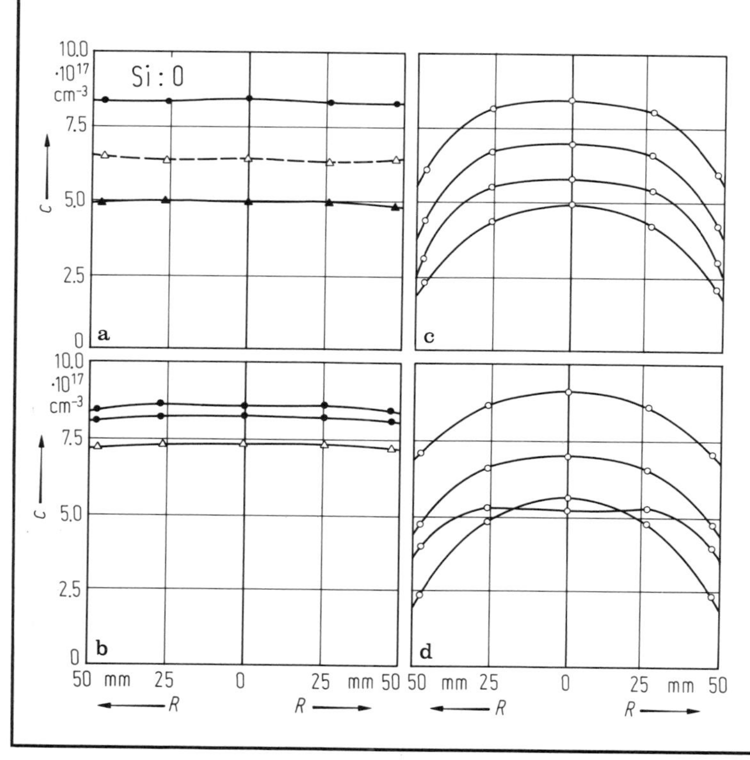

◀ Fig. 18. Si:O. Radial oxygen concentration c in CZ—Si-single-crystals vs. radial position R. The crystals in a) and b) were grown according to the melt flow characteristics in Fig. 28d of section 6.1.2, whereas the crystals in c) and d) were grown according to the melt flow characteristics which are situated between that of Figs. 28a and 28c of section 6.1.2.

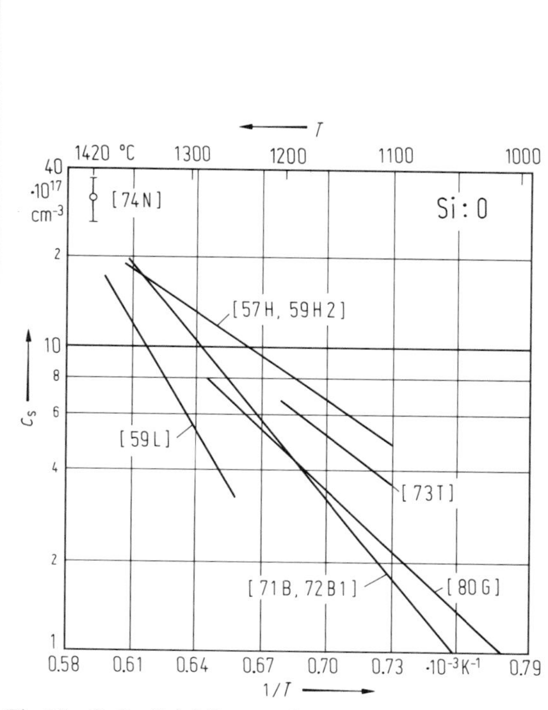

Fig. 19. Si:O. Solubility c_S of oxygen in solid silicon vs. reciprocal temperature $1/T$.

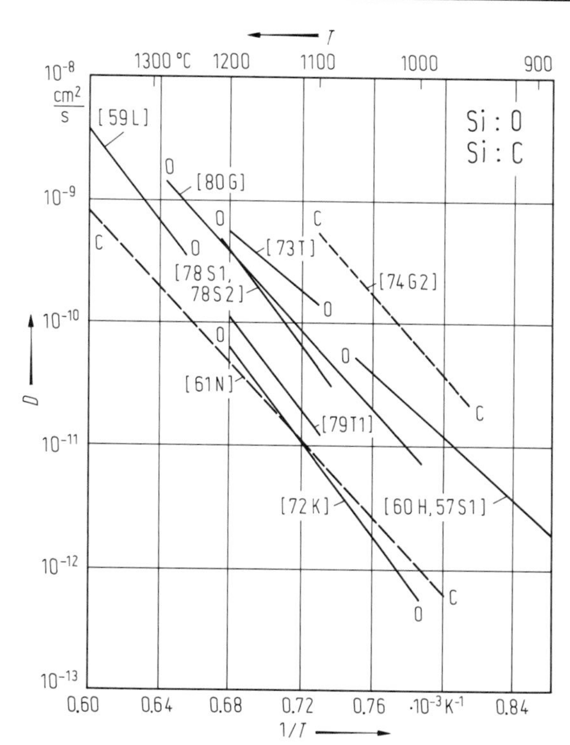

Fig. 20. Si:O and Si:C. Diffusion coefficients D of oxygen and carbon in solid silicon vs. reciprocal temperature $1/T$. [79T1] describe their values by the diffusion of Si self-interstitials during the precipitation of oxygen, (see also Fig. 29).

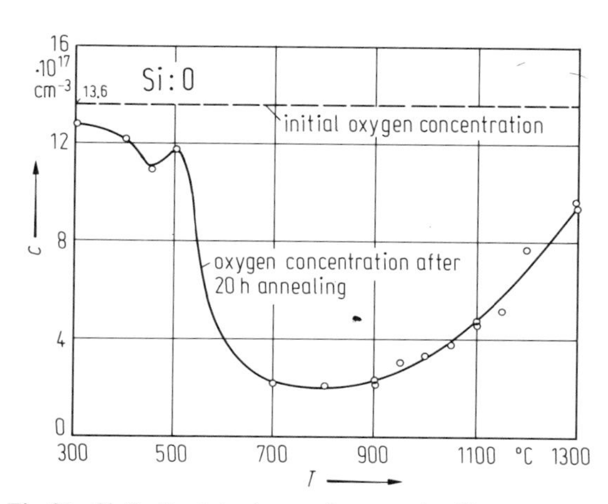

Fig. 21. Si:O. Precipitation c of oxygen in silicon vs. temperature T of annealing for 20 hours. The starting material had a dissolved oxygen content of $13.6 \cdot 10^{17}\,\mathrm{cm}^{-3}$ or 27.4 ppma determined by IR-absorption.

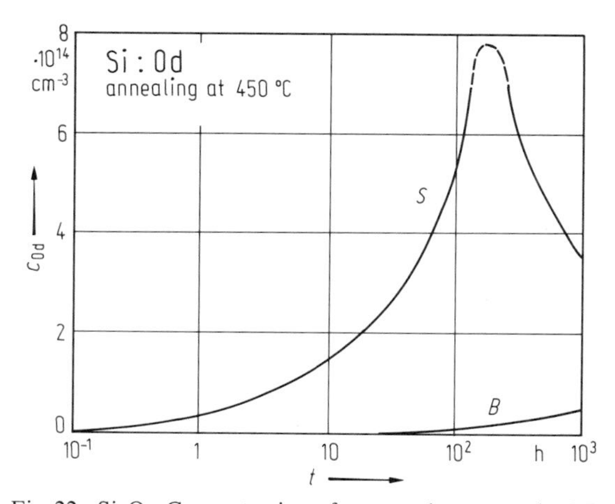

Fig. 22. Si:O. Concentration of oxygen donors c_{Od} in CZ-silicon samples vs. annealing time at 450 °C. The samples S were cut from the seed end of the crystal (oxygen rich), and the samples B from the bottom end (less oxygen) [77C1].

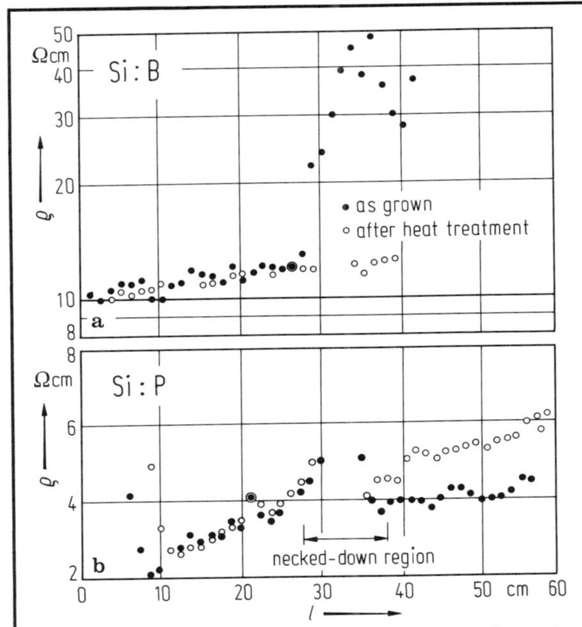

Fig. 23a, b. Si:O. Resistivity ϱ vs. distance l from the bottom of the crystal. The oxygen donor concentrations in as-grown crystals may be determined by measuring the axial resistivity variation before and after an annealing above 500 °C, e.g. at 650 °C. [73V1]. a) Boron-doped crystal, 5.8 cm diameter, b) phosphorus-doped crystal, 5.8 cm diameter. After 28 cm length, the diameter was necked down to 0.16 cm and then increased again for the final 22.9 cm of growth.

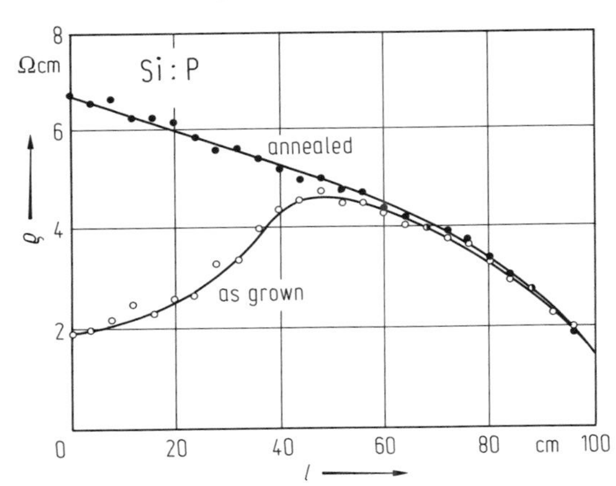

Fig. 25. Si:P. Resistivity ϱ vs. crystal length l. Oxygen donors in an as-grown phosphorus-doped CZ—Si-single-crystal. After annealing and a following rapid cooling oxygen donors are eliminated. In the second half of the crystal only a few donor centers are formed.

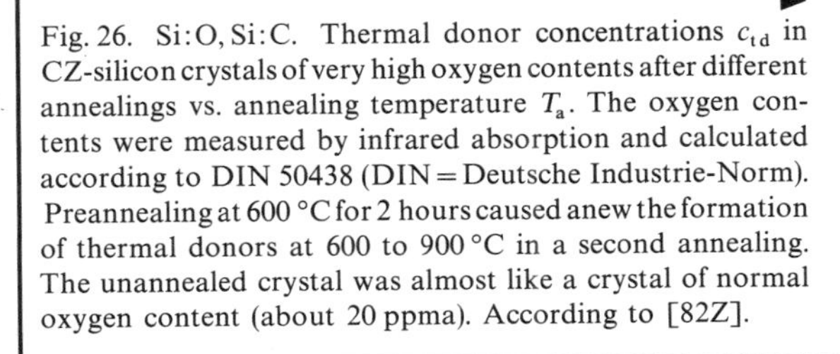

▶

Fig. 26. Si:O, Si:C. Thermal donor concentrations c_{td} in CZ-silicon crystals of very high oxygen contents after different annealings vs. annealing temperature T_a. The oxygen contents were measured by infrared absorption and calculated according to DIN 50438 (DIN = Deutsche Industrie-Norm). Preannealing at 600 °C for 2 hours caused anew the formation of thermal donors at 600 to 900 °C in a second annealing. The unannealed crystal was almost like a crystal of normal oxygen content (about 20 ppma). According to [82Z].

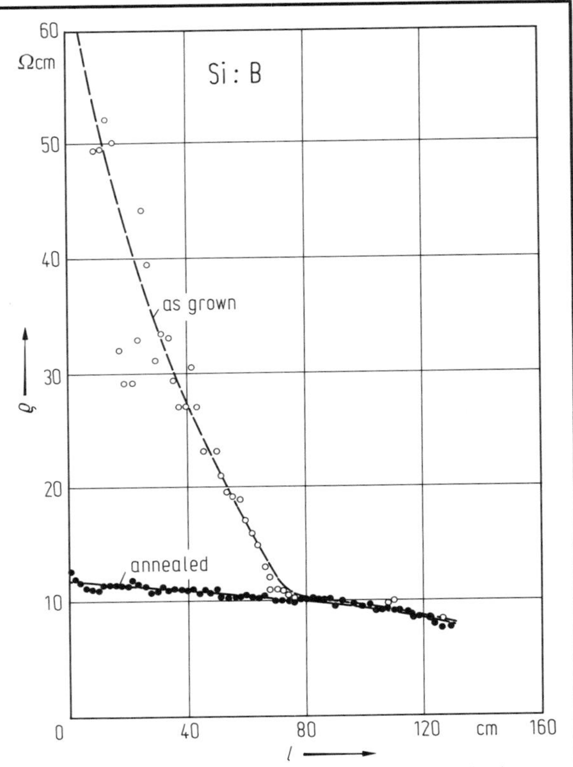

Fig. 24. Si:B. Resistivity ϱ vs. crystal length l. The boron-doping $(1.2 \cdot 10^{15}\ \mathrm{cm}^{-3})$ of the crystal is totally compensated by oxygen donors in the seed end of the as-grown crystal. After annealing the crystal for two hours at 650 °C and a following rapid cooling, the oxygen donors are eliminated. In the second half of the as-grown crystal only a few oxygen donor centers are formed.

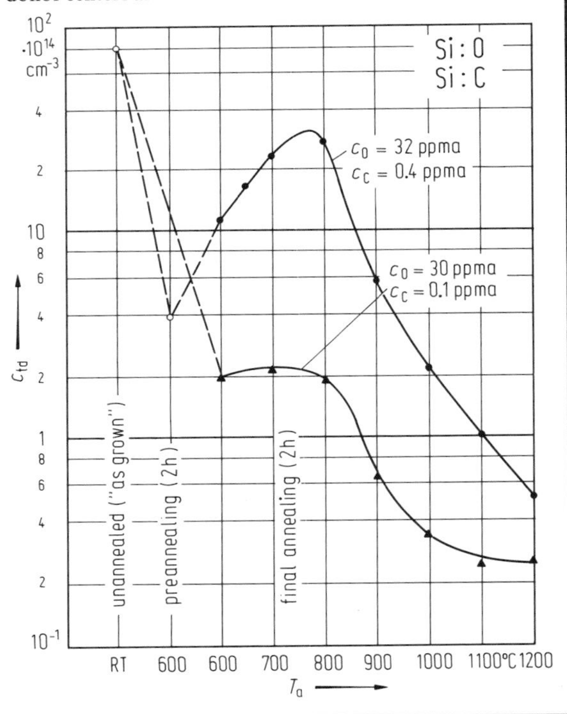

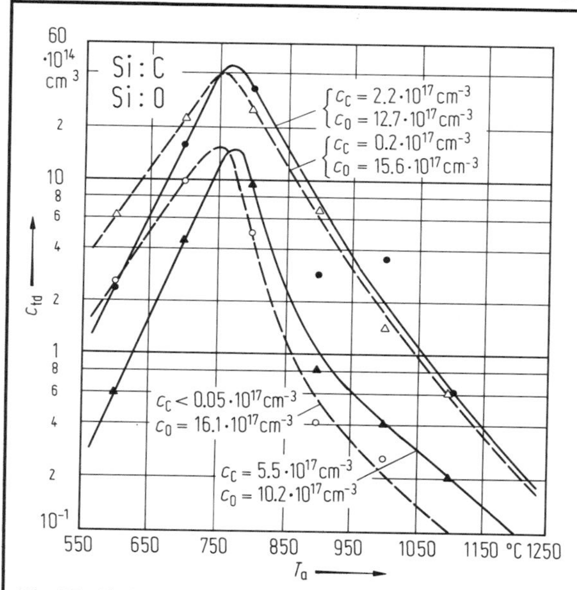

Fig. 27. Si:C, Si:O. Concentration of 770 °C thermal donors c_{td} in CZ-silicon crystals of various carbon and oxygen contents after two-step annealings vs. annealing temperature T_a. The first annealing is 2 h at 600 °C, the second annealing is 2 h at various temperatures from 600 to 1100 °C. Determination of carbon and oxygen contents was performed by infrared absorption, of donor concentrations by resistivity measurement [82Z].

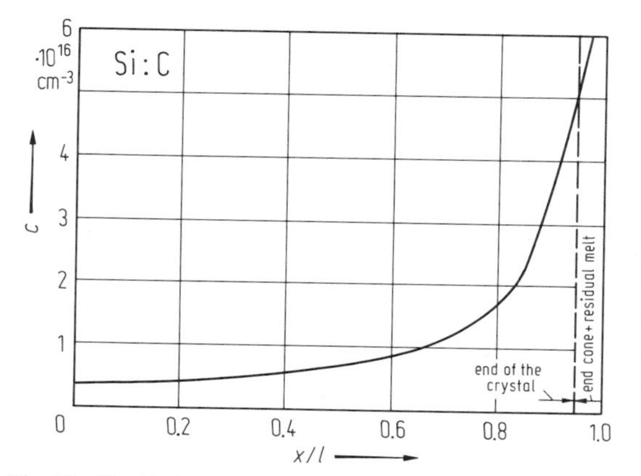

Fig. 28. Si. Carbon concentration c vs. axial position x/l of a typical CZ-silicon-single-crystal.

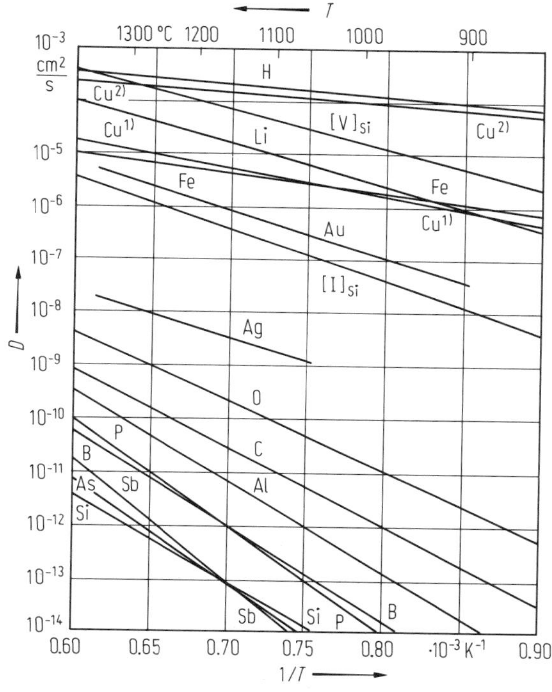

Fig. 29. Si. Diffusion coefficients D in solid silicon vs. reciprocal temperature $1/T$. Vacancies $[V]_{Si}$ after [78M1], self-interstitials $[I]_{Si}$ after [77S3], carbon after [61N], oxygen after Fig. 20 (average value), copper ([1]) and [2])) after [57S2, 64H] and the others after [75R2] (see also section 6.1.4.1).

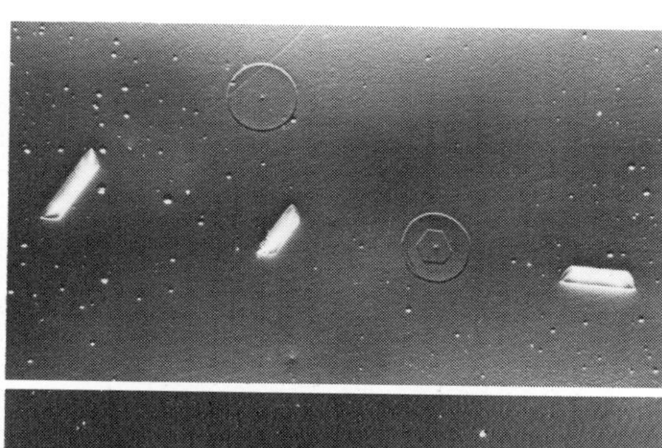

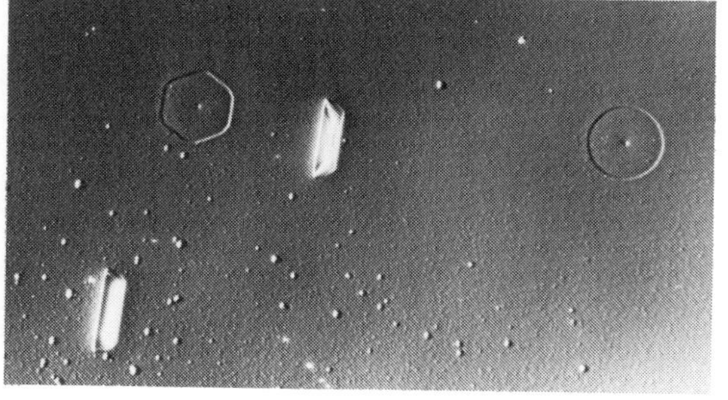

◄ Fig. 30. Si. Extrinsic stacking faults in CZ—Si-single-crystals after 20 h of annealing at 1200 °C preferentially etched in {111}-surfaces. Some of the stacking faults lying parallel to the etched surface are visible in their contours, most of them are of circular shape. The preferential etch reveals the partial dislocations of the stacking faults whereby the etch reacts on the strain field and/or the impurity cloud which surround the partial dislocations. Most of these stacking faults show the central precipitate (SiO_2-particle) as hillock in the surface.

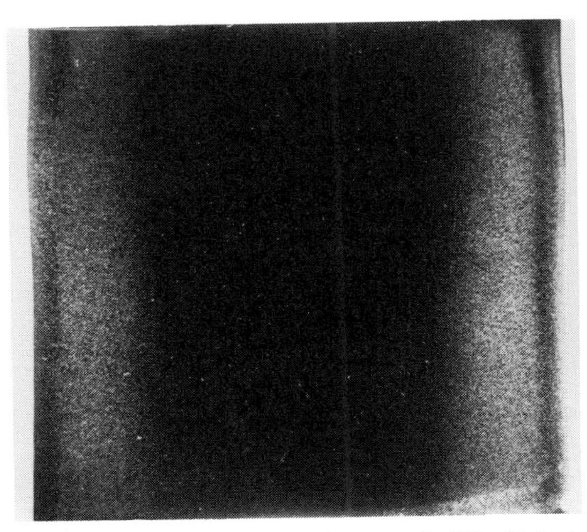

Fig. 31. Si. Axial section through a part of a CZ—Si-single-crystal showing precipitates without layer structure. They are probably precipitates of self-pointdefects (Si-interstitials) because they can be rapidly resolved and diffused out of thin wafers at high temperatures. Typically, they have their highest density near the crystal surface and the lowest (in most cases about zero) in the centre of the crystal. The defects are revealed by preferential etching.

Fig. 32. Si. Axial section through the seed end of a CZ—Si-single-crystal showing oxygen precipitates in a pronounced layer structure. In the area of the seed cone, the layers, i.e. the solidification front, are convex, in the cylindrical part of the crystal they become concave. The oxygen precipitates were generated by annealing the sample at 1000 °C for several hours and made visible by preferential etching.

►

Fig. 33. Si. Layers of oxygen precipitates in CZ—Si-single-crystals after annealing at 1000 °C for several hours. The pattern is made visible by preferential etching.

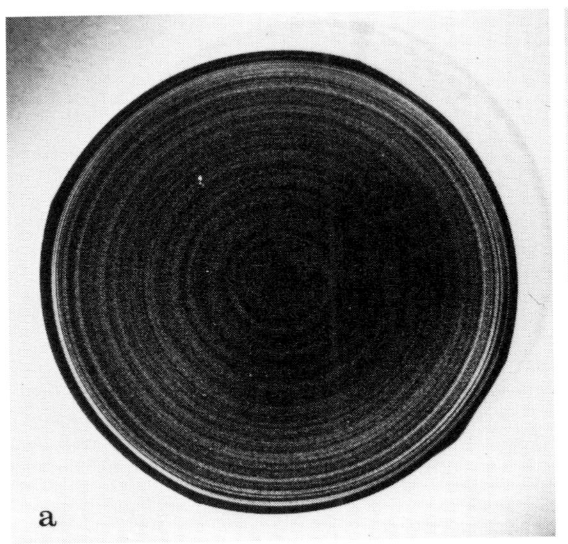

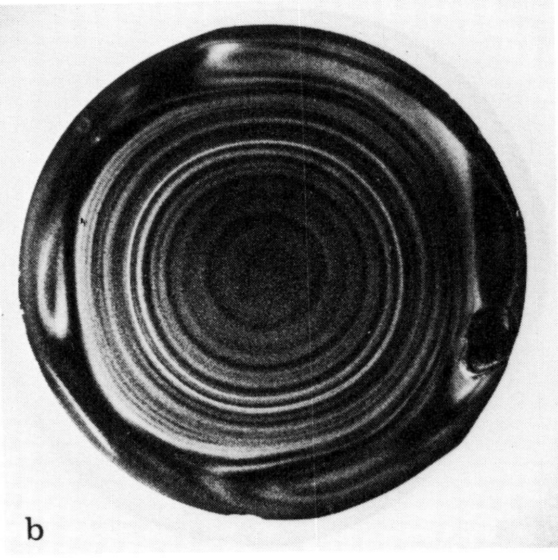

Fig. 34a, b. Si. Cross sections of CZ—Si-single-crystals taken near the seed end from two crystals containing extremely strong swirls (= pointlike defects arranged in circular or swirl like patterns). The pattern is made visible by preferential etching. Wafers a) and b) show typical patterns.

◀ Fig. 35. Axial section through a part of a crystal showing a transition from the dislocation free state to an area containing slip dislocations made visible by annealing at 1000 °C and preferential etching. The dislocation-free area shows concave layers of precipitates at high densities. In the lower part of the crystal the slip dislocations have dissolved the precipitates or prevented their formation.

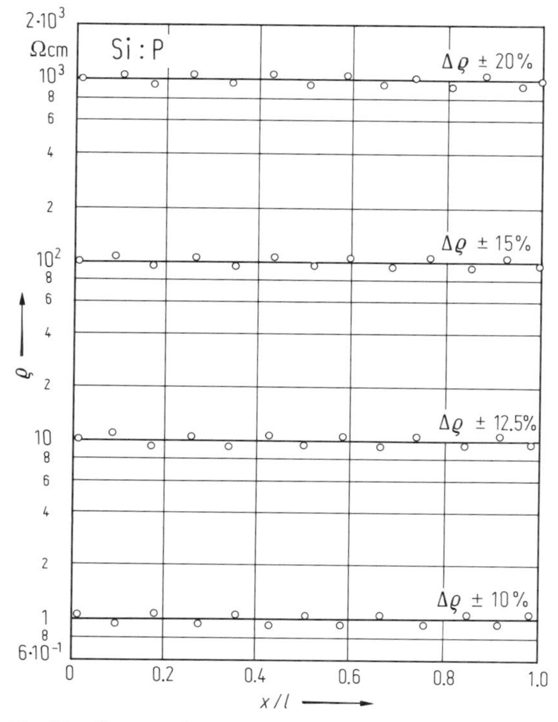

Fig. 38. Si: P. Axial resistivity variations $\Delta\varrho$ typical for phosphorus-doped crystals of 75 mm and 100 mm diameter at different resistivity levels ϱ vs. distance in relative crystal length x/l. Typical lengths are 100 to 150 cm for 75 mm crystals and up to 100 cm for 100 mm crystals, respectively. (By courtesy of H. Herzer, Wacker-Chemitronic).

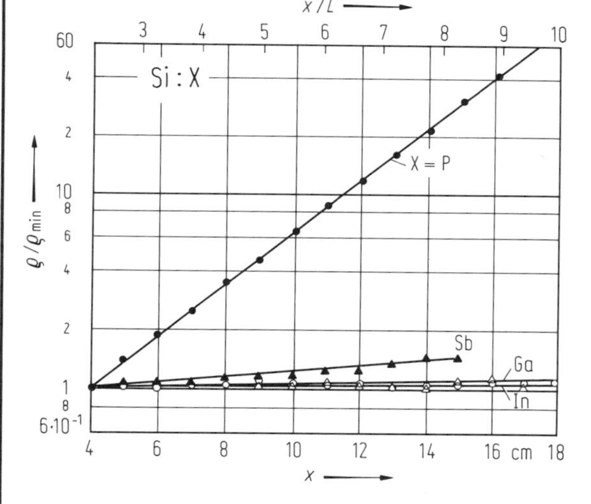

◀ Fig. 36. Si:X (X = P, Sb, Ga, In). Relative axial resistivity distribution ϱ/ϱ_{min} measured for pill-doped crystals of 35 mm diameter after one zone pass in argon. The profiles shown are for phosphorus-, antimony-, gallium-, and indium-doped crystals, respectively [81K]. (L = zone length).

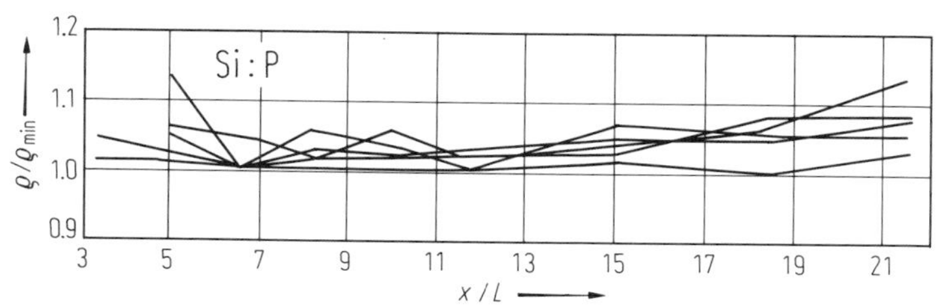

Fig. 37. Si: P. Relative axial resistivity profiles ϱ/ϱ_{min} measured for gas-doped crystals of 35 mm diameter vs. distance in zone length x/L. The profiles shown are for phosphorus-doped crystals.

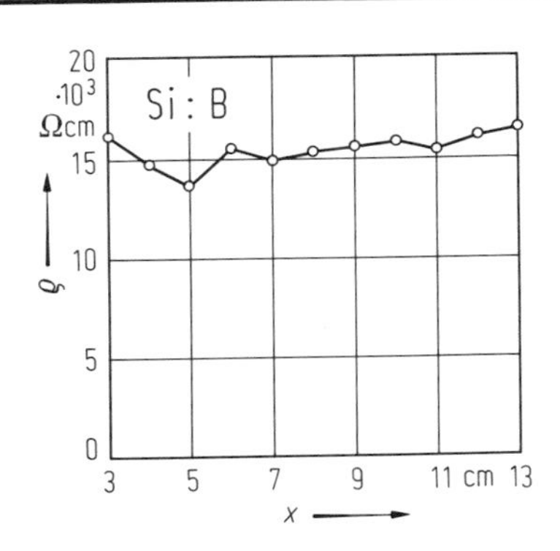

Fig. 39. Si:B. Axial resistivity profile of a high-resistivity p-type crystal measured after 6 zone passes [77K1].

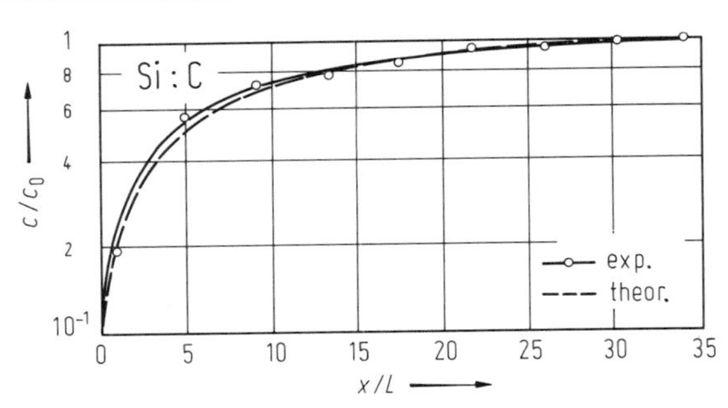

Fig. 40. Si:C. Relative axial carbon distribution c/c_0 after one zone pass vs. distance in zone length x/L. Experimental and theoretical profile the latter calculated by an effective distribution coefficient of $k=0.11$ [82K].

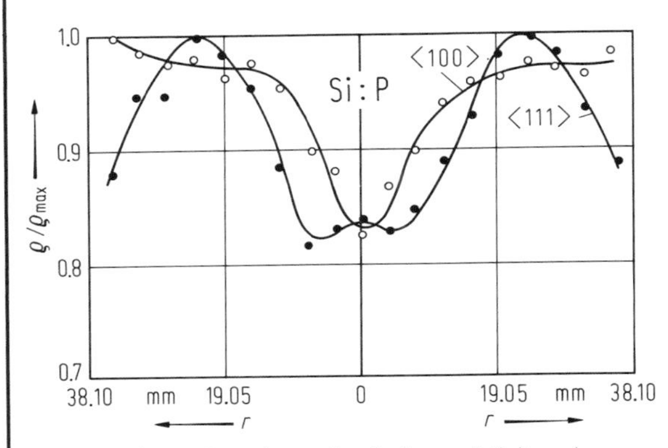

Fig. 41. Si:P. Experimental relative resistivity ϱ/ϱ_{max} vs. radial position r of phosphorus-doped crystals of 75 mm diameter and for resistivities typically between 1 and 100 Ω cm. Parameter is the crystal orientation ⟨100⟩, and ⟨111⟩, respectively. (By courtesy of H. Herzer, Wacker Chemitronic).

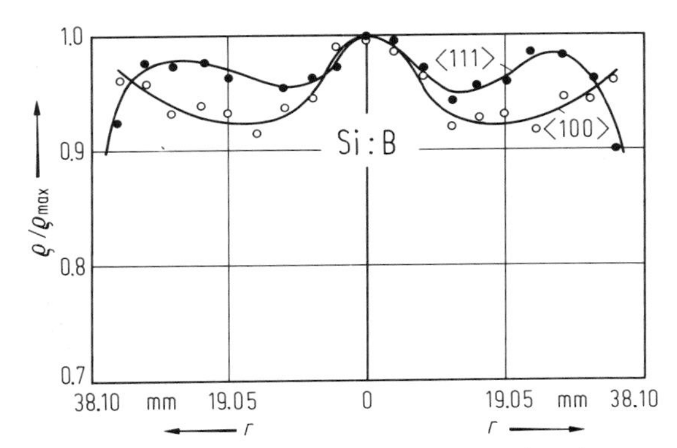

Fig. 42. Si:B. Experimental relative resistivity ϱ/ϱ_{max} vs. radial position r of boron-doped crystals of 75 mm diameter typical for resistivities between 1 and 100 Ω cm. Parameter is the crystal orientation ⟨100⟩, and ⟨111⟩, respectively. (By courtesy of H. Herzer, Wacker-Chemitronic).

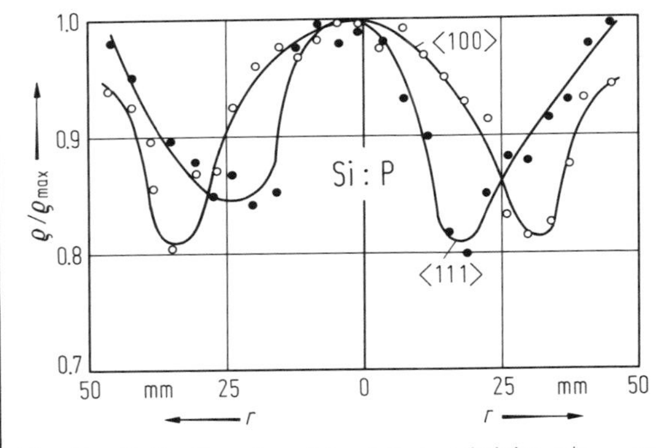

Fig. 43. Si:P. Experimental relative resistivity ϱ/ϱ_{max} vs. radial position r of phosphorus-doped crystals of 100 mm diameter for resistivities typically between 1 and 100 Ω cm. Parameter is the crystal orientation ⟨100⟩, and ⟨111⟩, respectively. (By courtesy of H. Herzer, Wacker-Chemitronic).

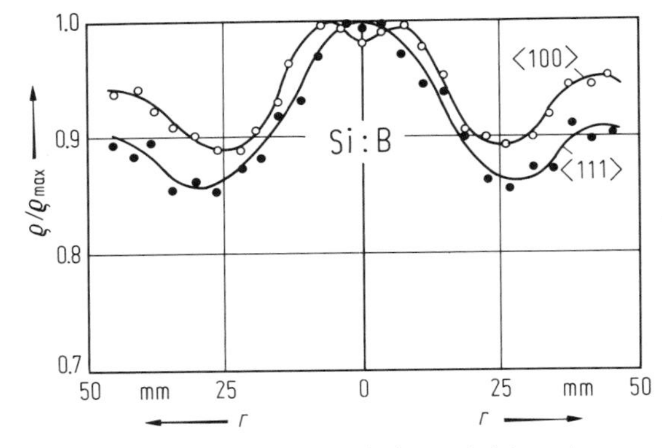

Fig. 44. Si:B. Experimental relative resistivity ϱ/ϱ_{max} vs. radial position r of boron-doped crystals of 100 mm diameter typical for resistivity levels between 1 and 100 Ω cm. Parameter is the crystal orientation ⟨100⟩, and ⟨111⟩, respectively. (By courtesy of H. Herzer, Wacker Chemitronic).

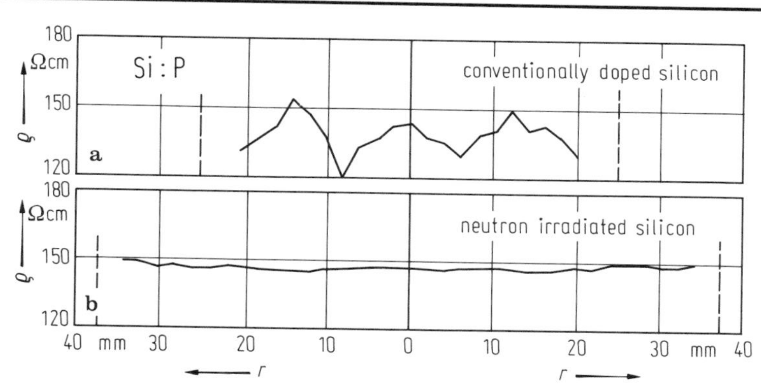

Fig. 45a, b. Si:P. Experimental resistivity ϱ vs. radial position r of phosphorus-doped crystals having a target resistivity of 145 Ω cm [76S]. a) Conventionally doped crystal; b) Neutron transmutation-doped crystal; Annealing; 4 hours at 1100 °C in vacuum. (Crystal diameters are indicated by dashed lines), (see also section 6.1.4.3).

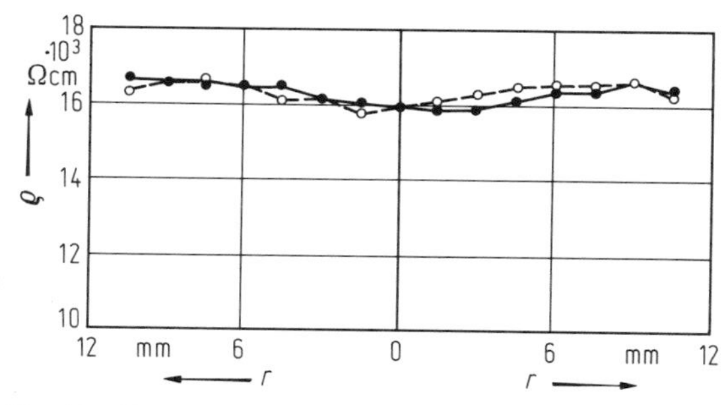

Fig. 46. Si. Experimental resistivity ϱ vs. radial position r of a high-resistivity p-type crystal of about 25 mm diameter [77K1]. The two curves are taken in two perpendicular directions, respectively.

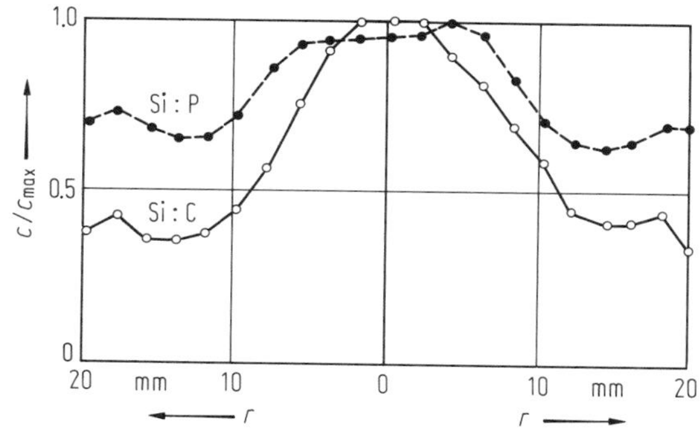

Fig. 47. Si. Comparison of the radial distribution of the relative phosphorus and carbon concentration c/c_{max} in a phosphorus-doped 50 mm-diameter crystal showing enhanced concentrations in the center [82K].

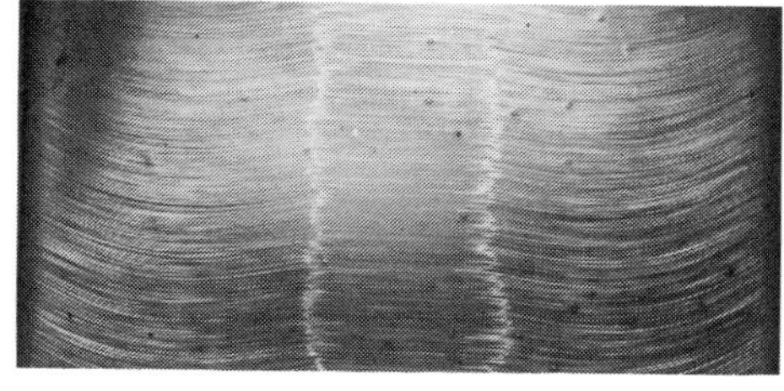

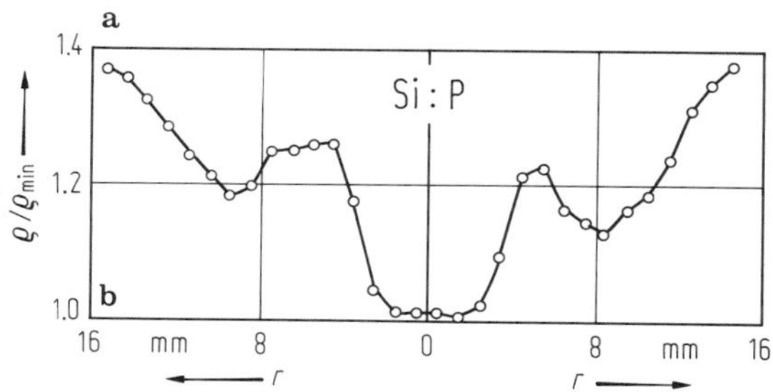

Fig. 48a, b. Si:P. Lengthwise cut of a ⟨111⟩-oriented phosphorus-doped crystal showing a striation pattern after preferential etching. a) central facets, b) the corresponding relative radial resistivity profile resulting in a central resistivity dip [81D3].

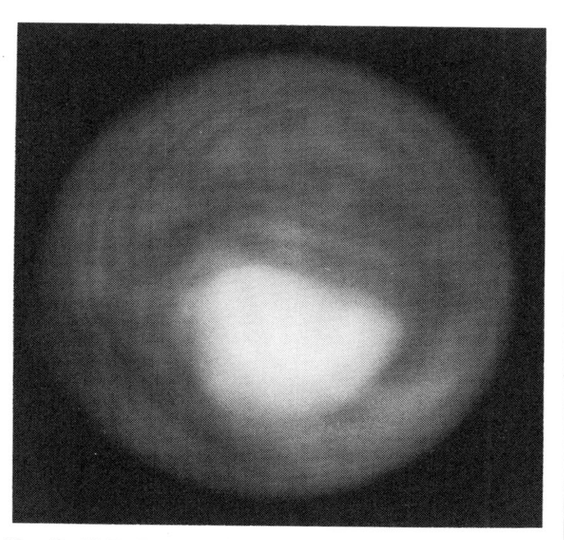

Fig. 49. Si:P. Pattern of the inhomogeneously incorporated phosphorus (bright part) in a crystal cross section having a marked phosphorus core as revealed by autoradiography [77H1].

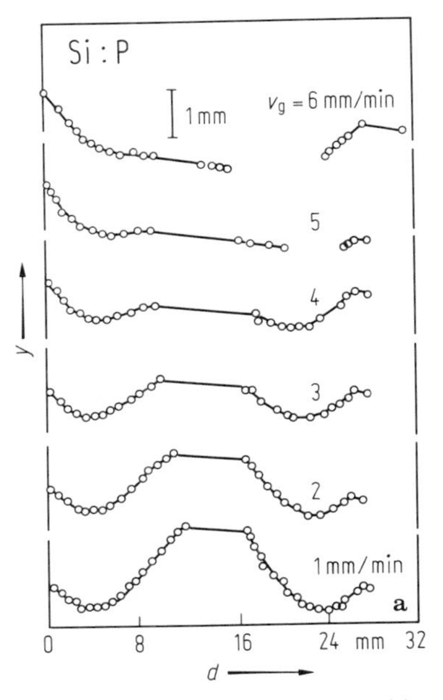

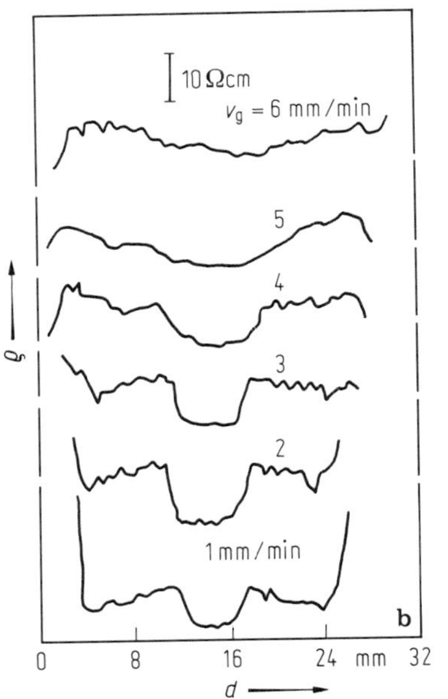

Fig. 50 a, b. Si:P. Interface shape in growth direction y (a), and resistivity ϱ (b) vs. diameter distance d of a ⟨111⟩-oriented phosphorus-doped crystal. Parameter at the curves is the growth velocity v_g [69C].

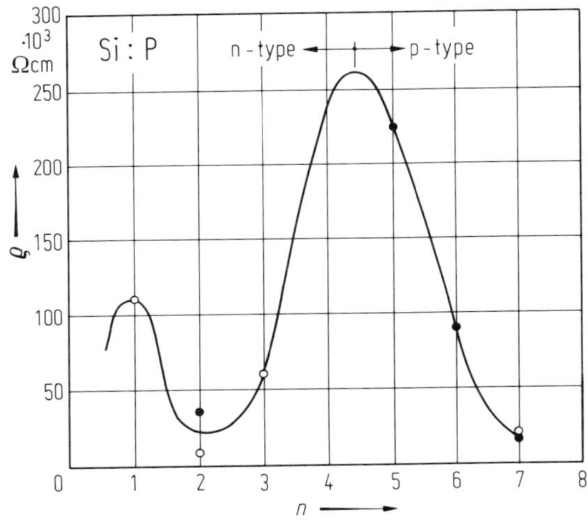

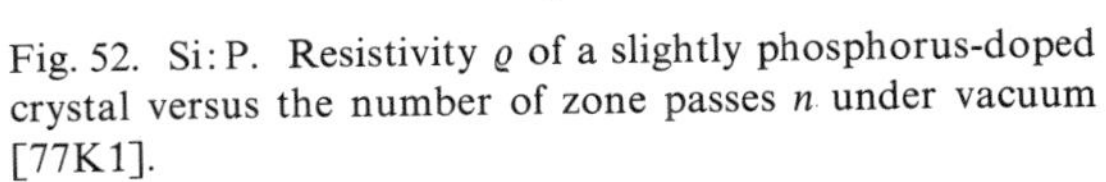

Fig. 52. Si:P. Resistivity ϱ of a slightly phosphorus-doped crystal versus the number of zone passes n under vacuum [77K1].

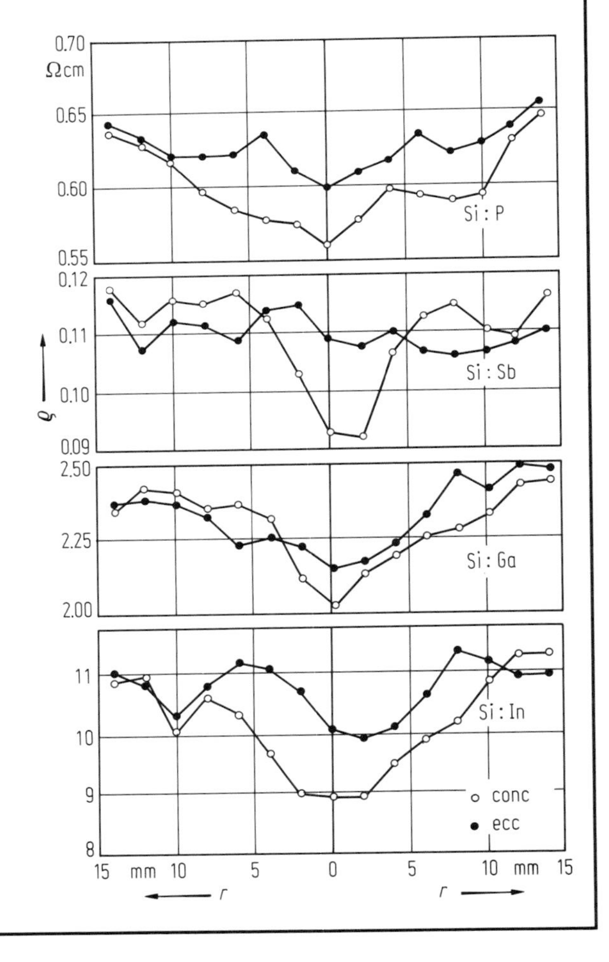

Fig. 51. Si:X (X = P, Sb, Ga, In). Experimental resistivity ϱ vs. radial position r of phosphorus-, antimony-, gallium-, and indium-doped ⟨100⟩-oriented crystals of 34 mm diameter. Parameter is the growth mode concentric (conc), and eccentric (ecc), respectively [81K].

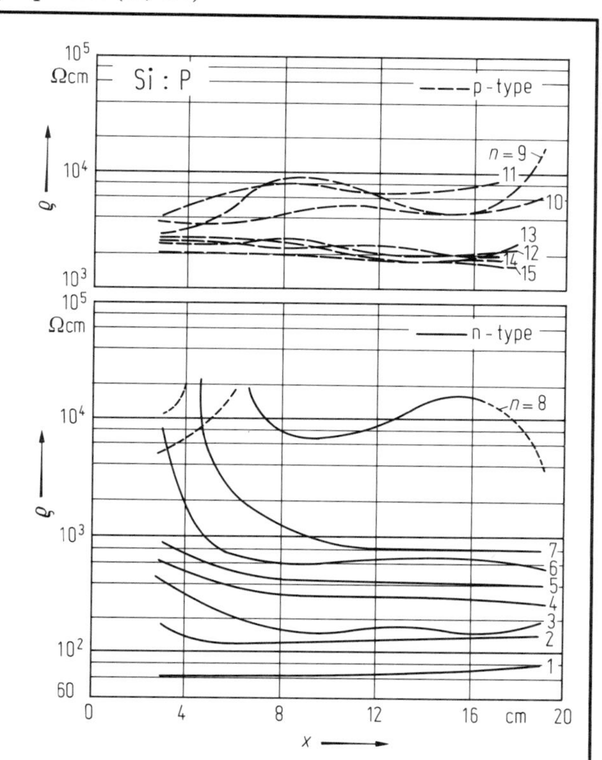

Fig. 53. Si:P. Experimental resistivity ϱ vs. axial position x of a predominantly phosphorus-doped crystal of 12 mm diameter. Parameter is the number of consecutive zone passes $n=1$ to 15 [81K].

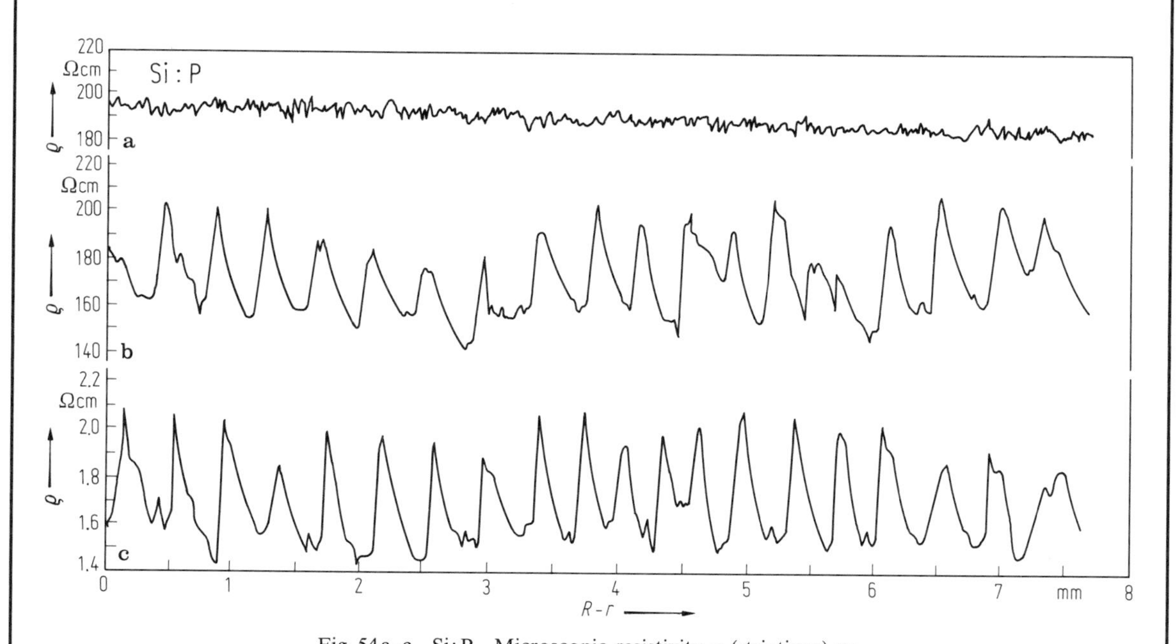

Fig. 54a–c. Si:P. Microscopic resistivity ϱ (striations) vs. radial position $(R-r)$ of phosphorus-doped 100 mm-diameter (2 R) crystals taken at the rim region. a) Neutron transmutation doped ⟨111⟩-oriented crystal showing a resistivity of 200 Ω cm. b) Conventionally doped ⟨111⟩-oriented crystal of 200 Ω cm resistivity. c) Conventionally doped ⟨100⟩-oriented crystal of 2 Ω cm resistivity. (By courtesy of H. Herzer, Wacker-Chemitronic).

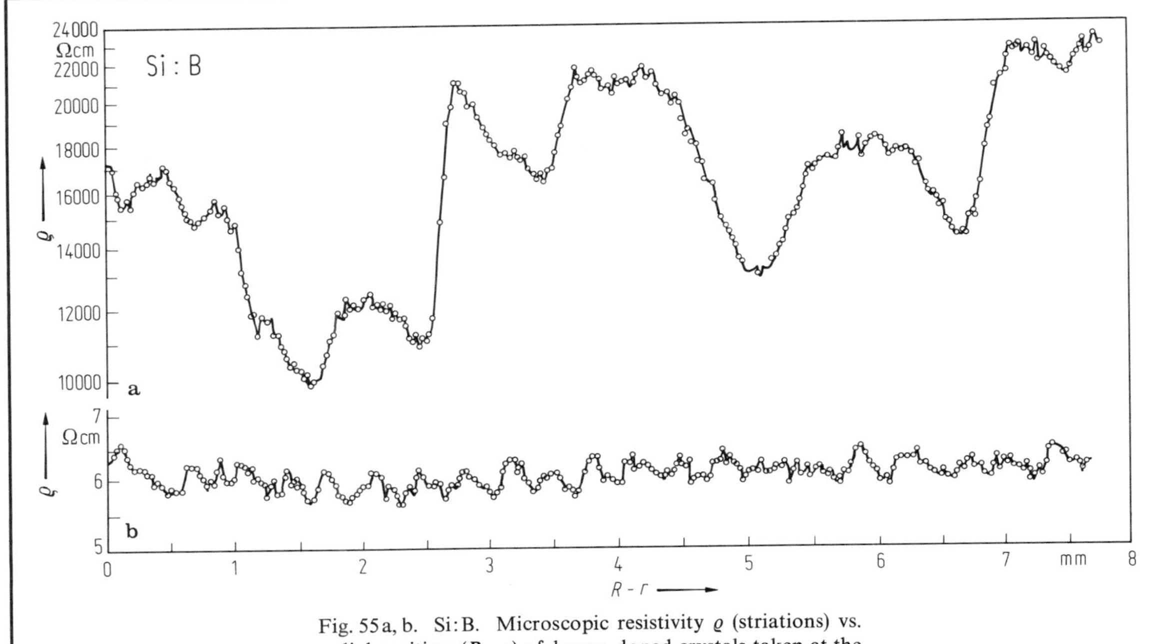

Fig. 55a, b. Si:B. Microscopic resistivity ϱ (striations) vs. radial position $(R-r)$ of boron-doped crystals taken at the rim region of the wafer. a) High-resistivity 33 mm-diameter (2 R) crystal, ⟨111⟩ oriented. b) Low-resistivity 100 mm-diameter (2 R) crystal, ⟨100⟩ oriented. (By courtesy of H. Herzer, Wacker-Chemitronic).

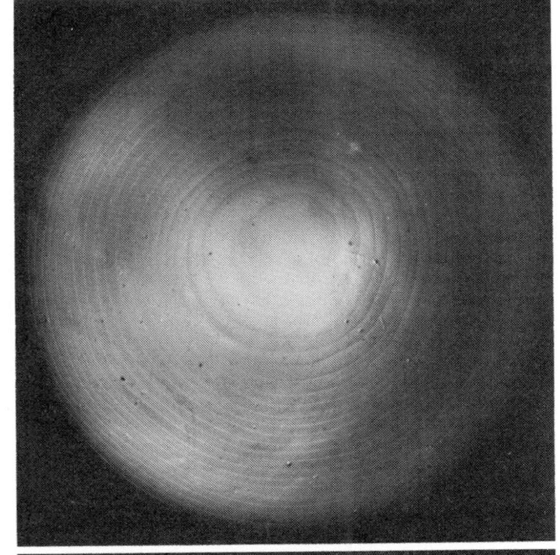

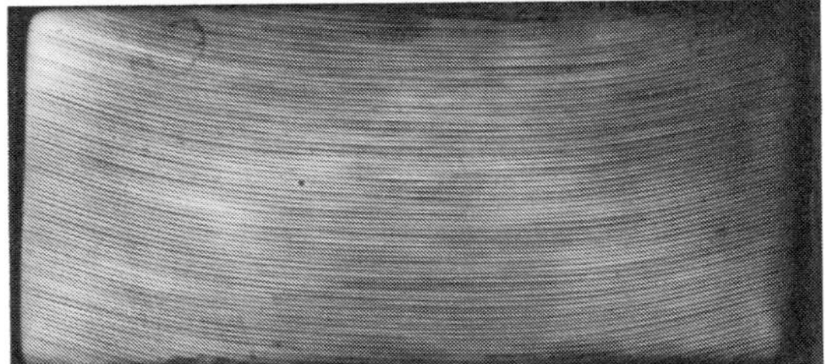

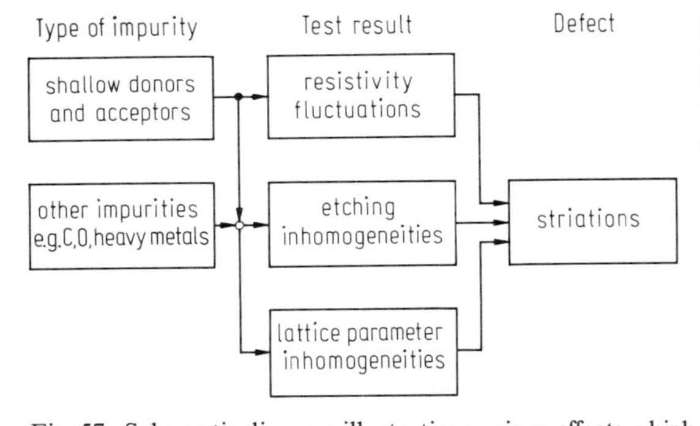

Fig. 57. Schematic diagram illustrating various effects which may exhibit striations in different tests.

◄
Fig. 56. Si. Striation pattern of a cross section (top) and a longitudinal cut (bottom) of a float-zone crystal revealed by preferential etching [81D3].

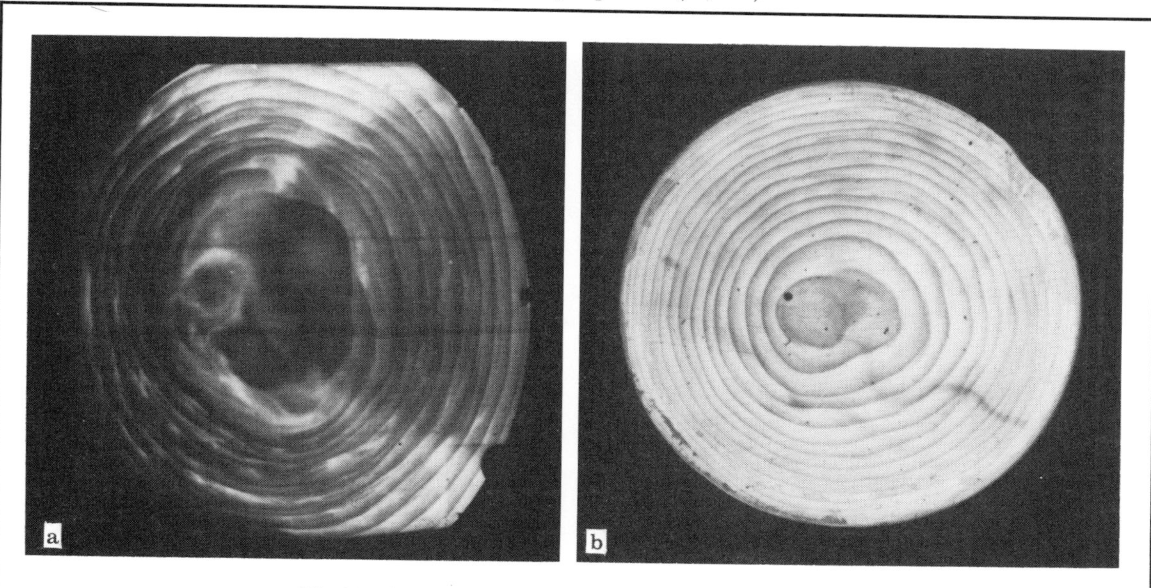

Fig. 58a, b. Si. Rotational striations revealed in carbon-rich homogeneously phosphorus-doped float-zone crystals by a) X-ray topography, b) Chemical etching.

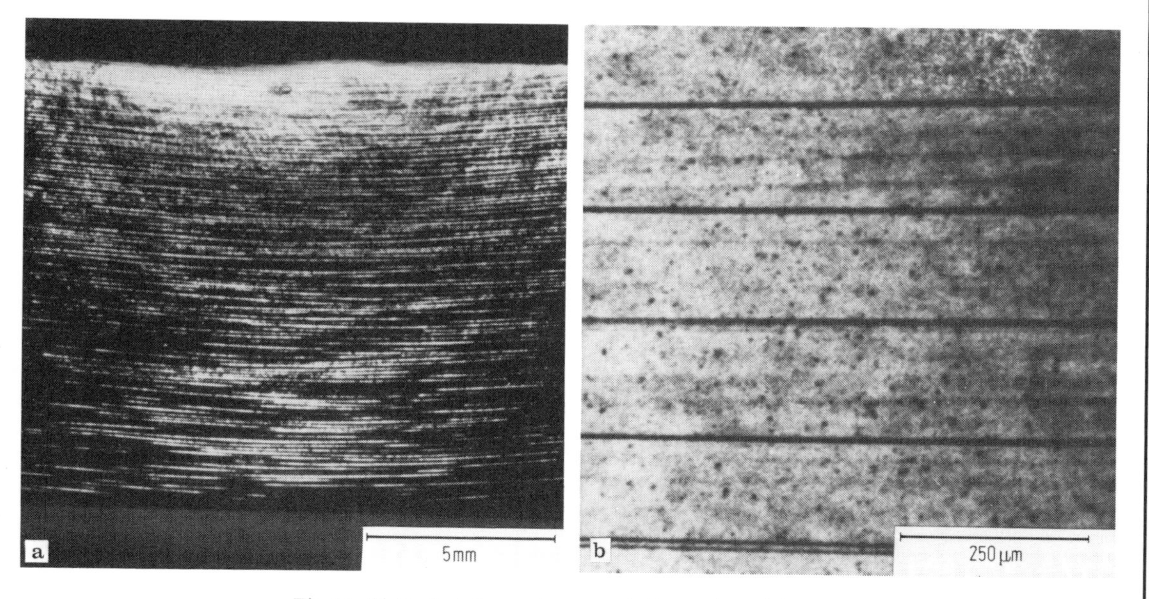

Fig. 59. Si: P. Striation pattern revealed by preferential etching on an axial section of a phosphorus-doped crystal. a) Overview. b) Magnified portion showing rotational striations (dark lines). (By courtesy of B.O. Kolbesen, Siemens).

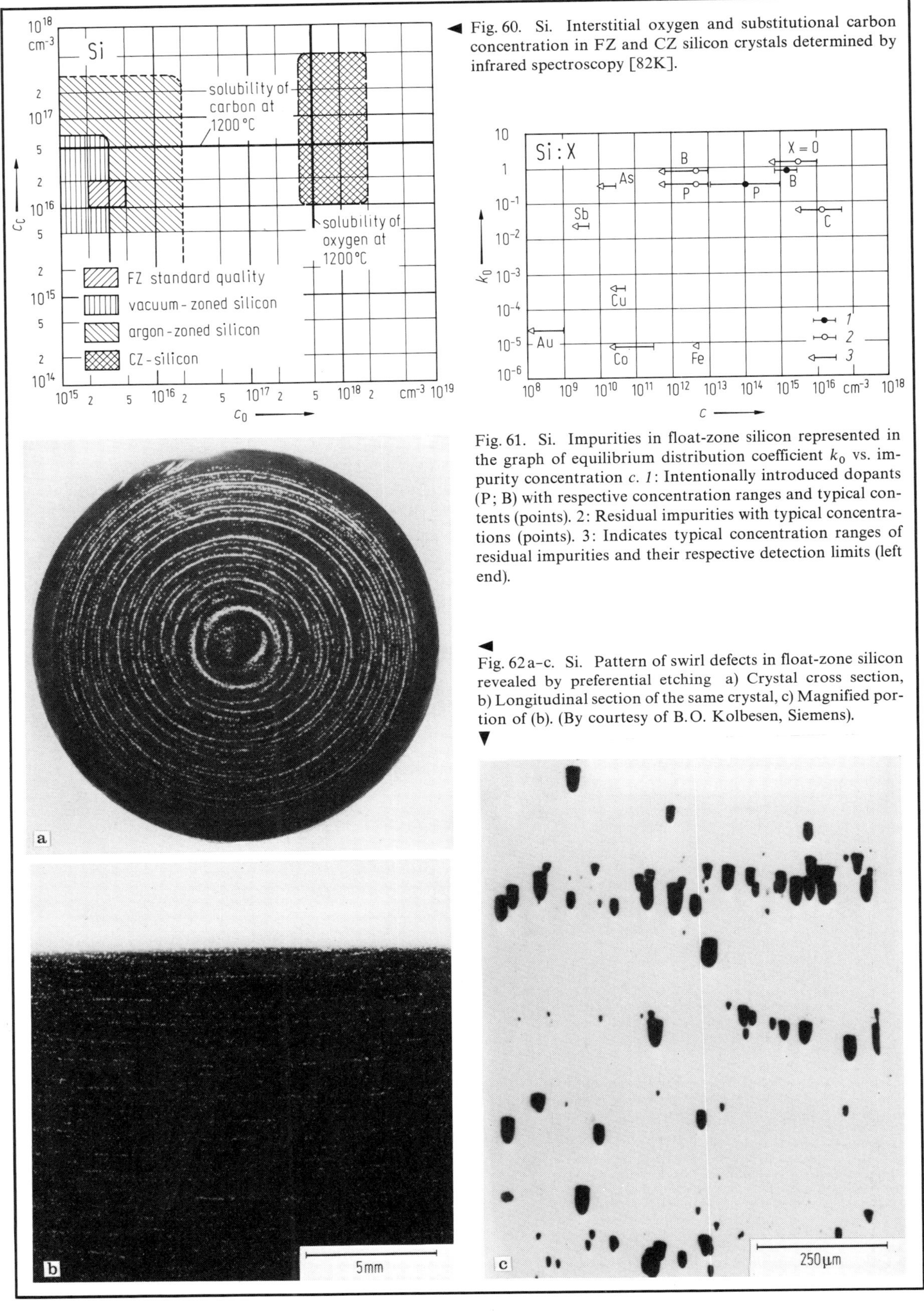

◄ Fig. 60. Si. Interstitial oxygen and substitutional carbon concentration in FZ and CZ silicon crystals determined by infrared spectroscopy [82K].

Fig. 61. Si. Impurities in float-zone silicon represented in the graph of equilibrium distribution coefficient k_0 vs. impurity concentration c. *1*: Intentionally introduced dopants (P; B) with respective concentration ranges and typical contents (points). 2: Residual impurities with typical concentrations (points). 3: Indicates typical concentration ranges of residual impurities and their respective detection limits (left end).

◄
Fig. 62a–c. Si. Pattern of swirl defects in float-zone silicon revealed by preferential etching a) Crystal cross section, b) Longitudinal section of the same crystal, c) Magnified portion of (b). (By courtesy of B.O. Kolbesen, Siemens).
▼

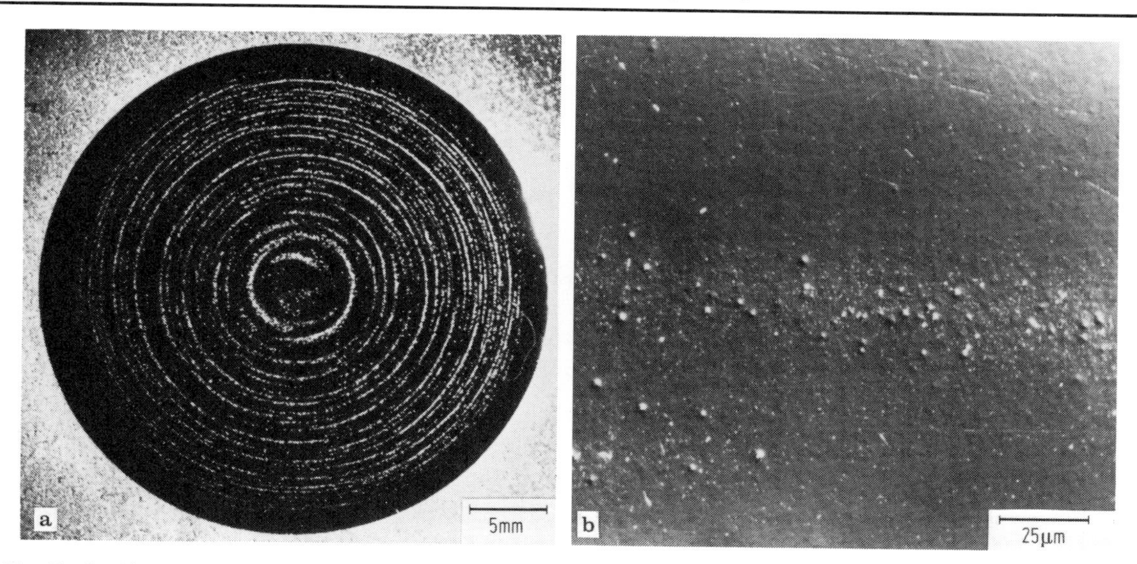

Fig. 63 a, b. Si. Pattern of swirl defects in float-zone silicon revealed by preferential etching [81F]. a) Crystal cross section, b) Magnified portion: Large dots correspond to *A*-swirl defects, small dots represent *B*-swirl defects.

For Fig. 64, see next page.

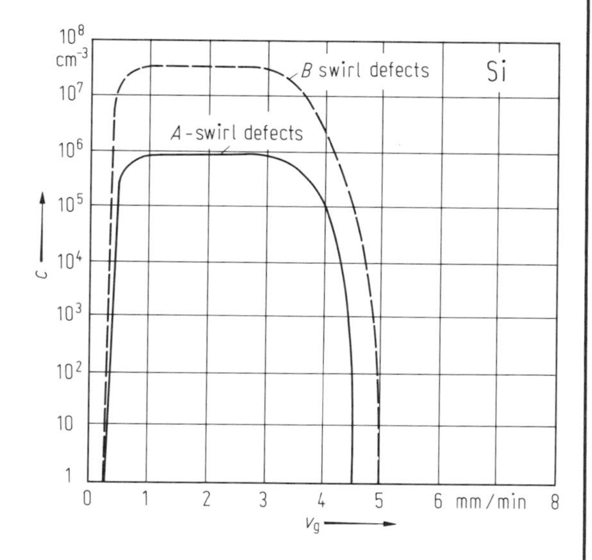

Fig. 65. Si. Experimental concentration c of *A*- and *B*-swirl defects vs. growth velocity v_g of a dislocation-free float-zone crystal [76R2].

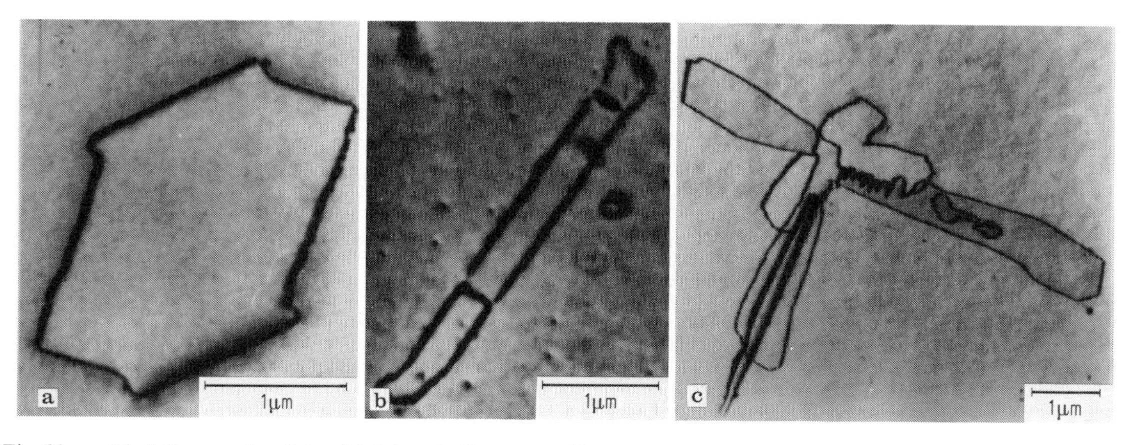

Fig. 66 a–c. Si. Micrographs of *A*-swirl defects in float-zone silicon revealed by high voltage transmission electron microscopy (650 keV). a) Single loop, b) Loop cluster, c) Complicated loop arrangement. (By courtesy of B.O. Kolbesen, Siemens).

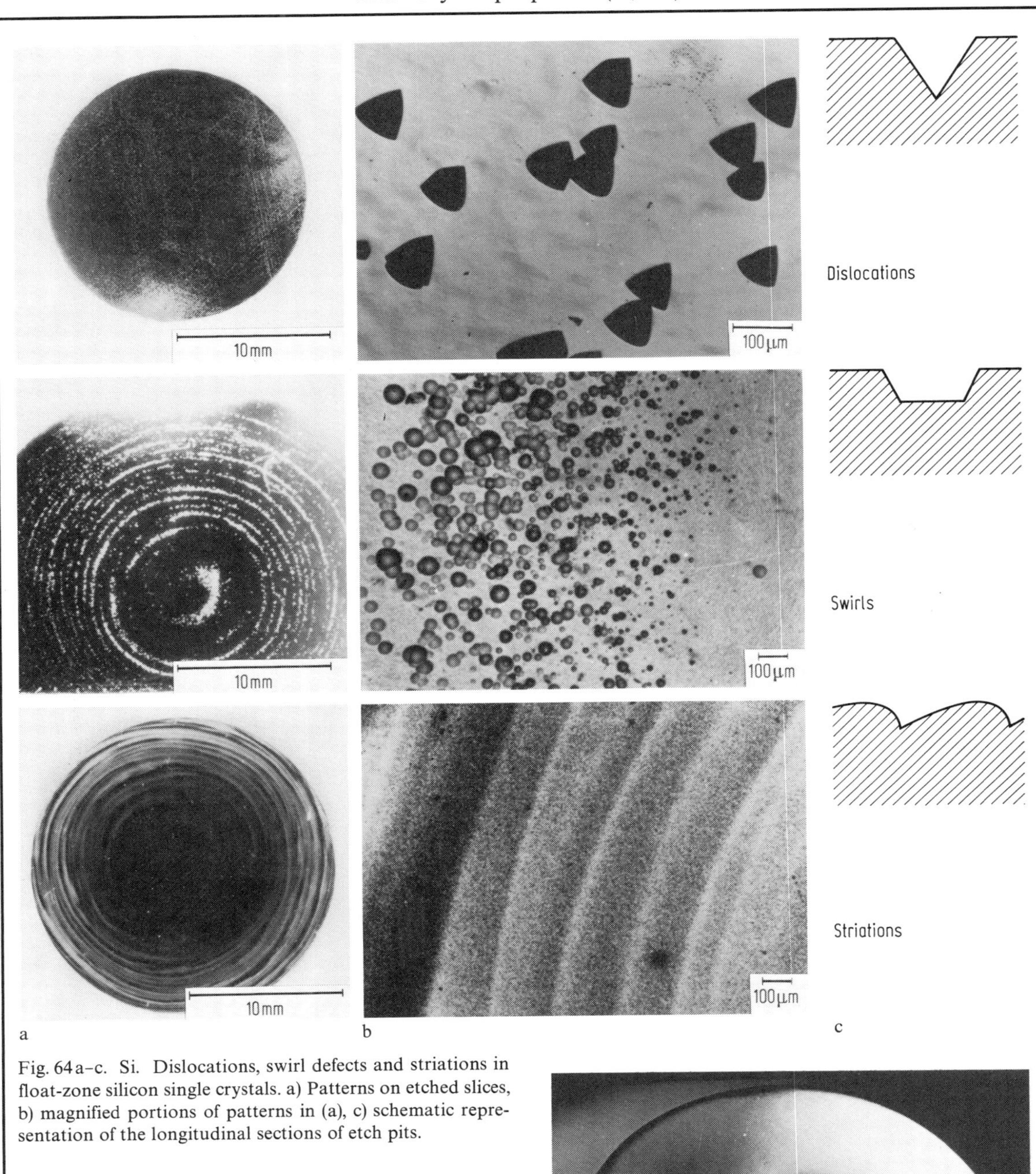

Fig. 64a–c. Si. Dislocations, swirl defects and striations in float-zone silicon single crystals. a) Patterns on etched slices, b) magnified portions of patterns in (a), c) schematic representation of the longitudinal sections of etch pits.

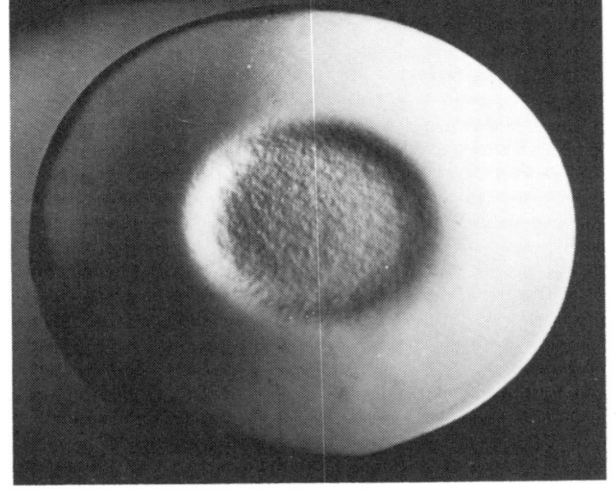

Fig. 67. Si. Etching depression on a polish-etched, dislocation-free float-zone silicon crystal [81K].

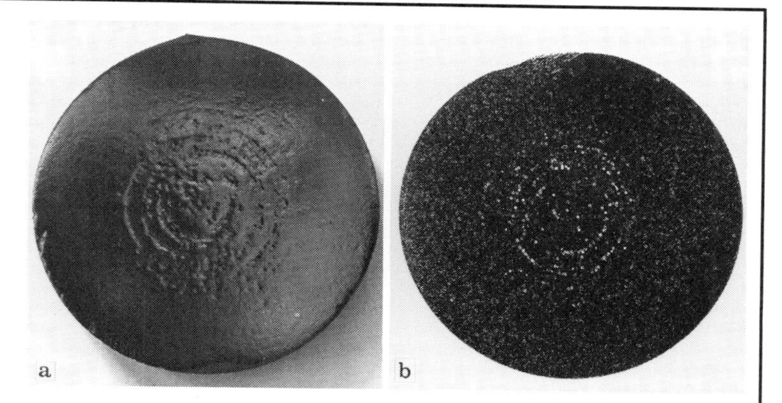

Fig. 68a, b. Si. Hydrogen defects on the cross-section of a dislocation-free float-zone crystal [81K]. a) Polish-etched silicon, b) Hydroxide-etched silicon.

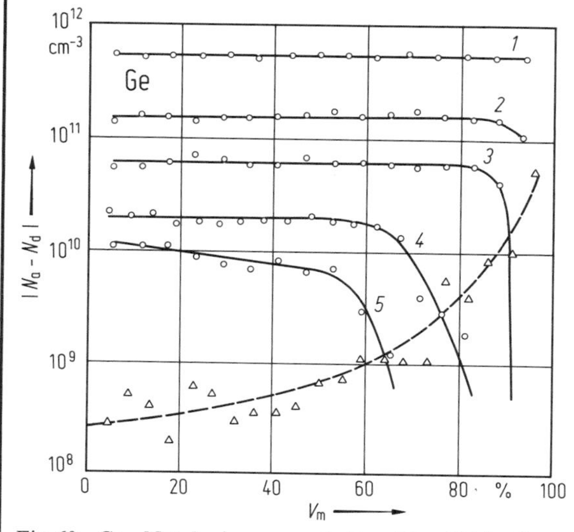

Fig. 69. Ge. Net doping concentration $|N_a - N_d|$ in Ge vs. crystal length as given by the melt volume V_m used for growing the ingot. The selection of the crystals *1*···*5* shows the common constant doping concentrations which have been shown by Fourier transform spectroscopy (FTS) as due to aluminum. The n-type portions of the crystals are not shown. The dashed line shows an extremely low n-type doping of a sample which apparently contains little aluminum [74H1].

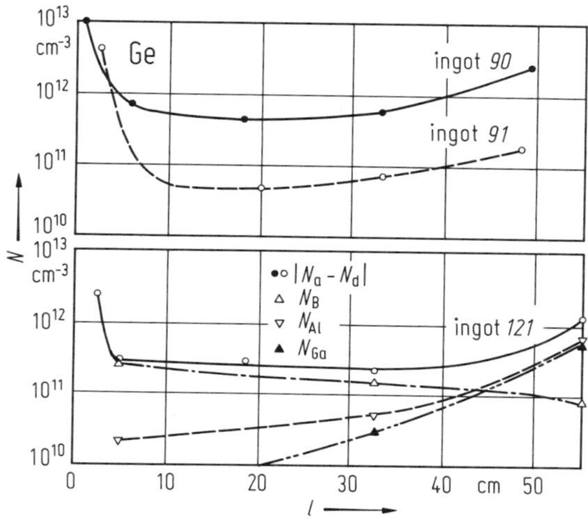

Fig. 70. Ge. Concentration N of impurities B, Al, Ga and net doping concentration $|N_a - N_d|$ vs. distance l along the Ge ingot. Ingot *90* has been refined in a solid graphite boat. Ingot *91* has been first refined in solid graphite followed by refining in a quartz boat smoked by carbon from burning butane. Ingot *121* was refined in a boat coated by carbon smoke followed by pyrolithic carbon. [78H3].

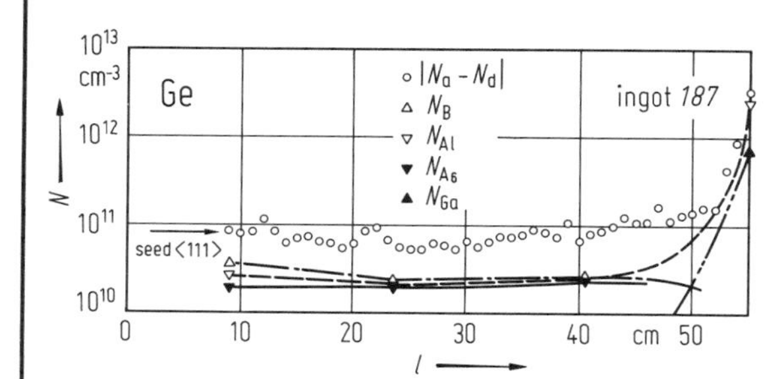

Fig. 72. Ge. Concentration N of impurities B, Al, A_6, Ga and net doping concentration $|N_a - N_d|$ vs. distance l along the Ge ingot. Ingot *187* is a single crystal refined by 12 passes in a pyrolithic carbon/silica-coated boat [78H3]. A_6 is a new acceptor, see [77H7].

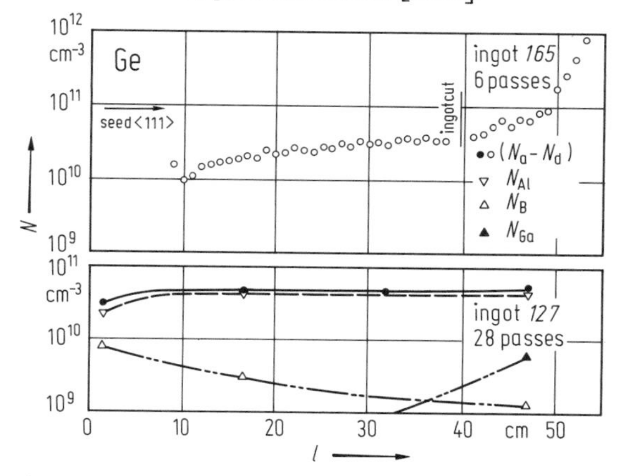

Fig. 71. Ge. Concentration N of impurities B, Al, Ga and net doping concentration $|N_a - N_d|$ vs. distance l along the Ge ingot. Ingot *165* is a single crystal refined in a silica smoked quartz boat. The net acceptor concentration is measured by the conductivity. The ingot *127* is polycrystalline refined in a silica-smoked quartz boat. [78H3].

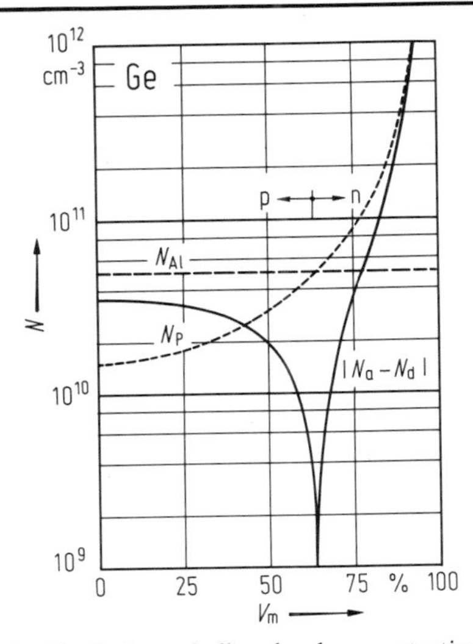

Fig. 73. Ge. Typical net shallow-level concentration $|N_a - N_d|$ vs. the fraction of frozen melt volume V_m along the growth axis of a HP–Ge single crystal. At the seed end (0% of melt frozen) the aluminum acceptor dominates yielding a p-type crystal. Near the tail end, the phosphorus concentration N_p exceeds the aluminum concentration N_{Al}, leading to an n-type crystal [81H2].

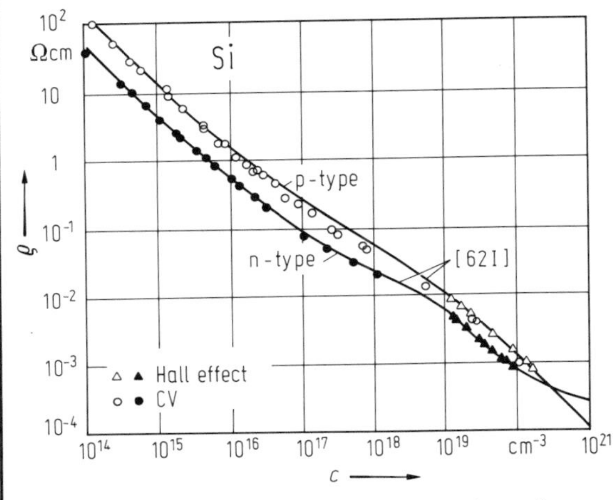

Fig. 74. Si. Resistivity ϱ vs. dopant concentration c of p-type and n-type silicon at 300 K. For both boron- and phosphorus-doped silicon, results were obtained by junction capacitance voltage (CV) and Hall effect measurements [78T2].

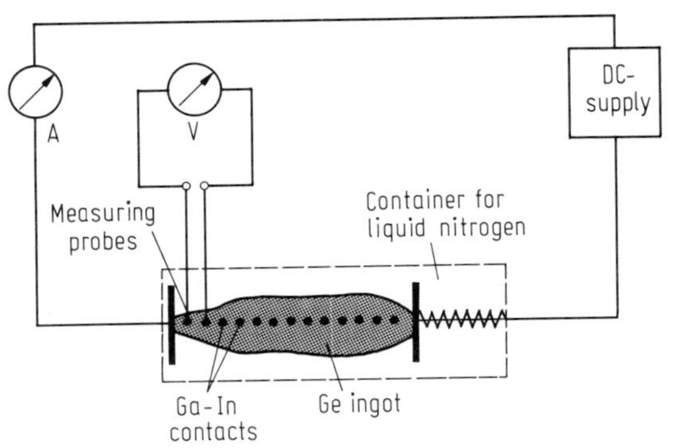

Fig. 75. Schematic drawing of the ingot conductivity measurement of HP–Ge [77G1].

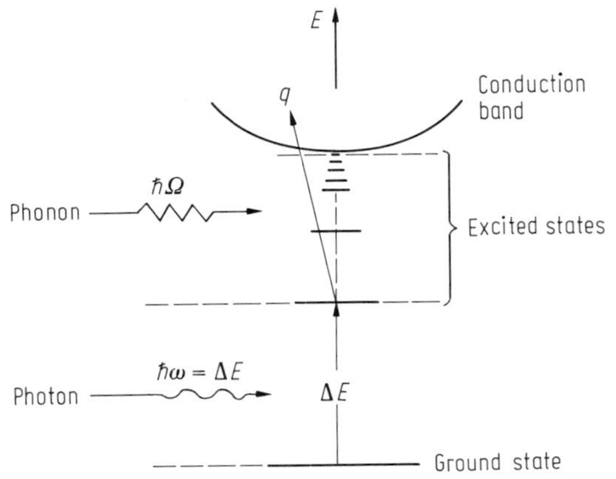

Fig. 76. Two-step excitation process of an impurity ground state which leads to the photothermal ionisation in PTIS (Photothermal infrared spectroscopy) [81H2]. An electron q is excited from the impurity ground state via an excited state into the conduction band by absorption of a photon of suitable energy $\hbar\omega = \Delta E$ and thermal phonons $\hbar\Omega$.

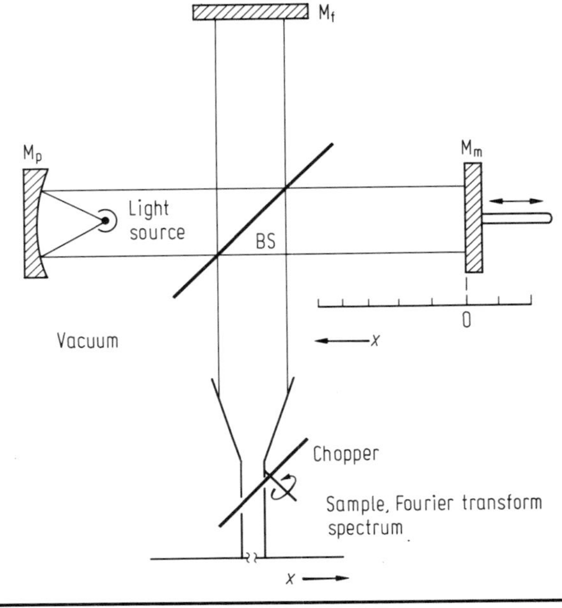

►

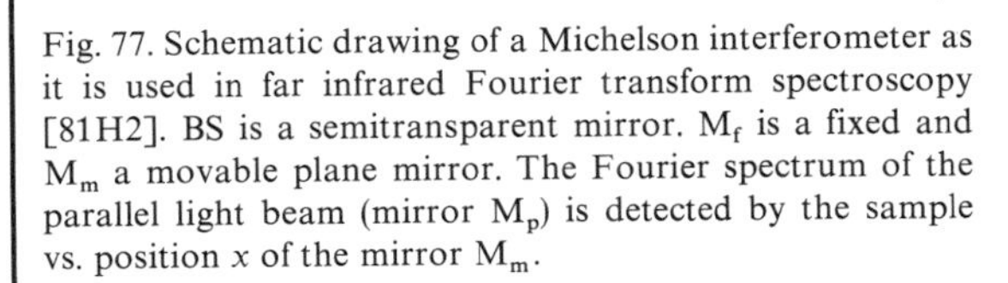

Fig. 77. Schematic drawing of a Michelson interferometer as it is used in far infrared Fourier transform spectroscopy [81H2]. BS is a semitransparent mirror. M_f is a fixed and M_m a movable plane mirror. The Fourier spectrum of the parallel light beam (mirror M_p) is detected by the sample vs. position x of the mirror M_m.

6.1.4 Device technology

6.1.4.0 Basic device structures

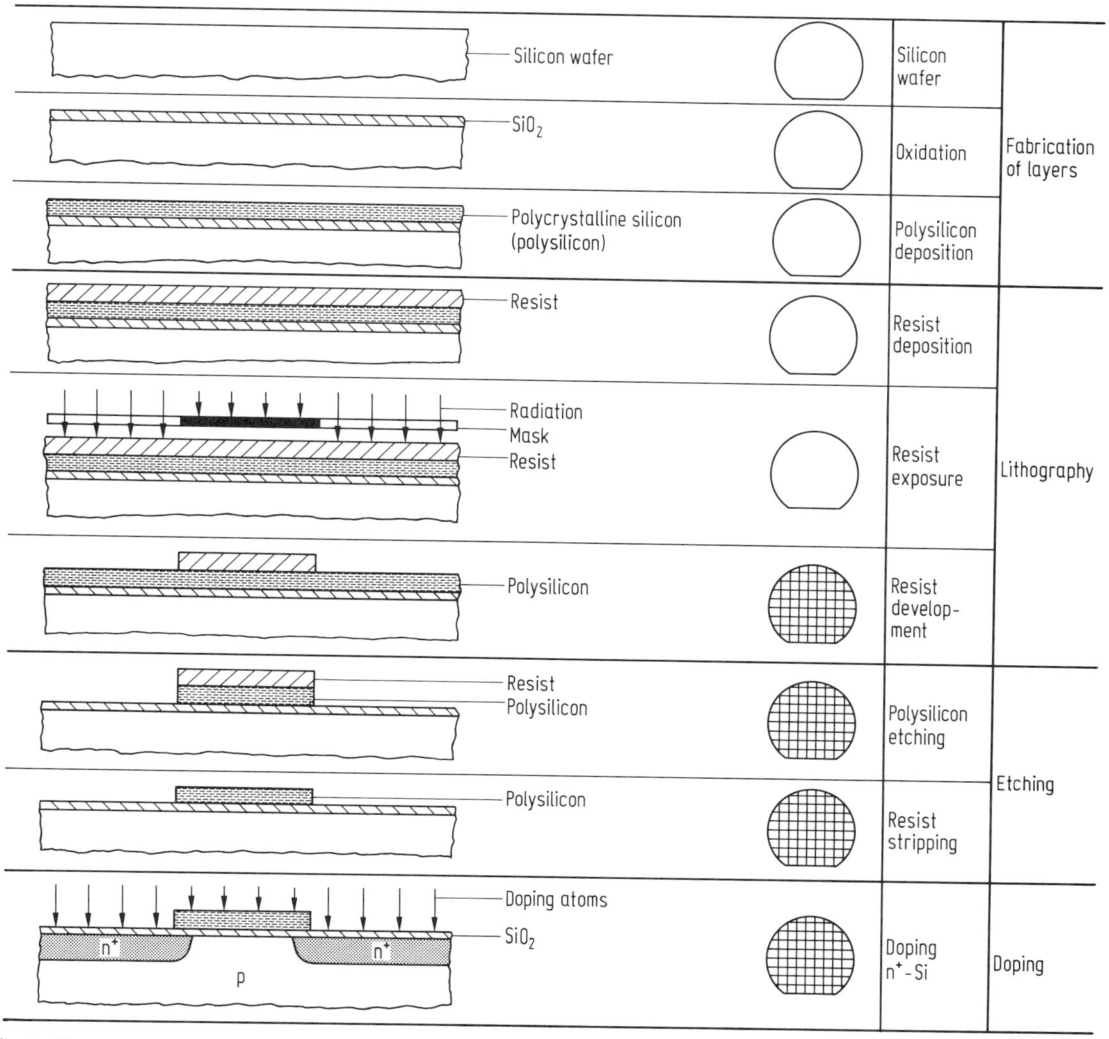

Structure labels	Process step	Category
Silicon wafer	Silicon wafer	Fabrication of layers
SiO_2	Oxidation	
Polycrystalline silicon (polysilicon)	Polysilicon deposition	
Resist	Resist deposition	Lithography
Radiation, Mask, Resist	Resist exposure	
Polysilicon	Resist development	
Resist, Polysilicon	Polysilicon etching	Etching
Polysilicon	Resist stripping	
Doping atoms, SiO_2, n^+, p, n^+	Doping n^+-Si	Doping

Fig. 1. Si.

continued

Fig. 1 (continued)

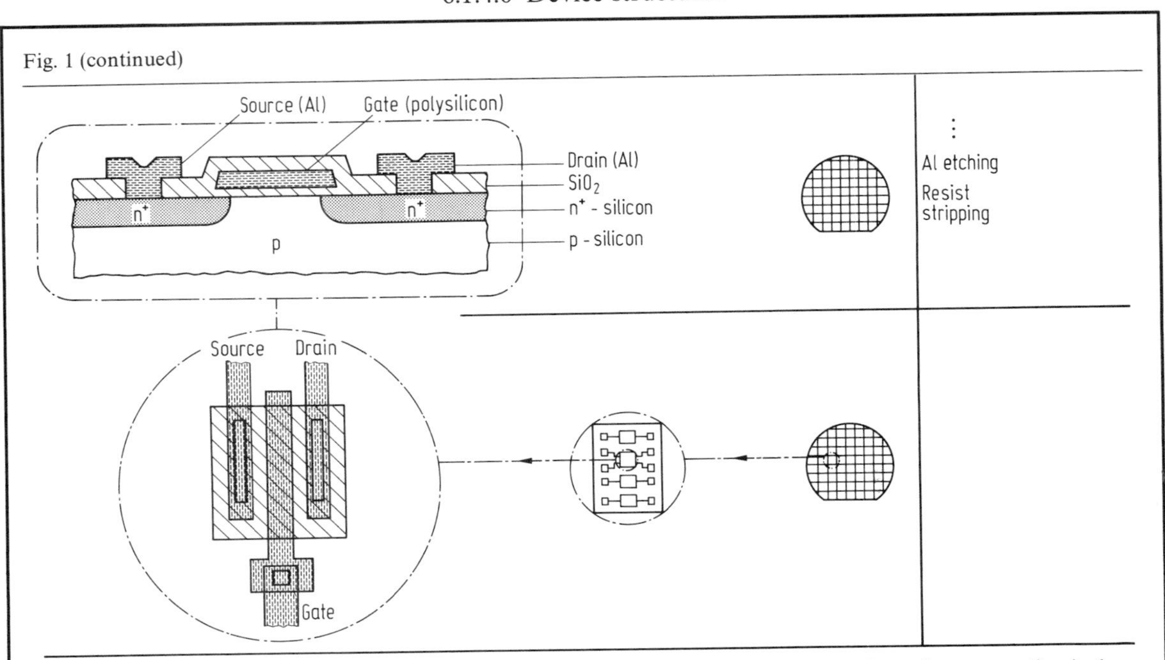

Fig. 1. Si. Typical steps for the fabrication of a MOS transistor having a polysilicon gate. The wafer cross-section is shown on the left. The top view of the wafer is shown in the center. The various process steps are explained on the right. The last sketch at the bottom shows the magnified portion of a selected chip.

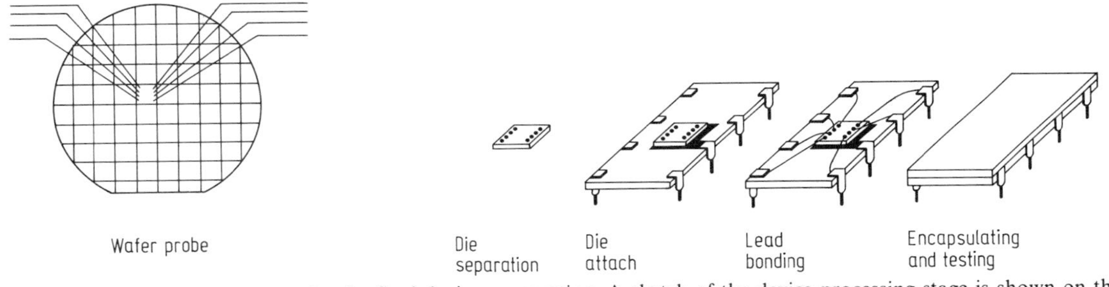

Fig. 2. Si. Typical process steps for the final device preparation. A sketch of the device processing stage is shown on the left. The processing steps involved are explained on the right.

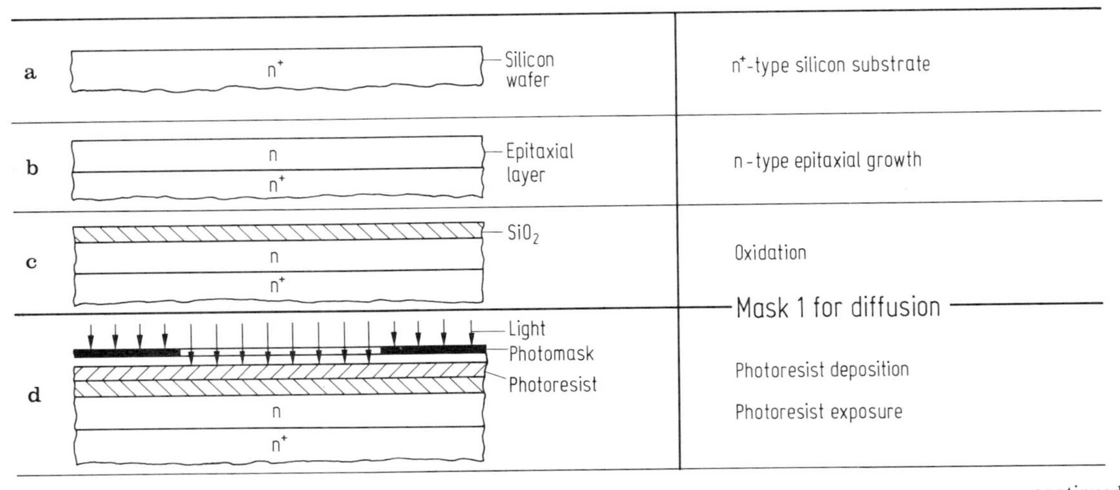

Fig. 3. Si.

continued

Fig. 3 (continued)

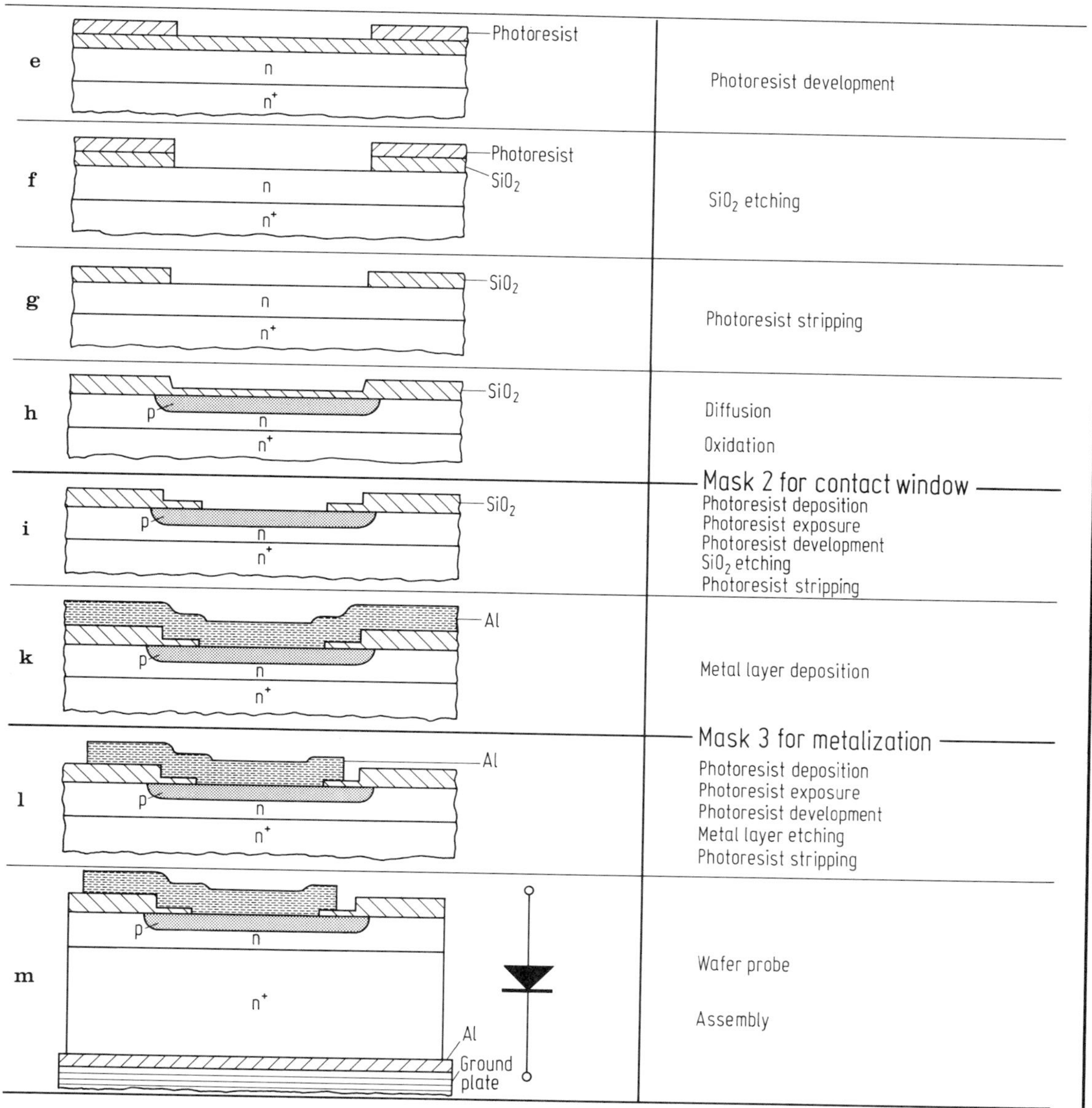

Step	Cross-section labels	Processing steps
e	Photoresist; n; n^+	Photoresist development
f	Photoresist; SiO_2; n; n^+	SiO_2 etching
g	SiO_2; n; n^+	Photoresist stripping
h	SiO_2; p; n; n^+	Diffusion Oxidation
		Mask 2 for contact window
i	SiO_2; p; n; n^+	Photoresist deposition Photoresist exposure Photoresist development SiO_2 etching Photoresist stripping
k	Al; p; n; n^+	Metal layer deposition
		Mask 3 for metalization
l	Al; p; n; n^+	Photoresist deposition Photoresist exposure Photoresist development Metal layer etching Photoresist stripping
m	p; n; n^+; Al; Ground plate	Wafer probe Assembly

Fig. 3. Si. Typical processing steps for the fabrication of silicon pn-diodes [72H2, 73M1, 75R1]. The cross-section of the wafer is shown on the left. The processing steps involved are explained in sequence on the right.

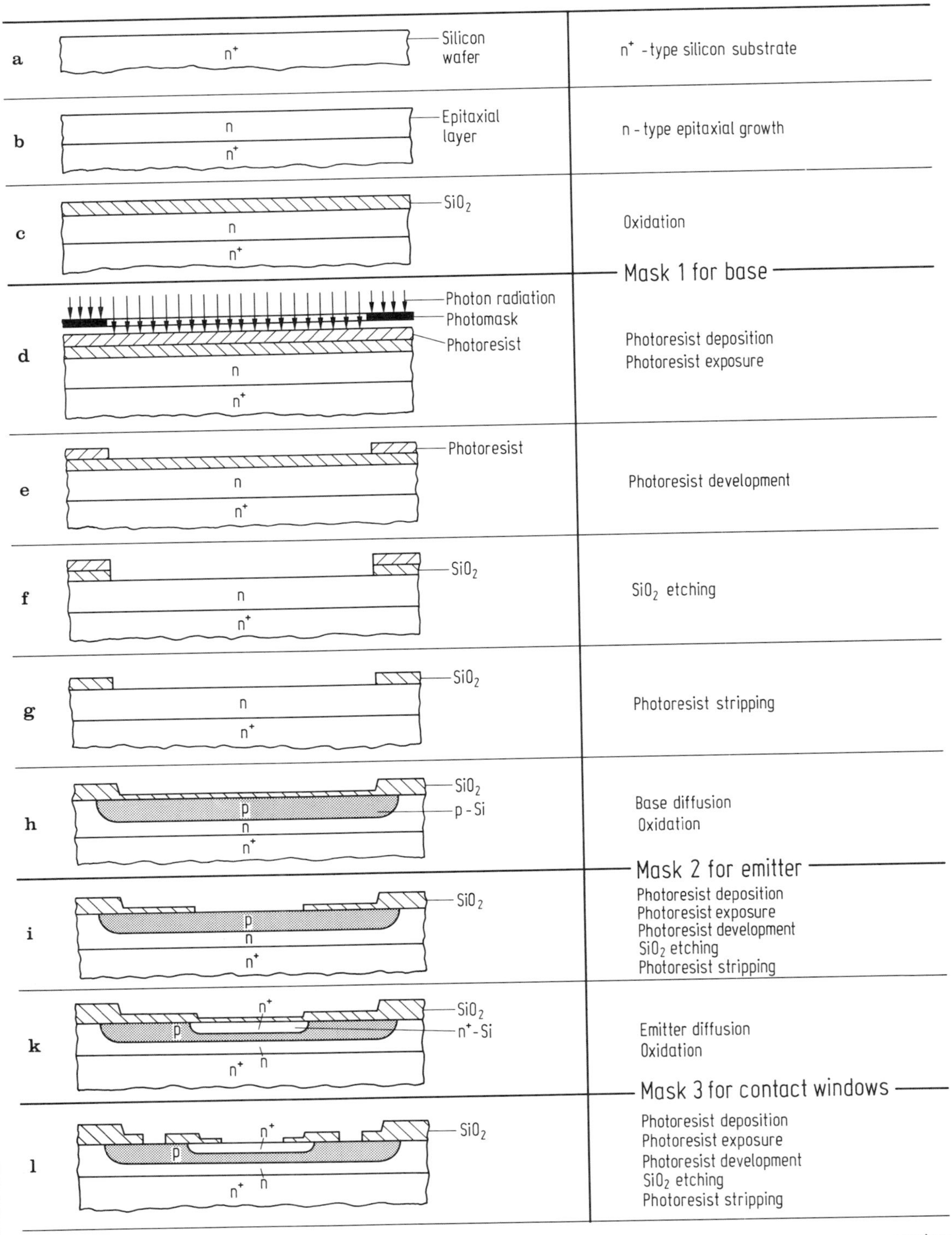

Step	Cross-section labels	Process
a	n^+; Silicon wafer	n^+-type silicon substrate
b	n; n^+; Epitaxial layer	n-type epitaxial growth
c	SiO_2; n; n^+	Oxidation
		Mask 1 for base
d	Photon radiation; Photomask; Photoresist; n; n^+	Photoresist deposition Photoresist exposure
e	Photoresist; n; n^+	Photoresist development
f	SiO_2; n; n^+	SiO_2 etching
g	SiO_2; n; n^+	Photoresist stripping
h	SiO_2; p-Si; p; n; n^+	Base diffusion Oxidation
		Mask 2 for emitter
i	SiO_2; p; n; n^+	Photoresist deposition Photoresist exposure Photoresist development SiO_2 etching Photoresist stripping
k	SiO_2; n^+-Si; n^+; p; n; n^+	Emitter diffusion Oxidation
		Mask 3 for contact windows
l	SiO_2; n^+; p; n; n^+	Photoresist deposition Photoresist exposure Photoresist development SiO_2 etching Photoresist stripping

Fig. 4. Si.

continued

Fig. 4 (continued)

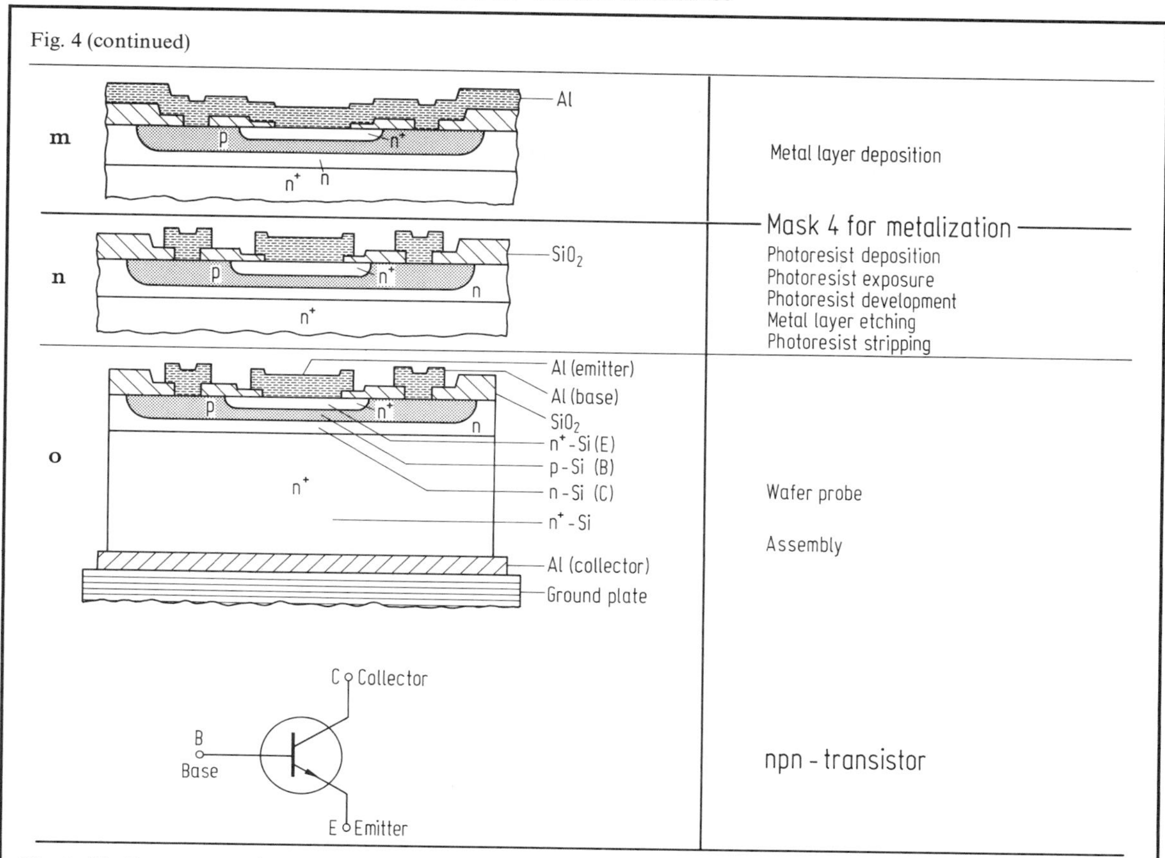

Fig. 4. Si. Process steps for the fabrication of silicon npn bipolar transistors [72H2, 73M1, 75M1, 78S1]. The wafer cross-section for each step a)···o) is shown on the left. The process applied is explained on the right. The device symbol is shown at the bottom.

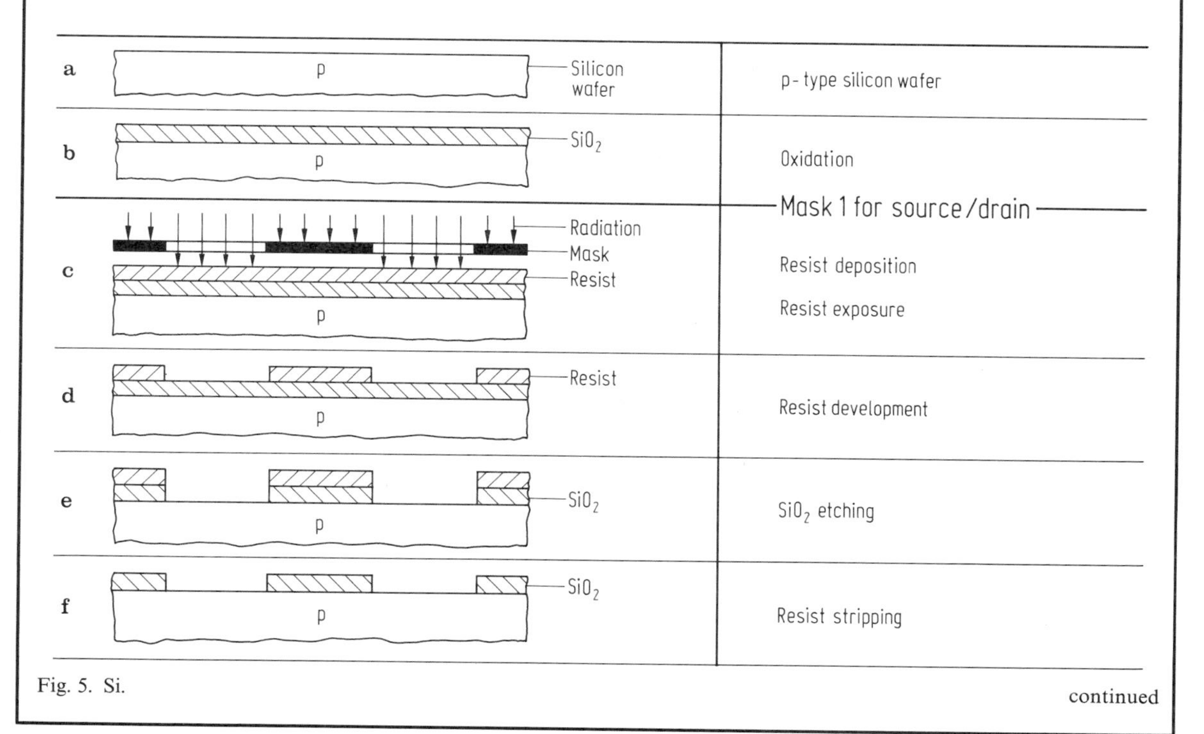

Fig. 5. Si.

continued

Fig. 5 (continued)

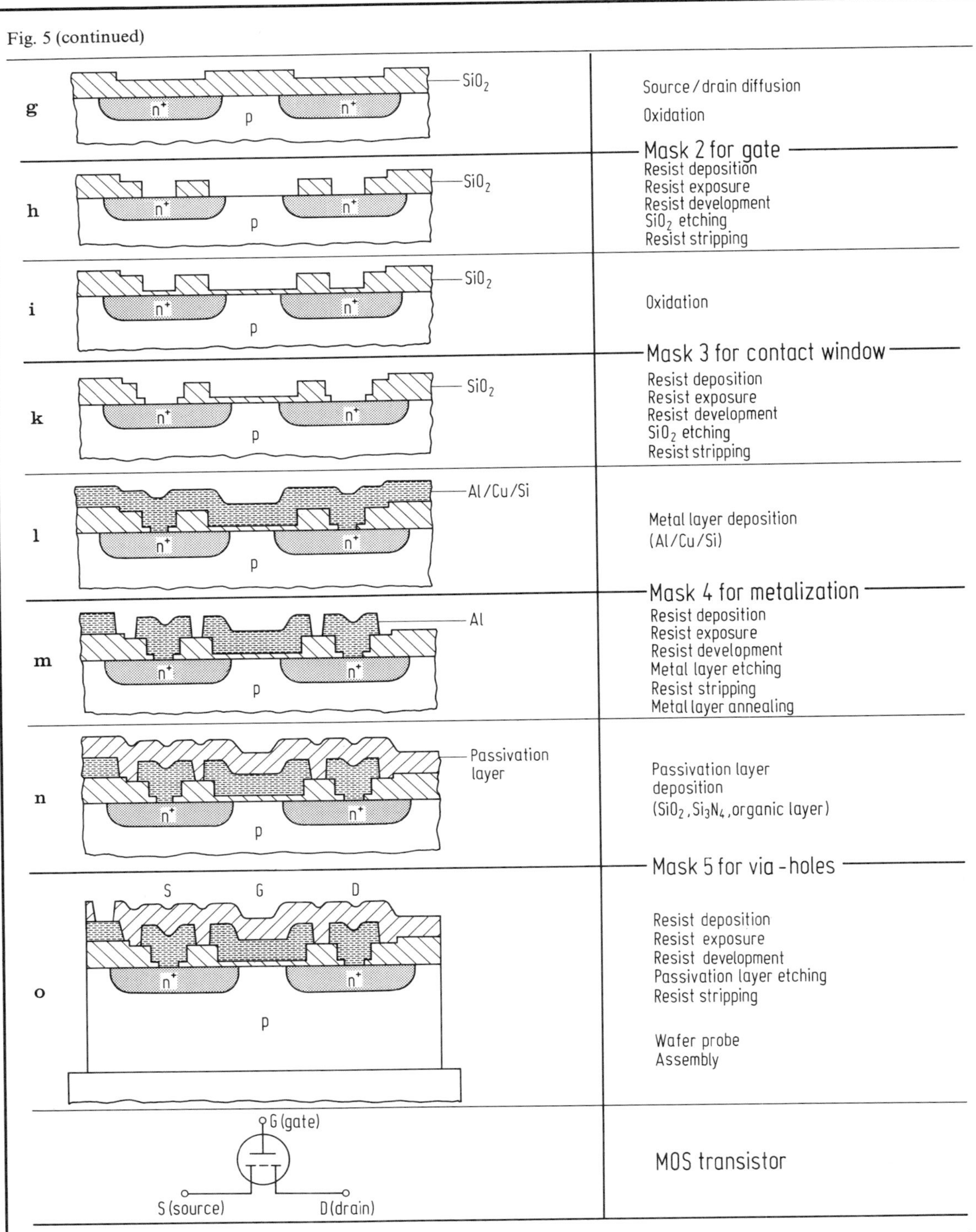

Fig. 5. Si. Process steps for the fabrication of the silicon metal gate n-channel MOS transistor [72H2, 73M1, 75R1]. The wafer cross-section for each step a)...o) is shown on the left. The process applied is explained on the right. The device symbol is shown at the bottom.

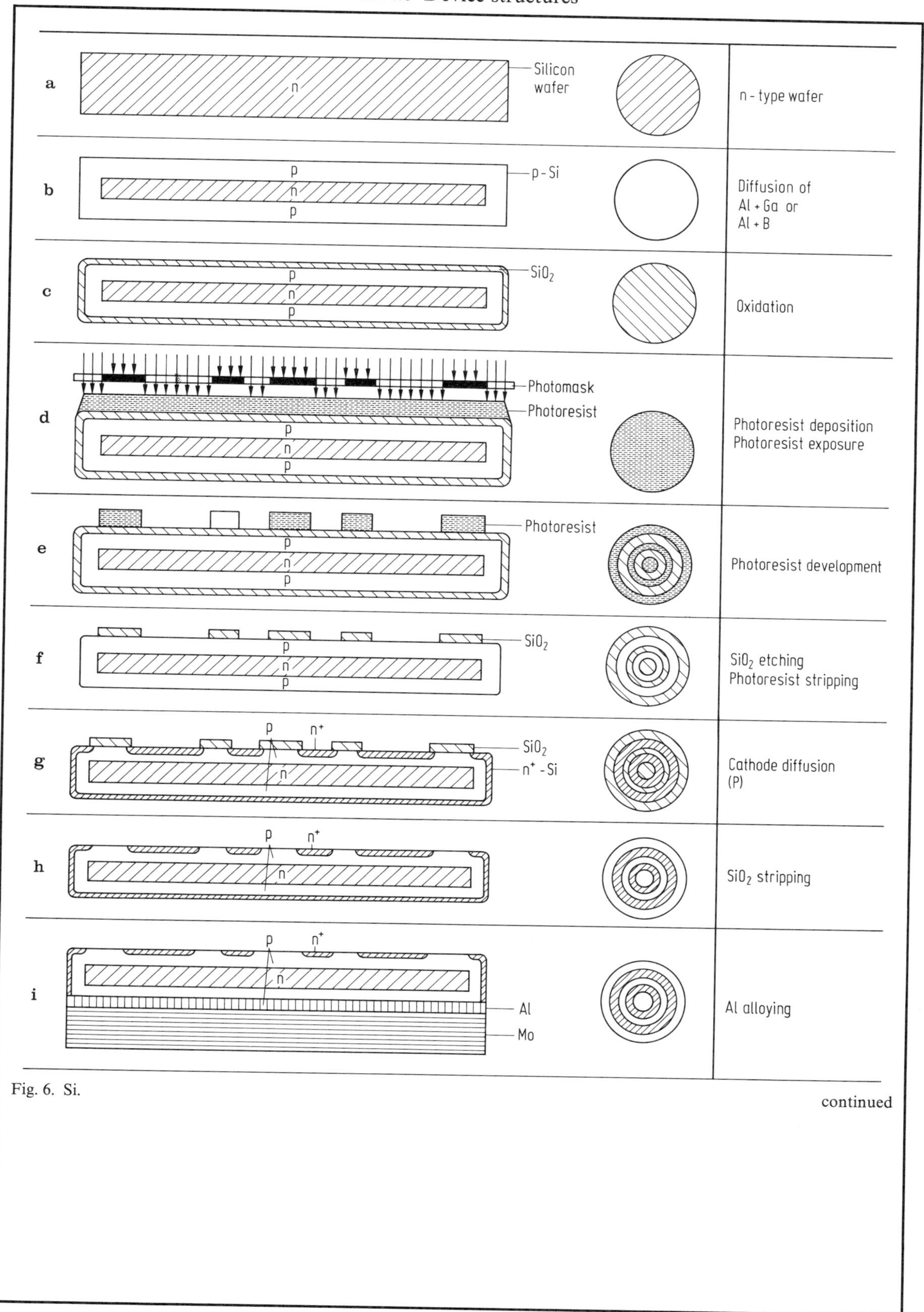

Fig. 6. Si.

continued

Fig. 6 (continued)

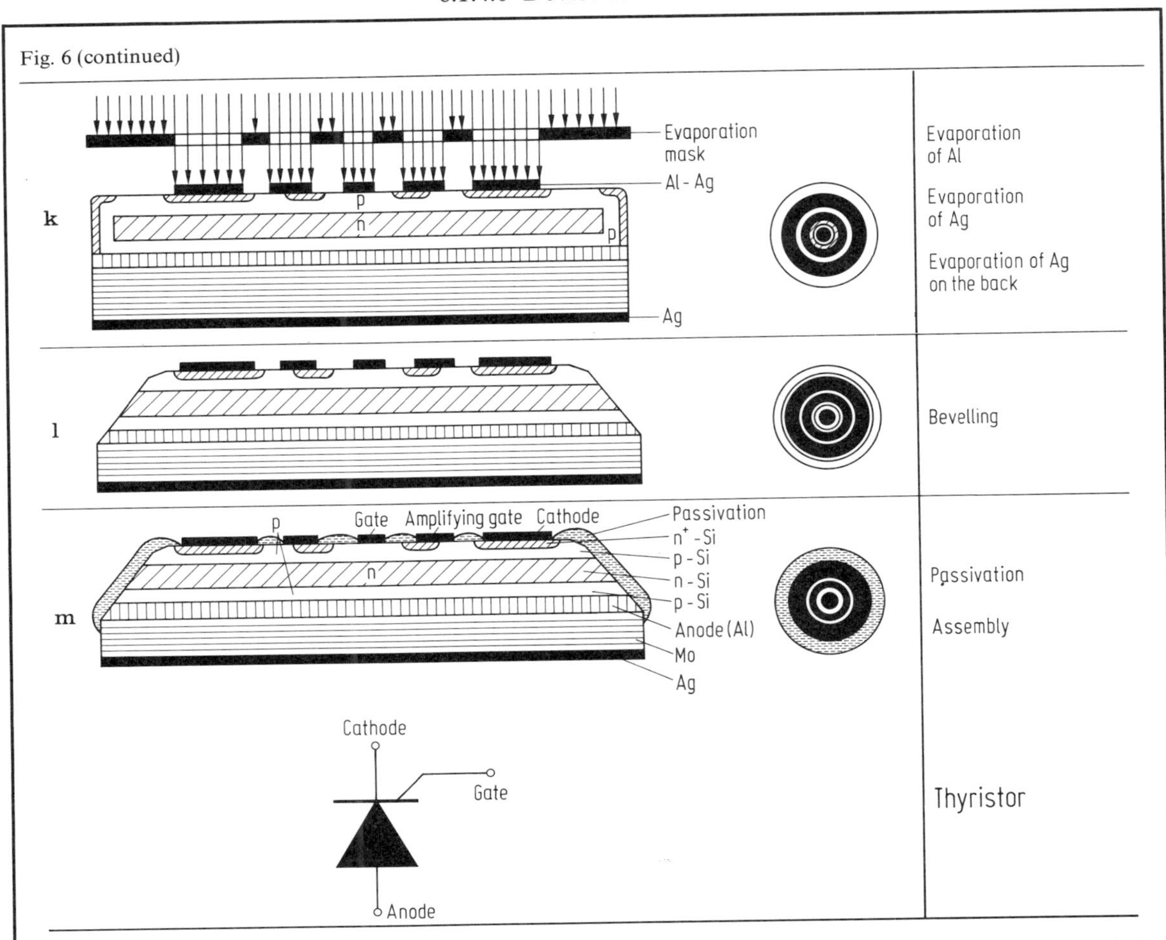

Fig. 6. Si. Process steps for the fabrication of a thyristor having an amplifying gate [68L1, 71H1, 73M1, 76H1, 79G1, 81P1]. The wafer cross-section after processing steps a)...m) is shown on the left. The top view of the wafer is shown in the center. The processing involved is explained on the right. The device symbol is given at the bottom.

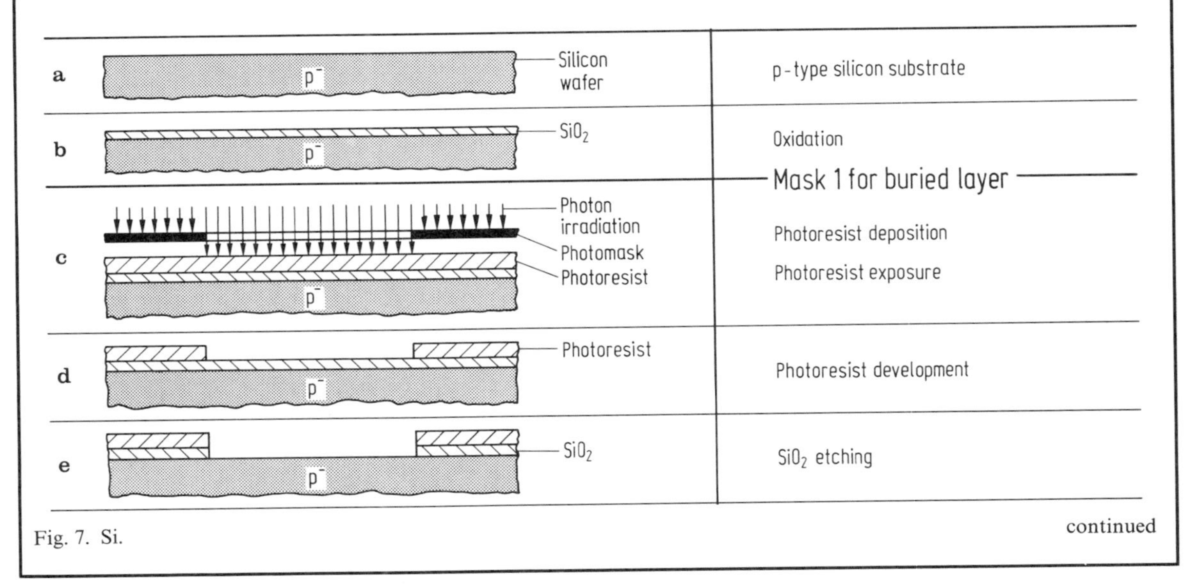

Fig. 7. Si.

continued

Fig. 7 (continued)

Step	Structure labels	Process
f	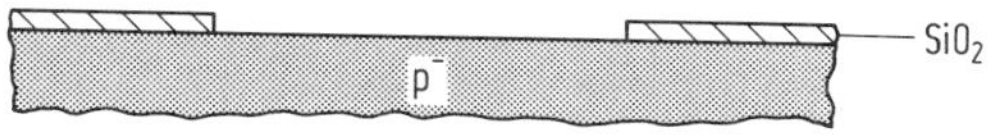 SiO_2; p^-	Photoresist stripping
g	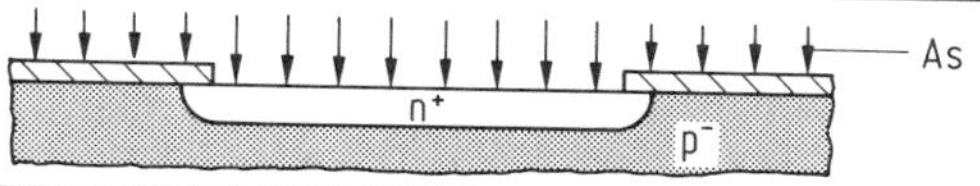 As; n^+; p^-	Buried layer diffusion
h	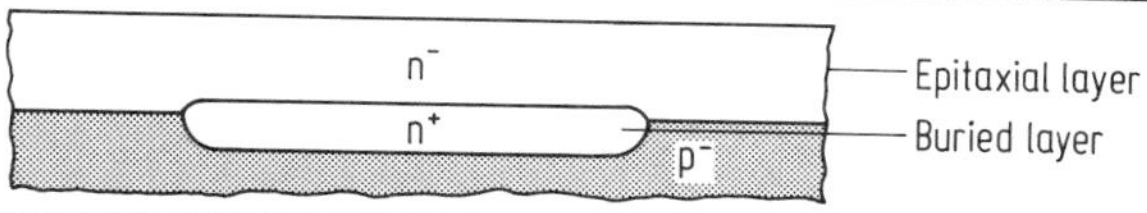 n^-; n^+; p^-; Epitaxial layer; Buried layer	Oxide stripping n-type epitaxial growth
i	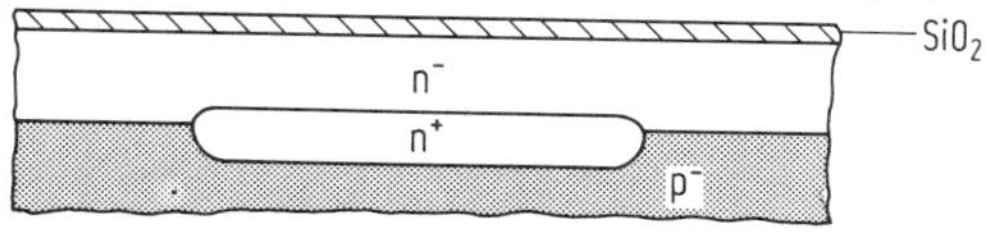 SiO_2; n^-; n^+; p^-	Oxidation
		Mask 2 for isolation
k	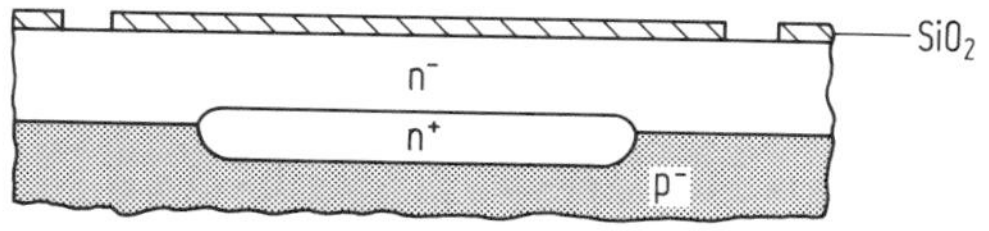 SiO_2; n^-; n^+; p^-	Photoresist deposition Photoresist exposure Photoresist development SiO_2 etching Photoresist stripping
l	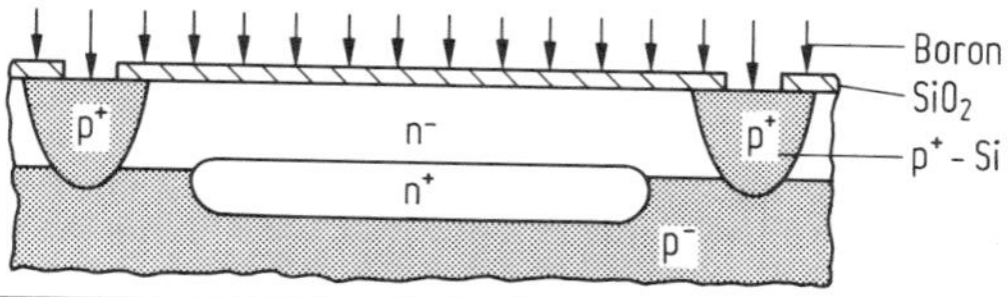 Boron; SiO_2; p^+-Si; p^+; n^-; n^+; p^-	Isolation diffusion
m	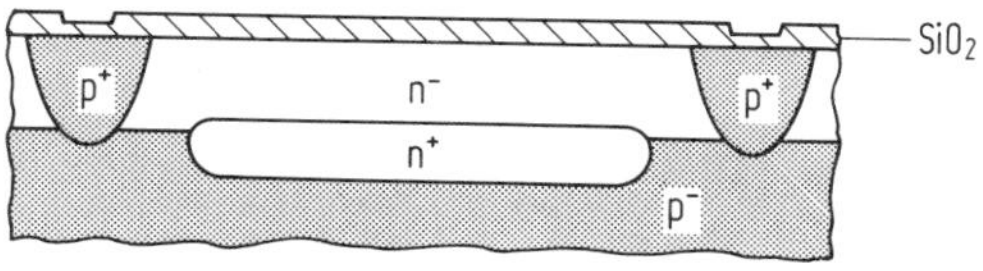 SiO_2; p^+; n^-; n^+; p^-	Oxidation
		Mask 3 for base
n	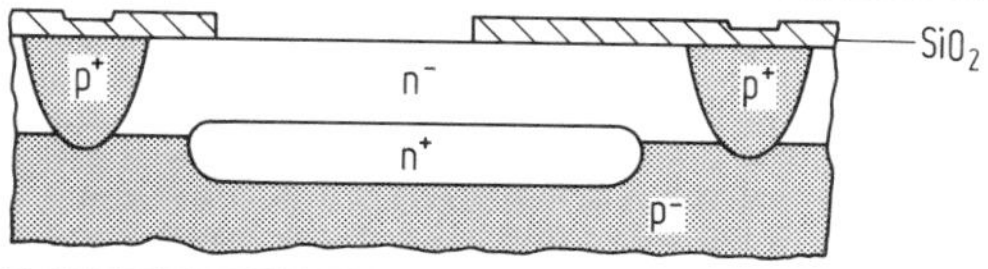 SiO_2; p^+; n^-; n^+; p^-	Photoresist deposition Photoresist exposure Photoresist development SiO_2 etching Photoresist stripping
o	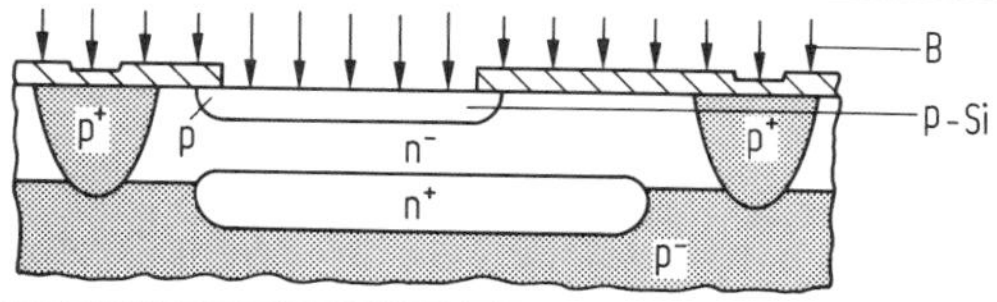 B; p-Si; p^+; p; n^-; n^+; p^-	Base diffusion
p	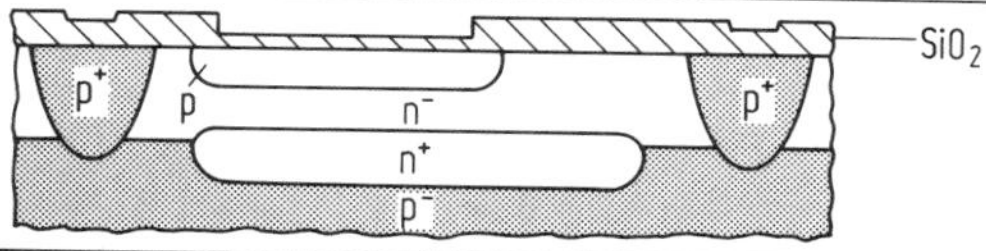 SiO_2; p^+; p; n^-; n^+; p^-	Oxidation
		Mask 4 for emitter and collector contact
q	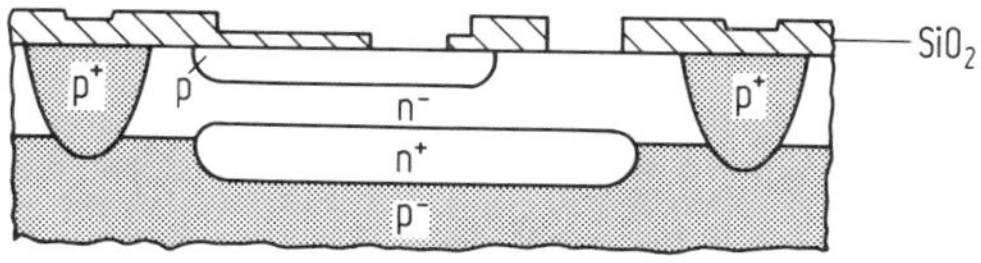 SiO_2; p^+; p; n^-; n^+; p^-	Photoresist deposition Photoresist exposure Photoresist development SiO_2 etching Photoresist stripping

Fig. 7. Si.

continued

Fig. 7 (continued)

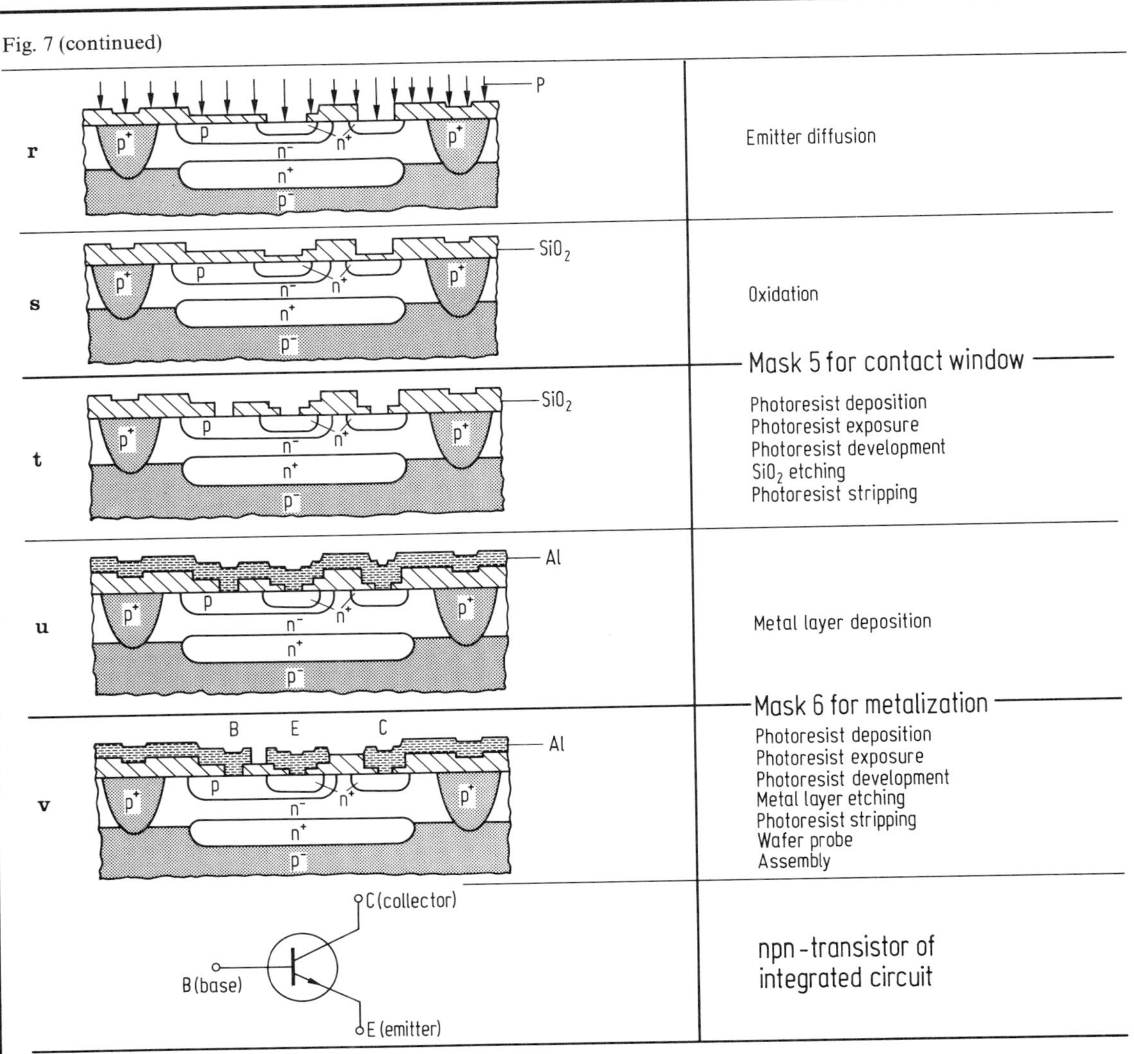

Fig. 7. Si. Process steps for the fabrication of bipolar integrated silicon circuits including junction isolation [72H2, 75R1, 80C1, 80R1]. The wafer cross-section after each processing step a)···v) is shown on the left. The process steps are explained on the right. The transistor symbol of an element in the circuit is shown at the bottom.

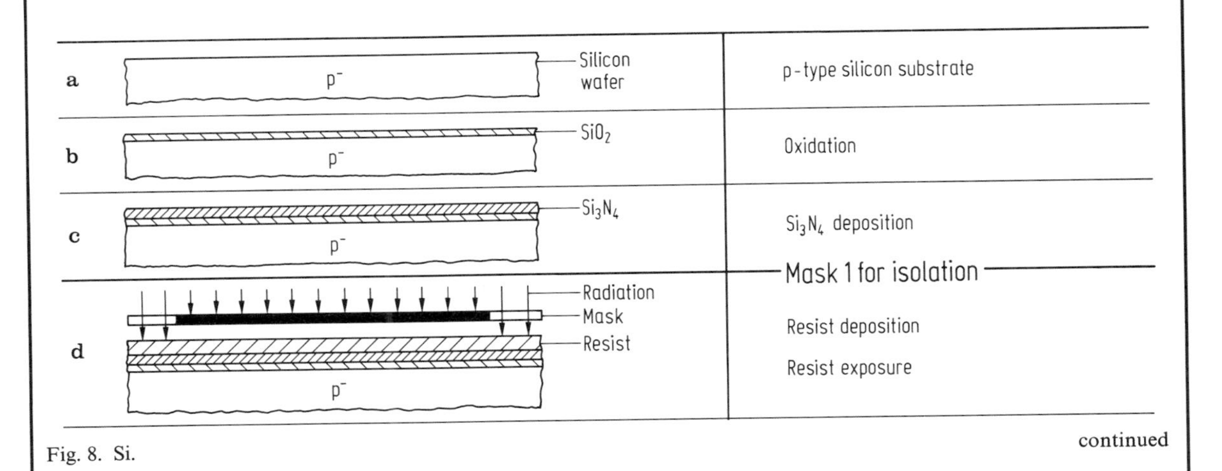

Fig. 8. Si.

continued

Fig. 8 (continued)

e	Resist development
f	Si_3N_4 etching
g	Ion implantation (p-type)
h	Resist stripping
i	Local oxidation
k	Si_3N_4 stripping Ion implantation (p-type)
l	Oxidation SiO_2 etching Reoxidation Polysilicon-1 deposition Polysilicon-1 doping (P or As)
	Mask 2 for polysilicon 1
m	Resist deposition Resist exposure Resist development Polysilicon-1 etching Resist stripping
n	Oxidation (isolation)
o	Polysilicon-2 deposition Polysilicon doping (P or As)
	Mask 3 for polysilicon 2
p	Resist deposition Resist exposure Resist development Polysilicon-2 etching Resist stripping Ion implantation (n^+)

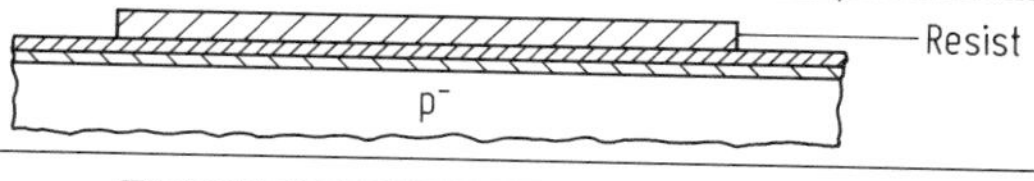

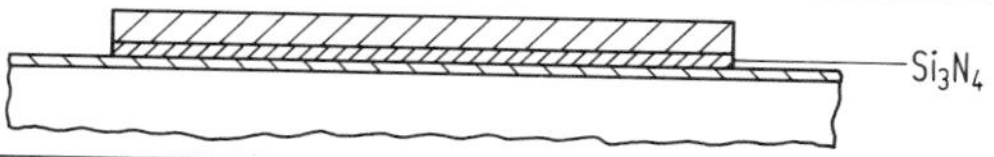

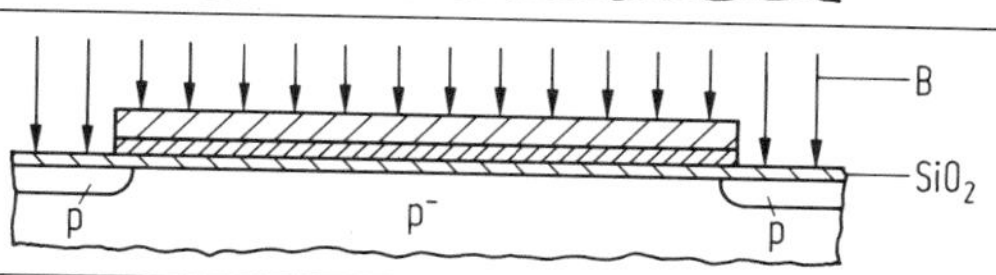

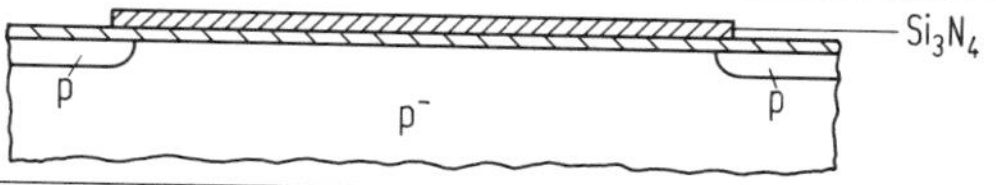

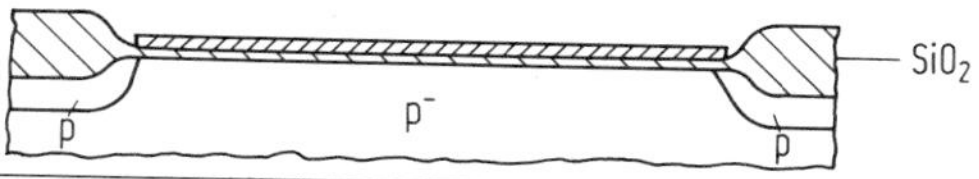

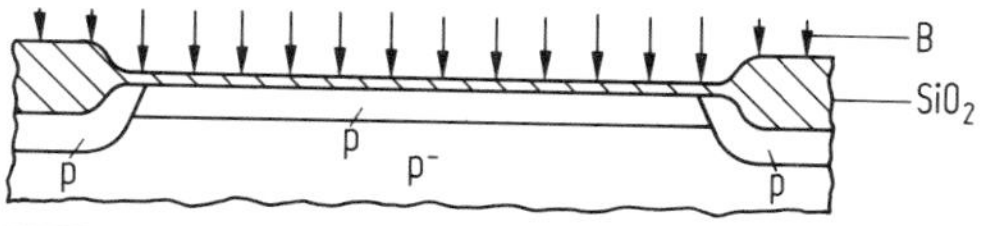

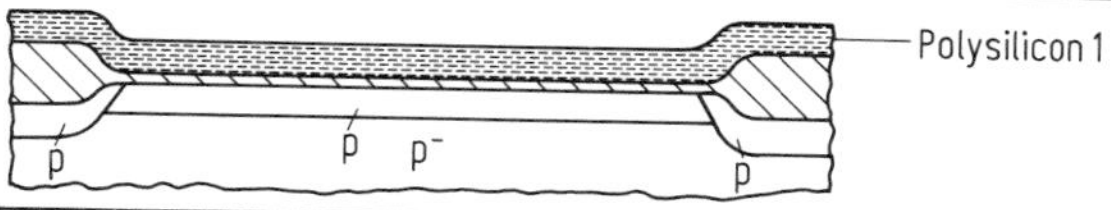

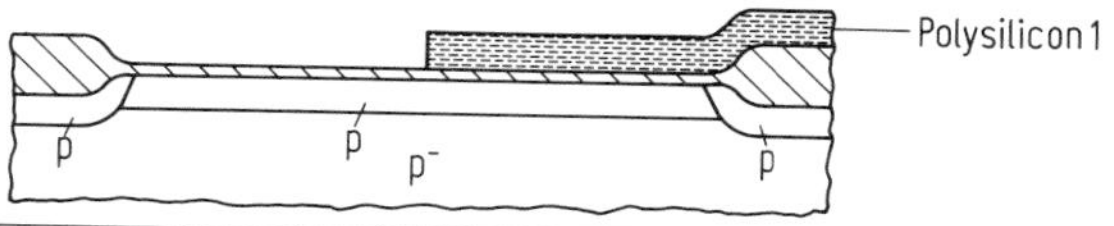

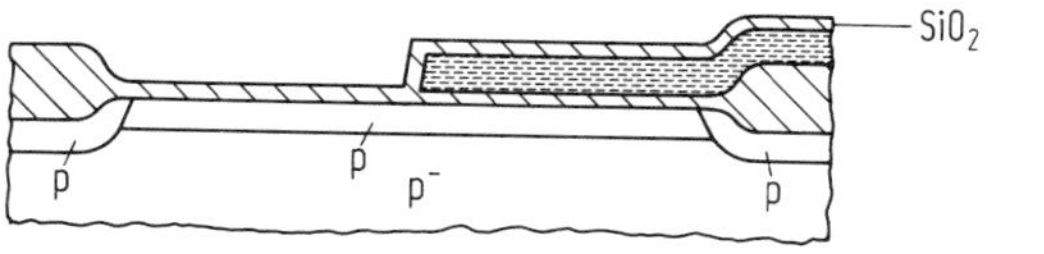

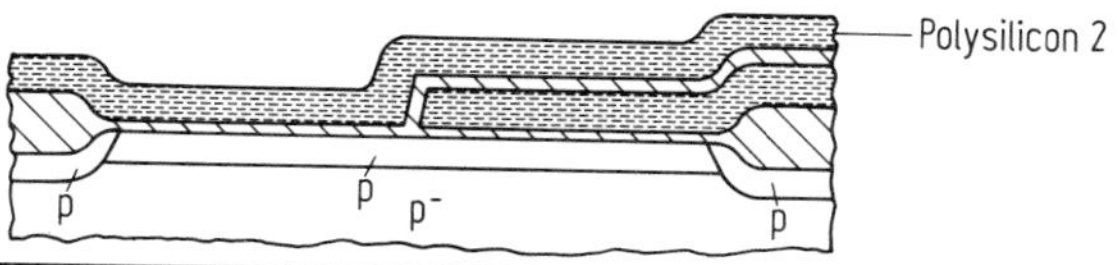

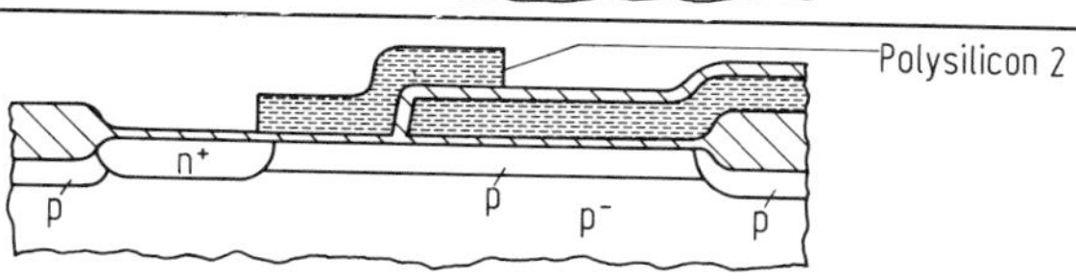

Fig. 8. Si.

continued

Fig. 8 (continued)

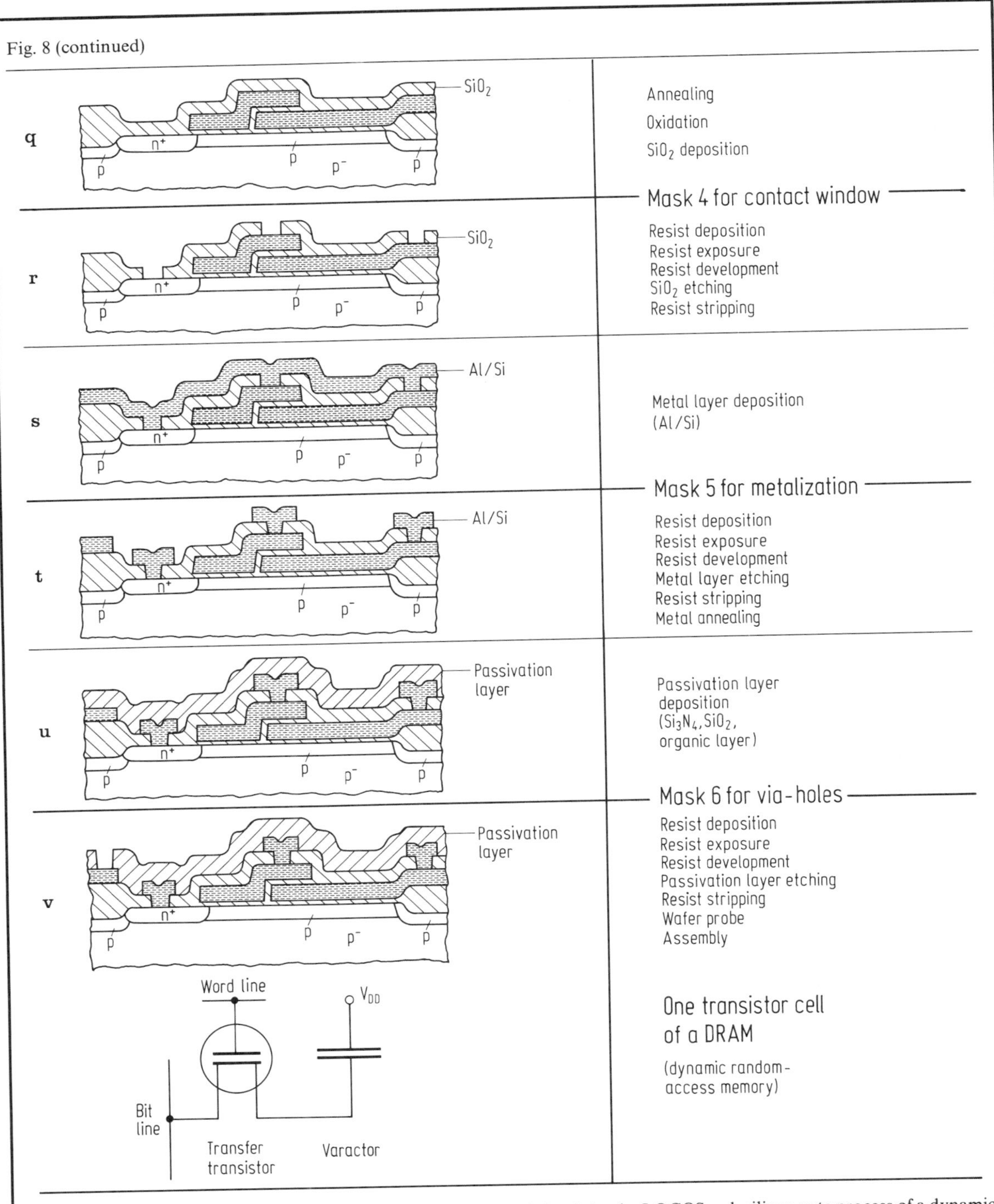

Fig. 8. Si. Process steps for the fabrication of an MOS integrated circuit by the LOCOS polysilicon gate process of a dynamic RAM [72H2, 75R1, 80F1]. The wafer cross-section after each processing step a)⋯v) is shown on the left. The process steps are explained on the right. The device symbol of an element of DRAM is given at the bottom.

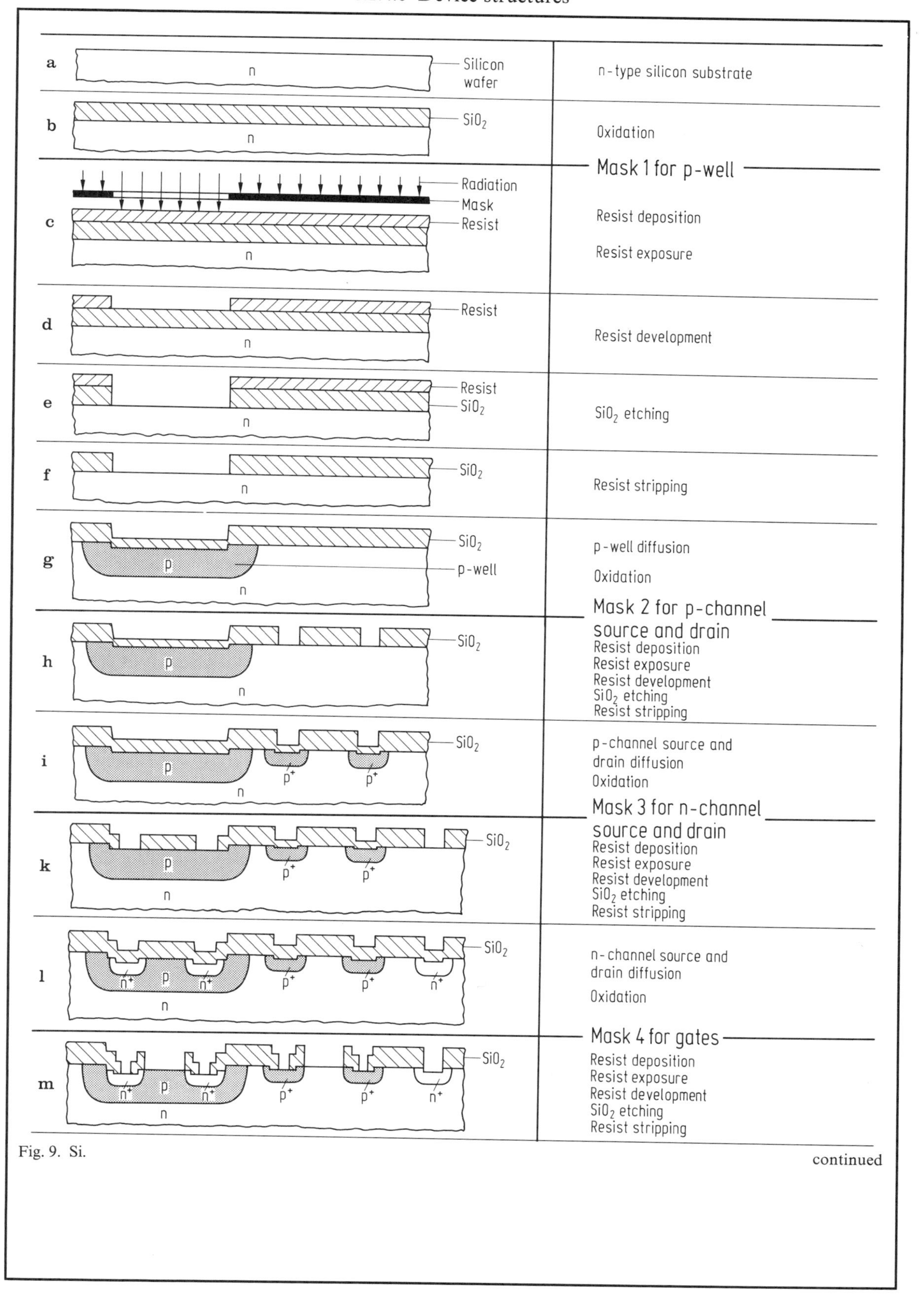

Fig. 9. Si.

continued

Fig. 9 (continued)

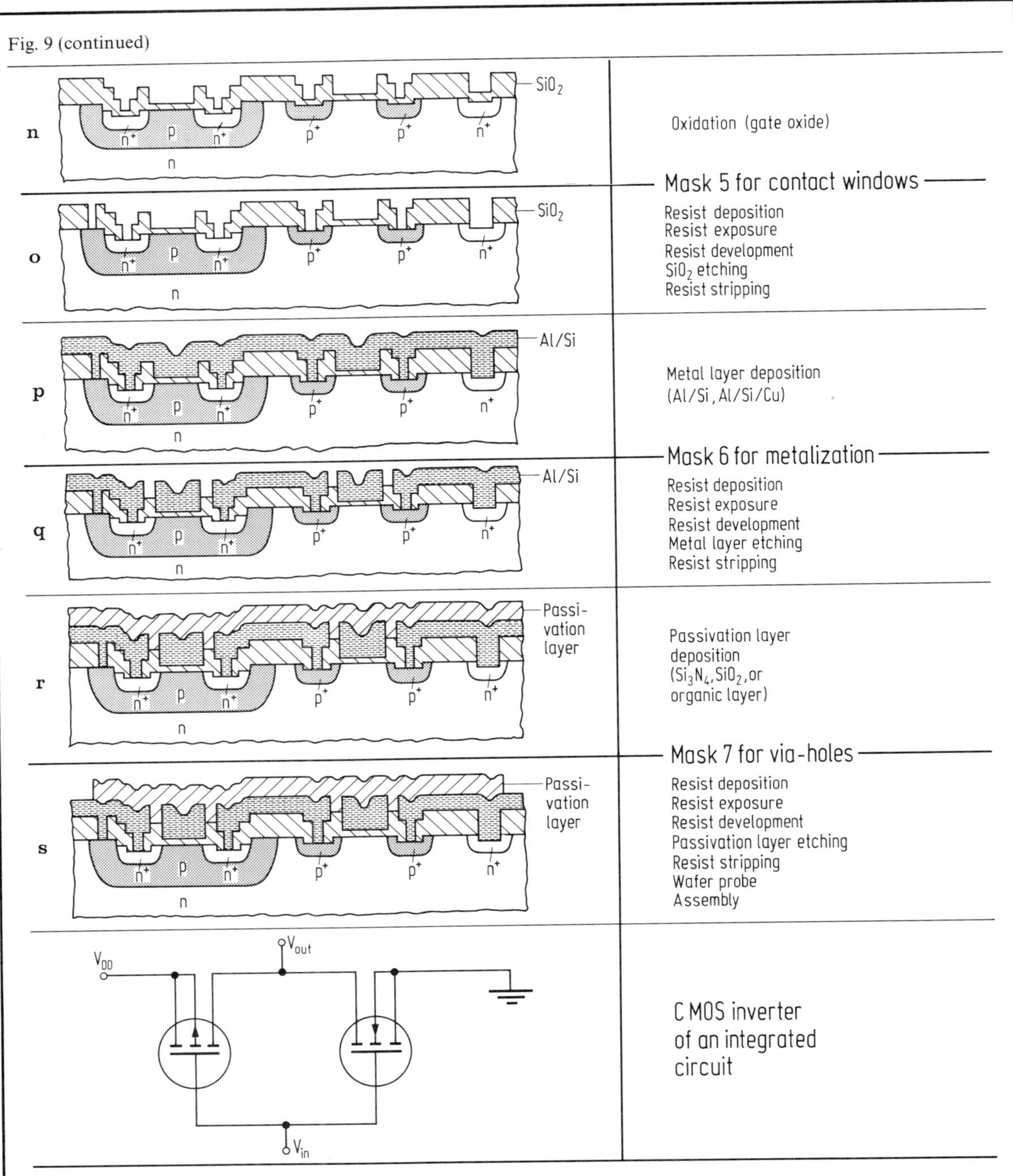

Fig. 9. Si. Process steps for the fabrication of CMOS integrated circuits [80C1]. The wafer cross-section after each processing step a)···s) is shown on the left. The process applied is given on the right. The device symbol of an inverter element is given at the bottom.

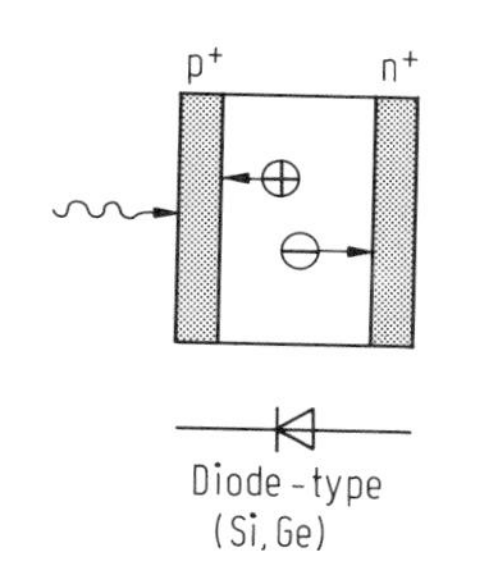

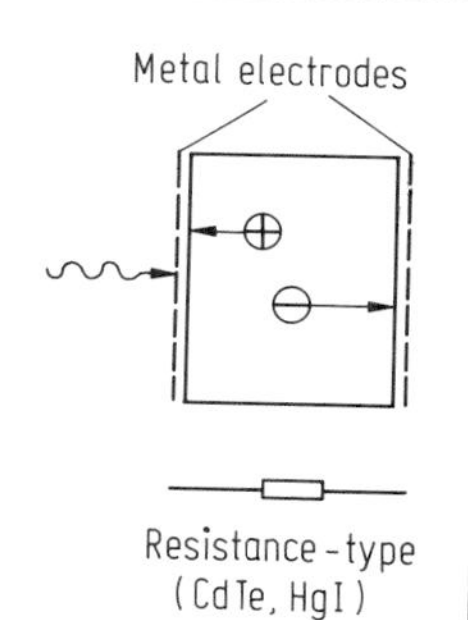

Fig. 10. Schematic drawing of the two fundamental structures and the basic mechanisms of semiconductor radiation detectors.

For Fig. 12, see next page.

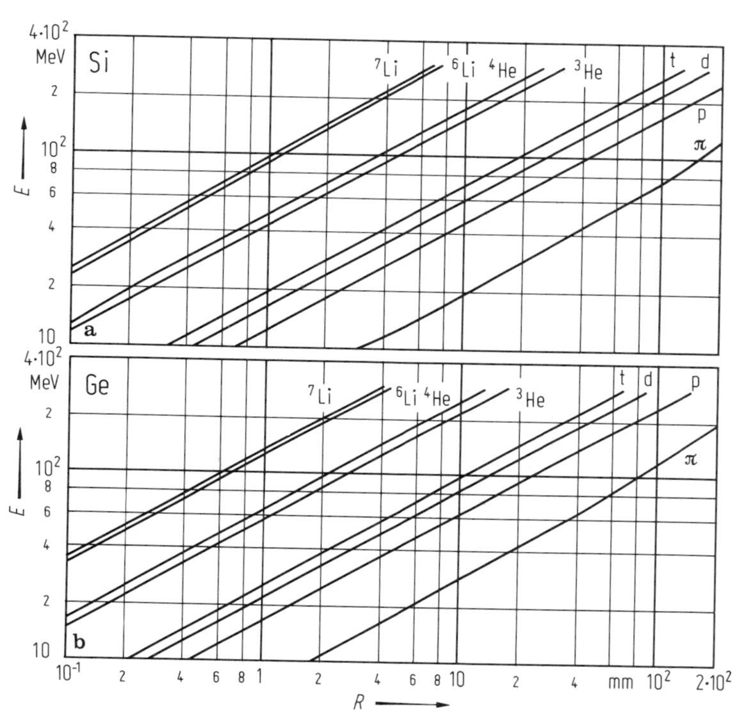

Fig. 11 a, b. Si, Ge. Penetration range R of pions (π), protons (p), deuterons (d), tritons (t), ^{3}He, ^{4}He, ^{6}Li, and ^{7}Li particles vs. energy E [82P1]. a) silicon, b) germanium.

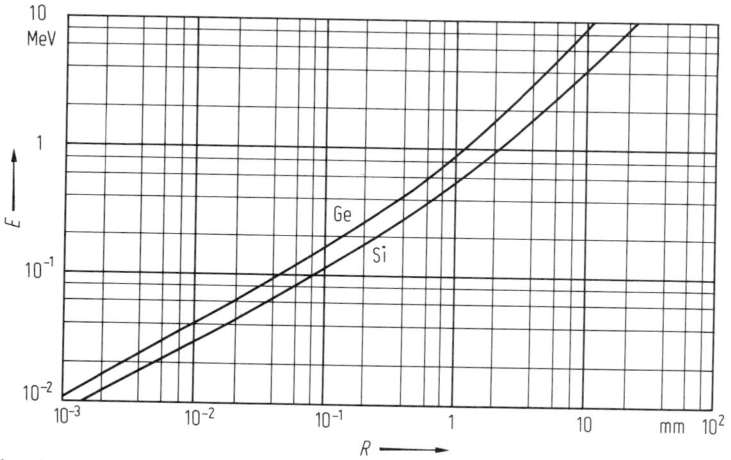

Fig. 13. Si, Ge. Extrapolated penetration range R of electrons in Si and Ge vs. energy E [74G1].

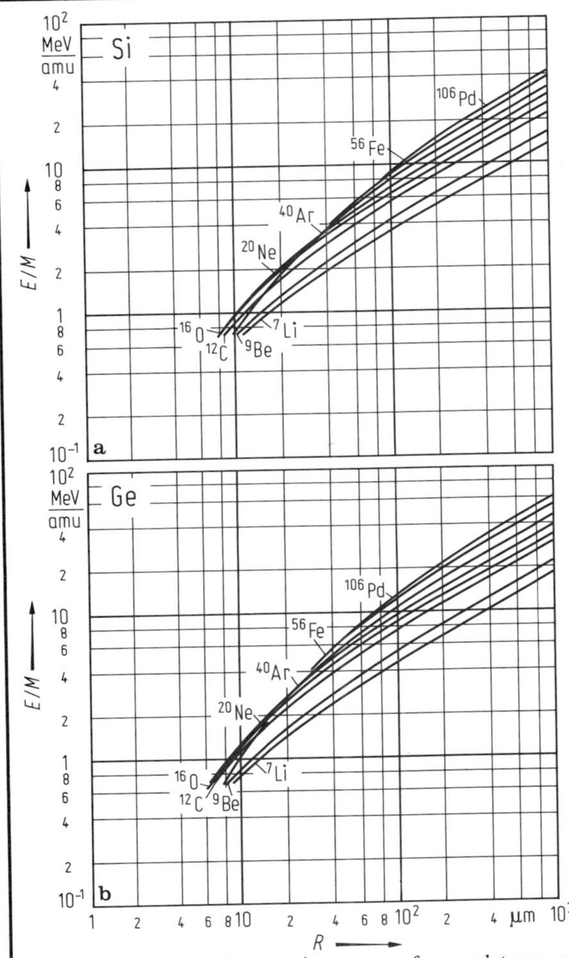

Fig. 12a, b. Si, Ge. Penetration range of several types of heavy ions vs. energy per mass unit E/M [74G1]. a) silicon, b) germanium.

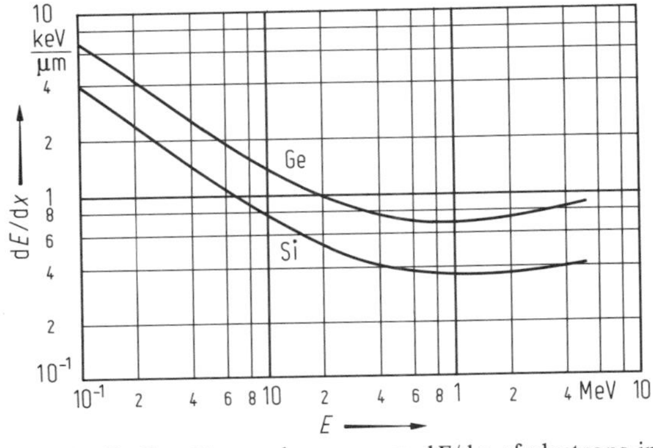

Fig. 15. Si, Ge. Energy loss per µm dE/dx of electrons in silicon and germanium vs. energy E [81H1].

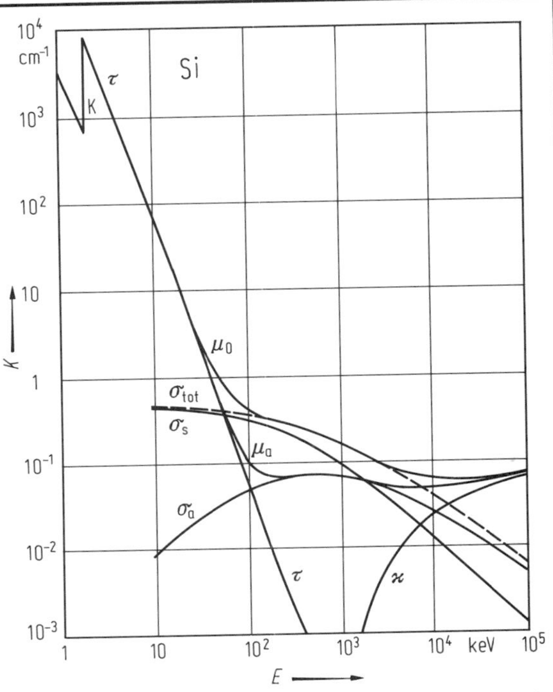

Fig. 16. Si. Absorption and scattering coefficients K in silicon vs. energy E of X- and γ-rays. τ: photoabsorption; σ_a: Compton-absorption; σ_s: Compton-scattering; κ: pair-production; $\mu_a = \tau + \sigma_a + \kappa$: energy-absorption; $\mu_0 = \mu_a + \sigma_s$: attenuation coefficient; absorption edge K at 1.838 keV.

Fig. 17. Ge. Absorption and scattering coefficients K in germanium vs. energy E of X- and γ-rays. ν: photoabsorption; σ_a: Compton absorption; σ_s: Compton-scattering; κ: pair-production; $\mu_a = \tau + \sigma_a + \kappa$: energy-absorption; $\mu_0 = \mu_a + \sigma_s$: attenuation coefficient; absorption edges K at 11.1 keV, L_I at 1.42 keV, L_{II} at 1.25 keV, L_{III} at 1.22 keV.

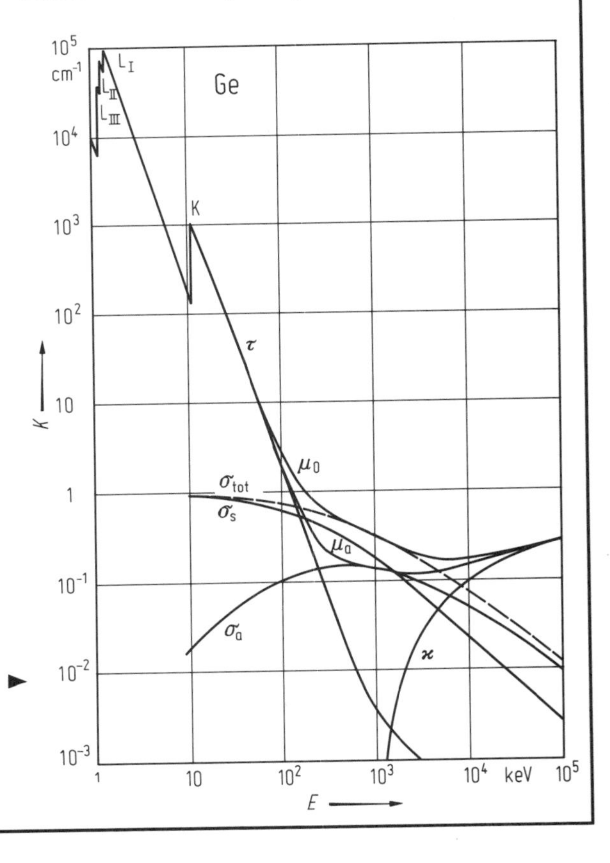

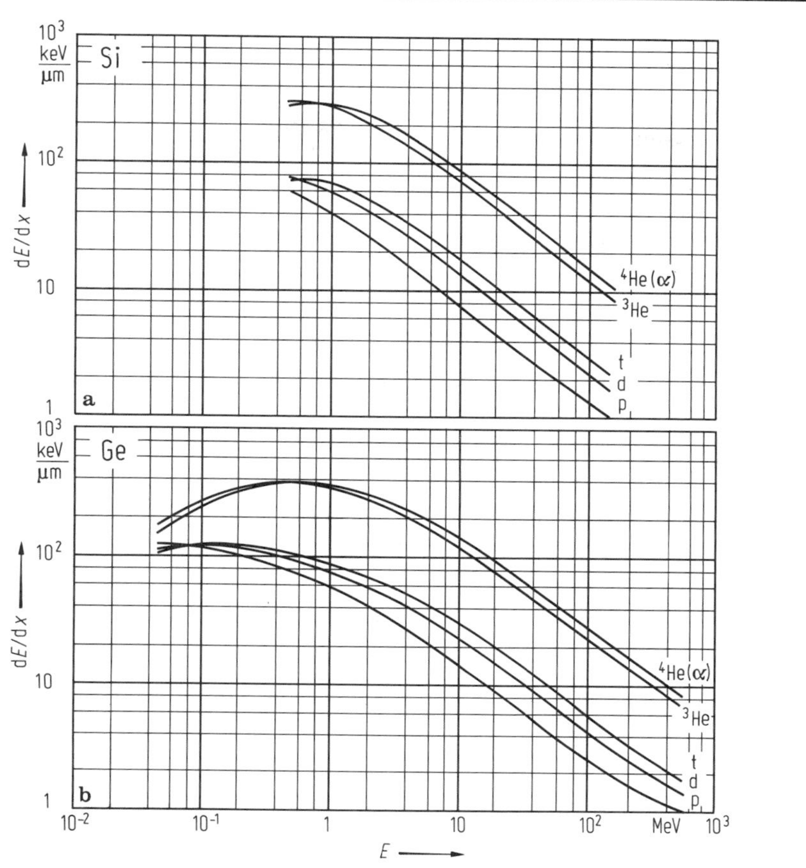

Fig. 14a, b. Si, Ge. Energy loss per µm dE/dx of protons (p), deuterons (d), tritons (t), ^{3}He, and ^{4}He (α's) in a) silicon [74G1] and b) germanium [66W1] as a function of energy E.

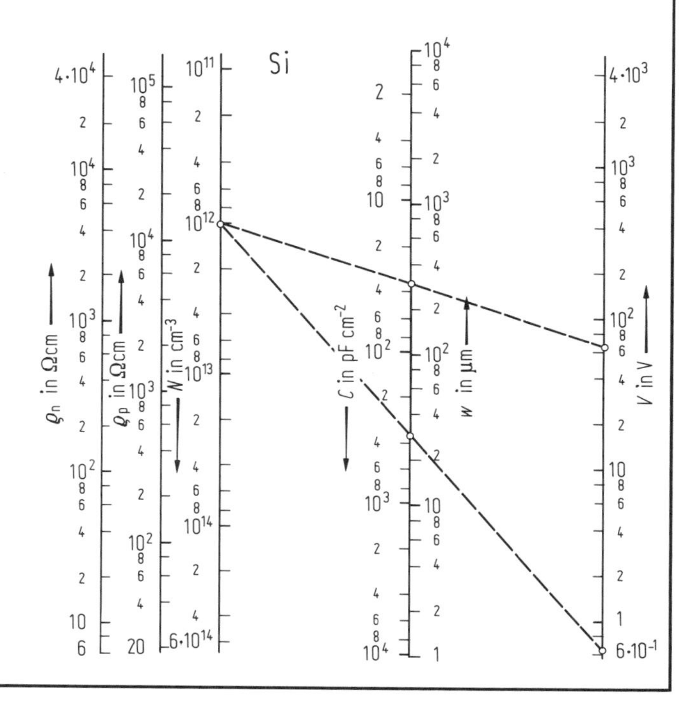

Fig. 18. Si. Nomograph to determine the width w and the capacitance C of the depleted layer in Si as functions of the resistivity ϱ and reverse bias voltage V. The data used are $\mu_n = 1400\ cm^2/Vs$, $\mu_p = 480\ cm^2/Vs$, and $\varepsilon = 12$. (N is the net-doping concentration.)

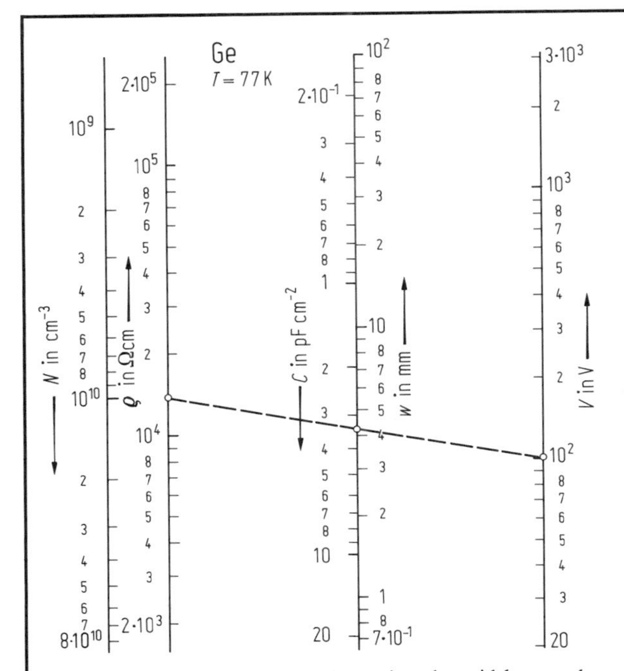

Fig. 19. Ge. Nomograph to determine the width w and the capacitance C of the depleted layer in HPGe at 77 K as functions of the resistivity ϱ and reverse bias voltage V. The data used are $\mu_n = \mu_p = 45000\ cm^2/Vs$ and $\varepsilon = 16$. (N is the net-doping concentration.)

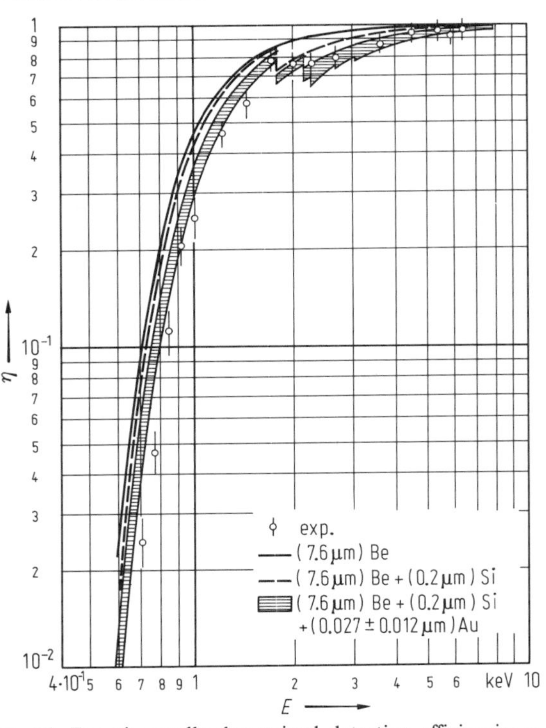

Fig. 20. Experimentally determined detection efficiencies η of a 3 mm Si(Li)-system and calculations of X-ray transmission efficiencies vs. X-ray energies for the layers Be, Be + Si, and Be + Si + Au [79K1].

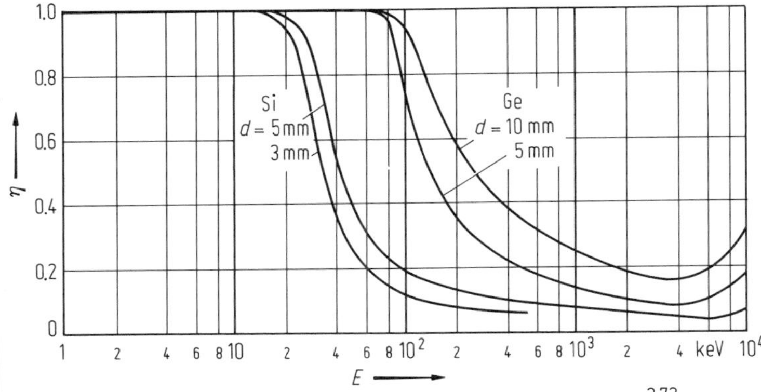

Fig. 21. Si, Ge. Fraction η of incident photons absorbed in Si and Ge of different thickness d vs. photon energy E [81H1].

For Fig. 22, see next page.

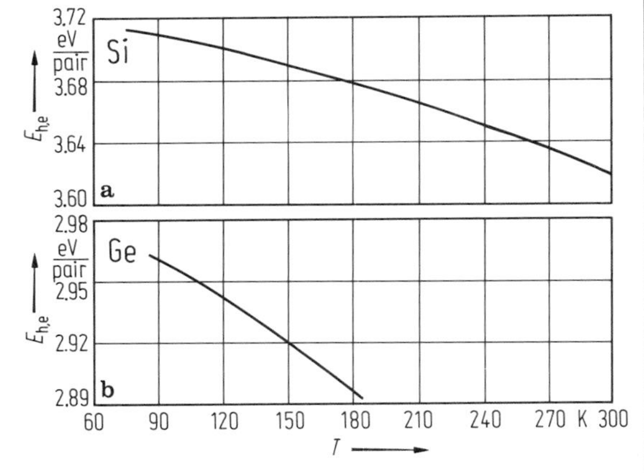

Fig. 23 a, b. Si, Ge. Ionisation energy $E_{h,e}$ vs. temperature T [68P1, 73S1] for a) silicon, b) germanium.

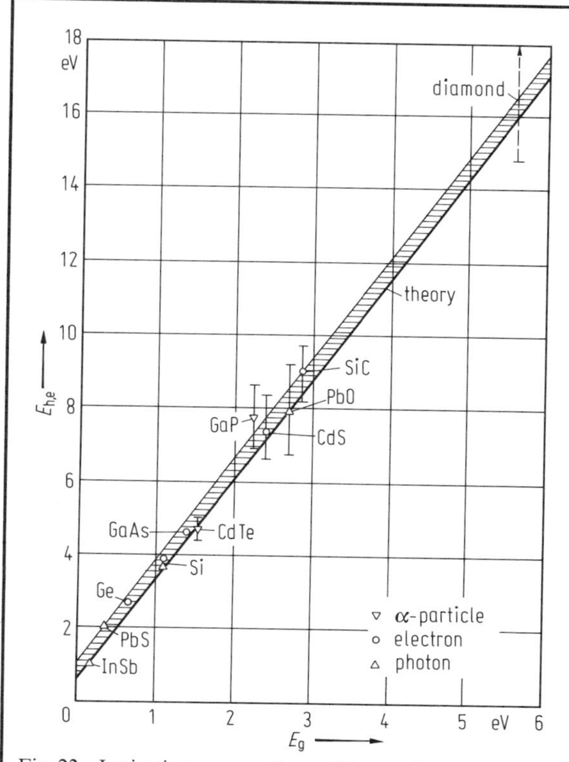

Fig. 22. Ionisation energy $E_{h,e}$ of X-rays, fast electrons or α-particles as a function of the bandgap E_g [68K1]. (Theory: $E_{h,e} = (14/5)\ E_g + r(\hbar\omega_R)$; with $0.5 \leqq r(\hbar\omega_R) \leqq 1.0$ eV).

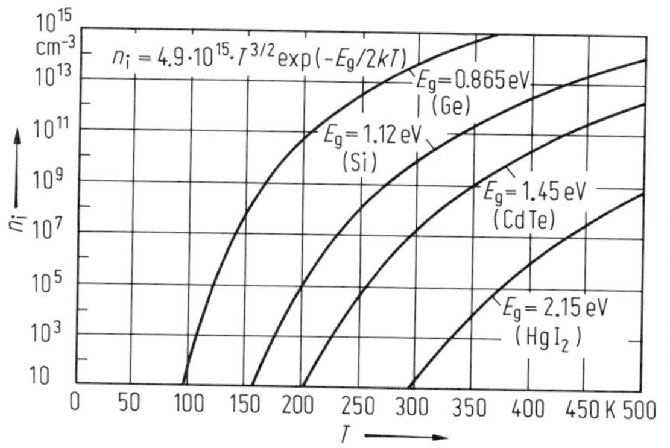

Fig. 25. Intrinsic carrier concentration n_i vs. temperature T for Ge, Si, CdTe, and HgI_2.

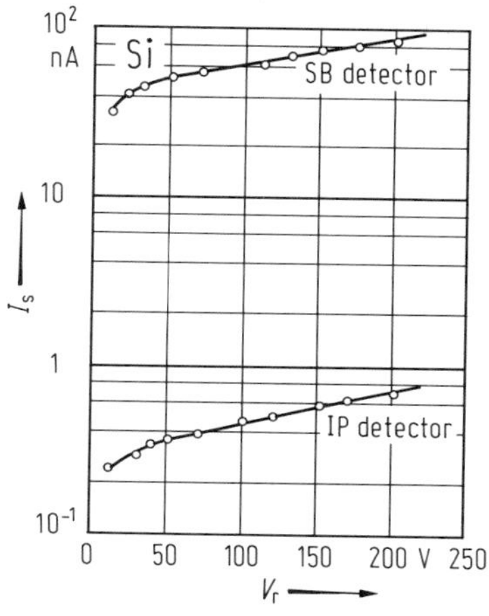

Fig. 26. Reverse current I_s of a surface barrier (SB) and a passivated ion-implanted (IP) silicon detector of the same size (25 mm² area, 300 μm thickness) vs. reverse bias voltage V_r [82K1].

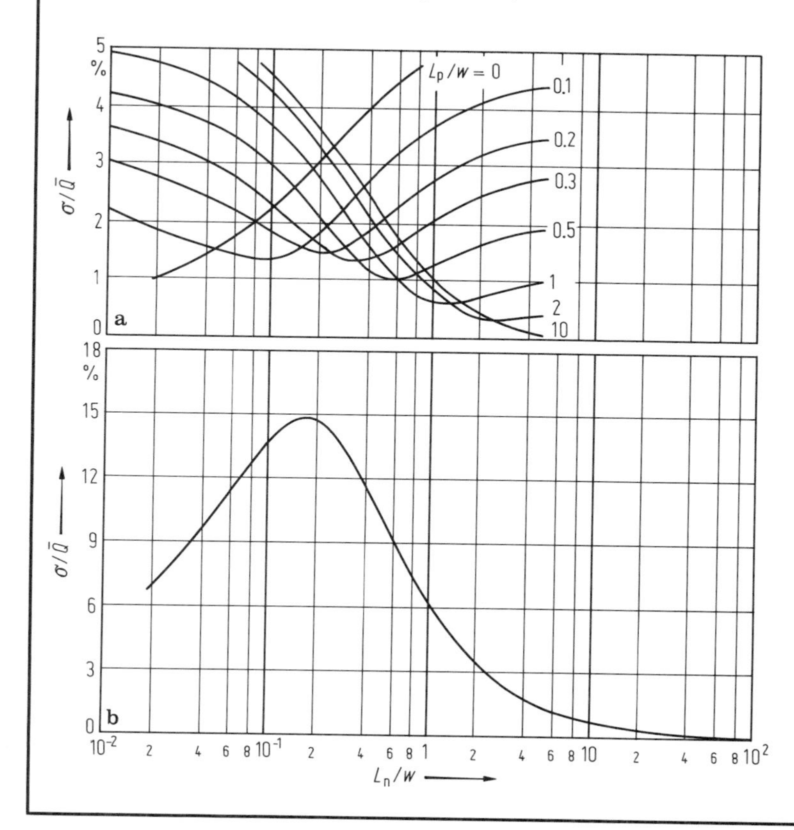

◄ Fig. 24a, b. Detector resolution $\sigma/\bar{Q}$ as a function of the electron drift length L_n/w [67D1]. a) Parameter: L_p/w as given, b) $L_n = L_p$. (For further explanation see text.)

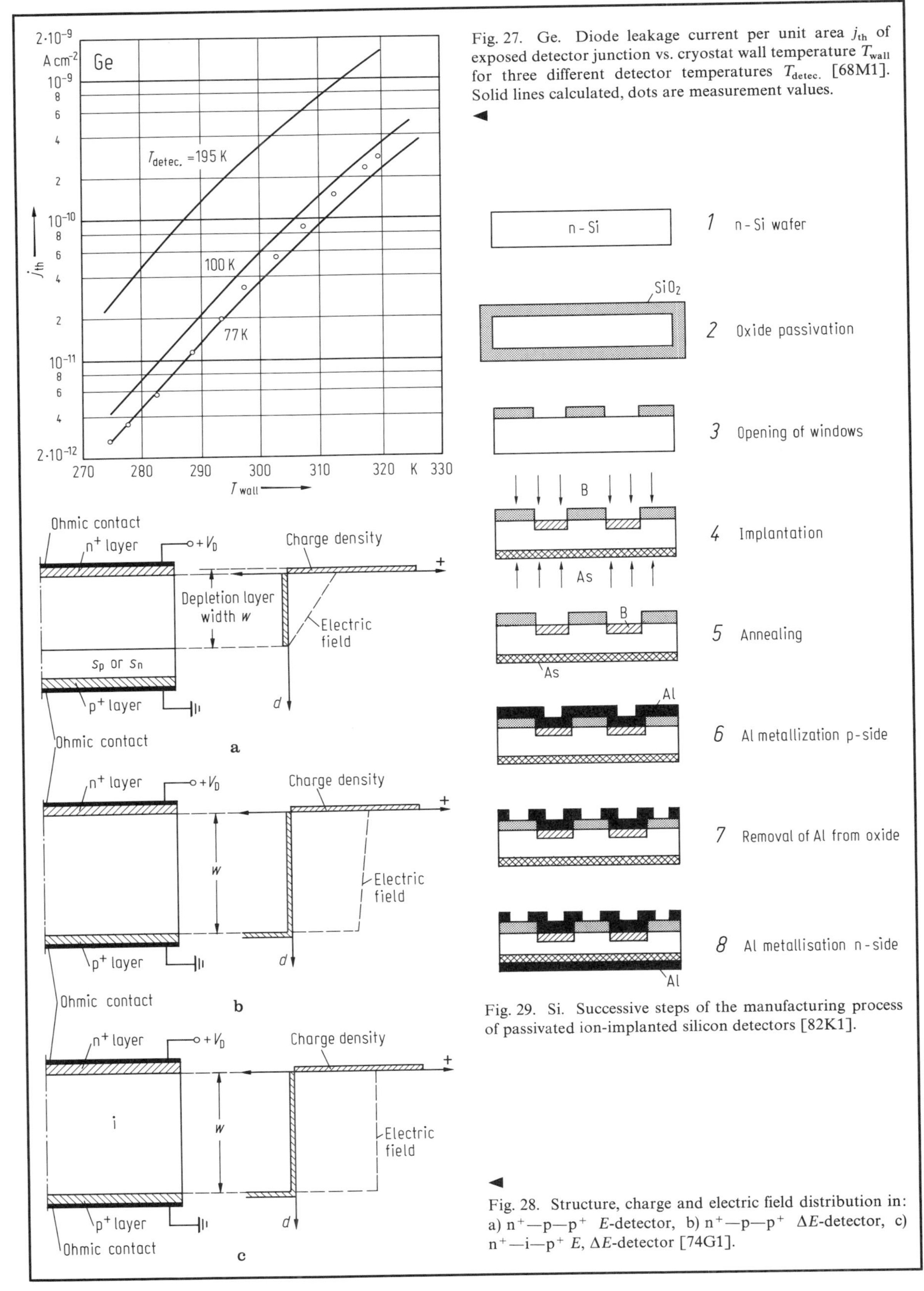

Fig. 27. Ge. Diode leakage current per unit area j_{th} of exposed detector junction vs. cryostat wall temperature T_{wall} for three different detector temperatures $T_{detec.}$ [68M1]. Solid lines calculated, dots are measurement values.

Fig. 29. Si. Successive steps of the manufacturing process of passivated ion-implanted silicon detectors [82K1].

Fig. 28. Structure, charge and electric field distribution in: a) n$^+$—p—p$^+$ E-detector, b) n$^+$—p—p$^+$ ΔE-detector, c) n$^+$—i—p$^+$ E, ΔE-detector [74G1].

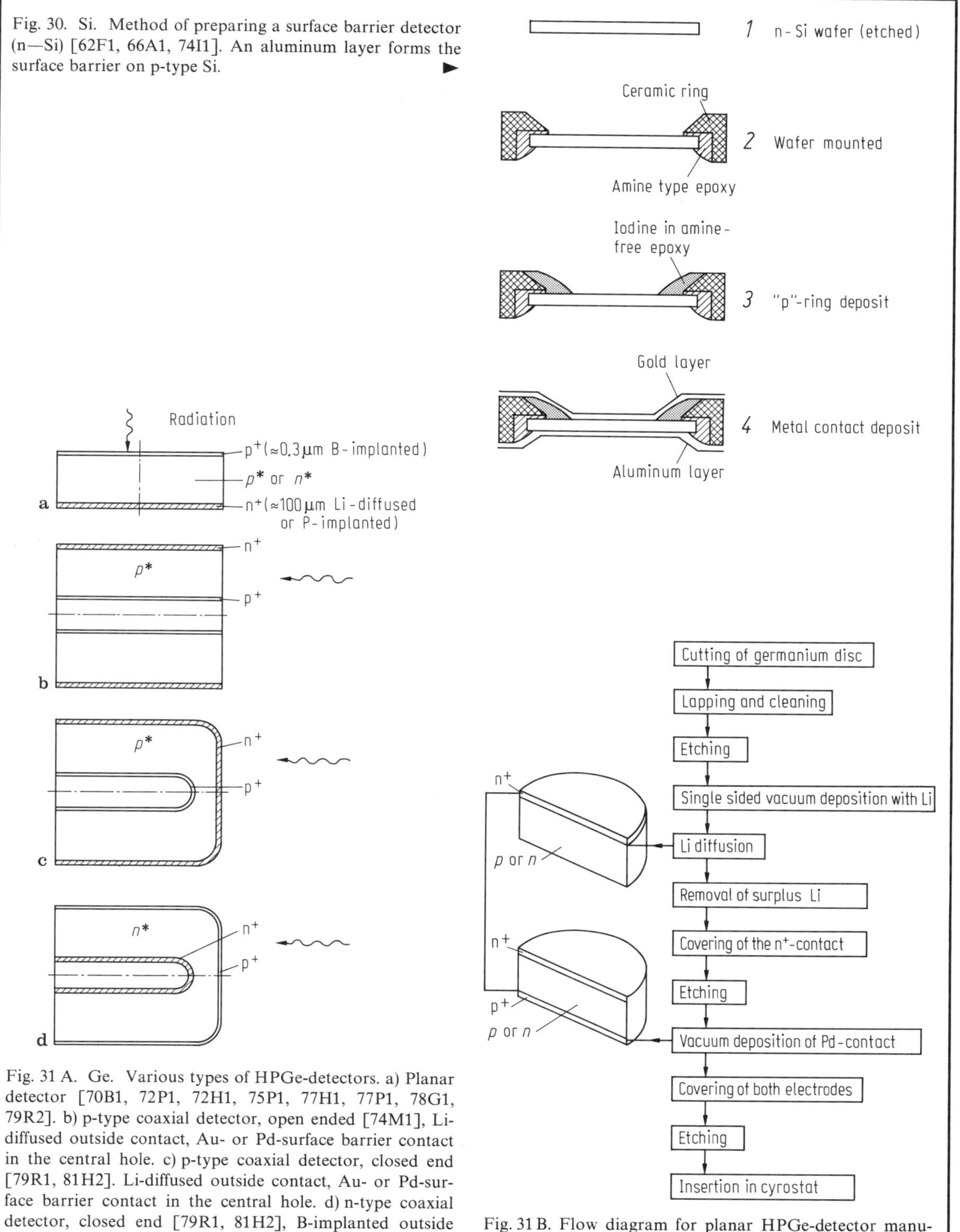

Fig. 30. Si. Method of preparing a surface barrier detector (n—Si) [62F1, 66A1, 74I1]. An aluminum layer forms the surface barrier on p-type Si. ►

Fig. 31 A. Ge. Various types of HPGe-detectors. a) Planar detector [70B1, 72P1, 72H1, 75P1, 77H1, 77P1, 78G1, 79R2]. b) p-type coaxial detector, open ended [74M1], Li-diffused outside contact, Au- or Pd-surface barrier contact in the central hole. c) p-type coaxial detector, closed end [79R1, 81H2]. Li-diffused outside contact, Au- or Pd-surface barrier contact in the central hole. d) n-type coaxial detector, closed end [79R1, 81H2], B-implanted outside contact, Li-diffused n^+-contact in the central hole. All coaxial detectors can be used as borehole detectors. (*) indicates *p* or *n* conduction type at the operating temperature 78K. (The diameters and the lengths of the coaxial detectors are up to 7.5 cm).

Fig. 31 B. Flow diagram for planar HPGe-detector manufacture [78G1]. Instead of a Pd-contact, a B-implanted layer can be employed. Phosphorus can be implanted instead of using Li-diffusion. In coaxial detectors, an axial hole is drilled into the cylindrical crystal. The electrodes are fabricated at the inner and the outer wall, respectively.

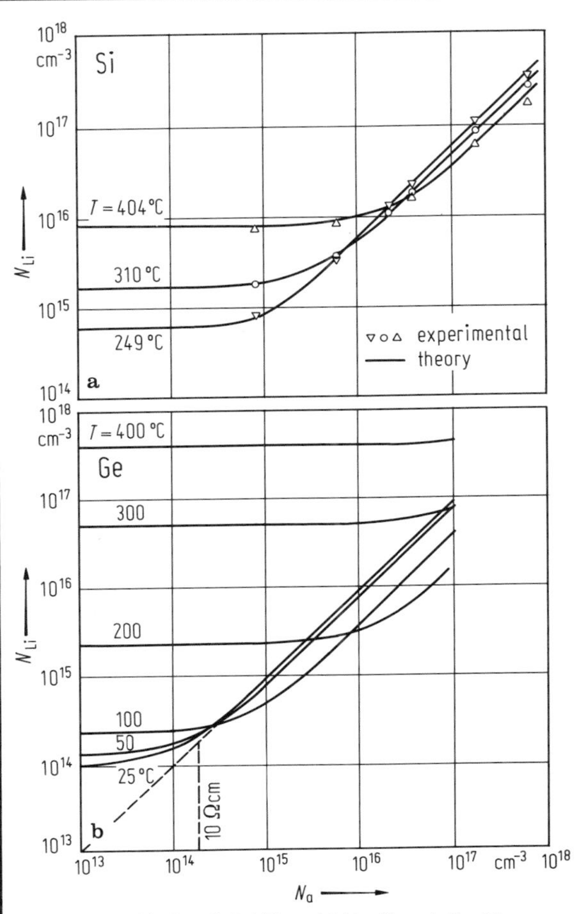

Fig. 32a, b. Si, Ge. Solubility of Li in Si and Ge. The concentration of lithium N_{Li} is shown vs. the acceptor concentration N_a present in the substrate. a) Silicon substrate, $N_a \triangleq$ boron concentration [68B1], b) germanium substrate, $N_a \triangleq$ gallium concentration [68B1].

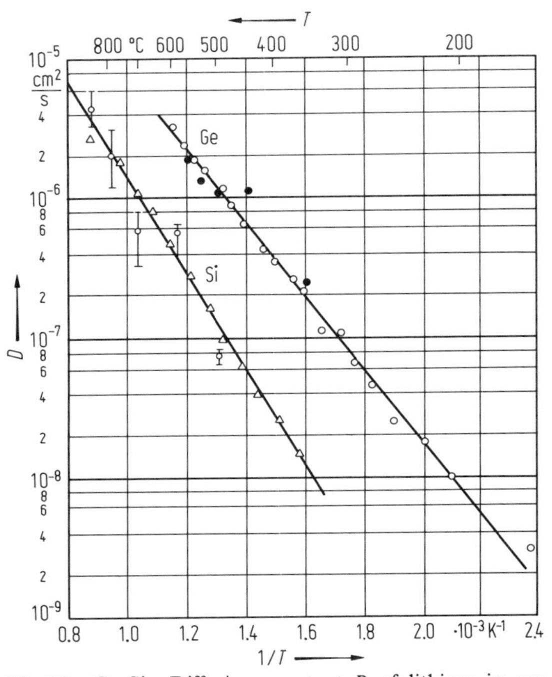

Fig. 33. Ge, Si. Diffusion constant D of lithium in germanium and silicon as a function of $1/T$ as determined from the measurement of ion drift in an electric field (open circles and triangles, respectively). Results of measurements by the p−n junction method are also shown (filled circles and vertical bars, respectively) [68B1]. ($D_{Ge} = 25 \cdot 10^{-4} \exp(-11800/RT)$; $D_{Si} = 23 \cdot 10^{-4} \exp(-15200/RT)$).

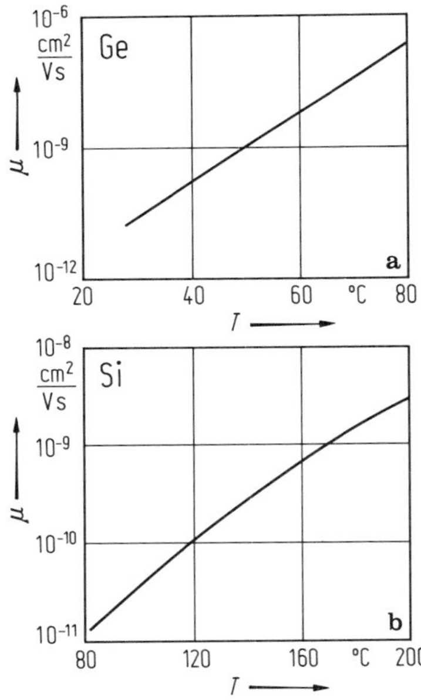

Fig. 34. Si, Ge. Mobility μ of lithium in germanium (a) and silicon (b) vs. temperature T in the range which is used for the drift process [68B1].

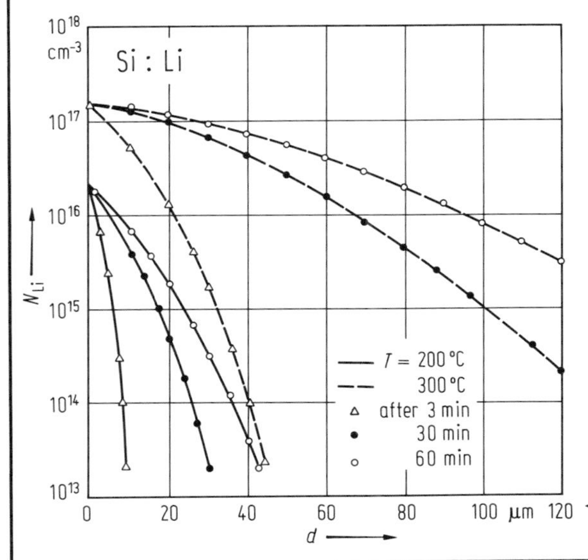

◄ Fig. 35. Si:Li. Calculated lithium concentration N_{Li} in silicon vs. depth d for various diffusion conditions [68B1].

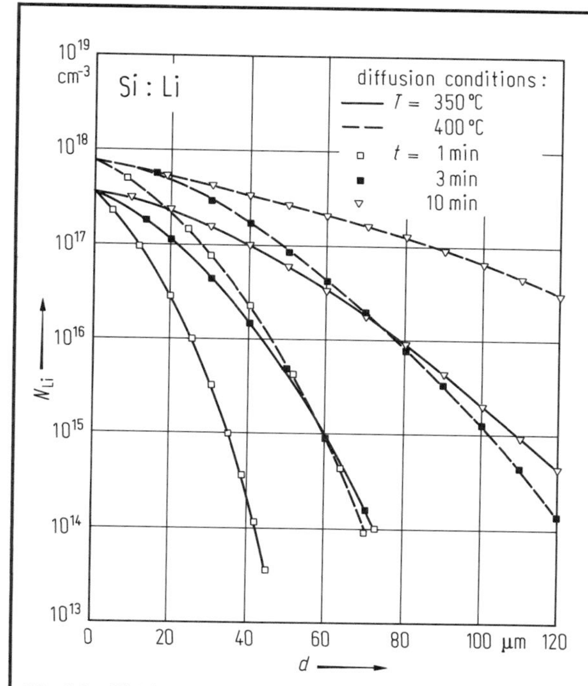

Fig. 36. Si:Li. Calculated lithium concentration N_{Li} in silicon vs. depth d, for various diffusion conditions [68B1].

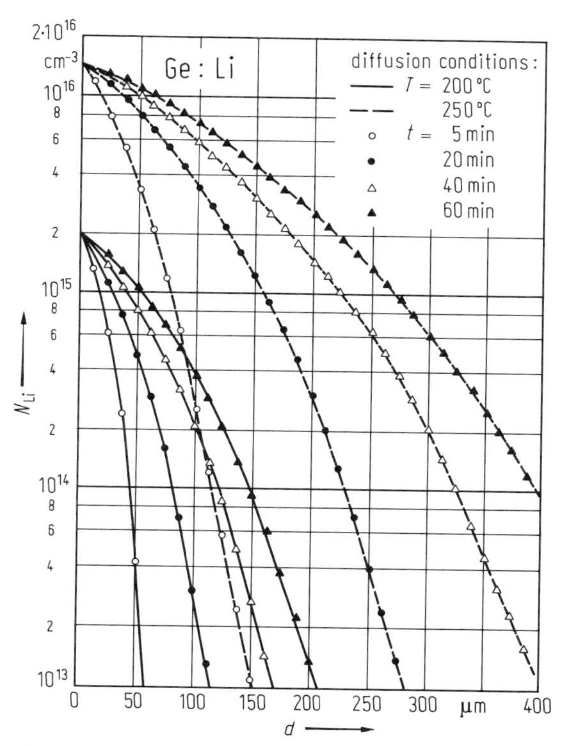

Fig. 37. Ge:Li. Calculated lithium concentration N_{Li} in germanium vs. depths d, for various diffusion conditions [68B1].

For Figs. 39, 40, see next page.

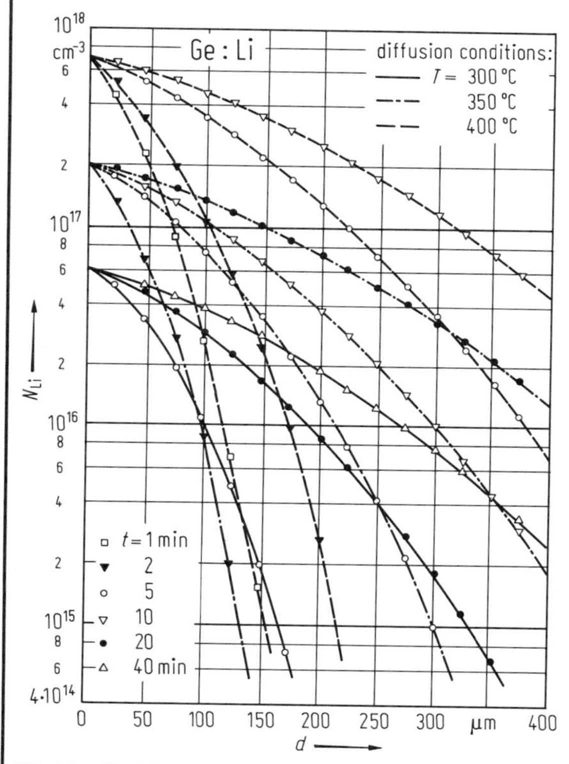

Fig. 38. Ge:Li. Calculated lithium concentration N_{Li} in germanium vs. depth d, for various diffusion conditions [68B1].

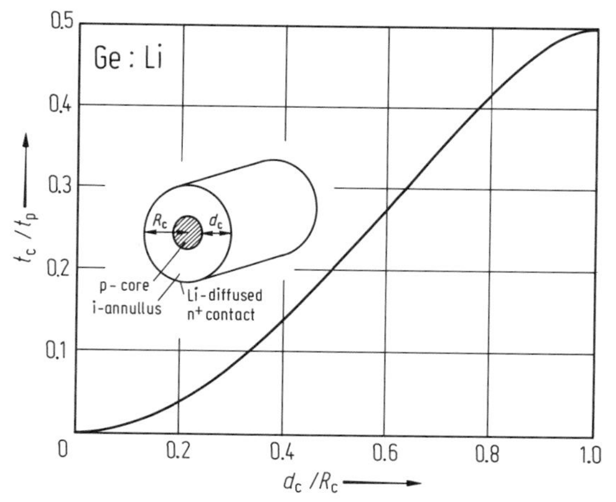

Fig. 41. Ge:Li. Ratio of drift times t_c/t_p in coaxial and planar Ge:Li diodes as a function of the ratio of drift depth, d_c/R_c, respectively. The insert shows the geometry of the coaxial cylindrical Ge:Li pin-diode [66T1].

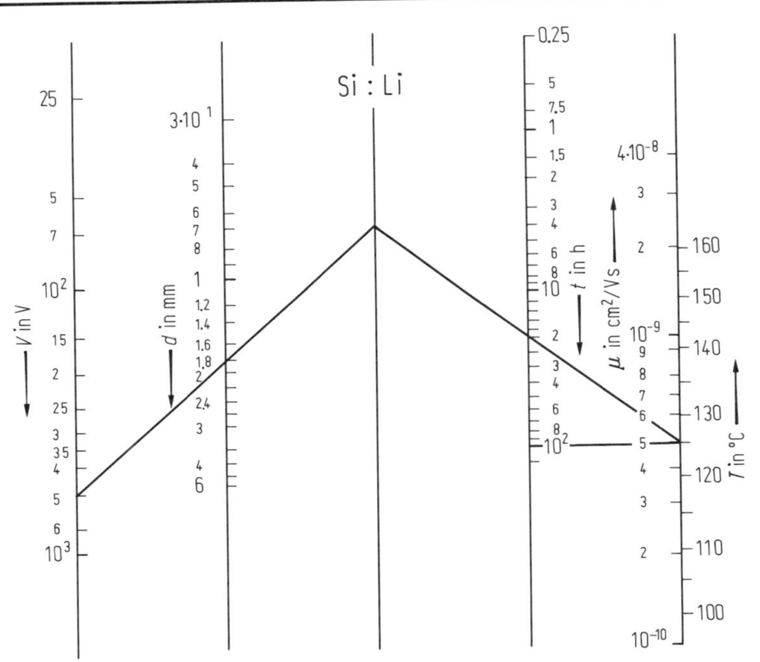

Fig. 39. Si:Li. Nomogram showing the compensated thickness d as a function of temperature T (or the mobility μ), bias voltage V and drift time t for Li in silicon. The straight lines give an example [68B1].

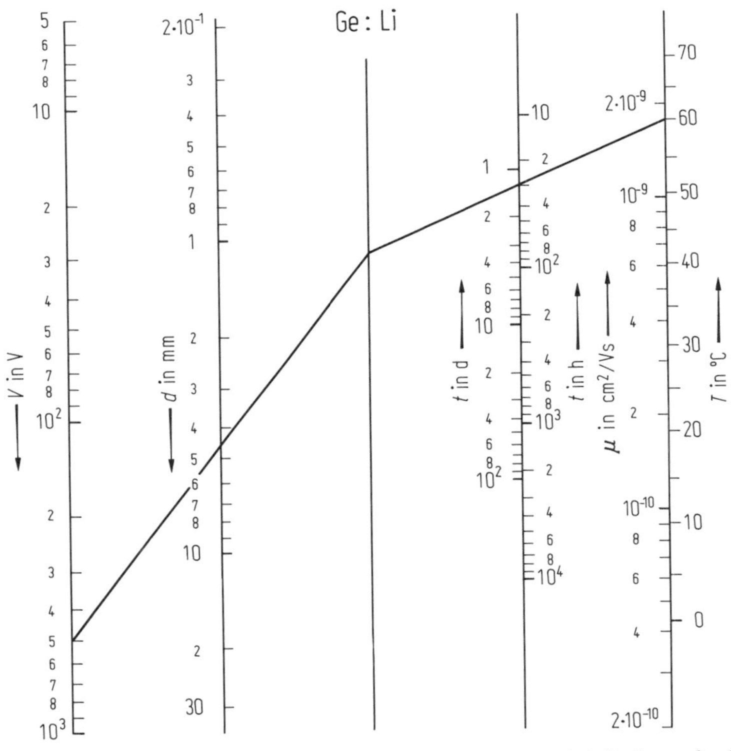

Fig. 40. Ge:Li. Nomogram showing the compensated thickness d as a function of temperature T (or the mobility μ), bias voltage V and drift time t for Li in germanium. The straight lines give an example [68B1].

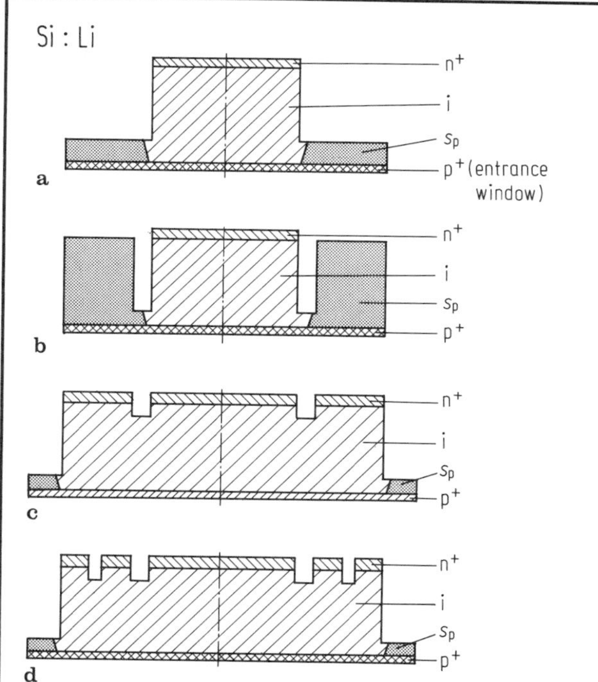

Fig. 42a···d. Si:Li. Various types of Si:Li-detectors. Cross section: a) Top hat [69L], b) grooved [78W], c) single guard ring [72J1], d) double guard ring [72J1]. (n^+: Li-diffused region, i: intrinsic region, Li-compensated, s_p: low p-doped region, p^+: Au-surface barrier, B-implanted or Al-alloyed.)

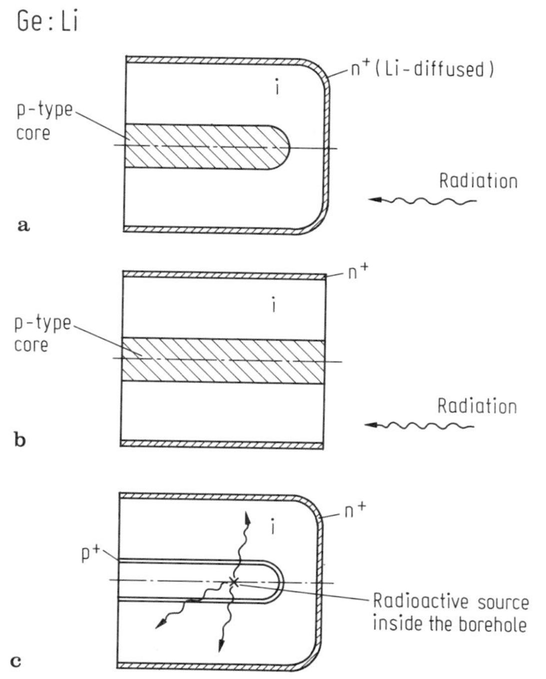

Fig. 43a···c. Ge:Li. Various types of coaxial Ge:Li-detectors (diameter and length up to 6 cm): a) closed ended and b) open ended [65M1, 66T1, 66T2, 71G1, 66M1]; c) bore hole or well type [70G2, 78B1].

For Fig. 44, see next page.

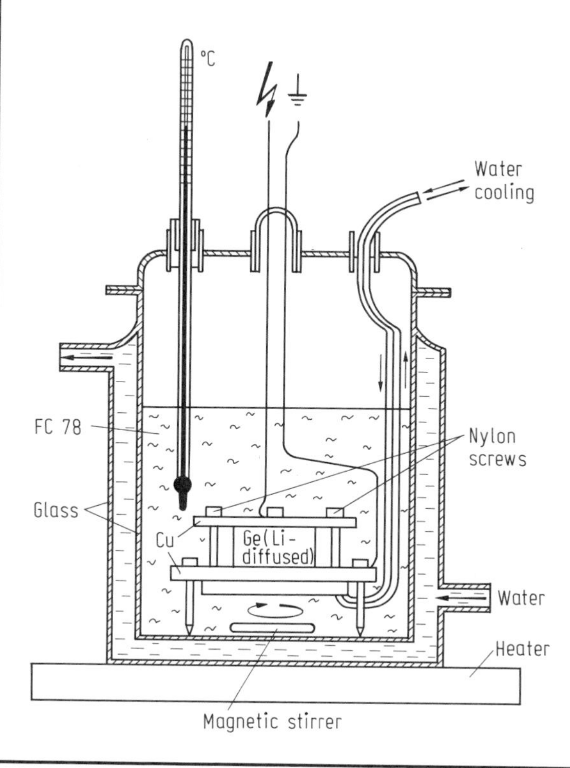

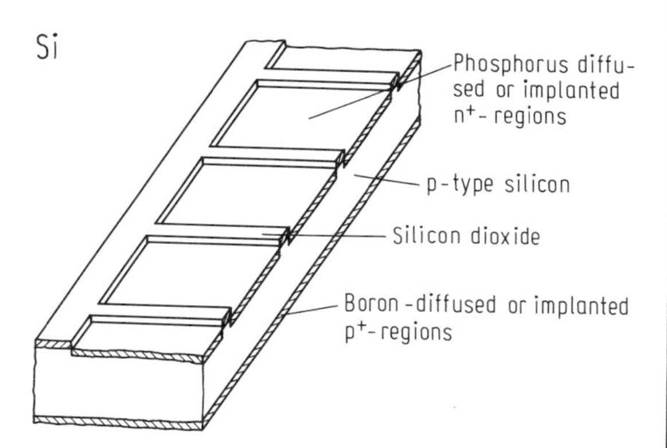

Fig. 46. Si. Cutaway view of a multielement detector with oxide passivation in a linear array [74G1].

◄ Fig. 45. Experimental set-up used for drifting Li in Ge at low temperatures (15···20 °C). (For drifting Li in Si there is no water cooling, the liquid is warmed up to 100···120 °C by the heater).

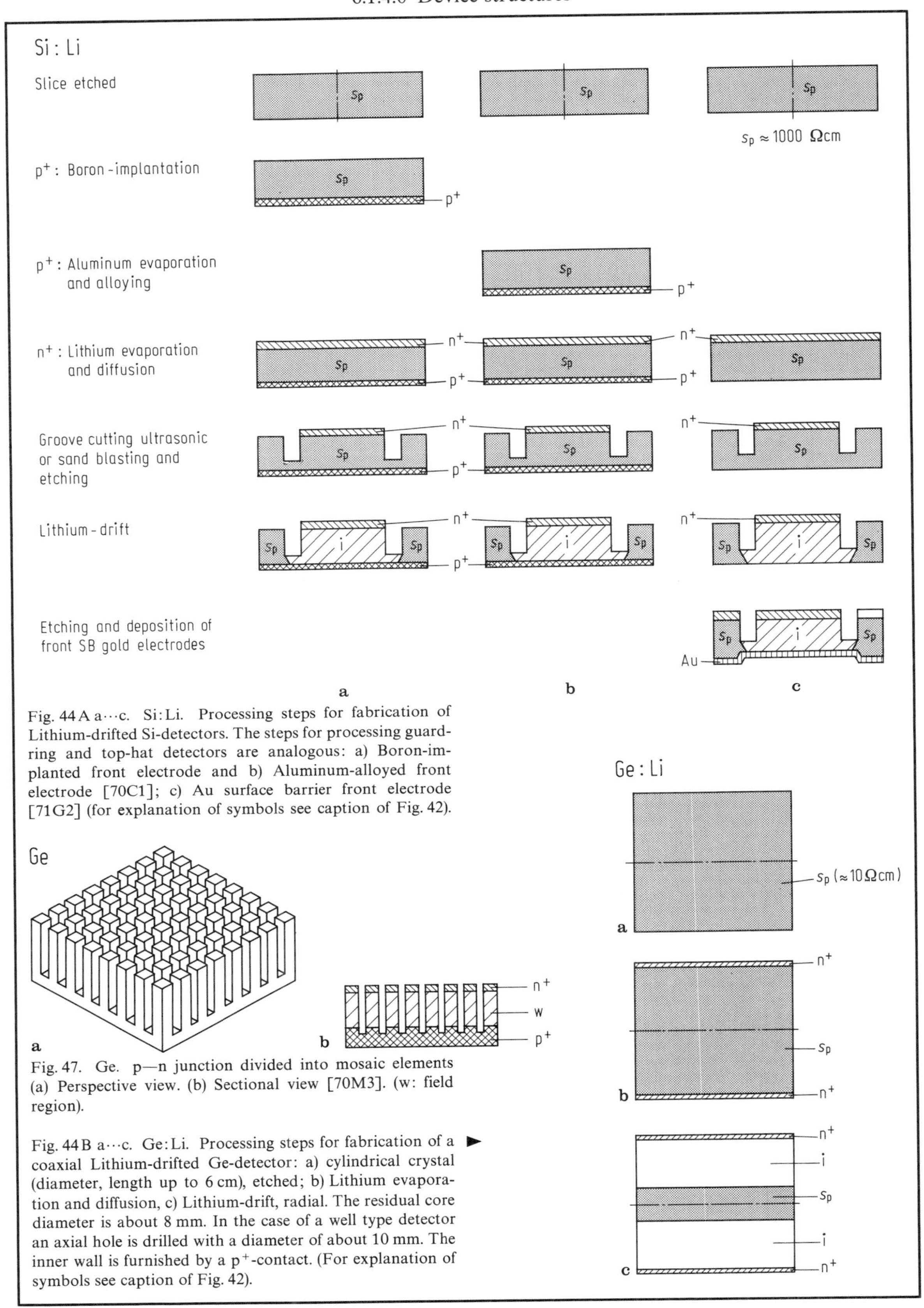

Fig. 44A a⋯c. Si:Li. Processing steps for fabrication of Lithium-drifted Si-detectors. The steps for processing guard-ring and top-hat detectors are analogous: a) Boron-implanted front electrode and b) Aluminum-alloyed front electrode [70C1]; c) Au surface barrier front electrode [71G2] (for explanation of symbols see caption of Fig. 42).

Fig. 47. Ge. p—n junction divided into mosaic elements (a) Perspective view. (b) Sectional view [70M3]. (w: field region).

Fig. 44B a⋯c. Ge:Li. Processing steps for fabrication of a coaxial Lithium-drifted Ge-detector: a) cylindrical crystal (diameter, length up to 6 cm), etched; b) Lithium evaporation and diffusion, c) Lithium-drift, radial. The residual core diameter is about 8 mm. In the case of a well type detector an axial hole is drilled with a diameter of about 10 mm. The inner wall is furnished by a p^+-contact. (For explanation of symbols see caption of Fig. 42). ►

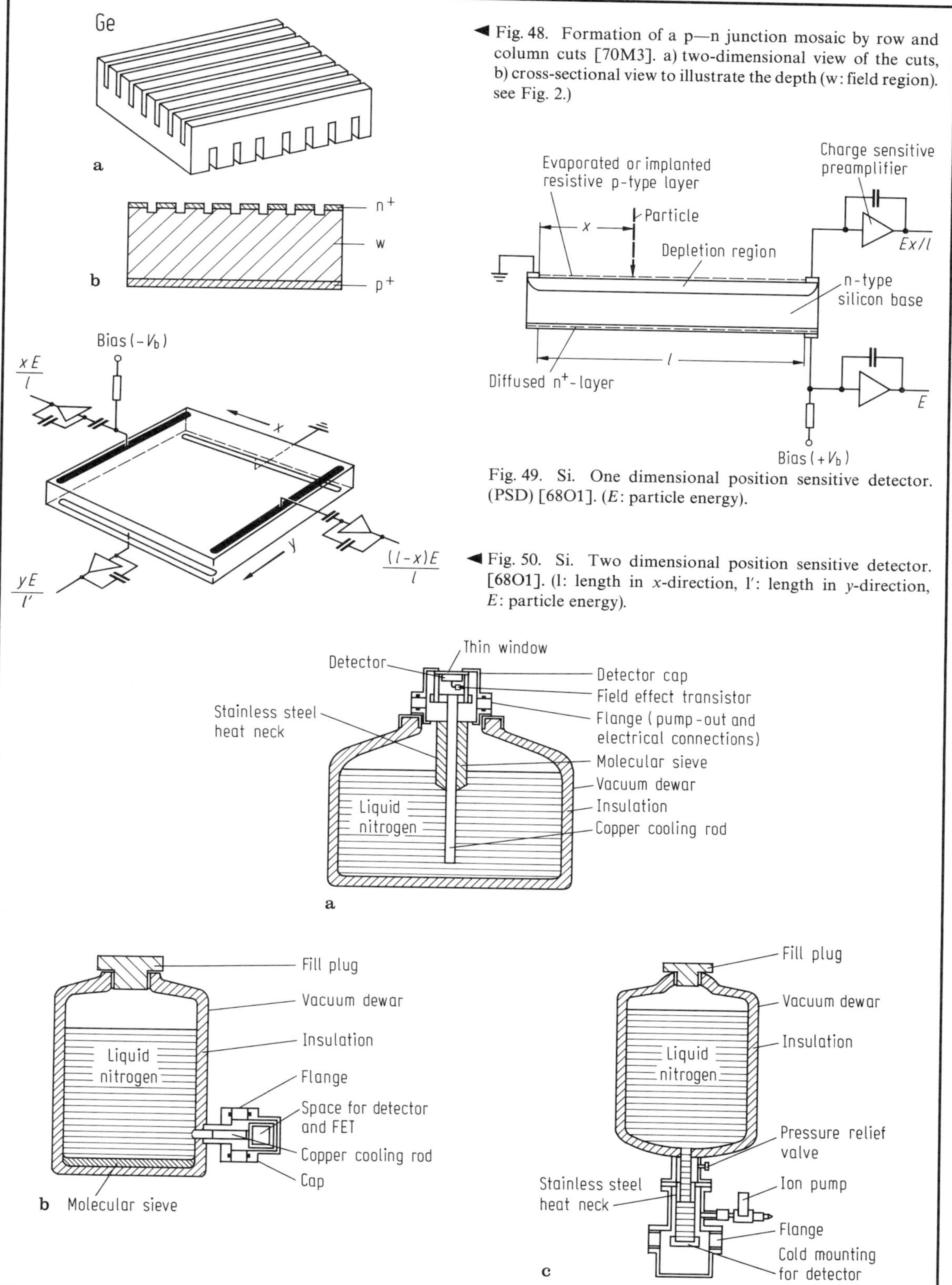

◄ Fig. 48. Formation of a p—n junction mosaic by row and column cuts [70M3]. a) two-dimensional view of the cuts, b) cross-sectional view to illustrate the depth (w: field region). see Fig. 2.)

Fig. 49. Si. One dimensional position sensitive detector. (PSD) [68O1]. (E: particle energy).

◄ Fig. 50. Si. Two dimensional position sensitive detector. [68O1]. (l: length in x-direction, l′: length in y-direction, E: particle energy).

Fig. 51 a···c. Typical detector housings and LN_2-reservoirs [74G1]. a) dipstick cryostat, b) side-locking cryostat and c) chicken-feeder cryostat.

6.1.4.1 Diffusion

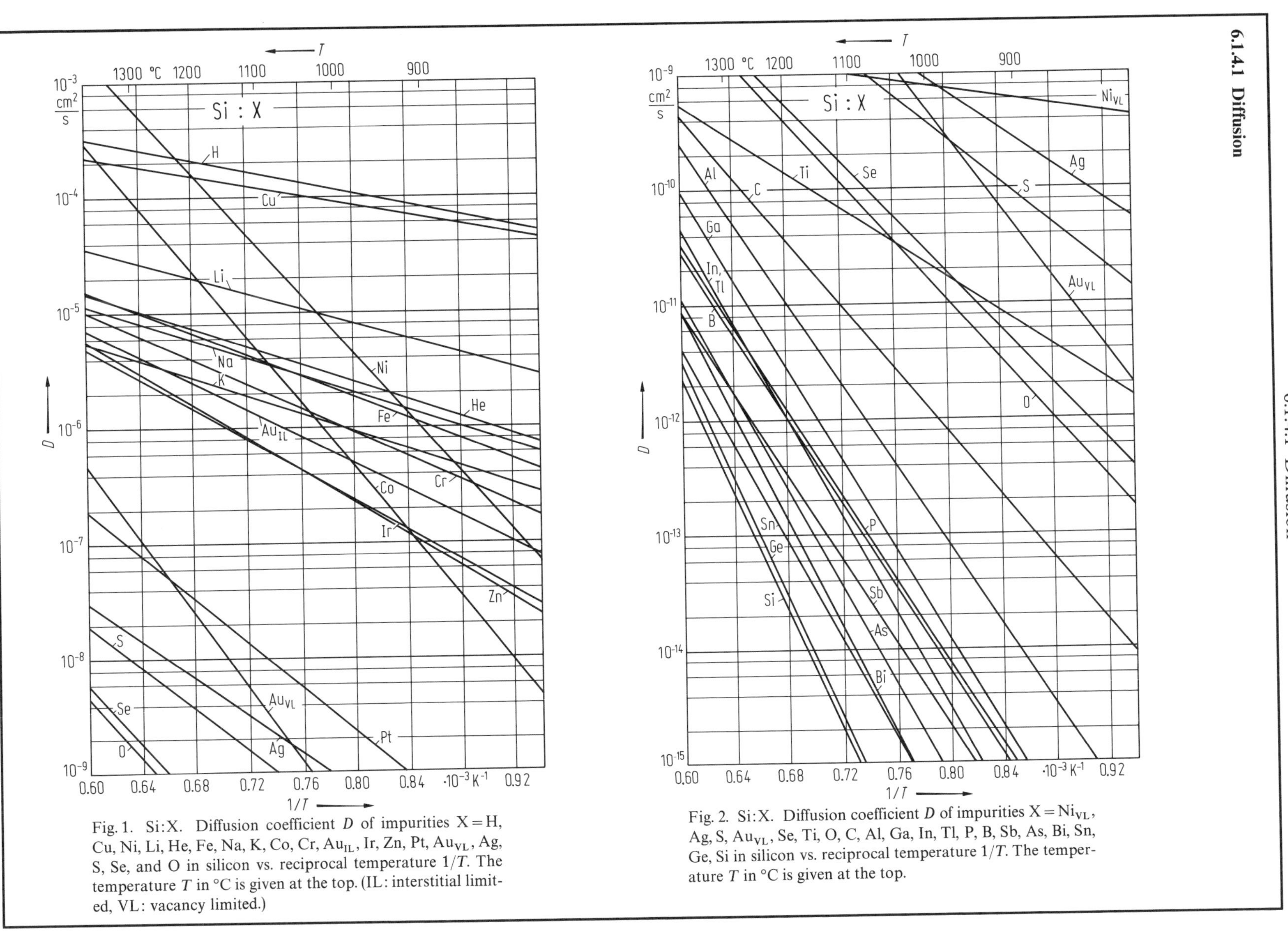

Fig. 1. Si:X. Diffusion coefficient D of impurities X = H, Cu, Ni, Li, He, Fe, Na, K, Co, Cr, Au_{IL}, Ir, Zn, Pt, Au_{VL}, Ag, S, Se, and O in silicon vs. reciprocal temperature $1/T$. The temperature T in °C is given at the top. (IL: interstitial limited, VL: vacancy limited.)

Fig. 2. Si:X. Diffusion coefficient D of impurities X = Ni_{VL}, Ag, S, Au_{VL}, Se, Ti, O, C, Al, Ga, In, Tl, P, B, Sb, As, Bi, Sn, Ge, Si in silicon vs. reciprocal temperature $1/T$. The temperature T in °C is given at the top.

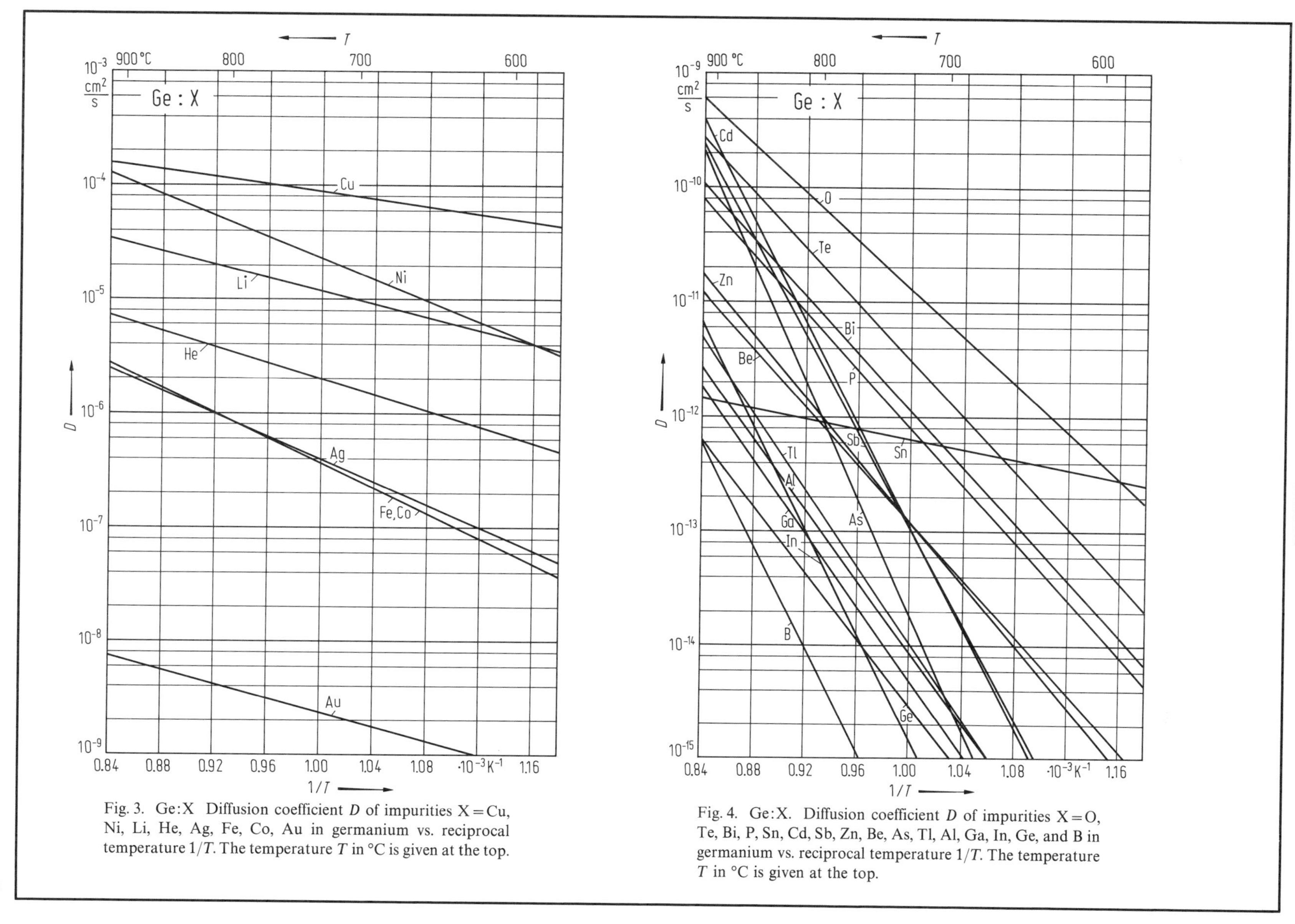

Fig. 3. Ge:X Diffusion coefficient D of impurities X = Cu, Ni, Li, He, Ag, Fe, Co, Au in germanium vs. reciprocal temperature $1/T$. The temperature T in °C is given at the top.

Fig. 4. Ge:X. Diffusion coefficient D of impurities X = O, Te, Bi, P, Sn, Cd, Sb, Zn, Be, As, Tl, Al, Ga, In, Ge, and B in germanium vs. reciprocal temperature $1/T$. The temperature T in °C is given at the top.

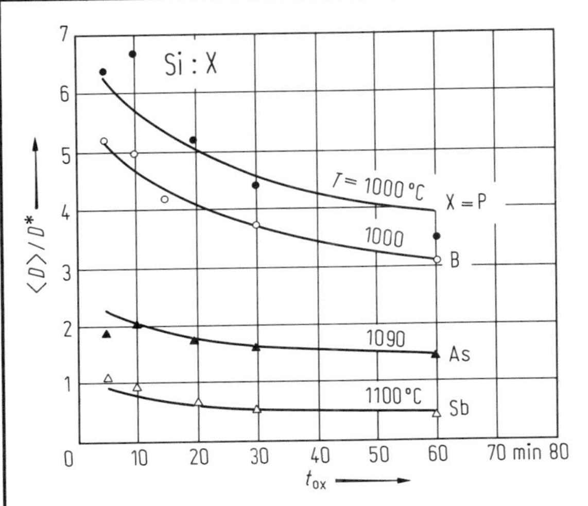

Fig. 5. Si:X (X = P, B, As, Sb). Normalized diffusion coefficient $\langle D\rangle/D^*$ vs. oxidation time t_{ox}. D^* is the intrinsic diffusion coefficient and $\langle D\rangle = \frac{1}{t}\int_0^t D\,dt$ is the time-averaged diffusion coefficient. Points are experimental data. Lines are fits of theoretical calculations. The temperatures are indicated [82A1].

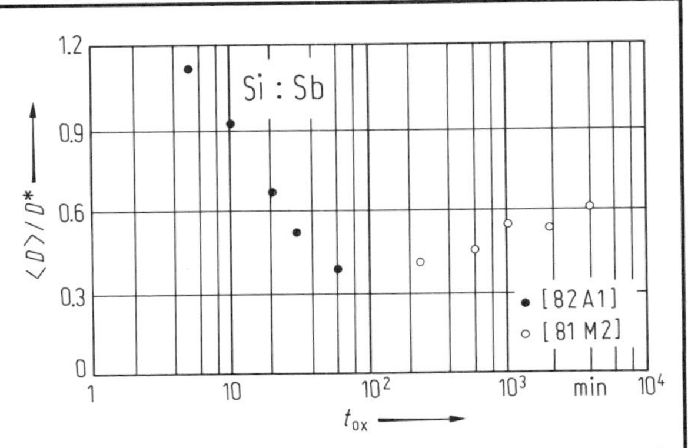

Fig. 6. Si:Sb. Normalized diffusion coefficient $\langle D\rangle/D^*$ of antimony in silicon vs. oxidation time t_{ox}. $\langle D\rangle$ and D^* are defined in Fig. 5 [82A1].

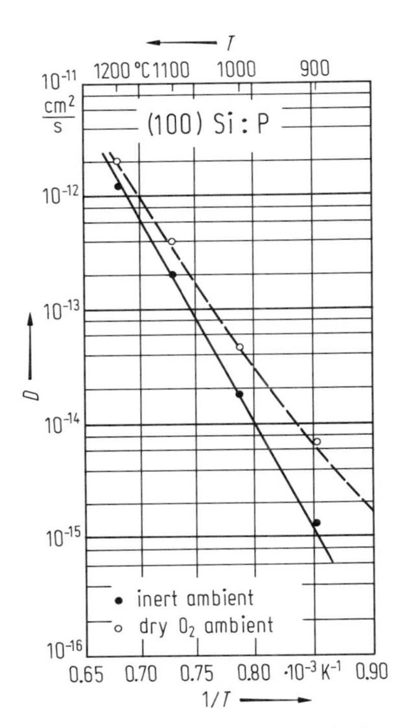

Fig. 8. Si:P. Diffusion coefficient D of phosphorus in silicon vs. reciprocal temperature $1/T$. Symbols and dashed curve are from [82A3, 78A1]. The wafer orientation is (100). The oxydation times are at 1000 °C, 6 h; at 1100 °C, 45 min; at 1200 °C, 10 min. The continuous line is from [62M2].

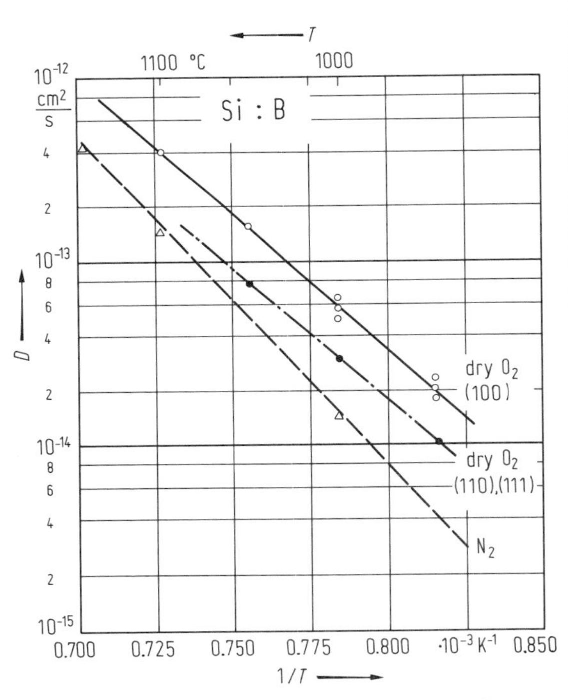

Fig. 7. Si:B. Diffusion coefficient D of boron in silicon vs. reciprocal temperature $1/T$. The influence of an inert ambient (intrinsic diffusion) and an oxidizing ambient on different crystal orientations is shown [82M]. The oxidation times are: (100) orientation: at 950 °C, 60, 130, and 200 min; at 1000 °C, 26, 60, and 100 min; at 1050 °C, 60 min; and at 1100 °C, 50 min.
(110), (111) orientation: at 950 °C, 130 min; at 1000 °C, 60 min; at 1050 °C, 60 min.

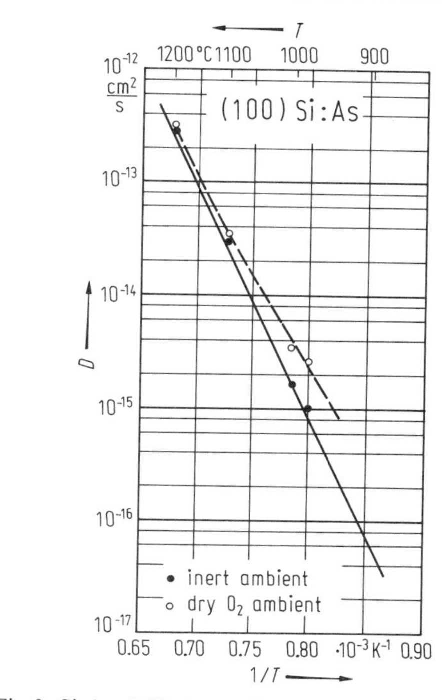

Fig. 9. Si:As. Diffusion coefficient D of arsenic in silicon vs. reciprocal temperature $1/T$. Symbols and dashed curve are from [78A1]. The wafer orientation is (100). The continuous line is from [71C2].

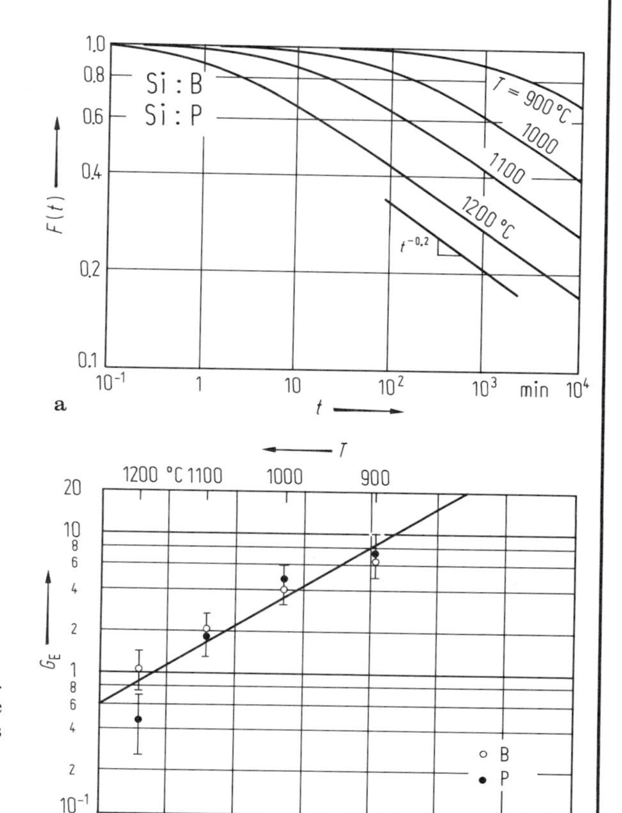

Fig. 10a, b. Si:B, Si:P. Correction factors G_E and $F(t)$ to determine the enhancement/reduction of the normalized diffusion coefficient of boron and phosphorus by $(\langle D\rangle/D^*)-1 = G_E \cdot F(t)$ [81L].
a) Diffusion reduction factor $F(t)$ vs. oxidation time in dry oxygen. Parameter is the oxidation temperature. b) Diffusion enhancement factor G_E vs. reciprocal oxidation temperature in dry oxygen.

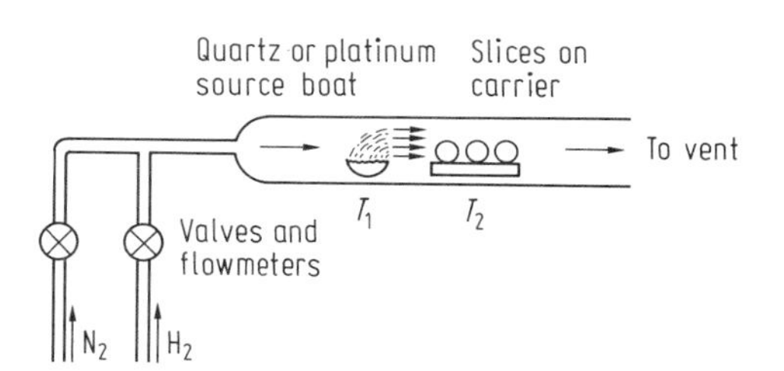

Fig. 11. Schematics of open tube diffusion in reducing (H_2) ambient. The temperature T_1 may be equal T_2 [68G1].

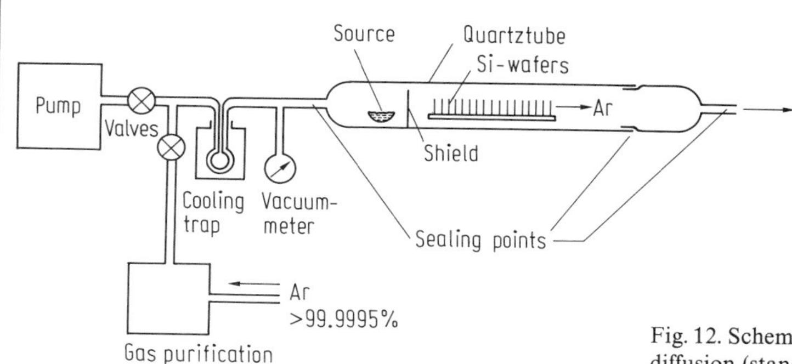

Fig. 12. Schematics of the crucible and apparatus of a closed diffusion (standard gallium diffusion) [75R].

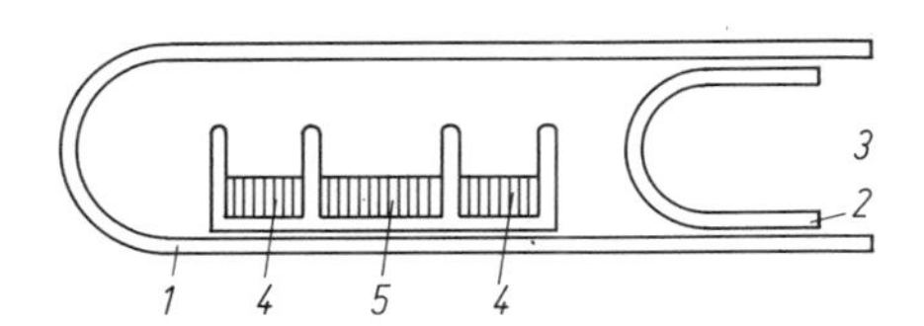

Fig. 13. Schematics of sealed tube diffusion. Doped silicon wafers are used as a diffusion source. Wafers and source wafers may also be arranged in an alternating sequence [73G2].
1 Capsule made of quartz tube, *2* sealing plug, *3* pump connection opening, *4* container for source wafers, *5* wafer carrier.

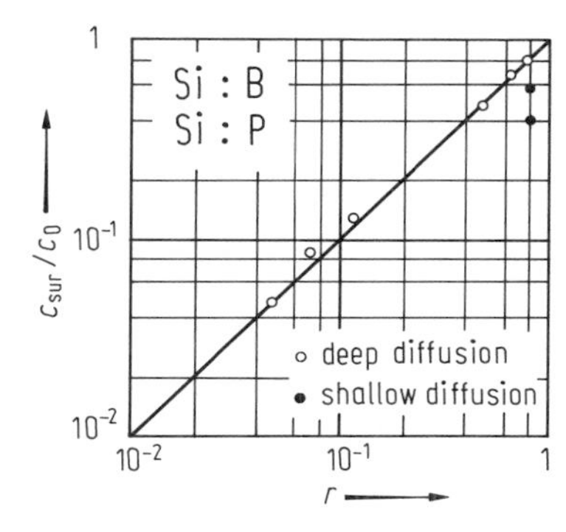

Fig. 14. Si:B, Si:P. Ratio of the surface doping concentration c_{sur} to the concentration c_0 in source wafers vs. the ratio r of the source area to sample silicon wafer area in close-tube diffusion. The straight line is theoretical; the points are experimental values of boron or phosphorus for deep and shallow diffusion [73G2].
Range of investigation:

	t min	c_0 cm^{-3}	c_{sur} cm^{-3}	r	T °C
B	60··· 1020	$2\cdot10^{19}$, $5\cdot10^{18}$	$2\cdot10^{18}$··· $1.5\cdot10^{19}$	0.75··· 0.95	1050··· 1200
P	60··· 1200	$5\cdot10^{18}$, $5\cdot10^{19}$	$2\cdot10^{18}$··· $5\cdot10^{18}$	0.75··· 0.9	1050··· 1200

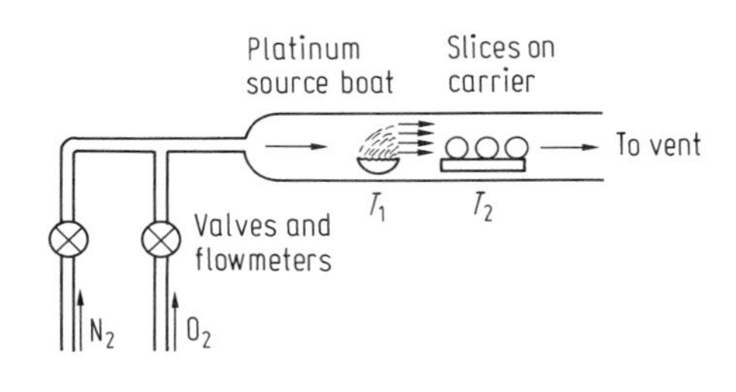

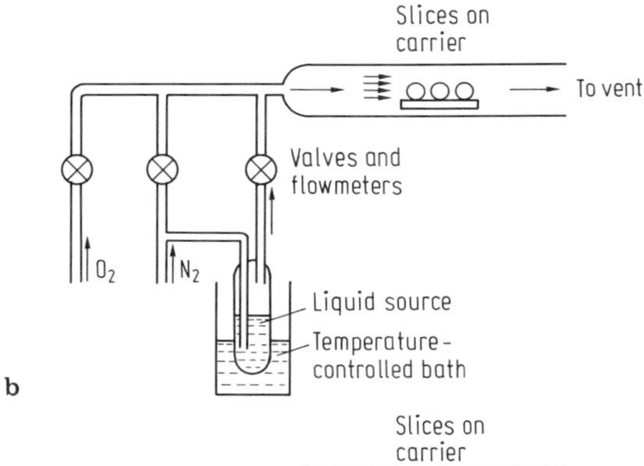

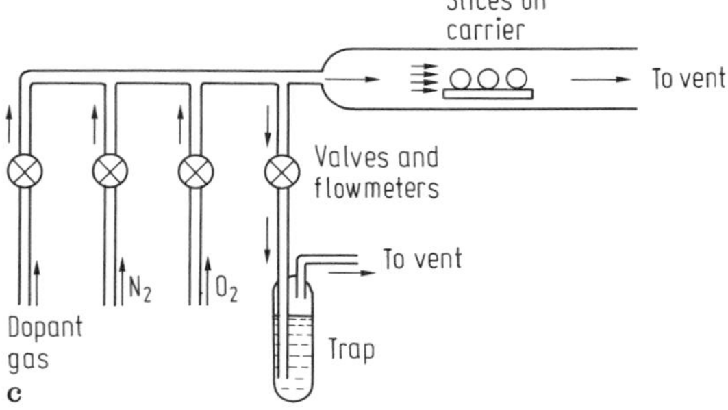

Fig. 15a···c. Schematics of open quartz tube diffusions [68G1] for a
a) solid source; in most cases the source temperature T_1 is different to the slice temperature T_2; b) liquid source; c) gaseous source.

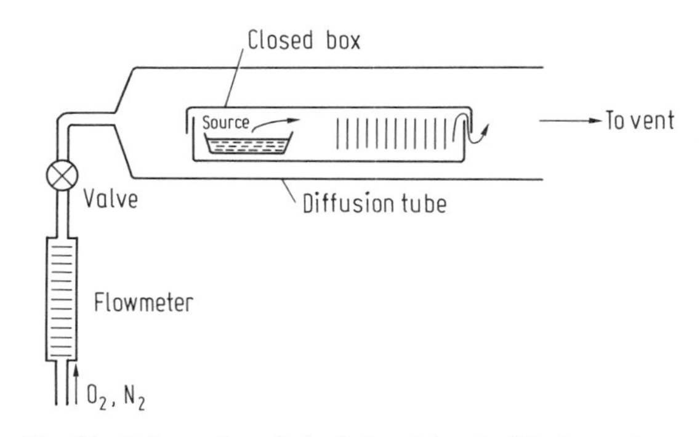

Fig. 16. Schematics of the "closed-box" diffusion using a box which contains the diffusion source material and the wafers.

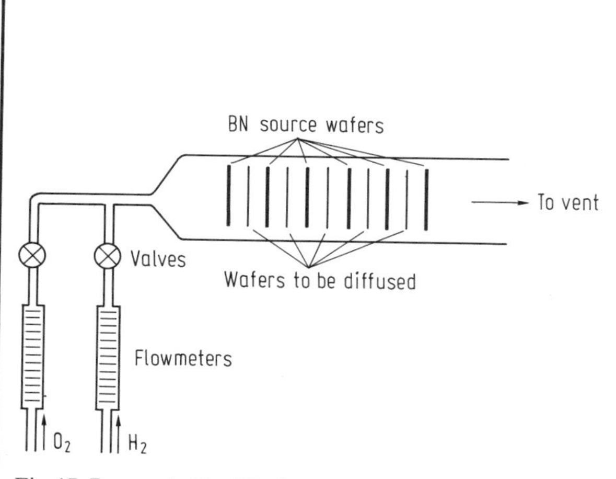

Fig. 17. Boron nitride diffusion system.

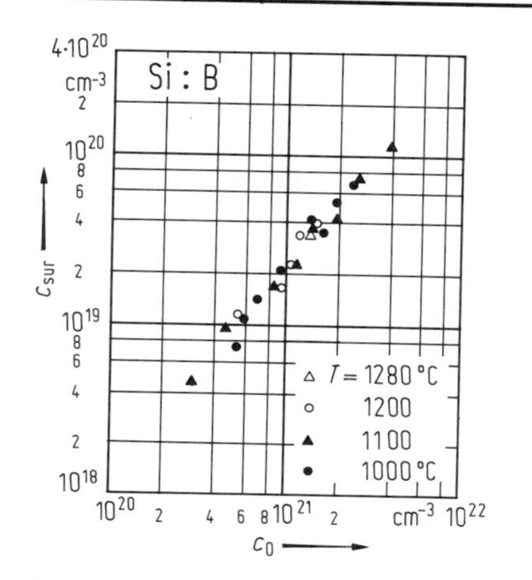

Fig. 18. Si:B. Boron surface concentration c_{sur} in silicon vs. boron concentration in the BSG source c_0 [69B].

For Fig. 19, see next page.

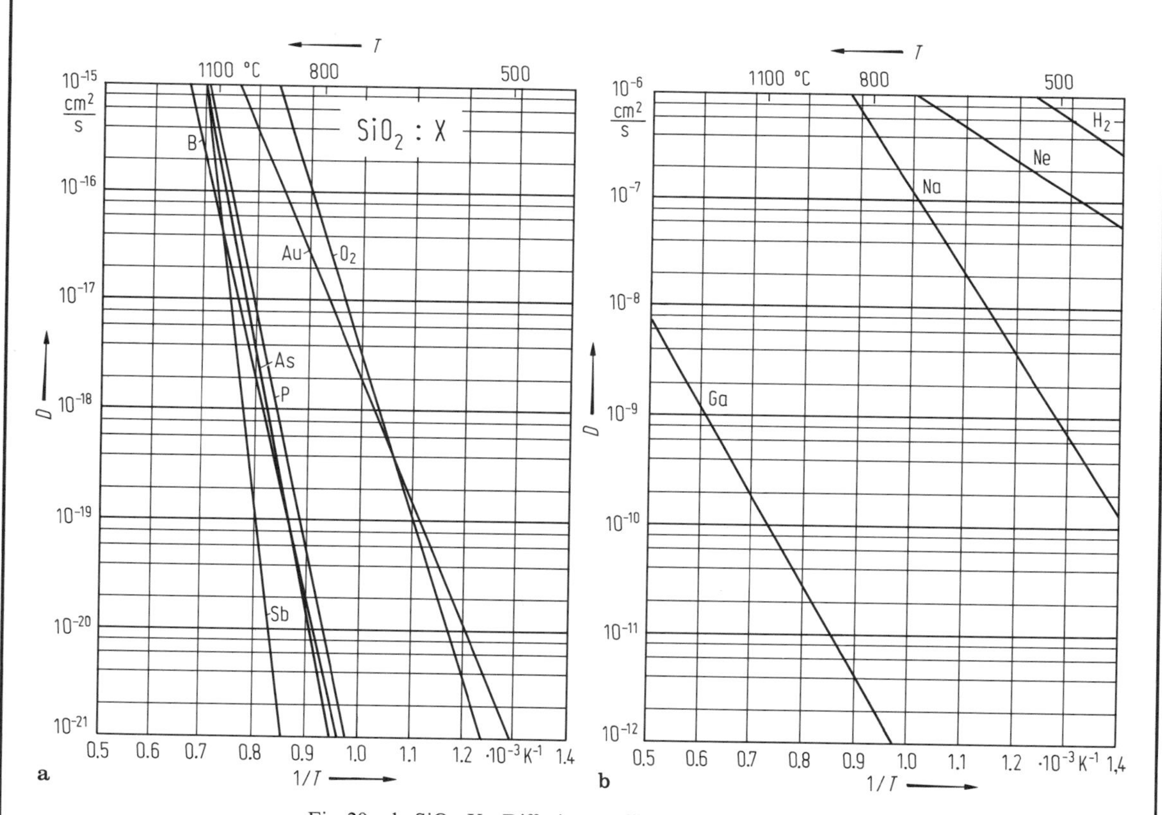

Fig. 20 a, b. SiO_2:X. Diffusion coefficient D of impurities X in SiO_2 vs. reciprocal temperature $1/T$. The temperature in °C is given at the top. a) X = B, P, As, Sb, Au, and O_2; b) X = Ga, Na, Ne, and H_2.

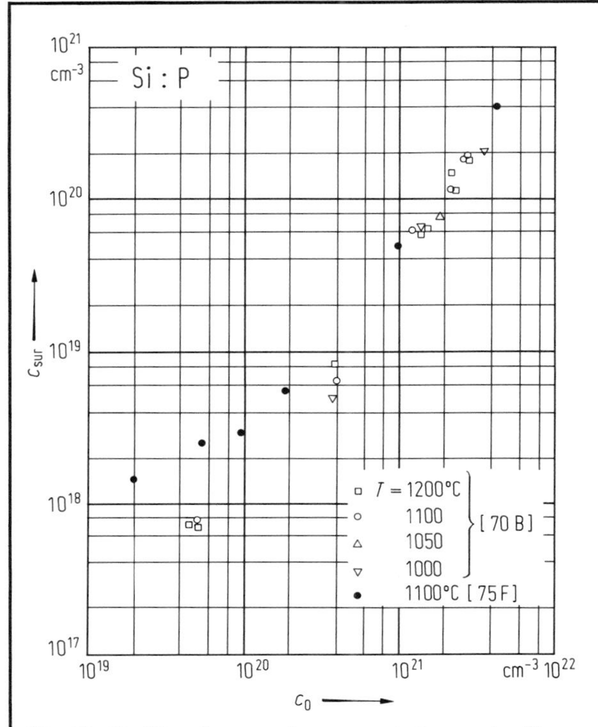

Fig. 19. Si. Phosphorus surface concentration c_{sur} in silicon vs. phosphorus concentration in the PSG source c_0.

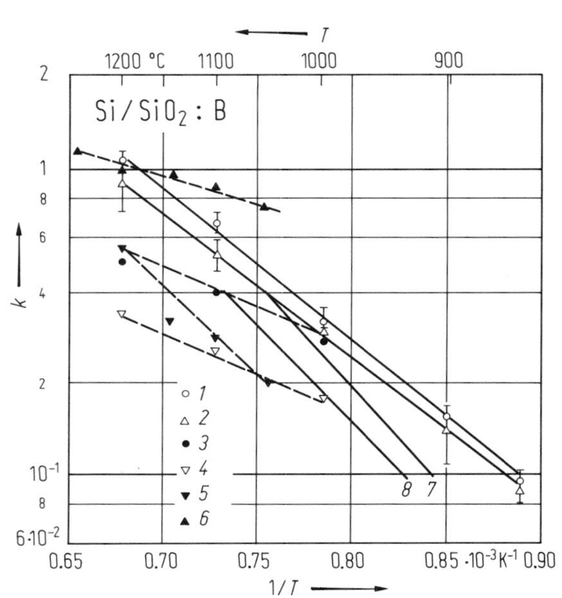

Fig. 21. Si/SiO_2:B. Distribution coefficient k of boron in the system Si/SiO_2 vs. reciprocal temperature $1/T$. Implanted boron was redistributed during oxidation. *1*: ⟨100⟩ orientation, dry O_2 [78A2]; *2*: ⟨111⟩ orientation, dry O_2 [78A2]; *3*: ⟨100⟩ orientation, dry O_2 [76C]; *4*: ⟨111⟩ orientation, dry O_2 [76C]; *5*: ⟨111⟩ orientation, steam oxide [74P]; *6*: ⟨100⟩ orientation, dry O_2 [75M]; *7*: ⟨110⟩ and ⟨111⟩ orientation, dry O_2 [82M]; *8*: ⟨100⟩ orientation, dry O_2 [82M]. [74P, 75M, 78A2]: Fitting of junction depth, sheet resistivity, and oxide thickness by using the process simulator SUPREM. [82M]: Fitting of profiles from the pulsed MOS C–V method to calculate profiles. [76C]: Direct measurements by SIMS.)

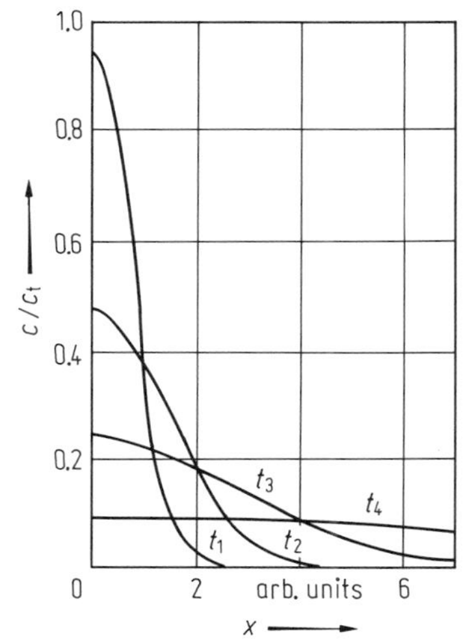

Fig. 23. Instantaneous source (Gaussian) diffusion profiles. The concentration normalized to the total concentration c/c_t is plotted vs. distance from the surface x, for different times: $t_1 < t_2 < t_3 < t_4$ [75R].

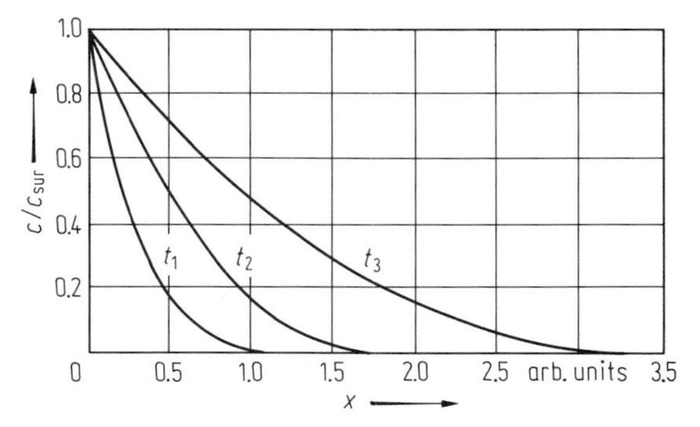

Fig. 22. Constant current source (erf.) diffusion profiles. The concentration normalized to the constant surface concentration c/c_{sur} is plotted vs. distance from the surface x, for different times: $t_1 < t_2 < t_3$ [75R].

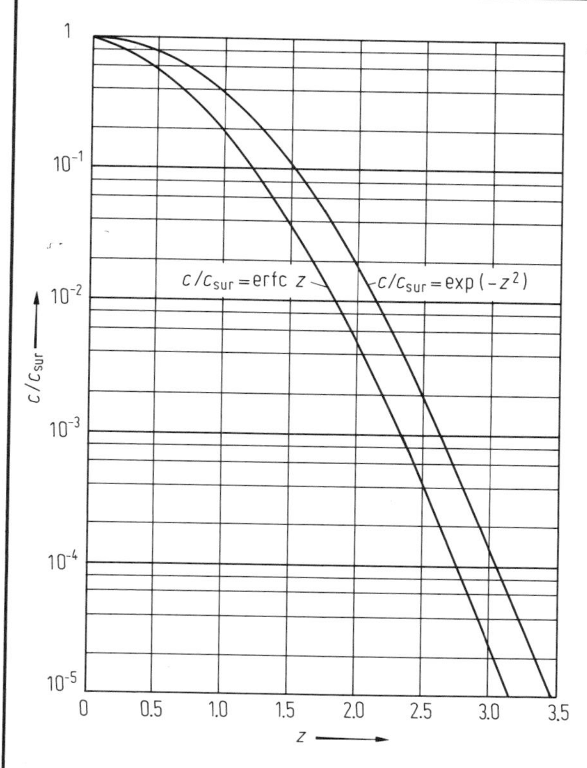

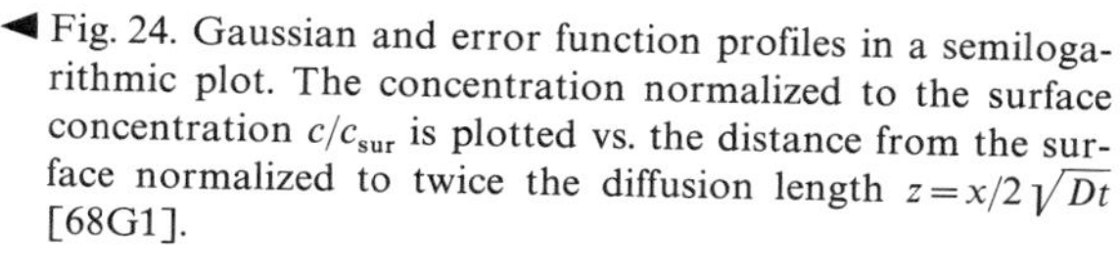
◀ Fig. 24. Gaussian and error function profiles in a semilogarithmic plot. The concentration normalized to the surface concentration c/c_{sur} is plotted vs. the distance from the surface normalized to twice the diffusion length $z = x/2\sqrt{Dt}$ [68G1].

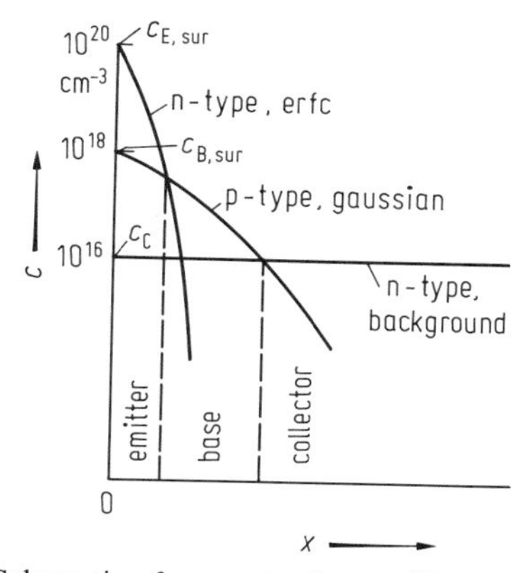

Fig. 25. Schematic of concentration profiles in double-diffused transistors. The homogeneous background doping concentration is present in the starting material.

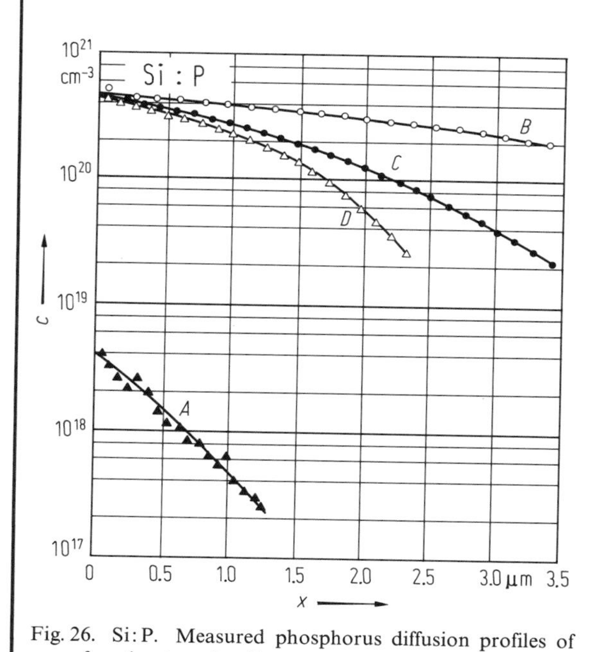

Fig. 26. Si:P. Measured phosphorus diffusion profiles of error function type in silicon, at $T = 1200\ ^\circ C$ [73M]. Phosphorus concentration c vs. depth x in Si. A: intrinsic diffusion 0.5 h; B: extrinsic diffusion 2.3 h; C: extrinsic iso-concentration diffusion 0.5 h; D: extrinsic diffusion into intrinsic Si 0.5 h.

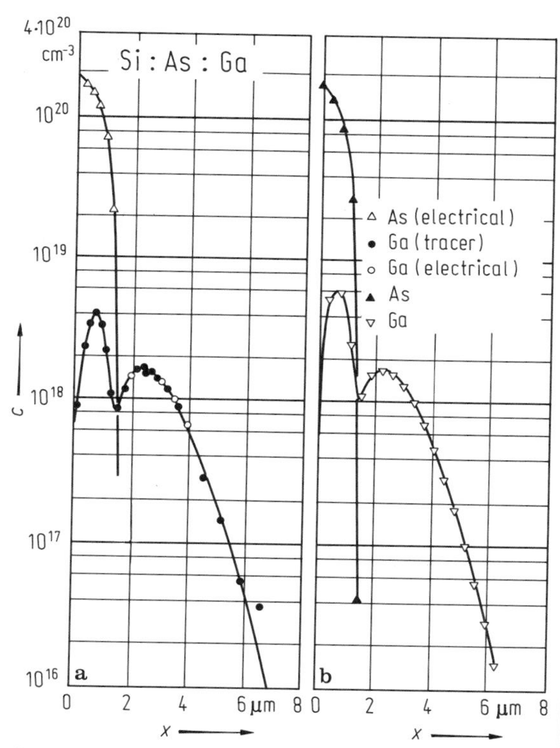

Fig. 27a, b. Si:As:Ga. Concentration c of As and Ga vs. depth x in Si [81M1].
a) 5 min/1075 °C Ga-diffusion followed by 15 min/1000 °C Ga drive-in diffusion followed by 15 min/1000 °C As diffusion in oxidizing ambient. b) Computer simulation of the experimental situation of a).

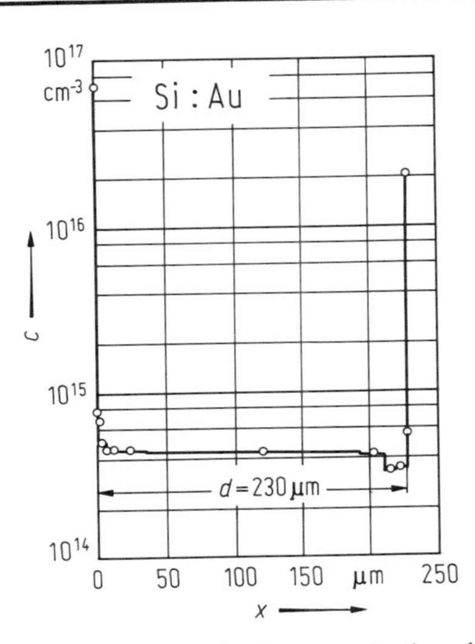

Fig. 28. Si:Au. Gold distribution, c vs. depth x observed in a 500 Ω cm, p-type, 230 μm thick silicon wafer after $3 \cdot 10^{12}$ cm^{-2} gold implantation and 4 hours of redistribution anneal at 1000 °C in argon ambient. Both surfaces were polished. The surface at $x = 0$ was implanted. Measurement is by radiotracer techniques [74S].

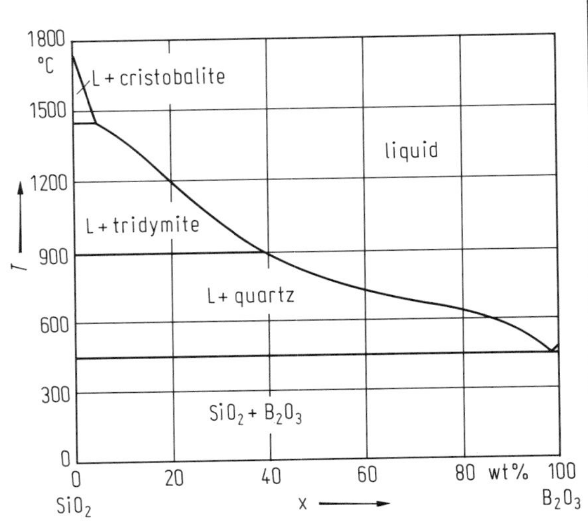

Fig. 29. SiO_2—B_2O_3. T-x phase diagram of the system SiO_2—B_2O_3 [65R].

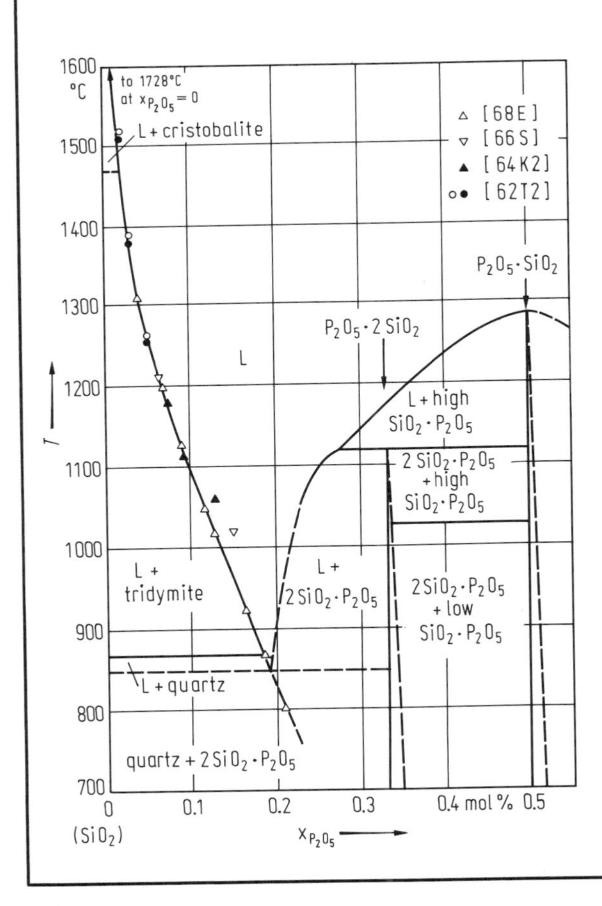

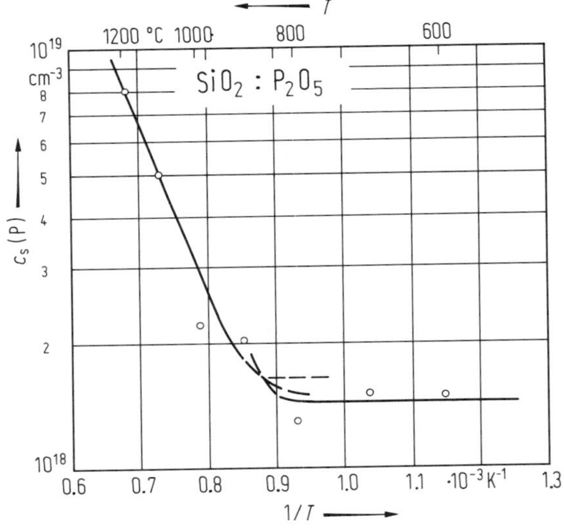

Fig. 31. SiO_2:P_2O_5. Concentration c_s(P) of phosphorus (solid solubility of P_2O_5) in SiO_2 vs. reciprocal temperature $1/T$ [68E].

◄ Fig. 30. SiO_2—P_2O_5. The T-x phase diagram of the system SiO_2—P_2O_5 [68E].

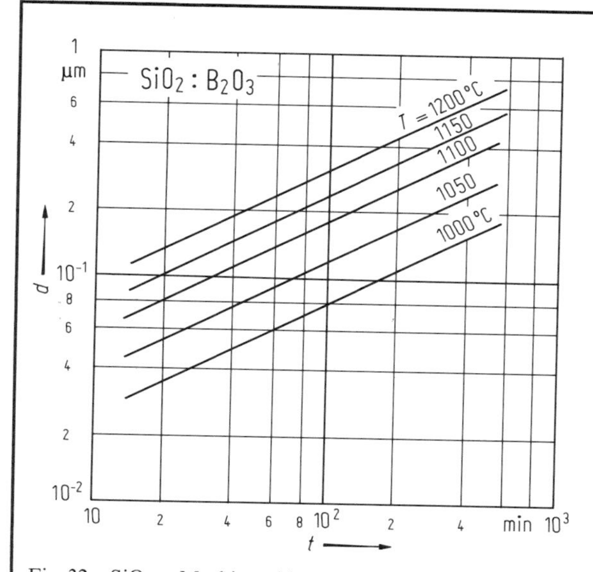

Fig. 32. SiO_2. Masking thickness d of thermal SiO_2 vs. time t of boron (B_2O_3 vapour) diffusion [68G1].

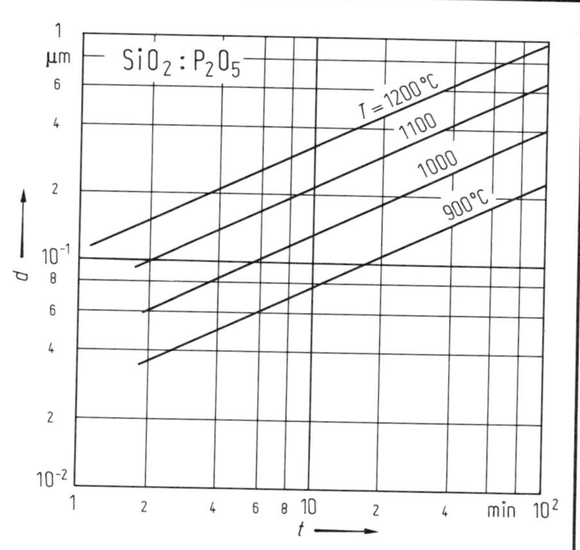

Fig. 33. SiO_2. Masking thickness d of thermal SiO_2 vs. time t of phosphorus (P_2O_5 vapour) diffusion [68G1].

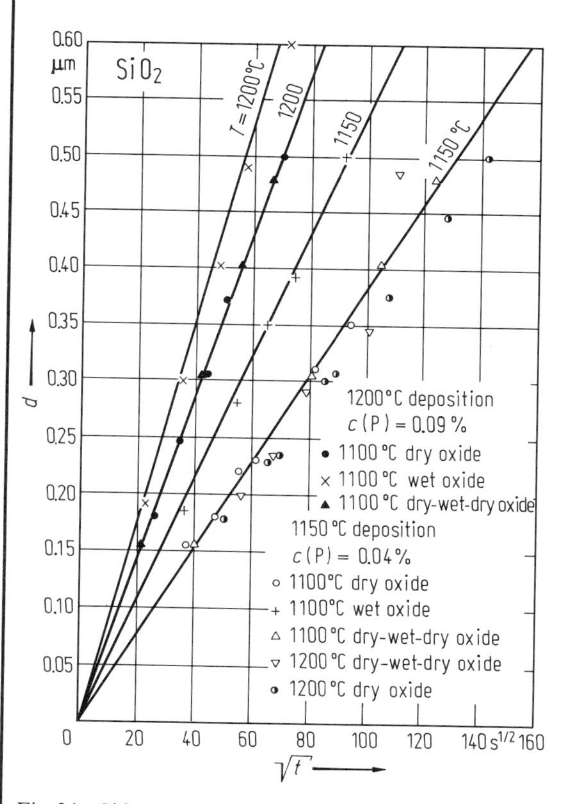

Fig. 34. SiO_2. Masking thickness d of thermal SiO_2 for phosphorus diffusion vs. square root of the deposition time $t^{1/2}$. The phosphorus diffusion is performed from a phosphine source under oxidizing conditions as indicated [75G].

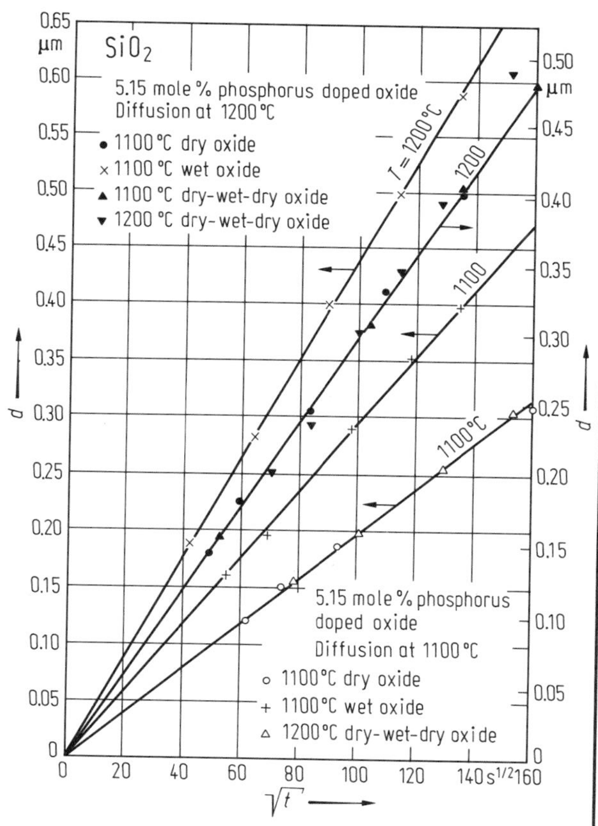

Fig. 35 SiO_2. Masking thickness d of thermal SiO_2 for phosphorus diffusion form a PSG source vs. square root of the annealing time $t^{1/2}$. The oxidizing conditions are indicated [75G].

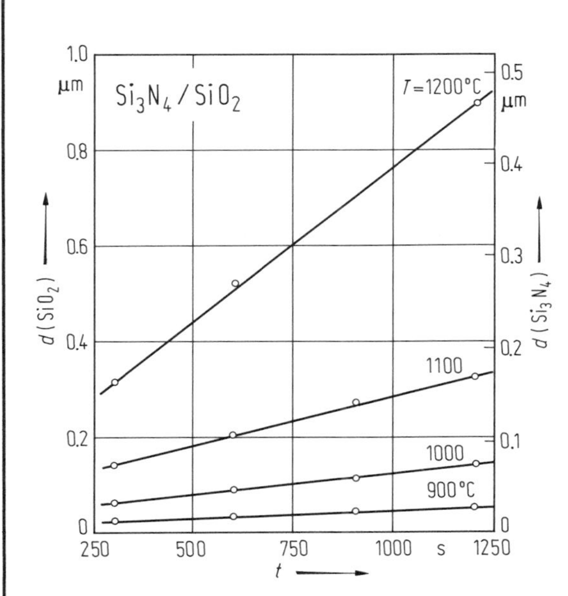

Fig. 36. Si_3N_4/SiO_2. Thickness d of Si_3N_4 converted into SiO_2 in an oxidizing ambient of 10 Vol % oxygen in nitrogen saturated with PBr_3 at 25 °C vs. reaction time t ($d_{Si_3N_4} = 0.52 \cdot d_{SiO_2}$) [71F1].

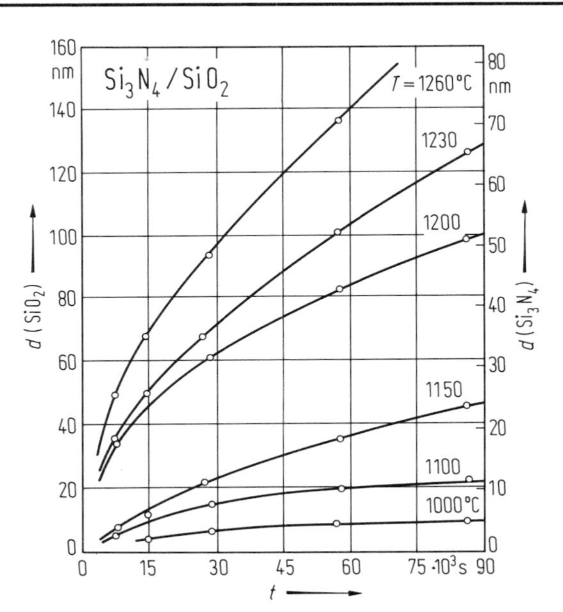

Fig. 37. Si_3N_4/SiO_2. Thickness d of Si_3N_4 converted into SiO_2 by dry oxygen vs. reaction time t($d_{Si_3N_4} = 0.52 \cdot d_{SiO_2}$) [71F1].

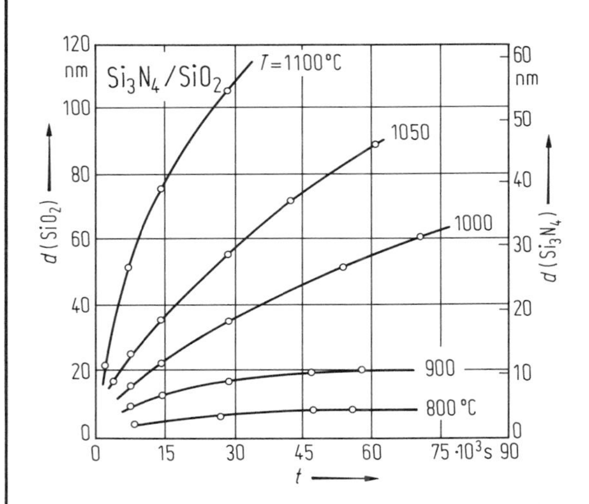

Fig. 38. Si_3N_4/SiO_2. Thickness d of Si_3N_4 converted into SiO_2 by wet (95 °C H_2O) oxygen vs. reaction time t ($d_{Si_3N_4} = 0.52 \cdot d_{SiO_2}$) [71F1].

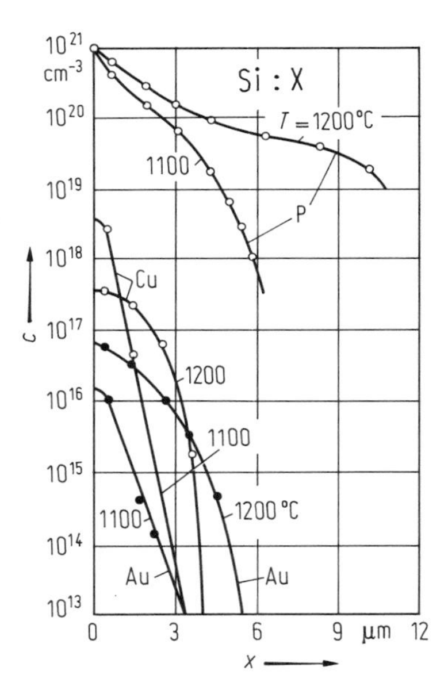

Fig. 39. Si:X (X = Au, Cu, P). Concentration of various dopants vs. distance x from the surface. Diffusion temperatures applied are 1100 °C and 1200 °C, respectively [68L].

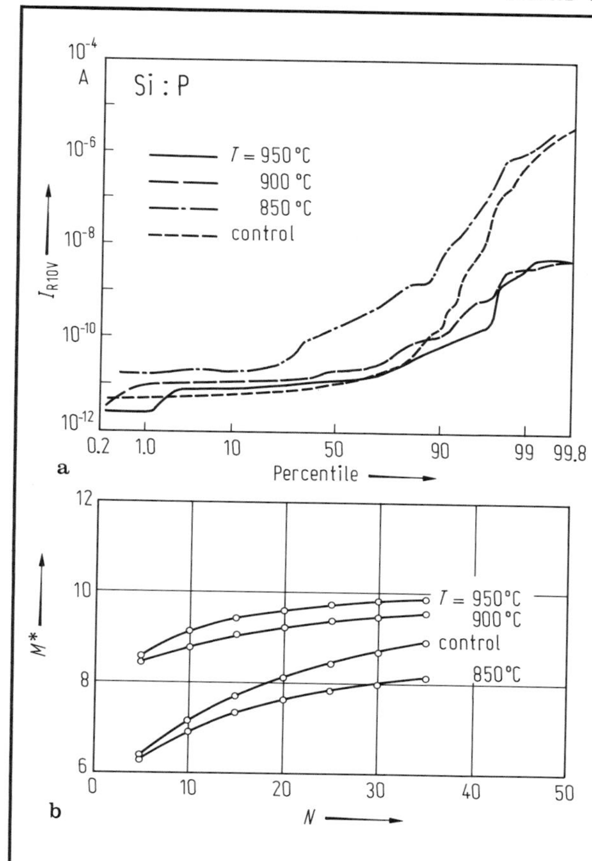

Fig. 40a, b. Si:P. a) Percentile of 140 or 210 pn-diodes vs. current I_{R10V} at 10 V reverse bias voltage. Parameter is the phosphorus diffusion gettering temperature (diffusion lasted 1 h) and a control sample. The pn-diodes (area = $2.5 \cdot 10^{-3}$ cm^{-2}) are fabricated on Si-wafers having the back surface saw damage removed by etching [82A2].

b) Figure of merit $M^* = -N^{-1}\left[\sum_{i=1}^{N} \log_{10}(I_{R10V})_i\right]$ of diodes in Fig. 40a vs. number N of worst devices [82A2].

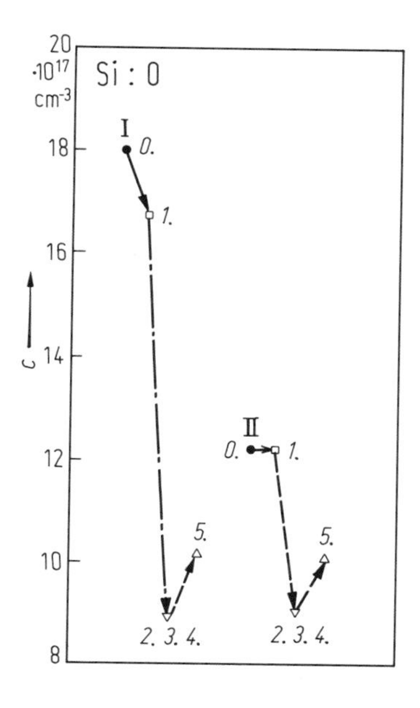

Fig. 41. Si:O. Interstitial oxygen concentration c after various MIG (multiple step intrinsic gettering) processing steps 1…5 [82S]. I = wafer having high initial oxygen concentration. II = wafer having low initial oxygen concentration. The processing steps are: *0.* as CZ grown; *1.* at 1230 °C for 3 h in dry N_2 (25% O_2); *2.* at 520 °C for 16 h in dry O_2; *3.* at 620 °C for 16 h in dry O_2; *4.* at 720 °C for 16 h in dry O_2; *5.* at 1140 °C for 2 h in wet O_2.

6.1.4.2 Ion implantation

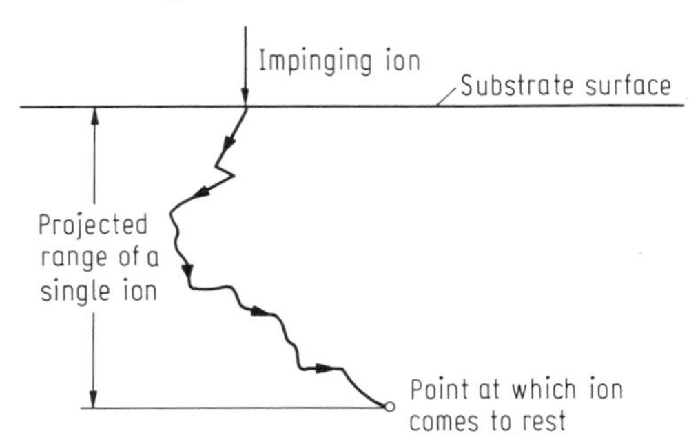

Fig. 1. Schematic path of implanted ion when slowed down. (R_p is the mean value of the projected ranges of many ions, see Fig. 2.)

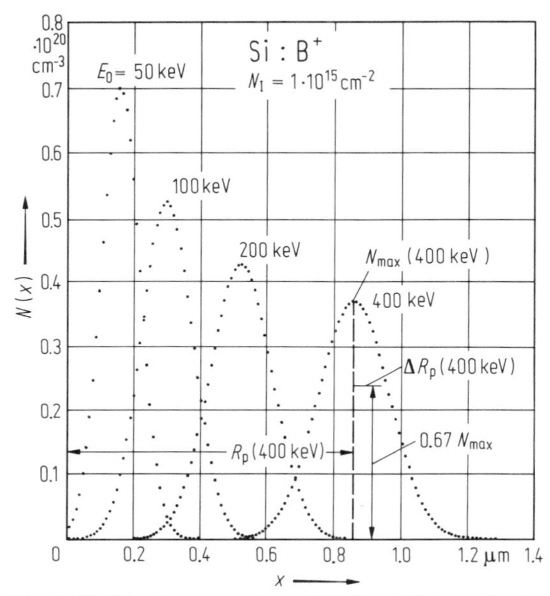

Fig. 2. Si:B⁺. Concentration of B⁺ ions $N(x)$ vs. depth x in Si for a constant implantation dose $N_I = 1 \cdot 10^{15}$ cm^{-2} and implantation energies E_0, varying as shown. (R_p, ΔR_p and N_{max} are labelled for the 400 keV ions.)

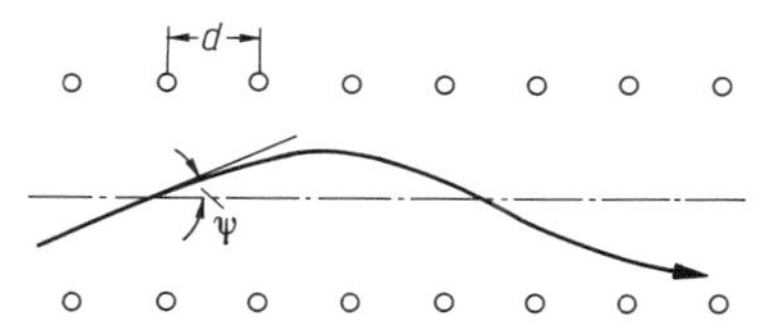

Fig. 3. Schematic trajectory of an ion in a single crystal having lattice constant d. The angle between ion beam and crystal axis is $\psi < \psi_c$.

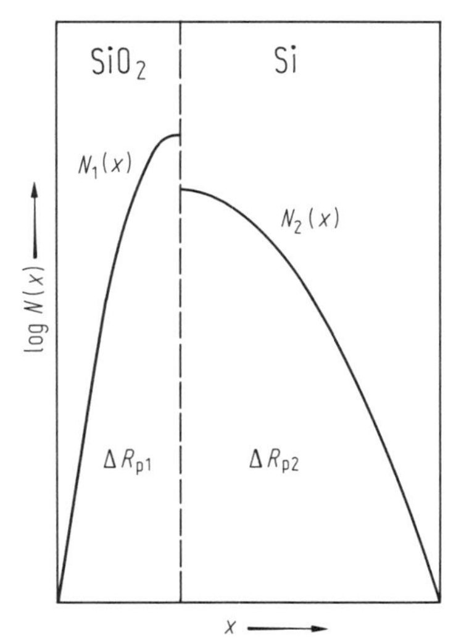

Fig. 4. Schematic concentration profile in a 2-layer structure e.g. SiO_2 (layer 1) on top of Si (layer 2).

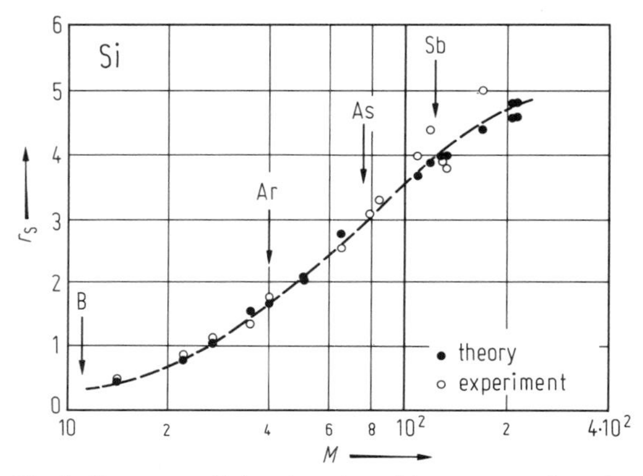

Fig. 5. Sputter coefficient (number of ions sputtered per incident ion) r_s of Si vs. mass number M of incident ion at 45 keV [69S1, 75A1].

Fig. 6. Lines of equal ion concentration, normalized to the maximum concentration, $N(x, y, z)/N_{max}$ under a mask edge for a B⁺-implantation at 70 keV in Si. Mask edge obtained by ion etching [77R1].

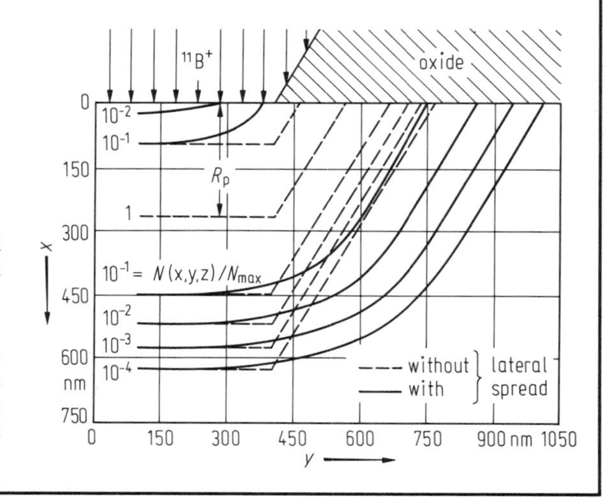

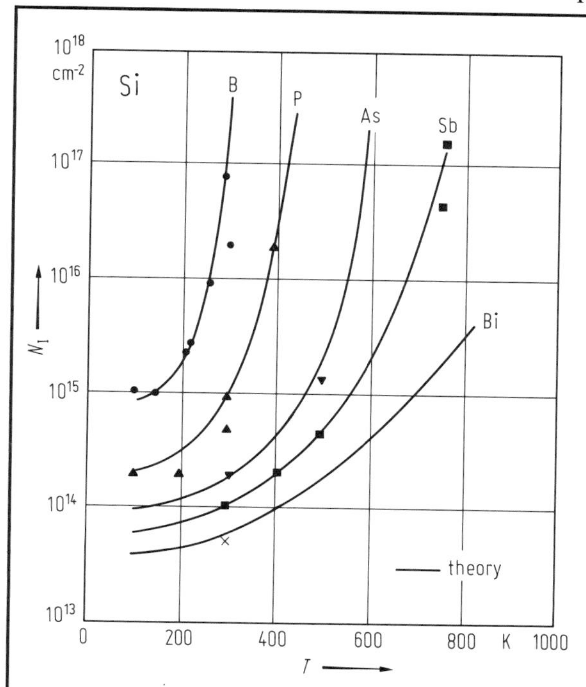

Fig. 7. Si. Theoretical and experimental values of the dose N_I, necessary to render the implanted layer in Si amorphous, vs. target temperature T, for different ion species [78R1].

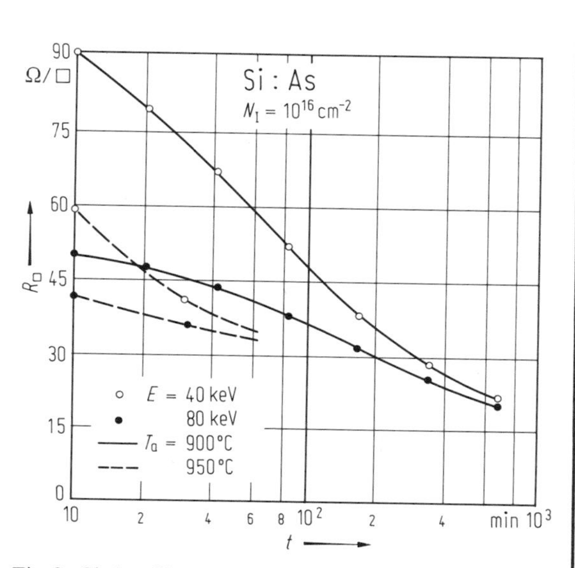

Fig. 8. Si:As. Sheet resistance $R_□$ vs. annealing time t at 900 and 950 °C for As^+ implanted at 40 and 80 keV and at dose $N_I = 1 \cdot 10^{16}$ cm^{-2} in Si [74M1].

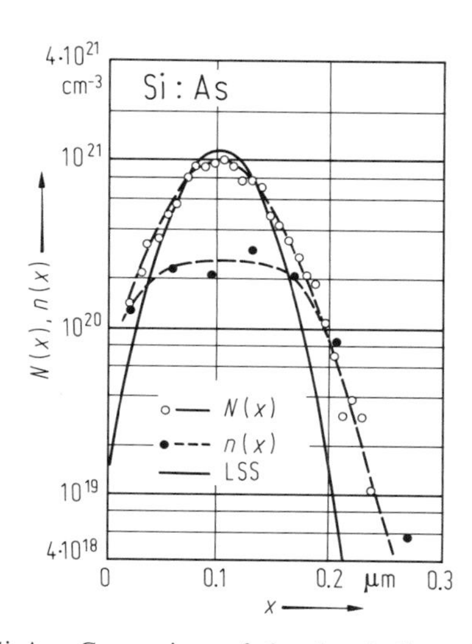

Fig. 9. Si:As. Comparison of the electrically active concentration $n(x)$, the implanted concentration $N(x)$, and the LSS profile [78R1]. As^+ implantation at $E = 180$ keV and $N_I = 1 \cdot 10^{16}$ cm^{-2}, annealing at 900 °C for 30 min.

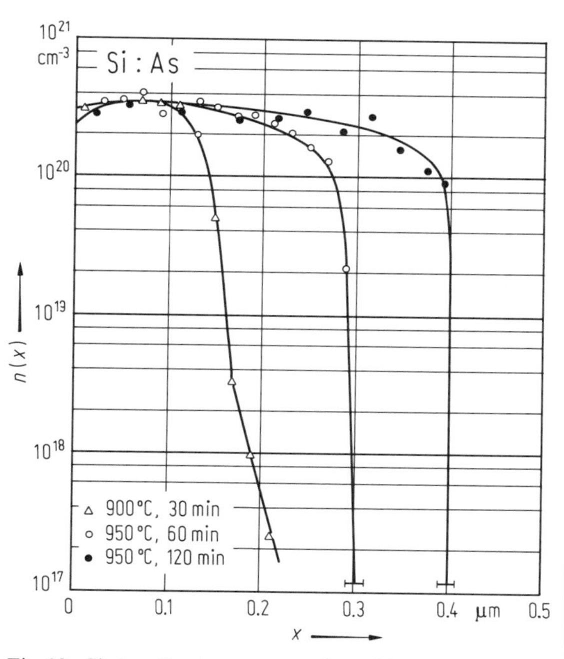

Fig. 10. Si:As. Carrier concentration $n(x)$ vs. depth x after As^+ implantation at $E = 120$ keV and $N_I = 1 \cdot 10^{16}$ cm^{-2} and after diffusion as shown [75R1].

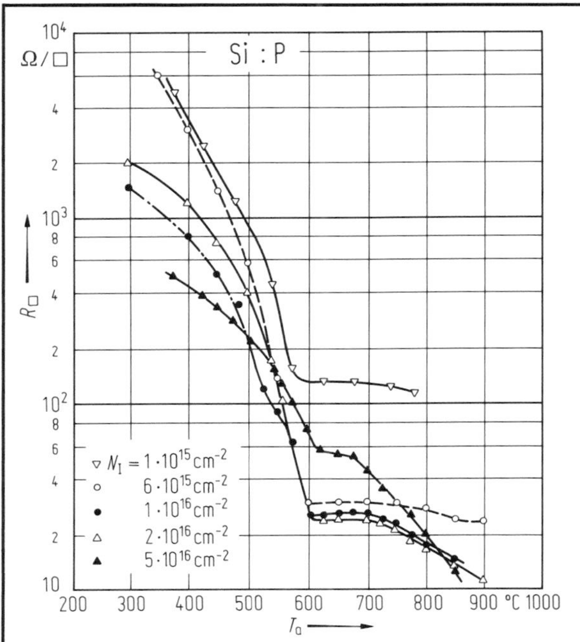

Fig. 11. Si:P. Sheet resistance $R_\square$ vs. annealing temperature T_a for P^+-implantation in Si at $E = 100$ keV. The implantation doses are between $N_I = 1 \cdot 10^{15}$ cm^{-2} and $N_I = 5 \cdot 10^{16}$ cm^{-2} [66S1].

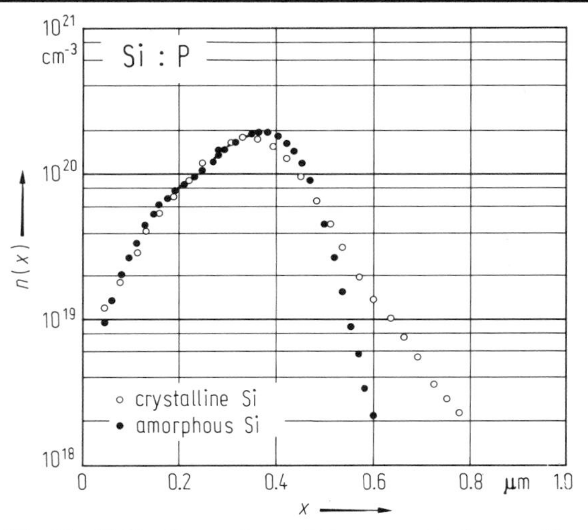

Fig. 12. Si:P. Comparison of the concentration profiles $n(x)$ of P^+ implanted into amorphous Si and into crystalline Si [79I2]. The implantation dose and energy are $N_I = 5 \cdot 10^{15}$ P^+/cm^2 and $E = 300$ keV, respectively.

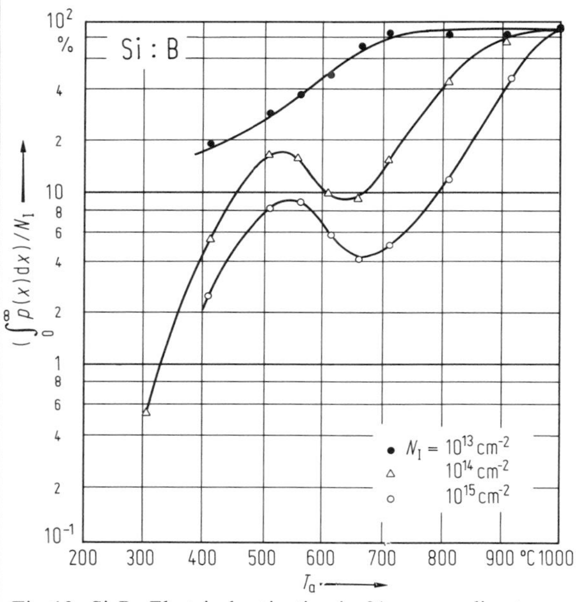

Fig. 13. Si:B. Electrical activation in % vs. annealing temperature T_a for B^+-implantation in Si at $E = 34$ keV. The implantation doses range between $N_I = 1 \cdot 10^{13}$ cm^{-2} and $N_I = 1 \cdot 10^{15}$ cm^{-2} [72R2].

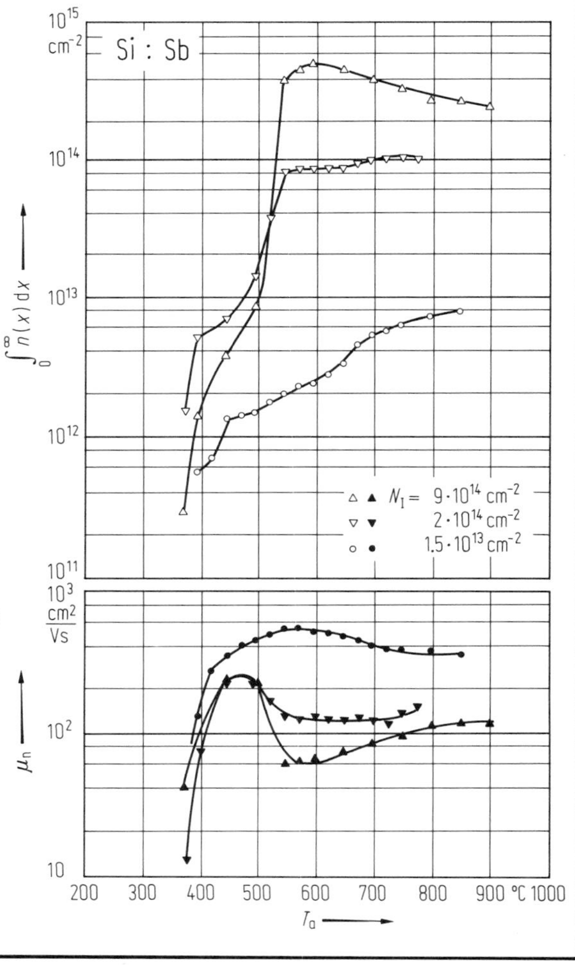

Fig. 14. Si:Sb. Integrated carrier concentration $\int_0^\infty n\,dx$ and mobility μ_n vs. annealing temperature T_a for Sb-implantation in Si at $E = 40$ keV and implantation doses between $N_I = 1.5 \cdot 10^{13}$ cm^{-2} and $N_I = 9 \cdot 10^{14}$ cm^{-2} [70J1].

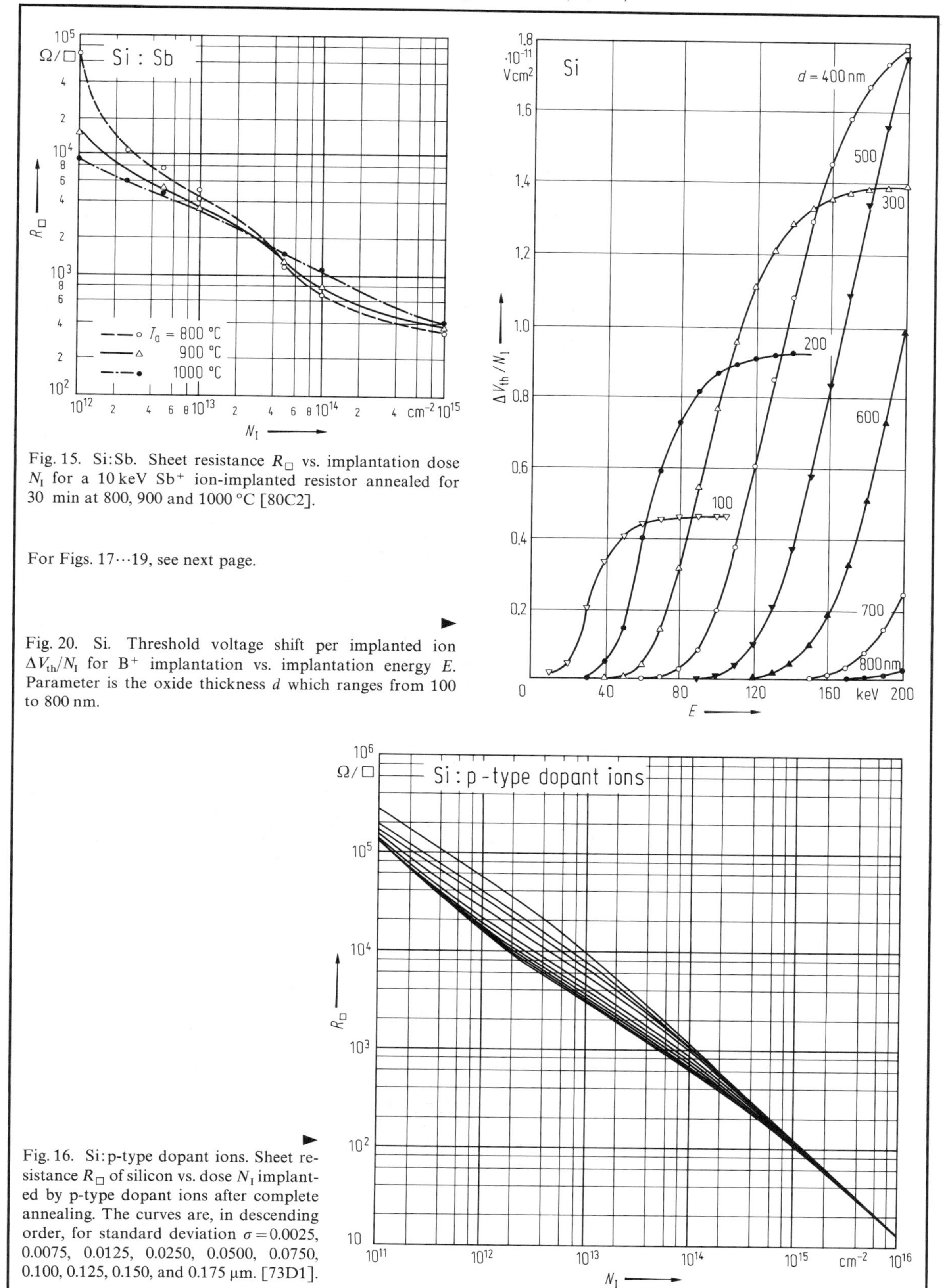

Fig. 15. Si:Sb. Sheet resistance $R_\square$ vs. implantation dose N_I for a 10 keV Sb^+ ion-implanted resistor annealed for 30 min at 800, 900 and 1000 °C [80C2].

For Figs. 17···19, see next page.

Fig. 20. Si. Threshold voltage shift per implanted ion $\Delta V_{th}/N_I$ for B^+ implantation vs. implantation energy E. Parameter is the oxide thickness d which ranges from 100 to 800 nm.

Fig. 16. Si:p-type dopant ions. Sheet resistance $R_\square$ of silicon vs. dose N_I implanted by p-type dopant ions after complete annealing. The curves are, in descending order, for standard deviation $\sigma = 0.0025$, 0.0075, 0.0125, 0.0250, 0.0500, 0.0750, 0.100, 0.125, 0.150, and 0.175 μm. [73D1].

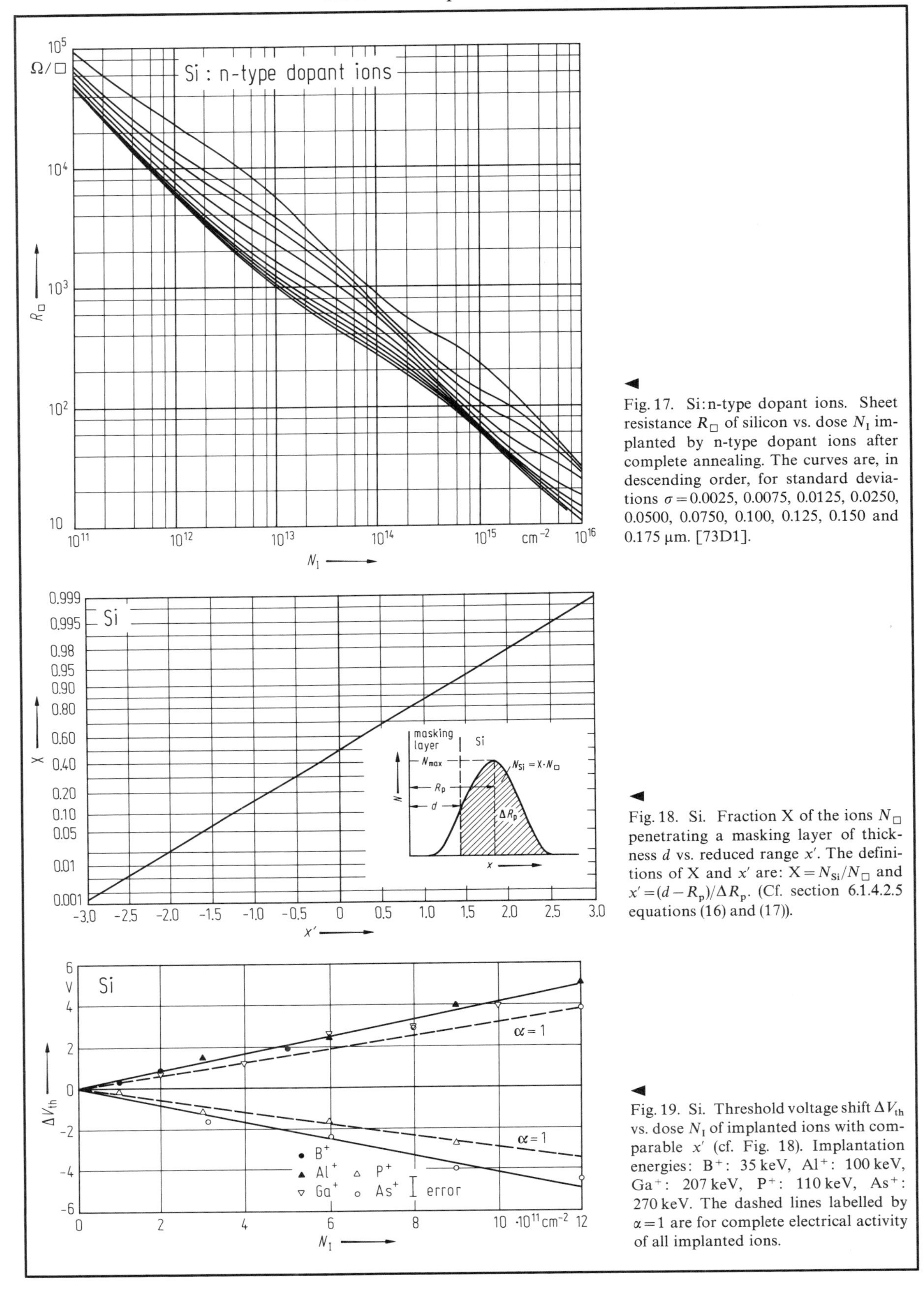

Fig. 17. Si: n-type dopant ions. Sheet resistance $R_\square$ of silicon vs. dose N_I implanted by n-type dopant ions after complete annealing. The curves are, in descending order, for standard deviations $\sigma = 0.0025$, 0.0075, 0.0125, 0.0250, 0.0500, 0.0750, 0.100, 0.125, 0.150 and 0.175 μm. [73D1].

Fig. 18. Si. Fraction X of the ions $N_\square$ penetrating a masking layer of thickness d vs. reduced range x'. The definitions of X and x' are: $X = N_{Si}/N_\square$ and $x' = (d - R_p)/\Delta R_p$. (Cf. section 6.1.4.2.5 equations (16) and (17)).

Fig. 19. Si. Threshold voltage shift ΔV_{th} vs. dose N_I of implanted ions with comparable x' (cf. Fig. 18). Implantation energies: B^+: 35 keV, Al^+: 100 keV, Ga^+: 207 keV, P^+: 110 keV, As^+: 270 keV. The dashed lines labelled by $\alpha = 1$ are for complete electrical activity of all implanted ions.

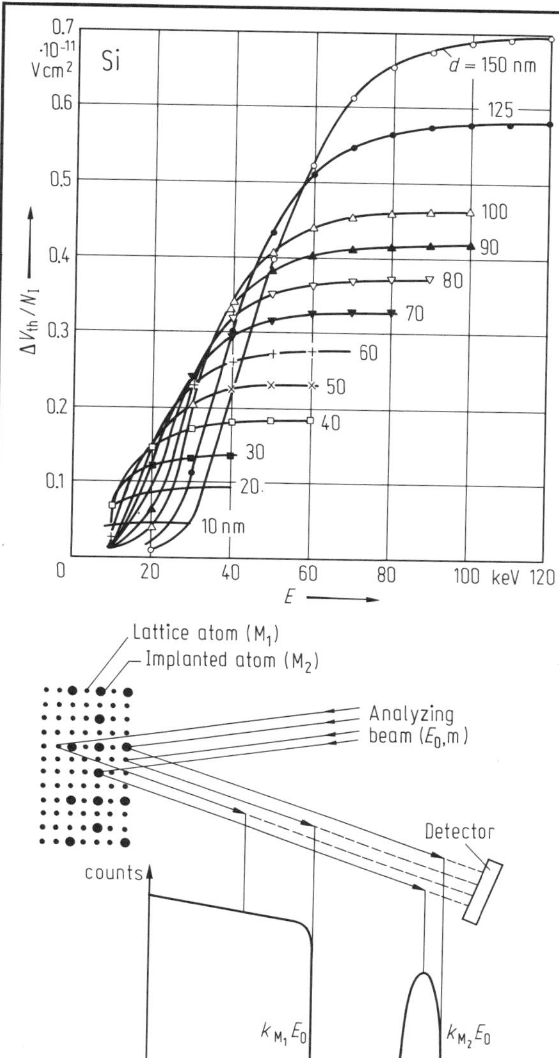

◄ Fig. 21. Si. Threshold voltage shift per implanted ion $\Delta V_{th}/N_I$ for B^+ implantation vs. implantation energy E. Parameter is the oxide thickness d which ranges from 10 to 150 nm.

Fig. 23. Relation of the "backscattering spectrum" (counts vs. particle energy) to the location of the backscattering event.

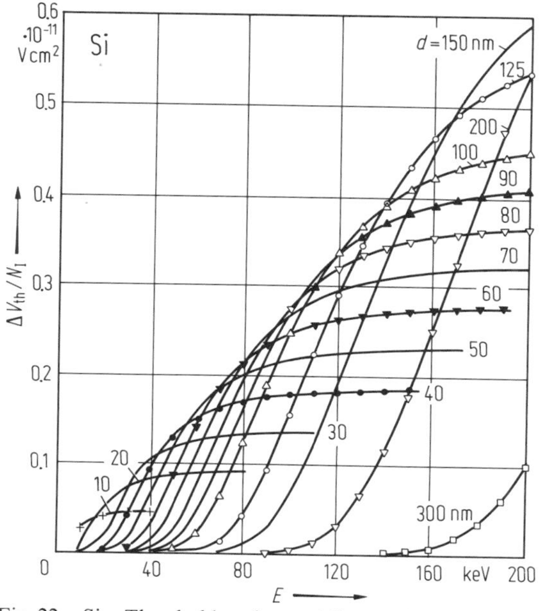

Fig. 22. Si. Threshold voltage shift per implanted ion $\Delta V_{th}/N_I$ for P^+ implantation vs. implantation energy E. Parameter is the oxide thickness d which ranges from 10 to 300 nm.

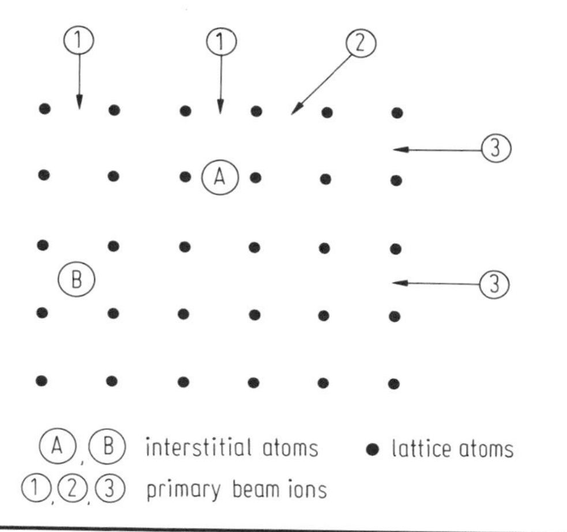

◄ Fig. 24. Location of interstitials by using the channeling effect. The arrows mark the orientation of the primary beam. Atoms at interstitial positions A and B can be identified by the channeling effect. For the direction of the primary beam 1: ions are backscattered from A and B, 2: ions are backscattered only from A, 3: ions are backscattered only from B.

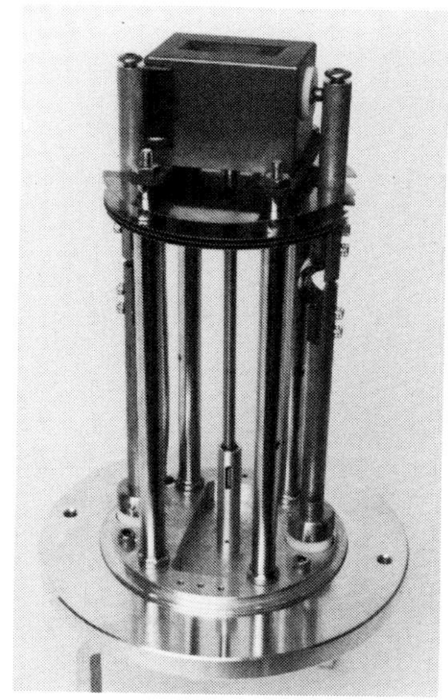

Fig. 25. Typical Freeman-type ion source (manufacturer Extrion-Varian).

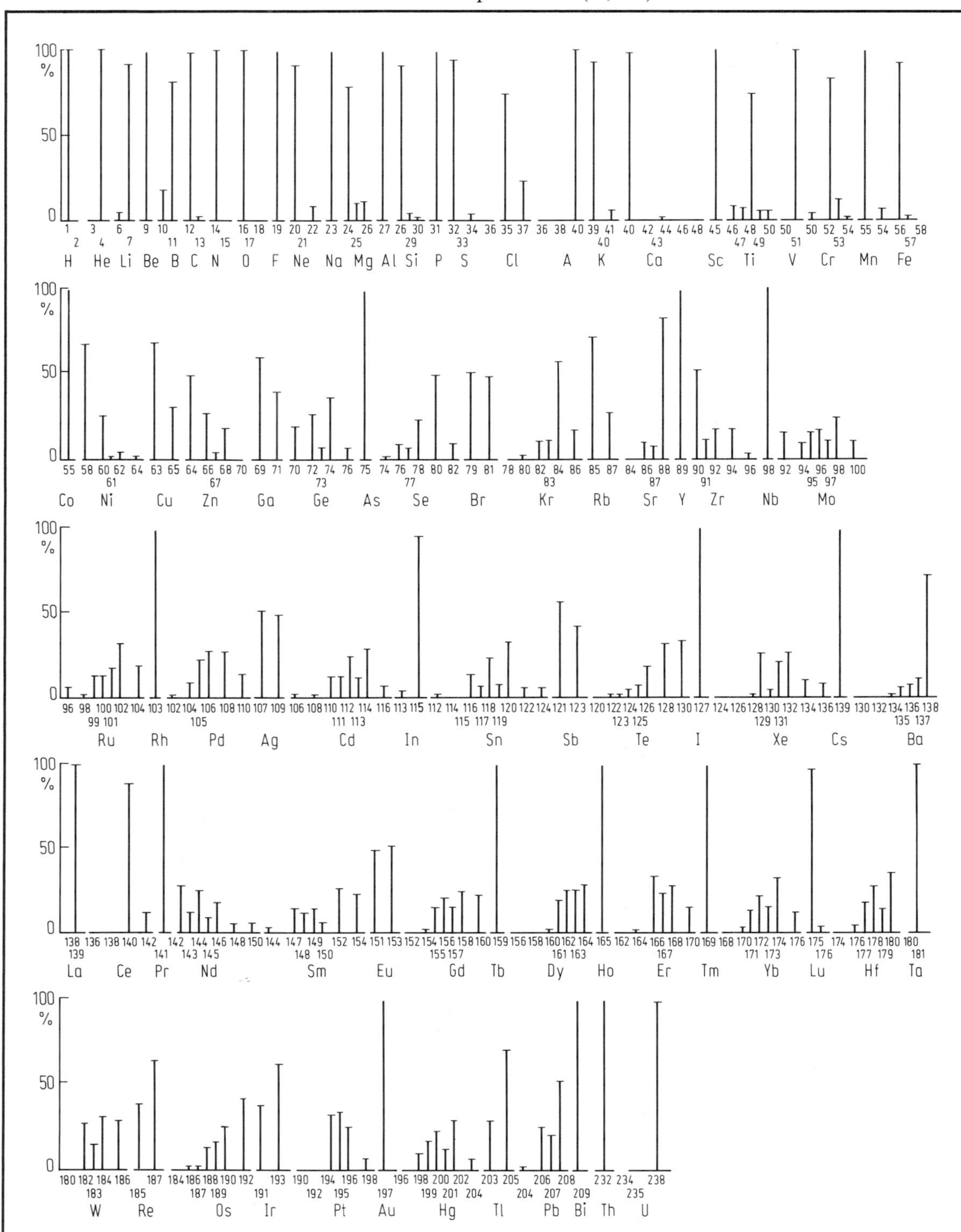

Fig. 26. Natural abundance of the isotopes [73D1]. Isotopes with a natural abundance less than 0.75 % appear only as numbers below the base line. 100 % isotopes have no top bar.

6.1.4.3 Nuclear transmutation doping

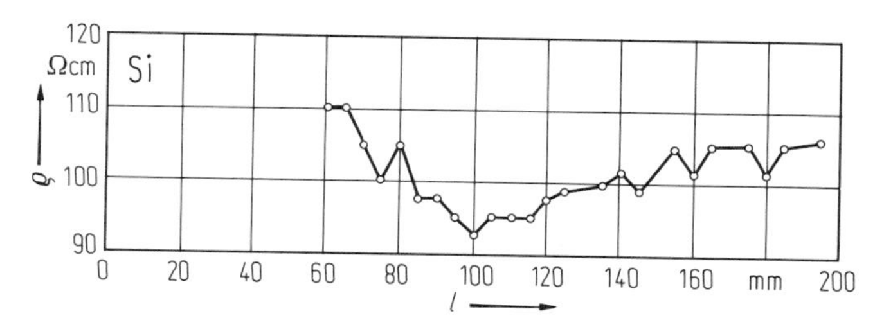

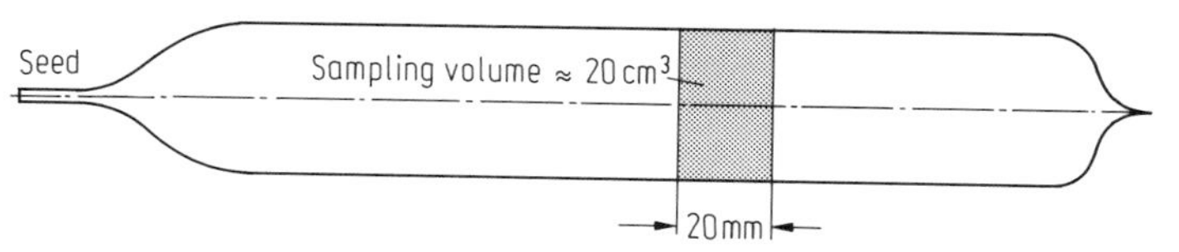

Fig. 1. Si. Typical axial distribution of the resistivity vs. distance l along a silicon crystal ingot. The measurement is performed by a two-point probe having 20 mm separation so that a microscopic average resistivity of a sampling volume of approximately 20 cm^3 is measured as indicated in the bottom sketch [74b1].

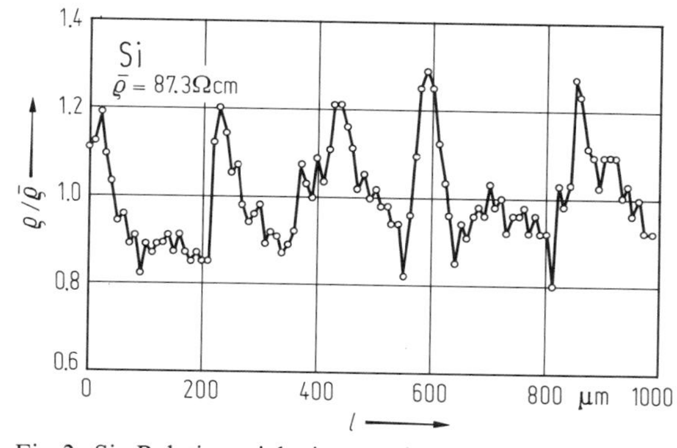

Fig. 2. Si. Relative axial microscopic resistivity $\varrho/\bar{\varrho}$ distribution vs. distance l from the seed of the silicon ingot. The average resistivity is $\bar{\varrho} = 87.3\ \Omega\text{cm}$ [74b1].

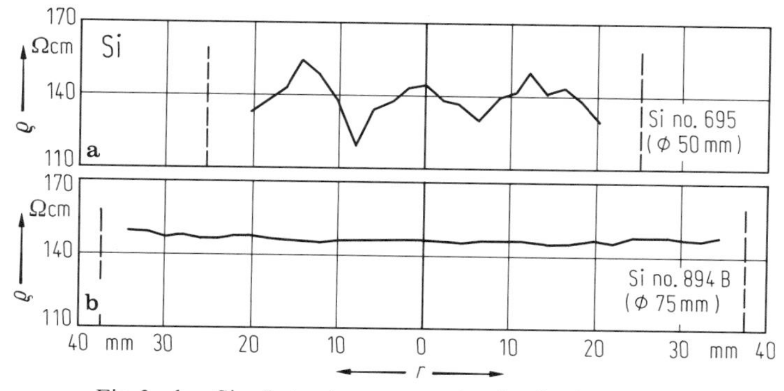

Fig. 3a, b. Si. Lateral macroscopic distribution of the resistivity ϱ vs. radial distance r from the center of a silicon wafer (4-point probe measurement) [76H1]; a) conventionally doped silicon, b) neutron-irradiated silicon.

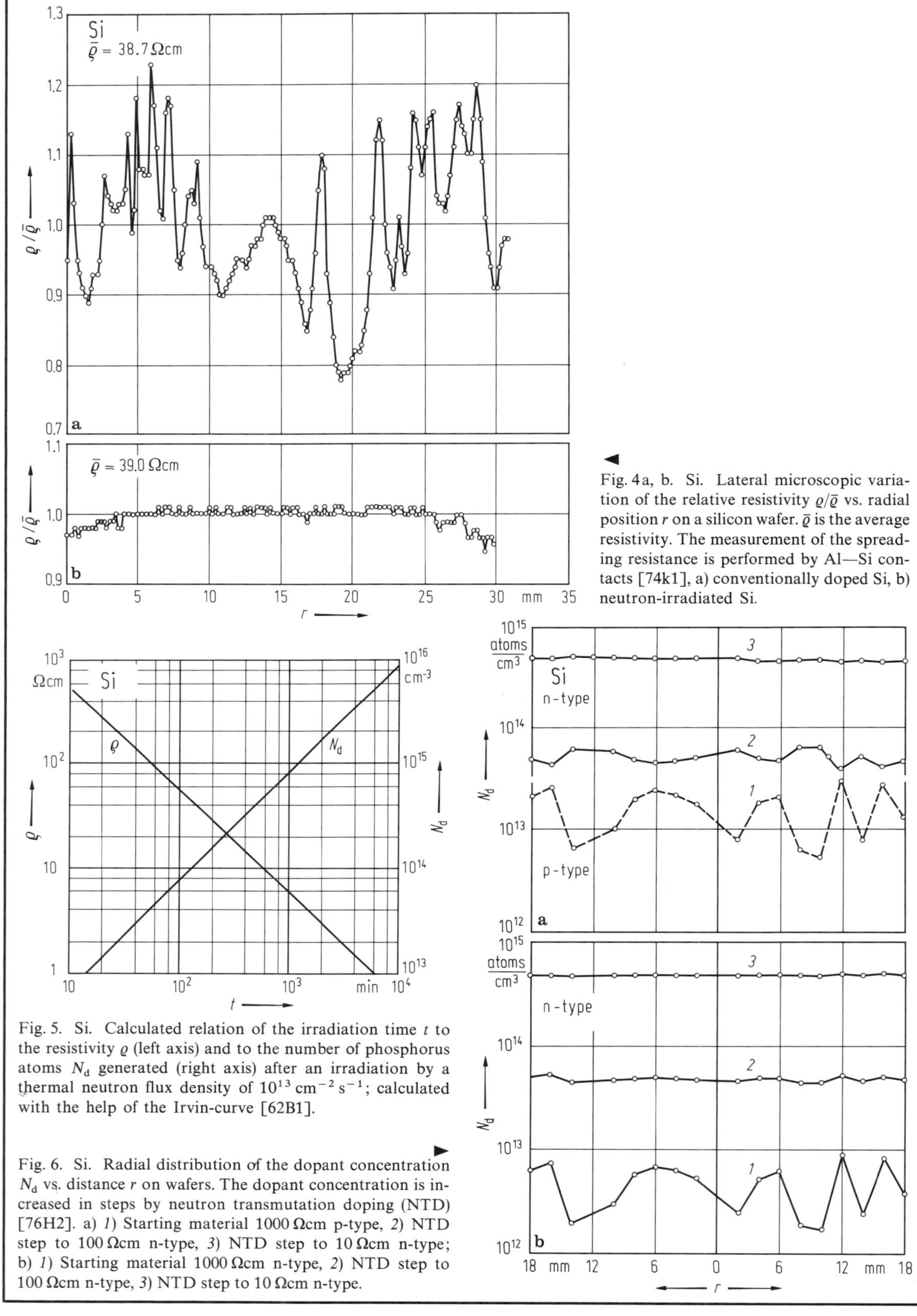

Fig. 4a, b. Si. Lateral microscopic variation of the relative resistivity $\varrho/\bar{\varrho}$ vs. radial position r on a silicon wafer. $\bar{\varrho}$ is the average resistivity. The measurement of the spreading resistance is performed by Al—Si contacts [74k1], a) conventionally doped Si, b) neutron-irradiated Si.

Fig. 5. Si. Calculated relation of the irradiation time t to the resistivity ϱ (left axis) and to the number of phosphorus atoms N_d generated (right axis) after an irradiation by a thermal neutron flux density of 10^{13} cm^{-2} s^{-1}; calculated with the help of the Irvin-curve [62B1].

Fig. 6. Si. Radial distribution of the dopant concentration N_d vs. distance r on wafers. The dopant concentration is increased in steps by neutron transmutation doping (NTD) [76H2]. a) *1*) Starting material 1000 Ωcm p-type, *2*) NTD step to 100 Ωcm n-type, *3*) NTD step to 10 Ωcm n-type; b) *1*) Starting material 1000 Ωcm n-type, *2*) NTD step to 100 Ωcm n-type, *3*) NTD step to 10 Ωcm n-type.

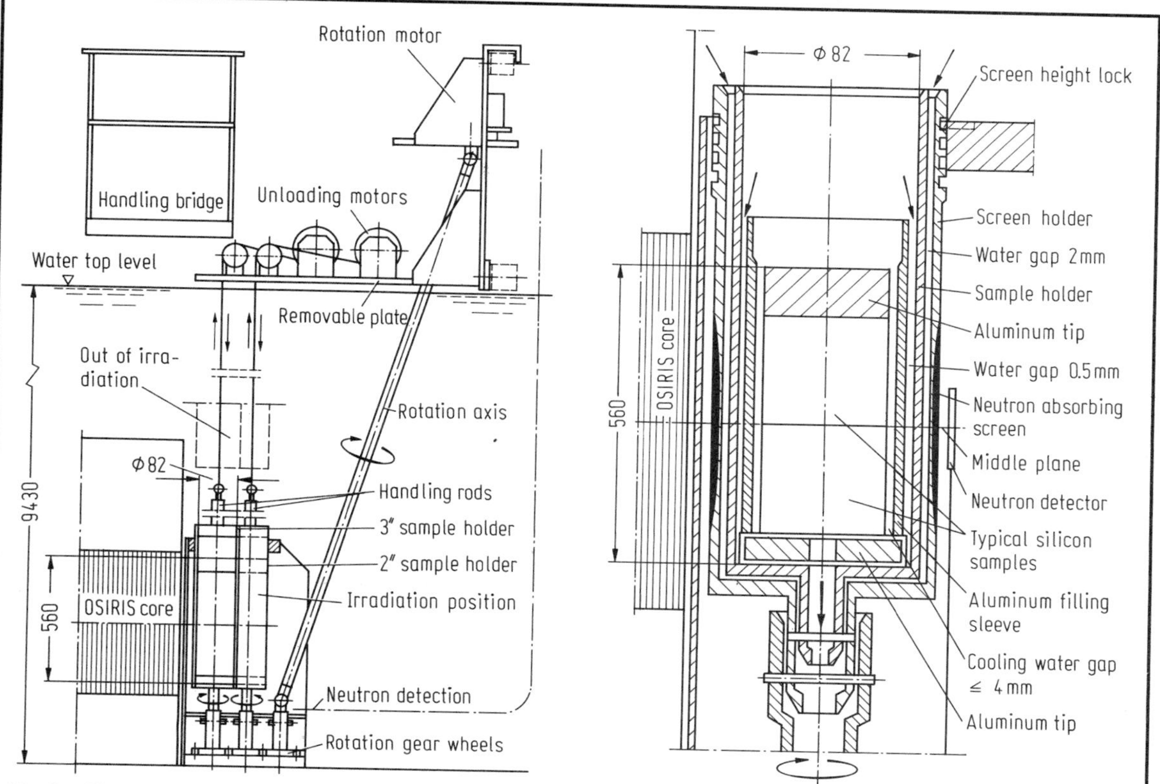

Fig. 7. Si. Diagram of an irradiation facility (reactor OSIRIS). Vertical section. [81b2]. (Measures given in mm).

Fig. 8. Si. Schematic axial section of the irradiated part of the irradiation-module and neutronic implications [81b2]. (Measures given in mm).

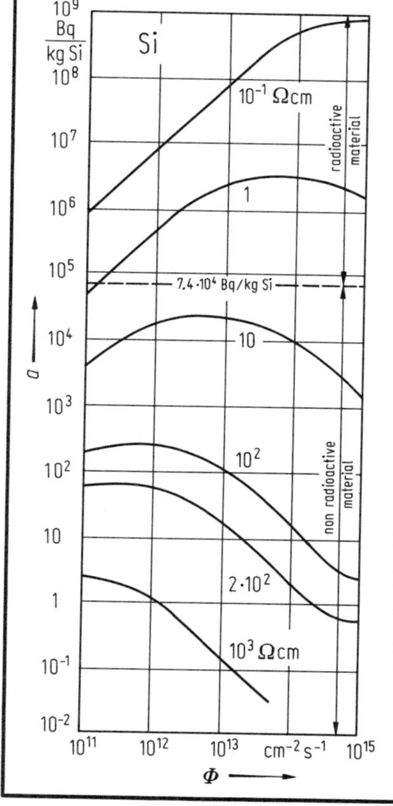

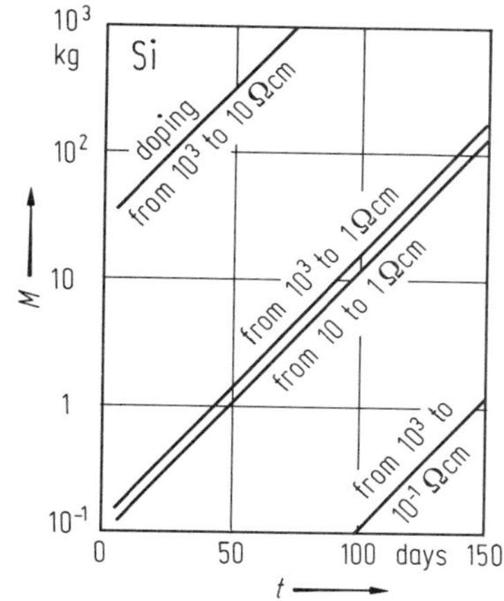

Fig. 10. Si. Decay time t of ^{32}P in neutron irradiated pure silicon for reaching the acceptable limit ($a \approx 3.7 \cdot 10^5$ Bq) vs. irradiated silicon quantity M for various resistivities. (The calculation is performed for a neutron flux density of $\Phi = 10^{14}$ cm^{-2} s^{-1}) [76H2].

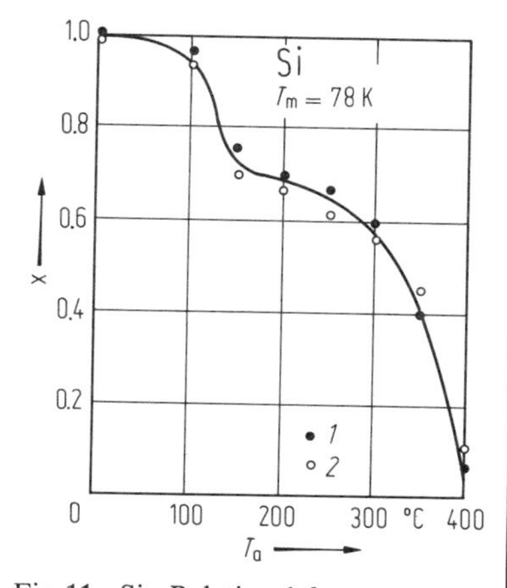

Fig. 11. Si. Relative defect concentration (mole fraction x) remaining as measured at $T_m = 78$ K by the van der Pauw technique vs. annealing temperature T_a after isochronal annealing of the samples *1* and *2* listed in Table 5 [78Y1].

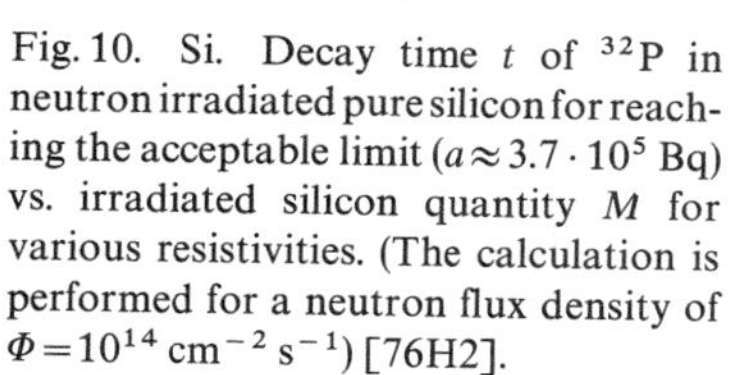

◄ Fig. 9. Si. Relationship of the neutron flux density Φ and ^{32}P radioactivity a for various resistivities [79h1, varied].

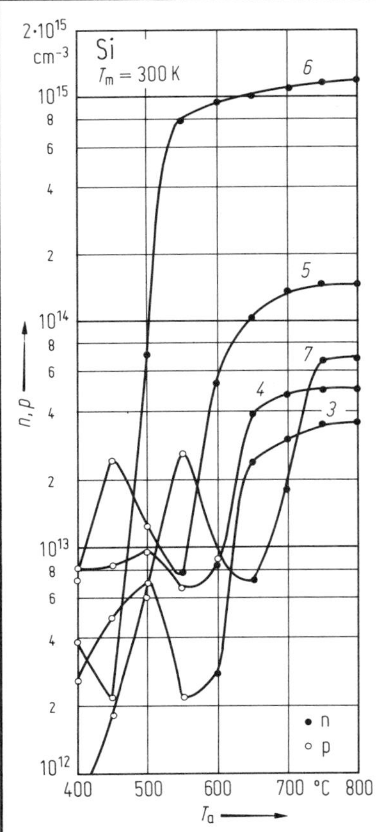

Fig. 12. Si. Recovery of the free carrier concentration n and p vs. annealing temperature T_a of isochronal annealing. The samples *3*···*7* as listed in Table 5 have received a light-to-moderate fast neutron fluence [78Y1]. (T_m: Temperature of measurements).

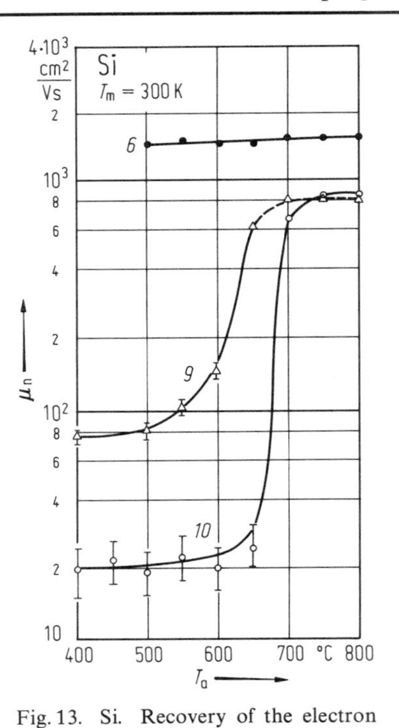

Fig. 13. Si. Recovery of the electron mobility μ_n vs. isochronal annealing temperature T_a of samples *6*, *9*, and *10* as listed in Table 5 [78Y1]. (T_m: Temperature of measurements).

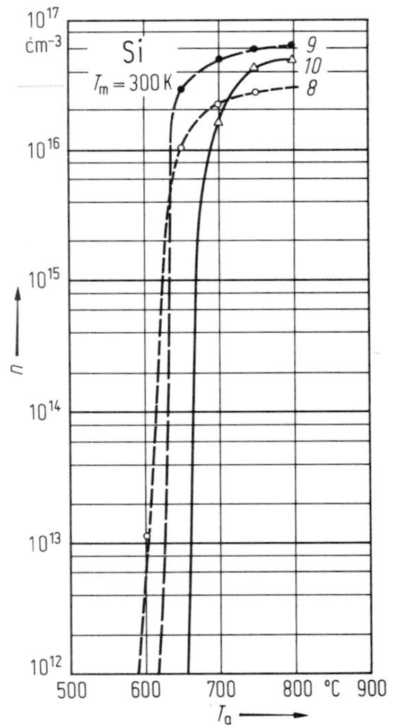

Fig. 14. Si. Recovery of the free carrier concentration n vs. annealing temperature T_a of the samples *8*···*10* as listed in Table 5 which have received a heavy fast-neutron fluence [78Y1]. (T_m: Temperature of measurements).

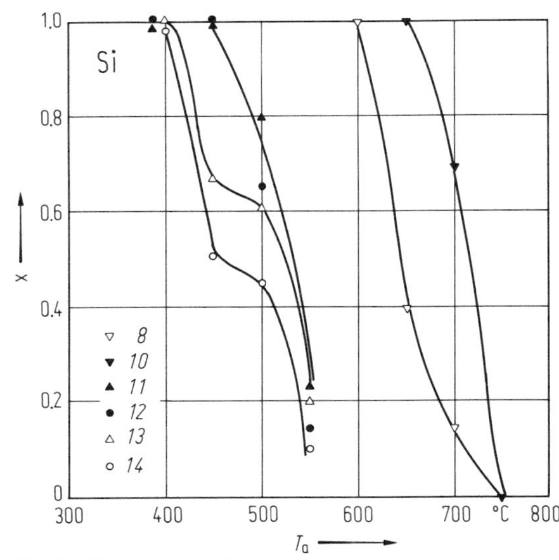

Fig. 15. Si. Isochronal annealing of the deep energy levels. Fraction x of deep levels unannealed vs. annealing temperature T_a of the samples *8* and *10*···*14* as listed in Table 5.

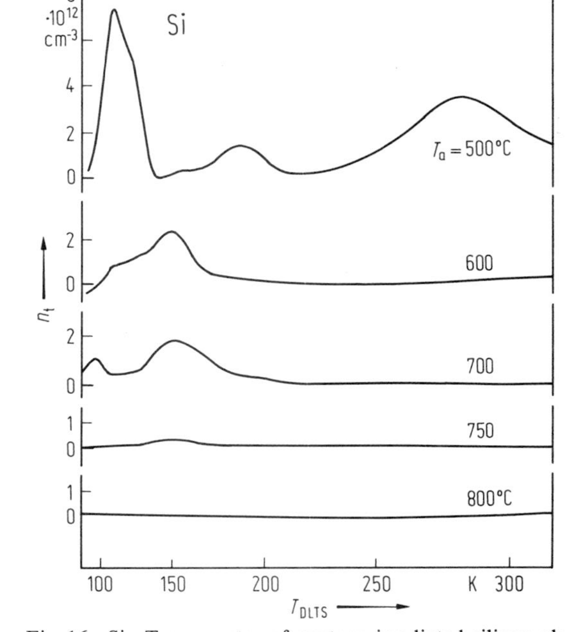

Fig. 16. Si. Trap spectra of neutron irradiated silicon obtained after 30 min. anneals in N_2 at various temperatures T_a. Trap concentration n_t vs. temperature during DLTS-measurement T_{DLTS} [78G1].

6.1.4.4 Silicon epitaxy

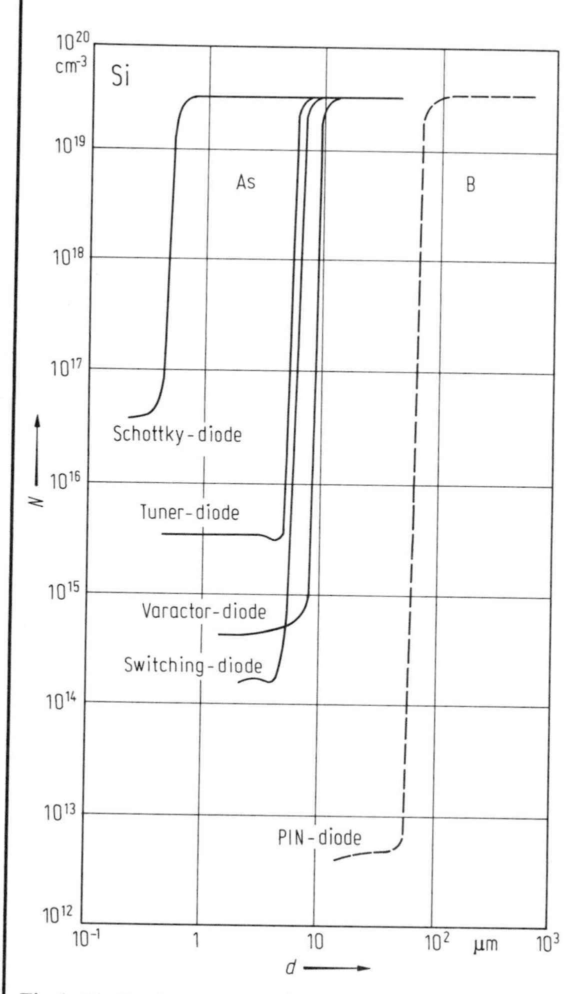

Fig. 1. Si. Doping concentration N vs. thickness d of homo-epitaxial silicon layers grown by VPE as they are typically used in diode fabrication. Layers doped with phosphorus (P) were grown on substrates heavily doped with arsenic (As) and boron (B), respectively.

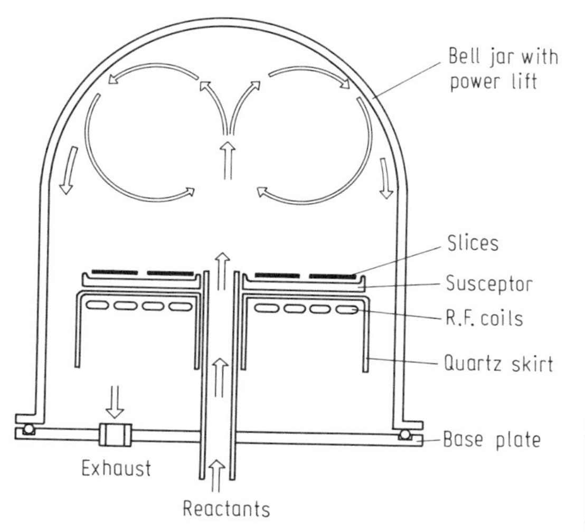

Fig. 2. Cross-sectional view of a *vertical reactor* for vapour-phase epitaxy. The arrows indicate the reactant gas flow direction.

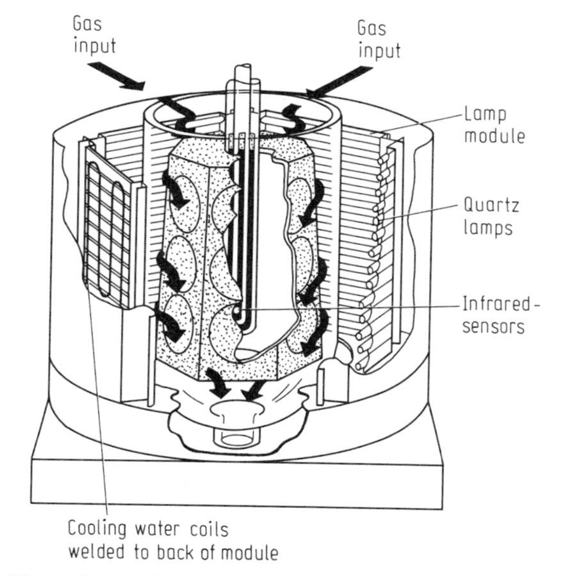

Fig. 4. Schematic drawing of a radiation-heated *cylinder reactor* for vapour-phase epitaxy.

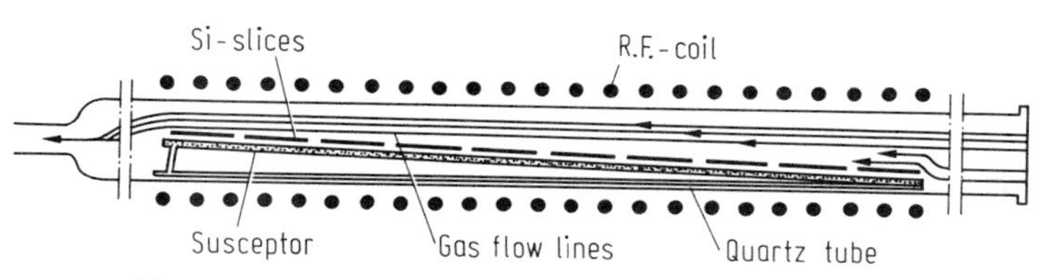

Fig. 3. Cross-sectional view of a *horizontal reactor* for vapour-phase epitaxy. The arrows indicate the flow lines of the reactant gas.

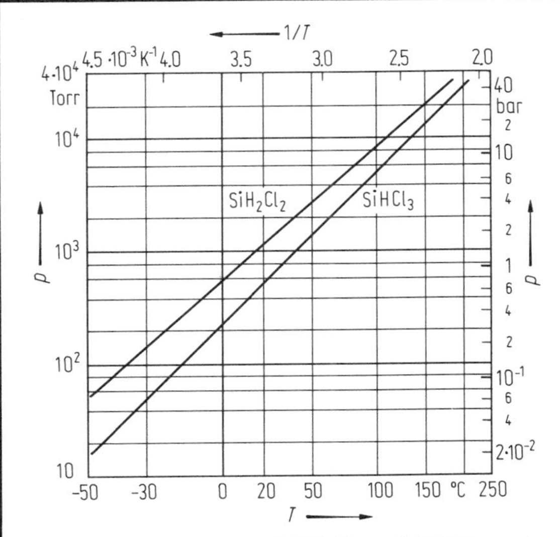

Fig. 5. Vapour pressure p of SiH_2Cl_2 and $SiHCl_3$ vs. temperature T [73D].

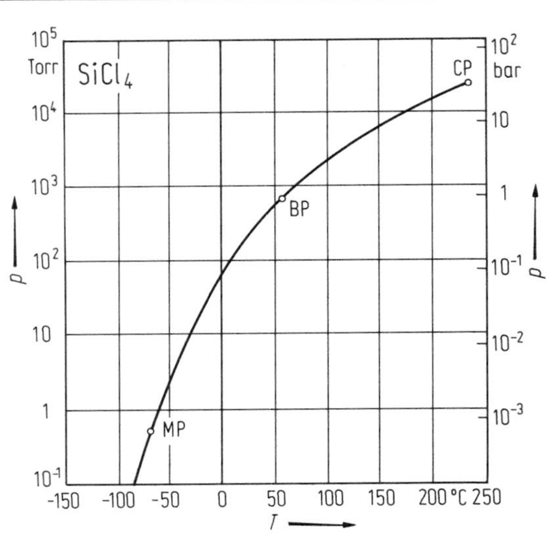

Fig. 6. $SiCl_4$. Vapour pressure p of $SiCl_4$ vs. temperature T [79Y]. The points marked MP, BP and CP indicate the melting point, boiling point, and critical point, respectively.

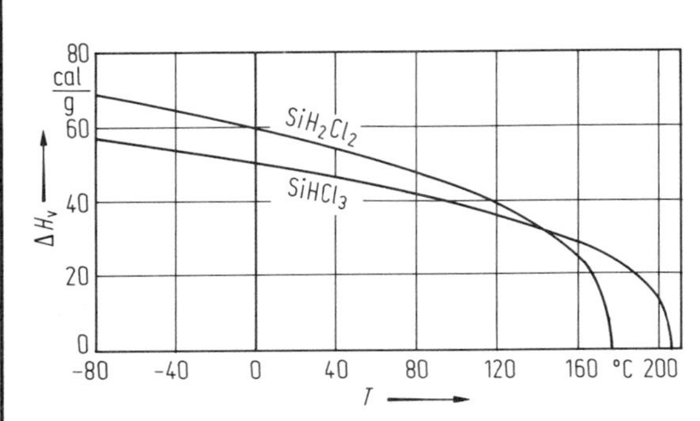

Fig. 7. Heat of vaporization ΔH_v of SiH_2Cl_2 and $SiHCl_3$ vs. temperature T [73D].

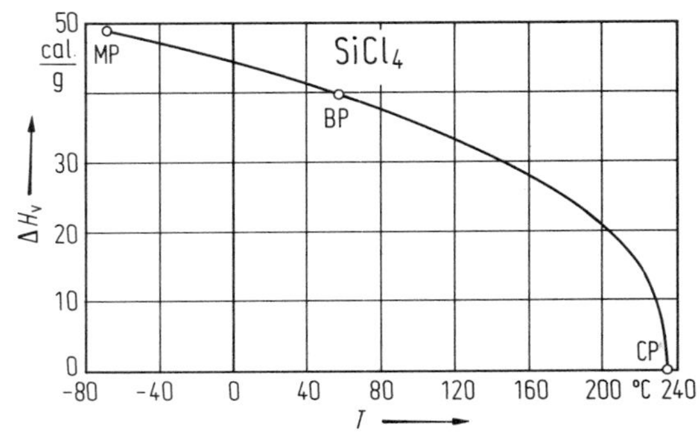

Fig. 8. $SiCl_4$. Heat of vaporization ΔH_v of $SiCl_4$ vs. temperature T. The points marked MP, BP and CP indicate the melting point, boiling point and critical point, respectively [79Y].

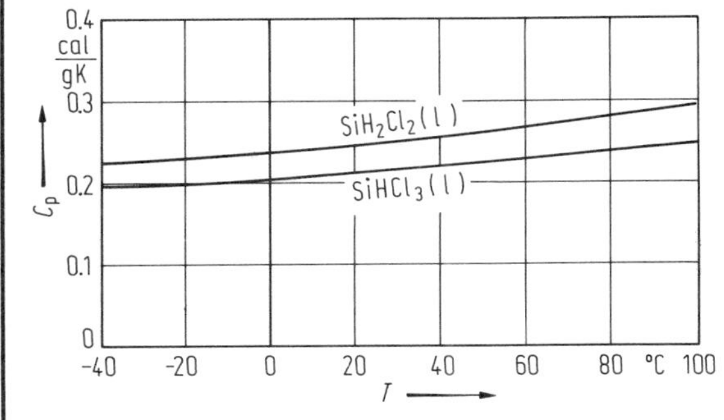

Fig. 9. Heat capacity C_p of liquid SiH_2Cl_2 and $SiHCl_3$ vs. temperature T [73D].

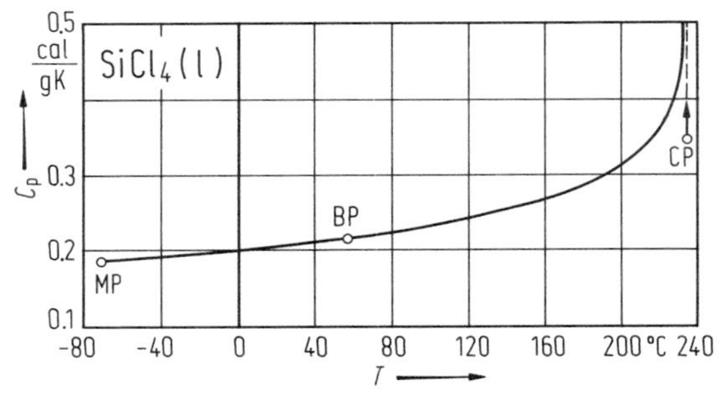

Fig. 10. $SiCl_4$. Heat capacity C_p of liquid $SiCl_4$ vs. temperature T. The points marked MP, BP and CP indicate the melting point, boiling point, and critical point, respectively [79Y].

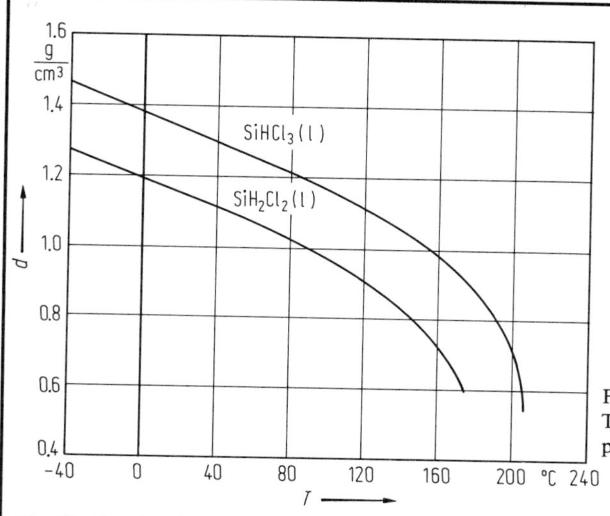

Fig. 11. Density d of liquid SiH_2Cl_2 and $SiHCl_3$ vs. temperature T [73D].

Fig. 12. $SiCl_4$. Density d of liquid $SiCl_4$ vs. temperature T. The points marked MP, BP, and CP indicate the melting point, boiling point, and critical point, respectively [79Y].

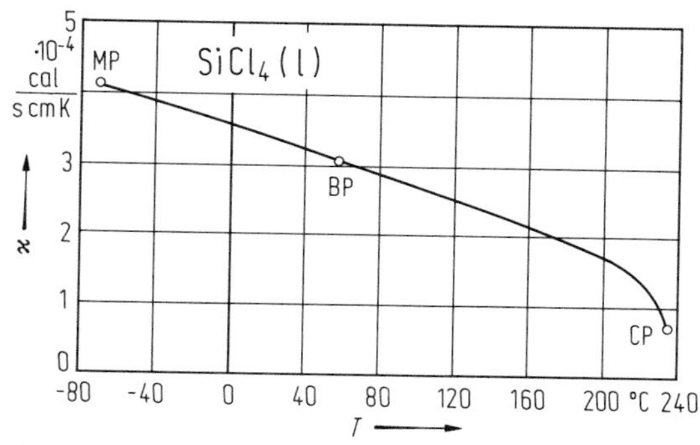

Fig. 13. $SiCl_4$. Thermal conductivity κ of liquid $SiCl_4$. The points marked MP, BP and CP indicate the melting point, boiling point, and critical point, respectively [79Y].

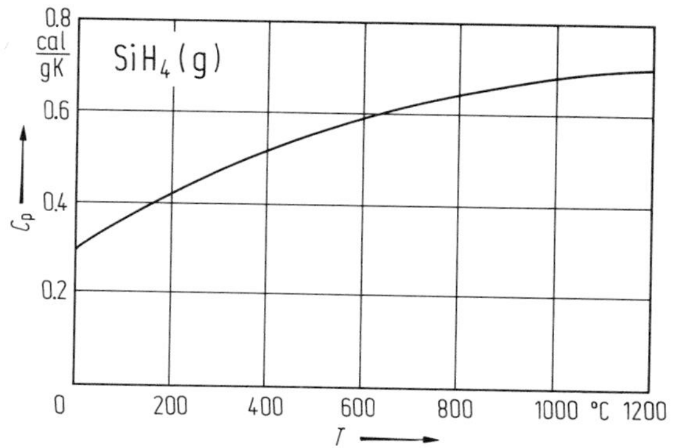

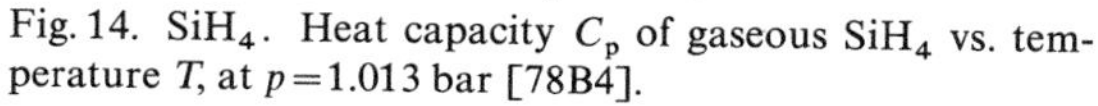

Fig. 14. SiH_4. Heat capacity C_p of gaseous SiH_4 vs. temperature T, at $p=1.013$ bar [78B4].

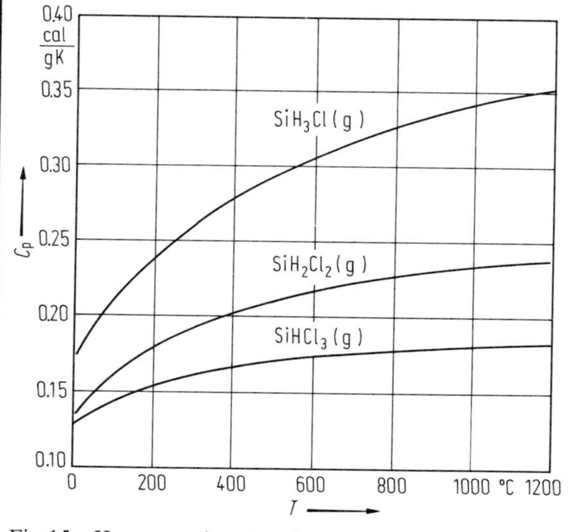

Fig. 15. Heat capacity C_p of gaseous SiH_3Cl, SiH_2Cl_2, and $SiHCl_3$ vs. temperature T, at $p=1.013$ bar [73D].

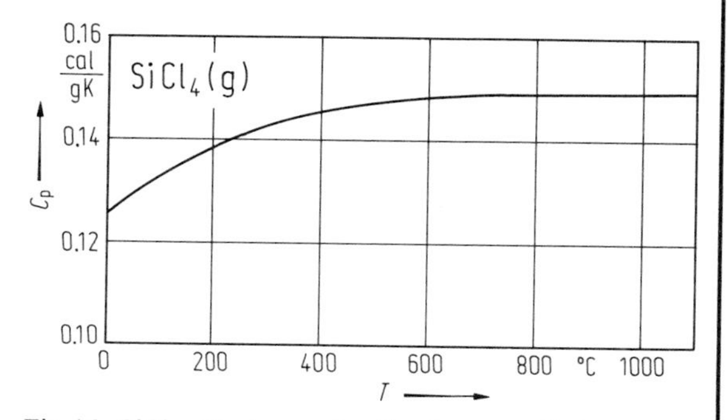

Fig. 16. $SiCl_4$. Heat capacity C_p of gaseous $SiCl_4$ vs. temperature T, at low pressure [79Y].

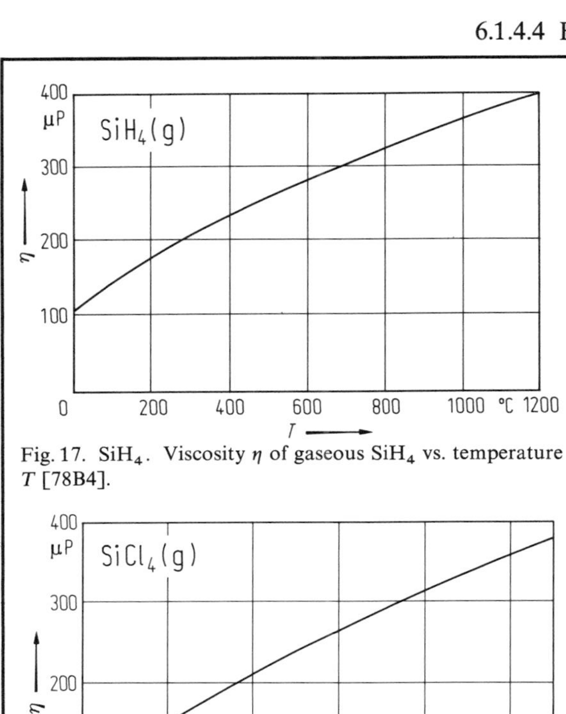

Fig. 17. SiH_4. Viscosity η of gaseous SiH_4 vs. temperature T [78B4].

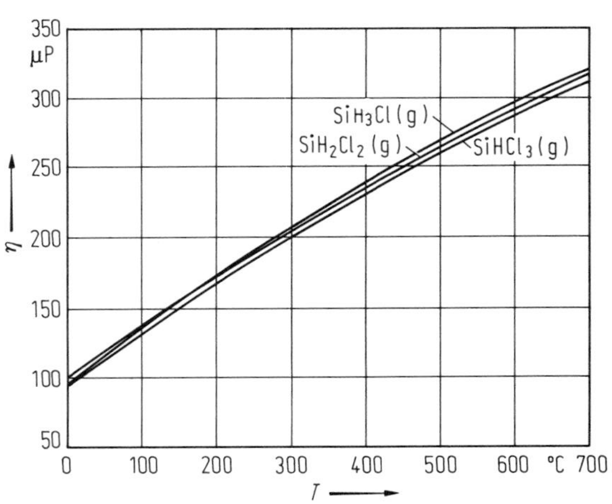

Fig. 18. Viscosity η of gaseous SiH_3Cl, SiH_2Cl_2, and $SiHCl_3$ vs. temperature T [73D].

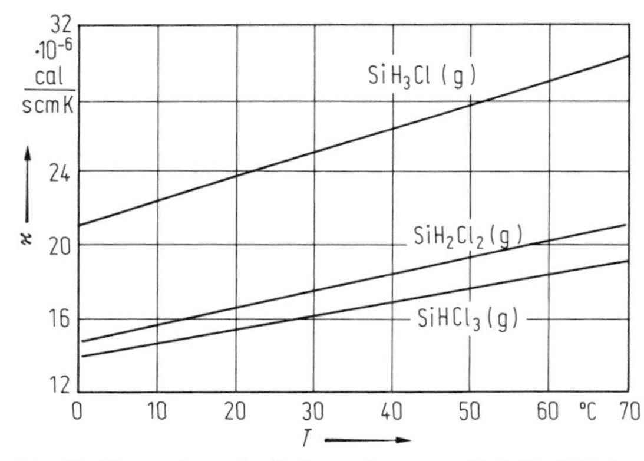

Fig. 19. $SiCl_4$. Viscosity η of gaseous $SiCl_4$ vs. temperature T [79Y].

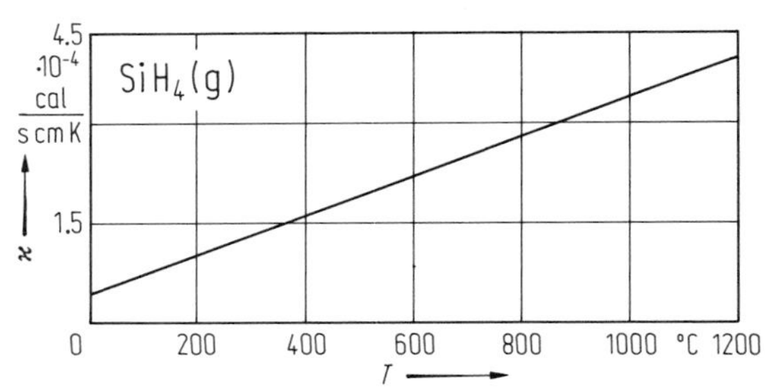

Fig. 20. SiH_4. Thermal conductivity κ of gaseous SiH_4 vs. temperature T [78B4].

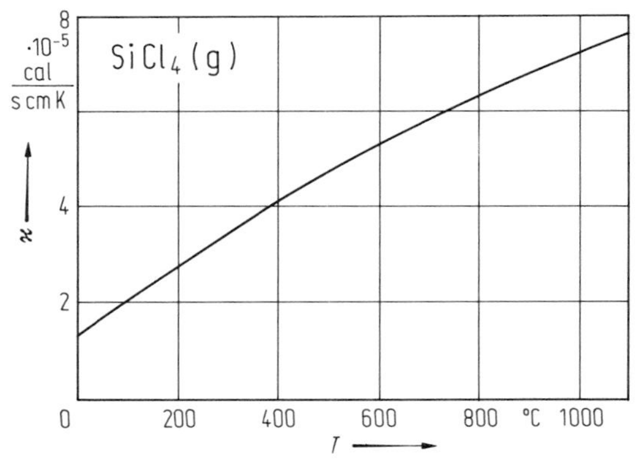

Fig. 22. $SiCl_4$. Thermal conductivity κ of gaseous $SiCl_4$ vs. temperature T [79Y].

Fig. 21. Thermal conductivity κ of gaseous SiH_3Cl, SiH_2Cl_2, and $SiHCl_3$ vs. temperature T [73D].

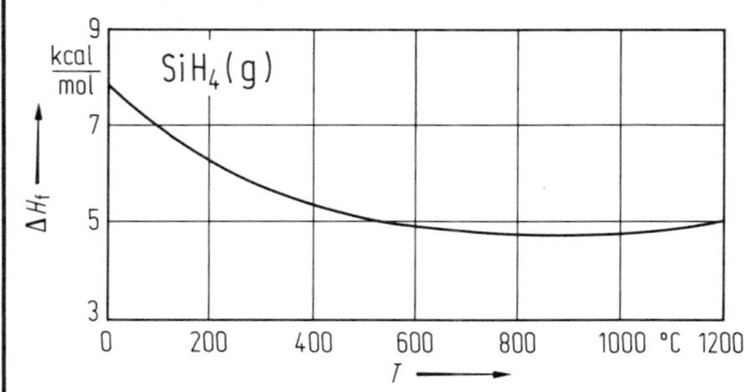

◄ Fig. 23. SiH_4. Heat of formation ΔH_f of gaseous SiH_4 vs. temperature T [78B4].

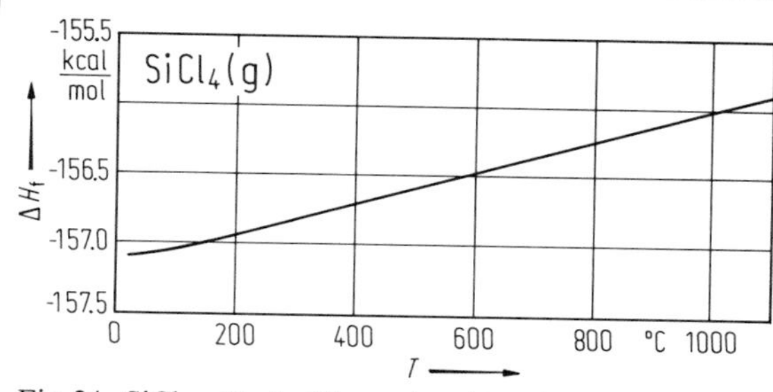

Fig. 24. $SiCl_4$. Heat of formation ΔH_f of gaseous $SiCl_4$ vs. temperature T [79Y].

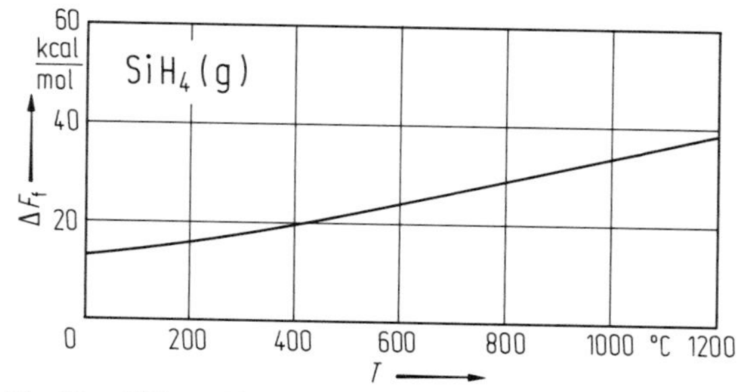

Fig. 25. SiH_4. Gibbs' free energy of formation ΔF_f of gaseous SiH_4 vs. temperature T [78B4].

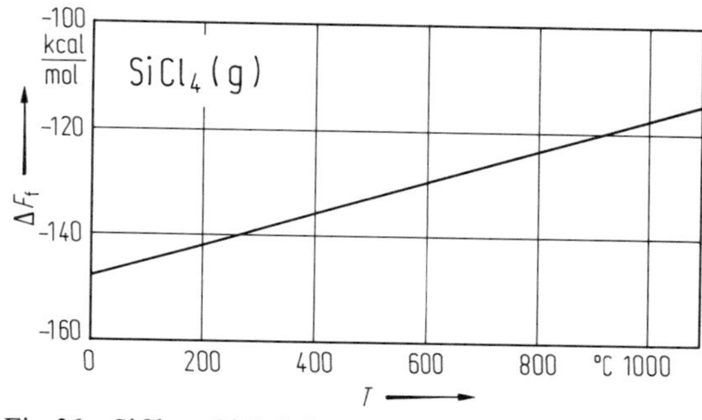

Fig. 26. $SiCl_4$. Gibbs' free energy of formation ΔF_f of gaseous $SiCl_4$ vs. temperature T [79Y].

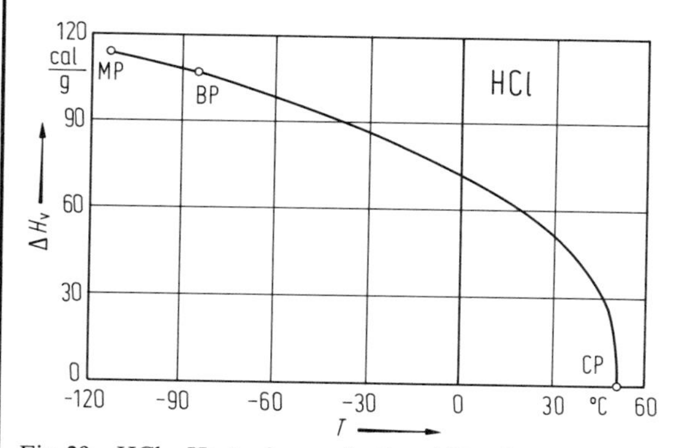

Fig. 29. HCl. Heat of vaporization ΔH_v of HCl vs. temperature T. The points marked MP, BP, and CP indicate the melting point, the boiling point, and the critical point, respectively [76A].

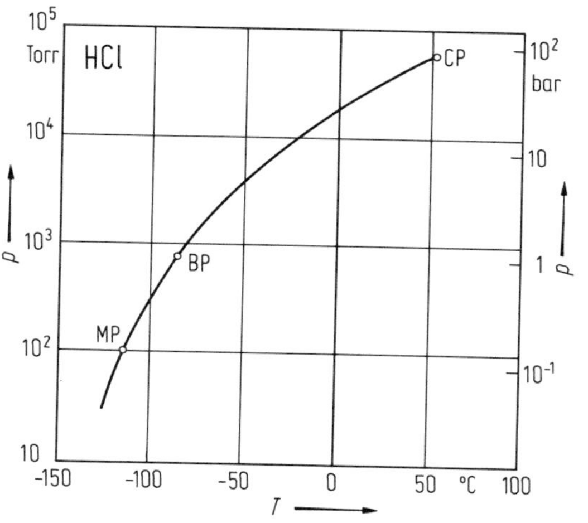

Fig. 27. HCl. Vapour pressure p of HCl vs. temperature T. The points marked MP, BP, and CP indicate the melting point, the boiling point, and critical point, respectively [76A].

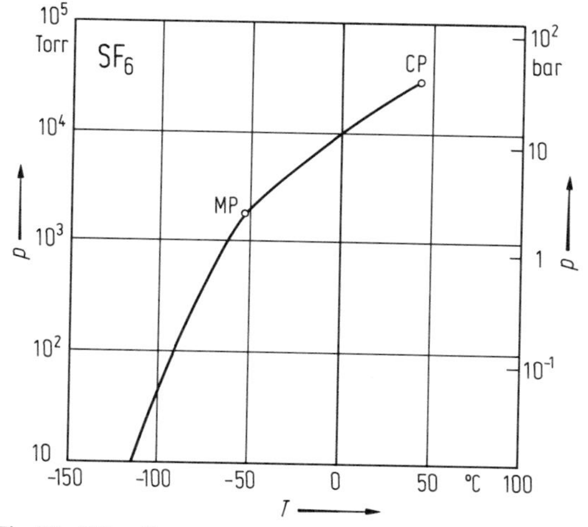

Fig. 28. SF_6. Vapour pressure p of SF_6 vs. temperature T. The points marked MP and CP indicate the melting point and critical point, respectively [75A].

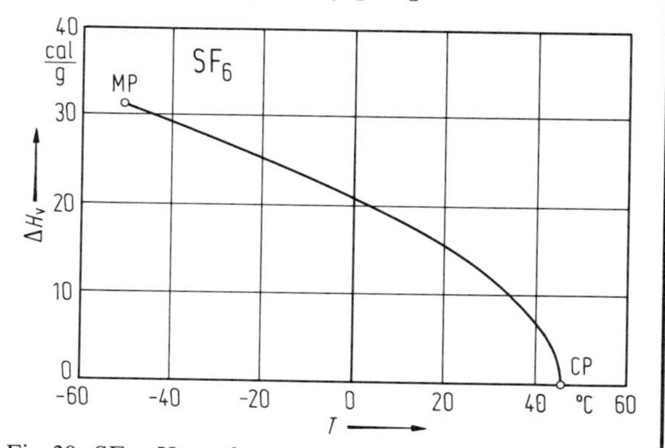

Fig. 30. SF_6. Heat of vaporization ΔH_v of SF_6 vs. temperature T. The points marked MP and CP indicate the melting point and the critical point, respectively [75A].

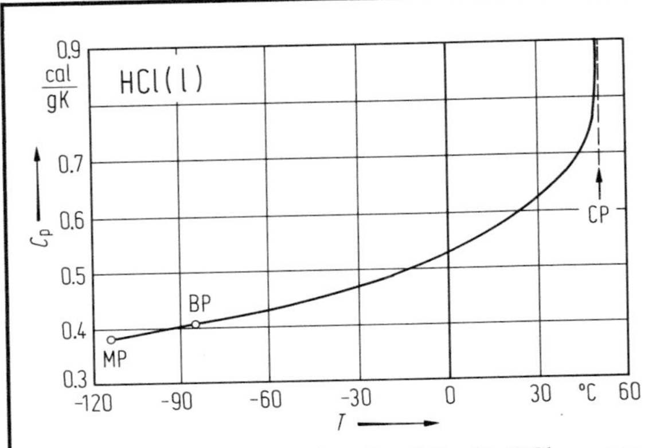

Fig. 31. HCl. Heat capacity C_p of liquid HCl vs. temperature T. The points marked MP, BP, and CP indicate the melting point, the boiling point, and the critical point, respectively [76A].

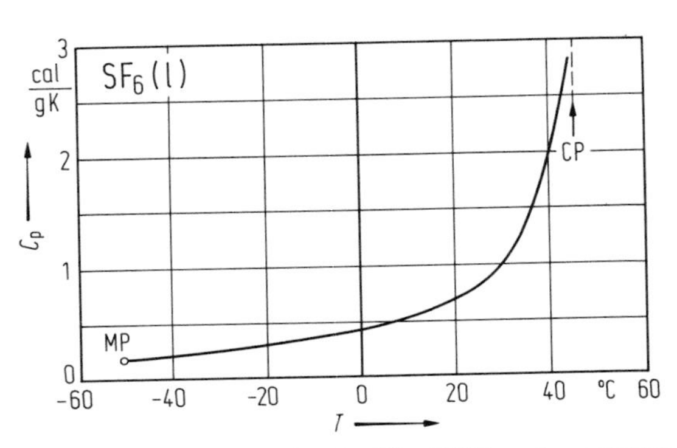

Fig. 32. SF_6. Heat capacity C_p of liquid SF_6 vs. temperature T. The points marked MP and CP indicate the melting point and critical point, respectively [75A].

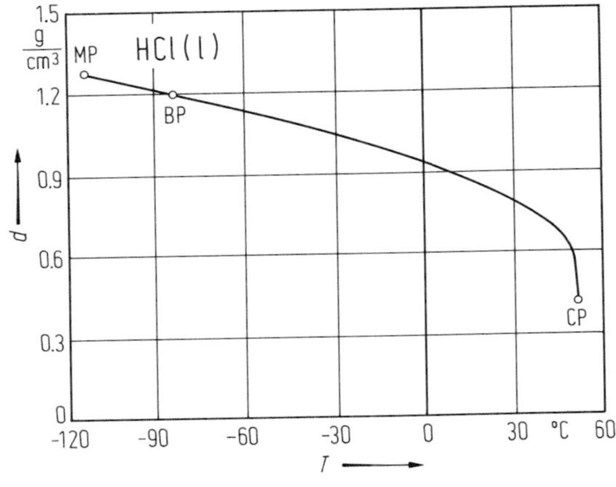

Fig. 33. HCl. Density d of liquid HCl vs. temperature T. The points marked MP, BP, and CP indicate the melting point, the boiling point, and the critical point, respectively [76A].

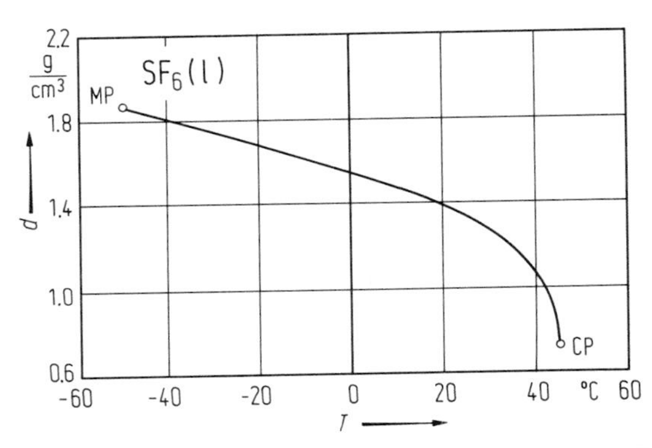

Fig. 34. SF_6. Density d of liquid SF_6 vs. temperature T. The points marked MP and CP indicate the melting point and the critical point, respectively [75A1].

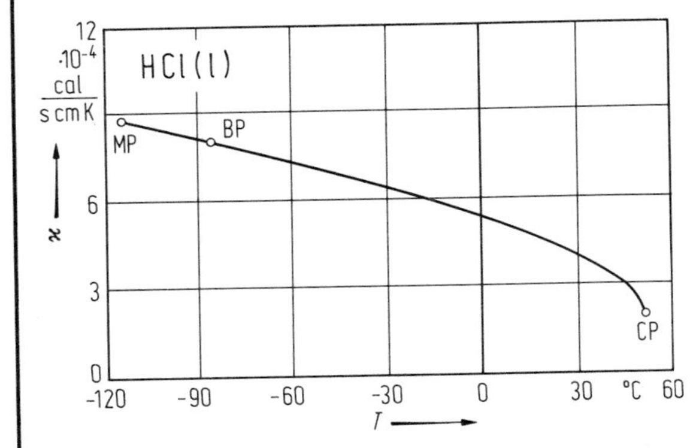

Fig. 35. HCl. Thermal conductivity κ of liquid HCl vs. temperature T. The points marked MP, BP, and CP indicate the melting point, boiling point, and critical point, respectively [76A].

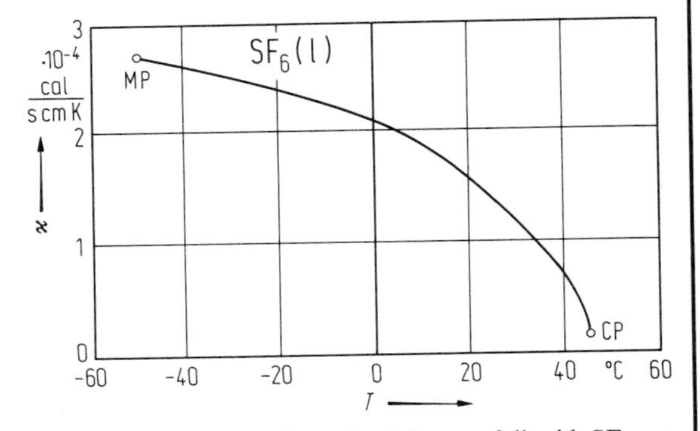

Fig. 36. SF_6. Thermal conductivity κ of liquid SF_6 vs. temperature T. The points marked MP and CP indicate the melting point and critical point, respectively [75A].

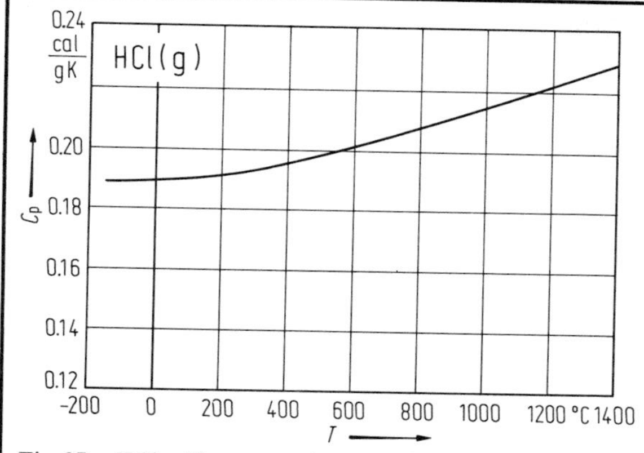

Fig. 37. HCl. Heat capacity C_p of vaporized HCl vs. temperature T at $p = 1.013$ bar [76A].

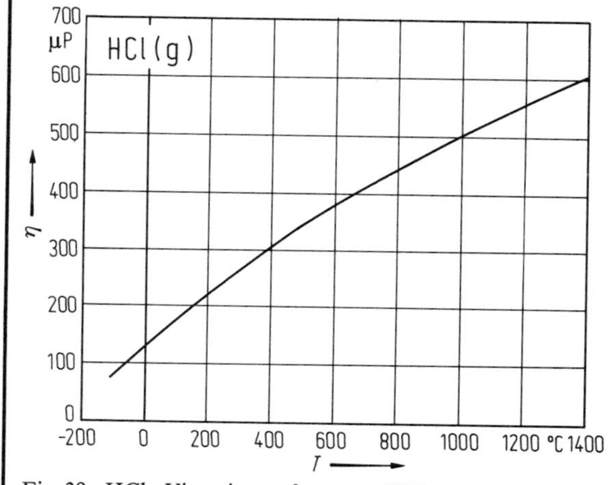

Fig. 39. HCl. Viscosity η of gaseous HCl vs. temperature T [76A].

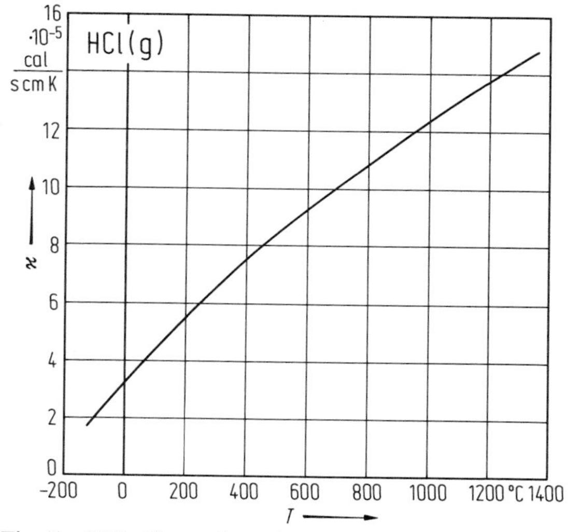

Fig. 41. HCl. Thermal conductivity κ of gaseous HCl vs. temperature T [76A].

Fig. 42. SF_6. Thermal conductivity κ of gaseous SF_6 vs. temperature T [75A].

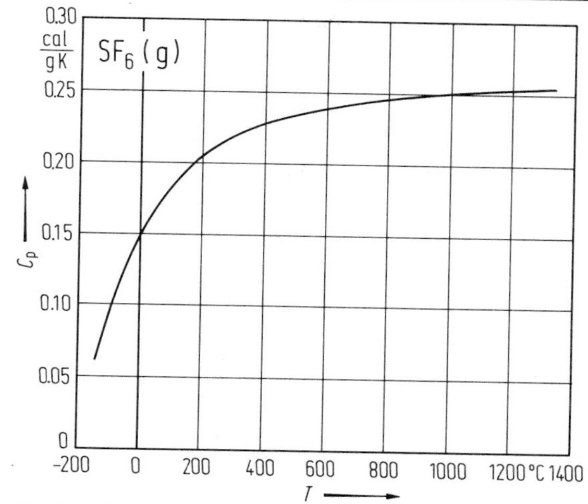

Fig. 38. SF_6. Heat capacity C_p of vaporized SF_6 vs. temperature T at $p = 1.013$ bar [75A].

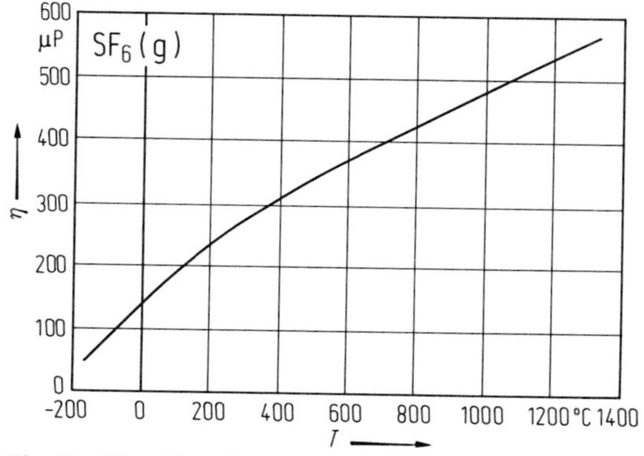

Fig. 40. SF_6. Viscosity η of gaseous SF_6 vs. temperature T [75A].

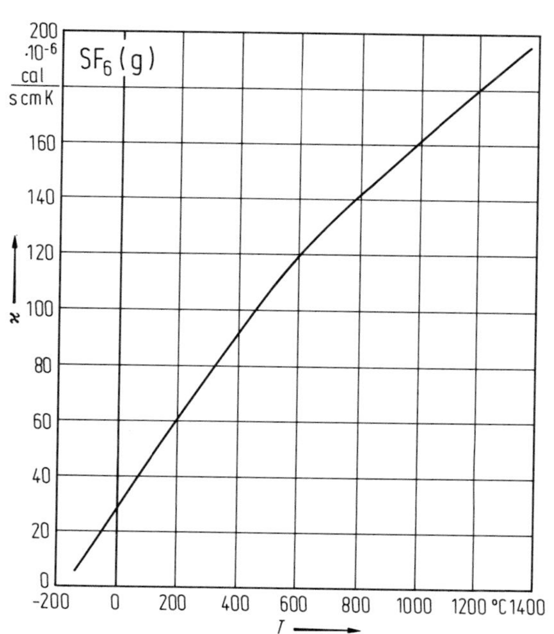

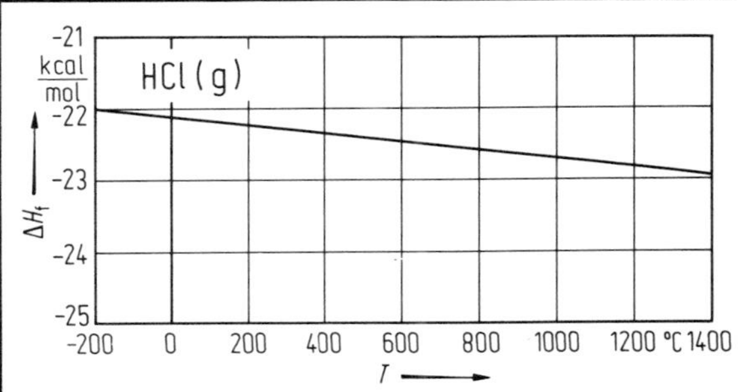

Fig. 43. HCl. Heat of formation ΔH_f of gaseous HCl vs. temperature T [76A].

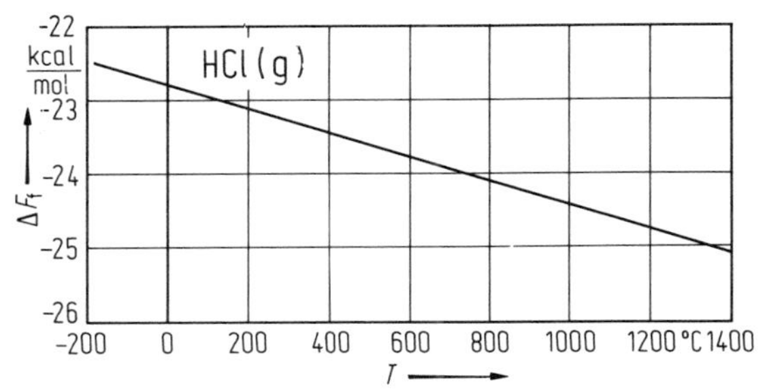

Fig. 45. HCl. Gibbs' free energy of formation ΔF_f of gaseous HCl vs. temperature T [76A].

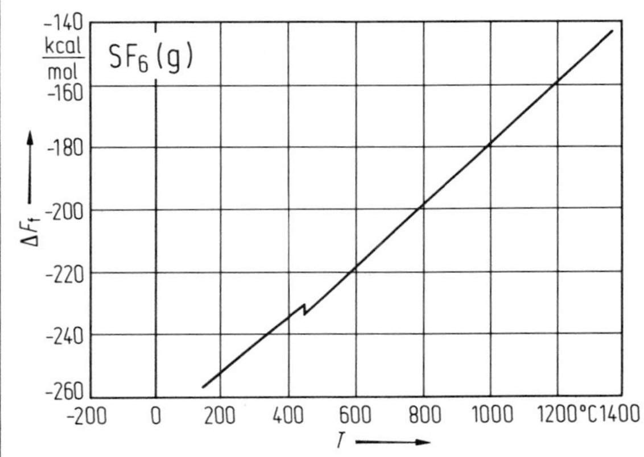

Fig. 46. SF_6. Gibbs' free energy of formation ΔF_f of gaseous SF_6 vs. temperature T [75A].

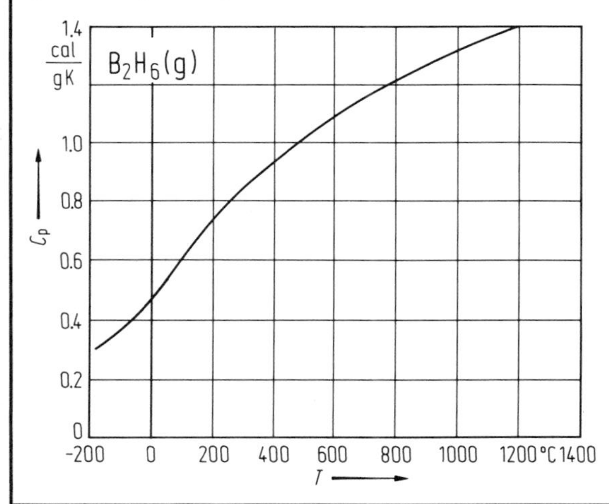

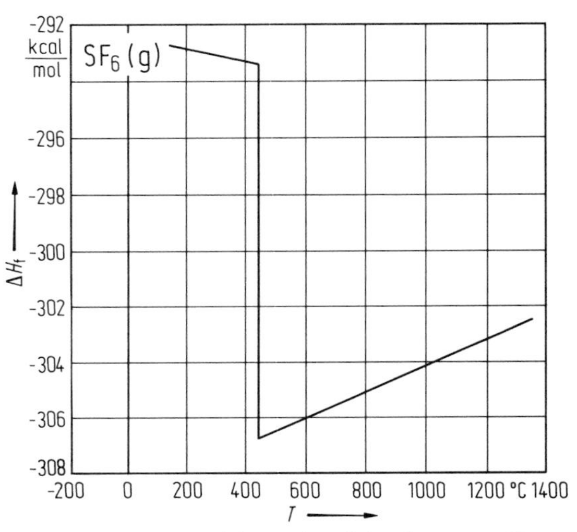

Fig. 44. SF_6. Heat of formation ΔH_f of gaseous SF_6 vs. temperature T [75A].

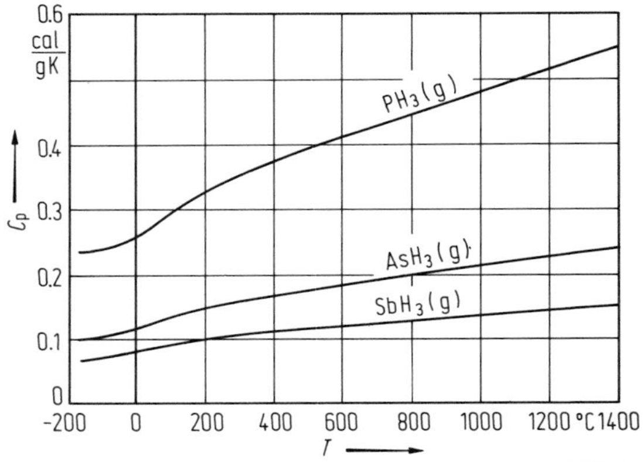

Fig. 47. Heat capacity C_p of gaseous PH_3, AsH_3, and SbH_3 vs. temperature T at $p = 1.013$ bar [74Y1].

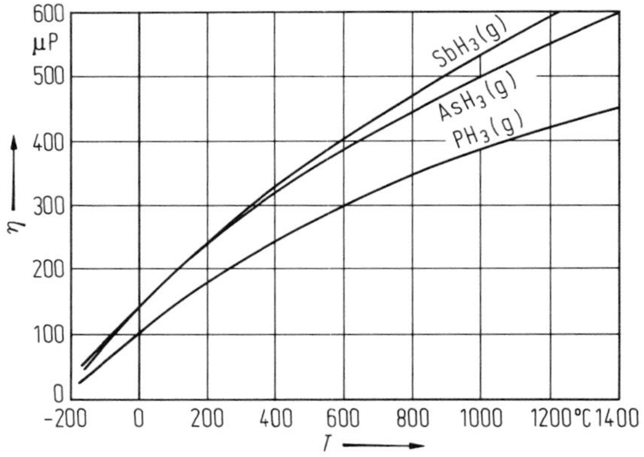

Fig. 49. Viscosity η of gaseous AsH_3, PH_3, and SbH_3 vs. temperature T [74Y1].

◀
Fig. 48. B_2H_6. Heat capacity C_p of gaseous B_2H_6 vs. temperature T at $p = 1.013$ bar [74Y2].

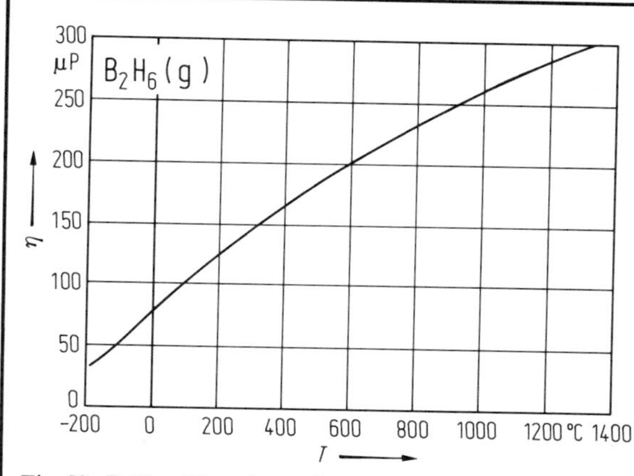

Fig. 50. B_2H_6. Viscosity η of gaseous B_2H_6 vs. temperature T [74Y1].

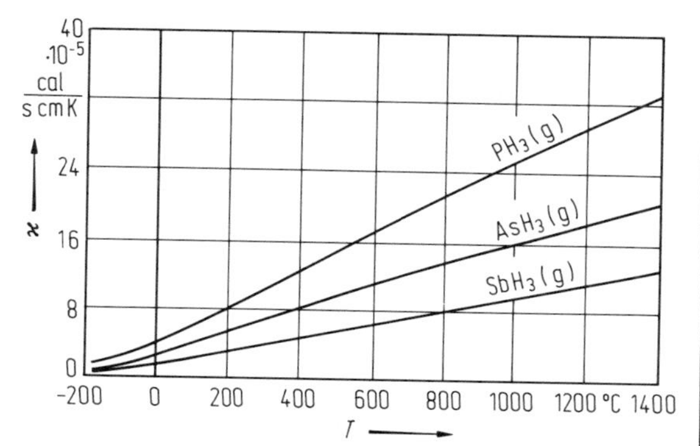

Fig. 51. Thermal conductivity κ of gaseous PH_3, AsH_3 and SbH_3 vs. temperature T [74Y1].

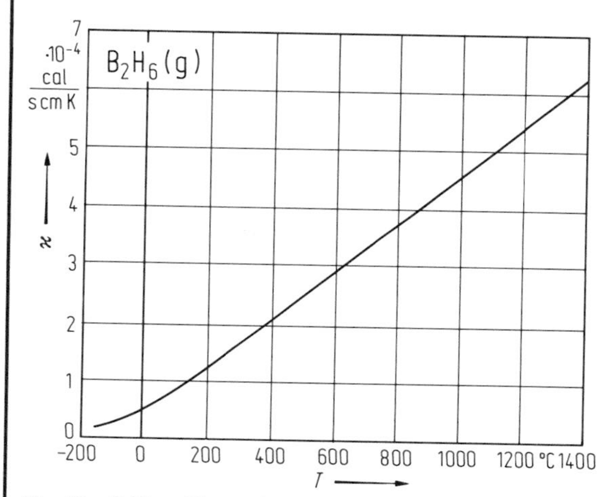

Fig. 52. B_2H_6. Thermal conductivity κ of gaseous B_2H_6 vs. temperature T [74Y2].

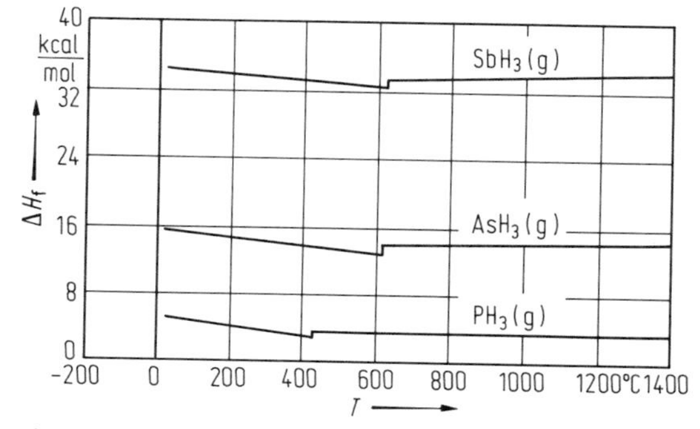

Fig. 53. Heat of formation ΔH_f of gaseous PH_3, AsH_3, and SbH_3 vs. temperature T [74Y1].

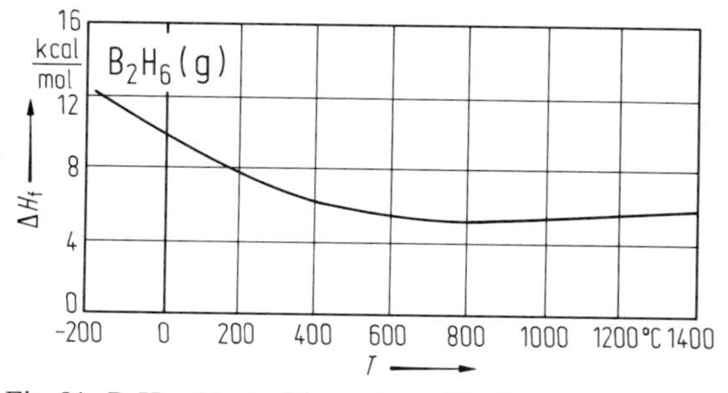

Fig. 54. B_2H_6. Heat of formation ΔH_f of gaseous B_2H_6 vs. temperature T [74Y2].

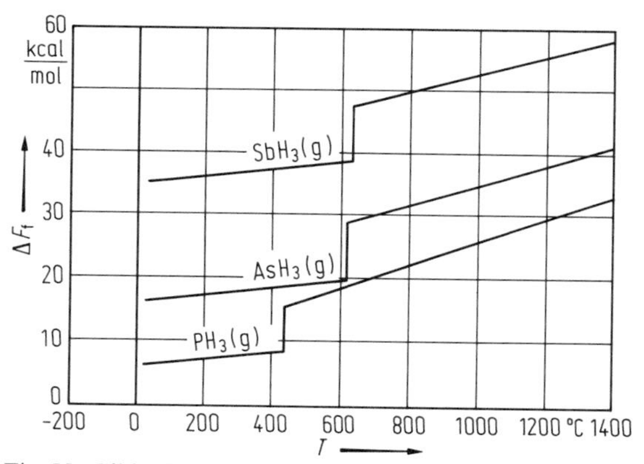

Fig. 55. Gibbs' free energy of formation ΔF_f of gaseous PH_3, AsH_3, and SbH_3 vs. temperature T [74Y1].

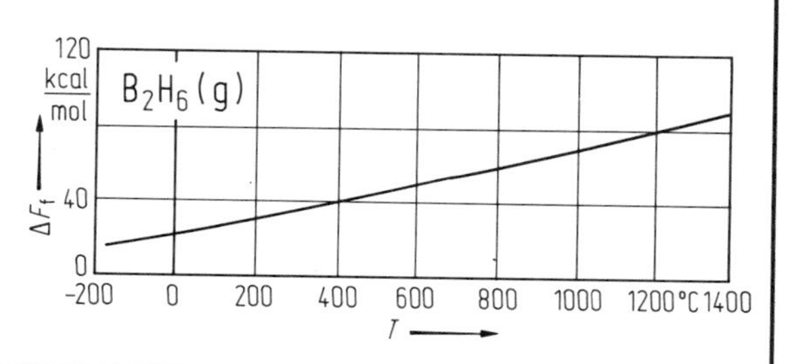

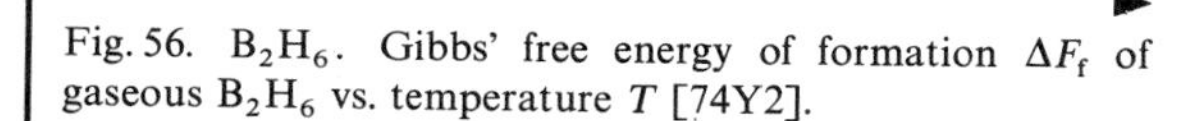

Fig. 56. B_2H_6. Gibbs' free energy of formation ΔF_f of gaseous B_2H_6 vs. temperature T [74Y2].

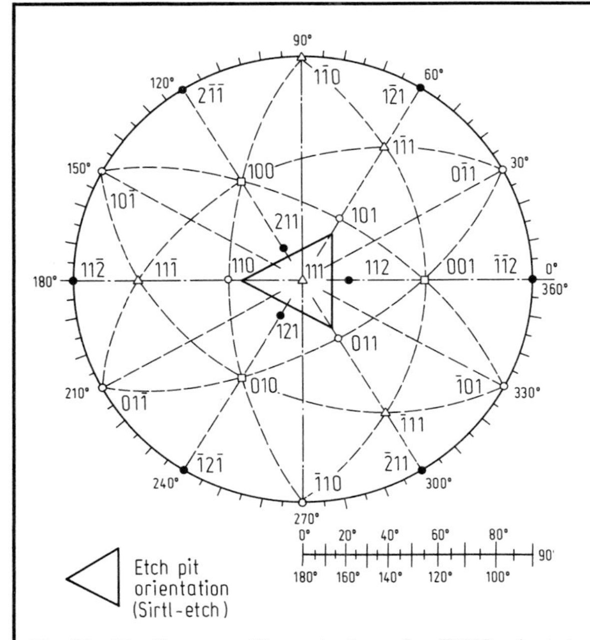

Fig. 57. Si. Stereographic projection of a [111]-oriented diamond cubic lattice with a polar net superimposed. The triangle marks the orientation of an etch pit obtained by the Sirtl etch.

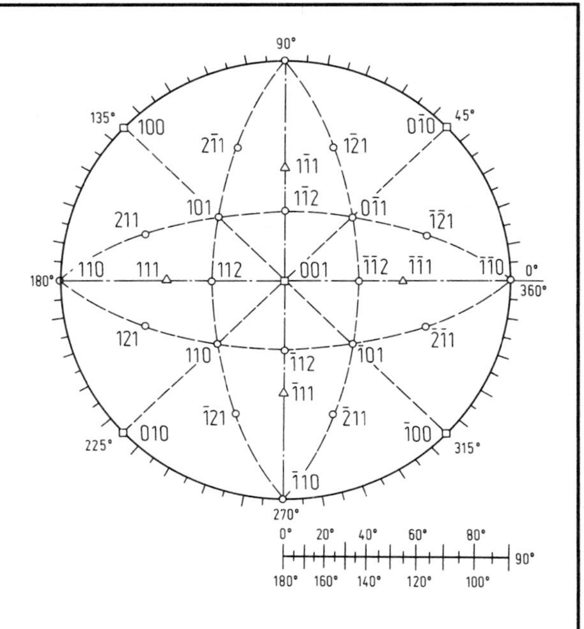

Fig. 58. Si. Stereographic projection of a [001]-oriented diamond cubic lattice with a polar net superimposed.

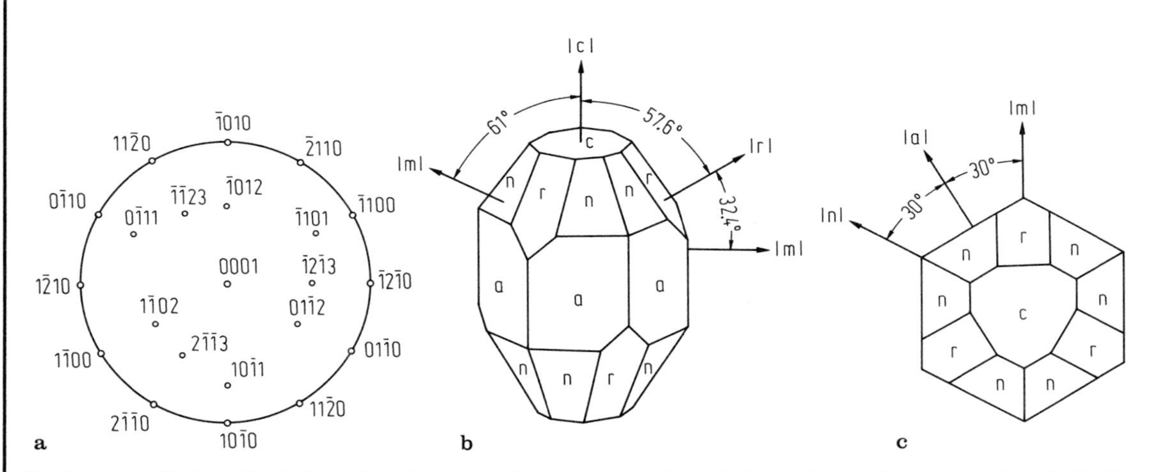

Fig. 59a–c. α-Al_2O_3. Crystallographic diagrams of sapphire:
a) stereographic projection of the [0001]-oriented lattice;
b) schematic drawing of a crystal facing the ($11\bar{2}0$) plane;
c) schematic drawing of a crystal facing the (0001) plane.
Mineralogical symbols are used to indicate the orientation of faces. The relation of the mineralogical symbols to structural indices are:

Structural indices	Mineralogical symbol
$1\bar{1}02$	r
$01\bar{1}2$	r
$\bar{1}012$	r
0001	c
$10\bar{1}0$	m
$11\bar{2}0$	a
$10\bar{1}1$	s
$2\bar{1}\bar{1}3$	n

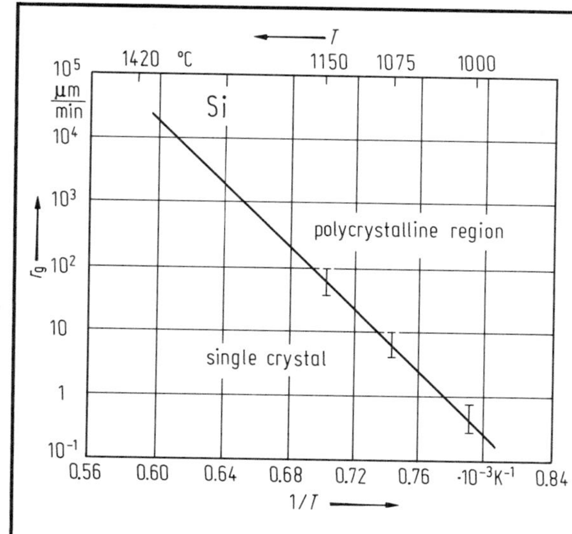

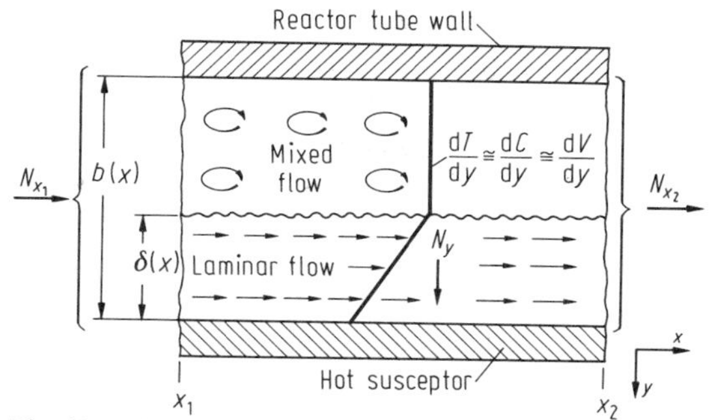

Fig. 61. Schematic drawing of the gas flow in a horizontal reactor [78B2]. (For further explanations see text.)

◀ Fig. 60. Si. Growth rate r_g of silicon on [111]-oriented silicon substrates vs. reciprocal temperature $1/T$. The line separates regions where polycrystalline and crystalline epitaxial layers are obtained [73B].

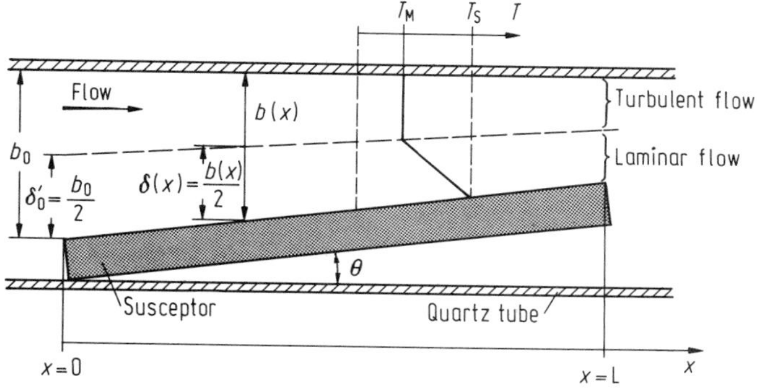

Fig. 62. Schematic drawing of a horizontal reactor with a tilted susceptor indicating the regions of laminar and turbulent flow [78B2]. (For further explanations see text.)

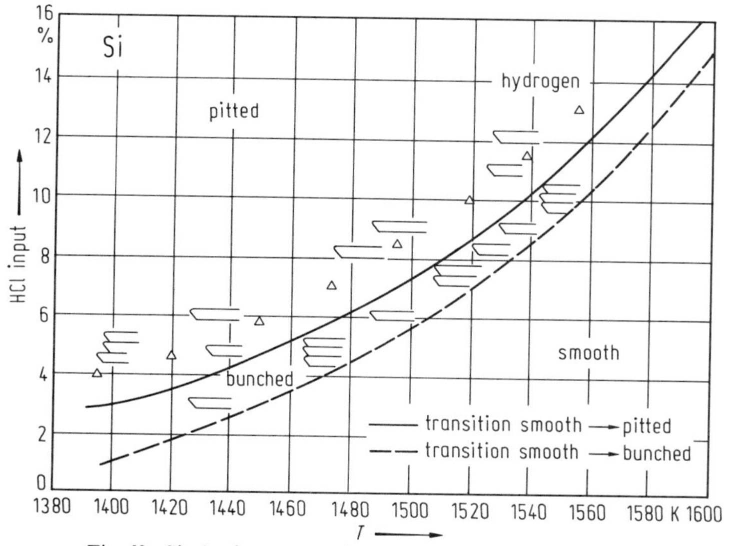

Fig. 63. Si. Surface properties influenced by the HCl input vs. temperature T. The solid line indicates the transition from a smooth to a pitted surface. The dashed line indicates the transition from a smooth to a bunched surface [78B3].

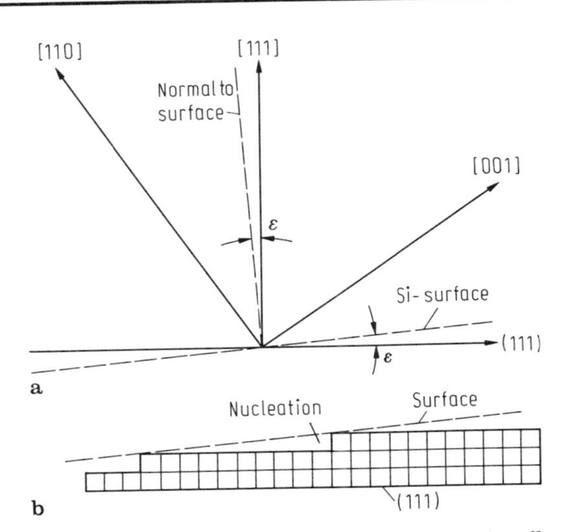

Fig. 64a, b. Si. Schematic drawing to illustrate the effect of a surface misorientation.
a) The vector normal to the surface of the slice is tilted out of the [111]-direction towards [110] by an angle ε.
b) Nucleation for epitaxial growth occurs at regular lattice steps in the surface which are caused by the misorientation.

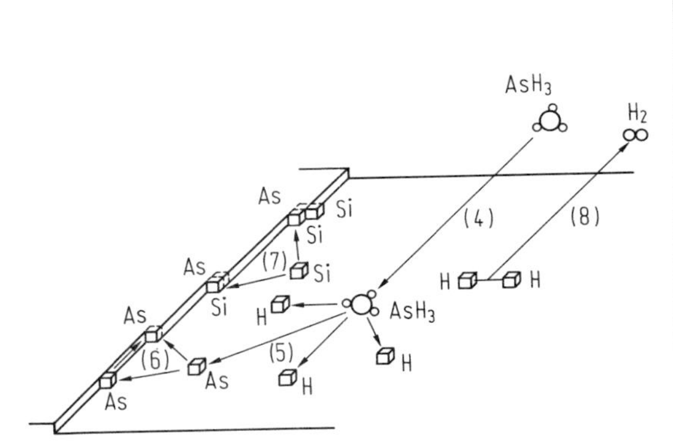

Fig. 65. Si. Schematic drawing of consecutive process steps (figures in parentheses) during doping of an epitaxial surface layer by deposition and decomposition of AsH_3 [79R1]. (For further explanations see text.)

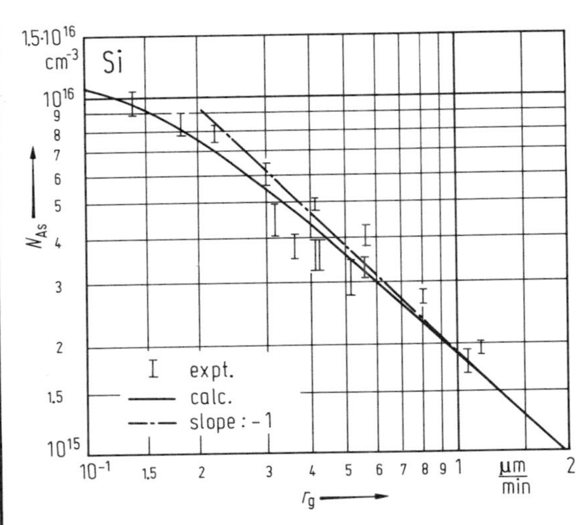

Fig. 66. Si. Arsenic dopant concentration N_{AS} vs. growth rate r_g of the uniformly doped epitaxial film [79R2].

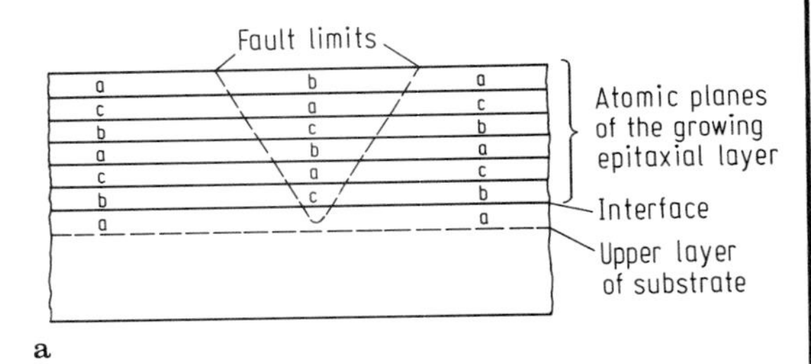

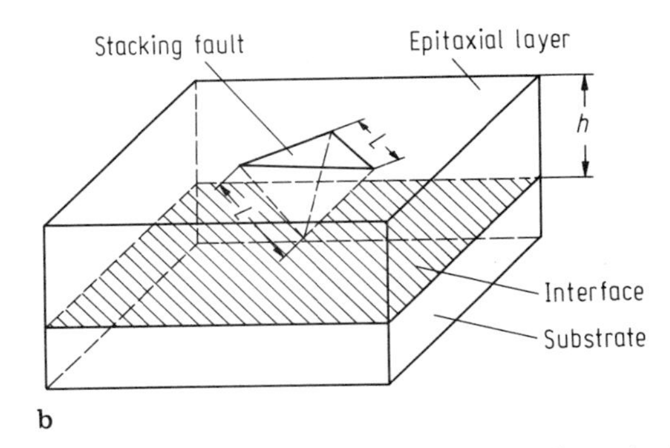

Fig. 67a, b. Si. Schematic drawing indicating the nucleation of a stacking fault in an epitaxial film.
a) Cross-sectional view of the layer stacking.
b) 3-dimensional view of the fault incorporation into the epitaxial film.

6.1.4.5 Fabrication of layers

(6.1.4.5.0···6.1.4.5.7 Fabrication of insulating films)

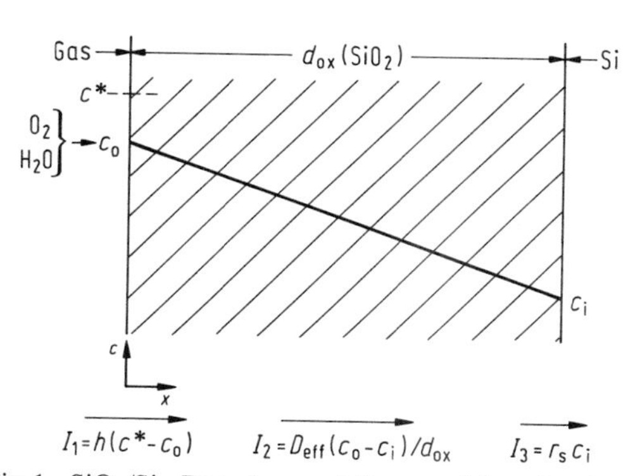

Fig. 1. SiO_2/Si. Boundary and flux conditions for the gas-SiO_2-Si system, schematically. For further explanations, see text.

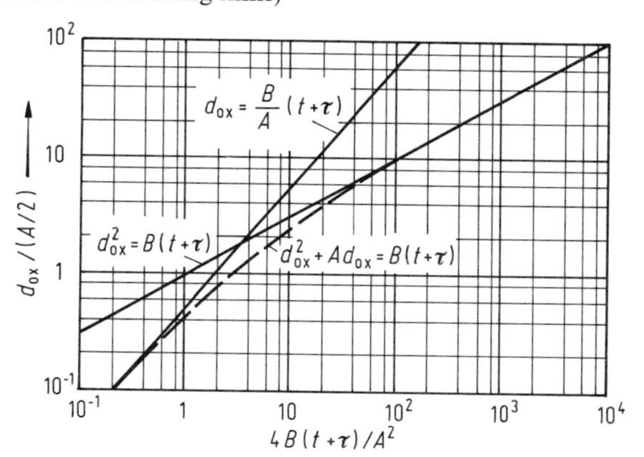

Fig. 2. SiO_2/Si. General relationship for thermal oxidation of silicon. The dashed curve is based on experimental data [65D3]. The normalized oxide thickness $d_{ox}/(A/2)$ is plotted vs. the normalized oxidation time $4B(t+\tau)/A^2$.

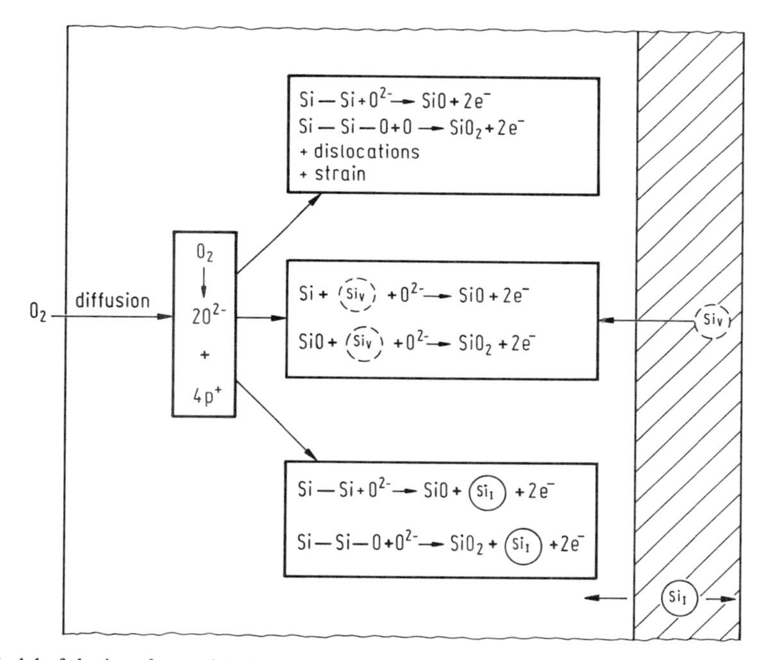

Fig. 3. SiO_2/Si. Model of the interface oxidation reaction. Possible sources of "free volume" supply [80T1, 80T2].

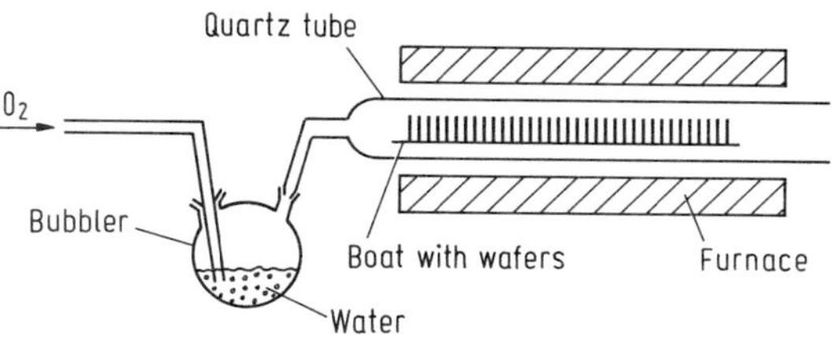

Fig. 4. SiO_2/Si. Apparatus for wet thermal oxidation (schematic).

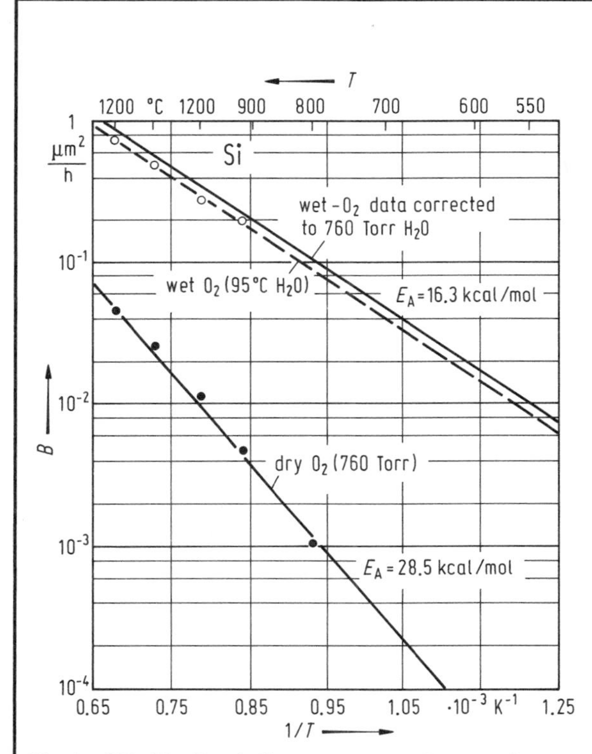

Fig. 5. SiO_2/Si. Parabolic rate constant B vs. reciprocal temperature $1/T$ [65D3].

Fig. 6. SiO_2/Si. Linear rate constant B/A vs. reciprocal temperature $1/T$ [65D3].

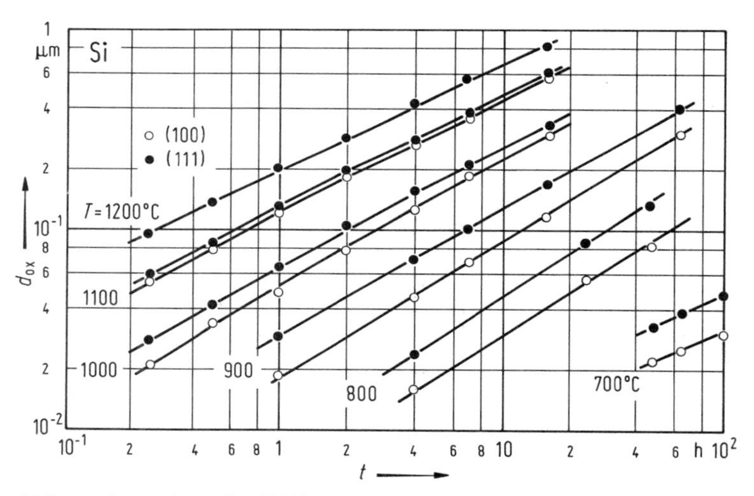

Fig. 7. SiO_2/Si. Oxide thickness d_{ox} vs. time t for ⟨111⟩ and ⟨100⟩ silicon [77D5].

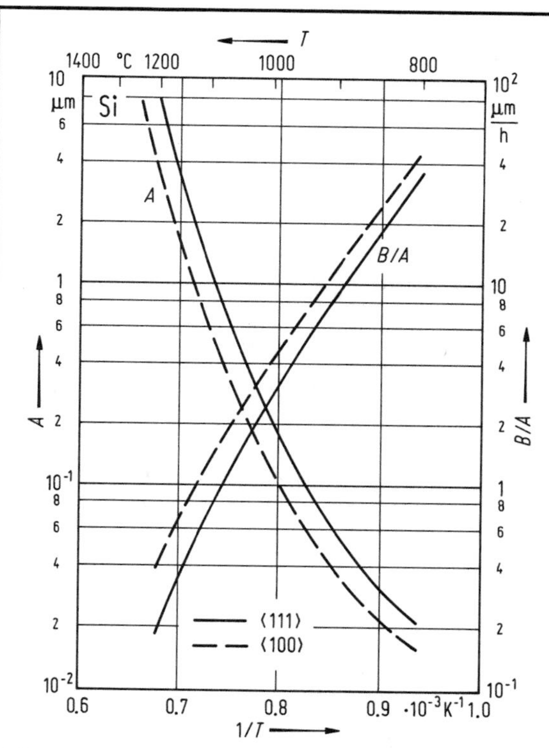

Fig. 8. SiO_2/Si. Oxidation rates A and B/A vs. reciprocal temperature $1/T$ for ⟨111⟩ and ⟨100⟩ crystal orientations [69w1].

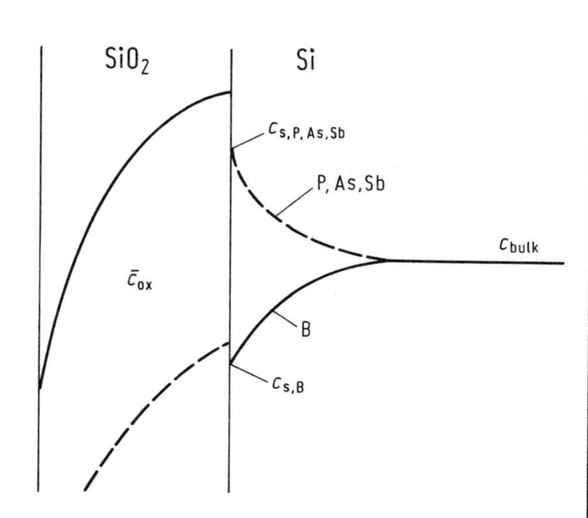

Fig. 9. SiO_2/Si. Redistribution of impurity doping profiles in the SiO_2/Si interface region [77D5]. (For further explanation see text).

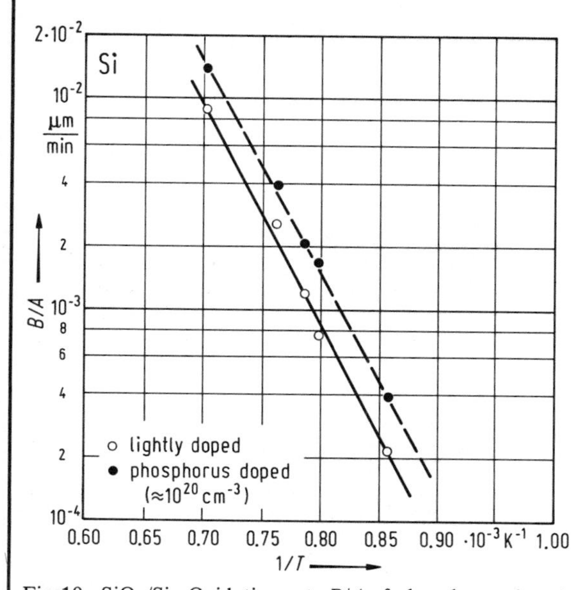

Fig. 10. SiO_2/Si. Oxidation rate B/A of phosphorus doped and undoped silicon vs. reciprocal temperature $1/T$ [78I2].

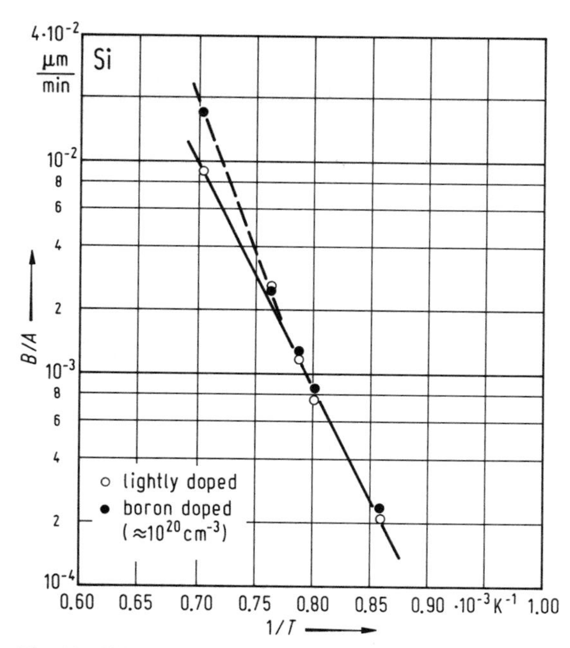

Fig. 11. SiO_2/Si. Oxidation rate B/A of boron doped and undoped silicon vs. reciprocal temperature $1/T$ [78I2].

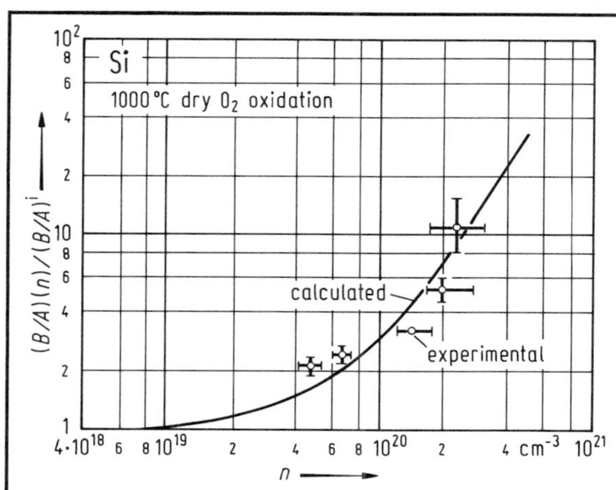

Fig. 12. SiO_2/Si. Normalized oxidation rate of doped substrates B/A (n) vs. doping concentration n (dry O_2 oxidation) [79H3]. $(B/A)^i$: = lightly doped substrate.

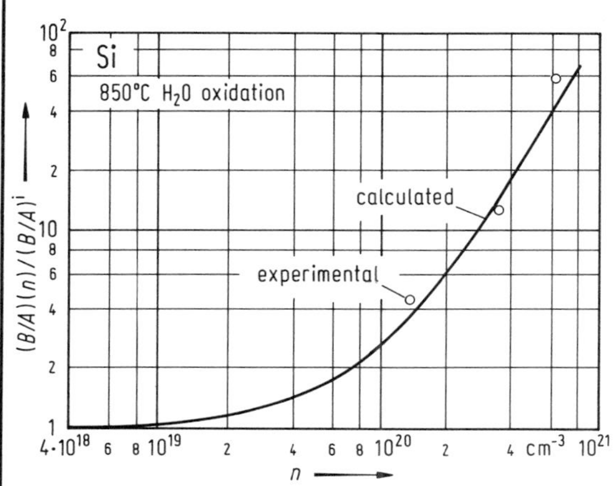

Fig. 13. SiO_2/Si. Normalized oxidation rate of doped substrates (B/A) (n) vs. doping concentration n (H_2O oxidation) [79H3]. $(B/A)^i$: = lightly doped substrate.

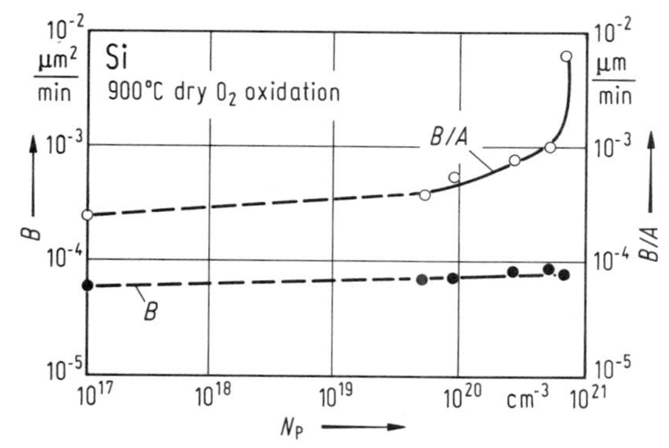

Fig. 14. SiO_2/Si. Linear and parabolic rate constants B and B/A vs. phosphorus doping concentration N_P (dry O_2-oxidation) [77D5].

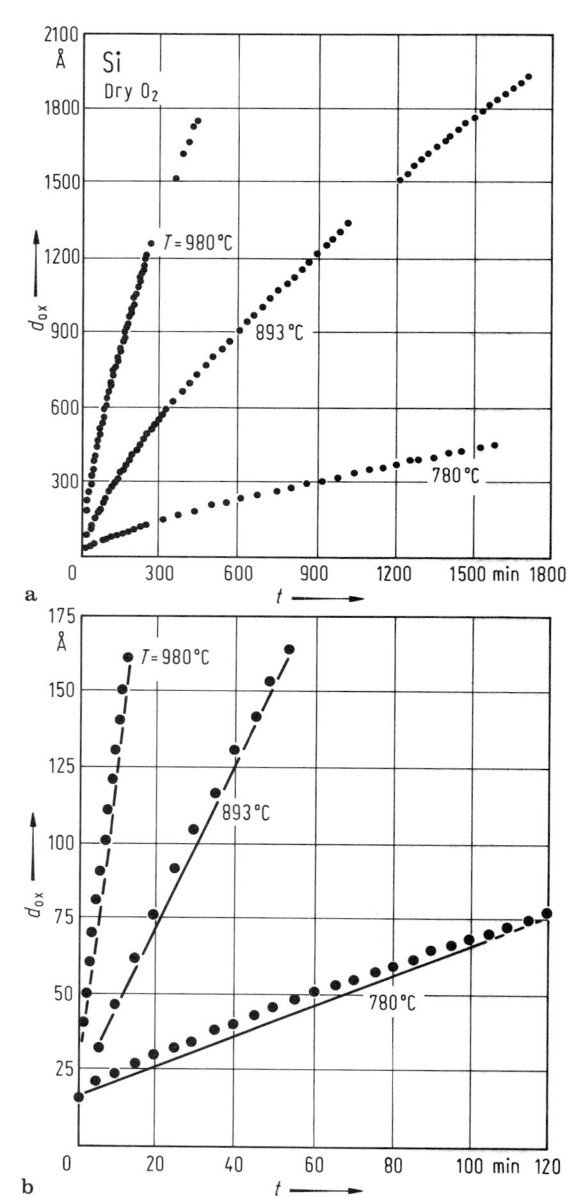

Fig. 15a, b. SiO_2/Si. SiO_2 film thickness d_{ox} vs. oxidation time t for 780 °C, 893 °C, and 980 °C oxidation temperatures of (100) Si in O_2, a) thick oxides, b) initial regime [76I1; 77I1].

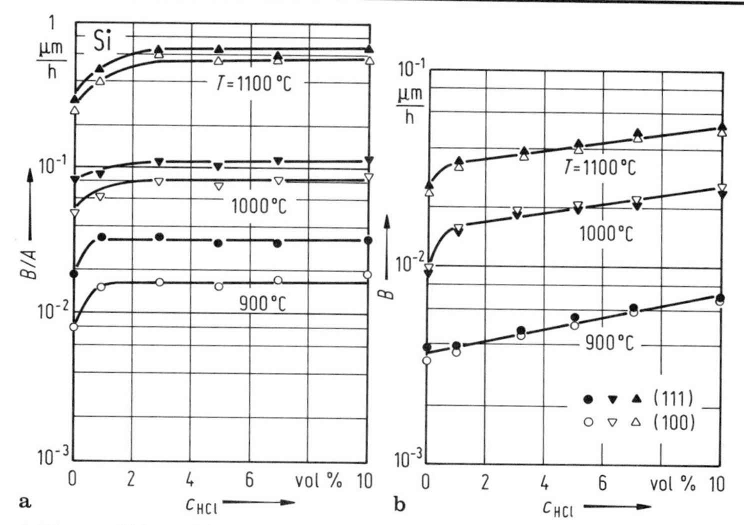

Fig. 16a, b. SiO_2/Si. a) Linear, B/A, and b) parabolic, B, rate constant vs. concentration c_{HCl} for O_2/HCl oxidations of (111) and (100) silicon at various temperatures [77D5].

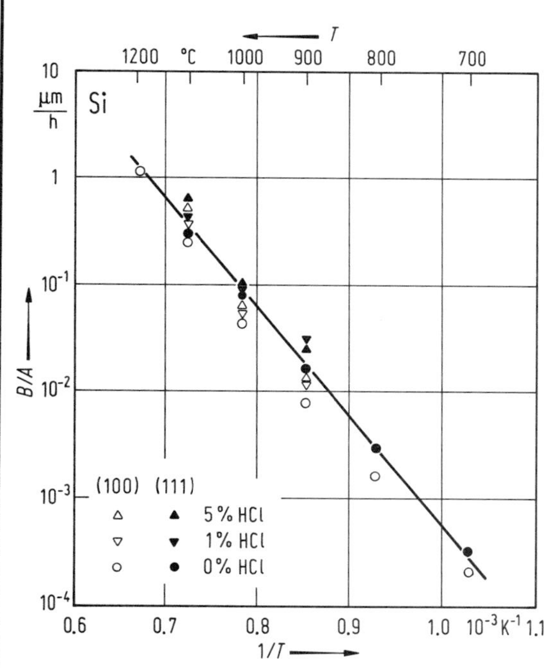

Fig. 17. SiO_2/Si. Arrhenius plot of the linear rate constant B/A vs. reciprocal temperature $1/T$ for O_2/HCl oxidations. The silicon orientation is (111) or (100) [77D5].

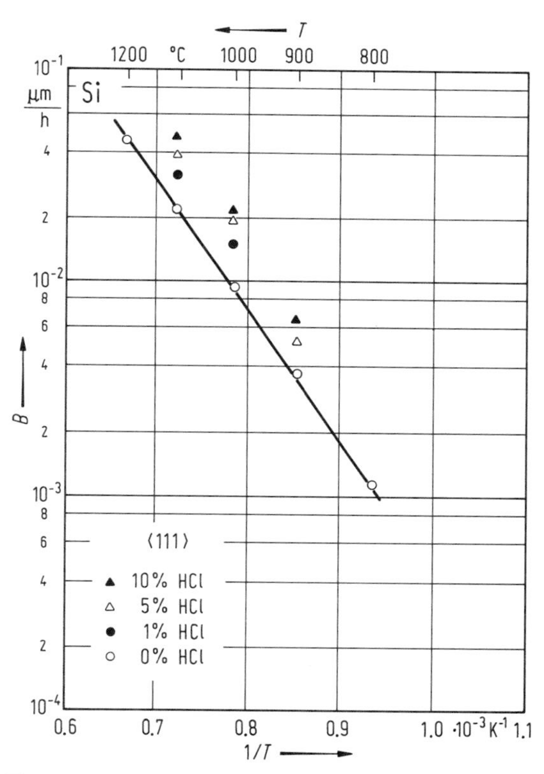

Fig. 18. SiO_2/Si. Arrhenius plot of the parabolic rate constant B vs. reciprocal temperature $1/T$ for O_2/HCl oxidations. The silicon orientation is ⟨111⟩ [77D5].

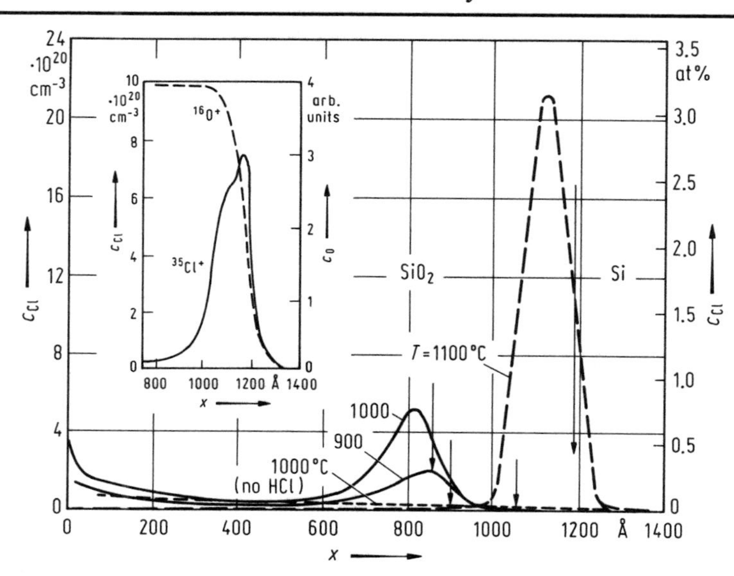

Fig. 19. SiO_2/Si. Chlorine concentration c_{Cl} vs. distance x into SiO_2 in thermally oxidized silicon. The oxides were prepared in a 5% HCl/O_2 mixture. Si—SiO_2 interfaces are indicated by arrows. The insert shows a profile of oxides prepared at 1100 °C in 3% HCl/O_2 ambient [78D4].

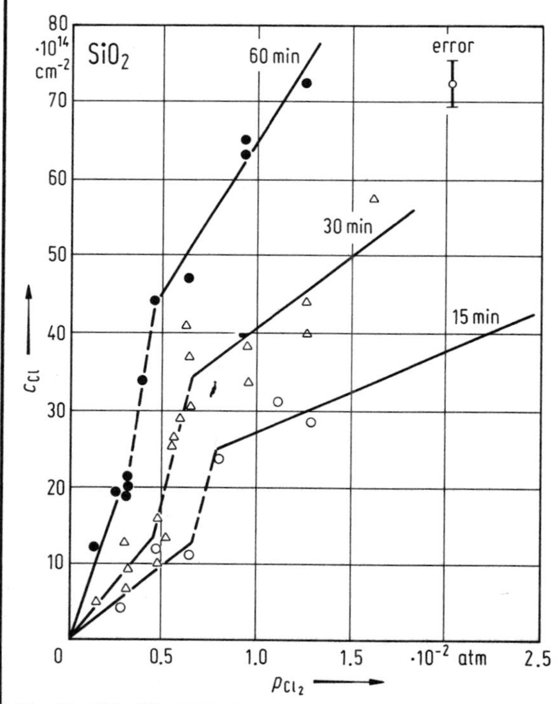

Fig. 20. SiO_2/Si. Chlorine concentration c_{Cl} of the oxide vs. p_{Cl_2} partial pressure for various oxidation times at 1200 °C [80M3].

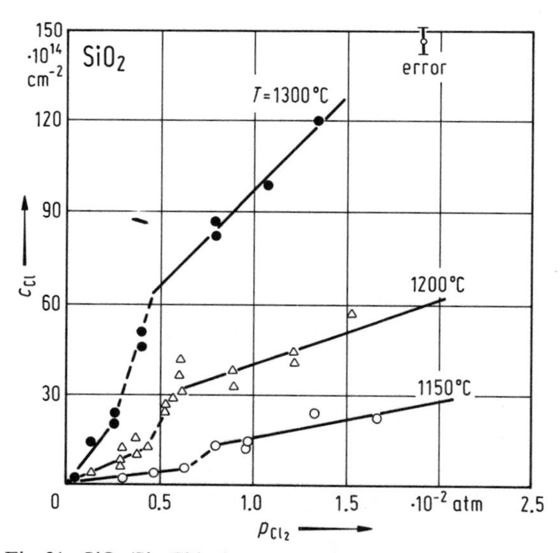

Fig. 21. SiO_2/Si. Chlorine concentration c_{Cl} in the oxide vs. p_{Cl_2} partial pressure for various oxidation temperatures and an oxidation time of 30 min [80M3].

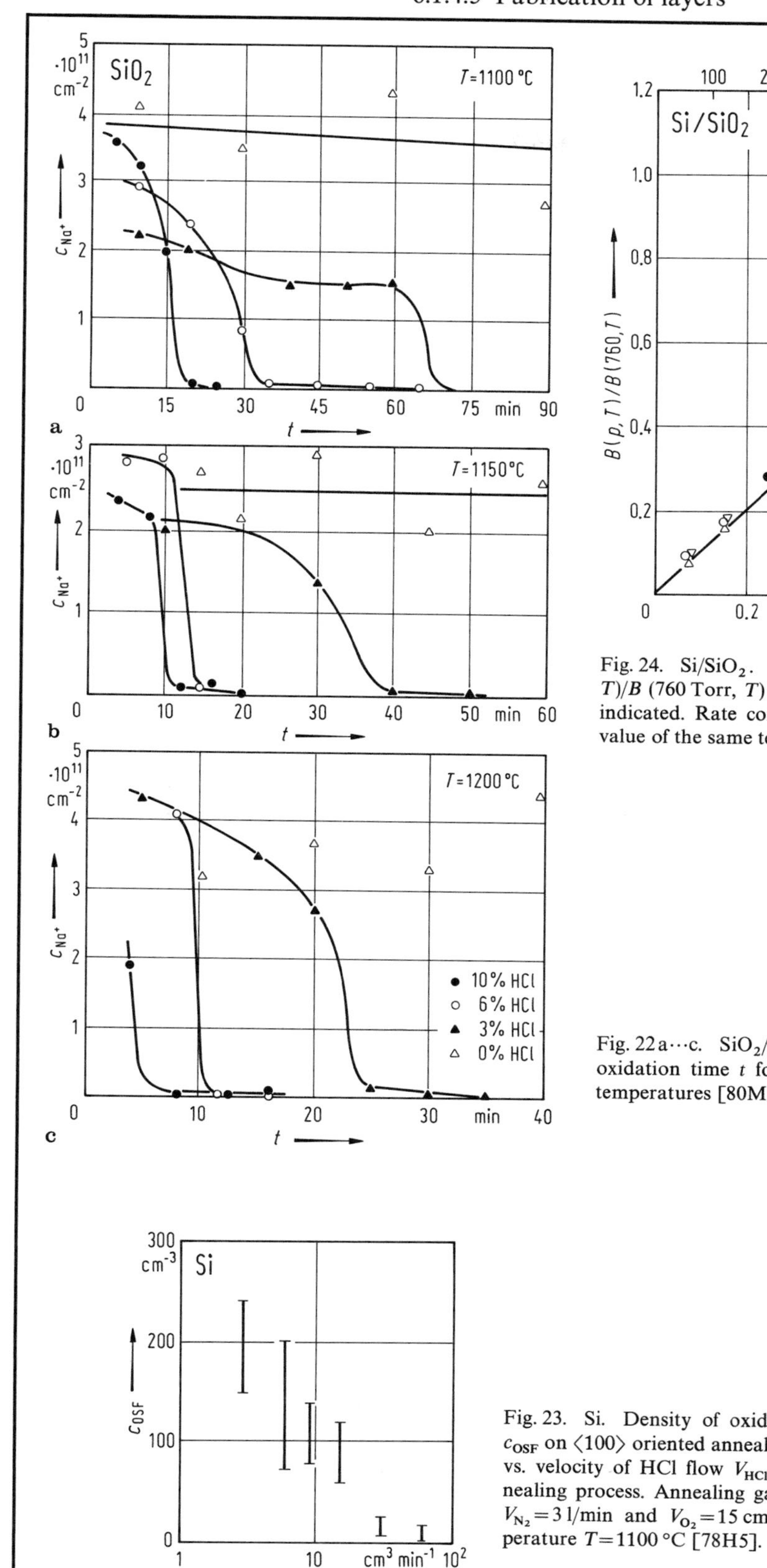

Fig. 24. Si/SiO_2. Parabolic oxidation rate constant $B(p, T)/B$ (760 Torr, T) vs. the partial pressure p of the oxidant indicated. Rate constants are normalized to the 760-Torr value of the same temperature [65D3].

Fig. 22a···c. SiO_2/Si. Mobile Na^+ concentration c_{Na} vs. oxidation time t for various HCl contents and oxidation temperatures [80M3].

Fig. 23. Si. Density of oxidation-induced stacking faults c_{OSF} on ⟨100⟩ oriented annealed and oxidized silicon wafers vs. velocity of HCl flow V_{HCl} during the preoxidation annealing process. Annealing gas mixture: $N_2/HCl/O_2$, with $V_{N_2}=3$ l/min and $V_{O_2}=15$ cm³/min; time $t=10$ min; temperature $T=1100$ °C [78H5].

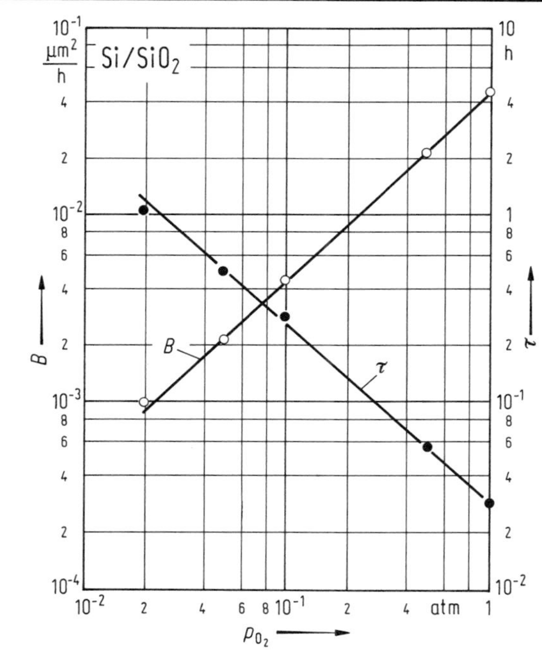

Fig. 25. Si/SiO_2. Parabolic oxidation rate constant B and the factor τ of the general oxidation equation vs. the oxygen partial pressure p_{O_2} for dry oxidation at 1200 °C. $\left(\tau = \frac{d_i^2 + A\,d_i}{B};\ d_i = \text{initial oxide thickness}\right)$ (see Fig. 2) [75H3].

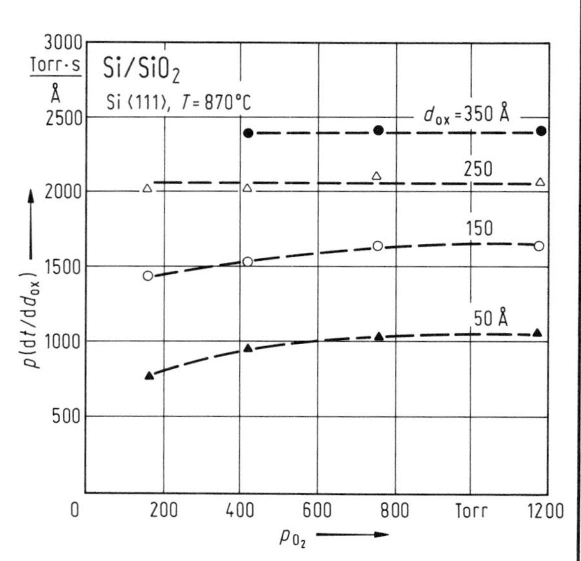

Fig. 26. Si/SiO_2. The inverse of the growth rate times pressure $p/(d\,d_{ox}/d\,t)$ vs. oxygen partial pressure p_{O_2} for ⟨111⟩ silicon at 870 °C. Parameter is the oxide thickness d_{ox} [75H2].

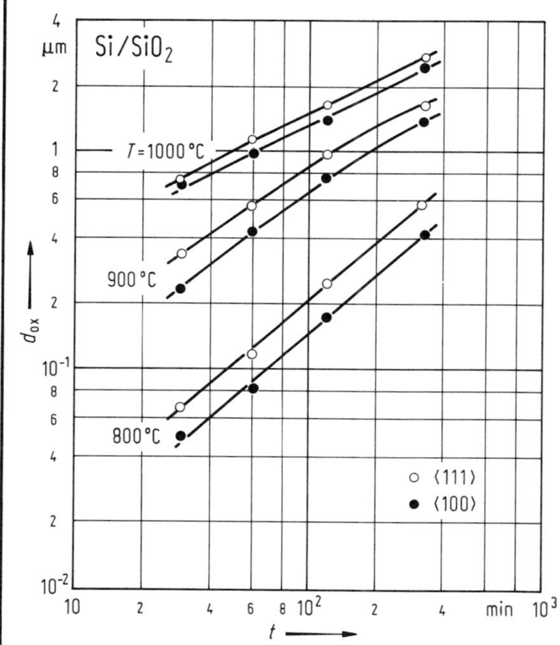

Fig. 27. Si/SiO_2. Oxide thickness d_{ox} vs. oxidation time t at high pressure (6.5 bar) steam oxidation ambient for p-type silicon at various temperatures T, and for two surface orientations: ⟨111⟩ and ⟨100⟩.

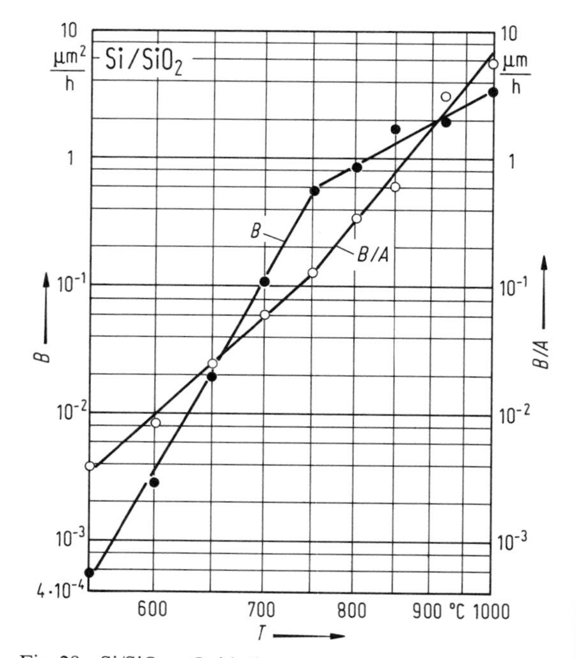

Fig. 28. Si/SiO_2. Oxidation rate constants B and B/A of steam oxidation vs. temperatures T at a pressure of $p = 9.81$ bar [81B1].

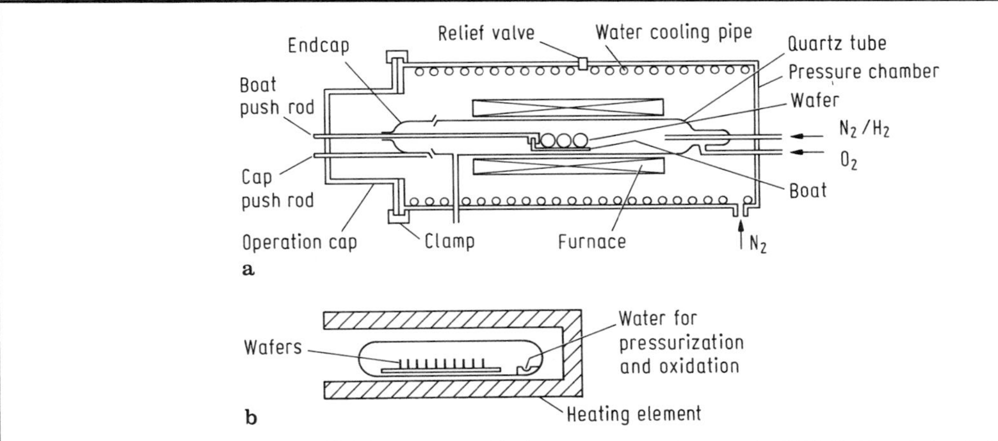

Fig. 29 a, b. Schematic diagrams of high-pressure oxidation apparatus: a) open tube [77T3], b) sealed tube [81O1].

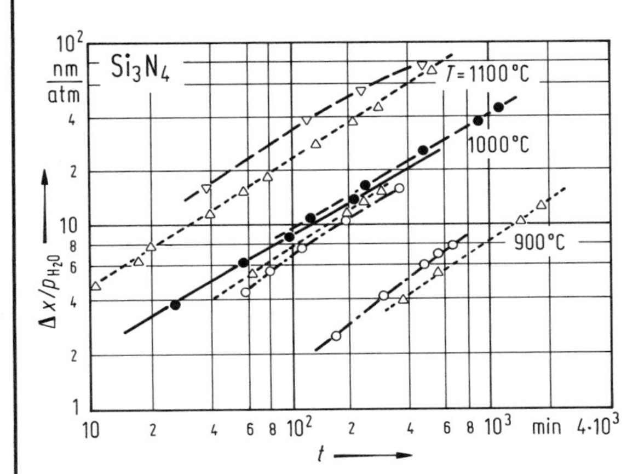

Fig. 30. Si_3N_4. Nitride consumption (converted layer thickness) normalized to oxidant pressure $\left(\frac{\Delta x}{p_{H_2O}}\right)$ vs. oxidation time t [81C2]. The data are collected from various references.

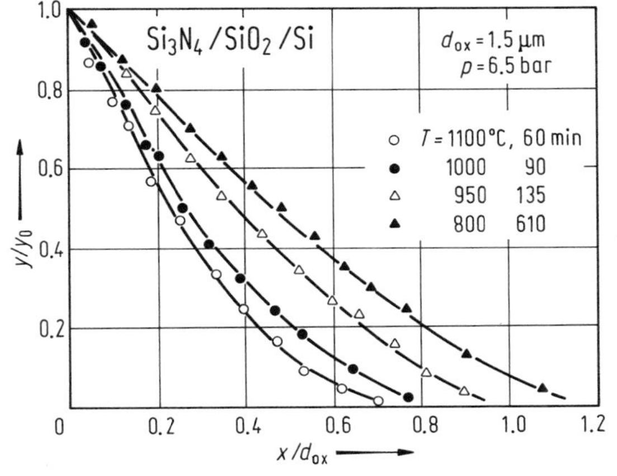

Fig. 31. $Si_3N_4/SiO_2/Si$. Shape of the lateral oxidation. Normalized vertical oxide thickness y/y_0 vs. the normalized lateral oxide thickness x/d_{ox}. Parameter is the oxidation temperature. (Denotation see Fig. 32) [78M3].

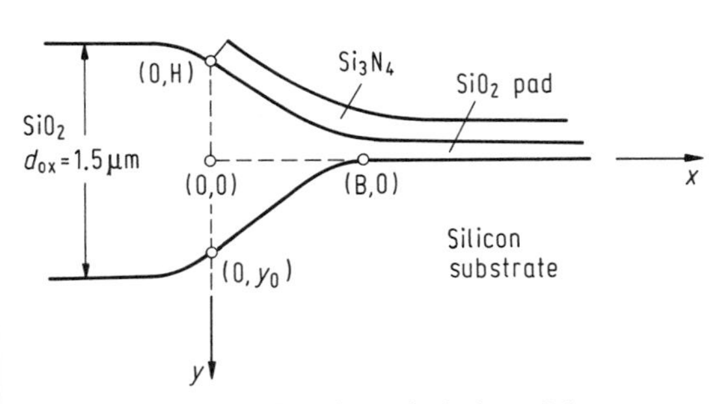

Fig. 32. $Si_3N_4/SiO_2/Si$. Schematical view of the cross section of a selective oxidation structure. The denotation is used in Fig. 31 [78M3].

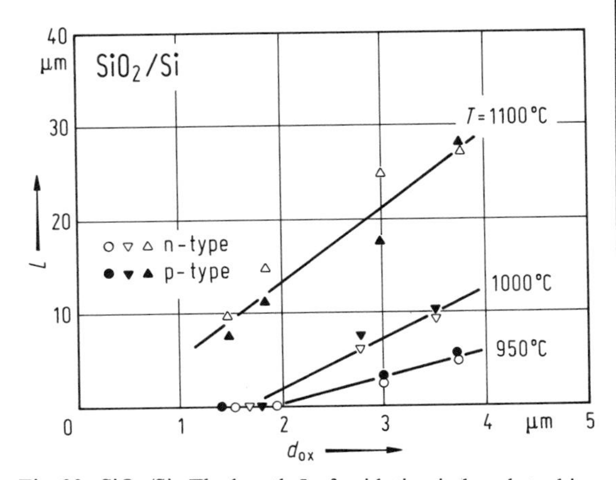

Fig. 33. SiO_2/Si. The length L of oxidation induced stacking faults for n-and p-type silicon vs. oxide thickness d_{ox} after high pressure ($p = 6.5$ bar) steam oxidation [77T1].

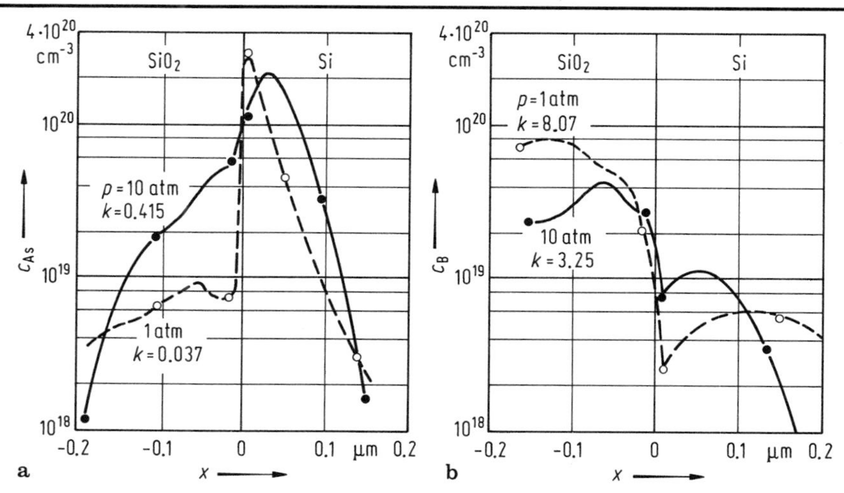

Fig. 34 a, b. SiO_2/Si. Influence of oxidant pressure p on the dopant segregation near the silicon/silicon oxide interface (the mean segregation coefficient k is indicated) [81C2]. a) Arsenic concentration c_{As} vs. position x; b) boron concentration c_B vs. position x.

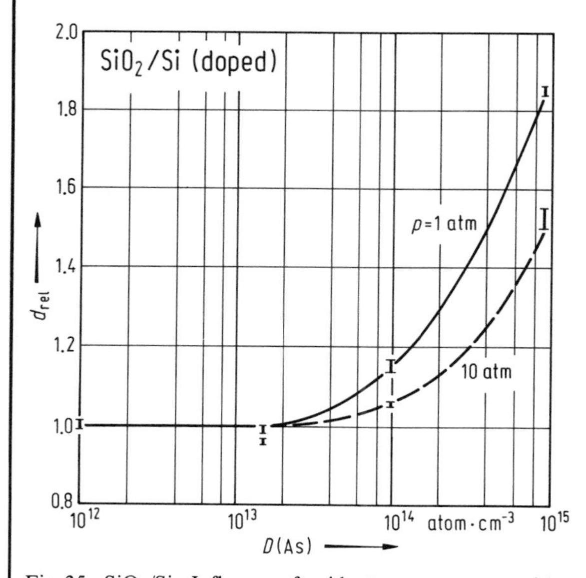

Fig. 35. SiO_2/Si. Influence of oxidant pressure p on oxidation depths of highly doped silicon. Normalized oxide-thickness d_{rel} on silicon vs. implantation dose D(As) of arsenic. Parameter is the oxidation pressure [80F1].

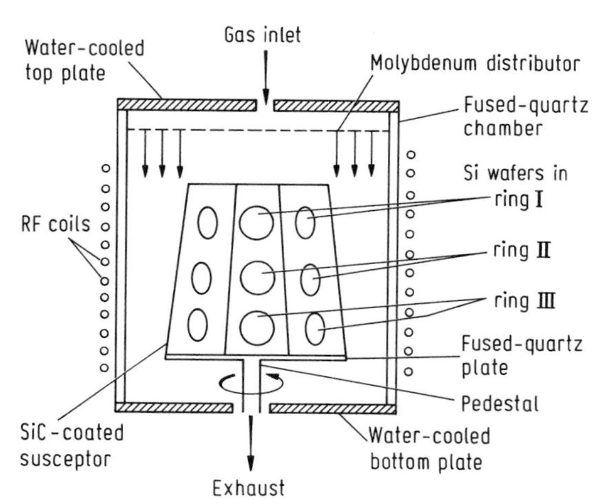

Fig. 38. Schematics of an RF-heated high-temperature CVD reactor [77G3].

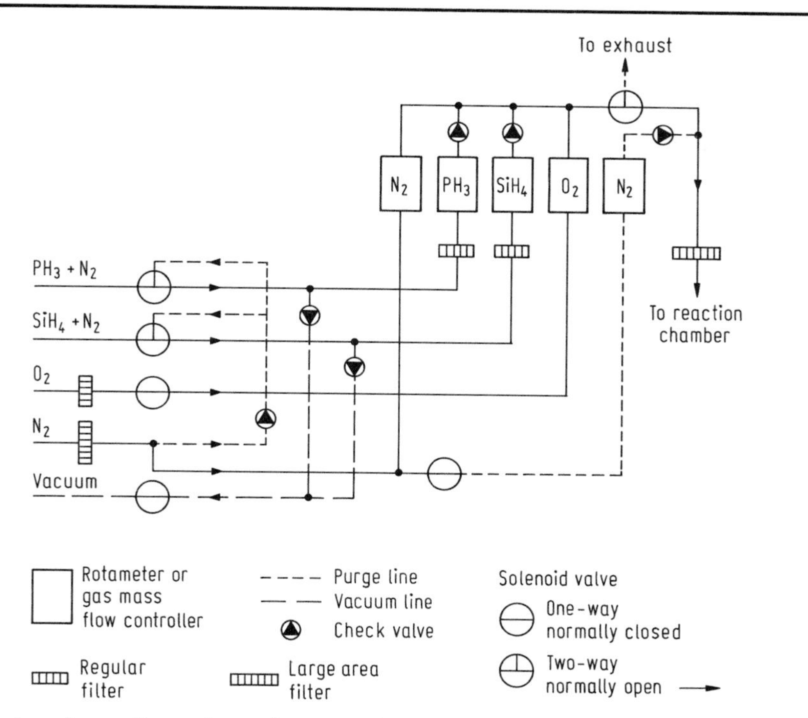

Fig. 36. Schematic view of a gas flow and metering system for an APCVD apparatus (shown gases are used for SiO_2 and PSG film deposition) [75K4].

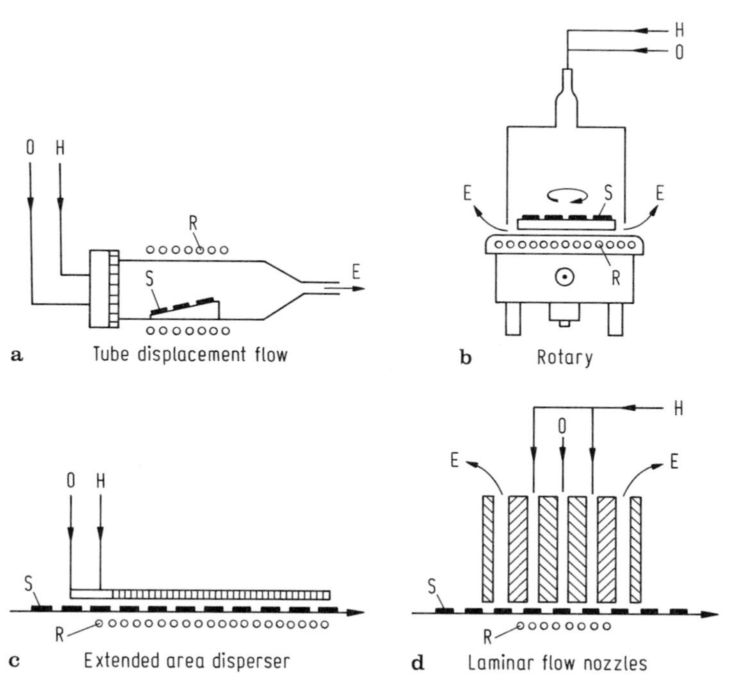

Fig. 37a⋯d. Schematics of basic low-temperature APCVD reactors. a) Horizontal tube displacement flow reactor, b) rotary vertical reactor, c) continuous reactor with gas disperser plate, d) continuous reactor with laminar flow nozzles. O, H: gas-inlets, E: exhaust gases, R: resistance heater, S: substrates; the arrow indicates the direction of substrate travel [75K4].

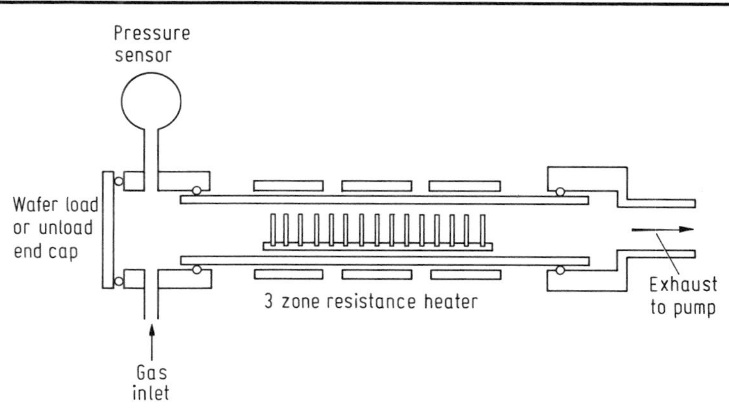

Fig. 39. Schematics of a low-pressure CVD reactor [79K1].

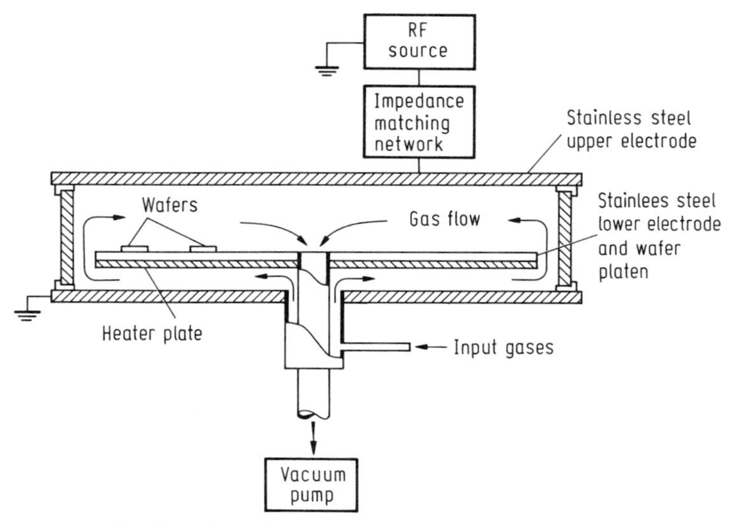

Fig. 40. Schematics of a planar radial-flow reactor for plasma-enhanced CVD [80E1].

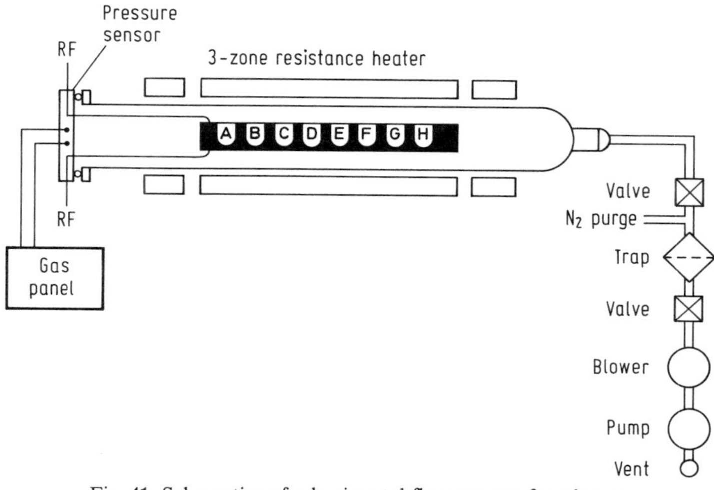

Fig. 41. Schematics of a horizontal-flow reactor for plasma-enhanced CVD [80R1]. (A⋯H chambers for substrates).

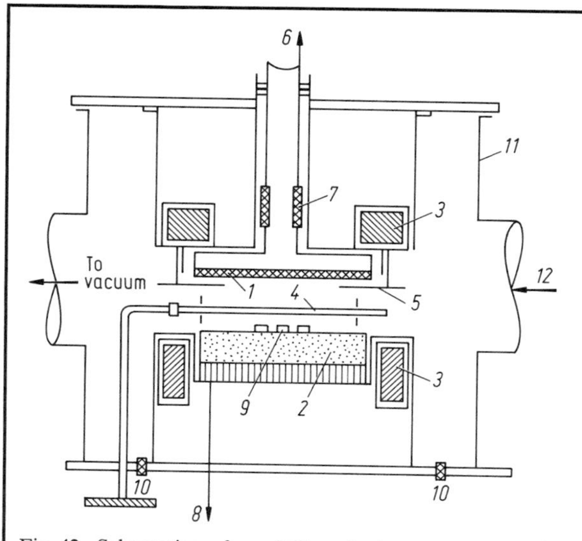

Fig. 42. Schematics of an RF-sputtering apparatus. The components are listed as follows: (1) cathode target: (2) anode substrate holder, (3) cathode and anode magnets (water-cooled); (4) shutter; (5) cathode shield; (6) RF power supply, (7) cathode isolation insulator; (8) substrate bias supply, (9) substrates, (10) anode isolation insulators, (11) vacuum chamber, (12) sputter gas [72V3].

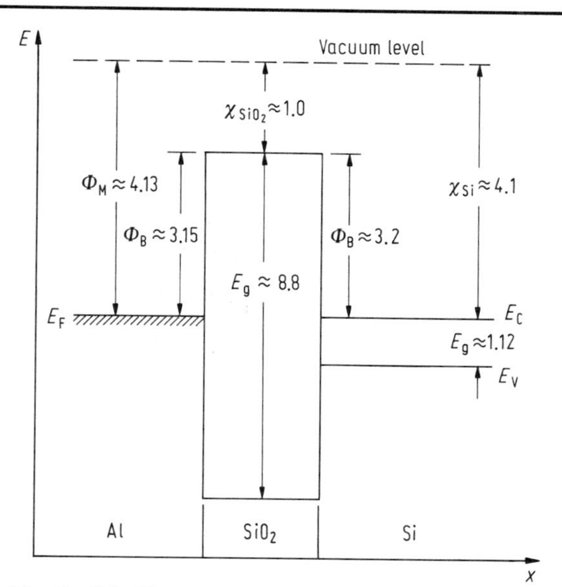

Fig. 43. SiO_2/Si. Schematic to illustrate the electron energy band diagram of the Al/SiO_2/Si system. (Energies are given in [eV]).

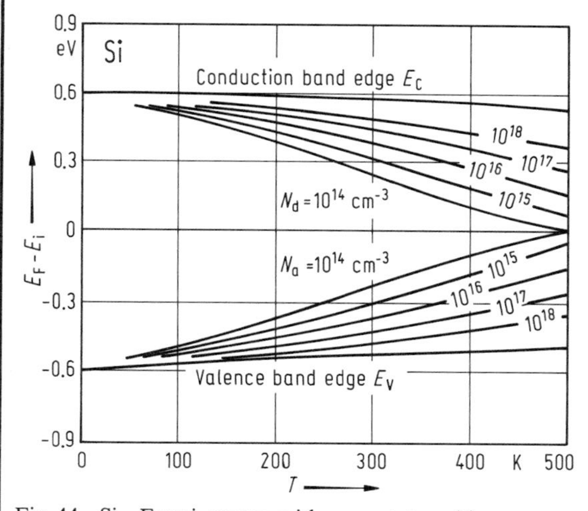

Fig. 44. Si. Fermi energy with respect to midgap energy $E_F - E_i$ vs. temperature T for silicon. Parameters are the doping concentrations N_a and N_d.

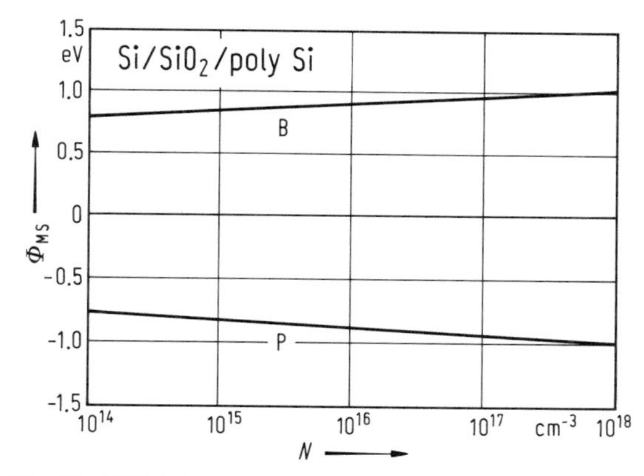

Fig. 45. Si/SiO_2/poly Si. Work function difference Φ_{MS} between degenerately doped poly-Si gate and silicon vs. doping density. B: the poly-Si is doped with 10^{20} cm^{-3} boron, and the Si-substrate is n-type. P: the poly-Si is doped with 10^{20} cm^{-3} phosphorus and the Si-substrate is p-type [82n, 74W].

For the figures of subsections 6.1.4.5.8···6.1.4.5.12 (Fabrication of metal layers) see p. 624 ff.

6.1.4.6 Lithography

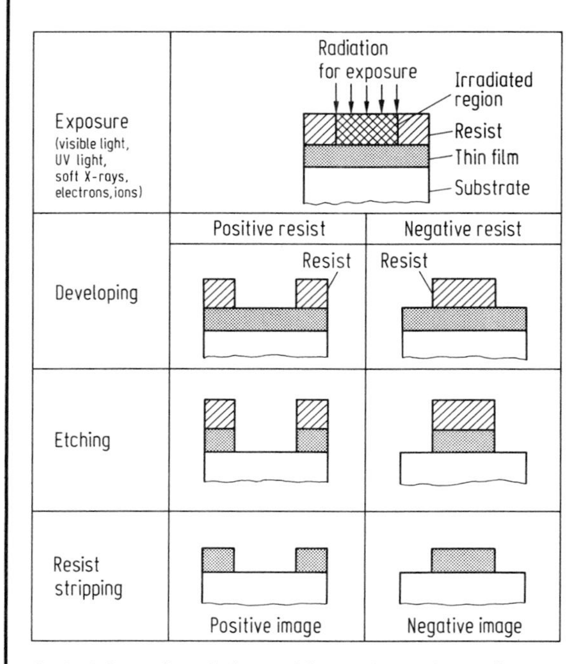

Fig. 1. Schematics of the positive and negative resist processing resulting in etched film patterns.

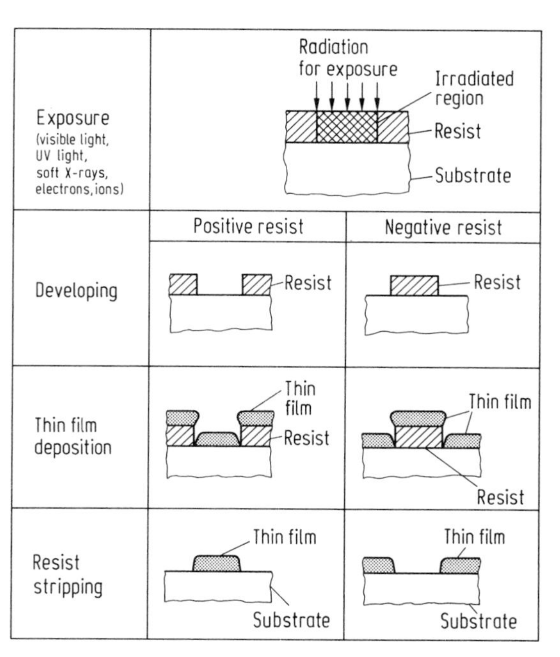

Fig. 2. Schematics of the positive and negative resist processing demonstrating the lift-off technique.

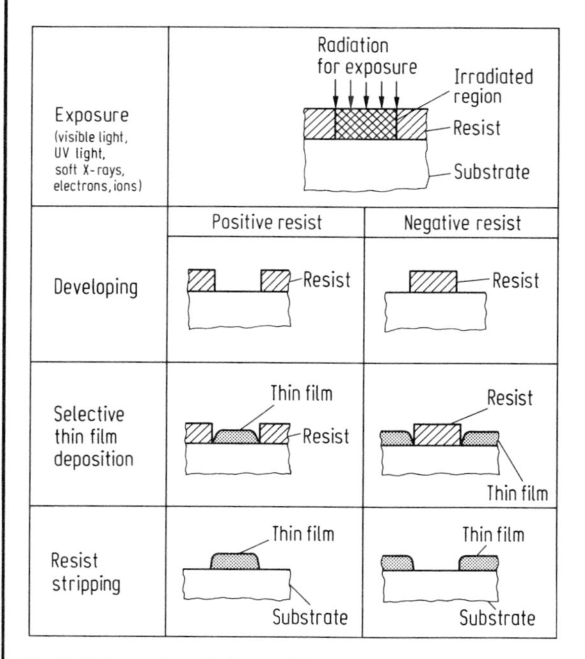

Fig. 3. Schematics of the positive and negative resist processing demonstrating the selective deposition of film patterns.

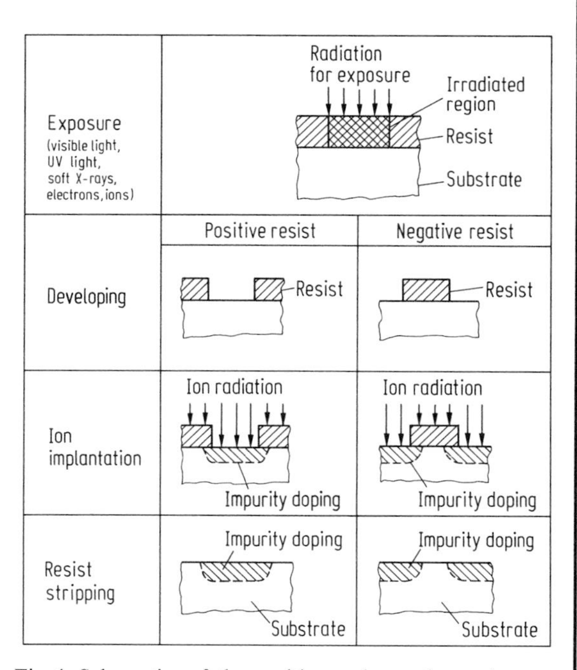

Fig. 4. Schematics of the positive and negative resist processing demonstrating the generation of impurity profiles by ion implantation.

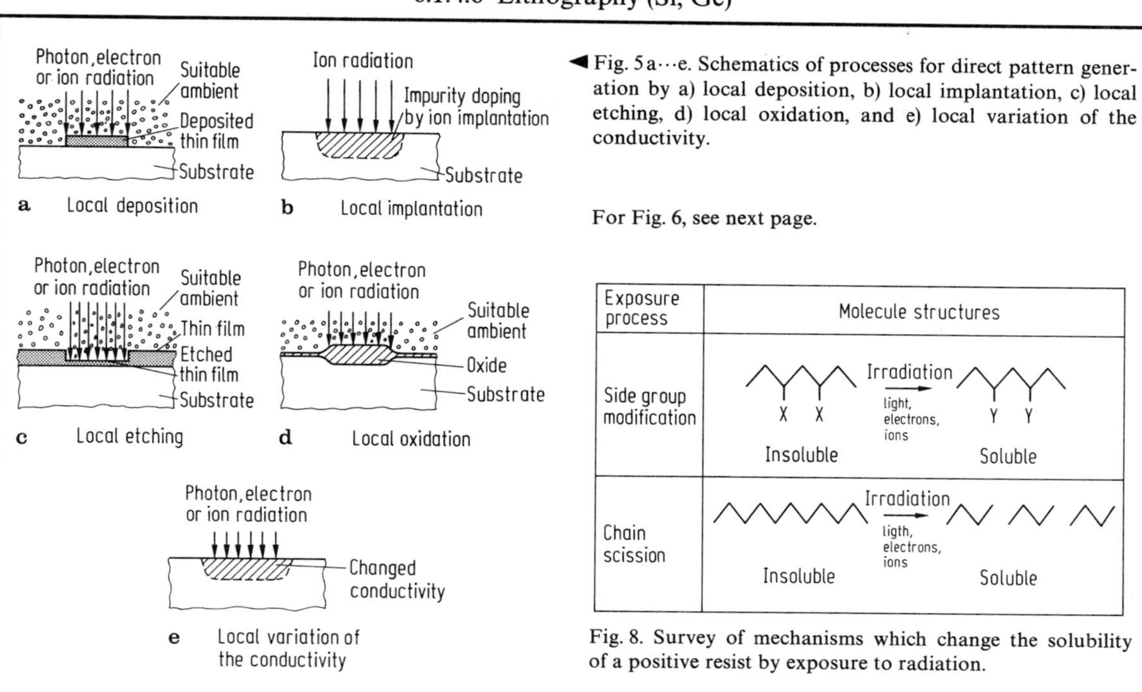

◄ Fig. 5a···e. Schematics of processes for direct pattern generation by a) local deposition, b) local implantation, c) local etching, d) local oxidation, and e) local variation of the conductivity.

For Fig. 6, see next page.

Fig. 8. Survey of mechanisms which change the solubility of a positive resist by exposure to radiation.

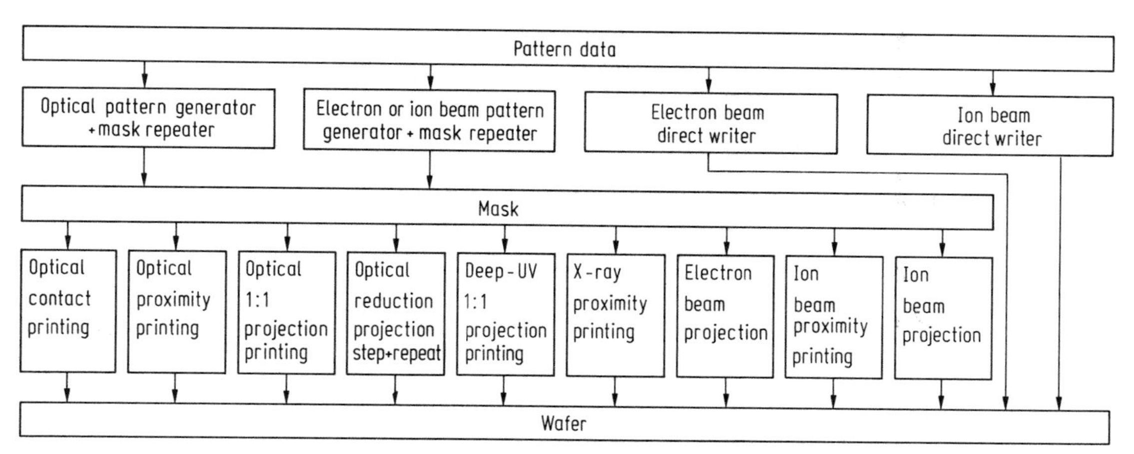

Fig. 7. Diagram for the demonstration of the various methods of how pattern data can be transferred onto a wafer by lithography methods.

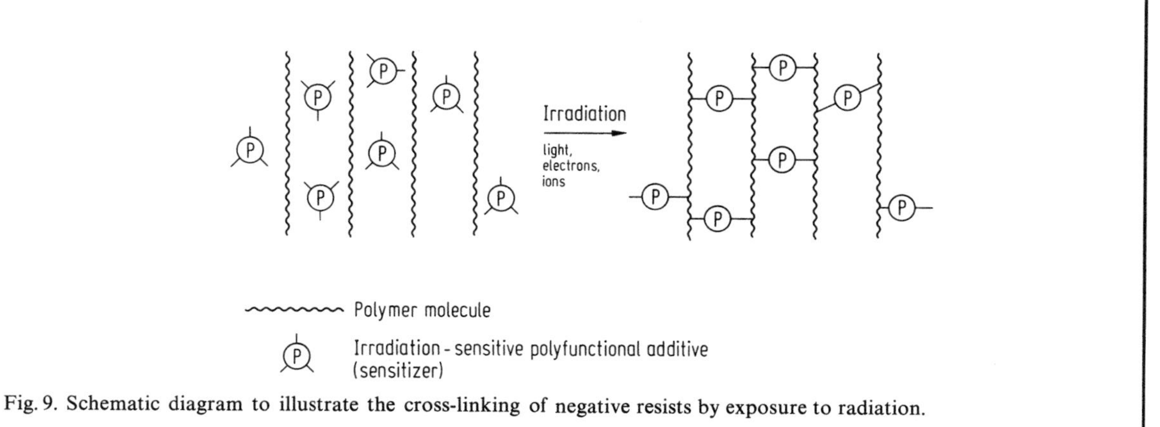

Fig. 9. Schematic diagram to illustrate the cross-linking of negative resists by exposure to radiation.

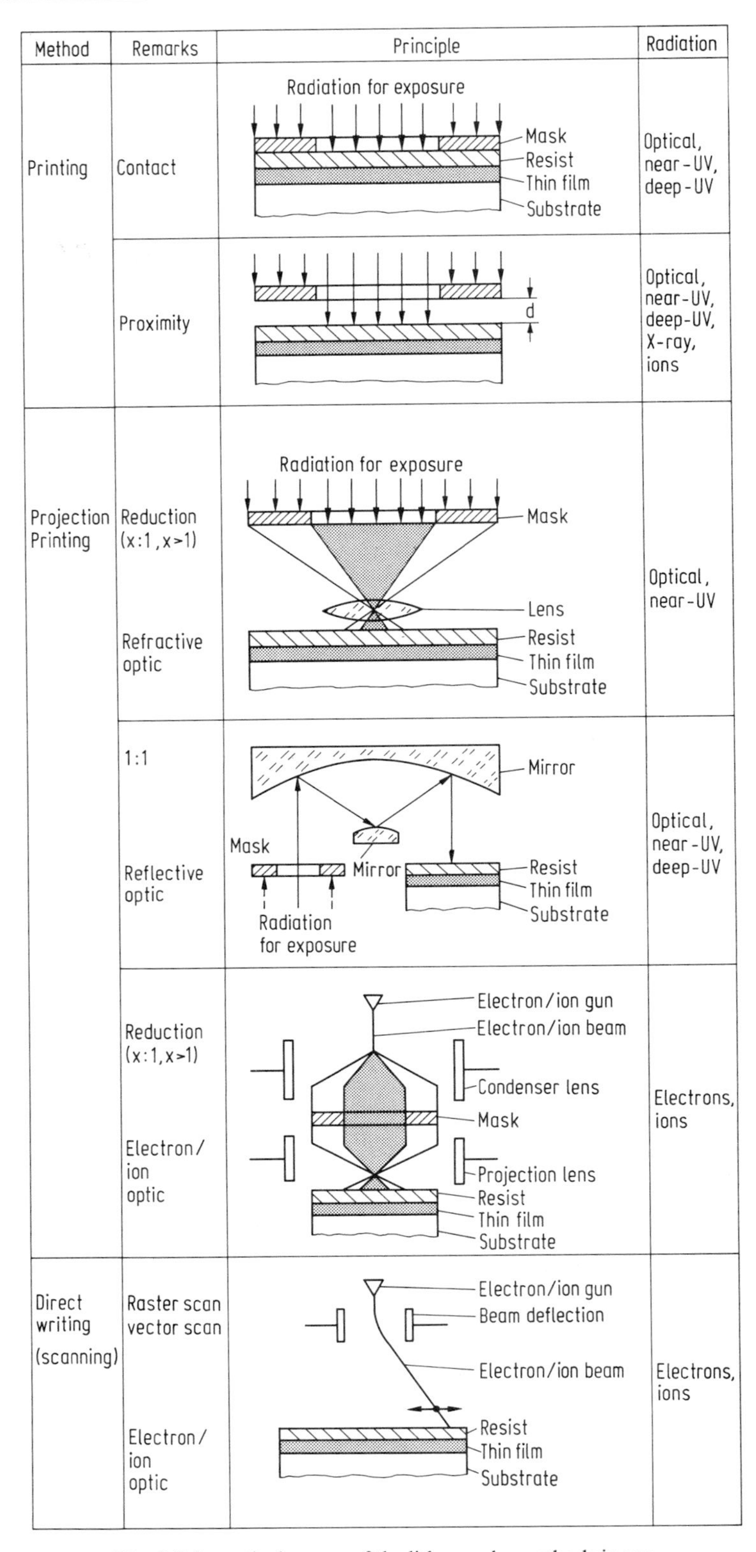

Fig. 6. Schematical survey of the lithography methods in use.

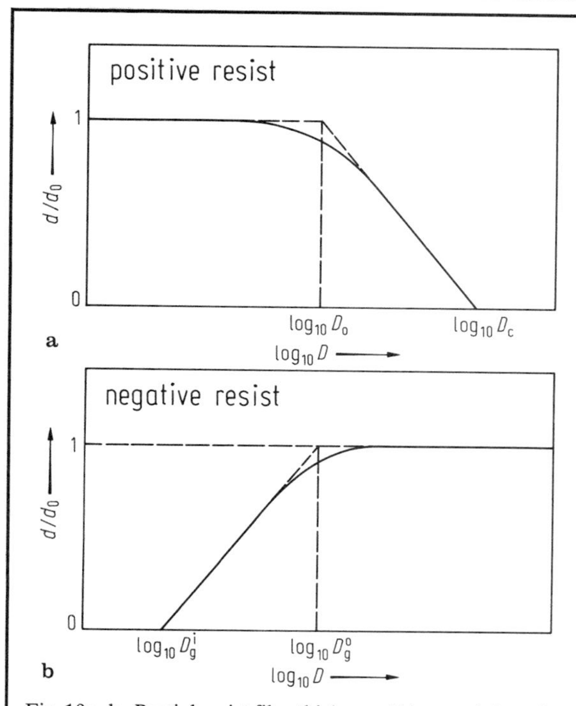

Fig. 10a, b. Partial resist film thickness d/d_0 remaining after development versus the exposure dose D in an arbitrary $\log_{10}$ scale for a) positive resist and b) negative resist. The doses D_0, D_c, D_g^i, and D_g^0 defined by the graphs are used to determine the contrast γ of the resist (Section 6.1.4.6.1).

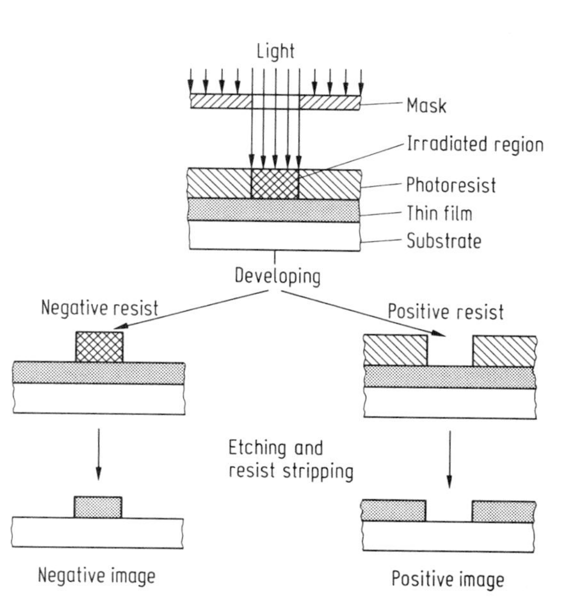

Fig. 11. Schematics of the processing sequence of negative (left) and positive (right) resist in order to obtain etched-film patterns.

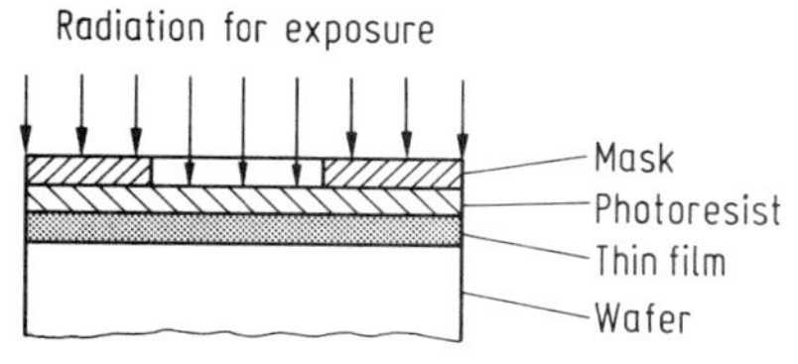

Fig. 12. Schematical drawing to demonstrate the contact printing.

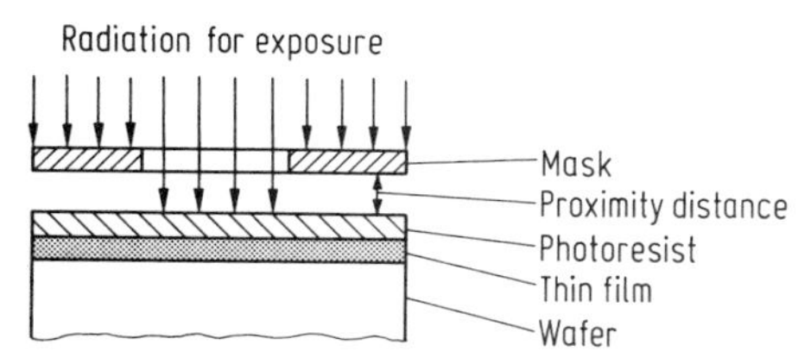

Fig. 13. Schematical drawing to demonstrate the proximity printing.

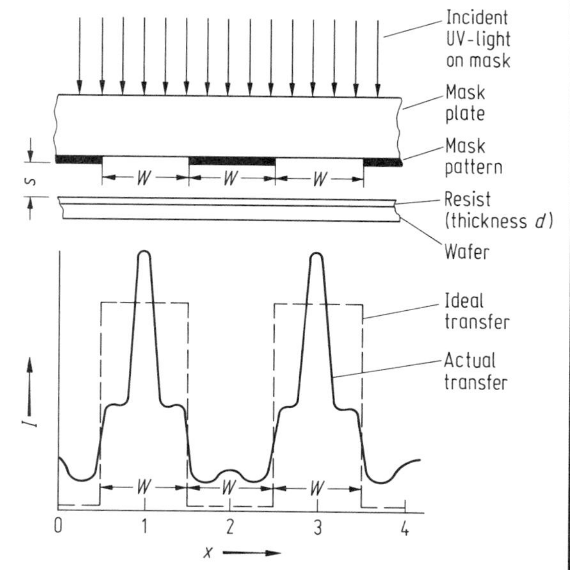

Fig. 14. Schematic representation of the pattern transfer by proximity printing using uv light [76W1]. The configuration of the mask and the wafer is shown in the top. The incident light intensity I versus position x is shown in the bottom graph. The minimum transferrable period is $2\,w_{min} \approx 3 \cdot [(s+d/2)\cdot\lambda]^{\frac{1}{2}}$.

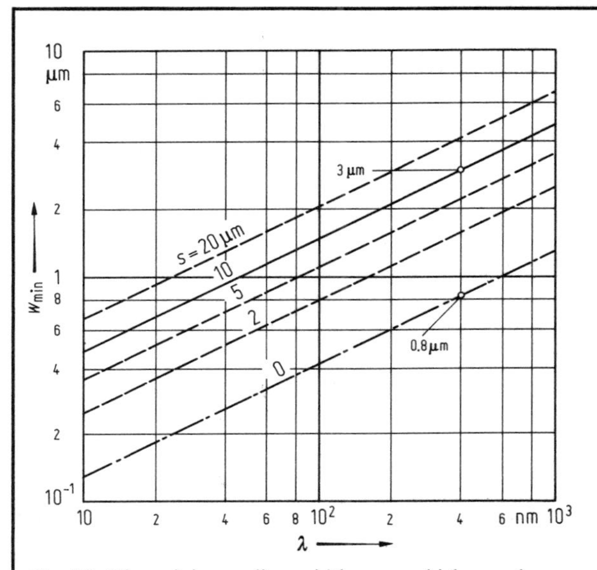

Fig. 15. The minimum line width w_{min} which can be replicated by proximity printing versus wavelength λ of the light used. The parameter s is the proximity distance from the wafer to the mask. The thickness of the photoresist is assumed to be 1.5 µm [76W1]. The points marked indicate the values $w_{min} = 3$ µm and $w_{min} = 0.8$ µm obtained for proximity printing and contact printing, respectively, at the commonly used wavelength $\lambda = 400$ nm.

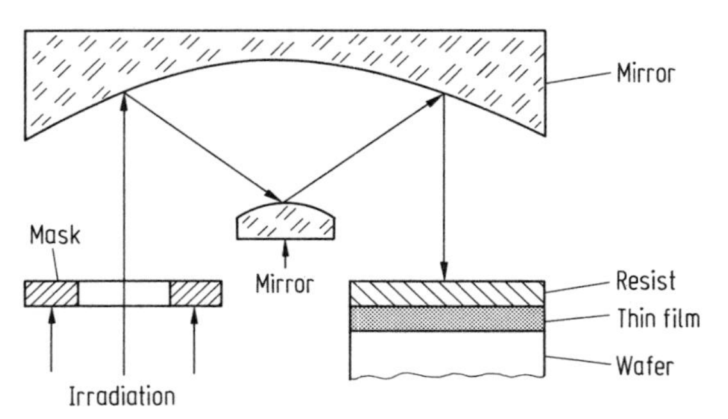

Fig. 16. Schematic drawing of the projection optics in a 1:1 reflective scanning projection system.

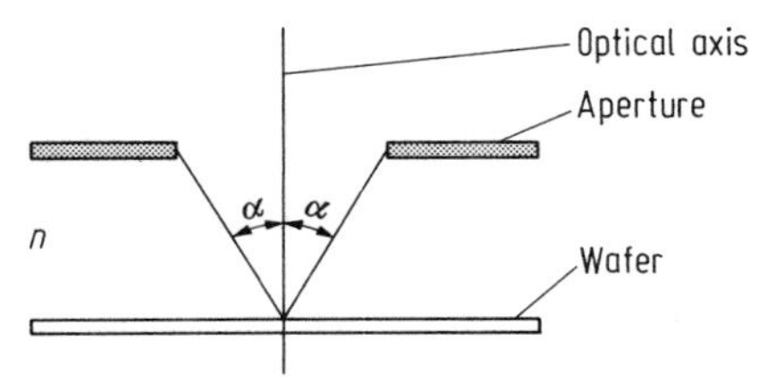

Fig. 18. Definition of the numerical aperture $NA = n \cdot \sin\alpha$; n is the refractive index of the medium between the aperture and the wafer and α is the angle indicated.

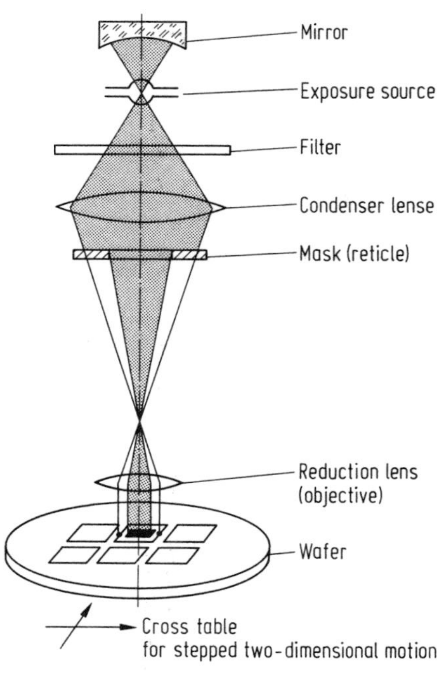

Fig. 17. Schematic drawing of a step-and-repeat optical reduction projection system.

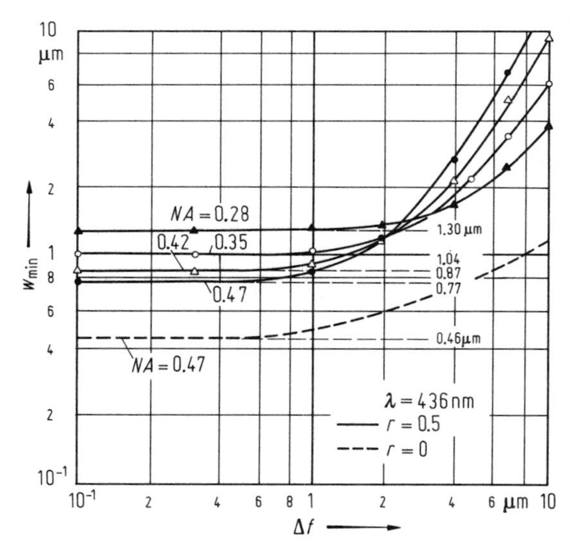

Fig. 19. Minimum line width w_{min} versus the degree of defocussing Δf. Parameter is the numerical aperture NA. A fixed wavelength λ and a coherence ratio r are assumed as indicated.

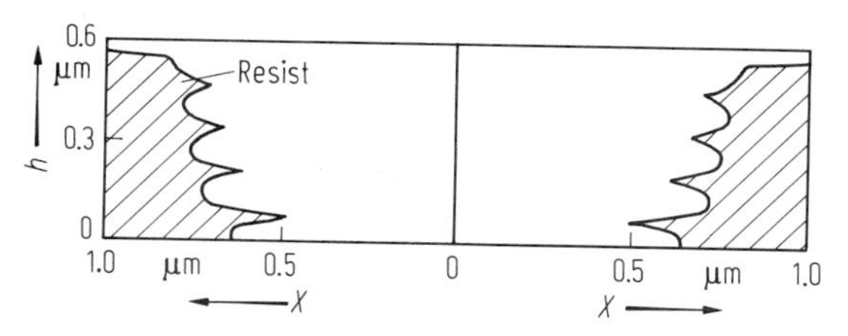

Fig. 20. Calculated edge profile for a nominal 1 μm line in positive photoresist (Shipley AZ 1350) exposed at $\lambda =$ 404.7 nm wavelength [75D4]. The height h of the edge is plotted versus distance x from the center line.

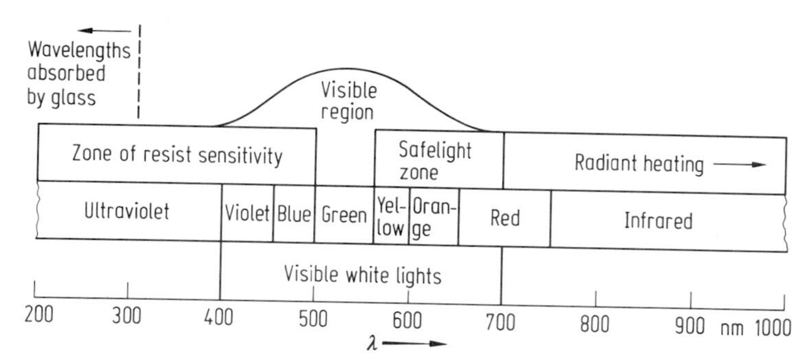

Fig. 21. Properties of photoresist versus wavelength λ [75D1].

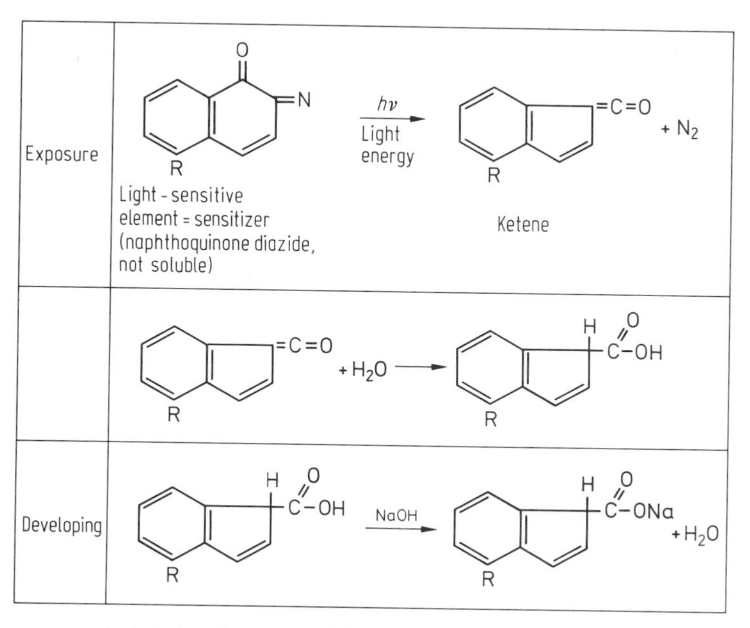

Fig. 22. Reactions of positive resists occuring during exposure to light (top) [44S1, 79P1] and during developing (bottom) [75D1, 44S1].

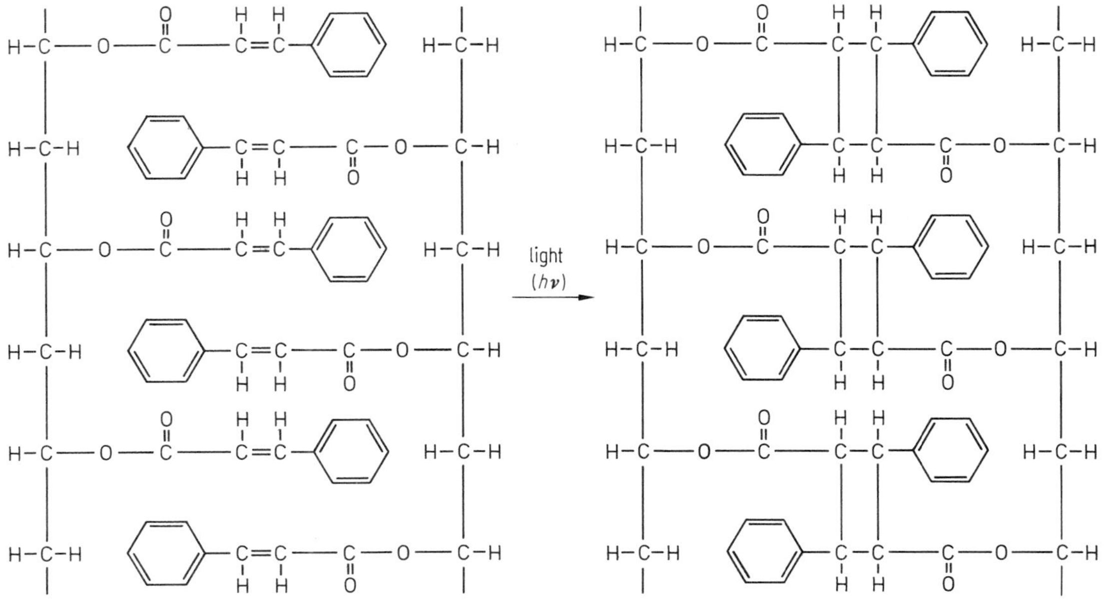

Fig. 23. Example of photochemical crosslinking in a negative photoresist for the case of polyvinyl cinnamate. The crosslinking occurs by α-truxillic dimerization [75D1].

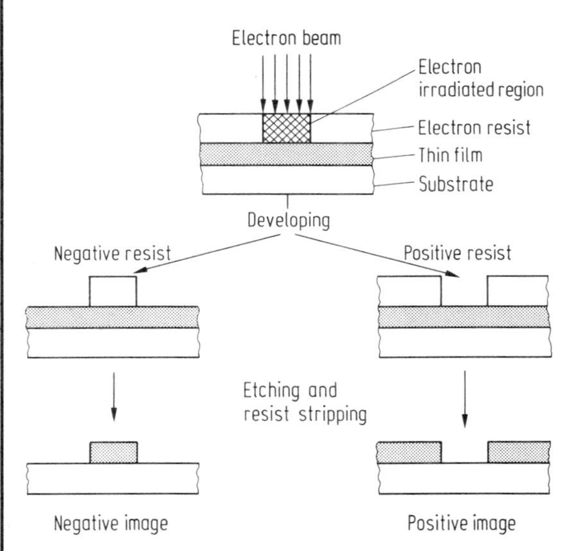

Fig. 24. Schematical drawing to demonstrate the processing steps in negative (left) and positive (right) resist in electron beam pattern generation.

►

Fig. 25a, b. Schematic drawings of electron beam projection systems; a) reduction projection system, b) 1:1 photocathode projection system.

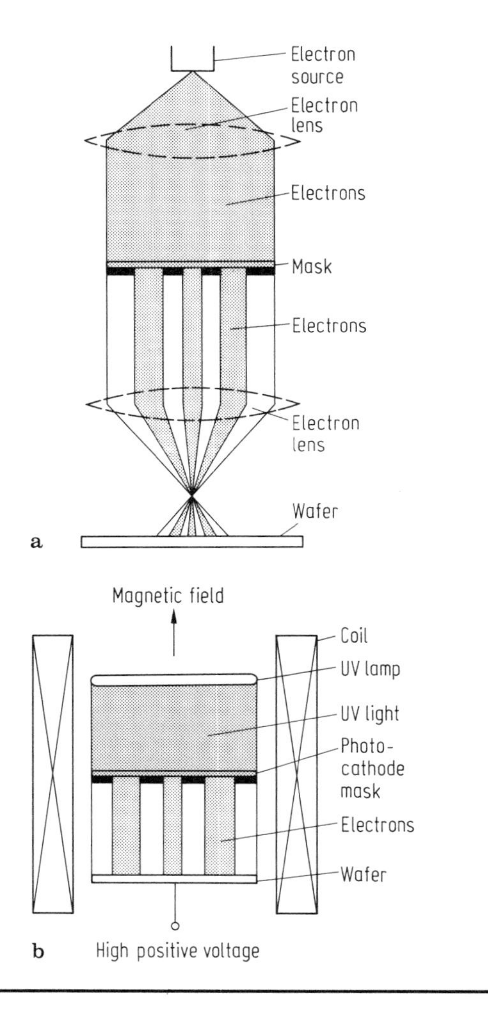

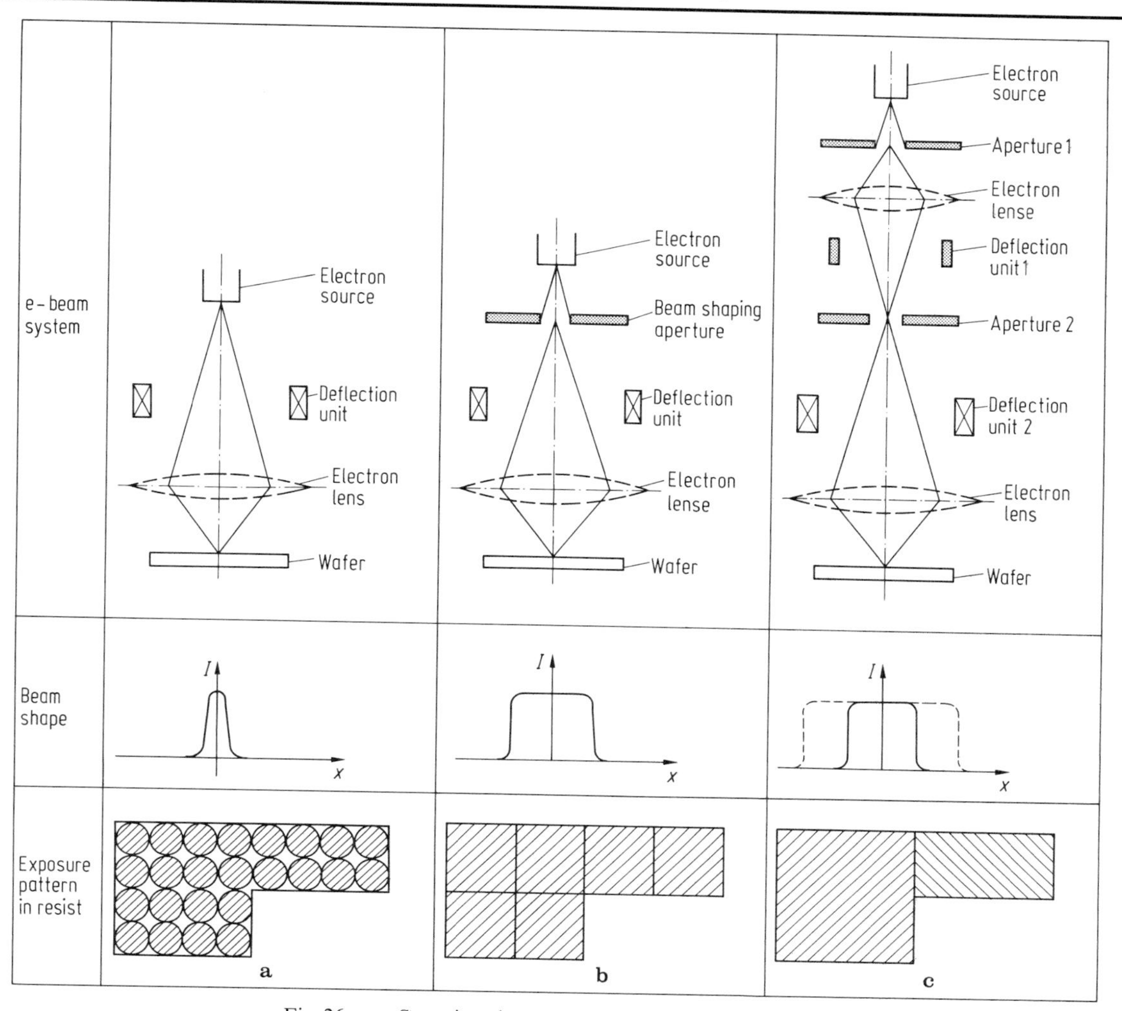

Fig. 26 a···c. Scanning electron beam systems using different spot-forming techniques. For each system, the e-beam imaging system is shown in the top, the beam shape-intensity I versus distance x – in the center, and the exposure pattern obtained in the resist in the bottom. a) source beam technique, b) shaped beam technique, c) variable shaped beam technique.

For Fig. 27, see next page.

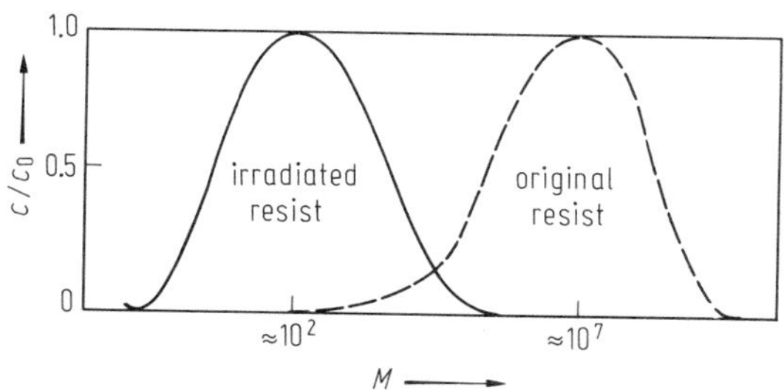

Fig. 28. Schematic representation of the concentration fraction c/c_0 of polymer molecules in the original and the irradiated positive resist versus the molecular weight M [79B1].

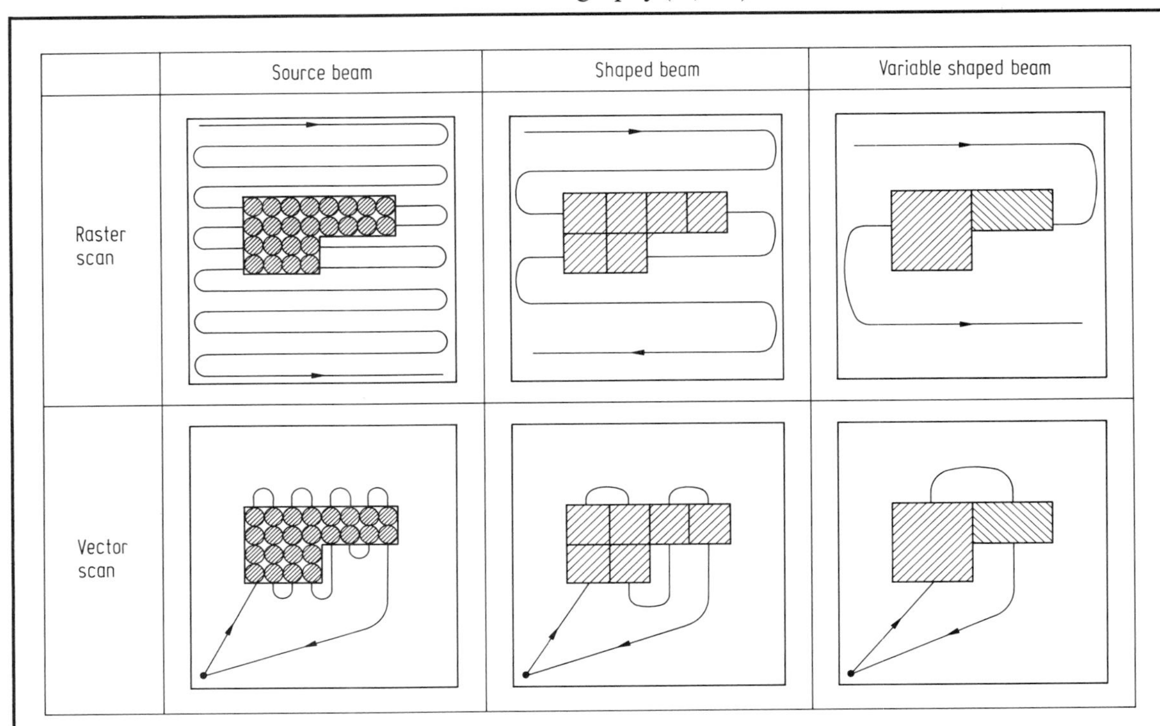

Fig. 27. Possible combinations of imaging and scanning techniques; the figures in the matrix demonstrate the beam scanning used and the exposure pattern obtained in the resist.

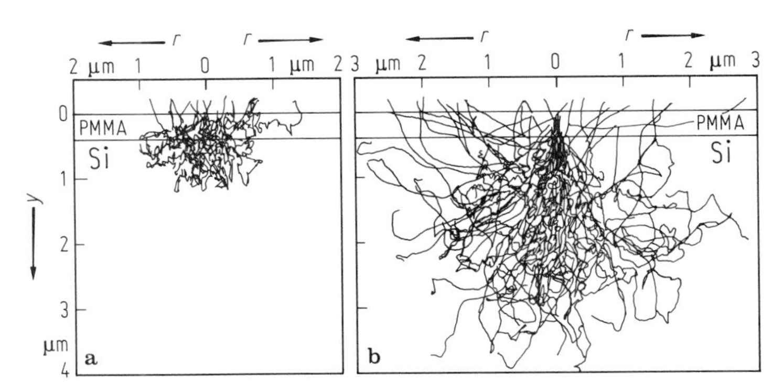

Fig. 29 a, b. PMMA/Si. Trajectories determined for 100 electrons in 400 nm thick PMMA resist film on silicon by a Monte Carlo calculation [75K1]. The electron energies assumed in the beam are a) 10 keV and b) 20 keV.

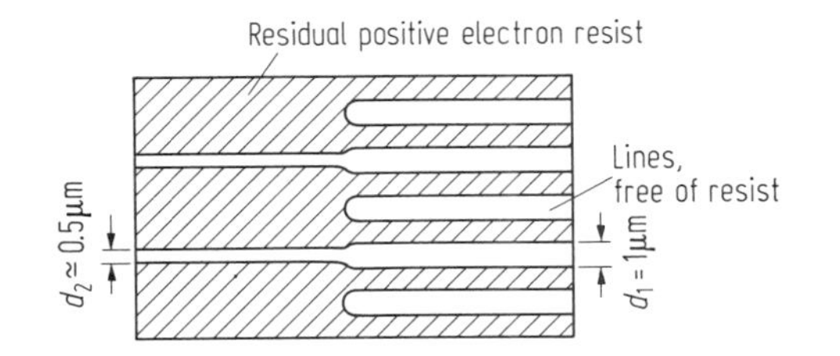

◀ Fig. 30. Schematic drawing to demonstrate the proximity effect. The pattern of fine lines is generated by identical electron beams in a positive electron resist. Backscattering of electrons from adjacent lines cause a higher exposure in the right part where lines are closely spaced, than on the left part where the lines are widely separated. The result is that the line width obtained after development on the right d_1 is larger than that on the left d_2, i.e. $d_1 > d_2$.

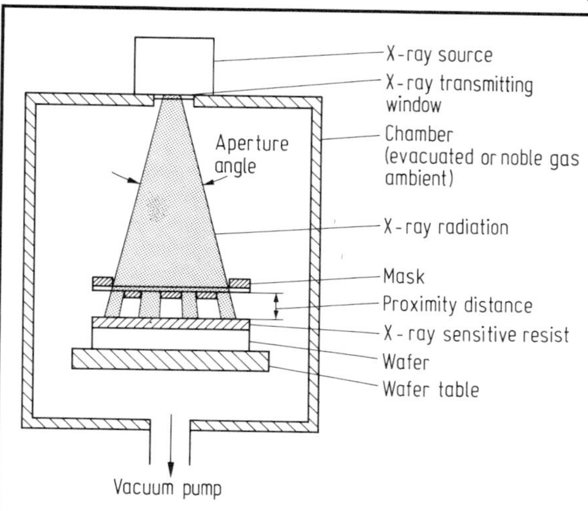

Fig. 31. Schematic drawing of an X-ray lithography system.

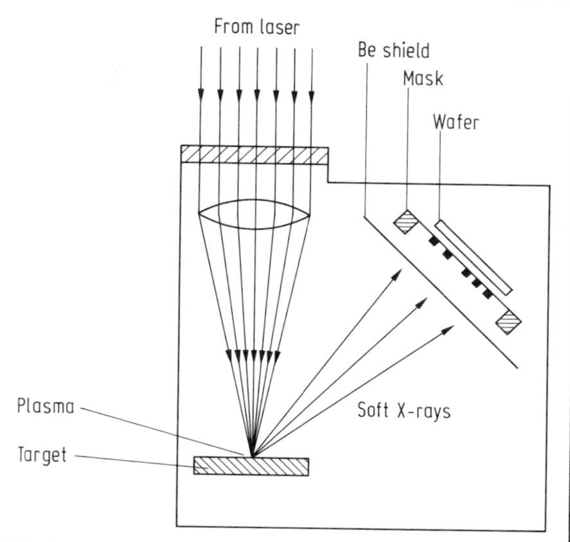

Fig. 34. X-ray lithography system using a laser-generated plasma as a soft X-ray source [78P1].

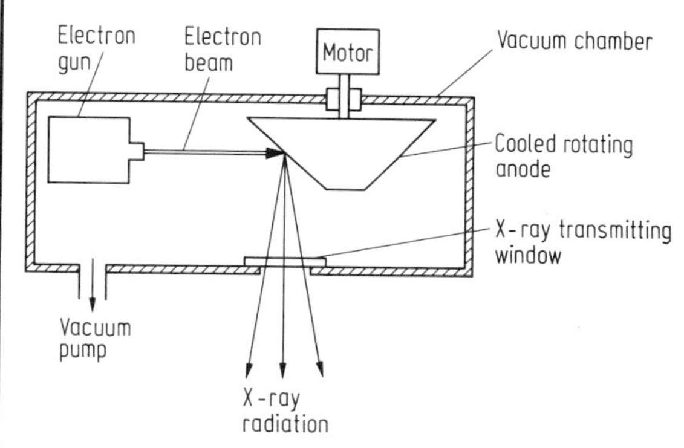

Fig. 32. Schematics of the design of an electron bombardment X-ray source.

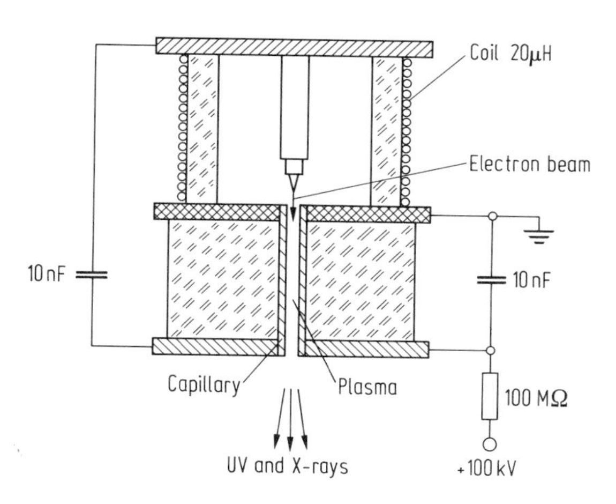

Fig. 35. Schematic drawing of an X-ray source in which a plasma generated by a spark discharge is intensified by an electron beam in order to increase the intensity of X-ray radiation [77M2].

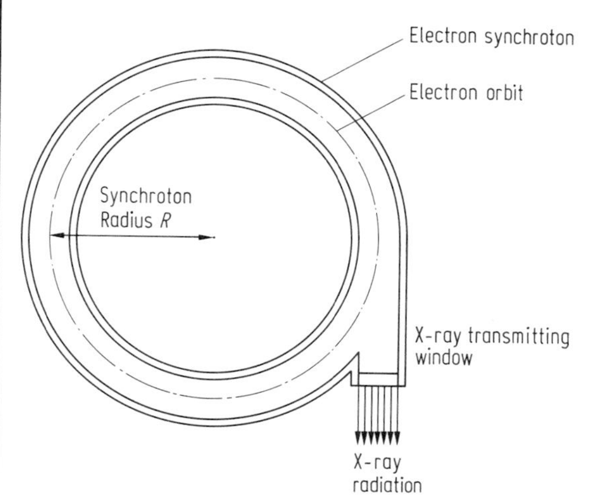

Fig. 33. Schematic drawing of a synchrotron used as a radiation source for X-ray lithography.

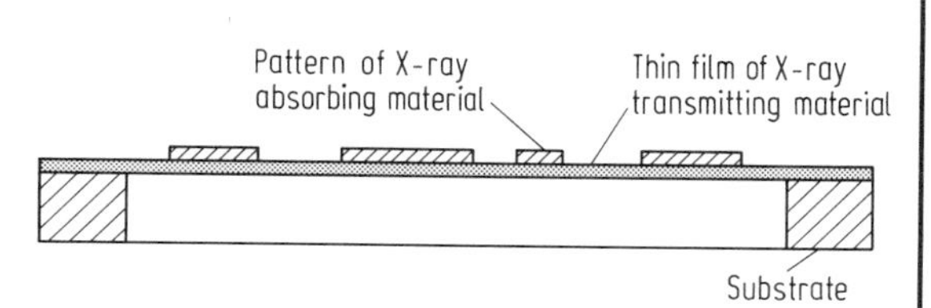

Fig. 36. Schematic drawing of a mask used in X-ray lithography.

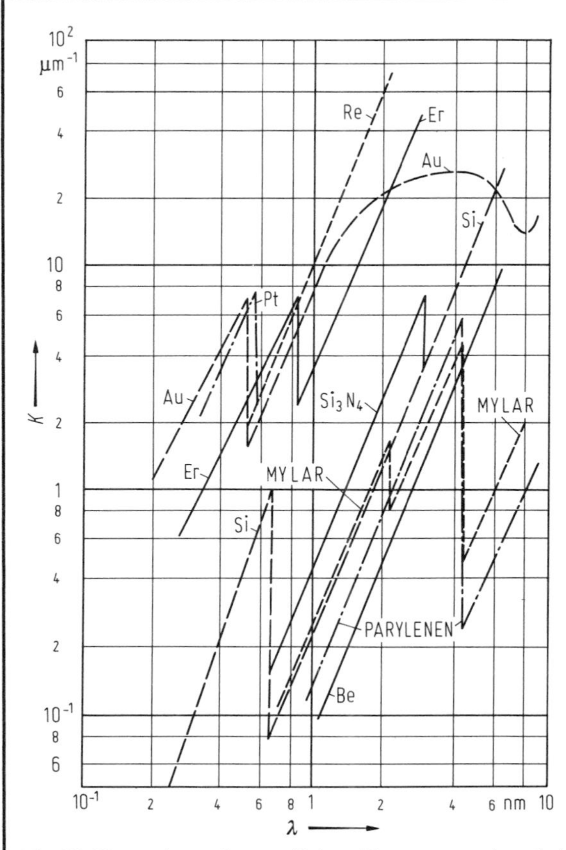

Fig. 37. X-ray absorption coefficient K versus wavelength λ for various materials of interest in X-ray lithography [77F1].

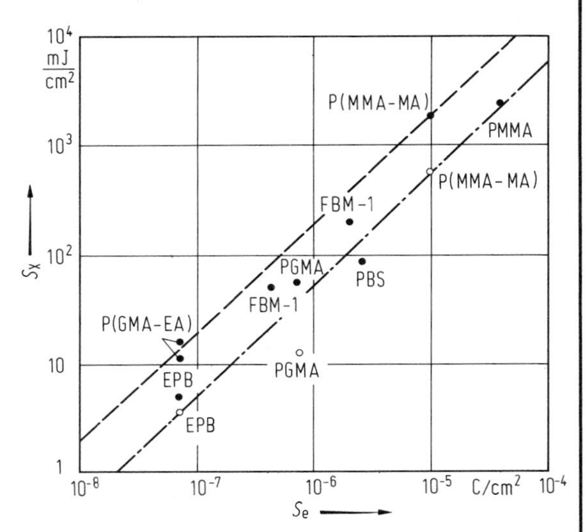

Fig. 38. Sensitivity S_x of resists to soft X-rays versus sensitivity S_e to a 20 keV electron beam [77M1] (full circles: X-rays of the L-line of a Mo-anode, open circles: X-rays of the K-line of an Al-anode).

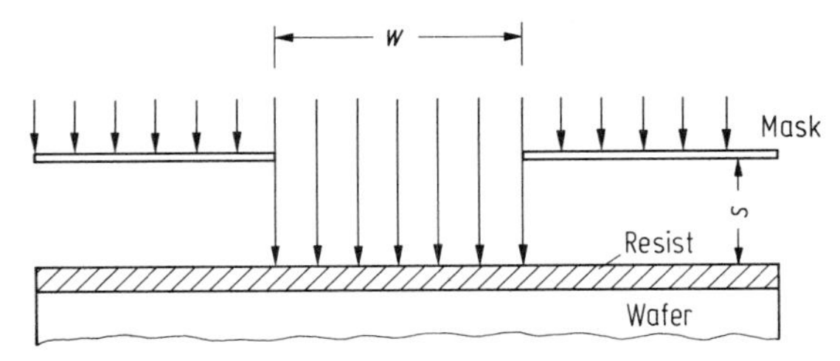

Fig. 39. Arrangement of mask and wafer in X-ray lithography. The proximity distance s and the line width w are defined for the discussion in Fig. 41.

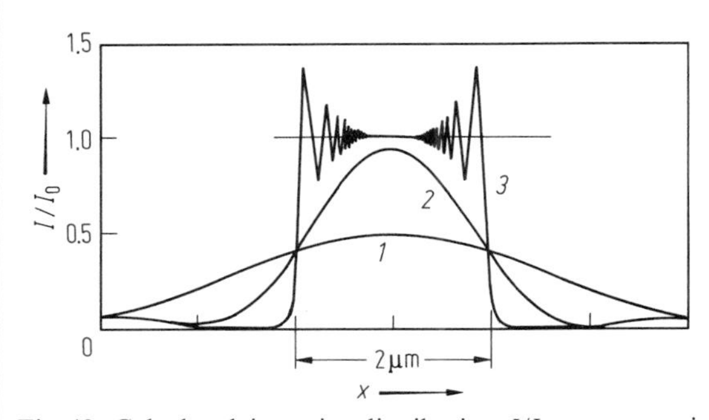

Fig. 40. Calculated intensity distribution I/I_0 versus position x behind a $w=2\ \mu m$ mask slit at a proximity distance of $s=20\ \mu m$ [80T1]. Parameter is the wavelength of the irradiation. 1) $\lambda=0.4\ \mu m$, 2) $\lambda=0.2\ \mu m$, 3) $\lambda=0.00083\ \mu m$ (Al-K radiation). The calculation is performed by the Fresnel integrals $C(v)$, $S(v)$ with $v_{1,2}=x_{1,2}(2/(\lambda s))^{\frac{1}{2}}$, $x_2-x_1=w=2\ \mu m$, and $I/I_0=0.5\{[C(v_2)-C(v_1)]^2+[S(v_2)-S(v_1)]^2\}$.

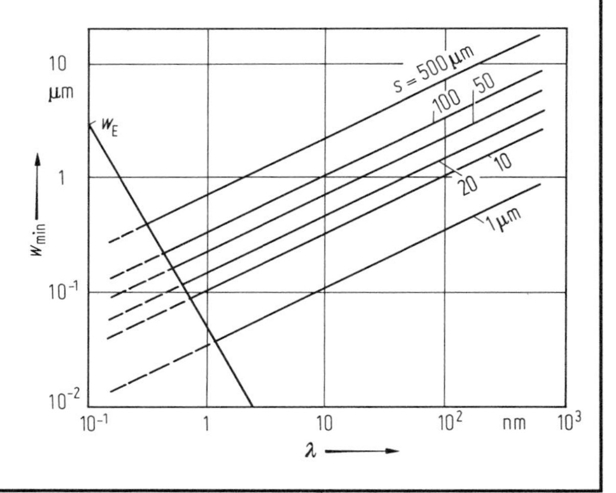

Fig. 41. Minimum resolvable width w_{min} versus wavelength λ as obtained in X-ray lithography due to Fresnel diffraction. Parameter is the proximity distance s of the mask. The straight line w_E marks the width limit due to the penetration range of photoelectrons in PMMA resist [80T1].

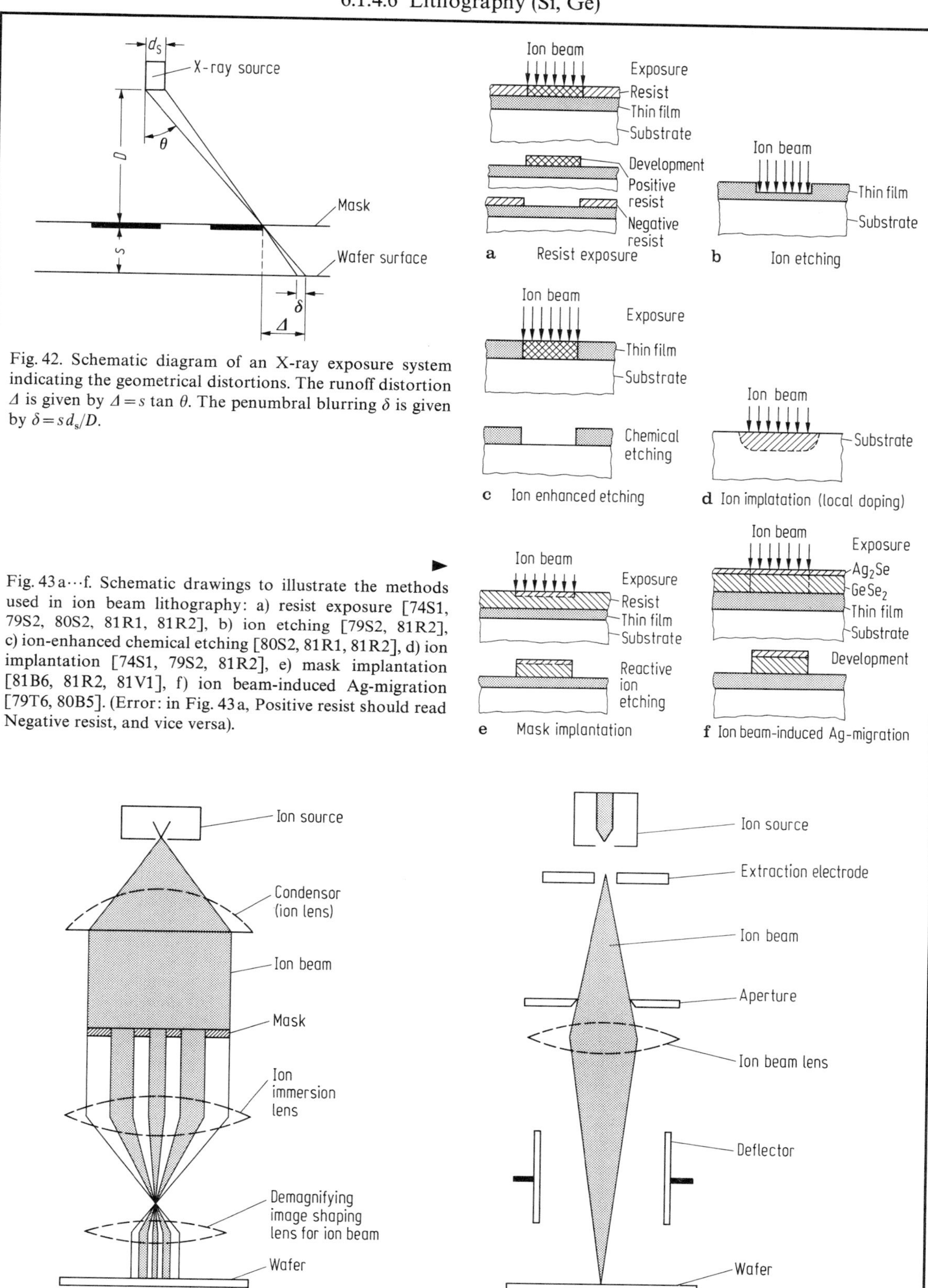

Fig. 42. Schematic diagram of an X-ray exposure system indicating the geometrical distortions. The runoff distortion Δ is given by $\Delta = s \tan\theta$. The penumbral blurring δ is given by $\delta = s d_s/D$.

Fig. 43a…f. Schematic drawings to illustrate the methods used in ion beam lithography: a) resist exposure [74S1, 79S2, 80S2, 81R1, 81R2], b) ion etching [79S2, 81R2], c) ion-enhanced chemical etching [80S2, 81R1, 81R2], d) ion implantation [74S1, 79S2, 81R2], e) mask implantation [81B6, 81R2, 81V1], f) ion beam-induced Ag-migration [79T6, 80B5]. (Error: in Fig. 43a, Positive resist should read Negative resist, and vice versa).

Fig. 44. Schematic drawing of an ion beam projection system [81R1].

Fig. 45. Schematic drawing of a scanning ion beam system [79S2].

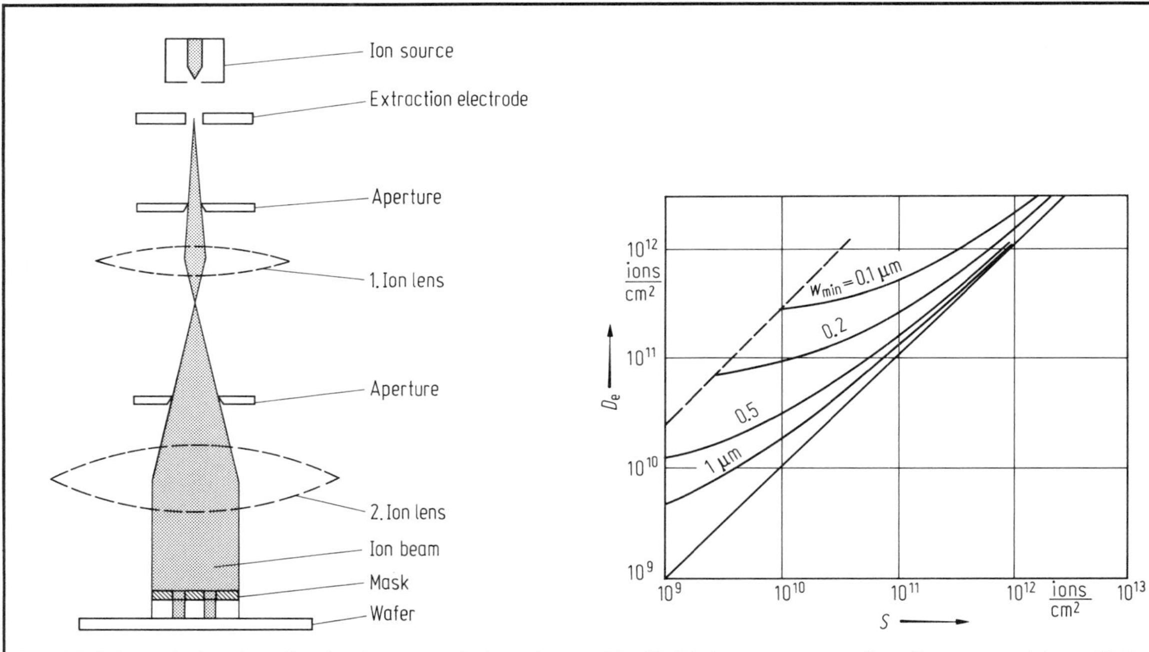

Fig. 46. Schematic drawing of an ion beam proximity printing system [81B12].

Fig. 48. Minimum exposure dose D_e versus resist sensitivity S. Parameter is the minimum line width w_{min} [81R2].

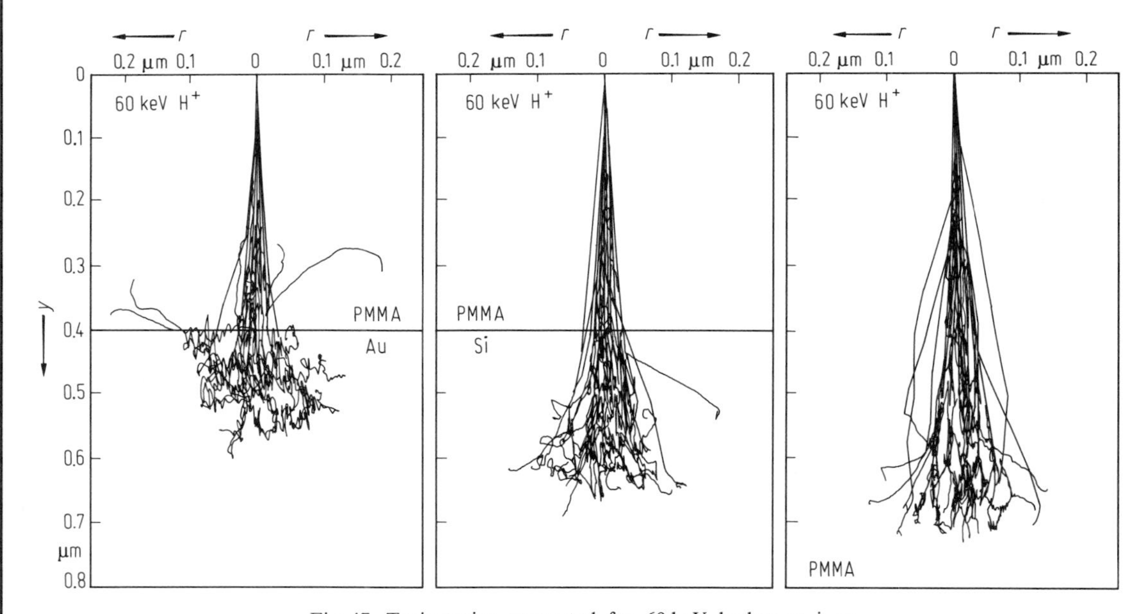

Fig. 47. Trajectories computed for 60 keV hydrogen ions traversing through PMMA, PMMA on gold, and PMMA on silicon by a Monte Carlo calculation [81K3].

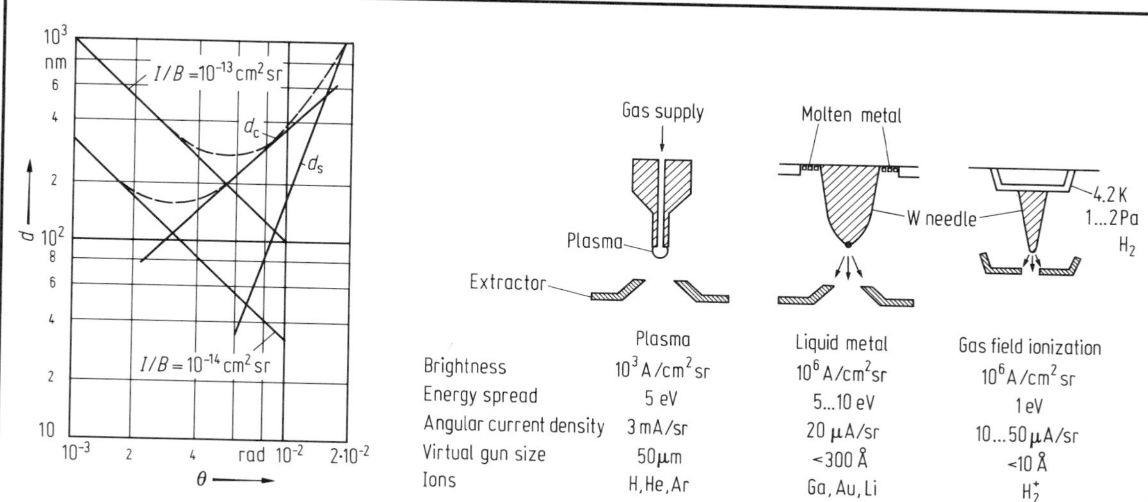

Fig. 49. Minimum spot diameter d versus convergence half angle θ. The lens bore diameter is 38 mm [81R2]. B is the ion source brightness, d_c is the chromatic aberration diameter, d_s is the spheric aberration diameter, $C_s = 17$ cm and $C_c = 19$ cm are the spherical and chromatic aberration coefficients, respectively. The calculation is performed for H^+ ions at energys of 50 keV, assuming that $\Delta E/E = 10^{-4}$.

Fig. 50. Comparison of plasma, liquid-metal, and gas field-ionization sources [81R2].

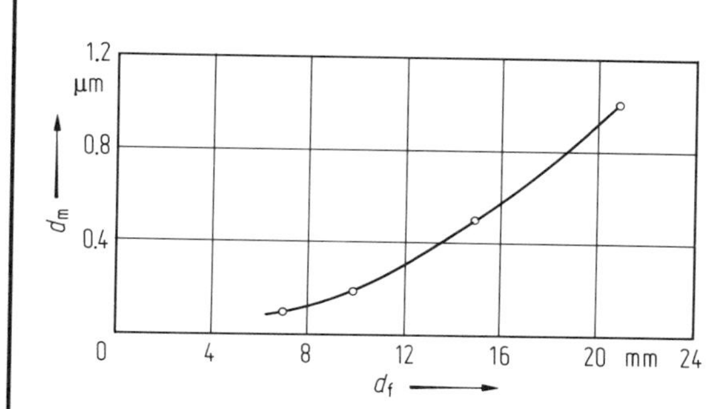

Fig. 51. Maximum diameter d_m of the aberration figure versus diameter d_f of the image field in the ion beam system. The diameter d_m equals the resolution width due to lens aberration [81S4].

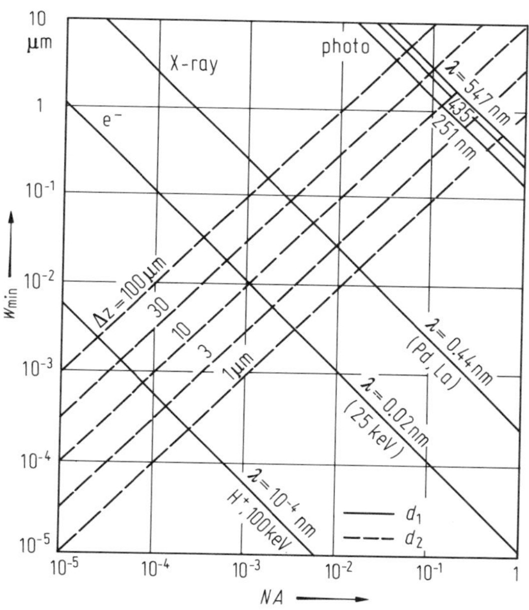

Fig. 52. Ultimate resolution width w_{min} versus numerical aperture NA of an aberration-free image-forming system. [81R6]. The lines marked d_1 and d_2 give the minimum distances which can be resolved by the Rayleigh criterion and the error disk when the image plane is out of focus by the distance Δz given. The resolution lines for e-beam lithography (marked e^-), X-ray lithography (marked *X-ray*), and photolithography (marked *photo*) are also shown for comparison. The conditions for each line are indicated in the graph.

6.1.4.7 Etching processes

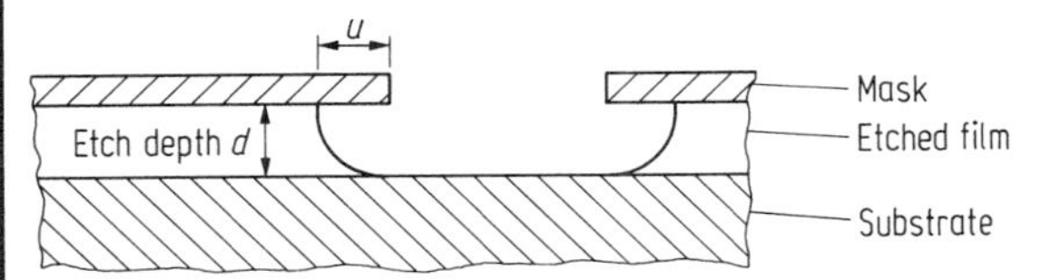

Fig. 1. Typical edge profile obtained by wet-chemical etching of a thin film using an etch mask. The etch factor f_e is defined by the ratio of the layer thickness d and the undercutting u: $f_e = d/u$.

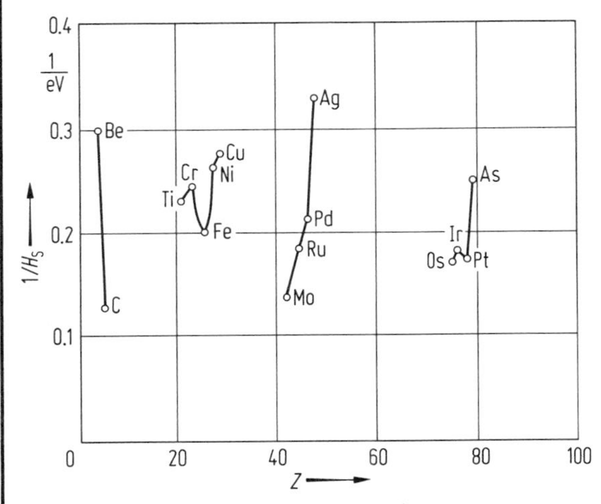

Fig. 2. Inverse heat of sublimation H_s^{-1} vs. element atomic number Z [76M2]. (Error: As should read Au).

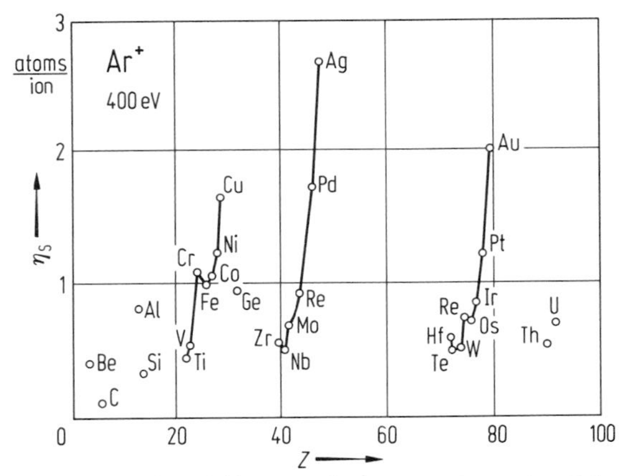

Fig. 3. Sputtering yield η_s vs. substrate atomic number Z for 400 eV argon ion bombardment [61L1]. (Error: Te should read Ta, and Re at $Z=44$ should read Ru).

For Fig. 4, see next page.

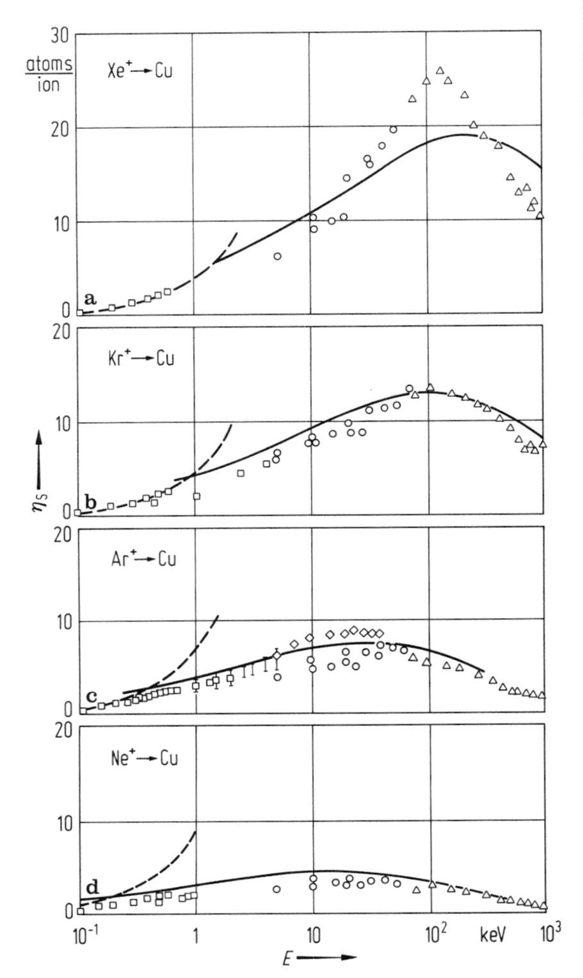

Fig. 5a⋯d. Sputtering yield η_s of copper vs. ion energy E. The dashed and continuous lines are calculations. The experimental points are taken from various references [69S1]. Bombardment by a) Xe^+ ions, b) Kr^+ ions, c) Ar^+ ions, d) Ne^+ ions.

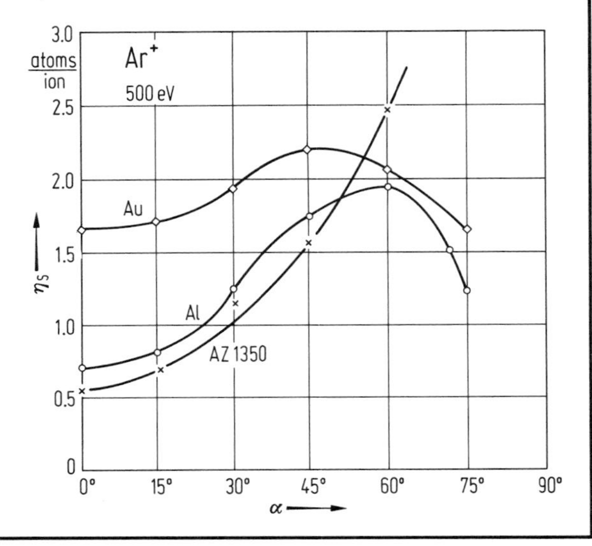

Fig. 6. Sputtering yield η_s of gold, aluminum, and Shipley photoresist AZ 1350 vs. angle α of incidence of the bombarding ions. The argon ion energy is 500 eV [76S1, 78M1].

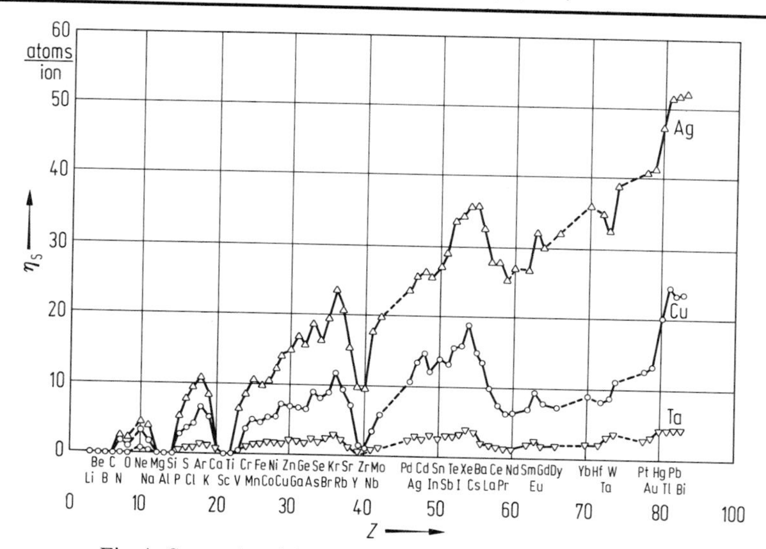

Fig. 4. Sputtering yield η_s of silver, copper, and tantalum vs. the atomic number Z of the bombarding ions. The sputtering yield is measured at an ion energy of 45 keV [61A2].

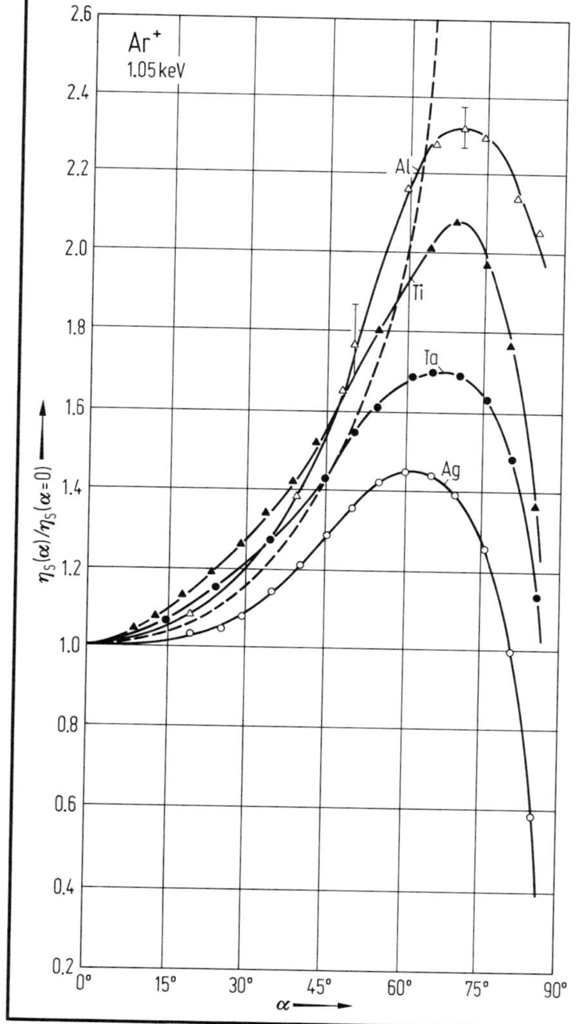

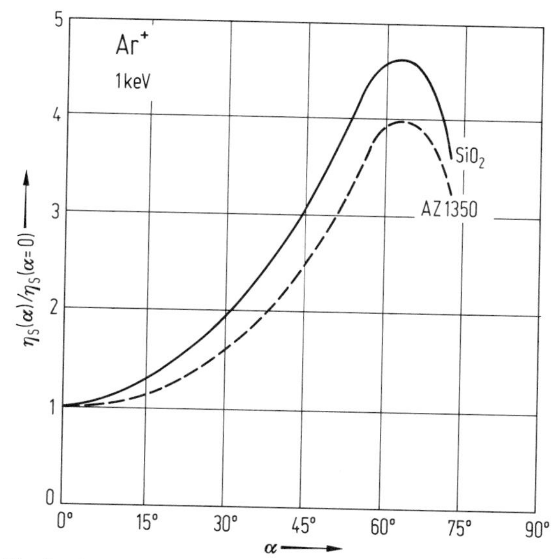

Fig. 8. Sputtering yield normalized to zero angle of incidence $\eta_s(\alpha)/\eta_s(\alpha=0)$ vs. angle of incidence α of the bombarding ions. Curves are shown for the target materials SiO_2 and Shipley photoresist AZ 1350. Bombarding ion is argon at energy 1 keV [76D2].

◄

Fig. 7. Sputtering yield normalized to zero angle of incidence $\eta_s(\alpha)/\eta_s(\alpha=0)$ vs. angle of incidence α of the bombarding ions. Curves are shown for the target materials Al, Ti, Ta, and Ag. Bombarding ion is Ar^+ at energy $E=1.05$ keV. The dashed line shows the function $1/\cos\alpha$.

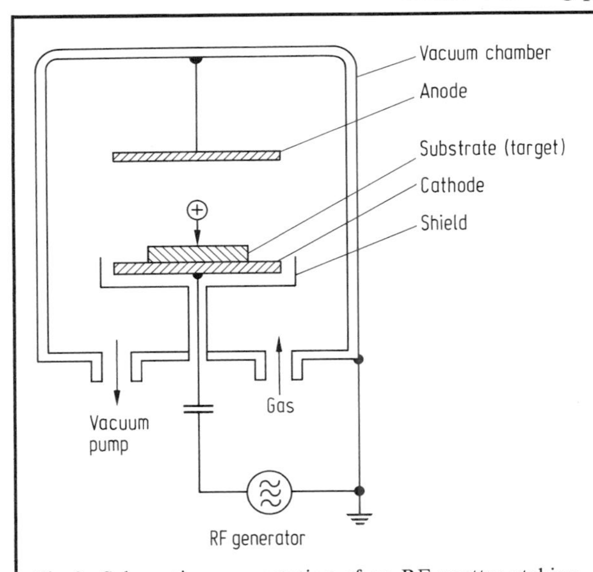

Fig. 9. Schematic representation of an RF sputter etching system.

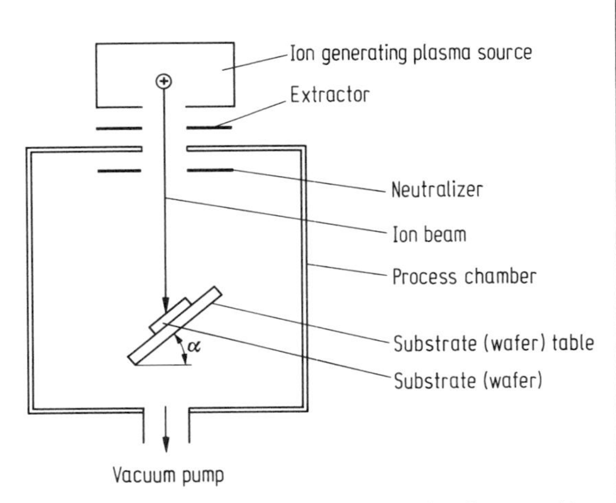

Fig. 10. Schematic representation of an ion beam etching system.

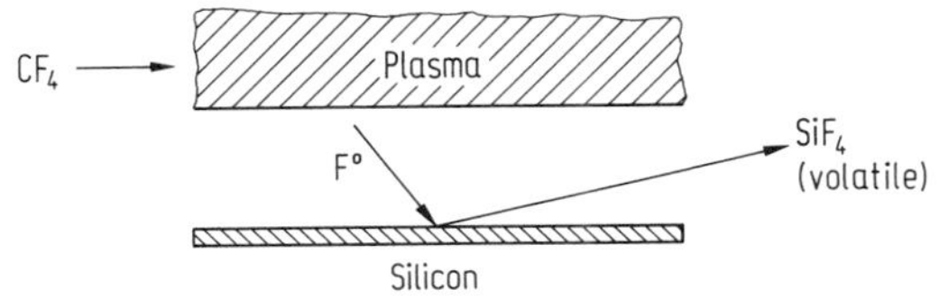

Fig. 11. Schematic representations of a reactive dry etching process by a spontaneous chemical reaction [71I1].

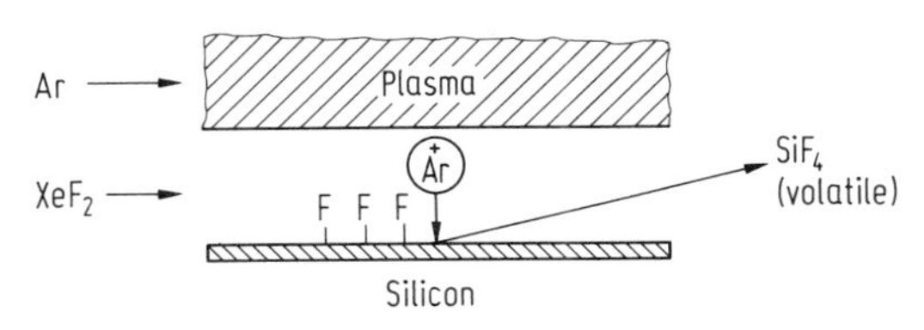

Fig. 12. Schematic representations of a reactive dry etching process by an ion-induced chemical reaction [79C4].

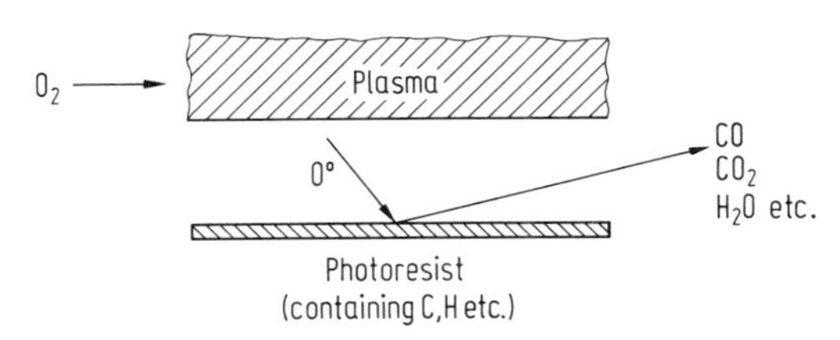

Fig. 13. Schematic representations of the plasma etching process of an organic photoresist [65H1].

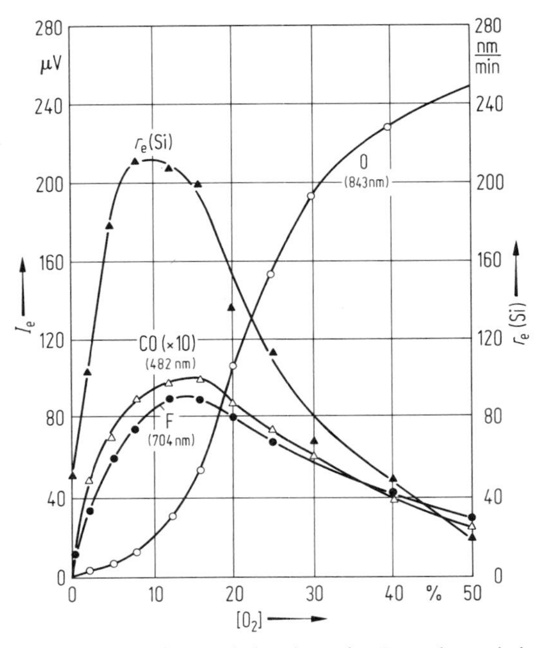

Fig. 14. Si. Optical emission intensity I_e at the emission lines of O (843 nm), of CO (482 nm) and F (704 nm) vs. the content of O_2 in the CF_4 gas [77H2]. The etch rate of Si, r_e(Si), is given on the right-hand scale. The etch rate of silicon increases with the fluorine content of the gas which is measured by the emission intensity.

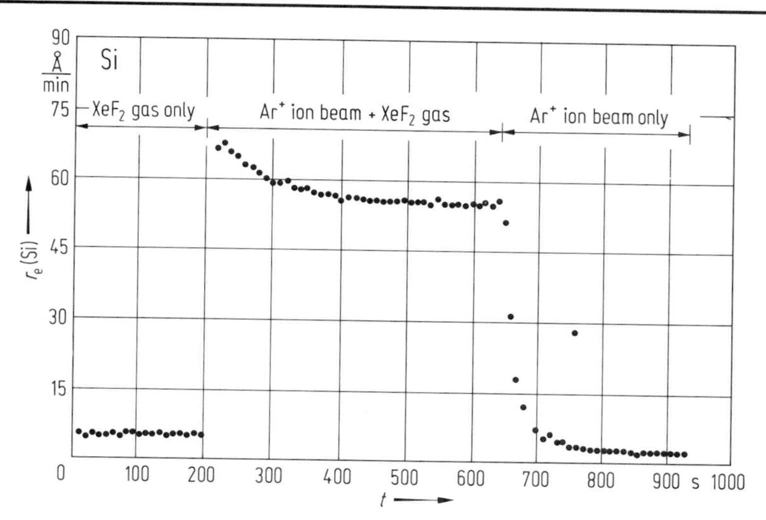

Fig. 15. Si. Etch rate of silicon r_e(Si) vs. time t. For the initial 200 seconds XeF_2 gas only is used for etching followed by combined etching of XeF_2 gas together with an Ar^+ ion beam. At the tail end etching is performed by the Ar^+ ion beam alone. The combined etching of the Ar^+ ion beam and the XeG_2 gas shows the largest etch rate.

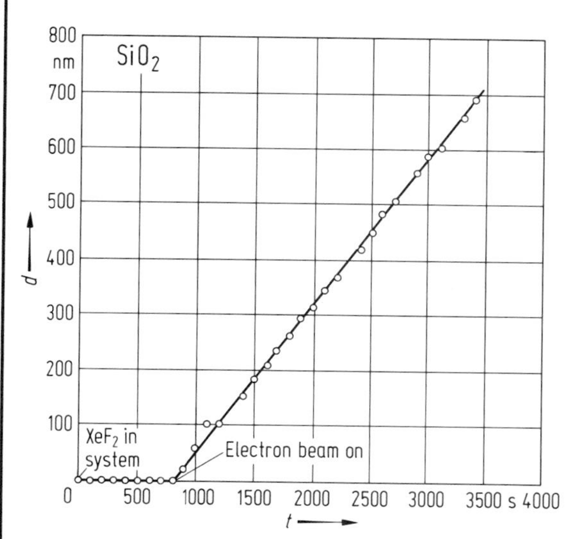

Fig. 16. SiO_2. Depth d of SiO_2 removed by etching vs. time t. For the initial 800 seconds XeF_2 alone is used for etching before the electron beam is applied. The total gas pressure (Xe ambient) in the system is $p=8\cdot10^{-2}$ Pa. The electron beam parameters are: $E=1500$ eV, $I=45$ μA, $j=50$ mA cm^{-2}. The combined effect of XeF_2 gas and electron beam leads to an etch rate $r_e(SiO_2)=200$ Å/min [79C4].

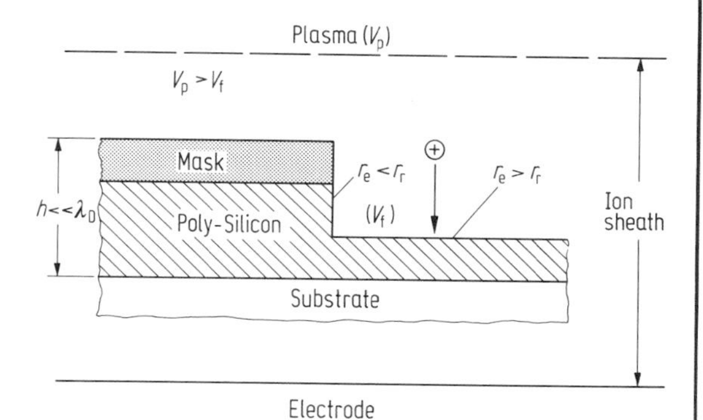

Fig. 17. Schematic representation of the model of Mogab and Levinstein [80M7] for anisotropic etching. V_p is the plasma potential, V_f is the floating potential of the target, r_e and r_r are the rates of the etching and the recombination reaction, respectively, and λ_D is the plasma Debye length. Etching occurs when $r_e > r_r$ in the irradiated region.

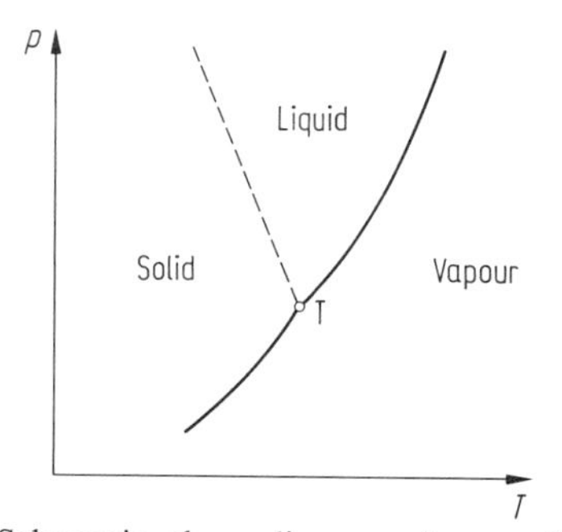

Fig. 18. Schematic phase diagram of a reaction product (T: triple point).

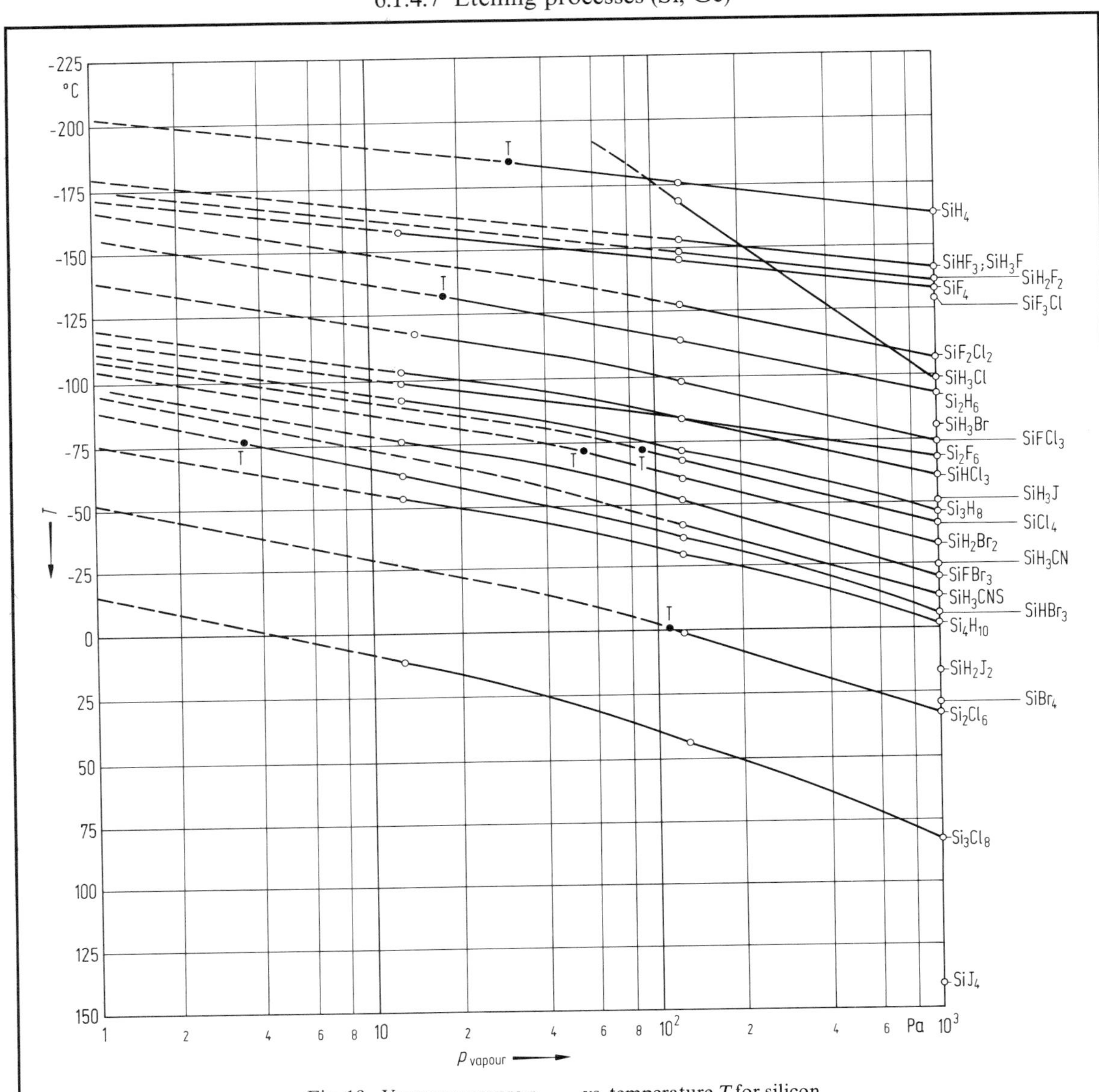

Fig. 19. Vapour pressure p_{vapour} vs. temperature T for silicon compounds [60G2, 65W1]. The point marked "T" indicates the triple point.

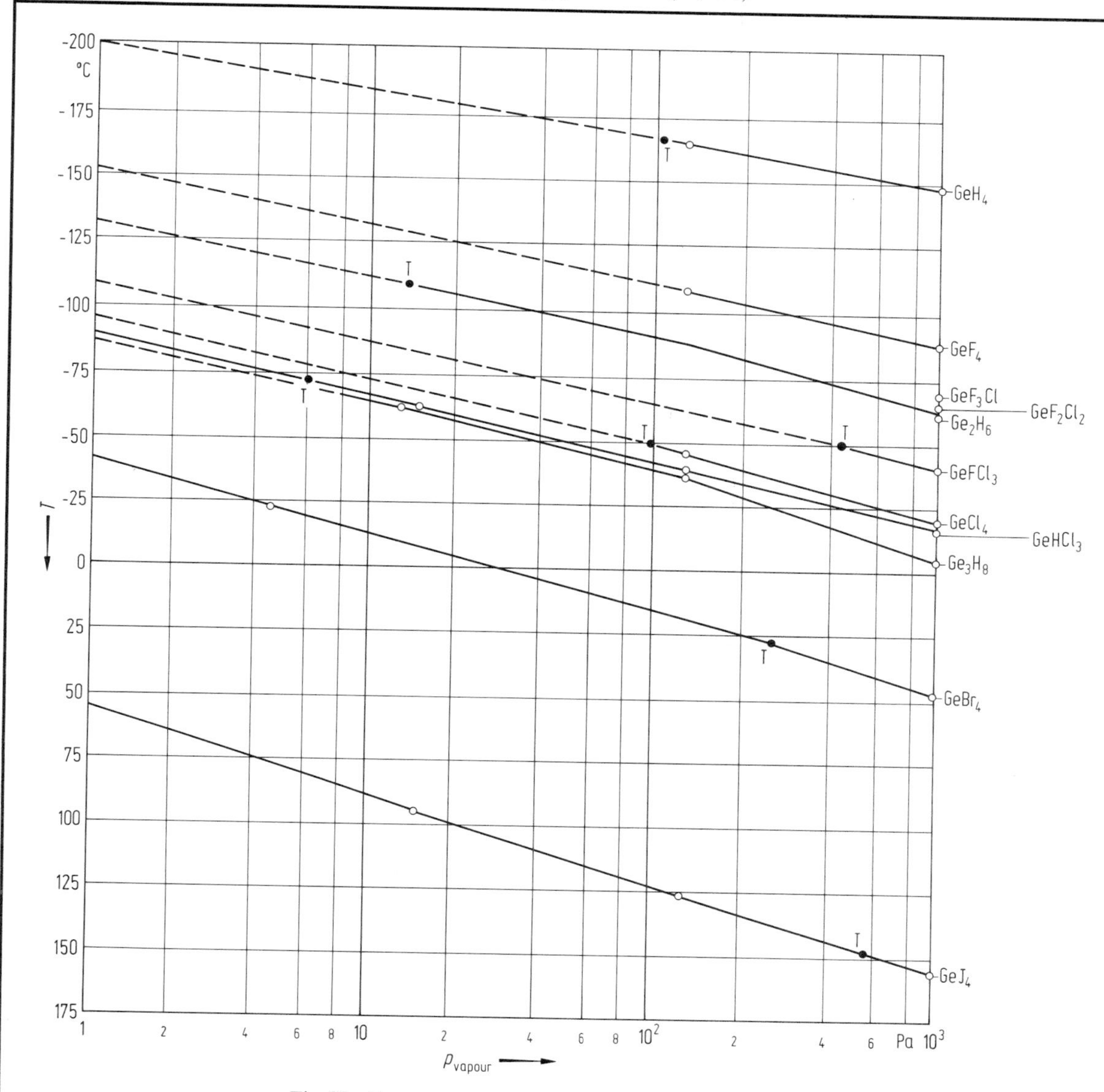

Fig. 20. Vapour pressure p_{vapour} vs. temperature T for germanium compounds [60G2, 65W1]. T indicates the triple point.

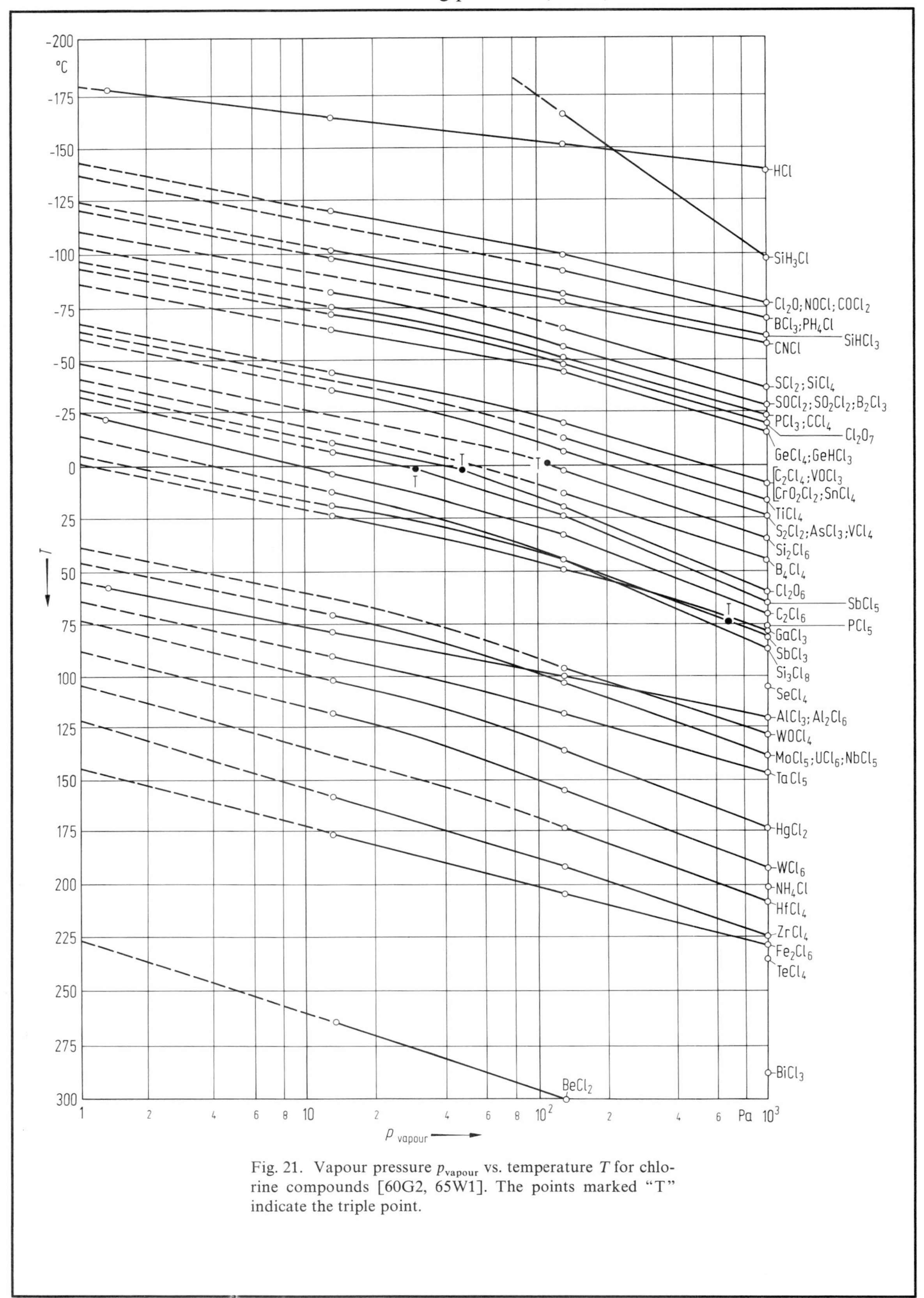

Fig. 21. Vapour pressure p_{vapour} vs. temperature T for chlorine compounds [60G2, 65W1]. The points marked "T" indicate the triple point.

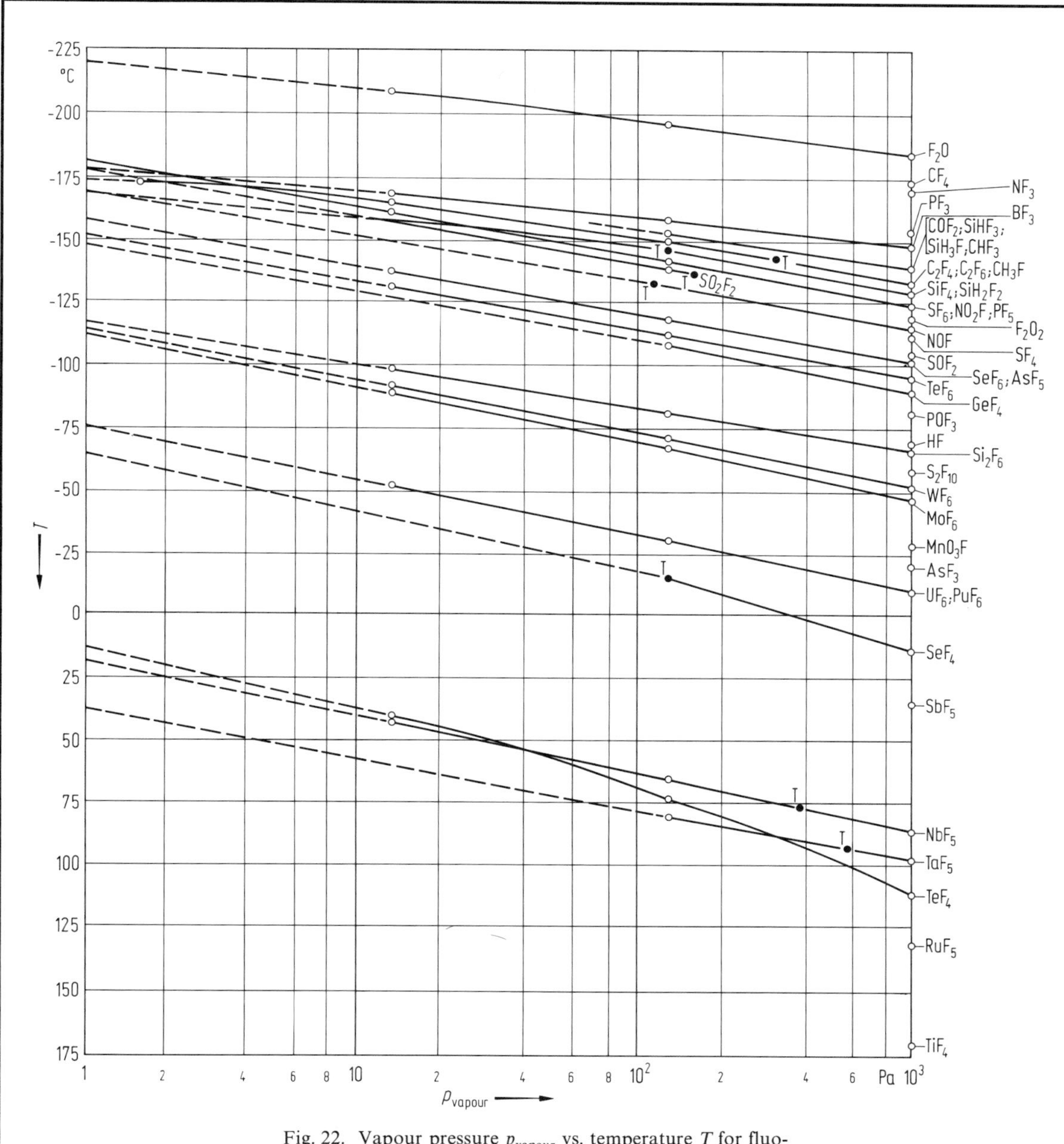

Fig. 22. Vapour pressure p_{vapour} vs. temperature T for fluorine compounds [60G2, 65W1]. The points marked "T" indicate the triple point.

For Fig. 23, see p. 566.

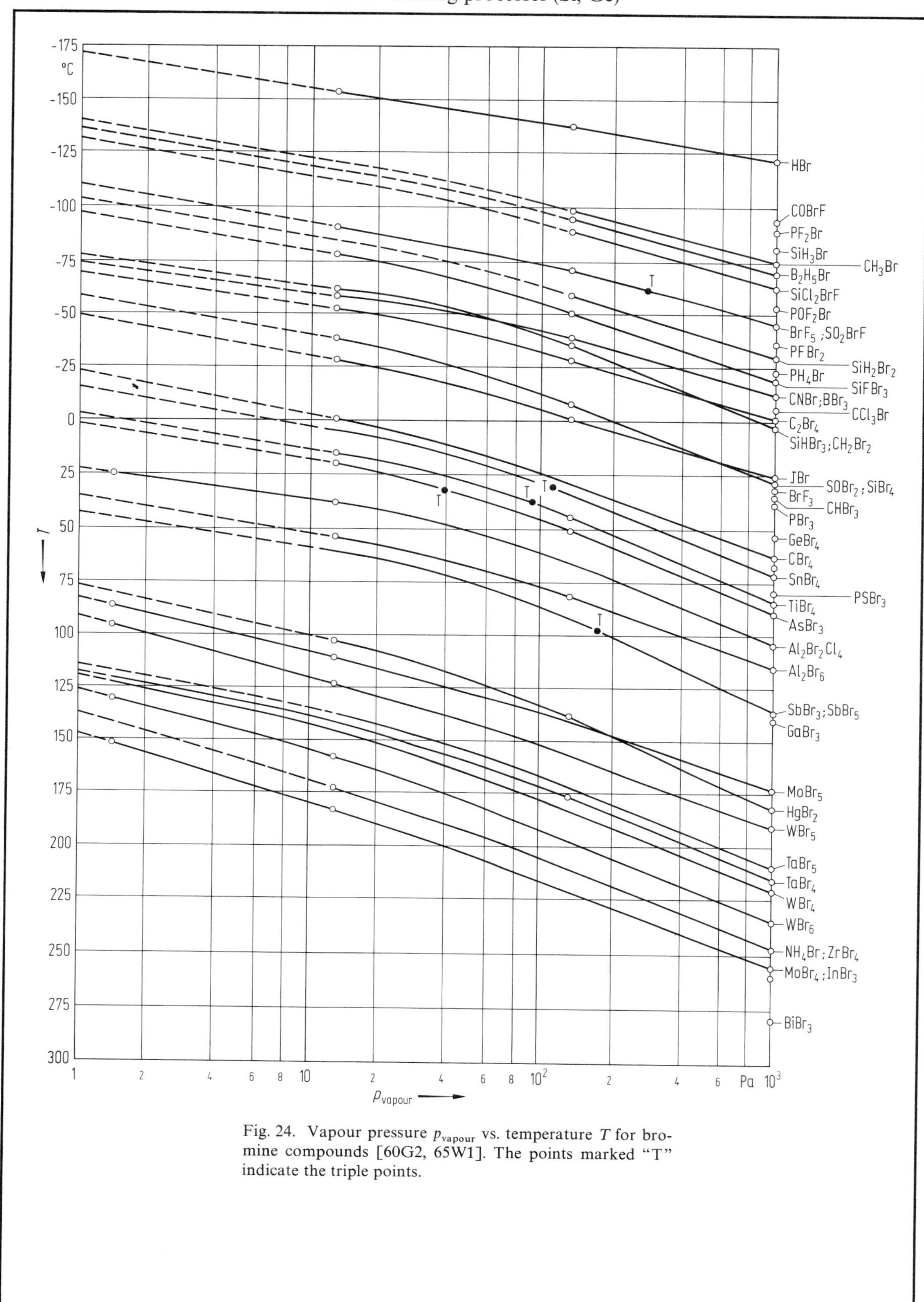

Fig. 24. Vapour pressure p_{vapour} vs. temperature T for bromine compounds [60G2, 65W1]. The points marked "T" indicate the triple points.

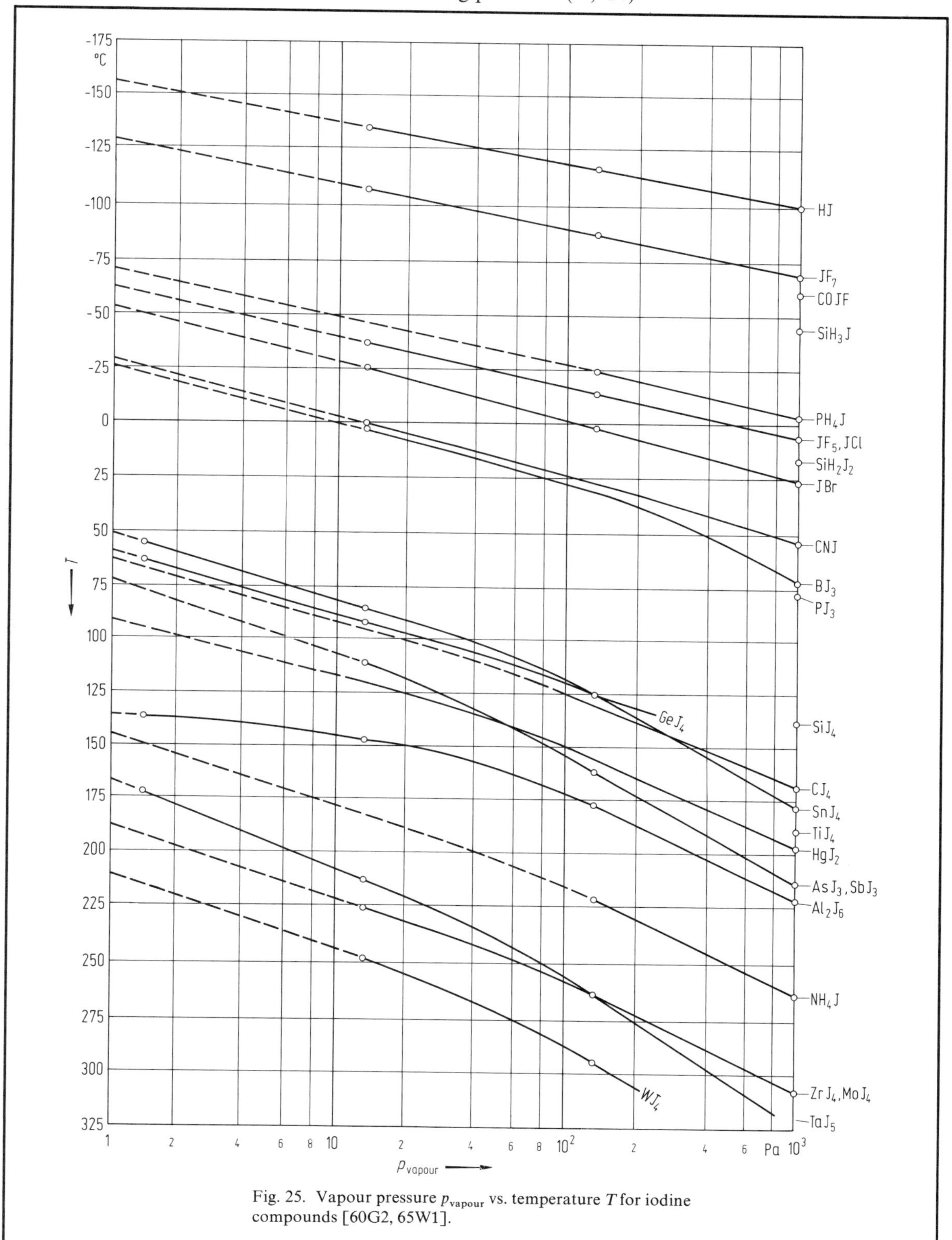

Fig. 25. Vapour pressure p_{vapour} vs. temperature T for iodine compounds [60G2, 65W1].

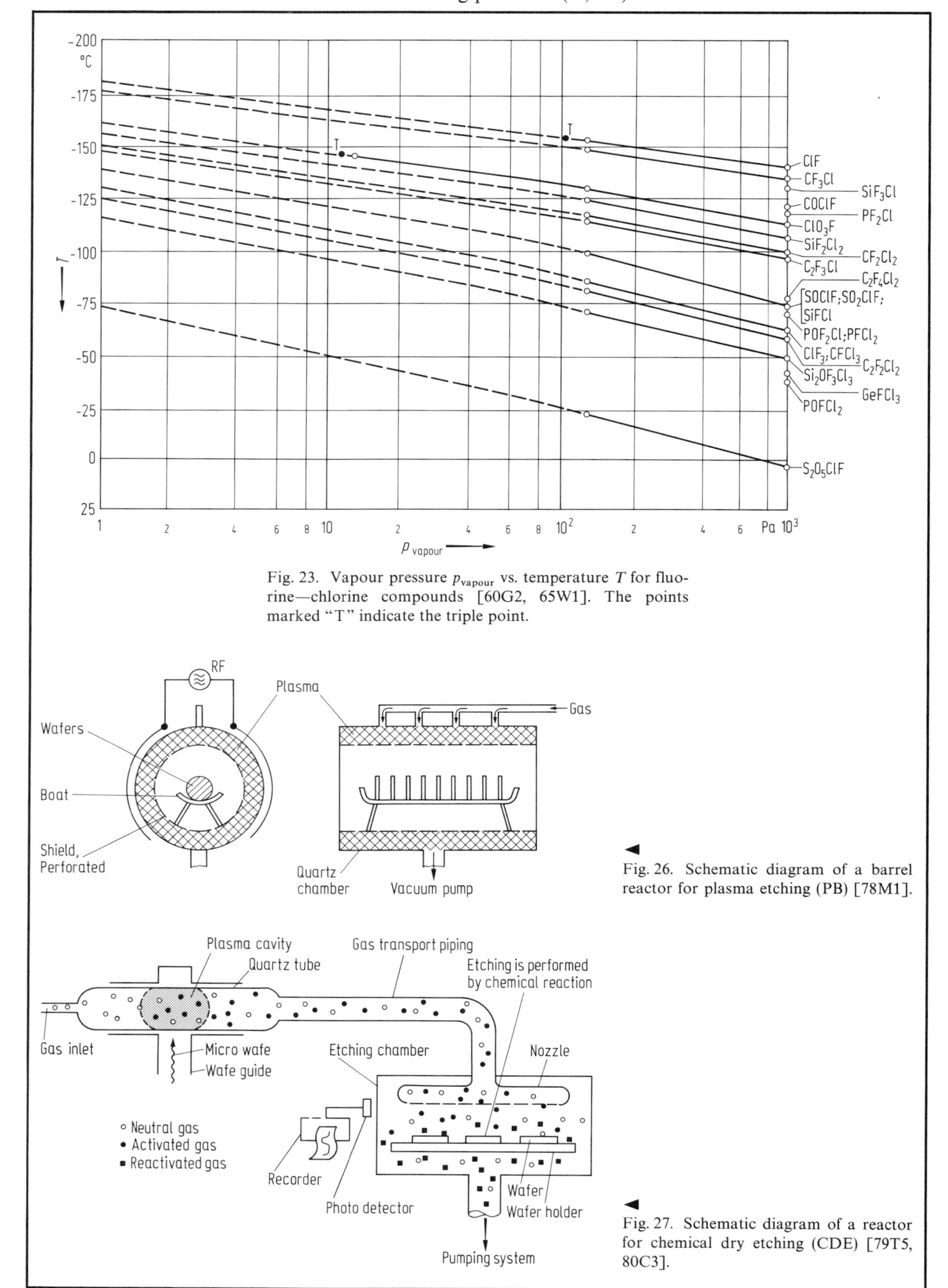

Fig. 23. Vapour pressure p_{vapour} vs. temperature T for fluorine—chlorine compounds [60G2, 65W1]. The points marked "T" indicate the triple point.

◄ Fig. 26. Schematic diagram of a barrel reactor for plasma etching (PB) [78M1].

◄ Fig. 27. Schematic diagram of a reactor for chemical dry etching (CDE) [79T5, 80C3].

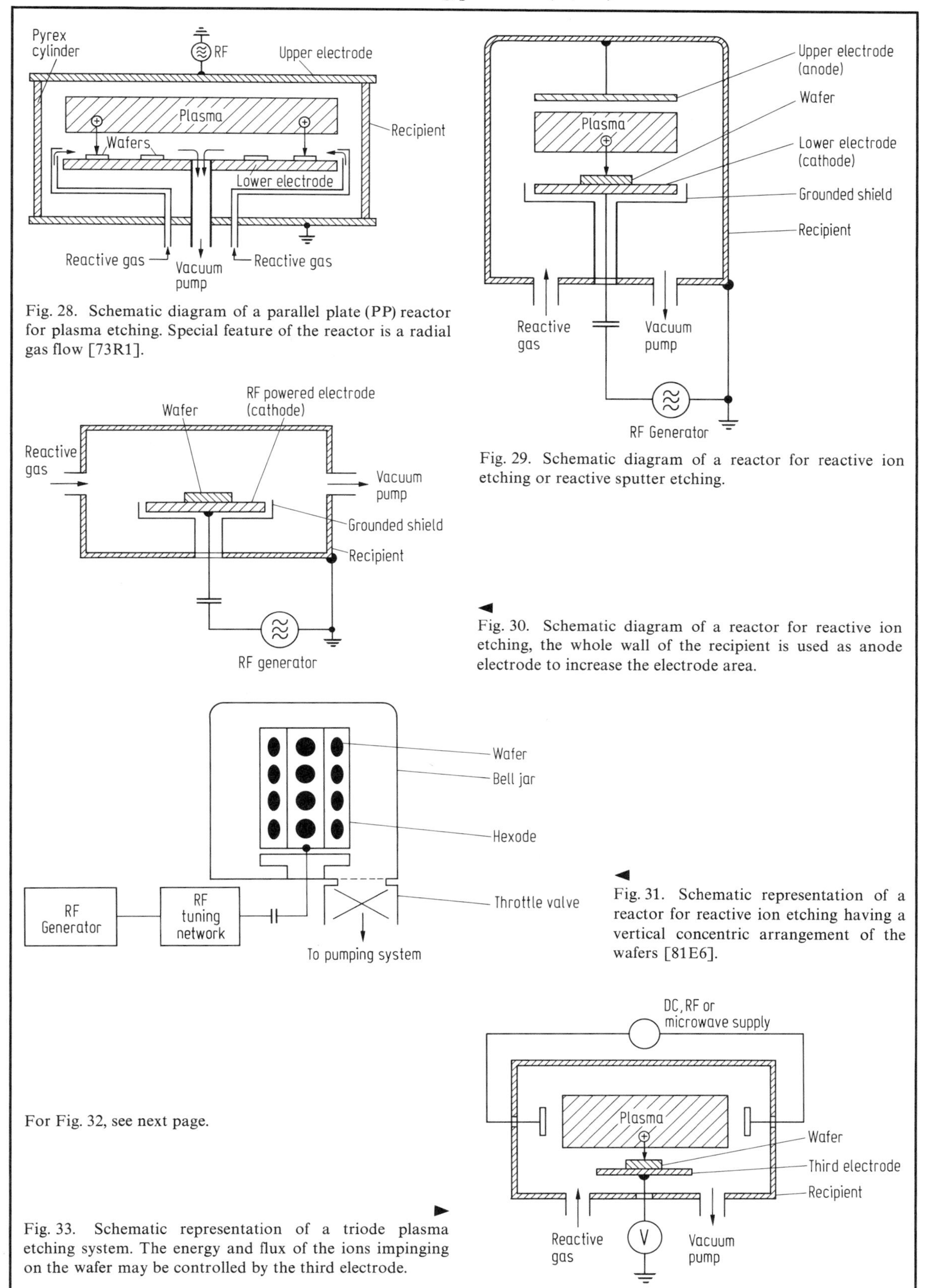

Fig. 28. Schematic diagram of a parallel plate (PP) reactor for plasma etching. Special feature of the reactor is a radial gas flow [73R1].

Fig. 29. Schematic diagram of a reactor for reactive ion etching or reactive sputter etching.

◄ Fig. 30. Schematic diagram of a reactor for reactive ion etching, the whole wall of the recipient is used as anode electrode to increase the electrode area.

◄ Fig. 31. Schematic representation of a reactor for reactive ion etching having a vertical concentric arrangement of the wafers [81E6].

For Fig. 32, see next page.

► Fig. 33. Schematic representation of a triode plasma etching system. The energy and flux of the ions impinging on the wafer may be controlled by the third electrode.

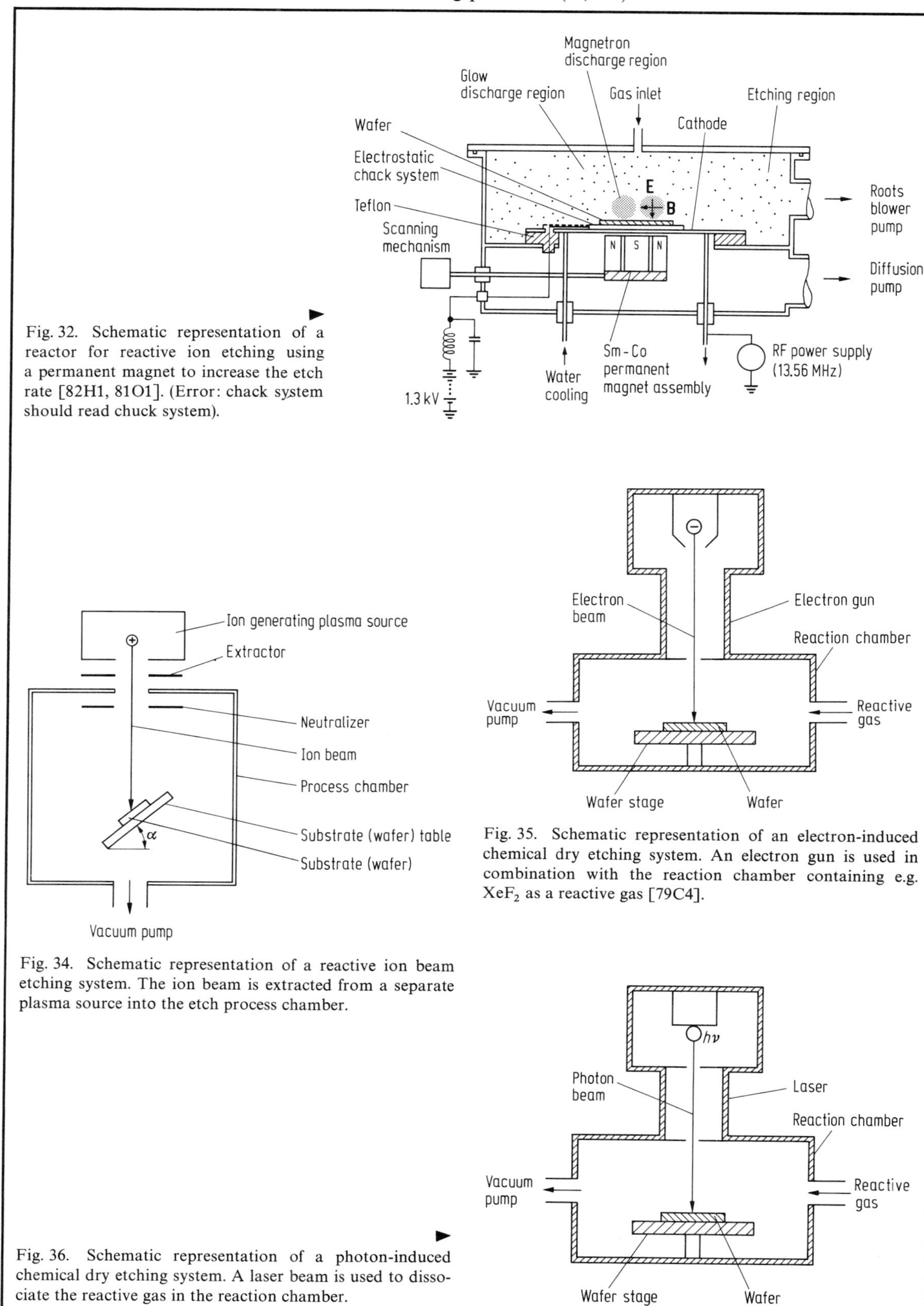

Fig. 32. Schematic representation of a reactor for reactive ion etching using a permanent magnet to increase the etch rate [82H1, 81O1]. (Error: chack system should read chuck system).

Fig. 35. Schematic representation of an electron-induced chemical dry etching system. An electron gun is used in combination with the reaction chamber containing e.g. XeF_2 as a reactive gas [79C4].

Fig. 34. Schematic representation of a reactive ion beam etching system. The ion beam is extracted from a separate plasma source into the etch process chamber.

Fig. 36. Schematic representation of a photon-induced chemical dry etching system. A laser beam is used to dissociate the reactive gas in the reaction chamber.

6.1.4.8 Final device preparation

(**Note:** if not stated otherwise all dimensions are given in mm)

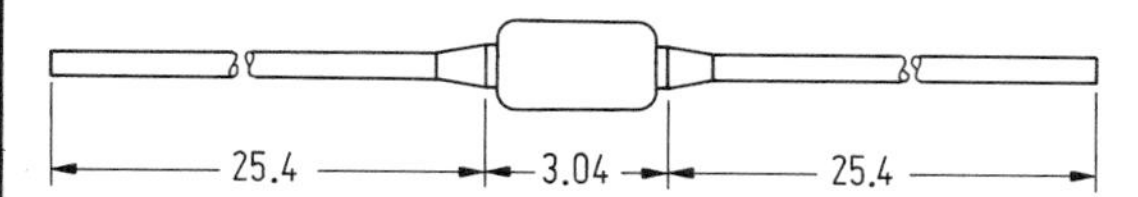

Fig. 1. Outline of a DO-34 package. (All dimensions are given in mm.)

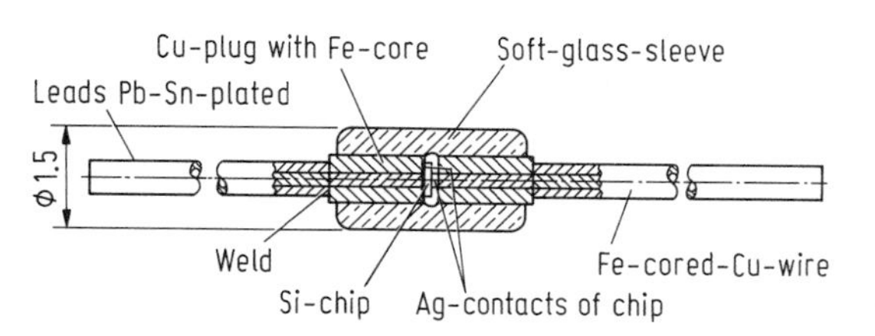

Fig. 2. Cross section of a DO-34 package.

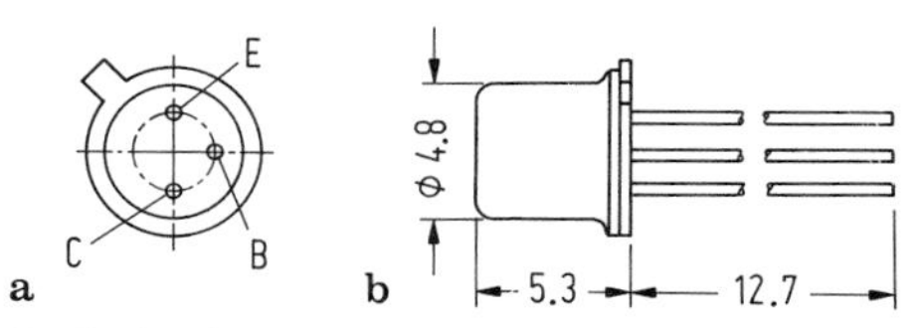

Fig. 3a, b. Outline of a TO-18 package, a) bottom view and, b) side view; E, B, and C are emitter, base, and collector leads, respectively. (All dimensions are given in mm.)

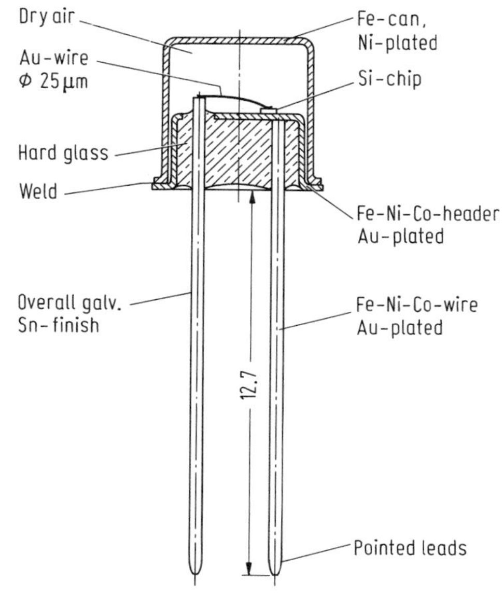

Fig. 4. Cross section of a TO-18 package for Si-transistors.

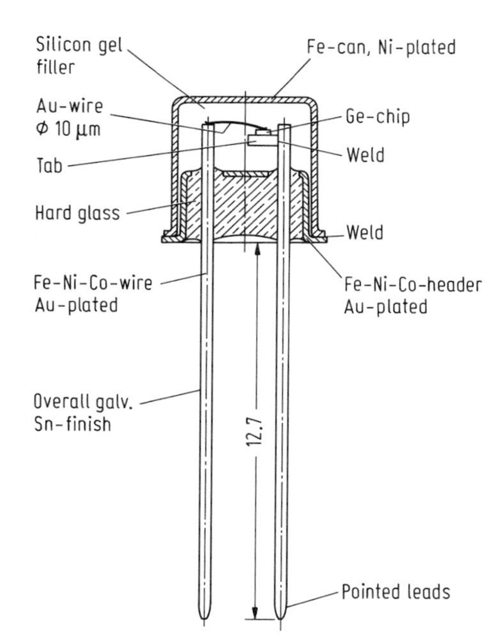

Fig. 5. Cross section of a TO-18 package for high frequency Ge-transistors.

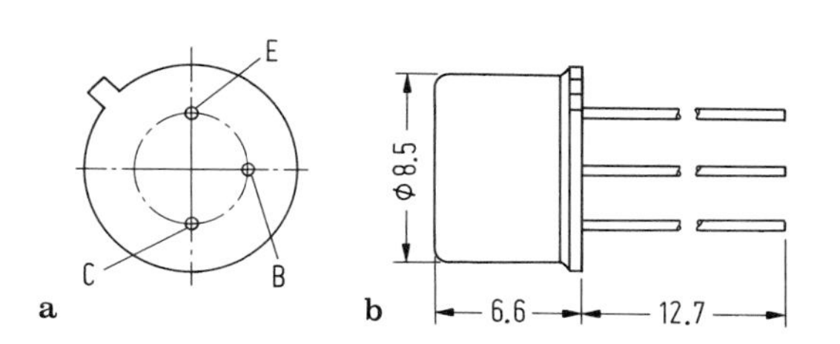

Fig. 6a, b. Outline of a TO-39 package. E, B, and C are emitter, base, and collector, respectively. a) bottom view and b) side view. (All dimensions are given in mm.)

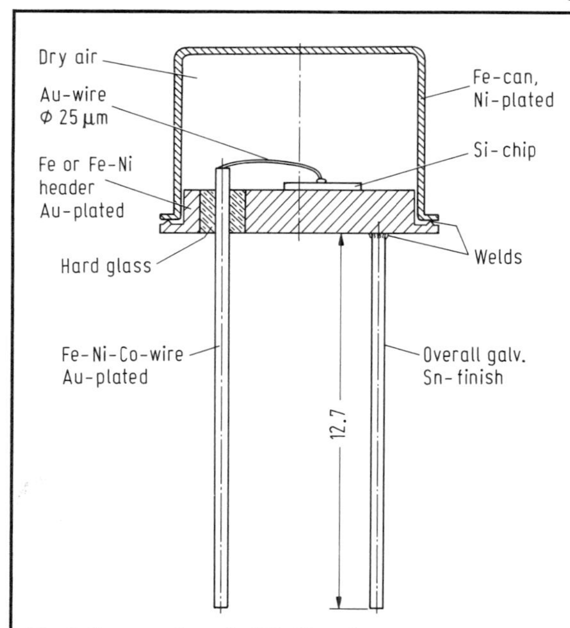

Fig. 7. Cross section of a TO-39 package.

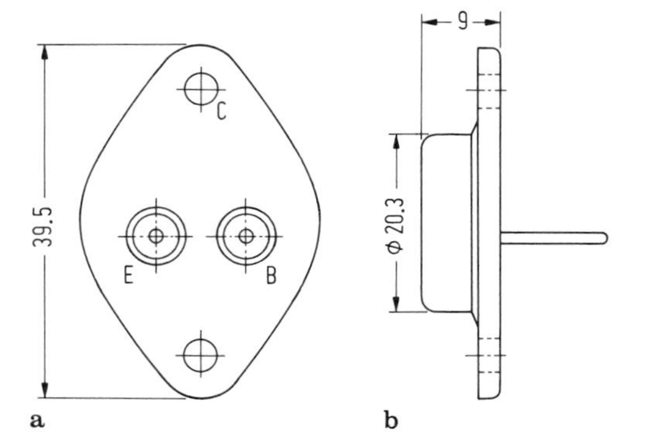

Fig. 8a, b. Outline of a TO-3 package. E, B, and C are emitter, base, and collector leads, respectively. a) bottom view, b) side view. (All dimensions are given in mm.)

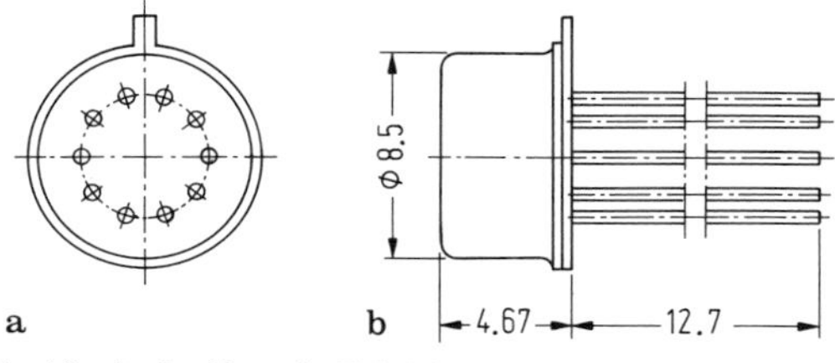

Fig. 10a, b. Outline of a TO-74 package. a) bottom view and, b) side view.

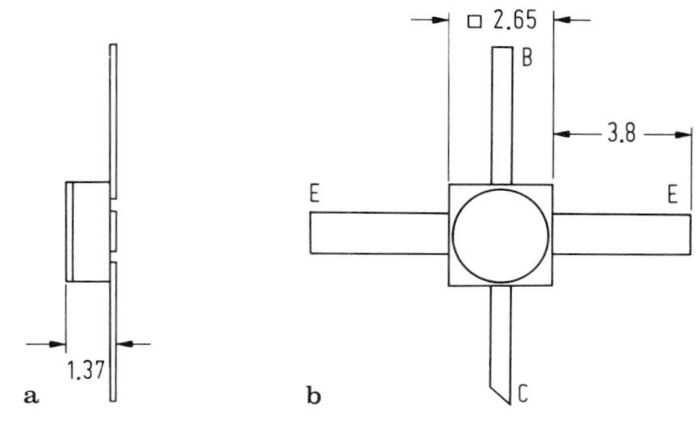

Fig. 9a, b. Outline of a ceramic package for high frequency transistors, JEDEC TO-120. E, B, and C are emitter, base, and collector leads, respectively a) side view, b) top view. (All dimensions are given in mm.)

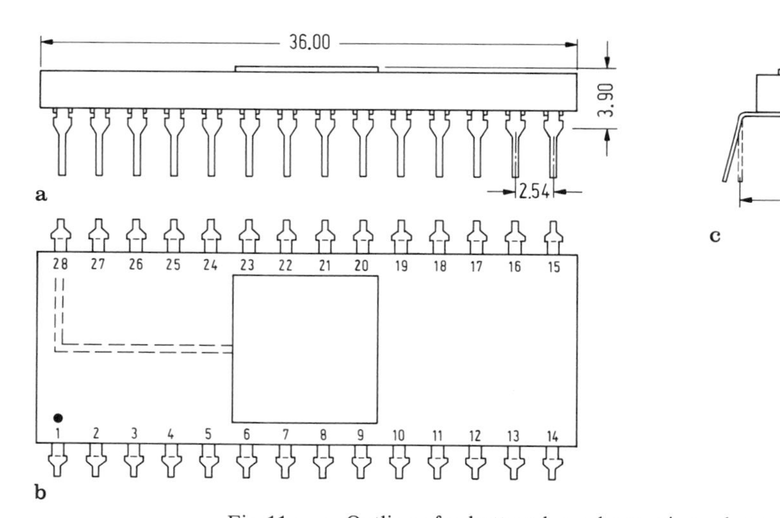

Fig. 11a···c. Outline of a bottom-brazed ceramic package, the "CERDIL". a) side view, b) top view, c) front side view. (All dimensions are given in mm.)

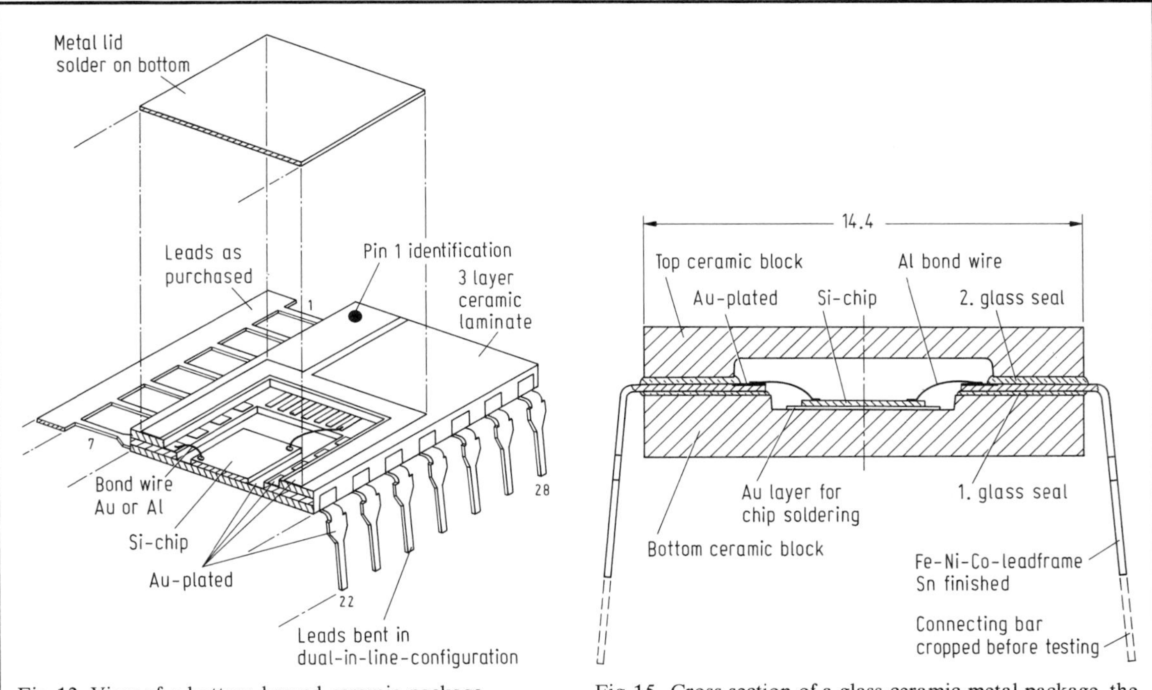

Fig. 12. View of a bottom-brazed ceramic package.

Fig. 15. Cross section of a glass-ceramic-metal package, the CERDIP.

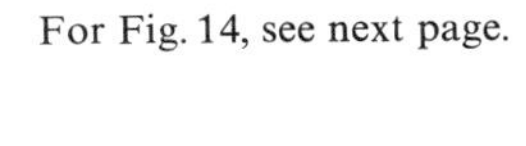

For Fig. 14, see next page.

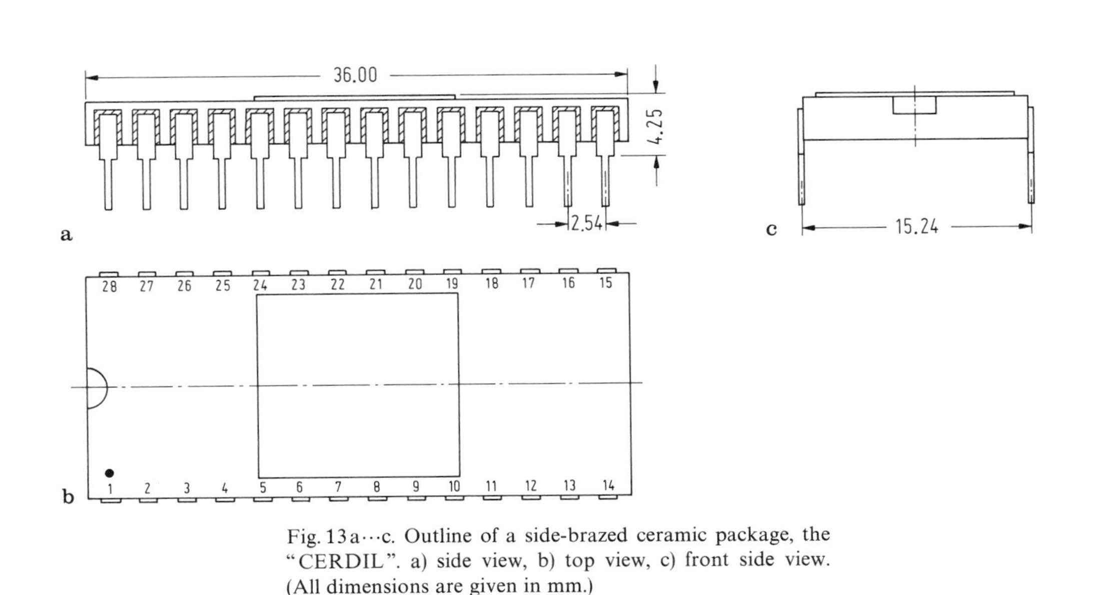

Fig. 13a···c. Outline of a side-brazed ceramic package, the "CERDIL". a) side view, b) top view, c) front side view. (All dimensions are given in mm.)

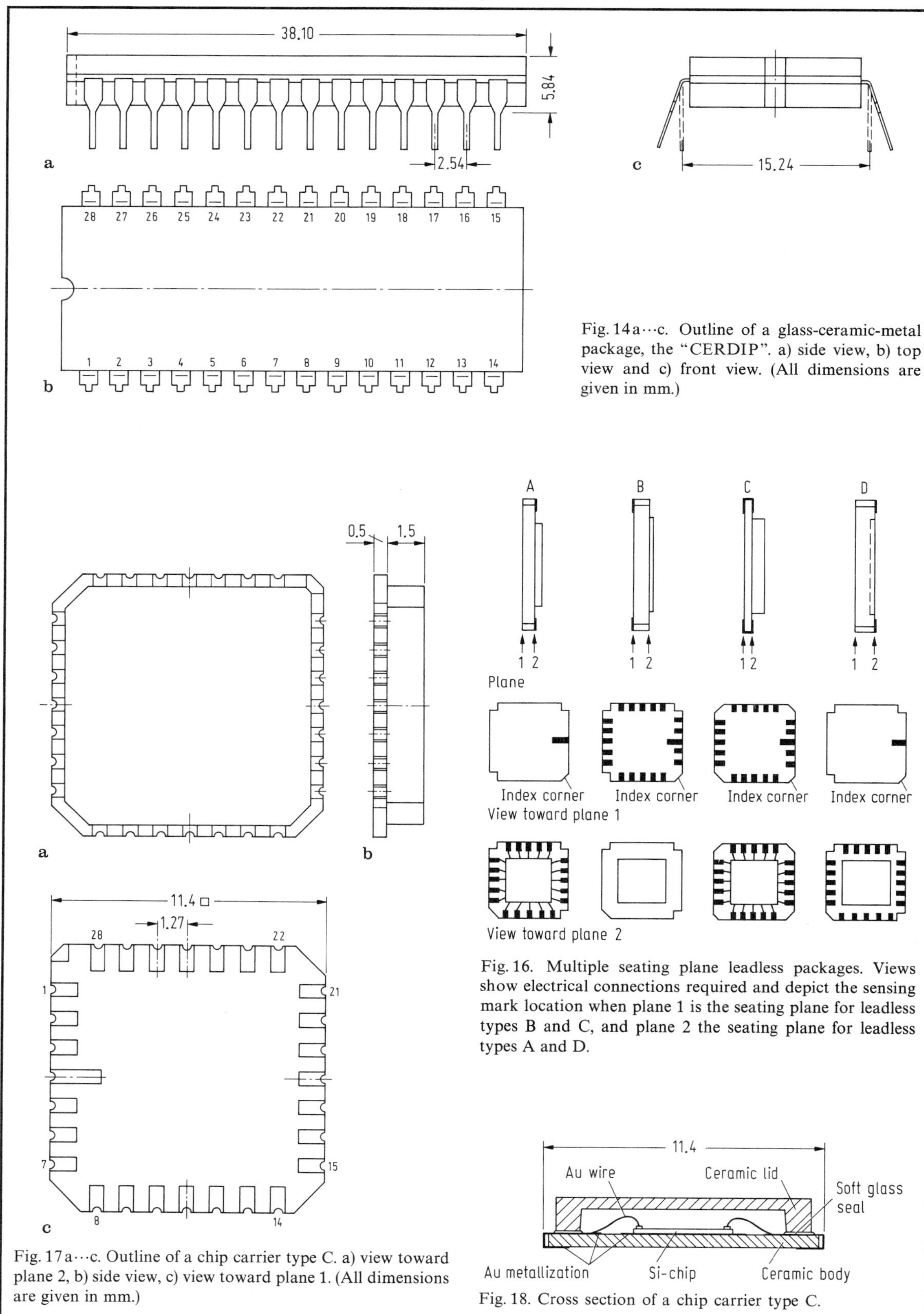

Fig. 14a···c. Outline of a glass-ceramic-metal package, the "CERDIP". a) side view, b) top view and c) front view. (All dimensions are given in mm.)

Fig. 16. Multiple seating plane leadless packages. Views show electrical connections required and depict the sensing mark location when plane 1 is the seating plane for leadless types B and C, and plane 2 the seating plane for leadless types A and D.

Fig. 17a···c. Outline of a chip carrier type C. a) view toward plane 2, b) side view, c) view toward plane 1. (All dimensions are given in mm.)

Fig. 18. Cross section of a chip carrier type C.

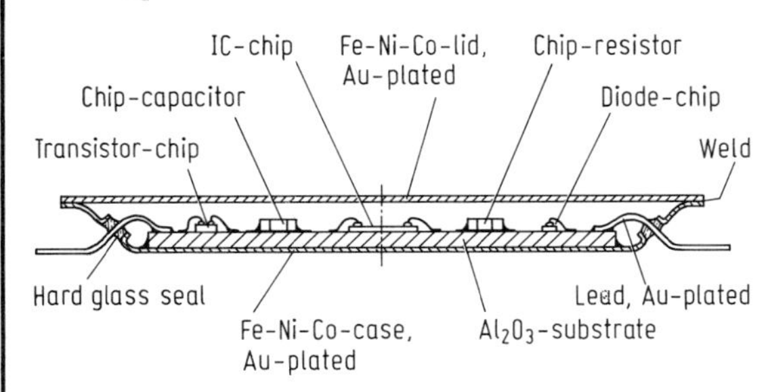

Fig. 19. Cross section of a hermetic hybrid package (not to scale).

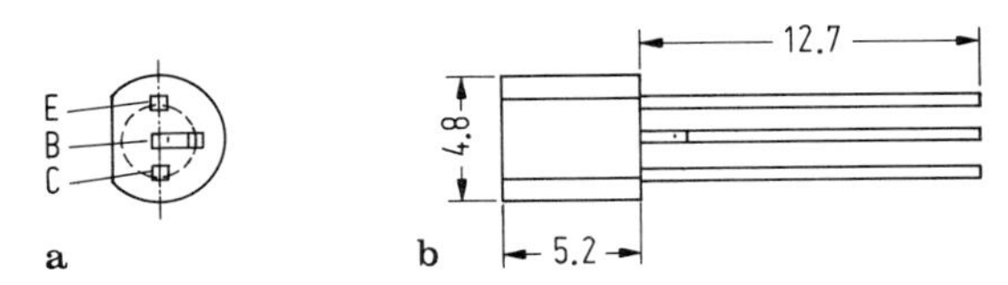

Fig. 20a, b. Outline of a TO-92 package. E, B, and C are emitter, base and collector, respectively. a) bottom view and b) side view. (All dimensions are given in mm.)

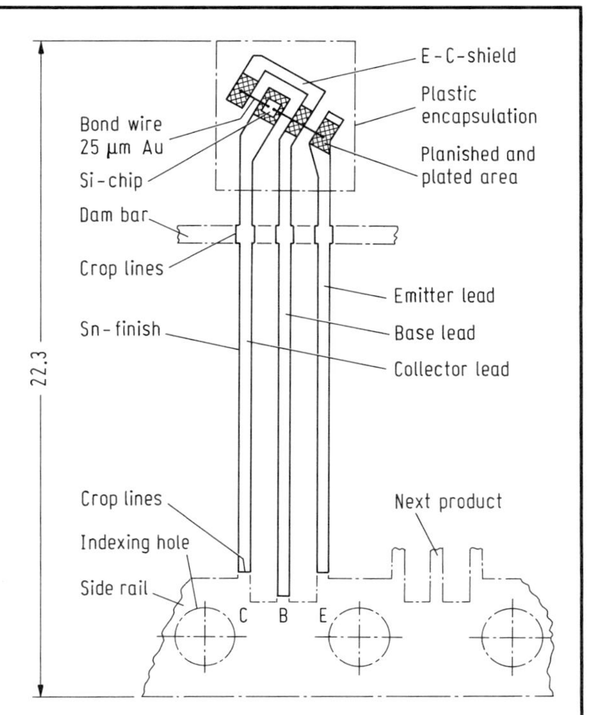

Fig. 22. Construction of a TO-92 package.

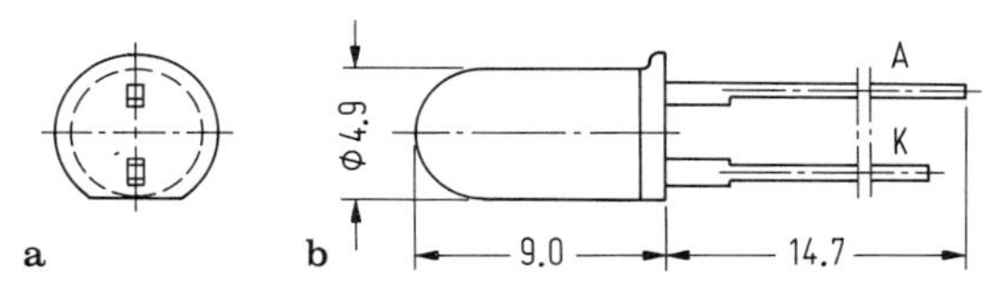

Fig. 21 a, b. Single-ended plastic package for light emitting diodes. A and K are anode and cathode, respectively

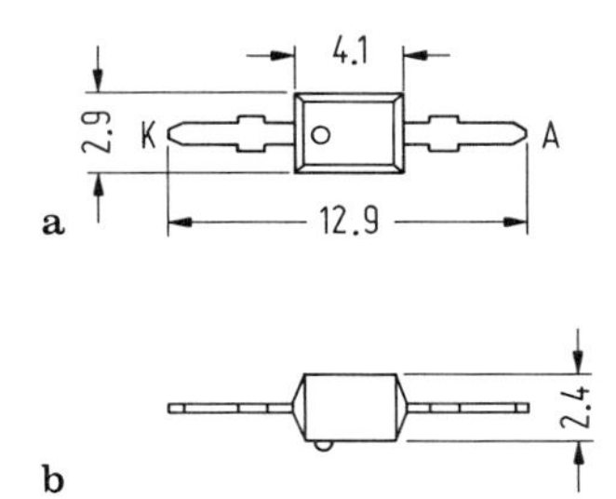

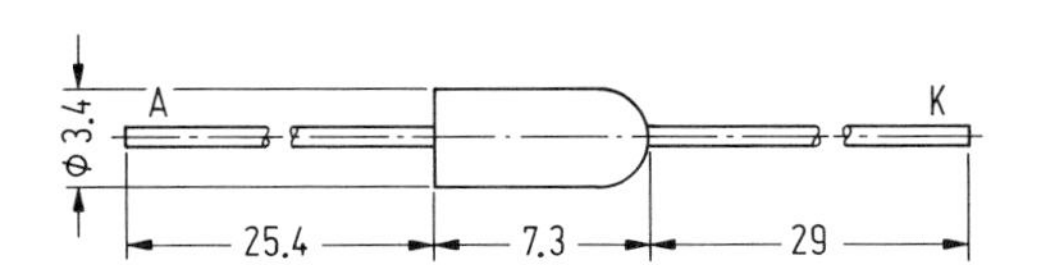

Fig. 23. Outline of a DO-14 package. (All dimensions are given in mm.)

Fig. 24a, b. Outline of a SOD-23 package for UHF varicap and switching diodes. K is the cathode and A is the anode. a) top view, b) side view. (All dimensions are given in mm.)

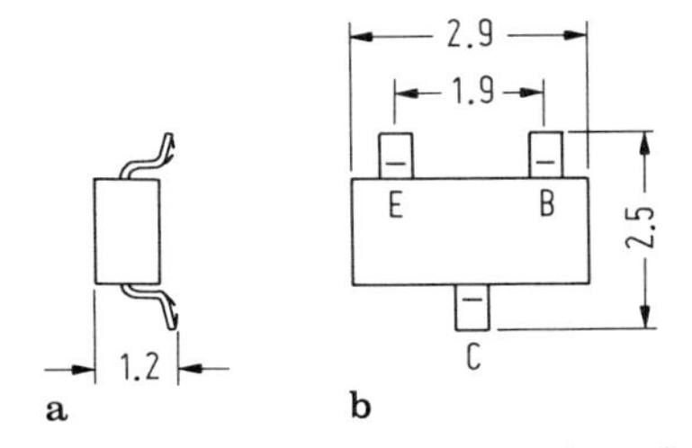

Fig. 25. Outline of a SOT-23 package for thickfilm applications. E, B, and C are emitter, base, and collector, respectively. a) side view, b) top view. (All dimensions are given in mm.)

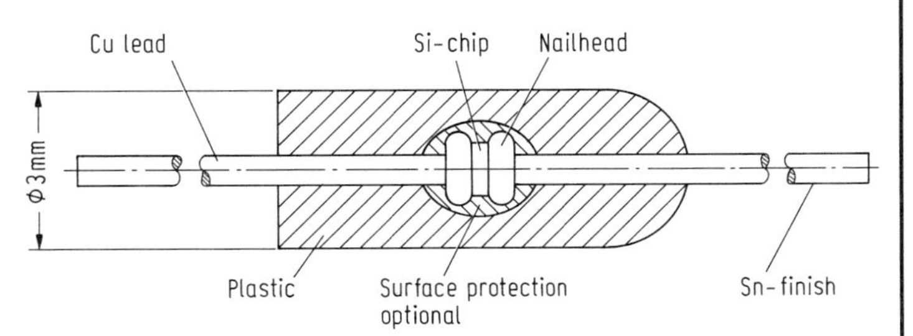

Fig. 26. Cross section of a DO-14 package for general purpose diodes.

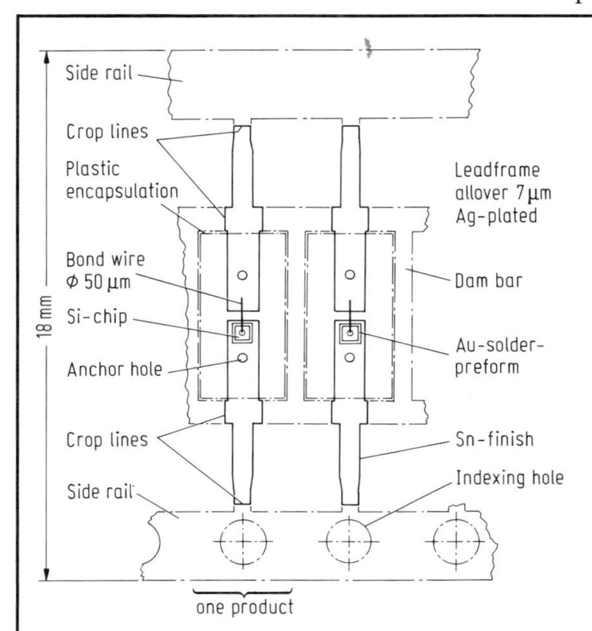

Fig. 27. Construction of a double-ended diode package, (SOD-23).

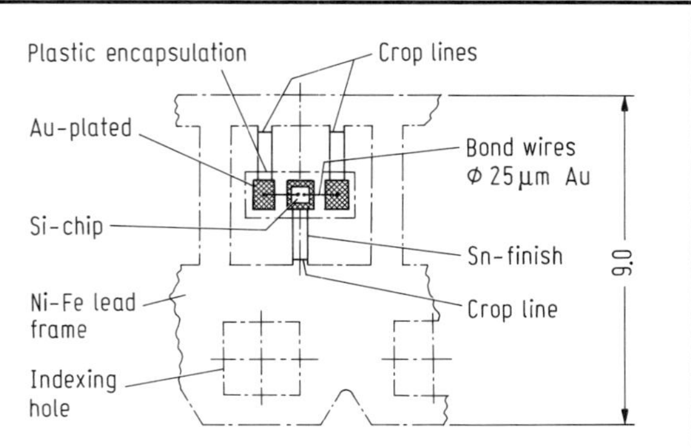

Fig. 28. Construction of a double-ended transistor package, (SOT-23).

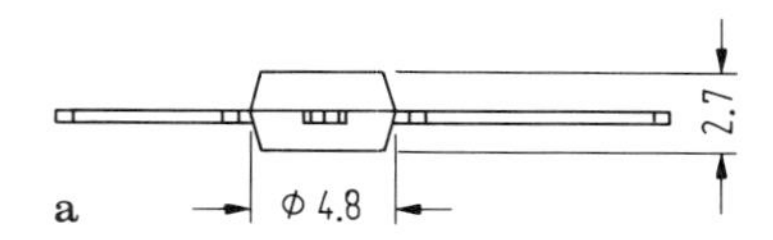

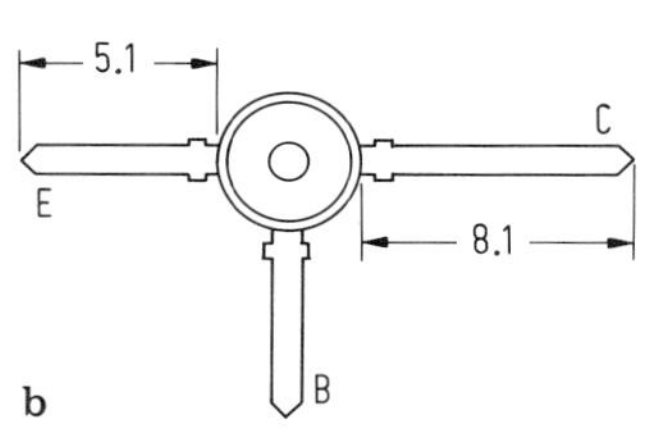

Fig. 29 a, b. Outline of a TO-119 package, a) side view, b) top view. E, B, and C are emitter, base, and collector, respectively. (All dimensions are given in mm.)

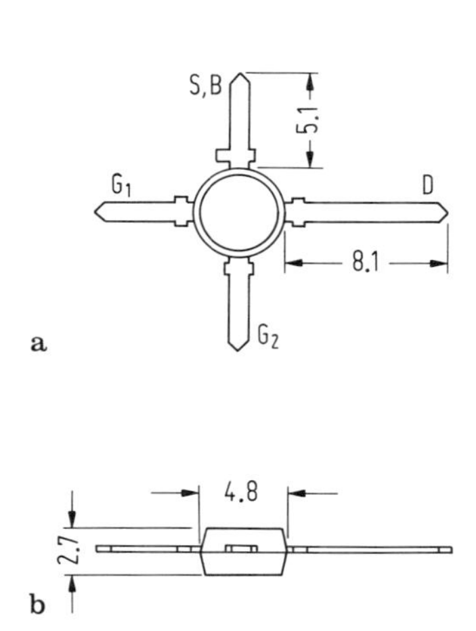

Fig. 30. Outline of a SOT-103 package, a 4-lead version of TO-119. S, D, G are source, drain, and gate leads, respectively. (All dimensions are given in mm.)

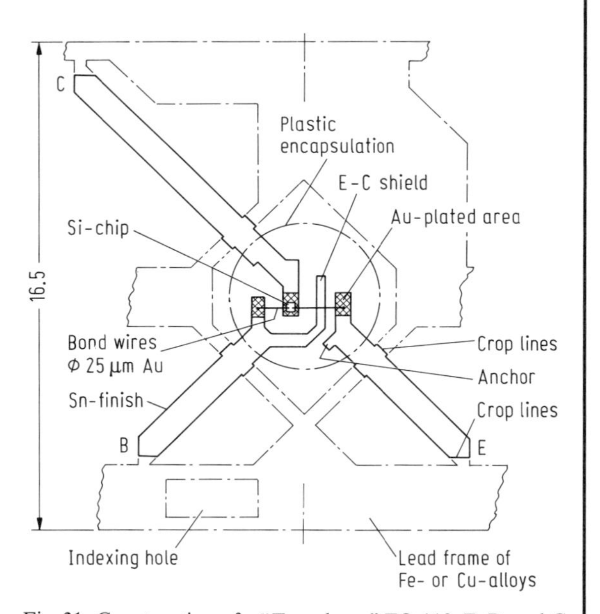

Fig. 31. Construction of a "T-package" TO-119, E, B, and C are the emitter, base, and collector leads, respectively.

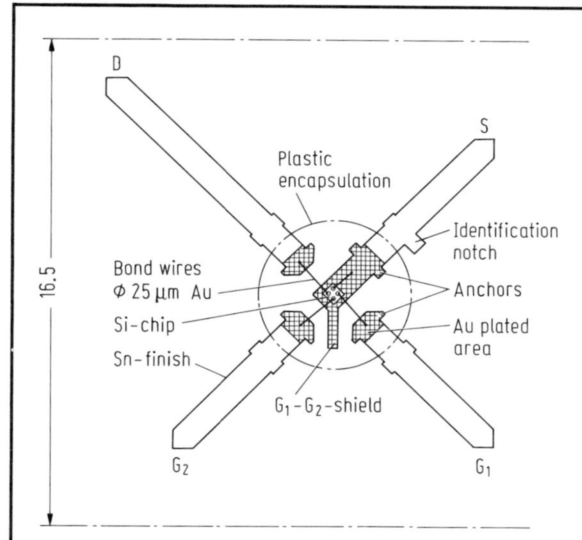

Fig. 32. Construction of an "X-package" SOT-103 for tetrodes; the lead frame details are analogous to Fig. 31. S, D, and G are source, drain and gate leads, respectively.

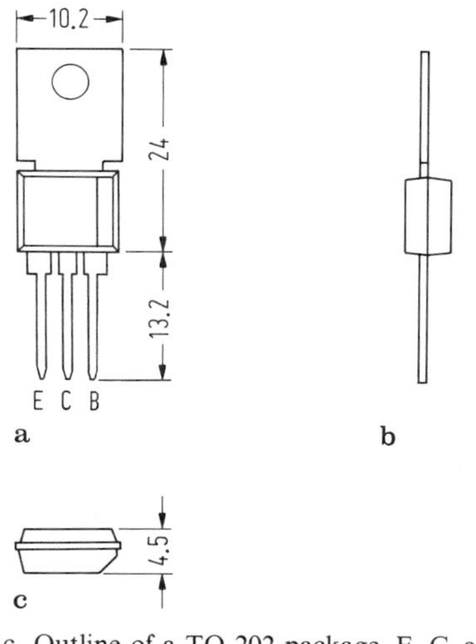

Fig. 34a···c. Outline of a TO-202 package. E, C, and B are the emitter, collector and base leads, respectively. a) top view, b) side view, c) front side view. (All dimensions given in mm.)

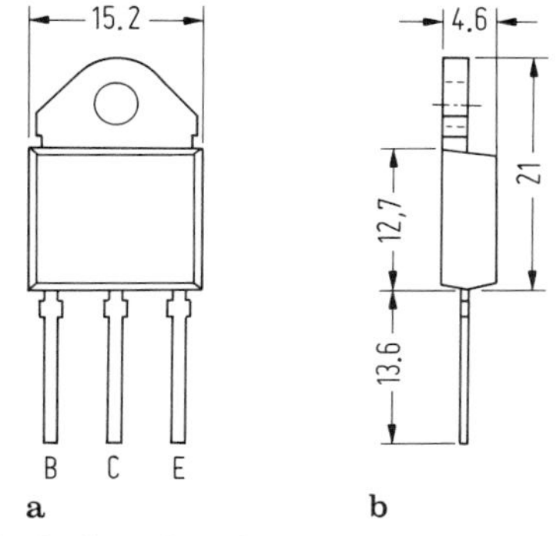

Fig. 33a, b. Outline of a SOT-93 package. B, C, and E are the base, collector, and emitter leads, respectively. a) top view, b) side view. (All dimensions are given in mm.)

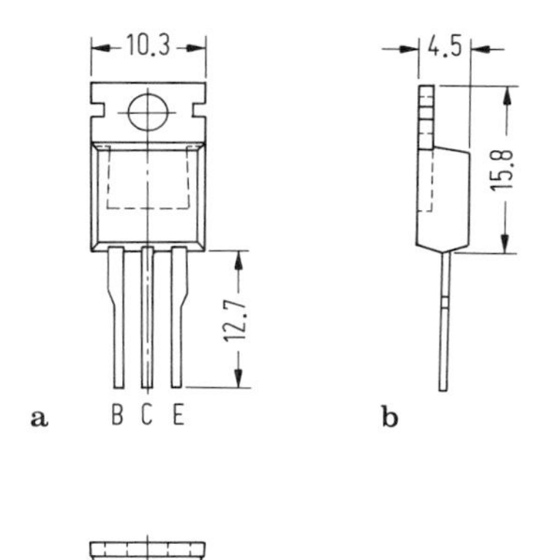

Fig. 35a···c. Outline of a TO-220 package for Si-power transistors. B, C, and E are the base, collector and emitter leads, respectively. a) top view, b) side view, c) front side view. (All measures are given in mm.)

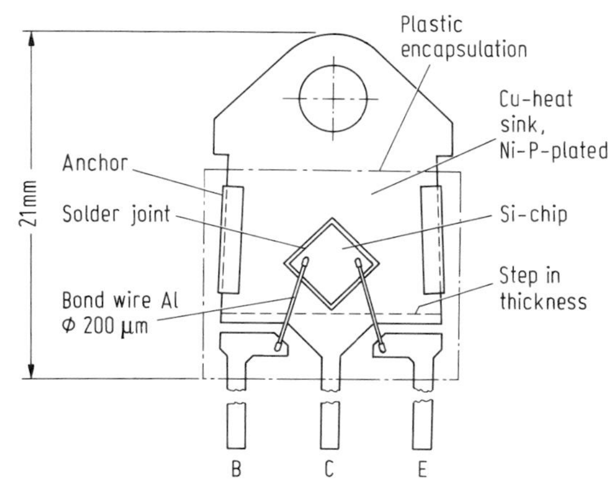

Fig. 36. Construction of a SOT-93 package.

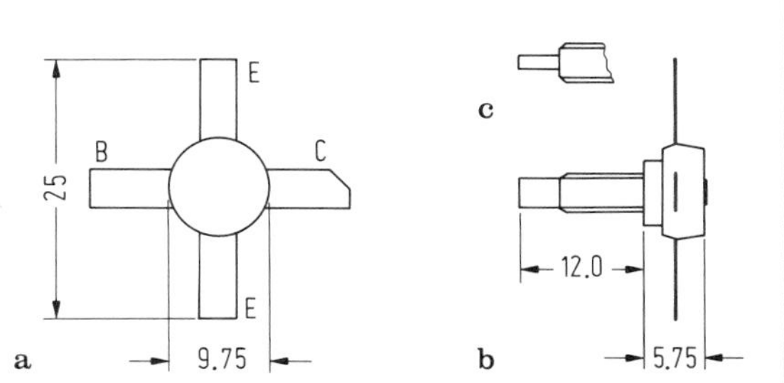

Fig. 37a···c. Outline of a TO-117 package. E, B, and C are emitter, base, and collector, respectively. a) top view, b) side view, c) detail thread tip. (All dimensions are given in mm.)

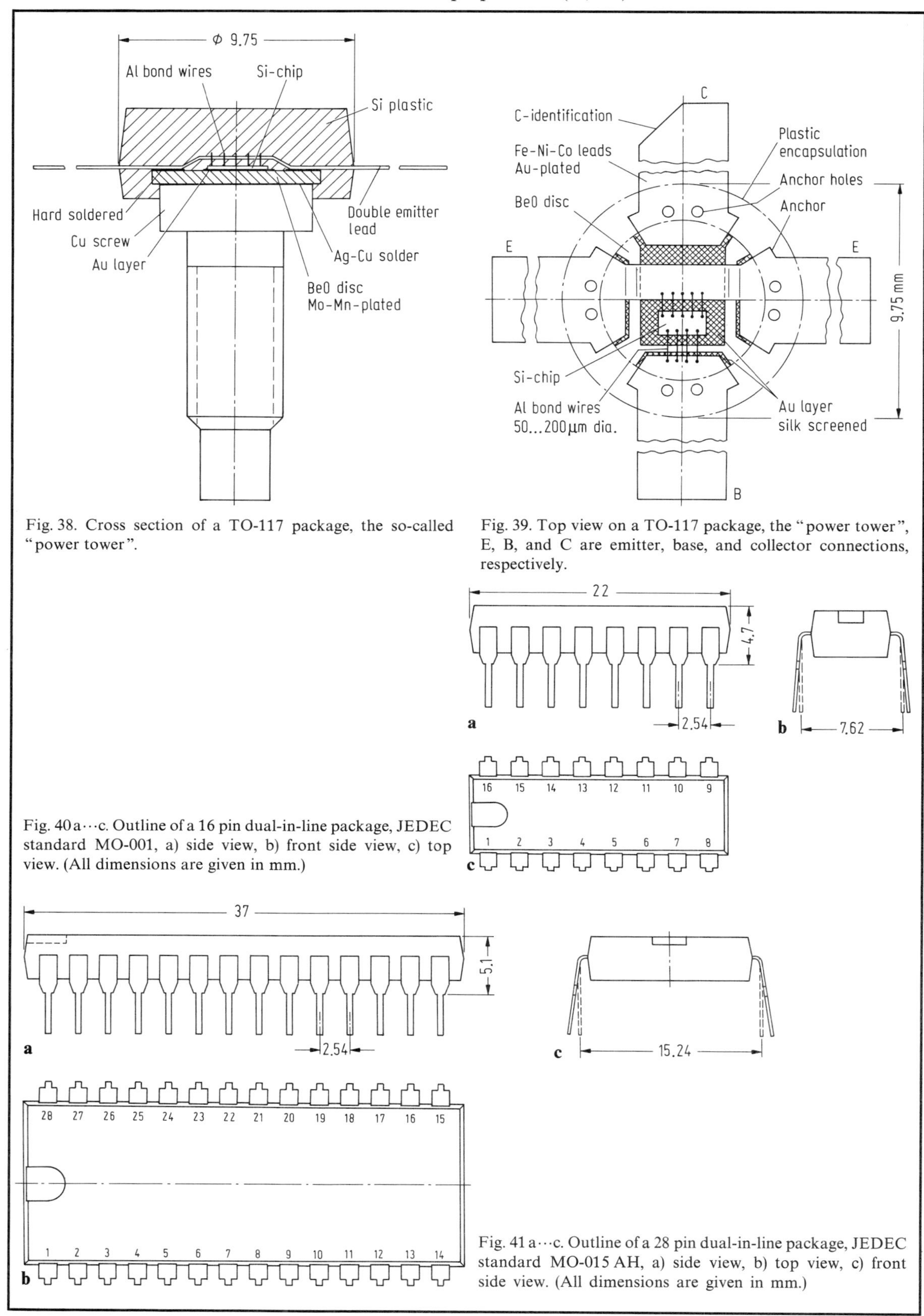

Fig. 38. Cross section of a TO-117 package, the so-called "power tower".

Fig. 39. Top view on a TO-117 package, the "power tower", E, B, and C are emitter, base, and collector connections, respectively.

Fig. 40 a···c. Outline of a 16 pin dual-in-line package, JEDEC standard MO-001, a) side view, b) front side view, c) top view. (All dimensions are given in mm.)

Fig. 41 a···c. Outline of a 28 pin dual-in-line package, JEDEC standard MO-015 AH, a) side view, b) top view, c) front side view. (All dimensions are given in mm.)

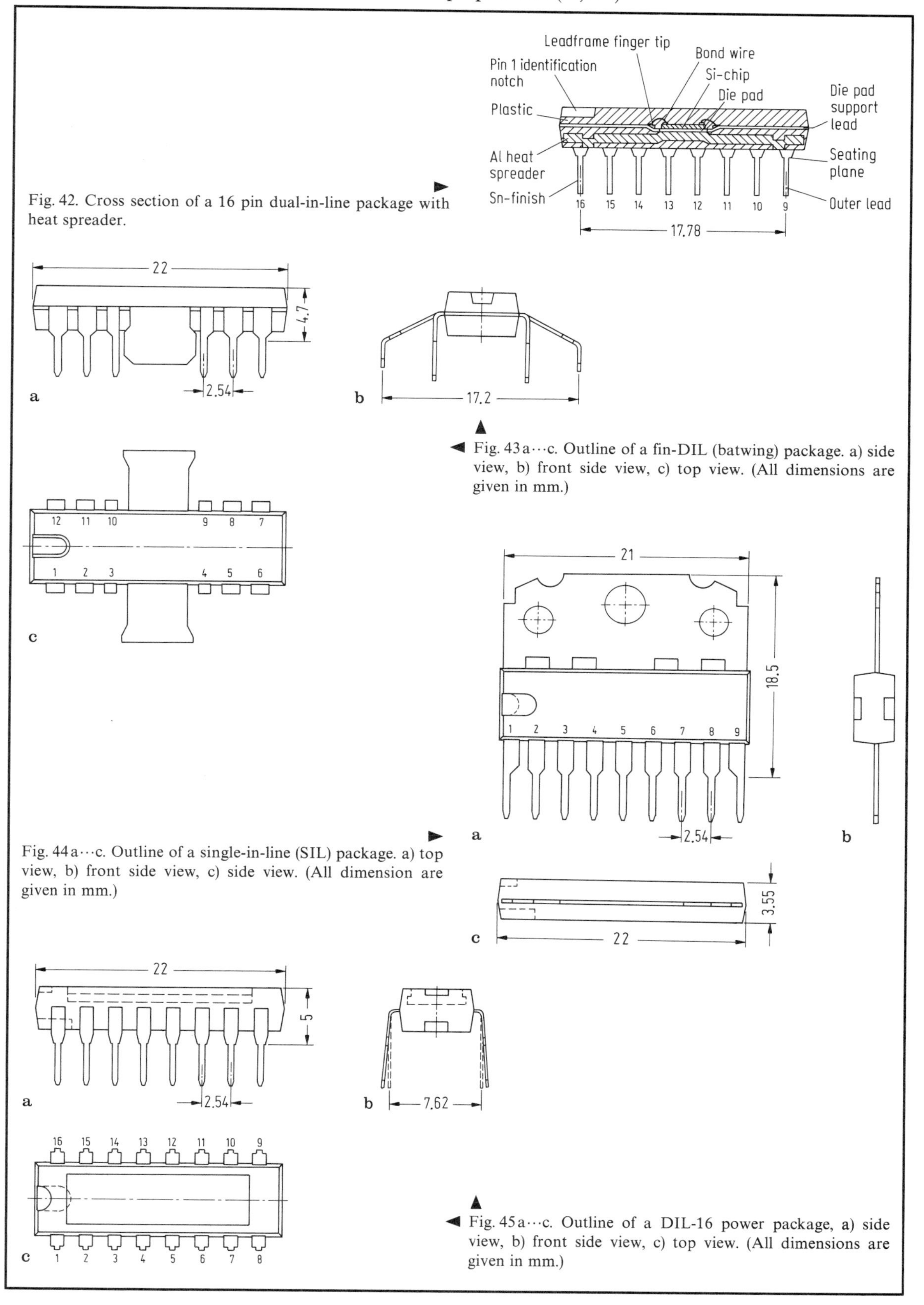

Fig. 42. Cross section of a 16 pin dual-in-line package with heat spreader.

Fig. 43a···c. Outline of a fin-DIL (batwing) package. a) side view, b) front side view, c) top view. (All dimensions are given in mm.)

Fig. 44a···c. Outline of a single-in-line (SIL) package. a) top view, b) front side view, c) side view. (All dimension are given in mm.)

Fig. 45a···c. Outline of a DIL-16 power package, a) side view, b) front side view, c) top view. (All dimensions are given in mm.)

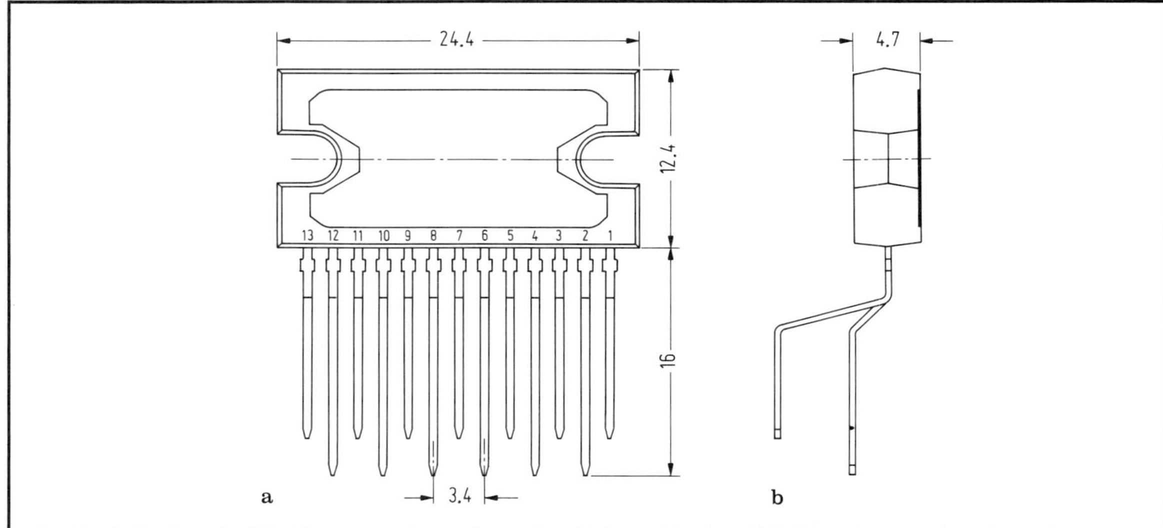

Fig. 46a, b. Outline of a SIL-13 power package. a) top view, b) front side view. (All dimensions are given in mm.)

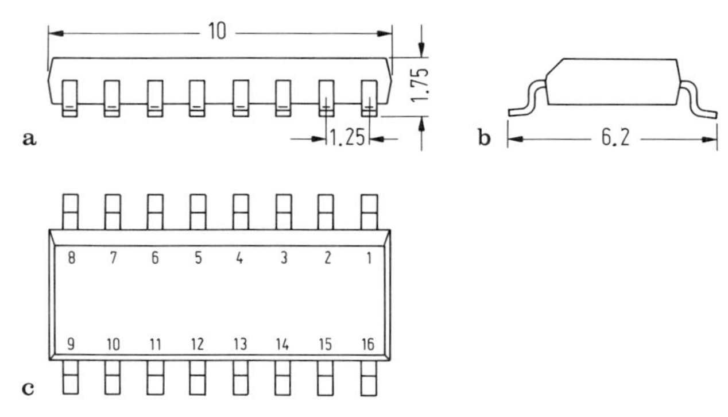

Fig. 47a···c. Outline of a SO-16 package a) side view, b) front side view, c) top view. (All dimensions are given in mm.)

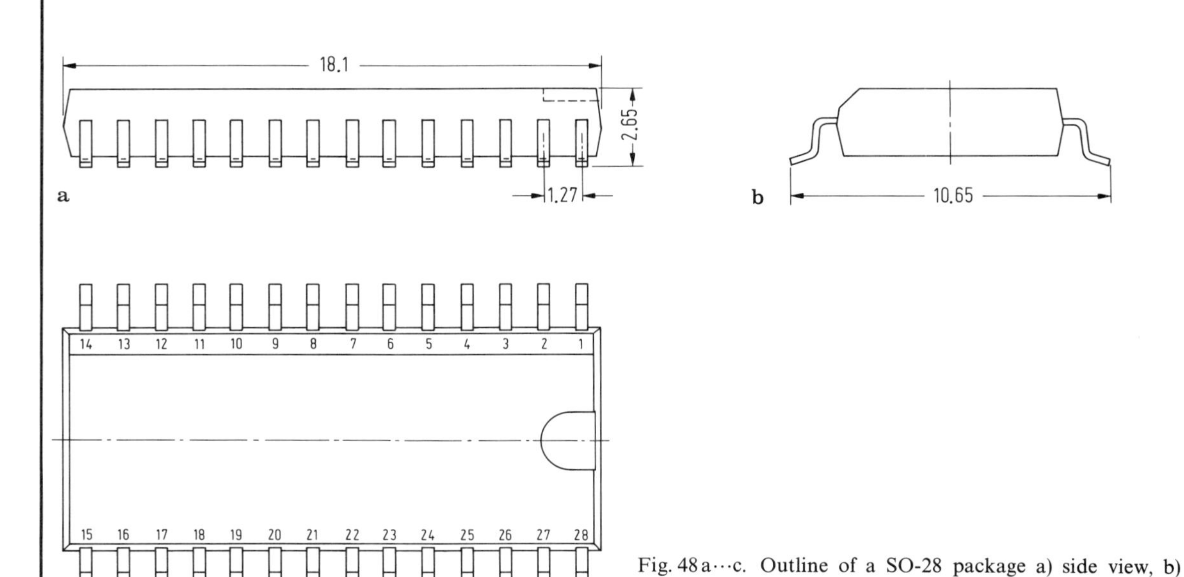

Fig. 48a···c. Outline of a SO-28 package a) side view, b) front side view, c) top view. (All dimensions are given in mm.)

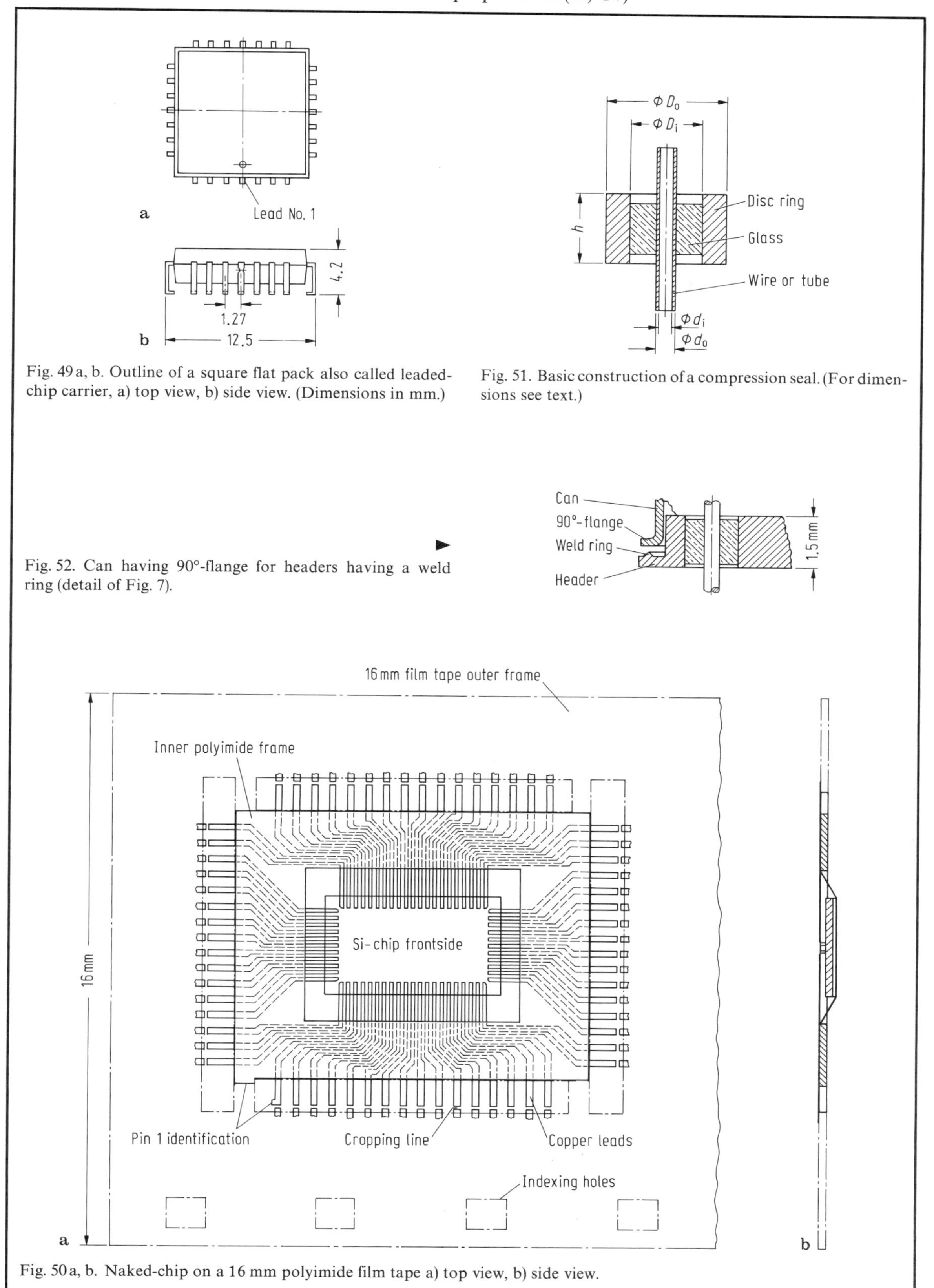

Fig. 49a, b. Outline of a square flat pack also called leaded-chip carrier, a) top view, b) side view. (Dimensions in mm.)

Fig. 51. Basic construction of a compression seal. (For dimensions see text.)

Fig. 52. Can having 90°-flange for headers having a weld ring (detail of Fig. 7).

Fig. 50a, b. Naked-chip on a 16 mm polyimide film tape a) top view, b) side view.

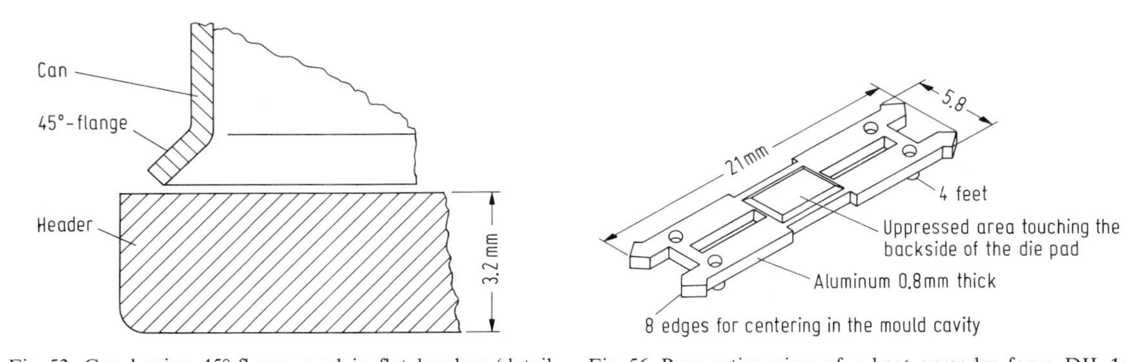

Fig. 53. Can having 45°-flange used in flat headers (detail of Fig. 8).

Fig. 56. Perspective view of a heat spreader for a DIL-16 package.

For Fig. 55, see next page.

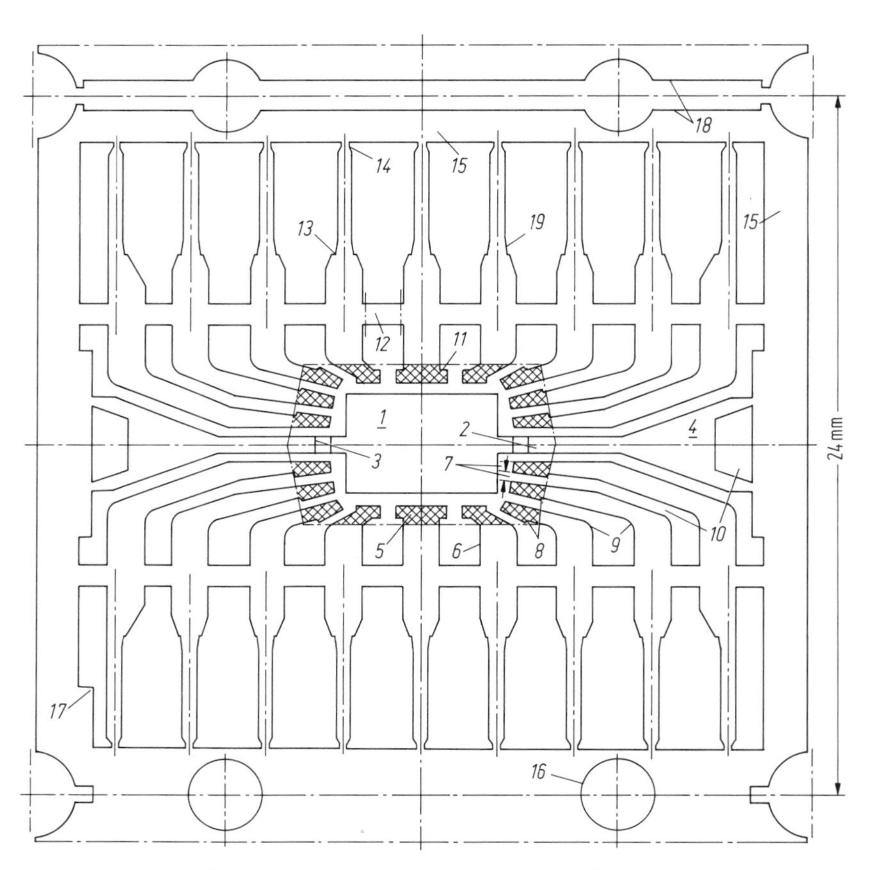

Fig. 54. Lead frame of an 18 lead dual-in-line plastic package: *1* die pad, *2* die pad support lead, *3* pad depression lines, *4* die support end, *5* inner lead tips, *6* leads, *7* spacing of slots and lead tips, *8* anchor steps, *9* rounded corners, *10* wide slots, *11* anchored lead tips, *12* dam bar, *13* seating shoulder, *14* pointed outer lead tip, *15* outer frame, *16* locating holes, *17* identification notch, *18* separation cut line, *19* conical lead part. (For further information see text.)

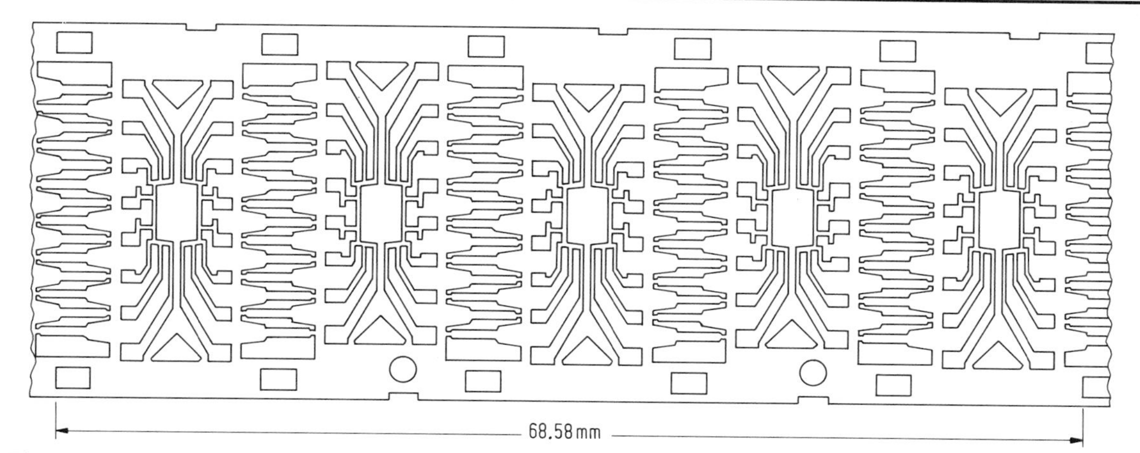

Fig. 55. Computer drawing of a material saving "interdigitated lead frame".

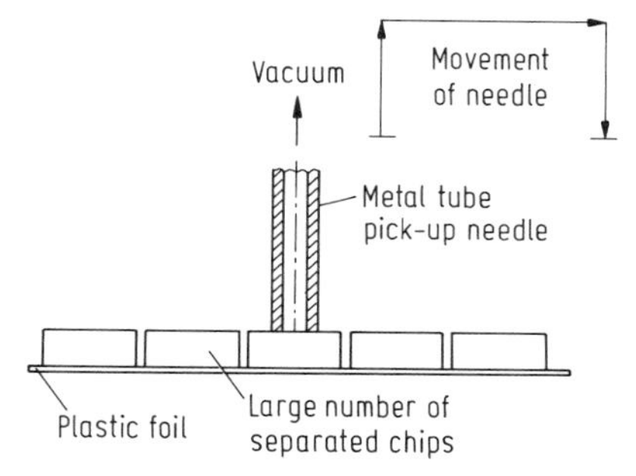

Fig. 57. Simple form of pick-up needle for chip bonding.

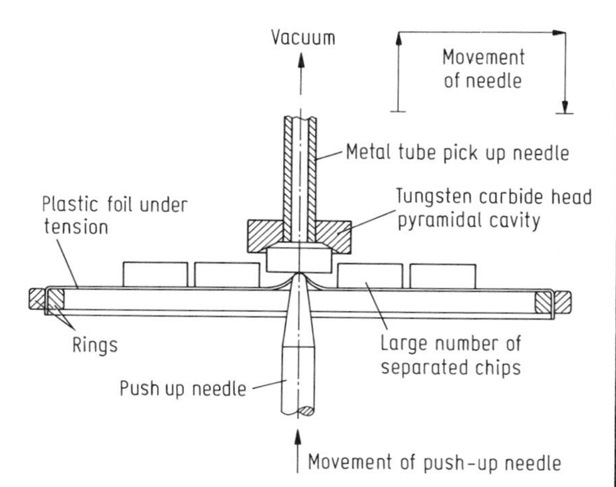

Fig. 58. Pick-up system having pyramidal needle.

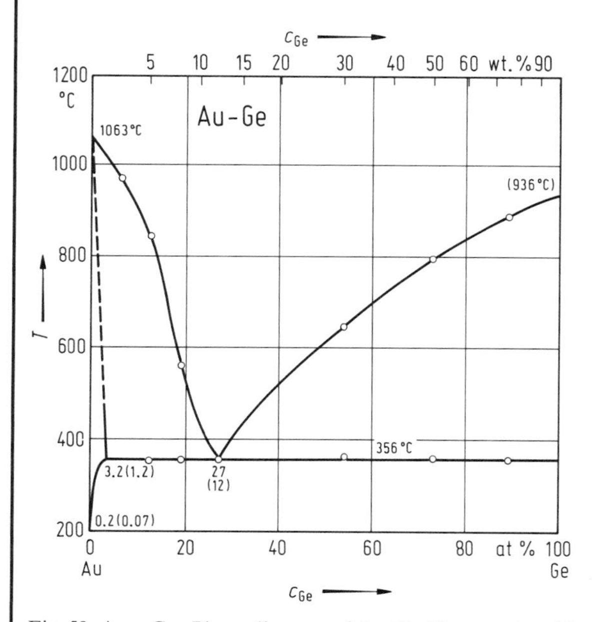

Fig. 59. Au—Ge: Phase diagram of Au-Ge. Temperature T vs. fraction of Ge, c_{Ge} in at % (wt %) [68H1].

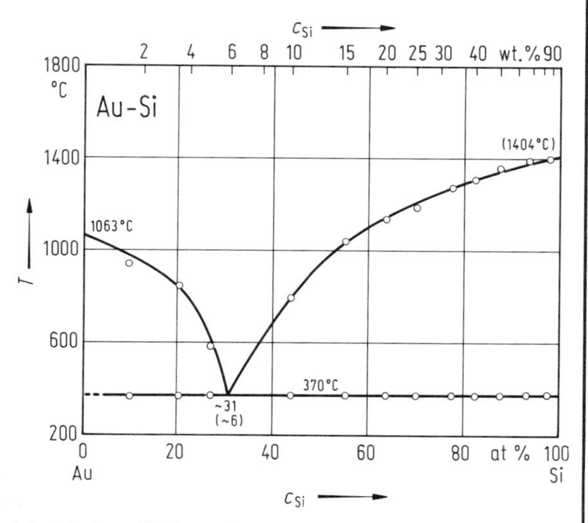

Fig. 60. Au—Si Phase diagram of Au-Si. Temperature T vs. fraction of Si, c_{Si} in at % (wt %) [68H1].

Fig. 61. Interconnection methods. The numbers and letters are explained in section 6.1.4.8.3.

		Method	Bondpad size	Wire diameter	Application
1	Wire bonding methods	Thermocompression wedge bonding 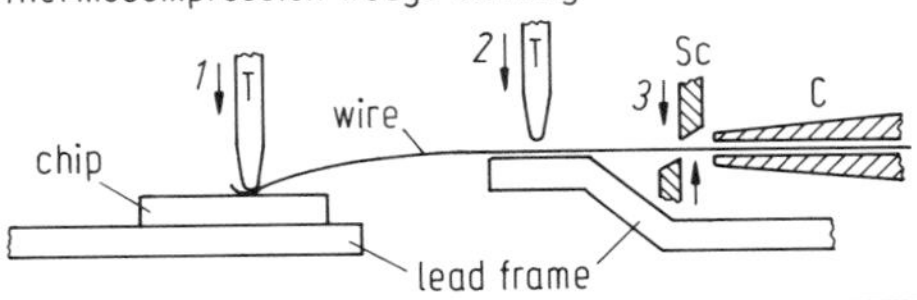 (1 T, 2 T, 3 Sc, C; chip, wire, lead frame)	≥ 13 × 33 μm²	7...10 μm Au	Ge-mesa-transistors
2	Wire bonding methods	Thermocompression stitch bonding 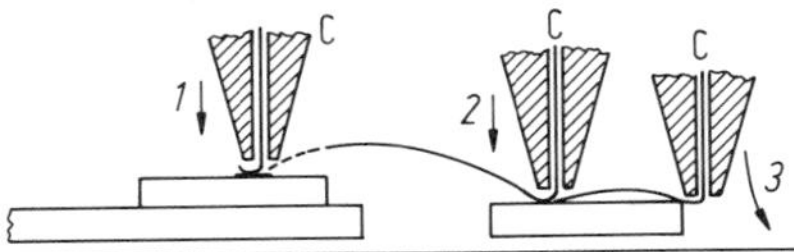 (1 C, 2 C, 3 C)	≥ 30 × 40 μm²	15...25 μm Au	Ge-semiplanar-transistors
3	Wire bonding methods	Thermocompression multiple wedge bonding 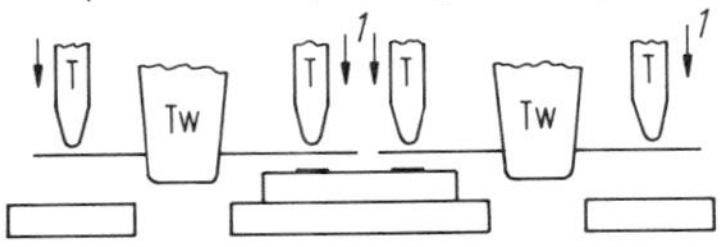 (T, Tw, 1)	≥ Φ 60 μm 40 × 50 μm²	25 μm Au 15 μm Au	Si-planar- Ge-semiplanar-transistors
4	Wire bonding methods	Thermocompression nailhead bonding 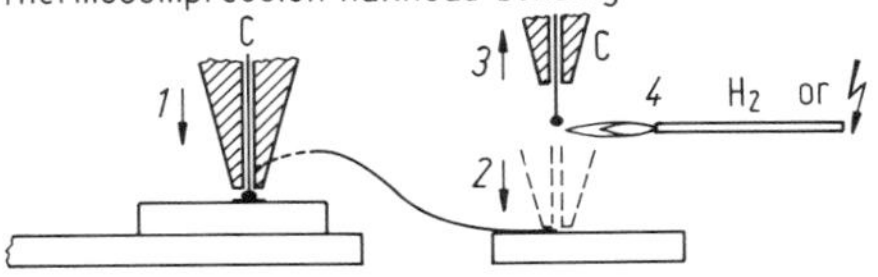 (1 C, 2, 3 C, 4 H_2 or ϟ)	Φ 120...170 μm	25...50 μm Au	Si-planar-transistors and integrated circuits
5	Wire bonding methods	Ultrasonic wedge bonding 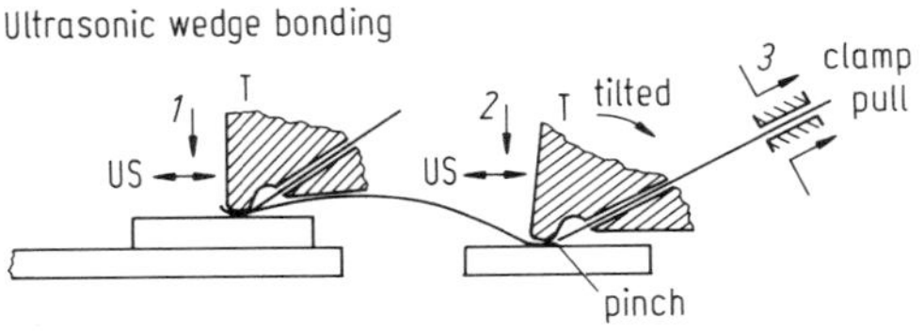 (1 T US, 2 T US tilted, 3 clamp pull, pinch)	≥ Φ 60 μm	25...500 μm Al	CERDIP, CERDIL, Thickfilm
6	Wire bonding methods	Ultrasonic and thermosonic nailhead bonding 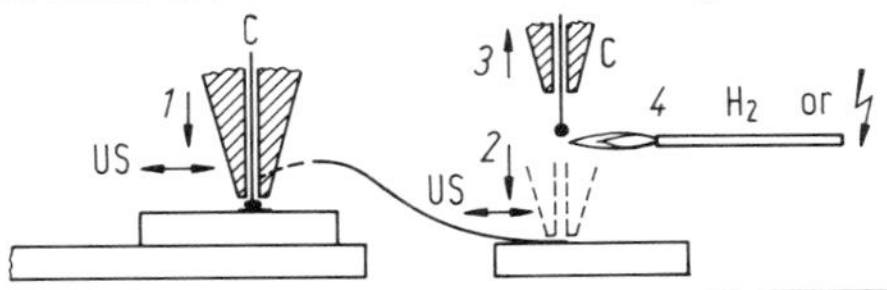 (1 C US, 2 US, 3 C, 4 H_2 or ϟ)	Φ 120 μm	25...38 μm Au	Integrated circuits. Thickfilm
7	Direct contact methods	Thermocompression beam lead bonding 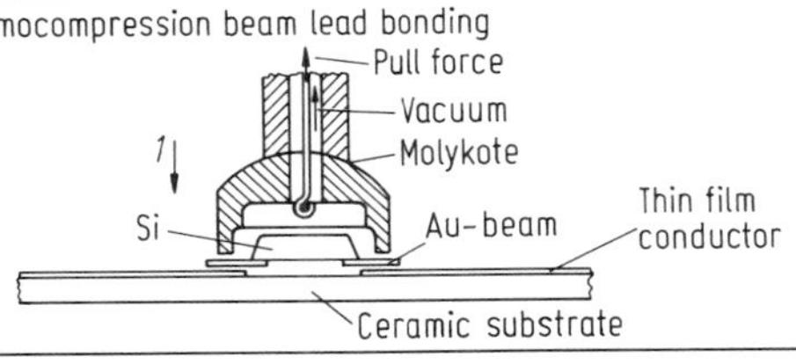 (1; Pull force, Vacuum, Molykote, Si, Au-beam, Thin film conductor, Ceramic substrate)	—	10 × 50 μm² Au-beams	High frequency High speed Thin film
8	Direct contact methods	Flip chip bonding with solder bumps 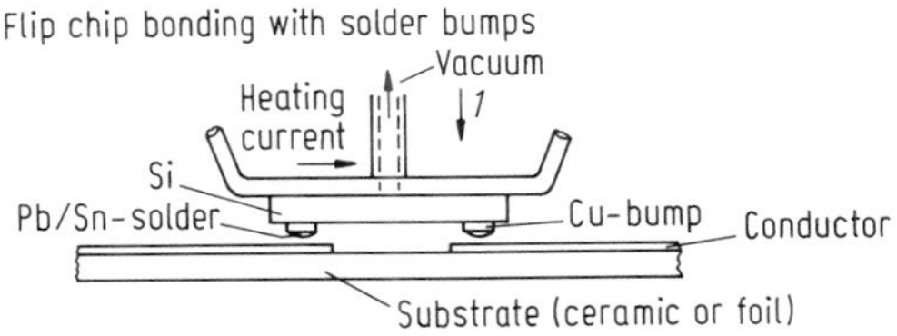 (1; Vacuum, Heating current, Si, Pb/Sn-solder, Cu-bump, Conductor, Substrate (ceramic or foil))	Φ 120 μm	—	Flip chips on foil and ceramic substrates

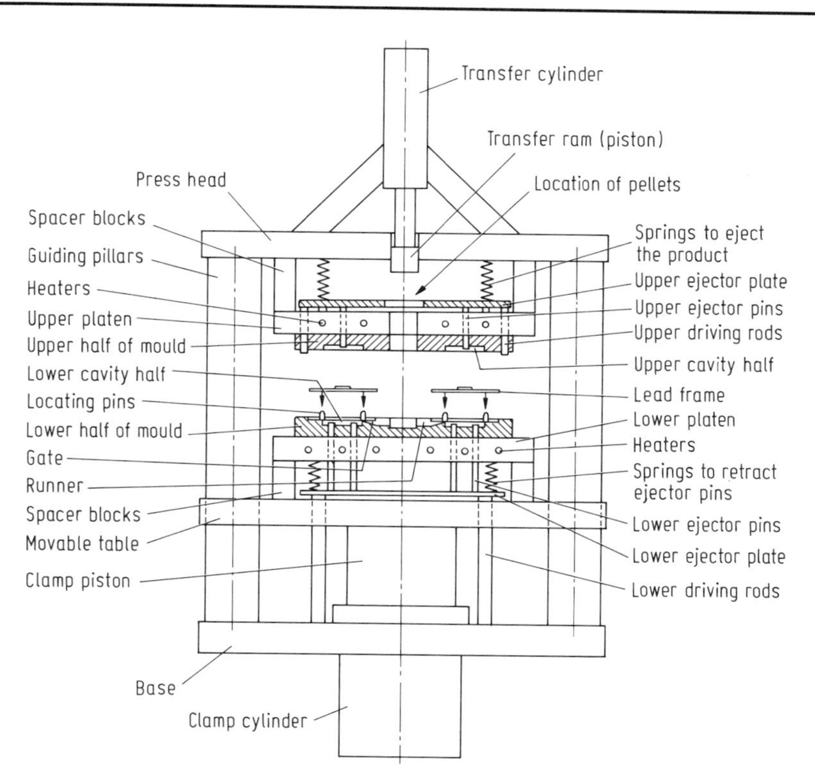

Fig. 62. Transfer moulding equipment. Function is explained in section 6.1.4.8.3.

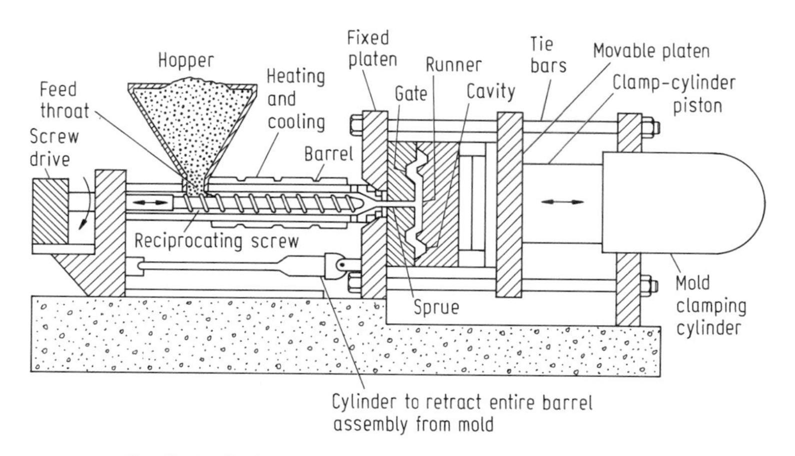

Fig. 63. Typical injection moulding machine (Dow Corning).

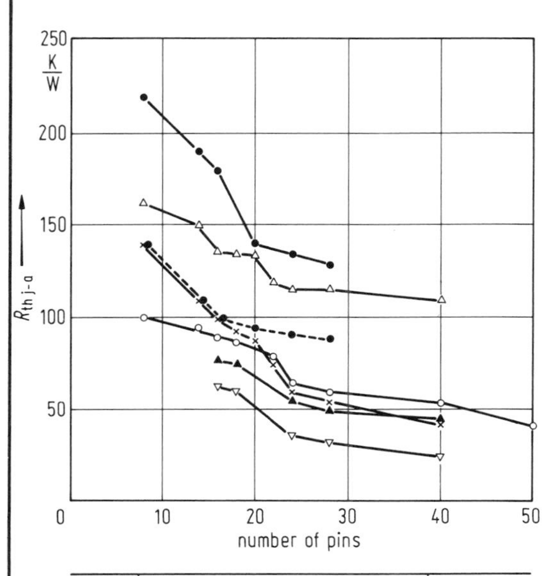

	package	outline: Fig. No.
●——●	SO on p.c. board	47, 48
●- - -●	SO on ceramic substrate	47, 48
△——△	Plastic DIL, Fe-Ni lead frame	40, 41
×——×	Cerdip	14, 15
○——○	Cerdil, side brazed	13
▲——▲	Plastic DIL, Cu-alloy lead frame	40, 41
▽——▽	Plastic DIL, Cu-alloy lead frame with heat spreader	40, 41, 42, 56

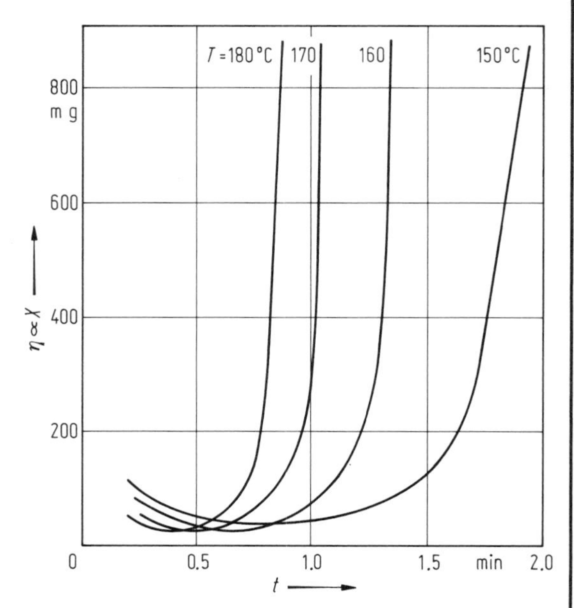

Fig. 65. Viscosity η as measured by the torque X of thermosetting plastic vs. moulding time t and tool temperature T as parameter.

◄

Fig. 64. Thermal resistance $R_{th_{j-a}}$ of small-outline- and dual-in-line packages on printed circuit board vs. number of pins.

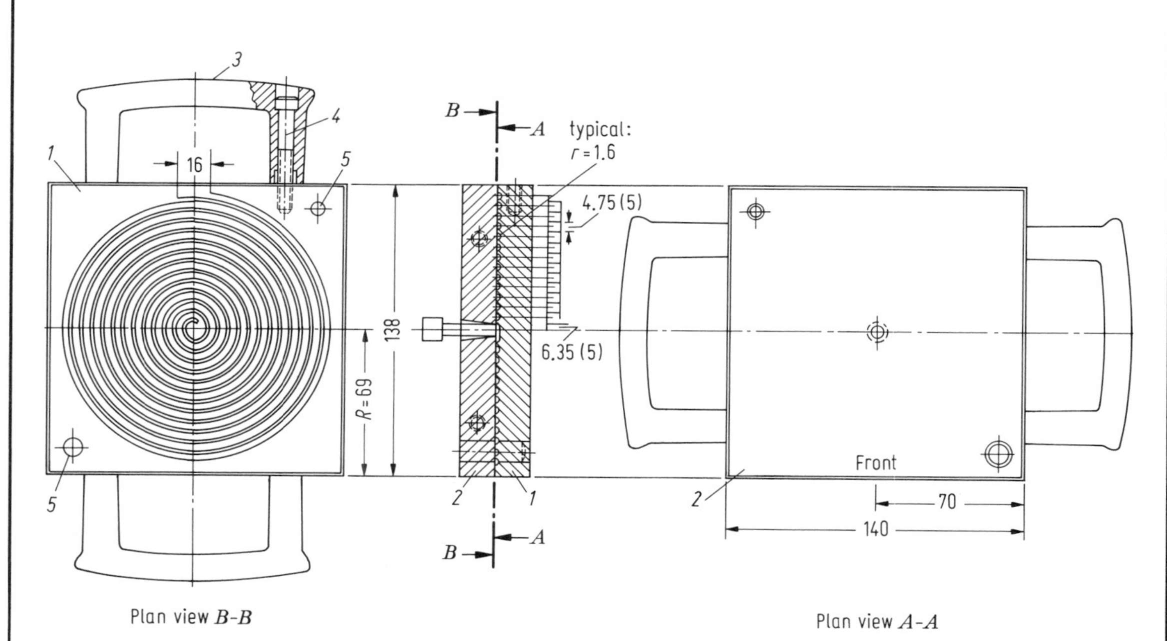

Fig. 66. Standard test tool "Emmi-Spiral" for testing of thermosetting plastics. *1*) spiral mould plate *2*) sprue plate *3*) Hull standard handles *4*) handle screws *5*) dowel pins. (Dimensions in mm.)

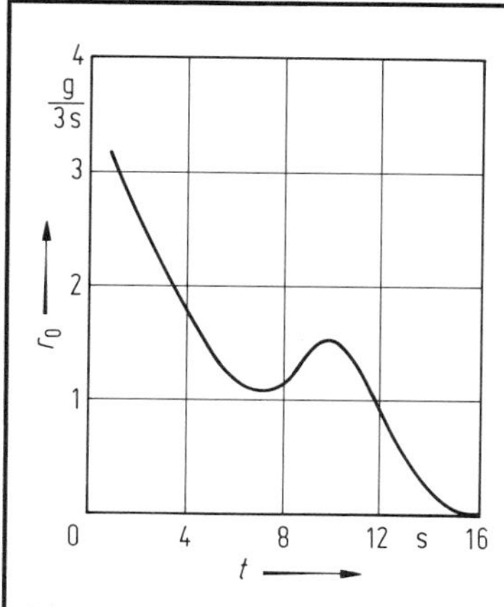

Fig. 67. Typical outflow rate of plastics r_0 vs. time t, measured in intervals of 3 s. For further explanations see section 6.1.4.8.2.

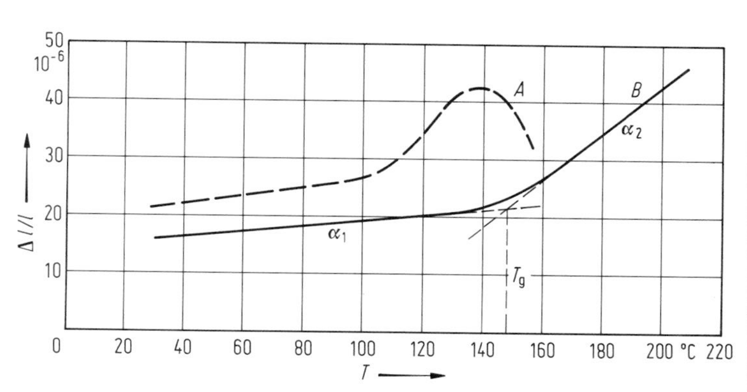

Fig. 68. Linear thermal expansion measurements of differently cured plastic: A: uncompletely cured plastic, B: well postcured plastic. (For further explanation see text.)

6.2 Silicon carbide

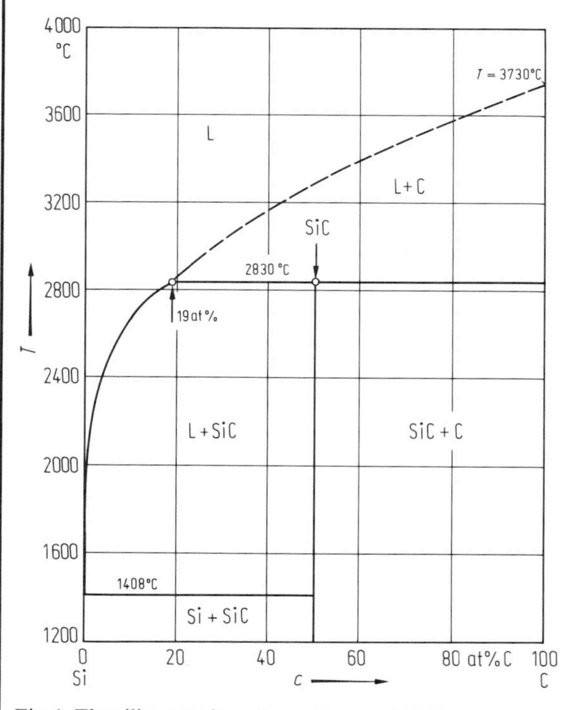

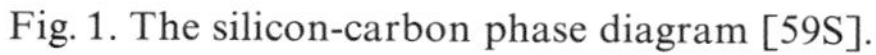

Fig. 1. The silicon-carbon phase diagram [59S].

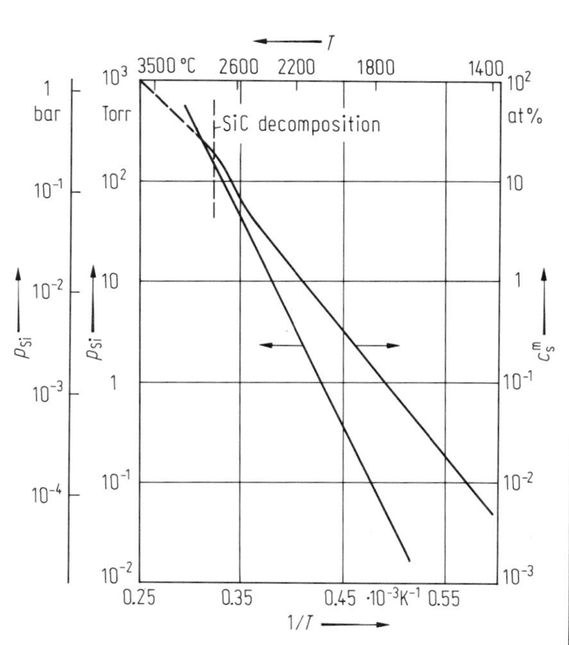

Fig. 2. Silicon vapour pressure p_{Si} (left scale) and solubility c_s^m of carbon in silicon (right scale) vs. reciprocal temperature T [59S].

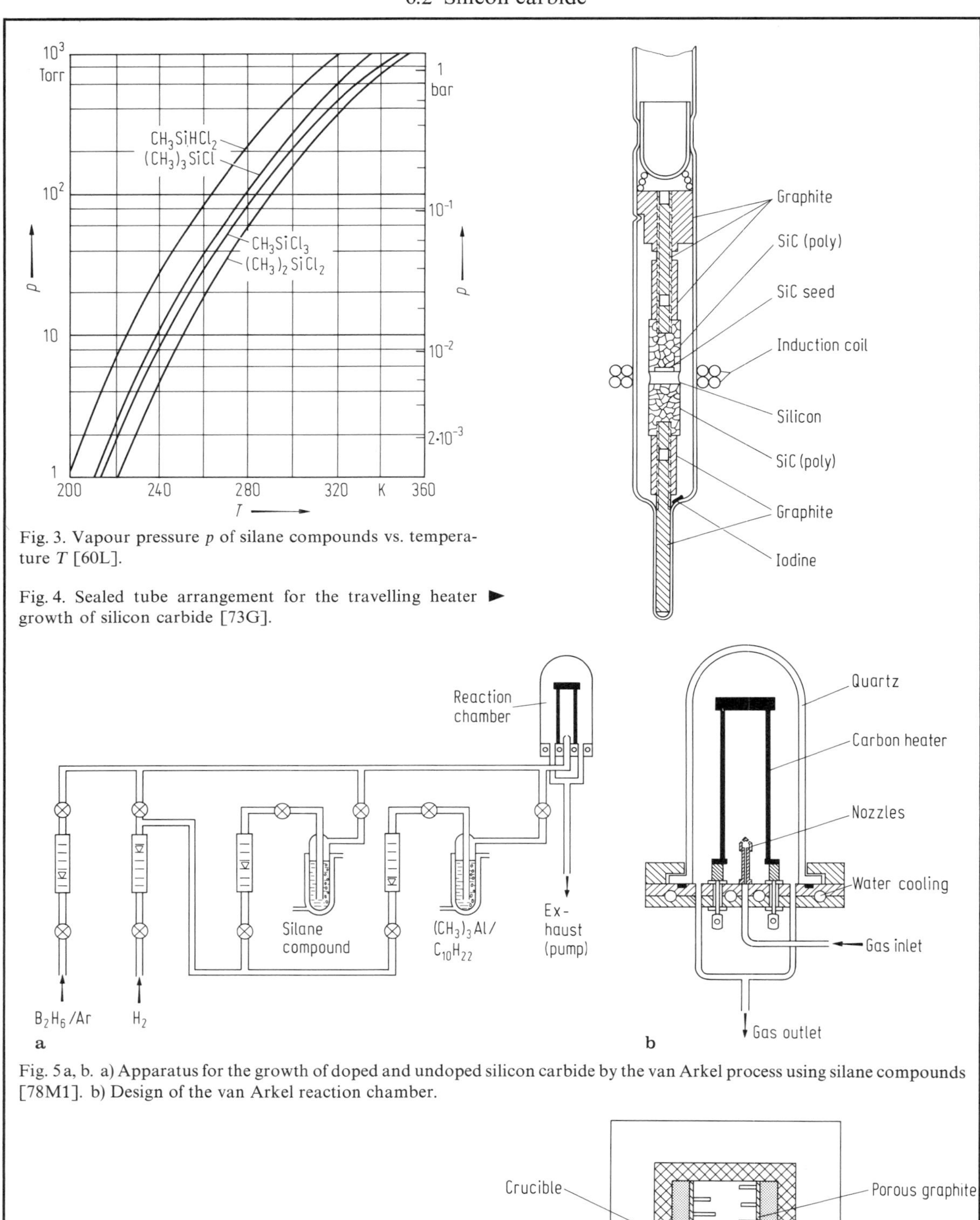

Fig. 3. Vapour pressure p of silane compounds vs. temperature T [60L].

Fig. 4. Sealed tube arrangement for the travelling heater ► growth of silicon carbide [73G].

Fig. 5a, b. a) Apparatus for the growth of doped and undoped silicon carbide by the van Arkel process using silane compounds [78M1]. b) Design of the van Arkel reaction chamber.

Fig. 6. Principle of silicon carbide crystal growth by the Lely technique. ►

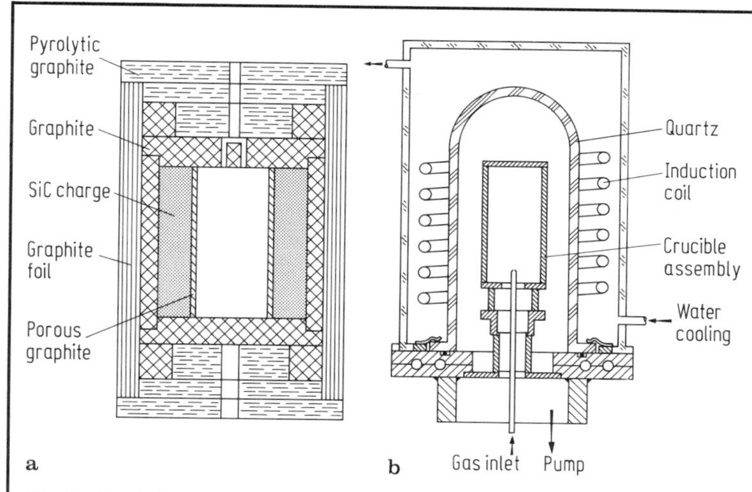

Fig. 7a, b. a) Crucible assembly for the growth of silicon carbide crystals. b) Apparatus for the growth of silicon carbide crystals (Lely technique).

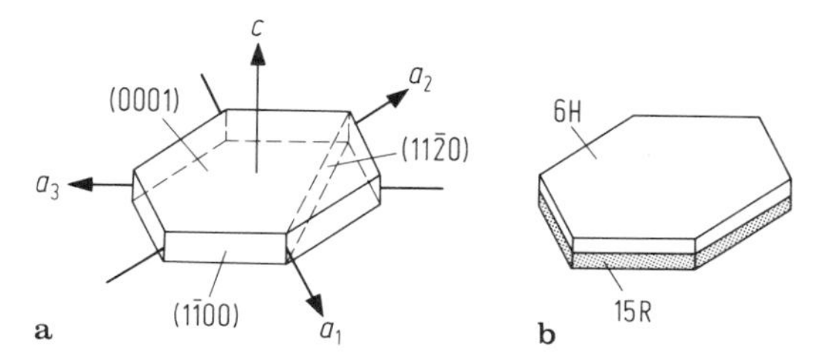

Fig. 8a, b. SiC. a) Shape and low index faces of silicon carbide crystals. b) Schematic drawing of silicon carbide crystal composed of two polytypes of SiC (6H and 15R).

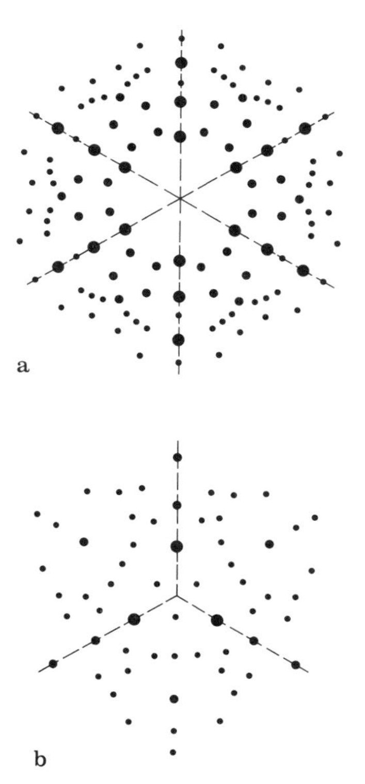

Fig. 9a, b. Idealized Laue transmission patterns of the most abundant polytypes of SiC. a) 6H; b) 15R.

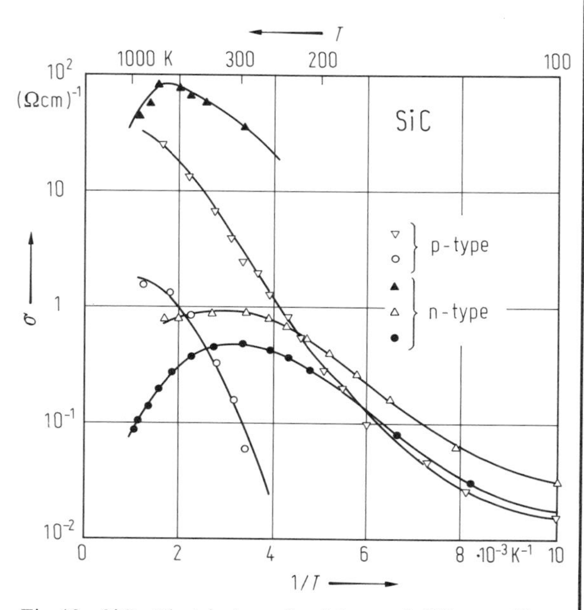

Fig. 10. SiC. Electrical conductivity σ of different silicon carbide crystals vs. reciprocal temperature T^{-1} [79B].

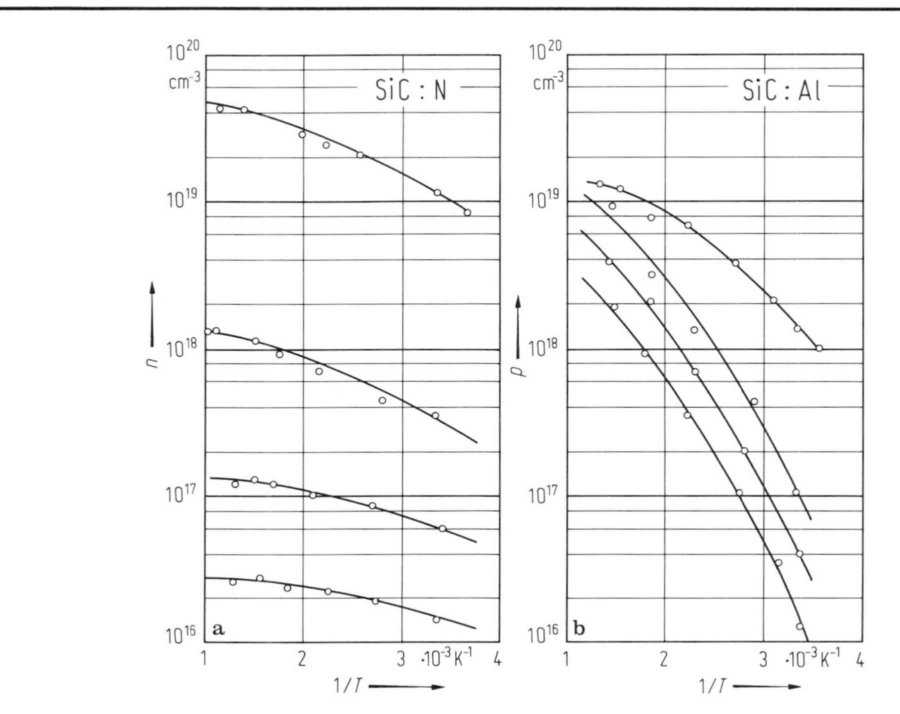

Fig. 11 a, b. SiC:N, SiC:Al. Carrier concentrations n, p of different silicon carbide crystals vs. reciprocal temperature T^{-1} [77P]. a) n-type crystals (N-doped); b) p-type crystals (Al-doped).

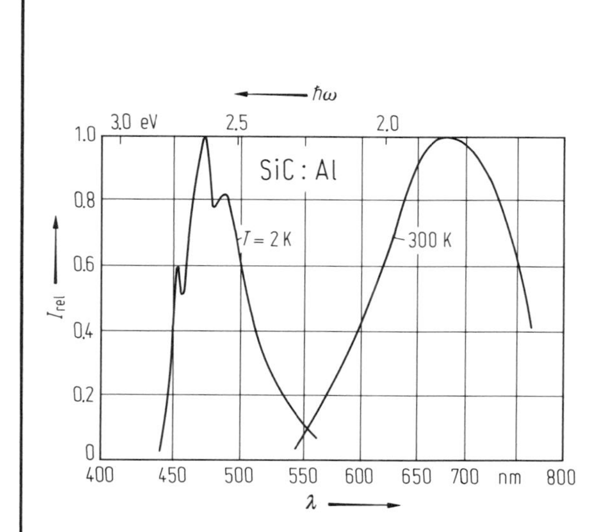

Fig. 12. SiC:Al. Photoluminescent intensity I_{rel} of Al-doped SiC crystal (6 H) vs. wavelength λ. Spectra are shown for sample temperatures of 2 K and 300 K, respectively. The hole concentration of the sample at room temperature is $p = 4 \cdot 10^{17}$ cm^{-3} [77P].

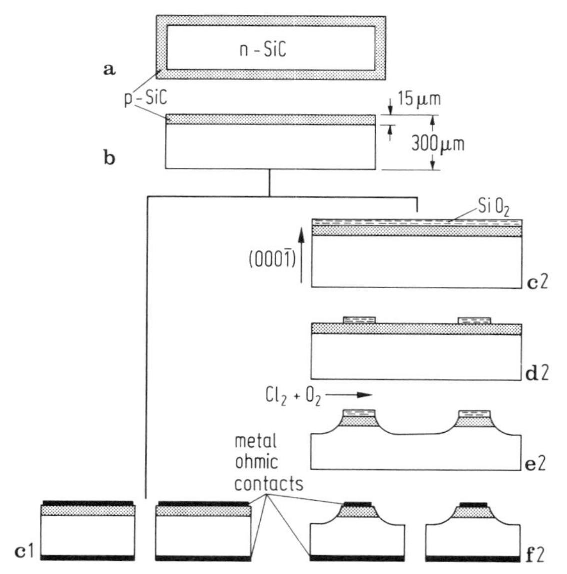

Fig. 13a···f. Process steps for the fabrication of SiC diodes.

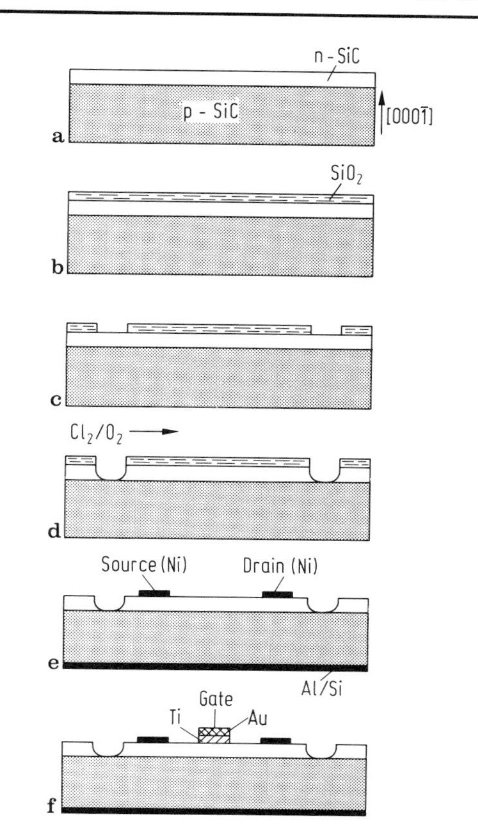

Fig. 14a···f. Process steps for the fabrication of SiC field-effect transistors.

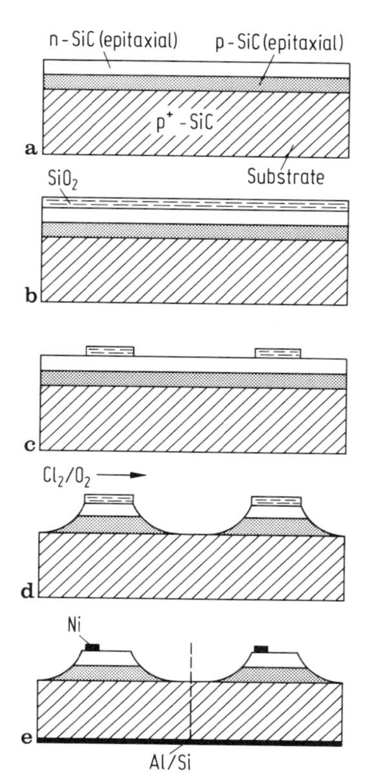

Fig. 15a···e. Process steps for the fabrication of blue-emitting diodes [77M].

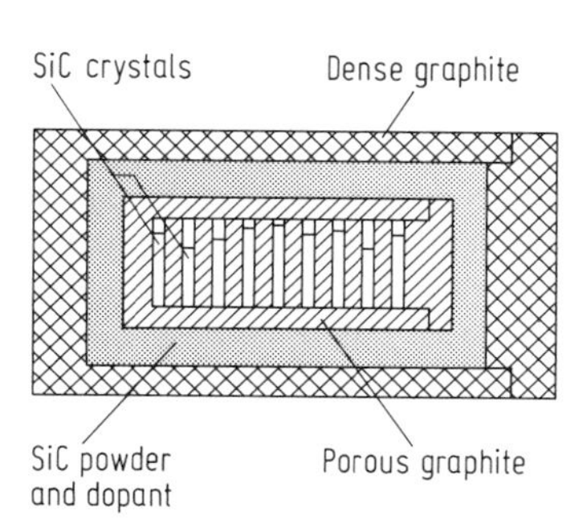

Fig. 16. Crucible for impurity diffusion into SiC [69B3].

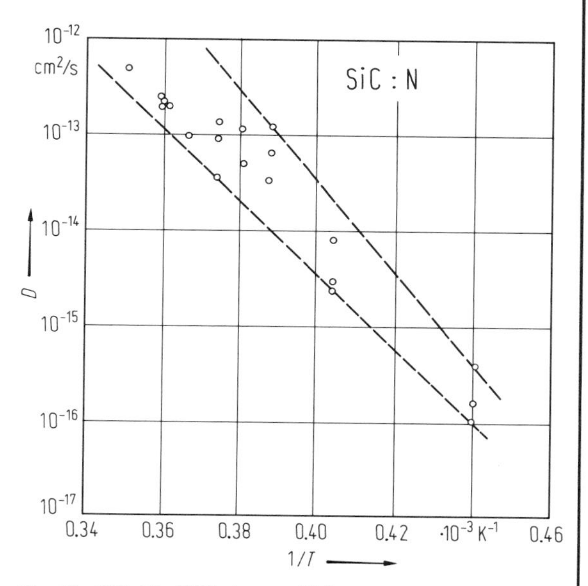

Fig. 17. SiC:N. Diffusion coefficient D of nitrogen in SiC vs. reciprocal temperature T^{-1} [66K2]. (Dashed lines represent lower and upper limits.)

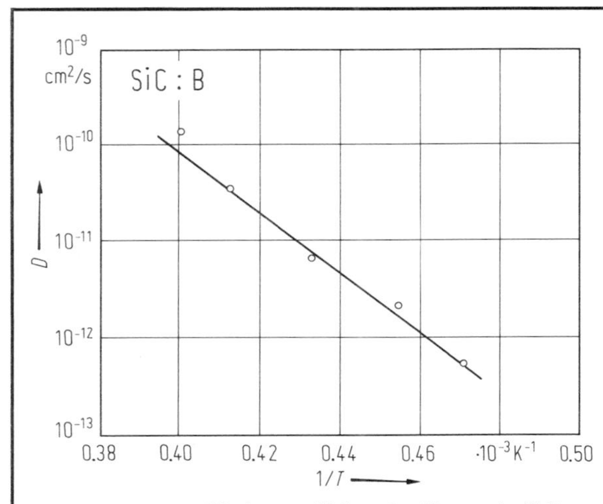

Fig. 18. SiC:B. Diffusion coefficient D of boron in SiC vs. reciprocal temperature T^{-1} [66V].

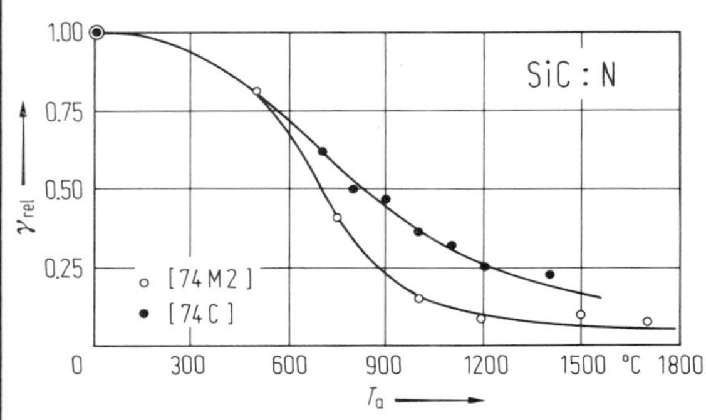

Fig. 20. SiC:N. Relative disorder γ_{rel} (back scattering analysis) vs. annealing temperature T_a for nitrogen-implanted (10^{15} cm^{-2}) SiC crystals.

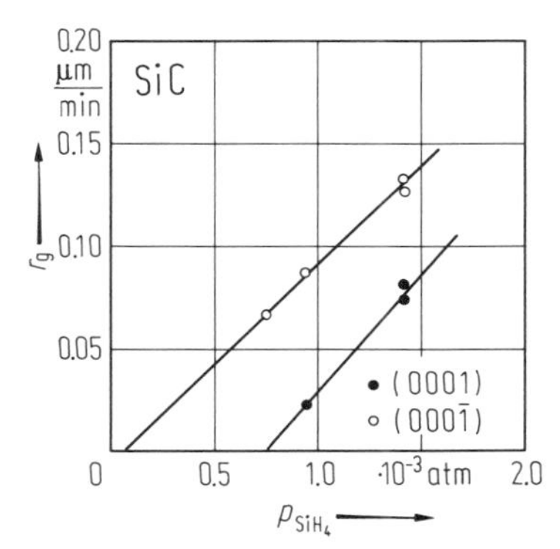

Fig. 22. SiC. Growth rate r_g vs. SiH_4 partial pressure p_{SiH_4} for SiC layers grown at 1650 °C; propane partial pressure $3 \cdot 10^{-5}$ bar [74W].

Fig. 23. SiC. Growth rate r_g vs. reciprocal temperature T^{-1} for SiC layers grown on (0001) surface; silane partial pressure $1.4 \cdot 10^{-3}$ bar; Si/C ratio is 5 [74W].

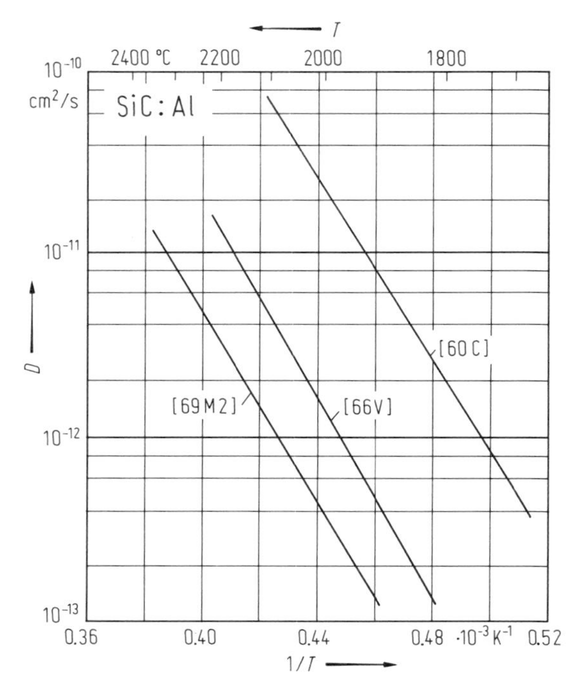

Fig. 19. SiC:Al. Diffusion coefficient D of aluminum in SiC vs. reciprocal temperature T^{-1}; from different authors.

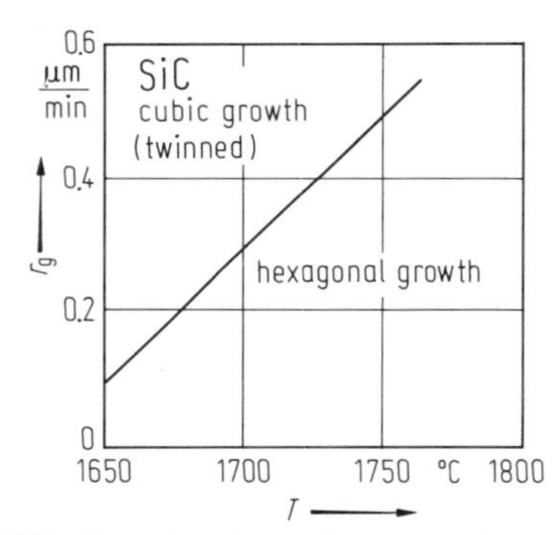

Fig. 21. SiC. Growth rate r_g of vapour phase epitaxy of SiC vs. temperature T [72G]. The solid line separates areas of cubic and hexagonal growth.

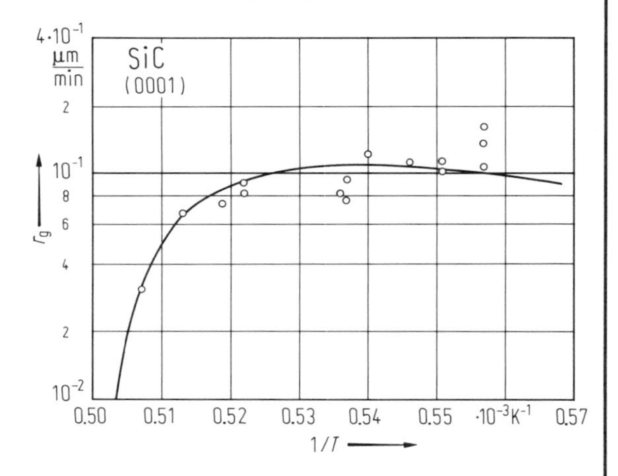

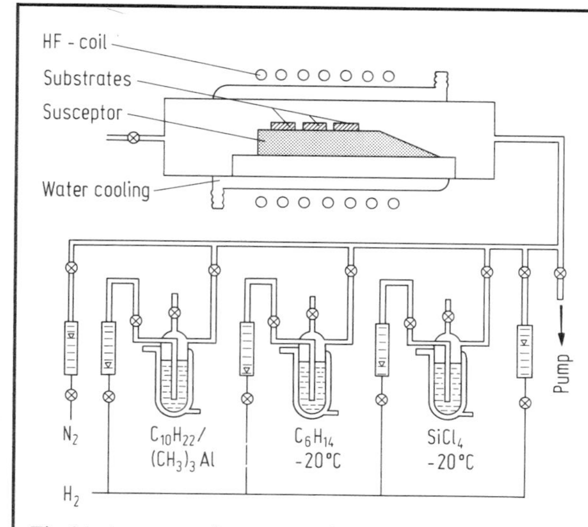

Fig. 24. Apparatus for vapour-phase epitaxy of SiC using liquid sources [77M].

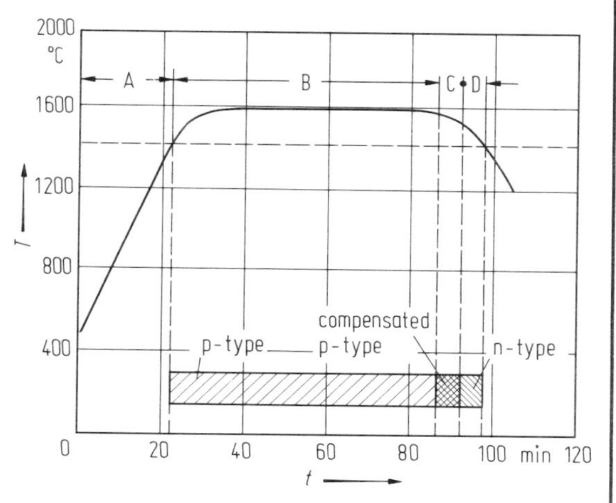

Fig. 26. Temperature T vs. time t program for the production of "overcompensated" LEDs.

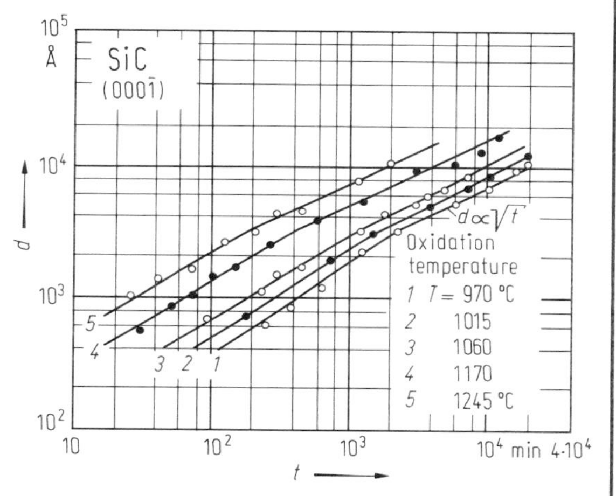

Fig. 27. SiC. Oxide thickness d on (000$\bar{1}$) face vs. oxidation time t; dry oxidation process [74H].

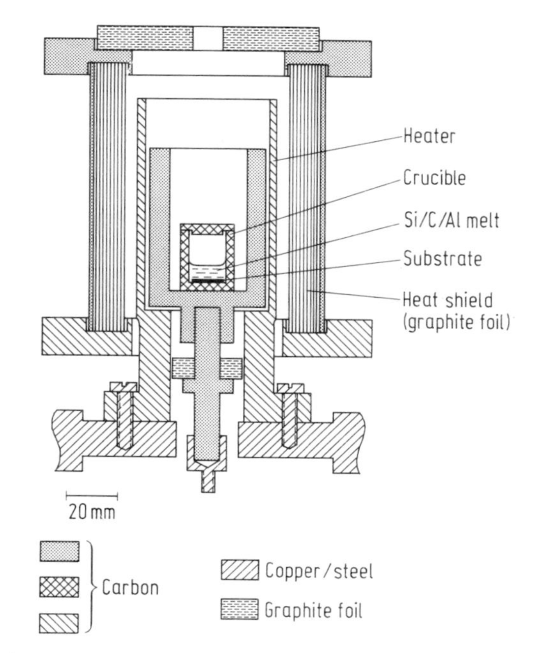

Fig. 25. Apparatus for liquid-phase epitaxy of SiC [77M].

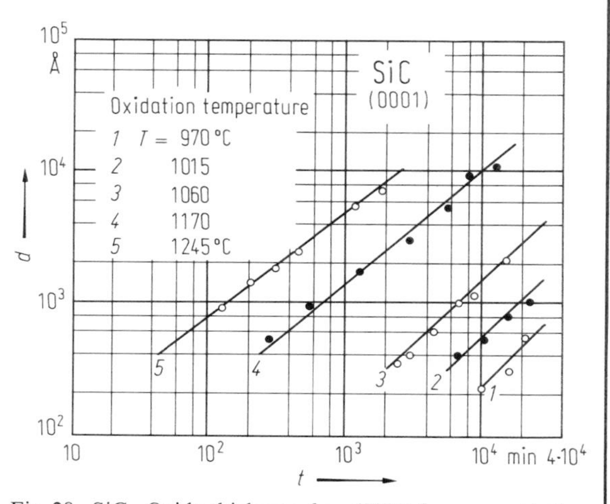

Fig. 28. SiC. Oxide thickness d on (0001) face vs. oxidation time t; dry oxidation process [74H].

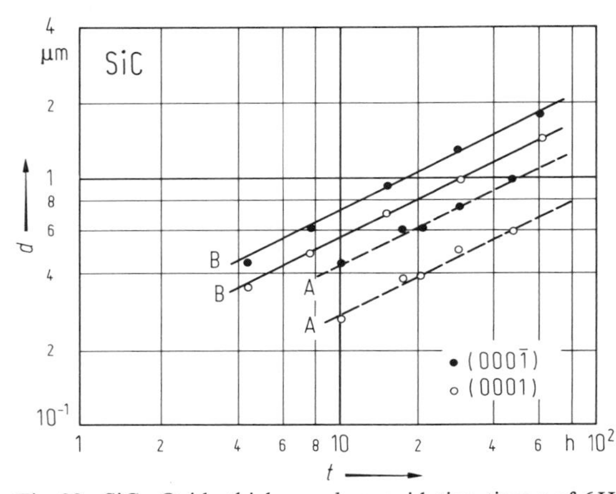

Fig. 29. SiC. Oxide thickness d vs. oxidation time t of 6H silicon carbide material at $T = 1200\,°C$ [67B2]. Moisture content : 10^2 ppm (curves A) and $2 \cdot 10^4$ ppm (curves B).

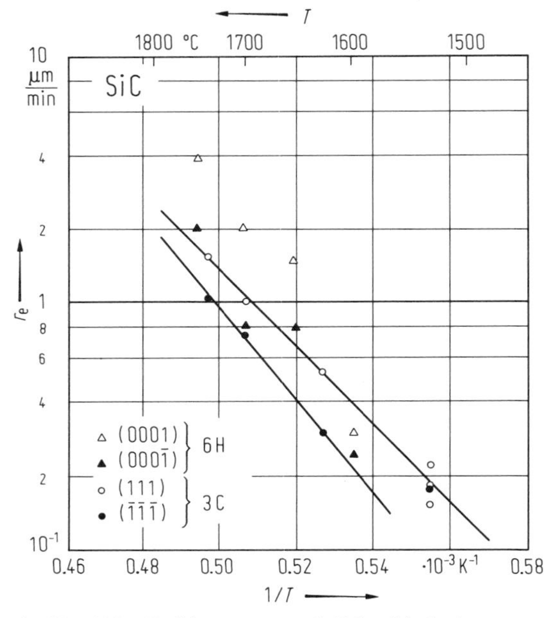

Fig. 30. SiC. Etching rate r_e of SiC with hydrogen vs. reciprocal temperature T^{-1} [69B1, 65C].

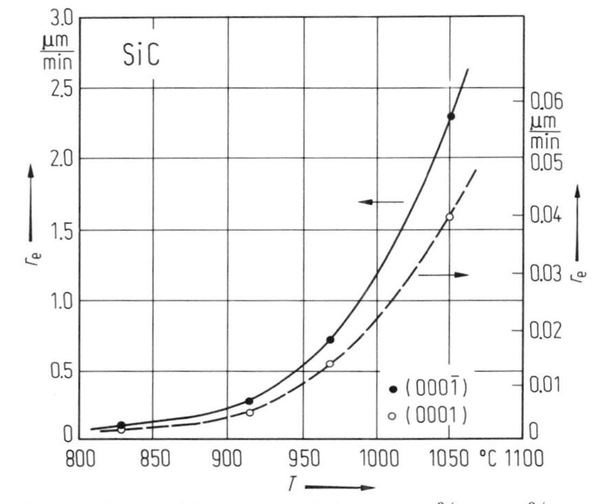

Fig. 31. SiC. Etching rate r_e of SiC with 68% Ar, 26% Cl_2, 6% O_2 vs. temperature T [69C].

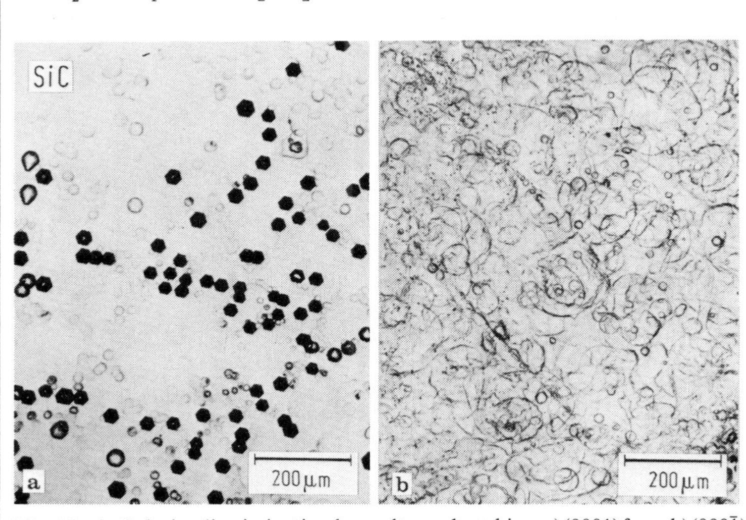

Fig. 33a, b. Polarity discrimination by molten-salt etching. a) (0001) face; b) (000$\bar{1}$) face.

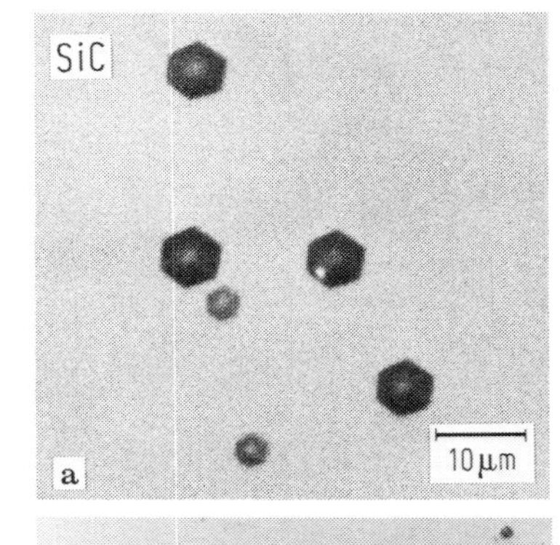

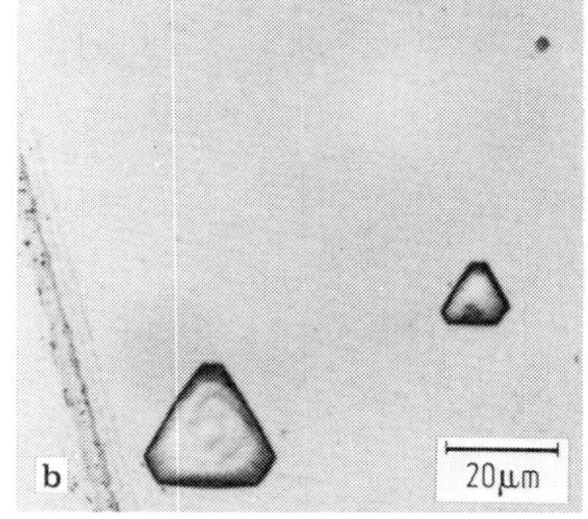

Fig. 32a, b. SiC. Etch pits generated by molten-salt etching. a) 6H crystal; b) 15R crystal.

6.1.4.5.8 Requirements on metal layers

Metal layers are employed in integrated circuits in three functions:
to contact device structures,
for interconnections of devices,
in contact pads for bonds to packages.

The basic requirements on metal layers are [74M1]:

A low-resistivity ohmic contact:
A low impedance must be achieved in contacts to n- as well as to p-type semiconductor regions. The current-voltage characteristics has to be linear.

A high conductivity:
In order to avoid a voltage drop along narrow and thin interconnect stripes, the conductivity of the metal layer must be as high as possible. A specific resistivity $\varrho \leqq 10^{-5}\ \Omega$ cm is usually required.

A good adhesion:
The metallization must exhibit good adhesion to silicon in the contact window and to the insulating SiO_2 layer.

A simple fabrication technology:
The fabrication of the metal layers should be achieved by a simple low-temperature process which does not affect the electrical characteristics of the device. The presence of the metal layer must not deteriorate the device.

A processing technology to fabricate fine structures in the layer:
The metallization must permit to fabricate fine contact and interconnect structures by lithography and etching techniques. The boundary conditions for these processes are that the substrate layers and the electrical device parameters are not affected by these processes. Contact bonds to the metal structures must also be feasable.

Besides these indispensible requirements, additional requirements must be met to achieve a high yield and a reliability for device fabrication:

Step and edge coverage:
Metal interconnections have to pass over steps and edges without fracture or change in cross section. Edges must be sloped by the lithography and etching processes of the planar technology. An optimization of the step height, the layer thickness, and the deposition process is necessary.

Mechanical stability:
Mechanical stability and resistance with respect to the following mounting processes is necessary.

Chemical stability:
Chemical reactions between the metal layer and the silicon, the SiO_2 insulator or the plastic encapsulation must be avoided.

Temperature stability:
High-temperature stress or temperature changes must not affect the electrical characteristics of the interconnects or contacts.

Current stability:
At high electrical current densities ($j \gtrsim 10^6$ A cm^{-2}), electromigration leads to interruptions of the metal interconnects. The material transport is caused by a momentum transfer of the conduction electrons to the lattice atoms. The magnitude of the effect is dependent on the type of the metal, the crystal structure, the current density, the line width of the metal, and the temperature [82G1].

Feasibility of multiple layer metallization:
In large scale integration, several layers of interconnections of devices by metal leads are required. The various layers are isolated by insulating films [81V1] (Fig. 1).

Al metallization

All the requirements listed are best met by aluminum. The aluminum metallization is deposited to a thickness of approximately 1 μm by sputtering or evaporation. For forming the contacts and for good adhesion, aluminum is deposited on unheated silicon substrates with a subsequent tempering at 500 °C for 30 minutes after structure-etching or is deposited on a heated (300···350 °C) substrate. The latter method improves the step and edge coverage by an increased surface mobility of the deposited metal atoms.

Problems of the Al metallization

Aluminum shows a low mechanical hardness. Al layers tend to form hillocks [82L2] during processing at elevated temperatures. These hillocks are local protrusions from the layer which are formed to relieve the internal strain.

Temperature treatments in the range 300 °C···500 °C are necessary for contact formation. At these temperatures, which are still well below the eutectic temperature (578 °C), a metallurgical reaction of Al with Si occurs which is caused by a noticable solubility of Si in Al and vice versa [83P1]. The equilibrium solubility for both cases is of the order of a few percent. The diffusivity of Si in the polycrystalline Al layer is high. Because of the high diffusivity, not only the amount of Si required for the formation of local Al—Si mixed crystals, but the large amount of silicon to saturate the whole adjacent metal stripe will be dissolved in the contact area. Especially at the edges of contact areas, the dissolution of silicon in metal layers is visible. It leads to alloyed metal regions, to irregular shape, especially for shallow device diffusions, and small device geometries. The dissolution of Si may lead to discontinuities in the metal conductor.

Aluminum is prone to electromigration because of its low atomic weight. If electromigration is a serious problem in special applications, other alternatives are preferred for the metallization.

Al—Si metallization

For this metallization, the Al layer is doped with silicon [83P1, 83P3] at a concentration which is above the saturation solubility of Si in Al so that reactions in the contact window are avoided. The concentration of the Si in the Al layer is quite critical in this case. Too low a Si concentration may lead to reactions in the contact window, too high a concentration leads to contact deterioration.

Since the vapour of Si and Al differs by several orders of magnitude, difficulties in the control of the evaporation process arise. The evaporation is performed out of two crucibles, one each at the appropriate temperature for Si and Al, or Al is flash-evaporated or sputtered from the specific alloy.

Besides the elaborate technology, the Al—Si alloy has the following disadvantages:

increased corrosion,

residual Si deposits after etching; Si deposits are difficult to remove and may cause problems in a multilayer metallization.

Bell metallization

The Bell metallization process was developed for high reliability beam lead transistors. The process is based on a multi-metal sandwich, in which various metals are deposited to solve several problems (Fig. 2).

The contact layer is formed by platinum silicide, which is fabricated by annealing of a Pt layer sputtered onto the Si substrate. Pt—Si establishes a uniform contact layer at a small vertical and lateral extension. The second titanium layer forms the adhesion to the silicide. The Au layer forms the interconnection lead. Since titanium and gold react to form an intermetallic compound at temperatures above 300 °C, a separating layer of platinum has to be deposited between the films. The platinum film may also be used as the base metal for an electroless deposition of a thick gold layer.

The multilayer sandwich structure of Ti/Pt/Au is frequently replaced by Ti—W/Au [78G1, 78N2]. Especially in bipolar ICs, the layer sequence Pt—Si/Ti—W/Al—Cu is employed [70C1]. For other interdiffusion barriers see section 6.1.4.5.10 (sputtering).

The advantages of multilayer metallizations are based on the well defined electrical contacts, the chemical resistance, and the stability against high current densities.

The deposition process, however, is elaborate. The complex layer sequence causes problems during the etching, especially when a special step profile is required in multilevel interconnects.

Future trends

Because a general solution does not exist for the fabrication of metal contacts and interconnects, the choice of the metallization employed must be adjusted to the specific application. In most applications, the trend is to deposit Al layers modified by Si, Cu additions or in simple sandwich structures, e.g. Ti—W/Al [82G1].

With the advancing reduction in device size in the MOS technology, metal silicides, e.g. molybdenum, tantal, titanium, and tungsten silicides are more and more applied. Metal silicides improve the electrical conductivity, when they are deposited on polysilicon, which is used in gate contacts or in low-level interconnects. The improved conductivity reduces the RC delay and thus the switching times (see also section 6.1.4.5.9 evaporation and sputtering and 6.1.4.5.12 resistivity).

Interconnects are passivated by deposition of an insulating film on top of the metal layer. The insulator cap improves the mechanical and chemical resistance as well as the electromigration effect.

6.1.4.5.9 Metal deposition

A survey of metallization processes is given in Table 1. The properties are listed in Table 2. In Table 3, various properties of metals frequently used in semiconductor technology are given. The important deposition processes are discussed in detail in the following subsections.

Table 1. Survey of deposition methods for metal layers.

Deposition Method	Process	Remarks	Ref.
I Vacuum deposition			
I.1 Physical vapour deposition (PVD)			
Evaporation (high vacuum, inert gas, reactive gas)	electron gun	mostly employed because of good flexibility with respect to deposition rates and materials, single layers or multilayer structures, coevaporation for silicides	72P1, 80H1
	induction heating	only suitable for vacuum setups having several chambers because the crucible must be heated and cooled slowly, radiation damage in MOS is avoided	79K1
	flash evaporation	suitable for deposition of alloy, risk of splashes commonly replaced by magnetron sputtering	
	resistance heating	poor flexibility with respect to choice of material and evaporation rate, risk of contamination by the source and residual gases, commonly replaced by the e-gun process	
	ion plating	not commonly used in Si technology	77S1, 79W1
Sputtering (high vacuum, inert gases, reactive gases)	magnetron sputtering DC, RF	generally suitable for metals, alloys, and admixtures Most important process for Al—Si, Al—Cu, Al—Si—Cu, employed in the cosputtering of silicides	79C1, 79H1, 81H1
	diode/triode sputtering	limited applications in Si technology, used for deposition of thin noble metal layers, low sputter rate, uniform target removal	
	ion beam deposition	only applied in research and development, applicable to all applications contamination by mask materials possible	77W1, 77S1
Molecular beam epitaxy (MBE)		single-crystalline layers, low defect density still in research stage for metals	77W1, 81F3
I.2 Chemical vapour deposition (CVD)			
Thermal reaction		industrial processes for Al and W, high substrate temperatures required, still increasing in importance	82C1, 82M1
Plasma assisted reaction		same properties as thermal reaction	82C2
II Plating			
II.1 Electroplating		fabrication of thick metal layers 0.1 μm···50 μm used in mounting and packaging of chips	79B1, 80M4, 81P1
II.2 Electroless plating		same as electroplating	

Table 2. Comparison of the three most important metallization processes.

Property	e-gun evaporation Unheated substrates	e-gun evaporation Heated substrate	Magnetron sputtering
Pressure range	$10^{-4}\cdots10^{-2}$ Pa	$10^{-4}\cdots10^{-2}$ Pa	$10^{-1}\cdots3$ Pa Ar
Mean free path	1 m···100 m	1 m···100 m	5 mm···0.1 m
Acceleration voltage	6···10 kV	6···10 kV	400···600 V
Source design	rigid, adjustments required	rigid, adjustments required	simple, rigid no adjustments
Source material efficiency	up to 40%	up to 40%	approx. 10%
Deposition of alloys	partly complicated, not reproducible	partly complicated, not reproducible	simple, reproducible
lock vacuum system	suitable under special conditions	suitable under special conditions	very suitable
Condensation rate	medium, 0.4 μm min^{-1} for Al	medium, 0.4 μm min^{-1} for Al	high 1···1.5 μm min^{-1} for Al
Kinetic energy/particles	0.3 eV/Al	0.3 eV/Al	3 eV/Al
Substrate temperature	low (≈100 °C)	high (≈400 °C)	adjustable by process high temperature possible
High step coverage	frequently unsymmetrical	frequently unsymmetrical	symmetrical
Steep step coverage	poor (microcracks)	good	good
Residual gas background	poor for single chambers	poor for single chambers	good for lock vacuum systems
Run-to-run reproducibility	poor	poor	good
Sensitivity to residual gases (H_2O, N_2, O_2)	not sensitive	sensitive	very sensitive
Grain size	small, irregular (≈1 μm)	medium, uniform (≈4 μm)	large, uniform (≈10 μm)
Reflectivity of Al—Si at 400 nm	>90%	60···80%	adjustable by process 95···45%
Reduction of electrical instabilities	simple, reproducible	difficult	simple
Electromigration-time of failure of Al	low	medium	high, very reliable

Evaporation

The metal deposition by evaporation is performed in a vacuum system. (Fig. 3). In an evaporation source which is described below, the metal to be evaporated is heated by electron bombardment, resistive heating, or inductive heating to a temperature, at which the vapour pressure exceeds 1 Pa. The metal vapour deposits on substrates, which are mounted on suitable substrate holders, e.g. planetary gearing. The substrate holder has to meet the boundary conditions for the process:

uniform deposition for a large number of substrates (typically better ±5%),
symmetrical step coverage, if required,
low risk of particle contamination.

The residual gas pressure during evaporation must be reduced, so that the contamination rate is low in relation to the deposition rate. Chemical or metallurgical reactions between the evaporation source and the metal must be avoided.

Electron beam gun

A schematic of the electron beam gun is shown in Fig. 4. A high electron current (typically up to 1 A) is generated in a filament (tungsten wire). The electrons are accelerated to the anode by a voltage of typically 10 kV. The focussed electron beam is deflected on to the surface of the metal to be vaporized by a magnetic field. The electron bombard-

Table 3. Properties of materials used to deposit conductive layers [68k1, 68b1]. (bcc: body centred cubic; fcc: face centred cubic; hcp: hexagonal close packed; tetr. bc: tetragonal body centred). (SiO_2 and Si_3N_4 only for comparison.)

Material	Density d (20 °C) $g\,cm^{-3}$	Atomic mass	Expansion coefficient α (0···100 °C) $10^{-6}\,K^{-1}$	Melting point T_m °C	Specific resistance ϱ (0 °C) $\mu\Omega\,cm$	Thermal conductivity κ (20 °C) $W\,m^{-1}\,K^{-1}$	Crystal structure	Atom radius nm	Diffusion constant prefactor D_0 $cm^2\,s^{-1}$	Activation energy of diffusion Q $kcal\,mol^{-1}$
Ag	10.5	107.87	19.7	960.8	1.50	428	fcc	0.144	0.40	44.1
Al	2.702	26.982	23.8	666.1	2.50	239	fcc	0.143	1.30	35.5
Au	19.29	196.967	14.3	1063	2.04	312	fcc	0.144	0.09	41.7
Co	8.9	58.933	13	1492	5.2		hcp	0.125	0.83	67.7
Cr	6.93	51.996	6.6	1900	15.0	90	bcc	0.125		
Cu	8.92	63.54	6.8	1083	1.55	395	fcc	0.128	0.20	47.1
Hf	13.3	178.49		2220	26.5					
Mg	1.74	24.312	26	650	3.94	171	hcp	0.160	1.10	32.2
Mo	10.21	95.94	5	2620	5.03	132	bcc	0.136	0.10	92.2
Nb	8.55	92.906	7.1	2500	23.3		bcc	0.143		
Ni	8.90	58.71	12.8	1453	6.58	81	fcc (hcp)	0.124	1.3	66.8
Pb	11.34	207.14	31.3	327.4	19.3	34.8	fcc	0.175		
Pd	11.4	106.4	11	1555	9.77		fcc	0.137		
Pt	21.45	195.09	9.0	1769.3	9.81	70.1	fcc	0.138		
Re	20.53	186.2		3180	18.9		hcp	0.138		
Rh	12.4	102.905	8.5	1960	4.3		fcc	0.134		
Sn	β7.28	118.69	2.7	231.9	10.1	65	tetr. bc	0.151		
Ta	16.6	180.848	6.5	3000	12.4	56	bcc	0.143		
Ti	4.52	47.90	9	1670	42	22	hcp	0.146		
V	5.96	50.942		1730	18.2		bcc	0.131		
W	19.3	183.85	4.3	3380	4.89	177	bcc	0.137		136
Zn	7.14	65.37	26.3	419.5	5.45	112	hcp	0.132	0.13	21.8
Zr	6.4	91.22	4.8	1855	40.5		hcp	0.158		
Si	2.32	28.086	2.3	1420		145	fcc (diamond type)	0.191		
(SiO_2)	2.56		0.5	1725		14.0				
(Si_3N_4)	3.44		2.8	≈1900		18.5				

ment generates a temperature of up to 3500 °C in an area of a few mm^2. The remainder of the melt remains relatively cool. For vaporizing special materials, the electron beam may be adjusted to the metal by defocussing of the beam and wobbeling of the beam position. The crucible is water-cooled to avoid or reduce a reaction of the metal with the crucible material.

The source is flexible with respect to the choice of the evaporation material. The crucible content usually is 10 g···100 g. Special e-guns and crucibles having a content of only a few grams are available for expensive evaporation materials, e.g. gold. For evaporation at high rates or of special materials, crucible liners made of tungsten, molybdenum or graphite are employed. In combination with automatic wirefeeders or rodfeeders, the e-gun source is suitable for multiple-chamber evaporation systems. Regular cleaning of the crucible, change of the filament (100 hours), and occasional change of the liner are required for routine maintenance.

Setups having 2 e-guns are applied for co-evaporation of alloys, e.g. Al—Si [78H1] or metal silicides used in gate electrodes and interconnections in Si-gate technology [83N1]. Deposition of layers for metal silicides is discussed in detail in section 6.1.4.5.12 (resistivity). Diffusion barriers (see Table 18), e.g. TiN, can be fabricated by reactive evaporation, e.g. of Ti in an N_2 ambient [82T1].

Induction-heated source

A boron-nitride-titanium-diboride crucible is used for the evaporation source in the induction-heated method (Fig. 5). The crucible is mounted inside a water-cooled RF coil, which is used to heat the crucible by an RF current at 50 kHz. The content of the crucible is heated by eddy currents. In contrast to the e-gun process, no radiation damage occurs in the silicon or the thermal oxide when the induction-heated source is used. However, the crucible must be heated and cooled very slowly [79K1].

Flash evaporation

The evaporation source material is fed continuously (wire feeder) or grain by grain onto a hot spot where it immediately vaporizes. The hot spot may be a boat fabricated of a metal having a high melting point, graphite, or metal oxide. The process is prone to splashing of particles because in contrast to the evaporation from a large molten source material supply, degassing cannot be achieved in a pre-evaporation process step. Flash evaporation was frequently applied for evaporating alloys before the introduction of magnetron sputtering.

Evaporation from resistive heater

Small boats or wire spirals are clamped between water-cooled electrodes and are heated by a high-current (100 A···1000 A) power supply. The evaporation source material is deposited into the boat prior to the heating or is suspended into the spiral by shaping it into a hook. Typical source materials to fabricate the boat or the spiral are tungsten, molybdenum, tantal, occasionally graphite is also used.

The advantage of this method is the low cost of the equipment. The disadvantages are the possible contamination by the boat material, the limited flexibility with respect to the choice and volume of the evaporation material, and the high running costs if the evaporation material reacts with the boat. This method is commonly replaced by the e-gun evaporation [72P1].

Ion plating

In the ion plating process, neutral particles are condensated on the substrate under simultaneous bombardment by highly accelerated (several keV) ions (Fig. 6). The layers obtained show good adhesion; however, there is a risk for causing radiation damage in the semiconductor [77W1].

Magnetron sputtering

A planar magnetron cathode is shown in Fig. 7. A system of strong magnets is located behind the cathode. In a region above the cathode, the electric and magnetic fields are perpendicular to each other and a high plasma current can be maintained. The plasma is contained in a ring of high ionization density by the magnet system. The ions accelerated in the cathode fall region impinge on the cathode and sputter off metal atoms from the cathode material. Power densities of 50 W cm^{-2} may be achieved.

Magnetron cathodes are designed for specific applications. Samples of various modifications are shown in Fig. 8.

The sputter yield of Al is shown in Fig. 9 as a function of the applied voltage. The sputter processes performed in the various voltage ranges are marked in Fig. 9 to differentiate the techniques. The magnetron sputtering operates in the range of the highest sputter yield. High sputter rates can be obtained because of the high sputter yield and the high plasma ion current densities (see Fig. 13). The radiation damage induced in the substrate, i.e. in sensitive devices, is low because of the low voltage applied and because the substrate is not necessarily exposed to an electron bombardment.

The sputter yield of various materials used to deposit conductive layers for Ar^+ ions at various energies is listed in Table 4. The sputter yield obtained for various sputter gases is surveyed in Table 5.

The deposition rates for magnetron sputtering are surveyed in Table 6. Results and variations are depicted in the Figs. 10···12.

For electron emission yields γ of molybdenum and Nickel see Figs. 28 and 29, respectively. For the energy and temperature dependence of the sputter yield for various ions see Figs. 25···27.

Table 4. Sputter yield η_s (in atoms/ion) of elements vs. Ar^+ ion energy [80C1, 82L1].

η_s	Ar^+ ion energy [eV]					
	100	200	300	400	500	600
Metal						
Ag	0.62	1.49	2.09	2.69	3.08	3.5
Al	0.12	0.36	0.69	0.85	1.18	1.23
Au	0.3	1.1	1.55	2.0	2.4	2.8
Be	0.04	0.19	0.29	0.39	0.49	0.5
C	0.02	0.04	0.09	0.1	0.12	0.15
Co	0.2	0.58	0.83	1.18	1.22	1.32
Cr	0.35	0.65	0.85	1.0	1.2	1.25
Cu	0.5	1.09	1.55	1.98	2.30	2.7
Dy	0.2	0.4	0.65	0.76	0.85	0.9
Er	0.14	0.3	0.42	0.65	0.75	0.8
Fe	0.22	0.49	0.75	0.95	1.15	1.22
Gd	0.13	0.37	0.55	0.7	0.8	0.92
Ge	0.25	0.5	0.75	0.95	1.19	1,24
Hf	0.12	0.31	0.45	0.6	0.7	0.76
Ir	0.1	0.45	0.7	0.85	1.0	1.21
Mo	0.12	0.38	0.55	0.75	0.85	0.95
Nb	0.05	0.2	0.37	0.47	0.6	0.7
Ni	0.3	0.66	0.95	1.24	1.41	1.69
Os	0.05	0.35	0.6	0.75	0.87	1.0
Pd	0.41	1.0	1.38	1.75	2.05	2.34
Pt	0.25	0.61	0.95	1.25	1.38	1.55
Re	0.12	0.35	0.59	0.72	0.85	0.95
Rh	0.16	0.5	0.8	1.05	1.25	1.42
Ru	0.12	0.38	0.62	0.87	1.06	1.25
Si	0.06	0.19	0.29	0.37	0.48	0.59
Sm	0.16	0.35	0.5	0.69	0.77	0.87
Sn	0.36	0.7	1.0	1.25	1.49	1.6
Ta	0.06	0.25	0.36	0.46	0.55	0,68
Th	0.1	0.29	0.37	0.5	0.62	0.7
Ti	0.1	0.22	0.29	0.37	0.42	0.55
U	0.16	0.35	0.51	0.7	0.8	0.9
V	0.12	0.25	0.37	0.5	0.6	0.65
W	0.1	0.25	0.37	0.48	0.6	0.62
Y	0.13	0.27	0.38	0.55	0.65	0.75
Zr	0.11	0.26	0.4	0.55	0.65	0.75

Table 5. Sputter yield η_s (in atoms/ion) of elements for different sputter gases at 500 V [78v1]*).

η_s	sputtering gas				
	He	Ne	Ar	Kr	Xe
Element					
Be	0.24	0.42	0.51	0.48	0.35
C	0.07		0.12	0.13	0.17
Al	0.16	0.73	1.05	0.96	0.82
Si	0.13	0.48	0.50	0.50	0.42
Ti	0.07	0.43	0.51	0.48	0.43
V	0.06	0.48	0.65	0.62	0.63
Cr	0.17	0.99	1.18	1.39	1.55
Mn				1.39	1.43
Mn			1.90		
Bi			6.64		
Fe	0.15	0.88	1.10	1.07	1.00
Fe		0.63	0.84	0.77	0.88
Co	0.13	0.90	1.22	1.08	1.08
Ni	0.16	1.10	1.45	1.30	1.22
Ni		0.99	1.33	1.06	1.22
Cu	0.24	1.80	2.35	2.35	2.05
Zr	0.02	0.38	0.65	0.51	0.58
Nb	0.03	0.33	0.60	0.55	0.53
Mo	0.03	0.48	0.80	0.87	0.87
Mo		0.24	0.64	0.59	0.72
Ru		0.57	1.15	1.27	1.20
Rh	0.06	0.70	1.30	1.43	1.38
Pd	0.13	1.15	2.08	2.22	2.23
Ag	0.20	1.77	3.12	3.27	3.32
Ag	1.0	1.70	2.4	3.1	
Ag			3.06		
Sm	0.05	0.69	0.80	1.09	1.28
Gd	0.03	0.48	0.83	1.12	1.20
Dy	0.03	0.55	0.88	1.15	1.29
Er	0.03	0.52	0.77	1.07	1.07
Hf	0.01	0.32	0.70	0.80	
Ta	0.01	0.28	0.57	0.87	0.88
W	0.01	0.28	0.57	0.91	1.01
Re	0.01	0.37	0.87	1.25	
Os	0.01	0.37	0.87	1.27	1.33
Ir	0.01	0.43	1.01	1.35	1.56
Pt	0.03	0.63	1.40	1.82	1.93
Au	0.07	1.08	2.40	3.06	3.01
Au	0.10	1.3	2.5		7.7
Pb	1.1		2.7		

*) Elements are ordered according to the atomic mass

Table 6. Deposition rates obtained by DC magnetron sputtering [78v1].

Cathode material	Sputter rate nm · min^{-1}	Efficiency $\frac{\text{nm min}^{-1}}{\text{W cm}^{-2}}$	Power density Wcm^{-2}	Source-substrate distance cm	Argon pressure Pa	Cathode dimensions cm
Al	550	52	10.5	5	0.53	9.21
	700	44	16			15.41
	1100	34	30	5	0.53	14
	1000	65	15.5			13.30
Au	2000	125	16			15.41
	1700	110	15.5			13.30
Cr	1000	63	16			15.41
	1000	65	15.5			13.30
Cu	1070	102	10.5	5	0.53	9.21
	2000	125	16			15.41
	2500	78	30	5	0.53	14
	1300	84	15.5			13.30
Ti	390	37	10.5	5	0.53	9.21

Configuration of the substrate holder and the magnetron cathode

Because a high sputter rate is only obtained in a plasma ring, the magnetron cathode cannot be arranged in a plane parallel configuration as in conventional diodes and the substrate has to be scanned to obtain a uniform deposition (except for special constructions of the magnetron cathode). Fig. 15 shows typical configurations and geometries used in magnetron sputtering.

In configuration 1 of Fig. 15, the substrates are mounted on a planetary gearing system. In the other configurations 2···5, the substrates are arranged at a short distance opposite to the magnetron cathodes and are scanned by a linear or rotary motion. The main difference of the configuration 1 and the other configurations 2···5 is the actual maximum condensation rate achieved, which is low for the configuration 1 and which is large for the short distance configurations 2···5. However, a good uniformity of the layer thickness is obtained for a large number of substrates by using the planetary gearing.

For all the configurations except no. 1, the substrates can be passed by the cathode once or several times. One or only a few passes are used for the Al deposition.

The magnetron cathode is especially suitable for the use in vacuum lock systems (Fig. 16), because no refills of the source material and no cleaning of the system between runs are required. Vacuum lock systems are necessary for Al deposition because of the sensitivity of Al layers to the residual gas content.

Sputter systems equipped with magnetron cathodes are commonly used today in semiconductor technology for metal deposition because of their flexibility. All the metal layers commonly applied (pure metals or alloys, Tables 3, 6, 7, and 18) can be deposited in single layers or in sandwich structures.

Magnetic target materials (Co, Ni) weaken the plasma-supporting magnetic field. Thin target layers or special magnet designs have to be used.

The metal combination Ti—W used as *diffusion barrier* is deposited by DC-magnetron sputtering from an alloy target. A number of diffusion barriers, e.g. TiN, are fabricated by reactive sputtering in a gas ambient defined by the material. The Table 18 lists a variety of diffusion barriers and their specific resistances.

Silicides fabricated of refractory metals (Table 7) replace more and more the poly-Si interconnects in the Si-gate technology when low-resistivity device connections are required. The silicides are deposited by DC magnetron sputtering. Three different methods are in use to fabricate silicides:

a metal layer is deposited on a Si layer deposited or otherwise fabricated prior to metal deposition (2-layer sandwich),

co-sputtering of the metal and the silicon from two different targets (multiple sandwich or admixture),

sputtering of the metal and the silicon from a target (compound target) which contains both materials as a mixture or is composed, e.g. by stripes of both materials. Compound targets are commercially available recently; but they are not yet sufficiently tested. The silicon content in compound targets must be larger than required for the silicide because silicon is more efficiently backsputtered especially for the case of bias sputtering, resulting in a reduced incorporation of silicon relative to the metal.

After deposition of the sandwich or admixture of the metal-Si compounds, the crystalline silicide layer is formed by annealing at 800 °C···1000 °C for approximately 1 hour in an inert or forming gas ambient. The resistivity is improved by the annealing (see section 6.1.4.5.12). The composition of the silicide, e.g. $MoSi_2$, Mo_5Si_3, or Mo_3Si for molybdenum and Si, is defined by the fractions of the metal and Si deposited from the sputter source. In semiconductor technology, silicides of the average composition "disilicide" are best suitable as gate electrodes (poly-Si/metal silicide sandwich (see Fig. 17a, b)) and as interconnections although these silicides do not show the lowest possible resistivity values. The silicides most commonly applied are $MoSi_2$, and WSi_2, (see Table 7).

Magnetron sputtering and diode sputtering may be used in combination with a DC or RF voltage applied to the substrate holder (bias sputtering). The bombardment of the substrate by ions or accelerated neutral particles during the deposition influences the layer properties (see section 6.1.4.5.12). The argon content of the layer is mainly increased (Fig. 24a, b).

Table 7. Resistivity and chemical reactivity of refractory metal silicides, [80M1, 80G1, 83M1, 83N1].

Silicide	Deposition mode	Sintering temperature °C	Resistivity μΩ cm	Soluble in	Insoluble in
$CrSi_2$	metal on poly-Si	700	600		
$MoSi_2$	co-evaporated	1000	40···60	$HF+HNO_3$	mineral acids, aqua regia
	compound sputtered		70···90	slowly in HF	aqueous alkali
	co-sputtered	1000	~100		
$NbSi_2$	metal on poly-Si	900	50	slowly in HF	
VSi_2	metal on poly-Si	900	50···55	HF	mineral acids except HF, aqua regia, aqueous alkali, $H_2SO_4+H_2O_2$
WSi_2	co-evaporation	1000	27	$HF+HNO_3$	mineral acids, aqua regia
	compound sputtered		50···70		
	co-sputtered	1000	70		
$TaSi_2$	compound sputtered		40···60	HF	
	co-sputtered	900	50···55		
$ZrSi_2$	metal on poly-Si	900	35···40	HF	mineral acids except HF, aqua regia, aqueous alkali, $H_2SO_4+H_2O_2$

RF diode sputtering

The common application of RF sputtering is the deposition of insulating films (see section 6.1.4.5.5) because the build-up of space charge inhibits the DC plasma discharge. A typical RF sputtering setup is shown in Fig. 18 [78V1].

RF diode sputtering is especially applied to the deposition of thin *noble metal layers*, e.g. PtSi. For contacts to shallow junctions, a very thin metal film is applied which reacts with the semiconductor silicon in a later annealing cycle. In order to obtain a uniform layer in the contact window, it is sometimes necessary to dilute the metal with either silicon or with a metal which does not form a silicide.

RF (and DC) bias sputtering may be applied to influence the layer properties (Figs. 19···24, see also previous section). According to König et al. [70K1], the ratio of the ion bombardment on the target and the substrate should be related to the 4-th power of the area ratio. Experiments favour a lower power ratio (≈1) [83H1].

Comparison of metallization methods

The three most important metallization processes are compared in Table 2 with respect to Al deposition. The features of electroplating and ion plating are surveyed in Table 8.

Because of the heating, sputter targets require special solders. Table 9 lists properties of target solders.

Target currents of various target materials used with DC diode sputtering are listed in Table 10. Power densities required for a temperature gradient 100 °C through a 1 cm thick cathode is listed in Table 11.
Energies and life times of metastable neutrals are surveyed in Table 12.

Table 8. Comparison of electroplating and ionplating [79W1].

A	Features of the process	Electroplating	Ion plating
	materials that can be deposited	pure metals, alloys	pure metals, alloys, some inorganic compounds
	non-conducting substrates	not possible without autocatalytic preplating	a wide range
	deposition rate (Å/second)	0 to 10^4	$10 \cdots 10^3$
	throwing power	limited	good
	substrate temperature	20···100 °C	50···400 °C depending on power levels and cathode cooling
	cost	moderate	high
	simplicity of equipment	simple in most instances	simple only for small batch processing
	gold recovery	costly	relatively simple
	pollution	problems – especially with cyanide	none

B	Properties of the deposits	Electroplated	Ion plated
	adhesion	moderate	excellent
	porosity	usually porous for thin coatings	generally non-porous if care is taken
	uniformity	good on flats, may show non-unifomity on edges	good or the front surface, thinner layers on the back. Rotation ensures uniformity
	purity	may contain gases and inclusions of inorganic or organic nature	limited only by purity of the source material
	resistance to tarnish and corrosion	excellent	excellent
	electrical conductivity	good for thick layers, moderate for thin layers	good
	optical reflectivity	good	excellent

Table 9. Target solders for sputtering targets (vapour pressure of Bi and Pb $<10^{-6}$ Pa, In and Sn $<10^{-8}$ Pa at 300 °C) [78v1].

Composition [wt%]				T [°C]	
Bi	In	Pb	Sn	Liquid	Solid
49	21	18	12	58	58
56	–	22	22	104	95
–	52	–	48	117	117
–	50	–	50	127	117
–	25	37.5	37.5	138	
58	–	–	42	138	
–	42	–	58	145	117
–	80	15 (Ag 5)	–	149	141

Composition [wt%]				T [°C]	
Bi	In	Pb	Sn	Liquid	Solid
–	99 (Cu 1)	–	–	153	153
–	100	–	–	157	157
–	12	18	70	174	150
–	70	30	–	174	160
–	–	37	63	182	
–	–	30	70	186	183
–	–	40	60	188	183
–	–	50	50	214	183
–	–	60	40	238	183

Table 10. Target current density j of different target materials with DC diode sputtering. Target voltage 2000 V. Gas pressure $p=9.3$ Pa. Cathode/anode spacing 4 cm [77W2].

Target material	j [mA/cm²]	Gas
Si	0.45	Ar
Mn_5Ge_3	0.55	Ar
GdFe	0.69	Ar
CoNi	0.49	Ar
Ni	0.46	Ar
CoNi	0.42	Ar
Ti	0.43	Ar
V	0.43	Ar
Ag	0.49	Ar
C	0.41	Ar
Ta	0.63	Ar
Ni	0.56	Ar+10% N_2
Au	0.39	Ar+10% N_2
Au	0.51	Ar+10% N_2
W	0.51	Ar+10% N_2
W	0.49	N_2
Ni	0.44	N_2

Table 11. Power density required for a temperature gradient 100 °C through a 1 cm thick cathode [78v1].

Cathode material	Power density W · cm⁻²
Ag	427
Cu	398
Au	315
Al	237
W	178
Si	149
Mo	138
Cr	94
Ni	91
In	82
Fe	89
Ge	76
Pt	73
Pd	72
Sn	67
Ta	58
Pb	35
Ti	22

Table 12. Energies and lifetime of metastable neutrals [78v1].

Species	Metastable energy [eV]	Lifetime [s]
H	10.20	0.12
H_2	11.86	long
He	19.82	very long
He	20.61	long
N	2.38	$6 \cdot 10^4$
N	3.58	13
N_2	6.16	0.9
N_2	8.54	$1.7 \cdot 10^{-4}$
O	1.97	110
O	4.17	0.78
O_2	0.98	very long
Ne	16.62, 16.71	long
Ar	11.55, 11.72	long
Kr	9.91, 9.99	
Xe	8.31, 8.44	

Ion beam sputtering

Ion bombardment of the target is used as the sputter source in ion beam sputtering (Fig. 30). The ions are generated in an ion source outside the deposition chamber containing the substrate. There is no plasma generated in the substrate chamber which is held at high vacuum ($10^{-3}\cdots10^{-4}$ Pa). The substrates are therefore not heated by secondary electrons. Because of the low substrate temperature and because sputtered metal atoms are not scattered by the sputter gas, ion beam sputtering is well suited for the photoresist lift-off process. The ion beam is usually neutralized before irradiation by a neutralizer filament (Fig. 30).

The deposition characteristics and the film properties obtained are surveyed in the Tables 13 and 14. Results are depicted in Figs. 31···34.

Table 13. Comparison of resistivity ϱ of ion beam sputtered films as deposited with other deposition techniques. Resistance measurements are made by four point probe [79K2].

Metal	ϱ [$\mu\Omega$ cm]		
	Ion-beam sputtered	RF sputtered	Evaporated
Au (3000 Å)	5.9	3.1	3.7
Al (4000 Å)	28.5	6.5	4.0
Ta (4000 Å)	225.8	218.1	233.7
NiFe (4000 Å)	27.5	17.0	18.0

Table 14. Room temperature stress (as deposited) in ion beam deposited metal films in comparison to other deposition techniques. Substrates are not heated during deposition [79K2].

	X [N m^{-2}]		
	Ion-beam sputtered	RF sputtered	Evaporated
Au	$1 \cdot 10^7$	$5 \cdot 10^7$	$4 \cdot 10^7$
Si	$6 \cdot 10^8$	$2 \cdot 10^8$	$4 \cdot 10^8$ *)
Al	$3 \cdot 10^7$	$6 \cdot 10^7$	$2 \cdot 10^7$
NiFe	$8 \cdot 10^8$	$3 \cdot 10^8$	$3 \cdot 10^8$ *)

*) Film thickness = 1000 Å; thickness of all other films = 4000 Å.

Conductive layers used in solar technology

The fabrication of conductive layers in the technology of solar cells requires processes which allow the fabrication of high volumes at low cost [80N2, 80M3]. Suitable processes are surveyed in Table 15.

Table 15. Metal deposition techniques applied in low-cost/high-volume photovoltaic systems [80M3, 80N2].

Deposition technique	Example	Remarks
Vacuum deposition (Evaporation/sputtering)	solar reflectors CdS—Cu_2S-films	magnetron sputtering for temperature-sensitive substrates (plastic)
Paints (printing and firing)	Ag-, Ag/Al-films organic or silicone binders	low coating costs, good environmental stability
Chemical processes (spray pyrolysis)	ITO-films CdS-, Cu_2S-films	low cost, high throughput, metal layers must be protected against corrosion by environment
CVD		application to single-crystal silicon solar cells (Figs. 35···40)
Electroplating	"black chrome" (chromium particles in chromiumoxide matrix), nickel films	widely used

6.1.4.5.10 Properties of metals used for deposition

Formation of layers and layer structure

During deposition, condensation of the metal occurs via impinging of atoms with occasional reevaporation, nucleation, and coalescence to formation of the thin film (see Fig. 41). A model of the microstructure was predicted by Mowchan et al. [69M1] and experimentally tested by Bunshah et al. [74B1]. The structure obtained for various deposition temperatures T in relation to the melting temperature T_m may be divided into three zones: (Fig. 42).

Zone 1: $T < 0.3\ T_m$
Porous layer, gaps at grain boundaries,

Zone 2: $T = 0.3\ T_m \cdots 0.45\ T_m$
Increasing nucleation and surface diffusion, density of bulk material; columnar structure with plane surface,

Zone 3: $T > 0.45\ T_m$
Recrystallization; increased grain size; equiaxial grain structure.

Thornton [74T1, 75T1] extended the model by including the effects of the argon pressure and sputter processes (Fig. 43). The Thornton model is confirmed by several experimental tests [81C1, 81C2, 82V1]. The important effects of the residual gases during evaporation and sputtering on the condensation, coalescence and diffusion are not taken into account by the model.

Diffusion and diffusion barriers

The diffusion coefficient D of metals in layers is dependent on the purity, the structure, the grain size, and the defect density. The exponential activation law

$$D = D_0 \, e^{-Q/RT}$$

is valid. The prefactor D_0 and the activation energy Q are listed in Table 16 for bulk metals. Lattice diffusion is usually prevalent in bulk metals. The diffusion coefficients in layers may be strongly affected by impurities (residual gases) and dependent on the temperature range. Grain boundary diffusion and grain-boundary-assisted bulk diffusion may be prevalent (Fig. 44).

Table 16. Prefactor D_0 and activation energy Q for various metals (bulk) [56E1].

Diffusion system	D_0 $cm_2 \, d^{-1}$	Q kcal mol^{-1}
Au in Au	$8 \cdot 10^5$	63
Cu in Au	50	27
Pd in Au	96	37
Pt in Au	107	37
Ag in Au	2510	38
Ni in Au	150	31
Au in Ag	46	30
Pd in Ag	0.55	20
Cd in Ag	4.2	22
Sn in Ag	6.7	21
Cu in Ag	5.1	25
Al in Cu	620	39
Mn in Cu	0.6	23
Ni in Cu	5.6	30
Zn in Cu	0.26	20
Pd in Cu	0.14	22

Diffusion system	D_0 $cm^2 \, d^{-1}$	Q kcal mol^{-1}
Sn in Cu	355	31
Pt in Cu	0.09	22
Au in Cu	0.6	22
Pb in Pb	$5.8 \cdot 10^5$	28
Sn in Pb	$3.5 \cdot 10^5$	26
Cd in Pb	$1.6 \cdot 10^2$	15
Ag in Pb	$6.5 \cdot 10^3$	15
Au in Pb	$3.0 \cdot 10^4$	14
Cu in Ni	90	35
Mo in W	54	80
Th in W	$3.3 \cdot 10^4$	90
Cu in Pt	$4.2 \cdot 10^3$	55.7
Ni in Pt	68	43.1

Diffusion barriers

Diffusion barriers are required to inhibit the indiffusion of Si into Al, which erodes the semiconductor contacts. Table 17 lists the diffusion coefficient D and the mean migration distance of Si in Al.

Table 17. Diffusion coefficient and mean migration distance $\sqrt{Dt}$ (t = 180 s) of the system Si in Al-films [79C1, 71M1].

T °C	D cm^2/s	$\sqrt{Dt}$ μm
560	$3.9 \cdot 10^{-8}$	26.5
496	$1.7 \cdot 10^{-8}$	17.5
441	$8.0 \cdot 10^{-9}$	12.0
394	$3.5 \cdot 10^{-9}$	7.93
352	$1.6 \cdot 10^{-9}$	5.36
315	$7.4 \cdot 10^{-10}$	3.65
282	$3.3 \cdot 10^{-10}$	2.44
227	$1.5 \cdot 10^{-10}$	1.64
203	$7.1 \cdot 10^{-11}$	1.13
181	$3.3 \cdot 10^{-11}$	0.77
162	$1.5 \cdot 10^{-11}$	0.52

The most frequently used diffusion barriers are Ti_x–W_{1-x} in the structures PtSi/Ti–W/Al and PtSi/Ti–W/Au, and Ti in the structures Ti/Al and PtSi/Ti/Pt/Au.

In the system Ti–W/Au, cracks and pinholes may occur in the Ti–W layer during annealing [78N2] (see also section 6.1.4.5.11). The system PtSi/Ti is thermodynamically unstable, because Ti also forms silicides. It is recommended to incorporate stable diffusion barriers consisting of oxides and mono-nitrides of transition metals, and also carbides, borides and silicides if they show a low electrical resistance [69N1, 75F1, 78N2, 82T1, 83E1, 83S1]. Table 18 lists resistivity values of these compounds. The position of various silicides in the periodic table are shown in Table 19.

Table 18. Measured resistivity ϱ at 20 °C of films used in diffusion barriers [78N2] ϱ_0^* is the specific resistivity of polycrystalline bulk material [68k1]. Variations (see also Fig. 47) are caused by the deposition parameters, e.g. due to residual gases.

Material	ϱ_0^* μΩ cm	ϱ μΩ cm
Cr	15.0	15.3···18.9
CrB_2		21.0···84
CrB		64···69
Cr_2B		52
Cr_3C_2		75
Cr_2N		81···84
CrN		640···2500
$CrSi_2$		91.4···1470
Hf	26.5	30.0
HfB_2		10···15
HfC		60···109.0
HfN		(56.5) (1200 K)
$HfSi_2$		45···50
Mo	5.03	5.2
Mo_2B_3		25.0
MoB_2		22.0···45.0
MoB		25.0···50.0
Mo_2B		40.0
Mo_2C		71.0···120
$MoSi_2$		21.0···60
MoSi		21.5
Mo_5Si_3		46.7
Mo_3Si		21.6···25
Nb	23.3	13.1···16.0 (?)
NbB_2		32.0···42.7
NbB		64.0
NbC		51.1···147.0
NbN		60.0···200.0
$NbSi_3$		6.3···52.0
$NbSi_2$		50
Ta	12.4	12.4···14.7
α-Ta		20.0
β-Ta		150.0
TaB_2		37.4···68.0
TaB		100.0
TaC		40.6···17.0
Ta_4N_5		>800
Ta_5N_6		≈500
TaN		135···1650
Ta_2N		186···245
$TaSi_2$		8.5···60.0
Ti	42.0	48.0···68.2
TiB_2		9.0···28.4
TiB		40.0
TiC		52.5···180.0
TiN		25.0
$TiSi_2$		13.0···39.3
Ti_5Si_3		350
Ti_xW_{1-x} *)		60···>150
V	18.2	26.0···26.6
VB_2		16.0
VB		30.0···35.0
VC		150
VN		85.0···200.0
VSi_2		9.5···66.5
V_5Si_3		114.5
V_3Si		203.5
W	4.89	5.03···5.5
W_2B_5		21.0···43.0
WC		19.6···54.0
W_2C		≈70···80.0
W_xN_y		1650
WSi_2		16.0···70.0
Zr	40.5	41.0···63.4
ZrB_2		7.0···38.8
ZrB		30.0
ZrC		56.6···75.0
ZrN		13.6···56.0
$ZrSi_2$		35.0···161
ZrSi		49.4

*) Depending on Ti concentration and deposition parameters.

Table 19. Positions of various silicides in the periodic table [80M1].

	IA	IIA	IIIA	IVA	VA	VIA	VIIA	VIII			IB	IIB	IIIB	IVB	VB	VIB	VIIB
1	H_4Si																
2	$Li_{15}Si_4$ Li_2Si												B_6Si B_4Si B_3Si	CSi	N_4Si_3	OSi O_2Si	F_4Si
3	NaSi	Mg_2Si												Si	PSi	S_2Si	Cl_4Si
4	KSi KSi_6	Ca_2Si CaSi $CaSi_2$	Sc_5Si_3 $ScSi_2$	Ti_5Si_2 TiSi $TiSi_2$	V_3Si V_5Si_3 VSi_2	Cr_3Si Cr_5Si_3 CrSi $CrSi_2$	Mn_3Si Mn_5Si_3 MnSi $MnSi_2$	Fe_3Si Fe_3Si_2 FeSi $FeSi_2$	Co_3Si CO_2Si CoSi $CoSi_2$	Ni_3Si Ni_2Si Ni_5Si_2 Ni_3Si_2 NiSi $NiSi_2$	Cu_3Si				As_2Si AsSi	Se_2Si	Br_4Si
5	RbSi $RbSi_6$	SrSi $SrSi_2$	Y_5Si_4 Y_5Si_3 YSi YSi_2	Zr_4Si Zr_2Si Zr_3Si_2 Zr_6Si_3 ZrSi $ZrSi_2$	Nb_4Si Nb_5Si_3 $NbSi_2$	Mo_3Si Mo_5Si_3 $MoSi_2$		RuSi Ru_2Si_3	Rh_2Si Rh_5Si_3 Rh_3Si_2 RhSi Rh_2Si_3	Pd_3Si Pd_2Si PdSi						Te_2Si TeSi	J_4Si
6	CaSi $CaSi_3$	BaSi $BaSi_2$	*) $LaSi_2$	Hf_2Si Hf_5Si_2 Hf_3Si_2 HfSi $HfSi_2$	$Ta_{4.5}Si$ Ta_2Si Ta_5Si_3 $TaSi_2$	W_5Si_3 WSi_2	Re_3Si Re_5Si_3 $ReSi_2$	OsSi $OsSi_2$ $OsSi_3$	Ir_3Si Ir_2Si Ir_3Si_2 IrSi $IrSi_2$ $IrSi_3$	Pt_3Si Pt_2Si PtSi							
			*) Ce_3Si Ce_2Si CeSi $CeSi_2$	$PrSi_2$	$NdSi_2$		$SmSi_2$		Gd_3Si_5 $GdSi_2$		Dy_3Si_5 $DySi_2$		Er_3Si_5		$YbSi_x$	Lu_2Si_5	
7				Th_3Si_2 ThSi $ThSi_2$		U_3Si_2 USi U_2Si_3 USi_2 USi_3	$NpSi_3$	PuSi $PuSi_2$									

Formation of silicides

Refractory metals which are frequently used in Si technology, the noble metals which are used in contact layers, and the transition metals, which are used in diffusion barriers, all form several silicide phases (Table 19). Several phases can exist at the same temperature; however, the formation starts in one phase and the next is formed with some delay by transformation of the first phase [79O1]. The increase in layer thickness follows a $\sqrt{t}$-law. The growth and interdiffusion of the phases can be observed by RBS (Rutherford back scattering) techniques (Fig. 46). The existence of several phases can be verified by X-ray diffraction (Table 20).

Table 21 surveys metallurgical data of noble metals and similar metals which start to form Me_2Si in the reaction. Table 22 lists transition metal silicides which start to form MeSi. Properties of refractory metals are listed in Table 23. They start the reaction by forming $MeSi_2$. Table 24 and 25 list electrical barrier heights and thermal expansion coefficients of silicides.

Fig. 48 depicts the volume change of silicide formation. Fig. 49 shows the reaction rate of Si—W.

Table 20. Comparison of MeV ^{4}He backscattering spectrometry and X-ray diffraction for analysis of metal silicide films (10 nm···1 μm thickness).

	MeV backscattering spectrometry	X-ray diffraction (glancing angle)	
		Read camera	Seemann–Bohlin
Film thickness	quantitative	qualitative	qualitative *)
Film composition	atomic ratios	phase identification	phase identification
Phase location in depth	concentration profile	no	yes, indirectly **)
Minimum phase thickness	20···50 nm	20···50 nm	≈10···20 nm
Typical sample dimension	$\approx\frac{1}{2}$ cm^2	$\approx\frac{1}{2}$ cm^2	>1 cm^2
Film structure:			
grain size	no	qualitative	quantitative
stress	no	no	yes
texture	no	variation of ring pattern intensity and discontinuity	indication from peak intensity
epitaxy	yes, by channeling	qualitative (single-crystal pattern)	qualitative (peaks absent)
ordering	no	yes, superlattice lines	yes, superlattice lines
Data acquisition	15···30 min	1···24 h	4···12 h
Additional limitations	requires lateral uniformity; not sensitive to light elements	qualitative only	intensity detection only along equator plane

*) Film thickness can be calculated from integrated intensity data provided that corrections such as absorption factor and temperature factor are known.

**) Relative intensity ratio of different phases will indicate the location of layers.

Table 23. Metallurgical properties of refractory metal silicides [79O1, 80M1].

Silicide	Silicide formation starts at °C	Activation energy eV	Growth rate	Moving species	T_m °C	Eutectic temperatures °C	Composition at lowest eutectic temperature at % Si
$CrSi_2$	450	1.7	t		1550···1590	1300···1390	87
$MoSi_2$	525	3.2	t	Si	1980···2050	1410···1900	97
VSi_2	600	2.9, 1.8	$t, \sqrt{t}$	Si	1750···1670	1375···1640	97
$NbSi_2$	650				1930···1950	1295···1850	95
WSi_2	650	3	$t, \sqrt{t}$	Si	2165	1400···2010	99.2
$TaSi_2$	650			Si	2200	1385···2100	99
$ZrSi_2$	700				1650···1700	1355	90

Table 21. Silicides formed by noble and near noble metals used in ohmic and Schottky contacts [79O1].

Silicide	Schottky-barrier ϕ_B eV	Formation temper-ature °C	Activation energy eV	Growth rate	Moving species	T_m °C.	Eutectic *) temper-atures °C
Co_2Si		350···500	1.5	$t^{1/2}$	Co	1332	1195···1286
CoSi	0.68	375···500	1.9	$t^{1/2}$		1460	1286···1310
$CoSi_2$	0.64	550				1362	1310···1259
Ni_2Si	0.7···0.75	200···350	1.5	$t^{1/2}$	Ni	1318	1265···964
NiSi	0.7···0.75	350···750	1.4	t(?)	Ni	992	964···966
$NiSi_2$	0.7	≦750				1100	
Ni_5Si_2		400				1282	1152···1265
Ni_3Si		450				1165	
Pd_2Si	0.74	100···300			Pd, Si	1398	1070···875
PdSi		800				1000	875···870
Pd_3Si		350				1070	820···1050
Pd_4Si		400				950	
Pd_5Si		650				835	825···822
Pt_2Si	0.78	200···500	1.5	$t^{1/2}$	Pt	1100	830···983
PtSi	0.87	300	1.6	$t^{1/2}$		1230	983···979
Pt_3Si		400				900	
Pt_2Si_3		Obtained by ion beam mixing					
Pt_4Si_2							

*) For compounds having a congruent melting point, the temperatures of the two closest eutectic points are given. The first temperature refers to the eutectic at the metal side.

Table 22. Silicides formed by transition metals [79O1].

Silicide	Formation temperature °C	Activation energy eV	Growth rate	Moving species	T_m °C	Eutectic *) temperatures °C
RhSi	350···425	1.95	$t^{1/2}$	Si		
Rh_2Si	400					
Rh_4Si_5	825···850					
Rh_3Si_4	925					
$HfSi_2$	550···700	2.5	$t^{1/2}$	Si	2200	
$HfSi_2$	750				1950	
TiSi	500				1920	
$TiSi_2$	600			Si	1540	1490···1330
MnSi	400···500				1275	1235···1145
$MnSi_2$	800				1150	
IrSi	400···500	1.9	$t^{1/2}$			
$IrSi_{1.75}$	500···1000				750	
$IrSi_3$	1000					

*) For compounds having a congruent melting point, the temperatures of the two closest eutectic points are given. The first temperature refers to the eutectic at the metal side.

Table 24. Schottky barrier heights (ϕ_B) of various silicides on n-type silicon [80M1]. (For pure metals see Table 43.)

Disilicides	ϕ_B eV	Other silicides	ϕ_B eV
$TiSi_2$	0.6	HfSi	0.53
$CrSi_2$	0.57	MnSi	0.76
$ZrSi_2$	0.55	CoSi	0.68
$MoSi_2$	0.55	NiSi	0.7···0.75
$TaSi_2$	0.59	Ni_2Si	0.7···0.75
WSi_2	0.65	RhSi	0.74
$CoSi_2$	0.64	Pd_2Si	0.74
$NiSi_2$	0.7	Pt_2Si	0.78
		PtSi	0.87
		IrSi	0.93
		Ir_2Si_3	0.85
		$IrSi_3$	0.94

Table 25. Linear thermal expansion coefficient α of silicides and constituents (see Fig. 45).

Silicide	α $10^{-6}\,K^{-1}$	Element	α $10^{-6}\,K^{-1}$
		Si	3
$TiSi_2$	10.5	Ti	8.5
$ZrSi_2$	8.6	Zr	5.7
VSi_2	11.2, 14.65	V	8
$NbSi_2$	8.4, 11.7	Nb	7
$TaSi_2$	8.8, 8.9	Ta	6.5
$MoSi_2$	8.25	Mo	5
WSi_2	6.25, 7.90	W	4.5

6.1.4.5.11 Intrinsic properties of metal layers

Thickness and uniformity of thickness

For evaporation in vacuum systems, the deposition by the vapour beam is directed and obeys a cosine-law. Magnetron cathodes exhibit a non-uniform deposition/sputter rate depending on the plasma density (Fig. 50). By moving the substrate (planetary gearing), a layer uniformity within ±6% can be obtained over the whole substrate holder and from run to run (Fig. 51) Table 26.

Table 26. Thickness uniformity, time for deposition of 1 μm Al film and deposition rate for various deposition processes.

Deposition process	Deposition configuration	Deposition rate		Time for deposition of 1 μm Al	Thickness uniformity over 100 wafers
		actual rate μm min^{-1}	average rate (over whole deposition time) μm min^{-1}	min	%
DC-magnetron sputtering	cylindrical substrate holder, planar magnetron cathode	1.5	0.17 (2 cathodes parallel)	6 (for 25 wafers 125 mm∅)	±6
	single wafer processing static deposition	1.5	1.5	0.67 (for 1 wafer)	±6
Electron beam gun evaporation	planetary system substrate holder (Knudsen type)		0.18	5.5 (for 25 wafers 125 mm∅)	±4···7

Stoichiometry of alloys

The two components of an alloy are co-evaporated from two e-guns in a *high-vacuum evaporation* process. The typical reproducibility of the stoichiometry in the evaporated layer is shown in Fig. 52 for Al—Si(1.5%). The maximum deviation from run to run is approximately 20% from the mean value. The same value is valid for Ti—W films.

For sputter deposition, the stoichiometric composition of films is dependent on the scattering of the alloy components on argon atoms in the ambient. For similar atom mass, e.g. Al and Si, the target stoichiometry is well reproduced within ±5%. For strongly differing atom masses, e.g. Ti and W, the stoichiometry is strongly dependent on the argon gas pressure. For a target composition 14% Ti—86% W, layer compositions in the range 3.5%···5.9% Ti are observed [80H2]. The Ti concentration can be reduced to 1/6 by applying a bias voltage of up to 300 V to the substrate [79H4] (see Fig. 53). The reduction is caused by preferential resputtering of Ti during deposition.

A non-uniform concentration may also be caused by diffusion in the layer during deposition [79C1] (see Figs. 54, 55).

For co-sputtering of alloys from different cathodes, additional variations (≈5%) are caused by the power instability of the two sputtering targets.

Adhesion of films on the substrate

No generally accepted methods exist to test the adhesion of films on the substrates (Si, SiO_2). Approximate measures may be obtained by shearing force tests of the layers. A qualitative measure is given in the heat of formation of oxides (Table 27). The adhesion is also dependent on the average mean energy when it impinges on the substrate (Table 28). This energy varies for the different vacuum deposition processes.

Table 27. Adhesion of metal films to SiO_2-substrates [69s1, 78H2].

Metal film	Metal oxide	Heat of metal-oxide formation kcal mol^{-1}	Shearing force 10^8 N m^{-2}	Adhesion to SiO_2
Ta	Ta_2O_5	−500		very strong
Al	Al_2O_3	−399	17.0	very strong
V	V_2O_3	−290		very strong
Cr	CrO_3	−270		very strong
Ti	TiO_2	−218	17.0	very strong
W	WO_3	−200		strong
Mo	MoO_3	−180	11.3	strong
Cu	Cu_2O	−40		weak
Ag	Ag_2O	−7		very weak
Au	Au_2O_3	+19		very weak

Table 29. Metal step coverage burnout testing [73G1].

Type of metal coverage	Burnout time (μs)		
Flat metal	0.5	1.0	5.0
45% coverage	0.38	0.78	4.6
30% coverage	0.14	0.32	2.1

Table 28. Mean energy $\bar{E}$ of particle impinging on the substrate for various vacuum deposition processes [77S1].

Deposition process	$\bar{E}$ eV
High vacuum evaporation	≈0.2
Sputtering (magnetron, diode)	2···20
Ion-beam sputtering	≈15···200
Ion plating	50

Table 30. Measures to improve the step coverage during sputter deposition.

Step profile	Measure
Steep, but not overhanging	increase of sputtering gas pressure
Overhanging step	high surface mobility of metal atoms
height < thickness of metal	high substrate temperature
Overhanging step heights > thickness of metal	bias sputtering

Step coverage

Because of the directionality of particle beams during evaporation, reductions of the layer cross-section and structural changes occur in the layer at steps (Fig. 56). Because these changes cause dramatic effects on the reliability of semiconductor devices, computer modelling is applied for the step coverage [74T2, 80N3] (see Figs. 57···59). Electrical measurement methods have been developed to characterize the step coverage [73G1]. Conductor stripes are melted by a short pulse so that the heat loss to the substrate is minimized. (Table 29). Visual inspection of the step coverage in cross-sections of cleaved or polished structures is performed by REM imaging [72E1, 82T2].

Poor step coverage is used to structure the metallization by the lift-off technique.

Surface active deposition methods, e.g. CVD are advantageous for good step coverage compared to vacuum evaporation. Improvements of the step coverage are obtained by planarization of steps [83R1]. Measures for improvement of the step coverage during sputter deposition are surveyed in Table 30.

Damage during deposition processes

Four causes of damages of the electrical properties by deposition processes are known:

X-ray and electron beam irradiation,
UV irradiation,
incorporation of mobile charges,
stress in deposited layers.

Most of the damages introduced during deposition of layers can be removed by annealing (Table 31). Exceptions are only the parameter changes caused by the incorporation of alkali ions [80F1] and by VUV irradiation which cause positive space charges and hole traps, respectively, in the gate oxide [80V1].

Table 31. Electrical damage observed after deposition of metal layers.

Deposition process	Electrical damage	Damage caused by	Annealing conditions	Ref.
e-gun evaporation of Al films on SiO_2	flatband voltage shift (Fig. 60) (⟨111⟩ and ⟨100⟩ Si-substrate, n- and p-type, 0.5···10 Ω cm resistivity)	trapped holes in SiO_2 close to Si—SiO_2 interface due to X-rays	30 min, 480 °C, N_2+H_2	78L2
e-gun evaporation of Al films on SiO_2	flatband voltage instability ⟨111⟩ Si-substrate	mobile charge concentration $Q_m=1.6\cdot10^{11}$ cm^{-2} (alkali ions)	not annealable	78H1
Al films evaporated on SiO_2 by induction-heated source	no defects detected after annealing, mobile ion concentration $Q_m=4\cdot10^{10}$ cm^{-2}		15 min, 450 °C, N_2	79F1
Magnetron-sputtered Al films on SiO_2	no defects detected after annealing, mobile ion concentration $Q_m=4\cdot10^{10}$ cm^{-2}		15 mm, 450 °C, N_2	79F1
Magnetron-sputtered Al—Si films on SiO_2	flatband voltage shift (≈0.6 V)	600 eV electrons	40 min, 450 °C, N_2+H_2	80V1
Magnetron-sputtered Al—Si films on SiO_2	flatband voltage shift	hole traps caused by vacuum ultraviolet radiation (VUV)	not annealed at temperatures up to 500 °C	80V1
RF-diode-sputtering of Al films on SiO_2	threshold voltage of transistor shifts towards positive voltages; dependence of threshold voltage on substrate voltage increased (Figs. 61···64)	positive fixed charges Q_{ox} and surface states $D_{it}(E)$ near Si/SiO_2-interface	20 mm, 450 °C, N_2+H_2	81S1
Magnetron-sputtered $TaSi_2$-films on n^+ doped poly-Si	no damage after sintering, MOS and IGFET parameters correspond to those of n^+ poly-Si gates, excellent V_{FB} stability (Fig. 65, Table 32)		30 min, 900 °C, H_2	80S2
RF triode-sputtered Ta-films on SiO_2	significant changes in recombination life time and capture cross sections (Table 33)	stresses in the Si/SiO_2 interface due to Ta-films	No annealing	78L3

Table 32. MOS- and IGFET parameters measured at $TaSi_2/n^+$-poly-Si gates [80S2].

Property	Data
Structure	2500 Å $TaSi_2$ on 2500 Å poly-Si (phosphorus-doped) on 500 Å SiO_2 on p(100) Si, $N_a \approx 1 \cdot 10^{16}\ cm^{-3}$
Sheet resistance	$2 \pm 0.3\ \Omega/\square$
Modified work function	3.30 V
Oxide fixed charge	$2 \cdot 10^{10}\ cm^{-2}$
Surface state density at midgap	$\lesssim 10^{10}\ cm^{-2}\ eV^{-1}$
ΔV_{FB} ($+2$ MV cm^{-1}, 250 °C, 15 min)	0.0 V
ΔV_{FB} (-2 MV cm^{-1}, 250 °C, 15 min)	≈ -0.1 V
Hot electron trapping efficiency	$N_{eff}/Q_{in} \approx 10^{-6}$ (at $Q_{in} = 10^{17}\ cm^{-2}$)
Lifetime	$\approx 14\ \mu s$
Median oxide dielectric strength $F_{50\%}$	$9\ (\pm 0.5)$ MV cm^{-1}
log normal sigma	0.05
Oxide defects (breakdown < 6 MV cm^{-1})	2.5 %
Oxide leakage (at < 6 MV cm^{-1})	$\lesssim 10^{-11}\ A$
Threshold voltage from C–V	0.8 V
IGFET threshold Voltage ($30 \times 30\ \mu m^2$)	0.9 V
Channel mobility	$635\ cm^2\ V^{-1} s^{-1}$
Short channel effect; (threshold voltage)	
5 µm	0.9 V
2 µm	0.75 V

Table 33. Flatband voltage V_{FB}, surface potential $\psi_{s0}(V=0)$, surface state charge density n_{it}, recombination life time τ_{m0}, capture cross section σ_p and stress X in p-type ⟨100⟩ silicon wafers after sputtering of 350 nm Ta-films.

Sputtering voltage [kV] *)	V_{FB} (V)	n_{it} ($\cdot 10^{11}\ cm^{-2}$)	ψ_{s0} (V)	τ_{m0} ($\cdot 10^{-2}$ s)	σ_p ($\cdot 10^{-16}\ cm^2$)	X **) kg mm^{-2}
0	1.8	7	0.15	1.6	3–4	0
2	2.45	7.1	0.18	8.7	0.75	0.89
4	2.76	7.8	0.25	16	0.4	
6	3.4	9.0	0.26	36	0.18	1.38

*) Peak to peak voltage.
**) Stress measurements performed by measuring the radius of curvature of bent Si wafers [78L3].

6.1.4.5.12 Residual gas-dependent properties of metal layers

Residual gases

Residual gas is still present in vacuum stations at high vacuum HV (10^{-1} Pa $> p >10^{-6}$ Pa) and at ultrahigh vacuum UHV ($p<10^{-6}$ Pa). Adsorption and desorption at surfaces in relation to the pump power defines the residual gas pressure, i.e. the sum of the partial pressures of the individual gases and vapours. Quadrupole spectrometers are generally used to analyze the residual gases. A typical spectrum is shown in Fig. 66. Hydrogen, water vapour, nitrogen, oxygen, and carbon dioxide and, in sputter setups, argon are usually observed, i.e.

$$p_{tot} = p_{H_2} + p_{H_2O} + p_{N_2} + p_{O_2} + p_{CO_2} + p_{Ar} + p_{NN},$$

p_{NN} is the sum of partial pressures of all residual gases not detected in the quadrupole spectrometer.

The base pressure in high-vacuum setups is at $\lesssim 3 \cdot 10^{-6}$ Pa. During deposition, the pressure increases by outgassing from the source material or by desorption from surfaces during temperature stressing. Spectra important for the quality of vacuum stations are shown in Figs. 67···82.

Sputter-deposited layers show a higher residual gas sensitivity than evaporated layers. The cause is the higher (one order of magnitude) kinetic energy of sputtered particles than that of evaporated particles (see Table 28). The high kinetic energy of particles leads to heating of the surfaces and thus increased desorption especially of water vapour from these surfaces. In the plasma of sputter setups, dissociation of desolved gases occurs. The incorporation of residual gas particles in layers leads to an increased electrical resistivity of the layers. Silicides fabricated by co-sputtering show a higher resistivity than those fabricated by co-evaporation (see Table 7).

When a deposition process is changed from evaporation to sputtering, the following parameters which affect the contamination by residual gases have to be considered [82w1]: sputter rate, leakage of the setup, optimization of the pump system, surface conditions in the sputter setup, purity of the sputter gases.

Electrical conductivity of thin films

The electrical resistance of thin metal films is more or less increased from the value of the bulk resistivity (Figs. 89 and 94). The increase is caused by various parameters, e.g. residual gas incorporation, substrate temperature, bombardment of ionized or neutral particles among others, during deposition. The increase is partially caused by impurities. The important contribution is the defect density.

Defects are reduced by annealing. The standard rule for the annealing temperature is $T_a \approx 2/3\,T_m$. The usual annealing procedure for aluminum layers is 30 min at 450 °C. The layer resistance after annealing for low contamination is at most 10···20% above the bulk value (Table 34 and 35).

Silicides of refractory metals and diffusion barriers of nitrides and carbides of transition elements have to be annealed at relatively high temperatures. The changes of the resistivity or the sheet resistances are depicted in Figs. 83···87. The lowest resistivity values of thin films reported are listed in Table 35. Resistivity values of diffusion barriers are presented in Table 18.

Table 34. Room temperature resistivity ϱ of e-gun evaporated aluminum films (thickness 0.8 μm) observed after 20 min annealing at various temperatures T_a.

T_a °C	ϱ μΩ cm
300	3.25
350	3.16
400	3.06
450	2.98
500	2.95

Table 35. Lowest resistivity values at room temperature (22 °C) reported for thin films of metals and silicides [78N2, 80G1, 80M1, 81H1, 83M1, 83N1].

Film *)	ϱ μΩ cm	Film	ϱ μΩ cm	Film	ϱ μΩ cm
Al	2.7	Cr	15.3	Zr	41
Al—Si(0.5)	2.8	Cu	1.6	$Ti_{0,1}$—$W_{0,9}$	70
Al—Si(1)	2.8	Hf	30.0	$CrSi_2$	600
Al—Si(1.2)	2.8	Mo	5.2	$MoSi_2$	40 (22?)
Al—Si(1.5)	2.9	Ni	6.7	$NbSi_2$	50
Al—Si(2)	2.9	Pd	10.0	VSi_2	50
Al—Cu(0.5)	2.8	Pt	10.2	WSi_2	27 (14?)
Al—Cu(0.5)—Si(1)	3.0	Sn	10.4	$TaSi_2$	40 (8?)
Al—Cu(4)—Si(1.5)	5.2	Ta	12.4	$TiSi_2$	13
Ag	1.6	Ti	48	$ZrSi_2$	35
Au	2.2	V	26		
Co	5.4	W	5.0		

*) The figures in parentheses give the contents in wt%.

According to Matthiessen's rule, the resistivity is composed of a temperature-dependent component $\varrho_M(T)$ specific to the pure metal without crystal defects and by a temperature-independent additional component, which is determined by the residual resistivity at low temperatures (T=4.2 K).

$$\varrho=\varrho_M(T)+\varrho_{4.2\,K}.$$

It is also common practice to define the resistivity ratio

$$RRR=\frac{\varrho_M(T)+\varrho_{4.2\,K}}{\varrho_{4.2\,K}}.$$

Only the residual resistivity at low temperature is affected by annealing. Measurements of the resistivity ratio RRR are used to characterize the microstructure of thin films. It yields information on the stress behavior and the expected lifetime of the film.

Among the residual gases, especially increased concentrations of water vapour and nitrogen increase the electrical resistivity (Fig. 88, Table 36). Oxygen causes an increase of the resistivity of Al layers only at relatively high concentrations (Fig. 89). The incorporation of nitrogen in W, Ni, and Au films is shown in Fig. 90.

Table 36. Influence of residual gas contamination during deposition on resistivity of magnetron-sputtered Al—Si(1.5%) films. Thickness 1 μm on p-type ⟨100⟩ Si-substrate. Base pressure of the sputter system was $9\cdot10^{-5}$ Pa [80N1].

Contaminant	Partial pressure of contaminant prior to sputtering Pa	Resistivity ϱ μΩ cm	Impurity concentration in the films (at.%)			
			H	C	N	O
None (Purity of argon 99,9999%)	$9\cdot10^{-5}$	3.02	0	0.1	0	0
Hydrogen	$5.3\cdot10^{-3}$	3.47	0	0.1	0.1	0.3
Nitrogen	$5.3\cdot10^{-3}$	6.58	0	0.4	1.6	0.5
Oxygen	$5.3\cdot10^{-3}$	4.25	0	0.9	0.1	1.4

The lowest electrical resistivity that may be achieved by *silicides* and *diffusion barriers*, e.g. Ti—W, Ti—N, is dependent on the stoichiometry. Results are shown for Mo—Si and Ti—Si films (Fig. 91), for Ti—N films (Fig. 92), and for Ti—W films (Fig. 93).

Bias sputtering affects the stoichiometry of alloys, gas incorporation, and microstructure of the films. Table 37 lists specific resistivities of bias-sputtered layers. The Figs. 24a, b and 53 show results.

Table 37. Resistivity of co-sputtered aluminum alloy films (with RF bias). Run No 1···6: Al sputtered with DC magnetron, Cu and Si, respectively, sputtered with RF diode; No 7···13: Al—Si(1%) sputtered with DC magnetron, Cu sputtered with RF diode.

Run No.	Composition of alloy films	DC power kW	RF power kW	RF induced bias voltage V	Thickness of films μm	Resistivity μΩ cm
1	Al—Cu(1.5%)	9	0.1	−25	1.10	2.97
2	Al—Cu(3.6%)	9	0.25	−25	1.10	3.08
3	Al—Cu(6.9%)	9	0.5	−25	1.10	3.19
4	Al—Cu(6.1%)	5	0.25	−50	1.20	3.24
5	Al—Si(1.8%)	9	1.0	−25	1.38	2.89
6	Al—Si(1.8%)	9	1.0	−50	1.30	2.87
7	Al—Si(1%)—Cu(2.4%)	6	0.1	−25	1.05	3.00
8	Al—Si(1%)—Cu(2.9%)	5	0.1	−35	1.15	2.99
9	Al—Si(1%)—Cu(1.5%)	9	0.1	−50	1.15	3.10
10	Al—Si(1%)—Cu(3.9%)	9	0.25	−25	1.14	3.11···3.13
11	Al—Si(1%)—Cu(3.8%)	9	0.25	−40	1.15	3.11
12	Al—Si(1%)—Cu(3.7%)	9	0.25	−50	1.18	3.12
13	Al—Si(1%)—Cu(4%)	9	0.25	−50	1.86	3.10···3.13

Contact resistance

Four different types of contacts occur in semiconductor technology:

the *ohmic contact* shows a linear current-voltage characteristic

the *Schottky barrier contact* is used in fast switching device structures. Only majority charge carriers determine the current flow.

metal-poly Si contacts occur in the MOS Si-gate technology. In some cases, poly Si is also used as a diffusion barrier.

via contacts are contacts between different metal layers in the multi-layer metallization of complex ICs.

The continuous reduction of contact areas in the VLSI technology increases the importance of contact resistances. Special structures using four point probes or contact chains [69M2, 70B1, 70Y1, 80R1] are employed to measure contact resistances. Measurement results are not always comparable.

For contacts to crystalline n- and p-type silicon, Al—Si alloys, platinum silicide or diffusion barriers, e.g. Ti—N, are preferentially used. The ohmic contact Al—Si is obtained by the metallurgical reaction during sintering (Fig. 95).

Aluminum spiking is caused at the contact by the temperature-dependent solution of Si in Al (Table 38). Alloys of Al and Si are employed to avoid contact spiking (see section 6.1.4.5.9).

Contact resistances of Al and other metals to Si are listed in Tables 39 and 40. Measurements are shown in Figs. 96 and 97.

Table 38. Solubility c_s of silicon in aluminum [58h1].

T °C	c_s wt. %
350	0.17
375	0.22
400	0.29
425	0.37
450	0.48
475	0.63
500	0.80
525	1.03
550	1.30
577 (eutectic)	1.65

Table 40. Specific contact resistance R of aluminum, molybdenum, and chromium contacts on p-type silicon for different surface dopant concentration c_{sur} [75R1, 71C1, 69K1, 70B1, 69s1].

c_{sur} cm^{-3}	R [$\Omega\,\mu m^2$]		
	Al	Cr	Mo
$1.5\cdot10^{20}$	$1.2\cdot10^2$	$1.2\cdot10^2$	$3\cdot10^2$
$6.0\cdot10^{19}$	$4\cdot10^2$	$4.0\cdot10^2$	$5.5\cdot10^2$
$1.0\cdot10^{19}$	$2.3\cdot10^3$	$3.0\cdot10^2$	$4.7\cdot10^4$
$1.0\cdot10^{18}$	$6\cdot10^3$		
$6.0\cdot10^{17}$	$1.1\cdot10^4$	$1.5\cdot10^4$	$1\cdot10^5$
$9.0\cdot10^{16}$	$3.8\cdot10^4$	$4.8\cdot10^4$	$4.5\cdot10^5$
$1.5\cdot10^{16}$	$1\cdot10^5$	no ohmic contact	

Table 39. Specific contact resistance R of aluminum, platinum silicide, chromium, titanium nitride, and molybdenum on n-type silicon for different surface dopant concentrations c_{sur} [81W1, 75R1, 71B1, 70B1, 69K1, 69s1]

c_{sur} cm^{-3}	R [$\Omega\,\mu m^2$]				
	Al	Pt	Cr	TiN	Mo
$3.5\cdot10^{20}$	10				
$1.0\cdot10^{20}$	190	150	120	$3.5\cdot10^3$	800
$1.0\cdot10^{19}$	$8\cdot10^4$	$8\cdot10^2$	$8\cdot10^3$	$1.5\cdot10^4$	$6\cdot10^4$
$5.0\cdot10^{18}$	no ohmic contact	$1.3\cdot10^3$	$2\cdot10^4$	$3\cdot10^4$	$6\cdot10^6$
$7.0\cdot10^{17}$	no ohmic contact		no ohmic contact	$6\cdot10^4$	

High contact resistances cannot only be caused by interfacial layers, e.g. oxides or polymerization layers after etching (section 6.1.4.7), but also by precipitation of aluminum, i.e. p-type doped Si from supersaturated Al/Si films. The p-type Si precipitate leads to a blocking characteristic of contacts on n-Si. The precipitation can occur in the form of small crystallites or as epitaxial layer. In order to avoid this contact failure, the Si content of the alloy must be minimized and the growth condition for epitaxial films must be avoided, e.g. by the use of low deposition rates and temperatures above the epitaxial growth limit.

Measurement techniques to characterize interfacial layers are AES (Auger electron spectroscopy), SIMS (secondary ion mass spectroscopy), and RBS (Rutherford backscattering).

AES and RBS measurements showed, that the *partial pressure of* O_2 during platinum evaporation affects the resulting uniformity of the silicide more than the oxide layer formed during etching of the surface [80C1]. No oxygen effects are visible at partial pressures $<1.3 \cdot 10^{-6}$ Pa. An oxygen partial pressure of $1.3 \cdot 10^{-5}$ Pa deteriorated the growth, but a nitrogen partial pressure of $1.3 \cdot 10^{-5}$ Pa had no influence on the formation of the silicide.

The oxygen content in the interface is reduced by the pumping time (Table 41) prior to deposition and by the deposition rate (Table 42).

Schottky barrier heights obtained are listed in Tables 24 and 43.

Contact resistances between different metal films are listed in Table 44. In-situ sputter cleaning of the substrate is recommended prior to the deposition of the next metallization layer.

Table 41. Oxygen content in the interface Si/Al—Si layer and in the Al—Si layer itself for different times of vacuum pumping prior to DC-magnetron sputtering of Al—Si(1 %) films in a vacuum lock sputter system. Measurements of oxygen content are performed by AES [83S1].

Time of pumping min	Oxygen content in the interface arb. units	Oxygen content in Al—Si(1 %) films atomic ratio O/Al
1	5.0	
5	3.0	$2.8 \cdot 10^{-4}$
15	2.8	$2.0 \cdot 10^{-4}$
120	2.8	$1.5 \cdot 10^{-4}$

Table 42. Oxygen content in the interface Si/Al—Si layer for different condensation rates during DC-magnetron sputtering of Al—Si(1 %) films in a vacuum lock sputter system. Measurements of the oxygen content are performed by AES [83S1].

Condensation rate of films μm min^{-1}	Oxygen content in the interface arb. units
0.5	4.8
0.7	3.0
1.0	2.8

For Table 43 see p. 618.

Table 44. Contact resistance of aluminum or aluminum alloy layers to various conductive layers after annealing.

1st layer	2nd layer	Contact resistance Ω	Contact window size μm^2	Annealing	Ref.
Poly-Si *)	Al pure	6	12×10	5 min, 450 °C	78N1
Poly-Si *)	Al—Si(1 %)	1	4×4	30 min, 450 °C	83N1
Poly-Si *)	Al—Si(2 %)	1.5	12×10	5 min, 450 °C	78N1
Poly-Si *)	Ti/Al 0.25 μm/0.70 μm	0.5	12×10	5 min, 450 °C	78N1
$MoSi_2$	Al—Si(1 %)	≦0.1	4×4	30 min, 450 °C	83N1
Mo_3Si	Al—Si(1 %)	≦0.1	4×4	30 min, 450 °C	83N1
Al—Si(1 %)	Al—Si(1 %)	≦0.02	4×4	no anneal **)	83S2
Al pure	Ag	≈0.02	2.5×100	no anneal	69s1

*) n^+ doped poly-Si, thickness 300 nm, resistivity 17···25 Ω/□.
**) Sputter etch was performed prior to sputtering the 2nd layer.

Table 43. Schottky barrier heights Φ_B of various metals to n-type and p-type silicon [75R1]. (For silicides see Table 24.)

Metal	Φ_B [eV]	
	n-Si	p-Si
Ag	0.55	0.54
Al	0.50	0.58
Au	0.81	0.34
Cu	0.69	0.46
Ni	0.67	0.51
Pb	0.41	0.55

Table 45. Median time to electromigration failure MTF for different aluminum metallizations at $j = 1 \cdot 10^6$ A cm^{-2}, $T = 250$ °C, line width = 2 µm [83P3].

Material	MTF h
Al pure	10
Al—Si(2 %)	≈10
Al—Cu(0,5 %)	≈10
Al—Cu(4 %)	≈100
Al—Cu(0,5 %) "bamboo structure"	800
multi-sandwich *)	7000

*) structure consisting of a diffusion barrier layer between Al layers.

Electromigration

An electric current in a conductive layer leads to forces (electric field) and momentum transfer from electrons to metal ions which cause a diffusion process. The mobility of metal atoms leads to growth of the grains and formation of voids. In areas where the current leaves the metal film, pile-up of the metal causes the formation of hillocks and whiskers. The formation of voids increases the local current density further until eventually the metallic film is interrupted. The medean time to failure MTF, i.e. the time elapsed after which 50% of the samples tested fail, is given by the relation

$$\mathrm{MTF} = M j^{-n} \, e^{Q/kT}.$$

The prefactor (M) is dependent on the microstructure of the film, the length and width of the stripe, and the physical properties of the film, the substrate (adhesion, heat conductivity etc.) and possible protective layers. j is the current density; n is a material-dependent power coefficient. Values reported are $n = 1 \cdots 3$ [67C1], $n = 1.7$ [81D1], $n = 2$ [69B1, 69B2] $n = 6 \cdots 7$ [70B2].

Q is the activation energy for diffusion. Values reported for Al metallization are $Q = 0.4 \cdots 0.6$ eV. Diffusion causing electromigration predominantly occurs in grain boundaries. In order to reduce the diffusion in grain boundaries and to increase the MTF, additions of Cu(4%), Ti(0.2%), Mg(1.5%) are alloyed into the Al—Si compound.

The effect of *the microstructure* on the MTF is described for Al films by the parameter δ

$$\delta = \frac{\Delta}{\sigma^2} \log \left(\frac{I_{111}}{I_{200}} \right)^3,$$

Δ = average grain size,
σ = standard deviation of grain size distribution,
I_{111}/I_{200} = ratio of X-ray diffraction intensity in ⟨111⟩ and ⟨200⟩ reflex.

Good correlation is found between the parameter δ and the MTF. The following properties are observed:

MTF increases with the reducing of the variety in the grain size. It is therefore dependent on the deposition rate, temperature, and the residual gas.

MTF increases with increasing average grain size.

MTF is dependent on the width of test stripes. A minimum occurs at approximately 2 µm width for Al films. The position is a function of the grain size. For a "bamboo" structure of the grains, i.e. arrays of long grains having grain boundaries perpendicular to the film surface, the longest MTF values are observed in single layers (Table 45).

MTF is increased by passivation of the test stripes, e.g. by glass passivation or an anodic oxide.

MTF is increased by pulsed (<1 µs) current relative to constant current of the same magnitude because heating of the conductor is reduced [73S1].

Studies of electromigration are usually performed on plane surfaces. Conclusions on the MTF of conductors on ICs are not straightforward because the effect of the step coverage is not included in the tests. Since standardized steps are not yet possible, the electromigration near steps can only be described qualitatively.

Reflectivity and hillock growth

A reflectivity $R > 90\%$ was used as a measure of layer quality for many years for Al metallization. A negative side effect of the high reflectivity is observed in recent years. The reflectivity in neighbouring steps of small device structures changes the photoresist line width in lithography (Fig. 98).

The reflectivity is reduced by the presence of residual gases (Fig. 99, Table 46). The reflectivity of LPCVD films is shown in Fig. 37.

A reduced reflectivity may be caused by *hillocks*. Hillocks are small crystallites which protrude from the metal surface. The growth occurs by grain boundary diffusion under compressive stress. Hillock formation may occur during deposition, by strong temperature changes during fabrication of the layer, or during operation of the device. Hillocks are undesired, because photoresist or insulating films are weakened or even interrupted. The formation of hillocks is reduced by the same process steps, i.e. alloy additions, which reduce the grain boundary diffusion causing the electromigration. Alloy additions used are Cu, Ti, Mg, Sn, Mn (Table 47). The tendency to hillock formation is also reduced by anodic oxidation of the Al film or by a cap of Ti—W—N [82L2].

Table 46. Influence of residual gas contamination during deposition on the reflectivity of magnetron-sputtered Al—Si(1.5%) films [80N1]. Thickness 1 μm on p-type ⟨100⟩ Si substrate [80N1].

Contaminant	Partial pressure of contaminant prior to sputtering Pa	Reflectivity at 400 nm %	Impurity concentration in the films [at.%]			
			H	C	N	O
None (Purity of argon 99.9999%)	$9 \cdot 10^{-5}$	90.5	0	0.1	0	0
Hydrogen	$5.3 \cdot 10^{-3}$	90.0	0	0.1	0.1	0.3
Nitrogen	$5.3 \cdot 10^{-3}$	1.8···10.7 *)	0	0.4	1.6	0.5
Oxygen	$5.3 \cdot 10^{-3}$	10···19.7 *)	0	0.9	0.1	1.4

*) A lower reflectivity is observed on the area of the substrate which has passed the deposition zone last, i.e. at the "trailing edge" of the substrate.

Table 47. Reduction of hillock density in Al films by alloying. Thermally evaporated films, thickness ≈5 μm [71S1].

Material of films	Hillock density *) cm^{-2}	Hillock density **) cm^{-2}	Mean diameter of hillocks **) μm
Al pure	$6 \cdot 10^5$	$1.2 \cdot 10^5$	6
Al—Sn(0.06 wt.%)	0	$2.1 \cdot 10^4$	3
Al—Mn(0.1 wt.%)	0	$3.4 \cdot 10^4$	8

*) After thermal cycling 200 °C→RT for 30 min.
**) After annealing at 400 °C for 24 h.

The *effect of hydrogen* on the reflectivity and hillock formation is still uncertain (compare Table 46 and Fig. 99); however, elevated partial pressures of nitrogen and water vapour reduce the reflectivity of Al films and increase the probability of hillock formation.

The *effect of oxygen* is uncertain. An optimum related to reliability of the Al films occurs at low partial pressures. A reduced reflectivity at low hillock density is observed after Al deposition at high substrate temperatures and high deposition rates.

Differences in reflectivity at the leading edge and the trailing edge of the substrate are caused by heating of the substrate during deposition [79M2, 80N1]. This effect can be avoided by preheating of the substrate.

Film hardness and bondability

The hardness of thin films (≲1000 nm) is determined by microhardness measurements (Knoop hardness). The hardness is an important feature because it affects the bondability of connecting wires to the package (Fig. 100). The film hardness is dependent on the deposition parameters (Fig. 21) and the residual gas content (Fig. 101).

Stress in conductive layers

Tensile or compressive stress at the interface of the metal film to the substrate consists of two components [69C1]:

intrinsic stress in the film caused by crystal defects,

thermal stress caused by differing thermal expansions of the film and the substrate.

The stress affects the

diffusion processes (hillock formation, corrosion),
electrical parameters (see 6.1.4.5.10),
adhesion of the film,
warping of the substrate.

A warping of the substrate causes problems during lithography. The intrinsic stress is correlated to the electrical resistivity of the film (Fig. 102).

Thermal stress can be varied by the substrate temperature and the film thickness.

Intrinsic stress shows a complex dependence on the deposition rate (Table 48), residual gas, substrate temperature, pressure of the sputter gas, substrate bias, the configuration of substrate and source (Fig. 103), and the film thickness (Figs. 104 and 105, Table 49). The same parameters affect the microstructure. In general, a low deposition rate is required, so that a strain-free lattice can be formed during deposition; however, the low deposition rate causes a high residual gas contamination which again is affected by the substrate bias.

Only the intrinsic stress can be reduced by annealing (Table 49).

The effect of the sputter gas on the microstructure and the stress is described by the Thornton model (Fig. 50). The following relations are observed:

For reduced Ar pressure, sputtered metal atoms and energetic neutral gas atoms (neutral gas ions backscattered from the target) lose less energy by collisions with argon atoms. The sputtered layer shows a dense packaging of atoms.

For reduced argon pressure, more argon gas atoms are implanted into the substrate (Fig. 106).

For increased argon pressure, the compressive stress changes at a critical pressure over into tensile stress (Fig. 107). This critical pressure increases with the atomic mass of the sputtered material (Fig. 107) and is reduced with increasing atomic mass of the sputter gas (Fig. 108).

A substrate bias generally shows the same effect on the film stress as a reduced sputter gas pressure.

The stress in sintered films is depicted in Figs. 109, 110, and 112, the Ar incorporation into sputtered films is shown in Fig. 111.

Table 48. Dependence of film stress X on deposition rate r_d for DC magnetron-sputtered chromium films of various film thicknesses d [77H2].

r_d nm min^{-1}	X [10^8 N m^{-2}]		
	d=0.1 μm	d=0.2 μm	d=0.3 μm
6	−22.5	−16.3	−13.9
60	− 8.0	−10.8	−11.5
600	+ 6.5	− 5.0	− 9.3

Table 49. Stress X (tensile) at RT in e-gun-evaporated aluminum films as deposited and after thermal cycling for various film thicknesses d [77S2]. (Evaporation without substrate heating.)

d μm	X *) 10^8 N m^{-2}	X **) 10^8 N m^{-2}
0.2	0.3	8.7
0.45	0.25	3.8
0.89	0.25	1.8
1.56	1.3	0.8

*) As deposited [77S2].
**) After temperature cycling:
RT → 450 °C···500 °C → RT, heating/cooling rate = 10 °C/min.

References for 6.1.4.5.8···6.1.4.5.12

Books and review articles

56h1 Holland, L.: Vacuum Deposition of Thin Films. London: Chapman and Hall **1956**.
58h1 Hansen, U., Anderko, K.: Constitution of Binary Alloys. New York, Toronto, London: McGraw-Hill Book Company **1958**.
68b1 Böhm, H.: Einführung in die Metallkunde. Mannheim, Wien, Zürich: Bibliographisches Institut **1968**.
68k1 Kohlrausch, F.: Praktische Physik **3**, Lautz, G., Taubert, R., (eds.). Stuttgart: B.G. Teubner **1968**.
69s1 Schwartz, B. (ed.): Ohmic Contacts to Semiconductors. New York: Electrochemical Society, Inc. **1969**.
70m1 Maissel, L.I., Glang, R., (eds.): Handbook of Thin Film Technology. New York, London: McGraw-Hill Book Company **1970**.
74c1 Chapman, B.N., Anderson, J.C., (eds.): Science and Technology of Surface Coating. London, New York: Academic Press **1974**.
76t1 Townsend, P.D., Kelly, J.C., Hartley, N.E.W.: Ion Implantation, Sputtering and their Applications. London, New York, San Francisco: Academic Press **1976**.
78v1 Vossen, J.L., Kern, W., (eds.): Thin Film Processes. London, New York, San Francisco: Academic Press **1978**.
80c1 Chapman, B.: Glow Discharge Processes. New York, Chichester, Brisbane, Toronto: John Wiley a. Sons **1980**.
82w1 Wutz, M., Adam, H., Walcher, W.: Theorie und Praxis der Vakuumtechnik. Braunschweig, Wiesbaden: Friedr. Vieweg u. Sohn **1982**.

Bibliography

48J1 Justi, E.: Leitfähigkeit u. Leitungsmechanismus fester Stoffe. Göttingen: Vandenhoeck u. Ruprecht **1948**.
54B1 Biondi, M.A., Chanin, L.M.: Phys. Rev. **94** (1954) 910.
56E1 Eder, F.X.: Moderne Meßmethoden der Physik, II. Berlin: Deutscher Verlag der Wissenschaften **1956**, 595.
57C1 Chanin, L.M., Biondi, M.A.: Phys. Rev. **107** (1957) 1219.
62S1 Stratton, R.: J. Phys. Chem. Solids **23** (1962) 1177.
63W1 Wobschall, D., Graham, J.R., Jr., Malone, D.P.: Phys. Rev. **131** (1963) 1565.
64F1 Frost, L.S., Phelps, A.V.: Phys. Rev. **136** (1964) 1538.
65C1 Carlston, C.E., Magnuson, G.D., Comeaux, A., Mahadevan, P.: Phys. Rev. **138** (1965) 759.
65M1 Maissel, L.I., Schaible, P.M.: J. Appl. Phys. **36** (1965) 237.
65M2 Mahagdevan, P., Magnuson, G.D., Layton, J.K., Carlston, C.E.: Phys. Rev. **140** (1965) 1407.
65T1 Takeishi, Y., Hagstrum, H.D.: Phys. Rev. **137** (1965) 641.
66H1 Hayden, H.C., Amme, R.C.: Phys. Rev. **141** (1966) 30.
67C1 Chabra, D.S., Ainslie, N.G.: Techn. Report 22.419, IBM Components Div., E. Fishkill Facility, New York: **1967**.
67W1 Winters, H.F., Kay, E.: J. Appl. Phys. **38** (1967) 3928.
68C1 Comas, J., Carosella, C.A.: J. Electrochem. Soc. **115** (1968) 974.
68H1 d'Heurle, F., Berenbaum, L., Rosenberg, R.: Trans. Met. Soc. AIME 242, **1968**, 502.
68V1 Vossen, J.L., O'Neill, J.J., Jr.: RCA Review 29, **1968**, 566.
69B1 Black, J.R.: Proc. IEEE, 57, **1969**, 1587.
69B2 Black, J.R.: IEEE Trans. Electron. Devices Ed. **16** (1969) 338.
69C1 Chopra, K.L.: Thin Film Phenomena. New York: McGraw-Hill **1969**, 177, 183, 269.
69H1 Heinicke, R., Eldridge, P.G., Pion, M.: Proc. Internat. Conf. on Microelectronics, Eastbourne, Engl., **1969**, 27.
69K1 Ki, D.K., Burgess, R.R., Coleman, M.G., Keil, J.G.: IEEE Trans. **16** (1969) 356.
69M1 Mowchan, B.A., Demchishin, A.V.: Fiz. Met. Metalloved **28** (1969) 653.
69M2 Murrmann, H., Widmann, D.: Solid State Electron. **12** (1969) 879.
69N1 Nelson, C.W.: Proc. Int. Symp. Hybrid Microelectronics, Sept. 29, Oct. 1, 1969, Dallas, Texas. Int. Soc. of Hybrid Microelectronics. Montgomory, USA, **1969**, 413.
70B1 Berger, H.: Modellbeschreibung planarer ohmscher Metall-Halbleiterkontakte, Diss., TH Aachen, **1970**.
70B2 Blair, J.C., Ghate, P.B., Haywood, C.T.: Appl. Phys. Lett. **17** (1970) 281.
70C1 Cunningham, J.A., Fuller, C.R., Haywood, C.T.: IEEE Trans. Reliab. **R-19** (1970) 182.
70D1 Dheer, R.R.: Proc. IEEE Electron. Comp. Conf. **1970**, 76.
70K1 Koenig, H.R., Maissel, L.I.: IBM J. Res. Dev. **14** (1970) 168.

70Y1 Yu, A.Y.C.: Solid State Electron. **13** (1970) 239.
71B1 Bhatt, H.J.: Appl. Phys. Lett. **19** (1971) 30.
71C1 Chang, C.Y., Fang, Y.K., Sze, S.M.: Solid State Electron. **14** (1971) 541.
71M1 Mc Caldin, J.O., Sankur, H.: Appl. Phys. Lett. **19** (1971) 524.
71S1 Sato, K., Oi, T., Matsumaru, H., Okubo, T., Nishimura, T.: Metall. Trans. **2** (March 1971) 691.
72B1 Berenbaum, L.: Appl. Phys. Lett. **20** (1972) 434.
72E1 Eberhardt, W., Kalus, C., Schleicher, L.: Siemens, Res. and Develop. Reports, **1**, 3 (1972) 315.
72P1 Porter, S.T., Hoffmann, V.E.: "Nepcon", Anaheim, Ca., **1972**, 209.
72P2 Patten, J.W., Mc Clanahan, E.D.: J. Appl. Phys. **43** (1972) 4811.
72W1 Winters, H.F., Kay, E.: J. Appl. Phys. **43** (1972) 794.
73G1 Gurev, H.S.: 11th. Ann. Proc. Reliab. Phys., **1973**, 245.
73H1 Heller, J.: Thin Solid Films **17** (1973) 163.
73H2 Hutchins, G.A., Shepela, A.: Thin Solid Films **18** (1973) 343.
73S1 Sigsbee, R.A.: 11th. Ann. Proc. Reliab. Phys. **1973**, 301.
74B1 Bunshah, R.F.: J. Vac. Sci. Technol. **11** (1974) 633.
74D1 Di Giacomo, G., Peressini, P., Rutledge, R.: J. Appl. Phys. **45** (1974) 1626.
74L1 Lau, S.S., Chu, W.K., Mayer, J.W., Tu, K.N.: Thin Solid Films **23** (1974) 205.
74M1 Murrmann, H., Schleicher, L.: Metalloberfläche, Angew. Elektrochemie **28** (1974) 121.
74T1 Thornton, J.A.: J. Vac. Sci. Technol. **11** (1974) 666.
74T2 Tisone, T.C., Bindell, J.B.: J. Vac. Sci. Technol. **11** (1974) 519.
75F1 Fournier, P.R.: U.S. Patent 3, 879, 746 (1975).
75F2 Fogelson, R.L., Ugai, Y.A., Akimova, I.A.: Phys. Met. Metalloved. **39** (1975) 212.
75R1 Ruge, I.: Halbleitertechnologie. Berlin, Heidelberg, New York: Springer Verlag **1975**.
75S1 Scoggan, G.A., Agarwala, B.N., Peressini, P.P., Brouillard, A.: 13th. Ann. Proc. Reliab. Phys. **1975**.
75T1 Thornton, J.A.: J. Vac. Sci. Technol. **12** (4) (1975) 830.
76A1 Airco Temescal: Physical Vapor Deposition. Berkeley, Ca., **1976**.
77H1 d'Heurle, F., Fiorio, R., Gangulee, A., Ranieri, V., Zirinsky, S.: Proc. of the Sev. Int. Vac. Congr. and 3rd Int. Conf. Solid Surfaces, Vienna 1977, III, 2123.
77H2 Hoffmann, D.W., Thornton, J.A.: Thin Solid Films **40** (1977) 355.
77K1 Kubovy, A., Janda, M.: Thin Solid Films **42** (1977) 169.
77K2 Koleshko, V.M.: Proc. 7th Intern. Vac. Congr. and 3rd Int. Conf. Solid Surfaces, Vienna, Austria **1977**, II 1871.
77M1 McLeod, P., Hartsough, L.D.: J. Vac. Sci. Technol. **14** (1977) 263.
77M2 Mat. Res. Corp., Orangeburg: Technical Publication No. 1002-PED-AAI-0577 R.
77S1 Schiller, S., Heissig, U., Goedicke, K.: Proc. 7th Int. Vac. Congr. and 3rd Int. Conf. Solid Surfaces. Vienna, Austria, Sept. **1977**, II, 1545.
77S2 Sinha, A.K., Sheng, T.T.: Thin Solid Films **48** (1978) 117.
77W1 Weißmantel, Ch.: Proc. 7th Internat. Vac. Congr. and 3rd Int. Conf. Solid Surfaces. Vienna, Austria, Sept. **1977**, II, 1533.
77W2 Winters, H.F., Coburn, J.W., Kay, E.: J. Appl. Phys. **48** (1977) 4973.
78A1 Aronson, A.J.: Solid State Technol. **21** (1978) 66.
78C1 Cardoso, J., Harsdorff, M.: Z. Naturforsch. **33a** (1978) 442.
78G1 Ghate, P.B., Blaire, J.C., Fuller, C.R., McGuire, G.E.: Thin Solid Films, **53** (1978) 117.
78H1 Hegner, F., Feuerstein, A.: Thin Solid Films **53** (1978) 141.
78H2 Hubregtse, J.: ATR Austral. Telecommun. Res. **12** (1978) 35.
78H3 Hecht, L.C.: J. Vac. Sci. Technol. **16** (2) (1979) 328.
78H4 Howard, J.K., White, J.F., Ho, P.S.: J. Appl. Phys. **49** (1978) 4083.
78L1 Lundquist, T.R.: J. Vac. Sci. Technol. **15** (1978) 684.
78L2 Lee, H.: IEEE Trans. Electron Devices **25** (1978) 795.
78L3 Lalevic, B., Murty, K., Suga, H., Weissmann, S.: Thin Solid Films **53** (1978) 153.
78M1 Morosanu, C., Soltuz, V.: Thin Solid Films **52** (1978) 181.
78N1 Naguib, H.M., Hobbs, L.H.: J. Electrochem. Soc. **125** (1978) 169.
78N2 Nicolet, M.A.: Thin Solid Films **52** (1978) 415.
78V1 Vossen, J.L., Cuomo, J.J.: Thin Film Processes, Vossen, J.L., Kern, W. (eds.). New York, San Francisco, London: Acad. Press **1978**, 32.
79B1 Bhar, T.N., Lamb, G.M.: J. Electrochem. Soc. **126** 9 (1979) 1514.
79C1 Class, W.H.: Solid State Technol. **22** (1979) 61.
79F1 Fuller, C.R., Ghate, P.B.: Thin Solid Films **64** (1979) 25.

79H1 Hartsough, L.D., Denison, D.R.: Aluminum-Silicon Sputter Deposition. Technical Report No. 79.01, Perkin-Elmer Ultek Division; Palo Alto, Ca. **1979**.
79H2 Hoffman, D.W., Thornton, J.A.: J. Vac. Sci. Technol. **16** (2) (1979) 134.
79H3 Hartsough, L.D.: Sputtering of Ti/W Films using RF Substrate Bias. Int. Conf. of Metallurgical Coatings, Aug. **1979**.
79H4 Hartsough, L.D.: Thin Solid Films **64** (1979) 17.
79K1 Kuljian, M.J.: Proc. 1979 joint automatic control conf. Denver, Colorado, USA, June **1979**, 266.
79K2 Kane, S.M., Ahn, K.Y.: J. Vac. Sci. Technol. **16** (2) (1979) 171.
79M1 Mukherjee, S.D., Morgan, D.V., Howes, M.J.: J. Electrochem. Soc.: Solid State Sci. and Technol., June **1979**, 1047.
79M2 McLeod, P.S., Hughes, J.L.: J. Vac. Sci. Technol. **16** (2) (1979) 369.
79O1 Ottaviani, G.: J. Vac. Sci. Technol. **16** (5) (1979) 1112.
79S2 Sim, S.P.: Microelectron. Reliab. **19** (1979) 207.
79W1 Williams, E.W.: Solid. State Technol. **22** (2) (1979) 80.
80B1 Balzers AG, Liechtenstein, Technical publication LLS 801, **1980**.
80C1 Crider, C.A., Poate, J.M., Sheny, T.T., Ferris, S.D.: J. Vac. Sci. Technol. **17** (1) (1980) 433.
80D1 Denison, D.R., Hartsong, L.D.: J. Vac. Sci. Technol. **17** (6) (1980) 1326.
80F1 Freeouf, J.L., Rubloff, G.W., Ho, P.S., Knon, T.S.: J. Vac. Sci. Technol. **17** (5) (1980) 916.
80G1 Geipel, H.J., Hsieh, N., Ishaq, M.H., Koburger, C.W., White, F.R.: IEEE Trans. Electron Devices **27** (1980) 1417.
80H1 Hoffman, D.W., Thornton, J.A.: J. Vac. Sci. Technol. **17** (1) (1980) 380.
80H2 Hill, M.: Magnetron sputtered Titanium-Tungsten films, Transactions Sputtering & Plasma Etching Conference, June **1980**, Selsdon Park Hotel, Surrey, England.
80M1 Murarka, S.P.: J. Vac. Sci. Technol. **14** (4) (1980) 775.
80M2 Murarka, S.P., Fraser, D.B., Sinha, A.K., Levinstein, H.J.: IEEE Trans. Electron Devices **27** (1980) 1409.
80M3 Mattox, D.M.: J. Vac. Sci. Technol. **17** (1) (1980) 370.
80M4 Missel, L., Duke, P., Montelbano, T.: Semiconductor Internat. **3** (2) (1980) 67.
80N1 Nowicki, R.S.: J. Vac. Sci. Technol. **17** (1) (1980) 384.
80N2 Nunoi, T., Nishimura, N., Nammori, T., Sawai, H., Suzuki, A.: J. Appl. Phys. **19** (1980) 67.
80N3 Neureuther, A.R., Chiu, H.T., Lin, C.: IEEE Trans. Electron Devices **ED-27** (1980) 1449.
80R1 Reeves, G.K.: Solid State Electron. **23** (1980) 487.
80S1 Sinha, A.K., Lindenberger, W.S., Fraser, D.B., Murarka, S.P., Fuls, E.N.: IEEE Trans. Electron Devices **27** (1980) 1425.
80S2 Sinha, A.K., Lindenberger, W.S., Fraser, D.B., Murarka, S.P., Fuls, E.N.: IEEE J. Solid State Circuits **15**, (1980) 490.
80V1 Vossen, J.L., O'Neill, J.J., Hughes, G.W., Taft, F.A., Snedeker, R.: J. Vac. Sci. Technol. **17** (1) (1980) 400.
80V2 Vaidya, S., Fraser, D.B., Sinha, A.K.: 18th Ann. Proc. Reliab. Phys. (1980) 165.
81B1 Balzers AG, eds., „Aufdampfmaterialien, Verdampfungsquellen, Targets, Hilfsmittel", Balzers, FL, **1981**.
81C1 Craig, S., Harding, H.L.: J. Vac. Sci. Technol. **19** (2) (1981) 205.
81C2 Craig, S., Harding, G.L.: J. Vac. Sci. Technol. **19** (3) (1981) 754.
81D1 Danso, K.A., Tullos, T.: Microelectronics Reliab. **21** (1981) 513.
81F1 Faith, T.J., Irven, R.S., O'Neill, J.J., Tams, F.J.: J. Vac. Sci. Technol. **19** (3) Sept./Oct. (1981) 709.
81F2 Faith, T.J.: J. Appl. Phys. **52** (1981) 4630.
81F3 Farrow, R.F.C.: J. Vac. Sci. Technol. **19** (2) (1981) 150.
81H1 Hoffman, V.E., Chang, H.M.: Solid State Technol. **2** (1981) 105.
81H2 Hecq, M., Hecq, A.: J. Vac. Sci. Technol. **18** (2) (1981) 219.
81H3 Huang, H.W., Baker, J.M., Serrano, C.M., Kircher, C.J.: J. Vac. Sci. Technol. **19** (1) (1981) 72.
81L1 Leybold-Heraeus GmbH: Technical publication No. 01-017. 1/2, Hanau **1981**.
81N1 Nagahuma, K., Nishitani, K., Ishii, M.: Jpn. J. Appl. Phys. **20** (1981) 1171.
81P1 Padmanabhan, K.R., Sorensen, G.: J. Vac. Sci. Technol. **18** (2) (1981) 231.
81S1 Serikawa, T., Yachi, T.: IEEE Trans. Electron Devices **28** (1981) 882.
81U1 UTI, Reference spectra, residual gases and common vacuum contaminants, Poster published by UTI, Sunnyvale, Ca., USA, **1981**.
81V1 Vossen, J.L.: Semiconductor Internat. Sep. **1981**, 91.
81W1 Wittmer, M., Studer, B., Melchior, H.: J. Appl. Phys. **52** (1981) 5722.
82C1 Cooke, M.J., Heinecke, R.A., Stern, R.G., Maes, J.W.C.: Solid State Technol. **24**, 12 (1982) 62.

82C2 Chu, J.K., Tang, C.C., Hess, D.W.: Appl. Phys. Lett. **41** (1982) 75.
82C3 Canali, C., Celotti, G., Fantini, F., Zanoni, E.: Thin Solid Films **88** (1982) 9.
82G1 Ghate, P.B.: Proc. IEEE, 20th Annual Reliability. Physics Symp. 1982, San Diego, CA., March **1982**, 292.
82L1 Leybold-Haereus GmbH, Vakuum-Verfahrenstechnik, Druckschrift Nr. 01-104.1/2.
82L2 Lim, S.C.P.: Semiconductor Internat. **5**, 4 (1982) 135.
82M1 Miller, N.E., Beinglass, I.: Solid State Technol. Dec. **1982**, 85.
82N1 Nowicki, R.S.: Solid State Technol. June **1982**, 127.
82N2 Naude, M.O., Pretorius, R., Marais, D.J.: Thin Solid Films **89** (1982) 339.
82P1 Petersson, C.S., Baylin, J.E.E., Dempsey, J.J., d'Heurle, F.M., La Placa, S.J.: J. Appl. Phys. **53** (1982) 4866.
82T1 Ting, C.Y.: J. Vac. Sci. Technol. **21** (1) (1982) 14.
82T2 Thomas, S.: Semicond Internat, April **1982**, 209.
82V1 Vincett, P.S.: J. Vac. Sci. Technol. **21** (4) (1982) 972.
83B1 Beinvogl, W., Hasler, B.: Solid State Technol., April **1983**, 125.
83E1 Eizenberg, M., Muraska, S.P., Heimann, P.A.: J. Appl. Phys. **54** (1983) 3195.
83H1 Horwitz, C.M.: J. Vac. Sci. Technol. A, **1** (1) (1983) 60.
83M1 Mc Lachlan, D., Avins, J.B.: Recent Advances in Thin Film Semiconductor Processing, Mat. Res. Corp., Orangeburg, New York, USA, **1983**, II, 1.
83N1 Neppl, F., Menzel, G., Schwabe, U.: J. Electrochem. Soc. **130** (1983) 1174.
83N2 Nava, F., Nobili, C., Ferla, G., Iannuzzi, G., Queirolo, G., Celotti, G.: J. Appl. Phys. **54** (1983) 2434.
83O1 Okuyama, K.: Jpn. J. Appl. Phys. **22** (6) (1983) 934.
83P1 Pramanik, D., Saxena, A.N.: Solid State Technol. Jan. **1983**, 127.
83P2 Pelleg, J., Murarka, S.P.: J. Appl. Phys. **54** (1983) 1337.
83P3 Pramanik, D., Saxena, A.N.: Solid State Technol. March **1983**, 131.
83R1 Rothman, L.B.: J. Electrochem. Soc. Solid State Sci. Technol. **130** (5) (1983) 1131.
83S1 Suni, I., Mäenpää, M., Nicolet, M.A., Luomajärvi, M.: J. Electrochem. Soc.: Solid State Technol., May **1983**, 1215.
83S2 Schleicher, L.: Recent Advances in Thin Film Semiconductor Processing; Mat. Res. Corp., Orangeburg, New York, USA **1983**, 2.

6.1.4.5 Fabrication of layers

(Subsections 6.1.4.5.8···12)

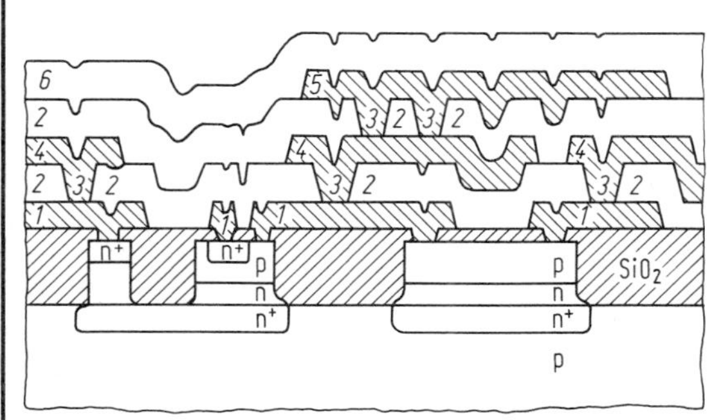

Fig. 1. Multilayer metallization. *1* first metal layer (Al—Si, Al—Si—Cu, Ti—W/Al—Cu); *2* intermetal isolator (SiO_2, Si_3N_4, Polyimide); *3* via contact; *4* 2nd metal layer (same material as *1*); *5* final metal layer; *6* glass passivation.

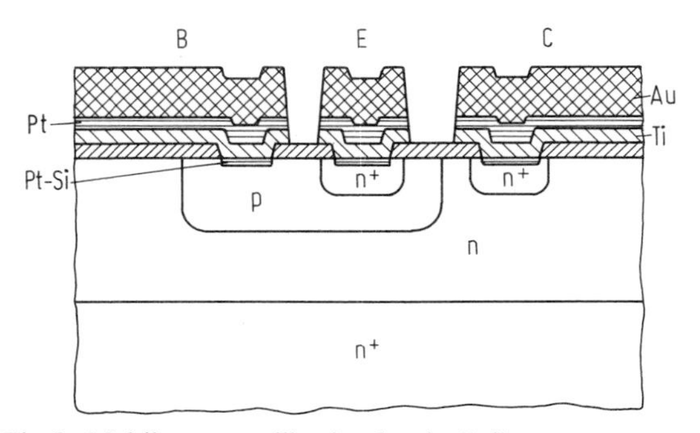

Fig. 2. Multilayer metallization by the Bell process.

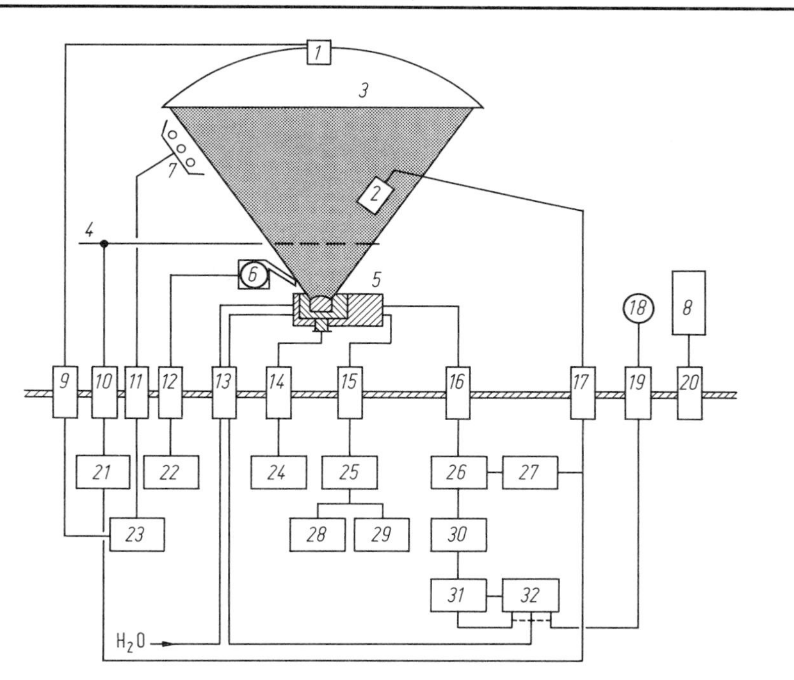

Fig. 3. Schematic of an e-beam evaporation system with auxiliary systems and controls [81L1]. *1* temperature sensor; *2* layer thickness sensor; *3* substrate holder; *4* shutter; *5* e-gun; *6* source material feeder; *7* heater; *8* residual gas analyzer; *9* T-sensor operating system; *10* shutter operating feed through; *11* heater power supply; *12* feeder mechanics; *13* coolant (H_2O) supply; *14* crucible mechanics; *15* e-beam deflection system; *16* high-voltage supply; *17* layer thickness measurement system; *18* pressure sensor; *19* pressure sensor system; *20* quadrupole residual gas analyzer system; *21* shutter operating control; *22* feeder control; *23* temperature control; *24* rotating crucible control; *25* x−y deflection; *26* cathode heater; *27* emission and evaporation rate control; *28* programmable beam deflection; *29* 2-crucible beam deflection; *30* high-voltage switch; *31* high-voltage power supply; *32* process control systems.

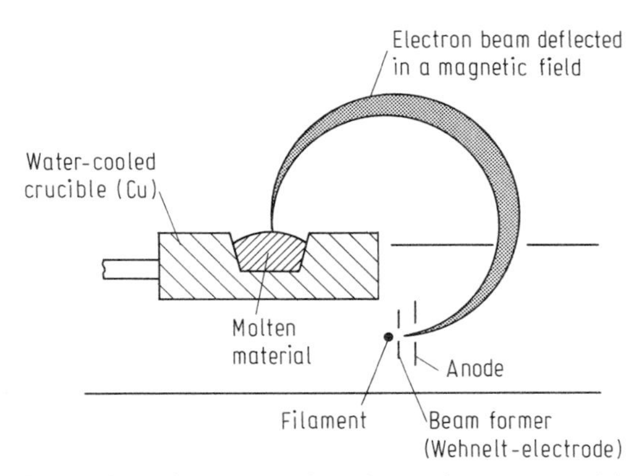

Fig. 4. Schematic drawing of an electron beam gun used in evaporation systems [76A1].

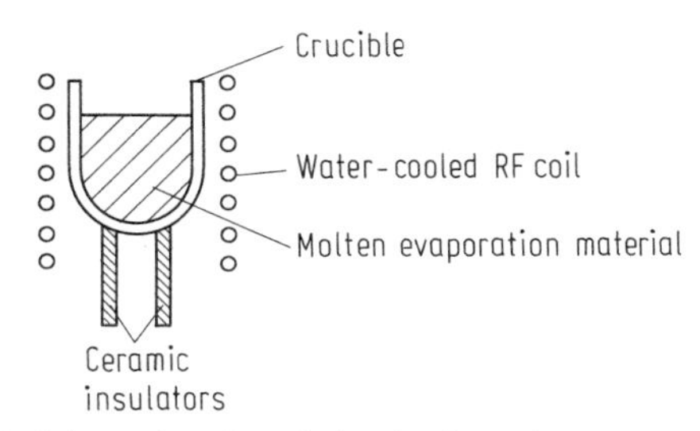

Fig. 5. Schematics of an induction-heated source.

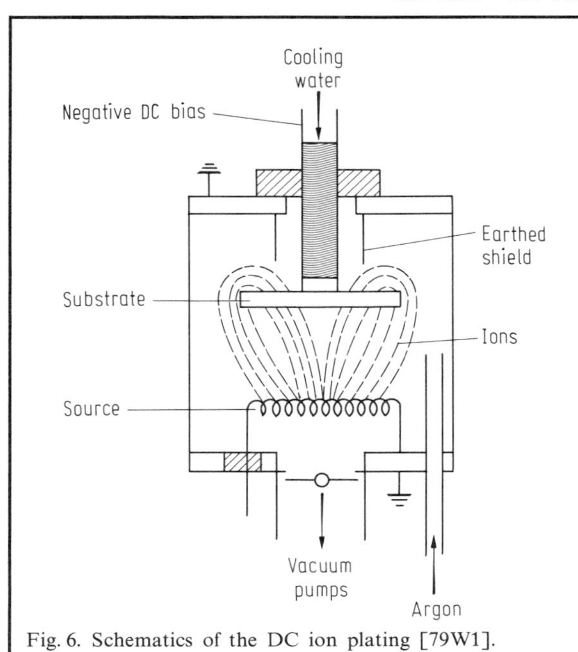

Fig. 6. Schematics of the DC ion plating [79W1].

Fig. 7. Rectangular planar magnetron cathode.

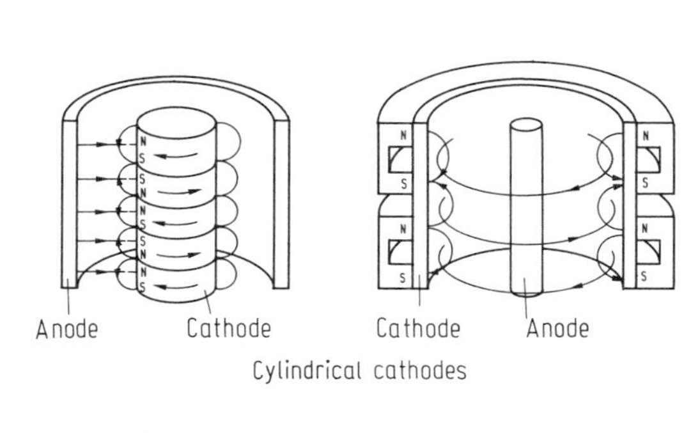

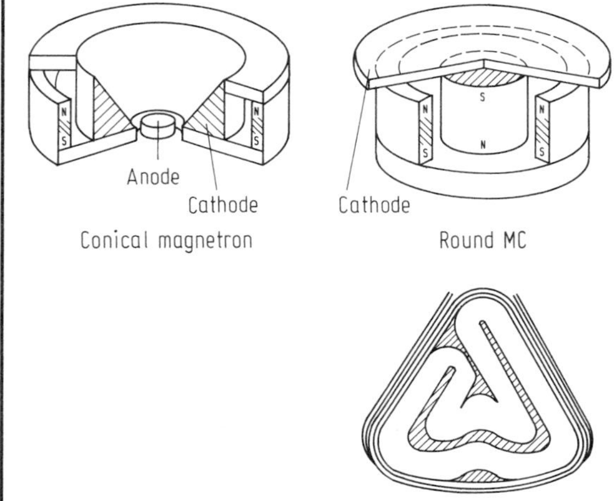

Fig. 8. Further designs of different magnetron cathodes (see Fig. 7).

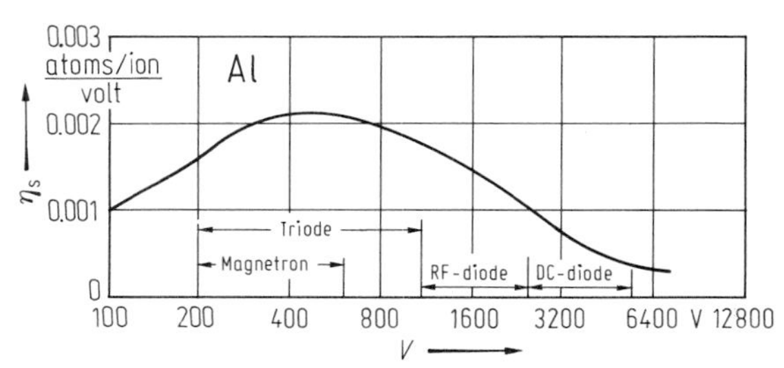

Fig. 9. Argon sputter yield for Al vs. voltage applied to indicate the efficiency of various processes [78A1].

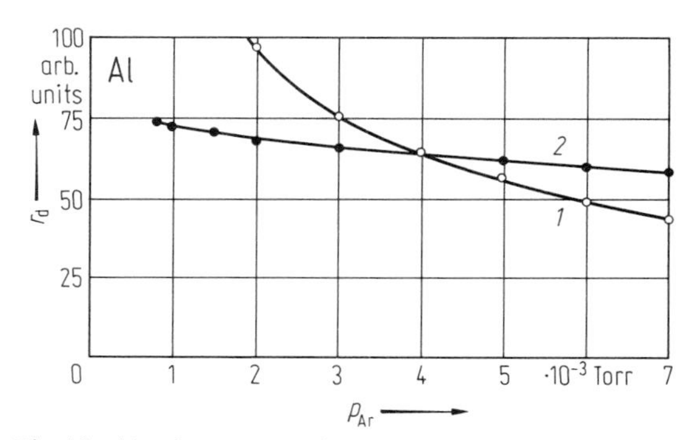

Fig. 10. Aluminum sputtering with s-gun magnetron; deposition rate r_d vs. Ar pressure p_{Ar} at *1)* 5 A constant current; *2)* 2.5 kW constant power [78V1].

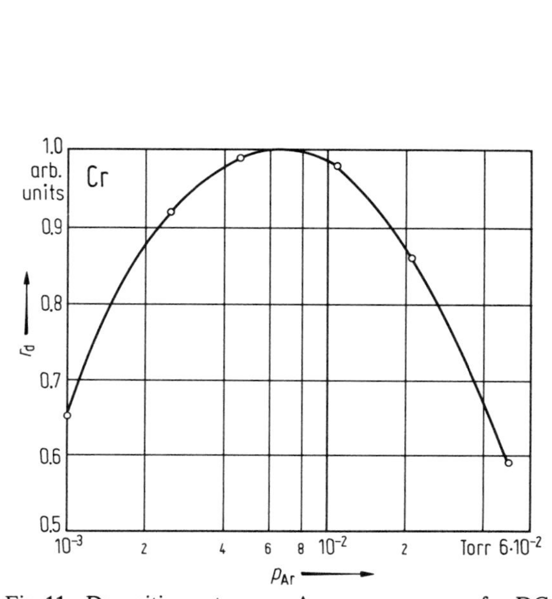

Fig. 11. Deposition rate r_d vs. Argon pressure p_{Ar} for DC magnetron sputtering of chromium (Cr cathode, 18 Wcm^{-2} power density, source/substrate distance 5 cm) [78V1].

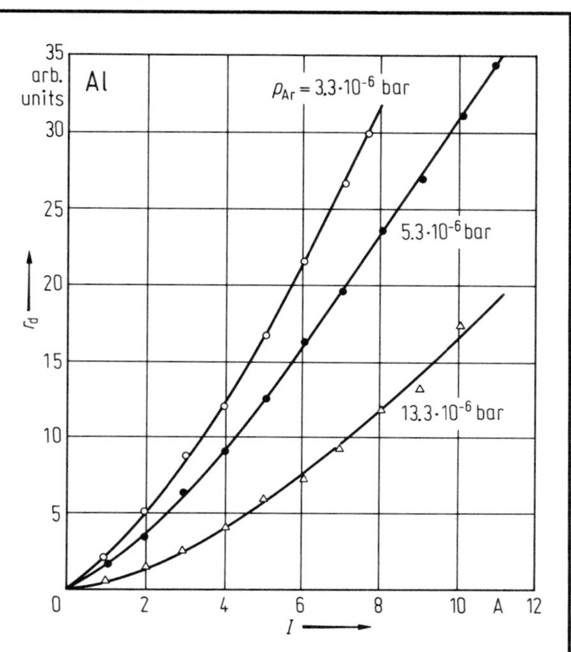

Fig. 12. Aluminum sputtering with s-gun magnetron; deposition rate r_d vs. cathode current I at three Ar pressures [78V1].

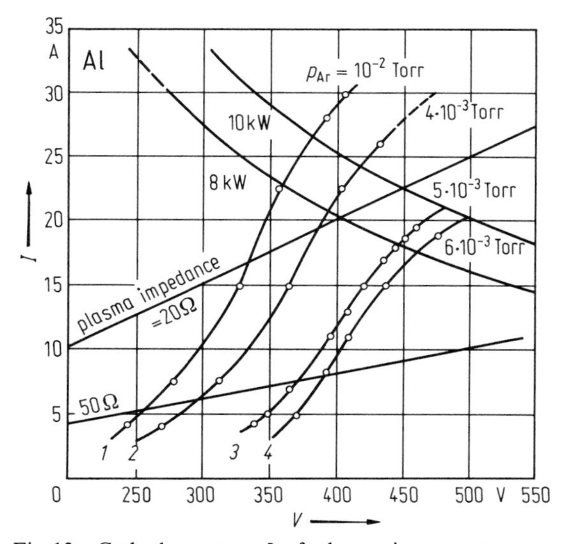

Fig. 13. Cathode current I of plasma in magnetron vs. voltage V applied at various Ar-pressures. The planar cathodes are fabricated of aluminum [78V1]. *1*, *2* are currents according to MRC-publication 1002-PED-AA1-0577 R [77M2] (cathode area: $12 \times 37.8\ cm^2$); *3*, *4* PE-DELTA target.

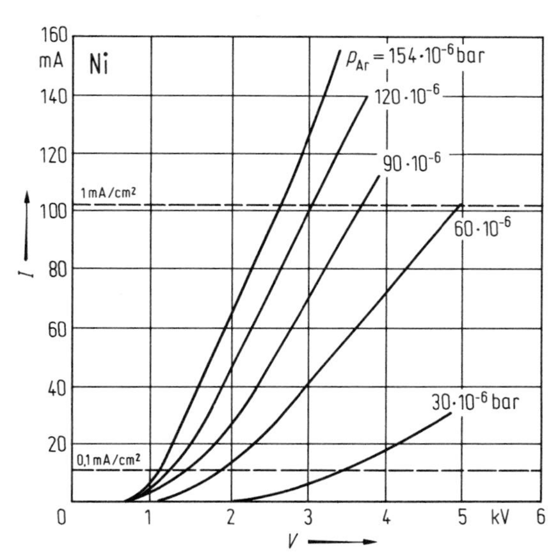

Fig. 14. DC diode sputtering; current I vs. voltage V of Ni cathode for 5 argon pressures [80c1].

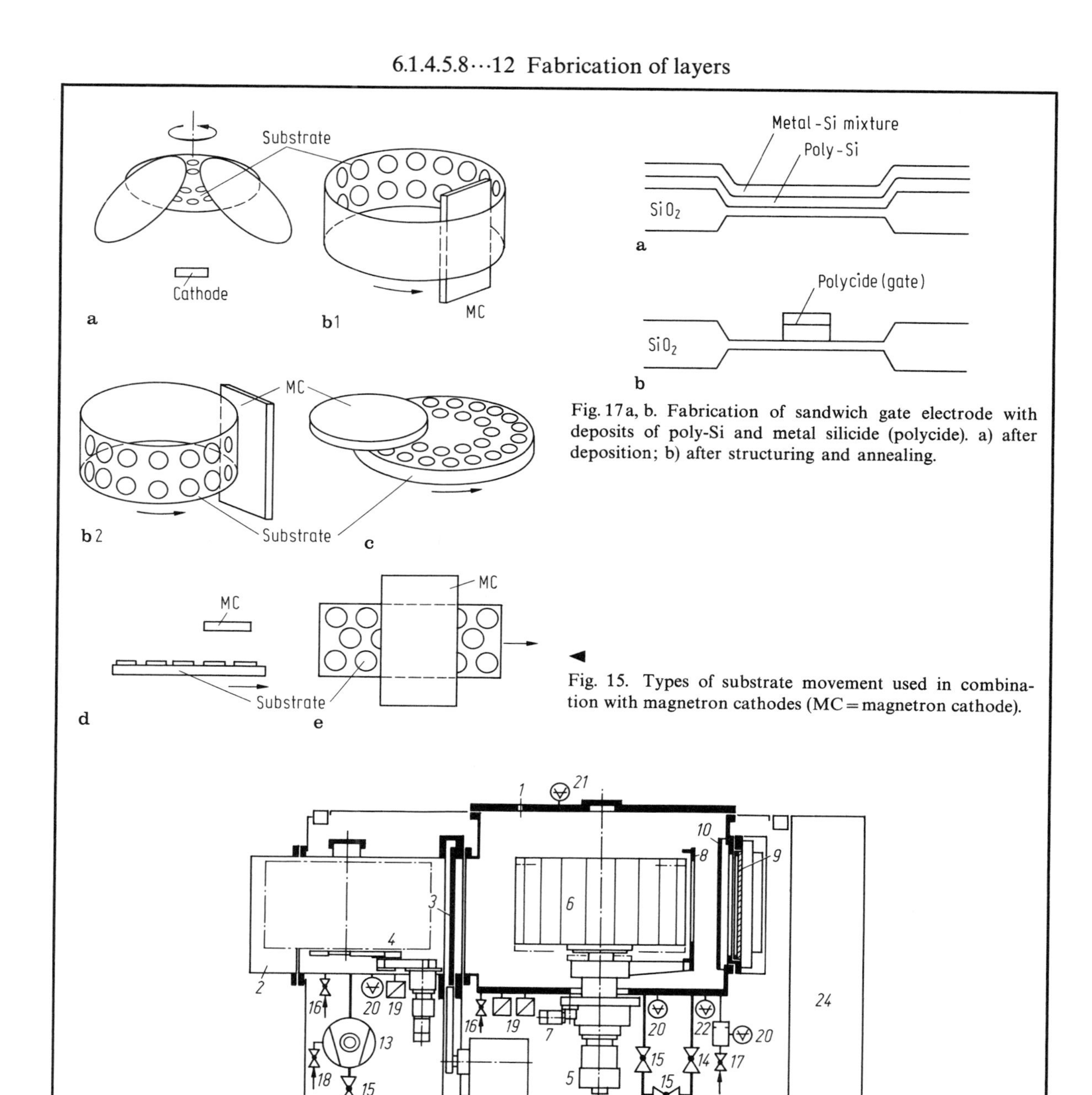

Fig. 17a, b. Fabrication of sandwich gate electrode with deposits of poly-Si and metal silicide (polycide). a) after deposition; b) after structuring and annealing.

◄

Fig. 15. Types of substrate movement used in combination with magnetron cathodes (MC = magnetron cathode).

Fig. 16. Lock vacuum system used in magnetron sputtering [80B1]. *1* processing chamber; *2* lock chamber; *3* lock valve; *4* transport mechanism; *5* substrate rotation motor; *6* substrate holder; *7* shutter motor; *8* shutter; *9* planar magnetron source; *10* mask box; *11* pump system; *12* cryo pump; *13* turbo molecular pump; *14···18* valves; *19* pressure sensitized switch; *20···22* vacuum gauges; *23* water/air pressure supply; *24* electrical control board.

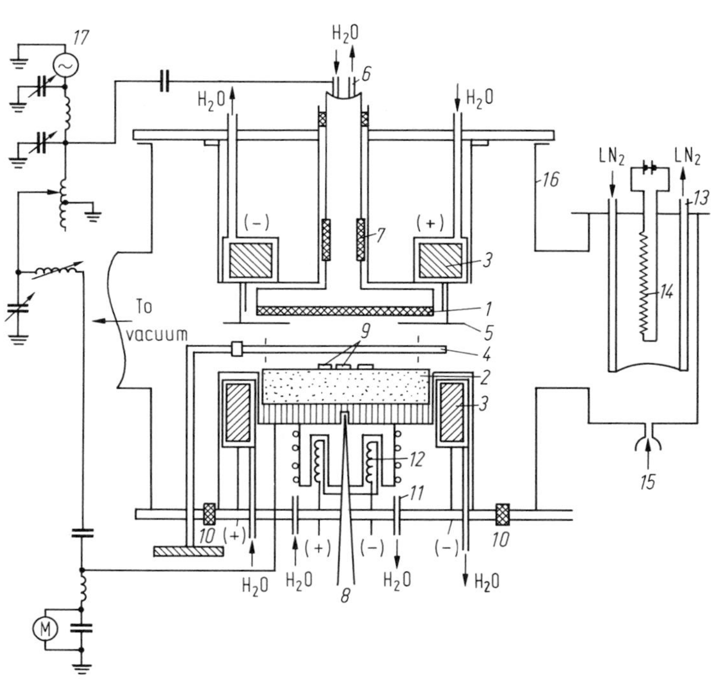

Fig. 18. Schematic of an RF sputtering apparatus [78V1]. The components are listed as follows: *1* cathode target; *2* anode substrate holder; *3* cathode and anode magnets (water-cooled); *4* shutter (SS = stainless steel); *5* cathode shield (SS); *6* cathode water cooling; *7* cathode isolation insulator; *8* substrate thermocouple; *9* substrates; *10* anode isolation insulator; *11* substrate cooling; *12* substrate heating; *13* liquid nitrogen-cooled SS shroud; *14* Ti sublimation filaments; *15* sputter gas; *16* SS vacuum chamber; *17* RF power supply matching network and substrate bias supply.

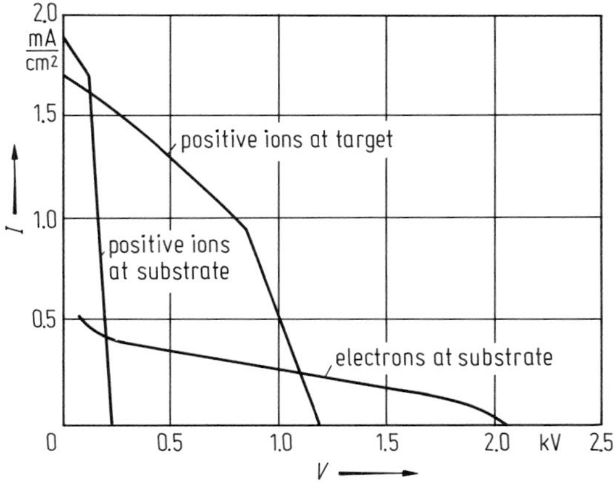

Fig. 20. Ion and electron current I in an RF diode sputter system at argon pressure $7 \cdot 10^{-6}$ bar vs. retarding potential [70K1].

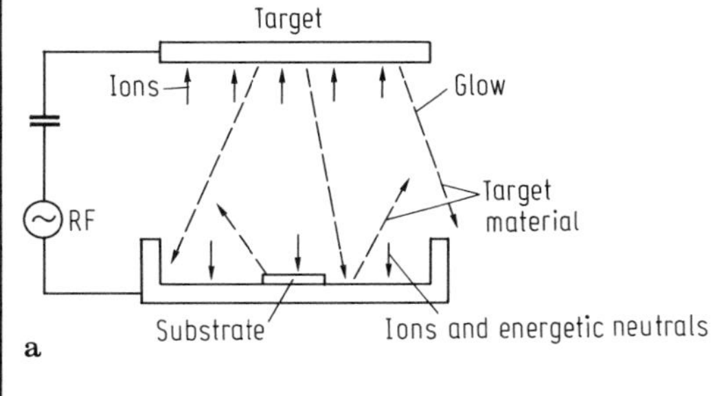

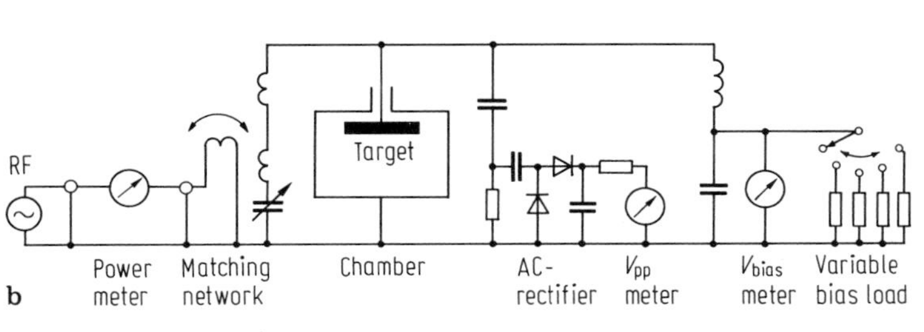

Fig. 19a, b. Schematics of the bias-sputter deposition [83H1]. a) basic principle; b) electrical circuit diagram.

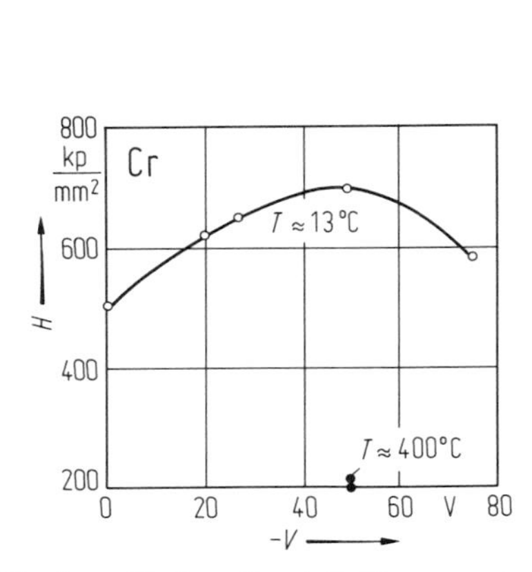

Fig. 21. Hardness H of sputtered Cr films vs. substrate bias voltage V at different temperatures (Hardness measurement: as deposited, 100 g loads) [72P1].

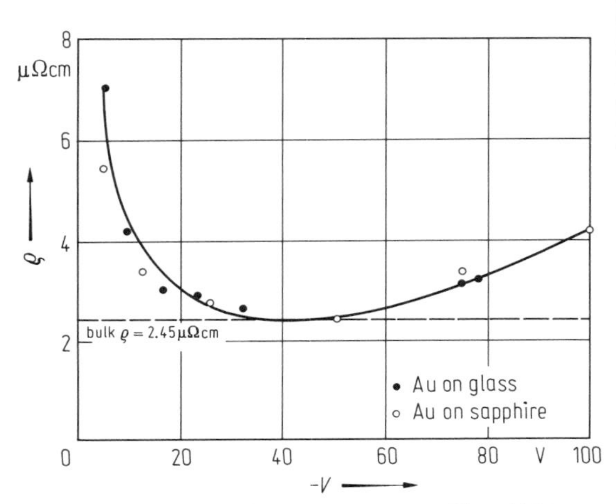

Fig. 22. Resistivity ϱ of DC sputtered Au films (thickness $d = 600$ nm) vs. substrate bias voltage V (RF induced) [68V1]. (Error: bulk $\varrho = 2.45$ μΩcm should read bulk $\varrho = 2.13$ μΩcm at $T = 22$ °C.)

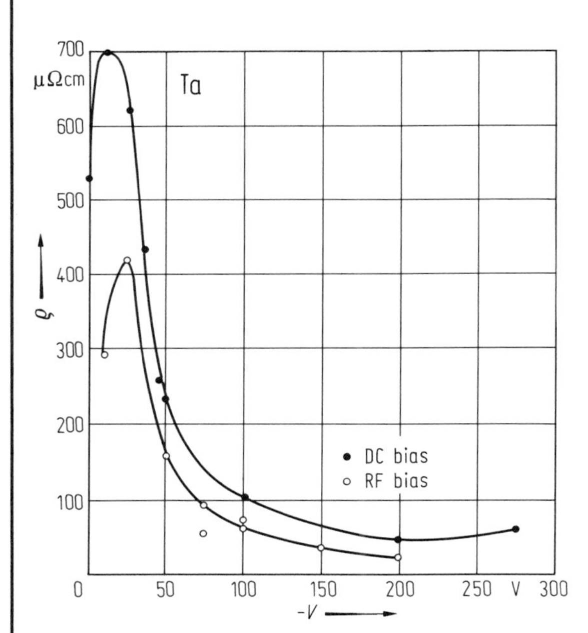

Fig. 23. Resistivity ϱ of DC-sputtered Ta films vs. substrate bias voltage V. (Target resistivity 17 μΩcm). DC bias [65M1]; RF bias [68V1].

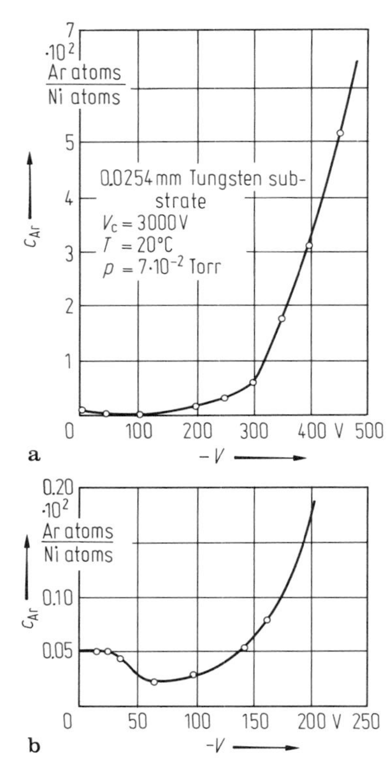

Fig. 24a, b. Argon concentration c_{Ar} in sputtered Ni films vs. DC-bias voltage V of the substrate. a) 0···500 V; b) detailed range 0···200 V.

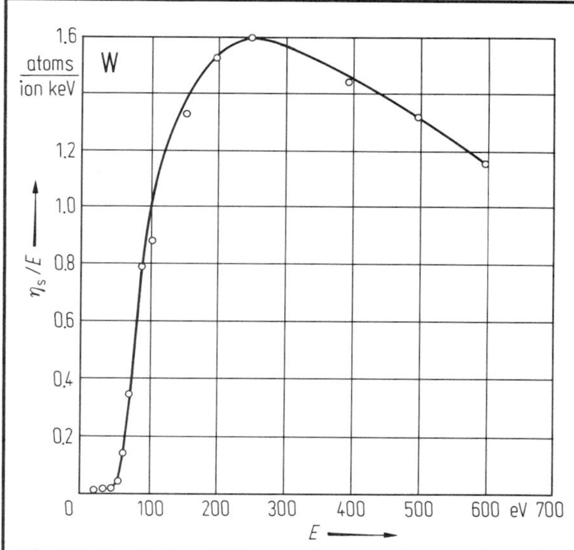

Fig. 25. Sputtering yield per unit energy input η_s/E vs. ion energy E (argon ions on tungsten target) [80C1].

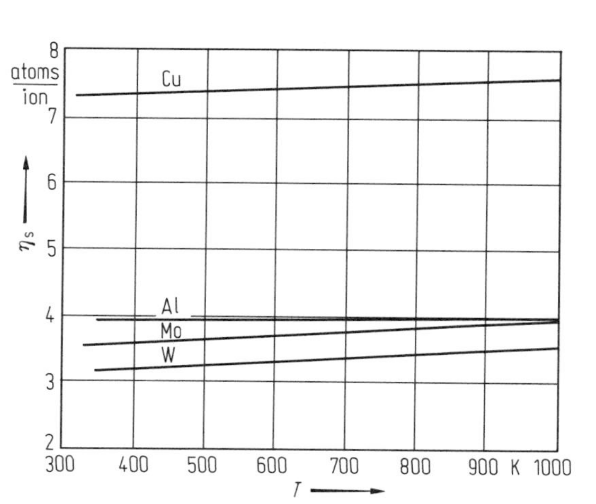

Fig. 27. Sputter yield η_s vs. target temperature T for sputtering of Al, Cu, Mo, W by 5 keV Ar^+ ions [65C1].

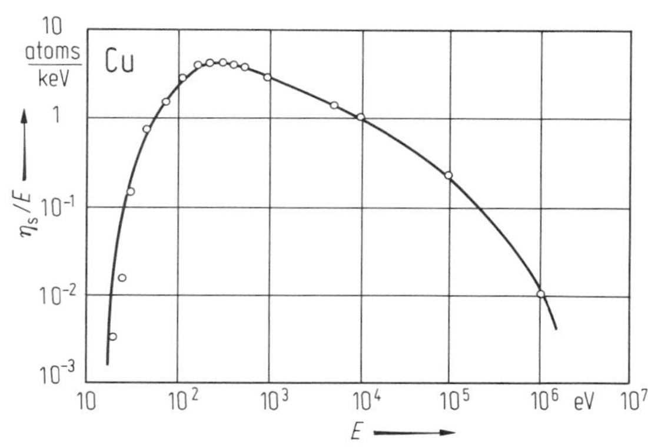

Fig. 26. Sputtering yield per unit energy input η_s/E vs. ion energy E (Xenon ions on copper target) [80c1].

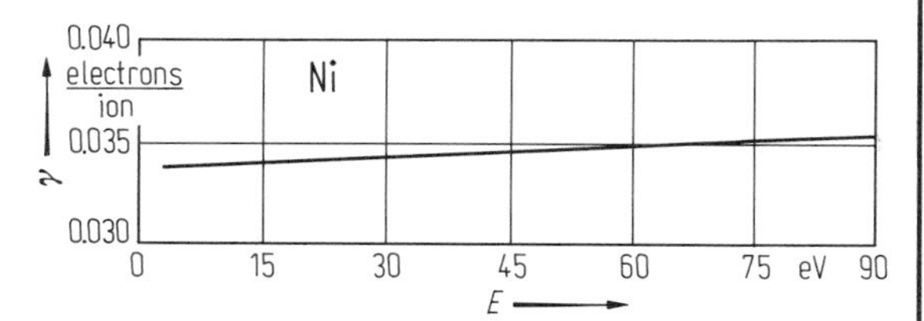

Fig. 29. Secondary electron emission γ vs. Ar^+ ion energy E for sputtering of Ni (111) surface [65T1].

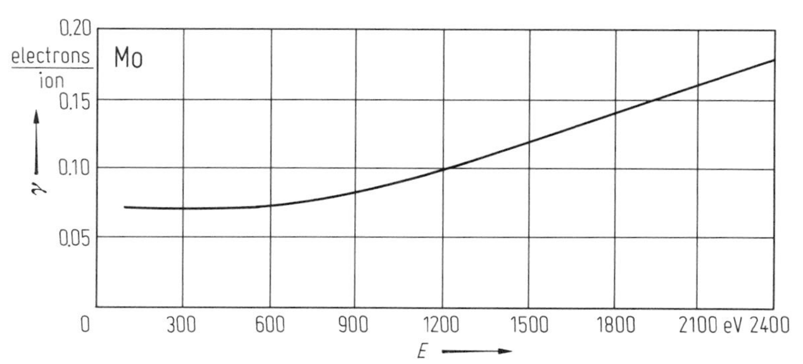

Fig. 28. Secondary electron emission γ vs. Ar^+ ion energy E for sputtering of Mo [65M2].

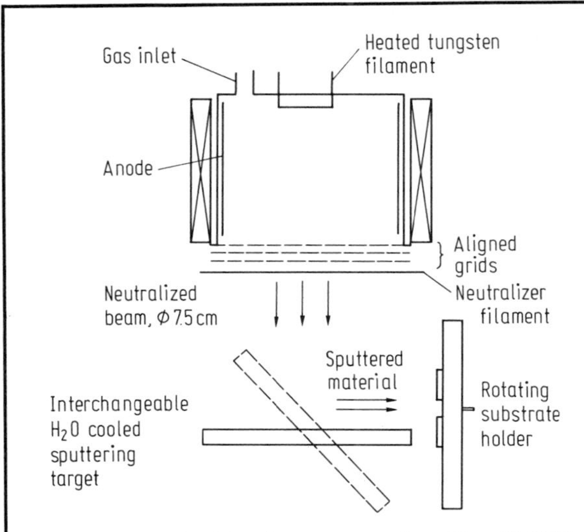

Fig. 30. Schematic of ion beam sputter deposition [79K2].

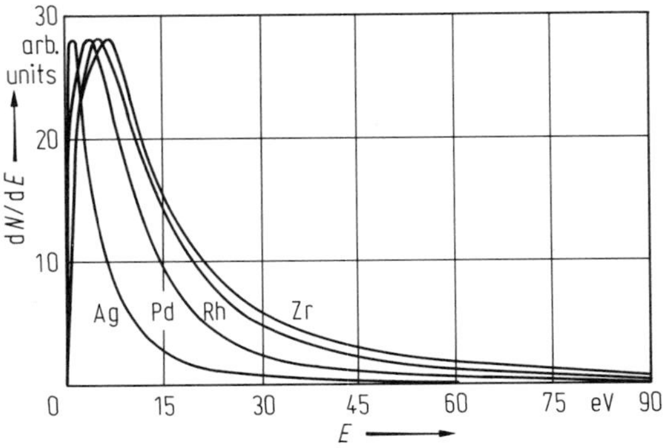

Fig. 31. Energy distribution of ion beam-sputtered atoms by Kr^+ ions of 1200 eV [70ml].

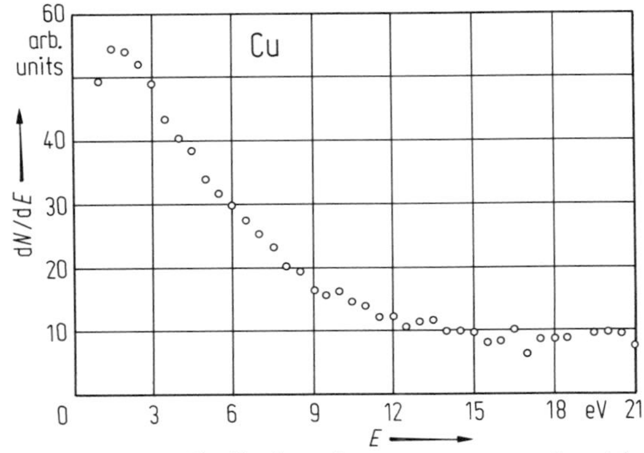

Fig. 32. Energy distribution of copper atoms sputtered by Ar^+ ions of 3 keV (45° incidence, 45° ejection) [78L1].

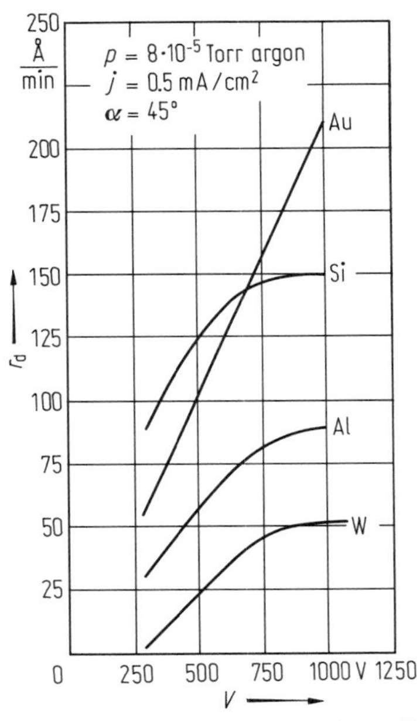

Fig. 34. Ion beam deposition rates r_d of Au, Si, Al, and W vs. accelerating voltage V [79K2].

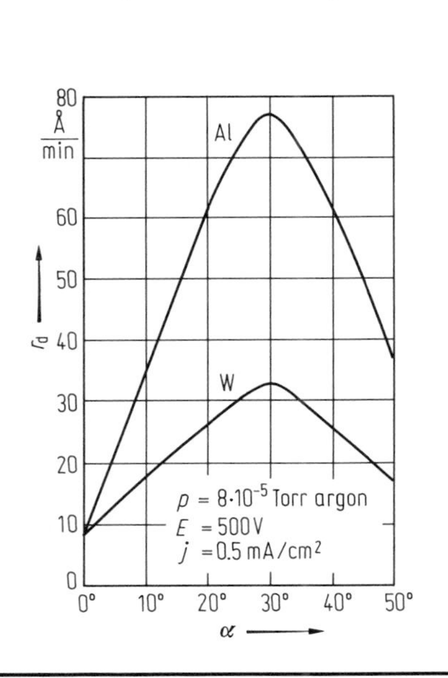

◄

Fig. 33. Ion beam deposition rate r_d of Al and W vs. target-to-beam angle α [79K2].

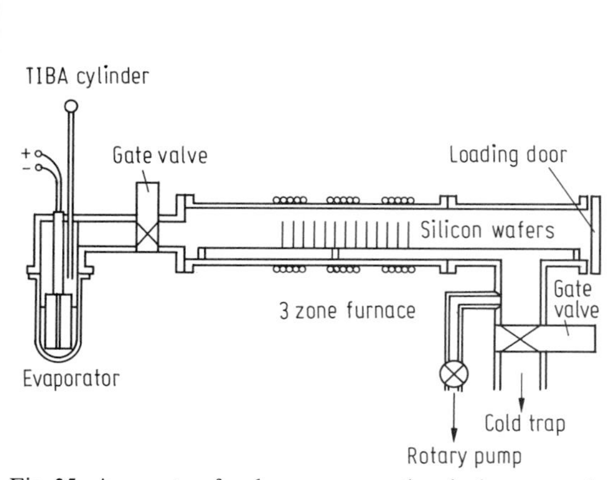

Fig. 35. Apparatus for low-pressure chemical vapour deposition LPCVD of aluminum and aluminum alloys [82C1]. (TIBA = tri-isobutyl aluminum)

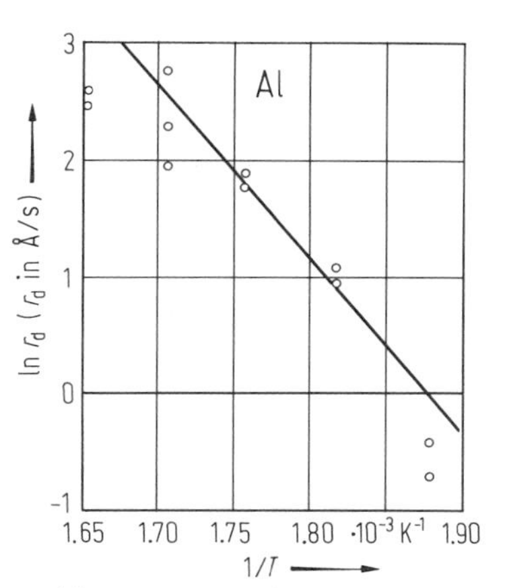

Fig. 36. Deposition rate r_d vs. reciprocal reactor temperature $1/T$ for LPCVD of Al [82C1].

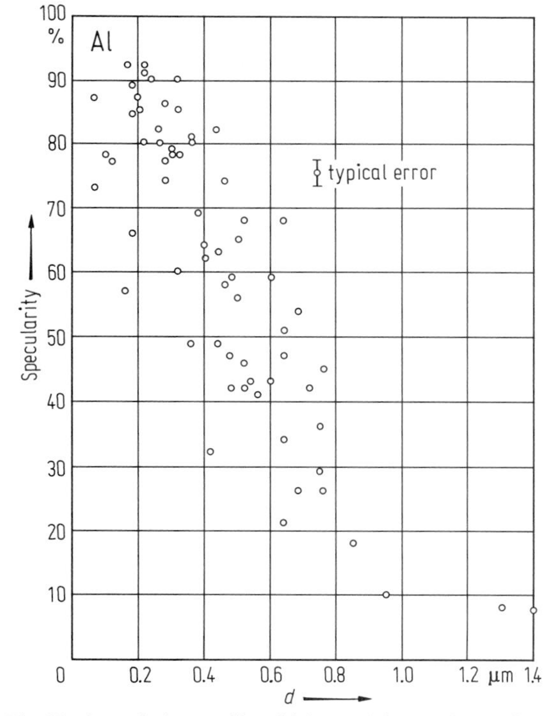

Fig. 37. Specularity vs. film thickness d for LPCVD of Al [82C1]. (Specularity 100% ≙ mirror; 0% = white paper).

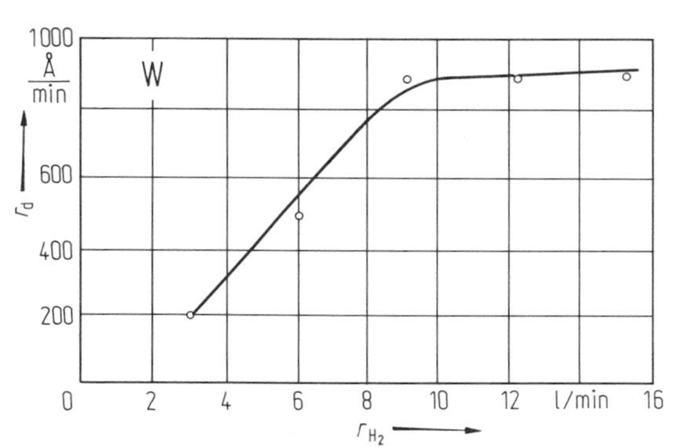

Fig. 38. Deposition rate r_d vs. hydrogen flow rate r_{H_2} for tungsten CVD [78M1]. (WF_6—H_2 system, $r_{WF_6} = 0.6$ cm^3 min^{-1}; $T = 600$ °C).

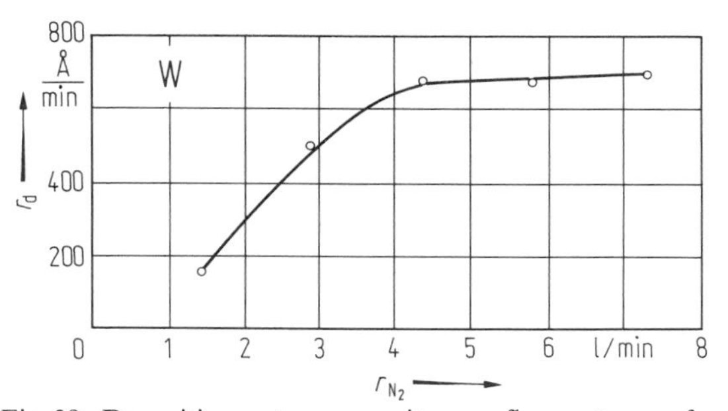

Fig. 39. Deposition rate r_d vs. nitrogen flow rate r_{N_2} for CVD of tungsten films [78M1]. (WF_6—N_2 system, $r_{WF_6} = 0.6$ cm^3 min^{-1}, $T = 540$ °C).

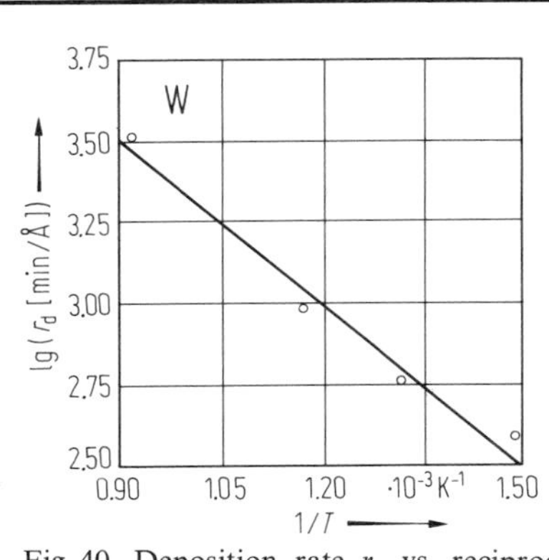

Fig. 40. Deposition rate r_d vs. reciprocal substrate temperature $1/T$ for CVD of tungsten films [78M1]. (WF_6—H_2 system, $r_{WF_6}=0.6\ cm^3\ min^{-1}$, $r_{H_2}=10.3\ l\ min^{-1}$).

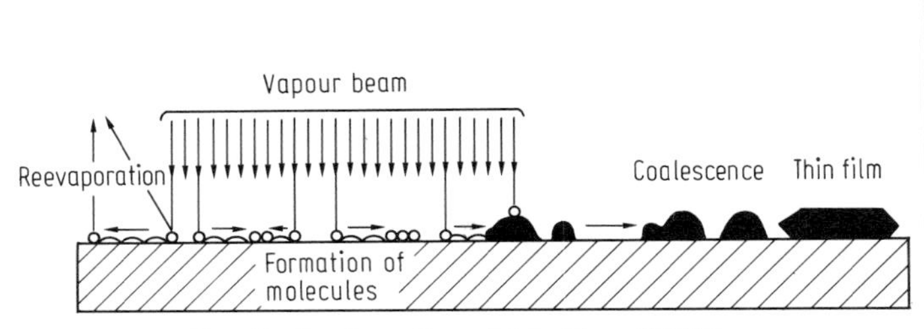

Fig. 41. Condensation of thin films [78C1].

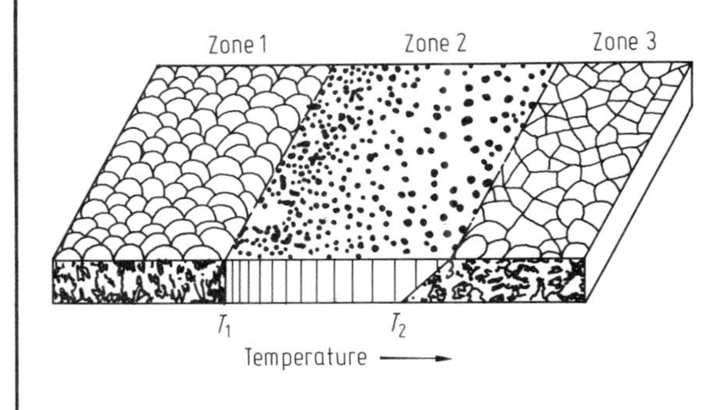

Fig. 42. Effect of the substrate temperature on the microstructure of evaporated films. Zone 1: $T<0.3\ T_m$; zone 2: $T=0.3\ T_m\cdots0.45\ T_m$; zone 3: $T>0.45\ T_m$. (T_m = melting temperature of the substrate) [69M1].

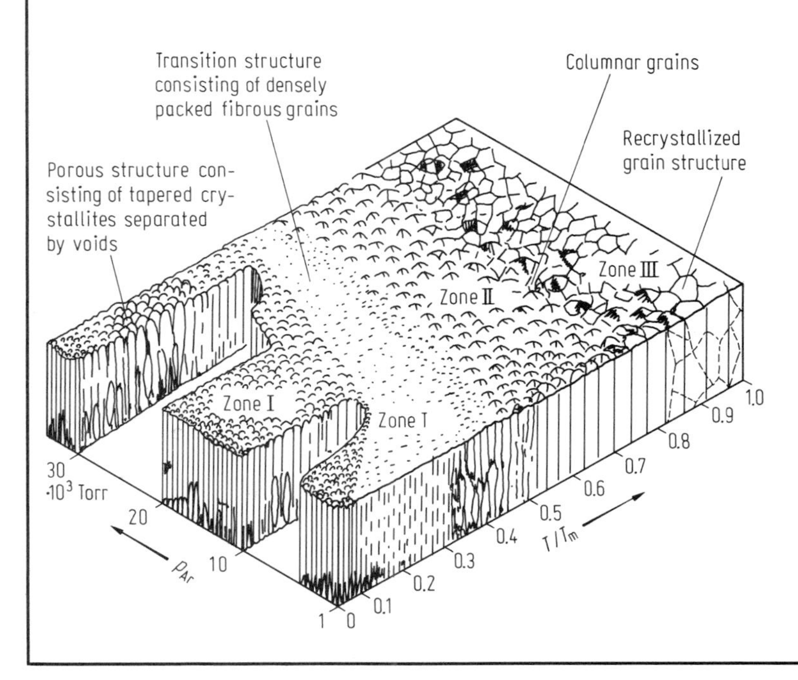

Fig. 43. Effect of the substrate temperature and the argon pressure on the microstructure of sputtered films [74T1, 75T1]. (Error: at the pressure axis the unit 10^3 Torr should read 10^{-3} Torr.)

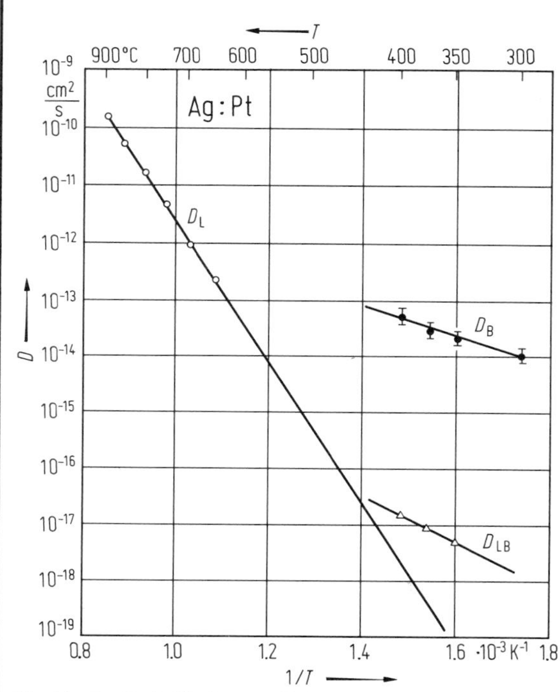

Fig. 44. Ag:Pt Diffusion coefficients D vs. reciprocal temperature $1/T$ for Pt in Ag [75F2, 79M1]. (D_L: lattice diffusion coefficient; D_B: grain boundary diffusion coefficient; D_{LB}: grain boundary-assisted bulk diffusion coefficient.)

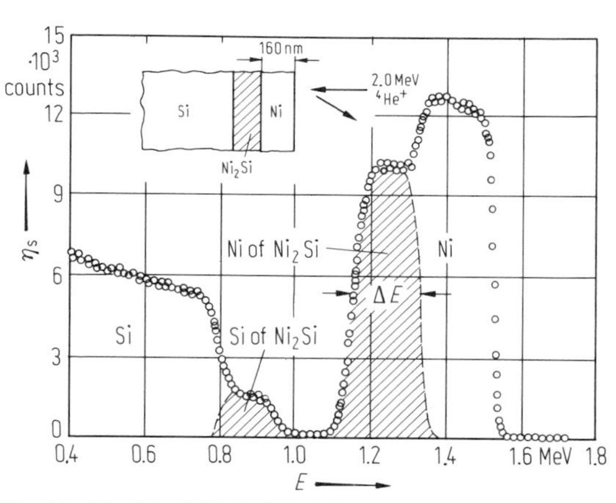

Fig. 46. Si—Ni. 2 MeV $^4He^+$ backscattering spectrum of secondary ions Si and Ni from 260 nm Ni e-gun evaporated on a Si wafer and annealed for 24 h at 250 °C [74L1].

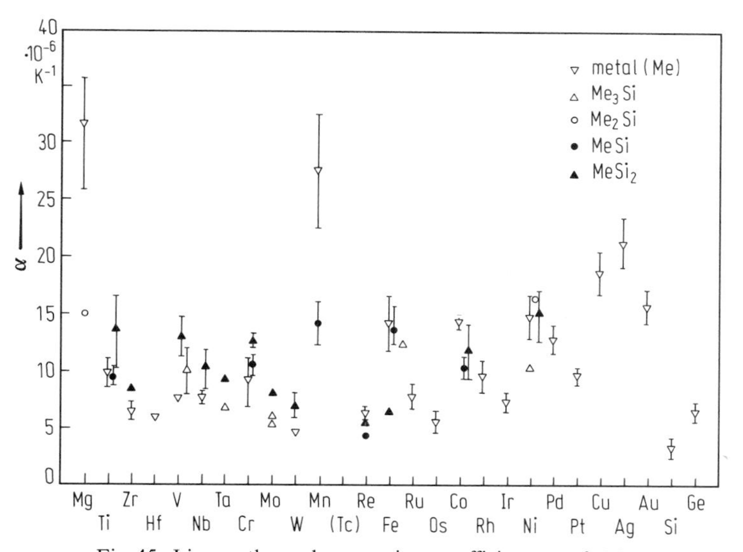

Fig. 45. Linear thermal expansion coefficient α of Mg, transition metals, Si, Ge, and silicides of metals [78N2].

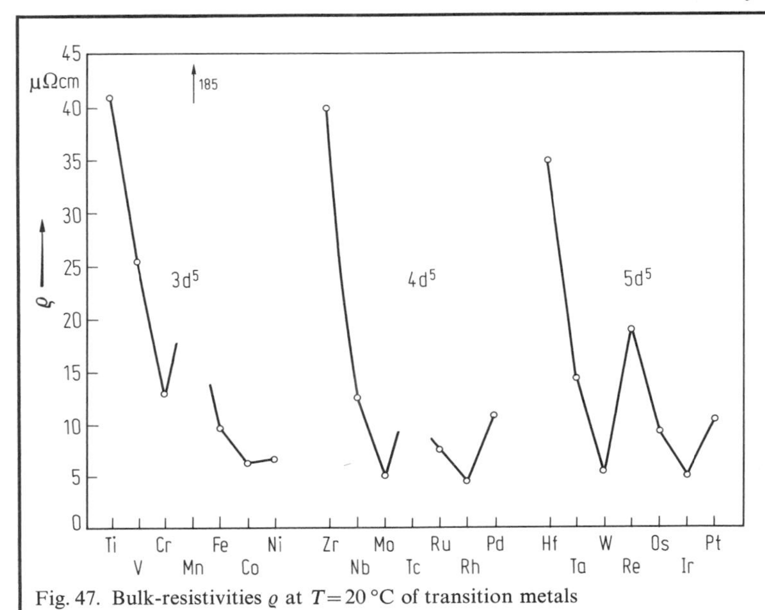

Fig. 47. Bulk-resistivities ϱ at $T=20\,°C$ of transition metals vs. configuration in the periodic table [80M1].

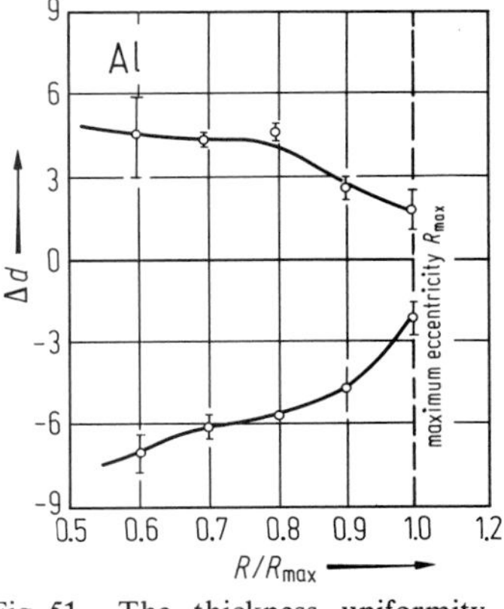

Fig. 51. The thickness uniformity Δd over the whole planetary system for electron beam gun-evaporated aluminum films vs. the position R/R_{max} of the e-gun. The deposition rate was 6 nm s^{-1} [80H1].

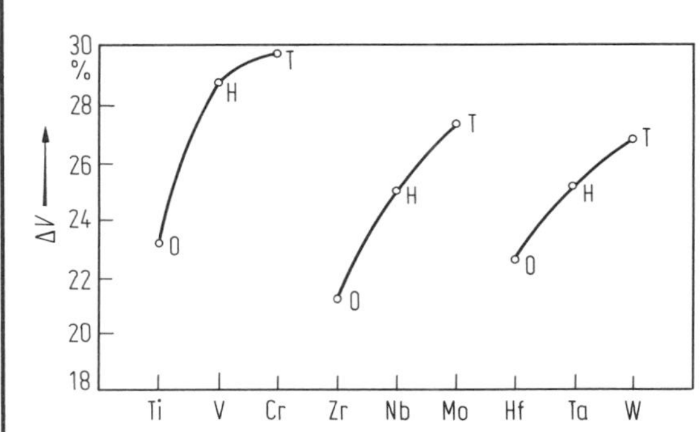

Fig. 48. Volume decrease ΔV during silicide formation (from metal on silicon reaction) [80M1]. (O, H, T stand for the orthorhombic, hexagonal, and tetragonal structures, respectively.)

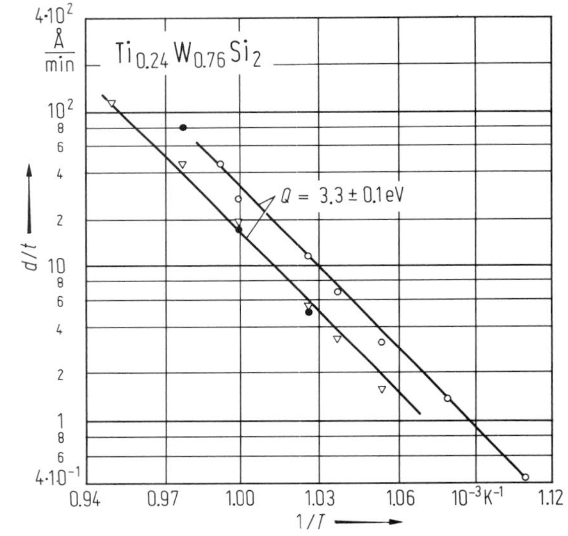

Fig. 49. Reaction rate (ratio of the produced layer thickness d of $Ti_{0.24}W_{0.76}Si_2$ and time t) d/t vs. reciprocal annealing temperature $1/T$ for the reaction between $Ti_{0.24}W_{0.76}$ and Si [83N2].

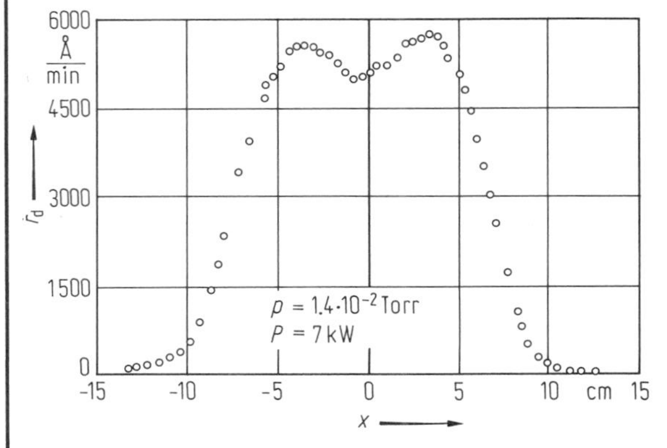

◄

Fig. 50. Deposition rate r_d of Al vs. distance x from the center line of the DC-magnetron cathode (INSET-Cathode), measured in wafer scan direction [79C1].

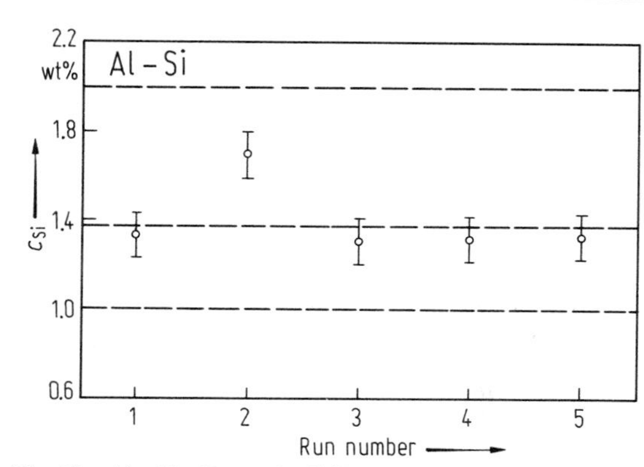

Fig. 52. Al—Si. Reproducibility of the silicon content c_{Si} in co-evaporated Al—Si(1.5 wt %) films from run to run over 5 runs [80H1].

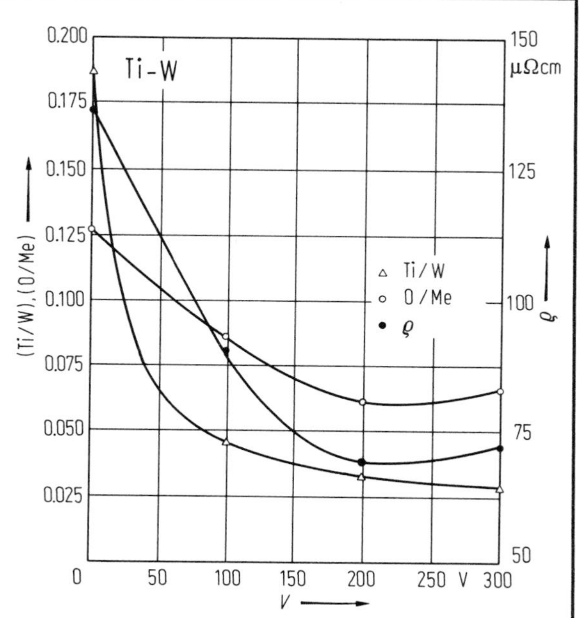

Fig. 53. Ti/W ratio, oxygen/metal ratio, and resistivity vs. RF-induced substrate bias voltage V of RF-sputtered Ti—W films [79H4].

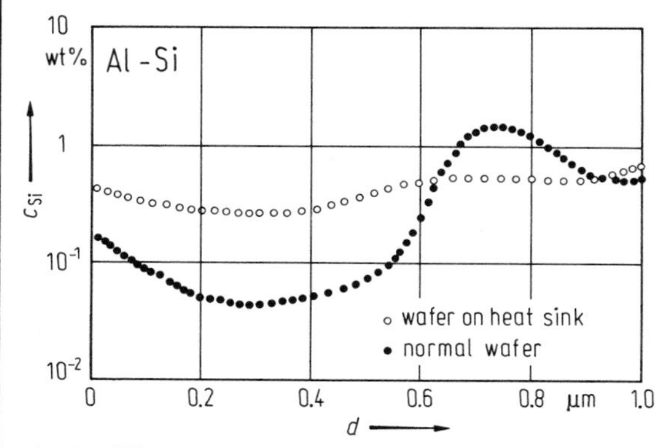

Fig. 54. Silicon concentration c_{Si} in as-deposited aluminum-silicon films (AES-profile) vs. film depth d for DC-magnetron sputter deposition from an Al—Si(0.5 wt %) target [79C1].

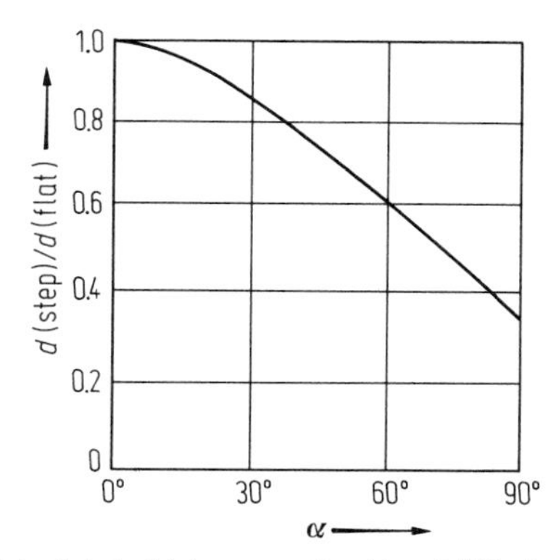

Fig. 56. Calculated thickness ratio d(step)/d(flat) vs. step angle α for a cosine distribution in the deposition process [79F1].

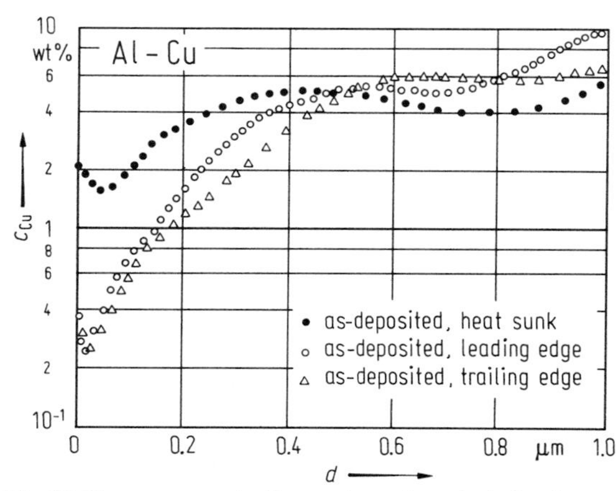

Fig. 55. Copper concentration c_{Cu} in as-deposited aluminum-copper films (AES-profile) vs. layer depth d for DC-magnetron sputter deposition from an Al—Cu(4 wt %) target [79C1]. (Film thickness = 1 μm).

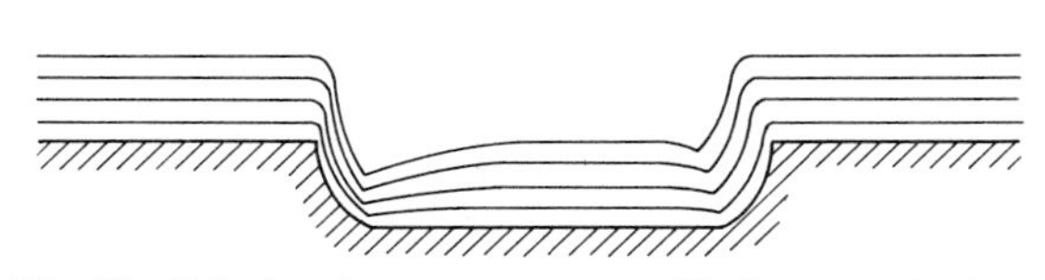

Fig. 57. Calculated step coverage profile for steps having different wall slopes [74T2].

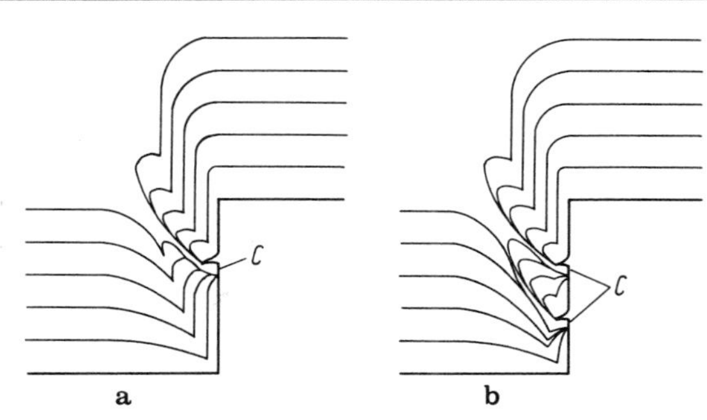

Fig. 58a, b. Schematic profiles for a step with surface protrusion [74T2]. a) effect of a single protrusion; "C" denotes area of no film grown; b) effect of two protrusions.

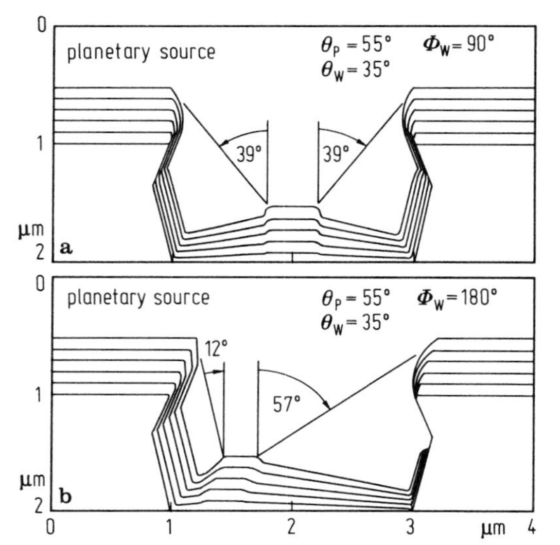

Fig. 59a, b. Step coverage during evaporation [80N3]. Simulation for 2 μm lines and spaces for 5 cm wafers located in the outboard planet position of an Airco-Temescal 1800 evaporator system. The symmetrical case is $\phi_W = 90°$ in a) and the asymmetrical case is $\phi_W = 180°$ in b).
(θ_P: angle between planet axis and dove axis; θ_W: angle between planet axis and the normal axis of the wafer; Φ_W: angle between the lines on the wafer and a line from the centre of the wafer to the centre of the planet.)

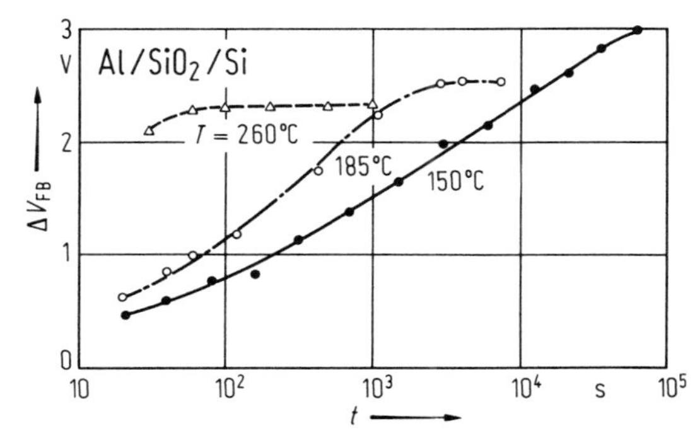

Fig. 60. Flatband voltage shift ΔV_{FB} vs. stress time t for 3 different stress temperatures [78L2]. MOS test capacitor with e-gun evaporated aluminum electrode before annealing. (Gate bias −65 V; $d_{ox} = 220$ nm; Si-substrate (100), n-type, 2 Ω cm).

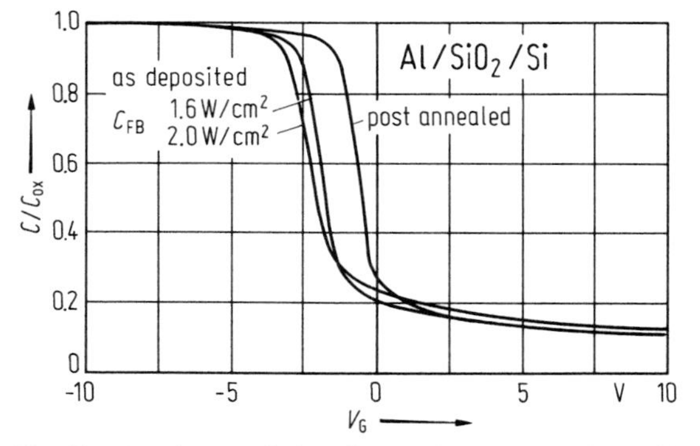

Fig. 61. CV characteristics of an MOS test capacitor with RF sputtered Al-gate, before and after annealing (20 min, 450 °C, $N_2 + H_2$). Normalized capacitance C/C_{ox} vs. gate voltage V_G: $C_{FB}/C_{ox} = 0.72$ indicates the flatband condition [81S1].

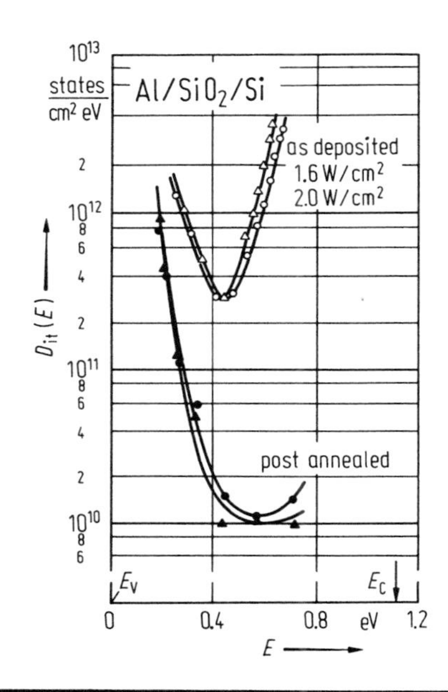

◄ Fig. 62. Distribution of surface states $D_{it}(E)$ over silicon bandgap of MOS capacitor with RF sputtered Al electrode before and after annealing (20 min, 450 °C, $N_2 + H_2$) [81S1].

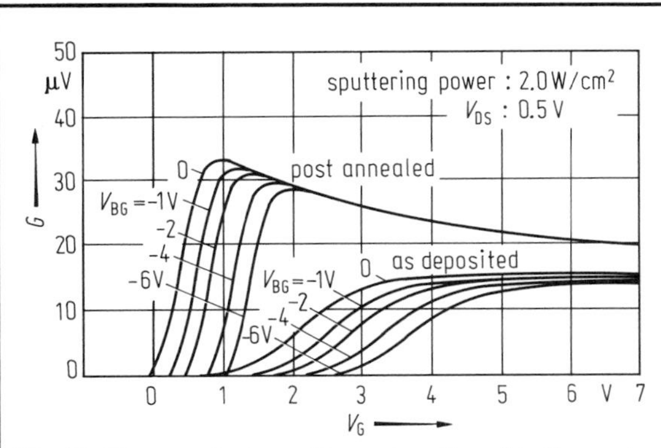

Fig. 63. Transconductance G vs. gate voltage V_G. Parameter is the substrate voltage V_{BG}. n-channel, poly-Si gate transistors with RF sputtered Al-metallization before and after annealing (20 min, 450 °C, N_2+H_2) Si-substrate: p-type, (100), 10 Ω cm [81S1].

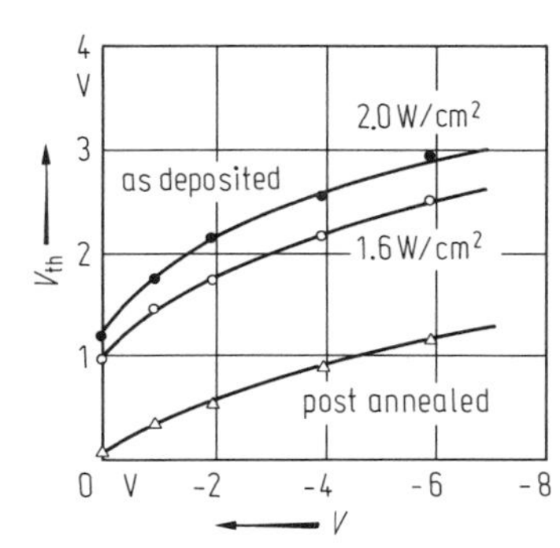

Fig. 64. Threshold voltage V_{th} vs. substrate voltage V of n-channel, poly-Si gate transistors with RF-sputtered Al-metallization before and after annealing (20 min, 450 °C, N_2+H_2) Si substrate: p-type, (100), 10 Ω cm [81S1].

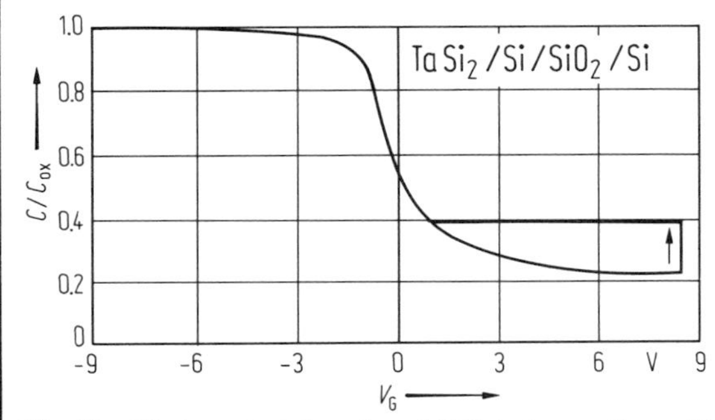

Fig. 65. CV characteristics of an MOS test capacitor with $TaSi_2/n^+$ poly-Si gate, identical characteristics as received and after stress: ±10 V, 250 °C, 15 min [80S2]. Gate structure: $TaSi_2/n^+$ poly-Si/SiO_2/p-Si (100); $d_{ox}=52.6$ nm; $n_{it}=2\cdot10^{10}$ cm^{-2}.

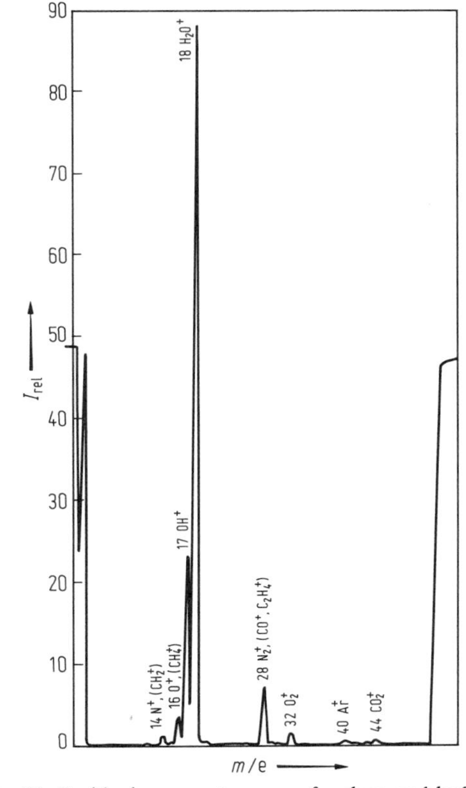

Fig. 66. Residual gas spectrogram of a clean and leak-tight vacuum system, $p_{tot}=1\cdot10^{-10}$ bar. Relative intensity I_{rel} vs. specific mass m/e in amu [83S2].

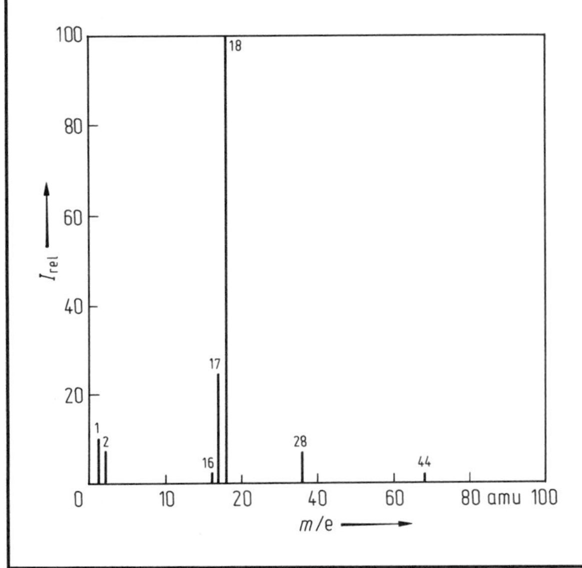

◄ Fig. 67. Residual gas spectrogram of an unbaked vacuum system [81U1]. Relative intensity I_{rel} vs. specific mass m/e.

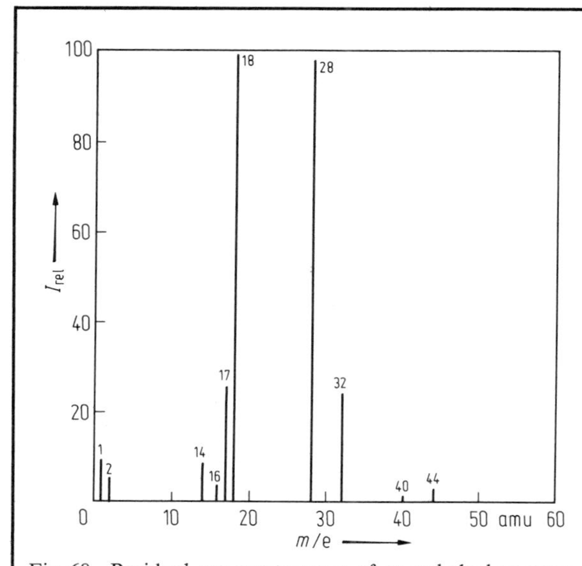

Fig. 68. Residual gas spectrogram of an unbaked vacuum system with an air-leak [81U1]. Relative intensity I_{rel} vs. specific mass m/e.

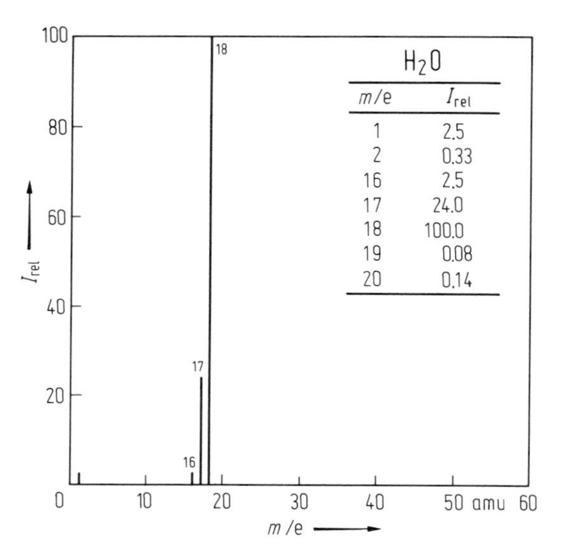

Fig. 69. Gas spectrogram of water vapour [81U1]. Relative intensity I_{rel} vs. specific mass m/e.

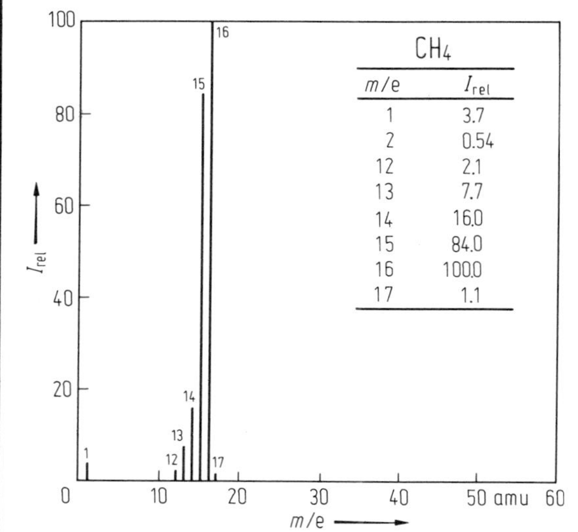

Fig. 70. Gas spectrogram of methane [81U1]. Relative intensity I_{rel} vs. specific mass m/e.

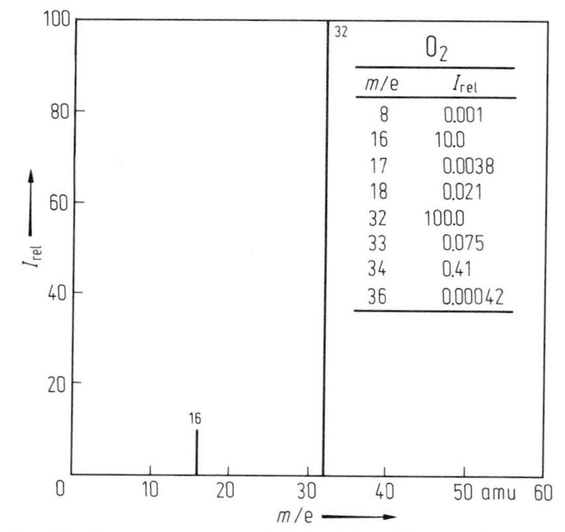

Fig. 71. Gas spectrogram of oxygen [81U1]. Relative intensity I_{rel} vs. specific mass m/e.

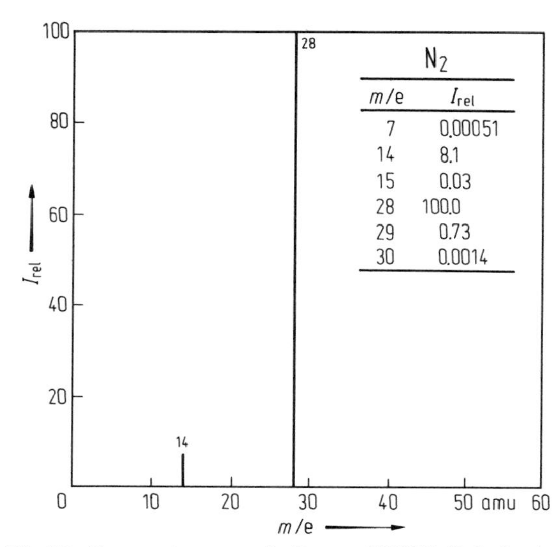

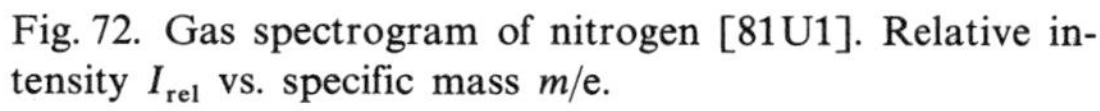

Fig. 72. Gas spectrogram of nitrogen [81U1]. Relative intensity I_{rel} vs. specific mass m/e.

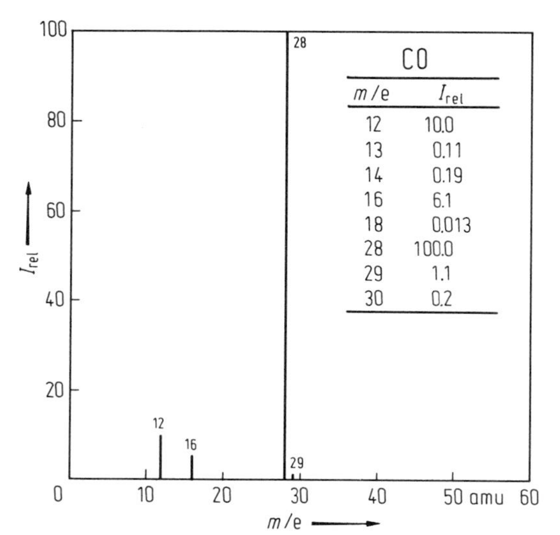

Fig. 73. Gas spectrogram of carbon monoxide [81U1]. Relative intensity I_{rel} vs. specific mass m/e.

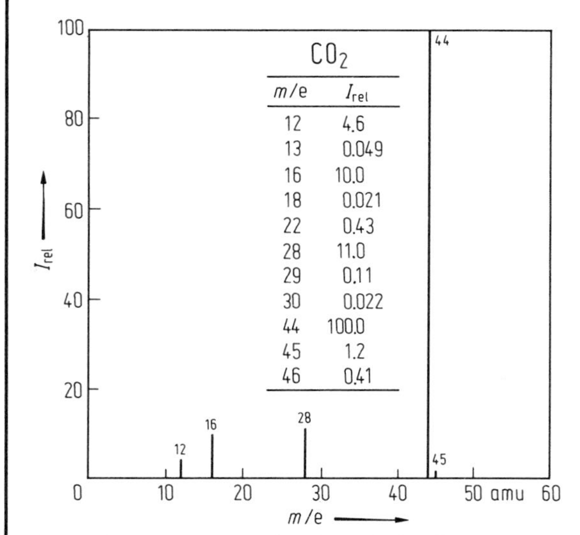

Fig. 74. Gas spectrogram of carbon dioxide [81U1]. Relative intensity I_{rel} vs. specific mass m/e.

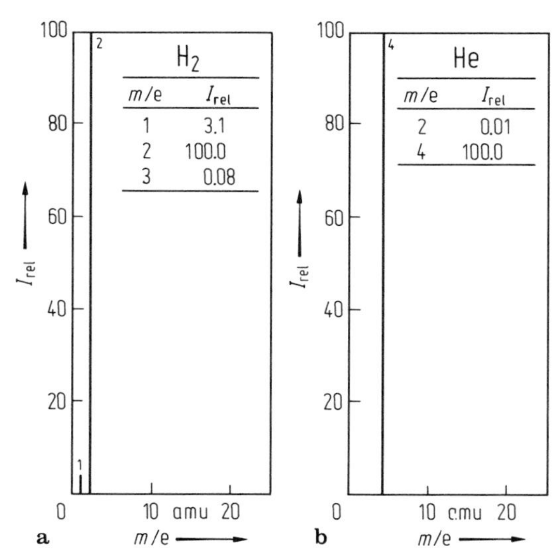

Fig. 75 a, b. Gas spectrogram of hydrogen (a) and helium (b) [81U1]. Relative intensity I_{rel} vs. specific mass m/e.

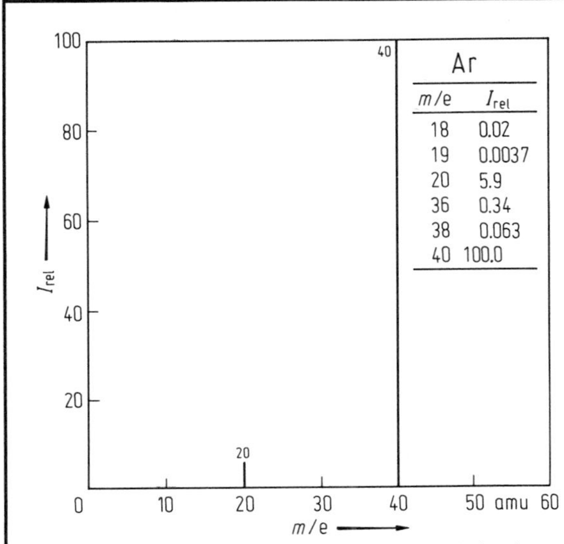

Ar	
m/e	I_{rel}
18	0.02
19	0.0037
20	5.9
36	0.34
38	0.063
40	100.0

Fig. 76. Gas spectrogram of argon [81U1]. Relative intensity I_{rel} vs. specific mass m/e.

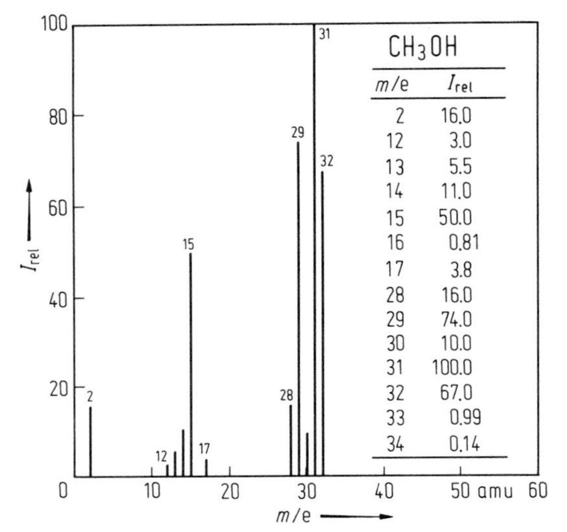

CH_3OH	
m/e	I_{rel}
2	16.0
12	3.0
13	5.5
14	11.0
15	50.0
16	0.81
17	3.8
28	16.0
29	74.0
30	10.0
31	100.0
32	67.0
33	0.99
34	0.14

Fig. 77. Gas spectrogram of methanol [81U1]. Relative intensity I_{rel} vs. specific mass m/e.

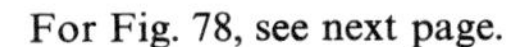

For Fig. 78, see next page.

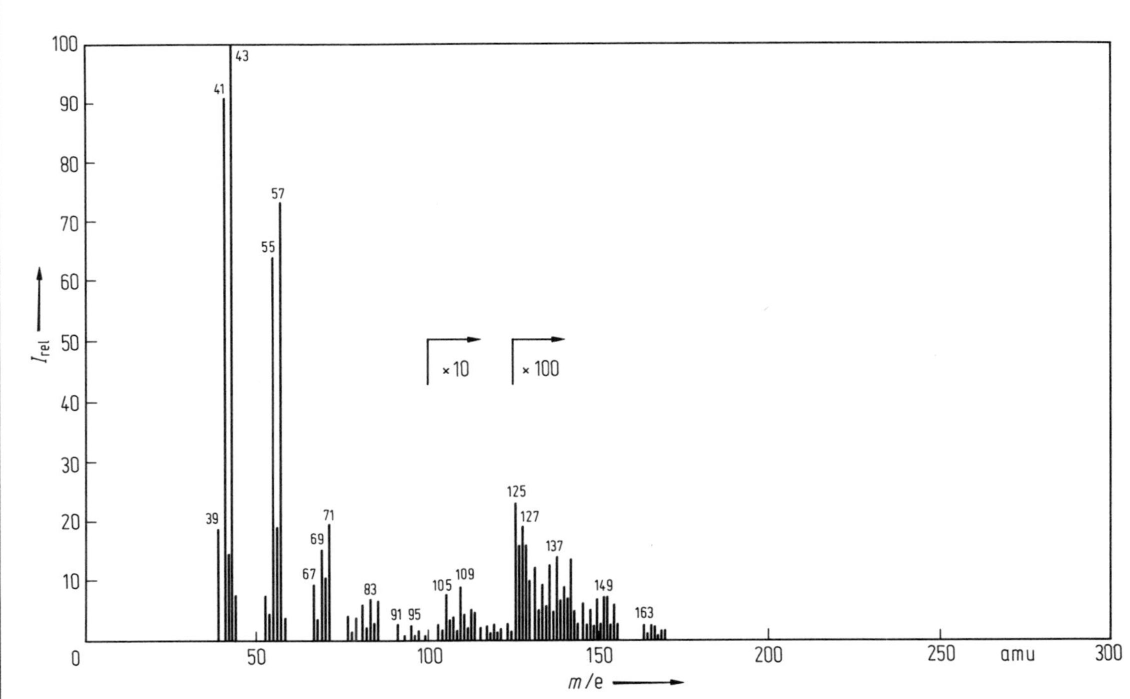

Fig. 79. Gas spectrogram of mechanical vacuum pump fluid [81U1]. Relative intensity I_{rel} vs. specific mass m/e.

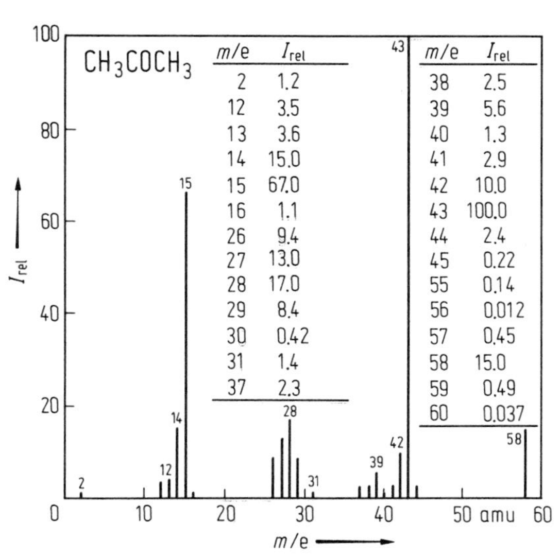

Fig. 78. Gas spectrogram of acetone [81U1]. Relative intensity I_{rel} vs. specific mass m/e.

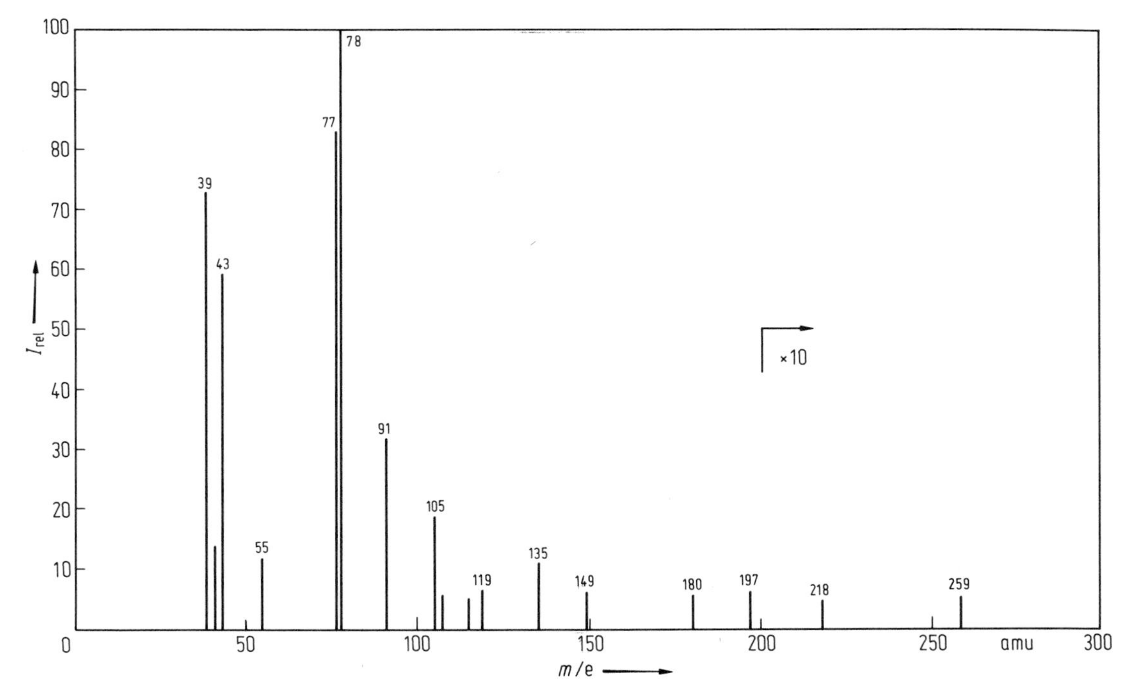

Fig. 80. Gas spectrogram of a diffusion pump fluid (silicone oil 705) [81U1]. Relative intensity I_{rel} vs. specific mass m/e.

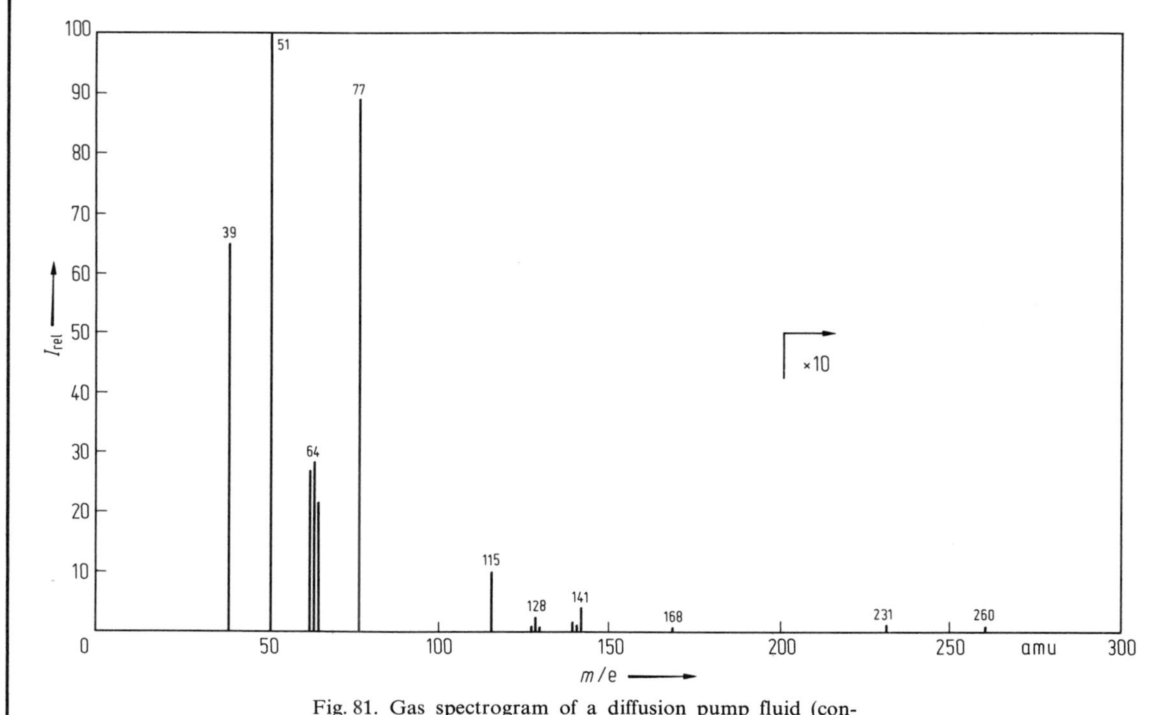

Fig. 81. Gas spectrogram of a diffusion pump fluid (convalex 10) [81U1]. Relative intensity I_{rel} vs. specific mass m/e.

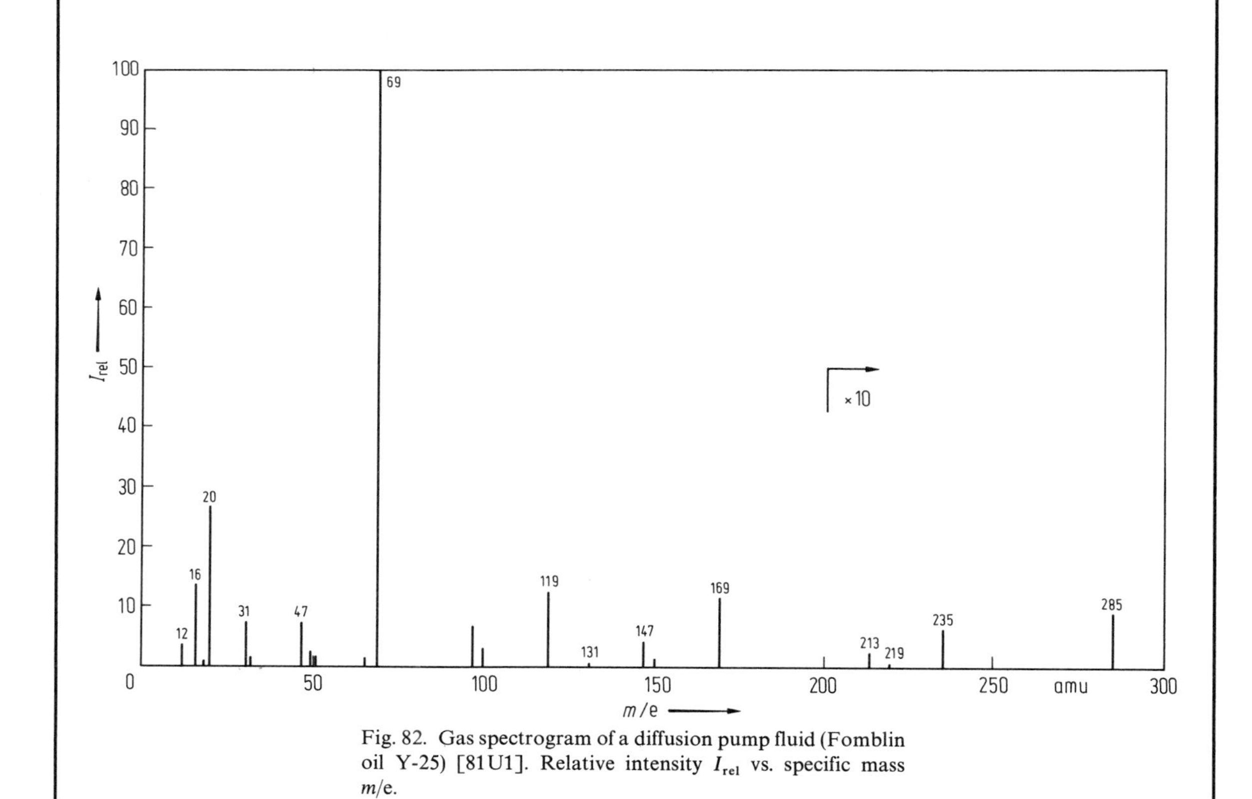

Fig. 82. Gas spectrogram of a diffusion pump fluid (Fomblin oil Y-25) [81U1]. Relative intensity I_{rel} vs. specific mass m/e.

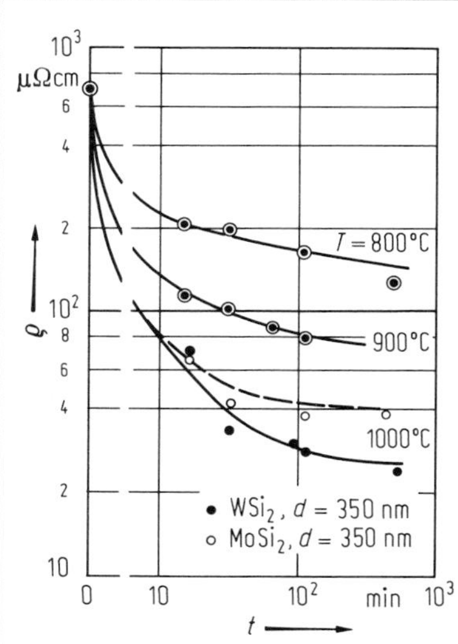

Fig. 83. Resistivity ϱ of $MoSi_2$- and WSi_2-films (350 nm, co-evaporated on SiO_2) vs. time t at various sintering temperatures in Ar [80G1].

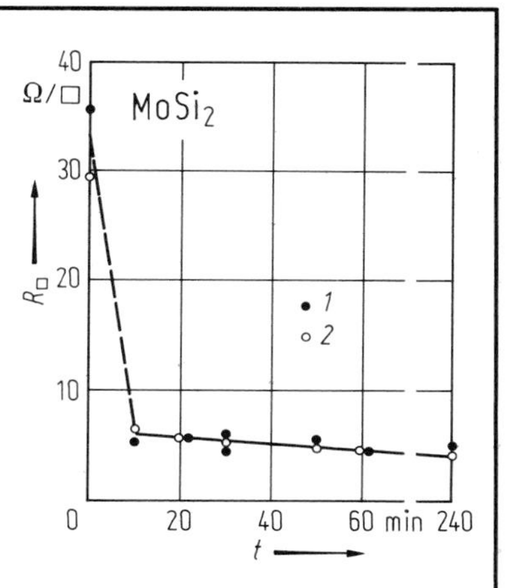

Fig. 84. $MoSi_2$. Sheet resistance $R_\square$ of $MoSi_2$ films vs. time t for sintering at $T = 900\,°C$, in N_2 [83N1]. *1* co-sputtered layer, 200 nm; *2* co-evaporated layer, 200 nm.

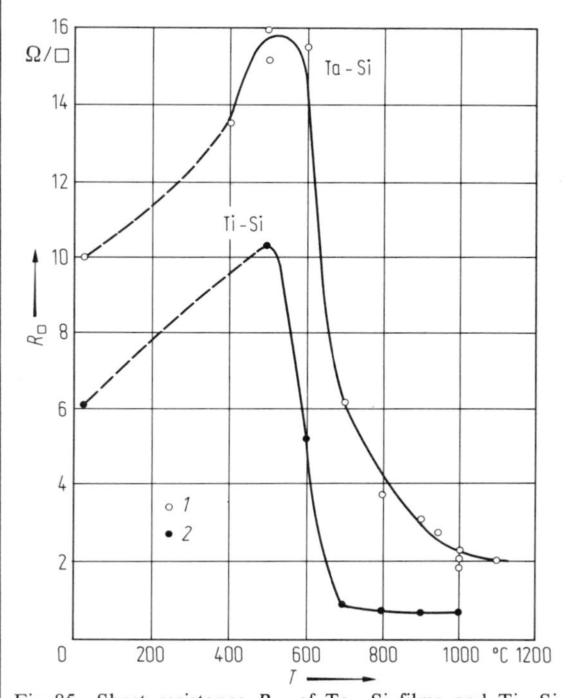

Fig. 85. Sheet resistance $R_\square$ of Ta—Si films and Ti—Si films vs. sintering temperature T [80M1]. *1* 100 nm Ta on doped 450 nm polysilicon; *2* 100 nm Ti on doped, 450 nm polysilicon.

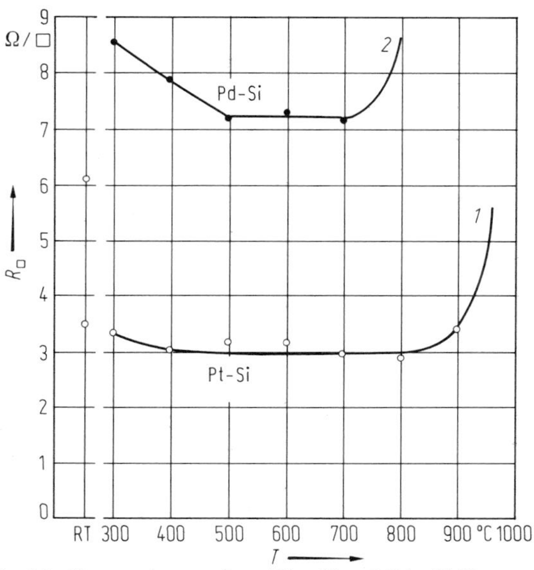

Fig. 86. Sheet resistance $R_\square$ of Pt—Si and Pd—Si films vs. sintering temperature T [80M1]. *1*) 50 nm Pt on (100) Si; *2*) 30 nm Pd on (100) Si.

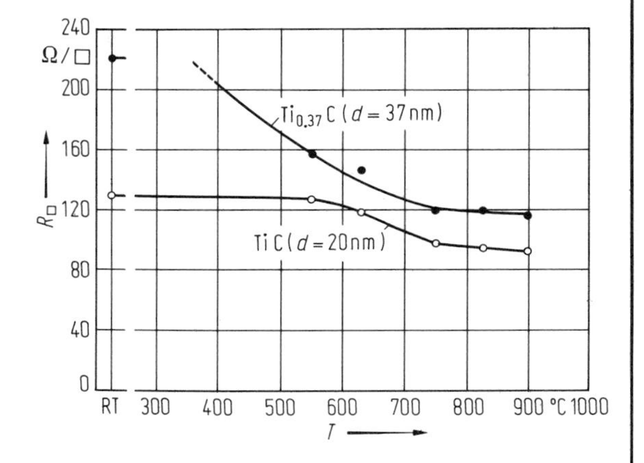

Fig. 87. Sheet resistance $R_\square$ for Ti_xC films on silicon vs. sintering temperature T (30 min anneal) [83E1].

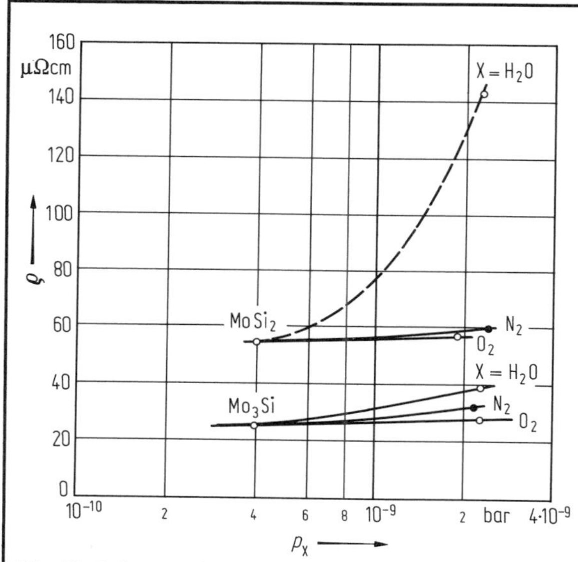

Fig. 88. Influence of the residual gas pressure during layer fabrication by co-evaporation of Mo and Si on the resistivity ϱ of Mo—Si. The resistivity of Mo-silicides is measured after sintering 60 min at 900 °C in N_2 [83N1].

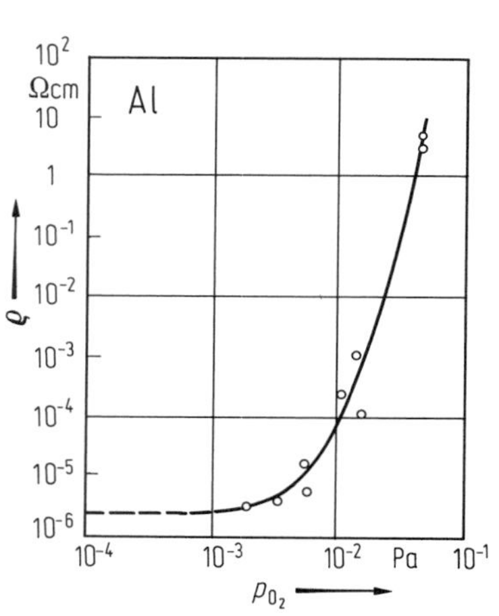

Fig. 89. Resistivity ϱ of aluminum films, magnetron-sputtered in an $Ar+O_2$ mixture vs. oxygen fill pressure p_{O_2} [81F1].

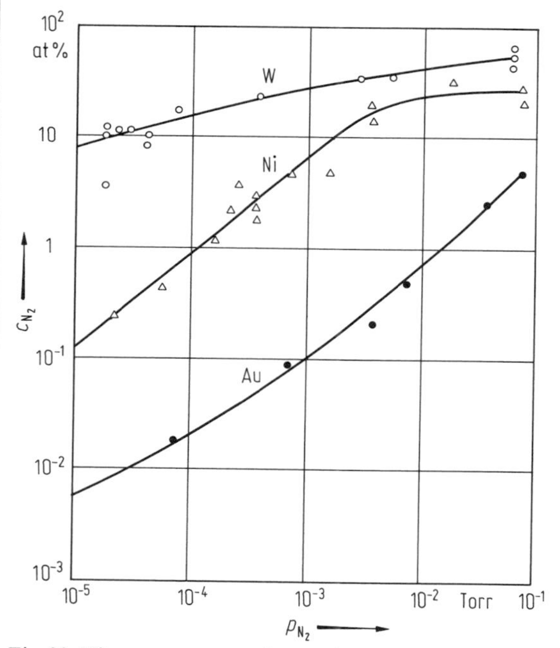

Fig. 90 Nitrogen concentration c_{N_2} in sputtered metal films vs. nitrogen partial pressure p_{N_2} during sputtering in $Ar+N_2$ gas [72W1]. Sputter conditions: DC diode sputtering. Target voltage 3000 V, no bias voltage, argon pressure $7\cdot10^{-2}$ Torr, no substrate heating, target-substrate distance $=4.5$ cm.

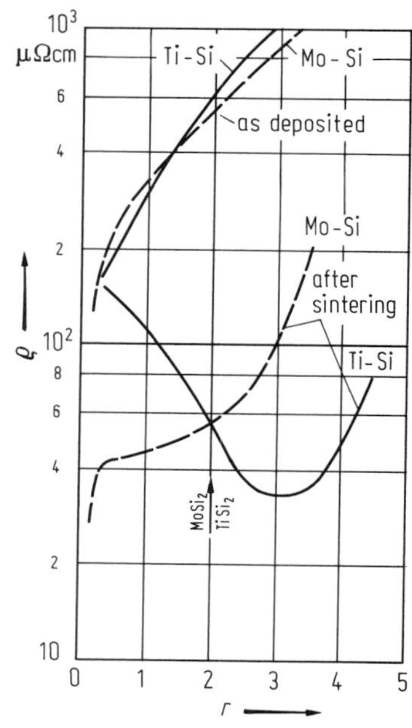

Fig. 91. Resistivity ϱ vs. nominal silicon/metal ratio r of co-sputtered Ti—Si films [80M1], and of co-evaporated Mo—Si films [83N1], as deposited, and after sintering 30 min at 900 °C, in H_2 for TiSi, respectively 60 min at 900 °C in N_2 for MoSi.

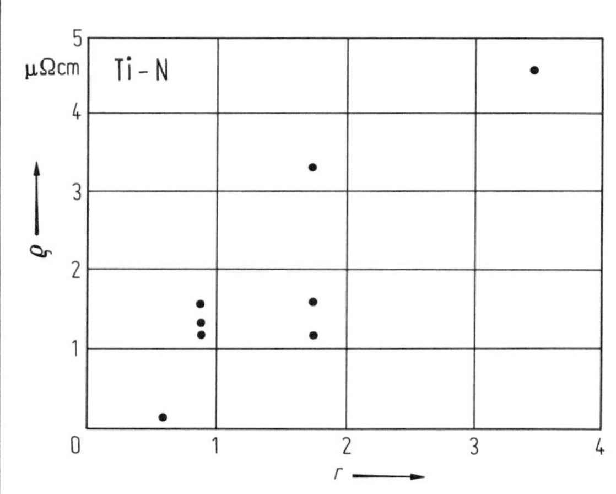

Fig. 92 Ti—N. Resistivity ϱ of Ti—N films (thickness 100 nm) vs. ratio r of N/Ti after annealing 30 min, at 900 °C in argon. The wide variation in resistivity is caused by the deposition parameters, probably the oxygen content [82T1].

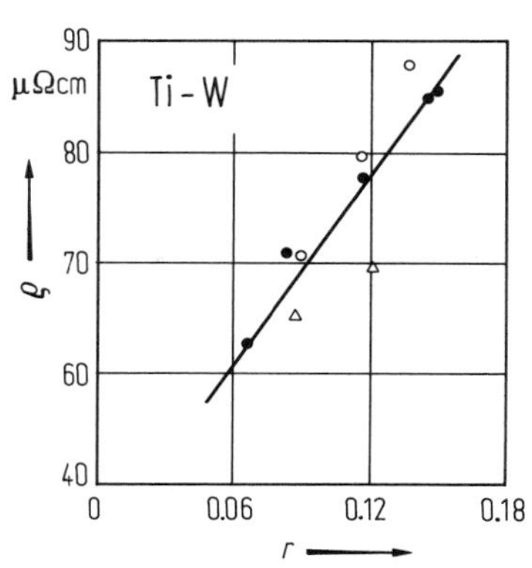

Fig. 93 Ti—W. Resistivity ϱ vs. Ti/W ratio r of DC magnetron (with RF bias) sputtered Ti—W films. The Ti/W ratio is determined from AES profiles [79H4].

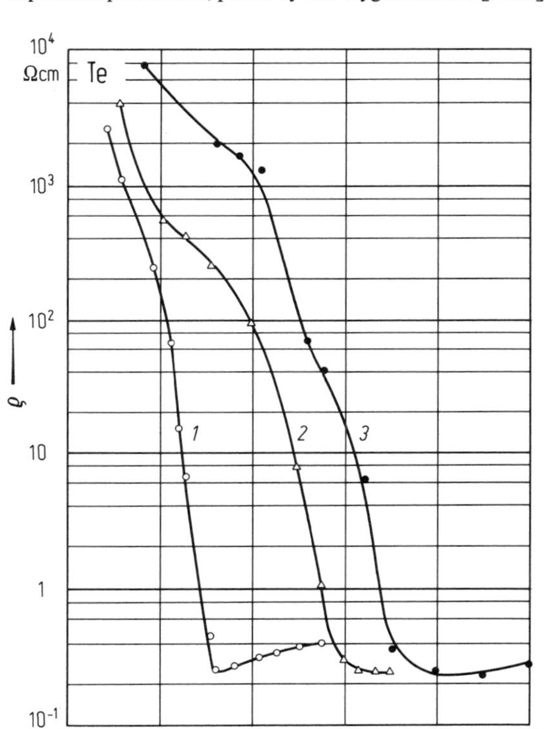

Fig. 94. Te. Resistivity ϱ vs. film thickness d for cold-evaporated tellurium films at different deposition rates, *1*) 110 nm min^{-1}; **2**) 300 nm min^{-1}; *3*) 630 nm min^{-1} [83O1].

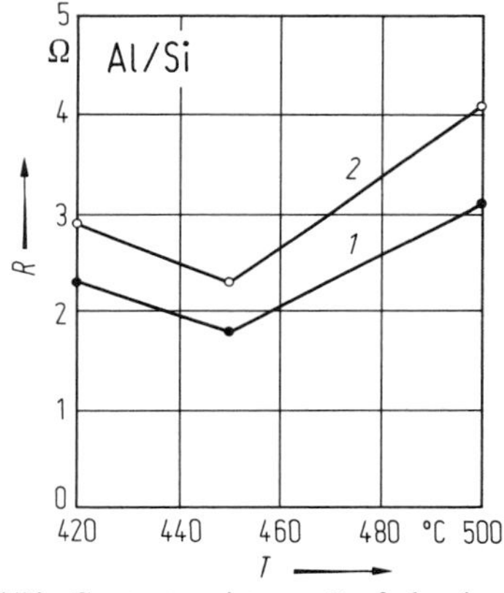

Fig. 95. Al/Si. Contact resistance R of aluminum contacts to p-type silicon, (100), $5 \cdot 10^{18}$ cm^{-3} surface doping, after 30 min sintering vs. sinter temperature T: *1* sintering in N_2; *2* sintering in $N_2 + H_2$ [78H3].

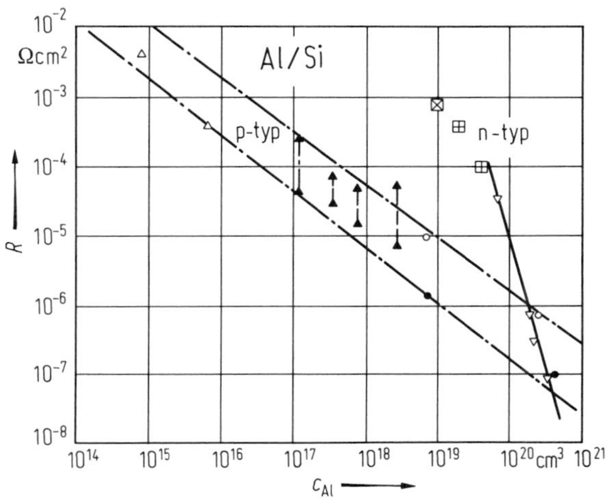

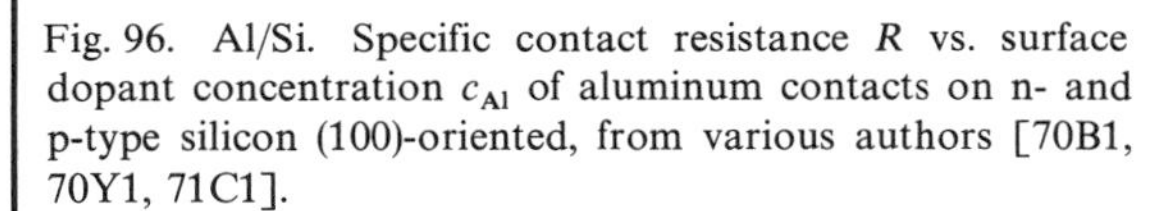

Fig. 96. Al/Si. Specific contact resistance R vs. surface dopant concentration c_{Al} of aluminum contacts on n- and p-type silicon (100)-oriented, from various authors [70B1, 70Y1, 71C1].

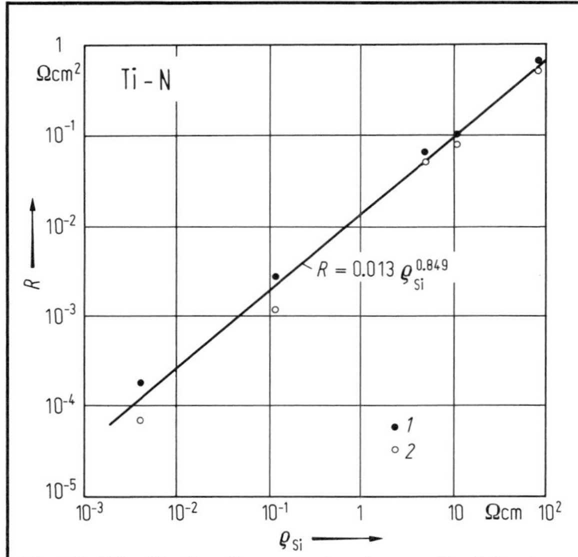

Fig. 97. Ti—N. Specific contact resistance R of titanium nitride vs. substrate resistivity ϱ_{Si} of n-type silicon [81W1]. *1* measured by method of Berger [70B1]; *2* measured by method of Reeves [80R1].

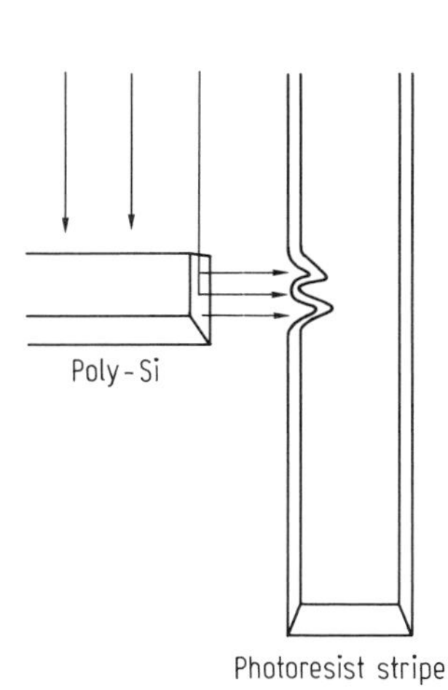

Fig. 98. Variation in linewidth of photoresist in the photolithographic process caused by reflection at opposite polysilicon walls on a wafer coated with aluminum [83S1].

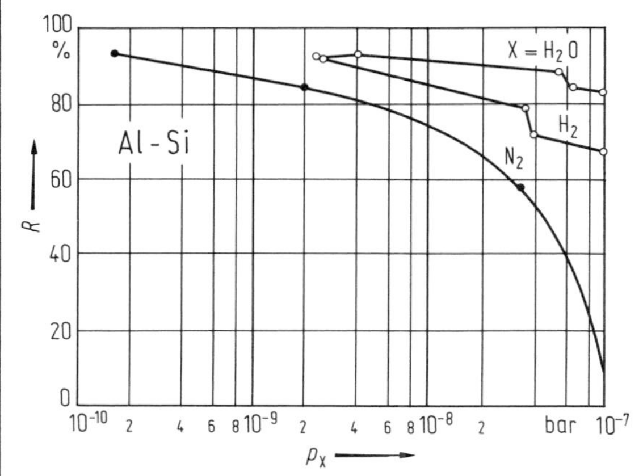

Fig. 99. Al—Si. Reflectivity R of DC-magnetron sputtered Al—Si(1 %) films (thickness ≈1 μm) vs. residual gas pressure p_X [79M2].

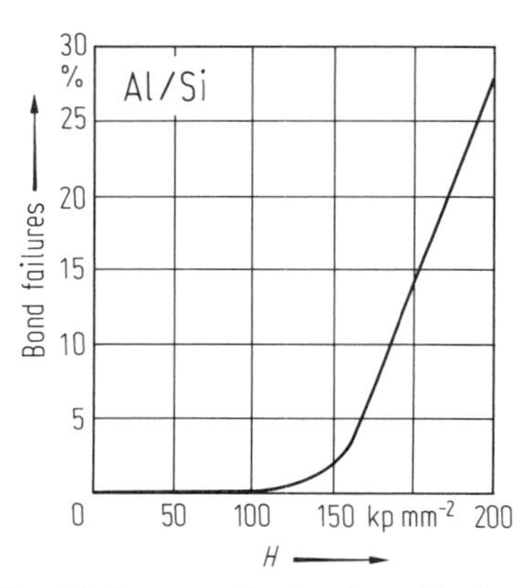

Fig. 100. Bond failures vs. film hardness H of magnetron-sputtered Al—Si films [79H1].

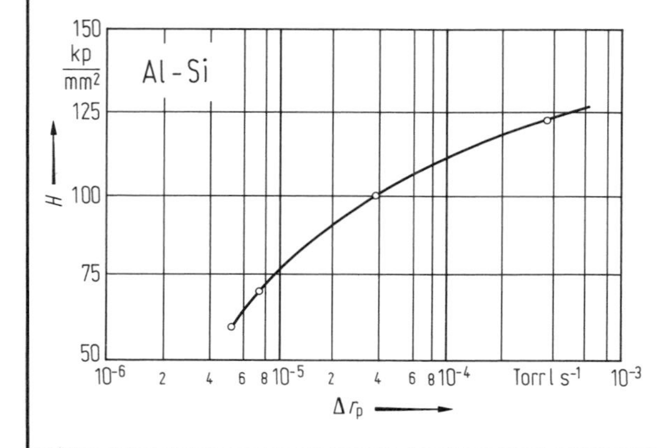

◄

Fig. 101. Microhardness of magnetron-sputtered Al—Si (1.5 %) layers vs. the rate of pressure increase Δr_p in the sputter chamber prior to sputtering. (The average hardness of evaporated Al—Si layers is 65 kp/mm²) [83S1].

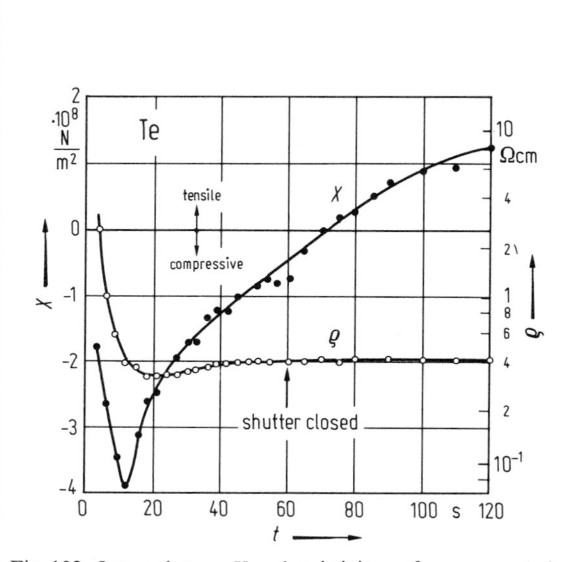

Fig. 102. Internal stress X and resistivity ϱ of an evaporated tellurium film (substrate temperature 100 °C, thickness 110 nm, deposition rate 110 nm/min), during and after deposition [83O1].

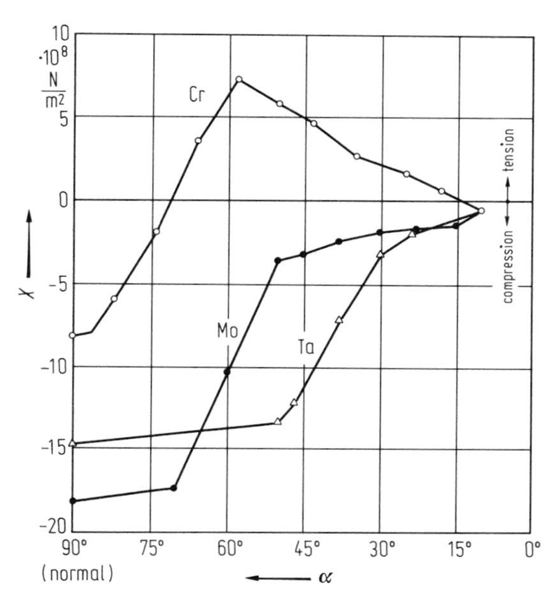

Fig. 103. Average internal film stress X vs. angle α of incidence at deposition for DC-magnetron sputtered films of chromium, molybdenum and tantalum [79H2]. Film thickness = 200 nm. Argon pressure during deposition = $1.3 \cdot 10^{-6}$ bar (≙0.13 Pa). Deposition rate (for normal incidence $\alpha = 90°$) = 24 nm min^{-1}.

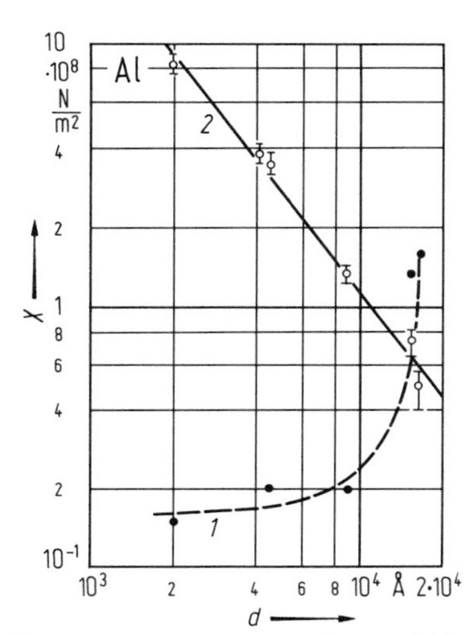

Fig. 104. Room temperature stress X vs. thickness d of e-gun-evaporated Al films (without substrate heating) [77S2]. *1* as deposited; *2* after thermal cycling: RT→ 450 °C···500 °C→RT, heating/cooling rate = 10 °C/min.

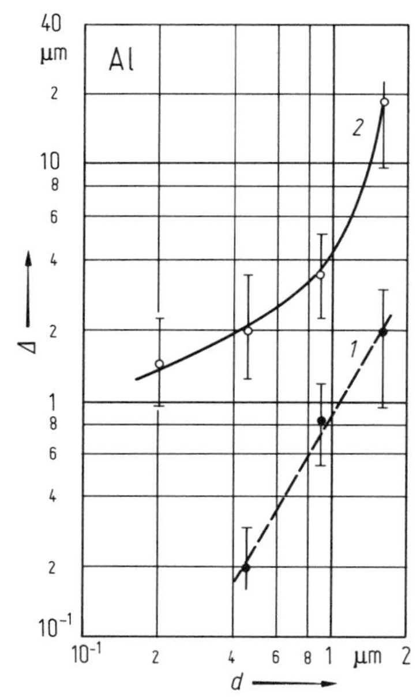

Fig. 105. Medium grain size Δ in e-gun-evaporated Al films vs. film thickness d [77S2]. *1* as-deposited; *2* after thermal cycling: RT→450 °C···500 °C→RT, heating/cooling rate = 10 °C/min; evaporation without substrate heating.

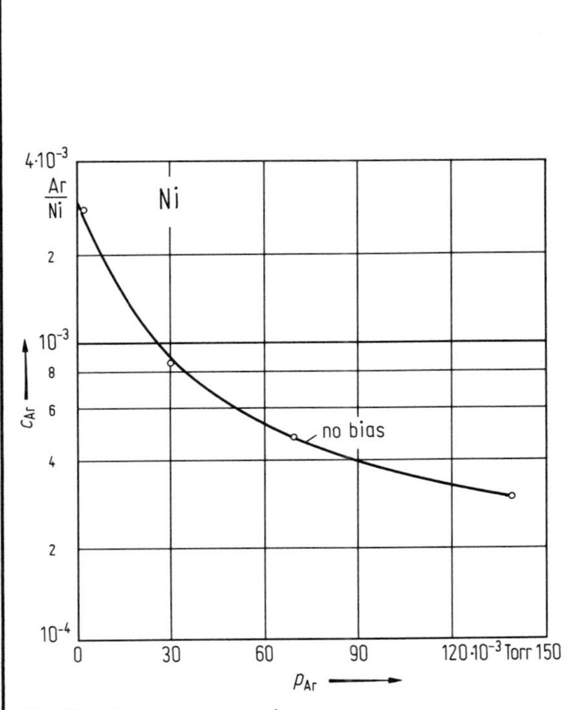

Fig. 106. Argon concentration c_{Ar} vs. argon pressure p_{Ar} during sputtering of Ni films. DC diode sputtering, target voltage 3000 V, substrate temperature 20 °C [67W1].

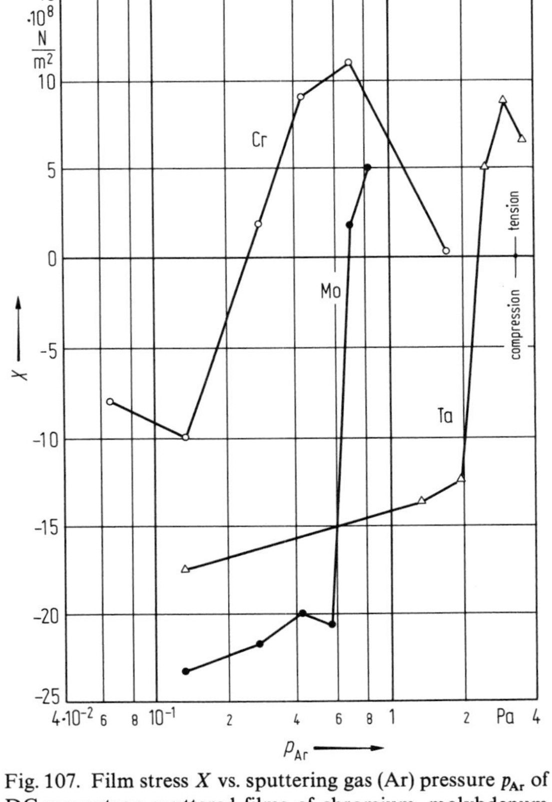

Fig. 107. Film stress X vs. sputtering gas (Ar) pressure p_{Ar} of DC-magnetron sputtered films of chromium, molybdenum and tantalum [79H2]. (Deposition rate = 60 nm min^{-1}; Film thickness = 200 nm; Substrate: glass, 25 mm × 25 mm × 0.15 mm).

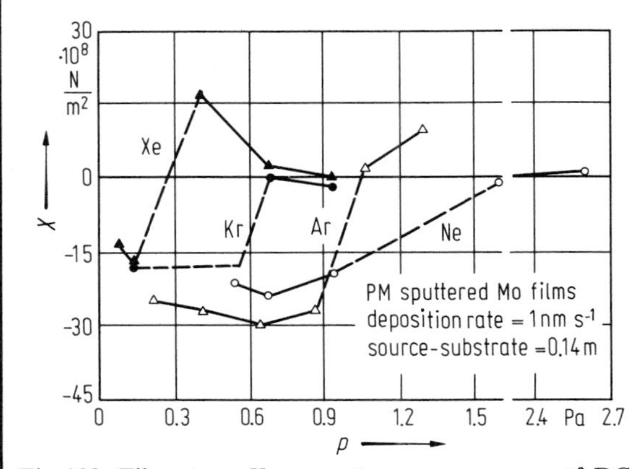

Fig. 108. Film stress X vs. sputter gas pressure p of DC-magnetron sputtered molybdenum films and various sputter gases: xenon, krypton, argon, neon [80H1]. (Deposition rate = 1 nm s^{-1}; distance source-substrate = 0.14 m).

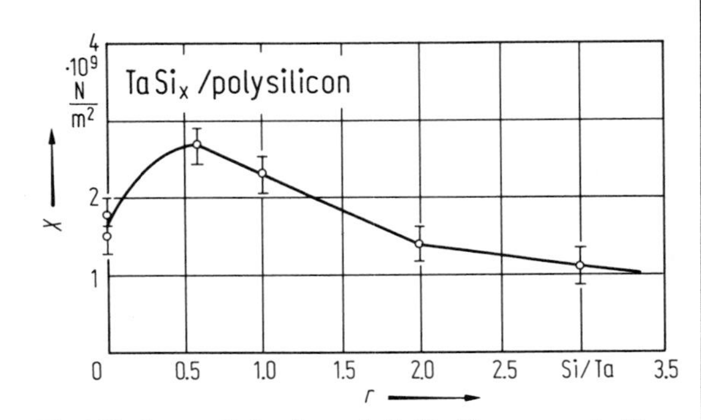

Fig. 109. Stress X in sintered $TaSi_x$ films on polysilicon vs. Si/Ta ratio r for co-sputtered $TaSi_x$ films [80M2].

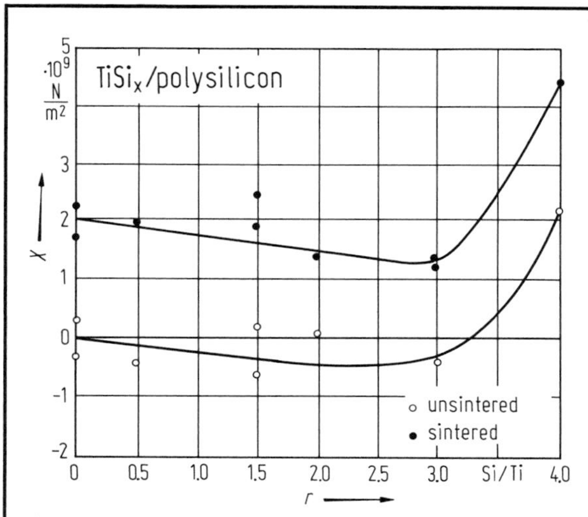

Fig. 110. Room temperature stress X in sintered and unsintered $TiSi_x$ films on polysilicon vs. Si/Ti ratio r for co-sputtered $TiSi_x$ films [80M2].

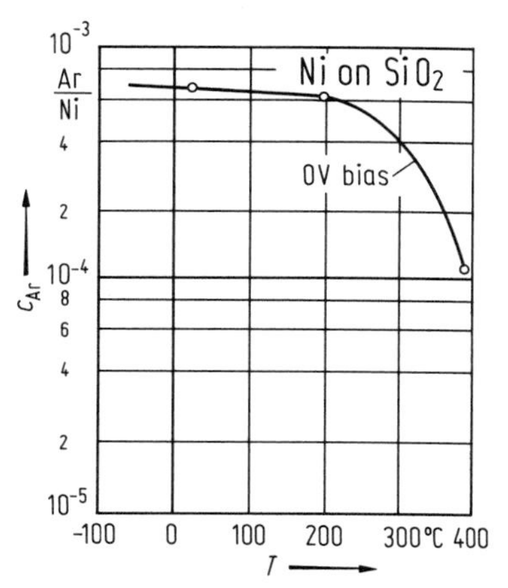

Fig. 111. Argon concentration c_{Ar} vs. substrate temperature T during sputtering of Ni films on SiO_2. DC diode sputtering, target voltage 3000 V, argon pressure $= 93 \cdot 10^{-6}$ bar [67W1].

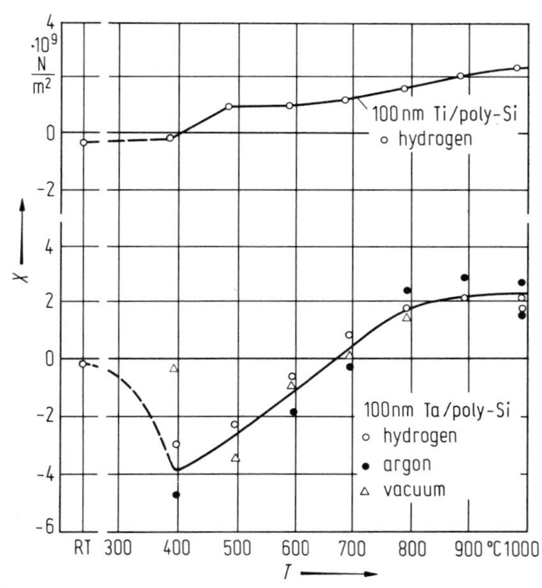

Fig. 112. Room temperature stress X vs. sintering temperature T of sintered Ta and Ti on polysilicon [80M1] for various sintering gases.